FUNGUS DISEASES OF TROPICAL CROPS

FUNGUS DISEASES OF TROPICAL CROPS

PAUL HOLLIDAY
Commonwealth Mycological Institute, Kew

CAMBRIDGE UNIVERSITY PRESS

Cambridge
London New York New Rochelle
Melbourne Sydney

Published by the Press Syndicate of the University of Cambridge
The Pitt Building, Trumpington Street, Cambridge CB2 1RP
32 East 57th Street, New York, NY 10022, USA
296 Beaconsfield Parade, Middle Park, Melbourne 3206, Australia

First published 1980

Printed in Great Britain by
the Alden Press, Osney Mead, Oxford

British Library Cataloguing in Publication Data

Holliday, Paul
Fungus diseases of tropical crops.
1. Tropical crops–Diseases and pests
2. Fungi, Phytopathogenic
I. Title
632′.4′0913 SB608.T8 79-41602

ISBN 0 521 22529 9

for
BETTY

'There's glory for you!' 'I don't know what you mean by "glory"', Alice said. 'I meant, "there's a nice knock down argument for you!"' 'But "glory" doesn't mean "a nice knock down argument"', Alice objected. 'When I use a word,' Humpty Dumpty said in a rather scornful tone, 'it means just what I choose it to mean, neither more nor less.'

LEWIS CARROLL (CHARLES LUTWIDGE DODGSON)

'I must begin with a good body of facts, and not from principle, in which I always suspect some fallacy.'

CHARLES DARWIN

'And he gave it for his opinion, that whoever could make two ears of corn or two blades of grass to grow upon a spot of ground where only one grew before, would deserve better of mankind, and do more essential service to his country, than the whole race of politicians put together.'

JONATHAN SWIFT

CONTENTS

Foreword	*page* ix
Acknowledgements	x
Introduction	xi
Abbreviations	xv
PATHOGENS	1
Epilogue	557
Appendix	
Hosts and pathogens	559
Lists	
Common disease names	577
Common host names	585
General Works	589
Addenda	591
Indexes	
Fungi	593
Fungus synonyms and conidial states	600
Other organisms	607

FOREWORD

by N. F. ROBERTSON, CBE, BSc, MA, PhD, Dip. Agr. Sci., FRSE,
Professor of Agriculture and Rural Economy, and Principal of the East of Scotland College of Agriculture, University of Edinburgh

The progress of plant pathology depends on labours of many kinds. The identification of the pathogen, the description of the responsible organism, the symptomatology, the aetiology of the disease, the epidemiology leading to the development of control measures and their integration into cultural practice are all required for an understanding of any disease syndrome. It is the hope of every plant pathologist that he can encompass all these tasks in a skilful fashion in the course of his professional career. But he is especially hopeful that in one of these areas he will, by his insight, bring about such an advance in the understanding of a plant disease that it will be effectively controlled.

Plant pathology needs not only practitioners skilled in the disciplines described above but also scholars who will bring order out of each wave of new advances in knowledge and present them to their colleagues as a solid base from which investigation can go forward. This book is such a work of scholarship: the orderly presentation and fastidious selection are readily apparent to any who read the book. The care in description and bibliography, and the insight with which the individual diseases are described and analysed, will be apparent to those who have worked with them. So, too, will they appreciate the background derived from the author's wide experience as a successful tropical plant pathologist.

Tropical plant pathology differs from temperate pathology mainly in its *locus* and in the difficulties for the practitioner of working at a distance from the necessary literature. Periodically sourcebooks are produced which aim to support the field pathologist in the tropics. Those which are successful combine a systematic treatment with an intuitive selection which makes them appear to a generation of plant pathologists to have an inspired capacity to provide the answer. It is my hope and belief that this, Paul Holliday's, book will become such a source for modern tropical plant pathology.

ACKNOWLEDGEMENTS

It was my good fortune to have been most ably influenced by two dedicated teachers, Miss E. M. Blackwell (a mycologist) and Professor N. F. Robertson (a plant pathologist). To them I owe both my enjoyment of the professional path that I chose and the seeds of inspiration for this book. I am also most grateful to my teacher at Cambridge for writing a foreword.

I wish first to thank Mr A. Johnston, Director of the Commonwealth Mycological Institute, for his kind permission to quote extensively from the publications of the Institute (CMI); particularly from the *CMI Descriptions of pathogenic Fungi and Bacteria*. I hope that adequate acknowledgement in the text on taxonomic sources has been made in each case (not only for CMI publications but also for all the others used). I am also indebted to Mr Johnston for the long loan of a set of the institute's abstracting journal in plant pathology (*Review of Applied Mycology* 1922–69; *Review of Plant Pathology* 1970–). Without the painstaking and consistent care of the many, traditionally anonymous, contributors to the review this book may not have been written.

I thank Dr I. A. S. Gibson and Dr J. M. Waller for advice and encouragement. Many helpful comments on fungal taxonomy came from my colleagues at CMI: Dr C. Booth (Assistant Director), Dr D. L. Hawksworth, Mrs J. A. Lunn, Dr J. E. M. Mordue, Dr E. Punithalingam, Dr A. Sivanesan and Dr B. C. Sutton (Principal Mycologist). For personal communications I thank Drs Booth, Sivanesan and Sutton. I am grateful to Dr D. N. Pegler (Herbarium, Royal Botanic Gardens, Kew) for his comments on the Agaricales. The library at the Royal Botanic Gardens, Kew, was of great value. The Directors of the Commonwealth Institutes of Entomology and Helminthology kindly had the names for insects and nematodes, respectively, checked. I thank Miss S. Daniels (CMI librarian) for help, and Mrs I. M. Ross and Mr P. Basu for translations.

For much preliminary typing, recording and checking references I owe a debt to: Mrs K. Franklin, Mrs E. Herriot, Mrs E. Holliday, Mrs S. North, Mrs E. Rainbow, Mrs M. S. Rainbow and Miss L. Thomson. Very many thanks are due to Mrs D. J. Harrison who readily undertook the formidable task of typing the final copy with speed and efficiency. I would also like to thank Miss L. M. Blackwell for help in the final preparation. Finally, I much appreciated the great care with which the members of the Cambridge University Press produced this book.

INTRODUCTION

This work is an attempt at a sourcebook. But it is neither an account of all described diseases nor one of all the fungi that have been reported to cause disease on tropical crops. The term pathogen is interpreted here as one that causes significant disease. Doubtless there are many tropical fungi which, locally, cause appreciable crop losses, but whose pathology has never been adequately investigated. The pitfalls of compression are obvious; any errors of omission, balance, selection, interpretation or assessment are regretted. My 4 aims have been to give a full description of fungus genera and species; a succinct account of the disease, emphasising those factors which are most important in the field; references adequate for access both to the bulk of the original work on the disease and the more recent, main literature on fungus morphology and taxonomy; an alphabetical arrangement to facilitate the identification of a given disease.

Crops and pathology

I have broadly followed J. W. Purseglove (*Tropical crops, Dicotyledons* Vols 1 and 2, 1968; *Monocotyledons* Vols 1 and 2, 1972) in the crops to be considered. Any name changes since these publications have not been adopted. Many temperate crops commonly grown in the tropics at high altitudes are omitted unless their main production is within the tropics. Some crops of temperate origin commonly or less commonly grown in the lowland tropics are considered; these are especially in the Cruciferae and Leguminosae. Some diseases of Coniferae and other timber trees are also included.

Tropical and temperate plant pathology differ neither in the inherent characteristics of the pathogen nor in the principles of disease investigation and control. The differences lie in weather and climate (see J. M. Waller, *Rev. Pl. Pathol.* 55: 185, 1976), crop characteristics, husbandry and socio-economic factors. In a tropical climate the carry over or dormant disease period (if one exists) is characterised particularly by dryness not low temperature. The amount, pattern and distribution of rain and moisture (including irrigation) are some of the dominant physical factors which affect disease build up and decline. Some tropical crops have both annual and perennial characteristics. More than one crop of an annual plant may be harvested each year. Most perennial crops renew their foliage only in part at any one time, i.e. in periodic flushes throughout the annual cycle. Leafless-ness in such crops is rare and nearly always short lived.

Introduction

Progress in agriculture reduces some pathogens and the diseases they cause to insignificance. Wherever such pathogens may still be considered important in the tropics it is probable that cultivation standards are inadequate. Characteristic still of much tropical agriculture is the wholly unsatisfactory imbalance in the degree of sophistication of our knowledge of diseases between crops. Where the product is a major component of world trade our knowledge is usually well advanced. Where it is of only local or regional importance our knowledge is often quite inadequate; furthermore, the advisory services are probably unsatisfactory or nonexistent. Therefore I have had to give too brief a treatment of some doubtless important, albeit local, diseases. The plant pathologist in the tropics can less afford the luxury of academic study than can his counterpart in temperate latitudes. The latter may only rarely experience a disease epidemic, a common (sometimes catastrophic) event in tropical crops. Tropical plant diseases will continue to take an often avoidable but heavy toll until more attention is given to factors which are usually out of control of the specialist in the field and laboratory.

Usage

Mycological taxonomic usage is generally that followed at the Commonwealth Mycological Institute, by *Ainsworth & Bisby's Dictionary of the fungi* (Ainsworth et al., 5th and 6th editions, 1961, 1971) and by *The fungi* Vols IVA and IVB (edited by G. C. Ainsworth, F. K. Sparrow & A. S. Sussman, 1973). The profusely illustrated books *Dematiaceous Hyphomycetes* and *More Dematiaceous Hyphomycetes* (M. B. Ellis, 1971, 1976) were the main taxonomic sources for this group. The books have keys to genera and some species. B. C. Sutton in *Mycol. Pap.* 141, 1977, gave the acceptable generic names for the Coelomycetes; his book (*The Coelomycetes*, 1980) will be the standard work for this group. The dictionary, and the books by Ellis and Sutton, provide fully adequate definitions of mycological terms. In pathology constant use was made of the invaluable *Host pathogen index to the review of applied mycology* Vols 1–40, 1922–61 (compiled by H. M. MacFarlane, 1968), and of the abstracts in this review and its continuation, *Review of Plant Pathology*, from 1970. The Federation of British Plant Pathologist's *A guide to the use of terms in plant pathology* (*Phytopathol. Pap.* 17, 1973) provided definitions of the diseased condition. Approved names of fungicides were taken from the British Crop Protection Council's *Pesticide Manual* (edited by H. Martin, 3rd and 5th editions, 1972, 1977). Crop names were taken from Purseglove (l.c.). For plant names reference was also made to *Willis' dictionary of the flowering plants and ferns* (H. K. Airy Shaw, 7th edition, 1966; now an 8th edition, 1973); *Pests and diseases of forest plantation trees* (F. G. Browne, 1968); *A dictionary of plants used by man* (G. Usher, 1974); and *Useful and everyday plants* (F. N. Howes, 1974). All other general texts used frequently are cited as appropriate and all are given at the end.

Pathogens

Where applicable these are given in the order as the perfect state; there are a few exceptions. Species morphology and some (not necessarily all) synonymy are taken from the CMI standardised description sheets or from another preferred source. For uniformity some very slight alterations have been made. These sheets are illustrated and give more details on synonymy, including the distinction between nomenclatural (obligate) synonyms and taxonomic (facultative) ones (see G. C. Ainsworth, Fungal nomenclature, *Rev. Pl. Pathol.* **52**: 59, 1973). But because of the inevitably uneven spread of up-to-date taxonomic work on pathogens it was not possible to give adequate descriptions in each case (see J. Walker on mutual responsibilities, *A. Rev. Phytopathol.* **13**: 335, 1975). Each genus is described as a separate entry. There follows an account that includes some mention of fungi which are not subsequently treated separately, usually because they cause only minor diseases or have been little studied as pathogens. Some important diseases affecting temperate crops may be mentioned here. Then follow the accounts of single species. Factors relevant to the disease(s) in the field are given more coverage than elementary histology, cultural and biochemical work. Where a pathogen causes important diseases on several crops an account of the disease is given separately (for each crop), following the description of the fungus. If a fungus name cannot be found in the main part of the book the indexes should be consulted.

Hosts (appendix)

The scientific and common host names head a list of the crop's main pathogens, i.e. those described separately, with the disease's common name and references to other mentions of the host. Where a pathogen causes significant disease on more than one crop its name is repeated under each. Plurivorous pathogens may also be given under several hosts but not under all. General literature on diseases of single crops or crop groups is given here.

References

For a full knowledge of the literature the reference lists should be consulted since not all the references given are mentioned in the text. Almost all the work cited has been examined in the original. I have been selective. Clearly much literature is elementary, ephemeral, routine, repetitive, of only local interest or peripheral to the disease situation. Original journal papers, monographs, reviews and books were the primary sources. The frequently secondary, preliminary or lesser sources of meetings, conferences, symposia, reports and surveys (including groups of pathogens and/or crops) were much less frequently searched. Some (often in vitro) work in biochemistry, genetics, physiology (and technique) has been omitted. Where such work is extensive it may be quoted as abstracts only. References are given in chronological order, and repetition of an author's name is represented by a dash. Repetition of an author's name plus 'et al.' is represented by two dashes. The *World list of scientific periodicals*, 4th edition, 1963–65, and the BUCOP Supplements, 1964–72, are guides to the title abbreviations which are not invariably those used in these lists; see also the *List of serial publications in the British Museum (natural history) library*, 2nd edition, 1975. The date given is that of the periodical sequence which is nearly always, but not invariably, the date of publication. After the citation there is an indication of contents; where a fungus or crop name is given the preferred one is usually used if this differs from that used by the author(s). The last component (omitted in the appendix) of each reference is the volume and page or abstract number to the abstract in the *Review of Applied Mycology* (1922–69) and the *Review of Plant Pathology* (1970–). Numbers preceded by a colon are page numbers whilst numbers preceded by a comma are abstract numbers. For title abbreviations of books on mycological taxonomy see *Mycologist's handbook* (D. L. Hawksworth, 1974).

Lists and indexes

Brief explanatory notes are given at the beginning of each list and index. Common disease names are sometimes notoriously unsatisfactory; some undistinctive ones have been omitted. I disagree with a slavish tendency to use a different common name for each symptom, i.e. where a pathogen causes disease effects on more than one plant organ. When giving common names more attention needs to be given to the syndrome as a whole. I am also not in favour of using a part of, or adaptation from, the pathogen's scientific name in the common name to distinguish a disease. Where it was considered necessary a choice of one common name was made for the list and the host appendix. But alternative names may be given in the text. Where there is no distinctive common name, for example, rust, the order is by the main host's common name.

Perhaps this book will go some way in reducing the undoubted truth contained in J. D. Bernal's remark: 'It is easier to make a scientific discovery than to learn whether it has already been made.' It is one which would have been useful to me as a raw recruit in the West Indies 30 years ago and in my later travels. I therefore hope that it will now be so to those who find themselves in similar situations and to those with experience.

P.H.

CMI, Kew; 1979

ABBREVIATIONS

av.	average
c.	circa (about, approximately)
cf.	compare
CMI	Commonwealth Mycological Institute
CMI Descr.	*CMI Descriptions of pathogenic Fungi and Bacteria*
conc(s).	concentration(s)
cv(s).	cultivar(s)
Demat. Hyphom.	*Dematiaceous Hyphomycetes*
diam.	diameter (in diameter)
e.g.	for example
et al.	and others
fide	according to
f. sp. (ff. sp.)	forma(e) specialis(es)
gen.	genus
gen. nov.	new genus
Ibid	the same place
i.e.	that is
l.c.	place (already) cited
LS	longitudinal section
Map	CMI distribution maps of plant diseases (and number)
max.	maximum
min.	minimum
opt.	optimum
PDA	potato dextrose agar
pH	hydrogen ion concentration
pp.	pages
q.v.	which see
race	physiologic race (of a fungus)
ref.	references
RH	relative humidity
sp., spp.	species (singular and plural)
sp. nov.	new species
ssp.	subspecies
stat. conid.	conidial (or imperfect) state
str(s).	strain(s)
temp(s).	temperature(s)
TS	transverse section
UV	ultraviolet
Univ.	University
var(s).	variety(ies)
var. nov.	new variety
μ	micron (10^{-3} mm)
$<$	less than
$>$	more than

(The usual abbreviations for metric measurements of length and symbols for chemical elements are used. All temperatures are in degrees centigrade and C is omitted.)

PATHOGENS

ACREMONIUM Link ex Fr., *Syst. Mycol.* 1: XLIV, 1821.
(synonymy in W. Gams, *Cephalosporium-artige Schimmelpilze* (Hyphomycetes): 38, 1971)
(Moniliaceae)

COLONIES slow growing, 25 mm diam. in 10 days, hyphae delicate, hyaline, without olive coloured pigment in the walls. ORTHOPHIALIDES formed in typical cases. CONIDIOPHORES basitonous (branches arising near the base), whorled or cymoid. PHIALIDES diminish gradually towards the tip. The tip is usually smooth, less frequently a small collarette (not funnel shaped) occurs. CONIDIA hyaline or pigmented, aseptate, catenuliform or in small mucilagenous heads. *Emericellopsis* and some *Nectria* spp. are perfect states (after W. Gams l.c.).

Gams described the mould fungi like *Cephalosporium*; the monograph contains a summary, keys and a glossary in English. *Cephalosporium* Corda can be replaced by the older generic name *Acremonium*, the type sp. of which has conidia in chains or heads. *Gliomastix* Guegen was merged with *Acremonium*; and, in an expanded form, retained as a separate section of the genus. *Paecilomyces* Bain. (monophialidic group) was transferred to *Acremonium*, which was defined as comprising slow or medium slow growing spp. producing aseptate (rarely 1 septate) phialospores in chains or heads, on thin-walled orthophialides or on basitonously branched conidiophores. Other spp. were placed in *Monocillium* Saksena, *Verticillium* (q.v.) and *Aphanocladium* Gams. *Acremonium zonatum* (Saw.) W. Gams (synonyms: *C. zonatum* Saw., *C. eichhorniae* Padwick, *C. fici* Tims & Olive) causes zonal leaf spot of coffee. The spots are irregular, 1–6 cm diam., greyish brown with dark brown concentric rings. *A. zonatum* has been recorded from widely differing plant genera (see Hawksworth). Samuels gave a key to the 12 spp. of *Nectria* and their *Acremonium* conidial states. Chesson et al. investigated the taxonomy through electrophoresis.

Some fungi referred to *Cephalosporium* in plant pathology literature apparently need further taxonomic study. *C. sacchari* (associated with sugarcane wilt) is discussed under *Gibberella fujikuroi* var. *subglutinans*. This fungus is said to be a *Fusarium*. The cause of black bundle of maize is usually attributed to *C. acremonium* (Reddy et al.), now referred to *Acremonium strictum* W. Gams. Gams considered that *A. kiliense* Grütz is probably the cause of black bundle but this may be in doubt. Late wilt of maize, described from Egypt (Samra et al. 1963), is caused by a fungus then named *C. maydis* Samra, Sabet & Hingorani. But this last organism apparently corresponds to the microconidial form of *Gaeumannomyces graminis* (Sacc.) Arx & Olivier (Gams; Domsch & Gams, *Fungi in agricultural soils*).

These 2 widespread soil fungi from maize have been studied in connection with the disease complex in this crop usually known as stalk rot. The earlier work on the pathogens concerned was analysed by J. J. Christensen et al. (*Monogr. Am. phytopathol. Soc.* 3, 1966). Sabet et al. (1970b) compared their saprophytic and pathogenic behaviour. Reddy et al. described the most typical symptoms of black bundle as blackened vascular tissue in the stem and sometimes in the leaf. As the season progresses the leaf and stem show a reddish colouration; there may be excessive tillering; there are barren stems and small ears. In late wilt there is a fairly rapid wilt from just before tasseling until shortly before maturity, beginning in the lower part of the plant. Leaves become dull green and dry. Stem vascular bundles and lower internodes become reddish brown (Samra et al. 1963). The fungus causing black bundle has been considered a weak parasite on maize (Harris; Sabet et al. 1970a). It is often associated with damage by insects, unfavourable growing conditions and is seedborne, probably internally. In late wilt young maize plants are readily infected but with age fewer plants become infected and none after *c.* 50 days from sowing. The roots are invaded and when these are wounded more plants, especially older ones, are attacked. The fungus is soilborne (persistence up to 180 days in deep soil, Samra et al. 1972) and a vascular pathogen. According to Payak et al. it can cause seed rot and prevent seedling emergence. Selection for resistance in Egypt showed it to be inherited in a dominant fashion (El Morshidy et al; Shehata). Some accounts of control of maize stalk rot will be found under the main pathogens involved: *Diplodia maydis, Gibberella fujikuroi* (and var. *subglutinans*) and *G. zeae*.

Chesson, A. et al. 1978. *Trans. Br. mycol. Soc.* **70**: 345 (electrophoresis of proteins of fungi like *Acremonium*).

El-Morshidy, M.A. et al. 1972. *Assiut J. agric. Sci.* **3**: 51 (resistance in late wilt; **55**, 3560).

Gams, W. 1971. *Cephalosporium-artige Schimmelpilze (Hyphomycetes)*, 262 pp., Gustav Fischer Verlag (monograph; **51**, 3191).

Gupta, B. M. et al. 1970. *Pl. Dis. Reptr* **54**: 989 (factors affecting infection in black bundle; **50**, 2905).

——. 1972. *Indian J. Mycol. Pl. Pathol.* **2**: 57 (resistance in black bundle; **52**, 2925).

Harris, M. R. 1936. *Phytopathology* **26**: 965 (pathogenicity, black bundle; **16**: 168).

Hawksworth, D. L. 1976. *C.M.I. Descr. pathog. Fungi Bact.* 502 (*Acremonium zonatum*).

Nag Raj, T. R. et al. 1962. *Curr. Sci.* **31**: 104 (*A. zonatum* on coffee; **41**, 656).

Payak, M. M. et al. 1970. *Indian Phytopathol.* **23**: 562 (general, late wilt; **50**, 3699).

Reddy, C. S. et al. 1924. *J. agric. Res.* **27**: 177 (general, black bundle; 3: 449).

Sabet, K. A. et al. 1970*a*. *Ann. appl. Biol.* **66**: 257 (pathogenicity, both maize fungi; **50**, 1764a).

——. 1970*b*. *Ibid* **66**: 265 (saprophytism, both maize fungi; **50**, 1764b).

——. 1972. *Phytopathol. Mediterranea* **11**: 10 (resistance & susceptibility in late wilt; **52**, 1522).

Salas, A. et al. 1967. *Turrialba* **17**: 292 (*A. zonatum* on coffee; **47**, 1560).

Samra, A. S. et al. 1963. *Phytopathology* **53**: 402 (general, late wilt; **42**: 611).

——. 1972. *Tech. Bull. Pl. Prot. Dep. Cairo* 2, 92 pp. (maize stalk rot complex in Egypt; **51**, 4813).

Samuels, G. J. 1976. *N.Z. Jl Bot.* **14**: 231 (perfect states of *Acremonium*; **56**, 2396).

Shehata, A. H. 1976. *Egyptian J. Genet. Cytol.* **5**: 42 (host gene effects in late wilt; **55**, 4064).

ALBUGO (Pers.) S. F. Gray, *Nat. Arr. Br. Pl.* Vol. 1: 540, 1821.
(Albuginaceae)
Uredo albugo Pers., 1801
Erysibe sensu Wallr., 1833
Cystopus Lév., 1847
Cystopus de Bary, 1863.

Parasitic and host family specific. MYCELIUM intercellular, widespread, particularly in leaves, bearing simple, globose, intracellular haustoria. SORI subepidermal, later erumpent, forming white to cream, mealy patches on all aerial parts of hosts. SPORANGIOPHORES crowded, clavate, hyaline, forming basipetal chains of sporangia joined by colourless connective links. SPORANGIA (zoosporangia) all alike or terminal sporangium larger than the others; globose elliptical, oblong cubical or truncated obovate, hyaline; wall smooth, equally thickened throughout or with an annular thickened band. SEXUAL ORGANS: antheridia and oogonia borne singly on short terminal or lateral mycelial branches within host tissue. OOGONIA globose, differentiated into periplasm and oosphere. ANTHERIDIA small, club shaped, arising near oogonia. OOSPORES thick walled, globose; epispore dark coloured, ornamented with warts, ridges or reticulations. Germination by zoospores (from S. D. Baker who upheld the validity of *Albugo* against *Cystopus*; see also Whipps et al. 1978a).

Biga re-examined the morphology of the sporangia or conidia (workers use both terms) and gave a key to the *Albugo* spp. Wilson described the North American spp., Savulescu the European (particularly Rumanian) spp., and Wakefield gave a historical note on the genus in her account of the spp. in South Africa. Members of the genus do not usually cause severe diseases. Those that have attracted attention as pathogens (besides the 3 spp. described individually) include: *A. occidentalis* G. W. Wilson on Chenopodiaceae including spinach (*Spinacia oleracea*), *A. bliti* (Biv.-Bern.) Kuntze on Amaranthaceae, and *A. tragopogonis* (DC.) S. F. Gray on Compositae including sunflower and salsify (*Tragopogon porrifolius*). The last sp. was described by Mukerji and the extensive synonymy given by Whipps et al. (1978a). The latter also described inoculations of Compositae with, and zoosporangia and zoospore behaviour in, *A. tragopogonis* (Whipps et al. 1978b & c). On sunflower most disease developed when inoculated plants (stomatal penetration) were kept at 20–25° in the post-penetration phase. Symptoms did not occur on such plants kept at 25° or above during the prepenetration and penetration stages (Kajornchaiyakul et al.).

Baker, S. D. 1955. *Trans. R. Soc. N.Z.* **82**: 987 (*Albugo* in New Zealand; **34**: 678).
Biga, M. L. B. 1955. *Sydowia* **9**: 339 (sporangial morphology & key; **36**: 63).
Doepel, R. F. 1965. *J. Agric. West. Aust.* Ser 4, **6**: 439 (control with zineb & maneb, *A. tragopogonis* on *Gerbera*; **45**, 471).
Kajornchaiyakul, P. et al. 1976. *Trans. Br. mycol. Soc.* **66**: 91 (infection of sunflower; **55**, 2833).
Khan, S. R. 1976. *Can. J. Bot.* **54**: 168 (electron microscopy of dividing somatic nuclei; **55**, 3948).
Mishra, M. D. et al. 1964. *Indian Phytopathol.* **16**: 333 (factors affecting sporangial germination in *A. bliti*; **43**, 3236).
Mukerji, K. G. 1975. *C.M.I. Descr. pathog. Fungi Bact.* 458 (*A. tragopogonis*).
Populer, C. 1966. *Bull. Rech. agron. Gembloux* **1**: 283 (sporangial dispersal in *A. tragopogonis*; **46**, 1900).
Raabe, R. D. et al. 1952. *Phytopathology* **42**: 448, 473 (*A. occidentalis* on spinach; **32**: 297, 419).
Săvulescu, O. 1946. Thesis 213, University of Bucharest (European *Albugo* spp.; **27**: 542).
Stevens, F. L. 1901. *Bot. Gaz.* **32**: 77, 157, 238 (gametogenesis & fertilisation).
Thakur, S. B. 1977. *Mycologia* **69**: 637 (percurrent proliferation of sporangiophores; **57**, 460).
Wakefield, E. M. 1927. *Bothalia* **2**: 242 (S. African *Albugo* spp.).
Whipps, J. M. et al. 1978a. *Trans. Br. mycol. Soc.* **70**: 285 (nomenclature of *A. tragopogonis*; **57**, 4353).
———. 1978b. *Ibid* **70**: 389 (interactions of Compositae with *A. tragopogonis* from *Senecio squalidus*; **57**, 4836).
———. 1978c. *Ibid* **71**: 121 (behaviour of zoospores & zoosporangia of *A. tragopogonis* in infection of *S. squalidus*).
Wiant, J. S. et al. 1939. *Phytopathology* **29**: 616 (*A. occidentalis* on spinach; **19**: 4).
Wilson, G. W. 1907. *Bull. Torrey bot. Club* **34**: 61 (N. American *Albugo* spp.).

Albugo candida (Pers. ex. Hooker) O. Kuntze, *Rev. Gen. Pl.* 2: 658, 1981.

(synonymy in Biga and Wilson, *Albugo* q.v.)

SORI on all parts of the host except the roots, white or rarely yellow, prominent and rather deep seated in tissues of the host, very variable in size and shape, often confluent and frequently producing marked distortion of the host. SPORANGIOPHORES hyaline, clavate, *c.* 15–17 × 35–40 μ. SPORANGIA similar, globular, hyaline with uniform thin walls, 15–18 μ. OOSPORES usually confined to the stems and fruits of the host, rarely in the leaves, chocolate coloured, *c.* 40–55 μ; epispore thick, verrucose or with low, blunt ridges which are often confluent and irregularly branched (after G. W. Wilson, *Albugo* q.v.; and see K. G. Mukerji, *CMI Descr.* 460, 1975).

A. candida is a widespread but minor pathogen on many genera of the Cruciferae; it is also found in the Capparidaceae and Cleomaceae. The diseases caused are called white rust or white blister. Butler (*Fungi and disease in plants*) gave a full description of the external and internal symptoms, and the pathogen's morphology; see Harper et al. for yield loss. The characteristically prominent, white sori (like blisters) are variable in size and shape and are formed mostly on the leaves and stems. On the latter they may be discrete or cover the whole stem. Distortion and hypertrophy of the host are frequent; the inflorescence axis may become many times its normal thickness and the leaves thickened, with rolled edges to the lamina. The floral parts persist and become distorted. Endo et al. made a more recent and detailed study of this white rust on horseradish (*Armoracia rusticana*); and see Verma et al. 1975b. Symptoms on roots seem rare but Fisher described root galls on radish. Oospores occur in the infected tissues. Petrie (1975) described oospores from seed samples; > 1500/g seed were recorded.

Both sporangia (presumably airborne) and oospores germinate indirectly by zoospores which penetrate the stomata. Light and dark have no differential effect on sporangial germination. The opt. temp. for zoospore release and germination is *c.* 15° and that for germ tube growth *c.* 20°. Temps. of 25° and over reduce both the rate and amount of release. Zoospore motility decreases with rise in temp.; at 10° 50% are active after 2 hours and at 20° only 5%. Disease development is reduced at temps. of < 10° and > 25°. Both young and old leaves can be infected but in the latter case the sori are smaller. Sporangia lose viability quickly, in 8–16 hours at 20° and RH 90%. Carry over is by mycelium in the host and by oospores. A period of dormancy before oospore germination does not appear to be necessary since the spores germinated 2 weeks after collection (Petrie et al. 1974). After dry storage for 21 years

43% of the oospores germinated in 4 days (Verma et al. 1975a). Work on host physiology after infection and fine structure has been described (*Rev. appl. Mycol.* **38**: 670; **41**: 492; **43**, 2759, 2760; **47**, 2310 2898; **48**, 635; *Rev. Pl. Pathol.* **55**, 1528; **56**, 900, 5479; **57**, 1466, 1889).

Forms from various hosts differ in pathogenicity and Togashi et al. proposed 2 groups within *A. candida* based on sporangial size: macrospora (18 × 20 μ) on *Brasscia* and *Raphanus* spp., microspora (14.5 × 15.5 μ) on *Capsella bursa-pastoris*, *Cardamine flexuosa*, *Draba nemorosa* and *Arabis hirsuta*. Pound et al. gave a review of the earlier work on pathogenic differences between isolates from various hosts. Using fungal collections from 6 hosts in 6 genera they considered each to be a distinct race; although, since most crucifers tested were susceptible to more than 1 race, these hosts were of little value as differentials. A high degree of heterogeneity was found in 115 cvs from 10 *Brassica* spp. when inoculated with races 1 and 2. Resistance is known in horseradish, brassicas and radish; in the last host Williams et al. found resistance to be controlled by a single dominant gene in 2 cvs; and see Petrie.

Akai, S. 1937. *Forschn Geb. PflKrankh. Kyoto* **3**: 71 (histology).
Davis, B. M. 1900. *Bot. Gaz.* **29**: 297 (fertilisation).
Eberhardt, A. 1904. *Zentbl. Bakt. ParasitKde* Abt. 2 **12**: 235, 426, 614, 714 (general).
Endo, R. M. et al. 1960. *Bull. Ill. agric. Exp. Stn* 655, 56 pp. (general on horseradish; **39**: 753).
Fisher, E. E. 1954. *Proc. R. Soc. Vict.* **65**: 61 (on radish roots; **33**: 459).
Harper, F. R. et al. 1974. *Phytopathology* **64**: 408 (yield loss).
Hiura, M. 1930. *Jap. J. Bot.* **5**: 1 (pathogenic strs; **9**: 573).
Hougas, R. W. et al. 1952. *Phytopathology* **42**: 109 (resistance in horseradish; **31**: 466).
Napper, M. E. 1933. *J. Pomol.* **11**: 81 (sporangial germination & pathogenic strs; **13**: 70).
Petrie, G. A. et al. 1974. *Can. J. Pl. Sci.* **54**: 595 (method for germinating oospores; **54**; 1043).
——. 1975. *Can. Pl. Dis. Surv.* **55**: 19 (oospores in seed samples; **55**, 2415).
Pound, G. S. et al. 1963. *Phytopathology* **53**: 1146 (races; **43**, 1185).
Sempio, C. 1940. *Riv. Patol. veg. Pavia* **30**: 29 (effects of external environment; **19**: 381).
Togashi, K. et al. 1934. *Bull. Imp. Coll. Agric. For. Morioka* 18, 88 pp. (morphologic & pathogenic strs; **14**: 1).
Vanterpool, T. C. 1959. *Can. J. Bot.* **37**: 169 (oospore germination; **38**: 582).
Verma, P. R. et al. 1975a & b. *Ibid.* **53**: 836, 1016 (oospore germination, infection & mycelial development in cotyledons; **55**, 951, 953).
Wager, H. 1896. *Ann. Bot.* **10**: 295 (structure & reproduction).
Williams, P. H. et al. 1963. *Phytopathology* **53**: 1150 (nature & inheritance of resistance in radish; **43**, 1186).

Albugo ipomoeae-aquaticae Sawada, *Descr. Cat. Formosan Fungi* **2**: 27, 1922.
(see Edie 1970 for note on confusion with *Albugo ipomoeae-panduratae* (q.v.) and error in Latin diagnosis by Ito et al.)

SORI hypophyllous and on stem of inflorescence, rounded to irregular, 1–3 mm diam; up to 7 mm long, dull white, becoming pulverulent; host sometimes distorted. SPORANGIOPHORES hyaline, clavate, 32–72 μ long, 18–23 μ wide. SPORANGIA catenulate, size uniform, globose to cuboid, hyaline, 16–23 × 18–26 (av. 19 × 21) μ; enclosed by a smooth, uniformly thickened membrane. OOGONIUM globose, subglobose or obovate with irregular patterns on the inner surface, 52–80 μ diam. when globose or 64–89 × 48–68 μ when otherwise shaped, containing one oospore. OOSPORE globose, hyaline, smooth 39–48 μ diam., the wall originally thin but becoming 6–8 μ thick after conjugation. (Sawada l.c., Eddie 1970 and Ito et al.)

A. ipomoeae-aquaticae, originally described from Taiwan, has recently been studied in detail as causing a serious disease of water spinach (*Ipomoea aquatica*) in Hong Kong. Earlier records of the pathogen on this host appear to have been erroneously attributed to the other and more widespread *Albugo* (*A. ipomoeae-panduratae*) on Convolvulaceae. *A. ipomoeae-aquaticae* has also been reported from India. On water spinach initial infection shows as chlorotic flecks; the white sori (5–6 mm diam.), which then develop, may cover the whole leaf. Necrotic areas form after the sporangia are shed. Cerebriform, stem galls occur on the stems (2–3 times the stem diam.), and plants that have root galls may be stunted. Bud galls develop after systemic infection via axillary buds which grow into thick shoots with compressed internodes, numerous, deformed lateral branches and systemically infected leaves that are distorted and reduced in size. These leaves differ in their symptoms from those infected by airborne sporangia. Sori form on the surface of bud galls but not on cerebriform galls. All older galls contain numerous oospores. Where reproductive structures originate from a bud gall systemic infection occurs in the inflorescence. Flower buds are swollen and aborted.

Opt. temp. for sporangial and zoospore germination is 25° or a little lower. At 20–25° sporangial germination is complete after 3 hours and at 15° or 28° it is much slower. Light and dark have no differential effect on germination of either spore. At RH 100% a fifth of the sporangia are viable after 24 hours; at 80% there is little or no germination after 12 hours. Most sori cease producing sporangia after 12 days. Most growth of the zoospore germ tubes is at 25° and they penetrate the stomata. Incubation at RH 100% and 25° for 8 hours gives 100% infection (40% after 2 hours). There is no infection at 15°; these opt. temps are higher than those previously reported for *Albugo* spp. Older, inoculated leaves form fewer, and less secondary sori, compared with younger ones. The sexual stage is common, particularly in systemically infected tissues. Root galls develop when plantings are made in soil infested with oospores. Attempts to infect other *Ipomoea* and *Convolvulus* spp. with the water spinach form of the fungus were unsuccessful. In Hong Kong all local cvs were susceptible and the only control is to avoid infected host debris and contaminated soil; destruction measures should include the roots.

Edie, H. H. 1970. *Trans. Br. mycol. Soc.* **55**: 167 (note on confusion with *Albugo ipomoeae-panduratae* & correction in Latin diagnosis; **50**, 540).

—— et al. 1970. *Ibid* **55**: 205 (factors affecting sporangial germination; **50**, 1036).

—— ——. 1970. *Agric. Sci. Hong Kong* **1**: 178 (general; **51**, 1032).

Ho, B. W. C. et al. 1969. *Pl. Dis. Reptr* **53**: 959 (host symptoms & fungal morphology; **49**, 1534).

Ito, S. et al. 1935. *Trans. Sapporo nat. Hist. Soc.* **14**: 11 (Latin diagnosis; **15**: 57).

Safeeulla, K. M. et al. 1953. *Cellule* **55**: 225 (morphology & cytology of *Albugo* on *Ipomoea aquatica* & *Merremia emarginata*).

Singh, H. et al. 1965. *Phytopathol. Z.* **53**: 201 (anatomy of infected *I. aquatica* & cytology of sex organs of pathogen as *A. ipomoeae-panduratae*).

Albugo ipomoeae-panduratae (Schw.) Swingle, *J. Mycol.* 7: 112, 1891.
(synonymy in Swingle l.c. and Wilson, *Albugo* q.v.)

SORI amphigenous or caulicolous, white or light yellow, prominent, superficial, 0.5–20 mm, rounded, often confluent and frequently producing marked distortions of the host. SPORANGIOPHORES hyaline, clavate, unequally curved at base, *c.* 15 × 30 μ. SPORANGIA short cylindric, similar or the terminal more rounded, hyaline; the membrane with an equatorial thickening, usually very pronounced, 14–20 × 12–18 μ. Oosporic sori separate from the sporangial, caulicolous, rarely on the petioles, 1–2 × 5–6 cm or even more, causing marked distortion of the host. OOSPORES light yellowish brown, 25–55 μ; epispore papillate or with irregular, more or less curved ridges (after Wilson, *Albugo* q.v.; see also K. G. Mukerji & C. Critchett, *CMI Descr.* 459, 1975).

Common and widespread on Convolvulaceae including *Ipomoea batatas* (sweet potato) and ornamental *Ipomoea* spp. *A. ipomoeae-panduratae* was described from *I. pandurata* and has recently been confused with a later described sp. (*A. ipomoeae-aquaticae* q.v.) which also occurs on Convolvulaceae. Leaves and stems of sweet potato are attacked; the whitish sori appear on the lower leaf surface beneath chlorotic spots. Distortion may occur and galls may form on the tubers. The anatomy of the 2 types of galls (bud and cerebriform on the internodes) on *I. pentaphylla* have been described in detail by Singh et al. Ciferri designated 2 forms, one on *I. batatas* and the other on *I. biloba*. Infection rarely ever seems to be serious enough to warrant control measures; copper fungicides were recommended against a severe attack on *I. horsfalliae.*

Ciferri, R. 1928. *Nuovo G. bot. ital.* **35**: 112 (forms on *Ipomoea batatas* & *I. biloba*; **8**: 11).

De Melo, J. L. 1947. *Bolm Agric. Pernambuco* **14**: 332 (on *I. batatas*; **27**: 347).

Drummond-Goncalves, R. 1955. *Biológico* **21**: 199 (on *I. horsfalliae* in Brazil; **35**: 459).

Kumar, K. et al. 1974. *Marcellia* **38**: 31 (gall anatomy on *I. sindica*; **55**, 6057.

Singh, H. 1963. *Curr. Sci.* **32**: 472 (growth on tissue culture; **43**, 912).

—— et al. 1966. *Can. J. Bot.* **44**: 535 (anatomy of galls formed by infection on *I. pentaphylla*; **45**, 2864).

Stevens, F. L. 1904. *Bot. Gaz.* **38**: 300 (oogenesis & fertilisation).

ALTERNARIA Nees ex Fr., Nees, *Syst. Pilze Schw.*: 72, 1816; Fries, *Syst. mycol.* 1: xlvi, 1821.
Macrosporium Fr., 1832
Rhopalidium Mont., 1846.
(Dematiaceae)

COLONIES effuse, usually grey, dark blackish brown or black. MYCELIUM all immersed or partly superficial; hyphae colourless, olivaceous brown or brown. STROMA rarely formed. SETAE and HYPHOPODIA absent. CONIDIOPHORES macronematous, simple or irregularly and loosely branched, pale brown or brown, solitary or in fascicles. CONIDIOGENOUS CELLS integrated, terminal becoming intercalary, polytretic, sympodial, or sometimes monotretic, cicatrised. CONIDIA catenate or solitary, dry, typically ovoid or obclavate, often rostrate, pale or mid-olivaceous brown or brown, smooth or verrucose, with transverse and frequently also oblique or longitudinal septa (M. B. Ellis, *Demat. Hyphom.*: 464, 1971).

Many spp. of this genus cause economically important diseases, mostly as necrotic lesions on leaves, stems, flowers and fruit. Ellis (l.c.) described 27 spp. The 4 he treated first usually form very long, sometimes branched chains of rather small conidia; 3 of these are described here: *A. citri*, *A. alternata* and *A. brassicicola*. The other spp. form either short or no chains, and are arranged by Ellis roughly according to how abrupt the transition is from spore body to beak, and how short or long the beak is relative to the length of the body. He gave the substrata upon which the 27 spp. occur most frequently. Further spp. (17) were described by Ellis in *More Demat. Hyphom.* Elliott, Joly and Neergaard have keyed and grouped *Alternaria* spp. in different ways; Wiltshire gave a detailed and fully illustrated account of a re-examination of the foundation spp., and reasons for the preferred generic name *Alternaria*; Groves et al. described some spp. as seedborne fungi; Simmons gave an illustrated comparison of the types of *Alternaria*, *Stemphylium* (q.v.) and *Ulocladium*; Rao gave an account of the genus from India; and Slifkin described the fine structure of the conidial wall.

Some spp. not described here are mentioned briefly. Two plurivorous spp. are *A. tenuissima* (Kunze ex Pers.) Wiltshire (synonyms: *Helminthosporium tenuissimum* Kunze and *Macrosporium tenuissimum* Fr.) and *A. longissima* Deighton & MacGarvie (F. C. Deighton et al., *Mycol. Pap.* 113, 15 pp., 1968). The former is extremely common and recorded on a very wide range of plants, usually as a secondary invader. *A. helianthi* (Hansf.) Tubaki & Nishihara (synonym: *H. helianthi* Hansford) causes leaf spots on sunflower (*Helianthus annuus*); in Brazil the disease has recently been said to be serious (Aquino et al.). *A. zinniae* M. B. Ellis is cosmopolitan on Compositae (Ellis; McDonald et al.) and causes a seedling blight of sunflower; it is seedborne (Gambogi et al.). *A. poonensis* Raghunath causes a general blight on coriander (*Coriandrum sativum*), *A. umbellifericola* Raghunath has been described from dill (*Anethum graveolens*) and fennel (*Foeniculum vulgare*); and Raghunath also compared *Alternaria* spp. on Umbelliferae. In work aimed at the biological control of the serious weed, water hyacinth (*Eichhornia crassipes*), *A. eichhorniae* Nag Raj & Ponnappa was described (see Maity et al. for toxic metabolite). *A. melongenae* Rangaswami & Sambandam was described from eggplant (*Solanum melongena*) causing a leaf spot and large lesions on the fruit; it also rots fruit of red pepper (*Capsicum annuum*). *A. cichorii* Nattrass occurs on endive (*Cichorium endivia*) and chicory (*C. intybus*; Garibaldi et al.; Sarasola).

Aquino, M. L. N. De, et al. 1971. *Revta Agric. Piracicaba* **46**: 151 (*Alternaria helianthi* on sunflower; **51**, 2751).

Baker, K. F. et al. 1950. *Pl. Dis. Reptr* **34**: 403 (on ornamentals with *Stemphylium*; **31**: 239).

Elliott, J. A. 1917. *Am. J. Bot.* **4**: 439 (taxonomic characters).

Ellis, M. B. 1972. *Mycol. Pap.* 131, 25 pp. (dematiaceous hyphomycetes, *A. zinniae* sp. nov.).

Gambogi, P. et al. 1976. *Seed Sci. Technol.* **4**: 33 (*A. zinniae* seedborne; **55**, 5798).

Garibaldi, A. et al. 1971. *Riv. Ortoflorofruttic. ital.* **55**: 350 (*A. cichorii*, resistance; **52**, 3900).

Groves, J. W. et al. 1944. *Can. J. Res.* Sect. C. **22**: 217 (seedborne *Alternaria*, 7 spp.; **26**: 405).

Joly, P. 1964. Le genre *Alternaria*. *Recherches physiologiques, biologiques, et sytématiques* 250 pp., Paul Lechevalier (monograph, 78 ref.; **44**, 1436).

——. 1964. *Rev. Mycol.* **29**: 348; translation in *Pl. Dis. Reptr* **51**: 296 by M. M. Mulik et al. (key to spp.; **46**, 2190).

McDonald, W. C. et al. 1963. *Phytopathology* **53**: 93 (*A. zinniae* on sunflower; **42**, 554).

Maity, B. R. et al. 1977. *Phytopathol. Z.* **88**: 78 (toxin from *A. eichhorniae*; **56**, 3947).

Nag Raj, T. R. et al. 1970. *Trans. Br. mycol. Soc.* **55**: 123 (*A. eichhorniae* sp. nov.; **50**, 469).

Neergaard, P. 1945. *Danish species of* Alternaria *and* Stemphylium, 560 pp., Oxford Univ. Press (monograph; **25**: 579).

Raghunath, T. 1963. *Pl. Dis. Reptr* **47**: 259 (on coriander; **42**: 624).

——. 1963. *Mycopathol. Mycol. appl.* **21**: 315 (*A. poonensis* sp. nov.; **43**, 2374).

——. 1966. *Ibid* **30**: 209 (*A. umbellifericola* sp. nov., *Alternaria* on Umbelliferae; **46**, 1305).

Rangaswami, G. et al. 1960. *Phytopathology* **50**: 489 (*A. melongenae*, effect of substrate on spore size; **41**: 117).

——— ———. 1960. *Mycologia* **52**: 517 (*A. melongenae* sp. nov.; **41**: 116).

——— ———. 1961. *Indian J. agric. Sci.* **31**: 160 (*Alternaria* on Solanaceae).

Rao, V. G. 1969. *Nova Hedwigia* **17**: 219 (*Alternaria* from India; **49**, 2780).

——— ———. 1971. *Mycopathol. Mycol. appl.* **43**: 361 (as above; **50**, 3515).

Sarasola, M. A. R. De. 1970. *Revta Investnes agropec.* Ser. 5 **7**: 21 (*A. cichorii*; **52**, 544).

Simmons, E. G. 1967. *Mycologia* **59**: 67 (comparison of types of *Alternaria*, *Stemphylium* & *Ulocladium*; **46**, 2394).

Slifkin, M. K. 1971. *J. Elisha Mitchell scient. Soc.* **87**: 231 (fine structure of conidial wall).

Tubaki, K. et al. 1969. *Trans. Br. mycol. Soc.* **53**: 147 (*A. helianthi* comb. nov.; **51**, 2750).

Wiltshire, S. P. 1933. *Ibid* **18**: 135 (foundation spp. of *Alternaria* & *Macrosporium*; **13**: 326).

Alternaria alternata (Fe.) Keissler, *Beih. Bot. Zbl.* **29**: 434, 1912.

Torula alternata Fr., 1832

Alternaria tenuis C.G. Nees, 1916/17.

(the reasons why the epithet *alternata* should be used instead of the more commonly accepted one *tenuis* were clearly stated by E. G. Simmons in *Mycologia* **59**: 73, 1967)

COLONIES usually black or olivaceous black, sometimes grey. CONIDIOPHORES arising singly or in small groups, simple or branched, straight or flexuous, sometimes geniculate, pale to mid-olivaceous or golden brown, smooth, up to 50 μ long, 3–6 μ thick with 1 or several conidial scars. CONIDIA formed in long, often branched chains, obclavate, obpyriform, ovoid or ellipsoidal, often with a short conical or cylindrical beak, sometimes up to, but not more than, one third the length of the conidium, pale to mid-golden brown, smooth or verruculose, with up to 8 transverse and usually several longitudinal or oblique septa, overall length 20–63(37) μ, 9–18(13) μ thick in the broadest part; beak pale, 2–5 μ thick. An extremely common saprophyte found on many kinds of plants and other substrata including foodstuffs, soil and textiles; cosmopolitan (M. B. Ellis, *Demat. Hyphom.*: 465, 1971; a perfect state was described by Bilgrami; see also Misaghi et al. for conidial morphology).

This fungus is very frequently mentioned in plant pathological literature as causing leaf and fruit spotting or fruit rots. In nearly all cases it is behaving as a wound parasite (Saad et al. 1969b found that both direct and stomatal penetration can occur) or invading a host that is physiologically or pathologically weakened. The pathogen causing brown spot of tobacco has been referred to as *A. alternata* (*A. longipes* q.v.; Lucas). Grogan et al. described a distinct pathotype causing a stem canker of tomato. Examples of pathogenicity are given for some crops: *Brassica*, citrus, cotton, *Cyamopsis tetragonoloba* (cluster bean), common bean, eggplant, red pepper, sunflower and tomato. D. Singh et al. described the internal infection of sunflower seeds. There has been recent work on a phytotoxin formed by *A. alternata* (*Rev. Pl. Pathol.* **54**, 2074, 5601; **55**, 622, 4281; **56**, **70**, 2623, 3885; **57**, 59).

Abawi, G. S. et al. 1977. *Pl. Dis. Reptr* **61**: 901 (on pods of common bean; **57**, 3178).

Bilgrami, R. S. 1974. *Curr. Sci.* **43**: 492 (perfect state; **54**, 2694).

Bose, A. B. 1942. *J. Indian bot. Soc.* **21**: 179 (sunflower leaf spot; **22**: 111).

Chand, J. N. et al. 1968. *Pl. Dis. Reptr* **52**: 145 (cluster bean leaf spot; **47**, 3639).

Courter, J. W. et al. 1965. *Ibid* **49**: 886 (field susceptibility in red pepper; **45**, 1272).

Giha, O. H. 1973. *Trans. Br. mycol. Soc.* **61**: 265 (in Sudan, leaf spots; **53**, 1747).

Grogan, R. G. et al. 1975. *Phytopathology* **65**: 880 (on tomato; **55**, 895).

Hartman, G. C. 1964. *Am. J. Bot.* **51**: 209 (nuclear division).

——. 1966. *Mycologia* **58**: 694 (cytology).

Howell, P. J. 1964. *Proc. int. Seed Test. Assoc.* **29**: 155 (UV & sporulation; **43**, 2841).

Joly, P. 1967. *Fruits* **22**: 89 (citrus fruit black rot; **46**, 2232).

Kamel, M. et al. 1971. *United Arab Repub. J. Bot.* **14**: 245, 255 (on cotton; **52**, 415, 416).

Kapoor, J. N. et al. 1958. *Indian J. agric. Sci.* **28**: 109 (eggplant leaf spot & fruit rot; **39**: 145).

Alternaria brassicae

Lucas, G. B. 1971. *Tob. Sci.* **15**: 37 (*Alternaria longipes* as a synonym; **51**, 1192).
McColloch, L. P. et al. 1952. *Phytopathology* **42**: 425 (on tomato fruit; **32**: 284).
Miller, J. W. 1969. *Ibid* **59**: 767 (cotton leaf blight complex; **48**, 3498).
Misaghi, I. J. et al. 1978. *Ibid* **68**: 29 (effects of environment & culture on conidial morphology).
Pearson, R. C. et al. 1975. *Ibid* **65**: 1352 (on tomato fruit; **55**, 3292).
Pinckard, J. A. et al. 1973. *Cotton Grow. Rev.* **50**: 115 (on cotton boll; **53**, 958).
Saad, S. et al. 1969a. *Phytopathology* **59**: 1530 (common bean leaf spot; **49**, 1210).
——. 1969b. *Ibid* **59**: 1773 (as above, penetration; **49**, 1211).
——. 1970. *Ibid* **60**: 903 (growth & nutrition; **49**, 3525).
Siddiqi, M. R. 1963. *J. Indian bot. Soc.* **42**: 260 (*A. alternata* in India; **43**, 1843).
Singh, D. et al. 1977. *Seed Sci. Technol.* **5**: 579 (in sunflower seed; **57**, 2217).
Singh, R. S. et al. 1966. *J. Indian bot. Soc.* **45**: 277 (growth in culture; **46**, 2233).
——. 1966. *Pl. Dis. Reptr* **50**: 127 (citrus fruit rot; **45**, 1794).
Thomas, H. R. 1944. *Phytopathology* **34**: 341 (tomato fruit spot; **23**: 364).
Vaartnow, H. et al. 1972. *Pl. Dis. Reptr* **56**: 676 (*Brassica campestris* leaf spot; **52**, 896).
Verneau, R. 1954. *Ric. Ossni. Divulg. fitopat. Campan Mezzogiorno* **12**: 51 (on red pepper fruit; **35**: 574).

Alternaria brassicae (Berk.) Sacc., *Michelia* **2**: 129, see also p. 172, 1880.

Macrosporium brassicae Berk., 1836.

(synonymy in S. P. Wiltshire, *Mycol. Pap.* 20, 1947) COLONIES amphigenous, effused, rather pale olive and hairy. MYCELIUM immersed, hyaline, smooth 4–8 μ wide. CONIDIOPHORES in groups of 2–10 or more, emerging through stomata, usually simple, erect or ascending, straight or flexuous, frequently geniculate, more or less cylindrical but often slightly swollen at the base, septate, mid-pale greyish olive, smooth up to 170 μ long, 6–11 μ wide, bearing one to several distinct conidial scars. CONIDIA solitary or occasionally in chains of up to 4, acropleurogenous, arising through small pores in the conidiophore wall, straight or slightly curved, obclavate, rostrate with 16–19 transverse septa and 0–8 (usually 0–3) longitudinal or oblique septa, pale or very pale olive or greyish olive, smooth or, infrequently, very conspicuously warted, 75–350 × 20–40 μ wide (in the broadest part), the beak about 1/3–1/2 the length of the conidium and 5–9 μ wide (M. B. Ellis, *CMI Descr.* 162, 1968).

The taxonomic confusion between this sp. and *A. brassicicola* (q.v.) was disentangled by Wiltshire in 1947 (l.c.). The fungus forms a rather pale brown lesion on leaves and has been called grey leaf spot, in contrast to the darker spot of *A. brassicicola*. The leaf spot caused by the latter sp. tends to lack the clear zonation, and the profuse sporulation on green tissue, which characterises *A. brassicae*. The small brown spots become irregular, increasing to > 1 cm diam. coalescing and leading to death of the leaves. Linear lesions form on the petioles. When external conditions are dry the lesions may darken and become black. Infected siliquae shrink and split prematurely; seed which is attacked is damped off. Cauliflower curd can be infected, and on cabbage in storage rapid spread and profuse sporulation can be destructive. On turnip leaves the lesions have been described as pale brown, papery, *c.* 0.5 cm diam. with a yellow halo; and roots may show symptoms after storage. On cabbage cankers below the head lead to stump rot.

The disease is very widespread (Map 353) and the fungus is confined to the Cruciferae. Degenhardt et al. described the effects of infection on yield in rape seed. Penetration appears to be through the stomata. Opt. temp. for linear growth in vitro is *c.* 24°, the rate of growth being probably lower than for *A. brassicicola*. Opt. temp. for conidial germination and infection may be a little lower. Spread is airborne or via host debris; microsclerotia and chlamydospores form from conidia (Tsuneda et al.). Infection can lead to growth of the mycelium into the seed and seed treatment is probably the most important single measure in control. Seed infection (see Jørgenson) not only increases loss in young plant growth but also the yield where brassicas are grown for oil. Hot water treatment for seed (Nielsen) or a thiram soak is effective. Field spraying should be designed so that infection of flowers and fruit is reduced to a min. Losses in storage can be severe. The disease is favoured by cooler conditions than the one caused by *A. brassicicola*. Skoropad et al. and Tewari et al. investigated the role of epicuticular wax in resistance.

Changsri, W. et al. 1963. *Phytopathology* **53**: 643 (comparison of *Alternaria brassicae*, *A. brassicicola*, *A. raphani*; **43**: 280).

Chupp, C. 1935. *Ibid* **25**: 269 (on turnip; **14**: 486).
Crosier, W. et al. 1940. *Proc. Assoc. Off. Seed Anal. N. Am.* 1939: 116 (hot water & chemical treatment; **19**: 636).
Degenhardt, K. J. et al. 1974. *Can. J. Pl. Sci.* **54**: 795 (effect on yield of rape seed; **54**, 4225).
Fajardo, T. G. et al. 1934. *Philipp. J. Agric.* **5**: 143 (general; **14**: 140).
Jørgensen, J. 1976. *Acta Agric. scand.* **26**, 109 (in seed of *Brassica alba*; **55**, 5986).
Louvet, J. 1958. *C.R. Acad. Agric. Fr.* **44**: 694 (epidemiology & control; **38**, 233).
—— et al. 1964. *Annls Epiphyt.* **15**: 229 (climate & control; **44**, 1723).
McDonald, W. C. 1959. *Can. J. Pl. Sci.* **39**: 409 (general on *B. napus*; **39**: 334).
Nielsen, O. 1936. *Tidsskr. Planteavl.* **41**: 450 (seed treatment; **16**: 83).
Skoropad, W. P. et al. 1977. *Can. J. Pl. Sci.* **57**: 1001 (field evaluation of role of epicuticular wax in resistance; **57**, 1464).
Tewari, J. P. et al. 1976. *Ibid* **56**: 781 (relationship between epicuticular wax & disease; **56**, 3285).
Tsuneda, A. et al. 1977. *Can. J. Bot.* **55**: 1276 (formation of microsclerotia & chlamydospores from conidia; **57**, 325).
Van Schreven, D. A. 1953. *Tijdschr. PlZiekt.* **59**: 105 (general including *A. brassicicola*; **33**: 129).
Weimer, J. L. 1924. *J. agric. Res.* **29**: 421 (general on *B. oleracea*; **4**: 324).
——. 1926. *Ibid* **33**: 645 (general; **6**: 202).

Alternaria brassicicola (Schw.) Wiltshire, *Mycol. Pap.* **20**: 8, 1947 (q.v. for synonymy).
Helminthosporium brassicicola Schweinitz, 1832
Macrosporium cheiranthi Fr. var. *circinans* Berk. & Curt., 1875
Alternaria circinans (Berk. & Curt.) Bolle, 1924
A. oleracea Milbraith, 1922.

COLONIES amphigenous, effused dark olivaceous brown to dark blackish brown, velvety. MYCELIUM immersed, hyaline then later brown or olivaceous, smooth, 1.5–1.7 μ wide. CONIDIOPHORES single or in groups of 2–12 or more, emerging through stomata, usually simple erect or ascending, occasionally geniculate, more or less cylindrical but often with slightly swollen base, septate, pale to mid-olivaceous brown, smooth, up to 70 μ long, 5–8 μ wide. CONIDIA mostly in chains, up to 20 or more, sometimes branched, acropleurogenous, arising through pores in conidiophore wall, straight, nearly cylindrical, usually tapering slightly towards the apex or obclavate, the basal cell rounded, the beak usually almost nonexistent, apical cell almost rectangular or like a truncated cone, occasionally better developed but then always short and thick, with 1–11 transverse septa and up to 6 longitudinal septa, often slightly constricted at the septa, pale to dark olivaceous brown, smooth or becoming slightly warted, 18–130 × 8–30 μ in the broadest part, beak 1/6 length of conidium, 6–8 μ wide (M. B. Ellis, *CMI Descr.* 163, 1968).

The symptoms have been confused with *A. brassicae* (q.v.). The leaf spotting is usually darker (black leaf spot) than in *A. brassicae* and less clearly zonate. Sporulation is generally sparse on the still vigorous leaves of cabbage and cauliflower but can become profuse on yellow or detached ones. On leaves of these hosts numerous small dark spots may appear simultaneously, later increasing in size. They have a purplish cast, a darker centre and appear lighter in colour on paler green leaves. Brown discolouration of cauliflower curds, and linear lesions on stems and petioles occur. Seedlings show damping off and shrunken (wire stem effect) stems. The dying parts of the flower become colonised and therefore an inoculum source close to the developing siliqua is formed.

The disease is very widespread (Map 457) and the fungus is confined to the Cruciferae. Penetration via the cuticle and the stomata has been reported. Growth opt. are a little higher than for *A. brassicae*, 25–27° in vitro and 25–30° for cauliflower infection. Spread is similar to that of *A. brassicae* and the fungus remains viable in the seed for several years; see Kilpatrick; Neergaard; Petrie; Schimmer for the fungus on seed. An RH approaching saturation for 18 hours and a temp. of 21–27° for at least 3 successive days can lead to an epidemic. Transmission of conidia by slugs has been reported. This disease may possibly be generally more serious than that caused by *A. brassicae*; control measures are similar. Maude found that the best control of pod (and hence seed) infection was given by Bordeaux; thiram gave control in seed but adversely affected germination (see also Domsch; Jouan et al.). For host reactions to infection see Braverman; Kilpatrick.

Braverman, S. W. 1971. *Pl. Dis. Reptr* **55**: 454 (reaction of broccoli & cauliflower introductions; **51**, 808).
———. 1977. *Ibid* **61**: 360 (reaction of Brussels sprout introductions; **57**, 831).
Campbell, R. 1968. *J. gen. Microbiol.* **54**: 381 (fine structure of conidia; **48**, 1107).

Campbell, R. 1969. *Arch. Mikrobiol.* **69**: 60 (as above; **49**, 2349).
——. 1970. *New Phytol.* **69**: 287 (fine structure of dormant conidia, conidial germination, hyphae & conidiophores; **49**, 3138).
Domsch, K. H. 1957. *Z. PflKrankh.* **64**: 65 (general; **36**: 442).
Hasan, S. et al. 1966. *Pl. Dis. Reptr* **50**: 764 (spread by slugs; **46**, 783).
Hemmi, T. et al. 1953. *Bull. Univ. Naniwa* Ser. *B* **3**: 93 (in vitro effect of temp., Cu & spore germination; **35**: 500).
Jouan, B. et al. 1972. *Annls Phytopathol.* **4**: 133 (ecology & control with *Mycosphaerella brassicicola*; **51**, 4449).
Kilpatrick, R. A. 1976. *Phytopathology* **66**: 945 (on seeds of *Crambe abyssinica* & virulence; **56**, 1687).
Maude, R. B. 1977. *Grower* **88**: 288 (fungicide treatment of seed crop; *Hort. Abstr.* **48**, 397).
Milbraith, D. G. 1922. *Bot. Gaz.* **74**: 320 (symptoms & morphology; **2**: 301).
Neergaard, P. 1941. *Gartnerdende* 28, 4 pp. (light & longevity in seed; **25**: 378).
Pace, M. A. et al. 1974. *Trans. Br. mycol. Soc.* **63**: 193 (effect of saprophytes on infection; **54**, 597).
Petrie, G. A. 1974. *Can. Pl. Dis. Surv.* **54**: 31 (infection from seed; **54**, 3030).
Rangel, J. F. 1945. *Phytopathology* **35**: 1002 (general; **25**: 244).
Schimmer, F. C. 1953. *Pl. Pathol.* **2**: 16 (on cauliflower seed; **32**: 529).

Alternaria burnsii Uppal, Patel & Kamat, *Indian J. agric. Sci.* 8: 61, 1938.

Mycelium on PDA 1.7–5.8 μ diam., septate, hyaline when young becoming light olive green. CONIDIOPHORES solitary or fasciculate, erect, simple or sometimes branched; light olive, septate, straight or somewhat irregularly bent, occasionally with a single terminal scar, geniculate, 3–5 celled (occasionally 8 celled on host), 15–52 × 1.9–3.7 μ. CONIDIA borne singly on the host but in chains of 2–10 in culture; most obovate or rarely obclavate with rounded base, tapering to apex which may be drawn into a septate or non-septate beak, often as long as the spore; septate with transverse, longitudinal and oblique septa; 1–9 (mostly 3–5) transverse septa, constricted at the septa (except in the region of the beak) olive green, 14–55 × 7–25 μ (without beak), 23–119 × 7–30 μ (with beak; B. N. Uppal et al. l.c.; Siddiqi, *A. alternata* q.v.).

The fungus was described from India; it occurs in Gujarat, Maharashtra and Rajasthan. Losses of up to 70% have been reported. It causes a blight of cumin (*Cuminum cyminum*). Very small whitish areas appear on the younger leaves and blossoms particularly; these become purplish with age, coalesce and the infected parts are killed. The disease becomes serious at flowering. Infection requires at least RH 90% for *c.* 3 days at 23–28°. The conidia are mostly dispersed in the forenoon. *A. burnsii* survives in seed and crop debris. The infected seeds are shrivelled and light in weight (Gemawat et al. 1972). Zineb or captafol sprays give some control and the seeds can be treated with methoxyethyl mercury chloride.

Gemawat, P. D. et al. 1969. *Indian Phytopathol.* **22**: 49 (fungicides; **49**, 817).
———. 1972. *Indian J. Mycol. Pl. Pathol.* **2**: 65 (epidemiology; **52**, 3400).
Patel, R. M. et al. 1971. *Indian Phytopathol.* **24**: 16 (fungicides, seed transmission & carry over; **51**, 2742).
Solanki, J. S. et al. 1973. *Indian J. Mycol. Pl. Pathol.* **3**: 196 (fungicides; **54**, 2924).
Uppal, B. N. et al. 1938. *Indian J. agric. Sci.* **8**: 49 (general; **17**: 486).

Alternaria carthami Chowdhury, *J. Indian bot. Soc.* 23: 65, 1944.

CONIDIOPHORES erect, simple, straight or flexuous, sometimes geniculate, septate, brown or olivaceous brown, paler near apex, up to 90 μ long, 5–8 μ thick, sometimes swollen at the base. CONIDIA solitary or in very short chains, straight or curved, obclavate, rostrate, sometimes constricted at the septa, pale or mid brown, smooth; body of conidium without beak, usually 60–110 μ long, 15–26 μ thick, with 7–11 transverse septa and up to 7 longitudinal or oblique septa, beak 25–160 μ long, 4–6 μ thick at the base tapering to 2–3 μ, with up to 5 transverse septa (M. B. Ellis & P. Holliday, *CMI Descr.* 241, 1970).

The fungus causes a widespread leaf spot of safflower. The symptoms appear before flowering on all parts but especially on the leaves. The brown to dark brown spots expand to 1 cm diam., the centre is lighter in colour and becomes surrounded by dark rings where sporulation occurs. Shotholes may develop, and coalescence of the spots leads to large, irregularly outlined lesions and break up of the lamina. Spotting is less severe on the petioles and stems. Infected flower buds do not open and shrivel. The opt. temp. for linear growth in vitro is 25–30°. The fungus is seedborne and probably penetrates the tissues below the testa. Infection reduces seed quality and causes pre-emergence death, seedling death and diseased seedlings. Infected seed show, but not invariably, dark sunken lesions. A rapid build up of the disease can follow the planting of infected seed (Irwin). The fungus has caused severe losses in Australia and USA and is probably a factor in the selection of disease resistant safflower cvs.

Chowdhury, S. 1944. *J. Indian bot. Soc.* **23**: 59 (morphology, culture & inoculations; **24**: 71).
Irwin, J. A. G. 1976. *Aust. J. exp. Agric. Anim. Husb.* **16**: 921 (in seed; **56**, 3654).

Alternaria citri Ellis & Pierce in Pierce, *Bot. Gaz.* **33**: 234, 1902.

COLONIES effused, olivaceous to black, in culture grey, olivaceous brown or black, sometimes zonate. CONIDIOPHORES simple or branched straight or flexuous, septate, pale to mid brown or olivaceous brown, up to 3–5 μ thick, with a terminal scar and sometimes one or two lateral ones. CONIDIA solitary or in simple or branched chains of 2–7, straight or slightly curved, variously shaped but commonly obclavate or oval, often rostrate, pale to mid or sometimes dark brown, or olivaceous brown, smooth to verruculose with up to 8 transverse and numerous longitudinal or oblique septa, constricted at the septa, 8–60 (42) μ long including beak when present, 6–24 (17) μ thick in the broadest part, beaks mostly 8 μ or less long, 2.5–4 μ thick, colourless or rather pale brown. Isolates of *A. citri* often become sterile after subculturing (M. B. Ellis & P. Holliday, *CMI Descr.* 242, 1970).

This fungus causes a common rot of orange (black rot), lemon and tangerine; and a leaf spot of rough lemon and Emperor mandarin; it is probably found generally on citrus fruit. In orange a firm, slow internal decay begins near the stylar end; there is no external sign of this but infected fruit tend to colour prematurely and may split. As maturity approaches the fruit may drop (Wager); the condition is particularly prevalent in the Washington navel orange. Tree ripe fruit tends to have more rot, and the condition may be aggravated by weather and host nutrient balance. On lemon 2 forms of rot have been described. A soft, central one in mature fruit, beginning at the button and spreading to the axis; the rind becomes translucent before the rot appears on the surface and the fruit becomes soft when decay is well developed. The other rot begins at one end and external spread is almost as extensive as that inside the fruit. On leaves of rough lemon circular or irregular necrotic spots with concentric zoning are produced. Leaf spotting, stem blight and fruit drop in mandarin also occur. Some forms of the pathogen may be rather limited in distribution (Bartholomew; Coit; Knorr). The fruit rotting isolates appear to be weak pathogens. Conidia become deposited in the navel and stylar ends or in any openings including the stem end. Infection is usually dormant until maturity approaches. Spread of the decay occurs on the tree before or at fruit ripening, and during transport and storage. Growth in vitro is opt. at 25°. Orange black rot is probably of only occasional importance, affecting mostly the industrial processing end; it may also predispose fruit to green and blue mould (*Penicillium* spp.). Picking should be prompt or delayed to allow infected fruit to fall.

In South Africa and USA (Florida) differences in vivo between the fruit and stem isolates have been reported (Doidge; Ruehle). In lemons a serious storage rot can develop from conidial infections in and under the button (receptacle and calyx). At 15–16° tree ripe, yellow fruit show decay after 1–1.5 months in storage; silvers (yellow with some green colour) after 2–2.5 months, and greens after 3–4 months. The number of sound, green buttons and the rate of fruit colour change indicates the max. safe storage period. The greater the deterioration of the button the greater the number of infected fruit. In storage the first external symptoms seldom indicate a decay rate of more than 1% of fruit. Fresh picked silver and green lemons show resistance to the growth of *A. citri* in the albedo, whilst ripe ones do not; this is lost during storage. Decay is delayed by treatment with 2,4-dichlorophenoxyacetic acid before storage. But when the fungus enters the albedo growth is not checked by this treatment or by fruit vigour.

Brown spot of Emperor mandarin (Kiely; Pegg), long known in Australia, was found in 1959 to be due to a strain of *A. citri*. Inoculations of young leaves resulted in necrotic lesions after 3 days, mature leaves were resistant. On orchard fruit symptoms appeared 18 hours after infection and drop occurred after 3 days. As the fruit develops it becomes more resistant. Cvs differ in susceptibility. An Australian isolate from rough lemon leaf spots was not pathogenic to Emperor mandarin. Isolates from the latter host cause infection through the secretion of a toxin, this being followed by penetration of the host. No direct penetration of the host cuticle was found. Preparations from the culture filtrates of the mandarin isolates caused necrosis when sprayed on the leaves and when shoots were placed in them. The toxin was heat labile and probably host specific.

Bartholomew, E. T. 1923. *Am. J. Bot*. **10**: 117 (symptoms; **2**: 406).
——. et al. 1923. *Ibid* **10**: 67 (physiology, lemons; **2**: 406).
—— ——. 1926. *Bull. Calif. agric. Exp. Stn* 408, 38 pp. (general; **6**: 287).
Bliss, D. E. et al. 1944. *Mycologia* **35**: 469 (morphology & taxonomy; **24**: 55).
Chand, J. N. et al. 1967. *Jnl Res. Ludhiana* **4**: 217 (pathogenicity; **47**, 195).
Coit, J. et al. 1919. *Univ. Calif. J. agric. Sci*. **3**: 283 (general).
De Wolfe, T. A. et al. 1959. *Proc. Am. Soc. hort. Sci*. **74**: 367 (lemon storage rot).
Doidge, E. M. 1929. *Bull. S. Afr. Dep. Agric. Sci*. 69, 29 pp. (general; **8**: 774).
Harvey, E. M. 1946. *Tech. Bull. U.S. Dep. Agric*. 908, 32 pp. (storage; **25**: 393).
Kiely, T. B. 1964. *Agric. Gaz. N.S.W*. **75**(2): 854 (brown spot, mandarin; **43**, 2616).
Knorr, L. C. et al. 1957. *Bull. Fla agric. Exp. Stn* 587 (general).
Pegg, K. G. 1966. *Qd. J. Agric. Anim. Sci*. **23**: 15 (brown spot, mandarin; **46**, 999).
Rattan, B. K. et al. 1968. *Jnl Res. Ludhiana* **5**: 513 (in vitro; **48**, 3491).
Ruehle, G. D. 1937. *Phytopathology* **27**: 863 (lemon leaf spot; **17**: 25).
Wager, V. A. 1939. *Bull. S. Afr. Dep. Agric. Sci*. 193, 18 pp. (fruit drop; **19**: 401).

Alternaria crassa (Sacc.) Rands, *Phytopathology* 7: 327, 1917.
Cercospora crassa Sacc., 1877
C. daturae Peck, 1882
Macrosporium daturae Fautrey, 1894
Alternaria daturae (Fautrey) Bubák & Ransjevič, 1909.

CONIDIOPHORES arising singly or in small groups, erect or ascending, straight or flexuous, sometimes geniculate, septate, pale or mid-pale brown, up to 90 μ long, 7–10 μ thick with 1 to several scars. CONIDIA usually solitary, occasionally in very short chains, obclavate, rostrate, beak generally greatly exceeding the length of the body of the spore, pale brown, smooth overall, 120–440 (250) μ long, 15–40 (22) μ thick in the broadest part; body up to 140 μ long with 7–10 transverse and usually several longitudinal septa; beak pale brown, septate, not branched, 4–8 μ thick at base, tapering to 2–2.5 μ. Cultures on PDA non-chromogenic in contrast to *A. solani* which usually stains the agar deep pink or yellow (M. B. Ellis & P. Holliday, *CMI Descr*. 243, 1970).

A. crassa causes a widespread leaf spot of *Datura stramonium* (Jimson weed, thorn apple) and other *Datura* spp. Irregular, straw coloured, zonate lesions occur on the leaves. These symptoms first appear on the lower foliage spreading upwards, and later in the growing season the seed pods develop dark sunken lesions. Heavily infected leaves are often shed. The pathogen was originally confused with *A. solani* (q.v.) but it does not infect Irish potato. More than 40% of the leaf area can be destroyed and complete defoliation can result. A method for obtaining sporulation on killed host tissue was described from Yugoslavia in 1966. In Israel *A. crassa* was found to be seedborne on *D. metel*. Infected seeds were grey, uninfected ones brown. Infection reduced seed quality, caused pre-emergence death and seedling blight. The mycelium occurs beneath the seed coat. Good control was obtained with a seed soak in thiram (24 hours at 30°).

Dimitrijević, B. 1966. *Zašt. Bilja* **17**: 315 (sporulation on host; **46**, 3014).
Halfon-Meiri, A. 1973. *Pl. Dis. Reptr* **57**: 960 (seedborne in *Datura metel*; **53**, 1839).
Grzybowska, T. et al. 1976. *Herba Pol*. **22**: 172 (on *D. inoxia* & control; **57**, 3563).

Alternaria cucumerina (Ellis & Everh.) Elliot, *Am. J. Bot*. 4: 472, 1917.
Macrosporium cucumerinum Ellis & Everh., 1895.

COLONIES amphigenous. CONIDIOPHORES arising singly or in small groups, erect, straight or flexuous, sometimes geniculate, cylindrical, septate, pale to mid brown, up to 110 μ long, 6–10 μ thick, usually with several well-developed conidial scars. CONIDIA solitary or occasionally in chains of 2, obclavate, rostrate, the beak longer, often much longer, than the body of the spore, pale to mid-golden brown, smooth to verruculose, overall 130–220 (180) μ long,

15–24 (10) μ thick in the broadest part, body with 6–9 transverse and several (sometimes many) longitudinal and oblique septa, beak pale brown, septate, not branched, 4–5 μ thick at the base rapidly narrowing to 1–2.5 μ (M. B. Ellis & P. Holliday, *CMI Descr.* 244, 1970).

This leaf spot of Cucurbitaceae occurs mostly on watermelon (*Citrullus lanatus*) and melon (*Cucumis melo*). The distribution is wide in both tropical and temperate countries. About 1958 it caused severe losses in USA (Florida) in melon. The spots begin as yellow brown flecks, 0.5 mm diam., usually on the upper surface of the crown leaves. They have a light green halo, with concentric ringing more common on the upper surface. The lesions coalesce and death of the leaves can be followed by spots (several cm diam.) forming on ripe fruit. These become covered with the dark, olive green, conidial masses; they may be sunken and also show concentric zonation. Fruit infection in the field may be aggravated by exposure to sunlight (Le Clerg). Stem and petiole infection does not apparently occur. The pathogen can also cause decay in transit and storage, especially of melons and squash (Guba). Symptoms develop more slowly on cucumber. Penetration may be through the cuticle and severe disease can probably occur over a wide temp. range, 20–32°. Linear growth in vitro is opt. at 27–30° with most sporulation at 27°. Ibrahim et al. found that mycelium in leaf tissue survived for a long period and that the fungus was not internally seedborne. But seed can become contaminated by conidia. Spraying with fungicides may be effective and resistance occurs in melon.

A. cyamopsidis Rangaswami & Venkata Rao was described from cluster bean (*Cyamopsis tetragonoloba*). The fungus was transferred to *A. cucumerina* as var. *cyamopsidis* (Rangaswami & Rao) Simmons (see Orellana et al.). Mathur reported on fungicide control and Jagtap on resistance.

Brisley, H. R. 1923. *Phytopathology* **13**: 199 (general; **3**: 49).

Guba, E. F. 1950. *Bull. Mass. agric. Exp. Stn* 457 (storage).

Ibrahim, A. N. et al. 1975. *Acta phytopathol. Acad. Sci. hung.* **10**: 309 (survival; **55**, 5997).

Jackson, C. R. 1959. *Phytopathology* **49**: 731 (general; **39**: 368).

—— et al. 1959. *Mycologia* **51**: 401 (morphology & taxonomy; **39**: 529).

Jagtap, R. P. 1975. *Res. J. Mahatma Phule agric. Univ.* **6**: 164 (resistance in cluster bean; **57**, 365).

Le Clerg, E. L. 1931. *Phytopathology* **21**: 97 (sun scorch; **10**: 431).

Mathur, R. L. et al. 1972. *Indian J. Mycol. Pl. Pathol.* **2**: 80 (fungicides on cluster bean; **52**, 3494).

Middleton, J. T. et al. 1946. *Pl. Dis. Reptr* **30**: 374 (on cantaloupe in USA, California).

Orellana, R. G. et al. 1966. *Mycopathol. Mycol. appl.* **29**: 129 (*Alternaria cucumerina* var. *cyamopsidis*; **45**, 3008).

Rangaswami, G. et al. 1957. *Indian Phytopathol.* **10**: 18 (*A. cyamopsidis* sp. nov.; **38**: 643).

Alternaria dauci (Kühn) Groves & Skolko, *Can. J. Res.* Sect. *C* **22**: 222, 1944.

Sporidesmium exitiosum Kühn var. *dauci* Kühn, 1855

Macrosporium dauci Rostrup, 1888

M. carotae Ellis & Langlois, 1890

Alternaria brassicae (Berk.) var. *dauci* Lindau, 1908

A. brassicae (Berk.) var. *dauci* Bolle, 1924

A. carotae (Ell. & Langl.) Stevenson & Wellman, 1944.

CONIDIOPHORES arising singly or in small groups, straight or flexuous sometimes geniculate, septate, pale or mid-pale olivaceous brown or brown, up to 80 μ long, 6–10 μ thick. CONIDIA usually solitary, occasionally in chains of 2, straight or curved, obclavate, rostrate with beak up to 3 times the length of the body of the spore, at first pale olivaceous brown, often becoming brown with age, smooth, overall 100–450 μ long, 16–25 (20) μ thick in the broadest part, with 7–11 transverse and 1 to several longitudinal or oblique septa, beaks often once branched, flexuous, hyaline or pale, 5–7 μ thick at the base tapering to 1–3 μ (M. B. Ellis, *Demat. Hyphom.*: 490, 1971; additional synonymy from Groves et al., l.c.).

A. dauci differs from the other *Alternaria* on carrot and other Umbelliferae (*A. radicina* q.v.) in its long beaked and larger conidia; the smaller conidia of *A. radicina* have virtually no beak. *A. dauci* causes a widespread (Map 352) leaf blight of carrot. Netzer et al. (1969) reported the occurrence of the fungus on 3 common wild hosts (*Daucus maximus*, *Caucalis tenella* and *Ridolfia segetum*) in Israel. Leaf blight has sometimes been studied with carrot black rot caused by *A. radicina*, for example, Maude (*A. radicina* q.v.) who described the aetiology of the 2 diseases and particularly their seedborne phases. Both fungi cause damping off and cannot be distinguished by symptoms alone at this stage. Penetration by *A. dauci*, which has not apparently been described adequately, occurs more commonly on the older leaves. The indefinite necrotic lesions appear at the margins and the entire leaf becomes shrivelled and killed. Unlike *A. radicina* the tap root is not generally attacked. The leaves may be destroyed just by the lesions on the petioles. Hawkins et al. briefly described an outbreak of the disease in UK. Affected plants occurred first in patches but later the disease became general and uniformly severe throughout the crop. Up to 7–8 outer leaves were dead and lying on the ground. The inner, more erect leaves looked ragged owing to the presence of irregular dark brown areas on the edges of the leaflets, with occasional dark flecks elsewhere on the leaves and petioles. In the centre there was a tuft of healthy green leaves giving the crop a characteristic appearance.

Spore germination has been reported to be best at 22–24° and 28°. Hooker found infection to be retarded at 16°; it increased with temp. to 24°. In Israel the pathogen remained viable (could sporulate) in carrot petioles after 3 months when these were kept dry. But in alternating wet and dry conditions viability was rapidly lost during this period. *A. dauci* colonising moribund foliage was viable for a longer period when the foliage remained on the soil surface than when it was buried in soil at 10 and 20 cm depths. In UK the pathogen did not apparently overwinter or could not attack a ware crop sown in April (Maude, *A. radicina* q.v.). In USA Strandberg (1977) found that RH 96–100% or free water was required for conidial formation (conidia were formed at 8–28°). The number of hours of foliar wetness during the night preceding conidial release was positively correlated with the observed number of conidia. Langenberg et al. found a diurnal periodicity for the conidia whose numbers increased after 0800 with a peak at *c.* 1300 hours.

Leach (and see Zimmer et al.) called the fungus a diurnal sporulator with 2 phases of photosporogenesis. An induction phase led to the formation of conidiophores and a terminal one when conidia were formed. The first was stimulated by near UV and operated at relatively high temps. The second was strongly inhibited by near UV and blue light, and operated at lower temps. Relatively high temps caused the formation of conidiophores in darkness, although only when the temp. was lowered did the conidia form, the amount of sporulation caused by high temp. induction was less than that caused by exposure to near UV.

Outbreaks of leaf blight seem most likely to arise from infected seed and treatment of seed is therefore very important; that with thiram is given under *A. radicina*. Saponaro reported treatment of seed with dichlone or soil with quintozene to be effective. Hewett described methods of testing for seed infection. For foliage spraying the following have given control; captafol, fentin acetate, fentin hydroxide, mancozeb, maneb, zineb and ziram (Carvalho et al.; Netzer et al. 1966; Roy; Saponaro); for resistance see Strandberg et al. 1972).

Carvalho, Y. De. et al. 1975. *Revta Ceres* **22**: 229 (fungicides; **56**, 2267).
Doran, W. L. et al. 1928. *Bull. Mass. agric. Exp. Stn* **245**: 271 (general; **8**: 352).
Hawkins, J. H. et al. 1959. *Pl. Pathol.* **8**: 76 (outbreak in UK; **39**: 141).
Hewett, P. D. 1964. *Proc. int. Seed Test. Assoc.* **29**: 463 (seed tests for infection; **44**, 1331).
Hooker, W. J. 1944. *Phytopathology* **34**: 606 (with *Cercospora carotae*, effect of temp.; **24**: 45).
Kotthoff, P. 1957. *Gesunde Pfl.* **8**: 106 (outbreaks in Germany; **36**: 370).
Langenberg, W. J. et al. 1977. *Phytopathology* **67**: 879 (weather & conidial periodicity; **57**, 837).
Leach, C. M. et al. 1966. *Photochem. Photobiol.* **5**: 621 (action spectra for light induced sporulation with *Pleospora herbarum*).
——. 1967. *Can. J. Bot.* **45**: 1999 (interaction of near UV & temp. on sporulation with *P. herbarum* inter alia; **47**, 1017).
Netzer, D. et al. 1966. *Pl. Dis. Reptr* **50**: 594 (fungicides with *Erysiphe heraclei*; **46**, 493.
—— ——. 1969. *Ann. appl. Biol.* **63**: 289 (persistence & transmission in Israel; **48**, 2677).
—— ——. 1970. *Can. J. Bot.* **48**: 831 (apparent heterocaryosis; **49**, 3539).
Roy, A. K. 1969. *Indian Phytopathol.* **22**: 105 (fungicides; **49**, 616).
Saponaro, A. 1963. *Boll. Staz. Patol. veg. Roma* Ser. 3 **21**: 163 (fungicide seed & foliage control with *C. carotae*; **44**, 306).
Strandberg, J. O. et al. 1972. *HortScience* **7**: 345 (sources of resistance; *Pl. Breed. Abstr.* **43**, 4658).
——. 1977. *Phytopathology* **67**: 1262 (conidial formation & dispersal; **57**, 4188).
Witsch, H. V. et al. 1955. *Arch. Mikrobiol.* **22**: 307 (effect of light on mycelium & sporulation; **35**: 704).
Zimmer, R. C. et al. 1969. *Phytopathology* **59**: 743 (interaction of light & temp., & effects on sporulation; **48**, 3221).

Alternaria longipes (Ellis & Everh.) Mason, *Mycol. Pap.* 2: 19, 1928.

Macrosporium longipes Ellis & Everh., 1892.

COLONIES amphigenous. CONIDIOPHORES arising singly or in groups, erect or ascending, simple or loosely branched, straight or flexuous, cylindrical, septate, rather pale olivaceous brown, up to 80 μ long, 3–5 μ thick, with 1 to several scars. CONIDIA sometimes solitary but usually in chains, obclavate, rostrate, pale to mid brown, smooth or verruculose, overall length 35–110 (69) μ, body of conidium 11–21 (14) μ thick in the broadest part, tapering gradually into the pale brown beak which is usually 1/3–1/2 the total length, 2–5 μ thick and often slightly swollen at the tip; there are 3–7 (usually 5–6) transverse septa and 1 to several longitudinal or oblique septa. (M. B. Ellis & P. Holliday, *CMI Descr.* 245, 1970.)

Brown spot of tobacco, the only host of economic importance, is an important disease in many tobacco areas and is widespread (Map 63). The pathogen is sometimes referred, more recently, to *A. alternata* (q.v.), particularly by workers in USA; see Lucas (*Diseases of tobacco*) for a discussion. Ellis (*Demat. Hyphom.*) gave the av. length of conidia for *A. alternata* as 37 μ and for *A. longipes* as 69 μ. In consequence the literature is confused. The symptoms first occur on the lower leaves as small watersoaked spots, becoming brown, orbicular, zonate, up to 3 cm diam., with the centre becoming lighter in colour. The lesion is surrounded by a yellow halo which is absent from spots formed on seedlings. Severe spotting leads to general leaf necrosis and death, and tends to increase during curing. Symptoms are much less conspicuous on young leaves. Necrotic lesions (halo absent) also occur on petioles, seed capsules and stems, in the last forming dark brown elongated depressions, running longitudinally.

The disease seems to be typically one of mature tobacco; the reasons for its absence (or presence as atypical symptoms) on seedlings are not clear (see Lloyd). The chlorotic halo is presumed to be due to a fungal toxin (see *Rev. Pl. Pathol.* **50**, 1968; **51**, 2844, 2855; **53**, 4111). Penetration is mainly through the cuticle (Stavely et al. 1971a). Infection builds up in wet, warm weather as the crop matures and becomes severe at 26–31°. Little or no infection occurs below 19°. (Stavely et al. 1970, indicated somewhat lower temps., infection being more severe at 20° compared with 24° and 16°. The opt. temp. for radial growth was 27°.) Norse (1973) gave an opt. for conidial germination and germ tube growth of 22.5°. The number of lesions/plant increased as wet periods did so from 8 to 96 hours (Stavely et al. 1975). The airborne conidia are more abundant on dry sunny days but spread through splash droplets probably occurs. Disease incidence is higher on plants topped low because this leads to heavier and coarser leaves. It is also more severe on land cropped for a second successive year and possibly on heavier soils. Unbalanced fertilising with major elements, giving relatively high nitrogen levels, tends to increase incidence; phosphorus and potassium deficient leaves are more susceptible. The pathogen is carried over in host debris and survival in dry soil in the absence of such debris is only a few weeks.

Lucas (l.c.) stated that there is no adequate control. Seed transmission in *A. longipes* seems unimportant. Control is based mainly on cultural measures: prompt harvesting, balanced fertiliser, deep ploughing after harvest, no successive cropping on the same land and fumigation of plant beds. The role of other hosts is not known and may be unimportant. Fungicides may only be partially successful. Some *Nicotiana* spp. show some resistance (e.g. Stavely et al. 1971b); the yellow halo tends to be absent in these and cvs exhibit differential susceptibility. One selection from the cigar var. Quinn Diaz (Beinhart 1000–1) was highly resistant in the field in USA (S. Carolina). Resistance is governed by a single factor, possibly of intermediate expression. (More recent ref. may call the pathogen *A. alternata*.)

Chaplin, J. F. et al. 1963. *Tob. Sci.* **7**: 59 (resistance).

Cole, J. S. et al. 1961. *Rhod. agric. J.* **58**: 354 (plant nutrients; **41**; 546).

De Fluiter, H. J. 1938 & 1939. *Meded. besoek. Proefstn* **59**: 1; **65**: 1. (general in Indonesia; **17**: 490; **19**: 497).

Hartill, W. F. T. et al. 1966. *Rhod. agric. J.* **63**: 132 (chemical control; **46**, 1327).

——. 1968. *Wld Crops* **20**: 54 (review; **48**, 922).

Hopkins, J. C. F. 1946. *Rhod. agric. J.* **43**: 114 (general; **25**: 529).

Lloyd, H. L. 1971. *Rhod. Sci. News* **5**: 6 (disease & host leaf age; **51**, 1904).

Maine, C. E. et al. 1971. *Tobacco* **172** (4): 69, 73 (damage & flue-cured tobacco quality; **51**, 669).

Norse, D. 1971. *Ann. appl. Biol.* **69**: 105 (lesion & epidemic development; **51**, 1903).

——. 1973. *Ibid* **74**: 297 (factors affecting conidial germination & penetration; **53**, 675).

Okura, M. et al. 1967. *Bull. Utsunomiya Tob. Exp. Stn* **5**: 1, 43 (ecology, overwintering, infection & control; **48**, 1308).
Riley, E. A. 1949. *Mem. Dep. Agric. S. Rhod.* 3, 32 pp. (general; **29**: 280).
Sobers, E. K. 1969. *Phytopathology* **59**: 202 (characteristics of single spore isolates; **48**, 1966).
Spurr, H. W. et al. 1972. *Ibid* **62**: 916 (tobacco leaf microflora, disease & fungicides; **52**, 1261).
—— ——. 1974. *Ibid* **64**: 738 (brown pigment formation in infected leaves; **54**, 223).
——. 1977. *Ibid* **67**: 128 (protective applications of conidia of non-pathogenic *Alternaria* spp.; **56**, 3712).
Stavely, J. R. et al. 1970. *Ibid* **60**: 1591 (effects of temp. & other factors on initiation of disease; **50**, 1969).
——. 1971a & b. *Ibid* **61**: 73, 541 (leaf age & infection; reaction of *Nicotiana* spp.; **50**, 3165; **51**, 670).
—— ——. 1975. *Ibid* **65**: 897 (post-inoculation leaf wetness & infection; **55**, 1459).
Tisdale, W. B. et al. 1931. *Ibid* **21**: 641 (general; **10**: 763).
Von Ramm, C. 1952. *Phytopathol. Z.* **45**: 391 (histology; **42**: 405).
—— et al. 1963. *Phytopathology* **53**: 450 (general; **42**: 633).
—— ——. 1963. *Pl. Dis. Reptr* **47**: 369 (resistance).
Wheeler, B. E. J. 1958. *Misc. Publs Commonw. mycol. Inst.* 15, 32 pp. (general in Malawi; **37**: 598).

Alternaria macrospora Zimm., *Ber. Land-u. Forstw. Dt.-Ostafr.* 2: 24, 1904.

Sporidesmium longipedicellatum Reichert, 1921
Alternaria longipedicellata Snowden, 1927.

COLONIES when on leaves amphigenous. CONIDIOPHORES arising singly or in groups, erect, simple, straight or flexuous, almost cylindrical or tapering slightly towards the apex, septate, pale to mid brown, smooth with 1 to several conidial scars, up to 80 μ long, 4–9 μ thick. CONIDIA solitary or sometimes in chains of 2, straight or curved, obclavate or with the body of the conidium ellipsoidal, tapering rather abruptly to a very narrow beak, which is equal in length to (or up to twice as long as) the body, mid to mid-dark reddish brown, usually minutely verruculose with 4–9 (normally 6–8) transverse and several longitudinal or oblique septa, often slightly constricted at the septa, overall length 90–180 (134) μ, 15–22 (17.7) μ thick in the broadest part, beak simple, pale 1–1.5 μ thick along most of its length (M. B. Ellis & P. Holliday, *CMI Descr.* 246, 1970).

A. macrospora causes a widespread leaf spot of cotton (*Gossypium* spp.) and possibly of other hosts. A necrotic spot, with a purplish halo, expands to about 1 cm diam., the centre becoming grey and cracked and the zonation is more clearly defined on the upper surface. Defoliation can be severe, especially where the peduncle becomes infected. Stem lesions begin as a small sunken spot which develops into a canker, the tissue splitting and cracking to cause a break. The glandular areas on the receptacle are also attacked and this can prevent the boll developing. Flowers and bolls may be shed, the latter becoming mummified and the fibre is infected. Penetration is probably through the cuticle. Infection of the glandular region of the receptacle possibly occurs through waterborne conidia passing through channels formed within the ridges at the junction of the 3 large bracts of the epicalyx. Leaf symptoms may only appear as the early bolls are approaching maturity. Linear growth in vitro is opt. at *c.* 28°. A high boll to leaf ratio and potassium deficiency predisposes the plant to infection. The pathogen was recovered from host debris after overwintering on the soil surface but not from such debris 12 cm below the surface. No recent control measures appear to have been described.

Biraghi, A. 1937. *Boll. Staz. Patol. veg. Roma* **17**: 475 (morphology; **17**: 674).
——. 1942. *Ibid* **22**: 181 (morphology; **26**: 14).
Ciccarone, A. 1948. *Riv. Agric. subtrop. trop.* **42**: 154 (symptoms & inoculation; **28**: 65).
Jones, G. H. 1928. *Ann. Bot.* **42**: 935 (general; **8**: 171).
Ling, L. et al. 1941. *Phytopathology* **31**: 664 (general; **20**: 573).
McDonald, D. et al. 1948. *Emp. Cotton Grs Corp. Progr. Rep.* 1946–47 (symptoms, development & effect of K; **27**: 472).
Nisikado, Y. et al. 1940. *Ann. phytopathol. Soc. Japan* **10**: 214 (two *Alternaria* spp. attacking cotton fibres; **20**: 461).
——. 1943. *Ber. Ōhara Inst. Landw. Forsch.* **9**: 238 (as above & control; **30**: 37).

Alternaria padwickii (Ganguly) M. B. Ellis, *Dematiaceous hyphomycetes*: 495, 1971.

Trichoconis padwickii Ganguly, 1947.

COLONIES effuse, thin. MYCELIUM partly superficial, partly immersed. SCLEROTIA spherical or subspherical, black, with reticulate walls, mostly 50–200 μ diam. CONIDIOPHORES up to 180 × 3–4 μ, often swollen to 5–6 μ at the apex, smooth except at the apex, which is often minutely echinulate. CONIDIA straight or curved, fusiform to obclavate and rostrate, the beak at least half and often more than half the length of the conidium, at first hyaline, later straw coloured to golden brown, smooth or minutely echinulate, often echinulate only near the scar, 95–170 (130) μ long, 11–20 (15.7) μ thick in the broadest part, 1.5–5 (2.7) μ thick in the centre of the beak, with 3–5 (most commonly 4) transverse septa in the body of the conidium and often 1 or more septa in the beak, sometimes constricted at the septa. The fungus grows well on PDA forming greyish colonies often deep pink or purple in reverse, sporulating freely, especially when exposed to near UV. Small, black sclerotia with very distinctly reticulate walls are often formed in old cultures. (M. B. Ellis & P. Holliday *CMI Descr.* 345, 1972.)

Stackburn, seedling blight and leaf spot of rice is widespread in S.E. Asia and parts of Oceania but not reported from Australia; it also occurs in Argentina, Costa Rica, Egypt, Ghana, Japan, Malagasy Republic, Nigeria and Surinam (Map 314). Necrotic spotting on the roots and coleoptile leads to death of seedlings. Leaf spots are circular or oval, up to 1 cm diam. with a dark margin; the centre becomes pale and bears the black sclerotia. Spotting occurs on the glumes, the kernels are invaded and become discoloured and shrivelled. Sclerotia are formed in all infected areas. Little is known of the life history of *A. padwickii*. In India a diurnal periodicity with a forenoon peak was found for the conidia. In vitro max. growth occurs at 28°. The pathogen probably survives between crops on the straw but spread is mostly through seed in which up to 75% infection has been found. Infected seed shows reduced germination (Mathur et al.). Provided healthy seed is used stackburn is not generally serious. Complete control of seed infection is given by mancozeb and ceresan (Dharam Vir et al.; see Padwick, *Manual of rice diseases* for a full account).

Dharam Vir et al. 1971. *Indian Phytopathol.* **24**: 343 (control in seed; **51**, 4000).

Ganguly, D. 1947. *J. Indian bot. Soc.* **26**: 233 (general; **27**: 447).

Mathur, S. B. et al. 1970. *Proc. 1st int. Symp. Pl. Pathol. 1966–67* IARI New Delhi: 69 (tests for seed infection).

——. 1972. *Proc. int. Seed Test. Assoc.* **37**: 803 (seedborne infection, distribution & damage; **53**, 534).

Sreeramulu, T. et al. 1966. *Indian Phytopathol.* **19**: 215 (diurnal periodicity in conidia; **46**, 3096).

Suryanarayana, D. et al. 1963. *Ibid* **16**: 232 (seedborne infection, hot water treatment; **43**: 2274).

Alternaria passiflorae Simmonds, *Proc. R. Soc. Qd* **49**: 151, 1938.

CONIDIOPHORES arising singly or in groups, simple or rarely branched, straight or flexuous, almost cylindrical or tapering towards the apex, septate, pale to mid-pale brown, smooth, with several conidial scars, up to 120 μ long, 6–10 μ thick. CONIDIA solitary on host, often in chains of up to 5 in culture, straight or slightly curved, obclavate or with the body of the conidium ellipsoidal tapering to the beak which is usually about the same length or longer than the body, pale to mid brown, smooth or occasionally minutely verruculose, with 5–12 transverse and a few longitudinal or obtuse septa, constricted at the septa, overall length 100–250 μ, 14–29 μ thick in the broadest part; beak simple or branched sometimes flexuous near base, pale, 1.5–4 μ thick (M. B. Ellis & P. Holliday, *CMI Descr.* 247, 1970).

Brown spot of passion fruit was first recorded in Australia in 1912 and later investigated in Queensland. It has been reported to be serious in Hawaii and New Zealand. The distribution is (Map 479): Australia, Canada (British Columbia), Fiji, Hawaii, Kenya, Malawi, New Zealand, Papua New Guinea, Tanzania, Uganda and Zambia. Besides *Passiflora edulis* the fungus occurs on other *Passiflora* spp. including *P. quadrangularis* and *P. foetida*. Infection is found on all the aerial parts of the vine. On the leaves a small necrotic spot enlarges to 1 cm diam., the lesion becomes lighter coloured in the centre and may be rounded or angular. Sporulation takes place mostly on the lower surface and continues after the leaf has fallen. Complete defoliation can occur. Spots appear on the fruit after it is half grown; the necrotic lesions are up to 2 cm diam. and have a dark green margin. They become wrinkled and the fruit shrivels up in a firm rot. Lesions 2–4 cm long, and usually associated with the leaf axils, form on the climbing stems which may be girdled and killed. Complete dieback of the plant can then result.

Infection becomes more severe as the summer develops, if it is sufficiently wet, and in Australia the disease can become severe in August to December. Conidial germination appears to decline slowly, being still 30% after 16 months. Linear growth in vitro is best at 24–28° and most spore germination occurs within similar limits. Carry over from season to season takes place either on the vine or the ground. Aragaki (*A. tomato* q.v.) described some characteristics of fungal isolates from passion fruit but referred them to *A. tomato*. Plantings should be well spaced, leaders trained systematically and laterals pruned at least once a year. The fungicides which have been recommended include copper, maneb, mancozeb and zineb.

Anon. 1952. *Agric. Gaz. N.S.W.* **63**: 373 (control; **32**: 668).
——. 1960. *Orchard N.Z.* **33**: 233 (fungicides; **40**: 235).
Brien, R. M. 1940. *N.Z. Jl Sci. Tech.* Sect. A **21**: 275 (general; **19**: 420).
Emechebe, A. M. et al. 1975. *Acta Hortic.* **49**: 281 (fungicides in Uganda; **55**, 3676).
Rosenberg, G. M. 1962. *Rep. Hawaii agric. Stn to June 1962* (hosts; **43**, 1519).
Simmonds, J. H. 1930. *Bull. Qd Dep. Agric. Ent. & Pl. Pathol.* 6, 15 pp. (general; **10**: 394).
Smith, W. P. C. 1939. *J. Dep. Agric. W. Aust.* **16**: 445 (symptoms, control; 19: 294).

Alternaria porri (Ellis) Cif., *J. Dep. Agric. P. Rico* **14**: 40, 1930.
Macrosporium porri Ellis, 1879.

CONIDIOPHORES arising singly or in groups, straight or flexuous, sometimes geniculate, septate, pale to mid brown, up to 120 μ long, 5–10 μ thick, with 1 or several well-defined conidial scars. CONIDIA usually solitary, straight or curved, obclavate or with the body of the conidium ellipsoidal, tapering to the beak which is commonly about the same length as the body but may be shorter or longer, pale to mid-golden brown, smooth or minutely verrucose, overall length usually 100–300 μ, 15–20 μ thick in the broadest part, with 8–12 transverse and 0 to several longitudinal or oblique septa, beak flexuous, pale, 2–4 μ thick, tapering. (M. B. Ellis & P. Holliday, *CMI Descr.* 248, 1970.)

Purple blotch of *Allium* spp. (including chive, garlic, leek, onion and shallot) has caused severe losses in onions in the southern USA, Puerto Rico, India (Punjab) and Kenya. The disease is widespread (Map 350). The first symptoms are small, white, leaf lesions which (under moist conditions) develop into elliptical, purplish areas with a yellow border, becoming several cm long. Sporulation on these lesions results in the formation of dark and light concentric zones. If conditions, after the initial appearance of symptoms, become dry (RH <70%) then the white fleck does not develop into the purple blotch. After 3–4 weeks the leaves collapse and infection can spread in the bulb causing a deep yellow to reddish watery rot. The scales become desiccated and dark; bulbs may be small or fail to develop. Conidial germination takes place in 45–60 minutes at 28–36°. Direct light is necessary for sporulation; 2 hours sunlight was effective. Germination, infection and symptoms have a broad opt. temp. of 21–30° (Angell; Bock; Boelema). Penetration is stomatal and probably also through the intact cuticle (see Fahim et al.; Yamada et al.). The airborne conidia increase with rain and wind, with a max. conc. in the forenoon and early afternoon (Meredith). Carry over is through crop debris (Pandotra 1965b). Control with anilazine, dicloran and mancozeb is effective. In Kenya Red Creole (hybrid and open pollinated) and Yellow Creole were very resistant, this being possibly due to a thicker cuticle. White Mexican and Burgundy Red show some resistance. In USA (Colorado) selections have been made among progeny from back crosses to Sweet Spanish (susceptible) hybridised with Italian Red 13–15 (resistant).

Angell, H. R. 1929. *J. Agric. Res.* **38**: 467 (general).
Bock, K. R. 1964. *Ann. appl. Biol.* **54**: 303 (general; **44**, 1732).
Boelema, B. H. et al. 1960. *Fmg S. Afr.* **43**: 15 (general; **47**, 964).
Brinkley, A. M. et al. 1944. *Bull. Col. Farm* **6**: 8 (resistance).
Fahim, M. M. 1966. *Trans. Br. mycol. Soc.* **49**: 73 (light & sporulation; **45**, 2313a).
—— et al. 1966. *Ibid* **49**: 79 (penetration; **45**, 2313b).
Fokkema, N. J. et al. 1974. *Phytopathology* **64**: 1128 (interactions with mycoflora on leaves; **54**, 1527).
Godfrey, G. H. 1945. *Pl. Dis. Reptr* **29**: 652 (fungicides; **25**: 149).
Meredith, D. S. 1966. *Ann. appl. Biol.* **57**: 67 (spore dispersal; **45**, 1596).
Nolla, J. A. B. 1927. *Phytopathology* **17**: 115 (general; **6**: 524).
Pandotra, V. R. 1964. *Proc. Indian Acad. Sci.* Sect. B, **60**: 336 (general; **44**, 1733).
——. 1965a. *Ibid* **61**: 326 (survival; **45**, 307).
——. 1965b. *Ibid* **62**: 229 (control; **45**, 1595).
Ponnappa, K. M. 1974. *Beih. nova Hedwigia* **47**: 547 (*Alternaria cepulae* sp. nov.; **54**, 3080).
Wasfy, E. H. et al. 1977. *Acta phytopathol. Acad. Sci. hung.* **12**: 277 (cell degrading enzymes; **57**, 3207).
Yamada, Y. et al. 1976. *Ann. phytopathol. Soc. Japan* **42**: 601 (fine structure of infection; **56**, 4298).

Alternaria radicina Meier, Drechsler & Eddy, *Phytopathology* **12**: 157, 1922.

Thyrospora radicina (Meier, Drechsler & Eddy) Neergaard, 1938

Stemphylium radicinum (Meier, Drechsler & Eddy) Neergaard, 1939

Pseudostemphylium radicinum (Meier, Drechsler & Eddy) Subram., 1961.

COLONIES effuse, dark blackish brown to black. CONIDIOPHORES arising usually singly from hyphae, simple or occasionally branched, straight or flexuous, cylindrical, septate, pale to mid brown or olivaceous brown, smooth up to 200 μ long, 3–9 μ thick, with 1 or several conidial scars. CONIDIA solitary or in chains of 2 or rarely 3, very variable in shape, often ellipsoidal, obclavate or obpyriform and 1 or several longitudinal or oblique septa, sometimes constricted at the septa, 27–57 (39) μ long, 9–27 (19) μ thick in the broadest part (M. B. Ellis & P. Holliday, *CMI Descr*. 346, 1972; *A. dauci* q.v.).

Black rot of carrot; *A. radicina* also attacks celery (*Apium graveolens*), parsley (*Petroselinum crispum*), parsnip (*Pastinaca sativa*) and dill (*Anethum graveolens*); widespread in Europe and N. America, also recorded in Argentina, Australia, India, Israel, Japan, Nigeria and Venezuela. The pathogen causes pre- and post-emergence damping off; primary lesions of older plants begin at the base of the petioles from where dark, usually shallow, lesions spread into the crown and sides of the roots. Secondary lesions develop below ground and are often coincident with cracks and splits. A dry, mealy rot may develop in storage. Necrotic lesions occur on the inflorescence and occasionally on the older leaves but the leaf blight is probably not as severe as that caused by *A. dauci*. The lower part of the flowering stem may be girdled. For behaviour of the fungus on carrot in the soil in Canada and UK see Benedict and Maude, respectively. Maude et al. (*Rev. Pl. Pathol.* **49**, 355d) found that the fungus was present in the soil after 6 years and then attacked carrots in their second year. *A. radicina* is seedborne (see De Tempe) and seed treatment for control is important. Complete elimination of seed infection, without damage, has been obtained with a thiram soak (24 hours at 30° in a 0.2% suspension, Maude).

Benedict, W. G. 1977. *Can. J. Bot.* **55**: 1410 (effect of soil temp.; **57**, 333).

Curren, T. 1968. *Can. J. Microbiol.* **14**: 337 (in vitro utilisation of C & N, thiamine requirement; **47**: 2342).

——. 1969. *Ibid* **15**: 1241 (pectolytic & cellulytic enzymes; **49**, 670).

De Tempe, J. 1962. *Proc. int. Seed Test. Assoc.* **27**: 773 (hot water treatment for determining seed weakness, methods for seed testing for *Alternaria radicina* inter alia; **43**, 401).

——. 1968. *Ibid* **33**: 547 (effects of light & moisture on seed infection in blotter medium, with *A. dauci*; **48**, 1421).

Grogan, R. G. et al. 1952. *Phytopathology* **42**: 215 (occurrence & pathological effects in California; **32**: 6).

Lang-De La Camp, M. M. 1966. *NachrBl. dt. PflSchutzdienst Berl.* **20**: 71 (infection & inoculation; **46**, 1810).

Lauritzen, J. I. 1926. *J. agric. Res.* **33**: 1025 (disease in storage; **6**: 269).

Maude, R. B. 1966. *Ann. appl. Biol.* **57**: 83 (general aetiology & thiram soak with *A. dauci*; **45**: 1969).

Neergaard, P. 1935. *Beretn. Nord. Jordbr. Forskn. Foren. Kongr. 1935* 7 pp., reprinted from (on celery, morphology, symptoms & inoculation; **15**: 275).

——. 1936. *K. VetHøjsk. Aarrskr. 1936* 42 pp., reprinted from (control of seed infection & comparison with *A. dauci*; **15**: 768).

Alternaria raphani Groves & Skolko, *Can. J. Res.* Sect. C **22**: 227, 1944.

Alternaria matthiolae Neergaard, 1945.

CONIDIOPHORES simple or occasionally branched, septate, olivaceous brown, up to 150 μ long, 3–7 μ thick, sometimes swollen slightly at the tip and usually with a single conidial scar. CONIDIA commonly in chains of 2–3, straight or slightly curved, obclavate or ellipsoidal, generally with a short beak, mid to dark golden brown or olivaceous brown, smooth or sometimes minutely verruculose, with 3–7 transverse and often a number of longitudinal or oblique septa: constricted at the septa, 50–130 (70) μ long, 14–30 (22) μ thick in the broadest part. CHLAMYDOSPORES formed abundantly in culture, sometimes in chains, at first aseptate, round, finally many celled and irregular, brown; conidiophores often develop from them (M. B. Ellis, *Demat. Hyphom.*: 474, 1971).

The fungus causes black pod blotch, seedling rot and leaf spot of radish; it also occurs on other crops in the Cruciferae. The 2 commoner *Alternaria* spp. on cruciferous plants are *A. brassicae* and *A. brassicicola* (q.v.); unlike these 2 spp. *A. raphani* forms chlamydospores and the beak of the conidium is much less long than that of *A. brassicae* (conidia 75–350 × 20–40 μ) but longer than the almost nonexistent beak of *A. brassicicola* (conidia 18–130 × 8–30 μ). The 3 spp. have been compared by Changsri et al. (*A. brassicae* q.v.) and Taber et al. The best temps. for in vitro growth are similar for all of them, with that for *A. brassicae* (20–25°) being somewhat lower than for the other 2 spp. (24–28°). Another sp. on Cruciferae is *A. cheiranthi* (Lib.) Bolle (synonyms: *Helminthosporium cheiranthi* Lib., *Macrosporium cheiranthi* (Lib.) Fr.), common on wallflower (*Cheiranthus* spp.); the conidia are mostly solitary, tapering to the apex which may be drawn out into a beak; 20–100 μ long, 13–32 μ thick in the broadest part.

On radish leaves the small chlorotic, slightly raised spots are 0.5 mm diam. becoming up to 1 cm diam., circular to elliptical with a thin papery centre showing dark sporulation; shotholes may occur. Mature fruit or pods bear necrotic lesions up to 1 cm diam.; these may be very numerous. Immature seeds may become infected, shrivelled and show dark areas; the seeds may not develop. Germination is low and damping off may take place. Comparable symptoms are found on *Brassica* and *Matthiola*. Black spotting and flecking occur on the petioles, and the round lesions on the pods have net like margins and sometimes chlorotic haloes. Lesions occur on leaves and stalks. Strs. of the pathogen are reported from Canada. Infection of the seed is internal and therefore chemically difficult to control. Hot water treatment, 50° for 10–44 minutes is effective. No other control measures seem to be necessary.

Atkinson, R. G. 1950. *Can. J. Res.* Sect. C **28**: 288 (general on *Brassica* & other Cruciferae; **30**: 208).
——. 1953. *Can. J. Bot.* **31**: 542 (survival & pathogenicity; **33**: 399).
Davis, L. H. et al. 1949. *Pl. Dis. Reptr* **33**: 432 (on *Matthiola incana*; **29**: 415).
McLean, D. M. 1949. *J. agric. Res.* **75**: 71 (general on radish; **27**: 344).
Taber, R. A. et al. 1968. *Phytopathology* **58**: 609 (temp. & nutrition comparison with *Alternaria brassicae* & *A. brassicicola*; **47**, 2997).

Alternaria ricini (Yoshii) Hansford, *Proc. Linn. Soc. Lond.* 1942–43: 52, 1943.

Macrosporium ricini Yoshii, 1929.

COLONIES when on leaves amphigenous. CONIDIOPHORES arising singly or in groups, erect, simple, straight or flexuous, almost cylindrical or rather thicker towards the base, septate, rather pale brown, smooth, with 1 or a few conidial scars, up to 80 μ long, 5–9 μ thick. CONIDIA solitary or occasionally in chains of 2, straight or curved, obclavate or with the body of the conidium ellipsoidal tapering rather abruptly to a very narrow beak which is equal in length to (or up to twice as long as) the body, pale to mid-golden brown or reddish brown, smooth, sometimes constricted at the septa, overall length 70–170 (140) μ, 13–27 (19) μ thick in the broadest part; beak simple, pale, 1–1.5 μ thick along most of its length (M. B. Ellis & P. Holliday, *CMI Descr.* 249, 1970).

A. ricini causes a seedling blight, leaf spot, inflorescence and capsule rot of castor (*Ricinis communis*). Heavy losses (up to 70–85%) in yield have been reported from USA (Mississippi) and have occurred in Texas and India. The pathogen is fairly widespread (Map 345). Leaf lesions are irregular in outline, variable in size but often quite large, brown, zonate with a yellow halo. Defoliation can be extensive. The inflorescence is attacked at any age and eventually develops a sooty appearance as sporulation appears under high RH. The capsules wilt suddenly and become purple dark brown, the seed is poorly filled and may become infected; normal dehiscence fails. Alternatively, a sunken area appears on one side of the capsule and eventually covers the whole of it, the seed is fairly well grown and dehiscence is normal. Seedlings which are infected have stunted cotyledons, become spotted and may be destroyed. The pathogen is seedborne, and infected capsules can give rise to seed which may have the coat, caruncle or endosperm infected. Seeds from such capsules give diseased seedlings. Sporulation in vitro is said to be poor but occurs at 20°. The opt. for linear growth is 28°. Spore germination is highest at 25–30°. Seed dressings have given only partial control. Little resistance is known and cvs from USA were generally susceptible in India. The Texas cv. Cimarron may have some resistance.

Bates, G. R. 1959. *Rep. Bot. Pl. Pathol. Minist. Agric. Rhod. & Nyasald* 1957–58: 26 (seed transmission; **39**: 211).
Pawar, V. H. et al. 1957. *Indian Phytopathol.* **10**: 110 (general; **37**: 499).
Poole, D. D. 1954. *Pl. Dis. Reptr* **38**: 218 (disease in USA, Texas; **33**: 631).

Singh, R. S. 1955. *J. Indian bot. Soc.* **34**: 130 (general).
Stevenson, E. C. 1945. *Phytopathology* **35**: 249 (symptoms & inoculation; **24**: 386).
——. 1947. *Ibid* **37**: 184 (seedling infection; **26**: 354).
Stone, W. J. 1959. *Pl. Dis. Reptr* **43**: 827 (disease in USA, Mississippi; **39**: 31).

Alternaria sesami (Kawamura) Mohanty & Behera, *Curr. Sci.* **27**: 493, 1958.
Macrosporium sesami Kawamura, 1931.

CONIDIOPHORES arising singly or in small groups, usually simple, straight or flexuous, almost cylindrical, septate, rather pale brown or yellowish brown, smooth with 1 or several conidial scars up to 100 μ long, 5–9 μ thick. CONIDIA solitary or sometimes in chains of 2, straight slightly curved, obclavate or with the body of the condidium ellipsoidal, tapering to the beak which is usually the same length or up to twice as long as the body, pale brown to mid-golden brown, smooth, with 6–11 transverse and several or many longitudinal or oblique septa, often constricted at the septa, overall length 90–260 μ, 14–33 μ thick in the broadest part, beak simple or branched, pale, 2–4 μ thick (M. B. Ellis & P. Holliday, *CMI Descr.* 250, 1970).

On sesame the fungus causes a fairly widespread (Map 410) seedling and mature plant blight, damping off and necrotic lesions on leaves, stems and capsules. On the leaves the brown, round or irregular spots are up to 2 cm diam. with concentric zonations on the upper surface, coalescent; defoliation can be severe. Stem and capsule lesions may be less conspicuous but the former can be watersoaked and spread almost the whole length of the stem. Infected seeds lead to reduced emergence and spread may have occurred mainly through seed. In vitro most isolates showed most growth at *c.* 25°. Field and glasshouse tests showed 10 strs, when inoculated to 11 cvs of sesame, to vary in the severity of disease symptoms caused. Venezuela 51 showed the least infection but under epidemic conditions this var. is not sufficiently resistant. Seed dressings will control the disease, at least in the early stages of growth, and delay infection of mature plants; rotation is advisable. Bordeaux and zineb gave control (Samuel et al.).

Berry, S. Z. 1960. *Phytopathology* **50**: 298 (general; **39**: 607).
Leppick, E. E. et al. 1964. *Pl. Prot. Bull. F.A.O.* **12**: 13 (distribution & spread by seed; **44**, 1196).
Samuel, G. S. et al. 1971. *Madras agric. J.* **58**: 882 (fungicides; **51**, 3484).
Thomas, C. A. 1959. *Phytopathology* **49**: 461 (seed dressing: **39**: 32).

Alternaria solani Sorauer, *Z. PflKrankh.* **6**: 6, 1896.
Macrosporium solani Ellis & Martin, 1882.

CONIDIOPHORES arising singly or in small groups, straight or flexuous, septate, rather pale brown or olivaceous brown up to 110 μ long, 6–10 μ thick. CONIDIA usually solitary, straight or slightly flexuous, obclavate or with the body of the conidium oblong or ellipsoidal tapering to a beak which is commonly the same length as, or rather longer than, the body, pale or mid-pale golden or olivaceous brown, smooth, overall length usually 150–300 μ, 15–19 μ thick in the broadest part, with 9–11 transverse and 0 or a few longitudinal or oblique septa; beak flexuous, pale, sometimes branched, 2.5–5 μ thick, tapering gradually (M. B. Ellis, *Demat. Hyphom.*: 482, 1971; and see Ellis et al. 1975).

The fungus is worldwide (Map 89) and causes diseases of the foliage (usually called early blight) of tomato and Irish potato. On the latter the common name of target spot has been used and a tuber rot is also caused. On the former a fruit rot (calyx or stem end rot) and a collar rot of young plants occurs. *A. solani* can infect eggplant, *Hyocyamus, Lycopersicon, Solanum* and other hosts (Solanaceae). On tomato the leaves show circular or angular spots (sometimes with a concentric zonation), dark, almost black, usually not >4–5 mm diam. and beginning on the lower leaves. There is a clear demarcation of the necrotic lesion which can have a chlorotic halo. Infection spreads up the plant; leaf decay and fall follow. The leaf symptoms can be confused with those caused by *Stemphylium solani* (q.v.). Lesions also form on the stem and cause breakage of the fruit stalk. On young plants the dark, shrunken, girdling stem lesions produce a collar rot and subsequent collapse; plants which survive are severely weakened. Older plants (>3 weeks) are quite resistant to this form of attack. The symptoms on the fruit, generally of less importance than those on the leaves, were described in detail by Teschner. The dark necrotic lesions usually begin at the calyx end, becoming somewhat sunken, leathery in appearance and spreading until much of the green or ripening fruit is involved. Penetration may also occur where fruit has become damaged or cracked. The nailhead spot type of symptom (*A. tomato* q.v.) is not caused (e.g. see Samuel; for losses in tomato see Basu 1974b).

Penetration of leaf and stem is cuticular. Opt. temp. for mycelial growth and conidial germination is *c.* 28° but it is *c.* 6° lower for sporulation and penetration; the latter taking place at about the opt. night temp. (20°) for the host. But the lesions expand most at temps below opt. for tomato (Waggoner et al. 1969). The effects of light and dark have been investigated, more recently by Douglas and Lukens. The pathogen is less sensitive to light at temps of <23°. In vitro, using a 16 hour light period, the number of conidia increased with a drop in temp. from 23° to 15°. In vivo there were 65% sporulating lesions under 16 hours light at 14° (comparable with the numbers under a 16 hour dark period at 14° and 24°), but at 24° only 1% of the lesions were sporulating. Survival between crops occurs on host debris; Basu (1971) described structures like chlamydospores that survived in soil (with or without host tissue) for at least 7 months. Conditions which reduce host vigour tend to increase disease and the plant becomes more susceptible when fruiting heavily.

Early blight on both crops has been studied intensively in the semi-arid conditions of parts of Israel by Rotem and others. The peak conc. for the air dispersed conidia is in the late forenoon, preceding the times when RH (1300 hours) and wind speed (1500 hours) are min. and max., respectively. Drier and more windy conditions tend to make the peak earlier in the day and wetter ones delay it. Dew provides moisture conditions adequate for infection and sporulation. Damage to the host by windblown sand increases the amount of infection. Plants damaged in this way give twice as many lesions/leaf than do undamaged ones; the effect decreases with increasing time between the initiation of such damage and inoculation. Disease incidence in the field is affected by fruit to foliage weight ratios; a low ratio being associated with a reduced incidence. Leaves kept continuously moist develop external mycelium but few conidia, under different light conditions. Interruption of the moist period by 1–3 days of dryness results in abundant sporulation. *A. solani* is very resistant to high temps and drought; the mycelium more so than the conidia. Max temp. for 24 hours survival in dryness is 88° (mycelium) and 58° (conidia). RH 14–38% is opt. for survival at most temps; an increasing RH is associated with a sharp decline in survival, a decreasing one with a slow decline. Conidia on leaves of plants which are in conditions that are unfavourable for infection remain infective for 8 weeks. Survival in host debris in semi-desert soil can be 8 months. Contamination of the seed can occur but whether actual penetration takes place seems in doubt (McWhorter; Massee; Samson).

Rotem (1966) examined single spore isolates from leaves of Irish potato, tomato and eggplant; he concluded that there was no justification for division into races (Asal; Henning et al.). Brian et al. (and see Pound et al.) obtained a highly phytotoxic metabolic product from *A. solani* (alternaric acid). If it was placed in the vascular system of tomato or Irish potato shoots necrotic legions similar to those caused by natural infection were induced. The water balance of the plant is disturbed.

Some resistance (or tolerance) has been found in tomato. Differences in reaction to infection have been described (e.g. Barksdale; Barksdale et al.; Locke). Brock found no resistance. Reynard et al. recorded segregation for resistance to the collar rot phase of the disease, it was controlled by a gene pair in a recessive fashion. Control is largely through spraying; the pattern of this depends on local conditions, including the incidence of other leaf diseases. Chlorothalonil, captafol, maneb, zineb, mancozeb and propineb are all effective (e.g. Abdel-Rahman; Johnson; Stevenson). Seed treatment should also be carried out (and see Basu 1974a; McCarter et al. for soil fumigation; Madden et al. for forecasting in USA).

Abdel-Rahman, M. 1977. *Pl. Dis. Reptr* **61**: 106 (interaction between ethephon & fungicides; **56**, 5191).

Andrus, C. F. et al. 1942. *J. agric. Res.* **65**: 339 (resistance to collar rot; **22**: 116).

Asal, U. M. 1975. *Rastit. Zasht.* **23**: 38 (pathogenic strs.; **54**, 5087).

Barksdale, T. H. 1969. *Phytopathology* **59**: 443 (resistance; **48**, 2572).

——. 1971. *Pl. Dis. Reptr* **55**: 807 (resistance; **51**; 1933).

—— et al. 1973. *Ibid* **57**: 964 (horizontal resistance, segregation; **53**, 2318).

—— ——. 1977. *Ibid* **61**: 63 (inheritance of resistance; **56**, 5197).

Basu, P. K. 1971. *Phytopathology* **61**: 1347 (chlamydospores as survival units; **51**, 2861).
——. 1974a & b. *Can. Pl. Dis. Surv.* **54**: 24, 45 (soil fumigation, losses; **54**, 2471, 2972).
Brian, P. W. et al. 1951. *J. gen. Microbiol.* **5**: 619 (alternaric acid, isolation & antifungal properties; **32**: 140).
—— ——. 1952. *Ann. appl. Biol.* **39**: 308 (phytoxicity of alternaric acid; **32**: 517).
Brock, R. D. 1950. *J. Aust. Inst. agric. Sci.* **16**: 90 (search for resistance; **30**: 292).
Charlton, K. M. 1953. *Trans. Br. mycol. Soc.* **36**: 349 (UV & sporulation in vitro; **33**: 567).
Coffey, M. D. et al. 1975. *Ann. appl. Biol.* **80**: 17 (effect on shoot growth of young plants; **54**, 5553).
Dillon Weston, W. A. R. 1935. *Trans. Br. mycol. Soc.* **20**: 112 (light & sporulation in vitro; **15**: 456).
Douglas, D. R. 1972. *Can. J. Bot.* **50**: 629 (effect of light & temp. on sporulation; **51**, 3771).
Ellis, M. B. et al. 1975. *C.M.I. Descr. pathog. Fungi Bact.* 475 (*Alternaria solani*).
Henning, R. G. et al. 1959. *Pl. Dis. Reptr* **43**: 298 (evidence for races; 38: 627).
Johnson, J. C. 1969. *Qd agric. J.* **95**: 369 (fungicides; **48**, 3666).
King, S. B. et al. 1969. *Am. J. Bot.* **56**: 249 (nuclear behaviour, septation & hyphal growth; **48**, 2215).
Kreutzer, W. A. et al. 1953. *Bull. Colo Exp. Stn* 402, 12 pp. (collar rot; **13**: 195).
Locke, S. B. 1949. *Phytopathology* **39**: 829 (resistance with *Septoria lycopersici*; **29**: 233).
Lukens, R. J. 1963. *Am. J. Bot.* **50**: 720 (photo-inhibition of sporulation; **43**, 383).
——. 1965. *Phytopathology* **55**: 1032 (reversal by red light of blue light inhibition of sporulation; **45**, 373).
——. 1966. *Ibid* **56**: 1430 (interaction of low temp. & light on sporulation; **46**, 1333).
—— et al. 1973. *Ibid* **63**: 176 (processes of sporulation & response to metabolic inhibitors; **52**, 3548).
McCallan, S. E. A. et al. 1944. *Contr. Boyce Thompson Inst. Pl. Res.* **13**: 323 (sporulation in vitro; **24**: 115).
McCarter, S. M. et al. 1976. *Phytopathology* **66**: 1122 (soil fumigation; **56**, 2216).
McWhorter, F. P. 1927. *Bull. Va Truck Exp. Stn* 59: 547 (general, symptoms; **7**: 411).
Madden, L. et al. 1978. *Phytopathology* **68**: 1354 (forecasting in USA).
Massee, G. 1914. *Kew Bull.* **4**: 145 (presence of mycelium in tomato seed).
Moore, W. D. et al. 1943. *Phytopathology* **33**: 1176 (disease on seedlings; **23**: 194).
Pound, G. S. 1951. *Ibid* **41**: 127 (effect of temp. on disease; **30**: 392).
—— et al. 1951. *Ibid* **41**: 1104 (formation of phytotoxic material; **31**: 305).
Rands, R. D. 1917. *Res. Bull. Wisc. agric. Exp. Stn* 42, 48 pp. (general, hosts).
Reynard, G. B. et al. 1945. *Phytopathology* **35**: 25 (inheritance of resistance to collar rot; **24**: 253).
Rotem, J. 1959. *Bull. Res. Coun. Israel* **7D**: 100 (effect of sand storms; **39**: 341).
—— et al. 1964. *Pl. Dis. Reptr* **48**: 211 (effect of dew).
——. 1964. *Phytopathology* **54**: 628 (effect of weather on conidial dispersal; **43**, 3019).
—— et al. 1965. *Israel Jnl agric. Res.* **15**: 115 (yield:foliage ratio & disease incidence; **45**, 1202).
——. 1965. *Agric. Met.* **2**: 281 (effect of sand & dust storms; **45**, 2608).
——. 1966. *Israel Jnl Bot.* **15**: 48 (variability in *A. solani*; **46**, 1923).
——. 1968. *Phytopathology* **58**: 1284 (behaviour in semi-desert conditions, thermoxerophytic; **48**, 604).
——. 1969. *Israel Jnl agric. Res.* **19**: 139 (effect of soil moisture; **49**, 983).
—— et al. 1969. *Trans. Br. mycol. Soc.* **53**: 433 (mycelial growth & sporulation, effects of leaf moisture & dryness; **49**, 1281).
Rowell, J. B. 1953. *Bull. Rhode Isl. agric. Exp. Stn* 320, 29 pp. (general).
Samson, R. W. 1942. *Phytopathology* **32**: 16 (no internal mycelium in tomato seed).
Samuel, G. 1932. *Ibid* **22**: 613 (fruit infection, not causing nailhead spot, *A. tomato*; **11**: 680).
Skrdla, W. H. et al. 1968. *Res. Bull. Ohio agric. Exp. Stn* 1009, 110 pp. (reaction of world collection of *Lycopersicon*; **48**, 1977).
Stall, R. E. 1958. *Am. J. Bot.* **45**: 657 (cytology).
Stevenson, W. R. 1977. *Pl. Dis. Reptr* **61**: 803 (fungicides; **57**, 2279).
Teschner, G. 1953. *Phytopathol. Z.* **21**: 133 (on tomato fruit & Irish potato tubers; **33**: 554).
Waggoner, P. E. et al. 1969. *Bull. Conn. agric. Exp. Stn* 689, 80 pp. (epidem: a simulator of the disease for a computer; **50**, 3399).
—— et al. 1975. *Phytopathology* **65**: 551 (slowing of conidial germination with changes between warm & cool temps; **54**, 5287).
—— ——. 1977. *Ibid* **67**: 1007 (changed metabolic pathways & conidial germination, **57**, 962).

Alternaria tomato (Cooke) G. F. Weber, *Bull. Fla agric. Exp. Stn* 332: 22, 1939.

Macrosporium tomato Cooke, 1883.

(the binomials *A. tomato* (Cooke) Jones, which undoubtedly refers to the saprophyte *A. fasciculata* (Cooke & Ellis) Jones & Grout, and *A. tomato* (Cooke) Brinkman, because it does not refer to the nailhead spot producing parasite as intended, should be dropped. *A. fasciculata* is probably a form of *A. alternata* q.v.)

SPOTS circular to oval or irregular, shallow to sunken, brownish black on foliage, tan to brown on fruit, seldom over 1 cm diam., amphigenous, irregularly scattered. HYPHAE hyaline, branched, irregularly septate, 4–10 μ diam., colonies greyish white. CONIDIOPHORES maculicole, dark brown, short, rigid, smooth, straight or curved, 3–8 septate, 7 μ diam., 75–100 μ in length, single or fasciculate, bases swollen. CONIDIA usually apical, dark, muriform, obclavate, $51.9 \times 14.7\ \mu$; 6–12 transverse and 2–8 longitudinal septa; terminating in an attenuate beak, $2\text{–}4 \times 82.2\ \mu$, sparingly septate, sometimes forked; conidia usually borne singly, occasional 2 spored chains. On foliage, stems and fruits of tomatoes, diagnostically characteristic spots on fruit (G. F. Weber l.c.).

The taxonomic position of this pathogen, which causes nailhead spot of tomato, has not been recently examined. Weber has fully discussed the disease and its confusion with the more important tomato pathogen *A. solani* (q.v.), *A. alternata* (a common saprophyte and weak pathogen) and blossom end rot (a physiological condition). He considered *A. tomato* to be a distinct fungus. Its distribution is uncertain but it may be fairly widespread; in USA (from where the only detailed descriptions have come) it is southerly and south eastern in distribution. Rosenbaum compared *A. tomato* with *A. solani*. He gave av. measurements for the spore body alone and with beak of: 55 μ and 153 μ (*A. tomato*); and 74 μ and 211 μ (*A. solani*). Joly (*Alternaria* q.v.) placed the fungus as a synonym of *A. tenuissima*.

Under natural conditions only tomato has been found to be attacked. In distinguishing the disease from that caused by *A. solani* (early blight) the diagnostic symptoms for *A. tomato* are those on the fruit; earlier symptoms on leaves and stems are virtually indistinguishable from those caused by *A. solani*. On very young fruit small grey black, slightly sunken spots form, and they may cover fruit, calyx and peduncle; if they cover a part of the fruit this becomes distorted with growth. On older fruit the slightly sunken spots are light brown becoming darker and usually not >5 mm diam. On mature coloured fruit the spots are rimose, sunken, with a narrow black border and a paler centre; the surrounding host tissue immediately beyond the spot may often be green. The fruit lesions may penetrate the seed cavity. Fruit infection is the most economically important phase of the disease. Rosenbaum was not able to produce the typical nailhead spots on green fruit with *A. solani* inoculum. Lesions are also formed on the leaves and stems, seedlings may be killed and defoliation occurs. The symptoms on the vegetative parts of the plant are described under *A. solani*.

Paulus et al. compared this pathogen with *Stemphylium solani* (q.v.). Both fungi showed max. in vitro growth at 24–28° (Weber gave 26–28° as opt. for conidial germination in *A. tomato*). Both caused more leaf infection at 26° compared with 14°, 18° and 22°; infection at 14° being always less than at the other temps. The amount of infection by *A. tomato* decreases with fruit age (Rosenbaum et al.). Up to 14 days old there was 100% infection, 21–35 days gave 23–85% and at 41–55 days there was no infection. Penetration of the intact cuticle and the stomata occur. Carry over is by conidia or infected host debris. Besides causing damage in the field infection leads to extensive damage during fruit transit and storage but with the use of resistant tomato cvs in USA nailhead spot is now of little importance. M. Aragaki (*Phytopathology* **51**: 803; **52**: 1227; **54**: 562, 565; **58**: 1041; *Mycologia* **65**: 1205) has described the effects of temp., light and chemicals on sporulation of *A. tomato*. But since his isolates apparently came from *Passiflora edulis* brown spot (caused by *A. passiflorae* q.v.) their identity seems in doubt.

Paulus, A. O. et al. 1955. *Phytopathology* **45**: 168 (effect of temp. on disease with *Stemphylium solani*; **34**: 680).
Ramsey, G. B. et al. 1929. *J. agric. Res.* **38**: 131 (disease in transit & marketing; **8**: 534).
Rosenbaum, J. 1920. *Phytopathology* **10**: 9 (general, comparison with *Alternaria solani*).
—— et al. 1920. *Am. J. Bot.* **7**: 78 (size of fruit & infection).
Weber, G. F. 1939. *Bull. Fla agric. Exp. Stn* 332, 54 pp. (monograph, 83 ref.; **18**: 766).

APHANOMYCES de Bary, *Jb. wiss. Bot.* 2: 178, 1860.

Hydatinophagus Valkanov, 1931.
(Saprolegniaceae)

HYPHAE delicate, rarely coarse, then not exceeding 20 μ diam., hyaline to light brown, sparingly branched or much branched and contorted. ZOOSPORANGIA filamentous, of variable length, isodiametric or rarely tapering toward the apex, formed from undifferentiated, vegetative hyphae, not proliferating internally, rarely with short side branches. Primary ZOOSPORES borne in a single row in the zoosporangia, encysting upon emergence at the orifice as in *Achlya* or, rarely, swimming prior to encystment as in *Leptolegnia*; primary zoospore cysts spherical, discharge poroid, papillate or, rarely, schistose; secondary zoospores reniform, laterally biflagellate. OOGONIA terminal on short or long branches, smooth walled or wall irregularly roughened or ornamented with sharp pointed spines or bluntly conical tubercles. ANTHERIDIA 1 to several, long cylindrical, clavate or short tuberous; antheridial branches simple, long cylindrical, clavate or short tuberous; antheridial branches simple or branched, diclinous, monoclinous or androgynous in origin; fertilisation tubes present or sometimes not visible; specialised gemmae lacking. OOSPORES single, rarely 2, hyaline or dark, contents homogeneous, finely granular, with or without a conspicuous oil globule; germination rarely observed, generally by the formation of a long, branched germ tube (from W. W. Scott).

The genus was monographed by Scott; he gave keys to 3 subgenera and to the spp. within each. The subgenus *Aphanomyces* contains the few plant pathogens. Dick commented on oospore structure as described by Scott. *A. cochlioides* Drechs. causes a seedling disease of beet (*Beta vulgaris*), particularly important in parts of USA (Papavizas et al. 1974, *A. euteiches* q.v.). *A. cladogamus* Drechs. causes damping off in various crops including vegetables. *A. brassicae* Pavgi & Singh causes a root rot of cabbage and cauliflower in India.

Dick, M. W. 1971. *Mycologia* **63**: 686 (oospore structure).

Drechsler, C. 1929. *J. agric. Res.* **38**: 309 (morphology of *Aphanomyces cochlioides* & *A. cladogamus* inter alia; **8**: 606).

——. 1954. *Sydowia* **8**: 334 (*A. cladogamus* on *Viola*; **34**: 370).

——. 1954. *J. Wash. Acad. Sci.* **44**: 212, 236 (morphology of *A. cladogamus*; **33**: 690, 689).

Fowles, B. 1976. *Mycologia* **68**: 1221 (factors affecting growth & reproduction in vitro; **56**, 3867).

McKeen, C. D. 1952. *Can. J. Bot.* **30**: 701 (*A. cladogamus* causing damping off; **32**: 360).

Scott, W. W. 1961. *Tech. Bull. Va agric. Exp. Stn* 151, 95 pp. (monograph, 150 ref.; **41**: 584).

Singh, S. L. et al. 1977. *Mycopathologia* **61**: 167 (*A. brassicae* sp. nov. on cauliflower; **57**, 1890).

Unestam, T. 1966. *Physiol. Pl.* **19**: 15 (chitinolytic, cellulytic & pectinolytic activity in vitro in 5 *Aphanomyces* spp.; **45**, 3289).

Aphanomyces euteiches Drechsler, *J. agric. Res.* 30: 311, 1925.

HYPHAE hyaline, 4–10 μ diam., not abruptly varying in width, in host cortex. SPORANGIA arising from conversion of vegetative mycelium, delimited by 1 or more septa, discharging through 1–4 tapering branches which, distally, are 4 μ diam. ZOOSPORES usually 3.5 × 30–50 μ; forming spherical cysts at mouth of sporangium, usually 8–11 μ diam.; diplanetic (but see Hoch et al. 1972), empty spherical wall with evacuation tube protruding, 1 × 2.5–3 μ. OOGONIA generally terminal, on a short lateral branch; subspherical, 25–35 μ diam.; when mature with wall of irregular thickness varying 1–5 μ (usually 1–2.5 μ). ANTHERIDIA diclinous, often branching, 8–10 × 15–18 μ, or when larger often more conspicuously arched, somewhat lobulate, with transverse septa. OOSPORES subspherical or more rarely ellipsoid, 18–25 (generally 20–23) μ, wall 1.2–1.8 μ thick; slightly eccentric internally; germinating without protracted resting period either directly (1–3 germ hyphae) or by single, unbranched sporangial filament (200–350 μ long) producing generally 13–18 zoospores, *c.* half delimited within the oospore wall (after C. Drechsler l.c.).

Scott (*Aphanomyces* q.v.) stated that this sp. has considerable similarity with two aquatic members of the genus, *A. laevis* de Bary and *A. helicoides* von Minden, but differs from them in having a greatly thickened oogonial wall with a characteristic, sinuous inner contour, complex zoosporangia and a parasitic habit. This author should be referred to for further data on morphology. Root rot of pea (*Pisum sativum*) occurs generally in USA (except N. New England and N. Pacific coast), Australia (Tasmania), Japan and Europe (Denmark, France, Norway, Sweden, UK and USSR; Map 78). *A. euteiches* has been reported from several leguminous genera (including *Lathyrus, Lupinus, Medicago, Melilotus, Phaseolus, Trifolium* and *Vicia*) but only causes serious disease in pea. The monograph by Papavizas et al. should be consulted for ref. The disease has been largely studied in USA where severe losses can be caused. Root symptoms, which are not particularly characteristic, are seen a few days after infection as watersoaked, cortical lesions becoming yellowish brown, and oospores are formed quickly (*c.* 8 days after infection). Spread through the roots from the tap root is rapid and part of the tap root separates readily from the rest of the root system when the plant is pulled up. The rot may extend a few cm above soil level and death of the whole plant can occur. Infected plants may show no above ground symptoms (if infected late) or they may be dwarfed with a poor crop.

The zoospores are attracted most strongly to the area behind the root cap (attraction seems to be non-specific) and penetrate within 2 hours. Most pea seedlings in infested soil become infected shortly before or after plant emergence. In vitro most zoospores form at 20–28° and vegetative growth is opt. at 28°. Presence or absence of light has no effect on zoospore production. Secondary swarming is inhibited at 32°. The development of infection and disease is similar at 20–28°, these are favourable temps; at 16° symptoms develop more slowly. High soil moisture and relatively low temps favour the disease. More oospores germinate near pea roots than near those of bean (*Phaseolus*), maize and soybean. The pathogen is carried over as oospores which provide the primary inoculum. Oospores from fresh and decomposed roots germinate only at pH 3.5–5.1 and germination is greater from the latter. Oospores are produced in vitro without added vitamins, being favoured by an acid reaction. In a method to obtain abundant oospores it was found that oospores retained infectivity after passage through water snails. These spores probably survive >2 years in the soil and root rot levels drop little in 3–4 years between pea crops; even after 6–8 years (e.g. Kotova) disease levels may still be high. Fields planted with crops other than legumes give greater disease reductions. There is little or no saprophytic development in the soil.

A. euteiches is sometimes associated with *Fusarium solani* f. sp. *pisi* (q.v.). Both pathogens are virtually restricted to the ploughed layer and where the subsoil is unfavourable disease tends to be higher. Adding *Pythium* (*P. debaryanum* and *P. ultimum* q.v.) to soil infested with *A. euteiches* has little or no effect on root rot. Plants infected with viruses tend to get more severe root rot caused by both *A. euteiches* and *F. solani* f. sp. *pisi*; and *Pratylenchus penetrans* can increase disease levels (e.g. Beute et al.; Farley et al.). Studies have been made on pectolytic activity, cytology, fine structure, the effect of chemicals in vitro and on the disease on peas growing in nutrient solutions. Some pathogenic variability occurs between single zoospore isolates. Races have been characterised in USA and Europe; and since pea lines did not have adequate resistance *Phaseolus* spp. have been used to differentiate races.

Control is largely cultural and a long rotation is necessary. Soils liable to poor drainage should be avoided. Organic amendments have shown some beneficial effects in glasshouse experiments. Soil treatment with fenaminosulf has been shown to be effective in the field but may be uneconomic. Pyroxychlor, when applied in the furrow in granular form, gave control (Pfleger et al.). No effective, highly resistant cvs have been developed.

Beute, M. K. et al. 1968. *Phytopathology* **58**: 1643 (increased disease in pea with viruses; **48**, 1388).

Drechsler, C. 1927. *J. agric. Res.* **34**: 287 (on tomato roots with *Plectospira*; **6**: 517).

Farley, J. P. et al. 1964. *Phytopathology* **54**: 1279 (increased disease in pea with viruses; **44**, 914).

Grau, C. R. 1977. *Ibid* **67**: 551 (effects of herbicides dinitramine & trifluralin; **56**, 5874).

—— et al. 1977. *Ibid* **67**: 273 (as above; **56**, 5271).

Hoch, H. C. et al. 1972. *Protoplasma* **75**: 113 (fine structure of zoospores, including encystment & germination; **52**, 623).

—— ——. 1973. *Can. J. Bot.* **51**: 413 (effects of osmotic water potentials during zoosporogenesis; **52**, 3219).

—— ——. 1975. *Ibid* **53**: 1085 (mechanisms in primary spore cleavage; **55**, 56).

Kotova, V. V. 1976. *Mikol. i Fitopatol.* **10**: 137 (oospore germination & survival; **55**, 4930).

Lewis, J. A. 1973. *Phytopathology* **63**: 989 (effect of mineral salts; **53**, 1187).

——. 1973. *Pl. Dis. Reptr* **57**: 876 (suppression of disease with Cu compounds; **53**, 3245).

——. 1977. *Ibid* **61**: 762 (suppression of disease with calcium minerals; **57**, 2339).

Marx, G. A. et al. 1972. *J. Am. Soc. hort. Sci.* **97**: 619 (a genetic study of tolerance; **52**, 910).

Mitchell, D. J. et al. 1973. *Phytopathology* **63**: 1053 (O_2 & CO_2 conc. effects on growth & reproduction with other soilborne pathogens; **53**, 2741).

Oyekan, P. O. et al. 1972. *Ibid* **62**: 369 (effect of *Pratylenchus penetrans* on disease; **51**, 4475).

Papavizas, G. C. et al. 1974. *Tech. Bull. U.S. Dep. Agric.* 1485, 158 pp. (monograph with *Aphanomyces cochlioides*, 349 ref.; **54**, 1915).

Pfleger, F. L. et al. 1976. *Pl. Dis. Reptr* **60**: 317 (fungicides; **56**, 1343).

Pueppke, S. G. et al. 1976. *Phytopathology* **66**: 1174 (pisatin & development of pathogen in pea; **56**, 1810).

Yokosawa, R. et al. 1977. *Ann. phytopathol. Soc. Japan* **43**: 501 (longevity & infectivity of *Aphanomyces* zoospores in soil; **57**, 4820).

Aphanomyces raphani Kendrick,
Phytopathology **17**: 43, 1927; *Bull. Purdue agric. Exp. Stn* 311, 1927.

HYPHAE 8–14 μ diam., coarse, hyaline, profusely branched at right angles and bearing short, conspicuous side branches. ZOOSPORANGIA moderately long, terminal or intercalary, much branched, differing from the vegetative hyphae only in the spiral twisting and marked tapering of the lateral branches. Primary ZOOSPORES spherical to elongate, numerous, encysting at the orifice upon emergence; primary zoospore cysts 8.8–12.7 μ diam.; secondary zoospores reniform, laterally biflagellate. OOGONIA terminal on short side branches, globose, thick walled with a smooth outer and an irregularly contoured inner surface; 32–44.9 μ diam. ANTHERIDIA 1–3, club shaped, occasionally with apical prolongations; antheridial stalk simple or branched, diclinous or monoclinous in origin; fertilisation tubes short or of appreciable length. OOSPORES hyaline, thick walled, 21.4–29 μ diam., contents granular with a large, central oil globule and small, conspicuous, refractive body (from W. W. Scott, *Aphanomyces* q.v.).

Black root of radish is a minor disease of scattered but widespread distribution: Australia (New South Wales, Queensland), Canada (British Columbia, Nova Scotia, Ontario, Quebec), Germany, Japan, New Zealand, South Africa, USA and Venezuela (Map 421). Humaydan et al. (1975) gave other hosts in *Brassica* and *Crambe*. Damping off, stunting, and root and hypocotyl discolouration are the commonest symptoms. Greyish blue, superficial lesions occur on the primary roots, apparently coinciding with the positions where the secondary roots emerge. The lesions darken and coalesce to form large necrotic areas; they may appear as bands of varying width (partially or wholly encircling the tap root) or as deeply penetrating strands. The rot becomes fissured and deformed due to unequal growth. Plants can be stunted and dark lesions may occur on hypocotyls, cotyledons and leaves. The rot is at first dry. Yokosawa et al. (1972) found that zoospore penetration of the hypocotyl led to most severe damping off; whilst exposure of the roots to infection led to little or no disease. Oospores are formed in the infected tissues and are said to require a 6 weeks resting period before germination. Most growth in vitro occurs at 23° or a few degrees higher (see Ogoshi et al.), and the disease is said to be more severe at rather high temps. In USA conventional soil sterilants are effective and long rotations (3–4 years) have been suggested. Cvs vary considerably in the degree of resistance; red globe types have effective resistance and in New Zealand out of 4 cvs tested French Breakfast was recommended (and see Wendland).

Boning, K. 1932. *Prakt. Bl. PflBau PflSchutz* **10**: 205 (general; **12**: 350).
Herold, F. 1952. *Phytopathol. Z.* **19**: 79 (general; **32**: 293).
Humaydan, H. S. et al. 1975. *Pl. Dis. Reptr* **59**: 113 (cruciferous hosts; **54**, 4228).
——. 1978. *Phytopathology* **68**: 377 (factors affecting in vitro growth and zoospore formation).
Kendrick, J. B. 1927. *Bull. Purdue agric. Exp. Stn* 311, 32 pp. (morphology & general; **7**: 4).
Ogoshi, A. et al 1972. *Ann. phytopathol. Soc. Japan* **38**: 130 (isolation, culture & morphology; **52**, 604).
Wendland, E. J. 1976. *Gemüse* **12**: 78 (control; **57**, 1470).
Wenham, H. T. 1960. *N.Z. Jl agric. Res.* **3**: 179 (general; **39**: 515).
Yokosawa, R. et al. 1972. *Ann. phytopathol. Soc. Japan* **38**: 284 (zoospore penetration; **52**, 2405).
——. 1974. *Ibid* **40**: 46 (taxis of zoospores to hypocotyl & role of host exudates; **53**, 4632).

ARMILLARIELLA Karst., *Hymenomycetes Fennici Acta flor. faun. Fenn.* **2**: 4, 1881.
Armillaria (Fr.) Quél., 1872 (non Kummer, 1871).
(a full description and synonymy was given by R. Singer, *The Agricales in modern taxonomy*: 259, 1975)
(Tricholomataceae)

Armillariella mellea (Vahl ex Fr.) P. Karst.,
Acta Soc. Fauna Flora fenn. **2**: 4, 1881.
Agaricus melleus Vahl. ex Fr., 1821
Armillaria mellea (Vahl. ex Fr.) Kummer, 1871.

Caespitose. CARPOPHORE clitocyboid, polymorphic. PILEUS 4–15 cm diam., fleshy, subglobose expanding to convex, finally sunken at the centre or umbonate; surface varying from honey colour to tawny or deep brown, with small, crowded, fibrillose squamules; margin entire, paler, striate. LAMELLAE adnate to subdecurrent, whitish to flesh pink moderately distant but with lamellulae of four lengths, moderately thick, tough, up to 9 mm wide; edge entire, concolorous. STIPE 5–15 × 1–2.5 cm, central, equal or slightly thickened towards the base, fistulose, fibrous fleshy; surface ochraceous yellow when young, darkening to brown with age, white above the annulus; often arising from black rhizomorphs. ANNULUS persistent, rather thick, membranous, whitish becoming more yellow in the outer portions, attached to the upper part of the stipe. CONTEXT white, woolly fleshy, consisting of thin walled, hyaline hyphae, 2–8 μ diam., inflating up to 15 μ diam., without clamp connections. SMELL and TASTE acrid. SPORE PRINT light cream colour. BASIDIOSPORES 7–12 × 5–7.5 (10.5 × 5.5) μ, short ovoid to ellipsoid, with or without a suprahilar depression, hyaline, inamyloid, with a thin or slightly thickened wall. BASIDIA 35–47 × 5–9 μ, elongate clavate, bearing four sterignata (up to 6 μ long). LAMELLA EDGE sterile, crowded with basidiole like CHEILOCYSTIDIA, 27–45 × 7–15 μ, fusoid, clavate to ventricose, hyaline, thin walled. PLEUROCYSTIDIA absent. HYMENOPHORAL TRAMA bilateral, with a broad mediostratum, and diverging lateral strata, consisting of hyaline hyphae, 3–7 μ diam., inflating to 15 μ diam. SUBHYMENIAL LAYER interwoven, up to 12 μ wide. PILEAL SURFACE a repent cutis, up to 30 μ thick, consisting of loose, unbranched chains of cylindrical elements, 15–45 × 6–8 μ, constricted at the septa, somewhat gelatinised. DEVELOPMENT hemiangiocarpous.

This may be distinguished from *A. montagnei* Singer which has a pure white spore print, a non-gelatinised pileal surface, and an almost regular to indistinctly bilateral hymenophoral trama. *Armillariella tabescens* (Scop. ex Fr.) Singer lacks a veil (D. N. Pegler & I. A. S. Gibson, *CMI Descr.* 321, 1972; and see Romagnesi; Singer).

A. mellea is worldwide, mostly on woody dicotyledons and conifers (Map 143; for hosts see Browne, *Pests and diseases of forest plantation trees*; Raabe 1962 & 1965; Wallace). Infection of monocotyledons and herbaceous plants is rare. The pathogen is commonly known as the honey, bootlace or shoestring fungus; the last 2 names refer to the characteristic rhizomorphs. It causes a root and butt rot. General accounts have been given by Gibson 1975; Pawsey; Peace, *Pathology of trees and shrubs*; Shaw et al.; Singh 1974a; Sokolov; Thomas. Thick, creamy white, fan like sheets of mycelium are found under the bark of the roots and stem base, and these are frequently accompanied by flattened, dark brown rhizomorphs (*forma subcorticalis*). These rhizomorphs may also be epiphytic on roots and grow out into the soil where their form is more cylindrical (*forma subterranea*). *A. mellea* may act as a lethal pathogen by invading and destroying the cambium of the root and lower stem, following which it may colonise and rot the woody tissues of the host after its death. Alternatively, it may cause a butt rot without attacking the living host tissues (Pegler et al. l.c.). In a lethal infection the plant dies rapidly and (in conifers and some dicotyledonous trees) the bark of the roots and stem base is cracked and covered with gum or resin exudate. Sporophores may be formed on the host in advanced stages of the disease. The butt rot is a typical soft, white rot with pronounced black zone lines (Campbell; and see Lopez-Real; Lopez-Real et al.).

The rhizomorphs of *A. mellea* have been studied in great detail (see Garrett, *Pathogenic root infecting fungi*). In some areas they can be free running through the soil for considerable distances, in others this habit is absent. The reasons for the differing behaviour of the fungus in different ecological conditions are not entirely understood. General accounts of particular aspects of the biology of the honey fungus have been given by: Fox; Garraway et al.; Redfern 1970, 1975; Swift 1970; and see Schönhar. (Further information on rhizomorph biology will be found in *Rev. appl. Mycol.* **44**, 1814; **48**, 3373; *Rev. Pl. Pathol.* **49**, 1303, 3652; **50**, 1118, 2767; **51**, 3160, 3783; **52**, 2162, 2757; **54**, 3157, 4358; **55**, 3078; **56**, 2373.) Some tropical crops on which the fungus has been studied are: cacao (Dade), coffee (Rayner), oil palm (Wardlaw), tea (Gadd), teak (Hocking) and tung (Wiehe). See also for Kenya forests and pines (Gibson 1960), Zimbabwe woodlands (Swift 1972), Scots pine (Rykowski), larch (Ono) and eucalyptus (Delatour; Marks et al.). *A. mellea* is extremely variable and there is some doubt as to whether *Armillaria fuscipes* Petch and *Armillariella tabescens* (*Clitocybe tabescens*) are valid (Pegler et al. l.c.; Rhoads 1945; Gibson et al. 1964). *A. tabescens* has been described on banana (Rhoads 1942), citrus (Rhoads 1948) and coffee (Dadant).

The distribution of diseases caused by *A. mellea* is patchy. Fox remarked that very little is known of the pathogen's behaviour in the lowland tropics, and that its distribution cannot be explained on the basis of temp. or hosts. Records of disease in the tropics have been usually montane but *A. mellea* was becoming serious at low altitudes on rubber (*Hevea brasiliensis*) in W. Africa (Fox). Infection by airborne basidiospores appears to be without significance (Rishbeth 1970). In temperate regions (i.e. excluding montane climates in tropical latitudes) infection of the root is by the characteristic rhizomorphs growing through the soil. Morrison and Redfern (1970, 1973) report on recent work on behaviour in soil. In tropical regions (montane and lowland) these free-running rhizomorphs are very sparse or non-existent, for example, in W., E. and central Africa, where most work has been done outside Europe and N. America. Disease spread is by root contact where the rhizomorphs are absent. Rishbeth (1968) tentatively suggested that the high temps of the tropics may limit initiation of rhizomorphs, since they were not initiated from woody inocula at >25° (and see Rishbeth 1978). Gibson (1961) grew isolates from temperate and tropical environments on malt agar, they were put in 3 groups on the basis of temp. opt. 22–24° (82 isolates), 25–27° (102), 27–29° (28). These results agreed with those of earlier workers. No correlation between opt. temp. for growth and the climate of origin was found. Boughey et al. attributed the absence of rhizomorphs in the high veld soils of Rhodesia to a toxin; see also Swift (1968) and Olembo.

The disease characteristically occurs in tree plantings established on recently exploited natural forest sites. Under natural conditions the rhizomorphs (if present) are only superficially attached to the plant. But the fungus becomes aggressive after clearing and planting, and under certain conditions. It has been overestimated in the role of a lethal pathogen, and aggressive attacks may only occur when the vigour of the host has declined. Gremmen recently described the secondary role played by *A. mellea*. Hardwood stumps provide more effective food bases (or centres of inoculum) than do those of conifers (Redfern 1975). In E. Africa *Acacia* and *Eucalyptus* spp. harbour the fungus and from which spread occurs to pine crops. Rishbeth (1972) found that rhizomorph formation was greater and lasted longer, after felling, on unpoisoned stumps of broad-leaved trees than on those of pines. Other fungi compete more efficiently with *A. mellea* in pine than in deciduous tree stumps. Several factors will affect the severity of an attack. The size of the active inoculum mass; young or unthrifty trees are more susceptible; moist soil with a high proportion of woody debris promotes disease; temps of 18–22° favour it; hosts differ in susceptibility (e.g. *Pinus elliottii* is less resistant than *P. patula*; Swift 1972). Felling and rhizomorph severing following soil disturbance caused by timber extraction or ploughing may cause a rapid increase in activity by the fungus (Redfern 1973). Shaw found that isolates from pine were pathogenic to *Pinus* spp. (seedlings) but those from hardwood trees were not highly pathogenic to pine.

Control in the forest is best achieved by avoidance; see Shaw et al. for control in conifers. High risk sites are generally discovered during the early stages of development and can be avoided. In second rotation crops it is generally best to avoid a succession of pines after hardwoods in areas where the fungus is likely to be active (Gibson 1975). Leach in E. Africa showed that if trees are ring barked and left to die their roots are not readily colonised by *A. mellea*. The method can be used in clearing forest sites where perennial crops such as tea are to be planted. Redfern (1968) in UK found that the colonisation of hardwood roots was not reduced by ring barking before felling or by frill girdling and poisoning. But the roots of treated trees (although fully colonised after 5 years) were less effective food bases than those of untreated trees. Swift (1970), in a discussion of control through ring barking, pointed out that the effect on rate of spread between stumps and within them should be distinguished. In the former case ring barking is probably effective, as Leach found. In the latter case the effects are more difficult to interpret. Here the initial effect is to stimulate fungal invasion; but infectivity is probably lowered because of the extended period between invasion and contact with the planted crop, and the rapid depletion of starch reserves. Rishbeth (1976) investigated the effects of chemical treatment and fungal inoculation of hardwood stumps on the activity of *A. mellea*. Bliss (1951) found that fumigation of citrus soils led to an increased development of *Trichoderma viride* (and see Aytoun) whose antagonism to *A. mellea* was considered to be the primary cause of the effectiveness of the fumigants. This work has been followed by that of Ohr et al.; Munnecke et al. Soil fumigation is used in California, USA.

Aytoun, R. S. C. 1953. *Trans. Proc. bot. Soc. Edinb.* **36**: 99 (relationship with *Trichoderma*; **32**: 497).

Bliss, D. E. 1946. *Phytopathology* **36**: 302 (soil temp. & disease development; **25**: 446).

Armillariella mellea

Bliss, D. E. 1951. *Ibid* **41**: 665 (soil fumigation; **30**: 608).

Boughey, A. S. et al. 1964. *Nature Lond.* **203**: 1302 (toxic factor in soil; **44**, 353).

Campbell, A. H. 1934. *Ann. appl. Biol.* **21**: 1 (black zone lines; **13**: 483).

Dadant, R. 1963. *Agron. trop.* **18**: 265 (on coffee; **42**: 683).

Dade, H. A. 1927. *Bull. Dep. Agric. Gold Coast* 5, 21 pp. (on cacao; **6**: 659).

Delatour, C. 1969. *Annls Inst. natn. Rech. For. Tunis* 2, 23 pp. (on *Eucalyptus*; **50**, 1396).

Fox, R. A. 1970. In *Root diseases and soil-borne pathogens* (editors Toussoun, T. A. et al.): 179 (*Armillariella mellea* inter alia on tropical, perennial crops).

Gadd, C. H. 1930. *Tea Q.* **3**: 109 (on tea; **10**: 275).

Garraway, M. O. et al. 1970. In *Root diseases and soil-borne pathogens* (editors Toussoun, T. A. et al.): 122 (rhizomorphs).

Gibson, I. A. S. 1960a. *E. Afr. agric. For. J.* **26**: 142 (in Kenya forests; **40**: 436).

——. 1960b. *Emp. For. Rev.* **39**: 94 (in Kenya on *Pinus*: **39**: 509).

——. 1961. *Trans. Br. mycol. Soc.* **44**: 123 (variation between isolates including temp.; **40**: 571).

—— et al. 1964. *Phytopathology* **54**: 122 (variation between isolates).

——. 1975. In *Diseases of forest trees widely planted as exotics in the tropics and southern hemisphere* Pt 1: 5.

Gremmen, J. 1976. *Ned. BoschbTijdschr.* **48**: 103 (significance of pathogen in dying trees; **55**, 4286).

Hocking, D. 1966. *Misc. Rep. trop. Pestic. Res. Inst.* (EAAFRO) 574, 4 pp. (on teak; **46**, 1348).

Kolbezen, M. J. et al. 1974. *Hilgardia* **42**: 465 (soil penetration by methyl bromide; **54**, 3681).

Leach, R. 1937. *Proc. R. Soc.* Ser. B **121**: 561 (parasitism & control; **16**: 564).

——. 1939. *Trans. Br. mycol. Soc.* **23**: 320 (ecology & biological control; **19**: 311).

Lopez-Real, J. M. 1975. *Ibid* **64**: 465 (morphology of zone lines; pseudosclerotia; **55**, 932).

—— et al. 1975. *Ibid* **64**: 473 (moisture content & zone lines; **55**, 933).

—— ——. 1977. *Ibid* **68**: 321 (zone lines in relation to gaseous atmosphere in wood; **57**, 319).

Marks, G. C. et al. 1976. *Aust. For Res.* **7**: 115 (spread in *Eucalyptus*; **56**, 3242).

Morrison, D. J. 1974. *Bi-mon. Res. Notes* **30**: 18 (effect of soil pH on rhizomorph growth; **54**, 1142).

——. 1976. *Trans. Br. mycol. Soc.* **66**: 393 (vertical distribution in soil; **56**, 72).

Munnecke, D. E. et al. 1970. *Phytopathology* **60**: 992 (dosage response to methyl bromide; **49**, 3661).

—— ——. 1973. *Ibid* **63**: 1352 (effects of soil fumigants & interaction with *Trichoderma*; **53**, 2811).

—— ——. 1976. *Ibid* **66**: 1363 (effect of heating or drying on *A. mellea* & *Trichoderma viride*, & survival in soil; **56**, 1943).

Ohr, H. D. et al. 1973. *Phytopathology* **63**: 965 (interaction with *Trichoderma* as modified by methyl bromide; **53**, 2193).

—— ——. 1974. *Trans. Br. mycol. Soc.* **62**: 65 (effects of methyl bromide on antibiotic formation; **53**, 2820).

Olembo, T. W. 1972. *Eur. J. For. Pathol.* **2**: 134 (effect of soil leachates on penetration & colonisation of conifer wood cylinders; **52**, 1711).

Ono, K. 1965. *Bull. Govt For. Exp. Stn Meguro* 179, 62 pp. (on larch, effects of topography & soil conditions; **45**, 635).

——. 1970. *Ibid* 229, 219 pp. (general ecology, especially soil conditions; **50**, 2001).

Pawsey, R. G. 1973. *Arboric. Assoc. J.* **4**: 116 (review; **52**, 3832).

Raabe, R. D. 1962. *Hilgardia* **33**: 25 (hosts; **42**: 447).

——. 1965. *Pl. Dis. Reptr* **49**: 812 (hosts).

——. 1966. *Phytopathology* **56**: 1241 (variation in culture; **46**, 883).

——. 1967. *Ibid* **57**: 73 (variation in pathogenicity; **46**, 1951).

——. 1972. *Mycologia* **64**: 1154 (as above in single spore isolates; **53**, 2496).

Rayner, R. W. 1959. *Kenya Coff.* **24**: 361 (on coffee; **39**: 228).

Redfern, D. B. 1968. *Ann. Bot.* **32**: 293 (ecology in UK, biological control; **47**, 2564).

——. 1970. In *Root diseases and soil-borne pathogens* (editors Toussoun, T. A. et al.): 147 (rhizomorph growth through soil).

——. 1973. *Trans. Br. mycol. Soc.* **61**: 569 (growth & behaviour in soil; **53**, 2086).

——. 1975. In *Biology and control of soil-borne plant pathogens* (editor Bruehl, G. W.): 69 (effect of food base on rhizomorph growth & pathogenicity).

Rhoads, A. S. 1942. *Phytopathology* **32**: 487 (*A. tabescens*; **21**: 497).

——. 1945. *Mycologia* **37**: 741 (comparison of *A. mellea* with *A. tabescens*; **25**, 186).

——. 1948. *Phytopathology* **38**: 44 (*A. tabescens* on citrus; **27**: 279).

Rishbeth, J. 1968. *Trans. Br. mycol. Soc.* **51**: 575 (rhizomorph growth; **48**, 39).

——. 1970. In *Root diseases and soil-borne pathogens* (editors Toussoun, T. A. et al.): 141 (basidiospores & infection).

——. 1972. *Eur. J. For. Pathol* **2**: 193 (formation of rhizomorphs from stumps; **52**, 3431).

——. 1976. *Ann. appl. Biol.* **82**: 57 (control by chemical treatment & inoculation; **55**, 3298).

——. 1978. *Trans. Br. mycol. Soc.* **70**: 213 (effects of soil temp. & atmosphere on rhizomorph growth; **57**, 4340).

Romagnesi, H. 1973. *Bull. Soc. mycol. Fr.* **89**: 195 (morphology of *A. mellea* inter alia; **55**, 421).

Rykowski, K. 1975. *Eur. J. For. Pathol.* **5**: 65 (infection of *Pinus sylvestris*; **55**, 421).
Schönhar, S. 1977. *Z. PflKrankh. PflSchutz* **84**: 304 (review, 1972–76; **56**, 5810).
Shaw, C. G. 1977. *Pl. Dis. Reptr* **61**: 416 (variation in pathogenicity; **57**, 810).
—— et al. 1978. *Eur. J. For. Pathol.* **8**: 163 (review, control in managed coniferous forests, 102 ref.).
Singer, R. 1970. *Schweiz. Z. Pilzk.* **48**: 65 (nomenclature, morphology & distribution; **49**, 3142).
Singh, P. 1974a. *Inf. Rep. For. Res. Centre* NX 126, 38 pp. (review; 106 ref.; **55**, 901).
—— et al. 1974b. *Eur. J. For. Pathol.* **4**: 20 (effect of disease on foliar nutrients & growth of some conifer spp.; **53**, 3203).
Sokolov, D. V. 1964. Moscow, Izdatel'stvo Lesnaya Promȳshlennost', 184 pp. (review, 22 pp. ref.; **45**; 1528).
Swift, M. J. 1968. *Trans. Br. mycol. Soc.* **51**: 241 (inhibition of rhizomorph development in soil; **47**, 3020).
——. 1970. In *Root diseases and soil-borne pathogens* (editors Toussoun, T. A. et al.): 150 (substrate colonisation relating to control by ring barking).
——. 1972. *Forestry* **45**: 67 (ecology in woodlands of Zimbabwe; **51**, 4431).
Thomas, H. E. 1934. *J. agric. Res.* **48**: 187 (infection, parasitism & resistance; **13**: 552).
Wallace, G. B. 1937. *E. Afr. agric. J.* **3**: 49 (tree hosts in Tanzania; **16**: 846).
Wardlaw, C. W. 1950. *Trop. Agric. Trin.* **27**: 95 (on oil palm; **30**: 155).
Wargo, P. M. 1975. *Physiol. Pl. Pathol.* **5**: 99 (lysis of *A. mellea* cell wall by enzymes from forest trees; **54**, 3496).
Wiehe, P. O. 1952. *E. Afr. agric. J.* **18**: 67 (on tung; **32**: 703).

ASCOCHYTA Lib., *Pl. Crypt. Ard.* (*Fasc. 1*): **8**, 1830.

(Sphaerioidaceae)

PYCNIDIA usually glabrous but sometimes hairy, mostly globose–subglobose or ampulliform to mammiform, sometimes more irregular in shape, separated or in small groups, usually subepidermal then erumpent with usually one, but sometimes more openings (pores) which may be simple, impressed, slightly protruding or papillate; wall mostly relatively thin, usually pseudoparenchymatic, the outer cells darker and more thick walled than the hyaline inner cells. CONIDIOGENOUS CELLS not distinctly differentiated from the inner cells, but recognisable by the cuspidate or somewhat elongated apex.

Under light microscopy the CONIDIA arise in basipetal succession as thin-walled protrusions at the apices of the conidiogenous cells. A kind of collar may be present or not evident at all. Under electron microscopy the CONIDIOGENOUS CELLS appear to be annellides producing from successively developing conidiogenous loci, thin-walled conidia seceding by a 3-layered septum. The wall of the first conidium arises as an outgrowth of the thin wall at the apex of the conidiogenous cell. The walls of successive conidia arise from the apical part of very short percurrent proliferations of the conidiogenous cell involving the basal layer of the 3-layered septum remaining after secession of the previous conidium. Percurrent growth, however, not only occurs at increasingly higher levels resulting in a series of annellations one above another encircling the apex of the conidiogenous cell, but also at *c.* the same level and then appears like a collar of periclinal annellations, the annellate collar. Conidiogenesis proceeds without evident production of mucilage.

CONIDIA hyaline or sometimes slightly coloured (yellow to pale brown), usually cylindrical to ellipsoidal or cymbiform, mostly twice or three times as long as wide, generally measuring between (5) 8–25 (29) × (2) 2.5–6 (8) μ. Conidia after secession often aseptate but then soon becoming 2 or occasionally 3 or even 4 celled. Under electron microscopy the cross-wall formation is characterised as distoseptation: production of a new inner conidial wall layer which concurrently by invagination divides the conidia in 2 or more cells; the invagination is initiated by the development of a septal plate. This distoseptation process is associated with abundant production of mucilage (from G. H. Boerema & G. J. Bollen, *Persoonia* 8: 136, 1975).

Ascochyta spp. can be confused with those of *Phoma* and *Phyllosticta*. *Ascochyta* spp. show annellidic ontogeny. Conidial septation is an essential part of conidial completion so that the conidia are always mainly 2 (or more) celled. *Phoma* spp. show phialidic ontogeny and the conidia are in principle aseptate, although secondary septation may occur. *Phyllosticta* spp. have aseptate conidia with an apical appendage. Boerema et al. and Brewer et al. described the differing formation of the conidia in *Ascochyta* and *Phoma*.

General studies (on taxonomy, morphology and host range) have been done by Mel'nik, for the Leguminoseae by Bondartzeva-Monteverde and

Sprague (1929), and for Gramineae by Sprague et al. *Ascochyta* spp. on Gramineae were very recently described by Punithalingam (1979). This monograph was not consulted for the account here. Crossan examined 200 isolates from 10 crop plants in inoculation experiments and in culture. It was considered that *A. abelmoschi* Harter and *A. gossypii* Woronich. (q.v.) should be tentatively considered synonymous with *A. phaseolorum*. But of the 4 other spp. also considered synonymous, 3 (*A. althaeina* Sacc. & Bizz., *A. capsici* Bond.-Mont. and *A. nicotianae* Pass.) were placed in synonymy with *Phoma exigua* Desm. (var. *exigua*) by Boerema et al. (1973a, *Phoma* q.v.). The fourth, *A. lycopersici*, is a synonym of *Didymella lycopersici* (q.v.).

A. abelmoschi causes a disease on fruit of okra (*Hibiscus esculentus*) and may also infect the leaves and stems. Bond considered that the same fungus infected the ornamental *H. rosa-sinensis*. Okra fruit show oval spots, with a brown to black margin and more or less zonate markings. A general rot may follow and pycnidia form both on the fruit and seed. The leaves show irregular spots with purple to black margins and grey centres, pycnidia are arranged more or less concentrically and shotholes form. Buds and flowers can also be attacked (Ellis). *A. boltshauseri* Sacc. from common bean was discussed by Sprague (1935) who compared it with *A. pisi* (q.v.) and *Mycosphaerella pinodes* (q.v.) and considered it distinct. *Stagonopsis phaseoli* Erikss. was the same as *A. boltshauseri*. *Stagonospora hortensis* Sacc. & Malbr. is considered to be the correct name with *A. boltshauseri* as a synonym (see Dingley).

A. caricae-papayae Tarr (for *A. caricae* Pat.) causes a fruit rot of papaw. The small, circular, watersoaked spots enlarge slowly and may reach 7–8 cm diam.; the lesions become sunken, brownish black, tough, with partially erumpent pycnidia. Several lesions may affect almost the whole of 1 fruit. Chowdhury infected uninjured fruit and found that more rot developed at 25° and 30° compared with 20° and 35° (at RH 100%). The diagnostic features and synonymy of *A. fabae* Speg. were given by Boerema et al. (l.c.); and see Punithalingam et al. This fungus is seedborne in broad bean (*Vicia faba*); infected seeds produce seedlings with leaf lesions and spread in the field was up to 10 m in an av. season (Hewett).

A. melongenae Padmanabhan causes leaf spots on eggplant; they were found on seedlings. The small brown spots, somewhat angular at first, enlarge to become circular or elliptical (or irregularly so) with a well-defined margin. The centres become papery and shotholes may be formed. The spots are mainly confined between the principal lateral veins. The fungus was compared with other fungi from Solanacae and referred to *Ascochyta*. *A. prasadii* Shukla & Pathak. causes leaf spots on hemp (*Cannabis sativa*); these are brick red to brownish, turning white and with pycnidia. Accounts of the minor leaf spots on *Sorghum* spp. have been given by Sprague, *Diseases of cereals and grasses in North America*; Sprague et al.; Singh et al.; Tarr, *Diseases of sorghum, Sudan grass and broom corn*. *A. sorghi* Sacc. has smaller conidia ($11–21 \times 1.5–4\ \mu$) than *A. sorghina* Sacc. ($20 \times 8\ \mu$). The larger pycnidia of the latter protrude conspicuously and are rough to the touch, hence the common name rough leaf spot. *A. sorghi* (for synonymy see Sprague et al.; Müller) infects other grasses but *A. sorghina* (see Singh et al. 1977) appears to be restricted to *Sorghum* spp. The characteristics of the elliptical leaf spots, apart from those of the pycnidia, caused by these 2 fungi are not particularly distinctive. Stout gave a key to 4 spp. on maize and Ou, *Rice diseases*, gave one (from K. Hara 1959) to 4 spp. on rice. Kanjanasoon described a collar rot of rice in Thailand caused by *A. oryzae* Catt. (see Watts Padwick, *Manual of Rice diseases*).

A. tarda Stewart was described from coffee in Ethiopia. The fungus attacks the young leaves and destroys the growing shoot. Necrotic lesions on the mature leaves are irregular, often beginning at the margins; they show concentric zonation in some cases, a watersoaked margin when expanding and resemble lesions caused by *Glomerella cingulata*. The opt. temp. for conidial germination and growth on PDA is 21–23°. Most germination occurs on the upper surface of leaves that are wounded and it is restricted at < RH 100% (Firman).

Aoki, J. 1955. *Forschn Geb. PflKrankh. Kyoto* **5**: 71 (*Ascochyta abelmoschi*; **35**: 268).

Boerema, G. H. et al. 1975. *Persoonia* **8**: 111 (conidiogenesis & conidial septation as differentiating criteria between *Phoma* & *Ascochyta*; **55**, 1676).

Bond, T. E. T. 1943. *Trop. Agric. Trin.* **20**: 67 (*A. abelmoschi*; **22**: 340).

Bondartzeva-Monteverde, V. N. et al. 1940. *Trudy̆ bot. Inst. Akad. Nauk. SSSR 1938* Ser. 2: 345 (on Leguminosae; **20**: 232).

Brewer, J. G. et al. 1965. *Proc. K. ned. Akad. Wet.* Ser. C **68**: 86 (conidial development in, & differentiation of, *Ascochyta* & *Phoma*; **44**, 3002).

Chowdhury, S. 1950. *Trans. Br. mycol. Soc.* **33**: 317 (*A. caricae*; **30**: 479).
Crossan, D. F. 1958. *Phytopathology* **48**: 248 (comparisons between spp.; **37**: 580).
Dingley, J. M. 1961. *N.Z. Jl agric. Res.* **4**: 336 (*Stagonospora hortensis* inter alia; **41**: 443).
Ellis, D. E. 1950. *Phytopathology* **40**: 1056 (*A. abelmoschi*; **30**: 302).
Firman, I. D. 1965. *Trans. Br. mycol. Soc.* **48**: 161 (*A. tarda*; **44**, 3055).
Harter, L. L. 1918. *J. agric. Res.* **14**: 207 (*A. abelmoschi* sp. nov.).
Hewett, P. D. 1966. *Pl. Pathol.* **15**: 161 (*A. fabae*; **46**, 1381).
——. 1973. *Ann. appl. Biol.* **74**: 287 (as above; **53**, 739).
Kanjanasoon, P. 1962. *Int. Rice Commn Newsl.* **11**: 22 (*A. oryzae*; **42**: 319).
Mel'nik, V. A. 1971. *Mikol. i Fitopatol.* **5**: 15 (taxonomy, literature survey & conclusions; **50**, 3516).
——. 1972. *Nov. Sist niz. Rast.* **9**: 159 (90 taxa).
——. 1975. Izdatel'stuo Nauka, Leningrad, 246 pp. (*Ascochyta* spp. key, 363 ref; **57**, 990).
Müller, E. 1952. *Phytopathol. Z.* **19**: 403 (*Ascochyta* inter alia on cereals; **32**, 245).
Padmanabhan, S. Y. 1948. *Indian J. agric. Sci.* **17**: 393 (*A. melongenae* sp. nov.; **28**: 112).
Punithalingam, E. et al. 1975. *C.M.I. Descr. pathog. Fungi Bact.* 461 (*A. fabae*).
——. 1979. *Mycol. Pap.* 142 (graminicolous *Ascochyta* spp.).
Shukla, D. D. et al. 1967. *Sydowia* **21**: 277 (*A. prasadii* sp. nov.; **48**, 1739).
Singh, D. S. et al. 1977. *Mycopathologia* **61**: 173 (*A. sorghina*, culture, infection & pycnidial formation; **57**, 2101).
Singh, R. S. et al. 1951. *Indian Phytopathol.* **4**: 45 (*A. sorghi*; **32**: 17).
Sprague, R. 1929. *Phytopathology* **19**: 917 (on Leguminosae; **9**: 273).
——. 1935. *Ibid* **25**: 416 (*A. boltshauseri*; **14**: 613).
——. et al. 1950. *Mycologia* **42**: 523 (on Gramineae; **30**: 80).
Stewart, R. B. 1957. *Ibid* **49**: 430 (*A. tarda* sp. nov.; **36**: 760).
Stout, G. L. 1930. *Ibid* **22**: 271 (*Ascochyta* inter alia on maize; **10**: 305).
Ullasa, B. A. et al. 1974. *Indian J. Mycol. Pl. Pathol.* **4**: 218 (*A. caricae* & perfect state; **55**, 3675).

Ascochyta gossypii Woronichin, *Vêst. tiflis. bot. Sada* **35**: 25, 1914.
Ascochyta gossypii Syd., 1916.

PYCNIDIA on all aerial parts of the host except the blossoms, globose, light brown to black, 120 × 150 (–180) μ diam., wall composed of 1–4 layers of angular pseudoparenchymatic cells; ostiole 30–40 μ wide. CONIDIA hyaline, straight or slightly curved, cylindrical to ovoid, 1 septate, not constricted at the septum, rounded at each end, 8 × 12 (–14)μ, formed from hyaline, obpyriform to flask-shaped phialides.

On *Gossypium* spp. and *Nicotiana tabacum*. Also by inoculation on *Althaea rosea*, *Capsicum annuum*, *Glycine max*, *Hibiscus esculentus*, *Lycopersicon esculentum*, *Phaseolus* spp., *Solanum melongena* and *Vigna unguiculata* (P. Holliday & E. Punithalingam, *CMI Descr*. 271, 1970.)

Wet-weather blight of cotton is widespread (Map 259). Small, round, whitish spots first appear on the cotyledons and lower leaves. They have a deep purplish brown border and enlarge; the centre becomes light brown, papery, bearing the pycnidia. The centre may fall out and severe spotting results in defoliation of the lower leaves. Stem infection leads to lesions sometimes several cm long, these show cracking and ragged edges. The flowers are not attacked, but mature lint is destroyed and, in the half-open bolls, may be discoloured grey with pycnidia. In severe attacks plants are killed. A similar leaf spot occurs on tobacco where the darker peripheral area may have concentric zonations. Leaf lesions may develop more frequently on the midrib causing it to fracture. Lenticular, dark spots are found on the stem.

Severe losses in cotton have occurred in S.E. USA. The disease is generally one of young plants and associated with periods of dull, wet weather. Although more spectacular the stem canker phase is probably less damaging than leaf infection. Direct penetration of the leaves can occur presumably but there is little information on the sites and manner of infection. The in vitro opt. temp. for growth is 20–25°, pycnidia being produced after 36 hours. Cultures from single spores show morphological variability. Spread probably takes place through rain splash. Seed infection may be internal and only partly controlled by treatment, but it is not considered serious. Most seedling infection arises from soil and plant debris-borne inoculum. Pycnidia with viable spores can overwinter. This species has been considered synonymous with *A. phaseolorum* (q.v.) and other *Ascochyta* spp. European and American isolates of *A. phaseolorum* differ in pathogenicity. *A. gossypii* has attacked *Lupinus angustifolius* but appears to be only weakly pathogenic on this host, though strongly so to cotton. Isolates of *A. gossypii* from tobacco show similar pathogenicity to this host and cotton. Control consists of seed treatment, rotation and deep ploughing to reduce infectivity of host debris.

Chippendale, H. G. 1929. *Trans. Br. mycol. Soc.* **14**: 201 (development of pycnidia in vitro; **9**: 240).
Elliot, J. A. 1922. *Bull. Ark. agric. Exp. Stn* 178, 18 pp. (general; **2**: 215).
Holdeman, Q. L. et al. 1952. *Pl. Dis. Reptr* **36**: 8 (on tobacco **31**: 356).
Smith, A. L. 1950. *Ibid* **34**: 233 (seedling blight; **30**: 105).
Wallace, M. M. 1948. *E. Afr. agric. J.* **14**: 10 (outbreak in Tanzania; **27**: 565).
Weimer, J. L. 1951. *Pl. Dis. Reptr* **35**: 81 (on *Lupinus angustifolius*; **30**: 470).

Ascochyta phaseolorum Sacc., *Michelia* 1: 164, 1878.

PYCNIDIA immersed, amphigenous, spherical to subglobose or sunken, honey yellow, 60–220 μ diam., wall up to 3 cells thick, pseudoparenchymatic, darker round the circular, slightly protruding ostiole which is 10–20 μ diam., finally up to 60 μ diam. CONIDIA straight or slightly bent at one or both ends, occasionally sigmoid, hyaline, 1 septate, apex rounded or flattened, base rounded to truncate, 13–16.5 × 3.5–5 μ, formed from hyaline, aseptate, cylindrical or globose phialides, 9 × 4–7 μ. (B. C. Sutton & J. M. Waterston, *CMI Descr.* 81, 1966; see G. H. Boerema, *Neth. J. Pl. Pathol.* **78**: 113, 1972.)

Primarily a leaf and pod spot of common bean but occurs naturally on a wide range of leguminous (and other?) hosts (Alcorn; Crossan, *Ascochyta* q.v.; Narita; Narita et al.). The symptoms, which can be confused with those caused by *A. boltshauseri* (*Ascochyta* q.v.), occur on leaves, petioles, calyx, pedical and pods. The leaf spot is a drab brown to grey, with a darker margin and sometimes an outermost border of light green to yellow. The lesions may become zonate, torn in the centre and be up to 2–3 cm diam., they bear the pycnidia. On the stems and petioles dark lesions may be up to 2 cm long and elongate ones can form on the veins of the lower leaf surface. When leaf attack is severe defoliation occurs. Infection of the floral remnants leads to a stem end rot of the pod which can also be infected at other points. Pycnidia occur at all the attacked sites.

The pathogen is widespread in both tropical and temperate areas. Presumably because the disease appears to be of only limited and/or local importance its biology is only sketchily known. It was reported to cause damage to black and green gram (*Phaseolus mungo* and *P. aureus*, respectively) in India (Punjab, 1955), and to be locally serious in Australia (Queensland, 1967) and the Netherlands. Infection of intact tissue is probably rare and, therefore, it normally occurs through wounds which include lesions caused by other pathogens or wind damaged leaf margins. Mineral deficient crops may also be less resistant. The disease seems more severe in relatively cool weather and, apart from spread through infected debris, conidia are disseminated through rain splash. The disease is also seedborne and is reported to persist in seed in a viable condition for > 2 years. In Queensland the pathogen can be well established in tropical legume pastures. Differences in pathogenicity between isolates occur and infection work on nearly 40 spp. from several families showed differences in the degree of susceptibility. Non-leguminous hosts may be important in spread. Control is largely cultural: healthy seed, field sanitation, protection against damage and maintenance of healthy growth. Zineb has given some control.

Alcorn, J. L. 1968. *Aust. J. biol. Sci.* **21**: 1143 (symptoms, pathogenicity & host range; **48**, 1395).
Ellis, D. E. 1952. *Pl. Dis. Reptr* **36**: 12 (symptoms; **31**: 365).
Ghani, A. A. et al. 1955. *Pakist. J. scient. Res.* **7**: 83 (general; **35**: 502).
Narita, T. et al. 1973 & 1975. *Rep. Soc. Pl. Prot. N. Japan* 24: 6; 26: 3 (isolates from several hosts similar to *Ascochyta phaseolorum*, synonyms & cross-inoculations; **54**, 2146; **56**, 3941).
——. 1976. *Ibid* 27: 12 (as above; **56**, 2935).
Pegg, K. G. et al. 1967. *Qd agric. J.* **93**: 321 (symptoms & control; **46**, 3276).
Sneep, J. 1945. *Tijdschr. PlZiekt.* **51**: 1–16 (general & control; **26**: 276).

Ascochyta pisi Lib., *Pl. Crypt. Ard.* (*Fasc.1*) No. 59, 1830.

Sphaeria concava Berk., 1841
Ascochyta pisicola (Berk.) Sacc., 1884.

PYCNIDIA on leaves and pods, globose, brown, immersed, later becoming erumpent, opening by apical ostioles 100–200 μ diam.; wall composed of 1–4 layers of elongated yellow brown, thin-walled cells.

CONIDIOPHORES (*phialides*) hyaline, short, obpyriform 6–14×3–8 μ, arising from the hyaline cells lining the inside of the pycnidium. CONIDIA hyaline, straight or slightly curved, 1 septate, slightly constricted at the septum, shortly cylindrical with rounded ends, 10–16×3–4.5 μ. Colonies on oat agar produce abundant pycnidia, spore exudate carrot red. On *Pisum*, *Lathyrus* and *Vicia*. (E. Punithalingam & P. Holliday, *CMI Descr*. 334, 1972.)

Leaf, stem and pod spot of pea (*P. sativum*) and other legumes is widespread (Map 273). The leaf lesion is somewhat light brown with a darker, frequently prominent, margin and pale centre. The stem lesions, rather sunken, are less abundant than those caused by *Mycosphaerella pinodes* (q.v.); see also *Phoma medicaginis* var. *pinodella*. *A. pisi*, which also causes damping off and dwarfing, is essentially an above-ground pathogen. Although a basal stem rot may be found, the characteristic foot rot syndrome caused by *M. pinodes* does not occur. Primary lesions often form on the first leaves. Pod infection can lead to aborted seed or a range of other damage to seed. Penetration is via the cuticle and stomata. Infection takes place over a wide temp. range (7–27°) but at lower temps symptoms are delayed. Linear growth in vitro is best at 20–28° and conidial germination good at 20–24°. Spread through seed (Dekker; Wharton et al.) is important; it also occurs from host debris and conidia are dispersed through water. The pathogen remains viable in seed for several years (e.g. Crosier; Wallen 1955). Effective spread in soil is unlikely since *A. pisi* has a low saprophytic ability and chlamydospores are rare or absent. Four races have been described from Canada (Gilpatrick et al.; Wallen 1957).

Detailed studies have been made on the effect of light (Leach; Leach et al.; Blakeman et al.; Trione et al.) on sporulation, leaf infection and conidial germination. Some isolates sporulate moderately in the dark but others do not. Near UV at 3131 °K is most effective in stimulating sporulation which is affected by wavelength, intensity and exposure time of light. An action spectrum for light-induced sporulation and a UV light-absorbing material which induced sporulation have been described. A detailed comparison of the development of lesions on detached leaflets, the roles of cell wall degrading enzymes and phytoalexins have been studied with *M. pinodes* (Uehara; Harrower; Heath et al.). Disease-free seed, which can be produced in dry areas, must be used. A 24 hours soak in captan or thiram at 30° or treatment with benomyl are effective (Maude; Maude et al.). Pea seed stored for 7 years can be free from infection. Resistance is controlled by dominant genes and resistant vars. are generally used. (For further ref. on this pathogen see *M. pinodes* with which it may occur in the field.)

Blakeman, J. P. et al. 1967. *Trans. Br. mycol. Soc.* **50**: 385 (effect of light on infection of leaf tissue with inter alia *Mycosphaerella pinodes*; **47**, 24).

Brewer, D. et al. 1953. *Can. J. Bot.* **31**: 739 (histology, ontogeny & resistant cvs; **33**: 650).

——. 1960. *Ibid* **38**: 705 (infection & temp., resistance; **40**; 258).

Crosier, W. 1939. *J. agric. Res.* **59**: 683 (occurrence & longevity in *Vicia* seed; **19**: 224).

Dekker, J. 1957. *Tijdschr. PlZiekt.* **63**: 65 (seed infection; **37**: 258).

Gilpatrick, J. D. et al. 1950. *Pl. Dis. Reptr* **34**: 383 (races; **31**: 268).

Harrower, K. M. 1973. *Trans. Br. mycol. Soc.* **61**: 383 (differential effects of phytoalexin on 2 races; **53**, 2369).

Heath, M. C. et al. 1969. *Ann. Bot.* **33**: 657 (leaf lesions & cell degrading enzymes with *M. pinodes*; **49**, 294).

—— ——. 1971. *Ibid* **35**: 451 (as above & phytoalexins; 50, 3275).

Leach, C. M. 1962. *Can. J. Bot.* **40**: 1577 (light & sporulation; **42**: 368).

—— et al. 1965. *Pl. Physiol.* **40**: 808 (action spectrum & light-induced sporulation; **45**, 732).

Ludwig, O. 1928. *Beitr. Biol. Pfl.* **16**: 464 (spore germination, infection & cytology; **8**: 416).

Lyall, L. H. et al. 1958. *Can. J. Pl. Sci.* **38**: 215 (inheritance of resistance; **37**: 616).

Maude, R. B. 1966. *Ann. appl. Biol.* **57**: 193 (seed infection & chemical control; **45**, 2660).

—— et al. 1970. *Ibid* **66**: 37 (as above; **50**, 1007).

Middleton, J. T. et al. 1947. *Phytopathology* **37**: 363 (disease free pea seed in dry areas; **26**: 475).

Skolko, A. J. et al. 1954. *Can. J. agric. Sci.* **34**: 417 (characteristics of seed infection & transmission; **34**: 200).

Trione, E. J. et al. 1966. *Nature Lond.* **212**: 163 (light-absorbing material & its effect on sporulation; **46**, 255).

Uehara, K. 1958. *Ann. phytopathol. Soc. Japan* **23**: 230 (phytoalexin; **9**: 362).

——. 1962. *Bull. Hiroshima agric. Coll.* **2**: 1 (identification & estimation of phytoalexin; **42**: 584).

Wallen, V. R. 1955. *Pl. Dis. Reptr* **39**: 674 (effect of storage on viability in pea seed; **35**: 502).

——. 1957. *Can. J. Pl. Sci.* **37**: 337 (races; **37**: 432).

Wark, D. C. 1950. *Aust. J. agric. Res.* **1**: 382 (inheritance of resistance; **30**: 400).
Weimer, J. L. 1947. *J. agric. Res.* **75**: 181 (resistance in *Lathyrus* & *Pisum* spp. with *M. pinodes*; **27**: 270).
Wharton, A. et al. 1970. *Proc. int. Seed Test. Assoc.* **35**: 173 (seed infection test with UV; **50**, 375).

Ascochyta rabiei (Pass.) Labrousse, *Rev. Pathol. vég. Ent. agric. Fr.* **18**: 228, 1931.
Zythia rabiei Pass., 1867
Phyllosticta rabiei (Pass.) Trotter, 1918
P. cicerina Prill. & Delacr., 1893.

PYCNIDIA on stems, leaves and seed pods, immersed becoming erumpent, globose, dark brown, 140–200 μ diam.; wall composed of 1–2 layers of elongated pseudoparenchymatic cells, ostiole 30×50 μ wide. CONIDIA hyaline, straight or slightly curved, 1 septate, some aseptate, slightly or not constricted at the septum, rounded at each end, $10\text{–}16 \times 3.5$ μ formed from hyaline, ampulliform phialides.

Kovachevski recorded *Mycosphaerella rabiei* (=*Didymella rabiei* (Kovachevski) Arx) on overwintered straw. According to him when ascospores were plated cultures producing pycnidia were obtained. As the perfect–imperfect state correlation needs further confirmation the chick pea pathogen is better retained here under *Ascochyta rabiei* (E. Punithalingam & P. Holliday, *CMI Descr.* 337, 1972).

Blight of gram or chick pea (*Cicer arietinum*) attacks all above ground parts of the plant; circular lesions on leaves and pods, and elongate ones on petioles and stems. The pycnidia form in concentric areas on these lesions and in severe attacks the whole plant is killed (see Grewal for an account of blight). *A. rabiei*, frequently destructive, is largely restricted to the Mediterranean region, S.E. Europe and S.W. Asia, and it is also reported from Tanzania (Map 151). Most conidial germination and growth occurs at 20–25° and the incubation on leaves is 7 days at 13–16° and 5 days at 18–25° (Zachos et al.). Kaiser in more recent work found the best growth and sporulation at 15–20°. Pycnidia with viable conidia formed on dry stem pieces at 10–30° (opt. 20°). The fungus survived >2 years in naturally infected tissue at 10–35°, RH 0–3% and on the soil surface. Viability was rapidly lost at RH 65–100% at soil depths of 10–40 cm.

Seed infection is important in spreading chick pea blight. Infection is carried both on and within the seed, and infected pods can show seed infestation of 50–80%. Seed formation, size and germination, and seedling growth are adversely affected (e.g. Halfon-Meiri; Luthra et al. 1932; Maden et al.; Sattar). Arif et al. found no evidence for the existence of races but Vir et al. (1974a) reported that there was evidence for 2. Field sanitation should be practised and seed treated with fungicides (e.g. Kaiser et al.); for fungicide applications on the plant see Vir et al. (1974b, captan and zineb). Sources of resistance occur (*Rev. appl. Mycol.* **21**: 120; **33**: 404; **43**, 1485; **47**, 375; *Rev. Pl. Pathol.* **51**, 3659; **55**, 4957). Vir et al. (1975) reported a single dominant gene controlling resistance in one cv.

Arif, A. G. et al. 1965. *W. Pakist. J. agric. Res.* **3**: 103 (no evidence for races; **47**, 2924).
Grewal, J. S. 1975. In *Advances in mycology and plant pathology* (editors S. P. Raychaudhuri et al.): 161 (review, 51 ref.).
Hafiz, A. 1952. *Phytopathology* **42**: 422 (resistance connected with number of hairs & secretion of organic acids; **32**: 300).
Halfon-Meiri, A. 1970. *Pl. Dis. Reptr* **54**: 442 (seed infection in Israel; **49**, 3059).
Kaiser, W. J. 1973. *Mycologia* **65**: 444 (factors affecting growth, sporulation, pathogenicity & survival; **52**, 4292).
—— et al. 1973. *Pl. Dis. Reptr* **57**: 742 (fungicide seed treatment; **53**, 1655).
Kovachevski, I. C. 1936. *Minist. agric. nat. Domains Sofia* 80 pp. (general, morphology, *Mycosphaerella rabiei* sp. nov., no Latin diagnosis; **15**: 700).
Labrousse, F. 1930 & 1931. *Rev. Pathol. vég. Ent. agric. Fr.* **17**: 174; **18**: 226 (symptoms, morphology & taxonomy; **9**: 697; **11**: 150).
Luthra, J. C. et al. 1932 & 1939. *Indian J. agric. Sci.* **2**: 499; **9**: 791 (general in India, Punjab; **12**: 264).
Maden, S. et al. 1975. *Seed Sci. Technol.* **3**: 667 (detection, site & spread of seedborne inoculum; **55**, 3381).
Sattar, A. 1933. *Ann. appl. Biol.* **20**: 612 (general in Punjab; **13**: 346).
Vir, S. et al. 1974a. *Indian Phytopathol.* **27**: 355 (races; **55**, 2979).
—— ——. 1974b. *Ibid* **27**: 641 (fungicides; **55**, 4956).
—— ——. 1975. *Euphytica* **24**: 209 (inheritance of resistance; **54**, 4272).
Zachos, D. G. et al. 1963. *Ann. Inst. Phytopathol. Benaki* **5**: 167 (general in Greece, role of ascospores; **44**, 299).

ASPERGILLUS Micheli ex Fr.; Micheli, *Nova Pl. Gen.*: 212–13, 1729; Fries, *Syst. Mycol.* 1: XLV, 1821.

(Dematiaceae)

COLONIES effuse, variously coloured, often green or yellowish, sometimes brown or black. MYCELIUM partly immersed, partly superficial. STROMA none. SETAE and HYPHODIA absent. CONIDIOPHORES macronematous, mononematous, often with a foot cell, straight or flexuous, colourless or with the upper part mid to dark brown, usually smooth, swollen at apex into a spherical or clavate vesicle the surface of which is covered by short branches or in some spp. by phialides; the branches are in 1 or several series and the terminal ones in the series always bear phialides. CONIDIOGENOUS CELLS monophialidic, discrete, several arising together at the ends of terminal branches or over the surface of the vesicle, mostly determinate, rarely percurrent, ampulliform or lageniform, collarettes sometimes present. CONIDIA catenate, dry, semi-endogenous or acrogenous, spherical, variously coloured, smooth, rugose, verruculose or echinulate, sometimes with spines arranged spirally, aseptate (M. B. Ellis, *Demat. Hyphom.*: 547, 1971).

The members of this genus are common moulds, inhabitants of soil and rhizosphere, frequent contaminants of cultures and stored seed, and components of physiological (including industrial) processes. They are not primarily plant pathogens, although they may often be associated with diseased conditions. The monograph by Raper et al. gave, besides taxonomy, accounts of the occurrence and significance of the individual spp. Further general accounts of the genus have been given by Christensen & Kaufmann, *Grain storage, the role of fungi in quality loss*; Sinha & Muir (editors), *Grain storage: part of a system*; Smith, *An introduction to industrial mycology*; Smith et al. The last work has a key to the groups of *Aspergillus*. Some later papers on the taxonomy and morphology of the genus are given, including that by Subramanian who gave a key to the perfect states and a list of spp. in each perfect genus. Recently Harman et al. described *Aspergillus* spp. on seed, particularly *A. ruber* (Konig, Spieckermann & Bremer) Thom & Church on pea (*Pisum sativum*); see 1974 for earlier ref. The effects of infection on stored seed include a reduced germination, for example, Ellis et al.; Harman et al. 1976. Work on seed invasion after harvest and in storage is outside the scope of this account, and ref. later than those given in the general works (already quoted) have been largely omitted.

Christensen, C. M. 1972. *Pl. Dis. Reptr* **56**: 173 (*A. glaucus* & *A. restrictus* on sunflower seed; **51**, 3488).

Christensen, M. et al. 1978. *Trans. Br. mycol. Soc.* **71**: 177 (key to *Aspergillus nidulans* group spp. & related *Emericella* spp.).

Ellis, M. A. et al. 1974. *Ibid* **58**: 332 (fungi from soybean seed; **53**, 3703).

Harman, G. E. et al. 1974. *Phytopathology* **64**: 1339 (pathogenicity & infection sites in stored seeds; **54**, 2747).

—— ——. 1976. *Can. J. Bot.* **54**: 39 (physiologic change in seed germination of pea & infection by *A. ruber*; **55**, 4362).

Hess, W. M. et al. 1969. *Mycologia* **61**: 560 (surface characteristics of conidia).

Kozakiewicz, Z. 1978. *Trans. Br. mycol. Soc.* **70**: 175 (phialide & conidium development; **57**, 4352).

Kulik, M. M. et al. 1969. *Mycologia* **61**: 1142 (characterising conidial colour by reflectance spectrophotometry; **49**, 2789).

Lazăr, V. 1972. *Rev. roum. Biol.* Ser. bot. **17**: 57, 357 (taxonomy & fine structure of conidia; **51**, 3213; **52**, 3579).

Lee, J. D. et al. 1977. *Trans. Br. mycol. Soc.* **69**: 137 (computerised classification; **57**, 457).

Locci, R. 1972. *Riv. Patol. Veg.* 4 8 (Supplemento) 172 pp. (monograph, fine structure of ascosporic Aspergilli; **53**, 846).

Nealson, K. H. et al. 1967. *Mycologia* **59**: 330 (electrophoretic survey of esterases, phosphatases & leucine aminopeptidases in mycelial extracts; **46**, 2638).

Raper, K. B. et al. 1965. *The genus* Aspergillus, Williams & Wilkins (monograph, 54 pp. ref., **45**, 392).

Singh, R. 1973. *Mycopathol. Mycol. appl.* **49**: 209 (morphology of foot cell).

Smith, J. E. et al. (editors) 1977. *Genetics and physiology of* Aspergillus. *Symp. Br. mycol. Soc.* 1, 522 pp. Academic Press (includes taxonomy, development & biodeterioration; **57**, 2826).

Subramanian, C. V. 1972. *Curr. Sci.* **41**: 755 (perfect states; **52**, 1457).

Aspergillus flavus Link ex Fr., *Syst. Mycol.* **3**: 386, 1832.

(see Raper et al., *Aspergillus* q.v., for probable synonyms)

COLONIES on Czapek and malt agar usually spreading, yellow green, reverse colourless to dark red brown; occasionally dominated by hard sclerotia, white at first, becoming red brown to almost black with age, 400–700 μ diam. CONIDIAL HEADS typically radiate, splitting into several poorly defined columns, rarely exceeding 500–600 μ diam. (mostly 300–400 μ), smaller heads occasionally columnar up to 300–400 μ. CONIDIOPHORES heavy walled, uncoloured, coarsely roughened, usually < 1 mm in length, 10–20 μ diam., just below the vesicle. VESICLES elongated when young, becoming subglobose to globose, varying from 10–65 μ diam., commonly 25–45 μ; both metulae and phialides present. METULAE usually 6–10 × 4–5.5 μ but sometimes up to 15–16 × 8–9 μ diam. PHIALIDES 6.5–10 × 3–5 μ. CONIDIA typically globose to subglobose, conspicuously echinulate, variable 3–6 μ diam., often 3.5–4.5 μ sometimes elliptical to pyriform at first and occasionally remaining so, then 4.5–5.5 × 3.5–4.5 μ (A. H. S. Onions, *CMI Descr.* 91, 1966).

A worldwide saprophyte in soil, particularly in the tropics; occurring also on decaying organic matter, in stored grain and seed; causing disease on young plants and in animals. Aflatoxins are toxic metabolic products of *A. flavus*; they cause disease in animals. Raper et al. (l.c.) gave an account of the occurrence and significance of the *A. flavus* group. The fungus is frequently described from groundnut (and *A. niger* q.v.) and cotton bolls. The *A. flavus* group has been widely studied, particularly with respect to enzymes, antibiotics, spoilage, deterioration and behaviour in soil. Sweet orange and grapefruit seed inoculated with strs of *A. flavus*, which caused albinism in maize, developed albinism in the seedlings. The fungus could be isolated from the seed coat but not from the remainder of the seedling. It was suggested that saprophytic fungal growth produces a metabolite which is taken up by the seedling and which inhibits photosynthesis (Durbin).

Maize ears inoculated at late milk and early dough were more susceptible to infection than when at silk or early milk. In the field wounding is probably required for infection (Rambo et al.). Hartill found that there was an increase in yellow mould of tobacco caused by *A. flavus* during curing at RH 90–100%. Incidence also tended to increase with higher dry bulb temp. Damage from yellowing was minimal at 29.5° or below. *A. flavus* has been isolated from soybean seed (found internally and externally), particularly at higher temps.; it causes damping off in this crop (Dhingra et al.; Saharan et al.). Olutiola, working with an isolate from cacao beans, reported that growth and sporulation in vitro were most at 35°.

Dhingra, O. D. et al. 1973. *Pl. Dis. Reptr* **57**: 185 (effect of temp. on recovery from soybean seed; **52**, 3486).

Durbin, R. D. 1959. *Ibid* **43**: 922 (albinism in citrus; **39**, 171).

Hartill, W. F. T. 1967. *Rhod. Zamb. Malawi J. agric. Res.* **5**: 61 (effects of temp. & RH on yellow mould of tobacco; **46**, 2824).

Olutiola, P. O. 1976. *Trans. Br. mycol. Soc.* **66**: 131 (growth & sporulation in vitro; **55**, 2541).

LaPrade, J. C. et al. 1976. *Phytopathology* **66**: 675 (aflatoxin production & fungal growth in maize hybrids).

Rambo, G. W. et al. 1974. *Ibid* **64**: 797 (on ears of maize; **54**, 1249).

Saharan, G. S. et al. 1973. *Phytopathol. Z.* **78**: 141 (on soybean seed; **53**, 3699).

ARACHIS HYPOGAEA

A. flavus causes pre- and post-emergence losses in groundnut. Chohan et al., referring to the disease as aflaroot, described it after seed inoculation and subsequent growth in sterilised soil. Carter in Nigeria found that where the testa was undamaged the emergence of white (susceptible) seed was only 50%, whilst that of coloured (resistant) seed was 95–98%. Damage to the testa decreased emergence. A seed dressing increased emergence of susceptible seed and restored emergence of damaged seed to that of undamaged seed. But the effect of complete removal of the testa was not counteracted by a seed dressing; such naked seed gave *c.* 10% emergence, with or without a dressing. El-Khadem described post-emergence disease symptoms as including stunted growth, lanceolate pale green leaves and inhibition of root growth. When kernels were injured before inoculation only pre-emergence rot was greatly increased. At 28–29° there were only slight, transient symptoms. Jackson found that surface disinfested, intact pods were attacked most readily at 26–38°. Under gnotobiotic conditions *A. flavus* penetrated the shell tissue but seed invasion was limited to the testa; colonisation of the embryos appeared to be limited (Lindsey). The behaviour of the fungus in soil (and above ground colonisation) has been extensively studied, for example see Griffin and Joffe. The presence of nematodes may increase the incidence of *A. flavus* on shells and kernels. Fungicide control has been described (e.g. Aujula et al.; Carter; Frank) and see Zambettakis et al. for var. differences in susceptibility.

Abdalla, M. H. 1974. *Trans. Br. mycol. Soc.* **63**: 353 (mycoflora of kernels; **54**, 2559).

Aujula, S. S. et al. 1975. *Jnl Res. Ludhiana* **12**: 52 (aflaroot, field occurrence & chemical control; **55**, 2039).

Bell, D. K. et al. 1971. *Phytopathology* **61**: 1038 (effects of *Meloidogyne arenaria* & curing time on infection of pods).

Carter, J. B. H. 1973. *Ann. appl. Biol.* **74**: 315 (effects of testa, damage & seed dressings on emergence; **53**, 1645).

Chohan, J. S. et al. 1968. *Indian J. agric. Sci.* **38**: 568 (aflaroot, general; **48**, 1412).

Diener, U. L. et al. 1965. *Pl. Dis. Reptr* **49**: 931 (pod invasion, **45**, 925).

El-Khadem, M. 1975. *Zentbl. Bakt. ParasitKde* Abt. 1 **130**: 245 (pre- & post-emergence losses; **55**, 2044).

Frank, Z. R. 1969. *Israel Jnl agric. Res.* **19**: 109 (seedborne inocula & control with *Rhizopus*; **49**, 1277).

Griffin, G. J. 1969. *Phytopathology* **59**: 1214 (spore germination in rhizosphere with *Fusarium oxysporum*; **49**, 609).

——. 1972. *Ibid* **62**: 1387 (spore germination & population in geocarposphere; **52**, 3491).

—— et al. 1974. *Ibid* **64**: 322 (population in soil with *Aspergillus niger*; **53**, 4208).

—— ——. 1976. *Ibid* **66**: 1161 (colonisation of aerial pegs with *A. niger*; **56**, 1824).

Jackson, C. R. 1965. *Ibid* **55**: 46 (temp., infection & growth with other fungi; **44**, 1729).

—— et al. 1968. *Oléagineux* **23**: 531 (pod invasion in presence of lesion nematodes; **47**, 3643).

Joffe, A. Z. 1968. *Mycologia* **60**: 908 (effect of soil inoculation on soil mycoflora; **48**, 664).

—— et al. 1968. *Pl. Dis. Reptr* **52**: 718 (effect of soil fungicides on fungal development in soil & kernels; **48**, 317).

——. 1969. *Phytopathol. Z.* **64**: 321 (effects on groundnut & other plants; **48**, 3771).

Ketring, D. L. et al. 1976. *Agron. J.* **68**: 661 (growing season & site effects on water uptake & drying of seed from resistant & susceptible genotypes; **56**, 4796).

Lindsey, D. L. 1970. *Phytopathology* **60**: 208 (host effects under gnotobiotic conditions; **49**, 2253).

Minton, N. A. et al. 1967. *Oléagineux* **22**: 543 (host invasion in presence of nematodes; **47**, 376).

—— ——. 1969. *J. Nematol.* **1**: 318 (host invasion in presence of *M. hapla*; *Helminth. Abstr.* **39**, 674).

Wells, T. R. et al. 1972. *Phytopathology* **62**: 1238 (host colonisation under gnotobiotic conditions; **52**, 3489).

Zambettakis, C. et al. 1977. *Oléagineux* **32**: 377 (var. differences in susceptibility; **57**, 1916).

GOSSYPIUM

A. flavus infection of cotton bolls (i.e. as one of the aetiological agents in the boll rot complex) has been largely investigated with respect to the formation of its toxic products (aflatoxins) and only a few selected ref. are given (see Ramey). A bright greenish yellow fluorescence is associated with the presence of the fungus in cotton seed and fibre, and this is used to detect *A. flavus* (Ashworth et al. 1966; Gardner et al.; Simpson et al.). Seed and fibre are protected from infection by the carpel wall in unopened bolls; and infection therefore probably only occurs in undamaged bolls as they open at maturity. Field infections are associated with imperfectly opened bolls (Ashworth et al. 1969a; Gardner et al.). Halloin detected *A. flavus* in cotton seed (and the embryo) after inoculation; it was the fungus most frequently isolated from the non-inoculated control seed. Damage by insects predisposes bolls to infection, for example, the pink bollworm (*Pectinophora gossypiella*). The fungus was isolated from insects infesting cotton, for example, lygus and stink bugs (*Lygus hesperus* and *Chlorochroa sayi*). Ashworth et al. (1969b) found that infection of fibre was rapid at 20–30° and max. seed infection was at 30–35°. Some forms of the fungus were most aggressive at 35°. Gilbert et al. found that seed infection increased as the duration of the daily max. temp. of 30° increased and/or as the number of diurnal max. temp. cycles of 30° increased.

Ashworth, L .J. et al. 1966. *Phytopathology* **56**: 1104 (association with a greenish yellow fluorescence of seed; **46**, 338).

—— ——. 1969a. *Ibid* **59**: 383 (time of infection & effect of fibre moisture; **48**, 2413).

—— ——. 1969b. *Ibid* **59**: 669 (effects of temp. & aeration on infection; **48**, 2997).

—— ——. 1971. *Ibid* **61**: 488 (relation of insects to infection; **50**, 3803).

Gardner, D. E. et al. 1974. *Ibid* **64**: 452 (geographic distribution & aflatoxin formation in infected seed; **54**, 1758).

Gilbert, R. G. et al. 1975. *Ibid* **65**: 1043 (effect of temp. on infection; **55**, 1814).

Halloin, J. M. 1975. *Ibid* **65**: 1229 (postharvest infection, with *Aspergillus niger* & *Rhizopus arrhizus*; **55**, 3190).

Kiyomoto, R. K. et al. 1974. *Ibid* **64**: 259 (boll rot & simulated pink bollworm injury; **53**, 3502).

McMeans, J. L. et al. 1975. *Crop Sci.* **15**: 865 (pink bollworm, infection & aflatoxin accumulation; **55**, 4735).

Marsh, P. B. et al. 1973. *Pl. Dis. Reptr* **57**: 664 (boll rot in relation to high aqueous extract pH of fibre; **53**, 964).

Ramey, H. H. 1974. *Phytopathology* **64**: 1451 (systems analysis of research objectives for control of aflatoxin formation; **54**, 2833).

Simpson, M. E. et al. 1971. *Pl. Dis. Reptr* **55**: 510 (distribution in USA crop of 1970; **51**, 376).

Stephenson, L. W. et al. 1974. *Ibid* **64**: 1502 (association with insects; **54**, 2270).

Aspergillus niger van Tieghem, *Annls Sci. nat. Botanique* Ser. 5 8: 240, 1867.

Sterigmatocystis niger van Tieghem, 1877.
(see Raper et al., *Aspergillus* q.v., for probable synonyms)

COLONIES with growth on Czapek agar rather restricted and on malt agar more rapid, 5–6 cm diam. in 2 weeks; with rather loose white to yellowish mycelium rapidly becoming black to dark brown on the upper surface with the development of conidia, but occasionally characterised by the presence of large white sclerotia *c.* 1 mm diam., usually with a plane even, slightly granular dusty surface, sometimes showing concentric zonation; reverse white to yellow. CONIDIAL HEADS globose, tangled or splitting into columns. CONIDIOPHORES smooth, hyaline or faintly brownish near the apex up to 3 mm, 15–20 μ diam. VESICLES globose or nearly so, up to 75 μ diam. but often quite small, fertile over the whole surface; metulae and phialides both produced but in some heads only phialides present. METULAE of varying lengths and sometimes septate, but when mature usually 20–30 μ long, in immature heads frequently shorter and occasionally longer ones are produced. PHIALIDES more uniform in length, usually $7–10 \times 2–3\ \mu$. CONIDIA more or less globose, often very rough or echinulate, mostly 4–5 μ diam., very dark (A. H. S. Onions, *CMI Descr*. 94, 1966).

A worldwide saprophyte in soil which occurs on decaying organic matter and causes disease in plants and animals (contamination from mouldy hay, straw, grain and other food). The fungus has been very widely studied in work on mould physiology and nutrition. It is frequently reported in investigations on rhizosphere fungi and on organisms causing plant storage rots. It may cause stalk rots and discolouration of cotton fibre. The most serious plant disease caused is crown rot (or collar rot) of groundnut (and *A. flavus* q.v.). All the ref. given are on this crop; some other ref. are given by Onions (l.c.). Crown rot is characterised in the field by wilt and death of seedlings accompanied by a hypocotyl rot. This rot is a yellow brown lesion with a profusely sporulating growth of *A. niger*; eventually the hypocotyl may become completely necrotic. In E. Africa Gibson (1953a) found the peak for death rate to be *c.* the seventeenth day, whereas that for hypocotyl lesions was *c.* 5 days later. Besides this a post-emergence seedling blight and a pre-emergence rot can be caused. Seeds below ground are covered in the black conidial masses. Late infection, i.e. beyond the seedling stage, may take place; lesions form on the stem below soil level, spread upwards to the branches and cause death. The pre- and post-emergence rots are, however, the more important forms of the disease. In India death of field plants was most at 50 days after sowing; most collar rot occurred at 13% and 16% soil moisture and 31° and 35° (Chohan 1965, 1969). The opt. temp. for fungal growth on agar was 37°; seedlings raised at 30–37° were most disposed to severe disease whose progress was most rapid at these temps, and (experimentally) there was most disease in light textured soil (Gibson 1953a). Seedlings were susceptible to *A. niger* under light of 500 foot candles or less but immune at 1500–5000 foot candles. Resistant seedlings were predisposed to infection by heat induced wounds or keeping in the dark for 2 days before inoculation (Ashworth et al.). Carry over is on crop (not necessarily groundnut) debris, in soil and on seed (mostly superficial infection).

Only undamaged seed of good quality should be used since there will be less risk that it has become infected after harvest or through damage during shelling. Damage to young plants should be avoided and seedling vigour maintained. Chohan et al. (1967) recommended rotation. Seed treatment with fungicides gives good control; thiram generally giving better results than organo-mercurials (e.g. Chohan et al. 1966). Captan is also effective. Gibson (1953b; and see Purss) found that, whilst seed dressings with thiram may reduce crown rot incidence by at least 50%, seed treatment with organo-mercurials generally increased disease. This effect was attributed to the selective toxicity of these mercury compounds operating to the advantage of *A. niger*. Mercury tolerant strs were isolated from soil in

Tanzania and these were more virulent than mercury sensitive ones. Chohan et al. (1970b) and Mathur et al. investigated the reaction of cvs to infection.

Agnihotri, J. P. et al. 1972. *Acta agron. hung.* **21**: 222 (fungicides; **53**, 2009).
Ashworth, L. J. et al. 1964. *Phytopathology* **54**: 1161 (epidemiology; **44**, 581).
Borut, S. Y. et al. 1966. *Israel Jnl Bot.* **15**: 112 (association with stored kernels; **46**, 3608).
Chohan, J. S. 1965. *Jnl Res. Ludhiana* **2**: 25 (general in India; **46**, 1806).
—— et al. 1966. *Ibid* **3**: 406 (fungicides; **46**, 2571).
—— ——. 1967. *Ibid* **4**: 536 (cultural control; **47**, 3287).
——. 1969. *Ibid* **6**: 634 (survival & soil factors; **50**, 1017).
——. 1970a. *Ibid* **7**: 567 (physiology of parasitism; **51**, 2088).
—— ——. 1970b. *Indian J. agric. Sci.* **40**: 546 (host resistance; **51**, 4556).
Gibson, I. A. S. 1953a. *Trans. Br. mycol. Soc.* **36**: 198 (general in E. Africa; **33**: 202).
——. 1953b. *Ibid* **36**: 324 (anomalous effect of organo-mercurial seed dressings; **33**: 519).
Griffin, G. J. et al. 1974. *Phytopathology* **64**: 322 (soil populations; **53**, 4208).
Mathur, R. L. et al. 1970. *Indian Phytopathol.* **23**: 143 (host resistance; **49**, 3525).
Purss, G. S. 1960. *Qd J. agric. Sci.* **17**: 1 (control; **40**: 447).
Sidhu, G. S. et al. 1971. *Jnl Res. Ludhiana* **8**: 211 (fungicides; **51**, 4557).

ASPERISPORIUM Maublanc, *Lavoura* 16: 212, 1913.

(Dematiaceae)

SPORODOCHIA punctiform, pulvinate, brown, olivaceous brown or black. MYCELIUM immersed. STROMA usually well developed, erumpent. SETAE and HYPOPODIA absent. CONIDIOPHORES macronematous, mononematous, closely packed together forming sporodochia, usually rather short, unbranched or occasionally branched, straight or flexuous, hyaline to olivaceous brown, smooth. CONIDIOGENOUS CELLS polyblastic, integrated, terminal, sympodial, cylindrical or clavate, cicatrised; scars prominent. CONIDIA solitary, dry, acropleurogenous, ellipsoidal, fusiform, obovoid, pyriform, clavate or obclavate, hyaline to brown or olivaceous brown, smooth or verrucose, with 0–3 transverse and sometimes 1 or more longitudinal or oblique septa (M. B. Ellis, *Demat. Hyphom*: 273, 1971; 6 spp. were described by Ellis in *More Demat. Hyphom.*, 1976).

Asperisporium caricae (Speg.) Maubl., *Lavoura* **16**: 212, 1913.

Cercospora caricae Speg., 1886
Fusicladium caricae (Speg.) Sacc., 1902
Scolecotrichum caricae Ellis & Everh., 1892
Epiclinium cumminsii Massee, 1898
Pucciniopsis caricae Earle, 1902.

MYCELIUM immersed, SPORODOCHIA hypophyllous, punctiform, pulvinate, dark blackish brown to black. STROMA well developed, erumpent. CONIDIOPHORES closely packed together and covering the surface of the stromata, unbranched or occasionally branched, straight or flexuous, hyaline to olivaceous brown, smooth, with several prominent conidial scars at the apex, up to 45 μ long, 6–9 μ thick. CONIDIA solitary, dry, ellipsoidal, pyriform or clavate, almost always 1 septate, hyaline to pale brown, distinctly verrucose, 14–26 $\times$ 7–10 μ (M. B. Ellis & P. Holliday, *CMI Descr.* 347, 1972).

Black spot, blight or 'rust' of papaw is a minor disease. Watersoaked spots on mature leaves become necrotic, usually circular and up to 4 mm diam., the dark conidial masses being conspicuous on the under surface. Abundant spotting causes defoliation and over 50% leaf fall can occur. Similar, conspicuous spots form on the fruit; they cause shallow lesions and no decay. Young leaves are not attacked. The fungus has been reported from central and S. America, West Indies, India, Kenya, Malawi, Zimbabwe, South Africa and USA (Florida and Texas; Map 488). Virtually nothing appears to be known of the biology of *A. caricae* which has not been grown in vitro. Penetration is stomatal and macroscopic symptoms are seen 8–10 days after inoculation. Incidence in Florida is seasonal, most infection occurring in late winter and spring. There is no confirmation of an early suggestion of seed transmission. Infection is said to reduce host vigour and it also causes fruit blemishes. Little control ever appears to be necessary apart from sanitary measures.

Stevens, H. E. 1939. *Proc. Fla. St. hort. Soc.* **52**: 57 (papaw diseases; **20**: 26).
Uphof, J. C. T. 1925. *Z. PflKrankh. PflPath. PflSchutz* **35**: 118 (general; **4**: 682).

Botryodiplodia theobromae

BOTRYODIPLODIA Sacc., *Sylloge Fung*. **3**: 377, 1884.
(Sphaerioidaceae)
(synonymy in B. C. Sutton, *Mycol. Pap*. 141: 26, 1977; see *B. theobromae* for a description)

Botryodiplodia theobromae Pat., *Bull. Soc. mycol. Fr*. **8**: 136, 1892.
Lasiodiplodia theobromae (Pat.) Griff & Maubl., 1909
Diplodia theobromae (Pat.) Nowell, 1923
D. gossypina Cooke, 1879
D. cacaoicola Henn., 1895
Macrophoma vestita Prill. & Delacr., 1894
L. tubericola Ellis & Everh. apud Clendinin, 1896
D. tubericola (Ellis & Everh. apud Clendinin) Taub., 1915
Botryodiplodia tubericola (Ellis & Everh. apud Clendinin) Petrak, 1923
B. gossypii Ellis & Barth., 1902
B. elasticae Petch, 1906
Chaetodiplodia grisea Petch, 1906
L. nigra Appel & Laubert, 1906
D. rapax Massee, 1910
D. natalensis Pole Evans, 1911
L. triflorae Higgins, 1916
D. ananassae Sacc., 1917
B. ananassae (Sacc.) Petrak, 1929.
(A perfect state *Physalospora rhodina* Berk & Curt. apud Cooke, *Grevillea* **17**: 92, 1900, has been described for *B. theobromae* but ascospores appear to play a very minor role in infection. For a description and illustration of the ascogenus state see Alvarez Garcia 1968; Stevens 1926; Voorhees; further synonyms in Zambettakis 1954.)

COLONIES on oat agar greyish sepia to mouse grey to black, fluffy with abundant aerial mycelium; reverse fuscous black to black. PYCNIDIA simple, or compound, often aggregated, stromatic, ostiolate, frequently setose, up to 5 mm wide. CONIDIOPHORES hyaline, simple, sometimes septate, rarely branched cylindrical, arising from the inner layers of cells lining the pycnidial cavity. CONIDIOGENOUS CELLS hyaline simple, cylindrical to subobpyriform, holoblastic, annelidic. CONIDIA initially aseptate, hyaline, granulose, subovoid to ellipsoid–oblong, thick walled, base truncate; mature conidia 1 septate, cinnamon to fawn, often longitudinally striate, (18–) 20–30 × 10–15 μ. PARAPHYSES when present hyaline cylindrical, sometimes septate, up to 50 μ long. PYCNIDIA on leaves, stems and fruits immersed, later becoming erumpent, simple or grouped, 2–4 mm wide, ostiolate, frequently pilose with conidia extruding in a black mass (E. Punithalingam, *CMI Descr*. 519, 1976).

B. theobromae is a plurivorous, wound and secondary pathogen, and a saprophyte, which is particularly common at relatively high temps. In isolation work from above-ground diseased tissue in the tropics it is an extremely frequent component of the fungal flora. It can cover a 9 cm Petri dish in <2 days, thus easily hindering the detection of any primary pathogen. Whilst sporulation occurs readily on host tissue the characteristic smoke grey, fluffy mycelium may be sterile in vitro. For sporulation in culture (light is essential) see, for example, Alasoudura; Ekundayo 1970; Ekundayo et al. 1969; Griffiths; Rao et al.; Uduebo 1974. The opt. temp. for growth both in vitro and in vivo is *c*. 30°. At 35° Uduebo (1974) found that vegetative growth diminished and there was no sporulation. The fungus is most important as a cause of postharvest fruit and tuber decay; for cellulytic and pectolytic enzymes see, for example, Arinze et al.; Pathak et al.; Umezurike; Wang et al. It is also commonly encountered as a secondary component in dieback complexes. Turner discussed the cacao dieback condition in detail and concluded that *B. theobromae* should only be regarded as a secondary invader. Only this fungus and *Calonectria rigidiuscula* (q.v.) merit consideration as pathogens in this particular dieback. *B. theobromae* may also be found affecting seed (e.g. Del Rosario; Srivastava 1964; Wilson). Only a few of the crops on which the fungus has been found causing damage, and on which some experimental work has been done, need be mentioned.

In citrus *B. theobromae*, commonly referred to as *Diplodia natalensis*, is one of the fungi causing stem end rot (*Diaporthe citri* q.v.) of the fruit. It is one of the most important postharvest diseases; the fruit rots caused by these 2 fungi are difficult to distinguish by symptoms only. The rind around the button end softens, becomes brown and the affected area spreads rapidly. A decay may appear at the stylar end as a result of the rapid spread of the internal decay. The invaded flesh has a bitter taste. Infection occurs, in undamaged fruit, from conidia lodged at the stem end. Actual penetration probably does not occur until natural openings develop in the separation layer between button and fruit at abscission (Brown et al. 1968; and see Littauer et al.; Nadel-Schiffman et al.). Decay mostly occurs 2–3 weeks after harvest. Fruit on the tree is not usually attacked unless injured or over ripe. Infection originates from dead wood. Conidia are washed on to the fruit and those lodging between calyx and rind escape desiccation. Brown found that released conidia could survive from one season to the next on the bark of citrus deadwood. The disease is checked by standard fungicidal treatment in the packing house. Brown et al. (1970) found that stem end rot was reduced in fruit from trees sprayed with benomyl the previous year. Applications to dormant trees in February reduced this rot in Hamlin oranges harvested in the following November and December. *B. theobromae* may also colonise weakened, woody tissue and cause a citrus dieback. Robinson dieback is a condition which occurs in Robinson tangerine. Young trees can be killed. Benomyl sprays can reduce the incidence of this dieback.

On banana an important in transit disease (finger rot), especially when transit times are >14 days from harvest to ripening, is caused by wound invasion (see Goos et al. for a general account and literature review). Infection often begins below the decayed flower. It results in a brownish black necrosis of the peel and softening of the pulp. Disease spread increases during ripening and the finger can become completely rotted (soft), with pycnidia forming in the wrinkled skin. Finger rot does not develop on fruit on the plant and is more common on fully mature fruit. Delay in placing harvested fruit at shipping temps increases disease incidence. Thiabendazole has given some control. *B. theobromae* is frequently reported causing a postharvest rot of mango; a rapidly spreading dark necrosis with a clearly defined margin begins at the stem end. Infection can occur via the exposed surface of the attached pedicel, the injured pedicel, abscission zone and wounded exocarp. Fruit can be completely rotted in 2–3 days. Quimo et al. described control in mango by hot water treatment (53° for 10 minutes).

The fungus is a very common invader of damaged cacao pods (brown pod) on the tree, and is one of the pathogens causing cotton black boll rot (commonly referred to as *Diplodia gossypina*). On pods the spread of the brown necrosis is so rapid that symptoms caused by other pod rotting agents (e.g. *Phytophthora palmivora* and *Crinipellis perniciosa*) can be completely masked. The dark, powdery spore masses develop on the pods which, when cut open, show the greyish black mycelium. When inoculated via deep wounds cotton bolls can be decayed in 4 days at 30° (9–11 days in uninjured bolls; McCarter).

B. theobromae is one of the organisms causing rots in tropical, edible tuber crops and recent work on cassava, sweet potato and yam is quoted. The rots occur mostly after harvest but on sweet potato (Java black rot) the disease can be severe in the plant bed on tubers and roots (Daines). Arinze et al. (1975) described the sweet potato storage rot as at first a dirty white and soft, then dark brown and finally becoming hard. Benomyl or captan treatment prevented rot. For cassava see Ekundayo et al. (1973); Pacca; and for yam see Adeniji; Ekundayo et al. (1972). A stem end rot of avocado is also caused. Ramsey et al. described a market disease of onion. A dry leathery rot with a grey to black discolouration of the outer dry scales was described. The condition only occurred on white cvs. In groundnut McGuire et al. found that heat injury predisposed the plants to infection (collar rot). The disease was more severe when the groundnut crop followed cotton than when planted after maize. *B. theobromae* overwintered and sporulated abundantly on old cotton bolls (and see Asuyama et al.). A disease on melon (*Cucumis melo*; Beraha et al.) causes stem gumming and blight which might be confused with the symptoms caused by *Didymella bryoniae* (q.v.) infection. On the crown or stem the lesions are watersoaked, oily, dark green and 1–5 cm long; a reddish or black gum-like exudate forms from older lesions. There is also a grey to black spongy postharvest decay of the fruit. There has been some mention of control measures in particular cases. In the field the association of *B. theobromae* with a diseased condition almost invariably indicates a predisposing factor. This may be poor physiological conditions in the crop; attack by another, possibly primary, pathogen; or animal or mechanical damage. Control of the potentially more serious rots that are caused after harvest must vary with local conditions. Sensible, cultural precautions at harvest and fungicidal treatment after it should reduce incidence.

Adeniji, M. O. 1970. *Phytopathology* **60**: 1698 (effects of moisture & temp. on yam decay organisms; **50**, 1550).
Alasoudura, S. O. 1970. *Mycopathol. Mycol. appl.* **42**: 153 (culture; **50**, 2154).
Alvarez Garcia, L. A. 1968. *J. Agric. Univ. P. Rico* **52**: 260 (perfect state; **48**, 1253).

Botryodiplodia theobromae

Arinze, A. E. et al. 1975 & 1976. *Int. Biodetn Bull.* **11**: 41; **12**: 15 (sweet potato, storage rot, cell degrading enzymes; **55**, 1004; **56**, 4967).

Arzee, T. et al. 1970. *Bot. Gaz.* **131**: 50 (avocado pedicel anatomy & localisation of mycelium; **49**, 3406).

Asuyama, H. et al. 1953. *Ann. phytopathol. Soc. Japan* **18**: 28 (on groundnut; **35**: 68).

Beraha, L. et al. 1976. *Pl. Dis. Reptr* **60**: 420 (on melon; **56**, 461).

Brown, G. E. et al. 1968. *Phytopathology* **58**: 736 (on citrus fruit, entry with *Diaporthe citri*; **47**, 3114).

—— ——. 1970. *Proc. Fla St. hort. Soc.* **83**: 222 (control by benomyl in citrus; **51**, 3367).

——. 1971. *Phytopathology* **61**: 559 (release & survival of conidia; **50**, 3752).

Daines, R. H. 1959. *Ibid* **49**: 252 (on sweet potato, temp. & fungicides; **38**: 622).

Del Rosario, M. S. E. 1954. *Philipp. Agric.* **37**: 623 (on maize kernels; **34**: 716).

Ekundayo, J. A. et al. 1969. *Can. J. Bot.* **47**: 1153, 1423; (pycnidial formation & light, pycnidial development & fine structure of conidium; **49**, 32, 387).

——. 1970. *Ibid* **48**: 67 (conidial germination; **49**, 1988).

—— ——. 1972. *Trans. Br. mycol. Soc.* **58**: 15 (on yam, pre-harvest rot; **51**, 3035).

—— ——. 1973. *Ibid* **61**: 27 (on cassava inter alia & benomyl control; **52**, 4304).

Frossard, P. 1964. *Fruits* **19**: 401 (on avocado; **44**, 776).

Goos, R. D. et al. 1961. *Mycologia* **53**: 262 (on banana, 43 ref.; **41**: 610).

Griffiths, D. A. 1967. *Mycopathol. Mycol. appl.* **33**: 273 (induced sporulation; **47**, 1441).

Halos, P. M. et al. 1970. *Philipp. Phytopathol.* **6**: 16 (on mango; **53**, 1477).

Haque, M. A. et al. 1973. *Bangladesh J. Bot.* **2**: 83 (on jute; **54**, 5446).

Littauer, F. et al. 1947. *Palest. J. Bot. Rehovot Ser.* **6**: 158 (on citrus, infection; **28**: 63).

McCarter, S. M. 1972. *Phytopathology* **62**: 1223 (on cotton, effects of temp. & boll injury; **52**, 2284).

McGuire, J. M. et al. 1965. *Ibid* **55**: 231 (on groundnut, heat injury; **44**, 2006).

Minz, G. 1946. *Palest. J. Bot. Rehovot Ser.* **5**: 152 (on citrus; **26**, 449).

—— et al. 1947. *Ibid* **6**: 165 (on citrus, infection).

Nadel-Schiffman, M. et al. 1947. *Ibid* **6**: 170 (on citrus, infection; **28**: 63).

Pacca, D. W. 1935. *Rodriguesia* **1**: 77 (on cassava; **15**: 278).

Pathak, V. N. et al. 1967. *Pl. Dis. Reptr* **51**: 744 (on mango, infection; **47**, 269).

—— ——. 1969. *Phytopathol. Z.* **65**: 164, 263 (on mango, pectolytic enzymes; **49**, 526, 807).

Quimo, A. J. et al. 1974. *Philipp. Phytopathol.* **10**: 16 (on mango, hot water treatment; **56**, 5124).

Ramsey, G. B. et al. 1946. *Phytopathology* **36**: 245 (on *Allium*; **25**: 485).

Rao, P. V. et al. 1978. *Trans. Br. mycol. Soc.* **70**: 121 (light dependent, synchronous pycnidial formation; **57**, 3303).

Roth, G. 1966. *Phytopathol. Z.* **57**: 201 (on citrus rootstocks; **46**, 1580).

Schiffmann-Nadel, M. et al. 1970. *Israel Jnl Bot.* **19**: 624 (on avocado; **50**, 1910).

Snow, J. P. et al. 1977. *Phytopathology* **67**: 589 (on cotton, infection fine structure; **57**, 178).

Srivastava, M. P. 1968. *Z. PflKrankh. PflPath. PflSchutz* **75**: 674 (on citrus; **48**, 1711).

Srivastava, S. N. S. 1964. *Indian Phytopathol.* **17**: 172 (on rubber seed; **44**, 1223).

Stevens, N. E. 1925. *Mycologia* **17**: 191 (life history & perfect state; **5**: 90).

——. 1926. *Ibid* **18**: 206 (perfect state inter alia; **6**: 126).

Turner, P. D. 1967. *Pl. Prot. Bull. F.A.O.* **15**: 81 (cacao dieback, review, 182 ref.; **47**, 1471).

Uduebo, A. E. et al. 1974. *Trans. Br. mycol. Soc.* **63**: 33, 45 (conidial germination & maturation; **54**, 322, 323).

——. 1974. *Can. J. Bot.* **52**: 2631 (temp., growth, sporulation & pigment; **54**, 2673).

——. 1975. *Ann. Bot.* **39**: 605 (conidial fine structure; **54**, 4380).

Umezurike, G. M. 1969. *Ibid* **33**: 451 (cellulytic enzymes; **48**, 2625).

Verral, A. F. 1942. *Phytopathology* **32**: 879 (in stained wood; **22**: 83).

Voorhees, R. K. 1942. *Bull. Fla agric. Exp. Stn* 371, 91 pp. (morphology, variation in vitro, taxonomy, including *Physalospora rhodina*, 86 ref.; **22**: 180).

Wang, S. C. et al. 1971. *Phytopathology* **61**: 1118 (on cotton, pectolytic enzymes; **51**, 1540).

—— ——. 1972. *Ibid* **62**: 560 (infection of cotton bolls; **51**, 4044).

—— ——. 1973. *Ibid* **63**: 1095, 1181 (on cotton, peroxidase activity & decay, conidial germination in presence of carbohydrates & phenolic compounds; **53**, 1827, 1831).

Wardlaw, C. W. 1932. *Ann. Bot.* **46**: (nomenclature & morphology).

Wergin, W. P. et al. 1973. *Devl Biol.* **32**: 1 (fine structure, conidial germination & septum formation; **54**, 3143).

Wilson, C. 1947. *Phytopathology* **37**: 657 (on groundnut fruit; **27**: 170).

Zambettakis, C. 1954. *Bull. Soc. mycol. Fr.* **70**: 219 (Sphaeropsidales, systematics; **35**: 237).

Zauberman, G. et al. 1975. *Phytopathology* **65**: 216 (respiration & ethylene evolution in infected citrus fruit; **54**, 3932).

BOTRYOSPHAERIA Ces. & de Not., *Comment. Soc. Crittogam. Ital.* 1: 211, 1863. (Botryosphaeriaceae)

PSEUDOTHECIA immersed or erumpent, usually on woody stems, 0.2 mm across or larger, globose, scattered or aggregated in large or small cushion like stromata, ostioles papillate. ASCI clavate, bitunicate, with numerous pseudoparaphyses, ASCOSPORES hyaline or nearly so, aseptate, usually an av. 18 μ long (similar fungi with smaller ascospores may be sought in *Guignardia*; from R. W. G. Dennis, *British ascomycetes*: 373, 1968; and see key by E. S. Luttrel, *The fungi* (editors G. C. Ainsworth et al.) Vol. IVA: 183, 1973).

B. obtusa (Schw.) Shoemaker (synonyms: *Sphaeria obtusa* Schw., *Physalospora obtusa* (Schw.) Cooke) causes canker and dieback of pomaceous fruits and grapevine in temperate regions.

Punithalingam, E. et al. 1973. *C.M.I. Descr. pathog. Fungi Bact.* 394 (*Botryosphaeria obtusa*).

Botryosphaeria ribes Grossenbacher & Duggar, *Tech. Bull. N.Y. agric. Exp. Stn* 18: 128, 1911.

ASCOSTROMATA in stems embedded in the cortex, later erumpent, sub-pulvinate, scattered, solitary, botyrose, black, up to 4 mm wide, individual ascogenous locules 170–250 μ diam., osiolate and darker around the neck. ASCI interspersed amongst filiform paraphyses, clavate, $100–110 \times 16–20$ μ, 8 spored, bitunicate. ASCOSPORES irregularly biseriate, hyaline, aseptate, ovoid, $17–23 \times 7–10$ μ. PYCNIDIA on stems, leaves and fruits, solitary or botyrose, stromatic, globose, 150–250 μ diam., with papillate ostioles darker around the neck region. Pycnidial wall many cells thick, composed of outer sclerotised cells and inner thin-walled cells lining the entire cavity. CONIDIOGENOUS CELLS holoblastic, hyaline and arising from the inner lining of the pycnidial cavity. MACROCONIDIA hyaline, fusoid, aseptate $17–25 \times 5–7$ μ. MICROCONIDIA (spermatia) hyaline, allantoid, $2–3 \times 1$ μ.

A saprophytic form which does not develop a purplish pink colour when grown on starch paste and has pulvinate ascostromata was designated by Grossenbacher and Duggar as *B. ribis* f. *achromogena*. Shear et al. (1924) refer to the pathogenic form as *B. ribis* f. *chromogena* (E. Punithalingam & P. Holliday, *CMI Descr.* 395, 1973).

B. ribis is plurivorous (see *CMI Descr.* 395 for list of host plant genera) particularly on trees; it also causes rots in fruit. It is widespread in temperate and tropical regions. Although a relatively minor pathogen, first attracting attention as a cause of cane blight in currant (*Ribes*) in USA, control measures may still be required in certain areas on certain hosts. Amongst tropical crops on which the fungus has been reported are: avocado, lemon (*Citrus limon*), pine and tung (*Aleurites fordii* and *A. montana*). In pines the killing of branch tips may lead to copious resin exudate. Spread in the wood kills the cambium and causes a dark purple brown stain in the wood with a black brown rot in the pith (Ivory); pycnidia form in the bark. In tung there is a wilt of branches and cankers occur as longitudinal bark splitting; spread to the trunk takes place. The external bark is dark brown becoming shrunken, the wood is brown to yellowish brown and there is brown streaking in living wood several cm beyond the brown lesions; pycnidia form on the dead wood (Webster et al.). Savastano described the disease on lemon. Frezzi reported severe cankers on the main branches and trunk of avocado in work which included other hosts. Horne et al. described the symptoms on avocado fruits. Light brown spots, vaguely bounded, darkening, circular, >1 cm diam., becoming soft, sunken; a watery rot develops in the flesh and the fruit shrivels.

Penetration is through wounds and lenticels, and infection might occur via lesions caused by other organisms as has been described in apple for fire blight (*Erwinia amylovora*; Drake). On *Rhododendron* 89% infection was obtained with inoculations done directly after wounding but only 13% when they were done 8 weeks after. Transmission can occur through pruning cuts (Schreiber). Cross inoculations suggest that different pathogenic strains do not occur, although an isolate from apple was less virulent to *Rhododendron* than one from the latter host. Wiehe described the infection pattern on tung where penetration of the fruit stalk was followed after *c.* 1–2 months by fruit fall, stem infection, twig blight and dieback of large branches. Pine dieback in Kenya is associated with adverse growing conditions (and see Anon. for apple). Smith et al. (1971) examined the effect of light on sporulation. In many cases control can probably be achieved by ensuring good growing conditions combined with other cultural measures, such as the removal and destruction of infected tissue. Bordeaux, thiram and zineb are effective and have been used for both the dieback and fruit rot forms of the disease. Control measures have fairly recently been described for apple in Australia (Anon.), where serious attacks have occurred in the coastal districts of New South Wales.

Anahosur, K. H. et al. 1972. *Caryologia* **25**: 327 (nuclear condition in ascospores & conidia; **52**, 2517).
Anon. 1970. *Agric. Gaz. N.S.W.* **81**(2): 98 (on apple; **49**, 3370).
Collins, R. P. et al. 1960. *Mycologia* **52**: 455 (pectolytic enzymes; **41**: 13).
Drake, C. R. 1971. *Pl. Dis. Reptr* **55**: 122 (on apple; **50**, 2353).
Frezzi, M. J. 1952. *Revta Invest. agric. B. Aires* **6**: 247 (in Argentina on avocado inter alia; **35**: 495).
Horne, W. T. et al. 1935. *Bull. Calif. agric. Exp. Stn* 594, 16 pp. (on avocado fruit; **15**: 238).
Ivory, M. H. 1967. *E. Afr. agric. For. J.* **32**: 341 (on pine; **47**, 344).
McClendon, J. H. et al. 1960. *Phytopathology* **50**: 258 (enzymes & rotting in apple fruit; **39**: 673).
Savastano, G. 1932. *Boll. Staz. Patol. veg. Roma* **12**: 245 (on lemon; **12**: 283).
Schreiber, L. R. 1964. *Pl. Dis. Reptr* **48**: 207 (on *Rhododendron*; **43**, 2296).
Shear, C. L. et al. 1924. *J. agric. Res.* **28**: 589 (on apple & *Ribes*; **4**: 178).
—— ——. 1925. *Mycologia* **17**: 98 (hosts in E. USA, morphology; **4**: 636).
Smith, C. O. 1934. *J. agric. Res.* **49**: 467 (host range, inoculations; **14**: 196).
Smith, D. H. et al. 1971. *Mycopathol. Mycol. appl.* **45**: 311 (effect of light on sporulation; **51**, 2188).
Stevens, N. E. et al. 1924. *J. agric. Res.* **27**: 837 (on temperate hosts; **3**: 725).
——. 1926. *Mycologia* **18**: 278 (as above; **6**: 106).
—— ——. 1929. *Ibid* **21**: 313 (in Hawaii; **9**: 344).
Webster, C. C. et al. 1950. *Pl. Dis. Circ. Nyasald* 1, 4 pp. (on tung; **29**: 339).
Wiehe, P. O. 1952. *Phytopathology* **42**: 521 (as above; **32**: 524).

BOTRYTIS Micheli ex Fr.; Micheli, *Nova. Pl. Gen.*: 212, 1729; Fries, *Syst. Mycol.* **1**: XLV, 1821. (Dematiaceae)

COLONIES effuse, often grey, powdery; under the low-power binocular microscope stout brown conidiophores are seen supporting glistening heads of pale conidia. MYCELIUM immersed or superficial. SCLEROTIA frequently formed both on natural substrata and in culture. STROMA none. SETAE and HYPHOPODIA absent. CONIDIOPHORES macronematous, mononematous, straight or flexuous, smooth, brown, branched, often dichotomously or trichotomously, with branches mostly restricted to the apical region forming a stipe and a rather open head; branches often markedly swollen at their ends to form colourless or pale conidiogenous ampullea, sometimes eventually collapsing in a concertina-like manner. CONIDIOGENOUS CELLS integrated, terminal on branches, polyblastic, determinate, inflated, clavate, spherical or subspherical, denticulate but often rather obscurely. CONIDIA solitary, acropleurogenous, simple, colourless or rather pale brown, smooth, aseptate or occasionally with a few conidia 1 or 2 septate, ellipsoidal, obovoid, spherical or subspherical. There is often also a phialidic state with small spherical or subspherical colourless phialoconidia (M. B. Ellis, *Demat. Hyphom.*: 178, 1971, 18 spp. described).

The genus was described by Hennebert (1973) who gave a synonymy, the recognised spp., names to be excluded, a synoptic key and a schematic chart for recognising *Botrytis* and like genera. Some spp. have perfect states in *Sclerotinia* (q.v.). Morgan examined 12 taxa, using data from 85 imperfect state characters (and see Morgan, *S. fuckeliana* q.v.). The addition of perfect state characters did not indicate any alteration in the relationship between taxa based on characters of the imperfect state. He gave a key to 12 *Botrytis* states. A comprehensive review was recently written by Jarvis. Several *Botrytis* spp. occur on *Allium*. Their taxonomy and relationships have been reviewed and discussed by Hennebert (1963). Of these the most important pathogen is *B. allii* (q.v.) on onion. This sp. is one of 3 causing the disease complex on onion bulbs known as neck rot. The others are *S. squamosa* (q.v.) and *B. byssoidea* Walker which is not described separately. The disease caused by *B. byssoidea* is called mycelial neck rot; the fungus differs in that conidial sporulation is sparse in the early stages of the disease and there is little or none in culture. The conidia are mostly $10–14 \times 6–9$ μ (length:breadth ratio 1.5–1.65), broader than those of *B. allii*, and smaller and less broad than those of *S. squamosa* which also has the characteristic concertina-like collapse of the conidiophores. This characteristic is absent in the 2 other spp. causing neck rot. *B. byssoidea* is widespread but not frequently recorded; it has been reported from Australia, Bulgaria, Denmark, Japan, Netherlands, New Zealand, Norway, Poland, UK and USA (Map 165). For other spp. on *Allium* see *Sclerotinia* and *S. porri*. The plurivorous *S. fuckeliana* (q.v.) can also cause disease in *Allium*. The dry conidia of *Botrytis* are probably spread mainly during the day. *Botrytis fabae* Sardina (chocolate spot of broad bean) was described by Ellis et al.

Ellis, M. B. et al. 1974. *C.M.I. Descr. pathog. Fungi Bact.* 432 (*Botrytis fabae*).
Hennebert, G. L. 1963. *Meded. LandbHoogesch. OpzoekStns Gent* **28**: 851 (on *Allium*, 65 ref.; **43**, 1512 p.).
——. 1973. *Persoonia* **7**: 183 (taxonomy & morphology; **52**, 3971).
Horiuchi, S. et al. 1978. *Bull. Chugoku Nat. agric. Exp. Stn* Ser. E **13**: 53 (identification by fine structure).
Jarvis, W. R. 1977. *Monogr., Res. Branch Agric. Can.* 15, 195 pp. (review, 1400 ref.; **57**, 2012).
Marsh, R. W. 1964. *Nature Lond.* **202**: 341 (conference on diseases, mostly temperate crops; **43**, 2498).
Menzinger, W. 1965. *Arch. Mikrobiol.* **52**: 178 (numbers & migration of nuclei; **45**, 2415).
——. 1966. *Zentbl. Bakt. ParasitKde* Abt. 2 **120**: 141, 179 (culture variability & taxonomy; **46**, 899).
Morgan, D. J. 1971. *Trans. Br. mycol. Soc.* **56**: 327 (numerical taxonomy; **51**, 133b).
Walker, J. C. 1925. *Phytopathology* **15**: 708 (*Botrytis byssoidea* and *B. squamosa* spp. nov.; **5**: 273).

Botrytis allii Munn, *Bull. N.Y. St. agric. Exp. Stn* 437: 396, 1917.

COLONIES dirty white bceoming smoky grey, dense felt like. CONIDIOPHORES short erect, septate, usually *c.* 0.5 mm (rarely >1 mm), single or in clusters; not often branched on host but tend to branch in culture, branches short, bearing conidial clusters. CONIDIA (macro-) hyaline or very slightly coloured, ellipsoid, often slightly tapering at both ends, 7–16×4–6 (av. 10–5) μ, attached to the vesicles in small clusters. SCLEROTIA dull black, firm, white inside with a dark outer cortical layer which is usually very finely wrinkled or striate, flattened or concave on side inseparably attached to host, strongly convex above; single sclerotia 1–5 mm diam., usually several form clusters or crusts which may be several cm diam.; sclerotia germinate by hyphae which form conidiophores. No apothecia produced (after M. T. Munn; and see Ellis et al.).

B. allii is the most important of the 3 spp. of *Botrytis* which cause neck rot of onion (*Allium cepa* var. *cepa*) bulbs. The least important is *B. byssoidea* (not described separately) and the third is *B. squamosa* (perfect state *Sclerotinia squamosa* q.v.). Some morphological differences between them (in the imperfect state) are given under *Botrytis*. The conidia are distinctly narrower than any other spp. of the genus on *Allium* (Ellis et al.). *B. allii*, which can also cause disease on garlic, leek and shallot, is widespread but mostly recorded in cooler regions (Map 169). The generally most damaging infection is on the bulb (which results in losses during storage); Walker (1926) called this form of the disease grey mould neck rot and he gave a full comparative account of these 3 neck rot fungi on *Allium*. Munn gave an earlier account of *B. allii*. For a full coverage of the literature on onion neck rot the ref. given under *S. fuckeliana* and *S. squamosa* should also be consulted. Netzer et al. (1967) described selective media for distinguishing between *B. allii* and *B. cinerea*, stat. conid. of *S. fuckeliana*. Hellmers in a full account of *B. allii* and other *Botrytis* spp. on *Allium* spp. reported *B. globosa* Raabe (perfect state *S. globosa* (Buchwald) Webster) on *A. ursinum* and found it to be pathogenic to onion bulbs.

Symptoms are most likely to be first seen on the bulbs after harvest. The infected area becomes soft and greyish (a cooked appearance), it may appear sunken and there is a definite healthy and diseased margin. The disease spread is more rapid down the scales (as the original site of infection is usually the neck) than it is from scale to scale. In section the host tissue looks watersoaked at the margin of the rot which tends to spread more quickly than the rot caused by *S. squamosa*. A dense grey sporulating mat eventually develops on the scales and is followed by the sclerotia. Such sporulation may be found in the field on the upper part of the neck and the base of the aerial green leaves. The sclerotia are most frequently crowded near the neck and the decayed bulb becomes mummified. The symptoms can first develop (but less frequently) from the base of the bulb or from wounds. The behaviour of, and the symptoms caused by, the pathogen on the leaves seem less clear. Segall et al. described the leaf spotting of onion leaves (called blast, blight, tip blight or tip burn) in USA and caused by *Botrytis* spp. The whitish leaf spots and flecks occurred when the foliage was at its max.; the leaves later began to wither from their tips. The spotting was caused by the conidia germinating on the leaf surface but no host penetration seemed to take place. *B. allii* was the commonest sp. involved. Hancock et al. (1963; *S. squamosa* q.v.) described (in USA) onion leaf spotting by *S. squamosa* and *S. fuckeliana*. The former caused deeper, elliptical lesions (1 × 1–3 mm) followed by withering of the leaf tip. The latter caused smaller (0.5 × 1–1.5 mm), superficial spots and no dieback. These 2 spp. were more prevalent than *B. allii* which was collected infrequently and caused much less spotting. Tichelaar in the Netherlands found that *B. allii* did not cause leaf spots. The fungus penetrated the leaves via the stomata (with appressoria) but, since the mycelium was restricted to the epidermal cells which lack chlorophyll, there were at first no symptoms. When the leaves became senescent the pathogen moved to the underlying parenchyma and the bulb. McKeen in Canada described infection of seedlings *c.* 5 weeks after transplanting to the field. Infection showed where one or more leaf blades arose from the central axis, slightly below soil level; a soft, watersoaked area developed a grey mycelial growth.

Hickman et al. in UK also described these fungi on onion leaves; in culture *S. squamosa* (q.v.) was the commonest. They described leaf dieback with water soaking at the junction of the withered area with the rest of the leaf; and also leaf spots. It would seem that, whilst *B. allii* is the most frequent cause of the bulb rot in onions, it is not associated with leaf spots (or less commonly so) and where *S. squamosa* is more frequent. In UK Maude et al. found that the major infection source was seed. Here *B. allii* persisted internally for 3.5 years at 10° and RH 50%. In onion crops the disease spread progressively in wet weather. Infection began at the leaf tips, and the fungus spread down the leaves to attack the bulb where it was deep seated at harvest. Thus infected seed was the main source of neck rot in store. In Israel Netzer et al. (1966) described an infection of the inflorescence by *B. allii* in onions being grown for seed. The umbel falls over and no seed is formed. Ellerbrock et al. also found inflorescence infection (with *S. fuckeliana* and *S. squamosa*).

Infection by *B. allii* takes place at a fairly low temp. opt. Walker (1926) found the best in vitro growth at 20–25° and most bulb decay at 15–20°. Hemmi et al. (1937) reported that decay was rapid at 13–25°. McKeen, working with seedlings, obtained most infection (and earlier deaths) at 15°, and progressively less at 20° and 25°. Doorn et al. gave a max. mycelial growth at 22–23°. The conidiophores can develop over a wide temp. range and at a fairly low RH. The fungus can penetrate through conidia germinating on the neck area around harvest time; the bulb base and wounds are infection points of less importance. Tichelaar found that neck rot developed equally in plants at ages: 17, 31, 45, 67 and 76 (days at inoculation). *B. allii* is readily carried over in host debris and organic matter generally in soil. Symptoms may only first appear in store, from infection that occurred in the field.

Recent work in UK showed that control of neck rot in the field and store should begin with healthy seed. Benomyl applied to seed reduced neck rot in store (Maude et al. 1977b). There should be efficient curing of onions so as to avoid neck infection with subsequent loss in storage. Drying of the senescent neck tissues prevents infection. Prolonged growth of the host should be avoided. Harvesting needs to be done in dry weather and field curing can be carried out if this weather persists. More efficient curing and storage is artificially achieved by forced air at 30° or more for 2–6 days (see Doorn et al.; Harrow et al.; Knoblauch; Vaughan et al. for onions and shallots). It may be advisable to avoid successive cropping (not > 2 crops); excessive nitrogen fertiliser may increase disease. Plant sanitation in the field and in storage houses is important. Storage should be at low temperatures (a little > 0° at *c.* 65% RH); where this is not possible very well ventilated conditions and cool temps are needed. Fungicides have been used as dips before planting: for example, dinocap, mancozeb and maneb (Elarosi et al.); captan and dicloran (Kaufman et al., fungicides applied to necks); benomyl (Vergniaud et al.). Sprays of benomyl, chlorothalonil and mancozeb controlled flower blight and increased seed yield (Ellerbrock et al.). Coloured cvs contract less infection than white ones but the spread of decay in both groups is similar after infection. Hatfield et al. and Owen et al. (*Colletotrichum circinans* q.v.) found that neck rot was more severe on the mild cvs than on the pungent ones. Meer et al. compared 3 inoculation methods. They found that seedlings of the cultivated onion showed little resistance whilst those of other *Allium* spp. were highly resistant. But leaves of older plants of most of these spp. were almost as susceptible as onion, although some ornamental spp. appeared very resistant. Some onion bulbs were very resistant.

Doorn, A. M. Van et al. 1962. *Versl. landbouwk. Onderz. RijkslandbProefstn* **68** (7); also as *Meded. Inst. Plziektenk. Onderz.* 299, 83 pp. (general; **43**, 319).
Elarosi, H. et al. 1965. *Alex. J. agric. Res.* **13**: 153 (fungicides at transplanting; **46**, 3292).
Ellerbrock, L. A. et al. 1977. *Phytopathology* **67**: 155 (*Botrytis* spp. causing flower blight & control; **56**, 5295).
Ellis, M. B. et al. 1974. *C.M.I. Descr. pathog. Fungi Bact.* 433 (*Botrytis allii*).
Harrow, K. M. et al. 1969. *N.Z. Jl agric. Res.* **12**: 592 (artificial curing; **49**, 902).
Hellmers, E. 1943. *Meddr. Vetløjsk. plantepat. Afd.* 25, 51 pp. (general in Denmark; **25**, 90).
Hemmi, T. et al. 1937. *Forschn Geb. PflKrankh. Kyoto* **3**: 234 (general; **17**: 789).
—— ——. 1939. *Ann. phytopathol. Soc. Japan* **8**: 309 (*Botrytis* spp. on stored onions; **18**: 430).
Kaufman, J. et al. 1967. *Pl. Dis. Reptr* **51**: 696 (fungicides; **47**, 380).
Knoblauch, F. 1958. *Tidsskr. PlAvl* **62**: 677 (curing in shallots; **38**, 439).
McKeen, C. D. 1951. *Scient. Agric.* **31**: 541 (on seedlings in Canada; **31**: 529).
Maude, R. B. et al. 1977a & b. *Ann. appl. Biol.* **86**: 163, 181 (seedborne infection & relation of disease in crop & in store, control; **57**, 887, 888).
Mayama, S. et al. 1977. *Phytopathology* **67**: 1300 (interference microscopy of penetration of epidermal cells; **57**, 4262).
Meer, Q. P. Van der. et al. 1970. *Euphytica* **19**: 152 (testing for resistance in *Allium* spp.; **50**, 421).
Munn, M. T. 1917. *Bull. N.Y. St. agric. Exp. Stn* 437: 363 (general).
Netzer, D. et al. 1966. *Pl. Dis. Reptr* **50**: 21 (on inflorescence in Israel; **45**, 1594).
—— ——. 1967. *Phytopathology* **57**: 795 (media to distinguish *B. allii* & *Sclerotinia fuckeliana*; **46**, 3291).
Owen, J. H. et al. 1950. *Ibid* **40**: 749 (variability in vitro; **30**: 211).
Priest, D. et al. 1961. *Ann. appl. Biol.* **49**: 445 (strs resistant to chlorinated nitrobenzenes; **41**, 208).
Segall, R. H. et al. 1960. *Phytopathology* **50**: 76 (*Botrytis* spp. on leaves; **39**, 452).
Tichelaar, G. M. 1967. *Neth. J. Pl. Pathol.* **73**: 157 (general; **47**, 966).
Vaughan, E. K. et al. 1964. *Tech. Bull. Ore. agric. Exp. Stn* 77, 22 pp. (curing & control; **44**, 2306).
Vergniaud, P. et al. 1972. *Pépiniéristes Horticulteurs Maraichers* **127**: 43 (use of benomyl; **53**, 356).
Walker, J. A. et al. 1975. *Trans. Br. mycol. Soc.* **65**: 335 (*Gliocladium roseum* on *B. allii*; **55**, 2053).
Walker, J. C. 1925. *J. agric. Res.* **30**: 365 (curing & control; **4**: 583).
——. 1926. *Ibid* **33**: 893 (general with *B. byssoidea* & *S. squamosa*; **6**: 267).

BREMIA Regel, *Bot. Ztg* **1**: 665, 1843.
(Peronosporaceae)

MYCELIUM intercellular; haustoria unbranched, clavate to globose. CONIDIOPHORES emerging from the stomata, where 2–3 have a common origin in a bulbous swelling of a single hypha; branching definitely dichotomous; each terminal branch broadened at its tip to form a shallow, saucer-shaped disk from the edges of which several short sterigmata arise each bearing a conidium. CONIDIA hyaline, with an apical papilla through which a germ tube is usually protruded. Distinguished by the characteristic disk-like projections (at the ends of the conidiophore branches) which bear the sterigmata and conidia on their margins (after H. M. Fitzpatrick, *The lower fungi Phycomycetes*: 219, 1930).

Bremia lactucae Regel, *Bot. Ztg* **1**: 665, 1843.

MYCELIUM hyaline, coenocytic, intercellular 5–12 μ, with globose or saccate (10×16 μ) haustoria. CONIDIOPHORES hyaline, emerging (2–3) through the stomata; 3–6 dichotomously branched, $200–1500 \times 9–11$ μ; clavate swelling at end of ultimate branches with 3–5 (usually 4) thin, digitiform sterigmata, each with a single conidium. CONIDIA ovoid–ellipsoid to spherical, $17–19 \times 18–21$ μ, hyaline with smooth, rather thick wall, attached at a small papilla; germination usually direct but zoospores have been described. OOSPORES spherical, light brown, 26–35 μ; see Humphreys-Jones; Ingram et al. 1975; Fletcher; Tommerup et al. for the occurrence of oospores in decayed stems, leaves, leaf debris, cotyledons and seedlings.

Lettuce downy mildew is worldwide but has not been reported from W. Africa or parts of southern Africa (Map 86). It is a disease of cool, damp weather and therefore not likely to be generally serious in the lowland tropics. Marlatt and Verhoeff gave full accounts. Ling et al., who described and proposed forms (5) for *B. lactucae*, stated that it had been recorded on 22 genera in the Compositeae. The form on the economic host (*Lactuca sativa*) also infects, in varying degrees, other spp. of *Lactuca* from some of which resistance has been obtained. Endive (*Cichorium endiva*) and chicory (*C. intybus*) are also infected but the forms from one plant genus do not apparently attack those of another. Leaf lesions begin as pale green to yellow areas, usually bounded by the veins, reaching 2–3 cm diam. and coalescing. The necrosis caused sometimes reaches the leaf edge, hence the name brown margin. Infection causes poor growth, excessive trimming, and loss in transport and marketing. Early attacks lead to stunting and death of seedlings.

Germination of the conidium is mostly direct but no extensive attempt appears to have been reported to determine whether, under certain conditions, the zoospores play a part in infection and spread. Sargent (1976) found that germination in distilled water was not affected by temps as low as 0° or as high as 21°, but at 25° <33% germinated and at 28° germination was nil. Viability was *c*. 48 hours at 28°. Other workers have various temp. opt. for germination (see Marlatt). Under moist conditions germination can begin in 1 hour from inoculation and by 3 hours germ tubes and appressoria are present.

Penetration of the leaf has been described as both stomatal and cuticular but the latter is most probable. There is little or no penetration at 25°. In seedlings penetration of the cotyledon is through the adaxial epidermis; the hyphae grow into the hypocotyl (after 90 hours), true leaves (108 hours) and the roots (114 hours). But further systemic spread of the developing plant does not occur. If there is no dark period immediately after inoculation penetration is inhibited. Sporulation is inhibited by light and max. sporulation occurred on seedlings in a dew chamber exposed to a min. 6 hours dark for 7 days after inoculation; continuous light over this period prevented sporulation (Raffray et al.). Conidiophores are formed early in the morning. In USA (California) infection of lettuce by *B. lactucae* was compared with that by *Erysiphe cichoracearum* q.v. (Schnathorst) under natural conditions. The former was more abundant where the temp. was an av. 13° and the RH an av. 88%. Carry over seems mostly as the asexual stage (but see Morgan who germinated the oospores) on lettuce; seed transmission is negligible, oospores are not abundant, soil and other hosts of no importance.

B. lactucae is pathogenically very adaptable and a combination of different major genes for resistance is unlikely to be a lasting control measure (Crute et al. 1976d). Races have been described in France, Germany, Israel, Netherlands, Sweden, UK and USA; they have also been compared (Blok et al.; Channon et al.; Crute et al. 1976; Dixon et al. 1973; Dixon 1976; Eenink; Globerson et al.; Johnson et al.; Netzer 1973; Sequeira et al.; Tjallingii et al.; Wellving et al.; Zink 1973; Zink et al. 1975; Zinkernagel). Fungicides (zineb and maneb) should be applied often up to planting out and then few more treatments are needed. No control is likely to be needed in high temp. regions (Anon.; Channon et al. 1967; Fletcher; Powlesland et al.).

Anon. 1973. *Advis. Leafl. Minist. Agric. Fish. Fd* 577, 4 pp. (advisory; **53**, 755).

Blok, I. et al. 1974. *Zaadbelangen* **28**: 138 (comparison of races; **55**, 992).

——. 1977. *Ibid* **31**: 57 (problems, new race; **57**, 1927).

Channon, A. G. et al. 1965. *Ann. appl. Biol.* **56**: 389 (2 races in UK; **45**, 928).

——. 1967. *Ibid* **59**: 355 (fungicides; **46**, 3297).

——. 1970. *Hort. Res.* **10**: 14 (comparison of races from the Netherlands & UK; **50**: 431).

——. 1971. *Ann. appl. Biol.* **68**: 185 (differentiation of 4 UK races; **51**, 965).

Cox, R. S. 1957. *Pl. Dis. Reptr* **41**: 455 (control with zineb; **37**: 65).

Crute, I. R. et al. 1976a,b,c. *Ann. appl. Biol.* **83**: 125, 173, 433 (genetical relationship between races & cvs, virulence gene combinations, behaviour on lettuce cvs & other Compositae; **55**, 2985, 4400, 4401).

——. 1976d. *Ibid* **84**: 287 (breeding for resistance; **56**, 982).

——. 1977. *Trans. Br. mycol. Soc.* **69**: 405 (host specificity; **57**, 3716).

——. 1978. *Ann. appl. Biol.* **89**: 467 (incomplete specific resistance).

Dawson, P. R. 1976. *Ibid* **84**: 282 (breeding for resistance; **56**, 980).

Dickinson, C. H. et al. 1974. *Ibid* **76**: 49 (effect of seedling age & development on infection; **53**, 3262).

Dixon, G. R. et al. 1973. *Hort. Res.* **13**: 89 (reaction of cvs & variation within races; **54**, 285).

——. 1973. *Ann. appl. Biol.* **74**: 307 (colonisation of adult plants; **53**, 754).

——. 1976. *Ibid* **84**: 283 (races with *Peronospora viciae* & *P. farinosa*; **56**, 981).

——. 1978. *Ibid* **88**: 287 (frequency & distribution of virulence factors in England; **57**, 3717).

Eenink, A. H. 1974. *Euphytica* **23**: 411 (in the Netherlands, races & resistance in wild *Lactuca* & lettuce cvs; **53**, 4680).

Faroqui, M. H. et al. 1966. *Mesopotamia Agric.* **1**: 29 (effect of temp. on conidial germination & sporulation; **46**, 1445).

Fletcher, J. T. 1976. *Ann. appl. Biol.* **84**: 294 (oospores, sporangial dissemination & fungicides; **56**, 984).

Globerson, D. et al. 1974. *Euphytica* **23**: 54 (inheritance of resistance; **53**, 3711).

Humphreys-Jones, D. R. 1971. *Pl. Soil* **35**: 187 (oospores in stem; **51**, 3663).

Ingram, D. S. et al. 1975. *Trans. Br. mycol. Soc.* **64**: 149 (occurrence of oospores; **54**, 5667).
——. 1976. *Ann. appl. Biol.* **84**: 299 (cytology of perfect state; **56**, 985).
Johnson, A. G. et al. 1977. *Ibid* **86**: 87 (genetics of race specific resistance; **56**, 5900).
——. 1978. *Ibid* **89**: 257 (as above).
Ling, L. et al. 1945. *Trans. Br. mycol. Soc.* **28**: 16 (host specialisation on Compositae; **24**: 398).
Marlatt, R. B. 1974. *Tech. Bull. Fla agric. Exp. Stn* 764, 25 pp. (review, 96 ref.; **54**, 4278).
Morgan, W. M. 1978. *Trans. Br. mycol. Soc.* **71**: 337 (germination of oospores).
Netzer, D. 1973. *Ibid* **61**: 375 (races in Israel; **53**, 2387).
—— et al. 1976. *HortScience* **11**: 612 (*L. saligna*, new source of resistance; **56**, 3354).
Ogilvie, L. 1944 & 1946. *Rep. hort. Res. Stn Bristol* 1943 & 1945: 90, 147 (other hosts & *Lactuca*, oospores on *Senecio vulgaris*; **23**: 470; **26**: 3).
Powlesland, R. 1954. *Trans. Br. mycol. Soc.* **37**: 362 (general, effects of temp. & RH; **34**: 697).
—— et al. 1954. *Ann. appl. Biol.* **41**: 461 (fungicides; **34**: 507).
Raffray, J. B. et al. 1971. *Can. J. Bot.* **49**: 237 (effect of light & dark on sporulation; **50**, 3336).
Sargent, J. A. et al. 1973. *Physiol. Pl. Pathol.* **3**: 231 (host penetration; **53**, 342).
——. 1974. *Trans. Br. mycol. Soc.* **63**: 509 (effect of temp. on germination, viability & fine structure; **54**, 3076).
——. 1976. *Ann. appl. Biol.* **84**: 290 (conidial germination; **56**, 983).
Schnathorst, W. C. 1962. *Phytopathology* **52**: 41 (effect of climate with *Erysiphe cichoracearum* in California; **41**: 565).
Sequeira, L. et al. 1971. *Ibid* **61**: 578 (inheritance of resistance, races in USA; **51**, 967).
Tao, C. F. 1965. *Acta phytophylac. sin.* **4**: 15 (in China; **44**, 2310).
Tjallingii, F. et al. 1969. *Zaadbelangen* **23**: 436 (resistance in 70 cvs to 4 races in the Netherlands; **50**, 2062).
Tommerup, I. C. et al. 1974. *Trans. Br. mycol. Soc.* **62**: 145 (oospores in cotyledons & seedlings; **53**, 3263).
Verhoeff, K. 1960. *Tijdschr. PlZiekt.* **66**: 133 (full account, 70 ref.; **40**: 197).
Virányi, F. et al. 1976. *Acta phytopathol. Acad. Sci. hung.* **11**: 173 (fine structure of penetration; **57**, 1533).
Wellving, A. et al. 1978. *Ann. appl. Biol.* **89**: 251 (virulence of populations in Sweden).
Whitaker, T. W. et al. 1958. *Proc. Am. Soc. hort. Sci.* **72**: 410 (history & development of resistance; **38**: 440).
Zink, F. W. et al. 1970. *Ibid* **95**: 420 (linkage of turnip mosaic virus susceptibility & resistance to *B. lactucae*; **50**, 430).
——. 1973. *J. Am. Soc. hort. Sci.* **98**: 293 (inheritance of resistance, races in USA; **52**, 4295).
——. 1975. *Phytopathology* **65**: 243 (comparison of races, reaction of resistant cvs to turnip mosaic virus; **54**, 4275).
Zinkernagel, V. 1975. *NachrBl. dt. PflSchutzdienst. Stuttg.* **27**: 185 (races in Germany; **55**, 3826).

CALONECTRIA de Not., *Comment. Soc. Crittogam. Ital.* Vol. 2: 447, 1867.

Scoleconectria Seaver, 1909 (in part).
(Hypocreaceae)

STROMA when present erumpent, pseudoparenchymatous. PERITHECIA scattered or caespitose, globose or oval, light or brightly coloured, usually red, ostiole papillate, perithecial wall pseudoparenchymatous. ASCI unitunicate, cylindrical or clavate, inoperculate, 6–8 spored; pseudoparaphyses evanescent, paraphyses absent. ASCOSPORES elliptical or fusiform, with 2 or more transverse septa, hyaline or lightly coloured (after J. M. Dingley, *Trans. R. Soc. N.Z.* 79: 403, 1952).

This genus is basically a *Nectria* with ascospores that have 2 or more septa. *C. theae* Loos (stat. conid. *Cylindrocladium theae* (Petch) Subram.; synonyms: *Cercosporella theae* Petch, *Candelospora theae* (Petch) Wakefield ex Gadd, *Cylindrocladium theae* (Petch) Alf. & Sob.) causes a minor leaf spot of tea. It has been called bird's eye spot. On young leaves the spots are brown or black and may coalesce. On mature leaves they are black, becoming grey or greyish white with a purplish margin. Tea bushes can be defoliated and spread from *Acacia decurrens*, used as a shade tree, has been described (and see Nishijima et al.). No special control measures seem to be required in Sri Lanka from where the fungus was first described. Peerally described the 5 *Calonectria* spp. whose hosts include tea (1947a). Besides *C. theae* (vesicle of sterile filaments narrowly clavate; conidia mostly 3 septate, $63–102.8 \times 5.1–6.8\ \mu$), there are *C. colhounii* Peerally (vesicle narrowly clavate; conidia 3 septate, $38.3–84.2 \times 3.4–5.7\ \mu$) and *C. quinqueseptata* Figueiredo & Namekata (vesicle narrowly clavate; conidia 1–6 (usually 5) septate, $59.8–104.6 \times 5.2–7\ \mu$ in vivo). *C. kytoensis* (q.v.), which also occurs on tea, has sterile filaments with a globose vesicle. *C. quinqueseptata* (stat. conid. *Cylindrocladium quinqueseptatum* Boedijn & Reitsma) causes several minor diseases: whitish leaf spots, circular with a reddish border on clove (*Eugenia caryophyllus*), damping off of *Eucalyptus*, and a nursery leaf spot of rubber (*Hevea brasiliensis*). In the last host immature leaves show chlorotic spots which become purplish brown with a yellow halo. On mature leaves the lesions are circular to subcircular up to 3 mm diam., with a whitish, papery centre and dark brown margin. The spots are raised and growing points may be attacked. The disease is severe in budwood nurseries only when flushing (after pruning) occurs in wet weather.

Anon. 1972. *Plrs' Bull. Rubb. Res. Inst. Malaya* 119: 55 (*Calonectria quinqueseptata* on rubber; **52**, 221).

Figueiredo, M. B. et al. 1967. *Arq. Inst. biol. S. Paulo* **34**: 91 (*C. quinqueseptata* on *Eucalyptus* spp. & *Annona squamosa*; **47**, 765).

Linderman, R. G. 1974. *Phytopathology* **64**: 567 (ascospore discharge from 3 *Calonectria* spp.; **53**, 4483).

Loos, C. A. 1950. *Trans. Br. mycol. Soc.* **33**: 13 (*C. theae* sp. nov.; **29**: 536).

——. 1951. *Tea Q.* **22**: 27 (*C. theae* on tea; **31**: 210).

Nishijima, W. T. et al. 1975. *Pl. Dis. Reptr* **59**: 883 (*C. theae* on *Acacia koa* & *Metrosideros collina*; **55**, 4300).

Peerally, M. A. 1973. *Trans. Br. mycol. Soc.* **61**: 89 (*C. colhounii* sp. nov.; **52**, 4208).

Peerally, A. 1974a. *Rev. agric. sucr. Ile Maurice* **53**: 57 (*Calonectria* spp. on tea; **54**, 1863).

——. 1974b. *C.M.I. Descr. pathog. Fungi Bact.* 423, 424, 430 (*C. quinqueseptata*, *C. theae* & *C. colhounii*).

Reitsma, J. et al. 1950. *Contr. gen. agric. Res. Stn Bogor* 109: 50 (*C. quinqueseptata* on clove; **30**: 431).

Webster, B. N. 1952. *Tea Q.* **23**: 70 (*C. theae* on *Acacia decurrens* & control; **32**: 454).

Calonectria crotalariae (Loos) Bell & Sobers, *Phytopathology* 56: 1364, 1966.

Calonectria theae var. *crotalariae* Loos, 1950
stat. conid. *Cylindrocladium crotalariae* (Loos) Bell & Sobers, 1966
Cercosporella theae Petch, 1917.

PERITHECIA on leaves and stems, superficial, scattered to gregarious, arising from a small erumpent stroma, orange to red, subglobose to oval, 280–450 μ wide, with small papillate ostiole. ASCI clavate, thin walled, $84.4–135.2 \times 12.6–20.4$ μ. ASCOSPORES hyaline, fusoid, 1–3 septate, $30.6–62.4 \times 6.4–8$ μ. CONIDIOPHORE BRANCHES arise laterally from a stipe; primary branches aseptate, 14.2–30.4 μ long; secondary branches aseptate, 12.6–24 μ long; tertiary branches aseptate, 8.4–12.6 μ long. PHIALIDES hyaline, 7.8–16.4 μ long. CONIDIA hyaline, cylindrical, 1–3 septate (mostly 3), $60.4–105.2 \times 4.8–7.2$ μ. STERILE FILAMENTS terminate in a subglobose to globose vesicle, 6.2–12.6 μ wide. CHLAMYDOSPORES and MICROSCLEROTIA formed in culture (A. Peerally, *CMI Descr.* 429, 1974).

The stat. conid. is distinguishable from other *Cylindrocladium* spp. by its 3 septate conidia and globose vesicles. The ascospores of *Calonectria crotalariae* are smaller than those of *C. theae* (*Calonectria* q.v.) and larger than those of *C. kyotensis* (q.v.). *C. crotalariae* causes black rot of groundnut (Bell et al. 1966) in USA; seedling collar rots of papaw (Nishijima et al.) and koa (*Acacia koa*; Aragaki et al.) in Hawaii; and a collar rot of *Crotalaria anagyroides* in Sri Lanka (Loos). Other plants, including *Eucalyptus* (Nishijima et al.), are infected. Rowe et al. (1973a) reported infection on soybean. Also in USA 5 cvs of flue cured tobacco were susceptible. Cotton (4 cvs) was not visibly affected but the fungus was isolated from the roots. Maize, wheat and rye were resistant (Rowe et al. 1973b).

Groundnut plants in the field show a chlorosis and wilt which is more extensive in the erect primary branches; tap roots and hypocotyls have a necrosis up to ground level and root tips are rotted. Brown to black lesions are formed on the pegs and pods. The disease has been described as primarily debilitative (Bell). Inoculated leaves developed circular, brown lesions, 0.5–1 mm diam., with chlorotic haloes of up to 2 mm diam. Perithecia develop in the stems 2–3 weeks after inoculation. Bell et al (1966) infected leaves of *Eucalyptus* spp. (reddish purple, irregularly circular spots, 0.5–3 mm diam.) and *C. spectabilis* (dark brown, circular to subcircular spots, 1–8 mm diam.). The pathogen is soilborne, and high germination of conidia and ascospores can occur in natural soil (Hwang et al. 1974). The groundnut hypocotyl is penetrated from infection cushions. Carry over (several months) takes place in infected stems and roots; from the latter Rowe et al. (1974b) described microsclerotia (av. 52.7×88.4 μ). Microsclerotia (but not ascospores or conidia) remained viable for 8 months in papaw tissue incubated in soil (Hwang et al. 1976). Most ascospores are discharged at *c.* 25° (a similar temp. is opt. for perithecial development and growth); forcible discharge takes place as the RH drops (see Bell; Phipps et al. 1977a for effects of temp. on disease).

Isolates vary in virulence to groundnut but there

is no evidence for the existence of races (Rowe et al. 1975b). The pathogenicity of *Calonectria crotalariae* was compared with that of *C. kyotensis* and *Cylindrocladium scoparium* (q.v.) to groundnut and soybean inter alia. *Calonectria crotalariae* was the most virulent and *C. kyotensis* the least so (Sobers et al.). Groundnut cvs differed in susceptibility but Spanish types were the least susceptible (Rowe et al. 1974a; and see Garren et al.; Phipps et al. 1977b). Disease intensity was reduced by broad spectrum fungicides. At planting these gave control for *c.* 65 days but at 135 days there was little difference between the treated and untreated (Bell et al. 1973). Except where stated the ref. mostly concern groundnut.

Aragaki, M. et al. 1972. *Pl. Dis. Reptr* **56**: 73 (on *Acacia koa*; **51**, 2881).
Bell, D. K. et al. 1966. *Phytopathology* **56**: 1361 (aetiology & morphology; **46**, 1159).
——. 1967. *Pl. Dis. Reptr* **51**: 986 (effects of soil temp. & plant age; **47**, 962).
—— ——. 1973. *Ibid* **57**: 90 (status in USA, Georgia, control; **52**, 3492).
Garren, K. H. et al. 1976. *Ibid* **60**: 175 (reaction of Virginia type cvs; **55**, 4952).
Griffin, C. J. et al. 1978. *Phytopathology* **68**: 887 (physical factors that affect microsclerotial recovery from soil).
Hanounik, S. B. et al. 1977. *Pl. Dis. Reptr* **61**: 431 (inoculum potential, Na azide treatment & host genotype; **57**, 361).
Hwang, S. C. et al. 1974. *Mycologia* **66**: 1053 (spore germination in soil; **54**, 2680).
—— ——. 1975. *Phytopathology* **65**: 1036 (medium for assessment in soil; **55**, 1667).
—— ——. 1976. *Ibid* **66**: 51 (biology of spores & microsclerotia in soil; **55**, 3466).
Johnston, S. A. et al. 1975. *Ibid* **65**: 649 (penetration of & growth in hypocotyl; **55**, 484).
Krigsvold, D. T. et al. 1977. *Pl. Dis. Reptr* **61**: 495 (spread in cultivation & soybean cropping; **57**, 362).
Loos, C. A. 1950. *Trans. Br. mycol. Soc.* **33**: 13 (*Calonectria theae* sp. nov. & var. *crotalariae* on tea & *Crotalaria*; **29**: 536).
Nishijima, W. T. et al. 1973. *Phytopathology* **63**: 553 (on papaw; **52**, 4162).
Phipps, P. M. et al. 1977a. *Ibid* **67**: 1104 (effect of soil temp. & moisture on disease; **57**, 3192).
—— ——. 1977b. *Pl. Dis. Reptr* **61**: 300 (sensitivity of groundnut cvs to inoculum density in soil; **56**, 5895).
Rowe, R. C. et al. 1973a. *Pl. Dis. Reptr* **57**: 387 (general in USA, North Carolina; **52**, 4289).
—— ——. 1973b. *Ibid* **57**: 1035 (susceptibility of crops in rotation; **53**, 2930).
—— ——. 1974a. *Ibid* **58**: 348 (control; **53**, 4206).
—— ——. 1974b. *Phytopathology* **64**: 1294 (formation & dispersal of microsclerotia; **54**, 3068).
—— ——. 1975a. *Ibid* **65**: 393 (ascospore formation & discharge; **54**, 5162).
—— ——. 1975b. *Ibid* **65**: 422 (variability in virulence; **55**, 483).
Sobers, E. K. et al. 1974. *Pl. Dis. Reptr* **58**: 1017 (comparison of pathogenicities, *Calonectria crotalariae*, *C. kyotensis* & *Cylindrocladium scoparium*; **54**, 2749).

Calonectria kyotensis Terashita, *Trans. mycol. Soc. Japan* 8: 124, 1968.
Calonectria uniseptata Gerlach, 1968
C. floridana Sobers, 1969
stat. conid. *Cylindrocladium floridanum* Sobers & Seymour, 1967.

PERITHECIA on leaves, stems and roots, scattered, arising from a small erumpent stroma, orange red, globose to oval, 309–450 μ wide; a papillate ostiole of small columnar cells. ACSI clavate, thin walled, hyaline, 8 spored, $70.4–143.8 \times 15.6–23.4$ μ. ASCOSPORES hyaline, 1 septate, $28.6–41.6 \times 6.2–8.2$ μ. CONIDIOPHORE BRANCHES arise laterally from a stipe; primary and secondary branches 0–1 septate, 10.2–26 μ long; tertiary branches aseptate, 7.8–10.2 μ. PHIALIDES hyaline, 7.8–15.6 μ. CONIDIA hyaline, cylindrical, 1 septate, $36.4–57.2 \times 2.6–4.6$ μ. STERILE FILAMENTS terminate in a globose vesicle; secondary sterile filaments present in most isolates. CULTURES on PDA dark reddish brown with MICROSCLEROTIA and CHLAMYDOSPORES (A. Peerally, *CMI Descr.* 421, 1974).

C. kyotensis occurs on temperate plants (including ornamentals), pines, tea and *Eucalyptus*. Several other *Calonectria* spp. have been described from tea; their distinguishing characteristics are noted under the genus. In Mauritius the fungus is associated with young, unthrifty (chlorosis and defoliation) tea plants and tea bushes in the field. Terashita (l.c.) compared *C. kyotensis* with 4 other *Calonectria* spp. including *C. theae* and *C. crotalariae* q.v.; and see Sobers (1969). Sobers et al. found that *C. kyotensis* was less virulent than *C. crotalariae* and *Cylindrocladium scoparium* (q.v.). In the glasshouse these 3 spp. were pathogenic to groundnut and soybean. The taxonomy of *C. scoparium* and *C. floridanum* was considered by Morrison et al. (1969a). The fungus attacking roots of conifer seedlings (Anderson et al.; Cox), originally referred to the former (oval to ellipsoid vesicles), was shown to be the latter (globose to ovate vesicles). *Calonectria kyotensis* is a soil inhabitant (rarely attacking aerial parts); it has the characteristics of a competitive soil saprophyte (Menge et al.). A technique for its isolation, as microsclerotia, was described by Morrison et al. (1969b). Soil drenches with benomyl have given control in temperate ornamentals. In tea such drenches (or mancozeb), artificial fertilisers and earthing up have given control of root rot of tea in Mauritius. In a small-scale test Weaver obtained control of microsclerotia with potassium azide.

Anderson, N. A. et al. 1962. *For. Sci.* **8**: 378 (on conifers in USA; **42**: 348).
Cox, R. S. 1954. *Tech. Bull. Del. agric. Exp. Stn* 301, 40 pp. (on conifer seedlings in USA).
Menge, J. A. et al. 1976. *Phytopathology* **66**: 862, 1085 (determination of inoculum potentials in cropped & chemically treated soils, effects of plant residues & chemical treatments on inoculum potential; **56**, 1766, 1941).
Morrison, R. H. et al. 1969a. *Mycologia* **61**: 957 (taxonomy of *Cylindrocladium floridanum* & *C. scoparium*; **49**, 1165).
—— ——. 1969b. *Pl. Dis. Reptr* **53**: 367 (isolation from soil; **48**, 2174).
Peerally, A. 1972. *Rev. agric. sucr. Ile Maurice* **51**: 115, 147 (decline of tea in Mauritius; **52**, 1249, 3818).
Sobers, E. K. 1969. *Phytopathology* **59**: 364 (morphology & comparison with *Calonectria crotalariae* & *C. ilicicola*; **48**, 2482).
——. 1972. *Ibid* **62**: 485 (morphology & pathogenicity; **51**, 3827).
—— et al. 1974. *Pl. Dis. Reptr* **58**: 1017 (pathogenicity with *C. crotalariae* & *Cylindrocladium scoparium*; **54**, 2749).
Weaver, D. J. 1971. *Ibid* **55**: 1094 (control with potassium azide; **51**, 2230).

Calonectria rigidiuscula (Berk. & Br.) Sacc., *Michelia* 1: 313, 1878.

Nectria rigidiuscula Berk. & Br., 1873
stat. conid. *Fusarium decemcellulare* Brick, 1908
Spicaria colorans De Jonge, 1909
Fusarium spicaria-colorantis Sacc. & Trott, 1913.

PERITHECIA on surface of stroma which arises from below the periderm of the host; globose, cream to yellow, roughly warted, 200–230 μ high × 190–300 μ diam. ASCI clavate, generally 4 spored, 70–100 × 12–14 μ. ASCOSPORES hyaline, ellipsoid to reniform with 3 transverse septa and faint longitudinal striations when mature, 22–28 × 7–10 μ. MICROCONIDIA in chains and resemble a *Spicaria*, ellipsoid, 10–12 × 3–4 μ with a flat circular scar at the base or at each end. MACROCONIDIA formed typically on sporodochia where they give rise to characteristic wedge-shaped masses, 50–65 × 5–7 μ, 7–10 septate, cylindrical, curved and narrowing apically to a point.

Cultures initially pale with white to cream floccose mycelium. After 3–5 days a rose pigmentation appears and this darkens with age. Pigment formation is affected by pH at isometabolic point and is most conspicuously developed in the dark. Diurnal fluctuations of light and temp. with opt. 12 hours light and 12 hours dark, and 25–30° favour sporulation in vitro. (From C. Booth & J. M. Waterson, *CMI Descr.* 21, 1964.)

Isolates are generally homothallic but those causing galls on cacao have been reported to be heterothallic or non-fertile. Plurivorous and common in the tropics, most frequently associated with woody tissue.The fungus has been reported from soil and from rice grain. It is usually a saprophyte and a weakly pathogenic, secondary invader of debilitated woody hosts in several genera; but most of the work done relates to infection of cacao. This crop fairly readily shows the stag-headed appearance of debilitated branches if the environment becomes unfavourable. The condition may be complex since diebacks can be initiated directly by pathogens or insects, or indirectly after soil or other cultural factors have become limiting. The patterns that emerge have been most studied in W. Africa and were reviewed by Kay. The associated organisms are the capsid bugs (mostly *Sahlbergella singularis* and *Distantiella theobroma*), cacao swollen shoot viruses, *Botryodiplodia theobromae* (q.v.) and *C. rigidiuscula*. *Phytophthora palmivora* (q.v.), a cause of cacao canker, may also be a factor.

Spread of any infection in artificially wounded but healthy cacao branch tissue is slow, and hardly greater than the formation of callus tissue. Under natural conditions infection by *C. rigidiuscula* is mainly via stem lesions caused by capsids. Microconidia are first seen as a whitish bloom, especially on these lesions on green stems. The macroconidia on pink to buff sporodochia form later on deady woody tissue as do the perithecia. In an active canker on stems a claret-coloured zone in the cortex and cambium markes the extent of the invasion with a clear line of demarcation. In dead wood there may be a black to grey zone which is an area colonised by *B. theobromae*. The healing of cankers caused by *P. palmivora* can be retarded through subsequent invasion by *C. rigidiuscula* which spreads most rapidly in the wood. Whilst *B. theobromae* can invade directly via capsid wounds it is less important than *C. rigidiuscula* which spreads in the host 5–6 times more quickly. Whilst combined insecticide and fungicide sprays help in controlling dieback, attention should also be given to the cultural and ecological conditions in which the crop is being grown. Some differential resistance may occur.

Various types of outgrowths (galls) on cacao have been reported, the earliest report may be from Guyana in 1905. The galls (sometimes called cushion or green point galls) are roughly hemispherical, 1–13 cm in diam. and consist of a compressed branch system which bears a large number of small green buds; they are attached to woody tissue by a short, central stalk. The discovery of large numbers of these galls on cacao in Central America in the early 1950s, and their occurrence in W. Africa, stimulated investigation. It was found that the condition could be transmitted to seedlings either mechanically or by insects, using gall tissue; but there was no systemic infection. *C. rigidiuscula* was isolated from galls and the isolates found to be pathogenic. Seedlings are highly susceptible, symptoms appearing after 12–14 days. Infection of cushions and pod stalks was in some cases successful but the incubation period was long and variable (3 weeks–5 months). After a variable time the gall dies; it becomes hard and woody with a crumbling, black surface. Dieback and gall isolates differ in pathogenicity, the latter being reported as heterothallic or unfertile. All isolates from naturally occurring perithecia in Central America were homothallic and did not cause galls. Perithecia of the heterothallic strs have not apparently been found in vivo. The gall forming strs have been designated *C. rigidiuscula* f. sp. *theobromae*. Nothing is known of the epidemiology of the gall disease, and spread in Ghana is slow. High infection rates have been reported for Central America but the economic effects are not clear. Some control, in inoculated seedlings, with mercury pastes has been described.

Brunt, A. A. et al. 1962. *Ann. appl. Biol.* **50**: 283 (aetiology of a gall disease in Ghana; **42**: 7).

Crowdy, S. H. 1947. *Ibid* **34**: 45 (general, pathogenicity & types of cacao dieback; **26**: 537).

Ford, E. J. et al. 1967. *Phytopathology* **57**: 710 (gall inducing pathogenic & non-pathogenic forms; **46**, 3391).

Griffiths, D. A. 1965. *Hort. Res.* **5**: 107 (cause of wilt in Bignoniaceae; **45**, 1392).

Hansen, A. J. et al. 1963. *Turrialba* **13**: 80 (pathogenicity of gall inducing forms; **43**, 972).

——. 1966. *Pl. Dis. Reptr* **50**: 229 (gall transmission by different isolates; **45**, 2447).

Helfenberger, A. 1966. *Cacao Turrialba* **11** (4): 1 (gall control in inoculated seedlings with Hg pastes; **46**, 3039).

Hutchins, L. M. 1958 & 1960. *Rep. 7th & 8th Int. Am. Cacao Conf. Colombia & Trinidad* (gall types & transmission; **39**: 402; **40**: 600).

Kay, D. 1961. *Tech. Bull. W. Afr. Cacao Res. Inst.* 8, 20 pp. (review cacao dieback; **41**: 220).

Owen, H. 1956. *Ann. appl. Biol.* **44**: 307 (pathogenicity in cacao dieback, & other hosts, association with *Botryodiplodia theobromae*; **36**: 10).

Reichle, R. E. et al. 1964. *Phytopathology* **54**: 1297 (heterothallism; **44**, 1484).

Thresh, J. M. 1960. *Emp. J. exp. Agric.* **28**: 193 (capsid bugs, swollen shoot virus & *C. rigidiuscula*; **40**: 94).

CALOSTILBE Sacc. & Syd., *Sylloge Fung.* **16**: 591, 1902.
(Hypocreaceae)

Samuels regarded *Calostilbe* as a synonym of *Nectria*. *Phaeonectria* (Sacc.) Sacc. was synonymous with *Calostilbe*. The latter has brownish ascospores; the brown colouration of ascospores was rejected by Samuels as a generic character in the Hypocreales. *Calostilbe* (associated with a synnematous conidial state) was considered equal to *Nectria* by Rogerson (*Sphaerostilbe* q.v.)

Samuels, G. J. 1973. *Can. J. Bot.* **51**: 1275 (genus *Macbridella* with notes on *Calostilbe, Herpotrichia, Phaeonectria* and *Letendraea*; **53**, 77).

Calostilbe striispora (Ell. & Ev.) Seaver, *Mycologia* **20**: 248, 1928.
Nectria striispora Ell. & Ev., 1893
Macbridella striispora (Ell. & Ev.) Seaver, 1909
Sphaerostilbe longiascus Möller, 1901
stat. conid. *Calostilbella calostilbe* Höhnel, 1919.

Yellow furfuraceous stromatic pustules *c.* 2 mm diam. develop on the surface of the host and several may coalesce. From these pustules the stipes of the stat. conid. develop; stipes are yellow orange, somewhat strap-like and up to 10 mm long and 350 μ diam.; when mature they have a reddish brown globose head up to 1 mm diam. CONIDIA 37–50 × 12–14 μ are formed from phialides, 3 septate with 2 large yellow brown central cells and 2 small crescent shaped, hyaline, apical cells. PERITHECIA develop later in dense groups on the same stromatic pustules as the conidial state; yellow, furfuraceous, broadly cylindrical but often laterally compressed and with a deep orange yellow, ostiolar papilla; 750–800 × 450–500 μ diam. ASCI 8 spored, clavate, 90–150 × 10–11 μ. ASCOSPORES 1 septate, 32–43 × 9–14 μ, hyaline, becoming light brown and longitudinally striate when mature. PARAPHYSES long filamentous, abundant (C. Booth & P. Holliday, *CMI Descr.* 392, 1973).

The only serious disease caused by *C. striispora* (bark rot) is the limited one on the immortelle trees used to shade cacao in Trinidad and parts of S. America. These are *Erythrina glauca* (the swamp or Bocare) and *E. poeppigiana* (the mountain or Anauca). Infection of banana, as bonnygate disease, has been described (see Wardlaw, *Banana Diseases*); and other crops attacked include cacao, rubber (*Hevea brasiliensis*), coffee and other *Erythrina* spp. The fungus has been reported from Colombia, Costa Rica, Ghana, Jamaica, Sierra Leone, Surinam, Trinidad, Venezuela and Zaire.

In Trinidad severe losses occurred many years ago in the wetter eastern area where the Bocare immortelle was used. Later, on one estate in Venezuela, 50% of these trees were attacked as against 18.7% for the Anauca immortelle. On this estate *Erythrina* comprised 97% of the permanent shade. Lesions can form on the trunk, main branches, buttress and main roots, and have been found 15 m above ground; but most attacks occur at or near soil level. A thick mat of flattened, white becoming purplish, rhizomorphs (seen within a month of inoculation) spreads in the cambial region. Spread is more rapid up and down the tree than laterally. Large areas of bark and underlying tissues are destroyed and mature trees can be killed in 10–12 months. Trees *c.* 10–12 years old appear to have more resistance than those of *c.* 20–30 years. Although the fungus is common as a saprophyte on cacao pods and can be inoculated to this host it does not seem to attack cacao growing under this *Erythrina* shade. The Anauca or mountain immortelle is somewhat resistant, and in areas where the Bocare has been killed out (and where other permanent shade is not used) the mountain immortelle has replaced it.

Baker, R. E. D. 1941. *Trop. Agric. Trin.* **18**: 96 (on *Erythrina* in Trinidad; **20**: 453).

Malaguti, G. et al. 1952. *Agron. trop.* **2**: 41 (on *Erythrina* in Venezuela; **33**: 414).

CERATOCYSTIS Ellis & Halst., emend.

Bakshi, *Mycol. Pap.* 35: 2, 1951.
(for the genera placed in synonymy or excluded see Hunt)
(Ophiostomataceae)

PERITHECIA produced singly or in clusters, non-stromatic, the bases brown to black with elongate black necks; perithecial centrum plectascaceous; paraphyses lacking. ASCI unitunicate, evanescent, 8 spored. ASCOSPORES hyaline, aseptate, exuded from the ostiole in a sticky matrix. Imperfect states include endogenously and exogenously produced conidia (after J. Hunt).

The genus, monographed by Hunt with a key to the spp., has been the subject of several more recent taxonomic studies. Wright et al. discussed the diagnostic characters for identification of the spp. Hoog divided the genus into *Ceratocystis* sensu stricto (stat. conid. in *Chalara*, *Chalaropsis* q.v. and *Thielaviopsis* q.v.) and *Ophiostoma* (stat. conid. in *Graphium*, *Sporothrix* and *Verticicladiella*); and see Weijman et al. The spp. from Canada (Manitoba) were placed in 4 groups (minuta, ips, fimbriata and pilifera) on ascospore characteristics (Olchowecki et al.). Eleven groups were distinguished by Spencer et al., using the nature of the proton magnetic resonance spectra of the mannose containing polysaccharides and the composition of the polysaccharides. Upadhyay et al. segregated a new genus from *Ceratocystis* and proposed 4 new form genera for the stat. conid.

Some *Ceratocystis* spp. are general pathogens; they are often associated with blue stain of timber trees and with bark beetles. Young trees can be killed through fungal colonisation of the cambium and growth into the sapwood (e.g. Basham; Mathre), and see general accounts by Hepting, *Diseases of forest and shade trees of the United States*; Leach, *Insect transmission of plant diseases*. But 2 important temperate spp. are very restricted in host range. One (*C. ulmi* (Buism.) C. Moreau) causes the devastating

Dutch elm disease (on *Ulmus*); see Booth et al.; Gibbs; Gibbs et al.; Laut et al. The other (*C. fagacearum* (Bretz) Hunt) causes oak wilt (on *Quercus*). *C. adiposa* (Butl.) C. Moreau causes black rot (a soft, watery rot) of sugarcane setts. It is less serious than a comparable disease on the same crop caused by *C. paradoxa* (q.v.), and can be controlled by a standard fungicide treatment (Byther; Sartoris). The ascospores of *C. adiposa* are half-moon shaped (6–7.5 × 3.5–5 μ), whilst those of *C. paradoxa* are ellipsoid (7–10 × 2.5–4 μ).

Basham, H. G. 1970. *Phytopathology* **60**: 750 (wilt of *Pinus taeda*, *Ceratocystis* spp.; **49**, 3497).

Booth, C. et al. 1973. *C.M.I. Descr. pathog. Fungi Bact.* 361 (*C. ulmi*).

Byther, R. S. 1971. *Pl. Dis. Reptr* **55**: 7 (*C. adiposa* on sugarcane; **50**, 2481).

Dart, R. K. et al. 1976. *Trans. Br. mycol. Soc.* **67**: 327 (classification of *Ceratocystis* & *Sporotrichum* based on their long chain fatty acids; **56**, 3452).

Davidson, R. W. 1958. *Mycologia* **50**: 661 (Ophiostomataceae spp. from USA, Colorado; **38**: 424).

——. 1966. *Mycopathol. Mycol. appl.* **28**: 273 (new *Ceratocystis* spp. from conifers; **45**, 2625).

Gibbs, J. N. et al. 1977. *For. Rec.* 115, 12 pp. (*C. ulmi*, general account, in UK; **57**, 3120).

——. 1978. *A. Rev. Phytopathol.* **16**: 287 (*C. ulmi*, intercontinental epidemiology, 79 ref.).

Griffin, H. D. 1968. *Can. J. Bot.* **46**: 689 (genus in Ontario, key to 60 spp.; **47**, 2621).

Hawes, C. R. et al. 1977. *Trans. Br. mycol. Soc.* **68**: 259, 267 (light & electron microscopy of conidium ontogeny in *Chalara* state of *Ceratocystis adiposa*; **56**, 4392, 4393).

Hoog, G. S. de 1974. *Stud. Mycol.* 7, 84 pp. (*Blastobotrys*, *Sporothrix*, *Calcarisporium* & *Calcarisporiella* gen. nov.; **54**, 740).

Hunt, J. 1956. *Lloydia* **19**: 1 (monograph, key; **36**: 63).

Laut, J. G. et al. 1976. Colorado Univ. (State Forest Service), 135 pp. (bibliography of *C. ulmi*, 1902 ref.; **56**, 3239).

Mathre, D. E. 1964. *Contr. Boyce Thompson Inst. Pl. Res.* **22**: 353, 363 (*Ceratocystis* spp. & bark beetles; pathogenicity of *C. ips* & *C. minor* to *Pinus ponderosa*; **44**: 884).

Olchowecki, A. et al. 1974. *Can. J. Bot.* **52**: 1675 (taxonomy of genus in Manitoba, keys; **54**, 244).

Reynolds, P. E. et al. 1972. *Trans. Br. mycol. Soc.* **59**: 1 (effect of temp. on growth; **52**, 596).

Roldan, E. F. 1958. *Philipp. J. Sci.* **87**: 37 (on rattan; **39**: 446).

——. 1962. *Ibid* **91**: 415 (as above; **42**: 625).

Sartoris, G. B. 1927. *J. agric. Res.* **35**: 577 (development & morphology of *C. adiposa*; **7**: 272).

Spencer, J. F. T. et al. 1971. *Mycologia* **63**: 387 (systematics of *Ceratocystis* & *Graphium*; **50**, 3520).

Upadhyay, H. P. et al. 1975. *Mycologia* **67**: 798 (prodromus for a revision of *Ceratocystis* & its conidial states; **55**, 1112).

Weijman, A. C. M. et al. 1975. *Antonie van Leewenhoek* **41**: 353 (subdivision of *Ceratocystis*; **55**, 1118).

Wright, E. F. et al. 1961. *Can. J. Bot.* **39**: 1215 (spp. nov.).

Ceratocystis fimbriata Ellis & Halsted, *Bull N.J. agric. Exp. Stn* 17: 14, 1890.
Sphaeronaema fimbriatum (Ell. & Halst.) Sacc., 1892
Ceratostomella fimbriatum (Ell. & Halst.) Elliott, 1923
Ophiostoma fimbriatum (Ell. & Halst.) Nannf., 1934
Encoconidiophora fimbriata (Ell. & Halst.) Davidson, 1935
Rostrella coffeae Zimm., 1900
O. coffeae (Zimm.) Arx, 1952.

PERITHECIA superficial or partly to completely immersed, brown to black, globose, 140–220 μ diam., unornamented, necks long, black, pale brown to subhyaline towards tip, tapering slightly, up to 900 μ, ostiolar hyphae hyaline, erect or moderately divergent. ASCOPORES elliptical with a gelatinous sheath forming a brim, giving a hat-shaped appearance, hyaline, non-septate, smooth 4.5–8 × 2.5–5.5 μ. CONIDIOPHORES slender, arising laterally from the hyphae, septate, phialidic, hyaline to very pale brown, up to 160 μ long, usually tapering towards the tip and producing a succession of conidia through the open end. CONIDIA cylindrical, truncate at the ends, hyaline, smooth walled, 11–25 (15) × 4–5.5 μ. A thick walled endoconidial form occurs. CHLAMYDOSPORES terminal in chains, obovate to oval, thick walled brown, 9–18 × 6–13 μ. Distinguished from *C. paradoxa* (q.v.) by its somewhat longer conidia and especially by the dark, unornamented perithecium and the hat shaped ascospores. (G. Morgan-Jones, *CMI Descr.* 141, 1967.)

Probably worldwide in distribution as non-pathogenic forms but restricted in its role as a pathogen. In the latter form it is widely distributed in the tropics and in N. America but appears to be absent from Europe, the Mediterranean and temperate Asia (Map 91). The fungus is largely a wound parasite usually attacking the trunks and branches of woody hosts. Over short distances spread is presumably aerial, although few studies have been reported. Over longer ones it is probably spread either by man or by insects. The disease (canker stain) it causes in London plane (*Platanus acerifolia*, Walter et al.) is a unique case of one being almost entirely spread by man in USA. This manner of spread is also important in rubber (*Hevea brasiliensis*). Insects may spread the fungus secondarily, i.e. after an initial infection has occurred by other means; such appears to be the case in cacao. Alternatively, insects have been reported to be responsible for direct transmission in deciduous stone fruit trees (Moller et al.). The fungus is plurivorous. Generally strs causing disease in any one host are not pathogenic, or only slightly so, to others. But more work is needed here. There is some evidence that disease producing strs might arise de novo from indigenous saprophytic forms.

The morphology, cytology and genetics have been fairly fully described. Differing morphological groups occur in vitro, but the sexual state seems quite uniform. The thick-walled spores survive through periods of dry, hot weather. All isolates require thiamine for the production of perithecia in vitro, where the linear growth was best at 24–27°. The numbers of conidia are not affected by the amount of thiamine except in so far as it affects the amount of mycelium which requires less than the sexual state. *C. fimbriata* has been considered to be basically homothallic. But single ascospores from perithecia in self-fertile cultures gave rise to both self-fertile and self-sterile strains. Sexuality is controlled by 2 major non-allelic factors. The absence of an incompatibility locus allows one mating type to be fertilised by any self-fertile type or the other mating type. Loss of both factors through mutation gives an asexual strain. Perithecial colour is controlled by a single factor independent of those controlling sexuality.

Andrus, C. F. et al. 1933. *J. agric. Res.* **46**: 1059 (morphology of reproduction).
—— ——. 1937. *Ibid* **54**: 19 (development of ascus; **16**: 494).
Barnett, H. L. et al. 1947. *Mycologia* **39**: 699 (thiamine & perithecial production; **27**: 405).
Chevaugeon, J. 1957. *Rev. Mycol. Paris* **22**, Suppl. Colon. 2: 45 (review, 42 ref.; **37**: 339).
Elliot, J. A. 1925. *Phytopathology* **15**: 417 (cytology of sexual state).
Gwynne-Vaughan, H. C. I. et al. 1936. *Ann. Bot.* **50**: 747 (cytology of sexual state; **16**: 494).
Lehman, S. G. 1918. *Mycologia* **10**: 155 (conidial formation).
Moller, W. J. et al. 1968. *Phytopathology* **58**: 1499 (insect transmission in deciduous stone fruit; **48**, 851).
Soave, J. et al. 1974. *Bragantia* **33**: XXI (effect of light on growth & sporulation; **54**, 4795).
Stiers, D. L. 1976. *Can. J. Bot.* **54**: 1714 (fine structure of ascospore formation; **56**, 596).
Walter, J. M. et al. 1952. *Ibid* **42**: 236 (spread & destruction in London plane; **32**: 43).
Webster, R. K. et al. 1967. *Can. J. Bot.* **45**: 1457 (morphological comparison & hybridisation in host strs; **47**, 766).
—— ——. 1967. *Mycologia* **59**: 212 (self-sterile & self-fertile strs; **46**, 2640a).
——. 1967. *Ibid* **59**: 222 (inheritance of sexuality and colony type: **46**, 2640b).

COFFEA spp.

The recent investigations on this canker disease, reported from Colombia and Venezuela in 1951, probably refer to the same condition first described by Zimmerman from Java on *C. arabica* in 1900. The disease also occurs in Guatemala, where losses (at 700–1500 m) have been severe, and probably Surinam (Van Emden et al.). Symptoms may first be seen as a leaf yellowing on the infected branch. The bark, where infection has occurred, becomes sunken. If it is scraped away, irregularly shaped necrotic lesions are exposed, penetrating to a depth of *c.* 1 cm. There is a fairly clear necrotic edge. Cankers can measure 45 cm vertically and if at the base of the trunk may extend into the roots. They occur usually on older trees and seldom on branches until the bark is 3 years old or more. A high percentage of 1-year-old trees were girdled and killed in artificial inoculations. Canker growth vertically was 2–3 cm/month and 1 cm/month laterally. Perithecia occur in the field if conditions are sufficiently moist. The disease is thought to be spread through wounds which result from cultural operations. The greater resistance of *C. canephora* and *C. liberica* compared with *C. arabica* has been attributed to higher concs of chlorogenic acid. In cross-inoculations strs from sweet potato and coffee were only pathogenic towards their respective hosts. Isolates may lose virulence with time in culture. Zuluaga reported on resistance.

Echandi, E. et al. 1962. *Phytopathology* **52**: 544 (chlorogenic acid & resistance; **42**: 123).
Fernandes-Borrero, O. 1964. *Cenicafé* **15**: 3 (pathogenicity & resistance; **45**, 2847).
Pontis, R. E. 1951. *Phytopathology* **41**: 178 (general; **30**: 366).
Snyder, W. C. et al. 1960. *Pl. Dis. Reptr* **44**: 566 (general; **40**: 106).
Szkolink, M. 1951. *Ibid* **35**: 500 (general; **31**: 237).
Van Emden, J. H. et al. 1959. *Surinam Landb.* **7**: 111 (with phloem necrosis; **39**: 579).
Zimmerman, A. 1900. *Bull. Inst. Bot. Buitenzorg* 4: 19 (in Indonesia, Java).
Zuluaga, V. J. et al. 1971. *Cenicafé* **22**: 43 (polyphenols & resistance; **52**, 2945).

HEVEA BRASILIENSIS

The gradual and systematic removal of the outer bark and cortex in the tapping panel of rubber provides an abnormal condition in the tree making the panel subject to fungal infections. Mouldy rot caused by *C. fimbriata* is one of the most serious of these. It occurs wherever the crop is cultivated; the fungus does not attack other parts of the tree. The first symptoms are a series of slightly sunken, discoloured spots, 0.5–2.5 cm above the tapping cut. These darken and often extend gradually, forming an irregularly outlined band parallel to the cut. This infected area becomes covered with a whitish mould (conidial stage) which darkens later. The fungus penetrates the thin layer of inner cortex (left over the cambium by the tapping process) and a few mm of the wood, the resulting damage to the cambium then detrimentally affects bark renewal. These bark wounds resemble those caused by bad tapping. Symptoms can first be seen 24 hours after infection, with rot appearing in 3 days.

The disease is essentially one of wet weather and particularly of moist environments in mature fields. In dry weather an attack may pass within a few days. Short-range spread probably occurs through wind and rain and possibly via insects. But spread by tappers on clothes and knives is important and can result in the infection of a single task in a few weeks. Tapping should cease on infected trees since infection only occurs on newly tapped bark. Change to third or fourth daily tapping reduces incidence. Up to 12 applications of the recommended fungicide (Anon.) should be made on the panel with a small pneumatic sprayer. The movement of tappers should be watched in relation to outbreaks of the disease and their knives given a disinfectant dip after each tapping. The usually dry, sunny mornings of Malaysia (W.) lessen any effect of the disease.

Anon. 1972. *Plrs' Bull. Rubb. Res. Inst. Malaya* 118: 3 (symptoms & control; **51**, 4303).
Martin, W. J. 1949. *Circ. U.S. Dep. Agric.* 798, 23 pp. (general; **29**: 118).
Olson, E. O. et al. 1949. *Phytopathology* **39**: 17 (rubber & sweet potato strs; **28**: 356).
Sanderson, A. R. et al. 1920–21. *Ann. appl. Biol.* **7**: 56 (general).
South, F. W. et al. 1925. *Bull. Dept Straits Settlements & Fed. Malay States* 37, 31 pp. (general; **5**: 184).

IPOMOEA BATATAS

The disease on sweet potato, commonly called black rot, is primarily one of the root tubers and often develops to the max. extent in storage and during marketing, from infection originating in the field. On tubers the first symptoms are brown, circular, slightly sunken, superficial spots *c.* 0.5 cm diam. They enlarge, becoming up to 5 cm across, greenish black, ashy mycelium develops and the black, long-necked perithecia appear. The rot is firm and generally shallow, a deeper rot usually being the result of invasion by other organisms. Infected tissue is bitter and after cooking the whole tuber has this flavour. In the seedbed diseased tubers or infection from soil or debris cause necrotic lesions on the young shoots. These may girdle the stems causing chlorosis and stunting; such infected plants will produce diseased tubers that will rot in storage. Any sort of wounding will increase the possibility of infection but it can probably also occur in apparently unwounded tissue. Disease development is most rapid at 25°, the opt. temp. for growth of the tuber lesions being very similar to that in vitro. At higher, more tropical temps incidence is reduced and the rot of shoots in seedbeds decreases as temp. increases.

The biochemical aspects of host specificity, pathogenic and non-pathogenic strs, resistance and susceptibility, and phytoalexins have been intensively studied, particularly by workers in Japan: *Rev. appl. Mycol.* **34**: 246, 248; **35**: 710; **39**: 440; **40**: 705; **41**: 329, 406, 614; **44**, 817; **45**, 585; **46**, 418, 2505; **48**, 3240. *Rev. Pl. Pathol.* **49**, 1246, 2281; **50**, 455; **52**, 1750, 1751, 4307; **54**, 1133; **55**, 2997, 3840, 3841, 5456, 5457; **56**, 510, 1377; **57**, 4739, 4740.

Control is through a combination of tuber and

seedbed treatment, and also general field and storage house sanitation. The tubers may be heat treated at 43° for 2–3 days or 39° for 5 days. Alternatively they may be dipped in 1% borax solution. Other chemical treatments do not appear to be effective except at temps above 30°. Daines (1971) found that dips in benomyl at 44–45° reduced cankers significantly compared with dips at 14°. Ferbam was effective at 49° or 54°. Both disease free seedbeds and fields should be used and rotation may be advisable. Tubers for storage should be carefully handled, inspected, placed in clean, disinfected bins after treatment by heat or chemicals and the correct storage temps and humidities maintained. Although some varietal differences in susceptibility have been found this does not have much practical significance.

Cheo, P. C. 1953. *Phytopathology* **43**: 78 (resistance; **32**: 693).
Cooley, J. S. et al. 1951. *Ibid* **41**: 801 (effect of temp.; **31**: 101).
Daines, R. H. 1959. *Ibid* **49**: 249 (effect of temp. on disease incidence; **38**: 622).
—— et al. 1962. *Ibid* **52**: 1138 (control by pre-bedding heat treatment; **42**: 339).
——. 1971. *Ibid* **61**: 1145 (fungicides; **51**, 1028).
Goto, K. et al. 1954. *Bull. Div. Pl. Breed. Cult. Tokai-Kinki agric. Exp. Stn* 1: 138 (soil spread & survival; **35**: 227).
Halsted, B. D. et al. 1891. *J. Mycol.* **7**: 1 (general & morphology).
Harter, L. L. et al. 1926. *J. agric. Res.* **32**: 1153 (soil temp. & moisture; **5**: 688).
Kushman, L. J. et al. 1949. *J. agric. Res.* **78**: 183 (effect of temp.).
Lauritzen, J. I. 1926. *Ibid* **33**: 663 (effect of temp. in storage).
Martin, W. J. et al. 1949. *Phytopathology* **39**: 580 (control in tubers with borax; **28**: 641).
—— ——. 1954. *Ibid* **44**: 383 (varietal reactions; **34**: 249).
Olson, E. O. 1949. *Ibid* **39**: 548 (strs; **29**; 53).
Person, L. H. et al. 1948. *Ibid* **38**: 474 (fungicides; **28**: 111).
Pinckard, J. S. et al. 1943. *Pl. Dis. Reptr* **27**: 151 (fungicides; **22**: 510).

THEOBROMA CACAO

The first report of cacao canker or wilt caused by *C. fimbriata* may be that from Costa Rica by R. A. Altson in 1925. Although this crop has been important for many years in parts of Central and S. America, and Trinidad, it was not until 20–30 years later that the disease began to cause noticeable losses, even though the fungus (in a non-pathogenic form?) had almost certainly occurred there earlier. In Trinidad (1932) A. K. Briant reported it to be the cause of wilt, stem and leaf spotting of *Crotalaria juncea*; in 1935 it was reported from cacao pods. But not until 1958 was it found to be causing severe disease in the island's cacao. There is a review by Saunders. The most noticeable symptoms are on the foliage; the leaves become pendulous, show irregular yellowing and browning, and eventually become necrotic remaining attached for several weeks. The pods dry up less rapidly. The site of infection in the wood reveals a brown to claret stain after tangential cutting. Wood sections show a blue grey (sometimes dark brown) stain. The discoloured wood decreases with increasing distance from the canker. Trunk infection leads to wilt of the whole tree but an attack may be confined to a single branch causing a partial wilt. Insect borings (1 mm diam.) are invariably associated with infection, and frass extruded from them can be found in the field.

Detailed studies on this disease, which appears so far to be confined to tropical America, have been done by Iton and others in Trinidad. Inoculation (through wounds) of 12–21-month-old clones produced symptoms in 4–16 days and death after 7–20 days. Sporulation does not occur extensively on the host surface, but forms on exposure of infected internal wood and also on the walls of the insect galleries. Two *Xyleborus* spp. attack the tree most frequently: *X. corniculatus* and *X. ferrugineus*. Their borings are mostly concentrated at ground level on the trunk and up to 50 cm above the soil; they also occur in the lateral roots. These insect populations are most advanced at ground level and progressively less so above and below this.

Chlamydospores and endoconidia are the most abundant in the field, and occur in the frass and on the insects, to which the ascospore masses also adhere. The sexual spores in frass, or from cut or broken wood and bark surfaces, are air dispersed. Although inocula, consisting of insects plus adhering spores, caused infections on surfaces of wood blocks, on living material these infections remain incipient. The *Xyleborus* spp. are apparently attracted to a tree which has already become infected and do not normally establish galleries in healthy trees. Other non-fungal organisms associated with the borers may also be a factor in spread. Initial infection presumably always occurs through wounds, mostly brought about by various cultural operations and, in Trinidad, by pruning in control of *Crinipellis perniciosa* (q.v.). This is followed by borer attack; the activities of the insects greatly increasing inoculum potential and also, presumably, infection rates in the field. The sudden appearance of this disease in Trinidad poses the question as to whether the pathogenic form was introduced in the 1950s from mainland tropical America (where it had earlier appeared) or whether it arose as a mutant in an already existing population. This in turn emphasises the need for detailed work both on the pathogenic forms in relation to their various hosts and on the saprophytic or quiescent stages.

The direction of control is not yet entirely clear. Until more has been done on what types (and intensity) of wounding lead to infection, recommendations involving cultural operations can have no proper basis. Chemical control of the borers is of doubtful value since they do not apparently initiate infection. Clones show varying degrees of resistance and this may be the most promising line of attack. But small scale tests may not correlate well with field performance.

Arbelaez, G. E. 1957. *Acta agron. Palmira* **7**: 71 (general in Colombia; **39**: 15).

Bartley, B. G. D. et al. 1966. *Rep. Cacao Res. Trinidad* 1965: 51 (resistance test & insect control; **47**, 1470b).

Capriles de Reyes, L. et al. 1966. *Agron. trop.* **16**: 273 (chlorogenic acid & resistance; **47**, 94).

Chong Gomez, L. 1962. *Turrialba* **42**: 752 (inoculation in Ecuador; **42**: 752).

Dominguez R. P. F. 1971 & 1972. *Revta Fac. Agron. Univ. cent. Venez.* **6** (2): 5; (4): 57 (breeding & selection for resistance; **51**, 4765; **53**, 2120).

Idrobo, M. S. et al. 1956. *Cacao en Colombia* **5**: 25, 37 (various aspects, including toxins, in Colombia; **37**: 221, 222).

Iton, E. F. 1959. *Rep. Cacao Res. Trinidad* 1957–58: 55 (general, outbreak in Trinidad; **39**: 685).

——. 1961. *Ibid* 1959–60: 47 (wind transmission; **41**: 377).

—— et al. 1961. *Ibid* 1959–60: 59 (biology of *Xyleborus* spp. & transmission; **41**: 377).

——. 1966. *Ibid* 1965: 44 (insect complex, attraction of infected trees for *Xyleborus* spp.; **47**, 1470a).

Malaguti, G. 1959. *Agron. trop. Maracay* **5**: 207 (in Venezuela; **37**: 271).

Naundorf, G. et al. 1956. *Cacao en Colombia* **5**: 29, 41 (spread & control; **37**: 222).

Saunders, J. L. 1965. *Cacao Turrialba* **10** (2): 7 (review, 44 ref.; **46**, 87).

Small, L. W. 1967. *Rep. Cacao Res. Trinidad* 1966: 40 (resistance test and insect control; **47**, 781a).

Soria V., J. 1973. *Turrialba* **23**: 231 (effect of plant age on symptom appearance; **53**, 100).

OTHER CROPS

In S. America a wilt of *Crotalaria juncea* (necrotic, elongated lesions on the stem) has been reported. Isolates differed in pathogenicity from those from sweet potato and coffee. Sporulation occurs on this host near soil level. In Brazil *C. anagyroides* and *C. goerensis* showed some resistance, as did some seedlings from the former sp. and from *C. juncea*.

On *Pimenta dioica* (pimento) infection of a wound in the wood causes an elliptical canker with a sunken, fissured, dry centre. The bark above the canker is often raised and when this is peeled off, dark brown pegs fitting into holes in the bark are found. The wood shows a necrosis which diminishes in extent at increasing distances from the site of infection. Immature trees are rarely attacked. Infection frequently begins on one limb, followed by a slow spread; a tree may die in a few months if the primary infection is below the crotch. Leaves wilt, become russet and remain attached. Dead bark developed fructifications of *Valsa eugeniae* (q.v.) and *Calonectria rigidiuscula* (q.v.) in Jamaica but neither fungus was pathogenic to pimento. Large branches broken in harvesting provide places of entry. Inoculations showed that canker elongation was much greater upwards. The rapid leaf death which results gave rise to the local name of fireblight. Insects do not appear to be associated with this disease and isolates did not attack cacao, coffee or sweet potato. Careful harvesting at the tips of fruiting branches and other measures to reduce wounding should exert some control.

In Brazil infection of mango causes a wilt of small crown branches, leaf chlorosis, fluting and spiralling of leaves around branches to cause characteristic flags. Association with an insect (*Xyleborus affinis*) has been described.

Costa, A. S. et al. 1935. *Phytopathol. Z.* **8**: 507 (on *Crotalaria* in Brazil; **15**: 155).

Davet, P. 1962. *Rev. Mycol. Paris* **26**: 225 (on *Crotalaria* in Ivory Coast; **41**: 700).

Leather, R. I. 1966. *Trans. Br. mycol. Soc.* **49**: 213 (on pimento in Jamaica; **45**, 2918).

Ceratocystis paradoxa

Malaguti, G. 1952. *Agron. trop. Maracay* **4**: 287 (on *Crotalaria* in Venezuela; **32**: 132).

Medeiros, A. G. 1967. *Fitopatologia* **2**: 29 (resistance in *Crotalaria*; **48**, 2418).

Viegas, A. P. 1960. *Bragantia* **19**: 163 (on mango in Brazil; **42**: 696; and see **46**, 2080).

Ceratocystis paradoxa (Dade) C. Moreau, *Rev. Mycol.* **17**: 22, 1952.

Ceratostomella paradoxa Dade, 1928

Ophiostoma paradoxa (Dade) Nannfeldt, 1934.

PERITHECIA partly or completely immersed, light brown, globose, 190–350 μ diam., ornamented with numerous stellate or coralloid, brown appendages, necks long, black, pale brown towards the tip, tapering, up to 1.4 mm, ostiolar hyphae hyaline, erect or moderately divergent. ASCOPORES ellipsoid, often with unequally curved sides, hyaline, aseptate, smooth, $7–10 \times 2.5–4\ \mu$. CONIDIOPHORES slender, arising laterally from the hyphae, septate, phialidic, hyaline to very pale brown, up to 200 μ long, tapering towards the tip and producing a succession of conidia through the open end. CONIDIA cylindrical to somewhat oval when mature, hyaline to mid brown, smooth walled, $6–24\ (13) \times 2–5.5\ \mu$. CHLAMYDOSPORES terminal in chains, obovate to oval, thick walled, brown, $10–25 \times 7.5–20\ \mu$. (G. Morgan-Jones, *CMI Descr.* 143, 1967.)

Like *C. fimbriata* (q.v.) this fungus is worldwide in distribution, but largely tropical and subtropical (Map 142). It is spread through soil and host debris and is plurivorous (see *CMI Descr.* 143). Petch fully described the early confusion over identity and Dade (1928) reported the perfect state. The fungus is heterothallic. The host is infected through wounded, cut or bruised surfaces. The evidence for strs based on host is slight, and isolates from one particular host are likely to be pathogenic towards others. Thiamine and biotin are required for growth, and ethyl acetate is formed; this causes the characteristic odour associated with *C. paradoxa*, especially in vitro. Chlamydospores are a resting stage and the host tissue is darkened by them.

Averna-Sacca, R. 1932. *Revta Agric. Piracicaba* **7**: 114 (general in Brazil; **13**: 127).

Dade, H. A. 1928. *Trans. Br. mycol. Soc.* **13**: 184 (perfect state; **8**: 227).

Ichinoe, M. et al. 1964. *Trans. mycol. Soc. Japan* **5**: 9 (fine structure of conidia; **44**, 1444).

Olutiola, P. O. 1976. *Mycologia* **68**: 1083 (cellulase enzymes in culture; **56**, 2869).

—— et al. 1977. *Ibid* **69**: 524 (growth & sporulation in culture; **57**, 247).

Petch, T. 1910. *Ann. R. bot. Gdns Peradeniya* **4**: 511 (early history).

Reddy, M. S. et al. 1964. *Indian Jnl expl Biol.* **2**: 211 (adaptation to Cu & Hg; **44**, 1386).

ANANAS COMOSUS

The first description of a disease was from pineapple where *C. paradoxa* causes a soft rot of the stem, leaf bases and fruit, and is sometimes referred to as water blister. Losses are most likely to be incurred where the fruit has to be transported some distance to the marketing area, and in handling, packing and storage. Symptoms on the fruit begin as a watersoaked area which develops into a soft, wet rot of the core and flesh; the colour changes to a vivid yellow, becoming grey and later black. Disintegration is complete. Less serious infection occurs on the leaves. A small yellow brown spot enlarges to become straw coloured, almost white, and conspicuous. The lesions are papery and may cause the leaf to wither. A basal rot can lead to death of the suckers.

Both green and ripe fruit are attacked. Most infection probably occurs through the cut stem after harvest. Rot develops rapidly at 23–29° and in vitro growth is greatest at *c.* 28°. No control of the leaf infection appears to be necessary. For the fruit careful handling is needed to reduce bruising and other damage at all stages after harvest. Basal rot can be reduced significantly by harvesting with a portion of the cut stem and treatment of the cut with a fungicide within 5 hours; 10% benzoic acid in ethanol or as a dry mix with kaolin has been effective. Strict hygiene must be observed in packing sheds. Cho et al. reported the control of pineapple propagation material with benomyl and thiabendazole (and see Frossard 1970, 1978).

Cho, J. J. et al. 1977. *Phytopathology* **67**: 700 (induction & chemical control; **57**, 231).

Chowdhury, S. 1945. *Indian J. agric. Sci.* **15**: 135 (symptoms, inoculation & morphology; **25**: 460).

Dickson, B. T. et al. 1931. *J. Aust. Coun. scient. ind. Res.* **4**: 152 (control; **11**: 192).

Frossard, P. 1964. *Fruits* **19**: 461 (temp. & pH effect in vitro; **44**, 2215).

——. 1970. *Ibid* **25**: 785 (fungicides; **50**, 3069).

——. 1978. *Ibid* **33**: 91 (storage temp. & fungicides; **57**, 5609).

McKnight, T. 1941. *Qd agric. J.* **55**: 180 (control by sanitation; **20**: 374).
Oxenham, B. L. 1953. *Qd J. agric. Sci.* **10**: 237 (leaf spot; **34**: 43).
Roldan, E. F. 1925. *Philipp. Agric.* **13**: 397 (general; **4**: 491).

PALMAE

Occurs on areca palm (*Areca catechu*) and coconut causing bleeding disease, and on oil palm as dry basal rot. An old disease on coconut reported *c.* 1906 from Sri Lanka and fully described from there by Petch. Provided any wounding of young palms (especially near the apical bud and base) is avoided the disease is of minor importance. Symptoms on coconut and areca are similar and usually begin as a small bleeding lesion at any point on the stem, the oozing liquid is reddish and becomes black. Decay of the internal cortex is yellow and later black; it can be much more extensive than the external symptoms suggest and large, hollow wounds can result. Serious infection can lead to reduced yield and death of young palms. Cook described a premature leaf and nut fall in coconut in Puerto Rico caused by the fungus (as the stat. conid. *Thielaviopsis paradoxa* (de Seynes) Hohnel). A leaf spot on coconut has also been reported from the Philippines. Since bleeding can have other causes isolation of the fungus may be necessary for diagnosis. Control is to avoid trunk damage and apply wound dressings.

Later (1959, see Robertson) a serious infection of oil palm was noted in Nigeria. Attack on 4–8-year-old trees was serious and yields reduced markedly. The external symptoms are a rot of the fruit or male inflorescence and rachis fracture in the lower leaves which become grey and brittle. Successively younger leaves may be affected, spear collapse and death can follow. Internally a dry rot with a well-defined margin is found at the stem base. Infection may occur via the roots and 1-year-old seedlings were killed 10–15 days after having their roots dipped in inoculum. Selection for resistance indicated that this may be controlled by a single gene.

Cook, M. T. 1924. *J. Dep. Agric. P. Rico* **8**(4): 12 (on coconut as *Thielaviopsis paradoxa*; **4**: 216).
Goberdhan, L. C. 1961. *J. agric. Soc. Trin.* **61**: 33 (general in Trinidad; **41**: 53).
Petch, T. 1909. *Circ. agric. J. R. bot. Gdns Peradeniya* **4**: 197 (general in Sri Lanka).
——. 1909. *Trans. Br. Mycol. Soc.* **3**: 108 (symptoms in Sri Lanka).
Protacio, D. B. 1965. *Philipp. J. Agric.* **25**: 67 (leaf spot in coconut; **45**, 1472).
Robertson, J. S. 1962. *Jl W. Afr. Inst. Oil Palm Res.* **3**: 339 (general; **41**: 612).
——. 1962. *Trans. Br. mycol. Soc.* **45**: 475 (general on oil palm; **42**: 336).
——. 1963. *Ann. Rep. W. Afr. Inst. Oil Palm Res.* 1962–63: 80 (resistance; **43**, 1111).
Salgado, M. L. N. 1942. *Trop. Agric.* **98**: 31 (physiological stem bleeding in coconut; **22**: 356).
Sundararaman, S. 1922. *Bull. agric. Res. Inst. Pusa* 127, 8 pp. (on coconut in India; **2**: 79).
—— et al. 1928. *Ibid* 169, 12 pp. (on areca palm in India; **7**: 629).

SACCHARUM

Wismer (1961) in a general review of pineapple disease of sugarcane stated that *C. paradoxa* is the principal cause of rots of cuttings (seedpieces or setts). A good early history was given by Cook. The ends of the cuttings are penetrated and the fungus spreads in the parenchyma, more rapidly in the internodes. Infected tissue is at first reddish, later black, with breakdown in the parenchyma causing the cutting to become almost hollow. Infection causes failure of the buds to germinate, one of the greatest causes of loss. If infected setts do develop the young shoots may dieback or be retarded in growth. If growing cane is damaged infection can result. As on other hosts the asexual state is commoner in the field.

Infection usually occurs via the soil through spores, mostly at 7–25 cm depth. When sterile discs of sugarcane stem were inoculated with soil the test could detect a conc. of only 6 spores/g soil. In vitro growth is opt. at *c.* 28° (and see Liu et al. 1972); but more ethyl acetate is produced at 20°. Infected sugarcane contains sufficient ethyl acetate to inhibit bud germination and the ester can be detected in the soil. Byther et al. found that rooting was inhibited when infected or healthy setts were suspended over cultures. Volatile substances from the cultures caused inhibition and a concomitant rise in the formation of ethylene by the plant. Ethylene in part causes this inhibition and the compound is formed under the stimulation of host invasion or volatile substances.

Control should aim at stimulation of germination; poor development of setts leads to increased infection and uneven stands. Larger cuttings are attacked less and not less than 3 nodes in a sett should be used. Mercurials are effective in protecting planted cuttings which may be treated as they are being planted, or dipped before in tanks where the fungicide is incorporated in the hot water treatment which stimulates germination. More recently benomyl, as a dip for setts, has been described, for example, for Hawaii (Hilton et al.), India (Muthusamy), Mauritius (Ricaud et al., where the hot dip was better than the cold one), South Africa (Bechet) and Taiwan (Wang et al.).

Aberdeen, J. E. C. 1969. *Aust. J. agric. Res.* **20**: 843 (conc. in soil; **49**, 833).
Bell, A. F. 1936. *Cane Grow. q. Bull.* **4**: 46 (effect on germination).
Bechet, G. R. 1978. *S. Afr. Sug. J.* **62**: 85 (systemic fungicides).
Boyd, H. W. et al. 1968. *Phytopathology* **58**: 839 (effect of cutting size & inoculum potential; **47**, 3201).
Byther, R. S. et al. 1974. *Can. J. Bot.* **52**: 761 (inhibition of rooting; **53**, 4108).
Chang, V. C. S. et al. 1974. *J. econ. Ent.* **67**: 190 (spread by nitidulid beetles in Hawaii; **53**, 4575).
Cook, M. T. 1932. *J. Dep. Agric. P. Rico.* **16**(2): 205 (early history, 30 ref.; **12**: 114).
Hilton, H. W. et al. 1971. *Hawaii. Plrs' Rec.* **58**: 159 (benomyl treatment; **52**, 2361).
Kiryu, T. 1939. *Rep. Govt Sug. Exp. Stn Taiwan* 6: 21 (physiology).
Kuo, T. T. et al. 1969. *Can. J. Bot.* **47**: 1459 (production of ethyl acetate & inhibition of buds; **49**, 550).
Liu, L. J. et al. 1972. *J. Agric. Univ. P. Rico* **56**: 162 (effects of temp. & moisture on development & pathogenicity; **52**: 3814).
——. 1973. *Ibid* **57**: 117 (compatibility, morphology, physiology, pathogenicity & in vitro sensitivity to fungicides, sugarcane & pineapple strs; **53**, 4532).
Muthusamy, S. 1973. *Sugcane Pathol. Newsl.* 10: 14 (systemic fungicides; **53**, 2302).
Ricaud, C. et al. 1974. *Rev. agric. sucr. Ile Maurice* **53**: 198 (benomyl treatment; **54**, 5540).
Steindl, D. R. L. 1970. *Sugcane Pathol. Newsl.* 5: 53 (control & stimulation of germination in setts; **50**, 930).
Story, C. G. 1952. *Cane Grow. q. Bull.* **15**: 92 (epidemic in Australia, Queensland; **32**: 100).
Wang, C. S. et al. 1973. *Pl. Prot. Bull. Taiwan* **15**: 134 (benomyl treatment; **53**, 4868).
Wismer, C. A. 1951. *Hawaii. Plrs. Rec.* **54**: 23 (control; **30**, 582).
——. 1961. In *Sugarcane diseases of the world* Vol. 1: 223, Elsevier (general, 62 ref.).

OTHER CROPS

These are: banana, cacao, coffee, custard or sugar apple (*Annona squamosa*), maize, *Pueraria*, sorghum and sweet potato. In banana a fruit, stem end rot and black head disease of rhizomes are caused. Fruit stalks decay rapidly, the tissue becoming black and soft with a sweetish smell. Infection spreads to the fingers producing an even discolouration of the skin and a wet decay; fingers may drop. In Australia differences between banana and sugarcane strs were reported, the former having a 24–26° opt. for growth and the latter (with those from pineapple) one of 29–30°. The banana str. attacked pineapple weakly and sugarcane not at all, whilst those from the last 2 hosts were only weak pathogens to banana (Mitchell). On wounded, nearly mature cacao pods a brown, rapidly spreading lesion is formed. It covers the pod in 5–7 days and causes a rot which allows the pod to be crushed easily.

Desrosiers, R. 1955. *Pl. Prot. Bull. F.A.O.* 3: 154 (on cacao pods; **35**: 883).
Joly, P. 1961. *Bull. Soc. mycol. Fr.* **77**: 219 (on banana; **41**: 399).
Mitchell, R. S. 1937. *J. Aust. Coun. scient. ind. Res.* **10**: 123 (differences between host strs.; **16**: 693).

CERCOSEPTORIA Petrak, *Annls Mycol.* 23: 69, 1928.

Septoriopsis Stevens & Dalbey, 1919
non *Septoriopsis* G. Fragoso & Paul, 1915.
(Dematiaceae)

MYCELIUM internal. STROMA present, substomatal. CONIDIOPHORES pale brown, smooth, septate, mostly simple, densely fasciculate. CONIDIOGENOUS CELLS integrated, sympodial and polyblastic or percurrent (pseudopercurrent); conidial scars unthickened and inconspicuous. CONIDIA pale brown, smooth, narrow (not $>3\ \mu$ wide), circular, tapering gradually from close to the base towards the apex but not obclavate, straight or curved, pluriseptate, not constricted.

The genus is in general very similar to many spp. which I have placed in *Pseudocercospora* and *Pseudocercosporella*, in which the conidial scars are also unthickened, but is characterised by the narrow, acicular conidia (from F. C. Deighton, *Mycol. Pap.* 140: 159, 1976).

Deighton (l.c.) described 4 spp, and referred to 6

more. *C. pini-densiflorae* (Hori & Nambu) Deighton (synonym: *Cercospora pini-densiflorae* Hori & Nambu) causes a blight of *Pinus* spp; it was described by Mulder et al.: STROMA amber to dark brown; intercellular usually filling stomatal cavity. Fascicles dense to very dense. CONIDIOPHORES dark brown, septate (sometimes not distinct), unbranched and slightly geniculate, up to 50 μ in length × 2.5–5 μ. CONIDIA pale olivaceous, long obclavate, straight or slightly curved with truncate or rounded base; 20–68 (mostly 40–50) × 2.5–4.5 μ.

The fungus causes yellowish brown to grey lesions, appearing generally towards the distal part of the needles. These coalesce to give complete needle necrosis and the needles are eventually cast. This is a disease of the later nursery stages and the first few years in a planting. It occurs in: Hong Kong, India, Japan, Malawi, Malaysia (W.), Zimbabwe, Taiwan, Tanzania, Vietnam and Zambia (Map 481). Ito has summarised the work done in Japan. Severe defoliation of susceptible pines can occur in young plantings. Conidia germinate at an opt. of 25° and the incubation period is possibly 6 weeks. This blight can be confused with the one caused by *Scirrhia pini* (q.v.) and is best distinguished by examining the conidia. Also *S. pini* generally (not always) causes a reddish tint to parts of the necrotic needle tissues; this is never seen in blight caused by *Cercoseptoria pini-densiflorae*. *Pinus* spp. differ in susceptibility; *caribaea*, *elliottii*, *patula* and *taeda* are reported to be effectively resistant (see Mulder et al.). Bordeaux gave control in Japan (Tokushige et al.), Vietnam (Uhlig) and Ivory found that benomyl, captafol, chlorothalonil and thiophanate were effective in Malaysia (W.).

C. theae (Cav.) Curzi (synonyms: *Septoria theae* Cavara, *Cercospora theae* Breda de Haan) is described by Deighton (l.c.). The fungus causes a very minor disease (bird's eye spot) of tea (see Kasai). The spots are circular, at first purple red, with an indefinite yellow green border, becoming white with a narrow purple red ring; not ⋗2–3 mm diam. *Cercoseptoria sesame* (Hansf.) Deighton (synonyms: *Cylindrosporium sesame* Hansford, *Cercospora sesamicola* Mohanty) occurs on sesame. It causes brown leaf spot. The spots are reddish or dirty brown, irregular, angular, sometimes indistinctly zonate, often confluent, raised margin, 2–10 mm diam. Seed treatment with agrosan and fernasan gave some control of infection in young plants (Schmutterer et al.). *Cercoseptoria cajanicola* M. S. Pavgi & R. A. Singh occurs on *Cajanus*. Ref. are on the pine pathogen except where shown.

Ito, K. 1972. *Bull. Govt For. Exp. Stn Meguro* 246: 21 (general in Japan, 44 ref.; **51**, 4421).

Ivory, M. H. 1975. *Commonw. For. Rev.* **54**: 154 (in Malaysia, W., fungicides inter alia; **54**, 5110).

Kasai, K. 1972. *Jap. agric. Res. Q.* **6**: 231 (on tea; **52**, 1251).

Mulder, J. M. et al. 1972. *C.M.I. Descr. pathog. Fungi Bact.* 329 (as *Cercospora pini-densiflorae*).

Orellana, R. G. 1961. *Phytopathology* **51**: 89 (on sesame; **40**: 620).

Schmutterer, H. et al. 1965. *Phytopathol. Z.* **54**: 193 (on sesame; **45**, 2205).

Suto, Y. 1971. *J. Jap. For. Soc.* **53**: 319 (sporulation in culture: **51**, 3595).

Tokushige, Y. et al. 1962. *Bull. Govt For. Exp. Stn Meguro* **135**: 15 (fungicides; **41**: 683).

Uhlig, S. K. 1977. *Arch. Phytopathol. PflSchutz.* **13**: 193 (fungicides: **57**, 1432).

CERCOSPORA Fresnius, *Beitr. Mykol.* 3: 91, 1863.

Virgasporium Cooke, 1875.
(Dematiaceae)

COLONIES effuse, greyish, tufted. MYCELIUM mostly immersed. STROMA often present but not large. SETAE and HYPHOPODIA absent. CONIDIOPHORES macronematous, mononematous, caespitose, straight or flexuous, sometimes geniculate, unbranched or rarely branched, olivaceous brown or brown paler towards the apex, smooth. CONIDIOGENOUS CELLS integrated, terminal, polyblastic, sympodial, cylindrical, cicatrised, scars usually conspicuous. CONIDIA solitary, acropleurogenous, simple, obclavate or subulate, colourless or pale, pluriseptate, smooth (M. B. Ellis, *Demat. Hyphom.*: 275, 1971).

This genus consists of 'a large and miscellaneous assortment of nearly 2000 spp. – about half of which are quite unlike the type of the genus *C. apii*' (Deighton et al. 1964). These authors gave a description of true *Cercospora* spp., morphologically similar to *C. apii*, and pointed out that relatively few of these are distinct. Although a few spp. are host limited there is now considerable evidence to contradict the validity of using a host as a criterion (often the only one) for their distinction. Deighton (1976) stated: 'The majority of spp. which have been included in

Cercospora

Cercospora fall into 2 distinct taxonomic categories; those in which the old conidial scars on the conidiophores are thickened to a greater or lesser degree, though always showing a distinct thickening even if this is only present as a rim; and those in which the scars are unthickened, being no thicker anywhere than the wall of the conidiophore. There are variations in the thickening of the scars in the first category, which may allow some further taxonomic division; but the distinction between "thickened" and "unthickened" scars is unambiguous. The hilum on the conidium is thickened or unthickened in correspondence with the scar on the conidiophore.' For a brief taxonomic account of the genus see Deighton (1976) who, with others, has transferred many *Cercospora* spp. to other genera (Deighton 1959, 1967, 1973, 1976, 1979). The conspicuous (thickened) scar is apparently a distinct characteristic of a true *Cercospora*.

Many spp. are identical with, or closely resemble, the type (lectotype) *C. apii* Fres., see Ellis, *Demat. Hyphom.*: 276, whose description of *C. apii* is: COLONIES grey, with small, scattered tufts of conidiophores. MYCELIUM immersed, hyphae colourless to pale olive, 2–4 μ diam., thickening to 8 μ and often forming hyphal knots or pseudostromata in the substomatal cavities. CONIDIOPHORES caespitose in groups of up to 30, usually emerging through stomata, unbranched, straight or flexuous, sometimes geniculate, often swollen slightly at the base and just below the apex, lower part olivaceous brown or brown, upper part pale, with conspicuous, widely spaced scars, usually 30–70 μ long but sometimes much longer, 3–4 μ thick near the apex, 5–9 μ at the base. CONIDIA straight or curved, narrow subulate, truncate at the base, 3.5–5 μ thick below the middle, tapering towards the apex to 1–2 μ, colourless, smooth, 9–17 septate, usually 60–200 μ long but sometimes longer; hilum conspicuous, refractive, appearing dark, 2.5–3.5 μ wide. Ellis (l.c.) listed 20 *Cercospora* spp., mainly on crop plants, which are probable synonyms of *C. apii* and 12 other spp. which are of the *C. apii* type.

The genus was monographed by Chupp; and there are regional accounts for India, Indonesia and Japan. There are reports in the literature on the difficulty of getting the spp. to sporulate in culture. This difficulty is probably more apparent than real. Several workers (see Goode for some ref.) have found that rigorous selection from colonies derived from single conidia demonstrates the existence of genotypes that sporulate adequately in vitro. The phytotoxin (cercosporin) was described by Fajola; Lynch et al. No attempt is made here to refer to many of the spp. described from tropical crops, most of which do not apparently cause serious diseases. Besides the 9 spp. described separately several are discussed under their perfect states. The following were given by Ellis (*More Demat. Hyphom.*) who described 107 spp.: *C. brassicicola* P. Henn, said to be destructive on mature leaves of Chinese cabbage in the Philippines (Lapis et al.); *C. corchori* Sawada on jute (Chowdhury 1948); *C. fusimaculans* Atk. on *Panicum maximum* (Pauvert et al.); *C. longissima* Cugini ex Trav. on lettuce (Szeto et al.); *C. sesami* Zimm. on sesame (Chowdhury 1944, 1945); and *C. traversiana* Sacc. which was reported to cause serious losses on fenugreek (*Trigonella foenum-graecum*) in Hungary (Vörös et al.).

The following spp. are not described by Ellis, but see Chupp. *C. angolensis* De Carvalho was reported by Brun to cause serious losses on citrus in central Africa. *C. anonae* Muller & Chupp damages custard apple in Brazil (Ponte). *C. hayi* Calpouzos causes brown spot of banana fruit (Kaiser et al.); it differs from the stat. conid. of *Mycosphaerella musicola* (q.v.) in the marked acute, attenuated conidial tip and the absence of sporodochia. Several spp. occur on soybean. Frog-eye leaf spot of this host is caused by *C. sojina* Hara (synonym: *C. diazu* Miura). The disease was described by Lehman, and several races have been differentiated (Athow et al. 1962; Ross). Roane et al. reported that increasing severity of *C. zeae-maydis* Tehon & Daniels, causing grey leaf spot of maize, was associated with the widespread practice of no tillage maize production in USA. Plants with maize dwarf mosaic virus were more susceptible to the fungus than virus-free plants. *C. vanderysti* on bean is referred to under *C. canescens* (q.v., and see Vieira et al.).

Assante, G. et al. 1977. *Phytochemistry* **16**: 243 (secondary metabolites, phytotoxins; **56**, 3439).
Athow, K. et al. 1952. *Phytopathology* **42**: 660 (*Cercospora sojina*, inheritance of resistance; **32**: 603).
—— ——. 1962. *Ibid* **52**: 712 (*C. sojina*, races; **42**: 169).
Boedijn, K. B. 1961. *Nova Hedwigia* **3**: 411 (in Indonesia; **41**: 373).
Brun, J. 1972. *Fruits* **27**: 539 (*C. angolensis*; **52**, 1146).
Chowdhury, S. 1944. *J. Indian hort. Soc.* **23**: 91 (*C. sesami*, physiology; **24**: 219).
——. 1945. *Indian J. agric. Sci.* **15**: 140 (*C. sesami*, control; **25**: 473).

——. 1948. *J. Indian bot. Soc.* **26**: 227 (*C. corchori*; **27**: 524).
Chupp, C. 1954. *A monograph of the fungus genus* Cercospora. Ithaca, New York; published by author, 667 pp. (**33**: 635).
De Carvalho, T. et al. 1953. *Bolm Soc. Broteriana* Ser. 2, **27**: 201 (*C. angolensis* sp. nov.; **35**: 364).
Deighton, F. C. 1959. *Mycol. Pap.* 71, 23 pp. (studies on *Cercospora* & allied genera 1. *Cercospora* spp. with coloured spores on *Phyllanthus* (Euphorbiaceae); **38**: 580).
—— et al. 1964. *Commonw. phytopathol. News* **10**: 49 (on *Cercospora*).
——. 1967. *Mycol. Pap.* 112, 80 pp. (studies on *Cercospora* & allied genera II. *Passalora*, *Cercosporidium* & some spp. of *Fusicladium* on *Euphorbia*; **47**, 450).
——. 1973. *Ibid* 133, 62 pp. (as above IV. *Cercosporella* Sacc., *Pseudocercosporella* gen. nov. & *Pseudocercosporidium* gen. nov.; **53**, 73).
——. 1976. *Ibid* 140, 168 pp. (as above VI. *Pseudocercospora* Speg., *Pantospora* Cif. & *Cercoseptoria* Petr.; **56**, 2397).
——. 1979. *Ibid* 144 (as above VII, new spp. & redispositions.
Fajola, A. O. 1978. *Physiol. Pl. Pathol.* **13**: 157 (phytotoxin cercosporin).
Goode, M. J. et al. 1970. *Phytopathology* **60**: 1502 (*C. citrullina*, sporulation in culture; **50**, 1541).
Hino, T. et al. 1976. *Jap. agric. Res. Q.* **10**: 215 (distribution in Brazil & Japan; **57**, 449).
Ibrahim, F. M. et al. 1974. *Mycopathol. Mycol. appl.* **52**: 141 (quantitative, morphological classification of 30 spp.; **53**, 2466).
Kaiser, W. J. et al. 1965. *Phytopathology* **55**: 977 (*C. hayi*; **45**, 518).
————. 1966. *Ibid* **56**: 1290 (*C. hayi*, dispersal & survival; **46**, 1052).
Katsuki, S. 1965. *Trans. mycol. Soc. Japan*, extra issue I, 100 pp. (in Japan; **44**, 3276).
——. 1966. *Ibid* 7: 101 (as above; **47**, 2620).
——. 1973. *Rep. Tottori mycol. Inst.* **10**: 561 (as above; **53**, 2882).
—— et al. 1975. *Trans. mycol. Soc. Japan* **16**: 1 (as above; **55**, 605).
Lapis, D. B. et al. 1974. *Philipp. Agric.* **58**: 167 (*C. brassicicola*; **56**, 1791).
Lehman, S. G. 1928. *J. agric. Res.* **36**: 811 (*C. sojina*; **7**: 760).
——. 1934. *Ibid* **48**: 131 (as above; **13**: 490).
Lynch, F. J. et al. 1977. *Trans. Br. mycol. Soc.* **69**: 496 (phytotoxin cercosporin; **57**, 3297).
Pauvert, P. et al. 1972. *Annls Phytopathol.* **4**: 245 (*C. fusimaculans* & *Helminthosporium* spp.; **52**, 2306).
Ponte, J. J. da 1973. *Revta Agric. Piracicaba* **48**: 121 (*C. anonae*; **54**, 516).
Probst, A. H. et al. 1958. *Phytopathology* **48**: 414 (*C. sojina*, inheritance of resistance; **38**: 50).
Roane, C.W. et al. 1974. *Pl. Dis. Reptr* **58**: 456 (*C. zeae-maydis*; **53**, 3946).
Ross, J. P. 1968. *Phytopathology* **58**: 708 (*C. sojina*, races; **47**, 2921).
Szeto, M. et al. 1975. *Agric. Hong Kong* **1**: 278 (*C. longissima*; **56**, 3788).
Vasudeva, R. S. 1963. IARI, New Delhi 245 pp. (in India; **45**, 3079).
Vieira, C. et al. 1965. *Revta Agric. Piracicaba* **40**: 3 (*C. vanderysti*; **44**, 2661).
Vörös, J. et al. 1972. *Acta phytopathol. Acad. Sci. hung.* 7: 71 (*C. traversiana*; **52**, 3022).
Yamamoto, W. et al. 1960. *Sci. Rep. Hyogo Univ. Agric.* Ser. agric. Biol. **4**: 41 (in Japan; **43**, 2545).

Cercospora canescens Ellis & Martin, *Am. Nat.* **16**: 1003, 1882.
C. vignicaulis Tehon, 1937.

Leaf spots subcircular to broadly irregular sometimes confluent, generally brown, pale tan to grey centre surrounded by a dark brown or reddish margin. STROMA slight. Fruiting amphigenous but more abundant on the lower surface of the leaf, also occurring in effuse patches on the stem, cotyledons and drying pods. CONIDIOPHORES in fascicles, sometimes dense, divergent, mostly straight, geniculate, rarely branched, uniform in colour, pale to medium brown, multiseptate, medium to large size conidial scar present on the rounded apex, width uniform, 20–175 (occasionally longer) × 3–6.5 μ. CONIDIA acicular or cylindric obclavate, hyaline, base truncate, apex acute, distinctly multiseptate, thickened hilum, straight or variously curved, 30–300 × 2.5–5 μ (J. L. Mulder & P. Holliday, *CMI Descr.* 462, 1975).

C. canescens causes leaf spot of *Phaseolus* spp.; also on *Lablab niger* (hyacinth bean), *Vigna* and *Voandzeia subterranea* (bambara groundnut). The symptoms are not particularly characteristic and often similar to those caused by other *Cercospora* spp. on these crops. The spp. were discussed briefly (Skiles et al.) in an account of *C. vanderysti* P. Henn. on common bean (bean grey spot) in Colombia. This fungus can be distinguished from *C. canescens* by the dense grey, cushiony growth over the leaf lesions. *C. canescens* is very widespread in warmer regions and recently was described as causing a severe leaf spot of bambara groundnut (Teyegaga et al.) in Ghana. On organic media most conidia are formed at 28° (less at 24° and 32°) and light increases their numbers. Conidial suspensions sprayed on susceptible vars of *P. aureus* (green gram, mung) showed no differences in host response up to the flowering stage, but lesion severity increased sharply from this stage onward. In work with cowpea (*Vigna unguiculata*) leaf diffusates both conidial germination and germ tube growth were inhibited on the surfaces of young leaves of susceptible cowpea but not on those of older leaves. Resistance has been described in green gram and cowpea; see Singh et al. for fungicides.

Mew, I. C. et al. 1975. *Pl. Dis. Reptr* **59**: 397 (culture, inoculation of & resistance in green gram; **55**, 470).
Ragunathan, A. N. 1969. *Madras agric. J.* **56**: 734 (culture; **50**, 410).
Rath, G. C. et al. 1973. *Indian J. Mycol. Pl. Pathol.* **3**: 204 (factors affecting disease severity & sporulation, resistance in green gram; **54**, 3568).
Schneider, R. W. et al. 1975. *Phytopathology* **65**: 63 (inhibition of conidial germination & germ tube growth by cowpea leaf diffusates; **54**, 4704).
Singh, D. V. et al. 1976. *Indian Phytopathol.* **29**: 337 (fungicides: **57**, 4699).
Skiles, R. L. et al. 1959. *Phytopathology* **49**: 133 (*Cercospora vanderysti*; **38**: 437).
Teyegaga, A. et al. 1972. *Trop. Agric. Trin.* **49**: 197 (conidial germination & survival; **51**, 4672).
Vakili, N. G. 1977. *Ibid* **54**: 69 (field screening for resistance in cowpea with *Mycosphaerella cruenta*; **56**, 2775).

Cercospora coffeicola Berkeley & Cooke, *Grevillea* **9**: 99, 1881.
Cercospora coffeae Zimm., 1904
C. herrerana Farn., 1911.

Sporulation amphigenous, mostly on the upper leaf surface. STROMATA up to 50 μ diam., globular, dark brown. CONIDIOPHORES in fascicles, 3–30, pale, medium brown, slightly paler towards the apex, occasionally branched, attenuated, slightly or strongly geniculate, septate, 20–275 × 4–6 μ; conidial scars distinct, thickened. CONIDIA hyaline, acicular to obclavate, apex acute, base truncate or subtruncate with conspicuous thickened hilum, straight or rarely curved, indistinctly multiseptate, 40–150 × 2–4 (5–7) μ. (J. M. Mulder & P. Holliday, *CMI Descr.* 415, 1974.)

Brown eyespot of coffee is found throughout the tropics (Map 59). Symptoms occur on leaves and berries. On the former the initially small chlorotic spots expand becoming deep brown, lighter in colour on the lower surface; the centre becomes grey white and is surrounded by a ring of dark brown tissue. Spots are 5–15 mm diam., sometimes with a yellowish halo which is most distinct on the upper leaf surface; dark sporulation is seen in the greyish area; leaves may be shed. On green berries there are brown, sunken, irregular or oval lesions, rarely > 5 mm long and sometimes with a purplish surround. Infection may penetrate the berry thus causing the pulp to stick to the seed or bean during fermentation.

Echandi and Siddiqi have both given accounts of this relatively minor disease of seedbeds and nurseries. But the practice of shade removal in coffee cultivation has apparently made it more prominent; shaded plants tend to have less disease. This effect of the shade factor suggests that adequate host nutrition would lessen infection. Incidence on unshaded plants was considerably reduced by applying increasing amounts of NPK. Three fertiliser levels both increased yield and reduced brown eyespot to 8.3, 5.9 and 2.2% compared with 21.8% for unfertilised plots (Fernández-Borrero et al.). Leaves are infected through the stomata, berries through wounds and probably other types of damage such as sun scorch. Conidial germination is > 80% at 15–30°. At 27° the germ tubes form after 2 hours but at other temps after 4–5 hours. The incubation period on leaves is relatively long, spots becoming recognisable after *c.* 38 days. the opt. temp. for in vitro growth is 24–25°; nutrient factors in culture have been investigated. Conidia are presumably windborne, mostly during the day. They can survive on leaves for up to 36 days. Carry over as mycelium in the host is likely and the possible role of any other host seems unimportant. The report of the existence of differential pathogenicity between isolates from leaves and berries has not been supported in later work.

Improvement in husbandry will presumably lessen the detrimental effects of the disease. Defoliation has been counteracted with 2,4-D (Valencia). There are numerous reports on the use of fungicides, for example, from Costa Rica, India, Malawi, Nicaragua and Salvador. Control has been given by benomyl, Bordeaux, tribasic copper sulphate and copper oxychloride; also by ziram, captafol and fentin acetate. Rodriguez et al. found copper superior to carbamates. The effectiveness of host resistance is not clear.

Abrego, L. 1966. *Bol. Inf. Inst. Salvador Invest. Cafe* **69**: 1 (fungicides; **46**, 3442).

Castillo, N. et al. 1977. *Tropenlandwirt* **78** (April): 64 (fungicides in Nicaragua; **57**, 3475).

Echandi, E. 1959. *Turrialba* **9**: 54 (general; **39**: 579).

Fernández-Borrero, O. et al. 1966. *Cenicafé* **17**: 5 (effects of fertiliser on disease incidence; **47**, 1133; and see **53**, 1398).

Lopéz-Duque, S. et al. 1969. *Ibid* **20**: 3 (epidemiology; **50**, 3775).

Nataraj, T. et al. 1975. *Indian Coff.* **39**: 179 (effect of shade; **55**, 1805).

——. 1976. *J. Coff. Res.* **6**: 81 (fungicides in India; **57**, 5513).

Rodriguez, R. A. et al. 1968. *Fitopatologia Bogotá* **3**: 5 (fungicides in Costa Rica; **52**, 412).

Siddiqi, M. A. 1970. *Trans. Br. mycol. Soc.* **54**: 415 (general; **50**, 100).

Sridhar, T. S. et al. 1968. *Riv. Patol. veg. Pavia* Ser IV **4**: 33, 41 (inoculations with berry & leaf isolates; attempted control with antibiotics; **47**, 2723a, b).

Subramanian, S. et al. 1966. *Ibid* **2**: 127, 133, 141, 147 (general; **46**, 2023 a, b, c, d).

——. 1967. *Phytopathol. Z.* **60**: 247 (nutrition in culture; **47**, 1561).

Valencia, A. G. 1972. *J. Coff. Res.* **2**: 15 (physiology of defoliation; **52**, 3327).

Venkataramaiah, G. H. et al. 1965. *Indian Coff.* **29**: 19 (fungicides in India; **45**, 1800).

Cercospora elaeidis Steyaert, *Bull. Soc. R. bot. Belg.* **80**: 35, 1948.

Stroma absent or very small. Fruiting hypophyllous. CONIDIOPHORES in fascicles (2–7), divergent, erect, geniculate, base swollen, dark brown, rough or finely echinulate, septate, conidial scars conspicuous, 185–250 × 5–7 μ. CONIDIA obclavate, base truncate, apex tapered, yellowish brown with the apical region paler or hyaline, septate (0–9), finely echinulate, hilum thickened, 125–187 × 6–10 μ (J. L. Mulder & P. Holliday, *CMI Descr.* 464, 1975).

Freckle of oil palm is restricted to Africa on this crop (Map 487); there is also a record from Australia (Northern Territory) on *Carpentaria acuminata*. In the nursery leaf infection shows as groups of dull brown spots, each rarely > 0.5 mm diam.; the spots coalesce, necrosis spreads and the surrounding tissue becomes orange; heavily infected areas of the lamina die. The final development of the symptoms is particularly noticeable on the older fronds. Necrotic spotting is also found on field palms and is usually most severe on older leaves. The spots have a small watersoaked halo which becomes yellow and then orange. Fully developed spots are round to oval, 3–4 mm diam. with the halo *c.* 10 × 4 mm; spots may coalesce and parts of the lamina become desiccated.

In some parts of Africa freckle appears to be a disease of the nursery but in others it may be serious on young (*c.* 3 years old) palms in the field. Nursery infection results in a severe check to growth and vigour. Penetration is through the stomata which are more abundant on the lower leaf surface; symptoms appear *c.* 20 days after inoculation. Conidiophores are formed at RH 81–100% at an opt. 27°; sporulation occurs only at RH > 93% and at temps < 32° (opt. 27°). Conidial germination requires a moisture film. Young potted palms were more susceptible when moisture conditions for the host were unfavourable (Moens et al.; Weir). Control measures appear to be restricted to the nursery and are largely through fungicides; benomyl, captan, mancozeb and ziram are recommended. Duff discussed failures to get adequate control with fungicides; these were attributed to the special form and development of the leaf which hinder the necessary surface distribution. In the nursery the disease may be reduced by wider spacing of seedlings. Techniques which allow min. of transplanting shock should reduce the level of field infection.

Duff, A. D. S. 1970. *Oléagineux* **25**: 329 (failure in fungicide control; loss in yield; **49**, 3433).

Kovachich, W. G. 1954. *Trans. Br. mycol. Soc.* **37**: 209 (inoculation, symptoms & control; **34**: 147).

——. 1956. *Ibid* **39**: 297 (symptoms on seedling & adult palms; **36**: 100).

Moens, P. et al. 1960. *Oléagineux* **15**: 609 (factors affecting freckle on potted palms; **40**: 179).

Rajagopalan, K. 1973. *Jnl Niger. Inst. Oil Palm Res.* **5**(18): 23 (fungicides; **53**, 1911).

Renard, J. L. et al. 1977. *Oléagineux* **32**: 43, 89 (fungicides; **56**, 3663, 4635).

Robertson, J. S. 1957. *Jl W. Afr. Inst. Oil Palm Res.* **2**: 265 (fungicides; **37**: 299).

Weir, G. M. 1968. *Jnl Niger. Inst. Oil Palm Res.* **5**(17): 41 (effects of RH & temp. on sporulation & conidial germination; **48**, 1270).

Cercospora kikuchi (Matsumoto & Tomoyasu) M. W. Gardner, *Proc. Indiana Acad. Sci.* **36**: 242, 1927 (1926).

Cercosporina kikuchii Matsumoto & Tomoyasu, 1925.

Leaf spots circular or irregular, often occurring

along the leaf margin or frequently covering the whole surface, tan to brown or dark violet with a brown or violet border, centre usually grey or tan. STROMA present, small. Sporulation amphigenous, also occurring on the stems, pods and seeds. CONIDIOPHORES in fascicles, divergent, occasionally 2–5 stalks, generally many, medium brown but paler towards the apex, unbranched, geniculate, prominent, conidial scars present, septate, 45–220 × 4–6 μ (generally longer in culture). CONIDIA hyaline, acicular, truncate base, apex tapered, multiseptate (0–22) straight or curved, thickened hilum, 50–375 × 2.5–5 μ (J. L. Mulder & P. Holliday, *CMI Descr.* 466, 1975).

C. kikuchi causes purple seed stain of soybean, also called purple blotch, purple stain or purple speck. Other legumes may be infected, i.e. common bean, cowpea (*Vigna*) and cluster bean or guar (*Cyamopsis tetragonoloba*). The most characteristic symptoms are the light to dark purple, sometimes pinkish, markings on the seed coat, the whole of which can be affected. Small cracks which may be quite wide form on the discoloured areas and they impart a dull appearance. Other *Cercospora* spp. may or may not cause this purplish staining but they are not apparently important in the field. The cotyledons are generally not discoloured. Reddish purple spots, becoming angular to irregular, up to 1 cm diam. occur on the leaves, and stems are also infected.

The pathogen is widespread with the main host. It is transmitted through the seed; Ilyas et al. isolated it from the seed coats, cotyledons and embryos; the purple staining is always associated with infected tissues. Pod infection is followed by fungal penetration of the unwounded seed coats (Jones; Lehman). When infected seed germinates the seedlings have darkened, shrunken cotyledons and a velvety, greyish white growth of conidiophores and conidia. Seedling germination may not be greatly reduced but infected seeds develop into weak seedlings and there may be a slow death (Lehman; Matsumoto et al.; Murakishi). Seed infection is lowest before plant maturity and can rise to 50% or more at maturity. Where maturity is delayed (and where the flowering period is longer) the plants tend to show more seed infection (Crane et al.). Chiu gave 20–24° as the temp. when conidiophores and conidia-like hyphae were most abundant. Matsumoto gave an opt. temp. for sporulation within 15–20°; at 20–25° the conidia quickly formed mycelium. This worker also reported that true chlamydospores formed on the pods at 25°, *c.* spherical, 5–8 (up to 12) μ diam. with an orange yellow membrane *c.* 1 μ thick. Murakishi reported most conidia on seed at 23–27° mycelial growth opt. at 27°. In USA (Mississipi) the fungus survived on the soil surface from January to April but on buried stems it had practically disappeared after 2 months (Jones 1968). The amount of purple seed stain was higher when tobacco ringspot virus was present in soybean (Crittenden et al.). Infected seeds can be more heavily infected with *Aspergillus niger*. Thiram or thiram + benomyl seed treatment gives some control and benomyl sprays a few weeks after planting have been suggested (Agarwal et al. 1974). Zineb and copper have also been used at flowering. Soybean cvs differ in their degree of susceptibility and early maturing ones may show some disease escape (and see Koyama et al.; Roy et al. 1976; Wilcox et al.).

Agarwal, V. K. et al. 1971. *Indian Phytopathol.* **24**: 810 (effect on seed germination & seed treatment; **52**, 1735).

—— ——. 1974. *Indian J. Mycol. Pl. Pathol.* **4**: 1 (fungicides; **54**, 4700).

Chiu, W. F. 1955. *Acta phytopathol. sin.* **1**: 191 (general, temp. & culture; **37**: 65).

Crane, J. L. et al. 1966. *Pl. Dis. Reptr* **50**: 464 (seed infection & plant development; **45**, 3447).

Crittenden, H. W. et al. 1966. *Ibid* **50**: 910 (with tobacco ringspot virus; **46**, 1156).

Deutschmann, F. 1953. *Phytopathol. Z.* **20**: 297 (in seed, culture & pigment; **33**: 462).

Ilyas, M. B. et al. 1975. *Pl. Dis. Reptr* **59**: 17 (in seed with *Diaporthe phaseolorum*; **54**, 3064).

Johnson, H. W. et al. 1962. *Phytopathology* **52**: 269 (on cluster bean; **42**: 69).

Jones, J. P. 1959. *Ibid* **49**: 430 (purple seed stain caused by *Cercospora* spp.; **39**: 68).

——. 1968. *Pl. Dis. Reptr* **52**: 931 (survival in the field; **48**, 1408).

Kilpatrick, R. A. et al. 1956. *Phytopathology* **46**: 201 (purple seed stain in legumes caused by *Cercospora* spp.; **35**: 861).

——. 1957. *Ibid* **47**: 131 (in seed; **36**: 569).

Koyama, T. et al. 1977. *Bull. Tohoku natn. agric. exp. Stn* 55: 235 (var. resistance; **57**, 354).

Lehman, S. G. 1950. *Bull. N. Carol. agric. Exp. Stn* 369, 11 pp. (general; **29**: 489).

Matsumoto, T. et al. 1925. *Ann. phytopathol. Soc. Japan* **1**(6): 1 (general, morphology including other *Cercospora* spp.; **4**: 714).

——. 1928. *Ibid* **2**(2): 65 (spore formation; **8**: 221).

Murakishi, H. H. 1951. *Phytopathology* **41**: 305 (general, effects of temp. **30**: 503).
Roy, K. W. et al. 1976. *Ibid* **66**: 1045 (pod inoculation effects; **56**, 2293).
——. 1977. *Ibid* **67**: 1062 (antagonism with other fungi on soybean; **57**, 1515).
Wilcox, J. R. et al. 1975. *Crop Sci.* **15**: 525 (inheritance of resistance; **55**, 3817).

Cercospora koepkei Kruger, *Ber. VersStn Zuckerrohr W. Java* **1**: 115, 1890.

Amphigenous but chiefly on the lower surface. STROMATA present, consisting of a few cells. CONIDIOPHORES fasciculate (2–20) divergent, brown or smoky brown, uniform, attenuated towards the tip, often tortuous, geniculate, multiseptate, rarely branched, 30–200 × 3.5–5 μ; conidial scars conspicuous, slightly thickened. CONIDIA hyaline, short cylindric obclavate, mostly straight, some slightly curved, septate (1–6, mostly 3), rarely constricted, base obconic or obconically truncate, apex obtuse, basal hilum thickened, 20–55 × 3–8 (mostly 35 × 5–8) μ. (J. M. Mulder & P. Holliday, *CMI Descr.* 417, 1974.)

C. koepkei, causing yellow spot of sugarcane, has recently caused epidemics in the wetter areas of N. Queensland, Australia (Egan) and is apparently a more important pathogen than these other members of the genus on the same host: *C. atrofiliformis* Yen, Lo & Chi, *C. longipes* (q.v.) and *C. vaginae* Krüger. *C. longipes* (brown spot) has longer conidia than the other spp. *C. atrofiliformis* infection (black stripe) causes narrow brownish black streaks, 5–36 cm long by 0.5–1.2 mm wide on the lamina. They begin as very small yellow spots and there is no definite centre to the spot or chlorotic halo. CONIDIOPHORES 2–6 (rarely 9) septate, 20–78 × 3.3–4.6 μ; CONIDIA 15–212 × 2–4.5 μ. *C. vaginae* infection (red spot) causes small, round, bright red spots on the upper leaf sheaths, sharply delimited, enlarging and coalescing to form irregular red patches (Abbott). CONIDIOPHORES 40–200 × 3–5 μ; CONIDIA 1–3 septate, 30–55 × 4–5 μ. Kiryu studied *C. vaginae* in culture (opt. temp. for growth 28°).

Yellow spot is widely distributed in S. and E. Asia; it also occurs in Australia, Oceania and Africa (Ghana, Kenya, Malagasy Republic, Mauritius, Reunion, South Africa, Tanzania and Uganda). The identity of the disease reported in Central America, West Indies and S. America (Argentina, Brazil, Colombia and Venezuela) is uncertain (Map 341). Irregularly outlined, yellow green spots form on the young leaves, up to 12 mm diam.; they may coalesce to cover large areas of the leaves. Reddish patches form as the leaf approaches maturity and infected fields appear a conspicuous rusty yellow. The leaves have a dirty grey growth of conidiophores and conidia on the lower surface and die prematurely. The breakdown of host resistance in Australia (causing considerable losses) has indicated that races of *C. koepkei* occur and thereby complicates selection for resistance. In Fiji noble canes have been found generally more resistant than hybrids. In S. India a diurnal periodicity for conidia was found, with a max. conc. in the forenoon. No natural hosts apart from *Saccharum* are known. Ricaud (1972) reported that in a spraying trial with benomyl there was a 65% control of infection and a 7% net increase in the av. profitable sugar index (but see Ricaud 1974).

Abbott, E. V. 1964. In *Sugarcane diseases of the world* Vol. 2: 49 (*Cercospora vaginae*).
Dignadice, P. B. et al. 1953. *Philipp. Agric.* **37**: 36 (general; **34**: 184).
Egan, B. T. 1970. *Sugcane Pathol. Newsl.* 5: 26 (probable existence of strs in Australia; **50**, 927).
——. 1971. *Ibid* 7: 14 (epidemic in Australia; **51**, 2823).
——. 1972. *Cane Grow q. Bull.* **35**: 86 (losses in Australia; **51**, 2824).
Hughes, C. G. et al. 1961. In *Sugarcane diseases of the world* Vol. 1: 357 (general).
Husain, A. A. et al. 1970. *Sugcane Pathol. Newsl.* 5: 25 (resistance in Fiji).
Kiryu, T. 1938. *Rep. Govt Sug. Exp. Stn Taiwan* 5: 53 (*C. vaginae* in culture; **17**: 839).
Prakasam, P. et al. 1967. *Indian J. agric. Sci.* **37**: 395 (sporulation in vitro; **47**, 1658).
Ricaud, C. 1972. *Sugcane Pathol. Newsl.* 8: 27 (trial with benomyl & yield assessment; **51**, 4320).
——. 1974. *Proc. 15th Congr. Int. Soc. Sugcane Technol.* Vol. 1: 354 (factors affecting disease, control & yield; **55**, 3730).
Roldan, E. F. 1938. *Philipp. J. Sci.* **66**: 7 (morphology of *C. vaginae* inter alia; **17**: 843).
Sreeramulu, T. et al. 1971. *Indian J. agric. Sci.* **41**: 655 (air dispersal & diurnal periodicity; **51**, 3537).
Yen, W. Y. et al. 1953. *J. Sug. Cane Res. Taiwan* 7: 1 (*C. atrofiliformis* sp. nov.; **36**: 729).

Cercospora longipes Butler, *Mem. Dep. Agric. India bot. Ser.* **1**: 41, 1906.

Sporulation chiefly hypophyllous. STROMATA

present. CONIDIOPHORES in fascicles, compact or divergent, medium brown to dark brown, slightly darker at the base, attenuated towards the apex, up to 200 (sometimes more) × 3–5 μ, rarely branched, septate, geniculate; conidial scars thickened. CONIDIA hyaline, obclavate, straight or mildly curved, distinctly multiseptate, base truncate, sometimes broader than the rest of the conidium, apex subacute 20–200 (mostly 100) × 3.5–8 μ (J. M. Mulder & P. Holliday, *CMI Descr.* 418, 1974).

C. longipes, causing brown spot of sugarcane, differs primarily from other *Cercospora* spp. on this host (*C. koepkei* q.v.) in its longer conidiophores and conidia. An account of this disease was also given by Butler, *Fungi and disease in plants*. It is a minor one and widespread (see *Proc. 14th Congr. Int. Soc. Sugcane Technol.*, 1971 for the distribution). The leaf spots are oval to linear, red brown and first appear on the older leaves; they enlarge up to 13 mm long, develop a chlorotic halo and the centre may become straw coloured. The spots are numerous and spread progressively upwards on the plant; severe attacks cause necrosis over large areas, leaf shed and death. No studies on physiologic specialisation or transmission have been reported. Resistance is known and new cvs should show some resistance to *C. longipes* when selected. Spread of infection through leaf debris on seedcane or setts can be prevented by a formalin dip at 52° for 20 minutes. Gupta et al. reported that copper oxychloride and mancozeb checked airborne infection, and hot water treatment of setts (52° for 2 hours) followed by a 20-minutedip in methoxyethylmercury chloride controlled sett infection.

Abbot, E. V. 1951. *Sug. Bull. New Orl.* **29**: 134, 139 (general in USA; **30**: 542).
——. 1964. In *Sugarcane diseases of the world* Vol. 2: 25 (general).
Gupta, M. R. et al. 1977. *Indian Sug.* **26**: 261 (fungicides; **57**, 1385).
Hsieh, W. H. et al. 1978. *Rep. Taiwan Sug. Res. Inst.* 79: 29 (general).

Cercospora nicotianae Ellis & Everhart, *Proc. Acad. nat. Sci. Philad.* **45**: 170, 1893.
Cercospora raciborskii Sacc. & Sydow, 1902.
(further synonymy in Johnson et al.)

Sporulation amphigenous, sometimes only on one leaf surface. STROMATA absent or composed of a few dark cells. CONIDIOPHORES fasciculate 2–7 or more, dense or divergent, brown at base, paler towards the apex, unbranched, multiseptate, undulate or mildly to abruptly geniculate, apex truncate or rounded, 20–600 × 4–4.5 μ, conidial scars large and conspicuous. CONIDIA hyaline, acicular, straight or variously curved, multiseptate, base truncate with thickened hilum and rounded apex, 35–300(150) × 3–4(5) μ. (J. M. Mulder & P. Holliday, *CMI Descr.* 416, 1974.)

Frog eye leaf spot of tobacco is widespread (Map 172). In inoculation studies Johnson et al. relegated *C. nicotianae* to synonymy with *C. apii* (*Cercospora* q.v.) and many other *Cercospora* spp. Sobers, in comparing *C. apii* (from celery) with *C. nicotianae*, found that only the latter was pathogenic to tobacco; but since he considered the 2 spp. similar morphologically, and in cultural characteristics, *C. apii* f. sp. *nicotianae* was proposed. Frog eye has been investigated intensively by Stephen in Zimbabwe, and has also been described in Australia (Hill), Indonesia (Jochems; Reitsma et al.), Nigeria (Alasoadura et al.), Sri Lanka (Park et al.) and Taiwan (Yen). The phase of the disease that shows during curing is called barn spot (also green spot, black barn spot, pole burn and leaf burn). In the seedbed and the field greenish brown leaf spots with somewhat indefinite margins form, more commonly on lower leaves. They become reddish brown to brown with the characteristic centre which turns progressively paler to a dingy grey, bleached and parchment like. The developed spot has a narrow, dark brown margin and a well defined outline, of 2–15 mm diam. Seedlings and the lower leaves of older field plants can be destroyed. Depending on growing conditions variations on spotting can occur; these can be larger, generally brown lesions with more irregular outlines, sometimes with zonate markings and chlorotic haloes. In the barn after harvest greenish or greyish spots form. On flue-cured leaves there are small, black, circular lesions which may be extremely numerous and coalesce to form irregular dark blotches. This barn spot phase of the syndrome can develop later in leaves which show no visible symptoms when reaped.

C. nicotianae is probably a comparatively weak pathogen which mostly attacks physiologically unbalanced plants. Host penetration is probably stomatal. The most in vitro growth was at 26–30° but sporulation was highest at 18° and was increased by light (Stavely et al. 1969; Yen). Poor cultural conditions in the seedbed predispose plants to infection; when infected seedlings are planted out high field incidence of frog eye can result and leads to an increase in the amount of barn spot. Nitrogen nutrition in the host is an important factor (Stephen 1958b). There are no particular studies on transmission. Spread is through host debris, possibly also soil, but neither seed transmission nor other hosts appear to be important in the epidemiology. Control must begin in the seedbed: destruction of tobacco plant debris, soil sterilisation, good fertiliser treatment (especially a correct nitrogen balance), promotion of vigorous growth and fungicide application. Bordeaux, nabam, mancozeb, maneb, benomyl and thiabendazole are effective. If weather conditions in the field seem particularly favourable for infection it may be necessary to apply fungicides here too, and infected leaves should be destroyed. For flue-cured tobacco a temp. of 38° (which must not be delayed) and RH 100% will largely eliminate barn spot. Resistance (apart from some earlier work by Stephen 1958c) has been described recently. It has been transferred to tobacco from *N. repanda* using *N. sylvestris* in an interspecific bridge cross (Stavely 1971; Stavely et al. 1972a, 1973; Wan et al.).

Alasoadura, S. O. et al. 1970. *Mycopathol. Mycol. appl.* **42**: 177 (general in Nigeria; **50**, 1965).

Fajola, A. O. et al. 1973. *Ann. appl. Biol.* **74**: 219 (fungicides, in Nigeria; **53**, 674).

Hill, A. V. 1936. *Res. Bull. Aust. Sci. Ind.* 98, 46 pp. (general; **15**: 612).

Hopkins, J. C. F. 1929. *Rhod. agric. J.* **62**: 817 (general; **9**: 140).

Jailloux, F. et al. 1970. *Phytiat. Phytopharm.* **19**: 107 (fungicides, in Guadeloupe; **51**, 666).

Jochems, S. C. J. 1931. *Meded. Deli-Proefstn Medan* Ser. 2 **72**, 38 pp. (general in Indonesia; **11**: 134).

Johnson, E. M. et al. 1949. *Phytopathology* **39**: 763 (synonymy, comparison with *Cercospora apii*; **29**: 175).

Mandelson, L. F. 1934. *Qd agric. J.* **41**: 132 (barn spot phase; **13**: 545).

Morgan, O. D. 1964. *Pl. Dis. Reptr* **48**: 693 (comparison with *C. kikuchi*; **43**, 3318).

Park, M. et al. 1937. *Trop. Agric.* **88**: 153 (control in seedbed; **16**: 565).

Reitsma, J. et al. 1947. *Chronica Nat.* **103**: 94 (general in Indonesia; **26**: 515).

Sobers, E. K. 1968. *Phytopathology* **58**: 1713 (comparison with *C. apii*; **48**, 1965).

Stavely, J. R. et al. 1968. *Ibid* **58**: 1372 (pH & nutrition in culture; **48**, 924).

———. 1969. *Ibid* **59**: 496 (effect of temp. in culture; **48**, 2555).

——. 1971. *Tob. Sci.* **15**: 132 (resistance in tobacco; **52**, 245).

—— et al. 1972a. *Phytopathology* **62**: 672 (resistance from *Nicotiana repanda*).

———. 1972b. *Ibid* **62**: 1392 (effect of infection of 4 major chemical constituents in cured leaves; **52**, 3067).

———. 1973. *J. Hered.* **64**: 265 (*N. repanda* as a source of resistance; **53**, 3611).

Stephen, R. C. 1955. *Bull. Tob. Res. Bd Rhod. Nyasald* 40, 36 pp. (general).

——. 1957. *Phytopathology* **47**: 663 (general in Zimbabwe; **37**: 312).

——. 1957. *Emp. J. exp. Agric.* **25**: 291 (effect of planting date on disease incidence; **37**: 185).

——. 1958a. *Rhod. agric. J.* **55**: 63 (disease incidence in flue cured seedbeds; **37**: 509).

——. 1958b. *Emp. J. exp. Agric.* **26**: 64 (effects of N, P, K on disease incidence; **37**: 421).

——. 1958c. *Ibid* **26**: 70 (resistance in *Nicotiana*).

Wan, H. et al. 1970. *A. Rep. Taiwan Tob. Res. Inst. 1970*: 46 (resistance from *N. repanda*; **51**, 664).

Yen, J. W. et al. 1956. *J. Agric. For. Taichung* **5**: 128 (general; **37**: 680).

Cercospora sorghi Ellis & Everhart, *J. Mycol.* **3**: 15, 1887.

Cercospora sorghi var. *maydis* Ell. & Ev.

Sporulation amphigenous. STROMATA lacking or 15–50 μ diam., globular, dark brown to black. CONIDIOPHORES in small scattered tufts 3–4 (up to 24) on the stroma, medium dark brown or olivaceous brown, paler towards the apex, width irregular, unbranched, multiseptate, 1–3 geniculate, 20–80 (–150) × 3–5.5(7.3) μ; conidial scars thickened on or near the subtruncate or rounded apex. CONIDIA hyaline, slightly coloured when old, acicular, obclavate to cylindrical, sometimes indistinct, multiseptate, straight to slightly curved, base truncate or obconically truncate, up to 300 μ long × 2–4 (rarely 5) μ wide. (J. M. Mulder & P. Holliday, *CMI Descr.* 419, 1974.)

Grey leaf spot of *Sorghum* spp., Sudan grass (*S. arundinaceum* var. *sudanesis*), Johnson grass (*S. halepense*) and broomcorn (*S. dochna*) is widespread (Map 338). The spots usually first form on the lower leaves and infection gradually spreads upwards in the plant; they are elongate, somewhat delimited by the veins and individually > 1 cm long and 3–5 mm wide. Coalescence can take place, to give larger stripes or patches and kill large areas of the leaf. The colour (shades of red to black, light brown or yellowish) of the lesions varies with that of the leaf. The deep coloured spots tend to have lighter coloured outer rings than the paler ones where the margins are darker. Colours darken with age and a faint zonation may be found. The field biology of *C. sorghi* is little known and the disease it causes appears to be serious only when the weather is wet. Losses have been reported from China and heavy infection from India. The conidia are presumably air dispersed and spread from host debris. Accounts of the infection of maize have not been substantiated. A different form of the fungus was described from Venezuela. No standard control measures have been described. Differences in reaction to infection amongst cvs were found in India (Madras).

Ciccarone, A. 1950. *Annali Sper. agr.* **4**: 281 (in Venezuela; **29**: 615).

Quebral, F. C. et al. 1959. *Philipp. Agric.* **43**: 271 (general; **39**: 575).

Rao, P. S. 1963. *Indian Phytopathol.* **16**: 344 (reaction of cvs; **44**, 127).

Ramakrishnan, T. S. 1931. *Mem. Dep. Agric. India bot. Ser.* **18**: 259 (general; **10**: 516).

CERCOSPORIDIUM Earle, *Muhlenbergia* 1(2): 15, 1901.

Berteromyces Cif., 1954.

(Dematiaceae)

MYCELIUM immersed. STROMA present, usually well developed. SETAE and HYPHOPODIA absent. CONIDIOPHORES macronematous, mononematous, usually simple, rarely branched, brown, septate or continuous, geniculate or not, densely fasciculate, the fascicles in many spp. incurved, particularly when old as a result of the formation of a thickened band along one side of the conidiophore. CONIDIOGENOUS CELLS integrated, terminal, polyblastic, sympodial, cicatrised, conidial scars always thickened and conspicuous, usually prominent, the old scars situated on rounded shoulders or on short protrusions (like pegs) but in some spp. lying more or less flat against the side of the conidiophore. CONIDIA solitary, dry, clavate, cylindrical, obclavate or broadly fusiform, with a conspicuous thickened hilum, more or less colourless or relatively pale brown, smooth to verrucose, 1–7 (mostly 1–3) septate (M. B. Ellis, *Demat. Hyphom.*: 279, 1971).

Deighton gave a revised description of the genus and accounts of the spp. (see also Ellis l.c. and *More Demat. Hyphom.*: 294, 1976). *Cercosporidium* is characterised by the fasciculate conidiophores with thickened bands, often incurved and in the rather broader conidia with less septa. Several fungi like *Cercospora* have been described from cassava. One is *Cercosporidium henningsii* (Allesch.) Deighton (synonym: *Cercospora henningsii* Allesch.; see also E. W. Mason, *Mycol. Pap.* 2: 32, 1928, who also gave *C. manihotis* P. Henn., *C. manihotis* P. Henn. apud Wildeman and *C. cassavae* Ell. & Ev. as synonyms). Another is *Phaeoramularia manihotis* (Stev. & Solh.) M. B. Ellis (synonyms: *Cercospora caribaea* Cif. and *Ragnhildiana manihotis* Stev. & Solheim). Ghesquière referred a *C. cassavae* to a perfect state in *Mycosphaerella*.

Cercosporidium henningsii has very pale olivaceous brown, smooth conidia, 3–8 septate, $30–85 \times 5–7\ \mu$. *P. manihotis* has conidia that are commonly in branched chains, very pale olivaceous, smooth, 1–3 septate, $15–45 \times 6–8\ \mu$ (*Phaeoramularia* q.v. was described by Ellis in *Demat. Hyphom.*: 307). Maduewesi (in Nigeria) referred to the 2 diseases of cassava as brown and white leaf spot, *C. henningsii* and *P. manihotis*, respectively. Brown leaf spot was predominant; both diseases were more severe on older leaves, and up to 22% of the leaf area could be destroyed. Cassava cvs differed in susceptibility. Castaño described white leaf spot: white, circular spots, occasionally angular, 1–2 mm diam., dark grey on lower surface, often with a violet margin. Golato et al. and Viégas recommended copper fungicides.

Castaño A. J. J. 1969. *Agric. trop.* **25**: 327 (white leaf spot; **49**, 907).

Deighton, F. C. 1967. *Mycol. Pap.* 112, 80 pp. (on *Cercospora* and allied genera II, *Passalora, Cercosporidium* & *Fusicladium* spp. on *Euphorbia*; **47**, 450).

Ghesquière, J. 1932. *Bull. Inst. r. Colon. Belg.* **3**: 160 (*Mycosphaerella* on cassava; **12**: 137).

Golato, C. et al. 1971. *Riv. Agric. subtrop. trop.* **65**: 21 (brown leaf spot; **51**, 1020).

Maduewesi, J. N. C. 1975. *Niger. J. Pl. Prot.* **1**: 29 (brown & white leaf spots; **54**, 5673).

Viégas, A. P. 1941. *Bragantia* **1**: 233 (as above; **20**: 445).

Viennot-Bourgin, G. et al. 1950. *Rev. Bot. appl. Agric. trop.* **30**: 138 (as above; **29**: 400).

CEROTELIUM Arthur, *Bull. Torrey bot. Club* 33: 30, 1906.
(correction *Ibid* 33: 513)
(Pucciniaceae)

PYCNIA subcuticular, applanate to conoid. AECIDIA subepidermal, cupulate and peridiate; AECIDIOSPORES orange yellow. UREDIA subepidermal, surrounded by incurved paraphyses developing from hyphoid peridium or none. UREDOSPORES usually hyaline and sessile with scattered germ pores. TELIA subepidermal, erumpent, waxy, becoming pulverulent at apex at maturity. TELIOSPORES aseptate, hyaline, developing in chains; telial chains laterally coalescent at the base to form short columnar structures; columns firm at base becoming pulverulent at apex, with mature teliospores germinating immediately. PROMYCELIUM external, 4 celled (M. J. Thirumalachar & B. B. Mundkur, *Indian Phytopathol.* 2: 81, 1949).

Sathe discussed the status of the genus and proposed that *Cerotelium* should be reserved for the rusts with peridiate uredia and telia as originally described by Arthur (l.c.). *C. fici* (Butler) Arth. causes fig (*Ficus carica*) rust; synonyms: *Kuehneola fici* Butler, *Uredo fici* Cast., *Physopella fici* (Cast.) Arth. and *U. moricola* P. Henn. The UREDOSPORES are globose, ellipsoid or obovoid, sometimes rather irregular and angular, 19–30 × 15–23 μ, echinulate with spines 1–3 μ apart, 0.5–1 μ high; pores usually very obscure but definitely 3(–4?) and equatorial in some of the deeper-coloured spores. TELIA are rare. The other fig rust *U. ficina* Juel (synonym: *P. ficina* (Juel) Arth.) can be distinguished from *C. fici* by the characteristic paraphyses, the larger (26–35 × 20–25 μ) uredospores, the higher (1.5–3 μ) spines and the different distribution of the pores which are apical (1) and basal (1–2) or occasionally apical and equatorial. The 2 spp. on fig have probably been sometimes confused in the literature (Laundon et al.).

Besides the common cultivated fig (which cannot be grown well in the wet lowland tropics) *C. fici* occurs on other *Ficus* spp. and *Morus* spp. It is widespread (Map 399) in warmer areas and can cause considerable damage to fig. The more recent reports on control with fungicides have come from Brazil (maneb and zineb), Egypt (Bordeaux and wettable sulphur), India (Bordeaux and aureofungin) and USA (Bordeaux). Sprays should be applied to the newly developing leaves.

Assawah, M. W. et al. 1965. *Alex. J. agric. Res.* **13**: 339 (fungicides in Egypt; **46**, 3151).
Boulos, Z. Y. et al. 1968. *Ibid* **15**: 333 (as above; **48**, 1862).
Filho, B. et al. 1970. *Seiva* **30**: 34 (fungicides in Brazil; **51**, 529).
Kezdorn, A. H. et al. 1961. *Agric. Handb. U.S. Dep. Agric.* 196, 26 pp. (fig growing in USA; **41**: 163).
Laundon, G. F. 1971. *C.M.I. Descr. pathog. Fungi Bact.* 281, 289 (*Cerotelium fici* & *Uredo ficina*).
Sathe, A. V. 1972. *Indian Phytopathol.* **25**: 76 (taxonomic status of *Cerotelium*; **52**, 2520).
Wani, D. D. et al. 1973. *Hindustan Antibiot. Bull.* **15**: 79 (fungicides in India; **53**, 1030).

CHALAROPSIS Peyronel, *Staz. sper. agr. ital.* 49: 585, 1916.
(Dematiaceae)

COLONIES effuse, greenish black or greyish black. MYCELIUM immersed and superficial. STROMA none. SETAE and HYPHOPODIA absent. CONIDIOPHORES semi-macronematous, flexuous, sympodially or irregularly branched, colourless, smooth. CONIDIOGENOUS CELLS monoblastic or polyblastic, integrated and terminal or discrete, determinate, cylindrical. CONIDIA solitary, acrogenous on short branches, simple, ellipsoidal, obovoid, spherical or subspherical, olive to dark blackish brown, aseptate, smooth or minutely spinulose (M. B. Ellis, *Demat. Hyphom.*: 61, 1971).

Only *C. thielavioides* Peyronel is of concern as a wound parasite, presumably spreading from soil; (CONIDIOPHORES erect or ascending, very variable in length, 4–9 μ thick; CONIDIA spherical or subspherical, olivaceous brown, mostly 14–19 μ diam.; *Chalara* conidiophores with stipes up to 70 × 5–7 μ, phialides cylindrical or lageniform, up to 60 μ long, base 5–6 μ wide, neck 3–4 μ thick, CONIDIA catenate, cylindrical with truncate ends, colourless, smooth, aseptate, 8–15 × 2.5–4.5 μ, Ellis: 63 l.c.). The fungus has been described as causing a root rot of elm (*Ulmus*) and walnut (*Juglans regia*); graft rots in *Rosa* and walnut; a rot of clamped and pre-packed carrot (*Daucus carota*); and associated with the root rot complex of poinsettia (*Euphorbia pulcherrima*), *Thielaviopsis basicola* q.v. Hennebert referred to 3 other spp. Bliss described *Ceratocystis radicicola* (Bliss) C. Moreau, with a *Chalaropsis* state, causing a root and trunk rot of date palm (*Phoenix dactylifera*). No particular control measures appear warranted apart from good plant hygiene; including the avoidance of damage, particularly to roots and after harvest.

Årsvoll, K. 1966. *Gartneryrket* **56**(6) 6 pp. (on washed carrot; **46**, 2574).
Baker, K. F. et al. 1946. *Phytopathology* **36**: 281 (on grafts of rose; **25**: 452).
Bliss, D. E. 1941. *Mycologia* **33**: 468 (on date palm; **21**: 13).
Boerema, G. H. 1959. *Versl. Meded. plziektenk. Dienst Wageningen* 134: 158 (on carrot in storage; **41**: 348).
Hammond, J. B. 1935. *Trans. Br. mycol. Soc.* **19**: 158 (on grafts of walnut; **14**: 408).
——. 1935. *J. Pomol.* **13**: 81 (on walnut; **14**: 801).
Hennebert, G. L. 1967. *Antonie van Leeuwenhoek* **33**: 333 (*Chalaropsis punctata*, a new hyphomycete).
Lamb, H. et al. 1935. *Phytopathology* **25**: 652 (on elm; **14**: 725).
Longrée, K. 1940. *Ibid* **30**: 793 (on grafts of rose; **20**: 63).
Schneider, R. et al. 1969. *NachrBl. dt. PflSchutzdienst. Stuttg.* **21**: 164 (on carrot in clamps; **49**, 1232).

CHOANEPHORA Currey, *J. Linn. Soc. (Bot)* **13**: 578, 1873.

(Choanephoraceae)

Both sporangia and conidia present, not infrequently arising from the same mycelium. SPORANGIUM terminal, usually pendent on the recurved end of an erect, unbranched sporangiophore, a definite columella which tends to be globose, and usually with many spores, although diminutive, few spored sporangia sometimes occur. SPORANGIOSPORES usually ovoid to fusiform, but occasionally varying to inequilateral or triangular, not striate like the conidia, with at both ends (and in spores of odd shape sometimes also at the side) a cluster of very fine radiating appendages. CONIDIOPHORE an erect hypha terminating in a capitate vesicle from which a few short branches emerge; these branches, usually without branching again, enlarging at their tips to form secondary vesicles which at maturity are covered with short sterigmata bearing conidia. CONIDIA resembling the sporangiospores in shape but non-ciliate, longitudinally striate and with a short hyaline appendage at the base; intercalary CHLAMYDOSPORES with more or less thickened walls borne on the mycelium; ZYGOSPORES observed in described spp. (from H. M. Fitzpatrick, *The lower fungi Phycomycetes*: 261, 1930).

The genus differs from the related genera *Blakeslea* Thaxt. and *Cunninghamella* Matr. in that the former has sporangiola, no conidia and longitudinally striate sporangiospores; and the latter has no sporangium, and sporangiola with echinulate conidia (Fitzpatrick, l.c.; and see Hesseltine).

Choanephora cucurbitarum (Berk. & Rav.) Thaxt. causes a soft (wet) rot in the field (and in storage) of blossoms and fruits; it can also attack stems and leaves, and may be found on seed. The fungus is common and occurs in soil. Its morphology was described by Wolf. The conidial state has a silvery, metallic lustre which is different from the comparable stages of the 2 other common fungal rots in which profuse mycelial growth also occurs on the substrate, i.e. *Sclerotinia fuckeliana* (*Botrytis* state) and *Rhizopus stolonifer* (q.v.). CONIDIA ($15–25 \times 7.5–11$ μ) and SPORANGIA ($35–160$ μ diam.) are black when mature; SPORANGIOSPORES $18–30 \times 10–15$ μ; ZYGOSPORES $50–90$ μ diam. (see Barnett et al. 1956). The fine structure of the perfect state in another sp. was described by Kirk. Amongst the crops on which this rot has been described are red pepper, *Colocasia esculenta* (taro), *Cucurbita* spp., *Hibiscus esculentus* (okra), common bean and *Vigna unguiculata* (cowpea).

In squash and red pepper infection frequently begins on the flowers; a dense, white growth forms and infection spreads to the young fruit. A soft, wet rot (spreading rapidly) is caused on older fruit. On cowpea infected pods show at first a watersoaked appearance; maturing pods near the ground are most commonly attacked. On okra *C. cucurbitarum* was pathogenic to flowers and wounded fruit but not to unwounded fruit and seedlings. On common bean leaf infection causes small irregular lesions, watersoaked at first, enlarging to become dry and dark with age; profuse sporulation occurs on both surfaces. In taro a wet rot of the leaves was described by Sinha (1940b) where *C. cucurbitarum* caused a secondary infection of the lesions due to *Phytophthora colocasiae*. Cuthbert et al. found that *C. cucurbitarum* caused disease in cowpea by infection via the feeding and oviposition punctures of *Chalcodermus aeneus*. Barnett et al. (1950) reported *c.* 25° as the opt. for formation of conidia; 31° was unfavourable but at this latter temp. many sporangia developed. No sporangia or conidia were formed at 34°.

Outbreaks of the disease may be found under wet, warm conditions in densely planted crops. Local conditions should determine what control measures are justified. Fungicide treatments for red pepper (mancozeb, thiram, zineb and ziram) and common bean were described by Chahal et al. and McMillan. Hot water treatment of okra seed (60° for 30 minutes) eliminated the fungus (Tai Luang Huan). Grover described a blossom rot of pumpkin (*Cucurbita maxima*) caused by *Cunninghamella echinulata* (Thaxt.) Thaxt. ex Blakeslee which also infected the young fruit and leaves. Most growth in culture, and most rot, took place at 28–32°. Conidial formation on the fruit was profuse at 24–32°. At 16° and 36° the fruit rot spread slowly.

Barnett, H. L. et al. 1950. *Phytopathology* **40**: 80 (effects of nutrition, temp. & light; **29**: 424).

—— ——. 1956. *Mycologia* **48**: 617 (factors affecting zygospore formation; **36**: 344).

Chahal, A. S. et al. 1974. *Haryana J. hort. Sci.* **3**: 190 (fungicides on red pepper; **56**, 946).

Christenberry, G. A. 1938. *J. Elisha Mitchell scient. Soc.* **54**: 297 (effect of light).

Cuthbert, F. P. et al. 1975. *J. econ. Ent.* **68**: 105 (on cowpea; **54**, 5165).

Dastur, J. F. 1920. *Ann. Bot.* **34**: 399 (on red pepper).

Grover, R. K. 1965. *Indian Phytopathol.* **18**: 257 (*Cunninghamella echinulata* on squash; **46**, 819).

Hesseltine, C. W. 1955. *Mycologia* **47**: 344 (Mucorales, keys & synonymy; **35**: 237).

Kirk, P. M. 1977. *Trans. Br. mycol. Soc.* **68**: 429 (zygospore fine structure in *Choanephora circinans*).

Lefebvre, C. L. et al. 1939. *Phytopathology* **29**: 898 (on cowpea; **19**: 133).

McMillan, R. T. 1972. *Pl. Dis. Reptr* **56**: 967 (on common bean; **53**, 324).

Sinha, S. 1940a. *Proc. Indian Acad. Sci.* Sect. B. **11**: 162 (on red pepper; **19**: 513).

——. 1940b. *Ibid* **11**: 167 (on taro; **19**: 514).

Tai Luang Huan et al. 1975. *MARDI Res. Bull.* **3**(2): 38 (on okra; **56**, 949).

Tiwari, V. N. et al. 1974. *Indian Phytopathol.* **27**: 611 (effect of light; **55**, 3923).

Wolf, F. A. 1917. *J. agric. Res.* **8**: 319 (on *Cucurbita*).

CLADOSPORIUM Link ex Fr.; Link, *Magazin Ges. naturf. Freunde Berlin* **7**: 37, 1815; Fries, *Syst. Mycol.* **1**: XLVI, 1821.
(S. J. Hughes in *Can. J. Bot.* **36**: 750, 1958, cites these synonyms: *Sporocladium* Chevallier, *Myxocladium* Corda, *Didymotrichum* Bonorden, *Heterosporium* Klotzsch in Cooke)
(Dematiaceae)

COLONIES effuse or occasionally punctiform, often olivaceous but also sometimes grey, buff, brown or dark blackish brown, velvety, floccose or hairy. MYCELIUM immersed and often also superficial. STROMA sometimes present. SETAE and HYPHOPODIA absent. CONIDIOPHORES macronematous or semi-macronematous and sometimes also micronematous; MACRONEMATOUS CONIDIOPHORES straight or flexuous, mostly unbranched or with branches restricted to the apical region forming a stipe and head, olivaceous brown or brown, smooth or verrucose. RAMO-CONIDIA often present. CONIDIOGENOUS CELLS polyblastic, usually integrated, terminal and intercalary but sometimes discrete, sympodial, more or less cylindrical, cicatrised, scars usually prominent. CONIDIA catenate as a rule but sometimes solitary especially in spp. with large conidia, often in branched chains, acropleurogenous, simple, cylindrical, doliiform, ellipsoidal, fusiform, ovoid, spherical or subspherical, often with a distinctly protuberant scar at each end or just at the base, pale to dark olivaceous brown or brown, smooth, verruculose or echinulate, with 0–3 or occasionally more septa (M. B. Ellis, *Demat. Hyphom.*: 308, 1971).

Ellis l.c. stated that there is no comprehensive monograph of the genus (some 500 spp. have been described) and he described and keyed out 15 of the common ones (further spp. were described by Ellis in *More Demat. Hyphom.*). *C. cladosporiodes* (Fresen.) de Vries (see Reddy for perfect state) and *C. herbarum* (Pers.) Link ex S. F. Gray are cosmopolitan on many kinds of substrata and may behave as secondary invaders. *C. oxysporum* Berk. & Curt. is widespread in the tropics on the dead parts of leaves and stems of herbaceous and woody plants. Two other common spp. are *C. macrocarpum* Preuss and *C. sphaerospermum* Penz. *C. spongiosum* Berk. & Curt. occurs on inflorescences of grasses, especially spp. of *Cenchrus* and *Setaria*. Spp. causing minor diseases, mostly as leaf spots, include: *C. colocasiae* Sawada (Bugnicourt) on *Colocasia*; *C. variabile* (Cooke) de Vries on spinach (*Spinacea oleracea*; Gambogi; Mathur et al.); *C. musae* Mason on banana (Martyn; Stover); *C. pisicolum* Snyder on pea (*Pisum sativum*); *C. vignae* Gardner on cowpea (*Vigna*); and *C. allii-cepae* (Ranojević) M. B. Ellis on onion (Ryan).

Aulakh, K. S. 1970. *Indian Phytopathol.* **23**: 573 (*Cladosporium oxysporum* on castor; **50**, 3087).

Bugnicourt, F. 1958. *Rev. Mycol.* **23**: 233 (*C. colocasiae*; **38**: 176).

Cladosporium cucumerinum

Gambogi, P. 1960. *Agricoltura ital.* **60** (NS 15); 385 (*C. variabile*; **40**: 645).
Gardner, M. W. 1925. *Phytopathology* **15**: 453 (*C. vignae*; **5**: 76).
Levkina, L. M. 1974. *Vestnik Moskovsko Universiteta* 6 **29**: 77 (keys; **54**, 1624).
Martyn, E. B. 1945. *Mycol. Pap.* 13, 5 pp. (*C. musae* inter alia; **25**: 220).
Mathur, R. L. et al. 1959. *Indian Phytopathol.* **12**: 161 (*C. variabile*; **40**: 449).
—— ——. 1965. *Ibid* **18**: 215 (fungicide control of *C. variabile*; **45**, 911).
Reddy, S. M. 1975. *Indian Phytopathol.* **28**: 286 (perfect state of *C. cladosporioides*; **56**, 1961).
Ryan, E. W. 1978. *Fm Fd Res.* **9**: 139 (*C. allii-cepae*).
Snyder, W. C. 1934. *Phytopathology* **24**: 890 (*C. pisicolum*; **14**: 71).
Stover, R. H. 1975. *Trans. Br. mycol. Soc.* **65**: 328 (*Cladosporium* on banana; **55**, 1885).
Strider, D. L. et al. 1963. *Pl. Dis. Reptr* **47**: 493 (host screening in *C. vignae*; **43**, 305).
Vries, G. A. de 1952 (1967 reprint). Thesis, University of Utrecht, 121 pp. (morphology, taxonomy & other characteristics; **33**: 116).

Cladosporium cucumerinum Ellis & Arth., *Bull. agric. Exp. Stn Indiana* **19**: 9, 1889.
Scolecotrichum melophthorum Prill. & Delacr., 1891
Macrosporium melophthorum (Prill. & Delacr.) Rostrup, 1893
Cladosporium cucumeris Frank, 1893.

COLONIES on natural substata and in culture effuse, rather pale greyish olive, velvety or felted; reverse on malt agar greenish black. HYPHAE sometimes spirally twisted, immersed hyphae usually with a slime coat. CONIDIOPHORES macronematous up to 400 μ long, 3–5 μ thick, sometimes swollen to 8 μ at the base and micronematous, pale olivaceous brown, smooth. RAMO-CONIDIA 0–2 septate, up to 30 μ long, 3–5 μ thick, smooth. CONIDIA formed in long, branched chains, mostly aseptate, occasionally 1 septate, cylindrical rounded at the ends, ellipsoidal, fusiform or subspherical, smooth to minutely verruculose, pale olivaceous brown, 4–25 × 2–6 (mostly 4–9 × 3–5) μ. (M. B. Ellis & P. Holliday, *CMI Descr.* 348, 1972.)

Cucurbit scab occurs especially on cucumber (*Cucumis sativus*), melon (*C. melo*), and pumpkin, squash and marrow (*Cucurbita pepo*). Symptoms are most severe on the young fruit where deeply sunken lesions are formed, up to 1 cm diam., with a gummy exudate. On older fruit infection is restricted by the host reaction and results in the formation of brown, corky scabs. Foliage infection (necrotic, watersoaked spots, sometimes with a gummy exudate) is not very destructive. Sporulation on the leaf tends to be sparse. At favourable temps the apical shoots of young plants are killed back.

An important disease which is widespread in N. America and Europe; it also occurs in parts of Africa, Asia and tropical America (Map 310). Scab has not been reported from Australasia. Penetration, with appressoria, appears to be always direct on stem, leaf, petiole and fruit. At 27° it is less frequent than at 23°. The opt. temp. for both in vitro and in vivo growth is *c.* 21° or a little lower; at 17° the growing tips of young plants are killed back rapidly. Symptoms diminish progressively as temp. rises above 21° and there is little disease at 30° (Pierson et al.; Strider et al. 1960; Van der Muyzenberg; Walker 1950). The effects of nutrients on growth, spore production in vitro and cell degrading enzymes produced by the pathogen have been described. The conidia are presumably air dispersed and *C. cucumerinum* survives between crops on host debris. Seedborne spread probably does not occur although treatment to prevent it has been recommended. No races have been described.

Control by rotation is in doubt and resistant cvs, supplemented by fungicides, are used. Some biochemical aspects of resistance (controlled in some cucumber lines by a single dominant gene) have been studied, and lignification of the parenchyma following infection is a possible defence mechanism. Although resistant cucumber and gherkin cvs are available less resistance seems to have been found in *Cucurbita*; in USA a few lines of *C. pepo*, *C. maxima* and *C. moschata* were resistant or tolerant but all commercial squash cvs were susceptible. Spraying with dithiocarbamates is satisfactory in several countries and early application before fruit formation is necessary (Biehn et al.; Dekhuijzen; Strider 1963). The systemic activity of chemicals against the pathogen has been studied in the Netherlands.

Andweg, J. M. 1956. *Euphytica* **5**: 185 (breeding for resistance in slicing cvs of cucumber; **36**: 83).
Behr, L. 1948. *Phytopathol. Z.* **15**: 92 (histology of infection; **28**: 436).
Biehn, W. L. et al. 1963. *Pl. Dis. Reptr* **47**: 241 (control with mancozeb; **42**, 647).

Crossan, D. F. et al. 1964. *Phytopathology* **54**: 1038 (good sporulation on Czapek with mannose, fructose or lactose; **44**, 585).
——. 1969. *Pl. Dis. Reptr* **53**: 452 (effect of rotation; **48**, 3234).
Dekhuijzen, H. M. 1964. *Neth. J. Pl. Pathol.* **70** suppl 1, 75 pp. (systemic action of dimethyl-dithiocarbamates; **43**, 2473).
—— et al. 1971. *Pestic. Biochem. Physiol.* **1**: 11 (pathogen resistance against 6-azauracil & 6-azauridinemonophosphate; **51**, 85).
Dekker, J. 1968. *Neth. J. Pl. Pathol.* **74** Suppl. 1: 127 (pathogen resistance against 6-azauracil & biosynthesis of RNA precursors).
El-Din Fouad, M. K. 1965. *Meded. LandbHogesch. Wageningen* **56**: 1 (histology of infection & chemically induced host resistance; **36**: 510).
Fuchs, A. 1965. *Neth. J. Pl. Pathol.* **71**: 157 (resistance, fungal & host polyphenoloxidases; **45**, 1604).
Hijwegen, T. 1963. *Ibid* **69**: 314 (lignification & host resistance; **43**, 1207).
Husain, A. et al. 1958. *Phytopathology* **48**: 316 (extracellular pectic & cellulytic enzymes; **37**: 752).
Kaars Sijpesteijin, A. et al. 1968. *Neth. J. Pl. Pathol.* **74** Suppl. 1: 121 (action of chemotherapeutant phenylthiourea).
Kubicki, B. et al. 1970. *Genet. Pol.* **11**: 53 (resistant cucumber; **50**, 1030).
Kuć, J. 1962. *Phytopathology* **52**: 961 (extracellular enzymes; **42**: 357).
Laborda, F. et al. 1976. *Can. J. Microbiol.* **22**: 394 (fine structure during pathogenesis; **55**, 4917).
Mahadevan, A. et al. 1965. *Phytopathology* **55**: 1000 (pectinase, unknown inhibitor in resistant cucumber; **45**, 323).
Morton, D. J. et al. 1967. *Pl. Dis. Reptr* **51**: 495 (effect of rotation; **46**, 2919).
Overeem, J. C. et al. 1967. *Phytochemistry* **6**: 99 (formation of perylenequinones in etiolated, infected cucumber; **46**, 2575).
Paus, F. et al. 1973. *Physiol. Pl. Pathol.* **3**: 461 (fine structure of infection & disease development; **53**, 1612).
Pierson, C. F. et al. 1954. *Phytopathology* **44**: 459 (histology of infection & temp.; **35**: 576).
Schultz, H. et al. 1939. *Gartenbauwissenschaft* **13**: 169, 605 (general on cucumber; **18**: 651).
Skare, N. H. et al. 1975. *Physiol. Pl.* **33**: 229 (formation of pectinase & cellulase on isolated host cell walls; **54**, 4239).
Strider, D. L. et al. 1960. *Phytopathology* **50**: 583 (effect of temp., pH & nutrients on growth in vitro; **40**: 139).
——. 1961. *Ibid* **51**: 765 (cell degrading enzymes; **41**: 351).
——. 1963. *Pl. Dis. Reptr* **47**: 418 (dithiocarbamate fungicides; **42**: 718).
Van Andel, O. M. 1969. *Neth. J. Pl. Pathol.* **75**: 151 (effect of chemotherapeutic amino acids on pathogencity & growth; **48**, 1532).
Van der Muyzenberg, E. W. B. 1932. *Tijdschr. PlZiekt.* **38**: 81 (general; **11**: 690).
Walker, J. C. 1950. *Phytopathology* **40**: 1094 (effect of temp. & host age on growth & disease, inheritance of resistance; **30**: 356).
—— et al. 1953 & 1955. *Ibid* **43**: 215; **45**: 451 (resistant pickling cucumber cvs; **33**: 135; **35**: 73).
——. 1958. *Pl. Dis. Reptr* **42**: 1337 (as above; **38**: 442).

CLAVICEPS Tul., *C.r. hebd. Séanc. Acad. Sci. Paris* 33: 646, 1851.
(Clavicipitaceae)

STROMATA (stalked), borne on a elongated black SCLEROTIUM, club shaped with globose heads. PERITHECIA completely sunk in the head. ASCI unitunicate, cylindrical; ASCOSPORES like threads. The sclerotium originates in the floral structures of monocotyledons. *Sphacelia* Lév. is the stat. conid. (after R. W. G. Dennis, *British ascomycetes*: 226, 1968).

The diseases caused by *Claviceps* spp. are called ergot which is a name for the sclerotia. Severe infection reduces yields. The *Sphacelia* imperfect state forms a secretion (honey dew) attractive to insects. Airborne ascospores or rain and insect-dispersed conidia in honey dew infect the ovaries, mostly of members of the Gramineae. The sclerotia, which may contain toxic alkaloids and hallucinogens, are formed in the inflorescences 2–3 weeks after infection. Classical ergotism (ergot poisoning), the St Anthony's Fire of the Middle Ages, was due to the contamination of bread by the rye ergot (*C. purpurea* (Fr.) Tul.); see A. Barger, *Ergot and ergotism*, Gurney & Jackson, 1931; F. J. Bové, *The story of ergot*, S. Karger, 1970. Brady studied the hosts (Cyperaceae, Gramineae and Juncaceae) of 28 recognised spp. The mechanism of honey dew formation was described by Mower et al. The widespread *C. purpurea* (Map 10), mostly on temperate cereals and grasses, was described by Sprague (*Diseases of cereals and grasses in North America*). It has been recorded on *Saccharum* (Rajendran; Singh). See also Dickens et al.; Dickson (*Diseases of field crops*); Western, editor (*Diseases of crop plants*).

C. gigantea Fuentes, Isla, Ullstrup & Rodriquez was described from maize at high altitudes in Mexico (Ullstrup). *C. maximensis* Theis was described from *Panicum maximum* (guinea grass). *C. microcephala* (Wallr.) Tul. causes ergot of *Pennisetum typhoides* (bulrush millet). *C. fusiformis* Loveless (described from Zimbabwe) also occurs on bulrush millet. Bhat and Siddiqi et al. supported Loveless (1967) in that ergot of this crop should be referred to *C. fusiformis* not *C. microcephala*. The disease is a serious one in India and for reviews on it see Nene et al.; Sundaram (1975). *C. paspali* Stev. & Hall, occurs on *Paspalum dilatum* (dallis grass); it was described by Sprague (l.c.). Luttrell recently reported on its life history (and see Cunfer et al.). The perfect state of sorghum sugary disease (*Sphacelia sorghi* McRae) was described in 1976 as *C. sorghi* P. Kulkarni, Seshadri & Hedge. General accounts were given by Sundaram (1970) and Tarr (*Diseases of sorghum, Sudan grass and broom corn*). Mantle considered that *S. sorghi* in Nigeria differed in certain respects from oriental collections. For hosts of *C. sorghi* see Chinnadurai et al. (1971). Futrell et al. (1966) infected sorghum with conidia from guinea grass, and maize was also infected with conidia from sorghum.

Bhat, R. V. 1977. *Curr. Sci.* **46**: 184 (identity of *Claviceps* sp. on bulrush millet; **57**, 1712).

Brady, L. R. 1962. *Lloydia* **25**: 1 (phylogenetic distribution; 153 ref.; **42**: 104).

Brown, H. B. 1916. *J. agric. Res.* **7**: 401 (*C. paspali*, life history & poisonous properties).

Chinnadurai, G. et al. 1970. *Z. PflKrankh. PflSchutz* **77**: 221 (*Sphacelia sorghi*; **50**, 657).

—— ——. 1970. *Phytopathol. Z.* **69**: 56 (effect of stigmatic exudates on parasitism of *S. sorghi*; **50**, 1768).

—— ——. 1971. *Madras agric. J.* **58**: 600 (*S. sorghi*, hosts; **51**, 2433).

Cunfer, B. M. et al. 1977. *Mycologia* **69**: 1137, 1142 (*C. paspali*, germination & survival of sclerotia; **57**, 3795, 3796).

Dickens, J. S. W. et al. 1974. *Advis. Leafl. Minist. Agric. Fish. Fd* 548 (*C. purpurea*; **54**, 3804).

Fuentes, S. F. et al. 1964. *Phytopathology* **54**: 379 (*C. gigantea* sp. nov.; **43**, 2599).

Futrell, M. C. et al. 1965. *Pl. Dis. Reptr* **49**: 680 (*S. sorghi*, infection & sterility in sorghum; **44**, 3342).

—— ——. 1966. *Ibid* **50**: 828 (*S. sorghi*, hosts & epidemiology; **46**, 991).

Kulkarni, B. G. P. et al. 1976. *Mysore Jnl agric. Sci.* **10**: 286 (*C. sorghi* sp. nov.; **56**, 2484).

Loveless, A. R. 1964. *Trans. Br. mycol. Soc.* **47**: 205 (honey dew state for identification; **44**, 67).

——. 1967. *Ibid* **50**: 15 (*C. fusiformis* sp. nov.).

Luttrell, E. S. 1977. *Phytopathology* **67**: 1461 (*C. paspali*, disease cycle & fungus host relationships; **57**, 4514).

Mantle, P. G. 1968. *Ann. appl. Biol.* **62**: 443 (*S. sorghi*; **48**, 1168).

Moreno, M. et al. 1972. *Fitopatologia* **5**: 7 (*C. gigantea*, effect of position & number of sclerotia on maize germination; **52**, 399).

Mower, R. L. et al. 1975. *Can. J. Bot.* **53**: 2826 (mechanism of honey dew formation).

Nene, Y. L. et al. 1976. *PANS* **22**: 366 (reviews, *C. microcephela* & *Sclerospora graminicola*; **56**, 693).

Rajendran, V. 1966. *Curr. Sci.* **35**: 472 (*C. purpurea* on *Saccharum*; **46**, 1716).

Siddiqi, M. R. et al. 1973. *Trans. mycol. Soc. Japan* **14**: 195 (*Claviceps* on bulrush millet; **53**, 1357).

Singh, S. 1976. *Z. PflKrankh. PflSchutz* **83**: 442 (*C. purpurea* on *Saccharum*; **56**, 1259).

Stewart, R. B. 1957. *Phytopathology* **47**: 444 (*C. paspali*, sclerotial morphology; **36**: 766).

Sundaram, N. V. 1970. in *Proc. 1st Int. Symp. Pl. Pathol.* 1966–67, New Delhi: 435 (*S. sorghi*).

——. 1975. In *Advances in mycology and plant pathology* (editors S. P. Raychaudhuri et al.): New Delhi (*C. microcephala*).

Thies, T. 1952. *Mycologia* **44**: 789 (*C. maximensis* sp. nov.; **32**: 487).

Ullstrup, A. J. 1973. *PANS* **19**: 389 (*C. gigantea*; **53**, 1352).

COCHLIOBOLUS Drechsler, *Phytopathology* **24**: 973, 1934

(Pleosporaceae)

PERITHECIA scattered, black, submembranaceous to subcoriaceous, smooth or covered more or less with flexuous vegetative filaments or with somewhat more bristling conidiophorous hyphae, globose, usually with evident paraboloid or short cylindrical ostiolar beak. ASCI subcylindrical, short stipitate, often becoming more or less distended, especially before dehiscence; bitunicate. ASCOSPORES 1–8, colourless or especially at maturity somewhat coloured, filamentous, with many septa, crowded, disposed in a strong helicoid arrangement. CONIDIOPHORES simple or somewhat branched, mostly olivaceous, septate, forming first conidium some distance from base, and after repeated subterminal elongation successive conidia at intervals later, often marked by geniculations. CONIDIA elongated ellipsoidal or sometimes somewhat fusoid, straight or curved, nearly colourless to deep olivaceous, with several septa, germinating by 2 polar germ tubes, one from the apex and the other from a zone immediately surrounding the basal scar (from C. Drechsler l.c.).

Luttrell et al. gave a key to 11 spp. of *Cochliobolus*. Ellis (*Demat. Hyphom.* and *More Demat. Hyphom.*) described the imperfect states of the genus in *Drechslera* (there are such states in other genera). *C. sativus* (Ito & Kuribayashi) Drechsler ex Dastur is widespread (Map 322) and mainly a pathogen of wheat and barley but it has been recorded on tropical crops (mostly Gramineae); see Luttrell and Shoemaker for cytology, morphology and taxonomy. The stat. conid. is *D. sorokiniana* (Sacc.) Subram. & Jain (synonyms: *Helminthosporium sorokinianum* Sacc. ex Sorok.; *H. sativum* Pammel, King & Bake; *H. acrothecioides* Lindfors, *H. californicum* Mackie & Paxton). The conidia are curved (in culture often straight), fusiform to broadly ellipsoidal, dark olivaceous brown, smooth, 3–12 (mostly 6–10) pseudoseptate, 40–120 (mostly 60–100) μ long, 17–28 (mostly 18–23) μ wide in the broadest part. *C. sativus* has been widely studied in the soil environment (see Chinn et al.; Old). Kline et at. crossed a pathogenic and a non-pathogenic strain and tested monoascospore cultures for pathogenicity on 6 grass spp. Recombinants were detected and pathogenicity to a given sp. was inherited simply (and see Berkenkamp). In Australia the fungus was described causing a rice leaf spot; the spots are dark brown 1 × 2 mm, occasionally large and sometimes occur as long brown streaks (Walker et al.). It has also been described from non-gramineous hosts (e.g. common bean, pea and lucerne) but appears to be of no economic importance on these. Ref. to the earlier literature on *C. sativus* and accounts of the pathogen are given by Dickson, *Diseases of field crops*, and Sprague, *Diseases of cereals and grasses in North America*.

C. victoriae Nelson causes a disease on oats (Map 267). It has been recorded on some tropical crops (see Tarr, *Diseases of sorghum, sudan grass and broom corn*; Sprague l.c.) but causes no serious disease on them. The stat. conid. is *D. victoriae* (Meehan & Murphy) Subram. & Jain (synonym; *H. victoriae* Meehan & Murphy). The conidia are slightly curved, broadly fusiform or obclavate to fusiform, pale or mid-pale golden brown, smooth, 4–11 (mostly 8–10) pseudoseptate, 40–120 (mostly 60–90) × 12–19 (mostly 14–18) μ. They are paler, thinner and narrower than those of *C. sativus*. *C. cynodontis* Nelson is cosmopolitan. It can cause browning on *Cynodon dactylon* (Bermuda grass) and has been recorded on many other plants. The stat. conid. is *D. cynodontis* (Marignoni) Subram. & Jain (synonym: *H. cynodontis* Marignoni). The conidia are mostly slightly curved sometimes almost cylindrical but usually broadest in the middle tapering towards the rounded ends, pale to mid-golden brown, smooth, thin walled, with 3–9 (mostly 7–8) pseudosepta, 30–75 (50) × 10–16 (13) μ; hilum 2.5–3 μ wide. *Cochliobolus cymbopogonis* and *C. geniculatus* have *Curvularia* (q.v.) states (see also *Drechslera*).

Berkenkamp, B. 1971. *Phytoprotection* **52**: 52 (*Cochliobolus sativus*, pathogenicity on cereals & grasses; **51**, 1584).

Chinn, S. H. F. et al. 1967. *Phytopathology* **57**: 580 (effect of substances and soil treatments on germination in *C. sativus*; **46**, 3001).

Drechsler, C. 1934. *Ibid* **24**: 953 (taxonomy, *Ophiobolus*, *Pyrenophora*, *Helminthosporium* and *Cochliobolus* gen. nov.; **14**: 124).

Gourley, C. O. 1968. *Can. Pl. Dis. Surv.* **48**: 34 (*C. sativus* on common bean; **47**, 3268).

Graham, K. M. et al. 1964. *Ibid* **44**: 113 (as above; **44**, 573).

Hodges, C. F. 1975. *Mycopathologia* **57**: 9 (effects of temp. & culture age on conidial germination in *Drechslera sorokiniana* & *Curvularia geniculata*; **55**, 2528).

Kline, D. M. et al. 1971. *Phytopathology* **61**: 1052 (inheritance of pathogenicity in *C. sativus*; **51**, 1585).

Luttrell, E. S. 1955. *Am. J. Bot.* **52**: 57 (taxonomy, stat. conid. of *C. sativus*; **34**: 324).

—— et al. 1959. *Mycologia* **51**: 195 (key to *Cochliobolus* spp. inter alia).

Nelson, R. R. 1960. *Phytopathology* **50**: 774 (*C. victoriae* sp. nov.; **40**: 220).

——. 1964. *Mycologia* **56**: 64 (*C. cynodontis* sp. nov.; **43**: 2239).

Old, K. M. 1967. *Trans. Br. mycol. Soc.* **50**: 615 (effects of natural soil on survival of *C. sativus*; **47**, 1021).

Pringle, R. B. 1976. *Can. J. Biochem.* **54**: 783 (formation of victoxinine by *C. sativus* & *C. victoriae*; **56**, 3958).

Renfro, B. L. 1963. *Pl. Dis. Reptr* **47**: 292 (*C. sativus* on forage legumes; **42**: 688).

Shoemaker, R. A. 1955. *Can. J. Bot.* **33**: 562 (biology, cytology and taxonomy of *C. sativus*; **35**: 288).

Subramanian, C. V. et al. 1964. *Trans. Br. mycol. Soc.* **47**: 613 (conidial types in *C. sativus*; **44**, 1441).

Tsuda, M. et al. 1975. *Trans. mycol. Soc. Japan* **16**: 324 (perfect state & life cycle, *Drechslera* & *Cochliobolus*; **55**, 5052).

Walker, J. et al. 1968. *Aust. J. Sci.* **31**: 82 (*C. sativus* on rice; **48**, 458).

Cochliobolus carbonum

Cochliobolus carbonum Nelson, *Phytopathology* **49**: 809, 1959.

stat. conid. *Drechslera zeicola* (Stout) Subram. & Jain, 1966

Helminthosporium carbonum Ullstrup, 1944.

Mature PERITHECIA black, ellipsoidal to globose, ostiolate, 355–550 μ high, 320–430 μ diam.; long brown setae are produced over the upper fourth of the perithecium; a well-defined ostiolate beak, subconical to paraboloid, 60–200 μ long, is usually produced. Mass of hyaline cells frequently covers beak apex. Locule with hyaline, filamentous pseudoparaphyses. ASCI cylindrical or clavate, straight or slightly curved, short stipitate, 162–257 × 18–27 (av. 196.6 × 22.6) μ. ASCOSPORES 1–8, coiled in a close helix, discharged by splitting of upper ascus wall; filiform or flagelliform, somewhat tapered at extremities; when mature typically hyaline, 5–9 septate. 182–306 × 6–10 (av. 245.4 × 7.8) μ; following discharge ascospore often surrounded by thin mucous envelope. CONIDIOPHORES arising singly or in small groups, straight or flexuous, mid to dark brown or olivaceous brown, up to 250 μ long, 5–8 μ thick. CONIDIA curved or sometimes straight, occasionally almost cylindrical but usually broader in the middle and tapering towards the rounded ends, 30–100 × 12–18 μ (mostly 60–80 × 14–16 μ), with 6–12 (usually 7–8) pseudosepta, often finally becoming dark or very dark brown or olivaceous brown, the end cells sometimes remaining paler than the middle ones; hilum not very conspicuous. Dark, thick-walled hyphae, similar to those which give the characteristic charred appearance to diseased maize ears, are formed also in culture, especially when the fungus is grown on wheat with tap water agar; these hyphae sometimes break up with segments. (Perfect state: from R. R. Nelson l.c.; stat. conid: M. B. Ellis & P. Holliday, *CMI Descr*. 349, 1972.)

A widely distributed leaf spot of maize: central and southern Africa, S.E. Europe (and Denmark), America (Argentina, Canada (Ontario), Colombia, Costa Rica, Guatemala, Salvador and E. USA), Asia (Cambodia, China and E. India), Australia (New South Wales and Queensland) and New Caledonia (Map 380). The disease was originally attributed to *C. heterostrophus* by Ullstrup (1941a & b) who later (1944) described the imperfect state as *H. carbonum* after the charred appearance of the affected ears. Considerably later Nelson (l.c.) found the perfect state. Leaf symptoms differ between the first 2 races described. Race 1 gives oval to circular, straw coloured, dry and papery lesions which become zonate with light to purplish brown margins; they are abundant, up to 1.5 × 2.5 cm and often coalescent. Race 2 gives oval, chocolate brown spots, sometimes irregular, up to 0.5 × 2.5 cm and less abundant than those of race 1. Both races cause ear rot, a black felt-like growth over the kernels and giving a charred appearance. Nelson et al. (1973) reported a new race (3; see also Castor et al.; Leonard).

C carbonum is heterothallic; Nelson and co-workers studied the genetic control of: compatibility, perithecial formation and inhibition, conidial morphology and arrangement, polygalacturonase production, pathogenicity to grasses and pathogenicity of interspecific hybrids with *C. victoriae*. Growth in vitro is best at 26°; more perithecia are formed at 24° than at 20° or 30°. A detailed account of penetration of the host was given by Murray et al. The pathogen is seedborne (seed infection is often deep seated) and airborne through the conidia. It (race 1) produces a host specific toxin (see *Rev. appl. Mycol.* **45**, 438; **47**, 803, 1516; *Rev. Pl. Pathol.* **49**, 1630, 2450, 2762; **50**, 1202; **52**, 2262, 3309; **53**, 512, 3468, 3469; **54**, 2215). Race 1 is more virulent and specialised than race 2; susceptibility to 1 is controlled by a monogenic recessive and resistance has been readily incorporated into maize cvs. Both races occur in both compatibility groups but most wild types in group A were race 1 and in group a they were race 2.

Amici, A. et al. 1968. *Riv. Patol. veg. Pavia* **4**: 67 (fine structure & effect of solanine; **48**, 33).

Castor, L. L. et al. 1976. *Pl. Dis. Reptr* **60**: 827 (reaction of maize inbreds to race 3; **56**, 3546).

Dalmacio, S. C. et al. 1976. *Phytopathology* **66**: 655 (semi-incompatibility; **56**, 82).

Dayal, R. et al. 1969. *Proc. natn. Inst. Sci. India* Pt B **35**: 453 (effect of temp & nutrients in vitro; **49**, 3785).

Fries, R. E. et al. 1972. *Can. J. Microbiol.* **18**: 199 (effect of maize leaf extracts on sexual reproduction; **51**, 3322).

Hale, M. G. et al. 1961. *Phytopathology* **51**: 234 (nutrition & parasitism; **40**: 677).

Jennings, P. H. et al. 1969. *Ibid* **59**: 963 (peroxidase & polyphenol activity; **49**, 132).

Kline, D. M. et al. 1969. *Ibid* **59**: 1133 (genetic control of symptoms on grasses; **49**, 182).

Leonard, K. J. 1978. *Can. J. Bot.* **56**: 1809 (polymorphisms for lesion type, fungicide tolerance & mating).

Mackenzie, D. R. et al. 1971. *Phytopathology* **61**: 458, 471 (inheritance of fungicide tolerance; **51**, 68).
Malca, I. et al. 1963. *Ibid* **53**: 341 (effect of infection on host organic acid content; **42**: 611).
—— ——. 1964. *Ibid* **54**: 663 (stimulation & depression of cell free carboxylating systems in relation to disease development; **43**, 3204).
Murray, G. M. et al. 1975. *Can. J. Bot.* **53**: 2872 (fine structure of penetration; **55**, 3160).
Nelson, R. R. 1959. *Phytopathology* **49**: 807 (perfect state; **39**: 304).
——. 1960. *Ibid* **50**: 158 (genetics of compatibility; **39**: 571).
—— et al. 1961. *Ibid* **51**: 1 (inheritance of pathogenicity; **40**: 747).
—— ——. 1963. *Ibid* **53**: 101 (genetic control of pathogenicity in *Cochliobolus carbonum* & *C. victoriae* hybrids; **42**: 527).
——. 1964. *Ibid* **54**: 876 (genetic control of perithecial formation; **44**, 367).
——. 1966. *Mycologia* **58**: 208 (genetic control of conidial morphology & arrangement; **45**, 2757).
—— ——. 1968. *Phytopathology* **58**: 1277 (genetic control of endopolygalacturonase formation; **48**, 370).
—— ——. 1969. *Ibid* **59**: 164 (genes for pathogenicity to grasses; **48**, 2224).
——. 1970. *Ibid* **60**: 1335 (as above; **50**, 1279).
——. 1970. *Can. J. Bot.* **48**: 261 (variation in mating capacities; **49**, 2347).
—— ——. 1971. *Pl. Dis. Reptr* **55**: 325 (pathogenicity to grasses; **50**, 3870).
—— ——. 1973. *Ibid* **57**: 822 (race 3; **53**, 1797).
Ullstrup, A. J. 1941a. *Phytopathology* **31**: 508 (races; **20**: 526).
——. 1941b. *J. agric. Res.* **63**: 331 (races; **21**: 71).
——. 1944. *Phytopathology* **34**: 214 (races, symptoms & morphology of stat. conid.; **23**: 293).
—— ——. 1947. *J. Am Soc. Agron.* **39**: 606 (linkage & inheritance of susceptibility in maize; **26**: 448).

Cochliobolus heterostrophus (Drechsler) Drechsler, *Phytopathology* **24**: 973, 1934.

Ophiobolus heterostrophus Drechsler, 1925
stat. conid. *Drechslera maydis* (Nisikado) Subram. & Jain, 1966
Helminthosporium maydis Nisikado, 1926.

PERITHECIA usually early erumpent, black, no differentiated sterile setae, ascigerous part subglobose or more frequently somewhat ellipsoidal, usually 0.4–0·6 mm diam., ostiolate beak well defined, subconical or paraboloid, *c*. 0.15 mm at base and 0.15 mm long. ASCI numerous, short stipitate, apex rounded, subcylindrical 160–180 × 24–28 μ with 1–4 (typically 4) spores. ASCOSPORES filamentous, fuligenous, thin walled, becoming 5–9 septate, usually constricted at septa, the delimited segments becoming somewhat swollen to diam. up to 9 μ; thrusting against apex and into stalk base in multiple, heterostrophic, helicoid arrangement with *c*. 4 turns to each spore, 130–340 μ long; discharged simultaneously often with mucous envelope; germinating promptly by producing indiscriminantly from any or all segments, either laterally or terminally, germ tubes up to 8 in number. CONIDIOPHORES arising in groups, often from flat, dark brown to black stromata, straight or flexuous, sometimes geniculate, mid to dark brown, pale near the apex, smooth, up to 700 μ long, 5–10 μ thick. CONIDIA distinctly curved, fusiform, pale to mid-dark golden brown, smooth, 5–11 pseudoseptate, mostly 70–160 (98) μ long, 15–20 (17.3) μ thick in the broadest part; hilum dark, often flat, 3–4.5 μ wide (perfect state: C. Drechsler, l.c.; stat. conid.: M. B. Ellis & P. Holliday, *CMI Descr*. 301, 1971).

C. heterostrophus (widespread, Map 346) causes southern leaf blight of maize and occurs on many spp. of Gramineae (including *Euchlaena mexicana* and sorghum; see Nelson et al. 1968; Ullstrup 1970b). The disease aroused considerable concern when a particular race of the fungus caused widespread and heavy damage to the maize crop in USA in 1970. The impact of this epidemic was described by Ullstrup (1970a, 1972; and see Kovács; Tatum). This race (called T) was first reported from the Philippines in 1961 and is now known to be widespread. It attacks maize lines which contain T (Texas) male sterile cytoplasm (Tcms). The disease outbreak led to a general reassessment of the genetic vulnerability of major crops (National Academy of Sciences, Washington, 307 pp., 1972; *Ann. N.Y. Acad. Sci.* 287, 1977). By 1970 the use of inbred lines and hybrids containing Tcms in USA was so widespread that 85% of hybrid maize seed was of this type (Ullstrup 1972). Thus the susceptibility of this cytoplasm to race T and the favourable weather led to the epidemic.

Numerous lesions are formed, mostly on the leaves and up to 2.5 cm long. They are at first elliptical, then longitudinally elongate, becoming rectangular as restriction by the veins occurs; cinnamon buff (sometimes with a purplish tint) with a reddish brown margin, occasionally zonate, coalescing and becoming greyish with conidia. Symptoms caused by race T show a less well defined, somewhat diffuse lesion with marginal chlorosis (leading to leaf collapse) and all parts of the plant can be attacked. Perithecia have been reported in the field at the junction of leaf sheath and blade (Schenck).

This recent outbreak of southern leaf blight in epidemic form led to intensive work on host pathogen relationships; particularly on fungal toxins, epidemiology and the genetics of resistance. Earlier, full studies on the disease (previously of minor importance in USA) were made by Nisikado; Nisikado et al.; Orillo; Yu. Penetration of the host is stomatal and cuticular (see Wheeler for fine structure). Race T attacks all parts of the plant. The opt. temp. for growth in vitro and conidial germination is near 30°, and for formation of perithecia *c.* 26–27°. With race T Nelson et al. (1973) found that conidial sporulation increased with increasing dew temps to 28°, and lesion size increased with increasing dew and colonisation temps. On potted maize plants Hyre (1974) found that no conidia were formed (after 16 hours at 24°) at $<$RH 93%, many were formed at RH 97–98%. Few lesions, caused by races O and T, formed on maize seedlings with $<$4 hours high RH. Their numbers increased, nearly exponentially, on plants held in a mist chamber for 4–8 hours; after 8 hours the rate of increase was greatly reduced (Larsen et al.). In USA (Florida) the disease became severe in field maize when temps were 20° or more and RH 100% for at least 6 hours on several consecutive nights (Schenck et al.). Gillespie, in Canada, derived a disease index from the number of hours each day that temps were $>$18° while the leaves were wet. Waggoner et al. described an epidemimetric model for forecasting outbreaks and disease progress.

In USA the fungus can overwinter in infected maize tissue on the soil surface, on sod or suspended above the soil, but not in debris buried at 5–20 cm (Ullstrup 1971). Several studies on the air dispersal of conidia have been reported (Aylor; Aylor et al.; Hyre 1973; Waggoner). The fungus occurs on seed and sporulates on seedlings from infected seed (Boothroyd; Kulik; Singh et al.). Seeds were not macroscopically affected but when planted the loss in plant stand correlated fairly well with the percentage of infected seed (White et al.). With infected (race T) seed wilting was 17–21% in seedlings begun at 18° and transferred to 24°. But it was 27–30% in those begun at 24° (Kommedahl et al.). Morgan reported sclerotia (up to 300 μ diam.) on the outer surface of the seed coat in ragdoll germination tests with race T. Fatima et al. recorded the fungus on seed of crops other than maize.

C. heterostrophus is heterothallic and the genetic control of sexuality and pathogenicity has been investigated (Nelson 1957, 1959; Nelson et al. 1969). The 2 distinct races (T and O) are separable on pathogenicity (and see Warren et al.). In USA race O causes little general damage; it attacks both Tcms and normal cytoplasms equally and rarely attacks ears. Race T is extremely virulent on Tcms maize and only mildly so on normal cytoplasm. Variations in host reaction to infection by race T appears to be controlled by nuclear genes. Race T attacks the ears and can be spread both on and in the kernels. In USA under natural conditions only teosinte (apart from maize) has been found to be susceptible. Other cytoplasms (both male sterile and normal) have been studied for their reactions to infection by these races (e.g. Calvert et al.; Lim et al. 1974; Smith; Smith et al.; Villareal et al.). Fukuki et al. (1972) found the best temp. for conidial sporulation by race O at 20–28° (continuous light) and 28° (dark); for race T it was 20° and 24°, respectively. Warren found that such sporulation by race T was more sensitive to temp. than that by race O. Both races produced most conidia at 30°, but T formed 5 times more conidia at 22.5° and twice as many at 30° compared with O. More lesions formed (and expanded more rapidly) at 30° than at 22.5° and 15°. Race T had a higher opt. temp. for max. sporulation. Nelson et al. (1971) classified the virulence of race T isolates as weak, intermediate and strong. An isolate from *Chloris gayana* did not infect maize and was considered to be a distinct str. (Prakash et al.).

Both races produce phytotoxins. Plants with Tcms cytoplasm are much more sensitive to the toxins formed by race T, and this toxin is specific to this cytoplasm. The toxin from race O is non-specific to maize cytoplasm (Lim et al. 1972; Wheeler et al.). Laughnan et al. investigated the effect of the toxins from both races on the germination and growth of maize pollen. In work on the progeny from crosses between races T and O Yoder et al. found that race T type pathogenicity was associated with the formation of T toxin, i.e. this toxin is necessary for this pathogenicity type. Several toxins from race T have been found (Karr et al.). The work on phytotoxins formed by race T is now very extensive and further ref. (including bioassay) are given in the *Rev. Pl. Pathol.* **46**, 3083; **48**, 3478; **51**, 306, 3318,

3979, 3981, 3982, 3983; **52**, 2258, 3306; **53**, 504, 1347, 2539, 4799; **54**, 1242, 2797, 4464, 4465, 5417; **55**, 715, 1228, 2214, 2687, 3156, 3556, 3557; **56**, 4492, 4984, 4985; **57**, 153, 573, 574, 575, 1162, 1705, 2919, 4447, 4448. For fine structure and physiology see: *Rev. Pl. Pathol.* **51**: 304; **52**, 3674; **53**, 1793, 2990, 2991, 2992, 3463; **54**, 4918; **55**, 2692, 5166; **57**, 152, 1159, 1602.

Control of race O is determined by several genes and can be done with resistant hybrids. The use of normal cytoplasm hybrids will give control of race T. But sources of male sterility resistant to T are known. A form of resistance ('chlorotic lesion') controlled by 1 recessive gene has been described (Craig et al.; Smith; Smith et al.). For the effects of the races of *C. heterostrophus* on maize hybrids and hybrids (including different cytoplasms) see *Rev. Pl. Pathol.* **49**, 3786; **50**, 649, 2257d, 3691; **51**, 308, 3302; **52**, 4049; **53**, 1348, 2999; **54**, 433, 3875, 5416; **55**, 717, 1768, 2215, 2690, 4681, 5718; **57**, 576, 5464. Schenck et al. reported fungicide control with mancozeb (and see Comstock et al.; Kumar et al.). Pritchard eradicated the fungus from maize seed with a steam air mixture for 17 minutes at 54–55°. Thiram+carboxin are effective for seed treatments (Kommedahl et al.; Lim et al. 1973).

Aist, J. R. et al. 1976. *Phytopathology* **66**: 1050 (fine structure & mechanics of conidium attachment; **56**, 2046).

Ayers, J. E. et al. 1976. *Pl. Dis. Reptr* **60**: 331 (yield losses; **56**, 1099).

Aylor, D. E. et al. 1974 & 1976. *Phytopathology* **64**: 1136; **66**: 537 (liberation of conidia; **54**, 1243; **56**, 66).

——. 1975. *Pl. Physiol.* **55**: 99 (force needed to detach conidia; **54**, 2798).

Beniwal, S. P. S. et al. 1974. *Phytopathology* **64**: 1197 (effect of maize dwarf mosaic virus; **54**, 2212).

Boothroyd, C. W. 1971. *Ibid* **61**: 747 (in seed; **51**, 310).

Burton, C. L. 1968. *Pl. Dis. Reptr* **52**: 847 (as market disorder; **48**, 451).

Calvert, O. H. et al. 1973. *Phytopathology* **63**: 769 (ear rot by race T; **53**, 927).

Comstock, J. C. et al. 1974. *Pl. Dis. Reptr* **58**: 104 (fungicides; **53**, 2590).

Contreras, M. R. et al. 1975. *Phytopathology* **65**: 1075 (histology & effects on position of epidermal nuclei; **55**, 3154).

Craig, J. et al. 1968. *Pl. Dis. Reptr* **52**: 134 (chlorotic lesion resistance; **48**, 142).

—— ——. 1969. *Ibid* **53**: 742 (inheritance of chlorotic lesion resistance; **49**, 463).

Crosier, W. F. et al. 1971. *Phytopathology* **61**: 427 (in seed; **50**, 3690).

Drechsler, C. 1925. *J. agric. Res.* **31**: 701 (symptoms, morphology & taxonomy; 5: 293).

Fatima, R. et al. 1974. *Seed Sci. Technol.* **2**: 371 (on seed of crops other than maize; **54**, 2796).

Fukuki, K. A. et al. 1972. *Phytopathology* **62**: 676 (temp. & light effects on races T & O; **52**, 704).

—— ——. 1973. *Mycologia* **65**: 705 (perithecial formation; **53**, 506).

Gillespie, T. J. 1972. *Can. J. Pl. Sci.* **52**: 671 (an index for disease activity; **52**, 706).

Gregory, L. V. et al. 1978. *Phytopathology* **68**: 517 (predicting yield losses).

Hill, J. P. et al. 1976. *Ibid* **66**: 873 (ecological forms of race T; **56**, 1583).

Hyre, R. A. 1970. *Pl. Dis. Reptr* **54**: 1131 (weather factors; **50**, 2257l).

——. 1973. *Ibid* **57**: 627 (effect of RH & air speed on conidial release; **53**, 2538).

——. 1974. *Ibid* **58**: 297 (effect of RH on conidial sporulation; 53, 3937).

Karr, A. L. et al. 1974. *Pl. Physiol.* **53**: 250 (4 toxins of race T; **53**, 4797).

Karr, D. B. et al. 1975. *Ibid* **55**: 727 (colorimetric determination of toxins & their appearance; **54**, 4468).

Kommedahl, T. et al. 1973. *Phytopathology* **63**: 138 (effect of temp. & fungicides on survival of maize from infected kernels; **52**, 3670).

Kovács, I. 1972. *Növenytermelés* **21**: 81 (review, race T epidemic in USA, 54 ref.; **51**, 3314).

Kulik, M. M. 1971. *Proc. Assoc. off. Seed Anal.* **61**: 123 (on seed; **52**, 2587).

Kumar, S. et al. 1976. *Sci. Cult.* **42**: 533 (fungicides, with *Setosphaeria turcica*; **56**, 4028).

Larsen, P. O. et al. 1973. *Pl. Dis. Reptr* **57**: 76 (effect of RH duration on lesion formation; **52**, 2924).

Laughnan, J. R. et al. 1973. *Crop Sci.* **13**: 681 (reaction of maize pollen to toxins; **55**, 4680).

Leonard, K. J. 1972. *Mycologia* **64**: 437 (formation of protothecia; **51**, 3824).

——. 1973. *Phytopathology* **63**: 112 (association of mating type & virulence; origin of race T population in USA; **52**, 3669).

——. 1977. *Ibid* **67**: 1273 (virulence, temp. opt., competitive ability of races; **57**, 4445).

Lim, S. M. et al. 1972. *Pl. Dis. Reptr* **56**: 805 (preliminary characterisation of toxins; **52**, 2590).

—— ——. 1973. *Ibid* **57**: 344 (seed treatment with fungicides; **52**, 3672).

—— ——. 1974. *Crop Sci.* **14**: 190 (effect of race T on grain yield; **55**, 2689).

Massie, L. B. et al. 1973. *Pl. Dis. Reptr* **57**: 730 (regression equations for predicting conidial sporulation in race T; **53**, 1346).

Morgan, O. D. 1971. *Ibid* **55**: 755 (sclerotia on seed; **51**, 1427).

Cochliobolus lunatus

Nelson, R. R. 1957. *Phytopathology* **47**: 191, 742 (heterothallism, compatibility; **36**: 583; **37**: 281).
——. 1959. *Mycologia* **51**: 18, 24, 132 (genetics of sexual process; **39**: 395).
—— et al. 1968. *Pl. Dis. Reptr* **52**: 879 (host range; **48**, 414).
—— ——. 1969. *Can. J. Bot.* **47**: 1311 (genes for pathogenicity; **49**, 69).
—— ——. 1970. *Pl. Dis. Reptr* **54**: 1123 (occurrence & distribution of race T; **50**, 2257i).
—— ——. 1971. *Ibid* **55**: 495 (distribution, race frequency, virulence & mating type in N.E. USA; **51**, 309).
——. 1973. *Ibid* **57**: 18 (world distribution of race T; **53**, 499).
—— ——. 1973. *Ibid* **57**: 145, 304, 971 (influence of climate on colonisation & sporulation, cross protection by race O against race T; **52**, 3307, 3673, 2160).
Nisikado, Y. et al. 1926. *Agric. Stud.* **8**: 56 pp. (temp & comparison with *Setosphaeria turcica*; **5**: 734).
—— ——. 1926. *Ber. Ohara Inst. landw. Forsch.* **3**: 221 (as above; **6**: 157).
——. 1927. *Ibid* **3**: 349 (as above; **6**: 547).
Orillo, F. T. 1953. *Philipp. Agric.* **36**: 327 (general; **34**: 32).
Payne, G. A. et al. 1978. *Phytopathology* **68**: 331, 707 (effect of maize nuclear genome on sensitivity to race T & its toxin, toxins formed by isolates of race T).
Prakash, O. et al. 1976. *Pl. Dis. Reptr* **60**: 355 (on *Chloris gayana*; **56**, 1168).
Pritchard, D. W. 1974. *Phytopathology* **64**: 757 (eradication from seed; **54**, 137).
Schenck, N. C. 1970. *Pl. Dis. Reptr* **54**: 1127 (perfect state in the field; **50**, 2257j).
—— et al. 1974. *Phytopathology* **64**: 619 (effect of temp. & moisture on disease development, & fungicides; **53**, 4415).
Singh, D. V. et al. 1974. *Seed Sci. Technol.* **2**: 349 (seed health testing; **54**, 2794).
Smith, D. R. 1970. *Pl. Dis. Reptr* **54**: 819 (2 races; **50**, 1762).
—— et al. 1973. *Crop Sci.* **13**: 330 (monogenic chlorotic lesion resistance; **55**, 4682).
——. 1975. *Phytopathology* **65**: 1160 (as above; **55**, 2691).
Sumner, D. R. et al. 1974. *Ibid* **64**: 168 (effects of tillage, planting date, inoculum & mixed populations on epidemiology; **53**, 3466).
Tatum, L. A. 1971. *Science N.Y.* **171**: 1113 (epidemic in USA; **51**, 305).
Ullstrup, A. J. 1970a & b. *Pl. Dis. Reptr* **54**: 1100, 1103 (history & hosts; **50**, 2257a & b).
——. 1971. *Ibid* **55**: 563 (overwintering of race T; **51**, 1423).
——. 1972. *A. Rev. Phytopathol.* **10**: 37 (epidemic in USA, 86 ref.).
Villareal, R. L. et al. 1965. *Philipp. Agric.* **49**: 294 (susceptibility & male sterile cytoplasm; **46**, 113).
Waggoner, P. E. et al. 1972. *Bull. Conn. agric. Exp. Stn* 729, 84 pp. (epimay, a simulator of the disease; **53**, 135).
——. 1973. *Phytopathology* **63**: 1252 (wind dispersal of conidia; **53**, 2161).
Wallin, J. R. et al. 1977. *Ibid* **67**: 1370 (temp., RH & sporulation of race T; **57**, 3443).
Warren, H. L. 1975. *Ibid* **65**: 623 (temp. effects on lesion development & sporulation; **55**, 198).
—— et al. 1977. *Mycologia* **69**: 773 (morphological & physiological differences between races O & T; **57**, 1161).
Wheeler, H. et al. 1971. *Pl. Dis. Reptr* **55**: 667 (phytotoxin; **51**, 1425).
——. 1977. *Physiol. Pl. Pathol.* **11**: 171 (fine structure of penetration; **57**, 1163).
White, D. G. et al. 1971. *Pl. Dis. Reptr* **55**: 382 (in seed; **50**, 3692).
Yoder, O. C. et al. 1975. *Phytopathology* **65**: 273 (segregation of pathogenicity & phytotoxins from crosses of race T & O; **54**, 3876).
Yu, T. F. 1933. *Sinensia Nanking* **3**: 273 (general; **13**: 366).

Cochliobolus lunatus Nelson & Haasis, *Mycologia* **56**: 316, 1964.

stat. conid. *Curvularia lunata* (Wakker) Boedijn, 1933

Acrothecium lunatum Wakker, 1989.

PERITHECIA black, ellipsoidal to globose, 410–700 μ in height, 300–530 μ diam; well defined ostiolate beak, subconical to paraboloid, 210–560 μ in height, 70–150 μ diam. A mass of hyaline cells usually cover the apex of the beak and frequently exude, coiled and like ribbons from the apex. Developing frequently from columnar stromata; locule with hyaline filamentous pseudoparaphyses. ASCI cylindrical or clavate, straight or slightly curved, short stipitate; 160–300 (180) × 10–20 (15) μ. ASCOSPORES 1–8, completely coiled to a nearly straight and parallel condition; coiling frequently evident apically with little or no coiling at the base. ASCOSPORES filiform or flagelliform and somewhat tapered at the ends, hyaline, 6–15 septate, 130–270 (190) × 3.8–6.5 (5) μ; heterothallic. COLONIES on natural substrata effused brown, blackish brown or black, hairy; on PDA greyish brown, dark blackish brown or black, usually velvety, sometimes floccose. MYCELIUM on natural substrata mostly immersed, of branched, septate, subhyaline to pale brown, smooth walled, 1.5–6 μ thick hyphae, sometimes up to 10 μ at point of origin of conidiophores; hyphae 1.5–4.5 μ thick in culture. STROMATA very rarely found on rice grains. CONIDIOPHORES arising singly or in groups, terminally and laterally on the hyphae, simple or branched, straight or flexuous, sometimes geniculate, pale brown, reddish brown or dark reddish brown, often paler near the apex, smooth walled, septate, on natural substrata up to 650 μ long, swollen at the base to 10–15 μ, 5–9 μ thick just above the basal swelling, 3–5 μ at the apex; in culture up to 400 μ long; 3–7 μ thick. CONIDIA acropleurogenous, 3 septate, almost always curved at the third cell from the base which is usually larger and often darker than the others, cell at each end subhyaline or pale brown, intermediate cells brown or dark brown, smooth walled, on natural substrata 20–32 (24.5) μ long, 9–15 (11.8) μ thick in broadest part, in culture 18–30 (22.4) 9–14 (11) μ. Common and widespread on many different substrata (perfect state: after R. R. Nelson & F. A. Haasis l.c., stat. conid.: M. B. Ellis, *Mycol. Pap.* 106: 33, 1966; and see Ellis et al.).

The fungus was also described by Lam-Quang-Bach who gave a tabular comparison with 4 other *Curvularia* (q.v.) spp. It is at most weakly pathogenic in the tropics. Often associated with black kernel of rice and discolourations in other cereal grains, for example, *Pennisetum typhoides* (bulrush millet). Leaf spots of sugarcane, *Cyamopsis tetragonoloba* (cluster bean) and *Rauvolfia serpentina* (medicinal) have been described. Byther et al. (*Drechslera rostrata* q.v.) reported on the role of *Cochliobolus lunatus* in causing a seedling blight of sugarcane. *C. lunatus* and *D. rostrata* were more pathogenic than *D. hawaiiensis* (*Drechslera* q.v.) and *Curvularia senegalensis* (Speg.) Subram. (*Curvularia* q.v.). Fungicides gave control. Such control was described for cluster bean and sorghum. *C. lunatus* may occur on non-cereal seed (e.g. Gadage et al.). Khurana et al. reported a synergistic relationship with *Meloidogyne javanica* in sugarcane seedling blight.

Agnihotri, B. S. et al. 1962. *Indian J. Sugarcane Res. Dev.* 7: 36 (culture, hosts; **43**, 221).

Chand, J. N. et al. 1968. *Indian Phytopathol.* **21**: 239 (on cluster bean; **48**, 1407).

Ellis, M. B. et al. 1975. *C.M.I. Descr. pathog. Fungi Bact.* 474 (*Curvularia lunata*).

Gadage, N. B. et al. 1977. *J. Maharashtra agric. Sci.* **2**: 239 (on cotton).

Ji, T. et al. 1975. *Indian Phytopathol.* **28**: 384 (fungicides on sorghum; **56**, 2482).

Khurana, S. M. P. et al. 1971. *Ann. phytopathol. Soc. Japan* **37**: 313 (with*Meloidogyne javanica* on sugarcane; **51**, 2828).

Lam-Quang-Bach. 1964. *Fich. Phytopathol. Soc.* 15, 7 pp. Suppl. to *Rev. Mycol.* **29**(4) (general, control on rice & *Gladiolus*; **44**, 1478).

Macri, F. et al. 1976. *Physiol. Pl. Pathol.* **8**: 325 (toxin; **56**, 61).

Martin, A. L. et al. 1940. *Bull. Tex. agric. Exp. Stn* 584, 14 pp. (black kernel on rice; **19**: 492).

Mathur, R. L. et al. 1960. *Proc. natn. Acad. Sci. India* Sect. B **30**: 323 (on bulrush millet; **41**, 225).

Singh, G. 1974. *Indian Phytopathol.* **27**: 234 (fungicides on cluster bean; **55**, 491).

Singh, S. 1969. *Sci. Cult.* **35**: 396 (on sugarcane; **49**, 2169).

Varadarajan, P. D. 1966. *Indian Phytopathol.* **19**: 298 (on *Rauvolfia serpentina*; **46**, 3526).

Vianello, A. et al. 1976. *Can. J. Bot.* **54**: 2918 (toxin; **56**, 3002).

Cochliobolus miyabeanus (Ito & Kuribayashi) Drechsler ex Dastur, *Indian J. agric. Sci.* **12**: 733, 1924.

stat. conid. *Drechslera oryzae* (Breda de Haan) Subram. & Jain, 1966
Helminthosporium oryzae Breda de Haan, 1900.

PERITHECIA globose to sunken globose; outer wall dark yellowish brown, pseudoparenchymatous; well-defined, smooth ostiole with rounded apex. Ostiolar beak 95–190 × 55–95 μ, remainder of perithecium 463–763 × 368–777 μ. ASCI cylindrical to long fusiform, 21–36 × 142–235 μ with 1–8 (mostly 4–6) ascospores. ASCOSPORES filamentous or long cylindrical, hyaline to light olive, coiled in a close helix with mucous envelope, 6–15 (mostly 9–12) septate, 6–9 μ × 250–469 μ. CONIDIOPHORES solitary or in small groups, straight or flexuous, sometimes geniculate, pale to mid brown or olivaceous brown, up to 600 μ long, 4–8 μ thick. CONIDIA usually curved, navicular, fusiform or obclavate, occasionally cylindrical and almost straight, pale to mid-golden brown, smooth, 6–14 pseudoseptate, 14–22 (17) × 63–153 (109) μ; hilum characteristic, minute, often protruding slightly, papillate. Perithecia formed in culture, opt. temp. *c.* 25°. Conidia formed on glumes and in culture on wheat straw and tap water agar under near UV light are usually larger and darker than those formed on leaf spots (P. Holliday & M. B. Ellis, *CMI Descr.* 302, 1971, as stat.

conid.; for perfect state see Ito et al.; Tsuda et al.; Ueyama et al.; 2 mating types occur).

The fungus occurs on rice (and other *Oryza* spp.; for example, see Dath et al.; Subramanian et al.) causing brown spot and seedling blight. An important disease which is fully described by Ou in *Diseases of rice* (q.v. for a bibliography). Although it has been said that brown spot is not generally a problem on the normal soils of S.E. Asia serious losses have occurred, and the disease was one of the factors in the Bengal famine of 1942–43 (Padmanabhan). The disease is widespread (Map 92). The characteristic, oval leaf spots are up to 1 cm long, at first usually brown, sometimes purplish, later forming white to grey centres; spots may coalesce and leaves wither; glumes may be spotted, becoming velvety with sporulation. Infected seed is shrivelled and discoloured; coleoptiles bear lesions which can also occur on the roots of seedlings. The disease is often particularly serious in the seedbed. The kernels are invaded and those in the flowering and milk stages are more susceptible than in the soft dough and mature stages. Mycelium can be found within the seed (where it can survive 3–4 years) including the endosperm. Glumes are directly penetrated; in leaves penetration is virtually always cuticular, through the motor or bulliform cells. Severe infection can take place before emergence.

The opt. temp. for conidial germination, infection and growth in vitro is 25–30°. At these temps infection occurs in *c*. 4 hours, at 35° in 6 hours and at 20–22° in 8–10 hours. Plants are most susceptible at flowering. Cloudy conditions with light rain, heavy dew, temps of 20–29° for several days can lead to severe outbreaks of rice brown spot. In Japan the disease can occur particularly on poorly drained (akiochi) fields or in those where nutrient conditions are unbalanced. It becomes severe when roots develop under adverse conditions. Other diseases may also affect severity. Some host resistance depends on the cuticle thickness and the degree to which the epidermal cells become silicified. Those with a higher degree of silicification being more resistant. Differentiation into races is not apparently clear cut; single conidial isolates show pathogenic variation. However, races in India and Japan have been described. Some recent ref. to fine structure, phytotoxin and the biochemical aspects of resistance are: *Rev. Pl. Pathol.* **51**, 1468; **52**, 405; **53**, 1374; **54**, 2811, 3289, 3909; **55**, 5183, 5186; **56**, 1597; **57**, 3464, 3465, 3960.

In control healthy seed is very important since seed infection is probably the source for most of the primary infection in crops (see Aulakh et al.; Kulik; for seed testing). Water and soil (physical and nutrient) conditions must be good. The more resistant cvs should be used. Hot water treatment of seed is possibly more effective than fungicides (e.g. Heaton). Chakrabarti described some recent work with fungicides (mainly only the more recent ref. are given).

Aulakh, K. S. et al. 1974. *Seed Sci. Technol.* **2**: 385 (testing seedborne infection; **54**, 1263).

Baldacci, E. 1948. *Ann. Acad. Agric. Torino* 90, 26 pp. (general; **28**: 239).

Bedair, F. A. et al. 1976. *Egyptian J. Genet. Cytol.* **5**: 433 (inheritance of resistance; **56**, 2061).

Chakrabarti, N. K. et al. 1975. *Pl. Dis. Reptr* **59**: 1028 (fungicides; **55**, 4096).

Chattopadhyay, S. B. et al. 1958. *Indian Phytopathol.* **10**: 130 (cuticle thickness & Si in resistance; **37**: 474).

Dath, A. P. et al. 1973. *Sci. Cult.* **39**: 394 (wild rice hosts; **53**, 3014).

Heaton, J. B. 1969. *J. Aust. Inst. agric. Sci.* **35**: 103 (general; conidial diurnal periodicity, seed treatment; **48**, 3486).

Imam Fazli, S. F. et al. 1966. *Phytopathology* **56**, 507, 1003 (kernel infection & its effect on yield; **45**, 2834; **46**, 317).

Inoue, Y. et al. 1966. *Bull. Tokai-Kinki natn. agric. Exp. Stn* 15: 179 (effect of soil conditions, akiochi fields, other diseases & control; **46**, 1571).

Ito, S. et al. 1927. *Ann. phytopathol. Soc. Japan* **2**(1): 1 (perfect state; **7**: 54).

Kulik, M. M. 1977. *Phytopathology* **67**: 1303 (seed tests for predicting field emergence, with *Alternaria padwickii*; **57**, 4458).

Lam, T. H. et al. 1973. *Phytopathol. Z.* **76**: 42 (viability & infectivity of hyphal fragments; **52**, 3688).

Leach, C. M. 1961. *Can. J. Bot.* **39**: 705 (UV light & conidial sporulation; **41**: 14).

Lindberg, G. G. 1971. *Phytopathology* **61**: 420 (toxin production by diseased *C. miyabeanus*; **50**, 3727).

Padmanabhan, S. Y. 1973. *A. Rev. Phytopathol.* **11**: 11 (environmental factors, famine in India, Bengal, 25 ref.).

Purkayastha, R. P. et al. 1974. *Trans. Br. mycol. Soc.* **62**: 402 (factors affecting leaf colonisation; **53**, 4809).

Subramanian, C. L. et al. 1973. *Curr. Sci.* **52**: 327 (wild rice hosts; **53**, 531).

Tsuda, M. et al. 1975 & 1976. *Ann. phytopathol. Soc. Japan* **41**: 447; **42**: 7 (perfect state & mating types; **55**, 3173, 5185).

Ueyama, A. et al. 1975 & 1976. *Ibid* **41**: 434; **42**: 1 (as above; **55**, 3172; 5184).

Watanabe, Y. et al. 1976. *Bull. Tokai-Kinki natn. agric. Exp. Stn* 29: 80 (ecology of panicle infection; **56**, 4039).
Yamamoto, M. et al. 1972. *Bull. Fac. Agric. Shimane Univ.* 6: 6 (pathogenicity to weeds in rice; **55**, 1791).
Yamamura, S. et al. 1978. *Can. J. Bot.* **56**: 206 (myochrome system & conidial development in a non-photoinduced isolate; **57**, 3793).

Cochliobolus nodulosus Luttrell, *Phytopathology* **47**: 547, 1957.
stat. conid. *Drechslera nodulosa* (Berk & Curt.) Subram. & Jain, 1966
Helminthosporium nodulosum Berk. & Curt. apud Sacc., 1886
H. leucostylum Drechsler, 1923.

PERITHECIA superficial, black, spherical, 276–414 μ diam., with long cylindrical beaks, 97–262 × 55–83 μ. ASCI 1–8 spored, cylindrical, straight, rounded at the apex, short stipulate, 120–193 × 14–17 μ, interspersed with pseudoparaphyses. ASCOSPORES hyaline, filiform, tapering slightly toward either end, spirally coiled within the ascus, 90–193 × 3.9–7.8 μ, 3–11 septate, more or less deeply constricted at the septa in the middle portion. CONIDIOPHORES sometimes solitary but frequently in quite large groups, often branched, flexuous, markedly geniculate with large, mid to dark brown scars, very pale straw coloured at first, becoming darker later, up to 150 μ long in the inflorescences but usually much shorter on leaves, 5–9 μ thick. CONIDIA straight, ovoid to obclavate, pale to mid-dark golden brown, 5–7 pseudoseptate, 43–78 (mostly 50–65) μ long, 16–19 μ thick in the broadest part; hilum 3–4 μ wide, usually with a hyaline zone just above it and then a dark zone (perfect state: from E. S. Luttrell (l.c.); stat. conid.: M. B. Ellis & P. Holliday, *CMI Descr.* 341, 1972).

Seedling blight, foot rot and leaf blight of finger millet (*Eleusine coracana* and *E. indica*) is widespread in E. and S. Africa; it has also been reported from India, Malaysia (Sarawak), Nigeria, Papua New Guinea, Philippines, Sierra Leone, Sri Lanka, Taiwan and USA (Map 454). Elliptical to long rectangular spots form on the leaves, becoming coalescent, straw coloured with narrow, red brown margins. Spikes are infected and may collapse; grain does not form or is undeveloped. Systemic infection has been described for *E. indica*. Many infected plants die in the seedling stage but others show the stunting caused by this type of infection. When older plants were inoculated systemic infection occurred in the new growth only, and such plants may show some tillers to be infected and stunted whilst others are normal and healthy.

C. nodulosus, which is heterothallic, causes a minor disease and also infects: *Echinochloa frumentacea, Panicum miliaceum, Pennisetum typhoides*, sugarcane, sorghum and maize. The perfect state has been reported on *Cassia siamea*. Besides infection of the seedlings, leaves are penetrated through the stomata and directly; most infection occurring at 30–32°. Growth in vitro is best at 30°. No races have been described but inoculation of *Eleusine coracana*, which is infected in India, was unsuccessful in USA. Resistant cvs are reported from India and seed treatment with ceresan gives some control.

Govindu, H. C. et al. 1970. *Proc. 1st. Int. Symp. Pl. Pathol., New Delhi* 1966–67: 415 (testing cvs for resistance).
Hegde, R. K. et al. 1969. *Indian Phytopathol.* **22**: 423 (behaviour in vitro; **50**, 1207).
Luttrell, E. S. 1957. *Phytopathology* **47**: 540 (systemic infection, symptoms, inoculation, possible races & taxonomy; **37**: 78).
Mitra, M. et al. 1934. *Indian J. agric. Sci.* **4**: 943 (morphology, infection, symptoms & temp.; **14**: 439).
Vidhyasekaran, P. 1971. *Indian Phytopathol.* **24**: 347 (loss of conidial sporulation in vitro; **51**, 3802).
Reddy, S. M. et al. 1970. *Sci. Cult.* **36**: 598 (perfect state on *Cassia siamea*; **51**, 408).

Cochliobolus setariae (Ito & Kuribayashi) Drechsler ex Dastur, *Indian J. agric. Sci.* **12**: 733, 1942.
Helminthosporium setariae Saw., 1912
stat. conid. *Drechslera setariae* (Saw.) Subram. & Jain, 1966.

PERITHECIA dark brownish, pseudoparenchymatous, flask shaped with ostiolar beak; bodies globose or short ellipsoidal, 240–500 × 220–315 μ; beaks well developed, paraboloid or cylindrical, 60–125 × 50–110 μ. ASCI numerous, fusiform, straight or slightly curved, widest somewhat below the middle, rounded at the apex, short stipulate at the base, hyaline, thin walled, 130–150 × 22–32 μ, with 1–8 ascospores. ASCOSPORES filiform, wavy, obtusely pointed at both ends, 5–9 septate, coiled in a close helix, 200–315 × 6–7 μ, hyaline or light olive. CONIDIOPHORES solitary or in small groups, straight or flexuous, sometimes geniculate, pale to mid brown or olivaceous brown, up to 200 μ long, 5–9 μ thick, sometimes swollen at the base to 11 μ. CONIDIA slightly curved or sometimes straight, fusiform or navicular, pale to mid-golden brown, smooth, 5–10 pseudoseptate, 45–100 (mostly 50–70) × 10–15 (mostly 12–14); (perfect state: from R. Sprague, *Diseases of cereals and grasses in North America*: 382, 1950; stat. conid.: from M. B. Ellis, *Demat. Hyphom.*: 443, 1971, and see Ellis et al.).

C. setariae (probably fairly widespread) has been mostly recorded on *Setaria italica* (foxtail millet) on which it causes a seedling blight and leaf spot. Some other hosts (by inoculation) are given by Sprague (l.c.). In USA it causes diseases on *Pennisetum typhoides* (bulrush millet) and *Panicum fasciculatum* (browntop millet). Further descriptions of the fungus were given by Ito and by Nisikado (*Drechslera* q.v.). Inoculation of plants of bulrush millet (1–13 weeks old) showed the disease to be most severe at 1 week. It became progressively less so up to 6–7 weeks and then increased in severity. More disease occurred at 21.2° and 15.6°, least at 37.8° and 32.2°; at 26.7° it was intermediate. A similar effect of age was reported for browntop millet. Resistance in young and old leaves increased from the seedling stage until the plants were 6–7 weeks old. It then decreased as heading began after 7 weeks. Under a long day regime (16 hours light) plants did not head and this second susceptible stage did not occur. On young plants leaf lesions are irregular buff or pale tan, usually 1–11 × 1–3 mm. At 8–9 weeks the lesions are brown flecks and stripes, 3–15 × 2–4 mm, often surrounded by chlorotic areas. At 4–7 weeks the lesions are smaller. The fungus is seedborne and seed treatment may be needed; organo-mercurials (e.g. agrasan) are effective. Some differences in disease reaction between cvs of foxtail millet have been described from India.

Carranza, J. M. 1967. *Revta Fac. Agron. Univ. nac. La Plata* **43**: 241 (general on foxtail millet inter alia; **48**, 1603).

Ellis, M. B. et al. 1975. *C.M.I. Descr. pathog. Fungi Bact.* 473 (*Drechslera setariae*).

Grewal, J. S. et al. 1965. *Indian Phytopathol.* **18**: 123 (on seed of foxtail millet inter alia; **45**, 1050).

Luttrell, E. S. et al. 1974. *Phytopathology* **64**: 476 (on browntop millet, effects of host age & photoperiod; **53**, 3952).

Mathur, S. B. et al. 1967. *Proc. int. Seed Test. Assoc.* **32**: 633 (on seed of foxtail millet inter alia; **47**, 2154).

Vidhyasekaran, P. et al. 1970. *Madras agric. J.* **57**: 455 (growth in vitro; **50**, 1771).

Wells, H. D. et al. 1967. *Crop Sci.* **7**: 621 (on bulrush millet, effect of host age; **48**, 2957).

——. 1967. *Phytopathology* **57**: 1002 (on bulrush millet, effect of temp. in vivo; **47**, 165).

Cochliobolus spicifer

Cochliobolus spicifer Nelson, *Mycologia* **56**: 198 1964 (as *C. spiciferus*).
stat. conid. *Drechslera* sp.
Brachycladium spiciferum Bain., 1908
Curvularia spicifera (Bain.) Boedijn, 1933
Helminthosporium spiciferum (Bain.) Nicot, 1953.

Mature PERITHECIA black, ellipsoid to globose, 350–650 μ diam., 460–710 μ high; well-defined ostiolate beak, subconical to paraboloid, 60–150 μ diam., 280–670 μ high. A mass of hyaline cells usually covers the apex of the beak of mature perithecia and frequently exudes in a coiled, ribbon-like way from the apex. ASCI cylindrical or clavate, straight or slightly curved, short stipitate, 130–260 (av. 188) × 12–20 (av. 16) μ, with 1–8 ascospores which are completely coiled in a close helix or varying to a nearly straight and parallel condition. ASCOSPORES filiform or flagelliform, somewhat tapered at the ends; typically hyaline, 6–16 septate, 135–240 (av. 182) × 3.8–7 (av. 5.3) μ. CONIDIOPHORES solitary or in small groups, flexuous, repeatedly geniculate with numerous well-defined scars, often torsive, mid to dark brown, up to 300 μ long or sometimes longer, 4–9 μ thick. CONIDIA straight, oblong or cylindrical, rounded at the ends, when mature golden brown except for a small area just above the dark scar which remains hyaline or very pale, smooth, constantly 3 pseudoseptate, 20–40 × 9–14 (mostly 30–36 × 11–13) μ; hilum 2–3 μ wide (perfect state after R. R. Nelson l.c., stat. conid. and synonymy from M. B. Ellis *Demat. Hyphom.*: 415, 1971).

The fungus is common, occurring on many plants, and widespread, mostly in tropical and subtropical regions. It can be associated with seeds, reducing germination and emergence; and cause leaf spots and root rots. The diseases may occasionally be severe enough to warrant treatment with fungicides. Examples of infection are given for: *Cynodon dactylon* (Bermuda grass), cotton, eggplant, sugarcane, tobacco and *Vicia faba* (broad bean). On the first host *C. spicifer* is associated with the condition known as spring dead spot (Wadsworth et al.).

Bedi, P. S. et al. 1969. *Jnl Res. Ludhiana* **6**: 647 (on cotton, fungicides; **50**, 688).
Chauhan, M. S. et al. 1973. *Indian J. Mycol. Pl. Pathol.* 3: 55, 169 (factors affecting disease on tobacco; **53**, 4875; **54**, 3475).
El Nur, E. et al. 1970. *Pl. Dis. Reptr* **54**: 405 (on broad bean; **49**, 3052).
Freeman, T. E. 1970. *Ibid* **54**: 358 (on seed of Bermuda grass; **49**, 3352).
Gudauskas, R. T. 1962. *Ibid* **46**: 498 (on Bermuda grass; **42**: 28).
Narendra, D. V. et al. 1978. *Proc. Indian Acad. Sci.* Sect. B **87**: 45 (seedling blight of sugarcane).
Nelson, R. R. et al. 1977. *Mycologia* **69**: 173 (dual compatibility; **56**, 3431).
Ruppel, E. G. 1974. *Ibid* **66**: 803 (factors affecting conidial size, **54**, 2670).
Sarode, M. S. et al. 1974. *J. Mahatma Phule agric. Univ.* 5: 123 (on eggplant; **55**, 2061).
Suryanarayana, D. et al. 1968. *Jnl Res. Ludhiana* **5**: 62 (on cotton; **48**, 175).
Wadsworth, D. F. et al. 1968. *Phytopathology* **58**: 1658 (on Bermuda grass; **48**, 1775).

COLEOSPORIUM Lév., *Annls Sci. nat.* Ser. 3 8: 373, xii, 1847.
(Melampsoraceae)

SPERMOGONIA subepidermal with paraphyses and flexuous hyphae. AECIDIA foliicolous, erumpent, with prominent tongue-shaped peridia composed of a single layer of verrucose cells dehiscing irregularly; AECIDIOSPORES ellipsoid or globoid with colourless, tesselate, superficially tuberculate wall. UREDIA erumpent, pulverulent, without peridia; UREDOSPORES globoid or oblong, catenuate, in wall structure resembling the aecidiospores. TELIA subepidermal, indehiscent except through weathering, flattened, waxy, becoming gelatinous on germination; TELIOSPORES sessile, in a single layer in lateral contact, aseptate, cylindroid, clavoid or prismatic with smooth, colourless walls, thin at the sides, strongly thickened and gelatinous above, becoming divided into 4 superposed cells, all of which, in autumn, can germinate as soon as they mature, each producing a long sterigma bearing a basidiospore; BASIDIOSPORES ovoid or ellipsoid, rather large, thin walled. Predominantly heteroecious (2 microforms on *Pinus* in N. America and E. Asia); spermogonia and aecidia on the needles of *Pinus*; uredia and telia on various families of Angiosperms, especially Compositae (from M. Wilson & D. M. Henderson, *British rust fungi*: 1, 1966).

Wilson et al. (l.c.) discussed the taxonomy of the genus; they fully described *C. tussilaginis* (Pers.) Berk. and its race or race groups. These needle rusts of pines are widespread in the N. Hemisphere. They are one of the pathogen groups that could be introduced to the tropics and S. Hemisphere because of the introduction of conifers from the north. One of them is *C. ipomoeae* (Schw.) Burrill (see Laundon et al.), the orange rust of sweet potato. Although this rust has been known on sweet potato in E. and central Africa for some years it has not yet been found on exotic susceptible pines in the region (Gibson et al.). *C. ipomoeae* (of no economic importance on sweet potato) occurs in S.E. USA, Bermuda, Mexico, W. Indies, S. America, Indonesia (Java), Malawi, South Africa, Tanzania and Uganda (Map 484). Browne (*Pests and diseases of forest plantation trees*) gave some account of: *C. asterum* (Diet.) Syd., *C. barclayense* Bagchee, *C. helianthi* (Schw.) Arth., *C. pinicola* (Arth.) Arth. and *C. tussilaginis*.

Gibson, I. A. S. et al. 1972. *PANS* **18**: 336 (notes from ODA plant pathology liaison officers; **52**, 662).
Hiratsuka, N. et al. 1975. *Rep. Tottori mycol. Inst.* 12: 1 (surface structure of *Coleosporium* spores; **55**, 2150).
Laundon, G. F. et al. 1971. *C.M.I. Descr. pathog. Fungi Bact.* 282 (*C. ipomoeae*).

COLLETOTRICHUM Cda. (as *Colletothricum*) in Sturm, *Deutschlands Flora* 3: 41, 1831–32.
(synonymy in B. C. Sutton, *Mycol. Pap.* 141: 44, 1977)
(Melanconiaceae)

Immersed mycelium branched, septate, hyaline to pale brown, intracellular. ACERVULI subcuticular, epidermal, subepidermal, separate or confluent, formed of hyaline or brown, thin or thick-walled pseudoparenchyma. SETAE present or absent, originating irregularly from the pseudoparenchyma, more or less straight, unbranched, tapered to an acute or obtuse apex, brown, smooth, thick walled, septate. CONIDIOPHORES septate, branched at the base, hyaline to pale brown, smooth, formed from upper cells of pseudoparenchyma. CONIDIOGENOUS CELLS enteroblastic, phialidic, discrete or incorporated on conidiophores, determinate, cylindrical, hyaline to pale brown, smooth, channel narrow, periclinal wall thickened, collarette sometimes distinct. CONIDIA hyaline, aseptate, more or less guttulate, cylindrical, long clavate, falcate, fusiform or muticate; on germination become pale brown, septate and form APPRESSORIA (B. C. Sutton, personal communication, September, 1974).

The genus, its spp. and synonymy were described by Arx; see also Duke; Edgerton; Hemmi; Stoneman; Wollenweber et al. In a revision of the fungi placed in *Gloeosporium* Desm. & Mont. (auct. non; type sp. a *Marssonina* Magn.), a heterogenous assemblage, nearly all the 735 spp. were redisposed, many to *Colletotrichum* (Arx 1957c, 1970). The presence or absence of setae to distinguish these 2 genera was rejected since the formation of these structures depends on external conditions (Frost). Duke pointed out that there was no essential difference between *Colletotrichum* and *Vermicularia* Tode ex. Fr. Arx reduced many of the described *Colletotrichum* spp. to synonymy. Sutton (*C. dematium* q.v.), who considered some taxonomic and morphological aspects of the genus, stated that the studies of Arx (1975a) were apparently done without adequate examination or citation of original material, and probably with little investigation of the biology of the spp. which were being considered. Ten spp. are fully described here, 2 under their perfect states in *Glomerella*. Because of the still confused taxonomy no attempt has been made to describe many other minor spp. which may or may not be accepted as distinct. Some are referred to below; they cause generally lesser diseases and their taxonomic status is uncertain.

The forms of *Colletotrichum* have a very wide range of behavioural patterns in nature; varying from saprophytes to specialised parasitic strs with a narrow host range. The conidia are produced in mucilagenous masses, often pinkish and fairly conspicuous on the typically sunken, irregularly outlined necrotic lesions (called anthracnose, see Jenkins) on fruits, leaves and stems. Sometimes lesions on fruits become raised growths; cankerous, scab-, scar- or wart-like in appearance. Some spp. cause latent, macroscopically invisible infection on fruits (see Verhoeff); these may develop into the characteristic anthracnose lesions during the ripening phase. Such lesions (particularly studied in *C. musae* q.v.; and see Simmonds) arise from direct penetration of the cuticle from the appressoria characteristic of the genus, and likened to chlamydospores by some. But lesions can also arise directly, i.e. without the intervening latent stage, when infection takes place through a wound in the host tissue. Branch dieback is frequently caused but root infection is less common. The spp. generally survive for long periods on plant debris in or on the soil. The commonest pathogen in the tropics is *C. gloeosporioides* (*G. cingulata* q.v.) and under this sp. some differences between it and some other *Colletotrichum* spp. are described very briefly

C. acutatum Simmonds was described from Australia. The diagnostic features of the conidia are the small size (11.1 × 3.1 μ, mostly 8.3–14.4 × 2.5–4 μ), the variability in length in the one collection, and the pointed ends. On PDA the cultural type is relatively constant; appressoria are rarely lobed and the germ pore is often poorly defined. Growth in vitro is opt. at 25–26.5° with a mean 4-day growth rate of 3.8 cm compared with 6.3 cm for *G. cingulata* and 5.1 cm for *C. dematium* (q.v.). *C. acutatum* infected fruits of papaw, apple, avocado pear, banana and mango when wound inoculated. Dingley et al. reported other hosts when investigating terminal crook disease of *Pinus* spp. (Gibson et al.; Gilmour). The fungus infects nursery pine plants (3–30 cm high); a crook develops at the end of the terminal shoot which becomes necrotic; the stem is rigid and thickened since elongation growth is retarded whilst growth in stem diam. is normal. The disease may spread from the tip of the leader to the ends of the small branches and a bushy stunt arises in severe cases; seedlings may recover in the nursery or in the first year after planting out (see Sang et al.). Fungicide control was described by Vanner et al.

Light leaf spot of *Brassica* spp. is commonly referred to as *Gloeosporium concentricum* (Grev.) Berk. & Br.; Thompson (supported by Arx 1957c) proposed its retention as *Cylindrosporium concentricum* Grev. and discussed the taxonomy. Hickman established cultures from a conidial suspension and these formed apothecia but a precise identification of this state was not given. Staunton et al. (1966) found an apothecial fungus, which agreed with Hickman's description, on decaying plant debris in the field. Inoculation with ascospores reproduced the disease with the typical imperfect state. Eventually, *Pyrenopeziza brassicae* Sutton & Rawlinson was described as the perfect state. The first leaf symptoms are white spots (< 1 mm diam.) between the veins and arranged concentrically in circles; as the spore masses from the acervuli become wetted a white film spreads over the leaf surface. As the lesion spreads the centre becomes dull, bleached, fawn colour and further white spots form beyond this centre. The very characteristic white conidial film forms particularly on the upper surface. On the lower surface the spots become blackened with age, and the disease appears as a series of concentrically arranged black dots often surrounded by one or more rings of white ones. Affected areas on veins and petioles become hard, brown and split longitudinally. Irregular growth of the lamina near cankered veins results in twisting and distortion of the leaves. Most growth in vitro is at 15–20°. The disease is a minor one and apart from crop hygiene no particular control measures are warranted (see Cabral; Kavanagh et al.; Smith; Staunton 1967).

Colletotrichum corchorum Ikata & Tanaka was described (no Latin diagnosis) from jute (*Corchorus capsularis*) by Ikata et al., the fungus is also described as *Colletotrichum corchori* Ikata & Yoshida. It has falcate conidia and may be a form of *C. dematium*. Jute anthracnose is severe in Bangladesh. The fungus is seedborne and can cause a seedling blight. On older plants necrotic, sunken lesions are formed on the stem, they may girdle it and cause stem break. Plants which remain standing at harvest, although infected, show shredding of the fibre at the cankerous lesions. The fibres extracted from such plants became speckled and knotted from adherent pieces of bark which resist the retting process; fibre quality is therefore poor. Islam et al. reported high resistance in lines of *Corchorus clitorius* and in the wild sp. *C. acutangulus*, whereas *C. capsularis* lines were susceptible in varying degrees. Fakir et al. found resistance in a cv. of *C. capsularis* to be inherited in an oligogenic fashion (and see Khan et al.; Purkayastha et al.).

Colletotrichum higginsianum Sacc. in Higgins was disposed as a synonym of *G. cingulata* by Arx (1957a) but may be distinct. The fungus was described as causing a leaf spot on turnip (*Brassica rapa*) and Chupp described a rot of turnip root (*Alternaria brassicae* q.v.); it infects several members of the Cruciferae. The leaf lesions appear as 1–2 mm diam. watersoaked spots which enlarge to *c.* 5 mm; the centre becomes pale greyish or straw coloured and may split. The spots are sometimes very numerous and may coalesce to form large irregular lesions. Elongate sunken spots occur on stems and petioles, and the siliquae can be attacked. The opt. temp. for growth is *c.* 28° and for conidial germination 25–30°. There is evidence for seed transmission and carry over in host debris. This turnip or brassica anthracnose has only been investigated in some detail by Scheffer who found that chloranil gave control. Another *Colletotrichum* on this host group is *C. brassicae* Schulz. & Sacc. which is reduced to synonymy with *C. dematium* by Arx (l.c.).

Anthracnose of cotton is described under *G. cingulata*. Another anthracnose of the same crop was described from India (*C. indicum* Dastur); this may be a distinct sp. Ramakrishnan considered that the fungus was the same as *C. capsici* (q.v.), Arx (l.c.) disposed it as a synonym of *C. dematium*, and Wilson considered it to be distinct. The disease has also been investigated by Bhide et al.; Ling et al.; Raynal. The fungus mostly attacks the bolls, seeds and seedlings. The watersoaked spots on the bolls develop into the typically sunken, irregular, necrotic areas; the bolls become mummified and fall. Seedlings can be killed through infection from seedborne inoculum and the fungus can occur within the seed coat. Germination of infected seed is poor. Reports of the opt. temp. for growth differ; this is *c.* 27°. Raynal (1971) in seedling inoculations reported *Gossypium hirsutum* to be more susceptible than *G. arboreum* and *G. herbaceum*. Mercurials have been recommended for control of seed infection. *C. tabacum* Boning (Map 307) can cause severe epidemics in Malawi (see Abington et al.). It generally infects tobacco at the seedling stage.

Abington, J. B. et al. 1974. *PANS* **20**: 315 (resistance to *Colletotrichum tabacum*; **54**, 2965).

Arx, J. A. von 1957a. *Phytopathol. Z.* **29**: 413 (taxonomy of *Colletotrichum*; **37**: 25).

——. 1975b. *Tijdschr. PlZiekt.* **63**: 171 (culture & infection; **37**: 26).

——. 1957c. *Verh. K. ned. Akad. Wet.* **51**, 153 pp. (revision of fungi placed in *Gloeosporium*; **37**: 26).

——. 1970. Revised edition of *Bibliotheca Mycologica* 24: 203 pp. J. Cramer (as above; **51**, 2233).

Bhide, V. P. et al. 1957. *Indian Cotton Grow. Rev.* **11**: 496 (control of seedborne infection of *C. indicum*; **38**: 145).

Blakeman, J. P. et al. 1977. *Physiol. Pl. Pathol.* **11**: 313 (stimulation of appressoria by phylloplane bacteria in *C. acutatum*; **57**, 2808).

Cabral, R. V. De G. 1940. *Broteria* Ser. trimest. **9**: 18 (*G. concentricum*; **19**: 449).

Dastur. J. F. 1934. *Indian J. agric. Sci.* **4**: 100 (*C. indicum*; **13**: 507).

Deighton, F. C. 1972. *Trans. Br. mycol. Soc.* **59**: 185 (inter alia *Didymariopsis*, synonym of *Colletotrichum*).

Dingley, J. M. et al. 1972. *N.Z. Jl For. Sci.* **2**: 192 (*C. acutatum*; **52**, 3091).

Duke, M. M. 1928. *Trans. Br. mycol. Soc.* **13**: 156 (*Colletotrichum* & *Vermicularia*; **8**: 268).

Colletotrichum

Edgerton, C. W. 1908. *Bot. Gaz.* **45**: 367 (physiology & development of some anthracnoses).
Fakir, G. A. et al. 1965. *Phytopathology* **55**: 749 (inheritance of resistance to *C. corchori*; **44**, 3373).
Frost, R. R. 1964. *Nature Lond.* **201**: 730 (setae formation; **43**, 2237).
Ghosh, T. 1957. *Indian Phytopathol.* **10**: 63 (*C. corchori*; **38**: 695).
Gibson, I. A. S. et al. 1969. *E. Afr. agric. For. J.* **35**: 135, (*C. acutatum*; **49**, 1806).
Gilmour, J. W. 1965. *Res. Leafl. For. Res. inst. (N.Z.)* 4 pp. (terminal crook of *Pinus radiata*; **45**, 1546).
Hemmi, T. 1920 & 1921. *J. Coll. Agric. Hokkaido Imp. Univ.* 9 (morphology & physiology of Japanese *Colletotrichum* spp.).
Hickman, C. J. et al. 1955. *Pl. Pathol.* **4**: 129 (*G. concentricum*; **35**: 802).
Higgins, B. B. 1917. *J. agric. Res.* **10**: 161 (*C. higginsianum*).
Ikata, S. et al. 1940. *Ann. phytopathol. Soc. Japan* **10**: 141 (*C. corchorum*; **20**: 465).
Islam, N. et al. 1964. *Trans. Br. mycol. Soc.* **47**: 227 (resistance to *C. corchori*; **44**, 146).
Jenkins, A. E. 1933. *Phytopathology* **23**: 389 (history of term anthracnose; **12**: 595).
Kavanagh, T. et al. 1965. *Irish J. agric. Res.* **4**: 112 (*G. concentricum*; **44**, 2649).
Khan, S. R. et al. 1975. *Physiol. Pl. Pathol.* **5**: 157 (role of fungal stimulant from jute as determinant of susceptibility to *C. corchori*; **54**, 3341).
——— ———. 1976. *Ibid* **9**: 273 (purification & properties of stimulant, see above; **56**, 3056).
Kulshrestha, D. D. et al. 1976. *Friesia* **11**: 116 (seedborne spp.; **56**, 5482).
Ling, L. et al. 1944. *Ann. Bot.* **8**: 91 (*C. indicum*; **24**: 148).
Messiaen, C. M. 1955. *Annls Epiphyt.* **6**: 285 (descriptions of some *Colletotrichum* spp.; **35**: 635).
Parberry, D. G. et al. 1978. *Trans. Br. mycol. Soc.* **70**: 7 (appressoria in *C. acutatum* & effect of substances associated with leaf surfaces).
Purkayastha, R. P. et al. 1975. *Physiol. Pl. Pathol.* **6**: 265 (phytoalexins & *C. corchorum*; **55**, 1274).
Ramakrishnan, T. S. 1947. *Proc. Indian Acad. Sci.* Sect. B **25**: 15 (*C. indicum* inter alia; **26**: 544).
Rawlinson, C. J. et al. 1978. *Trans. Br. mycol. Soc.* **71**: 425 (taxonomy & biology of *Pyrenopeziza brassicae* sp. nov.).
Raynal, G. 1970 & 1971. *Coton Fibr. trop.* **25**: 443; **26**: 243, 251, 395 (*C. inducum* culture, morphology, physiology & pathogenicity; **50**, 2288, 3800; **51**, 2535).
Sang, F. K. arap et al. 1973. *E. Afr. agric. For. J.* **39**: 41 (fungicides & *C. acutatum* on pine; **53**, 3208).
Scheffer, R. P. 1950. *Tech. Bull. N. Carol. agric. Exp. Stn* 92, 26 pp. (*C. higginsianum*; **30**: 132).
Simmonds, J. H. 1965. *Qd J. agric. Anim. Sci.* **22**: 437 (*Glomerella cingulata* var. *minor* & *C. acutatum* sp. nov. inter alia; **46**, 370).
Smith, H. C. 1948. *N.Z. Jl Sci. Technol.* Sect. A **30**: 83 (*G. concentricum*; **29**: 239).
Staunton, W. P. et al. 1966. *Irish J. agric. Res.* **5**: 140 (as above; **45**, 2652).
——. 1967. *Ibid* **6**: 203 (as above; **47**, 939).
Stoneman, B. 1898. *Bot. Gaz.* **26**: 69 (development of some anthracnoses).
Thomson, J. R. 1935. *Trans. Br. mycol. Soc.* **20**: 123 (taxonomy, *G. concentricum*; **15**: 474).
Vanner, A. L. et al. 1973. *Proc. 26th N.Z. Weed Pest Control Conf.*: 139 (fungicides & *C. acutatum* on pine; **56**, 2235).
Verhoeff, K. 1974. *A. Rev. Phytopathol.* **12**: 99 (latent infections by fungi, 76 ref.).
Wilson, K. I. 1961. *Indian Phytopathol.* **14**: 53 (*C. indicum*; **41**: 305).
Wollenweber, H. W. et al. 1947. *Z. ParasitKde* **14**: 181 (general on *Colletotrichum*, *Gloeosporium* & *Vermicularia*; **29**: 157).

LEGUMINOSAE

Many *Colletotrichum* spp. have been described on Leguminosae, mostly on temperate crops. They were discussed by Tiffany et al.; Massenot et al.; the latter gave a bibliography. Those with curved conidia are given under *C. dematium* (q.v.) and these seem to be more important in the tropics. Those with straight conidia are mostly placed in *C. trifolii* Bain apud Bain & Essary or *C. destructivum* O'Gara (synonym: *C. sativum* Horn). Both spp. were described by Tiffany et al. who gave the hosts. *C. destructivum* has larger conidia, 14–22 × 3–4.5 μ, compared with 10.5–13 × 3.5–4 μ for the other sp. Arx (1957a, *Colletotrichum* q.v.) maintained *C. destructivum* as a distinct sp. but treated *C. trifolii* as a more or less specialised form of *Glomerella cingulata* on legumes; *C. pisi* Pat. is placed in synonymy with *G. cingulata*. *C. lindemuthianum* (q.v.), also on legumes, is considered separately. Arx (l.c.) placed *C. glycines* Hori (curved conidia) in synonymy with *C. dematium* f. sp. *truncata*; a perfect state was described, *G. glycines* Lehman & Wolf, from soybean. However Tiffany et al. described a fungus which agreed with Lehman & Wolf's description but the conidia were straight and therefore differed from the earlier description. Soybean anthracnose is discussed under *C. truncatum*. Weimer attributed anthracnoses of *Lupinus angustifolius* and *Lespedeza* spp. to *G. cingulata*. This fungus also causes anthracnose of *Stylosanthes*; Ellis et al. recovered *G. cingulata* from surface-sterilised pods and seeds of *Stylosanthes* spp.; and see Baldion for resistance.

C. trifolii causes diseases (southern anthracnose) on clovers, lucerne and other legumes (e.g. see Kort). Sampson compared the fungus with *Kabatiella* (q.v.) *caulivora* on clover. Light to dark brown lesions are caused on stems and petioles; the former may be girdled. A crown rot phase on lucerne has also been described. Plants have wilted and yellow tops, often appearing bleached when old, varying numbers of tillers are affected. A characteristic symptom is an internal bluish black discolouration from the base of the tillers, through the crown and often into the tap root for up to 8 cm. This discoloured tissue tends to be dry and shredded, and merges into a marginal brown area of firm consistency. Sometimes an external dark lesion is present on the crown and upper portion of the tap root. Plants apparently only show a complete wilt when the crown is infected (Irwin). Miehle et al. found that there was most conidial germination and appressorial formation at relatively low temps (21° and 24°), and that the latter was more sensitive to temp. than the former. The exposure of germinated conidia to temps of >27° reduced the number forming an appressorium. In N.E. USA the fungus persists from one harvesting season to the next in debris on protected harvesting equipment. It survived a little >3 months in stems in the field (Carroll et al.). Anthracnose is important in predisposing plants to winter injury (Lukezic). Schietinger compared *C. destructivum* and *C. trifolii* on lucerne. This anthracnose is widespread in S.W. Germany and in some cases the second or third crop became heavily infected. Most isolates were of the *C. destructivum* type. Graham et al. found *C. trifolii* to be the main cause of lucerne anthracnose in mid Atlantic USA. Local isolates of *C. destructivum* and *C. truncatum* were weakly pathogenic on this crop. Both these fungi formed acervuli on lesions caused by *C. trifolii*. Repeated selection in USA has led to the development of cvs with high resistance, whose control has been reported to be due to a single dominant in one experimental population, or is more complex (Barnes et al.; Devine et al.; Ostazeski et al.). Irwin also found high levels of resistance in a few cvs which could be used in selection. Campbell et al. reported resistance conditioned by a complete dominant.

Pea anthracnose is attributed to *C. pisi* which may not be distinct from *C. gloeosporioides* (*G. cingulata*). The disease is a minor one and infection is said to follow that by other organisms, for example, *Mycosphaerella pinodes*. Hagedorn described an epidemic in USA (Wisconsin), the first outbreak in the area since 1942. Irregular spots formed on the leaves, tan with a dark brown border and 2–8 mm diam. There were elongated stem lesions, 8–12 mm long, and some stem girdling. The lesions were so numerous on the plants that the crop appeared generally brown. The outbreak occurred in fields where pea and Irish potato alternated. It was attributed to the fact that pea debris had not decomposed sufficiently since it had not been ploughed in and the weather had been dry. It is doubtful whether the pathogen is seedborne in pea. Except where stated the ref. relate to lucerne anthracnose.

Antonova, S. P. 1940. *Vest Zashch. Rast.* **5**: 133 (on pea; **21**: 115).

Baldión, R. et al. 1975. *Fitopatologia* **10**: 104 (*Glomerella cingulata* on *Stylosanthes*, resistance; **55**, 4174).

Barnes, D. K. et al. 1969. *Crop Sci.* **9**: 344 (effect on yield, stand & vigour, resistance; **50**, 3897).

Campbell, T. A. et al. 1974. *Ibid* **14**: 667 (inheritance of resistance; **55**, 1838).

Carroll, R. B. et al. 1977. *Pl. Dis. Reptr* **61**: 12 (winter survival in USA; **56**, 3604).

Chilton, S. J. P. 1943. *Mycologia* **35**: 13 (variation in vitro of *Colletotrichum destructivum*; **22**: 250).

Devine, T. E. et al. 1971. *Crop Sci.* **11**: 854 (resistance; **52**, 4124).

———. 1975. *Ibid* **15**: 505 (as above; **55**, 2777).

Ellis, M. A. et al. 1976. *Pl. Dis. Reptr* **60**: 844 (*G. cingulata* in seed of *Stylosanthes* spp.; **56**, 4098).

Graham, J. H. et al. 1976. *Phytopathology* **66**: 538 (occurrence & interaction of 3 *Colletotrichum* spp. on lucerne in mid Atlantic USA; **55**, 5810).

Hagedorn, D. J. 1974. *Pl. Dis. Reptr* **58**: 226 (on pea with *Peronospora viciae*; **53**, 2745).

Hancock, J. G. et al. 1965. *Phytopathology* **55**: 346, 356 (host cell degrading enzymes in *C. trifolii*, *Ascochyta imperfecta* & *Pleospora herbarum*; **44**, 2182a & b).

——. 1966. *Ibid* **56**: 1112 (pectate lyase formation by *C. trifolii* & pH; **46**, 363).

Irwin, J .A. G. 1974. *Aust. J. exp. Agric. Anim. Husb.* **14**: 197 (crown rot syndrome & resistance; **53**, 4021).

Jones, F. R. et al. 1921. *Phytopathology* **11**: 500 (on pea; **1**: 282).

Kort, J. 1955. *Versl. Meded. plziektenk. Dienst Wageningen* 129: 179 (on legumes; **37**: 99).

Lukezic, F. L. 1974. *Phytopathology* **64**: 57 (dissemination & survival; **53**, 2601).

Massenot, M. et al. 1973. *Annls Phytopathol.* **5**: 83 (general, on leguminous fodders, 160 ref.; **53**, 988).

Miehle, B. R. et al. 1972. *Can. J. Microbiol.* **18**: 1263

(conidial germination & appressorium formation in *C. trifolii*; **52**, 338).
Ostazeski, S. A. et al. 1969. *Crop Sci.* **9**: 351 (selection for resistance; **50**, 3898).
Ou, S. H. et al. 1945. *Phytopathology* **35**: 565 (on pea; **25**: 22).
Sampson, K. 1928. *Trans. Br. mycol. Soc.* **13**: 103 (on *Trifolium* with *Kabatiella cauliflora*; **7**: 583).
Schietinger, R. 1970. *Z. PflKrankh. PflSchutz* **77**: 26 (*C. trifolii* & *C. destructivum*; **50**, 148).
Tiffany, L. H. et al. 1954. *Mycologia* **46**: 52 (*Colletotrichum* spp. from legumes; **33**; 637).
Weimer, J. L. 1943. *Phytopathology* **33**: 249 (on *Lupinus* spp.; **22**: 299).
——. 1946. *Ibid* **36**: 524 (on *Lespedeza* spp.; **26**: 16).

Colletotrichum capsici (Syd.) Butler & Bisby in The fungi of India, *Imp. Coun. agric. Res. India. Sci. Monogr.* **1**: 152, 1931.
Vermicularia capsici Syd., 1913.

ACERVULI on fruits, leaves and stems rounded or elongated, attaining *c.* 350 μ diam., intra- and subepidermal, disrupting outer epidermal cell walls of the host. SETAE brown, 1–5 septate, rigid, hardly swollen at the base, slightly tapered to the paler acute apex, up to 250 μ long, 5–8 μ wide. CONIDIA hyaline, falcate with acute apex and narrow truncate base, aseptate, uninucleate, 16–30 × 2.5–4 μ, formed from aseptate, hyaline to faintly brown cylindrical phialidic conidiophores. COLONIES on PDA at first white, rapidly becoming grey. Aerial mycelium forms light to dark grey felt over surface of colony, sometimes showing diurnal zonation in density, with acervuli conspicuous for their dark setae on the thinner areas. Spore droplets are pale buff to salmon. Reverse dark; appressoria and their supporting structures develop in large numbers against Petri dish surfaces. On potato carrot agar mycelium is sparse and shows little pigment and few acervuli. APPRESSORIA formed on mycelium in old cultures or slide cultures, sepia brown, 6–25 × 4–10 μ, sometimes solitary but frequently associated with or borne on convoluted chains of paler brown chlamydosporic structures. Close to *C. dematium* (Pers. ex Fr.) Grove, a normally saprophytic sp., which Arx (1957c, *Colletotrichum* q.v.) regarded as a synonym. *C. capsici* shows exceptionally complex development of appressoria and their supporting hyphae in slide cultures; setae and stromatic tissues are also strongly developed on the host (J. E. M. Mordue, *CMI Descr.* 317, 1971).

C. capsici is widespread and has been mostly studied (as a pathogen) on *Capsicum* causing red pepper anthracnose. *Colletotrichum dematium* (q.v.) was fully described by Sutton who gave evidence to justify retention of its specific status. *C. capsici* is a general, unspecialised parasite which has been recorded from many plants in a wide range of families; but the diseases caused (on stems, leaves and fruits) are apparently all relatively minor or only locally serious. Ramakrishnan (1954) described a leaf spot of turmeric (*Curcuma domestica*) and Nair et al. (1973b) reported a dry rhizome loss of >60%. The elliptical spots, up to *c.* 5 × 4 cm, have greyish white centres with the black acervuli arranged concentrically; the margin is brown and there is an indefinite chlorotic halo. The spots are at first most conspicuous on the upper surface, they coalesce and most of the leaf can dry up, with fields presenting a parched appearance. Srinivasan reported a seedling blight of *Sesbania grandiflora*, cankers (6–10 mm long) were caused at the collar; and see Chopra et al. for cotton boll infection. Bhardwaj et al. described a leaf blight of the cover crop *Desmodium gangeticum*. Isolates of *C. capsici* (although possibly showing minor differences in morphology and pathogenicity) from one host will generally infect others. Desai et al. considered a form of the fungus to be specialised on cluster bean (*Cyamopsis tetragonoloba*), and Sowell described a form of *Colletotrichum dematium* from seed of this host. Other *Colletotrichum* spp. have been reported on red pepper (and see Verma).

Dastur in the earliest full description of anthracnose of red pepper described it as a dieback in India (and see Butler, *Fungi and disease in plants*). Lesions may begin at the top of the stem or at a wound on the stem. They are at first brown, becoming whitish and sharply delimited by a black line; spread occurs so that branches are killed. On the fruit, as it turns red, the somewhat elliptical lesions are greenish black or a dirty grey. The necrotic area, which may extend to affect the whole side of a fruit, becomes lighter in colour, straw coloured and enclosed by a dark border; it may appear papery. Where the lesions are restricted they can be sunken; developing acervuli may be concentric. In USA the disease is apparently restricted to the fruit; the fungus spreads to the central cavity and infects the seed (Grover et al.; Higgins 1930; Rai et al.). Seed taken from diseased fruit can give diseased seedlings. Smith et al. reported survival on the seed for at least 9 months. The opt. temp. for growth in vitro is 28–30° or a little less and at these temps (under high moisture conditions) the disease develops rapidly (Chowdhury; Misra et al.).

Removal and destruction of infected plant material should reduce the severity of attacks. But since the fungus is common in the tropics and other warm regions, occurring on several hosts, chemical treatment both of the seed and growing plant may be necessary. On turmeric captan and zineb gave control in India (Dakshinamurti et al.). On cluster bean Agnihotri et al. (1966) found that copper was better than organo-metal compounds. Copper, maneb, wettable sulphur and zineb have been used against infection of red pepper; and for seed treatment in this crop Grover et al., in pot experiments, found that thiram and quintozene were effective. Some resistance has been reported in India (Bansal et al.). Except where stated the ref. are to anthracnose on red pepper.

Agnihotri, J. P. et al. 1966. *Pl. Dis. Reptr* **50**: 921 (fungicides on cluster bean; **46**, 1154).

——. 1971. *Indian J. Mycol. Pl. Pathol.* **1**: 51 (formation of protopectinase; **51**, 3169).

Bansal, R. D. et al. 1969. *J. Res. Punjab agric. Univ.* **6**: 345 (reaction of cvs in N. India; **49**, 2276).

Bhardwaj, S. D. et al. 1968. *Indian For.* **95**: 873 (on *Desmodium gangeticum*; **49**, 2578).

Bhullar, B. S. et al. 1972. *Phytopathol. Z.* **75**: 236 (phenols in resistant & susceptible cvs; **52**, 3148).

Chatrath, M. S. et al. 1968. *Indian Phytopathol.* **21**: 464 (nuclei in conidia; **48**, 2821).

Chopra, B. L. et al. 1975. *J. Res. Punjab agric. Univ.* **12**: 1 (on cotton bolls; **55**, 1267).

Chowdhury, S. 1957. *Indian Phytopathol.* **10**: 55 (general in Assam; **38**: 646).

Dakshinamurti, V. et al. 1966. *Andhra agric. J.* **13**: 69 (fungicides on turmeric; **48**: 1890).

Dastur, J. F. 1921. *Mem. Dep. Agric. India bot. Ser.* **11**: 129 (general in Bihar; **1**: 195).

Desai, M. V. et al. 1955. *Indian Phytopathol.* **8**: 52 (*C. capsici* f. sp. *cyamopsicola*; **35**: 268).

Grover, R. K. et al. 1970. *Ibid* **23**: 664 (seed transmission & fungicides; **51**, 1009).

Higgins, B. B. 1926. *Phytopathology* **16**: 333 (anthracnose fungi in USA; **5**: 647).

——. 1930. *Bull. Ga. agric. Exp. Stn* 162, 10 pp. (general; **10**: 56).

Ling, L. et al. 1944. *Indian J. agric. Sci.* **14**: 162 (distribution & in China; **26**: 141).

Malabanan, D. B. 1926. *Philipp. Agric.* **14**: 491 (general; **5**: 402).

Misra, A. P. et al. 1960. *Indian Phytopathol.* **13**: 12 (growth in vitro; **42**: 183).

——. 1963. *J. Indian bot. Soc.* **42**: 74 (comparison of 2 isolates; **43**, 1212).

Nair, M. C. et al. 1973a. *Curr. Sci.* **42**: 362 (infection of turmeric & toxic metabolites; **53**, 648).

——. 1973b. *Ibid* **42**: 549 (effect on quality & yield in turmeric; **53**, 1498).

——. 1974. *Proc. Indian Acad. Sci.* Sect. B **80**: 222 (on turmeric, phytotoxin; **54**, 4072).

Rai, I. S. et al. 1966. *J. Res. Punjab agric. Univ.* **3**: 32 (variation in vitro & carry over; **46**, 1830).

Ramakrishnan, T. S. 1941. *Proc. Indian Acad. Sci.* Sect. B **13**: 60 (variation in isolate from *Carthamus tinctorius*; **20**: 317).

——. 1947. *Ibid* **25**: 15 (comparison of isolates from several plant hosts; **26**: 544).

——. 1954. *Indian Phytopathol.* **7**: 111 (on turmeric; **35**, 396).

Smith, R. W. et al. 1958. *Pl. Dis. Reptr* **42**: 1099 (general in E. USA; **38**: 176).

Solanki, J. S. et al. 1974. *Mycopathol Mycol. appl.* **52**: 191 (temp., conidial germination & appressorial formation; **54**, 43).

Sowell, G. 1965. *Pl. Dis. Reptr* **49**: 607 (*C. dematium* f. sp. *truncata* from cluster bean; **44**, 3215b).

Srinivasan, K. V. 1952. *Curr. Sci.* **21**: 318 (on *Sesbania grandiflora*; **32**: 406).

Subramanian, K. S. et al. 1971. *Madras agric. J.* **58**: 548 (fungicides; **51**, 1010).

Thirupathaiah, V. et al. 1972. *Phytopathol. Z.* **75**: 175 (sporulation & polygalacturonate transeliminase formation; **52**, 4301).

Verma, M. L. 1973. *Indian Phytopathol.* **26**: 28 (virulence of 4 *Colletotrichum* spp.; **53**, 4221).

Colletotrichum circinans (Berk.) Vogl., *Ann. Accad. Agric. Torino* **49**: 175, 1907.

Vermicularia circinans Berk., 1851

(see W. B. Grove, *British stem and leaf fungi* Vol. 2: 239, 1937).

Smudge of onion is widespread and described by Walker in *Plant pathology*. Besides *Allium cepa* var. *cepa* the disease affects *A. cepa* var. *aggregatum* (shallot) and *A. ampeloprasum* var. *porrum* (leek). Kondo et al. studied the pathogen on *A. fistulosum* (Welsh onion). Arx (1957a, *Colletotrichum* q.v.) placed the fungus as a f. sp. of *C. dematium*. The disease is restricted to the bulb scales and the lower parts of the unthickened leaves. Behr recorded a case of damping off in Germany. Somewhat circular spots, at first dark green becoming black, form on the bulbs (outer scales); the acervuli are borne on stromata that are occasionally arranged concentrically. The conidial masses are cream in colour. Infection of underlying scales leads to the formation of similar spots, sometimes with chlorotic borders. These deeper-seated lesions may arise from spots on the outer scales or, where such scales have been broken, the fleshy ones may be directly infected from the soil. In the absence of the black stromata, where the fleshy scales have dried before sporulation of the fungus, slightly raised yellowish lesions occur. *C. circinans* persists in the soil for several years in the absence of the main host. The usual direct cuticular penetration takes place and the disease has an opt. temp. of *c*. 26° at which there is most growth in vitro. The stromata are very resistant to desiccation. Moist conditions are necessary for the spread of onion smudge which reduces the market value of the crop.

It was noted very early in the history of the disease that coloured cvs of onions escaped infection, although growing near infected bulbs of white cvs. Subsequently the early classic work by Walker and his associates described one of the first examples of plant disease resistance which was due to specific chemical compounds. The resistance mechanism is an example of substances acting outside the plant. When conidia are placed in water drops on the dry scales of red or yellow bulb cvs their germination is abnormal and without the formation of appressoria, or does not occur at all. On dry colourless scales germination is normal. If the dry scales are removed from the coloured bulbs the coloured living fleshy scales are infected as readily as white bulbs. This resistance is directly associated only with the pigments in the dead cells of the dry scales. In such scales the exudation of catechol and protocatechuic acid into the infection droplets prevents normal germination of the conidia. The inheritance of this resistance is apparently identical with the inheritance of colour (Clarke et al.; Jones et al.; Rieman). Three gene pairs interact to give phenotypes (27 genotypes) with 3 resistance reactions: high (red and yellow bulbs), intermediate (pink and cream bulbs) and no resistance (white bulbs). The higher resistance of pungent cvs was investigated by Hatfield et al. and Owen et al. with *Botrytis allii* and *Aspergillus niger*. Where *C. circinans* is present there is no complete control against infection of white onion cvs. Bulbs should be protected from rain at harvest and cured promptly. Coloured cvs should be used where the fungus prevails, and those with few, easily sloughing off outer scales avoided.

Angell, H. R. et al. 1930. *Phytopathology* **20**: 431 (protocatechuic acid & resistance; **9**: 758).
Behr, L. 1963. *Zentbl. Bakt. ParasitKde* Abt. 2 **116**: 552 (damping off; **43**, 2149).
Clarke, A. E. et al. 1944. *Genetics Princeton* **29**: 569 (inheritance of onion bulb colour).
Hatfield, W. C. et al. 1948. *J. agric. Res.* **77**: 115 (resistance with *Aspergillus niger* & *Botrytis allii*; **28**: 46).
Johnson, B. 1932. *Am. J. Bot.* **19**: 12 (host penetration; **11**: 391).
Jones, H. A. et al. 1946. *J. agric. Res.* **72**: 259 (bulb pigment, inheritance & resistance; **25**: 380).
Kondo, A. et al. 1953. *Scient. Rep. Shiga agric. Coll.* 3: 19 (on Welsh onion; **35**: 264).
Link, K. P. et al. 1929. *J. biol. Chem.* **84**: 719 (protocatechuic acid & resistance; **9**: 284).
—— ——. 1933. *Ibid* **100**: 379 (catechol & resistance; **12**: 672).
Owen, J. H. et al. 1950. *Phytopathology* **40**: 292 (resistance with *A. niger* & *B. allii*; **29**: 550).
Rieman, G. H. 1931. *J. agric. Res.* **42**: 251 (inheritance of pigmentation in relation to resistance; **10**: 701).
Walker, J. C. 1921. *Ibid.* **20**: 685 (general; **1**: 278).
——. 1923. *Ibid* **24**: 1019 (resistance; **3**: 118).
—— et al. 1924. *Ibid* **29**: 507 (as above; **4**: 392).
—— ——. 1925. *Ibid* **30**: 175 (as above; **4**: 519).
—— ——. 1929. *Proc. natn. Acad. Sci. U.S.A.* **15**: 845 (as above; **9**: 284).

Colletotrichum coccodes (Wallr.) Hughes, *Can. J. Bot.* **36**: 754, 1958.
Chaetomium coccodes Wallr., 1833
Vermicularia atramentaria Berk. & Br., 1850
Colletotrichum atramentarium (Berk. & Br.) Tauben., 1916.

ACERVULI on stems and roots, rounded or elongated, *c*. 300 μ diam.; acervular tissue intra- and subepidermal, disrupting the outer epidermal cell walls of host. Occasional cells of acervulus develop as SETAE which are brown, septate, slightly swollen at the base then tapered to the often slightly paler tip. Setose SCLEROTIA common. CONIDIA cylindrical, ends obtuse, hyaline, aseptate, 16–24 (19) × 2.5–4.5 (3) μ, formed from aseptate hyaline cylindrical phialidic conidiophores. COLONIES on PDA sparse, aerial mycelium whitish. SCLEROTIA usually abundant, evenly distributed over surface; when young, greyish, glabrous, rapidly becoming dark and setose. ACERVULI formed near sclerotia or as separate aggregates of setose mycelium; solitary phialides often found on mycelium. CONIDIOPHORES sometimes septate and branched. Spore masses normally small, colourless to salmon orange. CONIDIA frequently shorter in proportion to their breadth than on host, many irregular forms occur. Reverse of colony grey, darker with age because of formation of appressoria. APPRESSORIA very readily formed in slide cultures, cinnamon buff, ovate or obclavate to elliptical, occasionally irregularly lobed, 5–14 (10.5) × 4–11 (6) μ, borne on hyaline thin-walled sigmoid supporting hyphae. Distinguished from conidial forms within the broad general concept of *Glomerella cingulata* (q.v.; e.g. *Colletotrichum phomoides*) chiefly by size and shape of conidia and frequency of sclerotium formation. (J. E. M. Mordue, *CMI Descr.* 131, 1967. A perfect state of *C. phomoides* (*G. phomoides* Swank) was considered by Arx et al. (*Glomerella* q.v.) to be a synonym of *G. cingulata*.)

C. coccodes is a widespread (Map 190), general root inhabitant which also infects the leaves and fruit of several crops; its hosts are mainly in the Cucurbitaceae, Leguminosae and Solanaceae; see Chesters et al. who also briefly discussed the nomenclature of the fungus, including the incorrect use of *C. phomoides* for the forms attacking tomato, particularly the fruit. The diseases caused are mainly anthracnose of tomato fruit, and black dot of roots of tomato and Irish potato. Red pepper and eggplant are amongst other crops that can be attacked. The diseases caused by this fungus were reviewed by Forlot. Tomato fruit can also be infected by other *Colletotrichum* spp. (e.g. Barksdale 1972b; Jakubczyk). Corky root disease (or brown root rot) of tomato in European glasshouse crops was long thought to be due in part to infection by *C. coccodes* (*Pyrenochaeta lycopersici* q.v.). Griffiths et al. examined the fine structure of conidial development and germination, and appressorial development.

On tomato roots a brown cortical rot develops; the cortex becomes cracked and loose, root growth slows down and there are few roots. Secondary, aboveground symptoms of yellowing stunting and reduced crops follow the root attack. Abundant sclerotia develop both superficially and within the roots. On tomato fruit the first symptoms are generally seen towards ripening or when ripe. The watersoaked spots become sunken, up to 1 cm diam.; the abundant dark sclerotia form just beneath the surface, they are close together and sometimes arranged concentrically; there is a soft decay. Necrotic spots, up to 2 mm diam. and with a chlorotic halo, can be caused on the leaves but seem to be rare in the field. Pantidou et al. (1956) described spotting on leaves of some Cucurbitaceae. Dunn et al. (1970) found that light brown to brown lesions could be caused on the hypocotyl and/or radicle of tomato seedlings which are killed.

The most growth in vitro is at 28–30° and the disease develops rapidly at these temps or a little lower. Cuticular penetration of the fruit (and infection through wounds) occurs; the fungus often forms latent infections which erupt into lesions as the fruits ripen in the field or during transport. Fulton found that susceptibility increased with fruit age and that lesions were formed most rapidly at 26–27°. Illman et al. stated that a period of low temp. storage partially overcame latency in fruit and caused symptoms when it was green. Tomato anthracnose, although widespread, takes different forms. Where the crop is grown under glass, infection of the fruit is absent or rare. In N. and E. USA the disease is mainly one of fruit (used for processing) in the field crop.

Davet, who has recently investigated the disease, was concerned with root infection in Lebanon where aerial infection of tomato was not found. Survival of *C. coccodes* is through the sclerotia and in plant debris, and there is virtually no free growth through the soil. Colquhoun found viability to persist for at least 11 months in moist or dry soil in commercial glasshouses in the absence of host plants. Gemeinhardt, working with Irish potato, showed that stems buried in field soil still had 68% viable sclerotia after 8 months. In natural soil, subjected to normal glasshouse watering, 53% of sclerotia were viable after 83 weeks burial; of 28 sclerotia retrieved from soil after a similar period 3 infected tomato. Conidia did not survive for >3 weeks (Blakeman et al.). Davet (1970a) reported fungal viability in 50% of fragments of tomato roots after 15 months in the field. In the root form of the disease infection probably arises from sclerotia, although roots can be attacked by waterborne conidia. The fungus can penetrate and develop in young roots but the disease is generally one of plants becoming mature; root colonisation increased with age (Davet 1970b, 1971a, 1972). Farley (1976) noted a decrease in the sclerotial soil population after a year in dry or moist conditions at 4° or 25°; or at 25° when moisture varied. Conidial populations were reduced in the first week by 89–96% in dry soil and by 28–45% in moist soil. After a year conidial populations were virtually nil.

Colletotrichum coccodes

Primary conidial inoculum from other hosts, including weeds, may be important in the epidemiology. Raniere et al. indicated the likelihood of increasing anthracnose on fruit in a crop with overhead irrigation. In a pot experiment Dunn et al. (1964) described the effects of infection of tomato plants by *C. coccodes*, *Heterodera rostochiensis* (Irish potato root eelworm) and *Rhizoctonia solani* (*Thanatephorus cucumeris*). Growth was checked in a combined inoculation with *C. coccodes* and *R. solani* but the former (with or without the eelworm) had no detrimental effect on growth. Roy found that the incidence of *C. coccodes* in tomato roots was decreased by adding *Aphelenchus avenae*.

C. coccodes is unlikely to be a serious primary invader of tomato roots unless it has built up to a high level in the soil and where drainage is faulty, soils heavy, temps too low or the roots damaged. Infection may be associated with that of other fungi such as *P. lycopersici*, *T. cucumeris* and *Thielaviopsis basicola* (q.v.). Intensive cropping with tomatoes (especially in glasshouses) requires soil sterilisation; in the field rotation with less susceptible vegetable crops is desirable. By these means, and by improving the physical conditions and nutrient status of the soil, harmful build up of inoculum is prevented. Since damping off can be caused partial sterilising treatments for seed and seedbed are needed. Later, for plants in the field, any measures to prevent contact of the fruit with soil should be taken; and soil treatment before transplanting (dazomet) reduces storage rot (Lockhart et al.). Regular field spraying to prevent fruit infection may be required; captan, maneb, nabam and ziram are effective. Resistance to fruit infection occurs in certain tomato lines, and it has been transmitted in crosses (no clear segregation ratios) done in USA (Barksdale 1970, 1971, 1972a; Barksdale et al. 1975; Robbins et al.).

Adeniji, M. O. 1966. *Niger. agric. J.* **3**: 24 (role of fungi in root rot; **51**, 688).

Barksdale, T. H. 1967. *Phytopathology* **57**: 1173 (light induced in vitro sporulation; **47**, 894).

——. 1970. *Pl. Dis. Reptr* **54**: 32 (resistance in *Lycopersicon*; **49**, 2181).

——. 1971. *Ibid* **55**: 253 (inheritance of resistance; **50**, 3179).

——. 1972a. *Ibid* **56**: 321 (as above; **51**, 4372).

——. 1972b. *Phytopathology* **62**: 660 (resistance in tomato fruit to *Colletotrichum* spp.; **52**, 860).

—— et al. 1975. *Pl. Dis. Reptr* **59**: 648 (breeding for resistance in tomato; **55**, 1468).

Bewley, W. F. et al. 1924. *Ann. appl. Biol.* **11**: 244 (general on roots; **4**, 70).

Blakeman, J. P. et al. 1966. *Trans. Br. mycol. Soc.* **49**: 227 (persistence in soil with *Didymella chrysanthemi*; **45**, 3066).

Campbell, W. P. et al. 1974. *Ibid* **63**: 19 (fine structure of sclerotial germination; **54**, 3140).

Chesters, C. G. C. et al. 1965. *Ibid* **48**: 573, 583 (taxonomy & variation; alternative hosts & fruit inoculation; **45**, 1204).

Colquhoun, T. T. 1941. *J. Dep. Agric. S. Aust.* **44**: 572 (on roots; **21**: 103).

Davet, P. 1970a. *Cah. ORSTOM Biol.* **12**: 83 (behaviour in soil; **50**, 1978).

——. 1970b. *Phytopathol. Mediterranea* **9**: 29 (general in Lebanon; **50**, 3578).

——. 1971a. *Ibid* **10**: 159 (sources of infection in Lebanon; **51**, 1261).

——. 1971b. *Rev. Mycol.* **35**: 307 (effect of temp. on sclerotia; **51**, 2863).

——. 1972. *Phytopathol. Mediterranea* **11**: 103 (root infection & plant age; **52**, 2747).

——. 1976. *Annls Phytopathol.* **8**: 25, 79 (cell degrading enzymes; **56**, 3229, 3230).

Dunn, E. et al. 1964. *Nature Lond.* **201**: 413 (relative effects of infection by *C. coccodes*, *Heterodera rostochiensis* & *Rhizoctonia solani*; **43**, 1416).

—— ——. 1970. *Pl. Pathol.* **19**: 196 (infection of seedlings; **50**, 1979).

Farley, J. D. 1972. *Phytopathology* **62**: 1288 (selective medium for soil assay; **52**, 2512).

——. 1976. *Ibid* **66**: 640 (soil survival; **55**, 5930).

Forlot, P. 1965. *Bull. Ec. nat. sup. agron. Nancy* **7**: 122 (review, 154 ref.; **45**, 1321).

Fulton, J. P. 1948. *Phytopathology* **38**: 235 (fruit infection; **27**: 391).

Gemeinhardt, H. 1957. *Phytopathol. Z.* **29**: 151 (on Irish potato, saprophytism & longevity; **36**: 719).

Griffiths, D. A. et al. 1972. *Trans. Br. mycol. Soc.* **59**: 483 (fine structure of conidial development; **52**, 1456).

—— ——. 1973. *Ibid* **61**: 529 (fine structure of conidial germination & appressorial development; **53**, 2066).

Hoffman, G. M. 1959. *Phytopathol. Z.* **35**: 31 (on *Cannabis sativa*; **38**: 478).

Illman, W. I. et al. 1959. *Can. J. Bot.* **37**: 1237 (general on fruit; **39**: 502).

Jakubczyk, H. 1961. *Acta agrobot.* **10**: 29 (*Colletotrichum* spp. from tomato fruit; **42**: 574).

Kendrick, J. B. et al. 1948. *Phytopathology* **38**: 247 (general; **27**: 392).

Lockhart, C. L. et al. 1963. *Can. J. Pl. Sci.* **43**: 503 (chemical & cultural control; **43**; 1145).

Loprieno, N. et al. 1962. *Phytopathol. Z.* **44**: 263 (pathogenic variability & cultural characters; **42**: 154).

Pantidou, M. E. et al. 1955. *Phytopathology* **45**: 338 (foliage as an inoculum source; **35**: 51).

——. 1956. *Pl. Dis. Reptr* **40**: 432 (foliage infection of *Cucumis* & *Cucurbita*; **36**: 283).
Raniere, L. C. et al. 1959. *Phytopathology* **49**: 72 (effect of overhead irrigation; **38**: 546).
Robbins, M. et al. 1970. *J. Am. Soc. hort. Sci.* **95**: 469 (inheritance of resistance; **50**, 280).
Roy, A. K. 1973. *Z. PflKrankh. PflPath. PflSchutz* **80**: 23 (effects of *Aphelenchus avenae* on infection by *C. coccodes* & *R. solani*; **52**, 4219).
Swank, G. 1953. *Phytopathology* **43**: 285 (perfect state of *C. phomoides*; **33**: 123).
Younkin, S. G. et al. 1944. *Phytopathology* **34**: 976 (foliage infection; **24**: 168).

Colletotrichum dematium (Pers. ex Fr.) Grove, *J. Bot. Lond.* **56**: 341, 1918.
Sphaeria dematium Pers. ex Fr., 1823.

ACERVULI 50–400 μ diam., black, abundant, gregarious, circular or longitudinally elongated, frequently confluent, at first covered by the cuticle and epidermis but later strongly erumpent with considerable stromatic development, exuding pale, smoke grey spore masses, and covered completely by many stiff, divergent setae. SETAE 60–200 μ long × 4–7.5 μ wide at the base, abundant, erect, rarely curved but occasionally irregular at the apex, divergent, smooth walled, 0–7 septate, usually 3 or 4, tapering to an acute apex, thick walled, vandyke brown through development, basal cells (apart from the colour) are differentiated from the stromatic tissue from which they arise. CONIDIA formed as simple swollen conidiophore apices. After the first spore has been liberated there develops in basipetal succession some phialospores without further increase in the length of the phialide itself. Conidia 18–26 μ × 3–3.5 μ, hyaline, aseptate, smooth walled, falcate, fusoid tapering gradually at both ends, apex acute, base truncate, guttulation irregular (from B. C. Sutton).

A full, illustrated account was given by Sutton who supported retention of the sp. in a comparison with *C. trichellum* (Fr. ex Fr.) Duke which had been placed in synonymy by Arx (1975a, *Colletotrichum* q.v.). The extensive synonymy for *C. dematium* (86 names given by Arx (l.c.)) requires re-examination (see Sutton). Gourley called *C. dematium* a well known coloniser of senescent tissue; the sp. was not considered to be parasitic by Sutton. It has been recorded on many plant hosts (temperate and tropical) and is sometimes reported as causing minor diseases such as leaf spots and diebacks. Wastie et al. compared the pathogenicity of *C. gloeosporioides* (*Glomerella cingulats* q.v.), *C. crassipes* (Speg.) Arx and *C. dematium* on leaves of *Hevea brasiliensis*. The first sp. is economically important as one cause of secondary leaf fall of rubber. *C. crassipes* had a low infectivity and *C. dematium* was intermediate, being much less pathogenic to older leaves. Attempts to infect leaves older than 7 days from bud burst were difficult except with *G. cingulata*. Chupp briefly described the *Colletotrichum* spp. on stems of solanaceous plants and concluded that these are *C. dematium* and *C. atramentarium*; the latter is a synonym of *C. coccodes* (q.v.). Spp. of the genus with curved and straight conidia have been described from *Piper betle* (betel, see Roy). In Brazil *C. crassipes* caused death of cacao seedlings (Ram et al.).

Besides *C. dematium*, other spp. with falcate conidia are *C. graminicola* (q.v.), *C. falcatum* (*G. tucumanensis* q.v.) and *C. curvatum* Briant & Martyn. The last sp. (placed as a synonym of *C. dematium* by Arx (l.c.) was described from seedlings of *Crotalaria juncea* (sunn hemp; see Mitra; Whiteside). The disease was called stem break in Zimbabwe. In the field lesions are caused in the leaf axils and a leaf wilt follows. Infection causes purplish brown leaf spots and elongated necrotic lesions on the stem which may break and collapse before maturity. The fungus is seedborne and may penetrate the seed coat; it was not detected in soil near buried infested crop debris. Several *Crotalaria* spp., including *C. angyroides* and *C. mucronata*, were reported to be resistant. Arx (l.c.) erected three ff. sp. in *C. dematium*: *spinaciae* Ellis & Halst., *circinans* Berk. and *truncata* Schw. Gourley considered *C. spinaciae* to be a synonym of *C. dematium*. The other 2 ff. sp. have not been always recognised and are often referred to as distinct spp.

Arya, H. C. et al. 1963. *Indian Phytopathol.* **16**: 234 (on *Bougainvillea*; **43**, 2290).
Briant, A. K. et al. 1929. *Trop. Agric. Trin.* **6**: 258 (*Colletotrichum curvatum* on sunn hemp; **9**: 186).
Chupp, C. 1964. *Mycologia* **56**: 393 (*Colletotrichum* spp. on tomato & Irish potato; **44**, 365).
Gourley, C. O. 1966. *Can. J. Pl. Sci.* **46**: 531 (pathogenicity on various hosts; **46**, 186).
Lele, V. C. et al. 1968. *Indian Phytopathol.* **21**: 349 (on *Rauvolfia serpentina*; **50**, 215).
Mitra, M. 1937. *Indian J. agric. Sci.* **7**: 443 (on sunn hemp; **17**: 41).
Pavgi, M. S. et al. 1965. *Phytopathol. Z.* **53**: 167 (on *Clitoria ternatea*; **45**, 529).
Ram, A. et al. 1968. *Indian Phytopathol.* **21**: 127 (on eggplant; **48**, 1019).

Colletotrichum graminicola

Roy, T. C. 1948. *J. Indian bot. Soc.* **27**: 96 (on betel; **28**: 487).

Saksena, H. K. et al. 1967. *Indian Phytopathol.* **20**: 67 (on groundnut; **47**, 961).

Sutton, B. C. 1962. *Trans. Br. mycol. Soc.* **45**: 222 (morphology & taxonomy of *C. dematium* & *C. trichellum*; **42**: 105).

Wastie, R. L. et al. 1970. *Ibid* **54**: 150 (on rubber; **49**, 2616b).

Whiteside, J. O. 1955 & 1957. *Rhod. agric. J.* **52**: 417; **54**: 327 (on sunn hemp; **35**: 677; **37**: 169).

Colletotrichum graminicola (Ces.) Wilson, *Phytopathology* **4**: 110, 1914.
Dicladium graminicolum Cesati, 1852.
(synonymy in Wilson; and see Arx 1957a, *Colletotrichum* q.v.).

ACERVULI on leaf and stem lesions, black, rounded or elongated, 70–300 μ diam.; outer epidermal wall of host normally penetrated by fine hyphae and not disorganised. Occasional cells of the acervulus develop as SETAE which are brown, slightly swollen at the base, then tapered to the rounded often slightly paler tip on which conidia are sometimes formed. CONIDIA falcate or spindle shaped, hyaline, aseptate, 19–29 (24) × 3.3–4.8 (4.2) μ, formed on aseptate, hyaline, cylindrical, phialidic conidiophores, 8–20 × 4–8 μ. COLONIES on PDA grey, even, compact, felty to woolly, occasionally tufted or sparse but showing no zonation. SCLEROTIA and immersed dendroid mycelial strands sometimes present. Acervuli formed in association with sclerotia or as separate aggregates of setose mycelium on surface of agar; smaller aggregates of mycelium with spore droplets on aerial mycelium. Spore masses colourless, grey or dirty pink to salmon orange. Conidia frequently broader and shorter than on the host; many irregular forms occur. Reverse of colony lilac to grey or greenish grey, darker with age because of the formation of appressoria. APPRESSORIA very readily formed in slide cultures, ochraceous tawny to cinnamon brown, irregularly obovate to clavate or ellipsoid, frequently lobed, 6–35 (15.6) × 4–25 (11.2) μ, with a single germ pore and borne terminally on thickened and pigmented supporting hyphae (J. E. M. Mordue, *CMI Descr.* 132, 1967).

C. graminicola has been considered as a conidial form of *Glomerella tucumanensis* (q.v.) or as distinct from *C. falcatum* the imperfect state. Mordue (l.c.) stated that the irregular or lobed appressoria with their thickened supporting hyphae and conidia generally slightly narrower in proportion to their length are also distinctive. Sutton (1968) considered that these 2 *Colletotrichum* spp. were distinct on examination of their appressoria. *G. graminicola* Politis (1975) was described as the perfect state, it most closely resembles *G. tucumanensis.* Le Beau (1950a) investigated the pathogenicity of nearly 600 *Colletotrichum* isolates from 18 grass spp., mostly *Saccharum* and *Sorghum.* One group was highly pathogenic to sugarcane only, another likewise to sorghum and a third from other spp. was not pathogenic to either of the 2 main host groups. Dale found that isolates from maize and sorghum only attacked their respective hosts (and see Bergquist 1973; Zwillenberg).

C. graminicola is plurivorous on Gramineae and widespread; it also occurs on some members of the Leguminosae. Hosts were given by Couch, *Diseases of turf grasses*, and Sprague, *Diseases of cereals and grasses in North America*. The main disease caused is anthracnose of *Sorghum* spp. (also called red stalk rot, red leaf spot and seedling blight). A full account was given by Tarr, *Diseases of sorghum, sudan grass and broom corn.* Minor diseases are also caused on maize, temperate cereals and turf grasses. The symptoms on sorghum vary with the sp. and cv., and sometimes with the geographical area. Circular to elliptical spots up to >0.5 cm diam. develop on leaves and leaf sheaths. They are tan orange red to blackish purple, with centres becoming grey to straw coloured with reddish borders; spots sometimes coalesce and they may show zonation. Leaf infection occurs at all stages of growth and may result in death of young plants. The stem in the maturing plant becomes discoloured (variation similar to that in leaves) and the internodes may rot to cause lodging. Pockets of infected tissue form in the internodes. Elliptical spots with pale centres and reddish purple borders are caused on the stem surface. The inflorescence can be attacked (head blight). The intensity of the syndrome depends very much on the degree of host resistance. The fungus has been reported to be seedborne, but neither seed infection nor that of adventitious roots seems to be of much practical importance. On maize the disease is probably restricted to one of old leaves in the field but infection of unwounded leaves of young plants can occur. The small brown spots become linear with straw coloured centres and black streaks form below the stem epidermis.

Politis (1973, 1976) described the fine structure of host penetration which is direct. The opt. temp. for growth in vitro is *c.* 28° and most conidial germination occurs at this temp. or a little higher (Bruehl et al.; Dale). Skoropad (on barley) found that appressoria formed at 15–35° but that most penetration of leaves was at 25–30°. Appressoria formed at 15 or 20° remained dormant until exposed to the higher temps. There is the expected splash dispersal of the conidia. Katsanos et al. investigated spread in the stem, and the effects of host injury and inflorescence removal on this. Some work on differences between isolates in pathogenicity to different spp. of Gramineae has been mentioned. In more recent work Wheeler et al. found that the fungus from maize infected sorghum but not some temperate cereals (and see Forgey et al.). They also reported that low light, as well as high RH, seemed to increase disease. Carry over takes place on crop residues and weeds, especially perennial grasses. On maize kernel infection can severely reduce emergence in the field. There are changes in susceptibility during growth. In USA these changes are reflected in the field where there are early summer leaf infections and late summer stalk infections; there is little or no disease in midsummer (Leonard et al.; Warren et al.). Survival in maize kernels was described by Warren et al.; and Naylor et al. found that survival of *C. graminicola* was relatively low in maize stalks on (or buried in) the soil.

Tarr (l.c.) has fully discussed the early work on host resistance (and its inheritance) which is probably the most effective approach to control. More recently Harris et al. reported good field resistance in >300 introductions of *S. bicolor* (and see Chohan 1967; Rodriquez et al.). Poneleit et al. found some resistance in maize commercial hybrids and inbred lines. In evaluations for resistance in maize some evidence for the existence of races was reported by Nicholson et al. (1976b). Bergquist et al. showed that a single sorghum selection with a loose head got less disease than one with a compact head. Where susceptible sorghum cvs are being grown it may be advisable to rotate with non-cereal crops and to dress seed with a fungicide.

Bergquist, R. R. 1973. *Pl. Dis. Reptr* **57**: 272 (general in Hawaii; **53**, 516).
—— et al. 1974. *Trop. Agric. Trin.* **51**: 431 (resistance; **54**, 436).
Böning, K. et al. 1936. *Phytopathol. Z.* **9**: 99 (general on maize; **15**: 494).
Bruehl, G. W. et al. 1950. *Tech. Bull. U.S. Dep. Agric.* 1005, 37 pp. (general on cereals & grasses; **29**: 608).
Chohan, J. S. 1967. *J. Res. Punjab agric. Univ.* **4**: 394 (variation in pathogenicity; **47**, 3099).
——. 1968. *Ibid* **5**: 220 (as above; **48**, 795).
Chowdhury, S. C. 1936. *Indian J. agric. Sci.* **6**: 833 (general on maize; **15**: 795).
Coleman, O. H. et al. 1954. *Agron. J.* **46**: 61 (inheritance of resistance; **33**: 478).
Dale, J. L. 1963. *Pl. Dis. Reptr* **47**: 245 (general on maize; **42**: 677).
Forgey, W. M. et al. 1978. *Ibid* **62**: 573 (evidence for physiologic specialisation).
Hammerschmidt, R. et al. 1977. *Phytopathology* **67**: 247, 251 (resistance in maize; host phenols & pigments; effect of light on lesion development; **56**, 4991).
Harris, H. B. et al. 1970. *Pl. Dis. Reptr* **54**: 60 (resistance in sorghum; **49**, 2057).
Hoof, H. A. Van. 1949. *Landbouw Buitenz* **21**: 267 (in Indonesia; **28**: 521).
Katsanos, R. A. et al. 1967. *Pl. Dis. Reptr* **51**: 957 (root injury & spread in stems; **47**, 806).
—— ——. 1968. *Ibid* **52**: 68 (cell death in stems as measure of susceptibility; **47**, 1535).
—— ——. 1969. *Phytopathology* **59**: 132 (relationship of living & dead cells to spread; **48**, 2358).
Kozar, F. et al. 1978. *Can. J. Bot.* **56**: 2234 (hyphal development & appressorium formation).
Kruger, W. 1965. *S. Afr. J. agric. Sci.* **8**: 881 (on maize; **45**, 1045).
Lapp, M. S. et al. 1976. *Can. J. Bot.* **54**: 2239 (a mathematical model of conidial germination & appressorial formation; **56**, 1437).
—— ——. 1978. *Trans. Br. mycol. Soc.* **70**: 221, 225 (nature of adhesive material of appressoria, site of appressoria on leaf surfaces; **57**, 4433, 4434).
Le Beau, F. J. 1950a. *Phytopathology* **40**: 430 (pathogenicity on sugarcane & sorghum; **29**: 535).
—— et al. 1950b. *Agron. J.* **42**: 33 (inheritance of resistance; **29**: 299).
—— ——. 1951. *Tech. Bull. U.S. Dep. Agric.* 1035, 21 pp. (general on sorghum).
Leonard, K. J. et al. 1976. *Phytopathology* **66**: 635 (effects of maize maturity & temp. on lesion development; **55**, 5720).
Narwal, R. P. 1973. *Agra Univ. J. Res.* **22**: 17 (silica & resistance in sorghum; **54**, 2803).
Naylor, V. D. et al. 1977. *Pl. Dis. Reptr* **61**: 382 (survival in maize; **57**, 577).
Nicholson, R. L. et al. 1976a. *Phytopathol. Z.* **87**: 324 (pectic enzymes & susceptibility in maize; **56**, 3001).
—— ——. 1976b. *Phytopathology* **66**: 86 (criteria for evaluation of resistance in maize; **55**, 3559).
Nishihara, N. 1975. *Ann. phytopathol. Soc. Japan* **41**: 171 (types of conidia & their pathogenicity; **55**, 1772).

Politis, D. J. 1973. *Physiol. Pl. Pathol.* **3**: 465 (fine structure of host penetration; **53**, 2164).
——. 1975. *Mycologia* **67**: 56 (*Glomerella graminicola* sp. nov.; **54**, 3193).
——. 1976. *Physiol. Pl. Pathol.* 8: 117 (fine structure of penetration of resistant oat leaves; **55**, 5156).
Poneleit, C. G. et al. 1972. *Crop Sci.* **12**: 875 (resistance in maize **53**, 141).
Rodriquez, E. et al. 1968. *Pl. Dis. Reptr* **52**: 164 (resistance of selected sorghum lines; **48**, 146).
Romano, J. et al. 1973. *Proc. Indiana Acad. Sci.* **83**: 351 (extracellular pectic enzymes; **55**, 203).
Skoropad, W. P. 1967. *Can. J. Pl. Sci.* **47**: 431 (effect of temp. on appressorium formation in barley; **47**, 140).
Sutton, B. C. 1966. *Can. J. Bot.* **44**: 887 (development of fructifications in *C. graminicola* & related spp.; **45**, 3303).
——. 1968. *Ibid* **46**: 873 (appressoria of *C. graminicola* & *C. falcatum*; **47**, 3744).
Warren, H. L. et al. 1975. *Phytopathology* **65**: 620 (kernel infection & seedling blight in maize; **55**, 721).
——. 1977. *Ibid* **67**: 160 (survival in maize kernels; **56**, 4990).
Wheeler, H. et al. 1974. *Phytopathology* **64**: 293 (pathogenicity on maize & other hosts; **53**, 3943).
Wilson, G. W. 1914. *Ibid* **4**: 106 (host range & synonymy).
Zwillenberg, H. H. L. 1959. *Phytopathol. Z.* **34**: 417 (general on maize; **38**: 474).

Colletotrichum lindemuthianum (Sacc. & Magn.) Br. & Cav., *Funghi Parass.* No. 50, 1889.

Gloeosporium lindemuthianum Sacc. & Magn., 1880.

ACERVULI on fruits, leaves and stems, rounded or elongated, *c.* 300 μ diam., intra- and subepidermal, disrupting outer epidermal cell walls of host. Occasional cells of acervulus develop as setae which are brown, septate, hardly swollen at the base and slightly tapered to the rounded paler apex, 4–9 μ wide and usually <100 μ long. CONIDIA hyaline, cylindrical with both ends obtuse or with base narrow and truncate, aseptate, uninucleate, 11–20 × 2.5–5.5 μ, formed from aseptate hyaline or faintly brown cylindrical phialidid conidiophores. CULTURE on PDA, growth slow, *c.* 6 cm diam. in 10 days at 22–24°. COLONIES at first grey, rapidly becoming dark with compact aerial mycelium. In old cultures sectors with whitish aerial mycelium sometimes occur. Acervuli with pale salmon spore masses most conspicuous in zone behind advancing edge and in centre of colony, elsewhere often more or less concealed by aerial mycelium; solitary phialides common; reverse of colony near black. APPRESSORIA infrequently formed from mycelium in old cultures or slide cultures, cinnamon brown, ovate to obclavate or faintly lobed, 6–18 × 4–9 μ, borne on hyaline thin-walled supporting hyphae. The sp. is most readily distinguished from the broad concept of conidial forms of *Glomerella cingulata* (q.v.) by its slow growth and dark pigmentation in culture (J. E. M. Mordue, *CMI Descr.* 316, 1971).

Anthracnose of common bean is widespread (Map 177). Infection also occurs on other *Phaseolus* spp. (Barrus), *Dolichos uniflorus*, *Lablab niger* (Ramakrishna), *Vicia faba* (Yu), *Vigna* and occasionally other Leguminosae. The most conspicuous symptoms are the dark brown to black, circular or oval lesions (up to *c.* 1 cm diam.) that form on the immature pods. The spots are sunken with raised edges and bear abundant acervuli. Infection spreads to the seed and penetrates the seedcoat. Somewhat elongate spots form on the leaves; at first these occur on the veins of the lower surface. The hypocotyl and young stem are also attacked; severe infection with the coalescence of lesions can lead to collapse of the plant. The cotyledons of plants grown from infected seed develop lesions. The roots are not usually attacked. Infected seeds have brown lesions. Initial infection by cuticular penetration has been described in detail by Mercer et al. (1971; and see Dey; Leach). The disease is most severe at moderate temps (17–22°) and under wet conditions. The waterborne inoculum originates from host debris of earlier crops (in which survival can be for 2–3 years) and from lesions on young plants grown from infected seed. Incipient, invisible infections of pods in the field may result in lesions developing during transport.

C. lindemuthianum shows great variability in pathogenicity and for a short review of the work on this and host resistance see Leakey et al. Allard (1972) considered that the different pathogenic races or forms could be grouped into 4 (and see Goth et al.). The dominant gene (from the Venezuelan cv. Cornell 49–242) confers wide resistance and has been used in selection in France and the Netherlands. Bannerot et al. (1971) demonstrated the existence of a second dominant gene (in 2 lines from Mexico) which conferred similar resistance. The variation in pathogenicity, increasing with the number of cvs used for differentiation (and the inheritance of resistance in some cases) has been investigated in many countries: Australia (Cruickshank; Walker; Waterhouse), Brazil (Augustin et al.; Kimati; Oliari et al.), France (Bannerot; Bannerot et al.; Charrier et al.; Fouilloux), Germany (Frandsen; Hoffman et al.; Krüger et al.; Peuser; Schnock et al.; Schreiber), Japan (Tochinai et al. 1952b, 1953), Malawi (Ayonoadu), Mexico (Yerkes et al.), Netherlands (Mastenbroek; Muller), Rumania (Roşca et al.), Sweden (Lundin et al.), Uganda (Leakey et al.), UK (Dixon et al.) and USA (Andrus et al.; Barrus; Burkholder; Cardenas et al.; Goth et al.; McRostie). The results from Uganda indicated that some isolates partially overcame the immunity derived from Cornell 49–242 (and see Hoffman et al.; Krüger et al.; Schnock et al.). There are many studies on the biochemical aspects (including phytoalexins) of this host and parasite relationship: *Rev. appl. Mycol.* **42**: 424; **48**, 1512, 2074; *Rev. Pl. Pathol.* **49**, 1212, 2231; **50**, 390, 2707; **51**, 3785; **52**, 4279; **53**, 3687; **54**, 611, 1056, 5142, 5633; **55**, 6011, 6012, 6015; **56**, 476, 3318, 4775, 5877; **57**, 5195.

Control, apart from resistance, is generally cultural. Of primary importance is the use of healthy seed. In USA the practice of growing the seed crop in dry areas has virtually eliminated the disease. Where this procedure is not possible seed must be taken from healthy crops and pods. General sanitation and a rotation of 2–3 years are also desirable. Chemical treatments of either the seed or standing crop have not been generally adopted, but captan, ferbam, zineb and ziram may have some control in the crops. Systemic fungicides have also been used (see Giroto; Meyer; Navarro et al.).

Allard, C. 1972. *Annls. Phytopathol.* **4**: 165 (present state of knowledge on resistance; **51**, 4487).

——. 1974. *Ibid* **6**: 359 (histology of resistance; **55**, 1542).

Andrus, C. F. et al. 1942. *Tech. Bull. U.S. Dep. Agric.* 810, 29 pp. (inheritance of resistance; **22**: 50).

Augustin, E. et al. 1971. *Pesqui. Agropec. Bras. Agron.* **6**: 265 (sources of resistance; **52**, 1729).

Ayonoadu, U. W. U. 1974. *Turrialba* **24**: 311 (races; **54**, 4253).

Bailey, J. A. et al. 1971. *Physiol. Pl. Pathol.* **1**: 435 (phytoalexin & races; **51**, 2050).

Bannerot, H. 1965. *Annls Amél. Pl.* **15**: 201 (infection with 6 races; **45**, 1254).

—— et al. 1968. *Ibid* **18**: 171 (inheritance of resistance; **48**, 2073).

—— ——. 1971. *Ibid* **21**: 83 (second gene for complete resistance; **52**, 529).

Barrus, M. F. 1921. *Mem. Cornell Univ. agric. Exp. Stn* **42**: 97 (general; **1**: 364).

Berard, D. F. et al. 1972. *Physiol. Pl. Pathol.* **2**: 123 (a cv. specific protection factor from incompatible interactions; **51**, 4489).

—— ——. 1973. *Ibid* **3**: 51 (relationship of genes for resistance to protection by diffusates from incompatible interaction; **52**, 3477).

Burkholder, W. H. 1923. *Phytopathology* **13**: 316 (the gamma strain; **3**: 110).

Cardenas, F. et al. 1964. *Euphytica* **13**: 178 (genetic system for reaction to 3 races; **45**, 3441).

Charrier, A. et al. 1970. *Annls Phytopathol.* **2**: 489 (15 races; **50**, 2043).

Cruickshank, I. A. M. 1966. *J. Aust. Inst. agric. Sci.* **32**: 134 (3 strs; **45**, 3440).

Dey, P. K. 1919. *Ann. Bot.* **33**: 305 (host penetration).

Dixon, G. R. et al. 1973. *J. natn. Inst. agric. Bot.* **13**: 87 (testing for resistance; **53**, 4916).

Edgerton, C. W. 1910. *Bull. La agric. Exp. Stn* **119**: 3–55 (general).

Elliston, J. E. et al. 1971. *Phytopathology* **61**: 1110 (induced resistance at a distance from site of inducing interaction; **51**, 872).

—— ——. 1976. *Phytopathol. Z.* **87**: 289 (compatible & incompatible interactions; **56**, 3316).

Fouilloux, G. 1976. *Annls Amél. Pl.* **26**: 443 (resistance sources & races; **56**, 2279).

Frandsen, N. O. 1953. *Z. PflKrankh. PflPath. PflSchutz* **60**: 113 (races; **32**: 658).

Giroto, R. 1974. *IDIA* 313/314: 29 (fungicides; **56**, 5280).

Goth, R. W. et al. 1965. *Pl. Dis. Reptr* **49**: 815 (4 races, alpha, beta, gamma & delta; **45**, 665).

Hegde, R. K. et al. 1968. *Indian Jnl exp. Biol.* **6**: 166 (serology in differentiation of races; **48**, 1397).

Hoffman, G. M. et al. 1974. *Z. PflKrankh. PflSchutz* **81**: 490 (races; **54**, 3562).

Kimati, H. 1966. *Anais Esc. sup. Agric. 'Luiz Queiros'* **23**: 247 (races; **47**, 954).

Krüger, J. et al. 1977. *Euphytica* **26**: 22 (resistance sources & races; **56**, 4275).

Leach, J. G. 1923. *Tech. Bull. Minn. agric. Exp. Stn* 14, 39 pp. (general; **3**: 109).

Leakey, C. L. A. et al. 1972. *Ann. appl. Biol.* **70**: 25 (races; **51**, 2974).

Lundin, P. et al. 1971. *Agri. Hort. Genet.* **29**: 30 (races; **51**, 4488).

McRostie, G. P. 1919. *Phytopathology* **9**: 141 (inheritance of resistance).

——. 1921. *J. Am. Soc. Agron.* **13**: 15 (as above; **1**: 365).

Mastenbroek, C. 1960. *Euphytica* **9**: 177 (breeding for resistance using dominant gene *Are* from Cornell 49–242; **40**: 258).

Mercer, P. et al. 1970. *Ann. Bot.* **34**: 593 (effect of orange extract & other additives on disease; **49**, 3522).

Mercer, P. et al. 1971. In *Ecology of leaf surface micro-organisms* (editors T. F. Preece & C. H. Dickinson): 381, Academic Press (initial infection).
—— ——. 1974. *Physiol. Pl. Pathol.* **4**: 291 (hypersensitivity; **53**, 4918).
Meyer, E. 1976. *Mitt. biol. BundAnst. Ld-u. Forstw.* 166, 135 pp. (resistance to benzimidazole derivatives; **55**, 3031).
Muller, H. R. A. 1926. *Meded. LandbHoogesch. Wageningen* **30**(1), 93 pp. (races; **6**: 322).
Navarro, A. R. et al. 1977. *Fitopatologia Colombiana* **6**: 87 (fungicides; **57**, 5197).
Oliari, L. et al. 1973. *Pl. Dis. Reptr* **57**: 870 (races; **53**, 3685).
Peuser, H. 1931. *Phytopathol. Z.* **4**: 83 (races; **11**: 277).
Ramakrishna, M. L. 1964. *Madras agric. J.* **51**: 506 (control in *Lablab niger*; **45**, 922).
Rosca, I. et al. 1975. *Problème Prot. Pl.* **3**: 207 (races; **56**, 4274).
Schaffnit, E. et al. 1925. *Zentbl. Bakt. ParasitKde* Abt 2 **58**: 176, 360, 481 (general; **4**: 456).
Schnock, M. G. et al. 1975. *HortScience* **10**: 140 (races; **54**, 4255).
Schreiber, F. 1932. *Phytopathol. Z.* **4**: 415 (breeding for resistance, races; **11**: 618).
Skipp, R. A. et al. 1972. *Physiol. Pl. Pathol.* **2**: 357 (visible fungal growth & host change on infection of susceptible & resistant hypocotyls; **52**, 2073).
—— ——. 1973. *Ibid* **3**: 299 (cross protection; **53**, 1626).
Tochinai, Y. et al. 1952a. *Mem. Fac. Agric. Hokkaido Univ.* **1**: 103 (overwintering; **32**: 114).
—— ——. 1952b. *Ann. phytopathol. Soc. Japan* **16**: 117 (races & varietal susceptibility; **32**: 464).
—— ——. 1953. *Ibid* **17**: 49 (infection of resistant & susceptible cvs; **33**: 133).
Walker, J. 1960. *J. Aust. Inst. agric. Sci.* **26**: 363 (races; **40**: 446).
Waterhouse, W. L. 1955. *Proc. Linn. Soc. N.S.W.* **80**: 71 (races).
Yerkes, W. D. et al. 1956. *Phytopathology* **46**: 564 (races; **36**: 295).
Yu, T. F. 1937. *Bull. Chin. bot. Soc.* **3**: (on broad bean; **18**: 568).

Colletotrichum musae (Berk. & Curt.) Arx, *Phytopathol. Z.* 29: 446, 1957.

Myxosporium musae Berk. & Curt., 1874
Gloeosporium musarum Cooke & Massee, 1887.

ACERVULI on fruit, stalks, petioles and occasionally leaves, usually rounded or occasionally elongated, erumpent, up to 400 μ diam., composed of epidermal and subepidermal, pale brown pseudoparenchyma, becoming subhyaline towards the conidiophore region, lacking SETAE. CONIDIOPHORES formed from the upper pseudoparenchyma, cylindrical, tapered towards the apex, hyaline, septate, branched and subhyaline towards the base, up to 30 μ long, 3–5 μ wide, each with a single terminal phialidic aperture. CONIDIA hyaline, aseptate, oval to elliptical or cylindrical, often with a flattened base, apex obtuse, variably guttulate, 11–17 × 3–6 (4.5) μ, produced in yellow to salmon orange masses. COLONIES on PDA with sparse to abundant white, grey or olivaceous floccose aerial mycelium; acervuli dark brown to black, abundant, irregularly scattered or sometimes confluent, produced throughout the culture, particularly in areas devoid of aerial mycelium and around the inoculation point; setae very rarely formed; conidium masses well developed, salmon orange; conidia also formed from separate phialides on the vegetative mycelium. On potato carrot agar aerial mycelium very sparse and acervuli not abundant, scattered irregularly throughout the culture. In all older cultures the reverse becomes greyish due to appressorium formation. APPRESSORIA readily formed from vegetative hyphae where they are navicular to ovate, becoming irregularly lobed, dark brown, 6–12 × 5–10 μ; and from germinating conidia where they are pale brown, subglobose to irregular, up to 8 μ diam. (B. C. Sutton & J. M. Waterston, *CMI Descr.* 222, 1970).

C. musae causes anthracnose of banana (*Musa*). It infects *M. textilis* (abaca) and hosts other than *Musa* but on these it causes no serious disease. The pathogen is widespread and causes a variety of diseased conditions (largely on the fruit) to which an assortment of common names have been applied. *C. gloeosporioides* (*Glomerella cingulata* q.v.) has also been isolated from banana (e.g. Ashby; Greene 1967a; Kaiser et al., *G. cingulata* q.v.). *C. musae* has slightly larger conidia and a higher opt. temp. for growth compared with *C. gloeosporioides* whose precise role in banana diseases is not clear.

Anthracnose on the fruit skin may usually be seen either on yellowing fingers (arising from latent infection on uninjured green fruit) or on green fruit (arising from non-latent infection through host injury) mostly after harvest. In the first case the black, circular specks enlarge to brown patches which coalesce, become sunken and bear the characteristic conidial masses. Much of the fruit may be affected, including the pulp, when temps are high or the fruit over-ripe. In the second case there are lenticular, slightly sunken lesions (up to 3 × 8 cm in the ripening room) and the necrosis has a chlorotic border. In ripening the pulp can be invaded and there is a watery rot. *C. musae* also causes fungal scald which is associated with controlled atmosphere packs. Reddish brown sunken spots form on green fruit, most commonly near the finger tips and on the bottom clusters in a box. The lesions enlarge during ripening. A rot of the fruit pedicel is caused by invasion through stalk damaged tissue; the conditions may arise through other fungi which may or may not be associated with *C. musae*. The disease has been given numerous names: black end, finger drop, finger stalk rot, neck rot, Santa Marta stem end rot and stem end. The pedicels become necrotic; this often begins in green fruit where the pedicels or necks have been injured by twisting or force exerted upwards or downwards, as may occur in transport. The whole pedicel is attacked and so weakened that the fingers drop. The conidial masses may be found.

C. musae is one of the several fungi which together cause crown rot. Amongst these are: *Botryodiplodia theobromae*, *Ceratocystis paradoxa*, *Gibberella fujikuroi* and *Verticillium theobromae*; also *Cephalosporium* sp. and *Fusarium*, section Gibbosum. A soft black rot begins on the crown at the cut surface and spreads towards the pedicels; external mycelium may form. The severity of the condition (which has become important with the tendency to ship in boxes as hands rather than as separate bunches) increases with time in transit. Main stalk rot is caused largely by *Ceratocystis paradoxa* but *Colletotrichum musae* may be associated with it. Reference should be made to specific texts for full accounts of the complex transport and storage rots of bananas (Feakin, *Pest control in bananas*; Stover, *Banana, plantain and abaca diseases*). Lesions on green fruit in the field are not often seen but can be caused by *C. musae* infection spreading from dead flower remains; infection of this type can also occur in ripening rooms. The tips of the fingers develop a dark rot. Cardeñosa-Barriga described infection of the floral bracts and other parts in a plantain cv. Agati described anthracnose on abaca, both on the leaves and fruit. Schlösser et al. described brown edge on banana leaves where characteristic necrotic bands occur along the leaf margin, especially in older leaves.

The conidia (abundant at 25–35°) are splash dispersed from the acervuli on banana debris. Most conidial germination occurs at 27–30° and these are the temp. limits for opt. growth. At humidities approaching saturation germination is rapid (in 4–12 hours) appressoria are formed and direct, cuticular penetration of the fruit in the field takes place. Infection via the appressorium develops slight, subcuticular spread and then becomes latent. This latency can last >5 months. These incipient infections may be abundant but only a small proportion grow into anthracnose lesions. This may occur as ripening approaches and damage to host tissue also results in further growth by the pathogen (Simmonds; Simmonds et al.). Apparently, there is no latency when infection takes place through fruit skin wounds. Goos et al. found that germinated conidia with appressoria were more resistant to a solution of 0.25% sodium hypochlorite than non-germinated ones (and see Greene 1966). At low RH, *c*. 30% or less, almost 90% of the conidia are viable after 60 days (Badger). Peacock found that infection of the stem end of green bananas reduces their green life. The amount of green life lost depended on the inoculum conc. used to begin infection and on the stage of maturation of the fruit. Disease development also varied with these 2 variables. Diseased fruit have greatly increased ethylene formation rates.

All the diseased conditions described and caused by *C. musae* and/or the other fungi mentioned can be reduced, if not completely controlled, by general cultural operations before and after harvesting bananas. Field sanitation requires that host plant debris in the field should be kept to a min.; this may include removal of the distal bud when hands are open, bracts, and leaves from the young stem. Harvesting should be at the correct time and the period that the fruit remains at ambient temps before refrigeration kept as short as possible. All handling must be such that bruising or wounding of any part of the bunch is avoided. In preparing hands or fingers clean cuts should be made; and, in dehanding the crown, tissue should be correctly trimmed or bevelled. Cleanliness in handling stations is also important. Besides those fungi already mentioned accounts of the other fungal diseases of banana fruit should be consulted, i.e. *Cercospora hayi*, *Deightoniella torulosa*, *Nigrospora sphaerica*, *Pyricularia grisea* and *Trachysphaera fructigena*.

Chemical control, where found necessary, is given by spraying or dipping fruit after harvest. Thiabendazole is generally used; benomyl is also very effective and thiophanate is an alternative (Frossard; Frossard et al.; Long; Rippon; Rippon et al.). Griffee reported resistance to benomyl in *Colletotrichum musae*. Griffee et al. concluded that postharvest treatment (one dip) with benomyl was to be preferred to preharvest sprays (4). Fungicides (e.g. chlorine, see Arneson) are also used as general sterilants in handling sheds. Main stalk rot is of no importance where fingers or hands in boxes are the exported article. Treatment of the cut end of the bunch stalk with a paint containing ferbam +2-mercaptobenzothiazole is effective.

Agati, J. A. 1925. *Philipp. Agric.* **13**: 337 (on abaca; **4**: 350).

Arneson, P. A. 1971. *Phytopathology* **61**: 344 (sensitivity of postharvest banana fungi to Cl; **50**, 3056).

Ashby, S. F. 1931. *Trop. Agric. Trin.* **8**: 322 (forms of *Colletotrichum* from *Musa*; **11**: 382).

Badger, A. M. 1965. *Phytopathology* **55**: 688 (effects of RH on *C. musae* inter alia; **45**, 160).

Cardeñosa-Barriga, R. 1963. *Turrialba* **13**: 88 (symptoms & control; **43**, 803).

Chakravarty, T. 1957. *Trans. Br. mycol. Soc.* **40**: 337 (latent infection; **37**: 175).

Frossard, P. 1971. *Fruits* **26**: 169, 819 (fungicides; **51**, 506, 4216).

—— et al. 1973. *Ibid* **28**: 195 (as above; **52**, 4158).

Goos, R. D. et al. 1962. *Mycologia* **54**: 353 (growth, conidial germination & survival; **42**; 396).

Greene, G. L. 1966. *Phytopathology* **56**: 1201 (effect of hypochlorous acid on conidia & appressoria; **46**, 389).

——. 1967a. *Physiol. Pl.* **20**: 580 (effect of 2-deoxy-D-glucose on respiration with *Glomerella cingulata*; **47**, 1415).

——. 1967b. *Turrialba* **17**: 231 (peel extracts & conidial germination).

—— et al. 1967c. *Ibid* **17**: 447 (tannins as the cause of latency; **47**, 1628).

Griffee, P. J. 1973. *Trans. Br. mycol. Soc.* **60**: 433 (resistance to benomyl & related fungicides; **52**, 4159).

—— et al. 1974. *Ann. appl. Biol.* **74**: 11 (disease incidence & fungicides; **53**, 4852).

Long, P. G. 1970. *Trop. Agric. Trin.* **47**: 9 (fungicides; **49**, 1737).

——. 1970. *Pl. Dis. Reptr* **54**: 93 (as above; **49**, 2136).

——. 1971. *Aust. J. exp. Agric. Anim. Husb.* **11**: 559 (as above; **51**, 1689).

Meredith, D. S. 1960. *Ann. appl. Biol.* **48**: 279, 518 (non-latent anthracnose & factors affecting it; **40**, 59, 317).

——. 1964. *Nature Lond.* **201**: 214 (appressoria on fruit; **43**, 1714).

Misra, A. P. et al. 1962. *Indian Phytopathol.* **15**: 11 (effects of temp. & RH; **42**: 38).

Müller, H. J. 1962. *Phytopathol. Z.* **44**: 381 (infection with *G. cingulata*; **42**: 267).

Peacock, B. C. 1973. *Qd J. agric. Anim. Sci.* **30**: 239 (effect of infection on the preclimacteric life of fruit; **53**, 4068).

Razakamanantsoa, S. 1966. *Fruits* **21**: 597 (effects of volatile products given off by ripening fruit; **46**, 1676).

Rippon, L. E. 1972. *Aust. J. exp. Agric. Anim. Husb.* **12**: 185 (fungicides; **51**, 4217).

—— et al. 1973. *Ibid* **13**: 465 (as above with *Nigrospora sphaerica*; **53**, 628).

Schlösser, E. et al. 1972. *Z. PflKrankh. PflSchutz* **79**: 530 (symptoms on leaves; **53**, 229).

Simmonds, J. H. et al. 1940. *Res. Bull. Aust. Inst. Sci. Ind.* 131, 63 pp. (comprehensive account; **19**: 551).

——. 1941. *Proc. R. Soc. Qd* **52**: 92 (*Colletotrichum* & latent infection in tropical fruits; **21**: 87).

——. 1963. *Qd J. agric. Sci.* **20**: 373 (latent infection & suggested interpretation of development in fruit; **43**, 2976).

Colletotrichum truncatum (Schw.) Andrus & W. D. Moore, *Phytopathology* **25**: 122, 1935.
Vermicularia truncata Schw., 1832.

The *Colletotrichum* spp. on Leguminosae are also very briefly discussed under the genus. In the present unsatisfactory state of their taxonomy it would appear that one might just assign them to either of the 2 groups, curved or straight conidia. Arx (1957a, *Colletotrichum* q.v.) placed *C. truncatum* as a f. sp. of *C. dematium* (curved conidia); amongst his synonyms for *C. dematium* f. sp. *truncata* is *C. glycines* Hori. *Glomerella glycines* Lehman & Wolf was described from soybean, having a stat. conid. with curved conidia and considered to be identical with *C. glycines*. Tiffany (1951a), investigating anthracnose of soybean, described the conidia of *G. glycines* as straight and therefore not like *C. glycines* but similar to *C. destructivum* (*Colletotrichum*, Leguminosae q.v.). The fungus which causes anthracnose of soybean (and some other legumes) is mostly described as *C. truncatum* or *C. dematium* f. sp. *truncata*; it was reported on beans (*Phaseolus lunatus* and *P. vulgaris*) by Andrus et al. (l.c.). Two other synonyms given by Arx (l.c.) are *C. caulicola* Heald & Wolf and *C. viciae* Dearn. & Overh. (see also Andrus et al. l.c.).

The fungus is primarily a pathogen of soybean. Besides *Phaseolus* spp. it has been described as causing anthracnose diseases on: *Acacia longifolia*, *Cyamopsis tetragonoloba* (cluster bean), *Desmodium* spp., *Dolichos uniflorus* (horsegram) and for other hosts on legumes see Tiffany (1951a) and Tiffany et al.

(*Colletotrichum*, Leguminosae q.v.). *C. truncatum* is seedborne in soybean, occurring in the seed coat but probably not in the cotyledons or embryo (Schneider et al.). A pre- and post-emergent seedling blight are the most serious forms of the disease. Brown sunken lesions develop on the cotyledons and from these the stem becomes infected. Symptomless establishment of the fungus (latency) in the stem cortex was described by Tiffany (1951b). A few days before flowering it spread to the lower stem, petioles, leaves, and developing pods and seeds. There were no external symptoms at this stage but at maturity acervuli formed on some pods. Lehman et al. described large necrotic lesions on the stem but found no lesions on the leaves. Iida found opt. temps for growth of 28–30° and 26–32° for conidial germination; Chung reported most appressoria at nearly 30°. The fungus persists in the seed and in crop debris. Many cvs have some resistance. Ahn et al. treated inoculated seed with captan, dichlone and thiram (amongst other fungicides) and reduced the amount of seedling blight. Chambers et al. and Cox described *C. truncatum* on *P. lunatus* (lima bean). The more important bean anthracnose is *C. lindemuthianum* (q.v.) which has straight (cylindrical) conidia.

Ahn, J. K. et al. 1970. *Jnl Pl. Prot. Korea* **9**: 21 (fungicides for seed treatment; **50**, 1522).
Chambers, A. Y. et al. 1962. *Bull. Tenn. agric. Exp. Stn* 338, 19 pp. (on lima bean; **42**: 227).
Chung, B. K. 1969. *Jnl Pl. Prot. Korea* **8**: 25 (appressorial formation; **50**, 1521).
Cox, R. S. 1950. *Tech. Bull. N. Carol. agric. Exp. Stn* 90, 28 pp. (on lima bean; **30**: 209).
Iida, W. 1951. *Forschn Geb. PflKrankh. Kyoto* **4**: 169 (temp., carry over & inoculation; **33**: 134).
Lehman, S. G. et al. 1926. *J. agric. Res.* **33**: 381 (*Glomerella glycines* sp. nov.; **6**, 76).
Ling, L. 1940. *Phytopathology* **30**: 345 (seedling blight; **19**: 512).
Merlo, P. A. 1969. *Revta Fac. Agron. Univ. nac. La Plata* **45**: 53 (on *Acacia longifolia*; **50**, 1387).
Schneider, R. W. et al. 1974. *Phytopathology* **64**: 154 (distribution in seed; **53**, 4662).
Sowell, G. 1965. *Pl. Dis. Reptr* **49**: 607 (on cluster bean; **44**, 3215b).
Staples, R. C. et al. 1976. *Arch. Mikrobiol.* **109**: 75 (appressorium formation & nuclear division; **56**, 4789).
Tiffany, L. H. 1951a. *Iowa St. Coll. J. Sci.* **25**: 371 (anthracnose complex on soybean; **30**: 598).
——. 1951b. *Phytopathology* **41**: 975 (spread in host & delayed sporulation; **31**, 267).
Wells, H. D. et al. 1963. *Pl. Dis. Reptr* **47**: 837 (on *Desmodium* spp.; **43**, 778).
——. 1971. *Ibid* **55**: 10 (on horse gram; **50**, 3014).

CORDANA Preuss, *Linnaea* 24: 129, 1851.

Preussiaster O. Kunze, 1891.
(Dematiceae)

COLONIES effuse, brown, greyish brown, or black, often thinly hairy. MYCELIUM mostly immersed. STROMA none. SETAE and HYPHOPODIA absent. CONIDIOPHORES macronematous, mononematous, straight or flexuous, unbranched, with terminal and usually also intercalary swellings, brown, smooth. CONIDIOGENOUS CELLS polyblastic, integrated, terminal and becoming also intercalary, sympodial, doliiform, spherical or subspherical, denticulate; denticles small, cylindrical. CONIDA solitary, acropleurogenous, simple, ellipsoidal, ovoid or pyriform, pale to dark brown, smooth, 1 septate, in some spp. with a thick dark band at the septum, often with a protuberant hilum (M. B. Ellis, *Demat. Hyphom.*: 199, 1971).

Hughes, S. J. 1955. *Can. J. Bot.* **33**: 259 (history of genus, *C. paucisepta* described inter alia).

Cordana musae (Zimm.) Hohnel, *Zentbl. Bakt. ParasitKde* Abt. 2 60: 7, 1923.

Scolecotrichum musae Zimm., 1902.

COLONIES hypophyllous, effuse, greyish brown, hairy. CONIDIOPHORES straight or flexuous, often nodose, pale to mid brown, smooth, up to 220 μ long, 4–6 μ thick, usually swollen at the base to 9–11 μ, terminal and intercalary swellings 6–8 μ diam. CONIDIA solitary on small pegs arising from terminal swellings which later become intercalary, obovoid or pyriform, 1 septate, sometimes slightly constricted at the septum, subhyaline to pale brown, smooth, 11–18 μ long, 7–10 μ thick in the broadest part, hilum protuberant (M. B. Ellis & P. Holliday, *CMI Descr.* 350, 1972).

Leaf blotch or spot of banana (*Musa*) is widespread in the tropics with the crop (Map 168). Small brown spots enlarge to an oval or sometimes a diamond shape with a darker, red brown margin and zonation becomes quite marked with age; a chlorotic halo is conspicuous, especially on the lower surface. Necrotic tissue may occur as strips from edge to midrib, and a marginal necrosis with an uneven, zig zag, chlorotic edge separating healthy from diseased tissue may develop. *C. musae* is a weak pathogen, often being found as a secondary invader of leaf lesions caused by other fungi (*Deightoniella torulosa*, *Khuskia oryzae*, *Mycosphaerella musicola* and *Pyricularia grisea*). Direct penetration with an appressorium has been described. The Horn plantain cv. is more extensively invaded. The conidia are air dispersed and in Jamaica a diurnal periodicity with a peak near 0700 hours occurs. Spore discharge has been described in detail. Fungicides used against *M. musicola* (maneb in an oil water emulsion) will give control but misting oil has been reported to increase the amount of spotting caused by *C. musae* in Samoa. Oil is phytotoxic to plantains but benomyl is effective.

Long, P. G. 1970. *Trop. Agric. Trin.* **47**: 229 (effect of misting oil; **49**, 3402).
Meredith, D. S. 1962. *Ann. Bot.* **26**: 233 (spore discharge; **41**: 666).
——. 1962. *Ann. appl. Biol.* **50**: 263 (spore dispersal; **42**: 37).
Stahel, G. 1934. *Trop. Agric. Trin.* **11**: 138 (host penetration with *Deightoniella torulosa*; **13**: 787).

CORTICIUM Fr.

'The name has long been used for a highly heterogeneous genus of resupinate Aphyllophorales characterised by a "smooth" hymenium and the absence of notable hymenial elements other than basidia and gloeocystidia, a wide circumscription . . . The process of segregating groups of structurally similar "*Corticium*" spp. into well-defined genera is not new, but has been accelerated recently . . .; it was shown that "*Corticium*" in its generally accepted sense is nomenclaturally invalid; residual spp. that cannot at present be referred to segregated genera are still retained in "*Corticium*" as a temporary measure.' (P. H. B. Talbot who also discussed *Hypochnus*, *Botryobasidium*, *Pellicularia* and *Thanatephorus*, in Parmeter, 1970: 22, *T. cucumeris* q.v.)

Corticium rolfsii Curzi, *Boll. Staz. Cat. veg. Roma* N.S. **11**: 365, 1932.

Sclerotial state: *Sclerotium rolfsii* Sacc., 1911.

COLONIES on PDA white, usually with many mycelial strands in the aerial mycelium. SCLEROTIA develop on colony surface, near spherical (slightly flattened below) mostly 1–2 mm across when fresh, shrinking slightly when dry, with smooth or shallow pitted shiny surface. Cells of primary hyphae at advancing edge of colony usually 4.5–9 μ wide and up to 350 μ long with 1 or more clamp connections at septa. Secondary hyphae arise immediately below the distal septum of cell and often grow adpressed to the primary hyphae; tertiary and subsequent branches are narrow (1.5–2 μ wide), have comparatively short cells and a wide angle of branching not closely associated with septation, and usually without clamps. Sclerotia show sharply differentiated rind with evenly thickened strongly pigmented walls, cortex with faintly pigmented and medulla with colourless, unevenly thickened walls. Cortex and medulla contain vesicles of reserve materials, rind cells do not. The hymenium is described on agar media by Curzi (l.c.) as white to ochraceous stromata sparse or aggregated in irregular masses; fructifications dense, forming crustose hymenial layer. BASIDIA clavate. BASIDIOSPORES smooth, hyaline, globose-pyriform, 4.5–6.75 × 3.5–4.5 μ. Dimensions given by other investigators are 2.01–5.61 × 1.67–3.67 to 18.9–56.9 × 4.4–7.3 for basidia, 1.64–2.51 × 1–1.72 to 5.8–12.4 × 3.6–8 for basidiospores and are partly dependent on media used. On host fans of white mycelium spread over organs near the soil surface. APPRESSORIA develop as swollen tips of short branches behind the advancing edge. Sclerotia often do not develop until after death of host. Hymenium development has been recorded on leaves.

Validity of the distinction from *S. delphinii* Welch, which has larger sclerotia, is controversial. Some unrelated spp. appear similar under field conditions; of these *Sclerotinia sclerotiorum* (Lib.) de Bary (q.v.) is probably the most important and is distinguished by its large irregularly shaped sclerotia and by the absence of clamps from its mycelium (J. E. M. Mordue, *CMI Descr.* 410, 1974).

C. rolfsii is an unspecialised parasite, a soil inhabitant and very widespread (Map 311), particularly in warm, wet areas. The numerous hosts were tabulated by Aycock whose monograph gave a full bibliography to the earlier literature (and see Mordue l.c.). There is a monograph on the perfect state by Goto. Many common names have been given to the diseases caused; southern stem rot is used here. The first symptoms are usually seen as a leaf yellowing and wilt. A cortical decay occurs on the stem at about ground level, the conspicuous mycelium appears and extends into the soil and on any organic debris. The characteristic light to dark brown sclerotia (like mustard seed) form. Mycelial mats may grow within leaf sheaths and spread over the leaf bases. Stem infection may be followed by infection of the upper root system, and of root and stem tubers. Fruits and leaves touching or near the soil may be directly attacked, and infection generally begins near or a little below soil level. Mycelial growth and the formation of sclerotia is high at 25–35° (opt. *c.* 30°). Infection is via the mycelium from germinating sclerotia. Direct penetration of the host occurs although entry is also through wounds. *C. rolfsii* has often been reported in association with other organisms in plant infection. Some recent examples are: stem girdling injury to soybean by *Spissistilus festinus* (Herzog et al.); synergism with *Meloidogyne incognita* on tomato seedlings (Shukla et al); predisposition by *M. arenaria* in soybean (Minton et al.). Outbreaks of southern stem rot have an erratic distribution. They are associated with warm and moist conditions. The fungus survives in the soil, and in organic matter, as sclerotia. The survival time is apparently very variable (a few weeks to >1 year). Sclerotial viability drops off with depth in the soil. Germination of sclerotia is most abundant at the soil surface. High amounts of organic matter in the soil presumably increase inoculum potential. The short-lived basidiospores are not an important infection source. The fungus has been isolated from seed. *Sclerotium coffeicolum* Stahel on coffee in Surinam was considered by the author to be closely related to *C. rolfsii*.

Considerable work has been done on the physiology of sclerotia; see Aycock and the more recent ref. in *Rev. appl. Mycol.* **46**, 1457; **47**, 3006, 3727; **48**, 1079, 1080, 1512, 2796, 3313; *Rev. Pl. Pathol.* **49**, 1274; **50**, 2718; **52**, 1420, 1789, 1790; **53**, 810, 1226, 4661; **54**, 327, 1607; **55**, 588, 4512; **56**, 65, 68, 3884, 4859; **57**, 433, 3803, 4223. See also: fine structure, **49**, 390; **53**, 4315; **55**, 3463; basidial formation, **46**, 67, 272, 885, 2388, 2670, 2996; **47**, 2604; **48**, 2226; selective isolation, **55**, 5061; deterimination of sclerotia in soil, **51**, 4701; **53**, 4730; **57**, 3315; cell degrading enzymes, **48**, 1511; **51**, 3782; **52**, 31, 597; **56**, 490.

Control of this disease relies mainly on cultural measures and soil fumigation where this can be justified on economic grounds. Deep and complete coverage of crop debris in cultivation before planting is necessary. Any competing weeds should be controlled. Quintozene, dazomet and other soil fumigants can be effective. McCarter et al. found metham sodium the most effective of 6 general purpose fumigants in control of *C. rolfsii* on tomatoes; a soil drench was better than soil injections. Chaudhuri et al. reported that quintozene did not eradicate the fungus, but when applied before planting it delayed onset of the disease. Fungicidal control of a rhizome rot of turmeric (*Curcuma domestica*) and a tuber rot of *Colocasia* has been described (Goyal et al.; Reddy et al.). Rodriquez-Kabana used potassium azide on nematodes and the fungus in groundnut; and see Sharma et al. for maize. Various factors may affect the level of control. Recent examples have been given by Backman et al.; Reynolds; and Smith. In groundnut and tomato a significant reduction in damage was given by *Trichoderma harzianum*. When fungicides were applied to control *Mycosphaerella arachidis* and *M. berkeleyi* the unsprayed plots had lower levels of infection by *C. rolfsii*. The reason for this effect was partly attributed to toxicity of the fungicide to *T. viride* which is a natural antagonist of *C. rolfsii*. The highest field levels of *C. rolfsii* were found for the fungicides with no toxicity to this pathogen, but which were toxic to *T. viride*. In common bean southern stem rot was more severe on unmulched compared with mulched plots. Baxter et al. protected tomato transplants with aluminium foil and plastic wrap. Drying sclerotia limits their survival. Dried sclerotia rot within 2 weeks in remoistened soil. Drying also stimulates germination in moist soil in the absence of hosts or another suitable substrate. Under natural conditions drying may be the main mechanism for sclerotial germination. This would explain the haphazard incidence of the disease, even in uniformly infested soil.

Avizohar-Hershenzon, Z. et al. 1969. *Phytopathology* **59**: 288 (mode of action of inorganic nitrogenous amendments; **48**, 2191).

Aycock, R. 1966. *Tech. Bull. N. Carol. Exp. Stn* 174, 202 pp. (monograph, 65 pp. ref.; **46**, 1518).

Backman, P. A. et al. 1975. *Phytopathology* **65**: 773 (effect of fungicides on groundnut, on the non-target pathogen; **35**, 1559).

——. 1975. *Ibid* **65**: 819 (system for growth & delivery of *Trichoderma*; **55**, 1560).

Baxter, L. W. et al. 1977. *Pl. Dis. Reptr* **61**: 341 (control by physical means; **57**, 787).

Boyd, H. W. et al. 1973. *Phytopathology* **63**: 70 (toxicity of crop residue; **52**, 3889).

Chambers, S. C. et al. 1977. *APPS Newsl.* **6**: 6 (soil fumigants; **57**, 332).

Corticium salmonicolor

Chaudhuri, S. et al. 1975. *Z. PflKrankh. PflSchutz* **82**: 674 (inhibitory activity of quintozene against sclerotia; **55**, 3033).

D'Ambra, V. et al. 1974. *Riv. Patol. veg.* Ser. IV **10**: 91 (effects of soil microflora on sclerotial germination, mycelial survival & saprophytic development; **54**, 4371).

Goto, K. 1952. *Spec. Bull. Tokai-Kinki natn. agric. Exp. Stn* 1, 82 pp. (monograph on perfect state; **33**: 184).

Goyal, J. P. et al. 1974. *Beih. nova Hedwigia* 47: 205 (fungicides against tuber rot in *Colocasia*; **54**, 3090).

Herzog, D. C. et al. 1975. *Environ. Ent.* **4**: 986 (disease & injury by *Spissistilus festinus* on soybean; **55**, 5428).

Javed, Z. U. R. et al. 1973. *Trans. Br. mycol. Soc.* **60**: 441 (germination of sclerotia; **52**, 3959).

Linderman, R. G. et al. 1969. *Phytopathology* **59**: 1366 (stimulation in soil by volatile components of lucerne hay; **49**, 1066).

——. 1973. *Ibid* **63**: 358, 500 (effect of volatile compounds from lucerne hay on soil microbial activity & growth of *C. rolfsii*, stimulation of sclerotial germination by volatiles, fungistasis & sodium hypochlorite; **52**, 3958; **53**, 422).

McCarter, S. M. et al. 1976. *Ibid* **66**: 910 (soil fumigants; **56**, 1273).

Minton, N. A. et al. 1975. *Pl. Dis. Reptr* **59**: 920 (*Meloidogyne* spp. & infection of soybean; **55**, 3815).

Reddy, G. S. et al. 1973. *Indian Phytopathol.* **26**: 24 (rhizome storage rot in turmeric; **53**, 4084).

Reynolds, S. G. 1970. *Trop. Agric. Trin.* **47**: 137 (effect of mulches in common bean; **49**, 2700).

——. 1975. *Agron. trop.* **30**: 245 (as above, season & yields; **55**, 2024).

Rodriquez-Kabana, R. et al. 1972. *Pl. Dis. Reptr* **56**: 362 (fungicides on groundnut; **51**, 4555).

Sharma, Y. R. et al. 1976. *Phytopathol. Mediterranea* **15**: 134 (fungicides on maize; **57**, 157).

Shukla, V. N. et al. 1970. *Bull. Indian phytopathol. Soc.* 6: 52 (*M. incognita* & infection of tomato seedlings; **53**, 1082).

Smith, A. M. 1972. *Soil Biol. Biochem.* **4**: 119 (drying & wetting sclerotia promotes biological control; **52**, 3954).

——. 1972. *Ibid* **4**: 125 (nutrient leakage promotes biological control of dried sclerotia; **52**, 3955).

Stahel, G. 1921. *Bull. Dep. Landbouw Surinam* 42, 29 pp. (*Sclerotium coffeicolum*; **1**: 14).

Thanassoulopoulos, C. C. et al. 1971. *Phytopathol. Mediterranea* **10**: 115 (in vitro survival of sclerotia; **50**, 3579).

Weerapat, P. et al. 1966. *Phytopathology* **56**, 640 (effect of soil temp. on resistance of rice seedlings; **45**, 3138).

Wells, H. D. et al. 1972. *Ibid* **62**: 442 (use of *Trichoderma* for control in tomato; **51**, 3798).

Corticium salmonicolor Berk. & Br., *J. Linn. Soc. (Bot.)* **18**: 71, 1783.
Phanerochaete salmonicolor (Berk. & Br.) Jühlich, 1975
Pellicularia salmonicolor (Berk. & Br.) Dastur, 1946
Botryobasidium salmonicolor (Berk. & Br.) Venkatarayan, 1950
stat. conid. *Necator decretus* Massee, 1898.

FRUITBODY extending several cm over surface of bark; resupinate, membranaceous, often about 500 μ thick, smooth, salmon pink when fresh, but cracking and fading to light cream or whitish when dry. HYPHAL SYSTEM monomitic, all hyphae without clamps, hyaline, smooth, distinct. Basal hyphae 6–10 μ diam., cells 30–150 μ long, walls up to 1.5 μ thick, sparsely branched, branching usually at a fairly wide angle. Hymenial branches narrower, shorter celled, thinnner walled and more richly branched than basal hyphae. HYMENIAL SURFACE even, composed of densely crowded basidia only. BASIDIA obovoid to broadly clavate when young, narrowly clavate to cylindrical when mature, thin walled, smooth, 30–55 × 5–10 μ with 4 slightly inwardly curved sterigmata *c.* 4–6 × 1.5–2 μ. SPORES broadly ellipsoidal with prominent apiculus, thin walled, smooth, not amyloid, dextrinoid or cyanophilous, 10–13 × 6–9 μ. Specimens are frequently found in which the pink membranous layer is developed over substantial areas, but basidia are lacking. SPORODOCHIA OF STAT. CONID. discoid, erumpent or superficial on bark, forming orange red pustules or conidiogenous cells. These fragment by separation at the septa to form hyaline, thin walled, aseptate ellipsoid conidia 10–18 (13) × 6–12 (8.5) μ. Although many spp. of resupinate basidiomycetes fructify on woody plants most of them are saprophytic and there is no evidence to suggest that they are likely to be confused with *C. salmonicolor*.

Donk (*Persoonia* 2: 217, 1962) mentioned *C. salmonicolor* as a likely member of a section of *Peniophora* with smooth hymenophore lacking cystidia. He suggested that if it were considered necessary to separate this section from *Peniophora*, *Phanerochaete* might be an appropriate disposition, but he did not make a transfer of *C. salmonicolor* to either *Peniophora* or *Phanerochaete*. The available evidence suggests that Jülich's combination will have to be taken up for this pathogen. However modern taxonomic criteria have been applied to relatively few tropical taxa in the group and reassessment may occur (J. E. M. Mordue & I. A. S. Gibson, *CMI Descr.* 511, 1976).

C. salmonicolor, which causes pink disease (sometimes called rubellosis), is worldwide in the tropics and has been recorded outside this region (Map 122). It is plurivorous and occurs mainly on woody perennials. Although the fungus can cause serious damage on several crops (e.g. cacao, citrus, coffee, tea and black pepper) it has only been studied in any depth as a pathogen of rubber (*Hevea brasiliensis*). Recently Seth et al. described high mortality in *Eucalyptus tereticornis* and *E. grandis* in high rainfall areas of S.W. India. On rubber in Malaysia control of pink disease is often needed (see the annual reports of the Rubber Research Institute of Malaysia). The early literature was largely covered by Hilton, Subba Rao and Tims. The distinctive salmon pink encrustation on the bark represents the advanced stage of the disease. The first symptoms appear to vary with the host. On rubber the initial stages of an attack occur on brown bark as latex drops and then silky white (cobweb like) mycelium on the surface. In black pepper the sterile, pink cushion stage (pustules <1 mm diam.) appear first on the young green stems. In tea Subba Rao found that the later pink encrustation was less common than the other symptoms. In citrus the sterile pustules may appear first, and there may be gumming (see Fawcett, *Citrus diseases and their control*). On rubber open wounds develop and the branch or stem dies. The cobweb stage dries up and the pink, sterile, mycelial masses appear as pustules and as a crust. The former erupt in lines through the bark which sloughs off and the pustules then disappear. The crust usually appears later (it may take several weeks to form) and is typically confined to the lower surface of the branch, although it may develop all round the branch. The crust becomes cracked. The conidia appear as orange red pustules scattered over the bark surface. The attacked branches die with the leaves remaining attached and long black streaks of coagulated latex appear.

Pink disease of rubber is largely an infection of young trees up to *c.* 7 years old, although older trees can be affected by persistent infections of the main fork. In young trees *C. salmonicolor* can cause extensive destruction of the upper stem and leading branches; if the main stem is involved the whole canopy can be lost. Precisely how natural infection occurs is not known. In artificial infections, only mycelium was successful. But in a recent report (1976) basidiospore infection was achieved on 5-year-old trees (Rubber Res. Inst. Malaysia). The basidiospores are wind dispersed and the conidia probably splash dispersed. The disease is one of wet weather (in W. Malaysia particularly where there are well marked rainy seasons); in dry weather spread ceases. Inland areas are more prone to infection than coastal ones. Rubber clones also vary in susceptibility.

Rubber trees 2–8 years old are those most likely to be severely damaged. Early recognition of infection is important, i.e. before stems and branches are killed, and then there should be prompt application of a fungicide. Pruning affected branches is not particularly useful and should be left until dry weather. Bordeaux sprays have been used but require repeated applications in wet weather. Recently brush on applications of other formulations have been shown to be more effective. A single such application of tridemorph (in prevulcanised latex) gave good control for up to 3 months (Anon.; Wastie et al. 1972; Yeoh et al.). In field trials with tridemorph in a rainproof formulation Wastie (1976) found that this treatment showed 3% diseased trees after 3 months compared with 80% such trees where Bordeaux was used.

Altona, T. 1926. *Tectona* **19**: 31 (on *Tectona grandis*; **5**: 457).

Anon. 1974. *Plrs' Bull. Rubb. Res. Inst. Malaysia* 130: 30 (advisory; **53**, 4867).

Dastur, J. F. 1941. *Indian J. agric. Sci.* **11**: 892 (on citrus; **21**, 487).

Hilton, R. N. 1958. *J. Rubb. Res. Inst. Malaya* **15**: 275 (on rubber; **38**: 623).

Ramakrishnan, T. S. et al. 1962. *Rubb. Bd Bull.* **5**: 120 (on rubber in India; **42**, 47).

Schwarz, M. B. 1925. *Meded. Inst. PlZiekt.* 68, 17 pp. (on *T. grandis*; **5**: 67).

Seth, S. K. et al. 1978. *Eur. J. For. Pathol.* **8**: 200 (on *Eucalyptus* spp. in India).

Subba Rao, M. K. 1936. *Bull. United Plrs' Assoc. S. India* 10, 25 pp. (on tea; **16**: 280).

Tims, L. C. 1963. *Pl. Dis. Reptr* **47**: 1055 (in USA; **43**, 1562).

Wastie, R. L. et al. 1972. *Proc. Rubb. Res. Inst. Malaysia Plrs' Conf.* 6 pp. (fungicides; **54**, 1858).

——. 1976. *Pesticides* **10**(2): 55 (as above; **56**, 2203).

Yeoh, C. S. et al. 1974. *Proc. Rubb. Res. Inst. Malaysia Plrs' Conf.* 7 pp. (as above; **54**, 1859).

Corynespora cassiicola

CORYNESPORA Güsow, *Z. PflKrankh.* **76**: 10, 1906.

(Dematiaceae)

COLONIES effuse, grey, olivaceous brown, brown, dark blackish brown or black, often hairy or velvety. MYCELIUM immersed or superficial. STROMA present in some spp. SETAE and HYPHOPODIA absent. CONIDIOPHORES macronematous, mononematous, straight or flexuous, unbranched, brown or olivaceous brown, smooth. CONIDIOGENOUS CELLS monotretic, integrated, terminal, percurrent, cylindrical or doliiform. CONIDIA solitary or catenate, dry, acrogenous, simple, obclavate in most spp., cylindrical in a few, subhyaline, pale to dark brown or olivaceous brown or straw coloured, septate or pseudoseptate, smooth in most spp., verrucose in only a few (M. B. Ellis, *Demat. Hyphom*: 372, 1971).

C. citricola M. B. Ellis (1957) occurs on lime (*Citrus aurantifolia*); see also M. B. Ellis, *Mycol. Pap.* 79, 82, 1961; 87, 93, 1963.

Ellis, M. B. 1957. *Mycol. Pap.* 65, 15 pp. (8 spp. *Corynespora*, key; **36**: 666).

——. 1960. *Ibid* 76, 36 pp. (10 spp. *Corynespora* inter alia; **40**: 29).

Wei, C. T. 1950. *Ibid* 34, 10 pp. (*Corynespora* including *C. cassiicola* & its synonymy; **32**: 342).

Corynespora cassiicola (Berk. & Curt.) Wei, *Mycol. Pap.* 34: 5, 1950.

Helminthosporium cassiicola Berk. & Curt., 1869
Cercospora melonis Cooke, 1896
Corynespora melonis (Cooke) Lindau, 1910
H. papayae H. Syd., 1923
Cercospora vignicola Kawamura, 1931
H. vignae, Olive, 1945.

COLONIES effuse, grey or brown, thinly hairy; viewed under the binocular dissecting microscope the conidiophores appear iridescent. MYCELIUM mostly immersed, no stroma. CONIDIOPHORES erect, simple or occasionally branched, straight or slightly flexuous, pale to mid brown, smooth, septate, monotretic, percurrent, with up to 9 successive cylindrical proliferations, 110–850 μ long, 4–11 μ thick. CONIDIA solitary or catenate, very variable in shape, obclavate to cylindrical, straight or curved, subhyaline to rather pale olivaceous brown or brown, smooth, 4–20 pseudoseptate, 40–220 μ long (up to 520 μ in culture), 9–22 μ thick in the broadest part, 4–8 μ wide at the truncate base; hilum often dark with a slight rim (M. B. Ellis & P. Holliday, *CMI Descr.* 303, 1971).

A widespread, plurivorous fungus which causes a disease usually called target spot; it is especially abundant in the tropics. Symptoms are commoner on leaves but can also be found on stems, roots, flowers and fruit. The necrotic lesions, becoming darker, are usually irregular, often with a wavy border and frequently zonate, up to 2 cm diam., shotholing and defoliation may occur. From Canada a serious infection of roots and hypocotyl of soybean seedlings was reported (Seaman et al.). On this host a soil temp. above 20° appears to arrest disease development; Canadian isolates grew best at *c.* 20° whilst those from south USA had an opt. near 28°. Resistance in soybean occurs. Although *C. cassiicola* has not attracted attention recently in Europe on *Cucurbitaceae*, serious outbreaks in Florida on *Cucumis sativus* (cucumber) have led to recent work on resistance (Strandberg). Resistance in this crop is controlled by a dominant gene (Abul-Hayja et al.). Olive et al. and Spencer et al. described the isolates from soybean and cowpea (*Vigna*) as more pathogenic to the host from which they were derived. On papaw the fungus causes a leaf, stem and fruit disease. Meléndez et al. described the watersoaked spots on the upper stem, petioles and leaves, and called the disease greasy spot. Norse also described it from Barbados. Stem break occurred and the spots developed into sunken lesions (up to 5 cm diam.). Mature fruit was seriously affected (and see Bird et al.).

On tomato a leaf and fruit spot (Blazquez) is caused. Resistance may be controlled by a recessive gene (Bliss et al.). *C. corynespora* affects the young and maturing leaves of rubber (*Hevea brasiliensis*). The yellowish brown spots enlarge to circular or irregular spots, papery, becoming pale grey, 1–8 mm diam. Severe defoliation in mature trees can occur and infection can be higher in unshaded nurseries. Thiram and benomyl have given good control in Malaysia in nurseries (Ananth et al.; Anon; Ramakrishnan et al.). Other hosts include: betel (*Piper betle*), cotton, sesame, tobacco and *Rauvolfia serpentina*. The conidia are air dispersed and a diurnal periodicity (forenoon max.) was described in India. Stone et al. found seed infection in soybean and sesame. Some control may be achieved on most hosts with Bordeaux and dithiocarbamates.

Abul-Hayja, Z. et al. 1978. *Pl. Dis. Reptr* **62**: 43 (inheritance of resistance in cucumber; **57**, 4681).
Ananth, K. C. et al. 1965. *Rubb. Bd Bull.* 8: 78 (on rubber in India, effect of shade & control; **45**, 202).
Anon. 1975. *Plrs' Bull. Rubb. Res. Inst. Malaysia* 139: 84 (advisory on rubber; **55**, 5336).
Bird, J. et al. 1966. *J. Agric. Univ. P. Rico* **50**: 186 (a decline of papaw; **46**, 1056).
Blazquez, C. H. 1977. *Pl. Dis. Reptr* **61**: 1002 (on tomato; **57**, 4141).
Bliss, F. A. et al. 1973. *Phytopathology* **63**: 837 (inheritance of resistance in tomato; **53**, 1086).
Boosalis, M. G. et al. 1957. *Pl. Dis. Reptr* **41**: 696 (general on soybean; **37**: 199).
Culp, T. W. et al. 1964. *Ibid* **48**: 608 (on sesame; **44**, 200).
Fajola, A. O. et al. 1973. *Ibid* **57**: 375 (on tobacco; **52**, 4214).
Hartwig, E. E. 1959. *Ibid* **43**: 504 (effect on yield in soybean; **38**: 724).
Jones, J. P. 1961. *Phytopathology* **51**: 305 (on cotton & cross inoculations to other hosts; **40**: 754).
Kawamura, E. 1931. *Fungi Nippon Fungological Soc.* **1**: 14 (on cowpea).
Meléndez, P. L. et al. 1971. *J. Agric. Univ. P. Rico* **55**: 411 (on papaw; **51**, 4228).
Meredith, D. S. 1963. *Ann. Bot.* **27**: 39 (conidial release; **43**, 2801).
Mohanty, N. N. et al. 1968. *Indian Phytopathol.* **21**: 275 (on betel; 50, 213).
Nair, M. C. et al. 1966. *Agric. Res. J. Kerala* **4**: 54 (conidial, diurnal periodicity; **46**, 745).
—— ——. 1967. *Sci. Cult.* **33**: 69 (stomatal penetration in tomato; **46**, 2525).
Norse, D. 1973. *Pl. Dis. Reptr* **57**: 404 (on papaw; **53**, 630).
Olive, L. S. et al. 1945. *Phytopathology* **35**: 822 (general, races; **25**: 150).
——. 1949. *Mycologia* **41**: 355 (taxonomy; **28**: 559).
Ramakrishnan, T. S. et al. 1961. *Rubb. Bd Bull.* **5**: 32 (on rubber & control; **41**, 330).
Reddy, M. A. R. et al. 1971. *Indian For.* **97**: 487 (on *Rauvolfia serpentina*; **51**, 1762).
Seaman, W. L. et al. 1965. *Can. J. Bot.* **43**: 1461 (full account on soybean; **45**, 1260).
Spencer, J. A. et al. 1969. *Phytopathology* **59**: 58 (races, temp. & growth; **48**, 2084).
Stone, W. J. et al. 1960. *Ibid* **50**: 263 (seedborne in soybean & sesame; **39**: 710).
Strandberg, J. O. 1971. *Pl. Dis. Reptr* **55**: 142 (resistance in cucumber; **50**, 2612).

CRINIPELLIS Pat., *J. Bot. Paris* 3: 336, 1889; em. Earle, *Bull. N.Y. bot. Gard.* 5: 414, 1909. (Tricholomataceae)

A full description is given in *The Agaricales in modern taxonomy*: 368, 1975, by Singer who also monographed the genus. The spp. are mostly saprophytic; some are horse hair blights (*Marasmius* q.v.); but there is one extremely important pathogen of cacao.

Singer, R. 1942. *Lilloa* 8: 441 (monograph, *Crinipellis* & *Chaetocalathus*; **23**: 411).

Crinipellis perniciosa (Stahel) Singer, *Lilloa* **8**: 503, 1942.
Marasmius perniciosus Stahel, 1915.

PILEUS 5–25 mm diam., thin, reviving, conico-campanulate expanding to convex and finally depressed to concave or umbilicate, sometimes umbonate; surface 'Eugenia Red' (M.2.5R/4.8/10.0), typically fading to 'Venetian Pink' (M.2.5R/7.8/4.0) and bleaching to whitish especially at the margin but retaining a dark reddish to black spot at the disk, radially fibrillose-striate with short strigose hairs densely crowded at the disk but scarce to absent elsewhere except in young specimens. LAMELLAE adnato-adnexed, white, rather thick, 1–2 mm broad, broadly spaced, 8–12 entire, with occasional lamellulae of 2 or 3 lengths and some intervening; edge entire, concolorous. STIPE $5–14 \times 0.5–1$ mm, short, often curved, cylindric with a small, sub-bulbous base up to 2 mm diam., solid to narrowly fistulose; surface initially light citron yellow at the base, white elsewhere, darkening to reddish brown or dark brown from the base upwards but generally remaining pale at the apex, pubescent to flocculose, insititious. CONTEXT very thin, 250–600 μ thick, white, inamyloid, of loosely woven, thin-walled hyphae, 2.5–9 μ diam. inflating to 32 μ diam.; ODOUR none; TASTE mild. Spore print pure white. SPORES $9–12.5 \times 4.5–6$ $(10.6 \pm 0.86 \times 5.3 \pm 0.35)$ μ, Q=2.02, ellipsoid to oblong ellipsoid, hyaline, inamyloid, thin walled, with few granular contents. BASIDIA $23–35 \times 7–10$ μ, clavate, bearing 4 sterigmata up to 4 μ long. LAMELLA EDGE sterile or heteromorphous, with crowded although at times inconspicuous cheilocystidia. CHEILOCYSTIDIA $26–50 \times 6.5–13$ μ, irregularly clavate to fusoid, often subcapitate or mucronate, rarely indistinctly digitate, hyaline, thin walled. PLEUROCYSTIDIA none. HYMENOPHORAL

TRAMA subregular, hyaline, of more or less parallel, inflated hyphae, 4–23 μ diam. Subhymenial layer well developed, interwoven, 20–26 μ wide. PILEIPELLIS a well-defined hypotrichial layer giving rise to thick-walled hairs, often in fasciculate clusters. Hypotrichial layer of inamyloid hyphae with a slightly thickened wall, regularly septate and forming short elements, 20–55 × 3–27 μ, sometimes almost isodiametric, often with a granular encrustation and then appearing crimson. Hairs numerous at pileal disk, scarce elsewhere, 25–175 (–250) × 3–9 μ, unbranched, with a thickened, dextrinoid wall up to 3 μ thick, typically with a crimson membrane pigment which may bleach in older specimens, narrowing gradually to an obtusely rounded apex, often with fine granular encrustation. STIPE HAIRS present, similar to those on the pileus but usually much shorter and scattered. CLAMP CONNECTIONS prominent on all hyphae (D. N. Pegler, *Kew Bull.* **32**: 731, 1978 as var. *perniciosa*).

Pegler, working with material investigated by H. C. Evans in Ecuador, recognised 3 vars: *perniciosa*, *ecuadoriensis* (Stahel) Pegler (synonym: *M. perniciosus* Stahel var. *ecuadoriensis* Stahel) and *citriniceps* Pegler. The first 2 vars have a pileus with pale to deep crimson tints. In var. *perniciosa* it is whitish, at least at the margin, with crimson tints confined to the centre and along the striae. Var. *ecuadoriensis* has a uniformly deep crimson red pileus which lacks the white areas. Var. *citriniceps* has a citron yellow pileus.

C. perniciosa var. *perniciosa* causes the extremely serious but geographically restricted witches' broom disease of cacao. Stahel (1915, 1919) determined the aetiology and fully described the disease which occurs in Bolivia, Brazil (N., centre and W.), Colombia, Ecuador, Grenada, Guyana, Peru, Trinidad and Tobago, Surinam and Venezuela (Map 37; and see Pegler for a Cuban record). The pathogen is indigenous on wild spp. of *Theobroma* in the forests of the Amazon and Orinoco river systems of S. America. Besides the primary host (*T. cacao*) it has been recorded on *T. bicolor*, *T. calodesmis*, *T. glauca*, *T. grandiflora*, *T. microcarpa*, *T. obovata* and *T. subincana*; also on the related genus *Herrania* (*H. albiflora*, *H. nitida*, *H. purpurea*). Recently, Evans, who described 2 pathotypes, found a form of the fungus to be endemic in the forests of Ecuador on lianes, shrubs and trees. A full account of the disease was given by Baker et al. (1957), a summary by Holliday (1970) and a later account by Thorold (*Diseases of cocoa*); see also the general work in Trinidad and Tobago by Baker et al., Bartley, Dale, Pound and Holliday.

Infection by the basidiospores causes the growing vegetative shoots to become hypertrophied; swollen and stunted stems and petioles, small leaves and abnormal proliferation of the axillary buds occur. These abnormalities, the witches' brooms, have been classified according to the particular branch type (vegetative brooms) on which they are borne (fan or chupon), and to their growth on the stem (terminal, lateral or grown through). They are at first green, later becoming necrotic and may be > 30 cm long. Cushion infection causes the flower and pod peduncles to become swollen, the young pod assumes a strawberry shape and fails to develop. Large, abnormal, vegetative shoots (cushion brooms) frequently form on the cushions. Pods are individually infected, becoming distorted with hard necrotic areas and few healthy beans.

The disease is characterised by a long period between initial infection by basidiospores (the only spore form) and fructification, 28 weeks or more for vegetative brooms. Infection occurs through breaking vegetative buds, young pods and cushions. More recently Cronshaw et al. found that penetration could occur on the stem, petiole and pulvinus of an unhardened flush. When a vegetative bud is infected the new shoot develops apparently without external symptoms, but after 5–6 weeks the hypertrophy becomes evident. The intercellular mycelium is confined to the witches' broom that is formed and there is no growth by the parasite from it into healthy mature tissue. After a further 3–4 weeks the brooms begin to go necrotic and dry out. It is only then, after another long interval, that sporophore formation begins. Sporophores are never produced on green, turgid brooms. The brooms remain attached to the tree for considerable periods, half of them for nearly 12 months. Sporophores also form on necrotic pods, but because these are usually removed, piled and allowed to rot on the ground they are less important as an infective source.

Spore deposition is heaviest at night at high RH and between 16–26°. The basidiospores are readily killed by desiccation and probably few survive beyond the day following the night they are shed. In Trinidad, where much of the study on this disease has been done, a dry season of 3–4 months markedly inhibits the formation of sporophores and hence infection. As the wet season settles in fructification begins and reaches a peak some 6 months later towards the end of the season. There is a strong positive correlation between the number of sporophores in 1 week and the number of brooms 5 weeks later; the time lag being the mean incubation time for the first symptoms. Flowering intensity is also positively correlated with cushion broom formation and self-incompatible clones tend to produce more cushion brooms. Natural spread is by the wind dispersed basidiospores and the disease is possibly most severe where there is a moderate annual rainfall of *c.* 200 cm. Infection builds up in the canopy and the trees become covered with hundreds of dry, necrotic brooms. Crop loss, through pod infection, can be over 80% under epidemic conditions. Seed infection can occur (Cronshaw et al.).

Growth on organic nutrient agar is readily achieved but is slow; a white mycelium is produced and under the influence of light a purplish pigmentation may appear. Sporophores have not been produced in vitro. The homothallic condition is thought to exist since a single spore culture forms clamp connections. The hypertrophies produced in the host may be due to the production of enzymes which have been demonstrated and which inactivate the host auxins.

Of the 2 pathotypes referred to the virulent form (var. *perniciosa*) occurs on cacao and the weakly or non-pathogenic form (var. *ecuadoriensis*) occurs on endemic plants in Ecuador and on *Theobroma*. The pathogenicity characteristics of the var. *citriniceps* do not appear to have been described. Highly resistant cvs developed in Trinidad, from material collected by Pound in S. America, were much less resistant when planted in Ecuador. This suggested the existence of races within the pathotype now referred to as var. *perniciosa*; this apparently remains to be demonstrated. Less, but significant, resistance occurs in selections made from the local Trinidad and Tobago cacao population (Imperial College selections). The inheritance of the high resistance of the S. American selections (to the Trinidad form(s) of the pathogen) is apparently complex.

Some control has been achieved by the physical removal of witches' brooms 2–4 times a year, depending on the rainfall pattern. The success of the method depends on high cultivation standards and the use of plant material with some resistance. Chemical treatments (e.g. Holliday 1960; Tollenaar) have not been adopted. They need further investigation and may become worth while with higher standards of cultivation. Cacao pods, in parts of the Americas, which develop in the dry season escape heavy infection from fungus pathogens. Edwards manipulated the timing of crop maturity by supplementing natural pollination during the early dry season. He reduced losses and found economic increases in yield.

Baker, R. E. D. et al. 1942. *Trop. Agric. Trin.* **19**: 207 (host susceptibility; **22**: 128).

——. 1943. *Ibid* **20**: 5 (epidemiology; **22**: 163).

——. 1943. *Ibid* **20**: 156 (host susceptibility; **22**: 471).

———. 1943. *Ibid* **20**: 188 (cushion & pod infection; **23**: 94).

———. 1943. *Mem. Imp. Coll. trop. Agric. Trin.* 7, 28 pp. (general; **22**: 242).

———. 1944. *Ibid* 8, 28 pp. (general; **24**: 9).

———. 1944. *Trop. Agric. Trin.* **21**: 170, 175, 196 (fan broom formation & pod loss; **24**: 10, 52).

———. 1954. *Rep. Cacao Res. Trin. 1953*: 8 (host range).

———. 1957. *Phytopathol. Pap.* 2, 42 pp. (monograph, 133 ref.; **37**: 151).

Bartley, B. G. D. 1959. *Rep. Cacao Res. Trin. 1957–58*: 49 (resistance tests; **39**: 684; and see subsequent reports).

— et al. 1959. *Ibid*: 53 (*Herrania* infection; **39**: 684).

Cronshaw, D. K. et al. 1978. *Ann. appl. Biol.* **89**: 193 (infection).

Dale, W. T. 1946. *Trop. Agric. Trin.* **23**: 217 (pod infection **26**: 383).

Delgado, J. C. et al. 1976. *Phytopathology* **66**: 717 (formation of arthroconidia in culture; **56**, 629).

———. 1976. *Can. J. Bot.* **54**: 66 (nuclear condition: **55**, 3504).

Dudman, W. F. et al. 1959. *Nature Lond.* **183**: 899 (growth substances; **38**: 681).

Edwards, D. F. 1978. *J. hort. Sci.* **53**: 243 (manipulation of crop maturity in Ecuador & pod diseases; **57**, 5372).

Evans, H. C. 1978. *Ann. appl. Biol.* **89**: 185 (pathotypes in Ecuador).

Holliday, P. 1950. *Proc. agric. Soc. Trin. Tob.* **50**: 393 (control; **32**: 305).

——. 1952. *H.M.S.O. Colonial* 286, 8 pp. (advisory).

——. 1954. *Rep. Cacao Res. Trin. 1953*: 56, 64 (host susceptibility & fungicides; **34**, 775).

——. 1954. *Trop. Agric. Trin.* **31**: 312 (control; **34**, 216).

——. 1955. *Rep. Cacao Res. Trin. 1954*: 50 (a resistance test; **35**: 427).

——. 1957. *Ibid 1955–56*: 48 (host susceptibility; **37**: 151).

——. 1960. *Trop. Agric. Trin.* **37**: 235 (fungicide; **40**: 215).

Holliday, P. 1970. *C.M.I. Descr. pathog. Fungi Bact.* 223 (*Crinipellis perniciosa*).
Krupasagar, V. et al. 1969. *Am. J. Bot.* **56**: 390 (auxin destruction; **48**, 2864).
Lems, G. 1963. *Surinam. Landb.* **11**: 138 (resistance field trial; **43**, 1584).
Lindelberg, G. et al. 1948. *Physiol. Pl.* **1**: 401 (enzymes; **29**: 45).
Pegler, D. N. 1978. *Kew Bull.* **32**: 731 (morphology & distribution of 3 vars).
Pound, F. J. 1938. *Rep. Dept Agric. Trin. Tob.* 58 pp. (exploration in S. America; **17**: 801).
——. 1940. *Trop. Agric. Trin.* **17**: 6 (resistance; **19**, 266).
——. 1943. *Rep. Dept Agric. Trin. Tob.* 14 pp. (exploration in Peru; **23**: 56).
——. 1943. *Proc. agric. Soc. Trin. Tob.* **43**: 55 (resistance; **22**: 346).
Stahel, G. 1915. *Bull. Dep. Landb. Suriname* 33, 26 pp. (description sp. nov.).
——. 1918. *Ann. Jard. bot. Buitenz.* Ser. 2, 15: 95 (infection & inflorescence morphology).
——. 1919. *Bull. Dep. Landb. Suriname* 39, 34 pp. (monograph).
Tollenaar, D. 1959. *Trop. Agric. Trin.* **36**: 177 (fungicides; **38**: 735).

CRISTULARIELLA Höhnel emend. Redhead, *Can. J. Bot.* **53**: 704, 1975. (Moniliaceae)

Foliar parasites producing epiphyllous propagules. PROPAGULES multicellular, subglobose, slightly flattened to disk or cone shaped, hyaline, becoming brown with age, internally branched. BRANCHES densely compacted, composed of clavate or subglobose cells formed by primary and secondary budding mainly in a radial direction away from an inflated central cell or an erect septate central hypha (a continuation of the propagulopore). PROPAGULOPHORE hyaline, erect (i.e. perpendicular to the substrate), septate, uniseriate, solitary, tapering acropetally. Propagules separated from the propagulophores at a constricted basal septum (from S. A. Redhead l.c.).

C. pyramidalis Waterman & Marshal has also been described by Latham (1969). The fungus causes zonate leaf spots on some temperate hosts; also on groundnut, lima bean (*Phaseolus lunatus*), common bean, kenaf (*Hibiscus cannabinus*) and tung (*Aleurites*). On kenaf severe defoliation has occurred in USA (Mississippi). The leaf lesions are initially circular, yellowish tan with a dark brown, well-defined margin; they expand to give the typical concentric zonation. Newly infected tissue is dark and oily in appearance; shotholes and tattered leaf margins develop. On groundnut the necrotic leaf spots are 2–13 mm diam., have light brown centres and a brown margin. Larger spots show zonations on both leaf surfaces. Latham (1974a) found most in vitro growth, sporulation and lesion development on pecan (*Carya pecan*) at 21°. No sporophores formed at <94% RH. The fungus was viable in 60-day-old leaf lesions. It forms a toxin (Kurian et al.).

Baniecki, J. F. et al. 1974. *Pl. Dis. Reptr* **58**: 421 (hosts; **53**, 3404).
Crandall, B. S. et al. 1972. *Ibid* **56**: 917 (on kenaf; **52**, 2629).
French, W. J. 1972. *Ibid* **56**: 135 (hosts; **51**, 3578).
Grand, L. F. et al. 1974. *Mycologia* **66**: 712 (sclerotia in nature; **54**, 1129).
Hirano, K. et al. 1975. *Tech. Bull. Fac. Hort. Chiba Univ.* **61**(23): 53 (on *Lagerstroemia indica*; **55**, 4148).
Kurian, P. et al. 1977. *Mycologia* **69**: 1203 (toxin; **57**, 3807).
Latham, A. J. 1969. *Phytopathology* **59**: 103 (on pecan; **48**, 2006).
——. 1974a & b. *Ibid* **64**: 635, 1255 (effects of temp. & moisture; **53**, 4887; **54**, 2987).
Neely, D. et al. 1976. *Pl. Dis. Reptr* **60**: 590 (hosts; **56**, 1761).
Pollack, F. G. et al. 1969. *Ibid* **53**: 810 (on kenaf; **49**, 770).
Redhead, S. A. 1975. *Can. J. Bot.* **53**: 700 (taxonomy & morphology of *Cristulariella*; **54**, 4832).
Smith, D. H. 1972. *Pl. Dis. Reptr* **56**: 796 (on groundnut; **52**, 1739).
Trolinger, J. C. et al. 1978. *Ibid* **62**: 710 (hosts).
Waterman, A. M. et al. 1947. *Mycologia* **39**: 690 (*C. pyramidalis* sp. nov.; **27**, 395).

CURVULARIA Boedijn, *Bull. Jard. bot. Buitenz.* III **13**: 120, 1933. *Malustela* Batista & Lima, 1960. (Dematiaceae)

COLONIES effuse, brown, grey or black, hairy, cottony or velvety. MYCELIUM immersed in natural substrata. STROMATA often large, erect, black, cylindrical, sometimes branched, formed by many species in culture, especially on firm substrata such as rice grains. CONIDIOPHORES macronematous, mononematous, straight or flexuous, often geniculate, sometimes nodose, brown, usually smooth. CONIDIOGENOUS CELLS polytretic, integrated, terminal, sometimes later becoming intercalary, sympodial, cylindrical or occasionally swollen, cicatrised. CONIDIA solitary, acropleurogenous, simple, often curved, clavate, ellipsoidal, broadly fusiform, obvoid or pyriform with 3 or more transverse septa. Pale or dark brown, often with some cells, usually the end ones, paler than the others, sometimes with dark bands at the septa, smooth or verrucose; hilum in some spp. protuberant. In many spp. occasional triradiate stauroconidia are formed at the same time as normal conidia (M. B. Ellis, *Demat. Hyphom.*: 452, 1971).

Ellis (1966; l.c.) described and keyed out the spp.; and see Somal; Upsher. They are saprophytes and at the most weak pathogens in the tropics, sometimes attacking the leaves of plants, particularly when young. Many are probably secondary invaders and *Curvularia* spp. are frequently associated with seed (Bugnicourt; Groves et al.; Rao et al.) and blights of cereal inflorescences. *C. lunata* (Wakker) Boedijn (*Cochliobolus lunatus* q.v.) is cosmopolitan. It seems likely that all the spp. in the tropics have many hosts and substrates; the commonest are: *Curvularia eragrostidis*, *C. geniculata*, *C. pallescens* and *C. verruculosa*. The following minor diseases are briefly described with the conidial measurements for 9 *Curvularia* spp.

C. andropogonis (Zimm.) Boedijn (45–66 × 18–28 μ) on *Andropogon* and *Cymbopogon* (Alam et al.; Sloof et al.).

C. clavata Jain (17–29 × 7–13 μ) on sorghum. Reported causing a maize leaf spot; small, necrotic, straw coloured spots or occasionally oval lesions (0.5–2 mm) with dark brown peripheral rings and surrounded by pale yellow translucent zones (Mandokhot et al.).

C. cymbopogonis (C. W. Dodge) Groves & Skolko (35–60 × 14–20 μ, in culture 28–40 × 12–18 μ) on Andropogoneae (synonym: *Helminthosporium cymbopogonis* C. W. Dodge as *cymbopogoni*). Described from *Cymbopogon citratus* and *C. nardus* (lemon and citronella grasses), and reported as pathogenic to seedlings of rice (Santamaria et al.). The perfect state is *Cochliobolus cymbopogonis* Hall & Sivanesan.

C. eragrostidis (P. Henn.) J. A. Meyer (22–23 × 10–18 μ, in culture 18–37 × 11–20 μ), plurivorous (synonyms: *Brachysporium eragrostidis* P. Henn., *Spondylocladium maculans* Bancroft, *Curvularia maculans* (Bancroft) Boedijn). Nelson (1956) described a leaf spot of maize; minute, straw coloured lesions 1–2 mm long (seldom > 5–7 mm) but may coalesce to form larger necrotic areas; separate lesions frequently have a reddish brown margin. Johnston described an oil palm seedling blight. The symptoms occur on the youngest leaves; at first as minute, round, translucent spots which develop a large brown, slightly sunken centre. The spots become darker brown, irregular in shape with a well-defined, narrow, translucent, yellow (becoming orange) halo. The mature spot (usually not > 7–8 mm diam.) within the halo has a narrow, raised, greenish brown, oily looking rim; within the rim the colour is red brown with 1–2 roughly concentric, lesser ridges; the centre is light brown. The disease is most severe a few months after transplanting to the field nursery; leaf and seedling death can occur. Dithiocarbamate sprays gave control. Infection can be severe on Malaysian yellow dwarf coconut seedlings. The small, circular, chlorotic spots expand becoming light brown. Captafol, captan and cycloheximide were promising for control (Chan).

C. geniculata (Tracy & Earle) Boedijn (28–48 × 8–13 μ, in culture 18–37 × 8–14 μ), plurivorous (perfect state *Cochliobolus geniculatus* Nelson, 1964; synonyms: *Helminthosporium geniculatum* Tracy & Earle, *Brachysporium sesami* Sawada).

C. pallescens Boedijn (17–32 × 7–12 μ) is plurivorous. Mabadeje described a leaf spot of maize and discussed references to *Curvularia* spp. on maize (and see Malaguti et al; Fajemisin for control).

C. penniseti (Mitra) Boedijn (29–42 × 13–20 μ) on *Pennisetum* and sorghum (synonym: *Acrothecium pennesiti* Mitra). On *P. typhoides* (bulrush millet) small yellowish spots form at first on the lower leaves, becoming oval longitudinally; coalescence of the spots leads to death of the leaves; leaf sheaths and ears can be spotted (Mitra; Patil et al. 1966b). On leaves of sorghum irregular, yellow spots become necrotic, coalesce and the leaf dries out; attacked ears are small (Patil et al. 1966a; see also Choudhary et al.; Singh et al.).

C. senegalensis (Speg.) Subram. (19–30 × 10–14 μ) was isolated from sugarcane (including seed) and Johnson grass (*Sorghum halepense*) by Yang. The fungus was pathogenic (opt. 25–30°) to seedlings of both plants.

C. verruculosa Tandon & Bilgrami ex M. B. Ellis (25–32 × 12–14 μ, in culture 20–35 × 12–17 μ) is plurivorous. On rice the leaf spots are pale yellow, becoming watery grey, translucent to light brown; leaves may dry out and brown or black lesions occur on the spikelets (Aulakh); also on tobacco (Prasad et al.).

Curvularia spp. are readily grown in culture and have frequently been studied in vitro for effects of temp., pH, nutrients, growth factors and light.

Agrawal, S. C. et al. 1969. *JNKVV Res. Jnl* 3: 67 (on sorghum; **50**, 658).

Alam, M. et al. 1976. *New Botanist* 3: 54 (on *Cymbopogon winterianus*).

Aulakh, K. S. 1966. *Pl. Dis. Reptr* **50**: 314 (or rice; **45**, 2836).

Bugnicourt, F. 1950. *Rev. gén. Bot.* **57**: 65 (on rice grain; **29**: 277).

Chan, C. L. 1974. *MARDI Res. Bull.* **2**(1): 19 (on coconut; **54**, 1395).

Choudhary, N. S. et al. 1974. *Nova Hedwigia* **25**: 255 (on bulrush millet; **54**, 1253).

Dodge, C. W. 1942. *Ann. Mo. bot. Gdn* **29**: 137 (on *Cymbopogon citratus* & *C. nardus*; **21**: 419).

Ellis, M. B. 1966. *Mycol. Pap.* 106, 57 pp. (*Curvularia* inter alia; **46**, 1228c).

Fajemisin, J. M. et al. 1976. *PANS* **22**: 234 (on maize; **55**, 5169).

Groves, J. W. et al. 1945. *Can. J. Res.* Sect. C **23**: 94 (on seed; **26**: 405).

Hall, J. A. et al. 1972. *Trans. Br. mycol. Soc.* **59**: 314 (*Cochliobolus cymbopogonis* sp. nov.).

Johnston, A. 1959. *Malay. agric. J.* **42**: 14 (on oil palm; **38**: 761).

Kranz, J. 1965. *Phytopathol. Z.* **52**: 202 (on pineapple; **44**, 3114).

Lele, V. C. et al. 1968. *Indian Phytopathol.* **21**: 66 (on citrus; **48**, 1183).

Mabadeje, S. A. 1969. *Trans. Br. mycol. Soc.* **52**: 267 (on maize; **48**, 2356).

Malaguti, G. et al. 1971. *Agron. trop.* **21**: 119 (on maize; **51**, 1435).

Mandokhot, A. M. et al. 1972. *Neth. J. Pl. Pathol.* **78**: 65 (on maize; **51**, 3985).

Mitra, M. 1921. *Mem. Dep. Agric. India bot. Ser.* **11**: 57 (on bulrush millet).

Nelson, R. R. 1956. *Pl. Dis. Reptr* **40**: 210 (on maize; **36**, 21).

——. 1964. *Mycologia* **56**: 777 (*Cochliobolus geniculatus* sp. nov.).

Patil, P. L. et al. 1966a. *Mycopathol. Mycol. appl.* **28**: 77 (on sorghum; **45**, 2097).

—— ——. 1966b. *Ibid* **28**: 348 (on bulrush millet; **46**: 314).

Prasad, S. S. et al. 1965. *Curr. Sci.* **34**: 542 (on tobacco; **45**, 605).

Rao, P. N. et al. 1954. *J. Indian bot. Soc.* **33**: 268 (5 spp. on cereal grains; **34**: 175).

Santamaria, P. A. et al. 1971. *Pl. Dis. Reptr* **55**: 349 (on rice; **50**, 3730).

Singh, R. R. et al. 1973. *Indian J. agric. Sci.* **43**: 895 (on bulrush millet; **54**, 4471).

Sloof, W. C. et al. 1947. *Chronica Nat.* **103**: 137 (on *Andropogon*; **27**: 44).

Somal, B. S. 1976. *Indian J. Mycol. Pl. Pathol.* **6**: 59 (key to *Curvularia* spp.; **57**, 1000).

Townsley, W. W. 1969. *Phytopathology* **59**: 523 (*C. geniculata*; **48**, 2790).

Upsher, F. J. 1975. *Int. Biodetn Bull.* **11**: 24 (general description of *Curvularia*, 43 ref.; **54**, 3754).

Yang, S. M. 1973. *Phytopathology* **63**: 1541 (*C. senegalensis*; **53**, 3605).

CYLINDROCARPON Wollenw.,

Phytopathology 3: 225, 1913.

Allantospora Wakker, 1896.

(Moniliaceae)

CONIDIA are slimy phialospores formed from phialides in basipetal succession, generally not adhering in chains. MICROCONIDIA, if present, hyaline, oval to ellipsoid, 0–1 septate. MACROCONIDIA always present, hyaline straight or curved, cylindrical to fusoid but with rounded ends and without *Fusarium* type foot cell, 1–10 transverse septa. PHIALIDES simple, with single apical pore bearing a collar, formed laterally on hyphae, terminally on simple lateral branches, or singly or in groups as termination to branches of penicillately branched CONIDIOPHORES. No continuation of conidiophore axis into a sterile appendage. CHLAMYDOSPORES present or absent, hyaline to brown, globose, formed singly, in chains or clumps, intercalary or terminal; or on lateral branches; or singly, or in chains in cells of the macroconidia. CULTURES white, beige, orange brown to purple, fluccose to felted. Sterile stromatic pustules or sporodochia present or absent. PERITHECIAL STATE where known a *Nectria* or, in one case, *Calonectria* (after C. Booth, *Mycol. Pap.* 104: 1, 1966).

The genus was revised by Booth who accepted 27 spp. and 6 vars., and gave a key. *Cylindrocarpon* spp. occur commonly in soil (see Domsch & Gams, *Fungi in agricultural soils*); most are saprophytes, weak or wound pathogens. *C. destructans* (*Nectria* q.v.) is plurivorous and causes diseases mostly on temperate crops. *C. lucidum* Booth (perfect state *N. lucida* Höhnel) has a wide range, generally on tropical trees. Lele et al. reported it as causing a dry root rot of trifoliate orange (*Poncirus trifoliata*) in the nursery on 6–18-month-old plants. *C. musae* Booth & Stover is commonly associated with the lesions on banana caused by the burrowing nematode (*Radopholus similis*); it was described as a wound invader. *C. panacis* Matuo & Miyazawa causes a dry root rot of ginseng (*Panax schinseng*) in Japan. The fungus causes round or irregular, blackish brown lesions on the main and lateral roots which are killed. *C. tenue* Bugn. occurs in the tropics and has been reported from cacao, coffee, cotton and tea. Agnihothrudu reduced *Gliocladiopsis sagariensis* Saksena to synonymy with *C. tenue*. These 2 spp. occur widely in the tropics: *C. tonkinense* Bugn. and *C. victoriae* (P. Henn. ex Wr.) Wollenw. (perfect state *N. jungneri* P. Henn.).

Agnihothrudu, V. 1959. *Trans. Br. mycol. Soc.* **42**: 458 (*Cylindrocarpon tenue*).
Booth, C. 1966. *Mycol. Pap.* 104, 56 pp. (monograph; **46**, 1228).
—— et al. 1974. *Trans. Br. mycol. Soc.* **63**: 503 (*C. musae* sp. nov.; **54**, 2905).
Evans, G. et al. 1966. *Ibid* **49**: 563 (antibiotics from *C. destructans*; **46**, 1196).
——. 1967. *Pl. Soil* **26**: 253 (phytotoxin from *C. destructans*; **46**, 2992).
Lele, V. C. et al. 1969. *Indian Phytopathol.* **22**: 497 (*C. lucidum* on trifoliate orange; **50**, 680).
Matuo, T. et al. 1969. *Trans. mycol. Soc. Japan* **9**: 109 (*C. panacis* sp. nov.; **49**, 229).
Subramanian, S. et al. 1968. *Pl. Dis. Reptr* **52**: 773 (*C. tenue* on coffee; **48**, 475).

CYLINDROCLADIUM Morgan, *Bot. Gaz.* 17: 191, 1892.

Candelospora Hawley apud Rea & Hawley, 1912
Tetracytum Vanderwalle, 1945.
(Moniliaceae)

Aerial mycelium well developed, at first white and cottony, later on a shade of brown in the centre; when submerged brown to red brown with numerous CHLAMYDOSPORES in chains or irregular; compact masses form sclerotial bodies. CONIDIOPHORES erect, dichotomously branched near apex, ultimate branches bearing phialides; main axis mostly forming a long, unbranched thread terminating in a globose to club shaped apex (or vesicle). In some spp. this structure, characteristic of the genus, is often only found in a limited number of conidiophores. CONIDIA solitary on the phialides, cylindrical, 1 to many septate, mostly adhering to a colourless substance. After germination the conidia may anastomose freely (after C. V. Subramanian, *Hyphomycetes*: 730, 1971).

Some *Cylindrocladium* spp. have perfect states in *Calonectria* (q.v.). *Cylindrocladium camelliae* Venkataramani & Venkata Ram has been reported as causing a minor root rot of tea and occurs on leguminous trees. *C. clavatum* Hodges & May is associated with a root disease of *Araucaria angustifolia*, *Eucalyptus saligna* and especially *Pinus* spp. in Brazil. Hodges et al. (1972) discussed the differences between *C. clavatum* and other members of the genus. Its conidia are 1 septate (av. 42.1 × 4.7 μ) and different in size from the 2 other spp. with 1 septate conidia (*C. parvum* Anderson, conidia 15–21 × 2–3 μ; *C. pteridis* F. A. Wolf, conidia 63–85 × 6.5–7.5 μ). All 3 spp. have clavate vesicles. Other described spp. of the genus with clavate vesicles have multiseptate conidia. *C. clavatum* was isolated from *Pinus* spp. where there were diseased centres of several dead and dying trees of up to 15 years old. The roots were pitch soaked and sometimes showed copious resin exudation. Needles of affected trees turned bright yellow, drooped and became brick red. *C. ilicicola* (Hawley) Boedijn & Reitsma was reported to have a perfect state, *Calonectria ilicicola* Boedijn & Reitsma (Boedijn et al.); it causes damping of *Eucalyptus* spp. Sobers described the host range and morphology of *Cylindrocladium pteridis* (synonym: *C. macrosporum* Sherbakoff). In Malaysia (West) a blight of the leading or secondary shoots (brown needle) of *Pinus* spp. was associated with *C. pteridis*. The condition occurred in nurseries and young field plantings (Ivory). A disease of pine caused by the same fungus in Brazil was described by Hodges et al. (1975). Deep reddish brown lesions, 2–5 mm long, form on the needles; heavily infected trees look scorched. Hunter et al. described the formation of microsclerotia, growth and sporulation of *Cylindrocladium* spp.

Boedijn, K. B. et al. 1950. *Reinwardtia* **1**: 51 (taxonomy; **30**: 345).
Figueredo, M. B. et al. 1963. *Arq. Inst. biol. S. Paulo* **30**: 29 (*Cylindrocladium ilicicola* on *Eucalyptus* spp.; **43**, 3028).
Hodges, C. S. et al. 1972. *Phytopathology* **62**: 898 (*C. clavatum* on *Pinus* spp.; **52**, 1289).
——. 1975. *Brasil Florestal* **6**(21): 8 (*C. pteridis* on *P. caribaea*; **55**, 2925).
Hunter, B. B. et al. 1976. *Phytopathology* **66**: 777 (formation of microsclerotia; **56**, 573).
——. 1978. *Mycologia* **70**: 614 (growth & sporulation in vitro).
Ivory, M. H. 1973. *PANS* **19**: 510 (*C. pteridis* on *Pinus* spp.; **53**, 3211).

Cylindrocladium scoparium

Peerally, A. 1974. *C.M.I. Descr. pathog. Fungi Bact.* 422 & 428 (*C. clavatum* & *C. camelliae*).
Sobers, E. K. 1968. *Phytopathology* **58**: 1265 (morphology & host range, *C. pteridis*; **48**, 609).
—— et al. 1972. *Proc. Fla St. hort. Soc.* **85**: 366 (*Cylindrocladium* spp. & hosts in Florida & Georgia, USA; **53**, 1716).
Terashita, T. 1969. *Trans. mycol. Soc. Japan* **9**: 113 (*C. camelliae* & *Calonectria hederae*).

Cylindrocladium scoparium Morgan, *Bot. Gaz.* **17**: 191, 1892.

Diplocladium cylindrosporum Ell. & Ev., 1900
Cylindrocladium pithecolobii Petch, 1917.

CONIDIOPHORES scattered over leaf or agar culture surface; up to 0.5 mm high, 7.5–8.5 μ diam. at stalk base; main axis continuing beyond the sporogenous zone to form the long sterile appendage characteristic of this genus, 120–150 μ long, 7 μ at the base and 5 μ just below the apical swollen cell (vesicle) which is hyaline, inequilaterally obovoid, 26–28 × 11–12 μ. Sporogenous part formed of 2 or more bifurcate lateral branches to the main stalk; primary branches have cells 22–44 × 5–7 μ and form secondary, sometimes tertiary, branches which have progressively smaller cells, each branch ending in 2 or more phialides, ovoid to doliiform and 7–12 × 3–4 μ. CONIDIA straight, cylindrical or slightly swollen in the upper cell, 1 septate, 50–60 × 4.5–6 μ, occasionally 3 septate. Cultures on PDA reddish brown from below, white margin irregular; aerial hyphae radially striate, white, becoming reddish brown with abundant conidiophores at first isolations. CHLAMYDOSPORES abundant in chains or clumps in older cultures which appear almost black from below; clumps are called MICROSCLEROTIA (from C. Booth & I. A. S. Gibson, *CMI Descr.* 362, 1973).

This worldwide fungus has been recorded from many hosts (both temperate and tropical) but has been particularly studied as a cause of damping off (seedling root rot or blight) of *Eucalyptus*, *Pinus* and other conifers. *C. braziliensis* (Batista & Cif.) Peerally (synonym: *C. scoparium* var. *braziliensis* Batista & Cif.) also occurs on *Eucalyptus*. It can be distinguished from *C. scoparium* by the smaller vesicles (6.2 × 2.6 μ) and conidia (24–38.2 × 2–2.8 μ). *C. clavatum* Hodges & May has also been isolated from *Pinus*. It differs from the other 2 spp. in having a typically clavate vesicle, the conidia are 36.4–59.8 × 3.2–5.2 μ (*Cylindrocladium* q.v.). *C. floridanum* (perfect state *Calonectria kyotensis* q.v.), which also attacks *Eucalyptus* and conifers, has a globose vesicle.

Viable microsclerotia were recovered in relatively high numbers from soil that had remained in clean fallow for 7 years; viability in soil was greatly reduced by drying. The fungus did not grow through non-sterile soil and did not move by root contact. Microsclerotia increased in soil when soybean was used as a cover crop and was lowest when maize was so used (Thies et al.). Symptoms on *Eucalyptus* in Brazil are: damping off; on leaves pale purple spots, darkening, a necrotic centre with a dark purple halo; defoliation and death of shoots; stem strangling of young plants at 20 cm above the collar and subsequent snapping off at this point; dieback of mature plants. On *Picea* penetration of the needles was stomatal, and microsclerotia developed in the substomatal chambers. They did so also in the root cortex. The fungus was recovered from roots of apparently healthy blue spruce grown for 6 months in infested, steamed soil (Bugbee et al.). On rice in Japan the fungus causes sheath net blotch. The symptoms occur on the leaf sheath near the water surface as yellow, watersoaked spots enlarging to 1–2 cm, oblong or elliptical; numerous dark brown lines form on the lesion and give it a net-like appearance; infected leaves turn chlorotic and die. Also from Japan Ogoshi described a chlorosis and dry root rot of groundnut from the beginning of flowering. The lesions on roots, pegs, pods and stems were brown black; the roots cracked and sometimes broke off. Normal, routine cultural and fungicide control in nurseries and young field plantings should be effective. Seed treatment has been recommended for *Eucalyptus* (Cruz et al. 1961) against *Cylindrocladium braziliensis* (ceresan or granosan M). Spp. of this tree differed in suceptibility. Bertus, in work on Australian native plants, found soil drenches with carbendazim or thiophanate methyl to be effective, and the latter as a foliar spray.

Aoyagi, K. 1962. *Ann. phytopathol. Soc. Japan* **27**: 147 (on rice; **42**: 196).
Batista, A. C. 1951. *Bolm Secr. Agric. Ind. Com. Est. Pernambuco* **18**: 188 (*Cylindrocladium braziliensis* on *Eucalyptus*; **31**: 524).
Bertus, A. L. 1976. *Phytopathol. Z.* **85**: 15 (on Australian native plants & control; **55**, 3111).
——. 1976. *Agric. Gaz. N.S.W.* **87**(5): 22 (as above; **56**, 3057).

Bugbee, W. M. et al. 1963. *Phytopathology* **53**: 1267 (on *Picea* seedlings; **43**, 1172).
Cruz, B. P. B. et al. 1960. *Arq. Inst. biol. S. Paulo* **27**: 96 (*C. braziliensis* on *Eucalyptus*; **41**: 414).
———. 1961. *Biológico* **27**: 106 (as above; **41**: 104).
Ogoshi, A. 1970. *Bull. natn. Inst. agric. Sci. Tokyo* Ser. C. **24**: 153 (on groundnut; **51**, 2084).
Peerally, A. 1974. *C.M.I. Descr. pathog. Fungi Bact.* 427 (*C. braziliensis*).
Terashita, T. et al. 1956. *Bull. For. Exp. Stn Meguro* 87: 33 (host range, physiology & culture; **36**: 470).
Thies, W. G. et al. 1970. *Phytopathology* **60**: 1662 (in USA, Wisconsin forest tree nurseries; **50**, 1990).
Weaver, D. J. 1974. *Can. J. Bot.* **52**: 1665 (growth & sclerotial formation with C & N sources, and *C. floridanum*; **54**, 3160).

CYTOSPORA Ehrenb. ex Fr. (as '*Cytispora*'), *Syst. Mycol.* 2(2): 540, 1823.

(synonymy in B. C. Sutton, *Mycol. Pap.* 141: 58, 1977)

(Sphaerioidaceae)

MYCELIUM immersed, septate, pale brown to hyaline, branched. CONIDIOMATA eustromatic, separate, subperidermal, conical, erumpent, and dark brown, multi-locular and convoluted, the locules enlarging, radiating from the centre, separated by walls of pale to dark brown textura angularis or intricata, but uniting in the ostiolar region; basal, lateral and upper walls of dark brown, thick-walled textura intricata, becoming very small celled textura angularis in the conidiogenous region. Ostiole single, circular, prominent, often surrounded by a furfuraceous ring of textura globulosa. CONIDIOPHORES hyaline, septate, branched irregularly at the base and above, smooth, mostly with apical conidia but sometimes acropleurogenous, formed from the inner cells of the locular walls. CONIDIOGENOUS CELLS enteroblastic, phialidic, determinate, integrated, straight, hyaline, smooth, occasionally formed as very small lateral branches immediately below transverse septa but more often as long distinct branches, collarette and channel minute. CONIDIA formed in distinct, variously coloured masses (as globose droplets or tendrils), hyaline, aseptate, thin walled, eguttulate, smooth, allantoid (B. C. Sutton, personal communication, August 1978).

Sheath rot of sugarcane is a minor disease caused by *C. sacchari* Butler (in *Mem. Dep. Agric. India bot. Ser.* 1(3): 31, 1906). The fungus is widespread on this crop (Map 256); it develops mostly on the sheaths but also occurs on cuttings, young shoots and stems, and stubble. The black stromata protrude above the surface, usually over the whole sheath area. The conidia (3–4 × 1–1.5 μ) are exuded in amber droplets. The infected rind becomes brown to black and the leaves can show drying from the tips downwards; severely diseased shoots may die. Growth of the fungus in culture is opt. at *c.* 31°. No control measures appear to be justified. Other spp. of *Cytospora* occur mostly on temperate woody hosts; some of their perfect state names are given under *Valsa*.

Cytospora spp. have been reported from *Eucalyptus* as causes of cankers. Of these *C. eucalypticola* Westhuizen is said to be the most damaging. STROMATA immersed, forming raised pustules in vertical rows, just below the epidermis of the bark, finally erumpent through longitudinal cracks, black, subglobose to somewhat oblong, 240–800 μ diam., ostiole inconspicuous, somewhat papillate, single or multiple in small crateriform opening; locules single or multiple, irregularly urceolate, often compressed, 30–250 μ diam., walls 18–60 μ thick, composed of compressed brown, sclerenchymatous cells. CONIDIOPHORES hyaline, slender, cylindrical to oblong fusiform, 8–12 × 1 μ, lining the pycnidial walls in a dense stand with occasional sterile branched paraphyses, somewhat curved, tapering towards the apex with 1 or 2 septa and projecting beyond the conidiophores, up to 25 × 2.5 μ. CONIDIA hyaline, allantoid, smooth, 3–4 × 0.7–1 μ; on *E. saligna* (from Westhuizen 1965a). Westhuizen (l.c.) gave reasons for supposing that *C. eucalyptina* Speg. is a synonym of *C. australiae* Speg.; *C. eucalypticola* differs from these 2 other fungi in its smaller conidia, short conidiophores and large pycnidial stromata. The locules of *C. eucalupticola* are most variable in size and irregularly disposed in the stroma. In *C. australiae* the locules are fairly regular in size and shape, and are radially disposed around a central columella.

C. eucalypticola is reported from Australia, Burma, Kenya, Malawi, Malaysia (Sabah), Pakistan, South Africa and Uganda. The leaves of infected trees droop and become coppery red; the condition spreads back from the younger to older branches and leaves fall. Longitudinal cracks form in the bark of the main stem at soil level. The inner bark turns brown, bark becomes detached, roots rot and the tree may die. Stress conditions (drought) apparently predispose the trees to attack.

Deightoniella torulosa

Abbot, E. V. 1938. *Proc. Int. Soc. Sug. Cane Technol.* 6th: 447 (*Cytospora sacchari*, general; **18**: 618).
Flores-Cáceres, S. 1963. *Revta téc. Col. Ingen. Agron. Méx.* **5**: 21 (*C. sacchari*, inoculation & host reaction; **46**, 1094).
Kiryu, T. 1937. *Rep. Govt Sug. Exp. Stn Taiwan* **4**: 172 (*C. sacchari*, in vitro culture; **17**: 840).
Luthra, J. C. et al. 1938. *Proc. Indian Acad. Sci.* Sect. B **8**: 188 (*C. sacchari*, general; **18**: 139).
——. 1940. *Ibid* **12**: 172 (*C. sacchari*, in vitro culture; **20**: 135).
Westhuizen, van der G. C. A. 1965a & b. *S. Afr. For. J.* **54**: 8, 12 (*C. eucalypticola* sp. nov.; disease on *Eucalyptus saligna*; **45**, 620).

DEIGHTONIELLA Hughes, *Mycol. Pap.* 48: 27, 1952, emend. D. Shaw, *Papua New Guin. agric. J.* **11**: 77, 1959.
(Dematiaceae)

COLONIES effuse, grey, brown or black, often hairy. MYCELIUM immersed in most spp., occasionally superficial. STROMA none. SETAE and HYPHOPODIA absent. CONIDIOPHORES macronematous, mononematous, torsive or flexuous, usually unbranched but in some spp. occasionally dichotomously branched, brown, smooth. CONIDIOGENOUS CELLS monoblastic, integrated, terminal, percurrent, cylindrical, doliiform, spherical or subspherical. After the first conidium has fallen the thin upper part of the conidiogenous cell tends to collapse and growth recommences inside and usually near its base. CONIDIA solitary, acrogenous, simple, clavate, cylindrical, doliiform, broadly ellipsoidal, obclavate, obpyriform, obturbinate, ovoid, or subspherical, pale to mid brown or golden brown, smooth or verruculose, with 0, 1, 2, 3 or more transverse septa or pseudosepta (M. B. Ellis, *Demat. Hyphom*: 163, 1971).

Keys to the spp. were given by Ellis (1957; l.c.). Besides the pathogen on *Musa* (*D. torulosa*) a leaf spot of *Saccharum* caused by *D. papuana* D. Shaw has been described and called veneer blotch because of the striking pattern that the symptoms develop into, particularly on the upper surface. The CONIDIOPHORES of *D. papuana* form a dense black pile, usually on the leaf under surface; they are simple, dark brown, 39–70 (90) × 6–8 μ, cylindrical throughout most of their length with twisted growth, 8–14 twists, base bulbous partly immersed in an epidermal cell. CONIDIA globose to nearly ovoid, aseptate, minutely echinulate, very pale olivaceous brown, 15–20 × 15–18 μ, no basal scar. Symptoms begin as a small oval leaf spot, light green to straw coloured with a distinct red brown border. The spot becomes flanked by 2 longitudinally elongated, similar lesions, one on each side; further lesions form, flanking the first pair, until sometimes 12 pairs or wings are formed on each side of the original spot. One lesion measured 61 × 1.2 cm. Lesions are outlined by a thin, dark red border 0.5–1 mm wide; the inner part finally becomes light brown.

Ellis, M. B. 1957. *Mycol. Pap.* 66, 12 pp. (*Deightoniella* spp., descriptions & key; **36**: 666).
Shaw, D. E. 1959. *Papua New Guin. agric. J.* **11** (1956): 77 (*D. papuana* on *Saccharum*; **38**: 624).

Deightoniella torulosa (Sydow) M. B. Ellis, *Mycol. Pap.* 66: 7, 1957.
Brachysporium torulosum Sydow, 1909
Helminthosporium torulosum (Sydow) Ashby, 1928
Cercospora musarum Ashby, 1913
H. nodosum Torrend, 1914.

COLONIES brown to black, effused. MYCELIUM mostly immersed in the substratum, composed of branched, septate, hyaline to pale brown, smooth-walled hyphae 2–5 μ thick. In epidermal cells of host plant hyphae sometimes swell to 10–15 μ, short branches grow through epidermal wall to form conidiophores; usually no septum cutting off swollen parts of mycelium from conidiophore basal part; these are joined by narrow isthmus. CONIDIOPHORES single or in small groups, simple or occasionally branched, straight or flexuous, septate, reddish brown 40–170 μ long, 6–10 μ thick, swollen into a bulb 13–16 μ diam. at apex, with up to 6 successive subglobose proliferations 13–16 μ diam. CONIDIA usually formed singly as blown out ends at the tips of the conidiophores and on successive proliferations through previous conidial scars; straight or slightly curved, obpyriform to obclavate, subhyaline to smoky olive, darker round the scar, 3–13 (usually 3–5) pseudoseptate, 30–150 (usually 35–70) × 15–25 μ (after M. B. Ellis, l.c.; and see Subramanian).

D. torulosa is a common inhabitant (Map 175) of plant debris from banana, particularly leaves, and causes relatively minor diseases of seedlings and older plants (pseudostem, leaves and fruit). The conidia are air dispersed, mostly in the forenoon and their numbers increase after rain. The violent discharge of the conidium has been described. Direct penetration of unwounded host tissue occurs. Control is largely cultural.

Seedling: A leaf necrosis, usually beginning at the tips of the lower leaves, spreads and often involves the pseudostem which develops a dark brown rot (reddish brow discolouration on the underlying leaf sheaths). The seedling damps off. Infection may begin at the base of the pseudostem or below ground and the outer tissues of rhizome and roots may be attacked. Emerging coleoptiles are also infected. Under less favourable conditions for damping off spread of the fungus is restricted; plant growth continues although it is slow and retarded. The usual steps against damping off should be taken; covering seedling with polythene reduces airborne spore inoculation; weekly spraying with captan has been recommended (Stover).

Leaf: In banana (black leaf spot) the spots 1–2 mm diam. increase in size, become oval and 2.5 cm or more in diam. with a black border and brownish centre. Spotting is commonest on the edges of the lamina of the oldest leaves. The lesions may coalesce to form bands of necrosis along the leaf edges. The spots can be confused with those caused by *Cordana musae* (q.v.). On abaca (*Musa textilis*) (brown leaf spot) the lenticular spots enlarge up to 20–25 cm long and 18–15 cm wide; they have brown centres (paler on the lower leaf surface), a well defined black border 1–2 mm wide surrounded by a bright chlorotic halo (Lopez et al.; Mendiola-Ela et al.). Where plantings are properly maintained the disease is of little consequence.

Pseudostem: Red sheath rot (also called trunk rot and pseudostem rot) has been described on abaca in the Philippines. Dark spots appear on the outer leaf sheaths; they are surrounded by a lighter area and a distinct brown ring; larger oval areas are formed by coalescence and the rot spreads inwards to the leaf sheaths successively. As a result older leaves fall prematurely, the pseudostem may fall and young plants (a few months old) are killed. The rot lowers fibre quality through discolouration. Good cultivation, sanitation (removal of infected debris) and prompt harvesting reduce the disease (Agati et al.; Calinisan et al.).

Fruit: Several forms of disease are found on banana fruit. One is called swamp spot and also speckle, but speckle is a term commonly used in describing banana diseases (*Veronaea* q.v. and *Mycosphaerella musae*). The spots are reddish brown to black up to 4 mm diam. but usually smaller; they have a dark green, watersoaked halo. Spots may be unevenly distributed both within the bunch and on the fingers; they tend to increase in severity as the bunch matures. Young fingers are more readily infected than older ones. The speckling on green fruit is mostly masked as ripening to yellow develops. Correct cultivation and covering the fruit bunches with perforated polythene bags prevent the disease becoming serious. Dithiocarbamate spraying can be carried out (Meredith 1961b,c,d; 1963; and see Temkin-Gorodeiski et al.). Black tip (also called tip rot and black end) is associated with *Verticillium theobromae* (q.v.). The peel of the finger tips becomes black and this rot, confined to the peel, may extend on the finger for 10 cm; the lesion has an irregular outline with a grey brown border. The ashy grey appearance of cigar end (*Trachysphaera fructigena* q.v.) is absent. At harvest the finger tip may rupture. The condition is uncommon and not important (Meredith 1961e; Ogilvie; Wardlaw). *D. torulosa* is one of the fungi associated with crown rot of bananas; this occurs after cutting the hands from the stem before packing because several fungi invade the freshly cut surface (see Stover, *Banana, plantain and abaca diseases*).

Agati, J. A. et al. 1934. *Philipp. J. Agric.* **5**: 191 (red sheath rot of abaca; **14**: 312).

Calinisan, M. R. et al. 1931. *Ibid* **2**: 223 (as above; **11**: 300).

Lopez, R. et al. 1953. *Turrialba* **3**: 159 (brown leaf spot of abaca; **33**: 353).

Mendiola-Ela, V. et al. 1954. *Philipp. Agric.* **38**: 251 (on abaca; **36**: 320).

Meredith, D. S. 1961a. *Ann. Bot.* **25**: 271 (conidial release; **41**: 242).

——. 1961b,c,d,e. *Trans. Br. mycol. Soc.* **44**: 95, 265, 391, 487 (swamp spot on banana, symptoms, pathogenicity, conidial germination, formation & dispersal; black tip on banana; **40**: 552; **41**: 97, 321).

——. 1961f. *Ann. appl. Biol.* **49**: 488 (conidial dispersal; **41**: 162).

——. 1963. *Ibid* **52**: 55 (control of swamp spot; **43**, 168).

Ogilvie, L. 1928. *Phytopathology* **18**: 531 (black tip; **7**: 731).

Stover, R. H. 1963. *Trop. Agric. Trin.* **40**: 9 (on seedlings; **42**: 396).

Subramanian, C. V. 1968. *C.M.I. Descr. pathog. Fungi Bact.* 165 (*Deightoniella torulosa*).

Deuterophoma tracheiphila

Temkin-Gorodeiski, N. et al. 1975. *Fruits* **30**: 613 (relation between benomyl treatment & fruit age in control of apical rots; **55**, 2319).
Wardlaw, C. W. 1932. *Trop. Agric. Trin.* **9**: 3 (black tip, morphology, histology & culture; **11**: 464).

DEUTEROPHOMA Petri, *Boll. R. Staz. Patol. veg. Roma* **9**: 396, 1929.
(Sphaerioidaceae)

PYCNIDIA membranous, astomatal, without stromata, at first fleshy and cellular inside, later full of sporules and mucilage, without any sporophore. SPORULES hyaline, very minute, rod shaped, continuous, derived from mucescent cells (L. Petri, Latin diagnosis, l.c.). The genus is monotypic. It has been considered to be a *Phoma*. B. C. Sutton (*Mycol. Pap*. 141: 63, 1977) gave the authors of the transfer to *Phoma* and continued: 'Whether the texture and structure of the pycnidial wall and the arrangement of the conidiogenous cells on it are sufficient to retain the single sp. in *Phoma* is debatable. For the moment however the generic name is maintained as a synonym of *Phoma*.'

Deuterophoma tracheiphila Petri, *Boll. R. Staz. Patol. veg. Roma* **9**: 396, 1929.
Bakerophoma tracheiphila (Petri) Cifferri, 1946
Phoma tracheiphila (Petri) Kantschaveli & Gikachvili, 1948.

PYCNIDIA on living stems aggregated within necrotic areas, globose to lenticular, black, 100–180 μ diam.; wall several cells thick, composed of outer sclerotial and deeply pigmented cells, and inner hyaline, thin walled, pseudoparenchymatic cells. CONIDIOGENOUS CELLS enteroblastic, phialidic, hyaline, simple, ampulliform, completely lining the inside of the pycnidial cavity. CONIDIA (phialospores) hyaline, aseptate, straight or curved, with apices rounded, 2–3 × 1 μ. IN CULTURE conidia are produced freely on hyphae. CONIDIOPHORES semi-macronematous, mononematous, simple, sometimes branched. CONIDIOGENOUS CELLS monophialidic, integrated, ampulliform to lageniform, determinate with well-defined collarettes. CONIDIA aggregated in slimy heads, semi-endogenous, simple, aseptate, straight with rounded ends, 2–2.5 × 1–1.5 μ, with a guttule at each end (E. Punithalingam & P. Holliday, *CMI Descr*. 399, 1973).

Mal secco of *Citrus* spp. (mainly lemon, *C. limon*) is a major disease of limited distribution (Map 155) in the Mediterranean and Black Sea areas: Algeria, Cyprus, France, Greece (including Crete and the Aegean Islands), Iraq, Israel, Italy (including Sardinia), Lebanon, Syria, Tunisia, Turkey and USSR (Georgia and Caucasus). Countries in the region that seem free from mal secco are Libya, Morocco, Portugal and Spain. The extensive investigations were reviewed some years ago by Chapot (1963) and Goidànich et al. (1953). The first, less specific, symptom of this vascular disease (tracheomycosis) is a wilt and death of the leaves. The pattern of these canopy symptoms varies, depending on the original site of infection. If this occurs towards the base of the trunk or on the main roots above ground collapse can be rapid; infection of one main root leads to wilt in one part of the tree. This type of infection kills the plant. When the pathogen invades the middle and upper branches the wilt is less severe. Cuts into the wood (before wilt and bark necrosis) reveal the characteristic orange yellow to reddish discolouration of the vascular tissue. The host epidermal covering over the pycnidia assumes an ashy appearance (Petri, 1930; Savastano et al.). External symptoms can be seen after injection of a susceptible stock with conidia.

Penetration takes place through wounds on the roots, trunk and branches. Natural openings may be penetrated but there appears to be little or no evidence for direct leaf entry. Penetration can occur in the zone where the fruits are attached to the peduncle. Such primary infection causes a premature, rapid yellowing of the fruits which wither and fall. Infection is waterborne, probably largely from host debris on the ground (branches, leaves and fruits); pycnidia form on pruning cuts. Trees are attacked in the cooler seasons of the year; probably less in autumn than in winter and spring. Growth in the host ceases during the summer at *c*. 30°; the best temps for mycelial growth, pycnidial formation and conidial germination are 14–20° (or somewhat higher, up to 25°, for germination). Injury to the tree through severe cold weather can predispose the tree to attack (Goidànich et al. 1949; Scaramuzzi et al.). Grasso described the conditions in Sicily which led to a dieback of a clementine cv. normally rather resistant to *D. tracheiphila*: hail and frequent rain in the previous autumn, a mild winter, young trees, high inoculum density and exposure. For cell-degrading enzymes, phytotoxins, inhibitors in the host and effects of virus infection see *Rev. appl.*

Mycol. **41**: 599; **46**, 2721; *Rev. Pl. Pathol.* **50**, 2950, 3746; **51**, 360, 3362; **53**, 3971, 4726; **56**, 4526, 5058.

No races have been described but see Salerno et al. (1966). Laviola reviewed control measures in Sicily where mal secco continues to be severe on lemon. The more susceptible spp. are lemon, sour orange (*C. aurantium*), citron (*C. medica*) and pummelo (*C. grandis*). Intermediate with some resistance or tolerance are: some lemon cvs, sweet orange (*C. sinensis*), mandarin (*C. reticulata*), lime (*C. aurantifolia*), trifoliate orange (*Poncirus trifoliata*) and kumquat (*Fortunella*). A resistant rootstock is generally recommended, see Chapot (1965) who described *C. volkameriana* as such a stock. Cleopatra mandarin and resistant or tolerant lemon cvs are used, such as Santa Teresa in Sicily and Lamas in Turkey (see Akteke et al.). Copper sprays are used and other fungicides recommended are dodine, folpet, nabam and ziram (Salerno et al. 1965, 1967). Recent work with benomyl, carboxin, thiabendazole and thiophanate has been promising (Salerno, et al. 1971; Scaramuzzi; Solel et al.; Somma et al.). Injury to the trees during cultivation must be avoided, and wherever possible cultivation should be done during the warmer times of the year. Infected branches should be pruned, the cut surfaces treated with an anti-fungal material and all infected debris removed.

Akteke, S. A. et al. 1977. *J. Turkish Phytopathol.* **6**: 91 (infection & resistance; **57**, 5506).

Carrante, V. et al. 1952. *Annali Sper. agr.* **6**: 323 (hybridisation for resistance in lemon; **32**: 125).

Catara, A. et al. 1972. *Annali Ist. sper. Agrumic.* **5**: 29 (hosts; **56**, 1126).

Chapot, H. 1963. *Al Awamia* **9**: 89 (review, 137 ref.; **43**, 3223).

——. 1965. *Ibid* **14**: 29 (resistant stock *Citrus volkameriana*; **45**, 1377).

Ciccarone, A. 1971. *Phytopathol. Mediterranea* **10**: 68 (taxonomy, name change to *Phoma tracheiphila*; **50**, 3745).

Demetradze, T. Ya. et al. 1970. *Subtrop. Kul't. 1970* (2): **90** (water movement in infected lemon; **50**, 1235).

Gassner, G. 1940. *Phytopathol. Z.* **13**: 1 (general; **20**: 298).

Goidànich, G. et al. 1947. *Atti Accad. Lincei Sci. nat.* Ser. 8 **3**: 395 (general; **27**: 318).

————. 1947. *Annali Sper. agr.* **1**: 141 (inoculation; **27**: 70).

————. 1948. *Ibid* **2**: 671 (conidial formation; **28**: 422).

————. 1949. *Ibid* **3**: 391 (effect of low temps in Sicily; **28**: 626).

————. 1953. *G. agr.* 3, 14 pp. (review, 47 ref.; **33**: 670).

Grasso, S. 1973. *Technica Agricola* **25**: 5 (conditions favouring infection; **53**, 2570).

Hohryakov, M. K. 1952. *Mikrobiologiya* **21**: 210 (reaction of *Citrus* & other spp.; **33**: 350).

Laviola, C. 1977. *Inftore fitopatol.* **27** (6–7): 45 (review on control, 22 ref.; **57**, 1199).

Perrotta, G. et al. 1976. *Riv. Patol. veg.* Ser. 4, **12**: 145 (fine structure of infection; **56**, 4527).

Petri, L. 1930. *Boll. R. Staz. Patol. veg. Roma* **10**: 63, 191, 353 (general; **10**: 181, 182, 593).

——. 1940. *Ibid* **20**: 81 (in Turkey; **25**: 444).

Ruggieri, G. 1940. *Ibid* **20**: 150 (fruit infection; **25**: 497).

——. 1948. *Annali Sper. agr.* **2**: 255 (general & control; **27**: 469).

Salerno, M. 1964. *Riv. Patol. veg.* Ser. 3, **4**: 289 (effect of temp. on fungal growth & conidial germination; **44**, 700).

—— et al. 1965. *Ibid* Ser. 4, **1**: 71 (fungicides; **45**, 453).

————. 1966. *Ibid* **2**: 303 (differences in pathogenicity; **46**, 2722).

————. 1967. *Tec. agric. Catania* **19**(3): 1; (4): 1 (on hosts other than *Citrus* & fungicides; **47**, 1883a & b).

————. 1971. *Phytopathol. Mediterranea* **10**: 99 (benomyl; **50**, 3744).

Savastano, G. et al. 1930. *Annali R. Staz. Sper. Agrumic. Fruttic. Acireale* 11: 37 pp. (general; **9**: 645).

Scaramuzzi, G. et al. 1964. *Riv. Patol. veg.* Ser. 3, **4**: 319 (effect of low temp. on disease; **44**, 2493).

——. 1970. *Inftore fitopatol.* **20**(14): 5 (systemic fungicides; **51**, 361).

Solel, Z. et al. 1972. *Phytopathology* **62**: 1007 (systemic fungicides; **52**, 1539).

Somma, V. et al. 1974. *Phytopathol. Mediterranea* **13**: 143 (benomyl; **55**, 5200).

Stepanov, K. M. 1950. *Dokl. Vses. Akad. sel'-khoz. Nauk* **15**: 39 (development of pycnidia in the field & control; **30**: 413).

Tramier, R. et al. 1963. *Rev. Pathol. vég.* **42**: 211 (in France; **43**, 2617).

DIAPORTHE Nits. emend. Wehmeyer, *The genus* Diaporthe *Nitschke and its segregates*, Ann Arbor, Univ. Michigan Press, 1933, by L. E. Wehmeyer.

(Diaporthaceae)

ASCOMATA (perithecia) immersed in substrate, erumpent through a pseudostroma mostly surrounding the ascomata which have more or less elongated perithecial necks. Pseudostroma distinct, often delimited by dark lines (marginal blackening at least at some points). ASCI unitunicate, clavate to clavate cylindric, loosening from the ascogenous cells at an early stage and lying free in the ascoma. PARAPHYSES broad, band like, present at first but disappearing with maturity. ASCOSPORES fusoid ellipsoid to cylindric, straight, inequilateral or curved, 1 septate, hyaline, sometimes with appendages, biseriate to uniseriate (after Wehmeyer l.c. and key by E. Müller & J. A. von Arx in *The fungi* (edited by G. C. Ainsworth et al.) Vol. IVA: 111, 1973).

The spp. occur mostly in wood and bark; they have conidial states in *Phomopsis* (q.v.) which are the forms more frequently encountered in the few pathogenic spp. The genus was monographed by Wehmeyer (l.c.) who considered related genera and gave a key to *Diaporthe* spp. which were divided into 2 sections: the Effusae in which the dorsal, blackened zones do not penetrate the bark and the Pustulatae in which the dorsal zones do dip into the bark between the perithecial clusters. A recent, fully illustrated description of a *Diaporthe* sp. and its *Phomopsis* state was given by Punithalingam et al. (and in *Guignardia* q.v.), *D. woodii* Punith., stat. conid. *P. leptostromiformis* (Kühn) Bubak. The *Phomopsis* state occurs abundantly on *Lupinus* spp. and causes lupinosis in sheep. *D. eres* Nits. is a complex whose forms have been associated, mostly in north temperate areas, with dieback and cankers in trees (see Browne, *Pests and diseases of forest plantation trees*). *P. manihotis* Swarup, Chanhan & Tripathi on cassava has a perfect state, *D. manihotis* Punith. *D. cubensis* Bruner is the name now preferred for a fungus causing a canker of *Eucalyptus* spp. in Brazil (*Endothia* q.v.).

Punithalingam, E. 1975. *Kavaka* 3: 29 (*Diaporthe manihotis* sp. nov.; 55, 5453).
—— et al. 1975. *C.M.I. Descr. pathog. Fungi Bact.* 476 (*D. woodii*).

Diaporthe citri Wolf, *J. agric. Res.* **33**: 625, 1926. (as *D. citri* Fawcett).
stat. conid. *Phomopsis citri* Fawcett, 1912
P. californica Fawcett, 1922
P. caribaea Horne, 1922
P. cytosporella Penzig & Sacc., 1887.

PERITHECIA single or in groups, immersed in black stromata which are covered by the bark, 125–160 μ diam.; beaks black, tapering, 200–800 μ in length. ASCI sessile, elongate clavate, 50–55 × 9–10 μ, apex thickened and pierced by a narrow pore. ASCOSPORES hyaline, 1 septate, constricted, elongate elliptical to spindle shaped, with 4 guttulae, 11.5–14.2 × 3.2–4.5 μ. PYCNIDIA on leaves, green and dead branches, bark, decaying fruit, scattered or clustered, immersed, later erumpent, black conical to lenticular, up to 600 μ diam., with ostioles. Pycnidial wall many cells thick, composed of outer sclerotised cells and inner thin pseudoparenchyma lining the cavity. CONIDIOGENOUS CELLS enteroblastic, phialidic, hyaline, simple, cylindrical to obclavate, arising from the innermost layer of pseudoparenchymatic cells lining the pycnidial cavity. A CONIDIA (phialospores) hyaline, aseptate, fusiform to ellipsoidal, biguttulate, with 1 guttule at each end, 6–10 × 2–3 μ. B CONIDIA (phialospores) hyaline, aseptate, filiform, curved and often strongly hooked, 20–30 × 0.5 μ. Although Wehmeyer (*Diaporthe* q.v.) did observe some difference between *D. medusaea* Nits and Fawcett's material of *D. citri*, Weymeyer cites *D. citri* as a synonym of *D. medusaea*; but see Yamato, 1976 (perfect state: Wolf l.c.; stat. conid.: E. Punithalingam & P. Holliday, *CMI Descr.* 396, 1973; F. E. Fisher made a comment on nomenclature, *Mycologia* **64**: 422, 1972).

Melanose of citrus and stem end rot of the fruit is widespread in citrus growing areas (Map 126). The disease is only of economic importance where the summers are wet; it is not a problem in areas with a climate of the Mediterranean pattern. On young leaves the symptoms begin as very small, dark, circular and sunken spots with chlorotic haloes. They increase in size (up to 1 mm diam.) as the leaf develops and become raised, brown, superficial spots which feel rough to the touch. Severe infection with numerous spots leads to leaf chlorosis, distortion and defoliation. Similar spots form on young shoots and fruit. The dark spotting can become very conspicuous on the developing fruit and coalescence of numerous lesions causes large, brown, scarred areas which become crusts and crack. Conidia, spread on the fruit by water, cause spot infections which develop in rings, lines or curves (tear drop patterns). In the field the first sign of stem end rot on immature fruit is a dark brown, reddish brown to almost black discolouration about the stem end. Orange and grapefruit may fall but tangerine tends not to. On lemon the first symptom is a pliable condition around the button, discolouration appears later as the natural yellow finally turns to buff. This fruit decay may first occur after harvest in the packed or stored condition where it can be more frequent than in the field. There is a softening around the button at first, the area becoming a drab brown; the peel remains pliable, not soft or easily ruptured by pressure. Internally decay spreads most rapidly in the central pith and the inner white part of the rind; the whole fruit is eventually affected.

D. citri is both a primary pathogen (cuticular penetration) of immature tissue (leaves, stems, pedicels and fruit), and a secondary invader of wounds and scars, and in certain other diseases or otherwise abnormal conditions. The opt. temp. for in vitro growth and formation of pycnidia is 24° or a little lower and this temp. is also favourable for infection. Entry to the fruit from the stem is prevented by the cuticle and wound periderm which separates healthy and necrotic tissue. Penetration of the button and albedo layer does not take place until maturity is approached; subsequently a rot develops in storage (Brown et al. 1968; Homma et al. 1969b). Inoculum comes from the waterborne conidia from pycnidia on the small, dead branches which are still attached. As these become more abundant during the life of the orchard, so both the incidence of melanose and the risk of stem end rot increase; the disease is rarely important in young plantings. The role of the ascospore, from branch debris on the ground, in the epidemiology is uncertain. In Japan a wetness period of 8–12 hours was sufficient for fruit and leaf infection at *c.* 25°. No fruit infection occurred at 17°. The fungus infected 1-year-old, dead or living twigs but not older material. Perithecia were formed on dead twigs (at >RH 95%) on the ground and on dead branches on the trees (Homma et al.; Inoue et al. 1972; Kuramoto et al.; Ushiyama). Blue light is required for conidial sporulation in vitro and Koizumi reported that the production of A conidia was favoured by light but that of B conidia was unaffected. Inoue et al. (1969) noted differences in pathogenicity between isolates. There is little difference in susceptibility between cvs but grapefruit is most severely affected.

Control should be directed at the fruit-disfiguring effects caused in the field and the rot after harvest. Pruning and other cultural operations should eliminate infected branch material so as to reduce inoculum. Fungicides are applied at the early stages of fruit formation (pre-blossom treatments are also sometimes recommended). Bordeaux (sometimes with additives) in 2–3 applications is still recommended; dithiocarbamates are not satisfactory (Kiely). In Florida captafol gave better control than copper when applied as a post-blossom spray (Moherek), and chlorothalonil has also been reported to be effective. In spite of good field control stem end rot may still cause losses after harvest; postharvest treatment with thiabendazole has been recommended (Brown et al. 1967). Whiteside compared copper and other fungicides, and their sites of action. Benomyl inhibited the formation of inoculum on dead twigs unlike other fungicides (and see Timmer).

Bach, W. J. et al. 1928. *J. agric. Res.* **37**: 243 (pathogenicity & histology; **8**: 99).

Bahgat, M. 1928. *Hilgardia* **3**: 153 (stem end rot; **7**: 713).

Brooks, C. 1944. *J. agric. Res.* **68**: 363 (as above; **23**: 386).

Brown, G. E. et al. 1967. *Pl. Dis. Reptr* **51**: 95 (postharvest treatment with thiabendazole; **46**, 1579).

—— ——. 1968. *Phytopathology* **58**: 736 (penetration of orange fruit; **47**, 3114).

Fawcett, H. S. 1911. *Bull. Fla agric. Exp. Stn* 107, 23 pp. (general, stem end rot).

—— 1912. *Phytopathology* 2: 109 (as above).

Floyd, B. F. et al. 1912. *Bull. Fla agric. Exp. Stn* 111, 16 pp. (general).

Homma, Y. et al. 1969a & b. *Bull. hort. Res. Stn Japan* Ser. B **9**: 85, 99 (factors affecting & mechanisms of infection & disease development; **50**, 678, 679).

Inoue, K. et al. 1969. *Bull. Citrus Exp. Stn Shizuoka* 8: 71 (differences in pathogenicity; **50**, 1238).

—— ——. 1972. *Ibid* 10: 76 (infection of dead twigs; **54**, 2249).

Kiely, T. B. 1967. *Agric. Gaz. N.S.W.* **78**(5): 309 (fungicide control with *Guignardia citricarpa*; **46**, 3438).

Koizumi, M. 1965. *Bull. hort. Res. Stn Okitsu* Ser. B **4**: 127 (effects of light, temp. & RH on conidial sporulation; **45**, 1378).

Kuramoto, T. et al. 1975. *Bull. Fruit Tree Res. Stn* Ser. B 2: 75 (environmental factors; **55**, 3186).

Moherek, E. A. 1970. *Proc. Fla St. hort. Soc.* **83**: 59 (fungicide control with *Elsinoë fawcettii*; **51**, 3366).

Sasaki, A. 1965. *Ann. phytopathol. Soc. Japan* **30**: 246 (fruit infection; **45**, 2117).

Timmer, L. W. 1974. *Pl. Dis. Reptr* **58**: 504 (fungicides on Texas grapefruit; **54**, 156).

Ushiyama, K. 1971. *Bull. Kanagwa hort. Exp. Stn* 19: 29 (infectivity of conidia; **53**, 2571).

——. 1973. *Ann. phytopathol. Soc. Japan* **39**: 120 (life history; **53**, 552).

Whiteside, J. O. 1975. *Pl. Dis. Reptr* **59**: 656 (copper & other fungicides; **55**, 2725).

Whiteside, J. O. 1977. *Phytopathology* **67**: 1067 (fungicides, sites of action; **57**, 1731).
Winston, J. R. et al. 1927. *Bull. U.S. Dep. Agric.* 1474, 62 pp. (general; **6**: 548).
Yamada, S. et al. 1962. *Bull. hort. Stn Tokai-Kinki* **6**: 108 (effect of temp, inoculum, infection & season; **42**: 262).
Yamato, H. 1971. *Ann. phytopathol. Soc. Japan* **37**: 355 (perfect state; **51**, 4012).
——. 1976. *Ibid* **42**: 56 (a *Diaporthe* nearer *D. medusaea* from citrus; **55**, 4720).

Diaporthe phaseolorum (Cooke & Ellis) Sacc., *Sylloge Fung.* **1**: 692, 1882.
Sphaeria phaseolorum Cooke & Ellis, 1878
Phoma phaseoli Desm., 1836
P. subcircinata Ellis & Everh., 1893
stat. conid. *Phomopsis phaseoli* (Desm.) Grove 1917.

ASCOSTROMATA scattered, crowded, botryose, up to 1 mm wide, black with long necks up to 500 μ, opening by narrow ostioles; wall several layers of cells thick, with a black border on the outside. ASCI 8 spored, clavate, sessile, 25–40 × 5–10 μ; wall thin, bitunicate, evanescent. ASCOSPORES biseriate, hyaline, ellipsoid, 1 septate, constricted at the septum, guttulate, 9–11 × 2.5–3.5 μ, slightly acute at the ends. PYCNIDIA on pods and stems, globose to irregular, immersed, becoming erumpent, opening by apical ostioles; cavity simple, sometimes convoluted; pycnidial wall multicellular, with heavily pigmented and sclerotised cells on the outerside. CONIDIOPHORES (phialides) cylindrical to obclavate, hyaline, simple or branched 6–10 × 2–3 μ. A CONIDIA hyaline, aseptate, fusiform to ellipsoid, 2 guttulate, with one guttule at each end of conidium, 6–8 × 2–2.5 μ. B CONIDIA hyaline, elongated, filiform, curved sometimes strongly hooked, 20–30 × 0.5–1 μ. An intermediate conidial form also occurs, 8–12 × 2–2.5 μ. Colonies on oat agar show white floccose aerial mycelium with abundant pycnidia distributed irregularly (E. Punithalingam & P. Holliday, *CMI Descr.* 336, 1972).

Four vars of *D. phaseolorum* have been distinguished on the basis of ascospore size, pathogenicity and host range. Var. *caulivora* Athow & Caldwell on *Glycine max* differs in its restriction to soybean and the relative rarity of the conidial state (Athow et al.; Hildebrand). Var. *batatatis* (Harter & Field) Wehmeyer is restricted to *Ipomoea batatas* (all other vars are generally on Leguminosae) and has ascospores 8–12 × 4–6 μ. Var. *sojae* (Lehman) Wehmeyer forms perithecia in culture in conspicuous diatrypoid stromata, has ascospores 11–11.5 × 3.5–4 μ and occurs on soybean, *Phaseolus lunatus*, other legumes and non-leguminous hosts. Var. *phaseolorum* occurs only on lima bean (*P. lunatus*), with ascospores 9.5 × 2.9 μ.

The fungus causes stem canker, pod and stem blight of soybean and other legumes, dry rot of sweet potato and pod blight of lima bean. The most serious forms of disease caused by this variable pathogen are those on soybean (var. *caulivora*). Seedlings can be damped off, be weak and stunted and show necrotic lesions on the cotyledons, hypocotyls and roots. In older plants lesions form on the petiole and stem, where they may become girdling cankers (characteristically red brown in var. *caulivora* in the early stages) leading to death. The fungus is internally seedborne; infected seed is reduced in size and discoloured. In the field perithecia predominate in var. *caulivora* and pycnidia likewise in var. *sojae*. On lima bean leaf lesions are up to 3 cm diam. with concentrically arranged pycnidia, shotholing may occur and infection of pods (with abundant pycnidia) may lead to destruction of the seed. Sweet potato tubers shrink and become mummified in storage; a dry rot and necrosis of sprouts occurs, and pycnidia form on tubers and sprouts.

In the form generally referred to as var. *sojae* the distribution is probably fairly widespread in the tropics and subtropics. Var. *batatatis* may occur more widely than has been reported and var. *phaseolorum* has been described from N. America, West Indies, and S. and E. Africa. Var. *caulivora* appears to be restricted to USA and Canada. Whilst vars *batatatis* and *phaseolorum* may be considered distinct the separation of vars *sojae* and *caulivora* has been questioned. It appears that a virulent form (or forms) occurs on soybean whilst others are only weakly parasitic on this host but can also attack other legumes and non-leguminous hosts, probably as wound parasites. Differences within var. *caulivora* have also been reported. Clearly further work is required on both physiologic and morphologic specialisation.

Most recent studies have been done in Canada (on soybean canker) and USA. The diseases caused are severe at *c.* 25°, and this temp. is opt. for growth in vitro and growth in soil. Isolates of var. *caulivora* do not always form conidia; light is required for sporulation. In soybean greater seed infection was found in less dense stands, since in this case there was a greater tendency for the lateral branches to break and fall close to the soil surface from which inoculum spreads. Seed infection reduces seed germination and a delay in harvesting increases the former (Chamberlain et al.; Kmetz et al.; Peterson et al.; Wallen et al.; Wilcox et al.). Vars *caulivora* and *sojae* were reported homothallic and heterothallic, respectively (Welch et al. 1948), but later workers have been unable to confirm this. In legumes the pathogen is seedborne and is generally spread in crop debris (on which it survives between seasons) and probably soil. Primarily a storage disease of sweet potato it is spread in the tubers and through the plant slips. It remains viable in soybean seed for 2 years, but thereafter there is a decrease in viability and a concurrent increase in germination.

No high resistance in soybean appears to be known but certain recommended vars in Canada contract less of the disease in the field. Fungicide treatment of soybean reduces seedborne infection (e.g. Bolkan et al.; Ellis et al.; Prasartsee et al.). In USA lima bean seed can be produced in areas free from the disease. Cultural measures including rotation should be practised, and mercurial dips for plant material in sweet potato are effective.

Athow, K. L. et al. 1954. *Phytopathology* **44**: 319 (vars *caulivora* & *sojae*, general; **34**: 126).
Bloss, H. E. et al. 1966. *Ibid* **56**: 92 (effect of amino acids & sugars on growth in vitro; **45**, 1584).
Bolkan, H. A. et al. 1976. *Fitopatologia Brasileira* **1**: 215 (fungicides on soybean; **56**, 3336).
Chamberlain, D. W. et al. 1974. *Pl. dis. Reptr* **58**: 50 (seed infection & treatment; **53**, 2755).
Daines, R. H. et al. 1960. *Phytopathology* **50**: 186 (effect of temp. on var. *batatatis* & control; **39**: 496).
Dunleavy, J. M. 1958. *Proc. Iowa Acad. Sci.* **65**: 131 (var. *caulivora*, inoculation & growth in soil; **39**: 205).
Ellis, M. A. et al. 1974. *Pl. Dis. Reptr* **58**: 760 (fungicides on soybean; **54**, 622).
Frosheiser, F. I. 1957. *Phytopathology* **47**: 87 (aetiology & epidemiology in vars *sojae* and *caulivora*, hosts; **36**: 447).
Harter, L. L. et al. 1913. *Bull. U.S. Dept agric. Bur. Pl. Ind.* 281 (general on var. *batatatis*).
—— ——. 1917. *J. agric. Res.* **11**: 473 (general on var. *phaseolorum*).
Hildebrand, A. A. 1956. *Can. J. Bot.* **34**: 577 (general, vars *sojae* & *caulivora* on soybean; **36**: 229).
Kmetz, K. et al. 1974. *Pl. Dis. Reptr* **58**: 978 (seed infection; **54**, 3062).
Kmetz, K. T. et al. 1978. *Phytopathology* **68**: 836 (as above).
Lehman, S. G. 1923. *Ann. Mo. bot. Gdn* **10**: 111 (general, var. *sojae* on soybean; **3**: 377).
Luttrel, E. S. 1947. *Phytopathology* **37**: 445 (var. *sojae*, symptoms, morphology & hosts; **27**: 55).
Peterson, J. L. et al. 1965. *Pl. Dis. Reptr* **49**: 228 (vars *sojae* & *caulivora* & soybean, seed germination; **44**: 2003).
Prasartsee, C. et al. 1975. *Ibid* **59**: 20 (fungicides on soybean; **54**, 3065).
Threinen, J. T. et al. 1959. *Phytopathology* **49**: 797 (hybridisation, mutants & vars *sojae* & *caulivora*; **39**: 364).
Timnick, M. B. et al. 1951. *Ibid* **41**: 327 (light, nutrients & sporulation in vitro; **30**: 503).
Tucker, C. M. 1935. *Mycologia* **27**: 580 (on red pepper; **15**: 277).
Wallen, V. R. et al. 1963. *Can. J. Bot.* **41**: 13 (seed infection in soybean; **42**: 425).
Welch, A. W. et al. 1948. *Phytopathology* **38**: 628 (heterothallism & homothallism; **28**: 106).
Whitehead, M. D. 1966. *Ibid* **56**: 396 (var. *sojae* on soybean & *Lotus corniculatus* **45**: 2667).
Wilcox, J. R. et al. 1971. *Pl. Dis. Reptr* **55**: 776 (effect of stem breakage on infection of soybean seed; **51**: 917).
—— ——. 1974. *Ibid* **58**: 130 (seedborne infection & harvest; **53**, 3701).

DIDYMELLA Sacc. ex Sacc., *Michelia* 2: 57, 1880.
Haplotheciella Höhnel, 1918.
(Venturiaceae)

PSEUDOTHECIA solitary, immersed or more or less erumpent at maturity, globular or flattened, without well developed stromata. Ostiole papilliform or in the form of a glabrous cylindrical collar. ASCI bitunicate, cylindrical, very slightly pedicillate, rounded at the apex, 8 spored. ASCOSPORES 1 septate, hyaline, distichous, septate in the middle or nearer the proximal end (rarely distally), ovoid to ellipsoid, constricted. PARAPHYSOIDS hyaline, filiform, simple (after R. Corbaz, *Phytopathol. Z.* **28**: 382, 1957; see L. Holm, *Taxon* **24**: 478, 1975 for nomenclature).

Corbaz described the genus and spp. (with a key to 19 spp.) and gave a list of excluded spp. The accepted spp. included: *D. applanata* (Niessl) Sacc. (causing spur blight of raspberry, Map 72) and *D. lycopersici* (q.v.). Little-known spp. on tropical crops are: *D. arcuata* Röder (on *Cannabis sativa*); *D. mangiferae* Batista (on *Mangifera indica*); and *D. sorghina* Saccas (on *Sorghum bicolor*, *Rev. appl. Mycol.* **33**: 534, 1954).

Didymella bryoniae

Batista, A. C. 1952. *Bolm Secr. Agric. Ind. Com. Est. Pernambuco* **19**: 165 (*Didymella mangiferae* sp. nov.; **32**: 684).

Corbaz, R. 1957. *Phytopathol. Z.* **28**: 375 (systematics & morphology, key; **36**: 667).

Röder, K. 1939. *Ibid* **12**: 321 (*D. arcuata* sp. nov.; **19**: 280).

Didymella bryoniae (Auersw.) Rehm, *Ber. d. Naturhist. Ver. in Augsberg* **26**: 27, 1881.
Sphaerella bryoniae Auersw., 1869
Didymosphaeria bryoniae (Auersw.) Niessl, 1875
stat. conid. *Ascochyta cucumis* Fautr. & Roum., 1891.
(full synonymy in *CMI Descr.* 332; frequently called *Mycosphaerella melonis* (Pass.) Chiu & Walker)

PSEUDOTHECIA on stems, leaves and fruit, globose, immersed becoming erumpent, black, 140–200 μ diam.; opening by apical, papillate ostiole. ASCI cylindrical to subclavate, short stipitate or sessile, 8 spored, 60–90 × 10–15 μ; ascus wall bitunicate. ASCOSPORES hyaline, biseriate, ellipsoid, ends mostly rounded, slightly constricted at the septum, guttulate, 14–18 × 4–7 μ. PSEUDOPARAPHYSES hyaline, septate and branched. PYCNIDIA on stems, leaves and fruit; solitary or gregarious, immersed, becoming erumpent, dark brown, 120–180 μ diam.; wall composed of 2–4 layers of yellow brown cells, slightly thicker walled on the outermost layer. CONIDIA hyaline, shortly cylindrical with rounded ends, guttulate, mostly one septate but a small percentage aseptate, 6–10 (–13) × 3–4 (–5) μ. On *Bryonia, Citrullus, Cucumis, Cucurbita, Luffa, Momordica* and *Trichosanthes* (E. Punithalingam & P. Holliday, *CMI Descr.* 332, 1972).

Gummy stem blight is an important disease of Cucurbitaceae and shows a variety of symptoms ambiguously referred to as leaf spot, stem canker, vine wilt and black fruit rot. The distribution is widespread (Map 450). Lesions on leaves and fruit usually begin as spreading watersoaked areas; in the former these may have a chlorotic halo, become light brown and irregular in outline; leaves can be destroyed. On fruit, dark, cracked, sunken lesions with an extensive rot beneath are formed. In the field the first symptoms may be plant collapse where girdling stem cankers lead to total loss. Infection also occurs on seedlings. But the main characteristic features are the gummy exudate on stem and fruit lesions, and the abundant pycnidia followed by pseudothecia. Brown etch or zonate ringspot of pumpkin (*Cucurbita moschata*) cv. Butternut, associated with *Fusarium* spp., was described by Johnson.

The most severe outbreaks of this disease have occurred outside the tropics and extensive development probably occurs at 20–28°. Fruit decay in watermelon (*Citrullus lanatus*) increases with temp. from 7° to 24°, thereafter it falls and is very limited at 29.5°. The max. growth in vitro has been reported at 26.7° and 20–24°. Penetration is cuticular in seedlings but in older tissue (especially fruit) infection is most likely through wounds and bruises. Infrequent stomatal penetration of watermelon fruit occurs. In cucumber (*Cucumis sativus*) spread takes place from the stub where the fruit is picked, thus resulting in a stem lesion; or the fruit may be attacked via the blossom end. Both spore types may serve as primary inoculum; conidia through water splash and aerially dispersed ascospores. The latter, in glasshouses in England, have a diurnal periodicity with a peak at 1800–2000 hours, and a night time peak has also been reported from USA. Reports on seed transmission seem conflicting; in cucumber, although seed could be inoculated successfully, no evidence for natural infection was found (Brown et al.). Transmission by pruning knives has been demonstrated. For survival in soil see Uspenskaya et al. *D. bryoniae* is homothallic. Although there is some variation in pathogenicity (and considerable variation occurs in vitro) no races have been described. Light is required for abundant formation of both spore states on organic media (Chiu et al.; Curren).

A field rotation of at least 18 months is necessary. Both chemical spray control (benomyl, chlorothalonil, captafol and dithiocarbamates; e.g. Hopkins; Kagiwata; Steekelenburg) and seed treatment can be effective. In the glasshouse sanitary measures are important. No high resistance appears to have been developed though certain watermelon vars from USA show some resistance which is also found in melon (*C. melo*). To avoid rot during transport and storage careful handling must be observed and fruit precooled to 10° or a little below, care being taken to avoid cold injury.

Brown, M. E. et al. 1968. *Pl. Pathol.* **17**: 116 (no evidence for natural seed transmission; **48**, 322).

Chiu, W. F. et al. 1949. *J. agric. Res.* **78**: 81, 589 (homothallism & variation in vitro; temp. effects on infection & growth in vitro; **28**: 614; **29**: 135).

Curren, T. 1969. *Can. J. Bot.* **47**: 2108 (light & sporulation in vitro; **49**, 2272).

——. 1969. *Ibid* **47**: 791 (production of pectic & cellulytic enzymes; **48**, 3232).
Fletcher, J. T. et al. 1966. *Ann. appl. Biol.* **58**: 423 (disease on cucumber, incidence, transmission, ascospore periodicity; **46**, 825).
Grossenbacher, J. G. 1909. *Tech. Bull. N.Y. agric. Exp. Stn* 9: 193 (general).
Hopkins, D. L. 1972. *Proc. Fla St. hort. Soc.* **85**: 108 (fungicides with *Pseudoperonospora cubensis*; **53**, 1994).
Johnson, G. I. 1976. *APPS Newsl.* **5**: 48 (on pumpkin with *Fusarium* spp.; **56**, 4259).
Kagiwata, T. 1967. *Bull. agric. Exp. Stn Kanagawa* 105: 40 (chemical control; **50**, 442).
Luepschen, N. S. 1961. *Pl. Dis. Reptr* **45**: 557 (decay & temp. with *Corticium rolfsii*; **41**: 198).
Prasad, K. et al. 1967. *Proc. Am. Soc. hort. Sci.* **91**: 396 (inheritance of resistance in melon; **47**, 1738).
Rankin, H. W. 1954. *Phytopathology* **44**: 675 (seed infection & control with *Colletotrichum lagenarium*; **34**: 511).
Schenck, N. C. 1962. *Ibid* **52**: 635 (penetration of & var. resistance in melon; **42**: 171).
——. 1968. *Ibid* **58**: 1420 (ascospore spread & pseudothecial survival; **48**, 672).
——. 1969. *Proc. Fla St. hort. Soc.* **82**: 151 (ascospore periodicity, with *P. cubensis*; **50**, 2621).
Sitterly, W. R. 1968 & 1969. *Pl. Dis. Reptr* **52**: 49; **53**: 417 (blossom end infection, control by rotation; **47**, 1737; **48**, 3233).
Steekelenburg, N. A. M. van 1978. *Neth. J. Pl. Pathol.* **84**: 27 (fungicides; **57**, 3691).
Svedelius, G. et al. 1978. *Trans. Br. mycol. Soc.* **71**: 89 (factors affecting infection of attached cucumber leaves).
Uspenskaya, G. D. et al. 1976. *Biol. Nauki* **19**(10): 76 (survival of spores in soil; **56**, 4757).
Wiant, J. S. 1945. *J. agric. Res.* **71**: 193 (general account, 37 ref.; **25**: 23).

Didymella chrysanthemi (Tassi) Garibaldi & Gullino, *Agricoltura ital.* **71**: 286, 1971.

Sphaerella chrysanthemi Tassi, 1900
Mycosphaerella ligulicola Baker, Dimock & Davis, 1949
Didymella ligulicola (Baker, Dimock & Davis) Arx, 1962
stat. conid. *Ascochyta chrysanthemi* F. L. Stevens, 1907
Phoma chrysanthemi Vogl., 1901.

PSEUDOTHECIA on greyish spots on old stems, less commonly on leaves and petals; erumpent, up to 225 μ broad with papillate ostioles. ASCI short stalked, broadly cylindrical, bitunicate, 8 spored, 45–90 × 8–12 μ. ASCOSPORES hyaline, 1 septate, fusoid, constricted at the septum, guttulate, 12–16 × 4–7 μ. PSEUDOPARAPHYSES present. PYCNIDIA immersed to erumpent, mostly on stems, also on petals, flattened, globose, ostiolate, scattered, up to 300 μ diam. CONIDIA oblong ellipsoid to ovoid, straight or somewhat curved, hyaline, 6–10 (–20) × 3–5 μ, with acute or obtuse ends (A. Sivanesan, personal communication, February 1978).

D. chrysanthemi causes ray blight of *Chrysanthemum*; it has been recorded on *C. cinerariaefolium* (pyrethrum). The fungus occurs in Europe, E. Africa, Australia, Canada, Israel, Japan, New Zealand, Papua New Guinea and USA (Map 406). The pathogen and disease on ornamental chrysanthemum have been studied in Germany (Sauthoff), Sweden (Nilsson), UK (Blakeman et al.; Chesters et al.; Hadley et al.) and USA (Baker et al.; Stevens). The symptoms are most conspicuous on the blossoms at various growth stages. They turn necrotic (straw coloured), often beginning at one side; a brownish necrosis spreads from the base to the tip of each ray flower. The peduncle may be infected and as a result the flower head droops. The head does not open if the bud is attacked. Leaf lesions are irregular, black blotches, 2–3 cm diam. Black girdling lesions may be found on the stems. No studies appear to have been done on infection of pyrethrum. Chesters et al. (1967) found that soilborne inoculum was mildly pathogenic to leaves and stems of dahlia, globe artichoke, *Rudbeckia*, sunflower and zinnia. But it was severely pathogenic to lettuce; on this host infection spread from the stem base and the lower leaves were destroyed. Successive passages through lettuce increased the virulence of *D. chrysanthemi* but virulence was reduced after a long period in culture.

Under wet conditions the spread of ray blight can be very rapid, the most favourable temps being 20–26°. Linear growth on PDA, formation of pycnidia and lesion development all have an opt. temp. of 26°. Most perithecia are formed at 21°. Lesion growth is most rapid at RH 100%, it is reduced to *c.* a half at RH 50% (McCoy et al. 1972). Conidia germinate in 3.5–4 hours and infection takes place after 6 hours. The total number of conidia increased with temp. (12–27°) but the conidial mass was little altered. Conidia formed at 26° germinated less at low RH, and showed a narrower temp. range for max. germination after 6 hours (opt. 25°), than did conidia formed at 15° (good germination at 15–30°; Blakeman et al. 1968a & 1969). These workers considered that the disease-producing potential of the larger number of conidia formed at 26° would be only realised when opt. conditions for dispersal and infection existed. The fungus can survive as epiphytic mycelium on the root surface of chrysanthemum cuttings throughout the life of a glasshouse crop. Symptomless, surface colonisation was induced in non-sterile soil following inoculation with mycelium and sclerotia or conidia. In plant material left in the open to dry the pathogen may enter the root and form pseudothecia (Chesters et al. 1966). It forms a loose type of sclerotium. Blakeman et al. (1966) found that all sclerotia were dead after 30 weeks' burial in natural soil. There was no survival of conidia after 3 weeks in natural soil. Sclerotia lost the capacity to infect chrysanthemum after 8 weeks in compost.

The disease has been spread by the movement of infected cuttings and the prevention of this is important in control. The fruitbodies of the fungus can develop and persist under very dry conditions. Such conditions favour perithecial formation. Fungicide control in France (chlorothalonil, mancozeb and zineb) and USA (benomyl and captan) has been described (Crane et al.; Grouet).

Baker, K. F. et al. 1949. *Phytopathology* **39**: 789 (general in USA; **29**: 215).

——— ———. 1961. *Ibid* **51**: 96 (cause & prevention of spread; **40**: 610).

Blakeman, J. P. et al. 1966. *Trans. Br. mycol. Soc.* **49**: 227 (persistence in soil; **45**, 3066).

——— ———. 1968a. *Ibid* **51**: 643 (pattern of asexual sporulation; **48**, 1208a).

———. 1968b. *Ann. appl. Biol.* **61**: 77 (effect of leaf washings on infection; **47**, 1905).

——— ———. 1969. *Ibid* **63**: 295 (significance of temp. during conidial sporulation; **48**, 2422).

Chesters, C. G. C. et al. 1966. *Ibid* **58**: 291 (survival on roots; **46**, 341).

——— ———. 1967. *Ibid* **60**: 385 (host range & variation in virulence; **47**, 1154).

Crane, G. L. et al. 1969. *Proc. Fla St. hort. Soc.* **82**: 379 (fungicides in USA; **50**, 2304).

Grouet, D. 1974. *Phytiat. Phytopharm.* **23**: 175 (fungicides in France; **54**, 3354).

Hadley, G. et al. 1968. *Trans. Br. mycol. Soc.* **51**: 653 (asexual sporulation & nutrition; **48**, 1280b).

McCoy, R. E. et al. 1972. *Phytopathology* **62**: 1188, 1195 (environment & reproduction; temp., RH & development; **52**, 1921, 1922).

——— ———. 1973. *Ibid* **63**: 586 (ascospore discharge; **53**, 1412).

Nilsson, L. 1963. *Växtskyddsnotiser* **27**: 8 (general in Sweden; **43**, 97).

Sauthoff, W. 1963. *Phytopathol. Z.* **48**: 240 (general in Germany; **43**, 1669).

Schadler, D. L. et al. 1975. *Phytopathology* **65**: 912 (toxin; **55**, 761).

Stevens, F. L. 1907. *Bot. Gaz.* **44**: 241 (aetiology).

Didymella lycopersici Klebahn, *Z. PflKrankh. PflPath. PflSchutz* **31**: 12, 1921.
Sphaeronaema lycopersici Plowright, 1881
Ascochyta lycopersici (Plowright) Brunaud, 1887
Phoma lycopersici (Plowright) Jaczewski, 1898
P. lycopersici Cooke, 1885
stat. conid. *Diplodina lycopersici* Hollós, 1907.
(Full synonymy in Boerema et al. 1973a, *Phoma* q.v.)

PSEUDOTHECIA subglobose, light brown with papillate ostioles, 120–200 μ diam. in culture (130–300 μ on host). ASCI cylindrical to subclavate, short stipitate or sessile, 8 spored, 50–70 × 8–9 μ in culture (70–95 × 6–10 μ on host), bitunicate. ASCOSPORES irregularly biseriate, 1 septate, ellipsoid, ends mostly obtuse, hyaline, slightly or not constricted at the septum, guttulate, 12–15 × 5 μ in culture (16–18 × 5.5 μ on host). PSEUDOPARAPHYSES filiform, hyaline and septate. PYCNIDIA on stem, leaves and fruit, solitary or gregarious, initially immersed becoming erumpent, dark brown, 140–200 μ diam. in culture (180–250 μ on host), wall composed of pseudoparenchymatic cells, slightly thicker walled on the outermost layer. CONIDIA hyaline, ellipsoid to ovoid, guttulate, aseptate or 1 septate, 6–10 × 2–3 μ.

According to Brooks and Searle *Phoma destructiva* is synonymous with *Diplodina lycopersici*, the pycnidial state of *Didymella lycopersici* Kleb., but Dennis and Knight (1960a) considered it distinct (from P. Holliday & E. Punithalingam, *CMI Descr.* 272, 1970; Boerema et al. (l.c.) accepted *P. destructiva* as distinct).

The fungus causes stem and fruit rot of tomato. With its low temp. requirements this disease has only caused serious losses (somewhat sporadic) in temperate areas, notably Europe. It is widely distributed (Map 324) and tropical countries in which *D. lycopersici* has been recorded are Barbados, Brazil, Brunei, Dominican Republic, Haiti, India,

Ivory Coast, Mexico, New Caledonia, Panama, Puerto Rico, Tahiti, Togo, Tonga, Trinidad, Uganda and Venezuela. A girdling canker, dark brown and sunken, develops at (or just above) soil level. Secondary cankers may form later higher up the stem and the plant collapses. The soft, outer, diseased tissues contain numerous pycnidia and in damp conditions conidia are extruded in slimy, pink masses. The perfect state is found only rarely. Infection can also occur on roots, leaves, flowers and fruit. Raicu et al. considered that eggplant isolates were the same fungus.

Penetration probably takes place directly, provided that the cuticle is not well developed but infection occurs more readily through wounds. The opt. temp. for mycelial growth in vitro and conidial germination is 19–20°; infection and spread are rapid at 19° or a little below; above this temp. infection falls off but can occur up to 28°. Infection arises via conidia from infected plant material in the soil where the pathogen can persist for *c.* 10 months; high moisture, organic matter and low temp. increase survival. Seed transmission is considered less important. Air dispersal by ascospores is of less consequence than splash and soil dispersal by conidia. Hyphae and pycnidia are found within the hairy seed coat. Conidia do not survive >9 months on the surface of seed (Derbyshire; Fisher; Knight et al. 1960b; Maud; Ogilvie; Phillips 1956a).

Host debris from the previous season's crop should be destroyed or very carefully composted (Phillips 1959). In moist conditions the fungus is destroyed at 35° in 3–6 days. All cultural implements, glasshouses and soil should be treated with the recommended sterilants and fumigants. In outdoor crops ploughing in of crop residues must be done as early as possible. Dithiocarbamate fungicides as soil drenches are effective. Channon described control with benomyl. There appear to be no resistant vars but one line of *Lycopersicon hirsutum* has shown high resistance.

Anon. 1968. *Advis. Leafl. Minist. Agric. Fish. Fd* 560 (advisory; **47**, 3572).

Brooks, F. T. et al. 1921. *Trans. Br. mycol. Soc.* **7**: 173 (*Phoma destructiva* is synonymous; **1**: 149).

Channon, A. G. 1972. *Hort. Res.* **12**: 89 (control with benomyl; **52**, 1267).

Dennis, R. W. G. 1946. *Trans. Br. mycol. Soc.* **29**: 11 (*P. destructiva* is distinct; **25**: 526).

Derbyshire, D. M. 1960. *Pl. Pathol.* **9**: 152 (in seed; **40**: 386).

——. 1961. *Proc. int. Seed Test. Assoc.* **26**: 61 (in seed).

Fisher, D. E. 1954. *Nature Lond.* **174**: 656 (infection of seed; **34**: 555).

Hack, J. et al. 1960. *Ann. appl. Biol.* **48**: 236 (effect of soil treatments on spore survival; **40**: 64).

Hickman, C. J. 1946. *J. Pomol.* **22**: 69 (source of infection of outdoor tomatoes; **25**: 371).

Knight, D. E. 1960a. *Trans. Br. mycol. Soc.* **43**: 519 (*P. destructiva* is distinct & evidence for heterothallism; **40**: 188).

—— et al. 1960b. *Ann. appl. Biol.* **48**: 245 (seed & other infection sources, resistance; **40**: 64).

—— ——. 1960c. *Ibid* **48**: 259 (fungicides; **40**: 64).

Liesau, O. F. 1932. *Phytopathol. Z.* **5**: 1 (general; **11**: 809).

Maude, R. B. 1962. *Ann. appl. Biol.* **50**: 104 (in seed; **41**: 546).

Ogilvie, L. 1945. *Gdnrs' chron.* **118**: 71 (in seed; **24**: 480).

Orth, H. 1939. *Zentbl. Bakt. ParasitKde* Abt. 2 **100**: 211 (general; **18**: 636).

Phillips, D. H. 1956a. *Trans. Br. mycol. Soc.* **39**: 319 (in seed; **36**: 138).

——. 1956b. *Ibid* **39**: 330 (soilborne infection; **36**: 138).

——. 1959. *Ann. appl. Biol.* **47**: 240 (destruction by composting; **39**: 50).

Raicu, C. et al. 1972. *An. Inst. Cerc. Prot. Pl.* **10**: 141 (stat. conid. from eggplant & tomato; **54**, 4155).

Schroevers, T. A. C. 1919. *Tidschr. PlZiekt.* **25**: 174 (general).

Small, T. 1943. *J. Minist. Agric.* **50**: 64 (general; **22**: 387).

Smith, H. C. et al. 1959. *N.Z. Jl Agric.* **99**: 577, 579, 581, 583 (inoculation & control by spraying; **39**: 626).

Verhoeff, K. 1963. *Neth. J. Pl. Pathol.* **69**: 283 (conditions for infection; **43**, 1148).

——. 1964. *Ibid* **70**: 149 (chemical control; **44**, 1257).

Williams, P. H. et al. 1953. *J. hort. Sci.* **28**: 278 (general; **33**: 266).

—— ——. 1957. *Ann. appl. Biol.* **45**: 304 (effect of soil treatments; **36**: 735).

DIPLODIA Fr. apud Mont., *Annls Sci. nat.* ser. 2, 1: 302, 1834.
(Sphaerioidaceae)

MYCELIUM immersed, branched, septate, dark brown. CONIDIOMATA pycnidial, separate or aggregated, globose, dark brown to black, immersed, unilocular, thick walled; wall of an outer layer of dark brown, thick-walled textura angularis, a median layer of dark brown thin-walled cells, and an inner layer of thin-walled, hyaline cells. OSTIOLE single, circular, central, papillate. CONIDIOPHORES hyaline, branched and septate above and at the base, smooth, cylindrical, formed from the inner cells of the pycnidial wall. CONIDIOGENOUS CELLS holoblastic, integrated or discrete, determinate, cylindrical, hyaline, smooth, forming a single apical conidium. CONIDIA at first hyaline, with a central guttule, thick walled, aseptate, smooth, later becoming dark brown and medianly 1 euseptate, apex obtuse, base truncate (B. C. Sutton, personal communication, August 1978, the genus requires revision).

Hewitt, W. B. et al. 1971. *Hilgardia* **41**: 77 (on *Diplodia* & like fungi; effects of pH, temp., light & vitamins on taxonomic characters; **51**, 3204).
Satour, M. M. et al. 1969. *Ibid* **39**: 601, 631 (as above; effects of C & N on growth & taxonomic characters; **49**, 1987a, b).
Webster, R. K. et al. 1969. *Ibid* **39**: 655 (as above; variation in *Botryodiplodia theobromae* from grapevine; **49**, 1987c).
———. 1971. *Ibid* **41**: 95, 107 (as above; effects of C:N ratio on growth, pycnida & conidia formation; and of natural substrates on variability in taxonomic characters; **51**, 3205, 3206).
———. 1974. *Ibid* **42**: 451 (as above; criteria for classification; **54**, 3189).

Diplodia maydis (Berk.) Sacc., *Sylloge Fung.* **3**: 373, 1884.

Sphaeria maydis Berk., 1847
S. zeae Schw., 1832
Diplodia zeae (Schw.) Lév., 1848
Hendersonia zeae (Schw.) Hazsl., 1872
Macrodiplodia zeae (Schw.) Petrak & Syd., 1923
Phaeostagonosporopsis zeae (Schw.) Woron., 1925
D. ? maydicola Speg., 1910
D. zeae-maydis Mechtij, 1962.

PYCNIDIA immersed, spherical to subglobose, dark brown to black, 150–300 μ diam., wall multicellular, darker round the circular, protruding, papillate ostiole, 40 μ diam. CONIDIA straight, curved or irregular, 1 (0–2) septate, smooth walled, pale brown, apex attenuated or rounded, base truncate, 15–34 × 5–8 μ, formed from hyaline, aseptate, cylindrical phialides 10–20 × 2–3 μ. Scolecospores have also been reported in culture (Johann, 1939; B. C. Sutton & J. M. Waterston, *CMI Descr.* 84, 1966).

Sutton described the *Diplodia* spp. on maize (morphology, taxonomy and nomenclature). *D. macrospora* Earle (synonyms: *Macrodioplodia macrospora* (Earle) Höhnel, *M. zeae* (Schw.) Petrak & Syd. var. *macrospora* (Earle) Petrak & Syd., *Stenocarpella zeae* Syd.) was also described by Sutton et al. It can be distinguished from *D. maydis* by the large, 0–3 septate conidia, 44–82 × 7.5–11.5 μ. *D. macrospora* is a minor, widespread (Map 227) pathogen on maize; the symptoms have been described by Eddins. The widespread *D. maydis* is one of the main causes of maize stalk rot, ear rot and seedling blight. The earlier literature on these disease conditions, caused by several organisms, was largely covered by Christensen et al. (*Gibberella zeae* q.v.). Other, mostly later, ref. will also be found under the other major fungal pathogens which can cause these diseases on maize (*G. fujikuroi* and its var. *subglutinans, G. zeae, Macrophomina phaseolina, Pythium arrhenomanes* and *P. butleri*). The maize stalk rot caused by *D. maydis* is probably the commonest in the USA maize belt. When infection takes place before plants reach full maturity the leaves become greyish green, the lower parts of the culm become brown and the pith becomes soft and disintegrates. The vascular bundles remain intact. Diseased stems break readily. Pycnidia form just beneath the surface of the lower stem internodes. In ear rot the infection usually spreads from the base of the ear and a complete rot may occur; pycnidia form in the husks, bracts and kernel pericarps. The ear can be covered with the mycelium. In USA the pycnidia form in late summer or autumn and mature in the following spring. Seedling blight can be caused by *D. maydis.* In seed tests Nwigwe found that the fungus caused a 5–37% loss in germination.

Some general accounts of the disease have been given by Clayton; Durrell; Eddins; Raleigh. In early growth the mesocotyl and crown are invaded (McNew); roots can be infected (Craig et al.). Localised infections of the stalk (mostly lower part), and silk, tip and shank of the ear occur. Growth takes place from the shank into the cob (Clayton). The fungus is favoured by high temperatures of *c.* 28–30°. Forms may differ in pathogenicity, temp. requirements and other characteristics. Maxwell et al. considered that the degree of virulence of a given isolate was important in discriminating between maize genotypes. Carry over takes place in seed and crop debris; the fungus can survive for 3 years in stem tissue. Several factors may increase disease incidence: excess nitrogen, soil moisture unfavourable for growth, large plant populations (e.g. Krüger et al.), partial defoliation, insect damage (e.g. Holbert et al.), root and leaf injury (Pappelis), and infection by other pathogens (e.g. Fajemisin et al.;

Michaelson). In South Africa Krüger et al. found that in seasons with av. rainfall the most stalk rot was caused by *D. maydis*; it was increased by phosphate but decreased by nitrogen. In dry seasons most such rot was caused by *G. fujikuroi* and it increased with high nitrogen. Healthy living host cells resist invasion by the pathogen which tends to spread only in senescent or dead cells (e.g. Pappelis et al.; Wysong et al.). Control is through healthy seed, balanced growth conditions (nitrogen especially should not be proportionately excessive) and the use of resistant hybrids; also Landis (*G. fujikuroi*, maize q.v.).

Bemiller, J. N. et al. 1965. *Phytopathology* **55**: 1237, 1241 (chemical content & tissue density in relation to resistance; **45**, 1047a & b).
———. 1968. *Ibid* **58**: 1336 (cellulytic activity; **48**, 791).
———. 1969. *Ibid* **59**: 674 (effects of phenolics & indole-3-acetic acid on cellulytic & pectolytic enzymes; **48**, 2944).
Calvert, O. H. et al. 1969. *Ibid* **59**: 239 (vivipary induced; **48**, 2354).
Clayton, E. E. 1927. *J. agric. Res.* **34**: 357 (ear & seed infection).
Craig, J. et al. 1961. *Phytopathology* **51**: 376, 382 (sugar trends, pith density & disease; root infection; **40**: 747).
Durrell, L. W. 1923. *Bull. Iowa agric. Exp. Stn* 77: 347 (general; 3: 207).
Eddins, A. H. 1930. *Phytopathology* **20**: 439 (general; **9**: 712).
Fajemisin, J. M. et al. 1974. *Ibid* **64**: 1496 (predisposition through infection by *Cochliobolus heterostrophus* & *Setosphaeria turcica*; **54**, 2791).
Johann, H. 1935. *J. agric. Res.* **51**: 855 (seed history & seed reaction to kernel rots; **15**: 290).
——. 1939. *Phytopathology* **29**: 67 (scolecospores; **18**: 307).
——. 1940. *Ibid* **30**: 979 (as above; **20**: 298).
Holbert, J. R. et al. 1935. *Ibid* **25**: 1113 (factors affecting infection & spread in host; **15**: 290).
Kang, M. S. et al. 1974. *Pl. Dis. Reptr* **58**: 1113 (effect of cob & shank inoculation on cell death in internodes; **54**, 2802).
Krüger, W. et al. 1965. *S. Afr. J. agric. Sci.* **8**: 703 (effect of fertilisers & plant population; **45**, 1046).
Loesch, P. J. et al. 1972. *Crop Sci.* **12**: 469 (effect of infection on stem quality; **53**, 140).
McNew, G. L. 1937. *Res. Bull. Iowa agric. Exp. Stn* 216: 191 (crown infection; **16**: 739).
Maxwell, J. D. et al. 1974. *Crop Sci.* **14**: 594 (balance between tester host resistance & isolate virulence in inbreds; **55**, 1231).
Michaelson, M. E. 1957. *Phytopathology* **47**: 499 (predisposition through infection by *Ustilago maydis*, inter alia, with *Gibberella zeae*; **37**: 39).
Murphy, J. A. et al. 1976. *J. Bact.* **127**: 1465 (fine structure & elemental composition of dormant & germinating conidia; **56**, 2048).
Nwigwe, C. 1974. *Pl. Dis. Reptr* **58**: 414 (effect on seed germination inter alia; **53**, 3459).
Pappelis, A. J. et al. 1963. *Phytopathology* **53**: 1100 (water content, living cells & spread of pathogen; **43**, 456).
———. 1970. *Ibid* **60**: 355 (effect of root & leaf injury on cell death & susceptibility; **49**, 2464).
———. 1973. *Pl. Dis. Reptr* **57**: 308 (parenchyma cell death & susceptibility in stem & ear; **52**, 3676).
Raleigh, W. P. 1930. *Res. Bull. Iowa agric. Exp. Stn* 124: 95 (infection; **9**: 773).
Sleper, D. A. et al. 1970. *Iowa St. J. Sci.* **45**: 197 (stem characteristics & their significance in resistance to stem break; **50**, 2908).
Stevens, N. E. et al. 1942. *Phytopathology* **32**: 184 (in vitro growth with biotin; **21**: 330).
Sutton, B. C. et al. 1966. *C.M.I. Descr. pathog. Fungi Bact.* 83 (*Diplodia macrospora*).
———. 1967. *Mycol. Pap.* 97, 42 pp. (Coelomycetes III; **44**, 3277).
Thompson, D. L. et al. 1971. *Pl. Dis. Reptr* **55**: 158 (correlations between stalk & ear rots; **50**, 2907).
Wysong, D. S. et al. 1966. *Phytopathology* **56**: 26 (relation of soluble solids & pith condition to stalk rot; **45**, 1366).

Diplodia pinea (Desm.) Kickx, *Fl. Flandres* **1**: 397, 1867.
Sphaeria pinea Desm., 1842
Botryodiplodia pinea (Desm.) Petrak, 1922
Macrophoma pinea (Desm.) Petrak & Syd., 1926
Sphaeropsis ellisii Sacc. 1884.
(further synonymy in *CMI Descr.* 273)

PYCNIDIA on bark, shoots, leaves and cone scales, solitary to gregarious, initially immersed becoming erumpent, black, ovoid, opening by apical ostioles; pycnidial wall multicellular with heavily pigmented cells around the neck region. CONIDIA yellowish brown eventually turning dark brown, wall rough, oblong to clavate, rounded at apex, basal end blunt, usually aseptate, later some become 1 septate, 30–45 × 10–16 μ. Distinguished from *D. sapinea* Fuckel, recorded on *Pinus*, by the often aseptate, large, rough-walled spores (E. Punithalingam & J. M. Waterston, *CMI Descr.* 273, 1970).

D. pinea on *Pinus* spp. and other conifers causes a shoot blight (also called tip and twig blight, stag head, dead or red top, whorl canker, bud wilt, dieback and seedling collar rot). The fungus, as a saprophyte causing blue stain of sapwood of fallen or freshly cut timbers, is widespread, especially in temperate areas (Map 459). Amongst the countries where the disease has been studied are: Argentina (Saravi Cisneros), Australia (Eldridge; Marks et al.; Millikan et al.; Purnell), Canada (Haddow et al.), Italy (Capretti), New Zealand (Birch; Chou; Gilmour), Portugal (Ferreirinha; Oliveira), Rumania (Prodan), Spain (Torres Juan) and USA (Peterson et al.; Schweitzer et al.; Slagg et al.; Waterman). In the nursery infection of seedlings (sometimes beginning on the youngest plants and usually in the upper parts) leads to dieback and death. In the field the young apical branch clusters show a brown or red brown necrosis of the needles which die and persist on the tree. Spread of the infection causes branch dieback, resin exudation, cankering, bending and curling of young shoots and death of parts of the crown. When needles are penetrated through the stomata small tan to reddish brown lesions form after a few days (Brookhouser et al.; Prodan). Chou (1978) described penetration of young stems through the junction of anticlinal epidermal walls and the cuticle. Infection may also be associated with wounds on the trunk and branches, and with pruning cuts. Gilmour (in New Zealand on 3–10-year-old *P. radiata*) found that infection was correlated with trunk wounding; incidence varied with the pruning tool used: secateurs 12%, axe 48% and slasher 68%. The disease is frequently associated with unfavourable growing and stress conditions, and physical damage to the tissues (wounds, hail, frost and insects). Birch, who gave an early, full account of the pathogen and the disease, described *D. pinea* as a saprophyte and a facultative parasite.

The conidia (water dispersed) germinate readily; >75% within 2 hours, with the most germination at 24° and germ tube growth at 28°. For infection of uninjured needles incubation for 12 hours at RH 100% and 24° was adequate (Brookhouser et al.). Young shoots are most susceptible. Haddow et al. described massive crown infection of *P. sylvestris* (Scots pine), particularly in trees >15 years old, associated with infestation by spittle bugs (*Aphrophora parallela*). Blight symptoms were caused when conidia were injected into the terminal shoots of Scots pine or were applied to a spittle bug infested branch. Kizikelashvili infected pine seedlings (1 year old) at 15–35° with most infection at 30°. Marks et al. (in *P. radiata*) found that injection of conidia gave high infection and the fungus spread rapidly in the pith. It can persist in infected tissue for over a year. Fast growing trees (*P. radiata*) were more severely affected than slower growing ones and the latter recovered more quickly. They found that a fresh wound remained susceptible to infection for 3–9 days under glasshouse conditions. A superficial wound that does not injure the cambium is less likely to produce lasting damage on infection than one that does. Rapid xylem growth exposes unsuberised tissue to infection. Wright et al. described the effects of the disease on standing merchantable wood. Chou (1977) found less dieback in trees that had reached 7–8 years of age but this was apparently not due to any increase in inherent plant resistance. Control measures depend very largely on local conditions. In some areas where the environment particularly favours the fungus some damage is unavoidable. Such sites could be used to produce lower-grade wood products, as infection does not kill trees or seriously reduce growth rates (Marks et al. 1970). Bordeaux reduced infection in USA (Nebraska; Peterson; Peterson et al.); and see Schweitzer et al. for benomyl.

Birch, T. T. C. 1936. *Bull. N.Z. For. Serv.* 8, 32 pp. (general; **16**: 148).

Brookhouser, L. W. et al. 1971. *Phytopathology* **61**: 409 (infection of pine needles & effects of temp. on conidial germination; **50**, 3223).

Capretti, C. 1956. *Annali Accad. ital. Sci. for.* **5**: 171 (general, 77 ref.; **36**: 436).

Chou, C. K. S. et al. 1976. *Eur. J. For. Pathol.* **6**: 354 (toxicity of monoterpenes from cortical oleoresin to conidia; **56**, 2701).

——. 1977. *Pl. Dis. Reptr* **61**: 101 (effect of tree age on infection; **56**, 5220).

——. 1978. *Physiol. Pl. Pathol.* **13**: 189 (penetration of young stems).

Eldridge, K. G. 1961. *Tech. Pap. For. Commn Vict.* 7: 16, 22 (factors affecting disease incidence & timber blue stain; **41**: 339).

Ferreirinha, M. P. 1955. *Estudos Inf. Dir. ger. Servs flor aquic* 46, 43 pp. (general; **35**: 131).

Foster, R. C. et al. 1968. *Aust. For.* **32**: 211 (histology; **48**, 959).

Gilmour, J. W. 1964. *Res. Leafl. N.Z. For. Serv.* 5, 4 pp. (wounding & canker; **44**, 889).

Haddow, W. R. et al. 1942. *Trans. R. Can. Inst.* **24**: 1 (association of disease with pine spittle bug; **21**: 398).

Kizikelashvili, O. G. 1968. *Trudy̆ Inst. Zashch. Rast. Tbilisi* **20**: 267 (conidial germination, growth & infection; **48**, 3188).

Marks, G. C. et al. 1969. *Aust. J. Bot.* **17**: 1 (pathogenicity; **48**, 3187).
——. 1970. *Ibid* **18**: 55 (resistance to infection; **50**, 316).
Millikan, C. R. et al. 1957. *Aust. For.* **21**: 4 (disease & general environment; **38**: 40).
Oliveira, A. L. F. 1944. *Agros Lisb.* **27**: 158 (morphology, growth in vitro & infection; **26**: 272).
Peterson, G. W. et al. 1968. *Pl. Dis. Reptr* **52**: 359 (disease damage & Bordeaux; **47**, 2873).
——. 1977. *Phytopathology* **67**: 511 (infection, epidemiology & Bordeaux; **56**, 5830).
Prodan, I. 1934. *Bul. Grăd. bot. Muz. bot. Univ. Cluj* **14**: 240 (needle infection; **14**: 483).
Purnell, H. M. 1957. *Bull. For. Commn Vict.* 5, 10 pp. (incidence, inoculation & fungal growth in vitro; **38**: 40).
Saravi Cisneros, R. 1950. *Revta Fac. Agron. Univ. nac. La Plata* **27**: 163 (general; **31**: 1).
Schweitzer, D. J. et al. 1976. *Pl. Dis. Reptr* **60**: 269 (control with benomyl; **55**, 5375).
Slagg, C. M. et al. 1943. *Phytopathology* **33**: 390 (disease in seedlings; **22**: 412).
Stahl, W. 1968. *Aust. For. Res.* **3**: 27 (host defence; **50**, 317).
Torres Juan, J. 1964. *An. Inst. For. Invest. Exp.* **36**: 103 (in Spain; **45**, 1545).
——. 1971. *Boln Serv. Plagas for.* **14**: 13 (in the Balearic Islands; **52**, 263).
Waterman, A. M. 1943. *Phytopathology* **33**: 1018 (general; **23**: 123).
Wright, J. P. et al. 1971. *Aust. For.* **34**: 107 (infection & loss of merchantable wood; **51**, 1992).
Young, H. E. 1936. *Qd agric. J.* **46**: 310 (infection, symptoms & pine hosts; **16**: 219).

DRECHSLERA Ito, *Proc. Imp. Acad. Japan* **6**: 355, 1930.
Bipolaris Shoemaker, 1959.
(B. C. Sutton, *Taxon* **26**: 591, 1977 on conservation against *Angiopoma* Lév.)
(Dematiaceae)

COLONIES effuse, grey, brown or blackish brown, often hairy, sometimes velvety. MYCELIUM mostly immersed. STROMA present in some spp. SCLEROTIA or PROTOTHECIA often formed in culture. SETAE and HYPHOPODIA absent. CONIDIOPHORES macronematous, mononematous, sometimes caespitose, straight or flexuous, often geniculate, unbranched or in a few spp. loosely branched, brown, smooth in most spp. CONIDIOGENOUS CELLS polytretic, integrated, terminal, frequently becoming intercalary, sympodial, cylindrical, cicatrised. CONIDIA solitary, in certain spp. also sometimes catenate or forming secondary conidiophores which bear conidia, acropleurogenous, simple, straight or curved, clavate, cylindrical rounded at the ends, ellipsoidal, fusiform or obclavate, straw coloured or pale to dark brown or olivaceous brown, sometimes with cells unequally coloured, the end cells then being paler than intermediate ones, mostly smooth, rarely verruculose, pseudoseptate. (M. B. Ellis, *Demat. Hyphom.*: 403, 1971, with key to, and descriptions of 47 spp., conidial states; some spp. with perfect states in *Cochliobolus*, *Pyrenophora* and *Setosphaeria* q.v.; and see Ellis, *More Demat. Hyphom.* 1976.)

The economically important spp. occur almost invariably on Gramineae (formerly under *Helminthosporium* q.v.); see Chidambaram et al.; Ito; Ito et al.; Misra; Richardson et al.; Shoemaker, 1959, 1962; Subramanian et al. *Drechslera* differs from *Helminthosporium* in that the conidiogenous cells develop sympodially and have scars (i.e. are cicatrised) where the conidia were borne. In the latter genus the conidiogenous cells are determinate, with the conidia developing through pores beneath the septa whilst the conidiophore tip is actively growing; conidiophore growth ceases when the terminal conidium is formed. The conidiophores are never geniculate and never have dark scars. In the former genus the first conidium not the last is formed at the apex of the conidiophore. After the first conidium has developed the conidiophore grows out laterally below the scar, sometimes splitting the side wall, pushing the scar to one side then growing on before forming another conidium at the newly constituted apex (Ellis 1974). Ellis (*Demat. Hyphom.*) gave a list of the common substrata for *Drechslera* spp.; and see Luttrel (1951). Some spp. which have been reported from tropical crops but are not described separately (either in the perfect or imperfect states) are:

D. australiensis (Bugnicourt) Subram. & Jain ex M. B. Ellis (synonym: *Helminthosporium australiense* Bugnicourt, invalidly published) on various hosts.

D. chloridis Alcorn on *Chloris gayana* (Rhodes grass); perfect state *Cochliobolus chloridis* Alcorn.

D. coicis (Nisikado) Subram. & Jain (synonyms: *H. coicis* Nisik., *Curvularia coicis* Castellani on *Coix*). The fungus causes a blight of Job's tears (*Coix lachryma-jobi*). It was placed in a new perfect state genus *Pseudocochliobolus* Tsuda, Ueyama & Nishihara as *P. nisikadoi* Tsuda, Ueyama & Nishihara. The genus differs from *Cochliobolus* particularly in the existence of basal columnar stromata and in ascospore arrangement with parallel to loose coiling in the ascus.

D. frumentacei (Mitra) M. B. Ellis (synonym: *H. frumentaceum* Mitra) on *Echinochloa* (as *Panicum frumentaceum*).

D. hawaiiensis (Bugnicourt) Subram. & Jain ex Ellis (synonym: *H. hawaiiense* Bugnicourt, invalidly published) on various hosts; perfect state *C. hawaiiensis* Alcorn.

D. miyakei (Nisikado) Subram. & Jain (synonym: *H. miyakei* Nisikado) on *Ergrostis*; Kidane et al. described the fungus on seed of *E. tef* (teff).

D. monoceras (Drechsler) Subram. & Jain (synonym: *H. monoceras* Drechsler) on *Echinochloa*.

D. musae-sapientum (Hansford) M. B. Ellis (synonym: *H. musae-sapientum* Hansford) on *Musa*.

D. sesami (Miyake) Richardson & Fraser (synonym: *H. sesami* Miyake) on *Sesamum indicum*; see Stone; Watanabe.

D. sorghicola (Lefebvre & Sherwin) Richardson & Fraser (synonym: *H. sorghicola* Lefebvre & Sherwin) on *Sorghum* (Ellis et al.). Chidambaram et al. gave the diagnostic features of 25 *Drechslera* spp.; and see Luttrel 1951.

Alcorn, J. L. 1976. *Trans. Br. mycol. Soc.* **67**: 148 (*Drechslera chloridis* sp. nov. & other *Drechslera* spp. on *Chloris*; **56**, 2109).

——. 1978. *Ibid* **70**: 61 (*Cochliobolus chloridis* & *C. hawaiiensis* spp. nov.; **57**, 3336).

Chidambaram, P. et al. 1973. *Friesia* **10**: 165 (characteristics & host range of seedborne spp.; **53**, 2116).

Drechsler, C. 1923. *J. agric. Res.* **24**: 641 (taxonomy, morphology, hosts, symptoms; **3**: 65).

Elazegui, F. A. et al. 1973. *Philipp. Agric.* **57**: 210 (*D. sorghicola*; **55**, 1236).

Ellis, M. B. 1974. *Trans. Br. mycol. Soc.* **62**: 225 (Dematiaceous Hyphomycetes in Britain).

—— et al. 1976. *C.M.I. Descr. pathog. Fungi Bact.* 491 (*D. sorghicola*).

Ghurde, V. R. 1966. *Trans. Br. mycol. Soc.* **49**, 241 (saprophytic ability of 3 spp.; **45**, 3068).

Ibrahim, J. M. et al. 1966. *Proc. R. Soc.* B **165**: 362 (application of numerical taxonomy; **46**, 1922).

Ito, S. 1930. *Proc. Imp. Acad. Japan* **6**: 352 (perfect states, *Drechslera* gen. nov.; **10**: 232).

—— et al. 1931. *J. Fac. Agric. Hokkaido (Imp.) Univ.* **29**: 85 (perfect states; **10**: 514).

Jiang, G. Z. 1959. *Acta phytopathol. sin.* **5**: 22 (spp. from China; **39**: 91).

Kafi, A. et al. 1966. *Trans. Br. mycol. Soc.* **49**: 327 (effect of nutrients on growth, sporulation & conidial characteristics in 5 spp.; **45**, 3083).

Kapoor, I. J. et al. 1969. *Curr. Sci* **38**: 120 (*D. australiensis* on tomato fruit; **48**, 2578).

Kenneth, R. 1958. *Bull. Res. Coun. Israel* Sect. D. **6**: 191 (spp. from Israel; **38**: 353).

Kidane, A. et al. 1978. *Pl. Dis. Reptr* **62**: 70 (*D. miyakei*, seed transmission; **57**, 5022).

Langdon, R. F. N. et al. 1971. *Search* **2**: 250 (the hilum as a taxonomic character; **52**, 621).

Luttrel, E. S. 1951. *Pl. Dis. Reptr* Suppl. 201: 59 (key to spp. in USA; **30**: 472).

——. 1963. *Mycologia* **55**: 643 (taxonomy with other genera; **43**, 943).

——. 1964. *Ibid* **56**: 119 (as above; **43**, 2238).

Misra, A. P. 1976. Tirhut Coll. Agric., Dholi, Muzaffapur, India, 289 pp. (in India; **56**, 78).

Mitra, M. 1923. *Mem. Dep. Agric. India bot. Ser.* **11**: 219 (on sorghum & maize in India; **3**: 29).

Nair, N. G. 1968. *Phytopathol. Z.* **61**: 331 (saprophytic activity in soil in 6 spp.; **47**, 3022).

Nelson, R. R. 1960. *Phytopathology* **50**: 375 (interspecific crosses; **40**: 30).

Nisikado, Y. 1929. *Ber. Ohara Inst. landw. Forsch.* **4**: 111 (spp. from Japan; **8**: 529).

Putterill, K. M. 1954. *Bothalia* **6**: 347 (spp. from South Africa; **34**: 323).

Rapilly, F. 1964. *Annls. Epiphyt.* **15**: 257 (taxonomic value of sporing structures; **44**, 1812).

Richardson, M. J. et al. 1968. *Trans. Br. mycol. Soc.* **51**: 147 (new combinations).

Shoemaker, R. A. 1957. *Can. J. Bot.* **35**: 269 (Atkinson's spp. from USA, Alabama).

——. 1959. *Ibid* **37**: 879 (nomenclature; **39**: 464).

——. 1962. *Ibid* **40**: 809 (taxonomy, keys & morphology).

Stone, W. J. 1959. *Phytopathology* **49**: 815 (on sesame; **39**: 333).

Subramanian, C. V. et al. 1966. *Curr. Sci.* **35**: 352 (taxonomy; **45**: 3491).

——. 1970. *Ibid* **39**: 2 (interpretation of conidial types; **49**, 2346).

Tarr, S. A. J. et al. 1968. *Trans. Br. mycol. Soc.* **51**: 771 (effect of N & pH in 5 spp. in vitro; **48**, 1071).

Thind, K. S. et al. 1967. *Proc. Indian Acad. Sci.* Sect. B **66**: 250 (trace elements; **47**, 2063).

—— ——. 1968. *Indian Phytopathol.* **21**: 190 (*D. australiensis*, growth in vitro; **48**, 1680).

Tsuda, M. et al. 1977. *Mycologia* **69**: 1109 (*Pseudocochliobolus nisikado* sp. nov.; **57**, 4017).

Watanabe, K. 1941. *Ann. phytopathol. Soc. Japan* **11**: 57 (on sesame; **30**: 344).

Drechslera gigantea (Heald & Wolf) Ito, *Proc. Imp. Acad. Japan* **6**: 355, 1930.

Helminthosporium giganteum Heald & Wolf, 1911.

CONIDIOPHORES brown, up to 400 μ long, 9–13 μ thick, sometimes swollen to 15 μ near the apex which often has a saucer like depression at the scar. CONIDIA straight, cylindrical, very pale, thin walled, with 3–6 (commonly 5) pseudosepta, usually 200–320 × 15–25 μ but occasionally up to 390 × 30 μ; 1–2 μ wide (M. B. Ellis, *Demat. Hyphom.*: 426, 1971).

Zonate eyespot (and the causal fungus) of grasses in America was described in detail by Drechsler; it is particularly common on *Cynodon dactylon* (Bermuda grass). The characteristics of the leaf spot differ with the host. The small brown spots increase in length and width, becoming generally elongate, with a central area lighter in colour and a narrow dark border when fully developed. The lesions may cover the whole width of the lamina, and the resulting necrotic areas show irregularly concentric brown, zonate markings. The plant genera attacked include *Cynodon*, *Echinochloa*, *Panicum* and *Pennisetum*. Meredith named other hosts including coconut and *Saccharum*, described the release of the conidium and the minor disease (eyespot) on banana leaves in Jamaica. Reddish brown spots appear first; these enlarge, become oval in the direction of the veins and the centre becomes dark brown fading to dull white or greyish. The spot has a narrow, well-defined dark brown margin, a pale green to yellowish halo and can be up to 8 × 16 mm. The spots originate mostly on the lower leaf surface and when a furled heart leaf is infected, on one side, a distinct line of spots is found on one half of the unfurled lamina. The airborne conidia are most abundant from 0800 to 1400 hours. The disease occurs on younger plants and only in areas where there is a nearby inoculum source, mostly *C. dactylon*; it is absent from well kept, mature, banana fields and no control is required.

Drechsler, C. 1928. *J. agric. Res.* **37**: 473 (full, general account; **8**: 384).
——. 1929. *Ibid* **39**: 129 (more grass hosts; **9**: 39).
Meredith, D. S. 1963. *Ann. appl. Biol.* **51**: 29 (on banana, general; **42**: 476).
——. 1963. *Trans. Br. mycol. Soc.* **46**: 201 (hosts & conidial release; **42**: 695).

Drechslera halodes (Drechsler) Subram. & Jain, *Curr. Sci.* **35**: 354, 1966.

Helminthosporium halodes Drechsler, 1923.

CONIDIOPHORES arising singly or in pairs, straight or flexuous, upper part often geniculate, brown, up to 150 μ long, 5–8 μ thick. CONIDIA straight or slightly curved, cylindrical to ellipsoidal with up to 12 (commonly 6–8) pseudosepta, end cells hyaline or very pale and cut off by thick, dark septa, intermediate cells golden brown, 30–100 (in culture usually 60–90) μ long, 11–20 μ thick in the broadest part; hilum distinctly protruberant. Frequently isolated from grasses, many other plants, soil and textiles (M. B. Ellis, *Demat. Hyphom.*: 409, 1971).

D. halodes differs from *D. rostrata* (q.v.) in the shape of the conidia which are also smaller in the former. Crops on which *D. halodes* causes minor leaf spot diseases include coconut, oil palm and sugarcane (target blotch). On young coconut leaves infection causes small, brown and sunken spots which later coalesce. Infection intensity is comparable on both leaf surfaces and no spotting occurs on mature leaves. Copper fungicides give some control. On the youngest and spear leaves of oil palm (*D. halodes* var. *elaeicola* Kovachich) small pale green, almost circular spots enlarge, become yellow with a central brown spot, turn necrotic becoming darker and coalescing so that large areas of the leaf become dried out; there are also chlorotic haloes to the brown spots (1–1.5 × 3–5 mm) which may eventually form into large chlorotic areas as well. Other leaf diseases of oil palm associated with *Drechslera* spp. have been described: *D. zeicola* (perfect state *Cochliobolus carbonum* q.v.), eyespot; *D. maydis* (perfect state *C. heterostrophus* q.v.), leaf rot. Captan gives good control.

Dupriez, G. et al. 1957. *Bull. Inf. I.N.E.A.C.* **6**: 205 (fungicide control on oil palm; **37**: 105).
Kovachich, W. G. 1954. *Trans. Br. mycol. Soc.* **37**: 422 (var. *elaeicola* on oil palm; **34**: 643).
——. 1957. *Ibid* **40**: 90 (other *Drechslera* spp. on oil palm; **36**: 586).
Lily, V. G. 1963. *Indian Cocon. J.* **16**: 97, 148 (symptoms, penetration & infection on coconut; **43**, 2386, 2387).
Priode, C. N. 1931. *Phytopathology* **21**: 41 (on sugarcane; **10**: 407).
Quillec, G. et al. 1975. *Oléagineux* **30**: 209 (on coconut; **54**, 5510).

Radha, K. 1965. *Sci. Cult.* **31**: 371 (effect of micronutrients in coconut **45**, 1124).
—— et al. 1961. *Indian Cocon. J.* **15**: 1 (on coconut; **41**: 537).
Subramaniam, L. S. 1936. *Indian J. agric. Sci.* **6**: 11 (on sugarcane; **15**: 465).

Drechslera heveae (Petch) M. B. Ellis, *Dematiaceous Hyphomycetes*: 451, 1971.
Helminthosporium heveae Petch, 1906.

CONIDIOPHORES solitary or in small groups, straight or flexuous, sometimes geniculate, pale to mid brown, up to 200 μ long, 6–8 μ thick; up to 10 μ thick, often much darker in culture. CONIDIA usually curved, navicular or fusiform, pale to mid golden or reddish brown, smooth, 6–11 pseudoseptate, mostly 90–130 μ long and 15–21 μ thick in the broadest part; hilum 3–5 μ wide. Conidia formed when the fungus is grown on wheat straw with tap water agar under near UV light are often up to 140 μ long, 22–28 μ thick and dark reddish brown (M. B. Ellis & P. Holliday, *CMI Descr.* 343, 1972).

Bird's eye spot of rubber (*Hevea brasiliensis*) is widespread in S.E. Asia, central and W. Africa, and central and S. America where rubber is grown (Map 270). Symptoms are most conspicuous on the leaves (varying with leaf age at the time of infection) but they also occur on green stems. On very young leaves the dark brown spots coalesce and sporulate heavily; there is leaf distortion and shedding. The upper part of a defoliated stem becomes swollen. On older leaves the initially chlorotic spots develop a straw coloured, papery centre with a distinct, red brown to black margin (up to 3–4 mm diam.) and a chlorotic halo; shotholes form. Conidia may be conspicuous on the under surface of the leaf in the centre of the lesion. Petioles show dark, elongated flecks.

This is a relatively minor disease of young plants and of concern only in the nursery; seedlings are generally more susceptible than buddings. Defoliation weakens stocks, thereby delaying budding and reducing the efficiency of its success; this is particularly so with green budding. Penetration is cuticular with appressoria and is complete in 6 hours. Leaves inoculated when up to 6 days old show severe distortion and large lesions; at 7–14 days there is moderate distortion and spotting and at 15–19 days spotting but no distortion. Older leaves are immune and young leaves of mature plants do not become infected. Light is required for infection and symptoms are almost completely suppressed under heavy shade. In potted seedlings high nitrogen increased the number of lesions but high calcium did not. The greater disease level with high nitrogen is possibly associated with a relative reduction in phosphorus and potassium conc. in the lamina. Nurseries should be sited in good soil conditions and properly maintained to ensure vigorous growth. If spraying is necessary good control has been obtained with weekly applications of carbamates, containing zinc or manganese, to the young leaves.

Anon. 1967. *Plrs' Bull. Rubb. Res. Inst. Malaya* **93**: 281 (advisory; **47**, 1256).
Bolle-Jones, E. W. et al. 1957. *J. Rubb. Res. Inst. Malaya* **15**: 80 (relation of susceptibility to host nutrient status; **37**: 506).
Hilton, R. N. 1952. *Ibid* **14**: 40 (general, 61 ref.; **32**: 397).
Turner, P. D. 1969. *Exp. Agric.* **5**: 32 (laboratory assessment of fungicides; **48**, 1289a).
Wastie, R. L. 1969. *Ibid* **5**: 41 (field assessment of fungicides; **48**, 1289b).

Drechslera incurvata (Ch. Bernard) M. B. Ellis, *Dematiaceous Hyphomycetes*: 450, 1971.
Helminthosporium incurvatum Ch. Bernard, 1906.

CONIDIOPHORES solitary or in small groups, straight or flexuous, sometimes geniculate, cylindrical to subulate, up to 500 μ long, 7–12 μ thick, often swollen to 16–20 μ at the base. CONIDIA typically curved, navicular or broadly fusiform, rather pale straw coloured, smooth, 8–13 pseudoseptate, 100–150 (130) μ long, 19–22 (20) μ thick in the broadest parts, hilum usually inconspicuous. Conidia formed when the fungus is grown on wheat straw with tap water agar and under near UV light are often broader (up to 24 μ) and darker (M. B. Ellis & P. Holliday, *CMI Descr.* 342, 1972).

A minor leaf spot of young coconut reported from S.E. Asia, Australasia, Oceania and also Jamaica, Seychelles and Sri Lanka. The spots are at first small, oval, brown; enlarging and becoming pale buff in the centre with a broad, dark brown margin. In severe attacks the edges of leaves become extensively necrotic. Severe attacks have been described from Malaysia and Sri Lanka; they can occur when young plants are overcrowded, heavily intercropped, under fertilised or given excessive shade. In Papua New Guinea and Sri Lanka applications of phosphate and potash to seedlings and 2–3-year-old palms, respectively, reduced the disease. Brown compared this leaf spot with the one caused by *Pestalotiopsis palmarum* (q.v.).

Brown, J. S. 1975. *Papua New Guin. agric. J.* **26**: 31 (isolation & inoculation with *Pestalotiopsis palmarum*; **55**, 5313).
Era, A. M. et al. 1972. *Araneta Res. J.* **19**: 137 (general, effect of irradiation; **53**, 1501).
Gallasch, H. 1974. *Papua New Guin. agric. J.* **25**: 38 (effects of fertiliser; **55**, 2333).
Johnston, A. 1962. *Inf. Lett. F.A.O. Pl. Prot. Comm. S.E. Asia* 15, 5 pp. (causing setback to seedlings in Sri Lanka; **41**: 734).
Keerthisinghe, J. K. F. 1962. *Ceylon Cocon. Plrs' Rev.* **2**: 17 (effects of fertiliser; **42**: 625).
Wiltshire, S. P. 1956. *Pl. Prot. Bull. F.A.O.* **5**: 6 (severe leaf spot in Malaysia; **36**: 308).

Drechslera rostrata (Drechsler) Richardson & Fraser, *Trans. Br. mycol. Soc.* **51**: 148, 1968.

Helminthosporium rostratum Drechsler, 1923.

CONIDIOPHORES solitary or in small groups, straight or flexuous, sometimes geniculate, mid to dark brown or olivaceous brown up to 200 μ long, 6–8 μ thick. CONIDIA straight or slightly curved, when mature obclavate, rostrate, 6–16 pseudoseptate, end cells often hyaline or very pale and cut off by thick, dark septa, intermediate cells golden brown, 40–180 μ long, 14–22 μ thick in the broadest part; hilum distinctly protruberant. Common on grasses and many other substrata, isolated from soil; cosmopolitan (M. B. Ellis, *Demat. Hyphom.*: 408, 1971).

Amongst the crops on which *D. rostrata* causes a leaf spot are sorghum, rice, maize, bulrush (pearl) millet, lemon grass, Bermuda (star) grass, sugarcane and papaw; see Whitehead et al. for gramineous hosts. The disease does not generally seem to be serious. The leaf spots are mostly small, oval, elongate, 1–2 × 2–5 mm (up to 1 cm long), the centre becoming pale grey or straw coloured with a darker border area. Spots may be brown or purplish and have chlorotic surrounds. A light red discolouration, developing to a black spot has been described on maize seed. Leaf spots may coalesce; on papaw leaf lesions begin as small yellowish, watersoaked spots, produced from the tips of the lower leaves. There is variation in lesion type between different hosts. No particular control measures seem to be required; sorghums, which vary in their reaction to infection, are more resistant than millet and maize.

Agarwal, G. P. et al. 1959. *Phyton* **13**: 45 (culture, effect of pH, C & N; **39**: 671).
Byther, R. S. et al. 1972. *Phytopathology* **62**: 120 (on sugarcane with *Cochliobolus lunata*, *Curvularia senegalensis* & *Drechslera hawaiiensis*; **51**, 3532).
Chattopadhyay, S. B. et al. 1959. *Pl. Dis. Reptr* **43**: 1241 (on rice; **39**: 308).
Honda, Y. et al. 1978. *Mycologia* **70**: 343 (photosporogenesis, temp. effects on sporulation & spore morphology).
—— ——. 1978. *Trans. mycol. Soc. Japan* **19**: 23 (photosporogenesis, light quality effects on sporulation & spore morphology).
Kafi, A. et al. 1963. *Trans. Br. mycol. Soc.* **46**: 549 (concentric zonation in culture; **43**, 1252).
Kucharek, T. A. 1973. *Phytopathology* **63**: 1336 (causing maize stalk rot; **53**, 2167).
Pal, M. et al. 1964. *Indian Phytopathol.* **17**: 188 (from sorghum; **44**, 2130).
Patil, L. K. et al. 1968. *Pl. Dis. Reptr* **52**: 784 (on papaw; **48**, 544).
Whitehead, M. D. et al. 1959. *Phytopathology* **49**: 817 (on Gramineae; **39**: 321).
Young, G. Y. et al. 1947. *Ibid* **37**: 180 (on Gramineae; **26**: 337).

Drechslera sacchari (Butler) Subram. & Jain, *Curr. Sci.* **35**: 354, 1966.

Helminthosporium sacchari Butler, 1913.
(*Cercospora sacchari* Breda de Haan and *H. ocellum* Faris may be synonyms of *D. sacchari*, see Martin)

CONIDIOPHORES arising singly or in small fascicles, often from stromata, straight or flexuous, mid to dark brown or olivaceous brown, paler towards the apex, up to 200 μ long, 5–8 μ thick (up to 700 × 10 μ in culture). CONIDIA slightly curved or occasionally straight, cylindrical or narrowly ellipsoidal, mid-pale to mid-golden brown, smooth, 5–9 (mostly 8) pseudoseptate, 35–96 (65) × 9–17 (13.8) μ; hilum dark, enclosed, 2–3 μ wide. Conidia in culture tend to be more curved than they are on natural substrata; groups of dark cells may be formed (M. B. Ellis & P. Holliday *CMI Descr.* 305, 1971).

D. sacchari causes eyespot and seedling blight of sugarcane, and leaf diseases of *Cymbopogon citratus* (lemon grass) and *Pennisetum purpureum* (elephant grass). Small reddish spots surrounded by a straw-coloured halo, develop on the leaves; very conspicuous on young leaves; becoming 5–12 × 3–6 mm, longer axis parallel to the main veins, coalescing. These spots may form chlorotic streaks running towards the leaf tip, 60–90 cm long. Germinating seed may be killed 12–14 days from sowing and severely diseased young sugarcane plants show a top rot. On lemon grass the leaf spots have pale centres with a dark purple border, 4–10 × 1.5–2 mm. Older leaves of elephant grass may be attacked most, severe outbreaks causing death.

The disease is widespread in the tropics and subtropics (Map 349). In sugarcane penetration is direct or via the stomata and infection decreases with leaf age. Opt. temp. for growth, which is reduced in darkness, is *c.* 26°. In a comparison with *Cochliobolus stenospilus* (invalid, *D. stenospila* q.v.) no infection by *D. sacchari* occurred at >25° whilst that by *D. stenospila* took place at 30° (Liu). Dispersal is presumably through airborne conidia and seedborne infection also takes place (Loveless et al.). A host specific toxin is produced on susceptible sugarcane cvs but not on resistant ones and the toxin can be used as a more rapid test for resistance which is the means of control. Temp. apparently affects the host's sensitivity to the toxin and susceptible clones become resistant in the warm, summer months. In Mauritius eyespot is more severe in the cooler months (Wiehe). A host lesion 32 cm long produces sufficient toxin to induce one of 4·5 cm on a susceptible leaf (Byther et al.; Dean et al.; Pinkerton et al.; Steiner et al.; Strobel; Strobel et al.). Resistance also occurs in elephant grass; isolates from this host were only mildly injurious to sugarcane and vice versa.

Bannerjee, S. K. 1967. *Sci. Cult.* **33**: 237 (light & growth; **48**, 3647).
Bourne, B. A. 1934. *Bull. Fla Exp. Stn* 267, 76 pp. (general & comparison with other sugarcane leaf fungi; **14**: 57).
——. 1941. *Phytopathology* **31**: 186 (on lemon grass; **20**: 229).
Breda de Haan, J. von. 1892. *Meded. Proefsta. Suikerr. W. Java* **3**: 15 (*Cercospora sacchari* sp. nov.).
Butler, E. J. et al. 1913. *Mem. Dep. Agric. India bot. ser.* **6**: 204 (*Helminthosporium sacchari* sp. nov.).
Byther, R. S. et al. 1972. *Phytopathology* **62**: 466 (use of toxin to test resistance).
—— ——. 1975. *Pl. Physiol.* **56**: 415 (effect of temp. on resistance; **55**, 1438).
—— ——. 1976. *Sugcane Pathol. Newsl.* 15–16: 54 (summer induced resistance; **56**, 378).
Cook, M. T. 1927. *J. Dep. Agric. P. Rico* **10**: 207 (general; **7**: 199).
Dean, J. L. et al. 1975. *Phytopathology* **65**: 955 (use of toxin to test resistance; **55**, 1939).
Faris, J. A. 1928. *Ibid* **18**: 135, 753 (*H. ocellum*, *H. sacchari* & *H. stenospilum*; **7**: 402; **8**: 134).
Halma, F. F. et al. 1925. *Ibid* **15**: 463 (growth & temp.; **5**: 55).
Liu, L. J. 1969. *Proc. int. Soc. Sug. Cane Technol.* 13th Congr. 1968: 1212 (infection & temp. by *Drechslera sacchari* and *D. stenospila*; **50**: 915p).
Loveless, A. R. et al. 1956. *Ann. appl. Biol.* **44**: 419 (seedling blight & seed transmission; **36**; 350).
Martin, J. P. 1961. In *Sugarcane diseases of the world* Vol. 1: 167 (review; 41 refs.).
Nishihara, N. 1967. *Bull. natn. Inst. Anim. Ind.* **15**: 33 (on elephant grass; **48**, 2361).
Parris, G. K. et al. 1941. *Phytopathology* **31**: 855 (resistance in elephant grass × merker grass; **21**: 21).
——. 1942. *Ibid* **32**: 46 (on elephant grass; **21**: 258).
——. 1950. *Ibid* **40**: 90 (*Helminthosporium* on sugarcane, lemon & elephant grasses; **29**: 477).
Pinkerton, F. et al. 1976. *Proc. natn. Acad. Sci. U.S.A.* **73**: 4007 (serinol as an activator of toxin formation; **56**, 5175).
Steiner, G. W. et al. 1971. *Phytopathology* **61**: 691 (partial characteristics & use of toxin; **51**, 628).
—— ——. 1976. *Ibid* **66**: 423 (comparison of toxin from different countries; **56**, 834).
Strobel, G. A. et al. 1972. *Ibid* **62**: 339 (cell fine structure in infected cells & treated with toxin; **51**, 4321).
—— ——. 1972. *Physiol. Pl. Pathol.* **2**: 129 (runner lesion formation & toxin; **51**, 3553).
——. 1973. *Proc. natn. Acad. Sci. U.S.A.* **70**: 1963 (biochemical basis of resistance; **53**, 4869).
——. 1974. *Ibid* **71**: 4232 (toxin binding protein in sugarcane; **54**, 4130).
—— ——. 1975. *Biochem. Genet.* **13**: 557 (deficiency of toxin binding in mutants of sugarcane; **55**, 4841).
Voorhees, R. K. 1938. *Phytopathology* **28**: 438 (on elephant grass; **17**: 753).
Wiehe, P. O. 1940. *Rev. agric. Maurice* **19**: 57 (effect of season; **20**: 134).

Drechslera stenospila (Drechsler) Subram. & Jain, *Curr. Sci.* **35**: 354, 1966.

Helminthosporium stenospilum Drechsler, 1928. (A *Cochliobolus* state was described in Japanese by T. Matsumoto & W. Yamamoto in 1936 but *C. stenospilus*, the name attributed to it, was published without a Latin diagnosis and is therefore invalid.)

CONIDIOPHORES solitary or in small groups, straight or flexuous, occasionally geniculate, mid to mid-dark brown, paler towards the apex, smooth, septate up to 200 μ long, 5–9 μ thick, sometimes swollen up to 13 μ at the base. CONIDIA mostly curved, cylindrical, ellipsoidal or broadly fusiform, rather dark olivaceous or golden brown, smooth, characteristically closely pseudoseptate, 70–135 (84) μ long, 14–22 (17) μ thick in the broadest part, with 6–14 pseudosepta; scar usually not very conspicuous. (P. Holliday & M. B. Ellis, *CMI Descr*. 306, 1971.)

D. stenospila is a relatively minor pathogen causing brown stripe of sugarcane. Although serious damage has been done the disease has become less important with the use of resistant cvs and improvements in growing conditions. Comparisons have been made between this sp. and *D. sacchari* (q.v.) by Parris (1950) and Faris; the former proposed an additional sp. on sugarcane (*Helminthosporium ocellum*) but this was considered by later workers to be identical with *D. sacchari*. Brown stripe has been reported from many sugarcane growing areas (Map 483). The linear spots turn reddish and elongate parallel to the long axis of the leaf. They are usually 2–10 mm long, often reach 25 mm and may be up to 75 mm but the streaks and runners characteristic of *D. sacchari* are absent. The spots may coalesce and form large necrotic areas.

Penetration is stomatal. Opt. temp. for mycelial growth is 28–32°, that for conidial germination being uncertain. Infection occurs over a wide temp. range, broader than that for *D. sacchari*. Spread is presumably through airborne conidia; no natural infection on other hosts occurs. The disease is more severe in dry weather and when growing conditions for the host are sub-optimal. No races have been described and resistance is reported as polygenic; cultural variants occur. The perfect state, already mentioned, has been referred to in Hawaii and Taiwan.

Adsuar, J. et al. 1966. *J. Agric. Univ. P. Rico* **50**: 73 (host resistance; **46**, 1095).

Ahmed, H. U. et al. 1976. *Sugcane Pathol. Newsl.* 17: 48 (on 176 clones; **56**, 3697).

Carpenter, C. W. 1931. *Proc. 50th A. Meet. Hawaiian Sugar Plrs' Assoc.*: 437 (perfect state reported; **11**: 4).

Drechsler, C. 1928. *Phytopathology* **18**: 135 (*Helminthosporium stenospilum* sp. nov.; **7**: 402).

Liu, L. J. 1967. *J. Agric. Univ. P. Rico* **51**: 334 (variability in vitro; **47**, 1264).

Martin, J. P. 1961. In *Sugarcane diseases of the world* Vol. 1: 129 (review, 9 ref.).

Matsumoto, T. et al. 1936. *J. Pl. Prot. Tokyo* **23**: 9, 107 (perfect state in Taiwan; **15**: 397).

Wang, C. K. 1950. *Rep. Taiwan Sug. Exp. Stn*: (in vitro studies, temp., penetration & pathogenicity; **30**: 544).

ELSINOË Racib. emend. Jenkins & Bitancourt, *Mycologia* **33**: 338, 1941.
Plectodiscella Woronich., 1914
Isotexis H. Syd., 1931.
(Myrangiaceae)

STROMATA embedded in host, scant or effuse, of more or less well-defined tissue, or of hyaline or pale yellowish pseudoparenchyma, internally as a more or less loose prosenchyma; originating in or below epidermis, becoming erumpent with well-defined exterior, often covered with dark layer (epithecium); occasionally linear and branched, following leaf veins. ASCI few to numerous, irregularly embedded in stroma, globose to pyriform, bitunicate; outer wall thin, elastic; inner wall thickened especially at apical region and with a more or less well-developed foveola, expanding upon rupture of outer wall; 1 to usually 8 ascospores. ASCOSPORES hyaline, 3 septate (rarely 4–5 septate), sometimes with a longitudinal or diagonal septum in 1 or more cells, the 2 upper cells often broader and shorter than the lower ones; usually germinating by sprout conidia but may form germ tubes, 1 from each cell (after A. E. Jenkins & A. A. Bitancourt l.c., conidial states in *Sphaceloma* q.v.).

Many spp. of *Elsinoë* have been described from tropical crops but very few cause important diseases and many have not been fully investigated. *E. batatas* (Saw.) Viégas & Jenkins on *Ipomoea batatas* has ascospores 7–8 × 3–4 μ; sweet potato scab is fairly widespread in S.E. Asia and Oceania, and it also occurs in Brazil and Mexico (Map 447). In 1961 the fungus was reported to be causing considerable damage to this crop in Indonesian New Guinea or W. Irian (Johnston). *E. canavaliae* Racib. occurs on *Canavalia ensiformis* and *C. gladiata* (jack and sword bean scab). A recent description of this fungus was given by Sivanesan et al. (1971; ascospores 9–12 × 2.9–3.5 μ). Early records of lima bean scab were attributed to this sp. but the cause of scab on *Phaseolus lunatus* is *E. phaseoli* (q.v.). *E. canavaliae* appears to have been recorded only from Indonesia, Philippines, Singapore, Sri Lanka, Uganda, Zaire and Zambia. *E. dolichi* Jenkins, Bitanc. & Cheo causes scab of *Lablab niger* (hyacinth bean scab; Jenkins et al. 1941, *Sphaceloma* q.v.); its ascospores are 7–13 × 3–5.2 μ. *E. heveae* Bitanc. & Jenkins occurs on *Hevea brasiliensis* (rubber scab), *E. mangiferae* Bitanc. & Jenkins on *Mangifera indica* (mango scab), and *E. piperis* Hansf. on *Piper nigrum* (black pepper scab). White rash of sugarcane (*Saccharum*) is caused by *E. sacchari* Lo, ascospores 8.6–10 × 3–3.3 μ; the leaf lesions eventually become whitish grey or chalk white and have the appearance of a fine irregular lacework extending over the entire affected area. Mung (green gram) scab on *Phaseolus aureus* (*E. iwatae* Kajiwara & Mukelar, ascospores 11–14 × 5–5.7 μ) was described from Indonesia (Mukelar et al.). Two spp. have been described from *Camellia sinensis* (tea scabs): *E. leucospila* Bitanc. & Jenkins (white scab, ascospores 13–16 × 5–7 μ) and *E. theae* Bitanc. & Jenkins (mottle scab, ascospores 10–14 × 3–7 μ). The former requires a leaf wetness period of 5–10 hours for infection which occurs at an opt. 16–24°; young leaves are the most susceptible (Fukuda et al.). *E. wisconsinensis* H. C. Green (see *Index of Fungi* 2: 428) was described from *Desmodium illoense* (see Mason et al.). Sivanesan et al. (1974) described *E. ampelina* Shear causing bird's eye rot (also spot anthracnose or black spot) of *Vitis* spp. (grapevine).

Bitancourt, A. A. et al. 1939. *Arq. Inst. biol. S. Paulo* **10**: 193 (*Elsinoë theae* sp. nov.; **19**: 369).

——— ———. 1946. *Ibid* **17**: 205 (*E. mangiferae* sp. nov.; **26**: 552).

——— ———. 1956. *Ibid* **23**: 41 (*E. heveae*; **37**: 110).

Cheo, C. C. et al. 1945. *Phytopathology* **35**: 339 (*E. dolichi* & *Sphaceloma ricini* inter alia; **24**: 441).

Durnerin, A. 1972. *Agron. Trop.* **27**: 740 (*E. piperis*; **52**, 470).

Fukuda, T. et al. 1977. *Study of tea* **52** (July): 1 (incidence of *E. leucospila* on tea; *Hort. Abstr.* **48**, 5043).

Jenkins, A. E. 1931. *J. agric. Res.* **42**: 1 (*E. canavaliae*; **10**: 577).

—— et al. 1943. *J. Wash. Acad. Sci.* **33**: 244 (*E. batatas*; **23**: 1).

——— ———. 1946. *Arq. Inst. biol. S. Paulo* **17**: 67 (*E. leucospila* sp. nov. & *E. theae*; **26**: 317).

——— ———. 1960. *Biológico* **26**: 111 (*S. sacchari*; **40**: 245).

Johnston, A. 1961. *Meded. Dienst. econ. Zak. Ned N. Guinea Landb.* Ser. 4, 55 pp. (preliminary plant disease survey in W. Irian; **41**: 360).

Kajiwara, T. et al. 1976. *Contr. Cent. Res. Inst. Agric. Bogor* 23, 12 pp. (*E. iwatae* sp. nov.).

Lo, T. C. 1957. *J. Agric. For. Taiwan* **6**: 70 (*S. sacchari*).

——. 1964. *Proc. biol. Soc. Wash.* **77**: 1 (*E. sacchari* sp. nov.; **43**, 3006).

Mason, D. L. et al. 1969. *Mycologia* **61**: 1124 (histology; **49**, 2900).

——— ———. 1978. *Phytopathology* **68**: 65 (fine structure of infection; **57**, 4998).

Mukelar, A. et al. 1976. *Contr. Cent. Res. Inst. Agric. Bogor* 24, 7 pp. (*E. iwatae*, fungicides; **57**, 4222).

Sivanesan, A. et al. 1971. *C.M.I. Descr. pathog. Fungi Bact.* 313 (*E. canavaliae*).

——— ———. 1974. *Ibid* 439 (*E. ampelina*).

Elsinoë australis Bitancourt & Jenkins, *Mycologia* **28**: 491, 1936.

stat. conid. *Sphaceloma australis* Bitanc. & Jenkins, 1936

S. fawcettii Jenkins var. *viscosa* Jenkins, 1933.

LESIONS on fruits, leaves and twigs, solitary or confluent, when fructicolous circular to irregular, raised, corky in texture and appearance, up to 4 mm diam., when foliicolous initially funnel-shaped pockets, later like scabs, smooth, glossy, up to 2 mm diam., amphigenous, especially along the midrib. Twig lesions resemble those on the leaf. ASCOMATA globose, separate or aggregated, pseudoparenchymatous, epidermal to subepidermal, up to 150 μ diam. ASCI in the upper part of the ascoma, elliptical to subglobose, 8 spored, 15–30 × 12–20 μ. ASCOSPORES hyaline, straight or curved, 1–3 septate, slightly constricted at the septa, but the upper middle cell may become longitudinally septate, 12–20 × 4–9 μ. ACERVULI similar in appearance to ascomata. CONIDIOGENOUS CELLS formed directly from the upper cells of the pseudoparenchyma or from 0–3 septate conidiophores, hyaline to pale brown, monophialidic to polyphialidic, terminal, integrated, determinate, 6–18 × 4–5 μ. CONIDIA hyaline, aseptate, 4–6 × 2–4 μ (A. Sivanesan & C. Critchett, *CMI Descr.* 440, 1974).

Sweet orange scab of *Citrus* spp., largely a South American disease, is found in Argentina, Bolivia, Brazil, Italy (Sicily), Paraguay and Uruguay (Map 55). This citrus scab should not be confused with the much more widespread, common citrus scab caused by *E. fawcettii* (q.v.) and the fairly widespread Australian citrus scab (or Tryon's scab) caused by *Sphaceloma fawcettii* var. *scabiosa* (q.v.). These 3 scabs have been reviewed by Brun, see *E. fawcettii* and the *CMI Descr.* 437, 438 and 440. The fruit lesions of *E. australis* are generally larger, less raised, smoother, more circular and not becoming warty or protuberant as is frequently found with *E. fawcettii*, or typically discoid or crateriform as in *S. fawcettii* var. *scabiosa*. The opt. temp. for growth in vitro is *c*. 26° for *E. australis* and *c*. 21° for *E. fawcettii*. *E. australis* is generally a disease of *C. sinensis*; other *Citrus* spp. and vars are attacked, sour orange (*C. aurantium*) is fairly resistant. A detailed description of sweet orange scab was given by Bitancourt et al. (1937a) who compared it with common citrus scab. Whilst the latter disease is frequent on leaves and green shoots, sweet orange scab is mainly a disfiguring disease on the fruit and may be absent from other tissues of the plant (leaf infection can occur). Young fruits may become distorted by the infection. Control is with copper fungicides applied before flowering, at late petal fall and a few weeks after fruit set. Ramallo used other fungicides including benomyl, captan and maneb.

Bitancourt, A. A. et al. 1937a. *J. agric. Res.* **54**: 1 (general; **16**: 451).

——. 1937b. *Rodriguesia* **2** (special no.): 315 (variability in culture; **17**: 27).

——. 1939. *Arq. Inst. biol. S. Paulo* **10**: 129 (culture, perfect & imperfect states; **19**: 339).

Jenkins, A. E. 1933. *Phytopathology* **23**: 538 (morphology & culture; **12**: 689).

Ramallo, N. E. V. de 1974. *Circ. Estac. exp. Agric. Tucumán* 193, 7 pp. (fungicides, with *Elsinoë fawcettii*; **54**, 4503).

Elsinoë fawcettii Bitancourt & Jenkins, *Phytopathology* **26**: 394, 1936.
stat. conid. *Sphaceloma fawcettii* Jenkins, 1925.

LESIONS on leaves, twigs and fruits. Leaf lesions amphigenous, initially semi-translucent dots, becoming papillate, cream to pale yellow or variously coloured, up to 3 mm diam. Scabs irregularly discoid to subglobose, confluent, mostly along the main veins, cinnamon to honey coloured, warty, deeply cracked and split when old. Fruits and young shoots may develop similar symptoms. ASCOMATA scattered, pulvinate, circular to elliptical, dark brown, up to 120 μ wide, composed of pseudoparenchymatous tissue containing several locules with asci. ASCI subglobose to elliptical, thick walled at the top, 8 spored, 12–16 μ diam. ASCOSPORES hyaline, oblong elliptical, 1–3 septate, usually constricted at the central septum, $10\text{–}12 \times 5\text{–}6\ \mu$. ACERVULI epidermal to subepidermal, separate or confluent, composed of hyaline to pale brown pseudoparenchyma. CONIDIOGENOUS CELLS formed directly from the upper cells of the pseudoparenchyma or from 0–2 septate conidiophores, monophialidic to polyphialidic, terminal, integrated, hyaline to pale brown, $12\text{–}22 \times 2\text{–}4\ \mu$. CONIDIA hyaline, aseptate, ellipsoidal, $5\text{–}10 \times 2\text{–}5\ \mu$ (A. Sivanesan & C. Critchett, *CMI Descr*. 438, 1974).

Common citrus scab (or sour orange scab) is widespread with this host group where moisture and temp. favour the disease (Map 125). Where these conditions occur, and where susceptible spp. and cvs are grown, the disease can be serious in the nursery and the field. Full, general accounts were given by Fawcett (*Citrus diseases and their control*), Jenkins (1931) and Winston. Brun, in a review, discussed common scab and the 2 other citrus scabs (*E. australis* and *Sphaceloma fawcettii* var. *scabiosa* q.v.). Fisher reviewed the work on common scab in USA (Florida) since 1953. The scabs caused by *E. fawcettii* differ from those caused by these other 2 fungi in that they are warty or protuberant, irregular, and deeply cracked or fissured. On leaves the rough, warty, scabs (like cork) are commoner on the lower surface, and can cover large areas. The colour of the lesions varies at first and can be rather bright (e.g. salmon buff or cream), later darkening to buff and dark olive. The leaves become distorted, puckered, stunted, show torn leaf margins, hollow conical outgrowths and may fall prematurely. Small branches are affected and can be killed. Fruits that are infected early in growth become misshapen and may drop. The colour around fruit scabs differs with *Citrus* sp. and cv.; the scabbing becomes extensive and more conspicuous with fruit growth; extensive, corky patches may develop.

Presumably both spore types cause infection. Plant tissues are most susceptible at emergence and become increasingly resistant with age. Leaves that are *c*. 1.5 cm in width and fruit of 2 cm diam. are immune. Transmission is probably mainly through water; and carry over occurs on leaves and small branches, fruit is less important. Severe outbreaks of common citrus scab only occur under fairly well-defined conditions, i.e. adequate moisture, favourable temps, a susceptible sp. or cv. and the presence of young plant tissue. The disease is not generally serious where the annual rainfall is < 130 cm and where there is a well-defined dry season. Temp. requirements for disease development are 15–23°, with an opt. of 20–21°. There is no infection at 14° or less and at 24.5° or more. Scab does not generally develop in regions which have a mean monthly temp. of 24° or more (Fawcett; Peltier et al. 1924 & 1926; Yamada). Conidia need a min. wet period of 2.5–3.5 hours for infection (Whiteside 1975). Thus the disease is largely one of the wet subtropics and cooler tropics. It is absent or unimportant in regions with the Mediterranean climatic type. Disease incidence is very subject to year by year (and local) variation in weather. Also any factor which stimulates a susceptible cv. to flush or set fruit during periods favourable for infection may also affect the extent to which disease develops. Most conidia are formed at RH 100% and the min. incubation is 5 days.

E. fawcettii is confined to a small number of genera and spp. in the Rutaceae. The degree and range of susceptibility in this group will vary with sp., cv. and climate. The most susceptible group includes: sour orange, rough and common lemon, and some tangelos. Moderately susceptible are: king orange, satsuma mandarin, tangerine, grapefruit, trifoliate orange (*Poncirus trifoliata*) and some tangelos. Those rarely attacked include sweet orange, oval kumquat (*Fortunella margarita*), Tahiti lime and pumelo (shaddock). Immune forms include citron, Mexican or Key lime, some sweet orange cvs and some kumquats. There may be degrees of susceptibility within a *Citrus* sp., for example, in grapefruit cvs (Peltier et al. 1923; Winston et al.). Whiteside (1978) described 2 biotypes of the fungus.

Control is with fungicides, both in the nursery and the field. In the latter case at least 2 applications (just before flushing and at late petal fall) are needed; a third may be applied *c.* 4 weeks after flowering. Copper has been widely used in the past (e.g. Ruehle et al.); ferbam (Fisher) and captafol (Moherek) are also effective. Good control with one application of benomyl has been reported; this treatment gave a much higher percentage of marketable fruit compared with copper (Hearn et al.). Spraying every 2 weeks in the nursery may be necessary. Whiteside (1974) found that captafol and benomyl were better than copper and ferbam.

Brun, J. 1971. *Fruits* **26**: 759 (review *Elsinoë fawcettii*; *E. australis* & *Sphaceloma fawcettii* var. *scabiosa*, 41 ref.; **51**, 2478).

Cunningham, H. S. 1928. *Phytopathology* **18**: 539 (histology; **7**: 714).

Fawcett, H. S. 1921. *J. agric. Res.* **21**: 243 (effects of temp.).

Fisher, F. E. 1970. *Citrus Veg. Mag.* **33**: 8 (work in USA, Florida, since 1953, control with ferbam; **50**, 1233).

Hearn, C. J. et al. 1971. *Pl. Dis. Reptr* **55**: 241 (control with Cu & benomyl; **50**, 3753).

Jenkins, A. E. 1931. *J. agric. Res.* **42**: 545 (general; **10**: 786).

——. 1936. *Archos Inst. biol. Def. agric. anim. S. Paulo* **7**: 23 (culture; **16**: 655).

Moherek, E. A. 1970. *Proc. Fla St. hort. Soc.* **83**: 59 (control with captalfol; **51**, 3366).

Peltier, G. L. et al. 1923. *J. agric. Res.* **24**: 955 (host susceptibility; **3**: 210).

—— ——. 1924. *Ibid* **28**: 241 (effects of environment; **4**: 476).

—— ——. 1926. *Ibid* **32**: 147 (effects of climate & weather on world distribution & prevalence with *Xanthomonas citri*; **5**: 420).

Ruehle, G. D. et al. 1939. *Bull. Fla agric. Exp. Stn* 337, 47 pp. (control with Cu; **20**: 460).

Whiteside, J. O. 1974. *Proc. Fla St. hort. Soc.* **87**: 9 (fungicides; **55**, 5746).

——. 1975. *Phytopathology* **65**: 1170 (epidemiology; **55**, 2243).

——. 1978. *Ibid* **68**: 1128 (pathogenicity of 2 biotypes).

Winston, J. R. 1923. *Bull. U.S. Dep. Agric.* 1118, 35 pp. (general; **2**: 364).

—— et al. 1925. *J. agric. Res.* **30**: 1087 (host susceptibility; **5**: 29).

Yamada, S. 1961. *Spec. Bull. hort. Stn, Tokai-Kinki natn. agric. Exp. Stn* 2, 56 pp. (epidemiology; **42**: 383).

Elsinoë phaseoli Jenkins, *J. agric. Res.* **47**: 788, 1933.

ASCOMATA subepidermal, often in erumpent punctiform groups, up to 140 μ wide, composed of pseudoparenchymatous tissue containing several locules, each with a single ascus. ASCI ovoid to subglobose, thick walled, 8 spored, 30–40 μ. ASCOSPORES oblong elliptical, basal part sometimes more obtuse than the apex, hyaline or pale yellow, transversely 3 septate, but one of the central cells may become longitudinally septate, somewhat constricted at the medium septum, 13–15 × 5–6 μ. Stat. conid. a *Sphaceloma*. CONIDIOPHORES aggregated in the centre of young lesions or developed on marginal zones of older lesions, hyaline to dark coloured, unbranched, blunt or pointed, continuous or 1 septate, 20 × 4 μ, bearing conidia at their tips, closely compacted in more or less effuse minute acervuli. CONIDIA variable, some minute spherical up to 4 μ diam., others ovoid or oblong–elliptical, guttulate, hyaline to greyish, 10×4 μ (A. Sivanesan & P. Holliday, *CMI Descr.* 314, 1971).

This relatively little known Central American disease, scab of lima bean (*Phaseolus lunatus*), was chiefly studied over 30 years ago when its severity attracted attention in Cuba and Puerto Rico. The disease occurs mainly in Central America and some of the West Indian islands. It is also present in Surinam, USA (S. Carolina) and E. Africa (Kenya, Malawi, Zimbabwe and Zambia; Map 194). It was first reported from Guatemala in 1890 and as recently as 1966 was severely damaging cowpea (*Vigna unguiculata*) in Surinam. It is possible that forms of the pathogen exist. Isolates from lima bean did not infect *P. vulgaris, Pisum sativus, Stizolobium deeringianum, Lablab niger, Canavalia ensiformis* and *C. gladiata*; the last 2 are hosts for *E. canavaliae* (*Elsinoë*, q.v.). But later the pathogen was reported attacking both *V. unguiculata* in Surinam and *Phaseolus vulgaris* in Zimbabwe; in the latter country it was referred to as *E. phaseoli* f. sp. *vulgare*.

The symptoms are most conspicuous on the pods where the red to brown, raised or sunken lesions are up to 1 cm diam., somewhat elongate, usually irregularly distributed and sometimes occurring on one valve; lesions do not penetrate the seed. Leaf lesions are up to 4 mm diam. or larger near the veins, more pronounced on the upper surface and vinaceous buff or liver brown. Elongate lesions which may be bordered by a purple brown band sometimes encircle the stem. Little of the biology of *E. phaseoli* is known. It has been detected both within the seed coat and on its surface. It survives on host debris and air dispersal may occur. More investigation is required particularly on host range, the nature of any physiologic variation, and the relationships with other *Elsinoë* spp. on Leguminoseae. Rotation, sanitation and seed treatment are control measures.

Bates, G. R. 1959. *Rep. Minist. Agric. Rhod. Nyasald 1957–58* pp. 26–33 (seed transmission; **39**: 211).

Bruner, S. C. et al. 1933. *J. agric. Res.* **47**: 783 (inoculation of other hosts, comparison with *Elsinoë canavaliae*; **13**: 344).

Jenkins, A. E. 1931. *Ibid* **42**: 13 (general; **10**: 578).

——. 1933. *Phytopathology* **23**: 662 (distribution, growth in vitro; **13**: 72).

McCubbin, W. A. 1933. *J. econ. Ent.* **26**: 625 (spread in Cuba; **12**: 742).

Sitterly, W. R. et al. 1958. *Pl. Dis. Reptr* **42**: 1309 (first record in USA, seed spread; **38**: 438).

Staples, R. R. 1957. *Rep. Minist. Agric. Rhod. Nyasald 1956–57* (designated *E. phaseoli* f. sp. *vulgare* on common bean; **38**: 295).

Van Hoof, H. S. 1963. *Surinaam. Landb.* **11**: 27 (field attack on cowpea; **42**: 717).

ENDOTHIA Fr., *Summ. veg. Scand.* Sect. Post.: 385, 1849.
(Diaporthaceae)

STROMA massive, yellowish or reddish. ASCOMATA (perithecia) immersed in a stroma or a pseudostroma, on twigs or woody stems, flask shaped with long slender necks. ASCI unitunicate, fusiform or clavate or nearly cylindrical, loosening from ascogenous cells at an early stage and lying free in the ascoma. ASCOSPORES 2 celled, with the septum in or near the middle, mostly fusiform, pyriform or ellipsoid, colourless or nearly so (after key by E. Müller & J. A. von Arx in *The fungi* (edited by G. C. Ainsworth et al.) Vol. IVA: 111, 1973; and R. W. G. Dennis, *British Ascomycetes*: 322, 1968).

E. havanensis Bruner, described from *Eucalyptus* spp., causes canker of *E. grandis* and *E. saligna* in Surinam, and in Brazil on these 2 spp. and *E. citriodora*. On the cankers numerous PYCNIDIA may be present, they have necks 1–2 mm long; the CONIDIA are aseptate, hyaline, 3.6×1.4 μ. The black PERITHECIA are identical in shape to the pycnidia; ASCI are 8 spored, subclavate or oblong fusoid, 45×6.5 μ; ASCOSPORES are 1 septate, hyaline, oblong-elliptic, 7.2×3.6 μ. On PDA cultures orange pycnidia are formed in concentric rings (Boerboom et al.). The first symptoms appear 14–18 months after planting. The base of the trunk swells slightly, longitudinal cracks form and reddish brown to black pycnidia develop on the bark. The cracks deepen, the base and young wood become dark red and these parts die off; further cankers form higher up the stem and may coalesce. A ruby coloured, water soluble gum is formed; it spreads on the trunk, turning it characteristically reddish. A callus forms around the site of infection and the outer layer of bark bulges; this layer is finally shed, leaving a canker. In TS there is a dark brown discolouration. Young trees (up to 3 years old) can be killed. Hodges et al. (1974b) considered that this eucalyptus canker fungus should be referred to *Diaporthe cubensis* Bruner (S. C. Bruner,

Endothia eugeniae

Informe Estac. Agron. Cuba 20: 745, 1918; see L. E. Wehmeyer, *Diaporthe* q.v.). In the more recent papers on the disease the pathogen is therefore called *D. cubensis.*

Endothia gyrosa (Schw. ex Fr.) Fr., which occurs on several tree spp. in N. America, was described causing a serious disease of *Liquidambar formosana* (from China and Taiwan, sweet gum) in Mississipi, USA. It occurs on *Eucalyptus* and has been reported from the Philippines and Sri Lanka (Snow et al.). *E. parasitica* (Murrill) P. J. & H. W. Anderson causes the highly destructive canker of some chestnut spp. (*Castanea*). It destroyed the stands of *C. dentata* in E. USA and has a restricted distribution in the northern hemisphere (Map 66).

Beattie, R. K. et al. 1954. *J. For.* **52**: 323 (50 years of *Endothia parasitica* in America; **34**: 113).
Boerboom, J. H. A. et al. 1970. *Turrialba* **20**: 94 (*E. havanensis* in Surinam; **50**, 1394).
Bruner, S. C. 1916. *Mycologia* **8**: 239 (*E. havanensis* sp. nov.).
Cardoso May, L. 1973. *Publicação Inst. Florestal* 2, 11 pp. (*E. havanensis* in Brazil; **53**, 1572).
Ferreira, F. A. et al. 1976. *Brasil Florestal* **7** (25): 13 (*Diaporthe cubensis*, effect on sprouting; **56**, 1756).
———. 1977. *Fitopatologia Brasileira* **2**: 225 (*D. cubensis*, resistance; **57**. 3643).
Golfari, L. 1975. *Brasil Florestal* **6** (23): 3 (effect of provenance on *D. cubensis* infection; **55**, 4291).
Hodges, C. S. et al. 1974a. *Ibid* **5** (18): 25 (*D. cubensis*, effect on sprouting; **54**, 5566).
———. 1974b. *Ibid* **5** (19): 14 (identification of fungus causing canker of *Eucalyptus* spp. in Brazil; **54**, 4647).
———. 1976. *Fitopatologia Brasileira* **1**: 129 (*Eucalyptus* canker; **56**, 4217).
Snow, G. A. et al. 1974. *Phytopathology* **64**: 602 (*Endothia gyrosa* in USA; **54**, 242).
———. 1975. *C.M.I. Descr. pathog. Fungi Bact.* 449 (*E. gyrosa*).

Endothia eugeniae (Nutman & Roberts) Reid & Booth, *Can. J. Bot.* **47**: 1059, 1969.
Cryptosporella eugeniae Nutm. & Roberts, 1952.

The fungus is visible under a hand lens on the host surface as clumps of protruding perithecial or pycnidial necks. In TS these are seen in an erumpent, conic stroma in the cortical tissue; it is broadest below and tapers above as it penetrates the layers of the periderm. PYCNIDIA in cavities in the stroma and superficially follow the same development as the later formed perithecia, but several pycnidial locules may fuse to form an irregular, diverticulate locule in which all the convoluted inner wall is lined with phialides. PHIALIDES simple or formed terminally on sparsely-branched conidiophores and interspersed with simple, elongate, sterile strands. PHIALOSPORES hyaline, oval, aseptate, $3.5–5 \times 1.2–2\ \mu$. PERITHECIA stromatic, clustered, frequently in rows in bark fissures and forming later in the same stroma as the pycnidia. The venters form at the base of the stroma, more or less below the pycnidia, and may have an irregular displacement of the neck or they may grow through the pycnidial locule; in either case they emerge at the apex of the stroma in a cluster or linearly, and grow to form protruding necks; $400–750 \times 100–150\ \mu$ diam. ASCI unitunicate, clavate, sessile, thin walled, $20–33 \times 4.6–5\ \mu$, 8 oblique, monostichous to distichous spores. ASCOSPORES oblong to ellipsoid, 2 celled, not appreciably constricted at the central septum, $6–8.5 \times 2–3\ \mu$; the widest point in the upper cell. PARAPHYSES absent (from C. Booth & I. A. S. Gibson, *CMI Descr.* 363, 1973).

Acute dieback of clove (*Eugenia caryophyllus*) formed part of the disease syndrome, whose cause was unknown for very many years, in Zanzibar and Pemba (Tanzania). The other fungus in this pathogenic complex is *Valsa eugeniae* (q.v. for further ref.). *Endothia eugeniae* (also reported from Malaysia, W.) causes a more serious dieback of the branches than does *V. eugeniae*, but does not generally attack the roots as does the latter fungus. The characteristic symptom is the dying back of a whole branch; the leaves dry up on the tree and become a conspicuous russet red. If (as often happens on young trees) the trunk is invaded the whole plant dies in a manner similar to that shown by sudden death (*V. eugeniae*); hence the obvious symptoms caused by the 2 fungi on the same host can be (and have been) confused. Both the pycnidia and perithecia are inconspicuous but both exude long, yellowish spore tendrils, sometimes in sufficient quantity to stain the bark a brownish orange. In a branch showing dieback, and split open at the junction between dead and living tissues, the diseased part is a dark red brown. Vessels are occluded as infection spreads. Spread down the branch eventually reaches the trunk; branches and trunks of young trees are girdled.

The conidia are probably splash dispersed and host penetration is invariably through wounds. It is not known whether the ascospores are forcibly discharged. Experimental infections that reproduce typical symptoms can rarely be caused in plants aged 18–36 months and not at all in young seedlings. Such infections of older trees become more certain with increasing age, and with saplings 7–10 years old failure is rare. The prevalence of acute dieback in Zanzibar was almost entirely attributable to destructive harvesting methods and the presence of much infected plant material in the groves. Control measures (which have apparently been quite successful) are removal of infected material (cutting back to healthy wood in branches and trunks) and subsequent treatment with fungicidal paints. Damage to trees during harvesting should be avoided.

Nutman, F. J. et al. 1952. *Ann. appl. Biol.* **39**: 599 (general in Zanzibar; **32**: 508).
——. 1971. *PANS* **17**: 147 (the clove industry, sudden death & dieback; **51**, 1751).

ENTYLOMA de Bary, *Bot. Ztg.* 32: 101, 1874.

Rhamphospora, D. D. Cunn., 1888.
(Tilletiaceae)

SORI as round, often bright coloured, slightly distorted swellings in the leaves, sometimes in stems also, permanently embedded in the tissues of the host. SPORES single, free, terminal or intercalary on fertile hyphal branches which do not always disappear through gelatinisation; hyaline to yellowish or reddish brown, of medium size. Germination by the formation of a short aseptate promycelium bearing a whorl of SPORIDIA usually conjugating in pairs and producing secondary sporidia or infection threads. CONIDIA often present, protruding through the stomata, hyaline, formed by the germination of the spores or on the mycelium (from B. B. Mundkur & M. J. Thirumalachar, *Ustilaginales of India*: 67, 1952).

Fischer (*Manual of the North American smut fungi*) said in 1953 that the genus needed critical revision. For the Indian spp. see Mundkur et al. (l.c.) and for the British ones Ainsworth & Sampson (*The British smut fungi, Ustilaginales*). An *Entyloma* sp. causing a leaf disease on *Phaseolus vulgaris* was reported from Central America and the Dominican Republic. From Brazil *E. vignae* Bat., Bezerra, Ponte & Vasconcelos (see *Index of Fungi* **3**: 407) was described causing a leaf disease of *Vigna unguiculata* (cowpea). Round or irregular spots, *c.* 2–12 mm diam., appear as a surface discolouration with diffuse edges, separate or confluent; at first brown becoming green grey as the leaf fades. The distribution of *Entyloma* on common bean and cowpea in tropical America was described by Vakili.

Ponte, J. J. Da. 1966. *Bolm Soc. cearense Agron.* **7**: 35 (*Entyloma vignae*; **47**, 2337).
—— et al. 1976. *Summa Phytopathol.* **2**: 50 (testing for resistance to *E. vignae*; **55**, 4955).
Schieber, E. et al. 1971. *Pl. Dis. Reptr* **55**: 207 (on common bean; **50**, 3296).
Vakili, N. G. 1978. *Pl. Prot. Bull. F.A.O.* **26**: 19 (distribution on common bean & cowpea & centre of origin).

Entyloma oryzae H. & P. Syd., *Ann. Mycol. Berl.* 12: 197, 1914.

SORI amphigenous, scattered, subcircular to more or less linear, beneath the epidermis forming a brown to blackish carbonaceous mass. SPORES olivaceous brown to dark brown, globose to subglobose, or polyhedral, firmly adhering in groups, 6–9 × 7.5–11.5 μ; wall smooth, 1–1.5 μ thick. On *Oryza sativa*. Macroscopically distinguished from the other rice smut, *Tilletia barclayana* (q.v.), in that infection is mainly restricted to the leaves (J. L. Mulder & P. Holliday, *CMI Descr.* 296, 1971).

Wherever it has occurred leaf smut of rice appears to have been considered of minor importance. The disease is fairly widespread in Asia; it also occurs in Australia and Papua New Guinea, in parts of America and in Africa (Egypt and Ghana); and in S. France (Map 451). Linear, rectangular or angular elliptical, leaden black leaf spots, not usually confluent, occur on both surfaces, 0·5–2 mm long and 0.5–1.5 mm broad. The epidermis covering the spots ruptures when wet. Where infection is heavy the leaves turn chlorotic and become split. In India (Uttar Pradesh) an increase in incidence was noted recently and 28 cvs were assessed for disease incidence. The early rice cvs had more disease than the late cvs; in USA little resistance was found in tests on 14 cvs and 9 breeding lines. Nitrogen applications, especially late ones, increased the severity of leaf smut which was more severe on heavy than on light soils. Details of the life history of *E. oryzae* are virtually unknown.

Singh, R. A. et al. 1966. *Indian Phytopathol.* **19**: 378 (reaction of cvs; **47**, 519).
Templeton, G. E. 1962. *Rice J.* **65**: 28 (effect of N fertiliser in USA, Arkansas; **42**: 462).
—— et al. 1963. *Arkansas Fm Res.* **12**: 4 (reaction of cvs; **43**, 463).

EREMOTHECIUM Borzi, *Nuovo G. Bot. ital.* **20**: 458, 1888.
Crebrothecium Routien, 1949.
(Spermophthoraceae)

VEGETATIVE CELLS multinucleate, sausage shaped and constricted near septa. CONIDIA or sprout cells present or absent. ASCI terminal or intercalary, cylindrical or flask shaped, many spored. ASCOSPORES acicular or semilunate (i.e. lunate with one end blunt), uninucleate, with or without a cytoplasmic appendage, usually germinating in the middle; liberated by deliquescence of the equatorial region of the ascus (from L. R. Batra, *Tech. Bull. U.S. Dep. Agric.* 1469: 10, 1973).

The related genus *Nematospora* (q.v.) has needle-shaped ascospores. Batra (l.c.) gave accounts of *E. cymbalariae* Borzi and *E. ashbyi* Guill. (synonym: *Crebrothecium ashbyi* (Guill.) Routien); both are rare. They can cause stigmatomycosis of cotton and citrus but do not cause any extensive damage to either crop. *E. cymbalariae* has terminal, fusiform ASCI, with 30–60 or more ascospores/ascus; ASCOSPORES are acicular, 19–25 μ long. *E. ashbyi* has intercalary, cylindrical ASCI, with (4–) 16–32 ascospores/ascus; ASCOSPORES semilunate, 16–25 μ long.

Guilliermond, A. 1936. *Rev. Mycol.* **1**: 115 (*Eremothecium ashbyi*; **15**: 719).
Marasas, W. F. O. 1971. *Bothalia* **10**: 407 (*E. ashbyi*, morphology & cotton staining; **51**, 390).
Mukerji, K. G. 1968. *C.M.I. Descr. pathog. Fungi Bact.* 181 (*E. ashbyi*).

ERYSIPHE Hedwig ex Mérat, *Nouvelle Flore de Paris* 1: 131, 1821.
Alphitomorpha Wallr.
(Erysiphaceae)

ASCOCARPS (cleistothecia) like thse of *Sphaerotheca*, i.e. with simple, hyphoid appendages but with several asci. CONIDIOPHORES on superficial hyphae with abundant CONIDIA borne in chains; base of the conidiophore straight or swollen. MYCELIUM superficial (partly from C. E. Yarwood, *The fungi* (edited by G. C. Ainsworth et al.) Vol. IVA: 71, 1973).

The imperfect state is *Oidium* Sacc. (q.v.). The perfect state differs from other genera in the powdery mildews (Erysiphaceae) in having an ascocarp with simple appendages and several asci. *E. graminis* DC. ex Mérat (on wheat, barley and other Gramineae) has been studied more intensively than any other powdery mildew (see Kapoor for morphology; it has relatively large ascocarps, > 150 μ diam. compared with other members of the genus). The powdery mildews form a very distinct group of obligate plant parasites. Their unique characteristics and economic importance has led frequently to general studies on the family, on several spp. and/or several genera. A brief account of the characteristics of the Erysiphaceae and the main, more recent, general papers are given here; see Spencer for a text on the powdery mildews. The conidium germinates on the host leaf surface in *c.* 2 hours and the short germ tube forms an appressorium from which the infection peg penetrates the plant cuticle and a haustorium (G. Smith) is formed in the epidermal cell. A fungal colony develops on the leaf surface through the growth of more germ tubes from which appressoria and haustoria form. The characteristic, superficial, whitish powdery growth is thus formed on the leaf; the ectoparasitic habit. *Leveillula taurica* (q.v.) is exceptional in having an endoparasitic habit and whose conidiophores (*Oidiopsis*) arise through the stomata. In other powdery mildews the conidiophores arise directly from the external mycelium. The conidia are formed in a few days from infection. Since the development of the perfect state is often erratic or rare an identification based on the conidial state may be useful (see Yarwood l.c. for keys and general taxonomy; Blumer; Boesewinkel; Junell). Particularly in the tropics ascocarps are absent or may be unknown (as in *Oidium heveae*; *Oidium* q.v.). Using spp. largely from temperate plant hosts Zaracovitis divided the conidia into 3 groups on the basis of their germination rate in vitro at 21° in a saturated atmosphere in the dark. These groups corresponded closely with the varying characteristics in the morphology of conidial maturation, germ tubes and appressoria. In a review Mount et al. discussed the primary development of powdery mildew. Wolfe et al. reviewed race changes.

Schnathorst reviewed the environmental relationships; and see Mount et al. The dry, hyaline, aseptate conidia are air dispersed, probably passively, and have a diurnal periodicity. Daytime peaks in numbers of conidia in the field have been shown for several economically important spp. (see also Cole; Manners; Pady). The confused literature on the external environment and powdery mildews (particularly on the effects of temp. and moisture on conidial germination, appressorial formation and infection) was very usefully discussed by Schnathorst. Some of his conclusions are as follows. Considerable variation in temp. requirements occurs; this is as wide between different pathogenic forms of the same sp. as it is between spp. As a group these fungi have an opt. temp. for germination and growth of *c.* 21°, lower than the av. opt. for plant pathogens. Germination is inhibited by free moisture, which also causes mycelial growth to be abnormal. Some mildews germinate best at RH 75–100%, others germinate best at saturation but can germinate at very low RH, and the third group germinate over a very wide range of RH. Although these fungi are generally favoured by high humidities they can tolerate dry climatic conditions, and the incidence of the diseases they cause tends to decrease as rain increases. Germination and appressorial maturation appear to be favoured by low illumination or darkness.

In temperate spp. ascocarp formation varies from very high to very low frequencies. In the tropics the perfect state is usually absent or unknown; here, clearly, carry over can take place as mildew colonies on volunteer plants (or in other situations), as it does sometimes in cool climates. Carry over may also occur as mycelium in dormant buds. There are few useful generalisations that can be made on factors affecting ascocarp formation. Heterothallism has been demonstrated in some spp. C. G. Smith suggested that lateness or irregularity in ascocarp formation in the field is due to the absence of the necessary mating types rather than to an unfavourable environment or the nutritive condition of the host. But Jackson et al. indicated that heterothallism may not be the only requirement for ascocarp formation, in work with *Sphaerotheca mors-uvae* (Schw.) Berk.; see these workers and C. G. Smith for literature on heterothallism and ascocarp formation.

The early literature on the Erysiphaceae was fully discussed by Yarwood. For genetics see Moseman; host range and distribution see Hirata; and major taxonomic works see Ainsworth et al. (in *Ainsworth and Bisby's Dictionary of the fungi*); Blumer; Junell. Some spp. in the following genera are considered: *Erysiphe*, *Leveillula*, *Microsphaera*, *Oidium*, *Phyllactinia* and *Sphaerotheca*.

Blumer, S. 1967. *True mildew fungi (Erysiphaceae). A book for the identification of the spp. occurring in Europe*, 436 pp., Gustav Fischer Verlag (general biology & accounts of taxa; **46**, 3007).

Boesewinkel, H. J. 1977. *Rev. Mycol.* **41**: 493 (identification by conidial characteristics; **57**, 3342).

Childs, J. F. L. 1940. *Phytopathology* **30**: 65 (diurnal cycle of spore maturation; **19**: 297).

Cole, J. S. 1976. In *Microbiology of aerial plant surfaces*, Academic Press: 627 (formation & dispersal of conidia in *Erysiphe cichoracearum*, *E. graminis* & *E. polygoni*).

Hammett, K. R. W. 1977. *N.Z. Jl Bot.* **15**: 687 (taxonomy of Erysiphaceae in New Zealand; **57**, 4827).

Hirata, K. 1966. Niigata Univ., Japan, 472 pp. (host range & geographical distribution, 1462 ref.; **45**, 3081).

——. 1968. *Trans. mycol. Soc. Japan* **9**: 73 (as above; **48**, 1594).

——. 1969. *Ibid* **10**: 47 (as above; **49**, 2344).

——. 1971. *Ibid* **12**: 1 (as above; **51**, 130).

——. 1976. *Ibid* **17**: 35 (as above; **56**: 589).

Jackson, G. V. H. et al. 1975. *Trans. Br. mycol. Soc.* **65**: 491 (ascocarp formation in *Sphaerotheca mors-uvae*; **55**, 2805).

Junell, L. 1965. *Ibid* **48**: 539 (nomenclature of 38 spp. in Erysiphaceae; **45**, 999).

——. 1967. *Symb. bot. upsal.* **19**(1), 117 pp. (Erysiphaceae of Sweden, monograph, keys to & descriptions of 82 spp., taxonomy of *E. cichoracearum* & *E. polyphaga*; **47**, 449).

Kapoor, J. N. 1967. *C.M.I. Descr. pathog. Fungi Bact.* 153 (*E. graminis*).

Manners, J. G. 1971. In *Ecology of leaf surface micro-organisms*, Academic Press: 339 (spore formation in *E. graminis* & *Puccinia striiformis*).

Moseman, J. G. 1966. *A. Rev. Phytopathol.* **4**: 269 (genetics, 150 ref.).

Mount, M. S. et al. 1971. In *Ecology of leaf surface micro-organisms*, Academic Press: 301 (primary development, 60 ref.).

Pady, S. M. et al. 1969. *Phytopathology* **59**: 844 (periodicity in sporulation in *E. cichoracearum*, *E. graminis* & *E. polygoni*; **48**, 3257).

——. 1972. *Ibid* **62**: 1099 (conidial release; **52**, 1484).

Salmon, E. S. 1900. *Mem. Torrey bot. Club* 9, 292 pp. (monograph on Erysiphaceae).

Schnathorst, W. C. 1965. *A. Rev. Phytopathol.* **3**: 343 (environmental relationships, 74 ref.).

Smith C. G. 1970. *Trans. Br. mycol. Soc.* **55**: 355 (formation of ascocarps in a controlled environment; **50**, 1686).

Smith, G. 1900. *Bot. Gaz.* **29**: 153 (haustoria).
Spencer, D. M. (editor). 1978. *The powdery mildews*, Academic Press.
Wolfe, M. S. et al. 1978. *A. Rev. Phytopathol.* **16**: 159 (patterns of race changes, 99 ref.).
Yarwood, C. E. 1957. *Bot. Rev.* **23**: 235 (review, 377 ref.; **36**: 657).
Zaracovitis, C. 1965. *Trans. Br. mycol. Soc.* **48**: 553 (attempts to identify by conidial characters; **45**, 1000).

Erysiphe cichoracearum DC. ex Mérat, *Nouv. fl. env. Paris* 1 ed. 2: 132, 1821.
Erysiphe scorzonerae Cast., 1845.

MYCELIUM usually well developed, evanescent but sometimes persistent and effused. CLEISTOTHECIA gregarious or scattered, globose becoming sunken or irregular, 90–135 μ diam., wall cells usually indistinct, 10–20 μ wide. APPENDAGES numerous, basally inserted, mycelioid, interwoven with mycelium, hyaline to dark brown, 1–4 times as long as diam. of ascocarp, rarely branched. ASCI 10–25, ovate to broadly ovate, rarely subglobose, more or less stalked, 60–90 × 25–50 μ. ASCOSPORES 2, very rarely 3, 20–30 × 12–18 μ. CONIDIA in long chains, ellipsoid to barrel shaped, very variable in size, 25–45 × 14–26 μ. (J. N. Kapoor, *CMI Descr.* 152, 1967 q.v. for further synonymy.)

The fungus is worldwide, particularly on Compositae but also on other plants. Junell described the taxonomy (1967, *Erysiphe* q.v.); Schmitt commented on the status of the name. *E. cichoracearum* can be recognised by the 2 spored asci and the basally inserted appendages which are much longer than the diam. of the ascocarp. Since it has conidia in long chains it has been confused with *Sphaerotheca fuliginea*, q.v. for distinguishing characteristics. *E. cichoracearum* forms well differentiated appressoria. Morrison (1960) confirmed earlier work in demonstrating a single gene mating system. An isolate from sunflower showed most conidial germination at 18–24°; free water on leaf disk surfaces inhibited germination and a high RH was favourable for it. In work on the perfect state on *Arctium lappa* (Burdock) it was found that ascospore discharge was greatest 6–24 hours after wetting dried cleistocarp material; that more ascospores were discharged at 20° than at 15°, 10° or 4°; and that cleistocarps subjected to weathering became progressively better sources of ascospores, their opt. germination temp. was *c.* 20° (Cook et al.; Cutter et al.; Wheeler et al.). In Brazil good control of the powdery mildew on okra (*Hibiscus esculentus*) was obtained with thiovit; maneb and dinocap were also effective (Perera et al.). Diseases caused by *E. cichoracearum* on lettuce and tobacco are described separately.

Cook R. T. A. et al. 1967. *Trans. Br. mycol. Soc.* **50**: 625 (overwintering of cleistocarps & ascospore infection on *Arctium lappa*; **47**, 982).
Cutter, E. C. et al. 1968. *Ibid* **51**: 791 (temp. & ascospore discharge from cleistocarps on *A. lappa*).
McKeen, W. E. et al. 1966. *Can. J. Bot.* **44**: 1299 (fine structure of haustorium & host parasite interface; **46**, 707).
Morrison, R. M. 1960. *Mycologia* **52**: 388 (mating types; **41**: 20).
——. 1964. *Ibid* **56**: 232 (conidial germination; **43**, 2524).
Pereira, C. A. L. et al. 1970. *Seiva* **30**: 26 (fungicides on okra; **51**, 1024).
Schmitt, J. A. 1955. *Mycologia* **47**: 422 (status of name *Erysiphe cichoracearum*; **35**: 330).
Wheeler, B. E. J. et al. 1973. *Trans. Br. mycol. Soc.* **60**: 117 (on *A. lappa*, cleistocarp dehiscence, ascospore germination & infection of cucumber; **52**, 3870).

LACTUCA SATIVA

The str. of *E. cichoracearum* (heterothallic) which causes powdery mildew of lettuce has been studied in California, U.S.A. This American str. appears to have originated in the Salinas Valley, California, possibly as a mutant from an endemic form on wild lettuce. The str. on cultivated lettuce is more sensitive to high temps (Schnathorst et al. 1958b). Little or no work appears to have been done elsewhere. The quickest conidial germination and germ tube growth occur at 18°. Over 24 hours the quickest germination is at 18–25°. Min. and max. temps for infection are 6–10° and near 27°. Most germination takes place at RH 95.6–98.2% and 25–30°); RH 100% is inhibitory. Most conidia are dispersed at 1200–1600 hours. The imperfect state overwinters but initial infections may be caused by ascospores from cleistothecia formed in the previous season. Ascospores infect lettuce and *Lactuca serriola* (lettuce str.; Schnathorst 1959b, 1960a); and see Schnathorst et al. 1958b for hosts. The powdery mildew and downy mildew (*Bremia lactucae* q.v., Schnathorst) lettuce diseases have been compared in the Salinas valley where temp. and RH affect their distribution. The former develops most at 17–19° and RH 77% or less; the latter at an av. 13° and av. RH 88%. Control of powdery mildew is through resistant cvs. In a cross between a resistant cv. and a susceptible wild lettuce Whitaker et al. showed control of resistance by a dominant gene.

Schnathorst, W. C. et al. 1958a. *Pl. Dis. Reptr* **42**: 1273 (susceptibility of cvs & hybrids; **38**: 175).
—— ——. 1958b. *Phytopathology* **48**: 538 (distribution, host range & origin; **38**: 175).
——. 1959a. *Mycologia* **51**: 708 (heterothallism; **40**: 645).
——. 1959b & c. *Phytopathology* **49**: 464, 562 (spread & life cycle; resistance; **39**: 142, 207).
——. 1960a & b. *Ibid* **50**: 304, 450 (effects of temp. & moisture stress; microclimate; **39**: 648, 762).
Whitaker, T. W. et al. 1941. *Ibid* **31**: 534 (inheritance of resistance; **20**: 556).

NICOTIANA TABACUM

Tobacco powdery mildew (also called white mould) is a widespread disease which can be severe in cooler climates (e.g. high altitudes in Africa and S.E. Europe). Lucas (*Diseases of tobacco*) gave a full account of the disease which does not occur in USA. Hopkins (*Tobacco diseases*) stated that the mildew first appears on the frills of the lower leaves as small, circular, white, powdery patches *c.* 0.5 cm diam. These become abundant on the upper surfaces of the bottom leaves and coalesce. If prolific conidial formation is not occurring the fungus may be detected by the pale yellowish colour of the ripening leaves. Such leaves have not the normal colour of ripeness but rather are devoid of pigment. Infected leaves (appearing 'falsely ripe') that are reaped and placed in barns quickly turn black. Rossouw (1959) found that the opt. temp. for conidial germination was 24–25° and for infection of tobacco it was 23.5°. Mildew did not occur where the av. daily max. and min. temps. were 25° or more. RH appears to have less effect on disease incidence than does temp. Although mildew may occur over a wide temp. range it appears to be only severe at moderate temps below 25°. The str. of *E. cichoracearum* on tobacco is virtually specific for this host (e.g. Rajendran).

The factors affecting the disease in the field have been widely studied in Zimbabwe by Cole and co-workers (see Cole 1978b). Leaves were resistant from their emergence until the main expansion was almost complete, some 6 weeks after transplanting. Upper leaves of intact plants had more natural infection than the corresponding leaves from topped plants. Susceptibility was apparently related inversely to free amino nitrogen and water content, and directly to insoluble carbohydrate/unit dry matter. Growth of colonies of the fungus was less on leaf disks from potassium deficient tobacco leaves than from those that were not deficient. In water culture experiments plants grown in media containing a low ratio of potassium to nitrogen produced the least susceptible leaves; their disease ratings were up to ×30 less than those of plants with full nutrients. Irrigation increased infection in all leaves; it increased the growth of the pathogen in dry weather, and the subsequent susceptibility of leaves that were still expanding but not yet infected. Irrigation increased the percentage of susceptible leaf area infected; in intact plants ×3 and in topped ones ×9. Topped plants had less infection than intact ones. Cole (1971) described the sporulation of powdery mildews, particularly *E. cichoracearum* on tobacco. Most conidia (which mature more rapidly in the light than in darkness) were trapped at 1300–1500 hours. Correlations of total conidia/day with temp. and RH were very variable. Cole (l.c.) concluded that in natural conditions the light dark cycle masks the responses to temp. and RH. Consequently, correlations of these parameters with hourly conidial catches appear to have no epidemiological significance.

Tobacco cvs differ in their resistance (see review by Lucas l.c.). In Bulgaria Palakartcheva found 16 *Nicotiana* spp. to be immune, 3 spp. were resistant and 2 spp. (including tobacco) were susceptible; and see Lamprecht; Lamprecht et al. Various fungicides including sulphur give some control. In Zimbabwe dinocap and salicylanilide were effective; in Japan thiophanate and benomyl gave protection in the glasshouse (Cole 1963, 1978a; Yamaguchi et al.). Stefanov used quinmethionate in USSR and benomyl has been used in Zimbabwe; see also Rajendran for India.

Cohen, Y. 1978. *Ann. appl. Biol.* **89**: 317 (protection by *Peronospora tabacina*).
Cole, J. S. 1963. *Rhod. Jnl agric. Res.* **1**: 65 (chemical control; **43**, 570).
——. 1964. *Ann. appl. Biol.* **54**: 291 (effects of N, P & K on susceptibility, growth & chemical composition of leaves in culture; **44**, 1672).

Cole, J. S. 1966a. *Ibid* **57**: 201, 435, 445 (effects of K deficiency on susceptibility, free amino N & carbohydrate content of leaves, effects of irrigation, conidial dispersal; **45**, 2601, 3219a & b).
——. 1966b. *Ibid* **58**: 61, 401 (susceptibility of leaves, stalk position, topped and intact plants in relation to free amino N & carbohydrate content, effects of inoculation methods & RH on conidial germination & hyphal growth on leaves; **45**, 3407; **46**, 1100).
—— et al. 1970. *Ibid* **66**: 239 (changes in leaf resistance induced by topping, cytokinins & antibiotics; **50**, 1373).
—— ——. 1970. *Trans. Br. mycol. Soc.* **55**: 345 (effects of light, temp. & RH on sporulation; **50**, 1966).
——. 1971. In *Ecology of leaf surface micro-organisms* (editors T. F. Preece & C. H. Dickinson): 323, Academic Press (sporulation).
——. 1976. *Trans. Br. mycol. Soc.* **67**: 339 (time lapse photography of conidial formation & release; **56**, 2658).
—— ——. 1978a. *Pestic. Sci.* **9**: 65 (efficiency of low & high volume pesticides; **57**, 5116).
——. 1978b. In *The powdery mildews* (editor D. M. Spencer): 447, Academic Press (90 ref.).
Lamprecht, M. P. 1969. *Agroplantae* **1**: 93 (improved yield, quality & disease resistance, with tobacco mosaic virus, in flue cured tobacco; **50**, 2491a).
—— et al. 1969. *Ibid* **1**: 99 (as above, with *Alternaria longipes*, in dark, air-cured tobacco; **50**, 2491b).
Oinuma, T. et al. 1977. *Bull. Morioka Tob. Exp. Stn* 12: 1 (breeding for resistance; 57, 1402).
Palakartcheva, M. 1974. *C.r. Acad. agric. Georgi Dimitrov* 7: 65 (resistance; **54**, 4140).
Rajendran, V. 1971. *Indian Phytopathol.* **24**: 684 (epidemiology & control; **52**, 2367).
Rossouw, D. J. 1959. *S. Afr. J. agric. Sci.* **2**: 19 (effects of temp. & RH; **39**: 127).
——. 1964. *Ibid* **7**: 151 (effects of climate; **43**, 2722).
Stefanov, D. 1972. *Rasteniev. Nauki* **9**: 103 (chemical control; **51**, 3541).
Yamaguchi, Y. 1973. *Bull. Hatano Tob. Exp. Stn* **73**: 377 (as above; **53**, 2312).

Erysiphe cruciferarum Opiz ex Junell, *Svensk bot. Tidskr.* 61: 217, 1967.

MYCELIUM amphigenous, white, dense, spreading and persistent. CONIDIA borne singly or in short chains, 30–40 × 12–16 μ, cylindric. CLEISTOTHECIA scattered, globose, at first yellow, becoming brown black with maturity, 95–125 μ diam. Cells of cleistothecia irregular, brown, 10–25 μ diam. APPENDAGES numerous, basally inserted, myceloid, rather narrow, hyaline to faintly coloured, seldom branched, often unequal in length, up to ×3 the diam. of cleistothecium. ASCI 3–12, usually *c.* 6–8, obovate, not stalked, 50–70 × 30–50 μ. ASCOSPORES ovoid, 2–7, 19–22 × 11–13 μ (T. J. Purnell & A. Sivanesan, *CMI Descr.* 251, 1970).

The perfect state is rare; see Sankhla et al. for a report. The taxonomy of the powdery mildew on Cruciferae was investigated by Junell and briefly referred to by Purnell et al. (l.c.). The former revised *E. communis* (Wallr.) Link sensu Blumer and divided it into 7 spp. including *E. betae* (Vañha) Weltzien and *E. cruciferarum*. *E. betae* (powdery mildew of sugar beet) may be distinguished on the basis of host specificity; see Kapoor for a description. *E. cruciferarum* (widespread) has longer appendages than *E. betae*. Crucifer powdery mildew is generally not serious; the few investigations have been done in temperate regions. Typical mildew symptoms occur on the leaves. Dixon, in a recent study, found that the fungus penetrated the buds of Brussel sprouts (*Brassica oleracea* var. *gemmifera*) deeply, and infected leaves and stems. Attempts to distinguish races have been made. Resistance in cabbage (*B. oleracea* var. *capitata*) and in cvs of Brussel sprouts has been described. Singh et al. controlled powdery mildew of Indian mustard (*B. juncea*) with dinocap.

Dixon, G. R. 1974. *Pl. Pathol.* **23**: 105 (infection & cv. reaction in Brussel sprouts; **54**, 3026).
Junell, L. 1967. *Svensk bot. Tidskr.* **61**: 209 (revision of *Erysiphe communis*; **46**, 2395).
Kapoor, J. N. 1961. *C.M.I. Descr. pathog. Fungi Bact.* 151 (*E. betae*).
Purnell, T. J. 1971. In *Ecology of leaf surface micro-organisms* (editors T. F. Preece & C. H. Dickinson): 269, Academic Press (effects of pre-inoculation washing of leaves on subsequent infections).
Sankhla, H. C. et al. 1967. *Pl. Dis. Reptr* **51**: 800 (perfect state; **47**, 364).
Singh, R. R. et al. 1976. *Indian J. Mycol. Pl. Pathol.* **6**: 73 (fungicides on mustard; **55**, 3783).
Walker, J. C. et al. 1965. *Pl. Dis. Reptr* **49**: 198 (inheritance of resistance in cabbage; **44**, 1991).

Erysiphe heraclei DC. ex St.-Am., *Fl. agen.*: 615, 1821.

Erysiphe umbelliferarum de Bary, 1870
E. heraclei DC., 1815
Alphitomorpha heraclei Wallr., 1819.

MYCELIUM well developed, much branched with

more or less distict lobed haustoria. CONIDIA usually single, rarely in short chains, cylindric, 34–46 × 14–20 μ. CLEISTOTHECIA at first globose becoming sunken, 85–120 μ diam. APPENDAGES numerous, basally inserted, mycelioid, brown, 1–2 times as long as the diam. of the ascocarp, 1 to many times irregularly branched. ASCI 3–8, rarely up to 10, obovate to subglobose, *c.* 55–70 × 30–45 μ. ASCOSPORES 3–5, rarely 6, ovate to elliptic, 20–28 × 10–15 μ (J. N. Kapoor, *CMI Descr.* 154, 1967, q.v. for further synonymy).

Powdery mildew of umbelliferous crops: carrot (*Daucus carota*), fennel (*Foeniculum vulgare*), parsley (*Petroselinum crispum*); and presumably it causes the disease on cumin (*Cuminum cyminum*) in India. *E. heraclei* is widespread and occurs on numerous genera in Umbelliferae. It is distinguished from other *Erysiphe* spp. by the numerous, short, irregularly branched appendages and elongate, cylindric conidia (Kapoor l.c.). On cumin in India, where the perfect state has not apparently been reported, the grey white lesions spread over the whole leaf surface, new leaves are infected as soon as they appear and other above ground parts of the plant are attacked. Uppal et al. found that many conidia germinate at 20–24° and that at 26.5–35° epidemics can arise on cumin. Specialised strs, by host, have been described, for example, Marras in Sardinia. On fennel in Italy the disease was found to be only severe in seedbeds. There is evidence for spread by seed and in Europe the cleistothecia have been found on seed (*Rev. appl. Mycol.* **28**: 510, 1949; Boerema et al.). Besides sulphur dust in India on cumin, dinocap and other fungicides gave control. In Israel tin compounds gave control on carrot; the highest yield being given by fentin acetate + maneb (and see Palti for an account of this fungus with special reference to carrot).

Boerema, G. H. et al. 1963. *Versl. PlZiekt. Dienst Wageningen* **138**: 184 (on carrot seed; **43**, 2185c).
Marras, F. 1961. *Studi sassar.* III 9(2), 12 pp. (host specialised strs in Sardinia; **43**, 635).
Mathur, R. L. et al. 1971. *Indian Phytopathol.* **24**: 796 (fungicide control on cumin; **52**, 1627).
Netzer, D. et al. 1966. *Pl. Dis. Reptr* **50**: 594 (fungicide control on carrot in Israel; **46**: 493).
Noviello, C. 1961. *Phytopathol. Z.* **42**: 167 (general on fennel; **41**: 323).
Palti, J. 1975. *Phytopathol. Mediterranea* **14**: 87 (on carrot, with *Leveillula taurica*, in Israel; **55**, 5989).
Uppal, B. N. et al. 1933. *Bull. Dep. Agric. Bombay* 169, 16 pp. (general on cumin; **12**: 393).

Erysiphe pisi DC. ex St.-Am., *Fl. agen.*: 614, 1821.

Erysiphe macropus Martius, 1817
Alphitomorpha pisi Wallr., 1819, pro parte
E. polygoni DC. emend. Salm., 1900, pro parte.

MYCELIUM very variable, persistent, thin effused and arachnoid, rarely dense or more often completely evanescent. CONIDIA formed singly, rarely in short chains, ellipsoid, 31–38 × 17–21 μ. CLEISTOTHECIA gregarious to scattered, globose, 85–126 μ diam. APPENDAGES very variable in number and length (10–30), basally inserted, mycelioid, sometimes knotty and frequently geniculate, rarely irregularly branched, brown, usually as long as diam. of ascocarp, at most × 2–3 as long. ASCI 3–10, usually 4–8, ovate to broadly ovate or subglobose, 50–60 × 30–40 μ. ASCOSPORES 3–5, rarely 6, 22–27 × 13–16 μ (J. N. Kapoor, *CMI Descr.* 155, 1967).

The pea (*Pisum sativum*; also on other legumes including *Lens, Lupinus, Medicago* and *Vicia*) powdery mildew is worldwide. It is distinguished from clover powdery mildew (*E. trifolii* Grev. (synonym: *E. martii* Lev.) see Kapoor) by the length and shape of the appendages. In *E. trifolii* these number *c.* 10–30, in part equatorially inserted, straight or bent, rarely flexuous, hyaline to brown, septate, × 2–12 as long as diam. of ascocarp (90–125 μ), a few dichotomously branched once or twice at the ends. *E. trifolii* occurs on numerous Leguminosae including *Trifolium, Lathyrus* and *Onobrychis* (common on these 3 genera). Stavely et al. (q.v. for ref.) studied the reaction of these plant genera (and others) to infection by powdery mildew; in clover 12 races were identified. There are forms differing in pathogenicity in both fungal spp. For powdery mildews on other legumes: common bean (*Phaseolus vulgaris*; see Dundas; Yarwood), green gram (*P. aureus*; see Soria et al.), black gram (*P. mungo*; see Singh et al. 1976) and fenugreek (*Trigonella foenum-graecum*; see Masih et al.).

On pea a yellow mottling of the leaf is followed by the characteristic white, powdery areas. The leaves become necrotic and infection can spread to the pods. Diseased plants can be dwarfed and killed. In India Uppal et al. considered that pea powdery mildew was most severe at 24–32°. Stavely et al. (1966b) found the opt. temp. for clover powdery mildew, i.e. conidial germination, mycelial growth and sporulation, to be *c*. 24°. The last 2 processes were inhibited at 28°. Initial growth was more rapid at RH 52% than at 75%, but subsequent growth and sporulation were similar at both humidities. For other aspects of this host and pathogen relationship see: Ayres; Gil et al.; Manners et al.; Oku et al.; Shiraishi et al.

In pea the fungus occurs within the seed. Both Crawford and Uppal et al. found that seed infection could be eliminated by hot water treatment (50–56° for 10–20 minutes). Seed should be obtained from sites free from infection. Amongst fungicides tested in India binapacryl, dinocap and quinomethionate reduced disease on pea and increased yields. Resistance (moderate) has been found in this crop (Ali-Khan et al.; and see Saxena et al.). The ref. are on pea except where stated.

Ali-Khan, S. T. et al. 1973. *Can. Pl. Dis. Surv.* **53**: 155 (resistance with *Mycosphaerella pinodes*; **53**, 4912).

Ayres, P. G. 1976. *J. exp. Bot.* **27**: 1196 (stomatal behaviour, transpiration & CO_2 exchange; **56**, 2740).

——. 1977. *Physiol. Pl. Pathol.* **10**: 139 (water stress & relations; **56**, 5872).

——. 1977. *Trans. Br. mycol. Soc.* **68**: 97 (leaf water potential & conidial formation; **56**, 3772).

Crawford, R. F. 1927. *Bull. New Mex. agric. Exp. Stn* 163, 13 pp. (general; **7**: 70).

Dundas, B. 1936 & 1941. *Hilgardia* **10**: 243; **13**: 551 (resistance in common bean; **16**: 363; **20**: 619).

Gil, F. et al. 1977. *Physiol. Pl. Pathol.* **10**: 1 (haustoria, fine structure & physiological properties; **56**, 4765).

Harland, S. C. 1948. *Heredity Lond.* **2**: 263 (inheritance of immunity in Peruvian forms of pea; **28**: 42).

Kapoor, J. N. 1967. *C.M.I. Descr. pathog. Fungi Bact.* 156 (*Erysiphe trifolii*).

Manners, J. M. et al. 1978. *Physiol. Pl. Pathol.* **12**: 199 uptake of ^{14}C photosynthates by haustoria; **57**, 4209).

Masih, B. et al. 1970. *Rajasthan J. agric. Sci.* **1**: 18 (fungicides on fenugreek; **53**, 4550).

Mathur, R. L. et al. 1971. *Indian J. Mycol. Pl. Pathol.* **1**: 95 (fungicides; **51**, 4474).

Oku, H. et al. 1976. *Ann. phytopathol. Soc. Japan* **42**: 597 (pisatin; **56**, 4269).

Saxena, J. K. et al. 1975. *Curr. Sci.* **44**: 746 (inheritance of resistance; **55**, 2425).

Shiraishi, T. et al. 1976. *Ann. phytopathol. Soc. Japan* **42**: 609 (pisatin; **56**, 4270).

———. 1977. *Phytopathol. Z.* **88**: 131 (pisatin; **56**, 4767).

Singh, S. D. et al. 1976. *Indian J. Mycol. Pl. Pathol.* **6**: 99 (fungicides on black gram; **57**, 1507).

———. 1977. *Indian J. agric. Sci.* **47**: 87 (fungicides; **56**, 5873).

Soria, J. A. et al. 1973. *Philipp. Agric.* **57**: 153 (fungicides on green gram; **54**, 5640).

Stavely, J. R. et al. 1966a. *Phytopathology* **56**: 309 (pathogenicity & morphology of *Erysiphe* on legumes & other plants; **45**, 2412).

———. 1966b. *Ibid* **56**: 940, 957 (effects of temp. & RH on *E. trifolii*, resistance reaction in *Trifolium*; **45**, 3569).

———. 1967. *Ibid* **57**: 193 (genetics of resistance to *E. trifolii* in *T. pratense*; **46**, 2054).

Uppal, B. N. et al. 1935. *Bull. Bombay Dep. Agric.* 177, 12 pp. (general; **15**: 338).

Yarwood, C. E. 1949. *Phytopathology* **39**: 780 (on common bean, effect of soil moisture & host nutrient conc.; **29**: 191).

Erysiphe polygoni DC. ex St.-Am., *Fl. agen.*: 614, 1821.

MYCELIUM white, dense, well developed on leaves and stems. CONIDIA borne singly or in short chains, cylindric or ovoid, 30–45 × 10–20 μ. CLEISTOTHECIA scattered, globose, at first yellow, becoming brownish black with maturity, 90–150 μ diam.; cells irregular, brown. APPENDAGES numerous basally inserted, narrow, more or less brown, simple or sometimes branched, interlaced with each other and with the hyaline mycelium, × 1–2 as long as the diam. of the ascocarp and forming a dense web around it. ASCI 3–12, obovate, 2–4 spored, 50–75 × 26–40 μ. ASCOSPORES elliptic to ovate, hyaline 20–30 × 10–12 μ (A. Sivanesan, *CMI Descr.* 509, 1976).

Salmon (*Erysiphe* q.v.) used *E. polygoni* in a collective sense and most reports in the literature are in this sense. Hammarlund divided the fungus into 26 ff. sp. Blumer restricted the name to forms infecting spp. in the Chenopodiaceae and Polygonaceae; and Junell (1967) restricted it to the latter family only (*Erysiphe* q.v.). Members of the *E. polygoni* complex have therefore now been dispersed as distinct spp. depending on the plant spp. and families that they infect. Sivanesan (l.c.) stated that a further morphological analysis may suggest that these spp. should be referred to as ff. sp. of *E. polygoni* rather than as distinct spp.

Hammarlund, C. 1925. *Hereditas* **6**: 1 (biology of some Erysiphaceae, 112 ref.; **4**: 431).

EXOBASIDIUM Woronin, *Ber. Naturf. Ges. Freiburg i.* B **4**: 397, 1867.
(Exobasidiaceae)

Parasitic on leaves, shoots, stems and flowers of dicotyledons causing hypertrophy and deformation or not; MYCELIUM intercellular forming a stroma beneath the epidermis; HYMENIUM continuous at maturity, on surface of infected part, effuse or compact, composed of basidia. BASIDIA emergent between epidermal cells, occasionally solitary and emergent through stomata; cylindrical to subclavate, projecting, 2–8 basidiospores on sterigmata. BASIDIOSPORES hyaline, smooth aseptate or septate (see Burt; McNabb).

E. vaccinii (Fuckel) Woron. is often collected from *Rhododendron* and *Vaccinium*, and other Ericaceae.

Burt, E. A. 1915. *Ann. Mo. bot. Gdn* **2**: 627 (Thelephoraceae of N. America IV *Exobasidium*).
McNabb, R. F. R. 1962. *Trans. R. Soc. N.Z.* **1**: 259 (genus in New Zealand including key).
Ramakrishnan, T. S. et al. 1949. *Proc. Indian Acad. Sci.* Sect. B **29**: 5 (genus in S. India; **28**: 362).
Savile, D. B. O. 1959. *Can. J. Bot.* **37**: 641 (the genus including key; **39**: 12).
Sundström, K. R. 1960. *Phytopathol. Z.* **40**: 213 (physiology & morphology; **40**: 406).

Exobasidium vexans Massee, *Kew Bull.* 138: 109, 1898.

MYCELIUM confined to the leaf blister, hyphae little >2 μ diam., sparingly septate, intercellular. HYMENIUM on convex surface of blister, lower leaf surface, dull, grey, becoming pure white, powdery. PARAPHYSES simple, apically rounded. BASIDIA projecting beyond sterile layer, subcylindric with 2 sterigma, a septum near the apex, 30–35 × 5–6 μ. BASIDIOSPORES ellipsoid, hyaline, at first aseptate, becoming 1 septate and readily dislodged from sterigmata, when mature 13–27 (18.7) × 4.3–6.5 (5.5 μ); thin walled but may develop thick walls (after Massee l.c., C. H. Gadd & C. A. Loos, *Trans. Br. mycol. Soc.* **31**: 229, 1948. The latter, agreeing with Sawada, pointed out that Massee was wrong to describe the 2 celled spores as conidia; see also Gadd et al. 1949).

E. vexans causes blister blight of tea, one of the crop's major diseases; there appears to be no other host. The disease moved out of the Assam valley (N.E. India) and caused severe losses near Darjeeling in 1908–9 (McRae). Mayne has reviewed its history. Since the product is the youngest leaves and buds the tea bush is kept perpetually in the vegetative phase by harvesting and pruning. A pathogen which causes defoliation of young leaves can therefore be particularly destructive. *E. vexans* has been reported from: Bangladesh, Cambodia, China, India, Indonesia, Japan, Malaysia (W.), Nepal, Sri Lanka, Taiwan, Thailand and Vietnam (Map 45). The first symptoms (on the leaves) are small, pale green, yellow or pink translucent spots; they are sometimes a deep red. The spot is circular and enlarges to 1–1.5 cm diam. (larger in rare cases). Its upper surface is smooth, shiny, pale green, yellowish or pinkish and becomes strongly concave. The lower, convex, surface, is dull, at first grey, becoming white and powdery. Sometimes the convex sporulating surface may be epiphyllous. Some 20 blisters may form on one leaf which may consequently become curled and distorted. Grey spots occur on the young stems which are completely penetrated by the fungus; thus the tissue is killed. The leaf blisters eventually become brown and shrunken with age. *E. reticulatum* Ito & Sawada causes a tea blight in Japan (Ezuka 1956, 1958). The symptoms were briefly described by Petch (*The diseases of the tea bush*). The leaf spot, which never forms a blister, is indefinite in shape, 2–3 cm or more in diam., brown, becoming darker with a network of dark lines. The hymenium (reticulate) forms along these lines when the epidermis splits. Studies on *E. vexans* have been done extensively in India and Sri Lanka, the earlier work in the latter was reviewed by Portsmouth; Venkata Ram (1970, 1975b) reviewed blister blight and control measures in S. India. Weille described the work on blister blight in Indonesia.

Infection is by the germinating basidiospore through the cuticle of young (up to 20–40 days old) leaf and stem tissue. At a high RH germination and appressorial formation take place in 6–16 hours. The first symptoms appear after 8–10 days and sporulation begins after 15–17 days. The white blister is readily seen after 18–21 days. The basidiospores are forcibly discharged; this stops if the water supply is interrupted. Discharge from a mature blister continues for 7–10 days. In Sri Lanka most airborne basidiospores were trapped during the S.W. and N.E. monsoons (May–June and November). There is a diurnal periodicity with a peak at 0001–0400 hours but high concs of basidiospores occur after rain during daylight. Diurnal liberation is almost entirely due to the diurnal fluctuations of RH. A few hours of sunshine kills most basidiospores; and *c.* 3.75 hours of sunshine/day over 5 days reduces the disease to insignificant levels (Loos; Portsmouth; Shanmuganathan et al. 1966; Visser et al.). Detailed epidemiological work in Sri Lanka suggests that accurate predictions of severe disease outbreaks can be made. Kerr et al. (1967b) stated that the accuracy of prediction is higher than for any other airborne plant disease. The epidemiological pattern is simplified, thus making predictions easier, by several factors. There is no seasonal change in host susceptibility; the susceptible young leaves are all at the top of the bush with little complication from microenvironments. Bushes are plucked weekly and therefore 52 disease assessments can be made each year (Kerr et al. 1969). The 2 most important factors to consider are the number of basidiospores landing on a leaf and the duration of leaf wetness; temp. can be virtually ignored. A simple method of prediction was devised for Sri Lanka. Because of the important local weather differences in tea areas such predictions need to be made for each area (Kerr et al. 1966, 1967a & b; Mulder et al.; Visser et al.).

Blister blight is controlled by regular spraying with copper fungicides. Rounds every 7–10 days (some 20–30 applications during the wet seasons) were soon found to be required in India and Sri Lanka. In India Venkata Ram (1971, 1972) found that combinations of nickel chloride with copper were advantageous. The superiority of copper fungicides over organic formulations was considered to be due to their better tenacity under monsoon conditions (Shanmuganathan et al. 1974). On St Coombs Estate (Sri Lanka) Silva et al. found that control was proportional to the quantity of fungicide used. In the first 2 years of the pruning cycle control was economically advantageous; but in the third and fourth years there was no increase in yield as a result of disease control. Phytotoxic effects may be a factor in the last 2 years. In seeking ways to reduce the number of spraying rounds systemic fungicides have been used and attempts at prediction made. In Sri Lanka accurate prediction could, theoretically, reduce the number of rounds to 4 in some years (Kerr et al. 1969). In N.E. India the systemic pyracarbolid (used also in Sri Lanka) in 2 fortnightly rounds was as effective as 4 weekly rounds of copper. In S. India fortnightly rounds with pyracarbolid and tridemorph showed promise. A schedule with calixin, copper and nickel chloride has also been described (Satyanarayana et al.; Shanmuganathan et al. 1978; Venkata Ram 1974, 1975, 1976).

Ezuka, A. 1955. *Bull. Tea Div. Tokai-Kinki agric. Exp. Stn* **3**: 28 (in vitro culture with *Exobasidium japonicum*; **35**: 723).

——. 1956. *Study of tea. Tea Div. Tokai-Kinki agric. Exp. Stn* **15**: 13 (ecology, temp. & RH, with *E. reticulatum*).

——. 1958. *Bull. Tea Div. Tokai-Kinki agric. Exp. Stn* **6**: 1; **38**: 225).

Gadd, C. H. et al. 1948. *Trans. Br. mycol. Soc.* **31**: 229 (on basidiospores; **28**: 145).

—— ——. 1949. *Tea Q.* **20**: 54 (general; **29**: 337).

Homburg, K. 1953. *Bergcultures* **22**: 345 (effect of climate; **33**: 324).

Kerr, A. et al. 1966. *Trans. Br. mycol. Soc.* **49**: 139 (epidemiology, sporulation; **45**, 2946a).

—— ——. 1967a. *Ibid* **50**: 49 (epidemiology, spore deposition & disease prediction; **46**, 2101).

—— ——. 1967b. *Ibid* **50**: 609 (epidemiology, disease forecasting; **47**, 1269).

—— ——. 1969. *Tea Q.* **40**: 9 (epidemiology; **49**, 244).

Loos, C. A. 1951. *Ibid* **22**: 65 (morphology & life history; **31**: 403).

McRae, W. 1910. *Agric. J. India* **5**: 126 (disease outbreak in Darjeeling).

Mayne, W. W. 1958. *Indian Coff.* **22**: 7 (history with *Hemileia vastratrix*; **38**: 145).

Mulder, D. et al. 1960. *Tea Q.* **31**: 56 (disease forecasting; **39**: 735).

Nozu, M. et al. 1975. *Bull. Fac. Agric. Shimane Univ.* **9**: 23 (fine structure of leaf gall; **56**, 381).

Portsmouth, G. B. 1961. *Outl. Agric.* **3**: 81 (review, Sri Lanka, 29 ref.; **40**: 764).

Reitsma, J. et al. 1949. *Bergcultures* **18**: 218–221, 223, 225, 227, 229, 231, 370–371, 373, 375, 377 (general in Indonesia; **29**: 58).

Satyanarayana, G. et al. 1975. *Two Bud* **22**: 87 (fungicides, pyracarbolid & Cu; **55**, 4843).

Sawada, K. 1922. *Trans. nat. Hist. Soc. Formosa* **59**: 7 (on basidiospores; **1**: 454).

Shanmuganathan, N. et al. 1966. *Trans. Br. mycol. Soc.* **49**: 219 (epidemiology, diurnal & seasonal spore periodicity; **45**, 2946b).

—— ——. 1974. *Pl. Dis. Reptr* **58**: 928 (fungicides, Cu & organic; **54**, 985).

—— ——. 1978. *PANS* **24**: 43 (fungicides, pyracarbolid; **57**, 4612).

Silva, R. L. et al. 1972. *Tea Q.* **43**: 140 (fungicides & crop loss; **53**, 2304).

Soepadmo, B. et al. 1974. *Menara Perkebunan* **42**: 111 (sunshine & disease forecasting in Indonesia; **54**, 4134).

Venkata Ram, C. S. 1970. *Bull. United Plrs' Assoc. S. India* 28: 53 (review, fungicide control, 42 ref.; **51**, 1868b).
——. 1971. *Ibid* 29: 4 (efficiency & Cu fungicide formulation; **52**, 3057).
——. *Plrs' Chron.* **67**: 259 (fungicides, Cu & nickel chloride; **52**, 3418).
——. 1974. *Ibid* **69**: 360 (systemic fungicides; **55**, 871).
——. 1975a. *Z. PflKrankh. PflSchutz* **82**: 65 (fungicides, systemic & Cu; **54**, 5069).
——. 1975b. In *Advances in mycology and plant pathology*, New Delhi: 211 (review, 66 ref.).
——. 1976. *Bull. scient. Dep. United Plrs' Assoc. S. India* 33: 70 (fungicides; **57**, 1390).
Visser, T. et al. 1961. *Ann. appl. Biol.* **49**: 306 (effects of sunshine & rain; **40**: 764).
Weille, G. A. De. 1960. *Neth. J. agric. Sci.* **8**: 183 (epidemiology, 37 ref.; **40**: 323).

FULVIA Ciferri, *Atti Ist. bot. Univ. Lab. crittogam. Pavia* Ser. 5, **10**: 245, 1954.
(Dematiaceae)

COLONIES effuse, velvety, buff to brown or purplish. STROMA present, pale substomatal. SETAE and HYPHOPODIA absent. CONIDIOPHORES macronematous, mononematous, caespitose, emerging through stomata, unbranched or occasionally branched, straight or flexuous, narrow at the base, thickening towards the apex, with unilateral nodose swellings which may proliferate as short lateral branchlets, very pale to mid-pale brown or olivaceous brown, smooth. CONIDIOGENOUS CELLS monoblastic or polyblastic, integrated, terminal becoming intercalary, sympodial, clavate or cylindrical, cicatrised. CONIDIA catenate, chains frequently branched, acropleurogenous, simple, cylindrical with rounded ends or ellipsoidal, very pale to mid-pale brown or olivaceous brown, smooth, 0–3 septate, hilum sometimes slightly protuberant (M. B. Ellis, *Demat. Hyphom.*: 306, 1971).

Fulvia fulva (Cooke) Ciferri, *Atti Ist. bot. Univ. Lab. crittogam. Pavia* Ser. 5 **10**: 246, 1954.
Cladosporium fulvum Cooke, 1883.

COLONIES hypophyllous, effuse, velvety, at first pale buff with whitish margin, later brown and finally often purplish; the upper surface of the leaf above infected areas is at first yellowish, later reddish brown. CONIDIOPHORES on leaves up to 200 μ long but usually 100 μ or less, 2–4 μ thick near the base broadening to 5–6 μ or up to 7–8 μ at the nodes. CONIDIA 12–47 × 4–10 μ (M. B. Ellis, *Demat. Hyphom.*: 307, 1971; see also Holliday et al.).

Leaf mould of tomato is a major worldwide (Map 77) disease. With an opt. temp. of 22–24° it is one that will be most severe in the tropics at cooler periods and when air moisture is high. In temperate countries leaf mould is almost invariably a disease of the glasshouse crop. The first and most readily observed symptoms are pale chlorotic spots (with indefinite margins) on the upper surface of the leaf. Already at this stage sporulation will probably have begun on the lower surface beneath the spots. This sporulation is downy, light grey, becoming buff to tawny brown or olive green. Infection can lead to defoliation as the spots increase in size. In work on the assessment of the disease and the effects of fungicides Smith et al. found that there was an interval of *c.* 6 weeks between the incidence of severe infection (colonisation of at least 50% leaf area) and decreases in yield. Flowers and fruits can be attacked but this phase of the disease is much less important than leaf attack. A black, somewhat hard rot spreads from the stem end and fruit may become lopsided (Gardner).

Leaf and sepal penetration is stomatal without appressoria; opt. temps for both conidial germination and in vitro growth are about the same as those for disease development. Leaf mould has been investigated almost entirely in temperate conditions. The importance of high RH for penetration, lesion growth and sporulation has long been recognised. Winspear et al. recently showed that in glasshouse crops disease incidence could be decreased (and yields increased) by limiting the periods of high RH. They found that there was less leaf mould at a constant 20° compared with a 20° (day) and 13° (night) temp. regime. Guba (1938) considered that the airborne conidia remained viable for 9–12 months under adverse conditions. Seed contamination has been described.

Control measures will depend in part on the method of cultivation, types of cv., distribution of races and probably the geographical locality. Ogilvie (*Diseases of vegetables*) discussed control with respect to glasshouse growing. The considerable breeding for resistance work in N. America and Europe has used oligogenic resistance in *Lycopersicon chilense*, *L. hirsutum*, *L. peruvianum* and *L. pimpinellifolium*; 5

resistance (R), dominant genes have been identified (e.g. Dijkman et al. 1971; Guba 1956; Kooistra; Kerr et al.; Termohlen). Each R gene controls resistance to certain races, and new races of *F. fulva* occur periodically. Hubbeling (1971) reported the existence of races 1.2.4 and 2.3.4. In Canada races which overcame resistance due to the R gene *Cf4* were designated 10, 11 and 12. Some accessions had high resistance to, or immunity from, races 10 and 11. An R gene in one accession gave immunity from races 6, 10, 11 and 12, and was designated *Cf5* (Kerr et al. 1971; Patrick et al.). For differentiation of races see also: Bailey; Bailey et al. 1964; Boukema et al.; Day; Hubbeling 1968; Kishi et al.; Langford 1937. Kaars Sijpesteijn et al. have described host parasite interactions; and see Lazarovits et al.; Wit.

Fungicide spraying is effective and may have to be used in the tropics where resistant cvs may be unavailable or agronomically unsuitable. Satisfactory results have been given by chlorothalonil, maneb, mancozeb, metiram and nabam+zinc sulphate; copper also gives some control and benomyl has been used with good effect in glasshouse tomatoes (Beaumont; Jones; Kankam; Pitblado et al.; Rogers; Wiggell).

Alexander, L. J. 1934. *Bull. Ohio Exp. Stn* 539, 26 pp. (resistance; **14**: 202).

——. 1942. *Phytopathology* **32**: 901 (races; **22**: 81).

Bailey, D. L. 1950. *Can. J. Res.* Ser. C **28**: 535 (races, differentiation, culture & population fluctuation; **30**: 250).

—— et al. 1962. *Can. J. Bot.* **40**: 1095 (physiology & resistance; **42**: 154).

—— ——. 1964. *Ibid* **42**: 1555 (race 10; **44**, 1255b).

Banga, O. 1941. *Meded. TuinbVoorlDienst* 24, 40 pp. (comparison of resistant cv. Vetmold with Dutch cvs; **25**: 528).

Beaumont, A. 1954. *Pl. Pathol.* **3**: 21 (Cu & other fungicides in UK; **33**: 767).

Bond, T. E. T. 1938. *Ann. appl. Biol.* **25**: 277 (penetration & histology, other hosts; **17**: 634).

Boukema, I. W. et al. 1975. *Euphytica* **24**: 99, 105 (uniform resistance, progenies from diallel crosses; **54**, 4152, 4153).

——. 1977. *Zaadbelangen* **31**: 17 (new race & manipulations of resistance genes; **57**, 786).

Curren, T. 1967. *Can. J. Bot.* **45**: 2125 (C & N nutrition; **47**, 1298).

Day, P. R. 1954. *Pl. Pathol.* **3**: 35 (race differentiation; **34**: 111).

Dijkman, A. Van et al. 1971. *Neth. J. Pl. Pathol.* **77**: 14 (biochemical mechanism for gene to gene resistance; **50**, 3181).

—— ——. 1973. *Ibid* **79**: 70 (gel electrophoresis of soluble proteins excreted by races & isolates; **52**, 3224).

—— ——. 1973. *Physiol. Pl. Pathol.* **3**: 57 (leakage of pre-absorbed ^{32}P from leaf disks infiltrated with high molecular weight products of incompatible races; **53**, 277).

Fajardo, T. G. 1937. *Philipp. J. Agric.* **8**: 163 (general; **17**: 140).

Gardner, M. W. 1925. *J. agric. Res.* **31**: 519 (fruit infection & seed transmission; **5**: 257).

Guba, E. F. 1936. *Phytopathology* **26**: 382 (resistance in USA; **15**: 690).

——. 1938. *Bull. Mass. agric. Exp. Stn* 350, 24 pp. (environment & disease; **18**: 142).

——. 1956. *Pl. Dis. Reptr* **40**: 647 (breeding for resistance in USA).

Hanusova, M. 1972. *Ochrana Rostlin* **8**: 201 (growth in vitro; **52**, 600).

Hasper, E. 1925. *Z. PflKrankh. PflPath. PflSchutz* **35**: 112 (general; **5**: 9).

Holliday, P. et al. 1976. *C.M.I. Descr. Pathog. Fungi Bact.* 487 (*Fulvia fulva*).

Hubbeling, N. 1966. *Meded. LandbHoogesch. OpzoekStns Gent* **31**: 925 (new races; **46**, 2369k).

——. 1968. *Meded. Rijksfac. LandbWetensch. Gent* **33**: 1011 (new race; **48**, 2704i).

——. 1971. *Meded. Fak. LandbWetensch. Gent* **36**: 300 (differentiation of new races; **51**, 1934).

Inoue, Y. 1937. *Forschn Geb. PflKrankh. Kyoto* **3**: 310 (penetration, culture, conidial germination & temp. **17**: 778).

Jones, J. P. 1973. *Pl. Dis. Reptr* **57**: 612 (fungicides; **53**, 678).

Kaars Sijpesteijn, A. et al. 1973. In *Fungal pathogenicity and the plant's response*: 437, Academic Press (host–parasite interactions).

Kankam, J. S. 1972. *Ghana J. agric. Sci.* **5**: 229 (fungicides maneb & mancozeb; **52**; 2372).

Kerr, E. A. et al. 1964. *Can. J. Bot.* **42**: 1541 (resistance from wild tomato spp.; **44**, 1255a).

—— ——. 1971. *Hort. Res.* **11**: 84 (resistance to new races; **51**, 1935).

Kishi, K. et al. 1976. *Ann. phytopathol. Soc. Japan* **42**: 497 (races; **56**, 2665).

Kooistra, E. 1964. *Euphytica* **13**: 103 (breeding for resistance in the Netherlands; **44**, 1253).

Langford, A. N. 1937. *Can. J. Res.* Ser. C **15**: 108 (races, inheritance of resistance; **16**: 571).

——. 1948. *Ibid* **26**: 36 (autogenous necrosis in immune tomato; **27**: 543).

Lazarovits, G. et al. 1976. *Can. J. Bot.* **54**: 224, 235 (histology with race 1, fine structure of host reactions to races; **55**, 3748, 3749).

Lowther, R. L. 1964. *Can. J. Bot.* **42**: 1365 (physiology & resistance; **44**, 1254).

Patrick, Z. A. et al. 1971. *Ibid* **49**: 189 (new races & resistance; **50**, 3182).
Persiel, F. 1967. *Z. PflZücht.* **57**: 325 (breeding for resistance).
Pitblado, R. E. et al. 1972. *Can. J. Pl. Sci.* **52**: 459 (fungicide benomyl; **52**, 493).
Rogers, I. S. 1962. *J. Agric. S. Aust.* **65**: 458 (fungicide nabam & Zn sulphate; **41**: 677).
Small, T. 1930. *Ann. appl. Biol.* **17**: 71 (effects of temp. & RH; **9**: 566).
Smith, P. M. et al. 1969. *Ibid* **63**: 19 (disease assessment & effect on yield; **48**, 1982).
Sprau, F. 1949. *Pflanzenschutz* **1**: 141 (control, seed treatment; **29**: 63).
Termohlen, G. P. 1960. *Tijdschr. PlZiekt.* **66**: 314 (races, breeding for resistance in the Netherlands; **40**: 710).
Wiggell, D. 1958. *Pl. Pathol.* **7**: 26 (fungicides in UK, nabam & Zn sulphate; **37**: 600).
Winspear, K. W. et al. 1970. *Ann. appl. Biol.* **65**: 75 (disease restriction by RH control in glasshouse with *Sclerotinia fuckeliana*; **49**, 2654).
Wit, P. J. G. M. de. 1977. *Neth. J. Pl. Pathol.* **83**: 109 (light & electron microscope study of infection; **56**, 5801).
Yordanov, M. et al. 1974. *C.r. Acad. Agric. Georgi Dimitrov* **7**(4): 69 (new sources of resistance; **54**, 4154).

FUSARIUM Link ex Fr., *Syst. Mycol.* **1** XLI (introduct.), 1821; Link *Mag. Ges. natuf. Freunde, Berlin* 3: 10, 1809.
(Moniliaceae)

The genus as originally defined by Link and validated by Fries was erected for spp. with fusiform conidia borne on a stroma and thus placed in the order Tuberculariales. Because stroma formation occurs only under certain conditions on the host, and because of the development of pure culture methods for spp. identification, little emphasis is now placed on the presence or absence of a stroma, and *Fusarium* is generally considered to belong to the moniliaceous Hyphomycetes. Basically 3 conidial forms may develop. The MICROCONIDIA, which in general are phialospores, belong in the genus *Acremonium* (*Cephalosporium*), although in some spp., for example *F. fusarioides*, microconidia are formed as blastospores. The MACROCONIDIA are fusoid, 1 or more septate, with a foot cell bearing a heel. This is the only distinct character separating the genus from *Cylindrocarpon* and other related genera. Resting spores (CHLAMYDOSPORES) are formed in many spp. They are globose with a thick wall, intercalary, solitary, in chains or clumps, or terminal on short lateral branches. They may also form from cells of the macroconidia. Perithecial states (where known) belong to the closely related genera *Nectria*, *Calonectria* and *Gibberella* or other genera in the Hypocreales (C. Booth, personal communication, April 1978).

The taxonomic treatment of this difficult genus follows that of Booth's (1971) monograph (see Toussoun et al. 1975b for a discussion of the systems of classification). Where a perfect state is known the disease(s) is discussed under this state. The most commonly encountered spp. were keyed out, with illustrations of diagnostic characters, by Booth (1977) and the present status of *Fusarium* taxonomy was discussed by Booth (1975). For the behaviour of these fungi, commonly associated with soilborne diseases, see Smith; Toussoun. Recent, general studies for different countries are by Carrera (Argentina), Gerlach et al. (Iran), Lim et al. (Malaysia), Matuo (Japan) and Sen Gupta (India); and see Joffe (1974b). The pathogenicity of *Fusarium* spp. has frequently been studied collectively on one crop or crop group; notably by Joffe and co-workers in Israel in recent years. A few other examples are given for avocado pear, cotton, maize and soybean. *F. semitectum* Berk. & Rav. is extremely common, particularly from tropical and subtropical regions. It is a secondary invader of plant tissue and can cause storage rots in several crops. *F. avenaceum* (Corda ex Fr.) Sacc. (*Gibberella* q.v.) and *F. culmorum* (W. G. Sm.) Sacc. (see Booth et al.) occur mainly in temperate areas. The latter occurs on maize in cooler climates.

Booth, C. et al. 1964. *C.M.I. Descr. pathog. Fungi Bact.* 25, 26 (*Fusarium avenaceum* & *F. culmorum*).
——. 1971. *The genus* Fusarium, 237 pp., Commonw. Mycol. Inst. (monograph, *c.* 500 ref.; **51**, 1181).
——. 1975. *A. Rev. Phytopathol.* **13**: 83 (status of taxonomy, 33 ref.).
——. 1977. Fusarium. *Laboratory guide to the identification of the major species*, 58 pp., Commonw. Mycol. Inst. (**56**, 4383).
Carrera, C. J. M. 1972. *Revta Investnes agrop. INTA* Ser. 5 **9**: 41 (in Argentina, 80 refs.; **55**, 5576).
Gerlach, W. et al. 1970. *Nova Hedwigia* **20**: 725 (in Iran, 45 refs.; **51**, 1191).
Joffe, A. Z. et al. 1967. *Fruits* **22**: 97 (on avocado pear; **46**, 2289).
—— ——. 1972. *Phytopathol. Mediterranea* **11**: 159 (distribution & pathogenicity on onion; **53**, 1659).

Fusarium oxysporum

Joffe, A. Z. 1972. *Pl. Dis. Reptr* **56**: 963 (on avocado pear, banana & citrus & pathogenicity; **52**, 2674).
——. 1973. *Pl. Soil* 38: 439 (from groundnut kernels & soil; **53**, 1144).
—— ——. 1974a. *Z. PflKrankh. PflSchutz* **81**: 196 (on field crops & pathogenicity to seedlings; **54**, 2141).
——. 1974b. *Mycopathologia* **53**: 201 (a taxonomic system, 35 refs.; **54**, 3757).
Knight, C. et al. 1977. *Phytopathol. Z.* **89**: 170 (*F. semitectum* rot of banana; **56**, 5734).
Lim, W. H. et al. 1977. *Malaysian appl. Biol.* **6**(1): 45 (from soil).
Matuo, T. 1972. *Rev. Pl. Prot. Res.* **5**: 34 (in Japan, 53 ref.; **52**, 3973).
Nirenberg, H. 1976. *Mitt. Biol. BundAnst. Ld-u. Forstw.* 169, 117 pp. (differentiation in section Liseola; **56**, 595).
Nyvall, R. E. 1976. *Mycologia* **68**: 1002 (on soybean; **56**, 3331).
Reddy, M. N. et al. 1972. *Phytopathol. Z.* **74**: 115 (taxonomic significance of isozyme patterns; **52**, 365).
Rintelen, J. 1967. *Ibid* **60**: 141 (on maize; **47**, 1111).
Sadasivan, T. S. et al. 1954. *J. Indian bot. Soc.* **33**: 162 (review, soilborne spp.; **34**: 179).
Saharan, G. S. et al. 1972. *Pl. Dis. Reptr* **56**: 693 (*F. semitectum* on soybean; **52**, 1330).
—— ——. 1973. *Indian J. Mycol. Pl. Pathol.* **3**: 222 (as above, fungicides; **54**, 3060).
Sen Gupta, P. K. 1974. *Nova Hedwigia* **25**: 699 (in India; **54**, 3735).
Smith, S. N. 1970. In *Root diseases and soilborne pathogens*: 28, Univ. Calif. Press (soil populations, 42 ref.).
Sparnicht, R. H. et al. 1972. *Phytopathology* **62**: 1381 (on cotton bolls, pathogenicity; **52**, 2951).
Toussoun, T. A. 1975a. In *Biology and control of soilborne plant pathogens*: 145, Am. Phytopathol. Soc. (suppressive soils, 40 ref.).
—— et al. 1975b. *A. Rev. Phytopathol.* **13**: 71 (variation & speciation, 28 ref.).
—— ——. 1976. *A pictorial guide to the identification of* Fusarium *spp.* 2nd edn, 43 pp. Pennsylvania State Univ. Press.
Zauberman, G. et al. 1977. *Phytopathol. Z.* **89**: 359 (on avocado pear; **57**, 1319).

Fusarium oxysporum Schlecht., *Flora berol* **2**: 139, 1824 emend. Snyder & Hansen pro parte *Am. J. Bot.* **27**: 64, 1940.

Fusarium angustum Sherb., 1915
F. bostrycoides Wollenw. & Reinkg., 1925
F. bulbigenum Cooke & Massee, 1913
F. conglutinans Wollenw., 1913
F. orthoceras Appel & Wollenw., 1910
F. vasinfectum Atkinson, 1892
F. lini Bolley, 1902
F. dianthi Prill. & Del., 1899.
(synonymy from C. Booth, *The genus* Fusarium: 130, 1971)

MYCELIUM delicate white or peach but usually with a purple tinge, sparse to abundant then floccose becoming felted and sometimes wrinkled in older cultures. MICROCONIDIA borne on simple phialides arising laterally on the hyphae or from the short sparsely branched conidiophores. Microconidia generally abundant, variable, oval to ellipsoid cylindrical, straight to curved, 5–12 × 2.2–3.5 µ. MACROCONIDIA, sparse in some strs, borne on more elaborately branched conidiophores or on the surface of sporodochia like *Tubercularia*. They are thin walled, generally 3–5 septate, fusoid–subulate and pointed at both ends, occasionally fusoid–falcate; macroconidia are found with a somewhat hooked apex and a pedicillate base; 3 septate, 27–46 × 3–5 µ; 5 septate, 35–60 × 3–5 µ; 6–7 septate, 50–60 × 3.5–5 µ. The 3 septate spore is most commonly found. Sterile stromatic pustules, pale or deep violet, form in some isolates and resemble perithecia of *Gibberella*. CHLAMYDOSPORES smooth and rough walled, generally abundant, terminal and intercalary, generally solitary but occasionally formed in pairs or chains.

F. oxysporum was the first of 9 spp. to be described in the section Elegans. By 1935, when Wollenweber and Reinking published their monograph of the genus, the number of spp., vars and forms included in Elegans had reached 40. This sp., apart from being the most economically important member of the genus *Fusarium*, is also one of the most labile and variable. This variation may be expressed as a series of morphological strs, and led to the acceptance or proposal of such names as *F. conglutinans*, *F. orthoceras*, *F. bulbigenum* and *F. vasinfectum*. Apart from the difficulty of separating these spp., especially after maintenance for some time on artificial media, it was gradually realised that the selective pathogenicity of the Elegans isolates was not necessarily linked to the morphological variant designated by a sp. name such as *bulbigenum* or *vasinfectum*. In fact almost any of these strs may carry the specific pathogenicity towards a particular host, the cultural appearance being more related to ecological and geographical factors than to true sp. characters. Consequently Snyder and Hansen in 1940 amended the description of *F. oxysporum* to agree with that of the section Elegans thus reducing all the other spp. in the section to synonymy. They then designated 25 forms of this sp. based on their pathogenicity patterns (from C. Booth, *CMI Descr.* 211, 1970, and see Booth (1971) l.c. for further morphology, taxonomy and a list of ff. sp.).

F. redolens Wollenw. (C. Booth & J. M. Waterston, *CMI Descr.* 27, 1964) is disposed (Booth 1971 l.c.) now as *F. oxysporum* var. *redolens* (Wollenw.) Gordon, *Can. J. Bot.* **30**: 238, 1952, with synonymy: *F. redolens* Wollenw., *F. redolens* var. *solani* Sherb., *F. solani* var. *redolens* (Wollenw.) Bilay. Var. *redolens* is somewhat intermediate in morphological characters between *F. oxysporum* and *F. solani* (q.v.). It can cause wilt and damping off, and is widespread in the soil of temperate regions. It has been associated with the wilt complex in pea (*Pisum sativum*) mainly caused by *F. oxysporum* f. sp. *pisi* (q.v.).

F. oxysporum has a worldwide distribution as a soil organism. Its survival, in the absence of a suitable substrate, is through chlamydospores and some aspects of this have been described by Park (1959) amongst others. He found no dispersal by continuous growth in soil or over organic matter. The fungus grew out a short distance from a colonised organic foodbase, then formed the resting spores which could colonise fresh pieces of organic material. A description of the germination of macroconidia and chlamydospores in the rhizosphere was given by Griffin. From this generalised type many pathogenic forms (ff. sp.) have arisen, and (in some cases) within these different races. An assessment of these ff. sp. was made by Armstrong et al. 1968 (and see Armstrong et al. 1975; Gordon). They designated 69 ff. sp. and 36 races produced by 12 ff. sp. In the field a characteristic symptom produced by a f. sp. is a wilt caused by colonisation of the vascular system via the roots (a tracheomycosis). The typical syndrome was described by Dimond who used *F. oxysporum* f. sp. *lycopersici* (q.v.) as his example. The ff. sp. vary in host specificity within each, some being rather more specific than others. Some of them are still inadequately known and the typical wilt should be clearly demonstrated before describing a new f. sp. In general one f. sp. causes disease in one particular crop but it may be secondarily pathogenic to others (Armstrong et al. 1969); i.e. the concept of primary and secondary hosts. Other hosts are described as non-susceptible carriers, from which the fungus can be isolated but in which it causes no disease (Armstrong et al. 1942, 1948, 1964; Davis 1963, 1966, 1967; Hendrix et al.). Probably the only consistent symptom common to all the wilt diseases caused by *F. oxysporum* is vascular discolouration; this is always moderately prominent although it can vary. Some of the ff. sp. produce the toxin fusaric acid (see Davis 1969, 1970; and under the ff. sp. described separately). Buxton (1959) reviewed mechanisms of variation.

Several factors have led to confusion in the work on the delimitation of both ff. sp. and races. Cultures in vitro exhibit variations in virulence with age, and loss of pathogenicity can be slight to complete. A particular f. sp. may invade a host in which it does not cause a true wilt, and saprophytic forms may colonise primary lesions resulting from infection by a pathogenic one (Nash et al.). Differential hosts for the designation of ff. sp. and races may be not only different vars within a host sp. but different spp., genus or even family. The early and inaccurate concept of highly selective pathogencity in *F. oxysporum* has probably resulted in the establishment of several ff. sp. that are now only recognised as races. Whilst some have several races (f. sp. *pisi* has 5 and f. sp. *vasinfectum* has 6), variability in this respect is apparently low in general. The 2 old and intensively investigated ff. sp. *lycopersici* and *cubense* each have only 2 characterised races. Banana cvs highly resistant to the common race 1 of the latter pathogen, and extensively grown in areas where the race occurs, show no evidence of breakdown in their resistance.

Continuous cropping of a susceptible host leads to a build up in inoculum and the creation of what has been termed a wilt sick soil. Soils have been described as favourable or unfavourable for wilt establishment. Chlamydospores from saprophytic isolates show a higher percentage germination compared with those from ff. sp. in unfavourable or wilt suppressive soils. In soils where wilt has occurred there is little difference, in this respect, between pathogens and saprophytes. Increasing soil nutrients tend to reduce any difference between the 2 groups and bacterial numbers increased more rapidly in wilt suppressive soils (S. N. Smith et al.). Since the pathogenic forms can exist (in very low numbers) almost indefinitely in their soil environment, in the absence of disease-prone hosts, normal rotational cropping is not a practical control measure. Resistance is generally the only form of control.

Fusarium oxysporum

Separate accounts of 20 of the ff. sp. and an end section listing another 22 follow this introduction.

Aist, J. R. et al. 1972. *J. cell. Biol.* **55**: 368 (fine structure & time course of mitosis; **52**, 1036).
Armstrong, G. M. et al. 1942. *Phytopathology* **32**: 685 (cross inoculations with ff. sp.; **22**: 65).
——. 1948. *Ibid* **38**: 808 (non-susceptible hosts as carriers; **28**: 189).
——. 1964. *Ibid* **54**: 1232 (*Lupinus* hosts for ff. sp. from several hosts; **44**, 2086).
——. 1968. *Ibid* **58**: 1242 (descriptive list of ff. sp. & races; **48**, 389).
——. 1969. *Ibid* **59**: 1256 (relationships between ff. sp. & pathogenicity for common hosts; **49**, 422).
——. 1975. *A. Rev. Phytopathol.* **13**: 95 (reflections on wilt fusaria, 30 ref.).
——. 1978. *Phytopathology* **68**: 19 (ff. sp. & races on cucurbits; **57**, 5178).
Buxton, E. W. 1955. *J. Gen. Microbiol.* **13**: 99 (differential tolerance to inhibition by Actinomycetes; **35**: 213).
——. 1959. In *Plant pathology – problems and progress*, 1908–1958: 183. Univ. Wisconsin Press.
Carlile, M. J. 1965. *J. gen. Microbiol.* **14**: 643 (factors affecting non-genetic variation).
Davis, D. 1963. *Phytopathology* **53**: 133 (selective pathogenicity in test tube culture; **42**: 525).
——. 1964. *Ibid* **54**: 290 (growth inhibition on root homogenates; **43**, 2530).
——. 1966. *Ibid* **56**: 825 (cross infection; **45**, 3510).
——. 1967. *Ibid* **57**: 311 (cross protection; **46**, 2210).
——. 1969. *Ibid* **59**: 1391 (pathogenicity & formation of fusaric acid; **49**, 940).
——. 1970. *Ibid* **60**: 111 (carbohydrate specificity for fusaric acid synthesis; **49**, 1919).
Dimond, A. E. 1959. *Trans. N.Y. Acad. Sci.* **21**: 609 (wilt syndrome; **39**: 50).
Garber, E. D. et al. 1961. *Am. J. Bot.* **48**: 325 (heterocaryons; **41**: 286).
Gordon, W. L. 1965. *Can. J. Bot.* **43**: 1309 (list of ff. sp.; **45**, 1700).
Griffin, G. J. 1969. *Phytopathology* **59**: 1214 (germination of chlamydospores & macroconidia; **49**, 609).
Griffiths, D. A. 1973. *Trans. Br. mycol. Soc.* **61**: 1, 7 (fine structure of chlamydospore wall & germination; **52**, 3951, 3952).
Hendrix, F. F. et al. 1958. *Phytopathology* **48**: 224 (other hosts; **37**: 593).
Joffe, A. Z. et al. 1974. *Phytoparasitica* **2**: 91 (in Israel; **54**, 4868).
Komada, H. 1972. *Bull. Tokai-Kinki natn. agric. Exp. Stn* 23: 144 (selective isolation from natural soil; **52**, 1027).
——. 1976. *Ibid* 29: 132 (on vegetable crops; 150 ref.; **56**, 4247).
Moore, H. et al. 1952. *Mycologia* **44**: 523 (physiology in vitro; **32**: 331).
Nash, S. M. et al. 1967. *Phytopathology* **57**: 293 (comparison, pathogens & saprophytes; **46**, 2187).
Olutiola, P. O. 1978. *Trans. Br. mycol. Soc.* **70**: 109 (growth, sporulation, pectic & cellulytic enzymes; **57**, 3799).
Papavizas, G. C. et al. 1966. *Phytopathology* **56**: 1269 (production of polygalacturonate trans-eliminase; **46**, 875).
Park, D. 1959. *Ann. Bot.* **23**: 35 (chlamydospore behaviour in soil).
——. 1961. *Trans. Br. mycol. Soc.* **44**: 119 (isolation from soil; **40**: 592).
——. 1961. *Ibid* **44**: 377 (morphogenesis, fungistasis & cultural staling; **41**: 83).
Robinson, P. M. 1972. *Ibid* **59**: 320 (isolation & characterisation of a branching factor; **52**, 1429).
Smith, S. N. et al. 1972. *Phytopathology* **62**: 273 (chlamydospore germination in soils favourable & unfavourable to wilt; **51**, 3183).
Smith, T. E. et al. 1943. *Ibid* **33**: 469 (pathogenicity of forms from several crops; **22**: 456).
Stevenson, I. L. et al. 1972. *Can. J. Microbiol.* **18**: 997 (fine structure & development of chlamydospores; **52**, 624).
Tuveson, R. W. et al. 1961. *Genetics Princeton* **46**: 485 (nuclear ratios in heterokaryons).
Winstead, N. N. et al. 1954. *Phytopathology* **44**: 153, 159 (enzyme production & toxic metabolites; **33**: 747, 748).

Fusarium oxysporum Schl. f. sp. batatas

(Wollenw.) Synder & Hansen, *Am. J. Bot.* **27**: 66, 1940.

Fusarium batatatis Wollenw., 1914
F. bulbigenum var. *batatas* Wollenw., 1931.

An important pathogen of sweet potato, especially in USA where it has caused heavy losses in the past. It is apparently of less importance in the tropics. Many plants can be infected (e.g. cabbage, cotton, cowpea, maize, okra, potato, sage, snap bean, soybean, tobacco, tomato and watermelon) but no external symptoms are caused except in the case of certain tobaccos. Interveinal yellowing of the leaves is followed by distortion and stunting, and the old leaves fall. There is extensive vascular necrosis which may appear purplish below soil level and the cortex can rupture. Infected tubers decay in storage but the fusaria which cause surface rots are probably different. In USA this sweet potato wilt occurs especially in the northern range of crop production, including the Pacific coast and western states. The disease has also been reported from China, Hawaii, India, Japan, Malawi, New Zealand and Taiwan. Infection occurs mainly through vascular wounds and not through intact tissue or callus of healed wounds. It takes place over a wide soil moisture range with an opt. temp. for infection of *c*. 30°; 32° is reported from Japan. Chlamydospores are formed (opt. 28°) within the central cell of the macroconidia and these occur in the parenchyma. After 11 months' storage about 90% germinated when immersed in water. If the tubers are stored at 13° growth is prevented and decay controlled (see Harter et al.; Nielsen 1969; Nielsen et al. for temp. effects). The pathogen is spread through plant material and by any means through soil. Physiological and in vitro work has been done in Japan. Two races (Armstrong et al.) are known. A complex relationship seems to exist not only between this f. sp. and its hosts (as carriers or in which it can cause wilt) but also with other *F. oxysporum* ff. sp. Besides causing disease in the primary host, race 1 attacks Burley tobacco and race 2 both Burley and flue-cured tobacco, resulting in a wilt.

Control is through the use of certified healthy plant material, fungicides and resistance. Fungicides are used generally as dip treatments of propagating material. They are only partly effective since relative host resistance and field conditions affect them. Resistant vars are now available in USA and the var. Tinian is widely used in breeding. Backcrosses to Tinian using moderately resistant F_1 material have given field-immune progeny. The inheritance of resistance is multifactorial (and see later work by Collins; Collins et al.; Jones, J. L. et al.).

Armstrong, G. M. et al. 1958. *Pl. Dis. Reptr* **42**: 1319 (races; **38**: 421).
Collins, W. W. et al. 1976. *Phytopathology* **66**: 489 (resistance; **56**, 509).
——. 1977. *J. Am Soc. hort. Sci.* **102**: 109 (resistance; **56**, 5309).
Diener, U. L. 1955. *Pl. Dis. Reptr* **39**: 918 (control; **35**: 632).
Fisher, K. D. 1965. *Phytopathology* **55**: 396 (hydrolytic enzyme & toxin formation; **44**, 2866).
French, E. R. et al. 1966. *Ibid* **56**: 1322 (production of macroconidia & chlamydospores in vitro; **46**, 1085).
Harter, L. L. 1914. *Ibid* **4**: 279 (general).
——. et al. 1927. *J. agric. Res.* **34**: 435, 915 (soil temp. & moisture, resistance; **6**: 749, 750).
Hemmi, T. et al. 1933. *Forschn Geb. PflKrankh. Kyoto* **2**: 314 (general; **13**: 123).
Jones, A. 1969. *J. Am. Soc. hort. Sci.* **94**: 207 (inheritance of resistance; **49**, 320).
Jones, J. L. et al. 1972. *Crop Sci.* **12**: 522 (resistance in tobacco; **53**, 274).
McClure, T. T. 1949. *Phytopathology* **39**: 876 (mode of infection; **29**: 285).
——. 1950. *Ibid* **40**: 769 (histology; **30**: 137).
Nielsen, L. W. 1969. *Ibid* **59**: 508 (storage temp., fungicides & resistance; **48**, 2694).
—— et al. 1974. *Ibid* **64**: 967 (postharvest temp. effects on wound healing & surface rot; **54**, 2589).
——. 1977. *Pl. Dis. Reptr* **61**: 1 (control with benomyl & thiabendazole; **56**, 3376).
Poole, R. F. 1924. *Bull. N.J. agric. Exp. Stn* 401 (general: **4**: 118).
Smith, S. N. et al. 1971. *Phytopathology* **61**: 1049 (inoculation density & soil types; **51**, 1030).
Steinbauer, C. E. 1948. *Proc. Am. Soc. hort. Sci.* **52**: 304 (resistance; **28**: 418).
Stuble, F. B. et al. 1966. *Phytopathology* **56**: 1217 (inheritance of resistance; **46**, 1086).
Watanabe, T. 1934. *Bull. Utsunomiya agric. Coll.* Sect. A **1**: 37 (in vitro; **14**: 254).
——. 1941. *Ibid* **3**: 43, 105 (general histology & physiology; **18**: 496; **19**: 191; **28**: 194).

Fusarium oxysporum Schl. f. sp. **carthami** Klisiewicz & Houston, *Phytopathology* 53: 241, 1963.

Safflower (*Carthamus tinctorius*) wilt was described from USA (California) and 6 other *Carthamus* spp. show some susceptibility. In the field plants are killed at early and late growth stages. Chlorosis begins on the lower leaves and spreads upwards; collapse of the host can occur on one side of the plant only. There is the characteristic necrosis in the vascular system; flowering heads can be distorted and seed yield reduced. The pathogen occurs within the seed and plants grown from infected seed seldom survive the seedling stage. Four races have been described. Material resistant to the races is available.

Ghosal, S. et al. 1977. *Phytopathology* **67**: 548 (control by mangiferin; **57**, 241).
Klisiewicz, J. M. et al. 1962. *Pl. Dis. Reptr* **46**: 748 (symptoms; **42**: 138).
—— ——. 1963. *Phytopathology* **53**: 241 (susceptibility of 7 *Carthamus* spp.; **42**: 566).
——. 1963. *Ibid* **53**: 1046 (seed transmission; **43**, 530).
—— ——. 1970. *Ibid* **60**: 83, 1706 (differentiation of races 1, 2 & 3; **49**, 2151; **50**, 1930).

Fusarium oxysporum

Klisiewicz, J. M. 1975. *Pl. Dis. Reptr* **59**: 712 (race 4; **55**, 1901).

Knowles, P. F. et al. 1968. *Crop Sci.* **8**: 636 (resistant cvs; **50**, 3088).

Fusarium oxysporum Schl. f. sp. **cepae** (Hanz). Snyder & Hansen, *Am. J. Bot.* **27**: 66, 1940.

Basal rot of onion (*Allium cepa*) is a relatively minor disease which has been mostly investigated in USA; it has been described from Brazil, Egypt, Hungary, India, Japan, Philippines and South Africa. The first field symptoms are chlorosis and dieback of the leaf tips; these are signs of earlier symptoms below soil level where a decay of the roots and stem plate has occurred. Fungal surface growth takes place at the base of the bulb and a slowly spreading, soft rot follows. The storage rot form of the disease is important and apparently healthy bulbs (when grown in infested soil) can develop basal rot when stored for 3 months at 24–29°. Damping off, in artificially infested soil, increases with temp. from 10–32°. This f. sp. is atypical since it causes a general rot and not the characteristic tracheomycosis.

The best growth in vitro occurs at 28° or a little lower, and most disease at a soil temp. of 28–32°. Onions, in storage, inoculated at 25–30° are killed; at 15° infection is hardly 20% after 30 days. Probably in most cases infection in the field occurs via the roots which are penetrated directly or through wounds. Root invasion leads to infection of the stem plate of the bulb. The population of the fungus in soil has been examined and this decreases in the absence of onion; the pathogen has been detected at a depth of 46 cm. Soils with a long history of basal rot have high populations but some organic soils do not show a high incidence of disease. Holz et al. (1976) attributed isolates from *Oxalis* spp. to this f. sp. These plants were in an onion field with a history of wilt and bulb rot, but in which onions had not been grown for 3 years. Seed transmission has been described. Infection has been investigated with *Pyrenochaeta terrestris* (q.v., Kehr et al.; Woolliams). Host reaction to the two fungi depends on susceptibility to either and resistance to one does not greatly modify susceptibility to the other.

No very high resistance in onion seems to have been found but there are cvs which get significantly less disease; these have been described as tolerant. Tolerance appears to be mostly expressed in the degree of bulb decay that occurs in storage and presumably also in the amount of infection of the bulb that takes place in the field. Spread of the pathogen from the stem plate is apparently resisted in some way by tolerant cvs. After several months in storage differences in the percentage of bulb decay between cvs are found. Of 25 lines tested (Abawi et al. 1971a) Japanese Bunching, Eastern Queen and Beltsville Bunching were highly tolerant. Onions should be correctly cured and stored. Rotation has been recommended and seed treatment with germisan; see also Ashour et al. for fungicides. Kawamoto et al. protected seedlings from damping off by treating seed in suspensions of *Pseudomonas cepacia*.

Abawi, G. S. et al. 1971a. *Pl. Dis. Reptr* **55**: 1000 (reaction of cvs to infection; **51**, 2097).

——— ———. 1971b. *Phytopathology* **61**: 1042, 1164 (pathogen in the soil, pathogenicity in tolerant & resistant cvs; **51**, 950, 2098).

——— ———. 1972. *Ibid* **62**: 870 (behaviour in soil, damping off & temp.; **52**, 1341).

Ashour, W. A. et al. 1973. *Agric. Res. Rev. Cairo* **51**: 153 (fungicides; **53**, 4685).

Holz, G. et al. 1974 & 1976. *Phytophylactica* **6**: 153; **8**: 89 (resistance & on *Oxalis* spp. in South Africa; **54**, 3586; **56**, 1835).

Kawamoto, S. O. et al. 1976. *Pl. Dis. Reptr* **60**: 189 (protection of seedlings with *Pseudomonas cepacia*; **55**, 6044).

Kehr, A. E. et al. 1962. *Euphytica* **11**: 197 (interaction with *Pyrenochaeta terrestris*, resistance & susceptibility; **42**: 67).

Link, G. K. K. et al. 1926. *J. agric. Res.* **33**: 929 (*Fusarium* spp. & onion bulb rot in Philippines; **6**: 268).

Lorbeer, J. W. et al. 1965. *Pl. Dis. Reptr* **49**: 522 (infection & reaction of 14 cvs; **44**, 2670).

Marlatt, R. B. 1958. *Ibid* **42**: 667 (bulb rot in storage & reaction of cvs; **37**: 619).

Palo, M. A. 1928. *Philipp. Agric.* **17**: 301 (general; **8**: 284).

Shalaby, G. I. et al. 1966. *Proc. Am. Soc. hort. Sci.* **89**: 438 (host penetration; **46**, 1809).

Szatala, O. 1964. *Ann. Inst. Prot. Plant. Hung.* **9** (1961–62): 301 (general, temp. & bulb rot in storage; **43**, 3075).

Takakuwa, M. et al. 1977. *Ann. phytopathol. Soc. Japan* **43**, 479 (in Japan, identification; **57**, 3727).

Walker, J. C. et al. 1924. *J. agric. Res.* **28**: 683 (general, effects of temp.; **4**: 202).

Woolliams, G. E. 1966. *Can. Pl. Dis. Surv.* **46**: 101 (host resistance with *P. terrestris*; **46**, 1808).

Fusarium oxysporum Schl. f. sp. **ciceris** (Padw.) Matuo & Sato (as *ciceri*), *Trans. mycol. Soc. Japan* **3**: 125, 1962.

Fusarium lateritium (Nees) Snyder & Hansen forma *ciceri* (Padw.) Erwin.

Wilt of gram or chick pea (*Cicer arietinum*) has been studied in India. There is a characteristic chlorosis, wilt and death of the plant, often beginning at the base, at various growth stages. An early symptom of leaf vein clearing can be seen before wilting begins. The following factors increase disease incidence in India: early sowing (latter part of September), sandy soil, low soil organic matter and possibly low pH. In pot tests most disease occurred at 25°. Organic amendments to the soil reduce incidence. Haware et al. found infection of seed and eradicated it with benomyl. Pathak et al. reported the control of resistance by a single recessive gene.

Chauhan, S. K. 1962. *Agra Univ. J. Res.* **11**: 285 (symptoms; **42**: 644).

——. 1962. *Vijnana Parishad anusandhan Patrika* **5**: 137 (effect of soil texture; **42**: 644).

——. 1962. *J. Indian bot. Soc.* **41**: 220 (effect of pH; **42**: 228).

——. 1962. *Proc. natn. Acad. Sci. India* Sect. B **32**: 385 (as above; **42**: 506).

——. 1962. *Ibid* **32**: 78 (variation between isolates; **42**: 426).

——. 1963. *Ibid* **33**: 552 (effect of soil temp.; **43**, 2776).

——. 1963. *Agra Univ. J. Res.* **12**: 143 (soil organic amendments; **45**, 920).

Gupta, M. N. 1960. *Proc. natn. Acad. Sci. India* Sect. B **30**: 365 (secretion of propectinase).

——. 1962. *Proc. Indian Acad. Sci.* Sect. B. **55**: 120 (as above; **41**: 638).

—— et al. 1966. *Mycopathol. Mycol. appl.* **29**: 193 (pectolytic enzymes; **45**, 3057).

—— ——. 1967. *Proc. natn. Acad. Sci. India* Sect. B **37**: 264 (cellulytic enzymes; **48**, 3757).

Haware, M. P. et al. 1978. *Phytopathology* **68**: 1364 (eradication from seed).

Misra, A. N. 1976. *Indian J. Mycol. Pl. Pathol.* **6**: 152 (resistance; **57**, 1922).

Padwick, G. W. 1940. *Indian J. agric. Sci.* **10**: 241 (morphology & cultural characteristics; **20**: 82).

—— et al. 1943. *Ibid* **13**: 289 (wilt & date of sowing; **23**: 325).

Pathak, M. M. et al. 1975. *Indian J. Fm Sci.* **3**: 10 (inheritance of resistance; **56**, 936).

Prasad, N. et al. 1939. *Indian J. agric. Sci.* **9**: 371 (general; **18**: 780).

Fusarium oxysporum Schl. f. sp. **conglutinans** (Wollenw.) Snyder & Hansen, *Am. J. Bot.* **27**: 66, 1940.

Fusarium conglutinans Wollenw., 1913

F. oxysporum var. *orthoceras* (Appel & Wollenw.) Bilay, 1955.

Yellows of cruciferous crops is caused by this f. sp. The particular plant hosts on which serious disease occurs depends on the race causing infection; Walker (*Plant Pathology*) and Subramanian have given accounts of the disease. Yellows may occur in any area; its known distribution is: America (Brazil, Canada, Central America, Cuba, Puerto Rico, Trinidad, USA); parts of Europe and USSR; Africa (Cameroon, Morocco, Zimbabwe, South Africa, Zaire); Asia (China, India, Iraq, Japan, Philippines, Thailand, Vietnam); Australasia (Australia, New Caledonia, New Zealand, Samoa; Map 54). The first symptom is a yellowing in the lower leaves; this then spreads upwards. The chlorosis is often more intense on one side of the leaf or plant and results in curling and distortion. The leaves become necrotic, brittle, fall prematurely and the plants are stunted. In section the vascular region shows as a dark, yellow brown area; it does not normally become black. There is no characteristic wilt.

Penetration is through the hypocotyl, root tip and at or near the zone of elongation; it is infrequent through root hairs. The characteristic spread in the vascular region takes place and mycelium occurs in the leaves, stem and apical meristem; no seed transmission appears to have been described. Spread and survival are as for other ff. sp. of *F. oxysporum*. Most disease occurs at relatively high temps (26–30°) and the most growth in vitro is at 26–28°. Walker (l.c.) stated that in general the disease temp. curve rises and falls roughly parallel with the growth curve of the pathogen, indicating that the effect of soil temp. on disease development is expressed mainly through its effect on the fungus. Little or no disease occurs at 17–18° unless the cv. is extremely susceptible. At temps very favourable for the disease polygenic resistance can be eroded. Proteins and oxidative enzymes in resistant and susceptible cvs have been described, and Heitefuss et al. considered that the low depolymerase activity and degradation of pectate to galacturonic acid may be factors in the absence of wilt in yellows. Fusaric acid was detected in vitro but not in vivo. Reyes et al. found that yellows symptoms were more severe when seedlings of *Brassica chinensis* (Chinese cabbage) were infected simultaneously with turnip mosaic virus and the fungus. Plants infected with the virus became more stunted when inoculated with the fungus. This interaction showed a greater effect at 28° compared with 14, 21 and 35°.

The disease has been most studied on cabbage (*B. oleracea* var. *capitata*) and Armstrong et al. (1966, 1974) distinguished 4 races. Races 1 and 2 (f. sp. *raphani*) are mainly separated by pathogenicity to cabbage and radish (*Raphanus sativus*), respectively. Both races infect other *B. oleracea* vars and also rape (*B. napus*), rutabaga (*B. napobrassica*), turnip (*B. rapa*), white mustard (*B. alba*) and stock (*Matthiola* spp.), but race 2 does not attack cabbage, cauliflower and Brussels sprouts. *Crambe* spp. were resistant to race 1 from cabbage and susceptible to race 2 from radish. Races 3 and 4 are separated on different cvs of stock (*M. incana*) and are of little or no importance on the vegetable hosts, some of which they can infect.

Control is through resistance; both polygenic and oligogenic (dominant) resistance have been used in breeding, often referred to as type B and A, respectively. The expression of polygenic resistance is affected by temp. and host nutrients. In hosts with this resistance the fungus penetrates to the xylem but grows there only sparsely and Peterson et al. found no spores in the vascular tissue of a radish cv. (Red Prince) with polygenic resistance. Initial penetration occurs in cvs with oligogenic resistance but fungal spread stops before the xylem is reached. Chupp and Sherf (*Vegetable diseases and their control*) list 23 resistant cvs, with resistance type and days to maturity.

Anderson, M. E. 1933. *J. agric. Res.* **47**: 639 (inheritance of & effect of temp. on polygenic resistance; **13**: 343).
—— et al. 1935. *Ibid* **50**: 823 (penetration & histology in resistant & susceptible cvs; **14**: 732).
Armstrong, G. M. et al. 1952. *Phytopathology* **42**: 255 (races of *Fusarium* causing wilt in Cruciferae; **32**: 54).
—— ——. 1966. *Ibid* **56**: 525 (characteristics of 4 races; **45**, 3109).
—— ——. 1974. *Pl. Dis. Reptr* **58**: 479 (wilt caused by races 1 & 2; **54**, 263).
Baker, K. F. 1948. *Phytopathology* **38**: 399 (on *Matthiola incana*; **27**: 476).
Blank, L. M. 1932. *Ibid* **22**: 191 (pathogenicity at low soil temps; **11**: 489).
—— et al. 1933. *J. agric. Res.* **46**: 1015 (inheritance of oligogenic resistance; **13**: 2).
——. 1934. *Ibid* **48**: 401 (comparisons between isolates; **13**: 557).
——. 1937. *Ibid* **55**: 497 (inheritance of oligogenic & polygenic resistance in one cv.; **17**: 218).
Gilman, J. C. 1916. *Ann. Mo. bot. Gdn* **3**: 25 (disease & temp.).
Heitefuss, R. et al. 1960. *Phytopathology* **50**: 198, 367, 370 (proteins in infected host, pectolytic enzymes & fusaric acid, oxidative enzymes in infected host; **39**: 514, 751).
Kendrick, J. B. 1930. *Hilgardia* **5**: 1 (general; **10**: 72).
—— et al. 1942. *Phytopathology* **32**: 1031 (race on radish; **22**: 160).
Melhus, I. E. et al. 1926. *Bull. Iowa agric. Exp. Stn* 235: 187 (general; **6**: 137).
Peterson, J. L. et al. 1960. *Phytopathology* **50**: 807 (penetration & spread in susceptible & resistant radish cvs; **40**: 327).
Pound, G. S. et al. 1953. *Ibid* **43**: 277 (inoculation, temp. & races 1 & 2, selection of radish resistant lines; **33**: 130).
Reyes, A. A. et al. 1972. *Ibid* **62**: 1424 (interaction with turnip mosaic virus; **52**, 3106).
Smith, R. et al. 1930. *J. agric. Res.* **41**: 17 (penetration & spread in resistant & susceptible cvs; **10**: 4).
Subramanian, C. V. 1970. *C.M.I. Descr. pathog. Fungi Bact.* 213 (*F. oxysporum* f. sp. *conglutinans*).
Tims, E. C. 1926. *J. agric. Res.* **33**: 971 (effects of soil temp. & moisture on disease; **6**: 265).
Tisdale, W. B. 1923. *Ibid* **24**: 55 (as above).
Walker, J. C. et al. 1928. *Ibid* **37**: 233 (survey of resistance in *Brassica oleracea*; **8**: 147).
—— ——. 1930. *Ibid* **41**: (effect of temp. on diseases in susceptible & resistant hosts; **10**: 4).
——. 1930. *Ibid* **40**: 721 (inheritance of oligogenic resistance; **9**: 694).
——. 1933. *Ibid* **46**: 639 (lines with oligogenic resistance; **12**: 608).
—— et al. 1945. *Am. J. Bot.* **32**: 314 (host nutrition & disease development; **24**: 484).
Winstead, N. N. et al. 1954. *Phytopathology* **44**: 159 (effect of culture filtrates on host, with *F. oxysporum* f. sp. *vasinfectum*; **33**: 748).

Fusarium oxysporum Schl. f. sp. **coriandrii** Narula & Joshi, *Sci. Cult.* **29**: 206, 1963.

Coriander (*Coriandrum sativum*) wilt has been described from India where severe losses have occurred in Madhya Pradesh and Rajasthan. The fungus causes pre- and post-emergence damping off, a wilt at any plant age or partial wilt with stunting; there is vascular browning and host sterility. Growth in vitro is good at 19–20° and severe disease occurs at 24–27°. Max. wilt takes place at 28°

and all cvs showed less disease at 20°. One cv. showed tolerance at 28° but none was resistant at this temp. The pathogen appears not to be seedborne. In Rajasthan more disease occurs in irrigated than in unirrigated fields.

Mall, O. P. 1968. *Indian Phytopathol.* **21**: 379 (general; **48**, 3107).
Srivastava, U.S. 1969. *Ibid* **22**: 406 (inoculum potential; **49**, 2953).
—— et al. 1971. *Indian J. agric. Sci.* **41**: 779 (effect of organic amendments; **51**, 3480).
——. 1971. *Indian Phytopathol.* **24**: 679 (edaphic factors & temp.; **52**, 1626).
——. 1972. *Indian J. agric. Sci.* **42**: 618 (factors, including temp., affecting wilt; **52**, 3399).

Fusarium oxysporum Schl. f. sp. **cubense** (E. F. Smith) Snyder & Hansen, *Am. J. Bot.* **27**: 66, 1940.

Fusarium cubense E. F. Smith, 1910
F. oxysporum Schl. var. *cubense* (E. F. Smith) Wollenw., 1935
F. oxysporum Schl. f. 3 Reink. & Wollenw., 1927
F. cubense E. F. Smith var. *inodoratum* Brandes, 1919
F. oxysporum Schl. f. 4 Wollenw., 1931.

Panama wilt of bananas (also called Panama disease, banana or fusarial wilt) is one of the devastating plant diseases; not only has it caused enormous losses but also has had disruptive social and political effects (see Carefoot and Sprott, *Famine on the Wind*). Nevertheless, as N. W. Simmonds (*Bananas*) has pointed out, this disease is essentially one of a single cv. (Gros Michel). Where banana export trades were based on it, as in Central America and Jamaica, Panama wilt became almost catastrophic; where they were not (Africa, Australia, Oceania and parts of the West Indies) the disease was of little or no importance. Only a brief account of this extensively studied disease is given here; reference should be made to the monograph by Stover (1962b), a short account by the same author (*Banana, plantain and abaca diseases*) and one by Wardlaw (*Banana diseases*). Summaries have also been given by Meredith (1970), Simmonds (l.c.) and Subramanian. The major exporting industries of Costa Rica, Guatemala, Honduras, Jamaica, Nicaragua and Panama (where max. production was variously reached between 1908 and 1956) collapsed, as far as the production of Gros Michel was concerned, between 1935 and 1959. Over 50 years some 40 000 hectares of bananas were destroyed or abandoned. The disease forced a change to the resistant (but commercially less desirable) cvs of the Cavendish group from 1945 in Jamaica and from 1955 in Central America.

F. oxysporum f. sp. *cubense* is widespread, and wherever susceptible bananas are cultivated the disease it causes is likely to occur (Map 31). Where it is not reported on such bananas it is unlikely that they have been cultivated for long or on a large scale. But large areas of Gros Michel are grown in N.W. Colombia (Santa Marta) and Ecuador. The disease may originate from corms imported from infested areas or by genetic change in widespread saprophytic forms. Outside *Musa*, *Heliconia* is the only systemicly infected host, and *H. caribaea* the only infected wild host (Waite 1963). *M. textilis* (abaca or Manila hemp) is susceptible. Although the pathogen has been recovered from the roots of several weeds its existence on them is largely saprophytic.

The first external symptoms are a chlorosis of the older leaves (at the junction of petiole with pseudostem) or their collapse while still green. In the first case chlorosis begins at the base of the petiole of 1 or 2 of the older leaves; it progresses inwards to the younger ones until only the youngest, partly unfurled leaf remains green. The emerging heart leaf commonly shows necrosis and the base of the pseudostem may split. All the leaves eventually collapse, leaving the pseudostem standing. Internally, the vascular discolouration (pale yellow, becoming dark reddish or almost black) is seen in the outer leaf sheaths, throughout the pseudostem, fruit bunch stalk (true stem) and rhizome where it is most conspicuous at the junction of stele with cortex. Almost the whole rhizome may be invaded. Suckers <1.5 m tall do not usually show external symptoms but they may be infected where the disease is spreading rapidly. There is no vascular discolouration of the fruit; fruit symptoms (premature yellowing of fingers and an internal, firm, dry, brown rot) are characteristic of moko wilt (*Pseudomonas solanacearum*) with which Panama wilt should not be confused. In abaca the internal symptoms are similar although externally they are less conspicuous; leaves dry up, become bunched and stunted. The fibre of decorticated pseudostems is yellow brown or reddish and of poor quality.

Initial penetration by the fungus is through secondary or tertiary roots; intact, healthy main roots are not directly attacked and direct invasion of the unwounded rhizome or pseudostem does not occur. In the advanced stages of infection mycelium spreads from the vascular tissue into the parenchyma. There is abundant sporulation in the host and conidia move in the transpiration steam. Beckman and others have described detailed investigations on the host's vascular response to invasion; the gel formation and vessel occlusion that takes place is disrupted in the susceptible host. Above ground symptoms in field plants are seen *c*. 2 months after initial infection. Nienhaus et al. described a heart rot of banana in the Lebanon caused by *F. oxysporum* which spread in the pseudostem via wounds.

Stover's monograph should be consulted for the very full analysis of the behaviour of *F. oxysporum* f. sp. *cubense* in the soil. Like other forms of the sp. there is a very long survival, although at extremely low levels, in the absence of disease. Large populations decline rapidly in disease free, natural soils and are not usually detectable after 6–8 months. Chlamydosores (abundant in diseased roots) are survival units. The pathogen fits more closely the criteria of a root inhabiting rather than a soil inhabiting fungus; except that all the evidence suggests that saprophytic adaptation to almost indefinite survival occurs through the colonisation of senescent or decaying non-host roots. Inoculum build up in the soil, and the rate of soil and water plant to plant spread, is affected (apart from host susceptibility) by the area of cultivation, climate and soil factors. There is also an indication that attack by *Radopholus similis* may increase wilt incidence or at least reduce the incubation time. In Honduras, over 14 years and on fertile irrigated land, the disease was more evident during the period of most rainfall, highest av. min. temps and max. plant growth rate. The soil characteristics which favour disease build up are: acidity (pH 6 and below), sandy and sandy loams, poor drainage, compaction, impervious subsoils and low fertility. Soils have been divided into those of long life (10–20 years before their abandonment) and short life (abandonment within 10 years). The cessation of Gros Michel cultivation was put off in Central America by the short sighted policy of continual shifting to new land, usually primary forest.

The attempts at control by methods other than host resistance led to the long investigations on the effects of soil factors on pathogen, host (susceptible and resistant) and their interaction. Eventually none of these methods proved effective and, apart from quarantine measures in particular cases, resistance (in the Cavendish bananas) is the only control. Banana cvs in world trade belong to the Eumusa series and are derived from the 2 wild spp., *M. acuminata* (genome A) and *M. balbisiana* (B). Two groups of cvs are derived solely from the former and the remaining 4 are of hybrid origin. *M. balbisiana* is immune and *M. acuminata* mostly so. Most banana cvs are resistant or highly resistant. The only important edible diploid (AA) group (Sucrier) is highly resistant. Gros Michel is triploid (AAA) and the other cvs in this genome are the Cavendish subgroup whose resistance is sufficient to reduce Panama wilt to negligible proportions. Simmonds (l.c.) listed the common cv. names in the 4 main divisions of the Cavendish bananas: Dwarf Cavendish, Giant Cavendish, Robusta and Pisang Masak Hijau. High susceptibility also occurs in the Silk cvs (AAB). Silk and Gros Michel are very susceptible to the common race (1) but resistant to race 2 which attacks Bluggoe cvs (ABB); Bluggoe is not susceptible to race 1. Abaca is susceptible to race 1 only and has some resistant cvs. Vakili found differential resistance to races 1 and 2 in seedlings and subspp. of *M. acuminata*, and indicated the presence of a single dominant factor for resistance in the edible diploid cv. Lidi (and see Waite 1977). There is no evidence that resistance in Cavendish may break down. Where such resistance may be eroded the cause is likely to be adverse growing conditions, see Stover et al. (1972) who found no evidence for a new race. Shepherd gave an account of banana breeding in which resistance to *F. oxysporum* f. sp. *cubense* is one consideration (see Stover 1962b for earlier ref.).

Beckman, C. H. et al. 1961. *Phytopathology* **51**: 507 (physical barriers in vascular tissue; **41**: 241).
——. 1962. *Ibid* **52**: 134, 893 (vessel occlusion in susceptible & resistant hosts; **41**: 665; **42**: 274).
——. 1967. *Ibid* **57**: 11 (origin & composition of vascular gel in infected *Musa*; **46**, 1675).
——. 1969. *Ibid* **59**: 837 (mechanics of gel formation by simulated plant cell wall membranes & perforation plates of vessels; **48**, 3590).
——. 1969. *Ibid* **59**: 1477 (plasticising of wall & gel induction in vessels; **49**, 1098).
——. 1970. *Ibid* **60**: 79 (distribution of phenols in roots of *Musa*; **49**, 2134).
——. 1974. *Ibid* **64**: 1214 (stabilisation of artificial & cell wall membranes by phenolic infusion & its relation to resistance; **54**, 2907).

Buxton, E. W. 1962. *Trans. Br. mycol. Soc.* **45**: 274 (parasexual recombination; **42**: 36).
——. 1962. *Ann. appl. Biol.* **50**: 269 (effect of root exudates from susceptible & resistant hosts on the pathogen; **42**: 36).
Castillo, B. S. et al. 1940. *Philipp. Agric.* **29**: 65 (infection of abaca; **19**: 707).
Deese, D. C. et al. 1962. *Phytopathology* **52**: 247 (formation of cellulytic enzymes in susceptible & resistant host tissue; **41**: 731).
Hansford, C. G. 1923. *Circ. Dep. Agric. Jamaica* 1, 28 pp. (general in Jamaica; **3**: 406).
——. 1926. *Ibid* 5, 35 pp. (as above; **6**: 42).
Leoncio, J. B. 1930. *Philipp. Agric.* **19**: 27 (on abaca; **9**: 785).
Meredith, D. S. 1970. *Rev. Pl. Pathol.* **49**: 539.
Nienhaus, F. et al. 1968. *Z. PflKrankh. PflPath. PflSchutz* **75**: 449 (aetiology of heart rot of banana in the Lebanon; **48**, 3591).
Page, O. T. 1961. *Can. J. Bot.* **39**: 545, 1509 (variation & production of polygalacturonase; **40**: 760; **41**: 213).
Ramakrishnan, T. S. et al. 1956. *Proc. Indian Acad. Sci.* B **43**: 213 (in India, reaction of cvs; **35**: 690).
Reinking, O. A. 1934. *Philipp. J. Sci.* **53**: 229 (distribution in Asia; **13**: 586).
Shepherd, K. 1968. *PANS* **14**: 370 (banana breeding in the West Indies; **48**, 536).
Stover, R. H. et al. 1961. *Can. J. Bot.* **39**: 197 (attempted cultural & chemical control; **40**: 480).
——. 1962a. *Ibid* **40**: 1467, 1473 (cultural variation & competitive saprophytic ability; **42**: 333).
——. 1962b. *Phytopathol. Pap.* 4, 117 pp. (monograph, 334 ref.; **41**: 610).
—— ——. 1972. *Pl. Dis. Reptr* **56**: 1000 (wilt in resistant dwarf cavendish; **52**, 4157).
Subramanian, C. V. 1970. *C.M.I. Descr. pathog. Fungi Bact.* 214 (*Fusarium oxysporum* f. sp. *cubense*).
Trujillo, E. E. 1963. *Phytopathology* **53**: 162 (histology, spread in xylem; **42**: 623).
—— et al. 1963. *Ibid* **53**: 167 (distribution in soil; **42**: 623).
Umali, D. L. et al. 1956. *Philipp. Agric.* **40**: 115 (reaction of abaca seedlings; **36**: 320).
Vakili, N. G. 1965. *Phytopathology* **55**: 135 (resistance to races 1 & 2 in *M. acuminata* seedlings & resistance inheritance in cv. Lidi; **44**, 1911).
Waite, B. H. 1954. *Pl. Dis. Reptr* **38**: 575 (on abaca in Central America; **34**: 151).
—— et al. 1960. *Can. J. Bot.* **38**: 985 (morphological & cultural characteristics; **40**: 318).
——. 1963. *Trop. Agric. Trin.* **40**: 299 (wilt of *Heliconia* spp., race 3; **43**, 522).
——. 1977. *Pl. Dis. Reptr* **61**: 15 (inoculation & natural infection with races 1 & 2; **56**, 3136).
Wallace, G. B. 1952. *E. Afr. agric. J.* **17**: 166 (in Tanzania; **32**: 324).
Ward, F. S. 1930. *Scient. Ser. Dep. Agric. Straits Settl. F.M.S.* 2, 26 pp. (in Malaysia; **10**: 254).
Zaroogian, G. E. et al. 1968. *Phytopathology* **58**: 733 (cell wall components in susceptible & resistant hosts; **47**, 3166).

Fusarium oxysporum Schl. f. sp. **cucumerinum** Owen, *Phytopathology* **46**: 156, 1956.

This f. sp. causes a fairly widespread wilt of cucumber (*Cucumis sativus*). Pre- and post-emergence damping off occur. Mature plants with runners, in USA (Florida), first show a wilt in a single crown branch and this is followed by a collapse of the whole plant. Vascular necrosis may extend into the vine for 6–8 nodes. A cortical decay, absent from older plants, occurs in seedlings. In England wilt symptoms have been described as beginning in the lower leaves, and the vascular system of the lower nodes becomes prominent, standing out as white lines. Isolates of *F. oxysporum* from melon (*C. melo*) may cause damping off in cucumber. The cucumber isolates do not cause damping off in melon, marrow (*Cucurbita pepo*) or pumpkin and squash (*C. maxima* and *C. mixta*). Mature cucumber is not attacked by isolates from these other hosts (Owen). In Florida wilt increased with successive field crops resulting in a 70% infection rate in the third consecutive season. Cucumber wilt may not be serious in the lowland tropics since relatively low temps are said to favour the disease (Van Koot). In Japan most infection occurred at a soil pH of 5.68–6.55 (Yoneyama 1954a). Armstrong et al. designated isolates from USA (Florida), Israel and Japan as races 1, 2 and 3, respectively.

In England (Fletcher et al.) 17 cvs were susceptible and in Japan pickling cvs were resistant but others were susceptible. Netzer et al. reported a dominant gene which conferred resistance. *C. ficifolia* has been used as a resistant stock. Soil fumigation is effective and Costache applied benomyl, carbendazim and thiophanate methyl around the stem bases.

Akai, S. et al. 1969. *Ann. phytopathol. Soc. Japan* **35**: 351 (fine struction of conidium; **49**, 2715).
Armstrong, G. M. et al. 1978. *Pl. Dis. Reptr* **62**: 824 (races).
Costache, M. 1972. *An. Inst. Cerc. agric. Prot. Pl.* **10**: 161 (fungicides in glasshouse; **54**, 4241).

Fusarium oxysporum

Fletcher, J. T. et al. 1966. *Pl. Pathol.* **15**: 85 (general; **45**, 3013).

Kobayashi, N. et al. 1969. *Forschn Geb. PflKrankh. Kyoto* **7**: 81 (fine structure of germinating conidium; **50**, 3352).

Komada, H. et al. 1974. *Bull. Veg. Ornament. Cr. Res. Stn* A 1: 233 (resistance; **55**, 3789).

Kosswig, W. 1955. Paul Parey, 148 pp. (*Fusarium* on cucumber; **35**: 265).

Netzer, D. et al. 1977. *Phytopathology* **67**: 525 (resistance; **56**, 5867).

Owen, J. H. 1955. *Ibid* **45**: 435 (general; **35**: 266).

Van Koot, Y. 1946. *Meded. TuinbVoorlDienst.* 42, 85 pp. (general in the Netherlands).

Yoneyama, S. 1974a & b. *Bull. Ibaraki-ken hort. Exp. Stn* 5: 67, 77 (resistance, histology; *Hort. Abstr.* **46**, 1131, 1132).

Fusarium oxysporum Schl. f. sp. **cumini** Patel, Prasad, R. L. Mathur & B. L. Mathur, *Curr. Sci.* **26**: 181, 1957.

Alavi reported field losses of 20–80% and stated that cumin (*Cuminum cyminum*) wilt occurs in Afghanistan, India, Iran, Pakistan, Syria and Turkey. The symptoms are a typical wilt with light brown vascular discolouration; wilt can be rapid without preliminary leaf chlorosis and often occurs about a month after sowing. Inoculation of other hosts has been unsuccessful. In vitro growth is best at 26–30°. In India (Rajasthan) wilt is most severe in the winter (13–19°) and in lighter soils. A summer fallow decreases disease incidence but no beneficial effects have been noted from mixed cropping, fertiliser treatment or green organic matter. No resistant cvs have been found. Singh et al., who found the fungus within the seed, reported that hot water treatment (54° for 15 minutes) was effective.

Alavi, A. 1969. *Iran. Jnl Pl. Pathol.* **5**: 31 (in Iran, **49**, 2952).

Joshi, N. C. et al. 1958. *Lloydia* **21**: 29 (symptoms, morphology & culture; **38**: 220).

Mathur, B. L. et al. 1964. *Indian J. agric. Sci.* **34**: 131, 273 (variation between isolates & inoculation of other plants; **44**, 1191, 2848).

——. 1964. *Indian Phytopathol.* **17**: 121 (effect of vitamins on fungal growth; **44**, 1192).

——. 1965. *Ibid* **18**: 335, 379, 381 (toxicity of culture filtrates & effects of chemicals on in vitro growth; **45**, 2198).

——. 1966 & 1967. *Proc. natn. Acad. Sci. India* Sect. B **36**: 33; **37**: 161 (factors affecting disease incidence, culture & morphology, pathogenicity to cvs; **46**, 1306; **48**, 3102).

——. 1967. *Indian Phytopathol.* **20**: 32, 42 (effects of cultural practices, toxicity of culture filtrates; **47**, 866, 867).

Patel, P. N. et al. 1963. *Pl. Dis. Reptr* **47**: 528 (symptoms, culture & morphology, growth in vitro; **43**, 179).

Singh, R. D. et al. 1972. *Phytopathol. Mediterranea* **11**: 19 (within seed & control; **52**, 1223).

Fusarium oxysporum Schl. f. sp. **elaeidis** Toovey, *Rep. Dep. Agric. Niger.* for 1948: 56, 1949.

This f. sp. causes an oil palm wilt of possibly great potential importance but which is still limited in its distribution: Cameroon, Ivory Coast, Nigeria and Zaire (Map 471; a record for Colombia is considered to need confirmation). In young palms the first symptoms are an extensive chlorosis in some of the central leaves (4–15th). This is followed by necrosis (giving a flat-topped appearance) and death within 1 year. In mature palms infection can cause a leaf necrosis and rapid breaking of the rachis. This acute form of the disease causes death in a few months. In the chronic form the progressive dying of the crown inwards may be very slow, with new (though smaller) leaves being produced. The vascular tissue in roots and stems becomes orange, darkening progressively to black. It is very characteristic of the disease that the internal necrosis is restricted to the xylem region. *Elaeis madagascariensis* and *E. melanococca* can be affected. The patch yellows disease of this palm is probably due to a different form of *F. oxysporum* (Gogoi 1949).

Locke et al. (1977) found that the fungus can infect seedlings when the roots have not been damaged. Renard considered the role of root injury in infection. Locke et al. (1973) isolated the pathogen from water used to wash oil palm seeds. Prendergast (1957) investigated the effects of fertilisers and soil on disease incidence. Replanting in diseased areas leads to an early recurrence of the disease and more wilt in young palms. Less vigorous trees tend to become diseased more frequently and wilt may increase in incidence between the ninth and twelfth years. High applications of potash may reduce incidence (Ollagnier et al.).

Bachy, A. 1970. *Oléagineux* **25**: 265 (review; **49**, 3432).

Fraselle, J. V. 1951. *Trans. Br. mycol. Soc.* **34**: 492 (pathogenicity; **31**: 282).

Gogoi, T. 1949. *Ibid* **32**: 171 (in vitro & patch yellows; **29**: 362).
——. 1950. *Ibid* **33**: 121 (in vitro & other fusaria; **29**: 507).
Guldentops, R. E. 1962. *Parasitica* **18**: 244 (review; **42**, 567).
Kovachich, W. G. 1948. *Ann. Bot.* **12**: 327 (histology; **27**: 523).
Locke, T. et al. 1973. *Trans. Br. mycol. Soc.* **60**: 594 (on seed; **52**, 4182).
—— ——. 1974. *Phytopathol. Z.* **79**: 77 (resistance testing method; **53**, 4552).
—— ——. 1977. *Ibid.* **88**: 18 (infection; **56**, 4152).
Ollagnier, M. et al. 1976. *Oléagineux* **31**: 203 (effect of K on disease; **55**, 5315).
Park, D. 1958. *Ann. Bot.* **22**: 19 (saprophytic status: **37**: 367).
Prendergast, A. G. 1957. *J. W. Afr. Inst. Oil Palm Res.* **2**: 148 (symptoms, hosts & soil conditions; **37**: 52).
——. 1963. *Ibid* **4**: 156 (inoculation method & resistance; **43**, 2382).
Renard, J. L. 1970. *Oléagineux* **25**: 581 (infection & root injury; **50**, 1348).
Trique, B. 1971. *Ibid* **26**: 163, 563 (pectinases & fusaric acid; **50**, 2418; **51**, 1780).
Wardlaw, C. W. 1950. *Trop. Agric. Trin.*, **27**: 42 (general; **30**: 156).

Fusarium oxysporum Schl. f. sp. **fabae** Yu & Fang, *Phytopathology* **38**: 587, 1948.

This wilt of broad or horse bean (*Vicia faba*) has been investigated in Canada, China, Egypt and Japan. The entire plant yellows and wilts, often at or near flowering. The upper leaves appear somewhat abnormally rigid. The reddish brown vascular discolouration may extend into the stem for a considerable distance. The pathogen has been described with *F. solani* f. sp. *fabae* which causes a cortical rot of the roots and both spp. can cause damping off. Growth in vitro is opt. at 24–26°. In Canada growing various vegetables for 4 years reduced incidence to nil (where broad bean had been grown for 5–6 years there were 66–68% diseased plants); soil fumigation with vapam or mylone reduced wilt to *c.* 2% but seed disinfection was ineffective. However in Egypt a hot water (60° for 10 minutes) treatment of seed reduced post-emergence losses for 33 days in infested soil.

Coulombe, L. J. 1961. *Can. Pl. Dis. Surv.* **41**: 191 (control, rotation, soil fumigation & seed treatment, **41**: 496).
Elarosi, H. et al. 1971. *Phytopathol. Mediterranea* **10**: 94 (hot water seed treatment; **51**, 890).
Ibrahim, I. A. et al. 1964. *Alex. J. agric. Res.* **12**: 221 (*F. oxysporum* & *F. solani* from *Lupinus* & potential danger to broad bean; **45**, 1420).
—— ——. 1965. *Ibid* **13**: 415 (with *F. solani* in Egypt; **46**, 3282).
Yamamoto, W. et al. 1955. *Sci. Rep. Hyogu Univ. Agric.* **2**: 53 (*F. oxysporum* & *F. solani*; **37**: 129).
Yu, T. F. 1944. *Phytopathology* **34**: 385 (general; **23**: 422).
—— et al. 1948. *Ibid* **38**: 331, 587 (general with *F. solani* f. sp. *fabae*; **27**: 550; **28**: 107).

Fusarium oxysporum Schl. f. sp. **lycopersici** (Sacc.) Snyder & Hansen, *Am. J. Bot.* **27**: 66, 1940.
F. oxysporum Schl. ssp. *lycopersici* Sacc., 1886
F. lycopersici Bruschi, 1912
F. bulbigenum var. *lycopersici* (Bruschi) Wollenw. & Reink., 1935.

This widespread wilt of tomato is one of the most studied of all the tracheomycotic or vascular plant diseases. There is a monograph by Walker and only some of the more recent references, including those published since his full account, are given here. Gäumann reviewed much of the work on fusaric acid and lycomarasmin; and see Wood, *Physiological plant pathology*. Subramanian gave a summary of the disease. On young, inoculated plants the first symptoms (within 24 hours) are a clearing of the ultimate leaf veins, giving a net-like appearance (Foster); after 72 hours the petioles begin to droop. Epinasty is also an early sign of wilt. In the field the first signs of disease may be a yellowing of the lower leaves; these die and symptoms appear on successively younger leaves. Leaflets on only one side of the petiole may be affected initially. Vascular browning is seen early in a TS of stem or petiole. Like other ff. sp. the fungus is soil and waterborne, and colonises the roots of weeds and cultivated plants. This latter phenomenon has recently been investigated by Katan who quantitatively estimated the amount of the pathogen in inoculated, non-disease prone hosts at 1–4% of that in susceptible tomato. Seed transmission can occur (e.g. Elliot et al.) and its present significance is as a factor in the spread of any new races. Besri reported a viability of the fungus in years in dried pulp fragments on the seed surface.

The environment affects development of the disease, both in susceptible cvs and in those with polygenic resistance. A temp. of 28–29° is opt. for fungal growth in vitro and the disease. No wilt occurs in infested soil at temps above 30° or below 20°, even if the air temp. is opt. If the soil temp. is opt., but that of the air too cool, root infection and spread to the stem takes place but no external symptoms appear and the plant continues to grow. Plants which are predisposed at the opt. temp. before inoculation get more disease than those predisposed at higher or lower temps. Disease severity is reduced with increasing nutrient concentration, and with high potassium and low nitrogen. Short day (6 hour/day) plants show more disease than long day (18) ones. Plants of low vigour grown in dry soil are predisposed to more disease than those of low vigour in very wet soil or of high vigour at opt. soil moisture levels. There are extensive studies on the behaviour of isolates which vary in virulence, and on their nutrient requirements in vitro. A foot and root rot disease of tomato caused apparently by a different pathotype of *F. oxysporum* has been described in Canada, Japan and USA (Jarvis; Jarvis et al.; Rowe et al.; Sato et al.; Sonoda). The symptoms are unlike those caused by the f. sp. and plant collapse occurs after fruit set. The roots become necrotic and there may be some discolouration of the vascular tissue. Isolates attack cvs which are resistant to races 1 and 2 of the f. sp. Failure to control this root rot in glasshouses in Ohio and Ontario by steam was apparently due to recontamination of the soil by air-dispersed microconidia from infested plant debris.

The intensive study of the cause of wilt in this disease, using tomato as a readily manipulated host tool, has arisen from the 2 conflicting theories of toxins and xylem blockage. Gäumann and his coworkers made the first attempt to characterise the apparently non-specific toxic principles which had been demonstrated in culture filtrates. One was the new compound lycomarasmin, a polypeptide; the other (fusaric acid) was already known and is also produced by the cotton wilt pathogen, *F. oxysporum* f. sp. *vasinfectum* (q.v.), and other ff. sp. These toxins cause wilt in tomato cuttings but do not produce the complete syndrome. Walker stated in 1971 that the relations between toxins and cell degrading enzymes in the disease are far from settled. But he concluded that wilt is caused by a functional failure in the xylem, induced in part by extracellular pectolytic and cellulytic enzymes of the pathogen, and that toxins have not been unequivocally implicated. More recent work supports this view (Beckman et al.; Duniway 1971b; Elgersma et al. 1972). The single, dominant gene type of resistance expresses itself in restricting the spread of the pathogen by the rapid development of gels and tyloses in the xylem. The morphological features of the stem xylem in resistant and susceptible tomato isolines affect spore distribution similarly and are therefore not factors in determining resistance or susceptibility. There is now also evidence for the existence of antifungal substances which may also play a role in the resistance mechanisms (Conway et al.; Drysdale et al.; Hammerschlag et al.; McCance et al.; Stromberg et al.).

The expected expression of the disease has been changed by previous or simultaneous inoculation with other ff. sp. of *F. oxysporum* (e.g. Homma et al.; Langton), by antibiotics and other compounds, and by nematode attack. Kawamura et al. demonstrated synergism with *Meloidogyne incognita*, and Hirano et al. with *Pratylenchus coffeae* and *P. penetrans*. Jones et al. (1976) and Orion et al. found little or no tendency for attack by root knot nematodes to break down resistance to the fungus (see also Bergeson et al.; Binder et al.; Cohn et al.; Jenkins et al.). Sidhu et al. (1974) reported breakdown of resistance to the fungus. Of 2 dominant genes segregating independently one was effective against the fungus and the other likewise against *M. incognita*. Plants susceptible to the nematode but resistant to the fungus showed a susceptible response to the latter when both infections took place.

There is very extensive literature on the biochemical aspects of this host pathogen relationship (amino acids, catechol, cell degrading enzymes, ethylene, heavy metals, nutrients, oxidases, pectic substances, phenols, phospholipids, photosynthesis, respiration and xylem flow). See Walker for earlier ref. and *Rev. Pl. Pathol.* **49**, 1785, 2325; **50**, 1980, 1981, 2501, 2502, 3489; **51**, 3555, 3556, 3557; **52**, 2371, 3826; **53**, 679, 1557, 1947, 4116, 4587; **54**, 994, 1875, 3483, 3485, 5083; **55**, 1466, 4277, 4525, 5352; **56**, 391, 4198; **57**, 1413, 3618, 4139, 5124.

Control is through resistance and (where economic) soil fumigation. Grafting on resistant stocks has also been investigated (e.g. Gindrat et al.; Okuda et al.). Only 2 races have apparently been differentiated. Walker should be consulted for an account of the development of cvs with polygenic or oligogenic resistance. Crill et al. have briefly reviewed the work in USA on resistance (see also: Adachi et al.; Bohn et al.; Cirulli et al.; Gabe; Giles et al.; Honma et al.;

Hutton et al.; Jones et al. 1974; Katan et al.; Laterrot). Race 2 was first reported in 1945; it did not spread as rapidly as race 1 but threatens the oligogenic resistance to the latter race which is now widely used (see Katan et al.). Single, dominant gene controlled resistance to both races is now known and has been incorporated into new cvs. For fungicides and soil fumigants see: Biehn; Fuchs et al.; Jones et al. 1971, 1972; Westeijn.

Adachi, T. et al. 1974. *Bull. Fac. Agric. Miyazaki Univ.* **21**: 97 (resistance; **54**, 3484).

Beckman, C. H. et al. 1972. *Phytopathology* **62**: 1256 (restriction of infection in near isogenic lines with oligogenic resistance; **52**, 4221).

Bergeson, G. B. et al. 1970. *Ibid* **60**: 1245 (*Meloidogyne javanica* & rhizosphere; **50**, 995).

Besri, M. 1977. In *Travaux dédiés à Georges Viennot-Bourgin* (Soc. Française Phytopathologie): 19 (on seed; **57**, 2652).

Biehn, W. L. 1973. *Pl. Dis. Reptr* **57**: 37 (use of benomyl; **52**, 3426).

Binder, E. et al. 1959. *Ibid* **43**: 972 (effect of *M. incognita*; **39**: 246).

Bohn, G. W. et al. 1940. *Res. Bull. Mo. agric. Exp. Stn* 311, 82 pp. (immunity in *Lycopersicon pimpinellifolium* & its inheritance; **19**: 501).

Buchenauer, H. 1971. *Phytopathol. Z.* **72**: 53 (effect of growth retardants; **51**, 1926).

Carrasco, A. et al. 1978. *Physiol. Pl. Pathol.* **12**: 225 (resistance).

Chambers, H. L. et al. 1963. *Phytopathology* **53**: 1006 (occlusion of xylem; **43**, 577).

Cirulli, M. et al. 1966. *Ibid* **56**: 1301 (races 1 & 2, pathogenicity & resistance; **46**, 1104).

Clayton, E. E. 1923. *Am. J. Bot.* **10**: 71, 133 (effect of temp. & soil moisture; **2**: 428, 477).

Cohn, E. et al. 1960. *Hassadeh* **40**: 1347 (effect of *M. hapla* & *M. incognita*; **40**: 564).

Conway, W. S. et al. 1978. *Phytopathology* **68**: 938 (distribution & growth in resistant & susceptible tomato).

Crill, P. et al. 1972. *Pl. Dis. Reptr* **56**: 695 (resistant cvs; **52**, 1265).

Dimond, A. E. et al. 1953. *Phytopathology* **43**: 663 (cause of epinastic symptoms; **33**: 507).

—— ——. 1953. *Ibid* **43**: 619 (water economy & wilt; **33**: 453).

Drysdale, R. B. et al. 1973. In *Fungal pathogenicity and the plant's response*: 423, Academic Press.

Duniway, J. M. et al. 1971a. *Phytopathology* **61**: 1377 (gas exchange in infected leaves; **51**, 2854).

——. 1971b. *Physiol. Pl. Pathol.* **1**: 537 (wilt & resistance to xylem flow; **51**, 1923).

Elgersma, D. M. et al. 1972. *Phytopathology* **62**: 1232 (restriction of pathogen growth in resistant, near isogenic lines; **52**, 4220).

Elliot, J. A. et al. 1922. *Ibid* **12**: 428 (seed transmission; **2**: 92).

Foster, R. E. 1946. *Ibid* **36**: 691 (initial symptoms of vein clearing; **26**: 83).

—— et al. 1947. *J. agric. Res.* **74**: 165 (effects of temp., soil moisture & major nutrients; **26**: 473).

Fuchs, A. et al. 1970. *Phytopathol. Z.* **69**: 330 (use of benomyl; **50**, 2677).

Gabe, H. L. 1975. *Trans. Br. mycol. Soc.* **64**: 156 (nomenclature of races; **54**, 2971).

Gäumann, E. 1957. *Phytopathol. Z.* **29**: 1 (fusaric acid as a wilt toxin, review, 93 ref.; **36**: 736).

——. 1957. *Phytopathology* **47**: 342 (as above, 88 ref.; **36**: 791).

Gerdeman, J. W. et al. 1951. *Ibid* **41**: 238 (pathogenicity of races 1 & 2; **30**: 493).

Giles, J. E. et al. 1958. *Aust. J. agric. Res.* **9**: 192 (*L. peruvianum* & resistance, including *M. javanica*).

Gindrat, D. et al. 1977. *Rev. suisse Vitic. Arboric.* **9**: 109 (cv. resistance & control by grafts; **56**, 5798).

Hammerschlag, F. et al. 1975. *Phytopathology* **65**: 93 (antifungal activity of host extracts; **54**, 4149).

Hirano, K. et al. 1972. *Tech. Bull. Fac. Hort. Chiba Univ.* **20**: 37 (infection with *Pratylenchus coffeae* & *P. penetrans*; **52**, 4222).

Homma, Y. et al. 1977. *Bull. Shikoku agric. Exp. Stn* 30: 103 (symptom suppression by prior inoculation with other ff. sp. of *Fusarium oxysporum* & *F. solani*; **57**, 296).

Honma, S. et al. 1971. *J. Am. Soc. hort. Sci.* **96**: 496 (synthesis of a jointless resistant tomato; **51**, 1925).

—— ——. 1972. *Euphytica* **21**: 143 (inheritance of resistance as affected by X gametophytic factor; **51**, 4366).

Hutton, E. M. et al. 1947. *J. Coun. scient. ind. Res. Aust.* **20**: 468 (inheritance of resistance; **27**: 454).

Jarvis, W. R. et al. 1975. *Can. Pl. Dis. Surv.* **55**: 25 (foot & root rot of tomato; **55**, 2365).

—— ——. 1976. *Pl. Dis. Reptr* **60**: 1027 (as above, susceptibility of *Lycopersicon* spp. & hybrids; **56**, 3223).

—— ——. 1977. *Ibid* **61**: 251 (as above, no interaction with *M. incognita*; **56**, 5797).

——. 1977. *Can. Agric.* **22**: 28 (as above, biological control; **56**, 5193).

Jenkins, W. R. et al. *Pl. Dis. Reptr* **41**: 182 (effect of *M. hapla* & *M. incognita*).

Jones, J. P. et al. 1971. *Phytopathology* **61**: 1415 (control with lime & soil fumigants; **51**, 2855).

—— ——. 1972. *Pl. Dis. Reptr* **56**: 953 (effect of mulching on efficiency of soil fumigant; **52**, 3427).

—— ——. 1974. *Phytopathology* **64**: 1507 (susceptibility of 'resistant' cvs; **54**, 2467).

Jones, J. P. et al. 1976. *Ibid.* **66**: 1339 (no erosion of resistance by *M. incognita*; **56**, 2660).
Katan, J. 1971. *Ibid* **61**: 1213 (symptomless host carriers of pathogen; **51**, 2853).
—— et al. 1974. *Phytoparasitica* **2**: 83 (race 2 distribution in Israel; **54**, 5082).
Kawamura, T. et al. 1967 & 1968. *Tech. Bull. Fac. Hort. Chiba Univ.* 15: 7; 16: 23 (effect of *M. incognita*; **47**, 3216; **49**, 567).
Keyworth, W. G. 1963 & 1964. *Ann. appl. Biol.* **52**: 257; **54**: 99 (reaction of resistant & susceptible cvs to inoculation of stems or rootstocks, effects on resistant scions to toxins formed in rootstocks; **43**, 1149; **44**, 240).
Langton, F. A. 1968. *Ann. appl. Biol.* **62**: 413 (effect of *F. oxysporum* f. sp. *pisi*; **48**, 1322).
Laterrot, H. 1972. *Phytopathol. Mediterranea* **11**: 154 (resistance selection; **53**, 1555).
——. 1976. *Annls Amél. Pl.* **26**: 485 (site of gene I_2 controlling resistance to race 2; **56**, 2662).
Ludwig, R. A. 1952. *Tech. Bull. Macdonald Coll. McGill Univ.* 20, 39 pp. (wilt & water flow; **33**: 56).
Mace, M. E. et al. 1971. *Phytopathology* **61**: 627, 834 (spore transport in resistant & susceptible isolines, expression of resistance above & below cotyledons; **51**, 681, 1924).
—— ——. 1972. *Ibid* **62**: 651 (phenols & histochemistry of vascular browning in isolines; **52**, 491).
McCance, D. J. et al. 1975. *Physiol. Pl. Pathol.* **7**: 221 (formation of tomatine & rishitin in inoculated plants; **55**, 3746).
Miller, P. M. et al. 1973. *Pl. Dis. Reptr* **57**: 267 (effect on disease of previous soil fumigation; **52**, 3827).
Okuda, T. et al. 1972. *Bull. Fukui agric. Exp. Stn* **9**: 61 (grafts for control; **54**, 1431).
Orion, D. et al. 1974. *Neth. J. Pl. Pathol.* **80**: 28 (effects of *M. incognita*, *M. javanica* & ethrel; **53**, 3623).
Ride, J. P. et al. 1971. *Physiol. Pl. Pathol.* **1**: 409 (chemical method for estimation of pathogen in host; **51**, 1922).
Rowe, R. C. et al. 1977. *Phytopathology* **67**: 1513 (foot & root rot, recolonisation of steamed soil; **57**, 4622).
—— ——. 1978. *Ibid* **68**: 1221 (foot & root rot, prevention of recolonisation).
Sato, R. et al. 1974. *A. Rep. Soc. Pl. Prot. North Japan* 25: 5 (foot & root rot; **55**, 2894).
Scheffer, R. P. et al. 1953. *Phytopathology* **43**: 116 (physiology of wilt; **33**, 122).
—— ——. 1954. *Ibid* **44**: 94 (distribution & nature of resistance; **33**: 691).
Sidhu, G. et al. 1974. *J. Hered.* **65**: 153 (genetics of resistance with *M. incognita*; **54**, 2463).
—— ——. 1977. *Nematologica* **23**, 436 (predisposition by *M. incognita*; *Hort. Abstr.* **48**, 5643).
Sonoda, R. M. 1976. *Pl. Dis. Reptr* **60**: 271 (foot & root rot in USA, Florida; **55**, 5925).
Stromberg, E. L. et al. 1977. *Phytopathology* **67**: 693 (antifungal activity of host extracts; **57**, 295).
Strong, M. C. 1946. *Ibid* **36**: 218 (effect of temp. & soil moisture; **25**: 423).
Subramanian, C. V. 1970. *C.M.I. Descr. pathog. Fungi Bact.* 217 (*F. oxysporum* f. sp. *lycopersici*).
Walker, J. C. 1971. *Monogr. Am. phytopathol. Soc.* 6, 56 pp. (monograph, 446 ref; **50**, 3174).
Wergin, W. P. 1972. *Phytopathology* **62**: 1045 (fine structure of microbodies in pathogenic & saprophytic hyphae; **52**, 1673).
Westeijn, G. 1973. *Neth. J. Pl. Pathol.* **79**: 36 (soil fumigation & glasshouse disinfection; **52**, 3073).

Fusarium oxysporum Schl. f. sp. **medicaginis** (Weimer) Snyder & Hansen, *Am. J. Bot.* **27**: 66, 1940.

Lucerne (*Medicago sativa*) wilt is a minor disease which occurs in USA, parts of Europe and S. America. One or more stems may become chlorotic and die whilst the remainder remain green for a longer time; the brown vascular discolouration is found in the roots and lower stem. Growth in vitro and infection have an opt. temp. of *c.* 25°. The pathogen caused wilt in 2 lupin cvs but not in 36 spp. or vars of other plants. No wilt occurred on lucerne when inoculated with 26 different wilt fusaria other than those from lucerne, cotton and *Cassia tora*. But in the field, when lucerne was planted in soil infested with the cotton *Fusarium*, fewer plants wilted than might have been expected from pot inoculation tests. Cvs (14) of lucerne showed highly significant differences in yield when grown for 5 years in soil infested with *F. oxysporum* f. sp. *vasinfectum* (q.v.) and root knot nematodes. But glasshouse tests with 8 cvs showed no high differences in resistance. Damage to the root system by *Sitona hispidulus* larvae and *Bradysia* increased infection by f. sp. *medicaginis*. The reaction of lines to infection with and without infection by *Corynebacterium insidiosum* has also been investigated (Frosheiser et al.; Hill et al.; Leath et al.).

Armstrong, G. M. et al. 1954. *Pl. Dis. Reptr* **38**: 221 (host relationships; **33**: 539).
—— ——. 1965. *Ibid* **49**: 412 (as above & reaction of lucerne cvs; **44**, 2830).
Chi, C. C. 1968. *Ibid* **52**: 939 (host nutrition & infection by *Fusarium oxysporum* & *F. solani*; **48**, 1220).
Frosheiser, F. I. et al. 1978. *Phytopathology* **68**: 943 (infection with *Corynebacterium insidiosum*).

Hill, R. R. et al. 1969. *Crop Sci.* **9**: 327 (relationship with *Sitona hispidulus* & *C. insidiosum*; **50**, 3903).
Leath, K. T. et al. 1969. *Phytopathology* **59**: 257 (effect of *Bradysia* sp.; **48**, 1787).
Marcley, M. D. 1970. *Pl. Dis. Reptr* **54**: 1061 (lucerne decline in W. Australia; **51**, 2627).
Weimer, J. L. 1928. *J. agric. Res.* **37**: 419 (general; **8**: 247).
——. 1929. *Ibid* **39**: 351 (additional hosts; **9**: 188).
——. 1930. *Ibid.* **40**: 97 (effects of temp. & soil moisture; **9**: 531).

Fusarium oxysporum Schl. f. sp. **melonis** Snyder & Hansen, *Am. J. Bot.*, **27**: 66, 1940.

A destructive wilt of melon (*Cucumis melo*) is caused by this f. sp. Infection occurs on seedlings (pre- and post-emergence damping off) and on older plants (leaf chlorosis, stunting and general wilt). Streaks up to 0.6 m long appear on the stems; they become necrotic and bear the salmon pink sporulating masses. In some cases stem cracks develop and form a brownish exudate. The vascular elements become an orange red and the fruit is much reduced in size. The disease has a widespread and scattered distribution (Map 496). It has been more often reported from temperate and subtropical regions. In the tropics it occurs in: Australia (Queensland), India (Tamil Nadu), Philippines, Zimbabwe, Saudi Arabia and Thailand.

The opt. temp. for in vitro growth (26° or more) may be higher than that for severe disease. Reid gave a soil temp. of 20° for max. disease; although disease incidence remained high at 30° it fell rapidly above this temp. Molot et al. (1975a) gave 18–22° for the most severe symptoms. Banihashemi et al. (1973a), studying saprophytic activity, found this to be more at 15° than at higher temps. The inoculum level in the soil is affected by soil type, the degree of resistance of the cv. being grown and major nutrients (McKeen et al.; Wensley et al.).

The ff. sp. from melon and watermelon (*F. oxysporum* f. sp. *niveum* q.v.) will each attack the seedlings of both hosts; but older plants are susceptible only to their respective f. sp. The fungus occurs in seed (Leach 1936). Breeding for resistance is complicated by the existence of several races, at least 4 have been differentiated. In races 1 and 2 resistance is inherited as single dominants (Banihashemi et al. 1975b; Benoit; Leary et al.; Messiaen et al.; Meyer et al.; Molot et al. 1974a, 1975a; Risser; Risser et al.). Bergeson found that infection by *Meloidogyne incognita* reduced the resistance in one cv. although that of others was not apparently affected. Besides control through resistance and soil fumigation the resistant stocks *Cucurbita ficifolia* and *Benincasa cerifera* have also been used (e.g. Benoit; Ruggeri); and see Wensley et al. (1970) for the use of benomyl.

Banihashemi, Z. et al. 1973a. *Trans. Br. mycol. Soc.* **60**: 205 (saprophytism in soil; **52**, 3953).
———. 1973b. *Pl. Soil* **38**: 465 (survival & soil temp.; **53**, 833).
———. 1975a. *Phytopathology* **65**: 1212 (behaviour in presence & absence of host plants; **55**, 2957).
———. 1975b. *Iran. J. agric. Res.* **3**: 41 (race 4; **55**, 455).
Benoit, F. 1974. *Tuinbouw berichten Belgium* **38**: 16 (cvs & resistance; **54**, 3036).
Bergeson, G. B. 1975. *Pl. Dis. Reptr* **59**: 410 (effect of *Meloidogyne incognita*; **55**, 450).
Bhaskaran, R. et al. 1972. *Phytopathol. Z.* **73**: 75 (preformed inhibitors & resistance; **51**, 4577).
Douglas, D. R. 1970. *Can. J. Bot.* **48**: 687 (effect of inoculum conc. on resistance; **49**, 3545).
Leach, J. G. 1933. *Phytopathology* **23**: 554 (general; **12**: 744).
——. 1936. *Ibid* **26**: 99 (seed penetration & soil temp.).
——. 1938. *Tech. Bull. Minn. agric. Exp. Stn* 129 (general).
Leary, J. V. et al. 1976. *Phytopathology* **66**: 15 (races; **55**, 3348).
McKeen, C. D. 1951. *Sci. Agric.* **31**: 413 (on melon & watermelon; **31**: 473).
—— et al. 1962. *Can. J. Microbiol.* **8**: 769 (cultural & pathogenic variation; **42**, 288).
———. 1965. *Ibid* **11**: 987 (effects of C & N amendments; **45**, 1980b).
Maraite, H. 1973. *Physiol. Pl. Pathol.* **3**: 29 (infection & changes in polyphenoloxidases & peroxidases; **52**, 3117).
Mas, P. M. et al. 1974. *Annls Phytopathol.* **6**: 237 (pre-immunising effect of a race against another; **54**, 5621).
Messiaen, C. M. et al. 1962. *Annls Amél. Pl.* **12**: 157 (inheritance of resistance; **42**, 430).
Meyer, J. A. et al. 1971. *Trans. Br. mycol. Soc.* **57**: 371 (inoculation with >1 race & symptom mitigation; **51**, 3016).
Millar, J. J. 1945 *Can. J. Res.* Ser. C **23**: 6, 166 (cultural & soil; **24**: 351; **25**: 249).
Molot, P. M. et al. 1974a. *C. r. hebd. Séanc. Acad. Sci. Paris* Ser. D **278**: 3327 (inoculation with >1 race; **54**, 3550).
———. 1974b. *Annls Phytopathol.* **6**: 245 (role of fusaric acid; **54**, 5622).
———. 1975a. *Ibid* **7**: 115 (effects of temp.; **55**, 5402).

Molot, P. M. et al. 1975b. *Ibid* **7**: 175 (effects of temp. on pre-immunisation; **55**, 5403).
Reid, J. 1958. *Can. J. Bot.* **36**: 393, 507 (general; **37**: 753).
Risser, G. et al. 1965. *Annls Amél. Pl.* **15**: 405 (races; **45**, 3015).
——. 1973. *Ibid* **23**: 259 (inheritance of resistance; **55**, 4350).
—— ——. 1976. *Phytopathology* **66**: 1105 (race nomenclature; **56**, 2729).
Rouxel, F. et al. 1977 *Annls Phytopathol.* **9**: 183 (effect of heat treatment on a disease in soil; **57**, 3170).
Rucceri, D. 1967. *Riv. Ortoflorofruttic. ital.* **51**: 328 (resistant rootstock; **47**, 716).
——. 1968. *Phytopathol. Mediterranea* **7**: 150 (as above; **48**, 2685).
Stoddard, D. L. 1947. *Phytopathology* **37**: 875 (host nutrition; **27**: 346).
Wensley, R. N. et al. 1962. *Can. J. Microbiol.* **8**: 818 (pathogenicity test; **42**: 288).
—— ——. 1963. *Ibid* **9**: 237 (pathogen populations & wilt potential of soil; **42**: 587).
—— ——. 1965. *Ibid* **11**: 581 (host nutrition, pathogen populations & wilt incidence; **45**, 1980a).
—— ——. 1966. *Ibid* **12**: 1115 (effect of host on pathogen populations & wilt; **46**, 1397).
——. 1969. *Ibid* **15**: 917 (biological & chemical control; **49**, 314).
—— ——. 1970. *Ibid* **16**: 615 (benomyl soil drenches; **50**, 444).

Fusarium oxysporum Schl. f. sp. **niveum** (E. F. Smith) Snyder & Hansen, *Am. J. Bot.* **27**: 66, 1940.

Fusarium niveum E. F. Smith, 1899
F. bulbigenum var. *niveum* (E. F. Smith) Wollenw., 1935.

This f. sp. causes a widespread wilt of watermelon (*Citrullus lanatus*) which is the only important host. Damping off, cortical rot and stunting of seedlings, and sudden or progressive wilt of older plants occur. Necrotic lesions form on the roots and browning, gum and tyloses are found in the vascular system. In mature plants wilt may be confined to a particular part, depending on which portion of the root system has been invaded from the soil. Chlorosis and stunting in maturing plants can occur and there may be a temporary recovery from wilt. Sporulation develops on dead stems in wet weather. *C. colocynthis* is said to be susceptible (Jotani).

Penetration is via the root tip meristematic zone and the epidermis of the zones of root elongation and maturation. It can also take place through the ruptures caused by new lateral roots. The opt. soil temp. for the disease in seedlings is 27°, but severe wilt can occur at 20–30° where the soil is heavy and where exposure to infection is as long as 30 days. Infection declines rapidly at over 30° and is absent above 33°. Sumner et al. investigated the effects of root knot nematodes. Nishimura reviewed the largely Japanese work on root exudates, xylem blockage and toxins (and see Hiroe et al.; Yoshii). In USA at least 2 races were described (Crall). Work by Netzer indicated that a highly virulent str. was present in Israel; the local isolates were pathogenic to resistant cvs from USA (e.g. see Barnes; Porter 1937). The inheritance of resistance is apparently complex with both dominant and recessive factors. Jones et al. found that wilt decreased with a rising soil pH. Hopkins et al. indicated that treatment with soil fungicides was effective where cvs were moderately resistant.

Barnes, G. L. 1972. *Pl. Dis. Reptr* **56**: 1022 (resistance reactions of cvs; **52**, 2776).
Bennett, L. S. 1936. *J. agric. Res.* **53**: 995 (inheritance of resistance; **16**: 85).
Cook, H. T. 1937. *Bull. Va Truck Exp. Stn* 97: 1513 (temp. & resistance; **17**: 290).
Crall, J. M. 1964. *A. Rep. agric. Exp. Stn Fla 1963–64* (races; **44**, 2954f).
Henderson, W. R. et al. 1970. *HortScience* **95**: 276 (inheritance of resistance; **50**, 445).
Hiroe, I. et al. 1956. *Ann. phytopathol. Soc. Japan* **20**: 161 (toxins; **36**: 511).
Hopkins, D. L. et al. 1975. *Proc. Fla St. hort. Soc.* **88**: 196 (soil fungicides; **57**, 2693).
Jones, J. P. et al. 1975. *Ibid* **88**: 200 (effect of liming & N; **57**, 2694).
Jotani, Y. 1954. *J. Jap. Bot.* **29**: 279 (on *Citrullus colocynthis*; **34**: 766).
Netzer, D. 1976. *Phytoparasitica* **4**: 131 (races & soil population; **56**, 4260).
Nishimura, S. 1971. *Rev. Pl. Prot. Res.* **4**: 71 (wilt mechanisms; **52**, 903).
Porter, D. R. 1928. *Res. Bull. Iowa agric. Exp. Stn* 112: 346 (general; **7**: 760).
—— et al. 1932. *Ibid* 149: 128 (general; **11**: 557).
——. 1937. *Bull. Calif. agric. Exp. Stn* 614 (breeding for resistance; **17**: 371).
Sleeth, B. 1934. *Bull. W. Va agric. Exp. Stn* 257, 23 pp. (in vitro & pathogenicity; **13**: 560).
Sumner, D. R. et al. 1973. *Phytpopathology* **63**: 857 (effects of root knot nematodes; **53**, 1992).

Sun, S. K. et al. 1977. *Pl. Prot. Bull. Taiwan* **19**: 257 (soil survival; **57**, 5754).
Taubenhaus, J. J. 1920. *Bull. Texas agric. Exp. Stn* 260 (general).
Walker, M. N. 1941. *Bull. Fla agric. Exp. Stn* 363 (soil temp.).
Welch, A. et al. 1942. *Phytopathology* **32**: 181 (inheritance of resistance; **21**: 362).
Wilson, J. J. 1936. *Res. Bull. Iowa agric. Exp. Stn* 195: 107 (general; **16**: 14).
Yoshii, H. 1933. *Bull. Sci. Falcultato Terkultura Kjusu Imp. Univ.* **5**: 313, 577 (infection & spread in host; **13**: 5, 419).
——. 1934. *Ibid* **6**: 1 (histology; **14**: 143).

Fusarium oxysporum Schl. f. sp. **passiflorae** Gordon apud Purss, *Qd J. agric. Sci.* **11**: 79, 1954.

This wilt of passion fruit (*Passiflora edulis*), described from Australia, has caused serious losses in New South Wales and Queensland. Mature vines can wilt and collapse within 24–48 hours after showing the relatively slight symptoms of slight paling of colour in the new leaf growth only. The wilt may be on one side of the plant and immature fruit may shrivel. Vascular browning is prominent up to 2 m above soil level. On seedlings the earliest symptom is a vein clearing 7–9 days after inoculation and the leaves absciss. No resistance has been found in the purple passion fruit (f. *edulis*) but it has been demonstrated in several *Passiflora* spp. and in the golden or yellow passion fruit (f. *flavicarpa*) which is an effective and acceptable rootstock. In 1961 Cox stated that 90% of vines grafted on to *flavicarpa* remained free from wilt when planted on infested soil. Progeny from the golden × purple cross and backcrossed to purple, show good resistance. The other resistant *Passiflora* spp. are apparently unsatisfactory as stocks.

Bastos, C. N. 1976. *Turrialba* **26**: 371 (toxin; **56**, 4127).
Cox, J. E. et al. 1961. *Agric. Gaz. N.S.W.* **72**: 314 (resistant rootstocks; **41**: 51).
Groszmann, H. M. et al. 1958. *Qd agric. J.* **84**: 341 (resistant rootstocks & seedling progeny).
Kiely, T. B. 1961. *Agric. Gaz. N.S.W.* **72**: 275 (general; **41**: 242).
McKnight, T. 1951. *Qd J. agric. Sci.* **8**: 1 (general & seedling inoculation; **32**: 575).
Purss, G. S. 1954. *Ibid* **11**: 79 (*F. oxysporum* f. sp. *passiflorae*; **35**: 203).
——. 1958. *Ibid* **15**: 95 (resistant rootstocks; **38**: 157).

Fusarium oxysporum Schl. f. sp. **phaseoli** Kendrick & Snyder, *Phytopathology* **32**: 1010, 1942.

Yellows of the common bean (*Phaseolus vulgaris*) and the scarlet runner bean (*P. coccineus*) is widespread; Brazil, Colombia, Costa Rica, Egypt, Netherlands, Peru, UK and USA. The characteristic yellowing symptoms usually occur first on the primary leaves, spreading upwards to the younger ones. The plant becomes conspicuously yellow before wilt and general collapse. As in other vascular wilts caused by ff. sp. of *F. oxysporum* the symptoms may be restricted at first to one side of the plant. As the leaves become chlorotic their margins may roll inwards and the leaflets droop. The vascular browning extends from the roots to the petioles of the youngest leaves and to the pods. The opt. temp. for growth in vitro is 28°. No internal seedborne transmission has been described although external contamination has. Besides the primary economic hosts only 2 *Lupinus* spp. were susceptible. Isolates from UK and USA had the same pathogenicity characteristics on 9 cvs of *P. vulgaris* (Armstrong et al.). Some cvs show little or no wilt in the field although their roots may be initially invaded.

Armstrong, G. M. et al. 1963. *Pl. Dis. Reptr* **47**: 1088 (inoculation of other plant genera & of *Phaseolus* with other wilt fusaria; **43**, 1472).
—— ——. 1964. *Ibid* **48**: 846 (comparison of UK and USA isolates; **44**, 919).
Cardoso, C. O. N. et al. 1966. *Anais Esc. sup. Agric. 'Luiz Querioz'* **23**: 273 (in Brazil; **47**, 953).
Davis, D. L. G. 1942. *Trans. Br. mycol. Soc.* **25**: 418 (general in UK on *P. coccineus*; **22**: 191).
Dongo D., S. L. et al. 1969. *Turrialba* **19**: 82 (tests on cv. resistance; **48**, 2661b).
Duque, S. L. et al. 1969. *Ibid* **19**: 71 (penetration & effect on host; **48**, 2661a).
Echandi, E. 1967. *Ibid* **17**: 409 (in Costa Rica, tests on cvs; **47**, 1723).
Kendrick, J. B. et al. 1942. *Phytopathology* **32**: 1010 (general; **22**: 192).
Youssef, Y. A. 1961. *Phytopathol. Z.* **41**: 353 (toxicity of culture filtrates; **41**: 190).

Fusarium oxysporum Schl. f. sp. **pisi** (van Hall) Snyder & Hansen, *Am. J. Bot.* **27**: 66, 1940.

Fusarium vasinfectum Atk. var. *pisi*. van Hall, 1903
F. orthoceras App. & Wr. var. *pisi* Linford, 1928

Fusarium oxysporum

F. oxysporum Schl. f. 8 Snyder & Walker, 1935
F. oxysporum Schl. var. *orthoceras* (App. & Wr.) Bilay, 1955.

Pea (*Pisum sativum*) wilt is widespread and has been investigated in N. America, W. Europe and Australia. The name near wilt has been used for the disease caused by race 2. In the Netherlands the disease called St John's is a complex, the syndrome of which was split by Buxton (1955) into 3 components: wilt, *F. oxysporum* f. sp. *pisi* and to a lesser extent *F. oxysporum* var. *redolens* (Wollenw.) Gordon; foot rot, the former pathogen; and thirdly, *F. solani* f. sp. *pisi* (q.v.) and *F. oxysporum* f. sp. *pisi* acting together. True wilt appears as a stunting and rolling under of the leaf margins and stipules; the upper parts become pale, develop a greyish bloom and chlorosis follows; the leaves become somewhat brittle and are readily removed; the upper part of the plant may wilt abruptly and dry up whilst still green. The vascular discolouration is pale yellow to deep orange brown. Severe symptoms commonly occur as the flowers are forming; such plants rarely bear full pods and death follows quickly.

Virgin et al. (1940) found that penetration occurred along any part of the seedling root and epicotyl; common entry points were the root tip and cotyledonary node. They reported the fungus within the seed coat; seed transmission had also been referred to earlier (Kadow et al.; Snyder 1932). Hepple found that seedlings in infested soil could become infected via the cotyledonary vascular bundles when the cotyledons had been decomposed by microbial action. Doling found that only when roots were damaged did wilt occur and suggested that the exposed stele was necessary for penetration by the pathogen. Nyvall et al. (1972), investigating the sites of infection with race 5, found that host invasion took place through wounds or the epidermis. In vitro growth (races 1 and 2) is best at *c.* 28°. Disease is most severe at this temp. but only slightly less so at 23°, and spread in the vascular tissues is rapid at 21–24° (MacNeill et al.; Schroeder et al.). Differences in the temp. opt. for races 1 and 2 have been reported. The fungus can be isolated from other legumes that do not show macroscopic symptoms. Pea, as a low temp. crop, presumably has a lower resistance at the high temps which favour the fungus. Schippers et al. found that germination of chlamydospores was highest in infested soil along the actively growing parts of pea roots (and see Whalley et al. 1976). In comparing a resistant with a susceptible cv. they detected no differences in germination on the root surface or in the rhizosphere, in mycelial growth on, or attachment to, the root surface. Buxton's (1957) findings that exudates from resistant roots reduced spore germination whilst those from susceptible ones increased it (in 3 races) could not be confirmed by Kommedahl or Whalley et al. (1973).

Buxton et al. (1959) and Perry found that inoculation with *F. solani* before *F. oxysporum* f. sp. *pisi* reduced the amount of wilt. The former colonised the epidermis and outer cortex of roots more rapidly and more extensively than did the latter. This effect on wilt also occurred, to a lesser degree, if both fungi were applied simultaneously. Worf et al. found that when *F. solani* f. sp. *pisi* spores were added to soil infested with *F. oxysporum* f. sp. *pisi* their development was restricted to chlamydospore formation until after planting peas. Very high populations of the latter caused a root rot indistinguishable from that caused by the former; lower populations reduced the amount of decay which occurred when *F. solani* spores were added (compared with *F. solani* alone) since such populations delayed development of *F. solani*. Kerr reported a marked interaction between *Pythium ultimum* (q.v.) and *F. oxysporum* f. sp. *pisi*; both together causing more severe disease than either alone. Synergistic effects with nematodes have also been described with *Rotylenchus robustus* (Labruyère et al.) and *Pratylenchus penetrans* (Oyekan et al.) amongst others. Jauch investigated cell degrading enzymes and Meiler a toxin.

Much of the work has been done with races 1, 2 and 3. Bolton et al. described race 4 and in 1970 Haglund et al. found a new race (5) which had become naturally established and caused losses in USA (N.W. Washington). These last authors reported on the reaction of 9 pea cvs to these 5 races. Armstrong et al. examined 19 isolates from 5 countries and their identification of the races did not wholly correspond with that of the original donor. They recognised 10 races but did not examine isolates of race 3. Kraft et al. re-examined the race situation in *F. oxysporum* f. sp. *pisi.* They concluded that the existing 11 race classification appeared to be based more on virulence differences than on true genetic differences in the host. These races should be grouped into either race 1 or race 2 types.

Control is through oligogenic, dominant, resistance (Wade; Hare et al.); some of the many resistant cvs are given by Ogilvie, *Diseases of vegetables*, and by Chupp and Sherf, *Vegetable diseases and their control*. More recent information is given in the reports of the Processors & Growers Research Organisation, UK (and see Buxton et al. 1960b; Crampton et al.; Cruickshank). Wilt has been reduced by the incorporation of chitin in the soil (Buxton et al. 1965; Khalifa), and Ebbels found that soil fumigation with chloropicrin and dazomet was effective without adversely affecting plant growth.

Armstrong, G. M. et al. 1974. *Phytopathology* **64**: 849 (race differentiation; **54**, 608).
Bolton, A. T. et al. 1966. *Can. J. Pl. Sci.* **46**: 343 (race 4; **45**, 3437).
Buxton, E. W. et al. 1954. *Pl. Pathol.* **3**: 13 (wilt in UK; **34**: 15).
——. 1955. *Trans. Br. mycol. Soc.* **38**: 309 (disease complex with *Fusarium solani* & *F. oxysporum* var. *redolens*; **35**: 411).
—— ——. 1955. *J. gen. Microbiol.* **13**: 99 (differential tolerance to inhibition by actinomycetes; **35**: 213).
——. 1956. *Ibid* **15**: 133 (heterokaryosis & parasexual recombination; **36**, 45).
——. 1957. *Trans. Br. mycol. Soc.* **40**: 305 (differential rhizosphere effects on 3 races; **37**: 126).
——. 1957. *Ibid* **40**: 145 (root exudate effects on 3 races; **36**: 629).
——. 1958. *Nature Lond.* **181**: 1222 (change from race 1 to 2 induced by root exudate; **37**: 615).
——. 1959. *Pl. Pathol.* **8**: 39 (resistant cvs; **39**: 66).
——. 1959. *Trans. Br. mycol. Soc.* **42**: 378 (interaction with *F. solani*; **39**: 520).
——. 1960a. *J. gen. Microbiol.* **22**: 678 (effects of root exudate on antagonism of rhizosphere micro-organisms; **40**: 69).
—— ——. 1960b. *Pl. Pathol.* **9**: 54 (resistant cvs; **40**: 137).
—— ——. 1965. *Ann. appl. Biol.* **55**: 83 (chitin soil amendment; **44**, 1998).
Crampton, M. J. et al. 1974. *N.Z. Jl Agric.* **128**(1): 12 (2 resistant cvs; **53**, 3681).
Cruickshank, I. A. M. 1952. *Ibid* **84**(2): 144 (resistant cvs; **32**: 114).
Doling, D. A. 1963. *Trans. Br. mycol. Soc.* **46**: 577 (effect of root damage; **43**, 1194).
Ebbels, D. L. 1967. *Ann. appl. Biol.* **60**: 391 (soil fumigation; **47**, 951).
Fleischmann, G. 1963. *Can. J. Bot.* **41**: 1569 (general; **43**, 1193).
Guy, S. O. et al. 1977. *Phytopathology* **67**: 72 (inoculum potential & biological control; **56**, 3771).
Haglund, W. A. et al. 1970. *Ibid* **60**: 1861 (race 5 & differentiation of races; **50**, 2575).
Hare, W. W. et al. 1949. *J. agric. Res.* **78**: 239 (inheritance of resistance; **28**: 558).
Hepple, S. 1960. *Nature Lond.* **185**: 333 (seedling infection; **39**: 450).
——. 1963. *Trans. Br. mycol. Soc.* **46**: 585 (as above; **43**, 1195).
Jauch, C. 1970. *Bull. Soc. mycol. Fr.* **86**: 839 (physiological & histochemical changes in infected host; **51**, 859).
Kadow, K. J. et al. 1932. *Bull. Wash. agric. Exp. Stn* 272, 30 pp. (general; **12**: 71).
Kerr, A. 1963. *Aust. J. biol. Sci.* **16**: 55 (interaction with *Pythium ultimum*; **42**: 507).
Khalifa, O. 1965. *Ann. appl. Biol.* **56**: 129 (chitin soil amendment; **45**, 287).
Kommedahl, T. 1966. *Phytopathology* **56**: 721 (root exudates & spore germination; **45**, 3242).
Kraft, J. M. et al. 1978. *Ibid* **68**: 273 (re-examination of races).
Labruyère, R. E. et al. 1959. *Nematologica* **4**: 336 (effect of *Rotylenchus robustus*; **40**: 573).
Linford, M. B. 1928. *Res. Bull. Wisc. agric. Exp. Stn* 85, 44 pp. (general; **8**: 214).
——. 1931. *Phytopathology* **21**: 791, 797, 827 (water loss in infected host, pathogenesis, effect of wound inoculation; **11**: 85, 86).
MacNeill, B. H. et al. 1959. *Can. J. Pl. Sci.* **39**: 483 (general; **39**: 450).
Meiler, D. 1970. *Phytopathol. Z.* **68**: 289 (toxins; **50**, 2040).
Nyvall, R. F. et al. 1972. *Phytopathology* **62**: 1419 (infection sites & race 5; **52**, 3122).
—— ——. 1976. *Ibid* **66**: 1093 (effect of plant age on wilt severity; **56**, 2738).
Oyekan, P. O. et al. 1971. *Pl. Dis. Reptr* **55**: 1032 (effect of *Pratylenchus penetrans*; **51**, 2045).
Perry, D. A. 1959. *Trans. Br. mycol. Soc.* **42**: 388 (effect of host colonisation by *F. solani* f. sp. *pisi*; **39**: 520).
Reddy, M. N. et al. 1975. *Physiol. Pl. Pathol.* **7**: 99 (malate dehydrogenase; **55**, 3351).
Roberts, D. D. et al. 1973. *Phytopathology* **63**: 765 (isolation of races from soil; **53**, 1621).
Schippers, B. et al. 1969. *Neth. J. Pl. Pathol.* **75**: 241 (germination of chlamydospores in rhizosphere & host penetration; **48**, 3724).
Schreuder, J. C. 1951. *Tijdschr. PlZiekt.* **57**: 175 (full account in the Netherlands; **31**: 471).
Schroeder, W. T. et al. 1942. *J. agric. Res.* **65**: 221 (effects of temp. & host nutrients in susceptible & resistant cvs; **22**: 87).
Snyder , W. C. 1932. *Phytopathology* **22**: 253 (seed transmission; **11**: 555).
——. 1933. *J. agric. Res.* **47**: 65 (variability in strs; **13**: 143).
——. 1935. *Zentbl. Bakt. ParasitKde* Abt 2 **91**: 449 (general; **14**: 613).

Fusarium oxysporum

Snyder, W.C. et al. 1935. *Ibid* **91**: 355 (general; **14**: 486).
Tuveson, R. W. et al. 1959. *Bot. Gaz.* **121**: 69, 74 (virulence of biochemical mutants, parasexual cycle; **39**: 756).
Virgin, W. J. et al. 1939. *J. agric. Res.* **59**: 591 (effects of temp. & moisture; **19**: 252).
——. 1940. *Ibid* **60**: 241 (penetration & spread in host, seed transmission; **19**: 510).
Wade, B. L. 1929. *Res. Bull. Wisc. agric. Exp. Stn* 97, 32 pp. (inheritance of resistance; **9**: 423).
Walker, J. C. 1931. *Ibid* 107, 15 pp. (resistant cvs).
Wells, D. G. et al. 1949. *Phytopathology* **39**: 771, 907 (testing for resistance, linkage between resistance factors; **29**: 191, 396).
Whalley, W. M. et al. 1973. *Ann. appl. Biol.* **73**: 269 (effects of root exudates on germination of conidia & chlamydospores; **52**, 3876).
——. 1976. *Trans. Br. mycol. Soc.* **66**: 7 (chlamydospore germination in soil near susceptible & resistant cvs; **55**, 4361).
Worf, G. L. et al. 1962. *Phytopathology* **52**: 1126 (interaction with *F. solani* f. sp. *pisi*; **42**: 354).

Fusarium oxysporum Schl. f. sp. tracheiphilum (E. F. Smith) Snyder & Hansen, *Am. J. Bot.* 27: 66, 1940.

Fusarium tracheiphilum E. F. Smith, 1899
F. bulbigenum Cooke & Masse var. *tracheiphilum* (E. F. Smith) Wollenw., 1931.

A fairly widespread disease of soybean and cowpea (*Vigna unguiculata*) is caused. In cowpea the leaves become flaccid and chlorotic and in young plants a quite rapid wilt leads to death. On older plants stunting and chlorosis precede leaf fall and a gradual wilt. The vascular tissue is necrotic, and the roots may be more severely diseased than the above-ground symptoms suggest. The lower stem may become swollen before any chlorosis occurs. In the woody soybean a general wilt is not usually found. The plants become stunted with the chlorotic leaves gradually falling, and death ensues later than in the case of cowpea. Internal stem necrosis is a conspicuous symptom. Although the fungus chiefly causes the typical vascular disorders of the group, there are reports of soybean pod infection from Japan.

The opt. temp. for conidial germination is 24°. The fungus can survive under low O_2 and is tolerant of CO_2. Flooding would therefore be of little value in control. Races have been reported (Armstrong et al.). Race 1 caused wilt in some cvs of cowpea and soybean but neither race 1 or 2 produced wilt in some other plant spp. Race 3 caused wilt in the cowpea cv. Arlington which was resistant to the other 2 races. This race also caused more severe disease in cvs partially resistant to race 2. Some isolates from cowpea have proved pathogenic to common bean and it has been suggested that isolates of *F. oxysporum* f. sp. *phaseoli* (q.v.) might be virulent towards cowpea. Nematodes complicate the expression of this disease. Soil infested with the fungus caused no wilt in the resistant cowpea cv. Grant but when *Meloidogyne javanica* was also present wilt occurred. The wilt susceptible soybean cv. Yelrado was killed more rapidly in plots with the fungus plus *Heterodera glycines* than with the former alone. The root knot resistant soybean cv. Jackson wilted in non-sterile soil with the fungus plus *H. glycines* but not with either pathogen alone (Ross; Thomason et al.). Kendrick showed that seed could be contaminated with the pathogen.

In USA this fusarial wilt is now generally controlled with resistant cvs. Breeding in cowpeas, using the cv. Iron which is resistant to the 3 races, showed that resistance is dominant, 2 genes being involved for races 2 and 3. In Nigeria Oyekan described the reaction of cowpea cvs to infection. Field treatment with nematicides has reduced wilt.

Armstrong, G. M. et al. 1950. *Phytopathology* **40**: 181 (races; **29**: 453).
——. 1963. *Pl. Dis. Reptr* **47**: 1088 (races, hosts; **43**, 1472).
Cromwell, R. O. 1917. *J. agric. Res.* **8**: 421 (general).
——. 1919. *Res. Bull. Neb. agric. Exp. Stn* 14, 43 pp. (general).
Kendrick, J. B. 1931. *Phytopathology* **21**: 979 (seed transmission; **11**: 220).
Liu, K. 1940. *Mem. Coll. Agric. Kyoto* **47**: 15 (general; **19**: 453).
Mackie, W. W. 1934. *Phytopathology* **24**: 1135 (inheritance of resistance; **14**: 208).
Orton, W. A. 1902. *Bull. U.S.D.A. Pl. Ind.* 17 (general).
Oyekan, P. O. 1977. *Trop. Grain Legume Bull.* 8: 47 (reaction of cowpea cvs; **57**, 4247).
Ross, J. P. 1965. *Phytopathology* **55**: 361 (with nematodes; **44**, 2298).
Singh, R. S. et al. 1955. *J. Indian bot. Soc.* **34**: 375 (general; **35**: 503).
Thomason, J. J. et al. 1959. *Phytopathology* **49**: 602 (with nematodes; **39**: 204).
Toler, R. W. 1966. *Ibid* **56**: 183 (physiology; **45**, 2016).

Fusarium oxysporum Schl. f. sp. **vasinfectum** (Atk.) Snyder & Hansen, *Am. J. Bot.* 27: 66, 1940.
Fusarium vasinfectum Atk., 1892.

This vascular wilt of cotton is widely distributed (Map 362). Ebbels gave a full review, with special reference to Tanzania, of the disease which has been extensively studied in East Africa, Egypt, India and USA. Brief, general accounts were given by Booth et al.; Santos; Steyaert. El Nur et al. described some of the work in Egypt and Kalyanasundaram et al. likewise for India; and see Bekker et al. (USSR). The earliest symptoms are seen on the cotyledons as a vein clearing and chlorotic mosaic followed by necrosis; they also occur on the first true leaves and the seedling may be killed. Older plants become stunted with fewer leaves and bolls; the leaf symptoms of chlorosis and death on such plants have been described as spreading both upwards from the base and downwards from the tip. The usual dark discolouration of the vascular areas occurs in the root and stem. Roots and hypocotyl are directly penetrated.

The fungus can penetrate the roots of many hosts which show no external symptoms and Wood et al., from Tanzania, reported inoculations on other hosts in the Malvaceae, Sterculiaceae and Tiliaceae; *Hibiscus cannabinus* and *H. panduraeformis* (both common in cotton areas) were invaded most (and see Ebbels). Grover et al. described a wilt of *H. esculentus* (okra); and see Armstrong et al. 1960. The long field survival typical of the sp. occurs and there is seed transmission, probably sometimes internal (Elliott; Lagière; Taubenhaus et al.). Opt. temps for spore germination, growth in vitro and development of wilt are *c.* 30°; disease can be severe a few degrees below 30° but falls off rapidly at >33°. Infected young plants, at a favourable temp., wilt in 7–12 days; incubation is appreciably longer in old ones. Host nutrient imbalance (especially low potash) and nematode root attack (Smith, A. L. 1953) tend to increase incidence. In USA wilt is most severe on sandy acidic soils but those on which it occurs in India are heavier and alkaline. There is recent work on field survival in the absence of cotton, saprophytic colonisation and chlamydospore germination (El-Abyad et al. 1973; Smith, S. N.; Smith, S. N. et al.).

Although severe wilt can occur in the absence of root knot nematodes, infection by *Meloidogyne incognita* increases the disease and the effect tends to be synergistic. Attack by this nematode will break down the resistance of some cvs; its effect appears to be largely one of physical damage to the root system, thus facilitating entry of the fungus (Martin et al.; Miles; Perry 1961; Smith, A. L. 1941; Wickens et al; Yang et al. 1976b). Other nematodes also affect infection by *F. oxysporum* f. sp. *vasinfectum.* In USA (Louisiana) high populations of *Rotylenchulus reniformis* increased wilt (Neal 1954). *Belonoliamus gracilis* broke down resistance to the fungus (Holdeman et al.); *B. longicaudatus* (in work with *M. incognita*) increased wilt in seedlings, although seedling emergence was not affected (Minton et al.). Using *Pratylenchus brachyurus* Michell et al. found more wilt when a susceptible cv. was attacked by both organisms simultaneously than when the nematode was added 2 weeks before the fungus or when the latter was used alone. No wilt occurred when both organisms were applied to a resistant cv. *P. brachyurus* reproduced on both cvs but the population was higher on the susceptible one where the fungus and nematode had been applied at the same time. (See also Yang et al. 1976a and Yassin for the effects of *B. longicaudatus* and *P. sudanensis.*)

In pot experiments Sabet et al. described an interaction between *Cephalosporium maydis* (causing late wilt of maize; *Acremonium* q.v.) and *F. oxysporum* f. sp. *vasinfectum.* The former does not cause a disease in cotton, on which it is just a root surface inhabitant. When it is applied before the *Fusarium* there is less wilt. This may be due to an increase in the number of lateral roots thereby increasing host tolerance to wilt. Also cotton roots with *C. maydis* on their surfaces may be an unfavourable substrate for the wilt pathogen. The latter in vitro produces a metabolite which is toxic to the former; this may explain the fact that *C. maydis* has little effect when placed in the soil after the pathogen. When *F. oxysporum* f. sp. *vasinfectum* and *Verticillium dahliae* (q.v.) were both present in the cotton root system the former seemed to penetrate the host more quickly (Al Shukri 1968). The cell degrading enzymes have been described by Lakshminarayanan and Mahadevan et al. who also studied fusaric acid. Bugbee (1970) reported on occlusion of vessels; he found that wilt was induced by both stem and root inoculations, and that the site of inoculation did not affect the differential reaction of cvs (and see Beckman et al.).

Control is through oligogenic resistance; it needs to be incorporated with resistance to nematodes and some selections have resistance to *Xanthomonas malvacearum* as well. Selection and the study of the inheritance of resistance can be complicated by the presence of nematodes (e.g. Smith, A. L. et al. 1960; Wickens). The first named workers in USA found that resistance in upland cotton (*Gossypium hirsutum*) is controlled by one major dominant gene plus modifiers; additional genes determining resistance to *M. incognita*. In sea island (*G. barbadense*) there are 2 dominant genes, additive in effect. Armstrong, G. M. et al. (1960) fully investigated the known differences in pathogenicity between the Egyptian, Indian and USA forms of the fungus and its relationship with other wilt fusaria. Since there are vars in *G. arboreum* (Asiatic cotton), *G. barbadense* and *G. herbaceum* (Asiatic cotton) susceptible to all forms the determination of races on the basis of vars of one *Gossypium* sp. is not possible. The only generalisation that could be made concerning an entire sp. as a differential host was that the commercial cvs of *G. hirsutum* tested were not susceptible to either the Egyptian or Indian forms. Of the races recognised 1 (USA), 3 (Egypt) and 4 (India) could be separated using the 2 cvs. Sakel and Rozi only; Sakel being susceptible only to races 1 and 3, and Rozi susceptible to races 3 and 4. The more recently described USA race 2 (Armstrong, J. K. et al. 1958) has a wider host range than race 1 and can be separated on the bases of its pathogenicity to the soybean cv. Yelredo and flue-cured tobacco. Races 3 and 4 do not cause wilt in other hosts as does race 1. Race 2 is probably restricted in its distribution. Wickens considered that isolates in Tanzania resembled those from USA (race 1) and were therefore distinct from races 3 and 4. Ibrahim described a fifth race from Sudan, distinguished by its pathogenicity to the var. Ashmouni (*G. barbadense*). Race 6 was described from Brazil (Armstrong, G. M. et al. 1978). Charudattan divided isolates from India, Italy, USA and USSR into 3 groups based on pathogenicity. Brinkerhoff et al. found that cotton lines developed for resistance to *X. malvacearum* (7 out of 9 lines) had higher percentages of plants resistant to *F. oxysporum* f. sp. *vasinfectum* than did 3 susceptible ones. For resistance see, for example, Kappelman; Kappelman et al.; Krasichkov et al.; Young P. A. Mercurials have been successful as seed treatments and soil sterilants on a small scale (see El Nur et al.). Abd-el-Rehim et al. found that hot water (60° for 10 minutes) treatment of cotton seed reduced pre- and post-emergence losses.

Abd-el-Rehim, M. A. et al. 1969. *Ann. appl. Biol.* **63**: 95 (hot water seed treatment; **48**, 1734).

Al Shukri, M. M. 1968. *Biológia Bratisl.* **23**: 819 (host penetration with *Verticillium dahliae*; **48**, 1189).

——. 1968. *Pl. Dis. Reptr* **52**: 910 (growth in vitro & in vivo with *V. dahliae*; **48**, 1190).

——. 1969. *Ibid* **53**: 126 (longevity in infected host with *V. dahliae* & *V. albo-atrum*; **48**, 1723).

——. 1969. *J. Bot. United Arab Repub.* **12**: 13 (predisposing environmental factors; **49**, 3306).

Arjunarao, V. 1971. *Proc. Indian Acad. Sci.* Sect. B **73**: 265; **74**: 16, 53 (effects of antagonistic micro-organisms; **51**, 1522, 1523, 1524).

Armstrong, G. M. et al. 1960. *Tech. Bull. U.S. Dep. Agric.* 1219, 18 pp. (races 1, 2, 3 & 4; relationship to other wilt fusaria; **40**: 416).

—— ——. 1978. *Pl. Dis. Reptr* **62**: 421 (race 6).

Armstrong, J. K. et al. 1958. *Ibid* **42**: 147 (race 2; **37**: 459).

Beckman, C. H. et al. 1976. *Physiol. Pl. Pathol.* **9**: 87 (vascular structure & pathogen distribution, with *Verticillium*; **56**, 721).

Bekker, E. E. et al. (editors) 1971 (disease in USSR; **51**, 378).

Booth, C. et al. 1964. *C.M.I. Descr. pathog. Fungi Bact.* 28 (*Fusarium oxysporum* f. sp. *vasinfectum*).

Brinkerhoff, L. A. et al. 1961. *Pl. Dis. Reptr* **45**: 126 (reaction of lines resistant to *Xanthomonas malvacearum*; **40**: 538).

Bugbee, W. M. 1970. *Phytopathology* **60**: 121 (vascular response to infection; **49**, 2080).

Charudattan, R. et al. 1966. *Phytopathol. Z.* **55**: 239 (hosts; **45**, 3538).

——. 1969. *Proc. Indian Acad. Sci.* Sect. B **70**: 139 (morphology & pathogenicity of strs; **49**, 1031).

Chopra, B. K. et al. 1970. *Phytopathology* **60**: 717 (effect of prometryne in soil on growth of pathogen; **49**, 3130).

Dharmarajulu, K. 1932. *Indian J. agric. Sci.* **2**: 293 (root penetration & spread; **11**: 781).

Ebbels, D. L. 1975. *Cotton Grow. Rev.* **52**: 295 (review, 170 ref.; **55**, 2248).

El-Abyad, M. S. et al. 1971. *Trans. Br. mycol. Soc.* **57**: 427 (germination, sporulation & growth in vitro; **51**, 2510).

—— ——. 1973. *Ibid* **60**: 187 (competitive saprophytic colonisation; **52**, 3704).

El-Gindi, A. Y. et al. 1974. *Potash Rev.* **23**: 1 (interrelationships with K nutrition & *Rotylenchulus reniformis*; **55**, 253).

Elliot, J. A. 1923. *J. agric. Res.* **23**: 387 (seed transmission).

El Nur, E. et al. 1970. *Tech. Bull. Gezira Res. Stn* N.S. 2, 16 pp. (general; **50**, 105).

Fahmy, T. 1928. *Bull. Minist. Agric. Egypt tech. scient. Serv.* 74, 106 pp. (general; **7**: 781).

——. 1931. *Ibid* 95, 30 pp. (genetics of resistance & selection; **11**: 178).
Fikry, A. 1932. *Ann. Bot.* **46**: 29 (general; **11**: 513).
Grover, R. K. et al. 1970. *Indian J. agric. Sci.* **40**: 989 (on okra; **51**, 3030).
Holdeman, Q. L. et al. 1954. *Phytopathology* **44**: 683 (inoculation with *Belonolaimus gracilis*; **34**: 453).
Ibrahim, F. M. 1966. *Emp. Cotton Grow Rev.* **43**: 296 (race 5; **46**: 334).
Kalyanasundaram, R. et al. 1958. *Phytopathol. Z.* **33**: 321 (work in India; **38**: 145).
Kappelman, A. J. 1971. *Pl. Dis. Reptr* **55**: 896 (resistant cvs; **51**, 1517).
——. 1971. *Crop Sci.* **11**: 672 (inheritance of resistance; **52**, 4085).
—— et al. 1973. *Ibid* **13**: 280 (resistance with *Meloidogyne* spp.; **55**, 4122).
——. 1975. *Pl. Dis. Reptr* **59**: 803 (resistance; **55**, 3191).
Krasichkov, V. P. et al. 1962. *Khlopkovodstvo* **12**: 35 (as above; **41**: 521).
Lagière, R. 1952. *Coton Fibr. trop.* **7**: 146 (seed transmission; **32**: 186).
Lakshminarayanan, K. 1956 & 1958. *Proc. Indian Acad. Sci.*, Sect. B **44**: 317; **47**: 78 (pectin metabolism; pectin methylesterase formation; **36**: 404; **37**: 479).
——. 1957. *Naturwissenschaften* **44**: 93 (in vivo detection of PME; **37**: 585).
——. 1957. *Physiol. Pl.* **10**: 877 (adaptive nature of PME; **37**: 585).
Mahadevan, A. et al. 1967. *Phytopathol. Mediterranea* **6**: 86 (proteases, transeliminases & fusaric acid in wilted plant; **47**, 3451).
——. 1970. *Plant disease problems, Proc. 1st Int. Symp. Pl. Pathol.* New Delhi: 707 (pectolytic enzymes).
Martin, W. J. et al. 1956. *Phytopathology* **46**: 285 (nematodes & wilt development).
Meyer, J. 1960. *Agricultura Louvain* Sér. 2 **8**: 203 (disease in Zaire; **40**: 108).
Michell, R. E. et al. 1972. *Phytopathology* **62**: 336 (effect of *Pratylenchus brachyurus* on wilt incidence; **51**, 4037).
Miles, L. E. 1939. *Ibid* **29**: 974 (effect of *M. incognita* on host susceptibility to the fungus; **19**: 146).
Minton, N. A. et al. 1966. *Phytopathology* **56**: 319 (effect of *B. longicaudatus* & *M. incognita* in 3 soils; **45**, 2502).
Mitra, M. et al. 1935. *Proc. Indian Acad. Sci.* Sect. B **2**: 495 (temp. & growth in vitro; **15**: 365).
Mundkur, B. B. 1936. *Ibid* **3**: 498 (resistance of American cottons in India; **15**: 799).
Neal, D. C. 1927. *Ann. Mo. bot. Gdn* **14**: 359 (general, early literature, also as *Tech. Bull. Mo. St.* 16, 87 pp., 1928; **7**: 320).
——. 1954. *Phytopathology* **44**: 447 (effect of *R. reniformis*; **34**: 226).
Perry, D. A. 1961. *6th Commonw. mycol. Conf. 1960*: 48 (effect of *M. incognita*).
——. 1962. *Emp. Cotton Grow. Rev.* **39**: 14 (general in Tanzania; **41**: 522).
Rao, K. R. et al. 1970. *Mycopathol. Mycol. appl.* **41**: 271; **42**: 299 (hydrolytic enzymes of 2 mutants; tyronase in pathogenicity & resistance; **50**, 2152, 2153).
Rao, M. V. et al. 1966. *Phytopathol. Z.* **56**: 393 (on seedling roots & inoculum potential; **46**, 1007).
—— ——. 1966. *Trans. Br. mycol. Soc.* **49**, 403 (inoculum potential; **46**, 128).
Rodriquez-Kabana, R. et al. 1970. *Phytopathology* **60**: 65 (effect of atrazine on growth; **49**, 1918).
Sabet, K. A. et al. 1966. *Ann. appl. Biol.* **58**: 93 (interaction with *Cephalosporium maydis*; **45**, 3537).
Santos, F. H. 1969. *Agron. angol.* **29**: 143 (review, 61 ref.; **50**, 3783).
Smith, A. L. 1941. *Phytopathology* **31**: 1099 (cvs & requirement of nematode resistance; **21**: 197).
——. 1953. *U.S. Dep. Agric. Yearbk.*: 292 (general & nematodes).
—— et al. 1960. *Phytopathology* **50**: 44 (inheritance of resistance & nematodes; **39**: 413).
Smith, S. N. et al. 1975. *Ibid* **65**: 190 (soil survival in absence of cotton; **54**, 4506).
——. 1977. *Ibid* **67**: 502 (chlamydospore germination in suppressive & conducive rhizosphere soils; **56**, 5467).
Steyaert, R. L. 1945. *Notes phytopathol. Inst. Nat. Etud. agron. Congo Belge* 2, 15 pp. (general; **26**: 244).
Subba Rao, N. S. 1965. *Indian Jnl expl Biol.* **3**: 193 (transpiration in infected host; **45**, 1384).
Subramanian, D. 1956. *Proc. Indian Acad. Sci.* Sect. B **43**: 302 (fusaric acid & trace element chelation; **36**: 404).
Taubenhaus, J. J. et al. 1932. *Science N.Y.* **76**: 61 (seed transmission; **11**: 713).
Wickens, G. M. et al. 1960. *Emp. Cotton Grow. Rev.* **37**: 15 (disease in Uganda; **39**: 580).
——. 1964. *Ibid* **41**: 172 (inoculation & selection for resistance; **43**, 3229).
Wood, C. M. et al. 1972. *Cotton Grow. Rev.* **49**: 79 (host range & survival; **51**, 2512).
Yang, H. et al. 1976a & b. *J. Nematol.* **8**: 74, 81 (interaction with nematodes & effect of *Trichoderma harzianum*; **56**, 232, 3047).
Yassin, A. M. 1974. *Sudan agric. J.* **9**: 48 (with *P. sudanensis*; **55**, 5758).
Young, P. A. 1943. *Bull. Tex. agric. Exp. Stn* 627, 26 pp. (cvs resistant to pathogen & *M. incognita*; **23**: 225).
Young, V. H. 1928. *Bull. Ark. agric. Exp. Stn* 226, 50 pp. (disease & soil temp.; **8**: 101).
—— et al. 1941. *Ibid* 410, 24 pp. (disease, fertiliser balance & potash deficiency; **21**: 74).

Fusarium oxysporum

Fusarium oxysporum, other formae speciales.

Some other ff. sp. reported from tropical crops and accepted by Armstrong et al. 1968 (*F. oxysporum* q.v.) are:

anethi Gordon, 1965; on dill, *Anethum graveolens* (Janson).

apii (Nels. & Sherb.) Snyder & Hansen, 1940; on celery, *Apium graveolens* (Bouhot et al.; Otto et al.).

cannabis Noviello & Snyder, 1962; on hemp, *Cannabis sativa*.

cassiae Gordon, 1965; on *Cassia tora* (Armstrong et al. 1966).

cattleyae Foster, 1955; on orchid, *Cattleya* spp.

coffeae (Garcia) Wellman, 1954; on coffee, *Coffea* spp.

eucalypti Arya & Jain, 1962; on *Eucalyptus* spp.

glycines Armstrong & Armstrong, 1965; on soybean, *Glycine max* (Dunleavy).

lactucae Matuo & Motohashi, 1967a; on lettuce, *Lactuca sativa* (not listed by Armstrong et al. 1968 (l.c.) or Booth 1971, *Fusarium* q.v.).

lagenariae Matuo & Yamamoto, 1967b; on bottle gourd, *Lagenaria siceraria*. (Kuniyasu; Kuniyasu et al.; Takeuchi et al.).

lathyri (Bhide & Uppal) Gordon, 1965; on grass pea, *Lathyrus sativus* (Prasad, T. et al.).

lentis (Vasudeva & Srinavasan) Gordon, 1965; on lentil, *Lens esculenta* (Kannaiyan et al.; Khare et al.).

melongenae Matuo & Ishigami, 1958; on eggplant, *Solanum melongena*.

perniciosum (Hept.) Toole, 1941; on *Albizia* spp. (Cappellini et al.; Gill; Griffin et al.; Hepting; Phipps et al.; Stipes et al.).

psidii Prasad, Mehta & Lal, 1952; on guava, *Psidium guajava* (Edward).

rauvolfiae Janardhanan, Ganguly & Husain, 1964; on sarpangandha, *Rauvolfia serpentina* (Janardhanan et al.).

ricini (Wr.) Gordon, 1965; on castor, *Ricinis communis* (Arruda et al.).

sesami (Zaprometoff) Castellani, 1950; on sesame, *Sesamum indicum* (Rivers et al.).

spinaciae (Sherb.) Snyder & Hansen, 1940; on spinach, *Spinacia oleracea* (Armstrong et al. 1976; Bassi et al.; O'Brien et al.; Reyes).

vanillae (Tucker) Gordon, 1965; on vanilla, *Vanilla fragrans* (Alconero et al.; Irvine et al.; Leakey; Theis et al.; Tucker).

voandzeiae Armstrong, Armstrong & Billington, 1975; on bambara groundnut, *Voandzeia subterranea* (Ebbels et al.).

zingiberi Trujillo, 1963; on ginger, *Zingiber officinale* (Haware et al., Teakle).

(*nicotianae* (Johns.) Snyder & Hansen; on tobacco, *Nicotiana tabacum* (Holdeman; Melendéz et al.; Porter et al.) is not accepted by Armstrong et al. 1968 l.c.). For *raphani* see *F. oxysporum* f. sp. *conglutinans* and Inque et al.

Alconero, R. 1968. *Phytopathology* **58**: 1281 (f. sp. *vanillae*, infection & histology; **48**, 559).

—— et al. 1969. *Ibid* **59**: 1521 (f. sp. *vanillae* with *Rhizoctonia solani*; **49**, 1107).

Arruda, S. C. et al. 1937. *Biológico* **3**: 232 (f. sp. *ricini*; **17**: 65).

—— ——. 1940. *Ibid* **6**: 144 (as above; **19**: 675).

Arya, H. C. et al. 1962. *Phytopathology* **52**: 638 (f. sp. *eucalypti*; **42**: 158).

Armstrong, G. M. et al. 1965. *Ibid* **55**: 237 (f. sp. *glycines*; **44**, 2004).

—— ——. 1966. *Ibid* **56**: 699 (f. sp. *cassiae*; **45**, 3110).

—— ——. 1975. *Mycologia* **67**: 709 (f. sp. *voandzeiae*; **55**, 1563).

—— ——. 1976. *Phytopathology* **66**: 542 (common hosts for ff. sp. *betae* & *spinaciae*; **55**, 5974).

Bassi. A. et al. 1978. *Pl. Dis. Reptr* **62**: 203 (f. sp. *spinaciae* seedborne; **57**, 5167).

Bouhot, D. et al. 1977. *Rev. Hort.* **181**: 49 (f. sp. *apii*; **57**, 1892).

Cappellini, R. A. et al. 1976. *Bull. Torrey bot. Club* **103**: 227 (f. sp. *perniciosum*, pectolytic enzymes; **56**, 2539).

Dunleavy, J. 1962. *Proc. Iowa Acad. Sci.* **68**: 106 (f. sp. *glycines*; **41**: 753).

Ebbels, D. L. et al. 1972. *Trans. Br. mycol. Soc.* **58**: 336 (on bambara groundnut; **51**, 4546).

Edward, J. C. 1960. *Indian Phytopathol.* **13**: 30, 168 (f. sp. *psidii*; **41**: 98; **42**: 136).

Foster, V. 1955. *Phytopathology* **45**: 599 (f. sp. *cattleyae*; **35**: 458).

Gill, D. L. 1958. *Pl. Dis. Reptr* **42**: 587 (f. sp. *perniciosum* & nematodes; **37**: 664).

——. 1959. *Nat. hort. Mag.* **38**: 105 (as above & resistance; **39**: 53).

——. 1968. *Pl. Dis. Reptr* **52**: 949 (as above in seed; **48**, 1329).

Griffin, G. J. et al. 1975. *Ibid* **59**: 787 (as above, populations; **55**, 1967).

Haware, M. P. et al. 1974. *Indian Phytopathol.* **27**: 236 (on ginger, fungicides; **55**, 1401).

Hepting, G. H. 1936. *Ibid* **20**: 177 (f. sp. *perniciosum*: **15**: 758).

Holdeman, Q. L. 1956. *Phytopathology* **46**: 129 (f. sp. *nicotianae*; **35**: 726).
Inque, Y. et al. 1964. *Res. Progr. Rep. Tokai-Kinki natn. agric. exp. Stn* 1: 6 (f. sp. *raphani*; **44**, 282).
Irvine, J. et al. 1964. *Phytopathology* **54**: 827 (f. sp. *vanillae*; **44**, 202).
Janardhanan, K. K. et al. 1964. *Curr. Sci.* **33**: 313 (f. sp. *rauvolfiae*; **43**, 3283).
——. 1969. *Pl. Sci.* **1**: 136 (as above, enzymes; **51**, 559).
Janson, B. F. 1952. *Phytopathology* **42**: 152 (f. sp. *anethi*; **31**: 576).
Johnson, J. 1921. *J. agric. Res.* **20**: 515 (f. sp. *nicotianae*; **1**: 321).
Kannaiyan, J. et al. 1974. *Indian J. Pl. Prot.* **2**: 80 (f. sp. *lentis*, fungicides; **55**, 2946).
Kendrick, J. B. et al. 1942. *Phytopathology* **32**: **1031** (f. sp. *raphani*; **22**: **160**).
Khare, M. N. et al. 1970. *Mysore Jnl agric. Sci.* **4**: 354 (f. sp. *lentis* & resistance; **51**, 2077).
——. 1971. *Rep. Dep. Pl. Pathol. JNKVV 1970–71* 59 pp. (as above; **51**, 2076).
Kuniyasu, K. et al. 1977. *Ann. phytopathol. Soc. Japan* **43**: 192 (f. sp. *lagenariae*, seed transmission; **57**, 1483).
—. 1977. *Ibid* **43**: 270 (as above; **57**, 2689).
——. 1978. *Bull. Veg. Ornament. Crops Res. Stn* A 4: 149 (as above & disinfection; *Hort. Abstr.* **48**, 10533).
Leakey, C. L. A. 1970. *E. Afr. agric. For. J.* **36**: 207 (f. sp. *vanillae*; **50**, 3097).
Matuo, T. et al. 1958. *Ann. phytopathol. Soc. Japan* **23**: 189 (f. sp. *melongenae*; **39**: 368).
——. 1962. *Trans. mycol. Soc. Japan* **3**: 120 (f. sp. *ciceri* inter alia; **42**: 104).
——. 1967a. *Ibid* **8**: 13 (f. sp. *lactucae*; **47**, 2580).
——. 1967b. *Ibid* **8**: 61 (f. sp. *lagenariae*; **47**, 1366).
Melendéz, P. L. et al. 1967. *Phytopathology* **57**: 286 (f. sp. *nicotianae*; **46**, 2312).
Noviello, C. et al. 1962. *Ibid* **52**: 1315 (f. sp. *cannabis*; **42**: 387).
O'Brien, M. J. et al. 1977. *J. Am. Soc. hort. Sci.* **102**: 424 (f. sp. *spinaciae*, resistance; **57**, 828).
——. 1978. *Pl. Dis. Reptr* **62**: 427 (as above).
Otto, H. W. et al. 1976. *Calif. Agric.* **30**(6): 10 (f. sp. *apii*; **56**, 457).
Phipps, P. M. et al. 1975. *Phytopathology* **65**: 504 (f. sp. *perniciosum*, fungicides; **55**, 412).
——. 1976. *Ibid* **66**: 839 (as above, histology; **56**, 1763).
Porter, D. M. et al. 1967. *Ibid* **57**: 282 (f. sp. *nicotianae*; **46**, 2312).
Prasad, N. et al. 1952. *Nature Lond.* **169**: 753 (f. sp. *psidii*; **31**: 390).
Prasad, T. et al. 1972. *Indian Phytopathol.* **25**: 423 (f. sp. *lathyri*, survival; **53**, 203).
Reyes, A. A. 1977. *Pl. Dis. Reptr* **61**: 1067 (f. sp. *spinaciae*; **57**, 4179).
Rivers, G. W. et al. 1965. *Ibid* **49**: 383 (f. sp. *sesami*; **44**, 2855).
Stipes, R. J. et al. 1975. *Phytopathology* **65**: 188 (f. sp. *perniciosum*; **54**, 4654).
Takeuchi, S. et al. 1978. *J. centr. agric. Exp. Stn* **28**: 49 (f. sp. *lagenariae*, seed transmission).
Teakle, D. S. 1965. *Qd J. agric. Sci.* **22**: 263 (f. sp. *zingiberi*; **45**, 2199).
Theis, T. et al. 1957. *Phytopathology* **47**: 579 (f. sp. *vanillae*; **37**: 178).
Toole, E. C. 1941. *Ibid* **31**: 599 (f. sp. *perniciosum*; **20**: 612).
Trujillo, E. E. 1963. *Ibid* **53**: 1370 (f. sp. *zingiberi*; **43**, 1100).
Tucker, C. M. 1927. *J. agric. Res.* **35**: 1121 (f. sp. *vanillae*; **7**: 669).
Vasudeva, R. S. et al. 1952. *Indian Phytopathol.* **5**: 23 (f. sp. *lentis*; **33**: 134).

Fusarium solani (Mart.) Sacc., *Michelia* **2**: 296, 1881, emend. Snyder & Hansen pro parte, *Am. J. Bot.* **28**: 740, 1941.
stat. conid. of *Nectria haematococca* Berk & Br., *J. Linn. Soc.* (*Bot.*) **14**: 116, 1873
Fusisporium solani Martius, 1842
Fusarium javanicum Koorders, 1907.
(*Hypomyces solani* used as the name for the perithecial state of *F. solani* is untenable)

Growth moderately rapid, covering agar slope in 4–6 days with a somewhat sparse, floccose, greyish white mycelium. Typically a bluish to bluish brown discolouration develops in the agar. MICROCONIDIA form abundantly in the aerial mycelium from elongated lateral phialides; hyaline, cylindrical, wedge shaped or allantoid, 9–16 × 2–4 μ and may become 1 septate. MACROCONIDIA develop in 4–7 days from branched and well developed conidiophores; cylindrical to falcate, often slightly wider towards the apex with a well-marked foot cell, 40–100 × 5–7.5 μ. Opt. temp. for growth and conidia formation 28°. CHLAMYDOSPORES globose to oval, smooth to rough walled, 10–11 × 8–9 μ, intercalary or terminal; formed more frequently in darkness than in light. PERITHECIA sparse to densely gregarious, superficial; generally with a sparse white byssus forming a pseudoparenchymatous stroma which may be reduced; irregularly globose, pale orange to ochre or light brown; appearing gelatinous with roughly warted to furfuraceous outer wall, when dry

130–200 (110–250) μ diam. ASCI cylindrical, becoming clavate; apex rounded with central pore, 60–80 × 8–12 μ. ASCOSPORES ellipsoid to obovate, 11–18 × 4–7 (av. 12.5–5.5) μ; hyaline becoming light brown, slightly constricted at the single central septum, developing longitudinal striations (stat. conid. C. Booth & J. M. Waterston, *CMI Descr*. 29, 1964; perfect state after C. Booth, *The genus* Fusarium: 46, 1971; and see Booth, *Mycol. Pap*. 74, 1960, for an account of nomenclature and the nectrioid perfect state).

This is not the place for a full account of *F. solani* which is a worldwide soil inhabitant and recorded on a very large number of plants and animals. Its parasitism is both more catholic and less specialised than that of *F. oxysporum*. Infection of the roots and collar region (not heavily suberised or cuticularised) leads to a general cortical rot, not to a tracheomycosis. It has been reported as causing cankers in woody stems and is frequently described in disease complexes with other soil fungi. The genetics of the perfect state have been described (Buxton et al.; Snyder et al.). Both homothallism and heterothallism occur; perithecia are found frequently in the wet tropics but are rare or absent in temperate areas; they require light for their formation in vitro (Curtis; Hwang). Cochrane and co-workers have conducted intensive studies on the biochemistry of spore germination; and see Griffin. They defined a medium in which chlamydospores formed in high yield (Cochrane, V. W. et al. 1971); these spores tend to develop in 7–14 days on low nutrient media. Cell wall degrading enzymes have been described by Griffiths et al.; and for fine structure see Eck and Tewari et al.

The ff. sp. proposed for *F. solani* number at least 18 (Booth 1971, l.c.) but only 3, which are described, are major pathogens, i.e. on cucurbits, pea (*Pisum sativum*) and common bean. Some other hosts on which disease has been described are given very briefly. *F. solani* can be one of the causes of damping off. Root or collar rots have been described in broad bean (*Vicia faba*; Ibrahim et al.; Yu et al.), some vegetables (Baldacci et al.; Govindaswamy), guava (Chattopadhyay et al.), citrus with *Tylenchus semipenetrans* (Gundy et al.), *Dalbergia sissoo* (Bakshi et al.) and *Samanea saman* (Bose et al.). Hocking described a collar and stem canker of mature teak, the wood developing a pink stain. Isolates of *F. solani* killed seedlings when inoculated into collar wounds, and pink stained zones formed. Brown reported the perfect state (*N. haematococca*) from stem cankers on *Maesopsis eminii*. Ofong found a higher canker incidence where growing conditions were unfavourable; similarly for a coffee wilt (Baker). Zauberman et al. concluded that infection of avocado fruit caused accelerated ripening and softening. For onion see Du Plessis and for papaw see Quimio. Chee (see also Anon.) described a panel necrosis of rubber (*Hevea brasiliensis*) associated with *Botryodiplodia theobromae* (q.v.). A severe stem canker was caused on a shade tree (*Albizia procera*) in tea (Bagchee). Synergism with *Pythium myriotylum* (q.v.) in a pod rot of groundnut was reported by Frank to take the form of a predisposition of the pods by *F. solani*. In sterilised soil pod inoculation with the latter pathogen followed by the former caused more diseased pods than vice versa. *F. solani* also causes a tuber rot of Irish potato. Elarosi described a synergistic action with *Rhizoctonia solani* in this condition and Murdoch et al. have recently done work on control with fungicides. Joffe et al. did pathogenicity tests on several crops in Israel. Apart from the usual precautions against fungal soil pathogens when growing from seed or in glasshouses it seems improbable that soil fumigation to protect mature plants in the field would be economic. Control measures may be affected by other pathogens.

Anon. 1971. *Plrs' Bull. Rubb. Res. Inst. Malaya* 113: 81 (on rubber; **51**, 1850).

Bagchee, K. 1954. *Indian For*. **80**: 246 (on *Albizia procera*; **33**: 510).

Baker, C. J. 1972. *E. Afr. agric. For. J*. **38**: 137 (on coffee; **53**, 174).

Bakshi, B. K. et al. 1959. *Indian For*. **85**, 310, 415 (on *Dalbergia sissoo*; **38**: 629; **39**: 53).

Baldacci, E. et al. 1953? *Quad. Pio Ist. agric. Carlo Gallini* 32 pp. (on red pepper; **34**: 508).

Bose, S. R. et al. 1961. *Curr. Sci*. **30**: 307 (on *Samanea saman*; **41**: 180).

Brown, K. W. 1964. *E. Afr. agric. For. J*. **30**: 54 (on *Maesopsis eminii*; **44**, 254).

Buxton, E. W. et al. 1962. *Trans. Br. mycol. Soc*. **45**: 261 (heterokaryons; **42**: 104).

Chattopadhyay, S. B. et al. 1967. *Bull. bot. Soc. Beng*. **21**: 107 (on guava; **47**, 3168).

—— ——. 1968. *Indian J. agric. Sci*. **38**: 65, 176 (on guava; **47**, 3514).

Chee, K. H. 1971. *Pl. Dis. Reptr* **55**: 152 (on rubber; **50**, 2477).

Cochrane, J. C. et al. 1963. *Phytopathology* **53**: 1155 (requirements for spore germination; **43**, 913).

———. 1971. *J. gen. Microbiol.* **65**: 45 (synthesis of macromolecules & polyribosome formation; **50**, 3488).
Cochrane, V. W. et al. 1963. *Am. J. Bot.* **50**: 806 (endogenous respiration & spore germination; **43**, 914).
———. 1963. *Pl. Physiol.* **38**: 533 (carbohydrate metabolism & spore germination; **43**, 915).
———. 1963. *J. Bact.* **86**: 312 (metabolism of ethanol & acetate; **43**, 916).
———. 1966. *Pl. Physiol.* **41**: 810 (anaerobic metabolism & enzyme activity).
———. 1971. *Mycologia* **63**: 462 (chlamydospore formation in vitro; **51**, 147).
Curtis, C. R. 1969. *Can. J. Microbiol.* **15**: 863 (light induced perithecial formation; **49**, 40).
Daniels, D. L. et al. 1976. *Physiol. Pl. Pathol.* **8**: 9 (pisatin inducing components in filtrates; **55**, 3801).
Du Plessis, S. J. 1932. *Annale Univ. Stellenbosch* Ser. A **10** (2), 17 pp. (on onion; **12**: 135).
Eck, W. H. Van 1976. *Can. J. Microbiol.* **22**: 1628, 1634 (fine structure in soil; **56**, 1030, 1031).
Elarosi, H. *Ann. Bot.* **21**: 555, 569 (Irish potato tuber rot & culture in vitro; **37**: 107).
Frank, Z. R. 1972. *Phytopathology* **62**: 1331 (pod rot in groundnut; **52**, 3490).
Govindaswamy, C. V. 1963. *J. Madras Univ.* Sect. B **33**: 203 (on vegetables; **44**, 1831).
Griffin, G. J. 1970. *Can. J. Microbiol.* **16**: 733, 1366 (exogenous C & N requirements for macroconidia & chlamydospore germination, & spore density; **50**, 1085, 2713).
——. 1973. *Ibid* **19**: 999 (exogenous C & N requirements for chlamydospore germination; **53**, 1214).
——. 1976. *Ibid* **22**: 1381 (physiology of chlamydospore formation; **56**, 1449).
Griffiths, D. A. et al. 1966. *Pl. Dis. Reptr* **50**: 116 (host colonisation & pectolytic enzymes, with *Calonectria rigidiuscula*; **45**, 1677).
———. 1967. *Mycopathol. Mycol. appl.* **33**: 17 (host reaction on infection; **47**, 1016).
Gundy, S. D. Van et al. 1963. *Phytopathology* **53**: 488 (on citrus with *Tylenchus semipenetrans*; **42**: 680).
Hocking, D. 1968. *Pl. Dis. Reptr* **52**: 628 (on teak; **47**, 3591).
Hwang, S. W. 1948. *Farlowia* **3**: 315 (perithecial formation; **28**: 229).
Ibrahim, I. A. et al. 1964 & 1965. *Alex. J. agric. Res.* **12**: 221; **13**: 415 (on broad bean; **45**, 1420; **46**, 3282).
Joffe, A. Z. et al. 1972. *Phytopathol. Z.* **73**: 123 (pathogenicity, culture & distribution; **51**, 3810).
Matuo, T. et al. 1973. *Phytopathology* **63**: 562 (identification of ff. sp.; **53**, 1244).
Meyers, J. A. et al. 1972. *Ibid* **62**: 1148 (induction of chlamydospore formation by abrupt removal of organic C substrate; **52**, 1787).
Murdoch, A. W. et al. 1972. *Ann. appl. Biol.* **72**: 53 (control of Irish potato tuber rot with thiabendazole; **52**, 1241).
Ofong, A. U. 1974. *Pl. Dis. Reptr* **58**: 463 (on *M. eminii*; **53**, 4144).
——. 1974. *E. Afr. agric. For. J.* **39**: 311 (on *M. emenii*, incidence of cankers; **54**, 1021).
Papavizas, G. C. et al. 1966. *Phytopathology* **56**: 1269 (polygalacturonate transeliminase formation; **46**, 875).
Quimio, T. H. 1976. *Kalikasan* **5**: 241 (on papaw; **56**, 2598).
Rado, T. A. et al. 1971. *J. Bact.* **106**: 301 (ribosomal competence & spore germination; **51**, 89).
Snyder, W. C. et al. 1954. *Phytopathology* **44**: 338 (spp. concepts, genetics & pathogenicity; **34**, 402).
Tewari, J. P. et al. 1975. *Can. J. Bot.* **53**: 2134 (fine structure of macroconidia; **55**, 2558).
Yu, T. F. et al. 1948. *Phytopathology* **38**: 587 (on broad bean; **28**: 107).
Zauberman, G. et al. 1974. *Ibid* **64**: 188 (on avocado fruit; **53**, 3565).

Fusarium solani (Mart.) Sacc. f. sp. **cucurbitae** Snyder & Hansen, *Am. J. Bot.* **28**: 740, 1941.
Fusarium javanicum Koord, pro parte.

Foot and root rot of *Cucurbita* spp.; *Cucumis* spp. are susceptible but may be less frequently attacked in the field. Plants show a sudden wilt and collapse due to a soft, cortical rot at the collar. The rot begins as light green, watersoaked areas and it does not spread far up the stem or along the roots. Sporulation can occur on the infected areas. Infected fruits show soft, watersoaked lesions which become light brown and then sometimes white to grey, the whole fruit developing a soft, watery, internal rot. The disease is favoured by fairly high temps and symptoms show quickly at 30°. Tousson et al. compared the 2 forms or races of the fungus in USA. Race 1 is the common root, stem or fruit infecting form, and race 2 which attacks the fruit only. Both occur within the seed to a high degree; viability of such seed is not impaired and germination is generally unaffected. After germination and establishment of both races in the soil their behaviour differs. Race 1 attacks at any time (stem and fruit), whilst race 2 infects the fruit as they are maturing and penetrates the seed. Race 1 is probably worldwide whilst 2 was found only in California and Oregon. A soil survival time of at least 18 months was reported from Australia (Conroy). In squash fields (USA, Georgia) Sumner found a soil survival time of 20 months (40 months in artificially infested soil). But survival seems less than might be expected by analogy with other *Fusarium* ff. sp. Nash et al. found that race 1 chlamydospores did not survive in soil as long as those of *F. solani* f. sp. *phaseoli*. When macroconidia were placed in the soil 30–50% were converted into chlamydospores in 1–8 weeks; most of these did not survive for >1 year. Isolates of f. sp. *cucurbitae* were much more uniform than those of f. sp. *phaseoli*. Lysis of macroconidia in amended soil was described by Schippers et al. (1972). In Israel fusarial wilt of cucurbits generally developed 3 months after sowing. It was particularly severe in fields of cucumber, melon and watermelon that had been sown with any of these crops in the 3 preceding years (Palti et al.).

Diaz-Polanco et al. found that squash (*Cucurbita maxima*) plants with squash mosaic and watermelon mosaic viruses were slower in exhibiting hypocotyl symptoms caused by *F. solani* f. sp. *cucurbitae* than those without these viruses. There was a similar, though lesser effect, by wild cucumber mosaic virus. The viruses induced a decrease in chlamydospore germination in the hypocotyl rhizospheres (and see Magyarosy et al.). The most important control measure, which can be completely effective, is the use of clean seed from healthy fruits. Toussoun et al. demonstrated a wide range of pathogenicity (race 1) to many cucurbits and their cvs but in the field *Citrullus lanatus* (watermelon) was not attacked. Besides the use of healthy seed a 4 year rotation will also reduce the disease caused by race 1. In glasshouse cultivation steaming may be required; for example, there was an outbreak in the Netherlands on *Cucurbita ficifolia*, the rootstock for *Cucumis sativus* (cucumber; Kerling et al.). Sumner reported that *Cucurbita pepo* (marrow) was more susceptible than *C. moschata* (pumpkin).

Conroy, R. J. 1953. *Agric. Gaz. N.S.W.* **64**:(12) 655 (general; **34**: 510).

——. 1953. *J. Aust. Inst. agric. Sci.* **19**: 106 (reaction of different spp. & cvs; **33**: 202).

Diaz-Polanco, C. et al. 1969. *Phytopathology* **59**: 18 (effects of virus infection of host; **48**, 2100).

Eck, W. H. Van et al. 1976. *Soil Biol. Biochem.* **8**: 1 (fine structure of developing chlamydospores in vitro; **55**, 3920).

Gries, G. A. 1946. *Bull. Conn. agric. Exp. Stn* **500**, 20 pp. (general on squash; **27**: 307).

Hancock, J. G. 1968. *Phytopathology* **58**: 62 (degradation of pectic substances in vivo; **47**, 1367).

—— et al. 1968. *Can. J. Bot.* **46**: 405 (Ca localisation in hypocotyls & its effect on pectate lyase & tissue maceration; **47**, 2583).

——. 1976. *Phytopathology* **66**: 40 (endo-pectate lyase in vitro & in vivo; **55**, 3790).

Joffe, A. Z. et al. 1970. *Mycopathol. Mycol. appl.* **42**: 305 (general & morphological comparisons in Israel; **50**, 2172).

Kerling, L. C. P. et al. 1967. *Neth. J. Pl. Pathol.* **73**: 15 on *Cucurbita ficifolia*; **46**, 2146).

Magyarosy, A. C. et al. 1974. *Phytopathology* **64**: 994 (effects of virus induced changes in soil around hypocotyls; **54**, 2531).

Nash, S. M. et al. 1965. *Ibid* **55**: 963 (comparative soil survival with *F. solani* f. sp. *phaseoli*; **45**, 686).

Old, K. M. et al. 1973. *Soil Biol. Biochem.* **5**: 613 (fine structure of chlamydospores formed in soil; **53**, 4729).

Palti, J. et al. 1971. *Phytopathol. Z.* **70**: 31 (fusarial wilts of cucurbits in Israel; **50**, 3350).

Prasad, N. 1949. *Phytopathology* **39**: 133 (in vitro & pathogenic differences between isolates; **28**: 436).

Schippers, B. et al. 1972. *Neth. J. Pl. Pathol.* **78**: 45 (chlamydospore formation & lysis of macroconidia in chitin amended soil; **51**, 3800).

——. 1972. *Ibid* **78**: 189 (as above in N amended soil; **52**, 1026).

—— ——. 1974. *Soil Biol. Biochem.* **6**: 153 (chlamydospore formation in vitro; **54**: 54).

Sumner, D. R. 1976. *Pl. Dis. Reptr* **60**: 923 (aetiology & control; **56**, 3301).

Toussoun, T. A. et al. 1961. *Phytopathology* **51**: 17 (life history & control of races 1 & 2; **41**: 77).

Fusarium solani (Mart.) Sacc. f. sp. **phaseoli** (Burk.) Snyder & Hansen, *Am. J. Bot.* **28**: 740, 1941.

Dry rot of common bean (*Phaseolus vulgaris*) is the disease. The name refers to the shrivelled appearance of the roots and bases of affected plants; this is accompanied by a reddish or reddish brown discolouration. The above ground parts of the plant are not directly attacked but they show the first clear symptoms of chlorosis in the lower leaves, slow growth and poor pod set. Once the soil has become heavily infested through repeated cropping the disease may be difficult to control. In severe attacks plants are killed and overall yields reduced by half. The pathogen penetrates stomata on the hypocotyl and wounded or intact roots. Mycelia form on the host surface before penetration but there are no appressoria. The whole of the cortex is invaded, spread being mostly stopped by the endodermis. Initial infection occurs more frequently via the hypocotyl (Chatterjee; Christou et al.). Maier (1961) reported most disease at a constant soil temp. of 24°

compared with 18, 28 and 32°. The most rapid development of root rot (at all temps) occurred during the first week after emergence and there was most disease after 3 weeks. Burke (1965a) found more root rot at 16° compared with 21, 27 and 32°. Nash et al. (1964) described surface contamination of seed. Pectolytic enzymes have been demonstrated in vitro and in vivo by Bateman et al.

This particular host and pathogen in soil relationship has been investigated in detail and some of this work was discussed by Toussoun (1970). Toussoun et al. (1961) found that there was a zone of chlamydospore germination close to the host surface. In soil most germination took place near germinating bean seed and root tips; mature roots had little effect; added organic matter also stimulated germination of chlamydospores (Schroth et al.). Nash et al. (1961–62) described the characteristics of these spores in which form the pathogen chiefly exists in the soil. Macroconidia in unsterilised field soil either germinated by a short germ tube which formed chlamydospores or were converted directly into chlamydospores. The count of pathogen propagules dropped markedly, and their distribution became less uniform, when bean was not grown for 1.5 years and the soil cropped to tomatoes. Cook et al. found 60% chlamydospore germination 16 hours after sowing bean seed. If adequate carbon and nitrogen was added to the soil in nutrients there was 40% germination in the same period. Ford et al. investigated the factors which induced chlamydospore germination. In extensive work Burke and others (1965b & c, 1966, 1968a, 1972) showed that in soil types with a history of high dry rot incidence the chlamydospores were both larger and more abundant compared with those with one of low dry rot incidence. They demonstrated the importance of lateral root infection and found that nitrogen did not greatly affect disease incidence. Fungal propagules were distributed throughout the ploughed layer but were seldom detected in subsoil (33–41 cm). In infested fields fewer roots entered the subsoil compared with non-infested ones. Miller et al. (1977) concluded that aggravation of the disease by low oxygen diffusion rates is the main cause of plant stunting and yield reduction; these result from temporary and excessive wetting of the soil in infested fields. Hutton et al. investigated the effects of *Pratylenchus penetrans* and *Meloidogyne* spp.

Thus cultural, soil factors affect the disease; soil compaction should be avoided, and everything done to promote deep and vigorous rooting. Long rotations can reduce the disease but those lasting 3 years are likely to fail eventually. Subsoiling (to reduce compaction) has been shown to increase yields especially in the less resistant cvs. It was more effective under the drill row than between rows (Burke 1968b; Burke et al. 1972). Natti et al. found that in furrow, sprays of benomyl were effective but that seed treatment with chemicals was not. Resistance has been found in *Phaseolus coccineus*, *P. lunatus* and *P. vulgaris*. Its inheritance is polygenic; both dominant and additive effects occur. Some resistant cvs have been developed in USA (Baggett et al.; Bravo et al.; Burke et al. 1967; Hassan et al.; Wallace et al.). Boomstra et al. discussed the strategy in breeding for resistance. In field trials Russell et al. found both benomyl and thiabendazole to be effective as seed treatments.

Adams, P. B. et al. 1968. *Phytopathology* **58**: 373, 378, 1603 (effect of cellulose amendment, fungistasis, control with coffee grounds; **47**, 2408b & c; **48**, 1398).

Alexander, J. V. et al. 1966. *Ibid* **56**: 353 (stimulation of chlamydospore formation by sterile soil extracts; **45**, 2406).

Baggett, J. R. et al. 1965. *Pl. Dis. Reptr* **49**: 630 (tests for resistance in *Phaseolus* spp.; **44**, 3210).

Baker, R. et al. 1965. *Phytopathology* **55**: 1381 (cellulose, N & effects on inoculum density in rhizosphere; **45**, 1255).

Bateman, D. F. et al. 1966. *Ibid* **56**: 238 (hydrolytic & trans-eliminative degradation of pectic substances; **45**, 2017).

Boomstra, A. G. et al. 1977. *J. Am. Soc. hort. Sci.* **102**: 186 (inheritance of, & breeding strategy for, resistance; **56**, 5282).

Bravo, A. et al. 1969. *Phytopathology* **59**: 1930 (inheritance of resistance; **49**, 1844).

Burke, D. W. 1965a, b & c. *Ibid* **55**: 757, 1122, 1188 (effects of plant spacing & soil types, immobility in soil, **45**, 293, 667, 913).

—— et al. 1966. *Ibid* **56**: 292 (importance of lateral root infection; **45**, 2298).

—— ——. 1967. *Bull. Wash. agric. Exp. Stn* 687, 5 pp. (N fertiliser & disease; **47**, 3267).

—— ——. 1968a. *Circ. Wash. agric. Exp. Stn* 490, 9 pp. (as above; **49**, 1213).

——. 1968b. *Phytopathology* **58**: 1575 (obstruction to root growth & disease; **48**, 985).

—— ——. 1972. *Ibid* **62**: 306, 550 (subsoiling & disease, effects of soil tillage & compaction on pathogen distribution; **51**, 4485, 4486).

Burkholder, W. H. 1919. *Mem. Cornell Univ. agric. Exp. Stn* 26: 1003 (general).

Fusarium solani

Byther, R. 1965. *Phytopathology* **55**: 852 (use of inorganic N & saprophytism; **45**, 43).
Chatterjee, P. 1958. *Ibid* **48**: 197 (penetration & histology; **37**: 618).
Christou, T. et al. 1962. *Ibid* **52**: 219 (penetration, invasion & spore formation in host; **41**: 751).
Cook, R. J. et al. 1965. *Ibid* **55**: 254, 1021 (effects of C & N compounds & host exudates on chlamydospore germination; **44**, 2291; **45**, 668).
Ford, E. J. et al. 1970. *Ibid* **60**: 124, 479, 1732 (induction of chlamydospore formation by soil substances, bacteria & C; **49**, 1970, 2770; **50**, 2742).
Griffin, G. J. 1964. *Can. J. Microbiol.* **10**: 605 (long-term effect of soil amendments on conidial germination; **44**, 634).
Hassan, A. A. et al. 1971. *J. Am. Soc. hort. Sci.* **96**: 623 (inheritance of resistance; **51**, 2051).
Hutton, D. G. et al. 1973. *Phytopathology* **63**: 749 (effect of nematodes; **53**, 732).
Ito, I. et al. 1975. *Mem. Fac. Agric. Hokkaido Univ.* **9**: 187 (behaviour in rhizosphere & underground tissue; **54**, 3564).
Lewis, J. A. et al. 1968. *Phytopathol. Z.* **63**: 124 (decomposition of tannins & lignins in soil; **48**, 986).
—— ——. 1977. *Phytopathology* **67**: 925 (effects of plant residues on chlamydospore germination & disease; **57**, 1501).
Maier, C. R. 1961. *Pl. Dis. Reptr* **45**: 960 (effects of soil temp. & crop residues; **41**: 496).
——. 1968. *Phytopathology* **58**: 620 (effect of N; **47**, 2909).
Maloy, O. C. et al. 1959. *Ibid* **49**: 583 (effect of crop rotation; **39**: 203).
——. 1960. *Ibid* **50**: 56 (comparison of mycelial & conidial types in vitro & in soil; **39**: 451).
Maurer, C. L. et al. 1964. *Ibid* **54**: 1425 (effect of chitin & lignin amendments; **44**, 2001a).
—— ——. 1965. *Ibid* **55**: 69 (effect of glucose, cellulose & inorganic N amendments; **44**, 2001b).
Miller, D. E. et al. 1974. *Ibid* **64**: 526 (effect of soil bulk density & water potential on disease; **53**, 4189).
—— ——. 1975. *Ibid* **65**: 519 (effect of soil aeration on disease; **54**, 5631).
—— ——. 1977. *Pl. Dis. Reptr* **61**: 175 (effect of temporary, excessive wetting on disease; **56**, 4776).
Mitchell, R. et al. 1962. *Proc. Soil Sci. Soc. Am.* **26**: 556 (control with chitin soil amendment; **42**: 451).
——. 1963. *Bull. Res. Coun. Israel* Sect. D **11**: 230 (lysis of pathogen in laminarin amended soil; **43**, 3149).
Nash, S. M. et al. 1961. *Phytopathology* **51**: 308 (characteristics of chlamydospores in soil; **41**: 74).
—— ——. 1962. *Ibid* **52**: 567 (plate counts of pathogen propagules in field soils; **42**: 65).
—— ——. 1964. *Ibid* **54**: 880 (spread with seed; **44**, 294).
Natti, J. J. et al. 1971. *Pl. Dis. Reptr* **55**: 483 (seed & soil chemical treatments; **51**, 868).
Papavizas, G. C. et al. 1968. *Phytopathology* **58**: 365, 414 (effect of soil C:N balance & saprophytic multiplication; **47**, 2408a, 2615a).
Russell, P. E. et al. 1977. *J. agric. Sci. Camb.* **89**: 235 (fungicides for seed; **56**, 5882).
Schroth, M. N. et al. 1961. *Phytopathology* **51**: 389 (effect of host exudate on chlamydospore germination; **41**: 74).
—— ——. 1962. *Ibid* **52**: 906 (effect of non-susceptible plants on survival; **42**: 286).
—— ——. 1963. *Ibid* **53**: 610, 809 (pathogen lesions, host exudates & chlamydospore germination; **42**: 715; **43**, 298).
Toussoun, T. A. et al. 1960. *Ibid* **50**: 137 (effect of N & glucose on pathogenicity to excised stems; **39**: 522).
—— ——. 1961. *Ibid* **51**: 620 (chlamydospore germination in unsterilised soils; **41**: 266).
—— ——. 1963. *Ibid* **53**: 265 (phytotoxic substances from plant residues & disease in excised stems; **42**: 584).
——. 1970. In *Root diseases and soil-borne pathogens*: 95. Univ. California Press.
Wallace, D. H. et al. 1965. *Phytopathology* **55**: 1227 (breeding for resistance; **45**, 914).
Weimer, J. L. et al. 1926. *J. agric. Res.* **32**: 311 (general; **5**: 462).

Fusarium solani (Mart.) Sacc. f. sp. **pisi** (Jones) Snyder & Hansen, *Am. J. Bot.* **28**: 740, 1941.

Fusarium martii var. *pisi* Jones, 1923.

Foot rot of pea (*Pisum sativum*) may be seen first as a withering of the lower leaves when the pods begin to form. The lower stem, tap root and sometimes main secondary roots, bear elongate, necrotic lesions variously described as reddish or purplish brown, chocolate brown or black. Discolouration may extend to the vascular region but this is usually limited; and the symptoms (primarily a cortical rot) need not be confused with the vascular wilt caused by *F. oxysporum* f. sp. *pisi* (q.v.) with which *F. solani* f. sp. *pisi*, as the less important of the 2 fungi, is often associated in a pathogenic complex. In severe infections the plants are killed. Basu et al. described loss conversion factors. The fungus can penetrate the stomata of the cotyledons; the first symptoms show after a week as necrotic flecks on parts of the seedling. The roots of 9-week-old plants can be almost completely invaded (Bywater). Under very humid conditions the aerial parts of older plants can be infected. Kraft (1969) reported chick pea (*Cicer arietinum*) as a host.

Walker (*Diseases of vegetable crops*) gave an opt. temp. for the disease of 27–30° and it occurs at soil temps of > 18°. The stimulation of chlamydospore germination by the host after seeds are planted leads to a max. germination after 20 hours (Cook et al. 1967). Bolton et al. have compared *F. solani* f. sp. *pisi* with *F. oxysporum* f. sp. *pisi*; and Kraft et al. (1974) with *F. solani* f. sp. *phaesoli*. Cook et al. (1968) recovered one isolate from wind blown soil > 1 km from a pea field and others from dust with pea seed. There appears to be no evidence for internal seed transmission. The pathogen has frequently been studied in association with other organisms. More severe disease can occur in infections with *Pythium ultimum* (q.v., Escobar et al.; Kraft et al.). In root rots associated with *Aphanomyces euteiches* (q.v.) Beute et al. found increased root rot caused by the 2 fungi in peas infected with pea common mosaic and bean yellow mosaic viruses. They considered that virus infection increased the permeability of root cell membranes and the resulting exudation led to an increase in substrates for fungal growth. Farley et al. similarly found more infection by these 2 fungi in virus infected pea. Short et al. reported more chlamydospore germination at 50% soil moisture than at 20%. The cv. type could affect such germination. (For phytoalexins and resistance, with *F. solani* f. sp. *phaseoli*, see *Rev. Pl. Pathol.* **53**, 418, 1131; **55**, 460, 1544, 3802; **57**, 850, 1496.)

Long rotations or new land may be advisable and also chemical seed dressings for control. Ogilvie (*Diseases of vegetables*) stated that pea cvs with round seeds are usually more susceptible than those with wrinkled ones, and that cvs with tough stems appear to be fairly resistant. Kraft et al. (1970) described resistance in some accessions. In 28 days the populations of the pathogen increased more in rhizosphere soil from the roots of susceptible plants than in those from resistant ones.

Basu, P. K. et al. 1976. *Can. Pl. Dis. Surv.* **56**:25 (loss conversion factors; **56**, 1809).

Bolton, A. T. et al. 1972. *Can. J. Pl. Sci.* **52**: 189 (comparisons with *F. oxysporum* f. sp. *pisi*; **51**, 4471).

Bywater, J. 1959. *Trans. Br. mycol. Soc.* **42**: 201 (infection & spread; **39**: 66).

Cook, R. J. et al. 1967. *Phytopathology* **57**: 178 (chlamydospore germination in soil; **46**, 2132).

—— ——. 1968. *Aust. J. agric. Res.* **19**: 253 (perfect state & dissemination; **47**, 2322).

Dorn, S. 1974. *Phytopathol. Z.* **81**: 193 (toxin; **54**, 4248).

Eck, W. H. Van 1978. *Can. J. Microbiol.* **24**: 65 (lipid body content & chlamydospore persistence in soil; **57**, 3316).

Hall, R. 1967. *Aust. J. biol. Sci.* **20**: 419 (proteins & catalase enzymes; **46**, 2637).

Harter, L. L. 1938. *Phytopathology* **28**: 432 (general; **17**: 787).

Jones, F. R. 1923. *J. agric. Res.* **26**: 459 (general).

Kerling, L. C. P. 1953. *Tijdschr PlZiekt.* **59**: 62 (effects of sand drift & water on infection; **33**: 63).

Knavel, D. E. 1967. *Proc. Am. Soc. hort. Sci.* **90**: 260 (resistance).

Kraft, J. M. 1969. *Pl. Dis. Reptr* **53**: 110 (inoculation of chick pea; **48**, 2085).

—— et al. 1970. *Phytopathology* **60**: 1814 (resistance; **50**, 2574).

—— ——. 1974. *Pl. Dis. Reptr* **58**: 500 (behaviour with *F. solani* f. sp. *phaseoli*; **54**, 1938).

Majumdar, M. et al. 1976. *Acta phytopathol. Acad. Sci. hung.* **11**: 45 (effect of temp.; **56**, 1342).

Reddy, M. N. et al. 1972. *Phytopathol. Z.* **74**: 55 (multiple molecular forms of enzymes in infected peas; **52**, 285).

Reiling, T. P. et al. 1960. *Phytopathology* **50**: 287 (soil indexing & effects on yield; **39**: 646).

Reinking, O. A. 1950. *Ibid* **40**: 664 (comparisons with *F. solani* f. sp. *phaseoli*; **30**: 92).

Reyes, A. A. et al. 1962. *Ibid* **52**: 1196 (growth response in rhizospheres of host & non-hosts; **42**: 308).

Schmidt, H. H. 1972. *Arch. PflSchutz.* **8**: 3 (effect of host & non-host plant residues & N on infection; **52**, 284).

Schuster, M. L. 1948. *Bull. Wash. agric. Exp. Stn* 499, 42 pp. (general, chemical treatment of seed; **29**: 133).

Shaykh, M. et al. 1977. *Pl. Physiol.* **60**: 170 (formation of cutinase; **57**, 341).

Short, G. E. et al. 1974. *Phytopathology* **64**: 558 (chlamydospore germination in spermosphere; **53**, 4648).

Yang, S. M. et al. 1968. *Ibid* **58**: 639 (cultural & pathogenic variation induced by UV; **47**, 2908).

Fusarium udum Butler, *Mem. Dep. Agric. India bot. Ser.* **2**(9): 54, 1910.

Fusarium oxysporum Schl. emend. Snyder & Hansen f. *udum* (Butler) Snyder & Hansen, 1940

F. udum Butler var. *cajani* Padwick, 1940

F. lateritium Nees emend. Snyder & Hansen f. sp. *cajani* (Padwick) Gordon, 1952

F. uncinatum Wollenw., 1917

F. lateritium Nees var. *uncinatum* (Wollenw.) Wollenw., 1931.

Cultures often with deep purple discolouration of agar, aerial mycelium felted or almost absent and usually with profuse development of pionnote sporodochia. CONIDIA initially produced on simple or verticillately branched CONIDIOPHORES, later formed from pionnote or normal SPORODOCHIA when they form pinkish or salmon-coloured masses. On the host they are usually formed from sporodochia. MICROCONIDIA aseptate, hyaline, ovoid–fusoid or curved, 6–11 × 2–3 μ. MACROCONIDIA hyaline, typically thin walled, 1–3 occasionally 5 septate, falcate with a distinct foot cell and an atypical cell of decreasing diam. towards the tip which may be curved or hooked; 15–30 (–46) × 2.5–3.5 μ. CHLAMYDOSPORES globose, intercalary in the mycelium, 8–10 μ diam. Perithecial state unknown (C. Booth, *The genus* Fusarium: 113, 1971, see also Booth 1978).

Vascular wilt of pigeon pea (*Cajanus cajan*) appears to be only of some importance in India. Although it is an old disease (see Butler, *Fungi and disease in plants*, for an early account) the details of the pathogen's life history in the field appear to be imperfectly known. Wilt symptoms can occur in the seedling stage or in maturing plants which may not flower or may drop their flowers prematurely. In inoculations Subramanian described epinasty after 10 days, followed by interveinal chlorosis, yellowing, turgor loss, leaf shrivelling and general wilt by the thirtieth day. In field plants the vascular discolouration can be traced in the stem for a considerable distance and plants are usually killed. The precise point where initial infection takes place and the opt. conditions for this and for disease development have not apparently been described. The disease can occur over a wide range of soil temps (17–29°). Differences in virulence between isolates have been reported (Sarojini); there is no indication that the pathogen is carried in the seed (Mohanty). Vasudeva et al. investigated the antagonism of other micro-organisms. As Butler (l.c.) implied rotations have probably reduced the severity of wilt in the important growing areas of central India. Cvs differ appreciably in their reaction towards infection. Subramanian, in 6 cvs, found the least resistant one to be colonised up the whole of the main stem, but in the most resistant one the fungus did not spread beyond the root region (and see Shaw; McRae et al.). Of 58 pigeon pea cvs Mukherjee et al. found that 9 were moderately resistant, and see Singh, D. V. No field scale attempts at fungicide control appear to have been described recently.

Booth, C. 1978. *C.M.I. Descr. pathog. Fungi Bact.* 575 (*Fusarium udum*).

Chadha, K. C. et al. 1965. *Indian J. agric. Sci.* **36**: 133 (interaction with pigeon pea sterility mosaic virus; **46**, 488).

Kaiser, S. A. K. M. et al. 1975. *Z. PflKrankh. PflSchutz* **82**: 485 (confirmation of host specificity, with nonpathogenic forms; **55**, 2981).

—— ——. 1976. *Trans. Br. mycol. Soc.* **67**: 33 (considered a f. sp. of *F. oxysporum*; **56**, 1485).

McRae, W. et al. 1933. *Scient. Monogr. Coun. agric. Res. India* 7, 68 pp. (effect of manure & resistant cvs; **13**: 345).

Maitra, A. et al. 1976. *Z. PflKrankh. PflSchutz* **83**: 742 (effect of infection on host & histology; **56**, 4293).

Mohanty, U. N. 1946. *Indian J. agric. Sci.* **16**: 379 (distribution in host; **28**: 46).

Mukherjee, D. et al. 1971. *Indian Phytopathol.* **24**: 598 (reaction of cvs; **52**, 941).

Mundkur, B. B. 1935. *Indian J. agric. Sci.* **5**: 609 (effect of temp. & host maturity, with *F. udum* f. sp. *crotalariae*; **15**: 771).

Padwick, G. W. 1940. *Ibid* **10**: 863 (morphology & culture; **20**: 496).

—— et al. 1940. *Ibid* **10**: 707 (cross infection tests with *Fusarium* isolates from *Cajanus cajan*, *Crotalaria juncea* & *Gossypium*; **20**: 232).

Sarojini, T. S. 1951. *Proc. Indian Acad. Sci.* Sect. B **33**: 49 (effect of soil micronutrients; **32**: 23).

Shaw, F. J. F. 1936. *Indian J. agric. Sci.* **6**: 139 (inheritance of resistance; **15**: 771).

Singh, D. V. et al. 1976. *Indian J. Mycol. Pl. Pathol.* **6**: 89 (reaction of cvs; **57**, 1919).

Singh, G. P. et al. 1962. *Curr. Sci.* **31**: 110 (formation of pectolytic & cellulytic enzymes; **41**: 690).

—— ——. 1968. *Indian Phytopathol.* **21**: 361 (as above; **48**, 3758).

——. 1968. *Indian Jnl Microbiol.* **8**: 95 (hydrolytic enzymes; **49**, 942).

—— ——. 1970. *Proc. natn. Acad. Sci. India* Sect. B **40**: 9 (phytotoxicity of culture filtrates; **51**, 899).

Singh, P. et al. 1965. *Ann. appl. Biol.* **55**: 89 (bulbiformin & seed protection; **44**, 2044).

Singh, R. S. 1974. *Indian Phytopathol.* **27**: 553 (effects of host & non-host seeds on growth & sporulation; **55**, 4532).

Subramanian, S. 1963. *Proc. Indian Acad. Sci.* Sect. B **57**: 134, 178, 259 (symptoms & infection, changes in host metabolism & effect of Mn; **42**: 716).

Vasudeva, R. S. et al. 1950. *Ann. appl. Biol.* **37**: 169 (effect of soil micro-organisms on wilt; **29**: 624).

—— ——. 1953. *Ibid* **40**: 573 (as above; **33**: 273).

Fusarium udum f. sp. **crotalariae** (Kulkarni) Subramanian comb. nov.

Fusarium vasinfectum Atk. *crotalariae* Kulkarni, *Indian J. agric. Sci.* **4**: 994, 1934

F. udum Butler var. *crotalariae* (Kulkarni) Padwick, 1940

F. lateritium Nees emend. Snyder & Hansen f. sp. *crotalariae* (Padwick) Gordon, 1952.

This f. sp. is distinguished by its pathogenicity to sunn hemp (*Crotalaria juncea*) in which it causes a vascular wilt. A form of *Fusarium* described from Brazil as attacking *Crotalaria* spp. may or may not be the same (*Rev. appl. Mycol.* **36**: 130, 1957). Uppal et al. gave an account of the disease in India, and it has also been described from USA and Trinidad. Seedlings can be killed after emergence, older plants collapsing in 2–3 weeks. In India wilting can be rapid in Sept.–Oct., becoming gradual in June–July. Mundkur found that higher temps (28–33°) favoured the wilt on sunn hemp compared with the one on pigeon pea (*F. udum*). Several other *Crotalaria* spp. are susceptible, for example, *anagyroides*, *intermedia*, *mucronata* and *zanzibarica*. Armstrong et al. used spp. from the genus to distinguish 3 races. The fungus may occur in the seed but apart from seed treatment there are no other specific control measures.

Armstrong, J. K. et al. 1951. *Phytopathology* **41**: 714 (races; **31**: 19).

Mitra, M. 1934. *Indian J. agric Sci.* **4**: 701 (inoculation; **14**: 144).

Mundkur, B. B. 1935. *Ibid* **5**: 609 (effect of temp. & maturity; **15**: 771).

Thorold, C. A. 1931. *Trop. Agric. Trin.* **8**: 176 (in Trinidad; **11**: 107).

Uppal, B. N. et al. 1937. *Indian J. agric. Sci.* **7**: 413 (general; **17**: 40).

Wright, J. et al. 1931. *Trop. Agric. Trin.* **8**: 151 (in Trinidad; **10**: 799).

GAEUMANNOMYCES Arx & Olivier, *Trans. Br. mycol. Soc.* **35**: 32, 1952.
(Diaporthaceae)

PERITHECIA single, dispersed, sunken in the substratum, globose, black, communicating with the exterior by a periphysis lined pore in a cylindrical, usually somewhat unilaterally inserted, beak; perithecial wall pseudoparenchymatous, light or dark brown. ASCI numerous, cylindrical, unitunicate, thin walled, stalked, with 8 spores and a refractive ring in the slightly thickened apical wall. ASCOSPORES parallel in the ascus, vermiculate, hyaline, continuous when young, with pseudosepta when ripe. PARAPHYSES filamentous and very delicate, evanescent (from J. A. von Arx & D. L. Olivier l.c., q.v. for taxonomy).

Crown sheath rot (a minor disease) of rice, also called brown or black sheath rot and Arkansas foot rot, is caused by *G. graminis* (Sacc.) Arx & Olivier var. *graminis* (synonyms: *Rhaphidophora graminis* Sacc., *Ophiobolus graminis* (Sacc.) Sacc., *O. oryzinus* Sacc., *Ophiochaeta graminis* (Sacc.) K. Hara, *Linocarpon oryzinum* (Sacc.) Petrak). The fungus was recently described by Walker (1973) who reviewed (1975) the recent taxonomy and pathology of the take-all diseases of Gramineae. *G. graminis* vars *avenae* (E. M. Turner) Dennis and *tritici* Walker cause these diseases on temperate cereals and grasses (see Walker 1973).

Var. *graminis* reduced from Walker (1973): PERITHECIA in lower sheaths, commonly oval, black, 200–400 × 150–300 μ, neck variable in length, 100–400 × 70–100 μ; ostiolar canal 25–40 μ wide, lined with hyaline, upward-pointing periphyses. ASCI 80–130 × 10–15 μ, with small, distinct apical ring, 2.5–3 μ diam. ASCOSPORES faintly yellowish in mass, hyaline to faintly tinted singly, slightly curved, 80–100 × 2.5–3 μ, 3–5 or more septate at maturity. PARAPHYSES hyaline, septate. HYPHAE a network on surface of rhizomes, leaf sheaths and culms, often as dark sheath several cm up the culm; runner hyphae brown, septate, 4–7 μ; branching hyphae paler brown forming HYPHOPODIA: (a) brown, lobed, terminal, 15–25 × 10–20 μ, on side branches up to 40 μ long; (b) simple unlobed, terminal or intercalary, 7–15 × 4–7 μ. Latter often united into irregular, sclerotic masses 200 μ or more diam. PHIALOSPORES formed in culture, hyaline, aseptate, curved, 4–7 × 1–1.5 μ. PHIALIDES slightly curved, hyaline or faintly tinted, clustered on sides of hyphae or terminally, 9–18 × 2–3 μ, collarette present. Hyphopodia usually develop on coleoptiles of cereal seedlings on culture plates. Distinguished from the vars *avenae* and *tritici* by the brown, lobed hyphopodia; its ascospores are similar to those of the latter var. and shorter than those of the former. Walker (1972) tabulated the perfect state measurements given by several workers for the fungus on rice. Wong et al. discussed the fungi like *Phialophora* associated with

G. graminis. Hornby et al. described a new sp. (*G. cylindrosporus* Hornby, Slope, Gutteridge & Sivanesan) from rotting roots of *Hordeum* and *Triticum*. It differs from the 3 vars of *G. graminis* mainly in the smaller ascospores, 37–69 × 3.2–5.6 μ.

G. graminis var. *graminis* occurs on several grass genera besides *Oryza*. It is probably widespread and, as *Ophiobolus graminis*, was confused with the wheat take-all fungus. Culture inoculation tests on wheat seedlings using ascospore isolates will often be necessary to determine the type of mycelium and hyphopodia associated with a particular perithecium. Simultaneous infections with more than one var. of *G. graminis* are possible. On rice the mycelium and hyphopodia of var. *graminis* could easily be confused with *Magnaporthe salvinii* (q.v.; Walker 1973). Rice plants develop a leaf sheath rot from the crown to well above water level. The disease is usually seen late in the season. Young plants can be killed but usually crown sheath rot occurs with plant maturity and severe damage is rare. Tillering may be reduced and there may be only one panicle. Lodging, incomplete development of the panicle and incomplete grain filling may occur. A black fungal sheath forms on the bases of tillers between the leaf sheaths and on the culm surfaces. Perithecia occur in the leaf sheaths. There is direct penetration (with appressoria) of the basal leaves or outer leaf sheaths. The fungus is presumably soilborne and seed infection was reported by E. Harris et al. (*Rev. appl. Mycol.* **42**: 727). Variation in pathogenicity between isolates may exist.

Deacon, J. W. 1974. *Pl. Pathol.* **23**: 85 (interactions between *Gaeumannomyces graminis* vars & *Phialophora radicicola* on Gramineae; **54**, 2857).
Hornby, D. et al. 1977. *Trans. Br. mycol. Soc.* **69**: 21 (*G. cylindrosporus* sp. nov.; **57**, 501).
Saccas, A. M. et al. 1954. *Agron. trop. Nogent* **9**: 7 (on rice; **33**: 444).
Speakman, J. B. et al. 1978. *Trans. Br. mycol. Soc.* **70**: 325 (invasion of grass & maize roots by *G. graminis* & *P. radicicola*).
Tullis, E. C. 1933. *J. agric. Res.* **46**: 799 (on rice; **12**: 655).
Walker, J. 1972. *Trans. Br. mycol. Soc.* **58**: 427 (type studies on *G. graminis* & related fungi; **51**, 4769).
——. 1973. *C.M.I. Descr. pathog. Fungi Bact.* 381, 382, 383 (*G. graminis* vars *graminis*, *avenae*, *tritici*).
——. 1975. *Rev. Pl. Pathol.* **54**: 113 (review, recent work on take-all diseases of Gramineae, 178 ref.).
Wong, P. T. W. et al. 1975. *Trans. Br. mycol. Soc.* **65**: 41 (germinating, phialidic conidia of *G. graminis* and fungi like *Phialophora* from Gramineae; **55**, 671).

GANODERMA Karst., *Rev. Mycol.* 3(9): 17, 1889.

(Ganodermataceae)

Polyporous, fibrous, coriaceous, elastic, sessile and eccentric dimidiate or stipitate, flesh coloured, brown or reddish. CAP covered with a rigid, shining, fragile crust, formed by the thickened extremity of the hyphae of the tissue. Subglobular BASIDIA with 4 sterigmata; cystidia slightly marked. BASIDIOSPORES yellowish brown, ovoid, verrucose, often with an internal droplet. TUBES elongated, brown. PORES small, rounded, whole. Lignicolous or growing near tree trunks. The cap tissue shows divergent, darker, sterile lines in European spp.; these lines are real conidia-bearing tubes in some exotic spp. (after Patouillard as quoted by D. N. Pegler & T. W. K. Young, *Kew Bull.* **28**: 352, 1973).

The taxonomy of *Ganoderma* rests on spore morphology and cutis anatomy. The spore has a hyaline, oviform *epispore* enclosing a coloured, ellipsoid *endospore* united by *echinules*. Spp. with other types of spores should be segregated from the genus. The echinules (at first long, later shorter) connect the endospore with warts on the epispore. The endospore, normally ellipsoid, may become flattened at the apex by the contraction of the echinules crowning it. The apex of the epispore is usually without echinules and collapses at spore discharge. The sporophore is made up of a cone of skeletal hyphae radiating from the base and branching upwards, to form the *cutis*. The cutis, in spp. with the most completely developed type, is composed of hyaline, parallel hyphae springing up from the branched extremities of the skeletal hyphae and swollen by a wax or a substance like it. Wax impregnates the cutis to a depth of *c.* 1 mm. Levels at which wax is deposited determine the type of cutis (from R. L. Steyaert, *Trans. Br. mycol. Soc.* **47**: 652, 1964); see Pegler et al. and Steyaert for taxonomy.

G. tornatum (Pers.) Bres. and other spp. were described by Steyaert (1975a & b). It appears to be the commonest sp. in the tropics and causes a heart and butt rot. The area of distribution overlaps that of *G. applanatum* (Pers. ex S. F. Gray) Pat. in N.W. India and Pakistan. The latter sp. (with a very wide host range of broad leaved and coniferous trees) occurs throughout the N. temperate zone. The southern limit is Florida (USA), Mediterranean, N. Iran, N. Pakistan and southern Himalayas. Childs described concentric canker of citrus caused by *G. applanatum*. The lenticular cankers, with the longer axis parallel to the trunk or branch long axis, show concentric cracking. The dead bark within the canker is scaly. *G. lucidum* ((W. Curt.) Fr.) Karst. occurs in the N. temperate zone and in central Africa above 1500 m. The name has been frequently misapplied in the tropics and the literature can be misleading (Steyaert 1975a); for *Ganoderma* spp. on oil palm see *G. boninense*.

Blackford, F. W. 1944. *Qd. J. agric. Sci.* **1**(4): 77 (*Ganoderma lucidum* on citrus; **24**: 411).
Childs, J. F. L. 1953. *Phytopathology* **43**: 99 (*G. applanatum* on citrus; **33**: 25).
Furtado, J. S. 1967. *Persoonia* **4**: 379 (some tropical *Ganoderma* spp. with pale context).
Pegler, D. N. et al. 1973. *Kew Bull.* **28**: 351 (basidiospore form in British spp. of *Ganoderma*; **53**, 2461).
Perreau, J. 1972. *Rev. Mycol.* **37**: 241 (spore ornamentation; **54**, 3183).
Steyaert, R. 1967. *Bull. Soc. r. Bot. Belg.* **100**: 189 (*Ganoderma* with emphasis on the European spp.).
Steyaert, R. L. 1972. *Persoonia* **7**: 55 (*Ganoderma* & related spp. mainly of the Bogor & Leiden herbaria).
——. 1975a. *Trans. Br. mycol. Soc.* **65**: 451 (concept & circumscription of *G. tornatum*; **55**, 2565).
——. 1975b. *C.M.I. Descr. pathog. Fungi Bact.* 443, 445, 447 (*G. applanatum*, *G. lucidum* & *G. tornatum*).

Ganoderma boninense Pat., *Bull. Soc. mycol. Fr.* **5**: 72, 1889.
Fomes lucidus (W. Curt.) Fr. forma *boninensis* Sacc., 1888
Ganoderma noukahivense Pat., 1889
F. lucidus (W. Curt) Fr. forma *noukahivensis* Cass., 1888
G. miniatocinctum Stey., 1967.
(In 1887 Patouillard in *Journal de Botanique* 169–170, when describing *G. neglectum* Pat., commented on collections of *Ganoderma* from the Bonin and Nouka-hiva Islands (Marquesas) and partially described them without establishing formal taxonomic distinctions.)

BASIDIOCARP (basidioma) very irregular, dimidiate to stipitate, vertically or horizontally pleuropode, flat to sub-ungulate, concentrically finely costulate or sulcate, up to 15 cm diam. and 5 cm thick. UPPER SURFACE sublaccate to shiny laccate, warm blackish brown, in growing specimen margined mars orange, white at the growing margin. PORE SURFACE white, sometimes with an orange margin. CUTIS *c.* 0.07 mm thick, usually underscored by a thin orange or yellow line. CONTEXT up to 1 cm thick at the base of the basidioma, 1–4 mm thick elsewhere, ochraceous buff. TUBE LAYER up to 4 cm thick, auburn. Cutis hymeniodermiform, with sphaeropedunculate to cuneate elements, 30–70 μ long, 6–12 μ wide at the sphaeroid apex, immersed in melanoid substances that must be saponified with KOH to enable them to be observed. PORES circular, 90–380 (155) μ diam., dissepiments to 10–140 (50) μ thick, distance between axes of pores 160–345 (200) μ. BASIDIOSPORES ellipsoid, gold yellow to brownish yellow, 8.5–13.5 (10.9) $\times$ 4.5–7 (5.9) μ with thin short echinules. Non-echinulate spores (apparently anomalous) may be observable, pale yellow, 8.5–13 (10.6) $\times$ 4.5–7 (5.85) μ (R. L. Steyaert, *CMI Descr.* 444, 1975).

Three *Ganoderma* spp., which have been implicated in the aetiology of basal stem rot of oil palm were described recently by Steyaert (1975). Besides *G. boninese* these are *G. tornatum* (Pers.) Bres. and *G. zonatum* Murrill.

G. tornatum: BASIDIOCARP (basidioma) sessile, dimidiate, up to 45 cm at its widest diam. and up to 10 cm thick at the base; UPPER SURFACE undulate, concentrically grooved, cracked by desiccation, melleus to cinnamon buff to olivaceous to clove brown, incurved margin, with horizontal grooves in old specimens, the last white in actively growing specimens; pore surface white, sometimes yellow, browning sometimes peripherally by bands 1–3 cm wide. CUTIS hard brittle, 600–700 μ thick, olivaceous. CONTEXT usually <1 cm thick, up to 4 cm in big specimens, with thin horny deposits of melanoid substances, usually visible as shiny striae near the tube layer in specimens cut vertically in the middle; other short striae can be seen dispersed in the context (no such striae are seen in *G. applanatum*). TUBE LAYER up to 6 cm thick in big specimens, umbrinus chestnut brown. Cutis of trichoderm type, similar to that of *G. applanatum* (*Ganoderma* q.v.). Pores circular, 90–200 (148) μ diam.; dissepiments 30–130 (64.5) μ thick, 135–285 (208) μ between axes of pores. Sizes of pores and dissepiments are affected by altitude; being *c.* 25% larger at 2500 m than they are at sea level. BASIDIOSPORES ovoid, yellow brown to brown, 6–12 $\times$ 4.5–8 μ, also increasing with altitude 6–9 μ from sea level to 8.5–11.5 μ at 600 m, exceptionally 13 μ above this. Non-echinulate basidiospores, ovoid, yellow 7–13 (7.7) × 4–7.5 (5.45) μ, usually bigger than the normal basidiospores of the same specimen but at times smaller. Occasionally a third type of basidiospore occurs with small warts instead of echinules (R. L. Steyaert, *CMI Descr*. 447, 1975; q.v. for synonymy).

G. zonatum: BASIDIOCARP (basidioma) sessile, dimidiate, reniform to irregular, tuberculate, up to 30–40 cm at its widest diam., up to 9 cm thick at the base; upper surface usually tuberculate sometimes sulcate, sub-shiny, victoria lake to bone brown or by long exposure saccardo umber, margin often tumid, pure white during growth; PORE SURFACE usually with an irregular margin, white. CUTIS very thin, at the most 20 μ, often underlined by a yellow layer. CONTEXT up to 1/3–1/2 of the basidioma thickness, bay. Tube layer 1/2–2/3 of the basidioma thickness, limit with context very irregular, especially in big specimens, sometimes parts enclosed in context tissue, walnut brown. Cutis hymeniodermiform with sphaeropedunculate very irregular elements, 70–80 μ long by 6–12 μ at the terminal sphaerule, very inflated with melanoid substances but abundantly immersed in it so that these have to be eliminated with KOH to observe the structures of the elements at the ends of the hyaline context hyphae. Pores regularly circular, 12–340 (210) μ diam.; dissepiments 10–160 (65) μ thick, distance between axes of pores 215–350 (275) μ. BASIDIOSPORES ellipsoid, golden yellow to light brown, 9–16 (12) × 5.5–9 (7) μ, with relatively long and thick echinules (R. L. Steyaert, *CMI Descr*. 448, 1975; q.v. for synonymy).

In addition to these 3 spp. on palms 2 others, also occurring on palms, were discussed by Steyaert (1967): *G. chalceum* (Cooke) Stey. and *G. xylonoides* Stey. Turner (*Phytopathol. Pap*. 14: 16, 1971) gave other spp. recorded on oil palm, these include *G. philippii* (q.v.). The relative importance of these fungi in this oil palm disease is not clear. Diseases on other palms (e.g. areca and coconut) have also been described. *G. boninense* occurs (mainly on palms) in S.E. Asia, Australia, Japan and the Pacific islands. *G. tornatum* (heart and butt rot) is found throughout the tropics and appears to be the commonest *Ganoderma* sp. in this climatic region. The distribution overlaps that of *G. applanatum* in N.W. India and Pakistan. *G. zonatum* occurs in tropical Africa on palms; it has been recorded from Florida, USA, and is probably found in S. America. These fungi are saprophytes or weak parasites which can build up massive inocula on debris of woody crops; the new crop then becomes infected.

On oil palm an internal dry rot at the stem base is caused. In TS a light brown rotting area is marked by darker bands of an irregular outline. Outside this area is a bright yellow transition zone between healthy and diseased tissue. In the brown tissue there are small cavities which contain white mycelium. Roots of diseased palms are friable, the cortex is necrotic and disintegrates. The first leaf symptom is an abnormal number of unexpanded spear leaves (not absolutely diagnostic); foliage is paler than in a healthy plant. Affected leaves die; necrosis begins in the older ones and extends upwards. Dead leaves droop from the point of attachment or fracture somewhere along the rachis; a skirt of dead leaves is the result. The trunk fractures at the base and the palm falls. Trees usually die within 6–12 months of the appearance of spear accumulation. The sporophores appear (at the stem base or occasionally on roots close by) after symptoms have been visible for some time or even after the palm has fallen (Turner & Bull, *Diseases and disorders of the oil palm in Malaysia*). For symptoms of the disease on *Areca catechu* (anabe-roga) and *Cocos nucifera* see Naidu et al.; Venkatarayan.

Basal stem rot had been considered to be a disease of old plams whose resistance had perhaps been lessened. But in Malaysia, since the last war, large numbers of oil palms of 5–15 years have been attacked. This epidemic situation was due to the spread of the disease after planting on old coconut land. After poisoning and/or felling the coconut palms their stumps and trunks become colonised by the fungus. Sporophores appear on stumps *c*. 2 years after felling, earlier on trunks. The fungus spreads to the stump root system which becomes the large inoculum source for subsequent invasion of oil palm roots. Four to 6 years may elapse before the oil palm shows symptoms. A similar build up of the pathogen(s) can occur in oil palm stands which are to be replanted with oil palm. But little infection of this crop takes place when it is planted on old rubber (*Hevea brasiliensis*) land (Navaratnam 1964; Turner 1965). The disease can also occur in oil palm planted in areas newly cleared from forest. Spread is entirely by root contact and there is no evidence for infection from airborne spores. Turner (1968b) found that wild palms may be possible sources of the disease.

The control measures recommended in Malaysia are as follows. In very old areas remove affected oil palms and accelerate replanting. In mature areas fell affected palms; excavate stump, bole and root mass, and disperse on ground surface; do not supply vacant points. In replants from rubber remove and destroy rubber wood (stumps left are potential breeding sites for rhinoceros beetle). In replants from oil palm or coconut fell and thoroughly excavate infected palms; no stumps to be left in the ground; destroy coconut wood by burning, oil palm trunks can be left in the interline or disposed of. In areas recently replanted from oil palm and coconut (<5 years old) remove remaining stumps; excavate bole and root tissue, and disperse on ground surface. In all cases where mature trees are slightly affected tree surgery may be effective (Turner 1968a); but only palms with dark green foliage are considered suitable for such treatment since this indicates restricted lesion development.

Dell, W. 1955. *Bergcultures* **24**: 191 (in Sumatra, Indonesia; **34**: 644).
Naidu, G. V. B. et al. 1966. *J. Mysore hort. Soc.* **11**: 14 (on areca palm; **48**, 3115).
Navaratnam, S. J. 1964. *Planter Kuala Lumpur* **40**: 256 (effect of previous crop; **44**, 1641).
—— et al. 1965. *Pl. Dis. Reptr* **49**: 1011 (inoculation of oil palm seedlings; **45**, 1126).
——. 1966. *Malaysian Agric.* **6**: 22 (inoculation of oil palm).
Steyaert, R. L. 1967. *Bull. Jard. bot. natn. Belg.* **37**: 465 (*Ganoderma* spp. on palms; **47**, 1945).
Turner, P. D. 1965. *Ann. appl. Biol.* **55**: 417 (effect of previous crop; **44**, 3127).
——. 1966. *Malaysian Agric.* **6**: 15 (control).
——. 1968a. *Planter Kuala Lumpur* **44**: 302 (control by surgery; **47**, 3531).
——. 1968b. *Ibid* **44**: 645 (wild palms as infection sources; **48**, 1893).
Venkatarayan, S. V. 1936. *Phytopathology* **26**: 153 (on areca & coconut palms **15**: 436).
Wijbrans, J. R. 1955. *Bergcultures* **24**: 112 (general in Sumatra, Indonesia; **34**: 450).

Ganoderma philippii (Bres. & P. Henn.) Bres., *Icongr. Mycol.* 21: tab. 1014, 1932.
Fomes philippii Bres. & P. Henn. ex Sacc., 1881
F. pseudoferreus Wakef., 1918
Ganoderma pseudoferreum (Wakef.) Over. & Steinm., 1925

BASIDIOCARP (basidioma) very variable but more often very flat and thin; upper surface army brown to bone brown, either with concentric grooves or knobbles, those with grooves sometimes with silvery speckles. PORE SURFACE usually cinnamon buff, sometimes lighter, rarely white except at the margin of growing specimens; up to 15–16 cm diam., up to 4 cm thick at the base. CUTIS from soft to horny, 30–120 μ thick as the basidioma gets older. CONTEXT from tawny close to cutis to russet towards tube layer, often with a very light brownish thin zone under the cutis; middle context usually with curved subvertical concentric lines, light and dark russet, from base towards cutis, interspersed with dark brown, slightly shiny horizontal deposits of melanoid substances. TUBE LAYER usually single but in old specimens several, barely distinguishable from each other. CUTIS of the anamixoderm type (intertwined hyphae with no free external ones). PORES round, 80–240 (125) μ diam., dissepiments 29–130 (55) μ thick; distance between axes 160–205 (180) μ. BASIDIOSPORES ovoid, 6–8.5 (7.5)×4–8 (5·2) μ, chamois. Sporulation is often not abundant and frequently lacking; non-echinulate spores 7–10 (8.33)×4–6 (5.1) μ, infrequent. RHIZOMORPHS red with white advancing edge, initially stranded and somewhat like a fan, fusing after the first few cm to form an encrusting sheath which becomes blackened with age. Black parts become red on moistening and rubbing, returning to black when dry (R. L. Steyaert, *CMI Descr.* 446, 1975).

G. philippii causes red root rot of rubber (*Hevea brasiliensis*) on which it has been largely studied with the 2 other general root pathogens *Phellinus noxius* (q.v.) and *Rigidoporus lignosus* (q.v.). It occurs on many woody and non-woody plant hosts in S.E. Asia through Indonesia to Papua New Guinea and New Caledonia (Map 98). Steyaert (l.c.) noted that the records for Africa are based on vegetative characters only and were not confirmed by corresponding sporophore collections. On rubber in Malaysia (W.) *G. philippii* is less important than *R. lignosus* which is the major root pathogen. The detection and recognition of these 3 fungi, causing red, brown and white root diseases has been described in detail in the *Plrs' Bull. Rubb. Res. Inst. Malaysia* 133: 111, 1974. Roots attacked by *G. philippii* have a red skin (rhizomorph) of mycelium to which soil adheres. The growing margin of the fungus is usually creamy white, the red colour forms some cm behind the advancing front. The inner surface of the rhizomorph is dirty white. In the early stages the rot caused is pale brown and hard; later it is pale buff, wet and spongy or dry depending on the soil conditions. Characteristically, the annular layers of the wood separate easily. The above-ground symptoms are not distinctive. In rubber experienced observers can detect root disease by the very early turning to the slightly off-green hue of the foliage of one or more parts of the tree, or of the whole tree if very young. The leaves develop a ripened appearance, in contrast to the deep green colour of the foliage in the unaffected branches, and soon turn yellow.

The disease spreads from massive, woody inocula by root contact. The fungus is relatively slow growing and disease may not be noticed until the trees are in tapping. The rhizomorphic skin rarely extends more than a few cm beyond the portion of the root penetrated by it. Evidence that infection can originate from airborne spores is not conclusive. Lim found that basidiospores germinated in water after passage through the larval gut of 2 spp. of Tipulid flies (*Limonia*), a member of the sporophore fauna. The spores were viable for up to 20 weeks after passage. The winged adult is an effective carrier of viable spores. Control of red root rot is largely that described for *R. lignosus* and some general comments made under this fungus also apply to *G. philippii.* Recent discussions on root disease control in rubber, with particular reference to Malaysia, have been given by Wastie, *PANS* **21**: 268, 1975, and in *Plrs' Bull. Rubb. Res. Inst. Malaysia* 134: 157, 1974. A prophylactic treatment against these root diseases relies on the fact that the external fungus growth does not sustain an infection in the absence of a food source. Removal of external growth ahead of the host tissue that has been penetrated prevents further fungus growth. Drazoxolon is recommended against red root. Diseased tissue is excised (but not superficial mycelium) and tar painted on the wounds. The remaining roots, and the collar and roots of neighbouring healthy trees, are treated to a depth and length of 25 cm with drazoxolon. Treated trees are reinspected after a year.

Corner, E. J. H. 1931. *J. Rubb. Res. Inst. Malaya* **3**: 120 (identification; **11**: 402).

Lim, T. M. 1977. *J. Rubb. Res. Inst. Malaysia* **25**: 89 (production germination & dispersal of basidiospores; **57**, 4117).

Overeem, C. Van 1925. *Bull. Jard. bot. Buitenz.* Ser. III **7**: 436 (general in Indonesia; **5**: 54).

Tan, A. M. et al. 1971. *Proc. Rubb. Res. Inst. Malaysia Plrs' Conf.*: 172 (collar protectant dressing; **54**, 1419).

Varghese, G. et al. 1973. *Malaysian agric. Res.* **2**: 31 (on tea; **54**, 219).

GEOTRICHUM Link ex Pers., *Mycol. eur.* 1: 26, 1822.

Coprotrichum Bonorden, 1851.
(Moniliaceae)

COLONIES spreading, creamy white, soft, somewhat yeast like in texture. VEGETATIVE MYCELIUM of hyaline, septate, branched, prostrate hyphae. CONIDIA (arthrospores) formed in chains by disarticulation of lateral hyphae at septa, sometimes intercalary in the broad vegetative hyphae also, usually maturing in rapid basipetal succession, at first cylindrical, later barrel shaped or ellipsoidal or even subglobose (from C. V. Subramanian, *Hyphomycetes*: 847, 1971, as *Geotrichum* Link ex Fr., *Syst. Mycol.* **1**: xlv, 1821; Caretta, 1963, gave a key to the spp., *G. candidum* q.v.).

Geotrichum candidum Link ex Pers. emend. Carmichael, *Mycologia* **49**: 823, 1957.

(synonymy in Caretta, 1963; Carmichael; Subramanian, *Hyphomycetes*: 847)

MYCELIUM hyaline, septate, more or less specialised into broad, radiating, vegetative hyphae which branch dichotomously, and narrower, lateral, sporulating hyphae which may also branch. On some media dichotomous branching may be reduced or absent. CONIDIA formed in chains by disarticulation at the septa of the lateral hyphae, sometimes intercalary in the broad vegetative hyphae as well; usually maturing in rapid basipetal succession; plentiful to scant, or almost absent in some strs on some media; chains may be aerial and erect, or decumbent, or flat on the medium surface, or under the surface; conidia cylindrical at first, later varying from barrel shaped to ellipsoidal or subglobose, 2–8 × 3–50 (commonly 3–6 × 6–12) μ. There are 1–4 (commonly 2) nuclei/spore (from Carmichael l.c., who described the morphology of the colonies on several media).

G. candidum is ubiquitous; a plant wound pathogen (fruit and vegetable rots), pathogenic to animals (including man), associated with food spoilage, isolated from water and waste products of wood and frequently encountered in air sampling. For further ref. to the literature see Butler and others, and El-Tobshy. A description of the fungus was also given by Tubaki who divided isolates into 3 groups based on culture growth characteristics. Butler (1960) and Butler et al. (1965), in work which included taxonomy, rejected a separate physiologic form (*parasitica*) for the isolates attacking tomato. But the form on citrus (synonyms: *Oidium citri-aurantii* Ferraris, *Oospora citri-aurantii* (Ferrar.) Sacc. & Syd.) was considered physiologically, although not morphologically, distinct. Butler et al. (1965) designated it the citrus race of *G. candidum*. It is sometimes referred to as *G. candidum* var. *citri-aurantii* (Ferrar.) R. Cif. & F. Cif. A perfect state, *Endomyces geotrichum* Butler & Peterson, has been described.

Isolates from citrus form a homogeneous group. They are highly virulent on lemon, require pyridoxine for growth, grow well in lemon juice at pH 2.2, and form slime in potato dextrose broth and lemon juice. Non-citrus isolates are weakly virulent to lemon, generally do not grow in lemon juice, do not form slime (but see Tubaki) and grow well without pyridoxine. Both forms of the fungus occur in citrus soils and on healthy lemon fruit on the tree; but the citrus form is generally unknown outside the citrus environment. The disease on citrus is called sour rot. On fruit the first symptoms are watersoaked, dark yellow, slightly raised areas on the rind. These become covered by a thin, creamy fungal growth. The plant tissue breaks down, softening quickly, becomes watery, leaks, and emits a sour odour; the resulting putrid mass attracts flies. Infection takes place through rind injuries (which may be caused by insects, e.g., Dadant) or through weakened areas around the buttons; subsequent spread is by contact. Control is through the prevention of fruit injury, the elimination of overmature fruit and the avoidance of lengthy storage. Certain fungicides in postharvest operations are effective (e.g., Bussel et al.; Kuramoto et al.; Wild et al.).

G. candidum has been described causing watery rots (in the field or in storage) of tomato, melon and carrot. Other crops may show some resistance to infection. Isolates from a variety of sources can infect tomato and carrot through wounds. Caretta (1961) found that other *Geotrichum* spp. were not pathogenic to tomato. Lewis et al. reported differences in lesion size amongst 15 crops inoculated through puncture wounds. Butler (1961) demonstrated that infection could be carried by *Drosophila melanogaster* (also *Rhizopus stolonifer* q.v.) and Elarosi et al. found more infection in tomato fruit invaded by *Phytophthora infestans*. No specific control measures can be described on crops other than citrus; these must relate to, and be affected by, the particular case of a given disease.

Bussel, J. et al. 1975. *Pl. Dis. Reptr* **59**: 269 (control in citrus; **54**, 3934).

Butler, E. E. 1960. *Phytopathology* **50**: 665 (pathogenicity & taxonomy; **40**: 249).

——. 1961. *Ibid* **51**: 250 (spread by *Drosophila melanogaster*; **40**: 632).

—— et al. 1965. *Ibid* **55**: 1262 (taxonomy, pathogenicity & physiology of citrus form; **45**, 1071).

—— ——. 1972. *Mycologia* **64**: 365 (*Endomyces geotrichum* sp. nov.; **51**, 3823).

Caretta, G. 1961. *Riv. Patol. veg. Pavia* 4, **1**: 324 (pathogenicity to tomato; **41**: 254).

——. 1963. *Atti Ist. bot. Univ. Lab. crittogam. Pavia* 5, **20**: 282 (taxonomy).

Carmichael, J. W. 1957. *Mycologia* **49**: 820 (taxonomy & culture).

Ceponis, M. J. 1966. *Pl. Dis. Reptr* **50**: 222 (on melon; **45**, 2679).

Ciferri, F. 1955. *Annali Sper. agr.* **9**: 5 (taxonomy of citrus form; **35**: 294).

Dadant, R. 1953. *Rev. Pathol. vég.* **32**: 87 (on citrus; **33**: 479).

Elarosi, H. M. et al. 1970. *Alex. J. agric. Res.* **18**: 233 (on tomato; **51**, 683).

El-Tobshy, Z. M. et al. 1965. *Phytopathology* **55**: 1210 (general pathogenicity to plants; **45**, 1072).

Harding, P. R. 1968. *Pl. Dis. Reptr* **52**: 433 (comparison with *Candida krusei* rot; **47**, 2720).

Hashimoto, T. et al. 1973. *J. Bact.* **116**: 447 (morphogenesis & fine structure of septa; **53**, 3371).

Kuramoto, T. et al. 1975. *Bull. Fruit Tree Res. Stn* B 2: 87 (occurrence & control on citrus in Japan; **55**, 3594).

Lewis, M. H. et al. 1966. *Pl. Dis. Reptr* **50**: 681 (plant tissue pH & susceptibility; **46**, 876).

Steele, S. D. et al. 1973. *Can. J. Microbiol.* **19**: 1507 (fine structure of hyphae; **53**, 2897).

Tubaki, K. 1962. *Trans. mycol. soc. Japan* **3**: 29 (from polluted water, morphology & culture).

Wells, J. M. 1975. *Phytopathology* **65**: 1299 (stimulation by low O_2 & high CO_2; **55**, 4279).

Wild, B. L. et al. 1976. *Pl. Dis. Reptr* **56**, 715 (control in citrus; 56, 715).

Wright, W. R. et al. 1964. *Ibid* **48**: 837 (on carrot; **44**, 950).

GIBBERELLA Sacc., *Michelia* **1**: 43, 1877. (Hypocreaceae)

STROMA, when present, erumpent, pseudoparenchymatous. PERITHECIA superficial, scattered or caespitose; globose, fuscous or black; ostiole distinct, usually papillate, wall pseudoparenchymatous, cell walls pigmented blue black. ASCI clavate or cylindrical, unitunicate, usually 8 spored, pseudoparaphyses evanescent; paraphyses absent. ASCOSPORES elliptical, oval sometimes fusiform, with 2 or more transverse septa, hyaline, sometimes lightly pigmented (after J. M. Dingley, *Trans. R. Soc. N.Z.* **79**: 405, 1952).

The taxonomy, morphology, culture and some other characteristics (including pathology) of 13 spp. and their *Fusarium* states have been fully described by Booth (*The genus* Fusarium). Four spp. and 1 var. are described separately here. *G. acuminata* Booth (stat. conid. *F. acuminatum* Ellis & Everhart) occurs on a wide variety of plants (and on insects) but is apparently a very minor pathogen. It has been recorded on maize and lucerne. On the latter it could cause a crown bud rot in Canada, as did *G. avenacea* Cook (Hawn). This latter fungus (stat. conid. *F. avenaceum* (Corda ex Fr.) Sacc.) has a worldwide distribution on many crop plants, but occurs chiefly in temperate regions, being often a serious root parasite on winter cereals and some legumes. It can cause damping off in many plants and is spread by seed, soil and crop debris. A form of the fungus associated with a root rot of broad bean was described from Japan as *F. avenaceum* (Fr.) Sacc. f. sp. *fabae* (Yu) Yamamoto.

G. baccata (Wallr.) Sacc. (stat. conid. *F. lateritium* Nees emend. Snyder & Hansen pro parte) has been also described by Booth (1971) elsewhere. Homothallic and heterothallic strs occur; the former seldom form mature perithecia in vitro but they can be obtained using wheat straw. The fungus has a wide host range and is reported as causing wilt, dieback and canker on woody plants. There are reports of it causing disease in citrus (e.g., Salerno); coffee, although in this crop it is sometimes confused with *G. stilboides* (q.v.), and several temperate tree crops. *F. lateritium* (Nees) emend. Snyder & Hansen f. sp. *pini* Hepting appears to be confined to pine on which it causes pitch canker in USA (cause now *F. moniliforme* var. *subglutinans, Mycologia* **70**: 1131). A copious pitch flow is the most noticeable symptom. The canker (usually sunken) retains the bark and the wood beneath is deeply pitch soaked; the latter symptom is diagnostic. Twig cankers may form at the base of needle clusters. Stems can be girdled and killed, and in some cases trunks also. The fungus can sometimes be seen if the bark is removed from the cankers (see Berry et al. who give the pine spp. on which pitch canker occurs; Bethune et al.). Matuo et al. found differences in pathogenicity between *F. lateritium* isolates from *Albizia*, *Morus* and *Robinia* and other isolates. They proposed the f. sp. *mori* (Desm.) Matuo & Sato for the form causing a disease on mulberry; and f. sp. *cerealis* Matuo & Sato for a str. which was pathogenic to these 3 plant genera but also to maize and wheat. In Nigeria *G. baccata* causes a leaf spot of the vegetable *Celosia argentea* (Afanide et al.).

G. cyanogena (Desm.) Sacc. (stat. conid. *F. sulphureum* Schlechtendahl; synonyms: *Sphaeria cyanogena* Desm., *Gibbera saubinetii* Mont., *Gibberella saubinetii* (Mont.) Sacc., *F. sambucinum* Fuckel f. 6 Wollenw.) is heterothallic. The fungus is frequent in soil, occurs in a potato storage rot and has been isolated from a groundnut rot originating in Gambia. Tyner found that it reduced the emergence of cucumber seedlings. It is often reported as *F. sambucinum* f. 6. *G. gordonia* Booth (stat. conid. *F. heterosporum* Nees) is heterothallic and particularly characteristic as a head blight of cereals and grasses. *G. intricans* Wollenw. (stat. conid. *F. equiseti* (Corda) Sacc.) has been isolated from many crops, both tropical and temperate. Joffe et al. found this fungus to be ubiquitous in soils in Israel; it was isolated from seeds of several crops including groundnut and was pathogenic (e.g., caused decay in wounded fruit of avocado pear; and see Winter et al.). Brian et al. obtained phytotoxic substances from *F. equiseti*. *G. pulicaris* (Fr.) Sacc. (stat. conid. *F. sambucinum* Fuckel) is heterothallic. It has been recorded on several crops, causing canker (e.g. hops), root and seedling rots (cereals) and a storage rot of potato, see Booth (1973).

The ascospore characteristics of the 7 mentioned *Gibberella* spp., from Booth (*The genus* Fusarium), are:

acuminata: smooth, hyaline, ellipsoidal to oval but becoming slightly constricted at the central septum, 13–20 × 6–7 μ, often remaining 1 septate but occasionally a second septum develops in the larger upper cell.

avenacea: hyaline, fusoid, constricted at the central septum, with a larger upper cell; often remaining 1 septate, 13–19 × 4–5 μ; others develop 1 or 2 more septa, 17–25 × 5–6.5 μ, such variations in size and septation occur in 1 perithecium.

baccata: smooth, hyaline, fusiform to ellipsoid or inequilaterally ellipsoid; occasionally, particularly in 8-spored asci, remaining 1 septate, 12–18 × 4.5–7.5 μ; more often becoming slightly curved, 3 septate, 13–18 × 5–8 μ.

cyanogena: ellipsoidal, straight or slightly curved, becoming slightly constricted at the 3 septa, 20–25 × 5–7 μ.

gordonia: hyaline, elliptical to fusoid, 1–3 septa, 15.6–18.5 × 4–4.5 μ.

intricans: hyaline, fusoid, 3 (rarely 1–2) septate, 21–33 × 4–5.5 μ; 4-spored asci have spores 22–40 × 4.5 μ.

pulicaris: fusiform to ellipsoid, curved and slightly constricted at the 3 transverse septa; 20–28 × 6–9 μ, some collections from hardwood trees (e.g., elm and oak) have spores 26–34 × 8–10 μ.

Afanide, B. et al. 1976. *Trans. Br. mycol. Soc.* **66**: 505 (*Gibberella baccata* on *Celosia argentea*; **56**, 511).

——— ———. 1976. *Mycologia* **68**: 1108 (*G. baccata* state; **56**, 3453).

Berry, C. R. et al. 1969. *For. Pest Leafl.* 35, 4 pp. (*Fusarium lateritium* f. sp. *pini*; **48**, 3684).

Bethune, J. E. et al. 1963. *J. For.* **61**: 517 (as above; **43**, 260).

Booth, C. 1971. *C.M.I. Descr. pathog. Fungi Bact.* 310 (*G. baccata*).

——. 1973. *Ibid* 385 (*G. pulicaris*).

Brian, P. W. et al. 1961. *J. exp. Bot.* **12**: 1 (phytotoxic compounds formed by *F. equiseti*; **40**: 456).

Hawn, E. J. 1959. *Can. J. Bot.* **37**: 1247 (*F. acuminatum* inter alia on lucerne; **39**: 473).

Joffe, A. Z. et al. 1967. *Israel Jnl Bot.* **16**: 1 (*G. intricans*; **47**, 1466).

Matuo, T. et al. 1962. *Trans. mycol. Soc. Japan* **3**: 120 (forms of *F. lateritium*; **42**: 104).

Salerno, M. 1959. *Tec. agric. Catania* **11**(4–5), 15 pp. (*G. baccata* on lemon; **40**, 607).

Snyder, W. C. et al. 1949. *J. agric. Res.* **78**: 365 (*Fusarium* spp. associated with *Albizia, Rhus* and *Pinus*, taxonomy & morphology; **28**: 603).

Tyner, L. E. 1945. *Scient. Agric.* **25**: 537 (*G. cyanogena* on Cucurbitaceae; **24**: 487).

Winter, W. E. et al. 1975. *Fitopatologia* **10**: 23 (*G. intricans* on seeds; **55**, 1167).

Yamamoto, W. et al. 1955. *Sci. Rep. Hyogo Univ. Agric.* Ser. Agric. **2**: 53 (*F. avenaceum* f. sp. *fabae* inter alia; **37**: 129).

Gibberella fujikuroi (Sawada) Ito apud Ito & Kimura, *Hokkaido Agric. Exp. Stn Rep.* 27: 28, 1931.

Lisea fujikuroi Sawada, 1919

Gibberella moniliforme (Sheld.) Wineland, 1924

stat. conid. *Fusarium moniliforme* Sheldon, 1904.

PERITHECIA usually occur only on dead plant material; superficial, dark blue, globose to conical, 250–350 μ high, 220–300 μ diam., outer wall rough. ASCI ellipsoid to clavate with 4–8 obliquely uniseriate to biseriate ascospores. ASCOSPORES hyaline, elliptical, often remaining 1 septate but occasionally 3 septate, 14–18 × 4.5–6 μ. CULTURES from below typically dark violet but occasionally lilac or vinaceous. Surface covered with rosy buff to vinaceous, floccose to felted mycelium which has a powdery appearance due to the formation of microconidia. MICROCONIDIA formed in chains, usually 1 but occasionally 2 celled, 5–12 × 1.5–2.5 μ, fusiform to clavate, slightly flattened at each end. MACROCONIDIA delicate, thin walled, straight or curved, somewhat dorsi-ventral narrowing at both ends, 3–7 septate; 3 septate, 25–36 × 2.5–3.5 μ; 5 septate, 30–50 × 2.5–4 μ; 7 septate 40–60 × 3–4 μ. CHLAMYDOSPORES absent, both in mycelium and conidia. Dark blue, irregularly globose sclerotia frequently formed; *G. fujikuroi* var. *subglutinans* (q.v.) for distinguishing characteristics (from C. Booth & J. M. Waterston, *CMI Descr.* 22, 1964; and see Booth, *The genus* Fusarium: 123, 1971).

The fungus is plurivorous, seed-, air- and soilborne, and economically most important on gramineous crops, but it also occurs on a very wide range of plants. Accounts of the diseases caused on maize, rice, sugarcane and other crops are given separately. *G. fujikuroi* is widespread in tropical and temperate zones (Map 102). It occurs frequently in disease complexes in the field and in storage. It causes damping off and lodging, and storage rots of fruits and vegetables. Under such circumstances its behaviour is usually that of a weak pathogen where certain predisposing conditions lead to disease. *G. fujikuroi* isolates generally have a higher opt. temp. for growth (25–30°) compared with 25° for those of its var. *subglutinans* (see Kingsland under *Zea mays*). The fungus may exist as homothallic and heterothallic forms. Hsieh et al. reported 3 mating groups (and see Wineland). But Voorhees (1933, *Z. mays* q.v.) obtained perithecia in cultures from single ascospores.

Gibberella fujikuroi

Hsieh, W. H. et al. 1977. *Phytopathology* **67**: 1041 (mating groups; **57**, 1003).
Matuo, T. et al. 1976. *Trans. mycol. Soc. Japan* **17**: 295 (morphology & physiology of isolates from several hosts; **57**, 1605).
Wineland, G. O. 1924. *J. agric. Res.* **28**: 909 (synonymy & perfect state; **4**: 162).

ORYZA SATIVA

G. fujikuroi causes the serious rice disease usually called bakanae (other names are foot rot, man rice, white stalk and seedling blight); losses of up to 70% were found by Heaton in Australia (Northern Territory) and 15–20% losses have been reported from several regions of Asia. Fairly recent accounts of bakanae have been given by Ramakrishnan (*Diseases of rice*), Ou (*Rice diseases*) and Hashioka (*Riso* **20**: 249); their bibliographies should be consulted. This summary is taken partly from those given by the last 2 authors. The conspicuous and characteristic symptom is the abnormal elongation of the plant in the seedbed and field. Affected seedlings (thin and yellowish) stand out, being markedly taller than normal plants. Such seedlings die, either before or after transplanting, if severely infected. In maturing fields infected plants have tall lanky tillers with pale green flag leaves and are conspicuous above the general level of the crop. Such plants usually die, the leaves drying up; or they may survive to maturity but bear only empty panicles. Some infected seedlings may not show the elongation symptoms but become stunted; others appear normal. Yamanaka et al. described 5 types of symptoms on inoculated seedlings. On dying plants sporulation of both stages occurs on the lower parts of the stem and extends upwards. The elongation symptom is due to the presence of gibberellin. The gibberellins are a family of natural plant hormones which are significant in the control of seed plant development. They are present in only trace amounts (even in the richest seed plant sources) and may not yet have been discovered except that they are formed by a few fungi. The discovery and commercial use of *G. fujikuroi* to produce gibberellin arose from the investigations of Kurosawa on this rice disease (see P. W. Brian, *Trans. Br. mycol. Soc.* **58**: 366, 1972). The elongation symptoms are caused only by some gibberellin forming strs of the fungus. The symptoms are most abundant at 35° (absent at 20°) and under moist soil conditions. The opt. temp. for fungal growth is 25–30° (Hemmi et al. gave 28–30° as the opt. temp. for macroconidial germination). Other strs form fusaric acid which causes stunting of the plant.

Bakanae is primarily seedborne; but symptoms developed on seedlings grown from healthy seed and sown in infested soil (see Watanabe). Rhizomorph like survival structures were observed when the fungus was grown on glass slides kept in soil. Ou (l.c.), quoting apparently unpublished work, considered that soil survival in the tropics is not long. In pot experiments it was found that ungerminated seed sown in infested soil developed more disease than did those which had already been germinated (72 hours after soaking; Rajagopalan et al.). The panicles are infected by the conidia and the seed becomes infected. Such seed may be discoloured, sometimes reddish, or appear healthy. Most infection occurs at flowering. In Japan Hino et al. found that most disease occurred in the seedbeds when flowering panicles had been inoculated in July compared with inoculations on those flowering in August and September. On seed germination the fungus becomes systemic, growing up within the plant but generally not penetrating the floral parts. Microconidia and mycelium occur particularly in the xylem. Survival occurs on crop debris as well as in the seed. Seed infection was increased by high av. temps and low insolation during flowering (Takeuchi). Whilst some strs of the fungus cause the characteristic elongation symptoms and others cause dwarfing a further group have no effect on seedling size. The phenomenon of elongation can be caused in barley, maize, sorghum and sugarcane.

Yu et al. recently described some ascospore characteristics. The spores were dispersed from the heading stage to harvest with a diurnal peak at 0100–0300 hours. But release in daylight occurred during rain. Seed could be contaminated by ascospores and develop into diseased plants. Watanabe et al, who confirmed the perfect state in Japan, found ascospores to be viable in perithecia which had been held for 9 months in the laboratory.

Rice vars differ in their reaction and Ou (l.c.) quotes work on resistance from India, Italy, Japan and the Philippines. Seed treatment with organic mercury gives some control, but Takeuchi found that 4 such treatments did not completely prevent bakanae; wet treatments were better than dry ones. Kitamura found that as the specific gravity of rice seeds decreased the percentage of non-viable seeds, incidence of bakanae and the detection of the pathogen in hulls and hulled rice increased. Seeds with a specific gravity of >1.2 soaked in benomyl at 250 ppm gave a disease control equal to that given by 2000 ppm. (The specific gravity of seeds from inoculated plants was rarely >1.25.) Selection by specific gravity did not eliminate all diseased seeds but was useful for increasing the efficiency of seed disinfection. Benomyl, with or without thiram, gives an effective seed treatment (Umehara et al.; and see *Pest control in rice*, PANS Manual No. 3).

Heaton, J. B. 1965. *Trop. Sci.* **7**: 116 (in Australia, general; **45**, 1371).

Hemmi, T. et al. 1941. *Ann. phytopathol. Soc. Japan* **11**: 66 (temp. & conidial germination; **30**: 342).

Hino, T. et al. 1968. *Bull. Chugoku agric. Exp. Stn* E-2: 97 (season, flower infection & seed transmission; **50**, 3719).

Kitamura, Y. 1975. *Proc. Kansai Pl. Prot. Soc.* 17: 32 (effect of seed selection by specific gravity on control; **55**, 2232).

Kurosawa, E. 1926. *Trans. nat. Hist. Soc. Formosa* **16**: 213 (on the filtrate of *G. fujikuroi*, English abstract in *Jap. J. Bot.* 3: 91, 1927).

Rajagopalan, K. et al. 1964. *Phytopathol. Z.* **50**: 221 (seed germination & infection; **44**, 131).

Singh, T. R. S. 1974. *Riso* **23**: 191 (role of gibberellic & fusaric acids, & pectic enzymes in disease; **55**, 4101).

Takeuchi, S. 1972. *Proc. Kansai Pl. Prot. Soc.* 14: 14 (climatic effect on seed infection & organic Hg control; **53**, 163).

Umehara, Y. et al. 1975. *Proc. Assoc. Pl. Prot. Hokuriku* 23: 75, 78 (control with benomyl & thiram; **56**, 1118, 1119).

Yamanaka, S. et al. 1978. *Ann. phytopathol. Soc. Japan* **44**: 57 (symptoms on seedlings; **57**, 5488).

Yu, K. S. et al. 1976. *Pl. Prot. Bull. Taiwan* **18**: 319 (ascospore dispersal & seed contamination; **56**, 5626).

Watanabe, Y. 1974. *Bull. Tokai-Kinki natn. agric. Exp. Stn* **27**: 35 (soil transmission; **54**, 2233).

—— et al. 1977. *Trans. mycol. Soc. Japan* **18**: 136 (perfect state & ascospore survival; **57**, 1178).

SACCHARUM

Accounts of the sugarcane diseases caused by *G. fujikuroi* have been given by Martin et al. (for pokkah boeng) and by Bourne (for sett or stem rot). In pokkah boeng chlorosis appears towards the base of the young leaves; there is leaf distortion (pronounced wrinkling, twisting and shortening). Leaf bases may be abnormally narrow. Irregular reddish areas form on the chlorotic zones, and these develop sometimes into lens-shaped holes with no definite arrangement or forming a ladder-like effect in a longitudinal manner. The leaf edge and tip may form irregular necrotic areas. Leaf sheaths can be affected like leaves. Symptoms occur on the stem as dark reddish streaks, as fine lines in the nodes but in the internodes as external and internal ladder-like lesions. The stem can be curved and distorted. A top rot may lead to death of the entire top of the plant; the tissue near the growing point becomes brown and soft. Infection is caused by conidia spreading through the spindle along the margin of a partially unfolded leaf. The incubation period is *c.* 1 month. The mycelium spreads to vascular bundles of the immature stem; blocked vessels lead to growth distortions and rupture, and the development of the ladder-like lesions. Reimers described pokkah boeng as a common disease of little consequence. It is not spread through setts. Most commercial cvs have some resistance. Very wet weather, late nitrogen applications or heavy watering after dry conditions may bring on the disease.

In infected setts (sett rot) a purplish red discolouration of the parenchyma and vascular bundles spreads from the cut ends inwards. The young roots redden, turn purplish and decay. Infected root eyes do not develop roots and buds may not germinate. Setts split longitudinally show numerous, dirty purple red bundles; and muddy, watersoaked or red brown discolouration in the surrounding parenchyma cells. In standing cane the rot (stem rot) can occur in places infested with the cane borer *Diatraea saccharalis*. The tissues in the stems are discoloured similarly to those in the setts. Stems may be infected after damage at harvest or by wind. Setts are infected from infested soil, through the cut ends, leaf scars, root eyes and wounded roots. They may also be internally infected before planting. The fungus spreads in the xylem and survives on plant debris. Sett rot, by reducing germination, may become important locally and the causal pathogen may be associated with red rot caused by *Glomerella tucumanensis* (q.v.). *Gibberella fujikuroi* has also been isolated from sugarcane affected by wilt; this disease is referred to briefly under *G. fujikuroi* var. *subglutinans*. Where sett rot is likely to occur the seed pieces should be treated with a fungicide and benomyl has shown promise. Lyrene et al. described the inheritance of resistance.

Bourne, B. A. 1961. In *Sugarcane diseases of the world* Vol. 1: 187 (general, 52 ref.).

Liu, L. J. et al. 1971. *J. Agric. Univ. P. Rico* **55**: 426 (reduction of sett germination & fungicides; **51**, 4327).
Lyrene, P. M. et al. 1977. *Phytopathology* **67**: 689 (pokkah boeng, inheritance of resistance; **57**, 276).
Martin, J. P. et al. 1961. In *Sugarcane diseases of the world* Vol. 1: 247 (general, 40 ref.).
Reimers, J. F. 1971. *Cane Grow. q. Bull.* **34**: 101 (note; **50**, 1954).

ZEA MAYS

G. fujikuroi is one of the main pathogens associated with the stalk rot complex of maize. For a full coverage of this diseased condition ref. given under *Diplodia maydis, Fusarium, G. zeae, Macrophomina phaseolina* and *Pythium arrhenomanes* should also be consulted; and see Christensen et al. (*G. zeae* q.v.) for literature up to mid 1964. The fungus also causes ear and kernel rot, and a seedling blight (including pre- and post-emergence death). The diseases caused have been most widely studied in USA but there is work from Canada, Egypt, Germany, India, Italy and South Africa; and see Edwards for Australia (*G. fujikuroi* var. *subglutinans* q.v.). In stalk rot the symptoms (not particularly characteristic) become evident as the plants become mature. There is no distinct colouration of the stem. The pith tissue becomes shredded and the vascular bundles remain intact. The leaves dry out and in damp weather white to pale pink conidial masses form on the leaf sheaths and culms. The weakened stem may break around the lower nodes. The symptoms of a kernel rot may be evident after husk removal; individual or groups of kernels become pinkish, red brown or grey in badly rotted kernels, with surface mycelium. But internally infected seed may show no external symptoms. In seed the fungus has been described as being: primarily confined to the pedicel and abscission layers (Sumner 1968a); seen in LS of mature seeds (Salama et al.); present in endosperm and embryo (Ragab et al.); seed transmitted especially in T male sterile lines (Warmke et al.); and see Kucharek et al.; Ooka et al. 1977a; Valleau; Voorhees.

In seedling blight dark brown, sunken lesions occur on roots and mesocotyl; plant stand and vigour are reduced and seedlings may be killed (Voorhees 1934). Futrell et al. found *G. fujikuroi* to be the primary pathogen from seedlings in poor stands. Seedlings grown with the fungus developed shorter roots with fewer laterals; the leaf blade was attacked and it sometimes died before emerging from the coleoptile. There was evidence that these effects were caused by a toxin produced by the pathogen (and see Scott et al.). Sumner (1968a) found no differences in the field between plants grown from infected and uninfected seed; and considered that air- and soilborne inoculum was more significant (see Ooka et al. 1977b). Likewise Kucharek et al. who found that the amount of seed infection at sowing may be so low that it could be disregarded as a source of root and stalk infection.

The pathogen in USA appears to be more important in the western reaches of the maize belt and in some of the warm, dry, inland valleys of California. Kingsland gave a higher opt. for growth (25–30°) compared with *G. fujikuroi* var. *subglutinans* (25°) which grew less rapidly at 35°. Edwards (1941b, l.c.) reviewed the early, somewhat conflicting, work on the effects of temp. and described the effect of this variable in Australia on both fungi. The work is briefly summarised under the var. *subglutinans*. *G. fujikuroi* is probably more serious at somewhat higher temps compared with those more favourable for maize diseases caused by *G. zeae*.

Besides infecting maize via the seed penetration takes place through the silk region (Koehler); the stem (from conidia deposited between the stalk and leaf sheath and through wounds); and via roots, although this last point of entry is probably of little practical importance (but see Whitney et al. who, from Canada, described root colonisation by *G. fujikuroi* inter alia). Foley (1962) described systemic infection. Although variation in pathogenicity has been described (Leonian) this appears to be of little or no significance in the field. The fungus survives in maize debris. 'Thickened hyphae' have been found and called survival structures. Survival was best at 30 cm soil depth, 5–35% soil moisture and 5–10° soil temp. (Nyvall et al.). Nath et al. reported on a synergistic effect with *Hoplolaimus indicus*. Palmer et al. (1974) described the effects of *Meloidogyne incognita* and other nematodes. The control of root worm (*Diabrotica longicornis*) reduced the incidence of roots infected by *Fusarium* spp. Fajemisin et al. found that maize leaf infections by *Cochliobolus heterostrophus* and *Setosphaeria turcica* led to more root rot caused by *G. fujikuroi*. Kruger described stalk and root rots in the S. African maize triangle. Amongst the main parasitic fungi involved were *Diplodia maydis* and *G. fujikuroi*. The regional distribution of the fungi was described.

Control measures, apart from considerations of

local conditions, for stalk rot are similar to those for the other fungi causing this disease, i.e. good seed, full season resistant hybrids, and well-balanced husbandry and soil fertility. Lunsford et al. found that additive gene action and maternal effects were more important than dominant gene action in the inheritance of resistance to seedling blight caused by *G. fujikuroi*. The systemic insecticide carbofuran reduced stalk rot (mainly caused by *D. maydis, G. fujikuroi* and *G. zeae*); and Landis therefore concluded that this was due to a reduction in the number of insect and nematode wounds. Piglionica et al. described heat treatment of seed.

Djakamihardja, S. et al. 1970. *Pl. Dis. Reptr* **54**: 307 (breeding, resistance; **49**, 3246).

Fajemisin, J. M. et al. 1974. *Ibid* **58**, 313 (effects of infection by *Diplodia maydis, Cochliobolus heterostrophus* & *Setosphaeria turcica*; **54**, 135).

Foley, D. C. 1962. *Phytopathology* **52**: 870 (systemic infection; **42**: 260).

——. 1969. *Ibid* **59**: 620 (stalk deterioration; **48**, 2941).

Futrell, M. C. et al. 1969. *Pl. Dis. Reptr* **53**: 213 (effects of infection on young plants; **48**, 1670).

Ikenberry, R. W. et al. 1967. *Iowa St. J. Sci.* **42**: 47 (cellulase activity in maize stem; **47**, 157).

Kingsland, G. C. 1959. *Diss. Abstr.* **19**: 2728 (aetiology & epidemiology, temp. effects; **38**: 740).

Koehler, B. 1942. *J. agric. Res.* **64**: 421 (host penetration & symptoms; **21**: 367).

Krüger, W. 1970. *Phytopathol. Z.* **67**: 259, 345; **68**: 1, 334 (distribution of fungi causing maize stalk rot in S. Africa, effects of cultural practices on disease; **49**, 3783a, b, c; **50**, 1753).

——. 1974. *Z. Acker-u. PflBau* **139**: 172 (effects of sowing date, crop density & var.; **54**, 2213).

Kucharek, T. A. et al. 1966. *Phytopathology* **56**: 983 (kernel infection & stalk rot incidence; **46**, 116).

Landis, W. R. 1971. *Pl. Dis. Reptr* **55**, 634 (effect of insecticide carbofuran; **51**, 1434).

Leonian, L. H. 1932. *Bull. W. Va Univ. agric. Exp. Stn* 248, 16 pp. (pathogenicity & variability; **11**: 569).

Lunsford, J. N. et al. 1975. *Phytopathology* **65**: 223 (maternal effects in inheritance of resistance; **54**, 3878; and see **56**, 687).

McKeen, W. E. 1953. *Can. J. Bot.* **31**: 132 (fungi causing root & stalk rot; **32**: 479).

Mortimore, C. G. et al. 1964. *Can. J. Pl. Sci.* **44**: 451 (as above & sugar levels in maize; **44**, 688).

—— ——. 1965. *Ibid* **45**: 487 (stalk rot in relation to plant population & grain yield; **45**, 441).

Nath, R. P. et al. 1974. *Indian J. Nematol.* **4**: 90 (synergism with *Hoplolaimus indicus*; **55**, 720).

Nyvall, R. F. et al. 1968. *Phytopathology* **58**: 1704 (hyphae as survival structures; **48**, 1669).

—— ——. 1970. *Ibid* **60**: 1233 (saprophytism & survival in maize; **50**, 650).

Ooka, J. J. et al. 1977b. *Ibid* **67**: 1023 (wind & rain dispersal; **57**, 1167).

—— ——. 1977a. *Pl. Dis. Reptr* **61**: 162 (kernel infection; **56**, 4493).

Palmer, L. T. et al. 1969. *Phytopathology* **59**: 1613 (root infecting *Fusarium* spp. & root worm; **49**, 1357).

—— ——. 1974. *Ibid* **64**: 14 (root infecting *Fusarium* spp. & nematodes; **53**, 4417).

Piglionica, V. et al. 1976. *Phytopathol. Mediterranea* **15**: 98 (heat treatment of kernels; **57**, 155).

Ragab, M. M. et al. 1970. *Agric. Res. Rev. Cairo* **48**: 141 (pathology & histology; **51**, 312).

Salama, A. M. et al. 1973. *Phytopathol. Z.* **77**: 356 (seed transmission; **53**, 2163).

Scott, G. E. et al. 1970. *Pl. Dis. Reptr* **54**: 483 (toxin; **49**, 3784).

Sumner, D. R. 1968a & b. *Phytopathology* **58**: 755, 761 (ecology, in seed; effect of soil moisture; **47**, 3095a & b).

Valleau, W. D. 1920. *Bull, Ky agric. Exp. Stn* 226, 51 pp. (infection).

Volker, P. 1975. *C. r. hebd. Séanc. Acad. Sci. D* **280**: 713 (root penetration & spread; **55**, 201).

Voorhees, R. K. 1933. *Phytopathology* **23**: 368 (general; **12**: 564).

——. 1934. *J. agric. Res.* **49**: 1009 (histology; **14**: 437).

Warmke, H. E. et al. 1971. *Pl. Dis. Reptr* **55**: 486 (occurrence in seed as related to T male sterile cytoplasm, with *C. heterostrophus*; **51**, 311).

Warren, H. L. et al. 1975. *Agron. J.* **67**: 655 (effect of nitrification inhibition on disease incidence & yield; **55**, 4060).

Whitney, N. J. et al. 1961. *Can. J. Pl. Sci.* **41**: 854 (fungi causing root & stalk rot; **41**: 301).

Younis, S. E. A. et al. 1969. *Indian J. Genet. Pl. Breed.* **29**: 418 (inheritance of resistance; **51**, 3323).

OTHER CROPS

As with some other gramineous crops *G. fujikuroi* is one organism in a stalk rot complex in sorghum. Outwardly, the symptoms resemble those caused by *Macrophomina phaseolina* (q.v.). In cvs where the leaf sheaths completely envelop the stems no infection is apparent until plants wilt and dry. On the exposed stem there is a brown or reddish purple discolouration of the vascular bundles on the periphery of the culm; and similarly in the stem pith and bundles. The pith rots and leaves loose, vascular strands. Tall and medium cvs lodge. Brown to reddish purple spots, with a pinkish growth, occur (but rarely) on the leaves. An attacked inflorescence becomes reddish and the grain is infected. The fungus can constitute a high proportion of the mycoflora of sorghum seed. It is found on the seed surface, in the testa and embryo. Seeds can be shrivelled and discoloured; they may germinate poorly or not at all. Infection of the seedling (via seed) causes root necrosis and shoot blight (Futrell et al.; Gourley et al.; Mathur, S. B. et al.; Mathur S. K. et al.). Isolates from maize, sorghum and sugarcane were not host specific for pathogenicity. Those from sugarcane caused the symptoms of pokkah boeng (*G. fujikuroi, Saccharum* q.v.) on sorghum and maize (Priode). Bain found that in sorghum seedlings the incidence of *Sclerospora sorghi* was higher when they later developed blight caused by *G. fujikuroi*. Sorghum stalk rot, caused primarily by *G. fujikuroi*, was reduced under ecofallow (reduced tillage) when the crop was grown in a winter wheat grain sorghum fallow rotation (Doupnik et al.). Control measures for this seed- and soilborne sorghum disease are in part obvious (healthy seed or seed disinfection) and in part will depend on local conditions. Wet conditions will aggravate the disease.

Oxenham described fruitlet core rot of pineapple in Australia (Queensland). Internal browning of a fruitlet develops immediately beneath the floral cavity, and in some cases spreads to the core. The lesions are firm. The rot may be indicated by a failure of the fruitlets to colour at maturity. The commonest fungi associated with the disease were *G. fujikuroi* and *Penicillium funiculosum* Thom behaving as wound pathogens. Infection could occur at any stage in fruit development but fleshy decay was most rapid in near ripe or ripe fruit. Rohrbach et al. reported on the susceptibility of cvs.

The work in Israel on infection of banana (black heart), citrus and avocado pear was reviewed by Joffe et al. They found *G. fujikuroi* to be widespread in soil but it rarely exceeded 5% of the total *Fusarium* population. Little pathogenic specialisation was found between isolates. The best growth in vitro was at 24–30°. In black heart apparently normal fingers, when cut, show a dark brown discolouration from the tip end and along the placenta to the stem end. Some infected fingers may become ripe prematurely or be smaller than normal. The fungus can penetrate banana fruit via the nectary. Control was achieved by removal of pistils not later than 14 days after the bunch had shot, to be followed at once by a zineb spray. Spraying alone did not control the rot (Temkin-Gorodeiski et al. 1971b).

Khurana et al. and Summanwar have given accounts of the work on malformation of mango in India. In adult trees the inflorescence is transformed into a compact mass of sterile flowers; apical or axillary buds develop abnormal, numerous vegetative shoots. Wound inoculations of 8–10-month-old mango seedlings with *G. fujikuroi* reproduce the disease. The fungus may enter the plant through the injuries caused by mites (*Aceria mangiferae*) from which it has been isolated. The disease is not a systemic one. Pruning malformed shoots and treatment with the insecticide diazinon+captan has given control. Among other crops on which *G. fujikuroi* causes disease are: cotton (boll rot), *Cyamopsis tetragonoloba* (blight of cluster bean), pea (wilt), pine (damping off) and sunflower (disease complex). Cole et al. reported that the fungus formed a toxin (moniliformin) which inhibited the growth of wheat coleoptiles; it caused necrosis and growth abnormalities when applied to maize, sorghum and tobacco.

Bain, D. C. 1973. *Phytopathology* **63**: 197 (association with *Sclerospora sorghi* on sorghum; **52**, 3679).
Belliard, J. 1972. *Coton Fibr. trop.* **27**: 243 (on cotton; **52**, 132).
Cole, R. J. et al. 1973. *Science N.Y.* **179**: 1324 (toxin; **52**, 2261).
Doupnik, B. et al. 1975. *Phytopathology* **65**: 1021 (cultural control on sorghum; **55**, 1776).
Futrell, M. C. et al. 1967. *Pl. Dis. Reptr.* **51**: 174 (on sorghum seed in Nigeria; **46**, 2007).
Gourley, L. M. et al. 1977. *Ibid* **61**: 616 (effects on sorghum seedling development; **57**, 1169).
Gupta, S. C. et al. 1967. *Proc. natn. Acad. Sci. India* Sect. B **37**: 241 (pectic enzymes, causing pea wilt; **48**, 3329).
Joffe, A. Z. et al. 1973. *Mycopathol. Mycol. appl.* **50**: 85 (general, in soil, on banana, citrus & avocado; **53**, 98).
Khurana, A. D. et al. 1973. *Pesticides* **7**(1): 12 (on mango; **53**, 3112).
Mathur, S. B. et al. 1967. *Proc. int. Seed Test. Assoc.* **32**: 639 (on sorghum seed; **47**, 2153).
Mathur, S. K. et al. 1975. *Seed Sci. Technol.* **3**: 683 (as above; **55**, 3161).
Orellana, R. G. 1973. *Pl. Dis. Reptr* **57**: 318 (on sunflower; **52**, 3789).
Oxenham, B. L. 1953. *Qd J. agric. Sci.* **10**: 237 (on pineapple; **34**: 43).
——. 1962. *Ibid* **19**: 27 (as above; **42**: 136).
Prasad, N. et al. 1952. *Curr. Sci.* **21**: 17 (on cluster bean; **31**: 557).
Priode, C. N. 1933. *Phytopathology* **23**: 672 (on maize, sorghum & sugarcane; **13**: 59).
Rohrbach, K. G. et al. 1976. *Ibid* **66**: 1386 (susceptibility

of pineapple cvs with *Penicillium funiculosum*; **56**, 4126).
Summanwar, A. S. 1970. In *Proc. 1st int. Symp. Pl. Pathol.* 1966–67: 627, IARI, New Delhi (on mango).
Temkin-Gorodeiski, N. et al. 1964. *Pl. Dis. Reptr* **48**: 134 (on banana; **43**, 2005).
——. 1971a. *Israel Jnl Bot.* **20**: 91 (on banana, role of nectary; **51**, 505, journal name error in *Rev. Pl. Pathol.*).
——. 1971b. *Phytopathol. Mediterranea* **10**: 223 (on banana, control; **52**, 2331).
Yamamoto, M. et al. 1965. *J. Jap. For. Soc.* **47**: 30 (on *Pinus* spp.; **44**, 2907).

Gibberella fujikuroi var. **subglutinans**

Edwards, *Agric. Gaz. N.S.W.* **44**: 896, 1933.
stat. conid. *Fusarium moniliforme* var. *subglutinans* Wr. & Reink., 1925.

Available records suggest perithecia are rare in nature. Those found have been on dead plant material and are similar in size and appearance to those of *G. fujikuroi* (q.v.). ASCI usually 8 spored; ASCOSPORES 12–15 × 4.5–5 μ. CULTURES similar to those of *G. fujikuroi* but MICROCONIDIA are not formed in chains and are more oval. MACROCONIDIA thin walled, narrowing at both ends, 3–5 septate, 32–50 × 3–3.5 μ. This var. is heterothallic; the 2 mating types grown together on wheat or rice straw under suitable conditions will form perithecia. Distinguished from *G. fujikuroi* by having thinner ascospores, microconidia not in chains, macroconidia smaller and less septate, and in culture by a more intense colouration on steamed rice, with a faster growth rate and lower opt. temp. (25°) for growth (from C. Booth & J. M. Waterston, *CMI Descr.* 23, 1964; and see Booth, *The genus* Fusarium: 127, 1971).

The fungus is plurivorous, seed- and soilborne, and economically most important on gramineous crops; but it has been recorded on plants in many other plant families and its distribution is worldwide (Map 191). Booth et al. (l.c.) considered *G. fujikuroi* var. *subglutinans* to be a weak parasite, attacking seedlings or invading plants growing under adverse or stress conditions. In maize it is one of the less important fungi in the stalk rot complex (*Diplodia maydis, G. fujikuroi* and *G. zeae* q.v.), and in maize ear and kernel rot. Infected kernels may be pinkish to reddish brown, with conspicuous surface mycelium; or with a lustreless pericarp surface and shrunken. Kernels may show no external symptoms but on plating pinkish white mycelium of *Gibberella* spp. may form on the surface. The symptoms on the maturing stalk are not particularly characteristic. Ullstrup described the fungus in USA and seedling inoculations. Edwards described poor germination and seedling blight in Australia caused by internally infected maize grain. Infection of grain could be shown in the field by inoculating ears at all stages from pollination to the nearly mature stage. Husk injury and moist conditions were important factors in enabling infection to become established. Two strs of the pathogen were reported. For both *G. fujikuroi* and its var. *subglutinans* high infection occurred at 16° and 20°; satisfactory infection took place only if the air temp. was not > 20°. Defective germination and pre-emergence death were higher at 16°, whilst lesions on emerging seedlings were more abundant at 20°. No seedling blight occurred at 28° or 32° (Edwards 1941b). Control should follow the general measures taken against maize stalk rot, seedling blight and kernel rot: resistance, balanced host growth and seed treatment.

Wilt of sugarcane is a disease whose precise aetiology apparently remains obscure; it seems unlikely that a single organism and/or other factor is the cause of wilt. Particularly in the early literature the main pathogen was described as *Cephalosporium sacchari* Butler. *G. fujikuroi* has also been isolated from sugarcane with wilt; see Ganguly for an account of the earlier work. Booth (*The genus* Fusarium: 129) stated that the disease situation is confused due to the frequent mis-identification of the var. *subglutinans* as *F. moniliforme*; and that wilt of sugarcane reported to be caused by *C. sacchari* is also probably caused by *G. fujikuroi* var. *subglutinans*. Material said to be infected with *C. sacchari* has always yielded a str. of *F. moniliforme* var. *subglutinans*. Gams (*Acremonium* q.v.) transferred *C. sacchari* Butler & Hafiz Khan to *F. sacchari* (Butler & Hafiz Khan) W. Gams. Raja et al. gave a brief review of sugarcane wilt, which is important in India, and described the condition as baffling. The disease reduces germination, dries up young shoots and causes wilt of the cane stem when half grown. Wilt can continue until harvest and the leaves become chlorotic. The pith shows a diffuse purple or dirty red colour which is mainly longitudinally streaky. Pith cavities contain abundant conidia. The organism(s) involved are soil inhabitants; penetration takes place through leaf scars, root eyes and injuries to the underground stem. Singh et al found that wilt symptoms could be produced in sugarcane by inoculation with *A. furcatum* F. & R. Moreau ex W. Gams and *A. terricola* (Miller, Giddens & Foster) W. Gams. Control measures lie through selection for resistance and fungicide treatment of sets.

G. fujikuroi var *subglutinans* has also been isolated from sorghum but, as with some other crops, its relative importance vis à vis *G. fujikuroi* is not clear. Hean described a wilt of sunn hemp (*Crotalaria juncea*) caused by the var. *subglutinans*. Plants which were several weeks old became yellowish and death occurred 3–4 days after the first obvious field symptoms. Sometimes brown streaking was found on the stem surface at about ground level. The fungus caused damping off but seed transmission tests were negative. In mango it causes extensive malformations of young inflorescences and prevents fruit development (see Chattopadhyay et al.). Ochoa et al. found the fungus to be associated with a spear rot of oil palm.

Chattopadhyay, N. C. et al. 1977. *Acta phytopathol. Acad. Sci. hung.* **12**: 283 (degradation of cellulose & lignin in mango inflorescences; **57**, 3039).

Edwards, E. T. 1935. *Sci. Bull. Dep. Agric. N.S.W.* 49, 68 pp. (general; **15**: 359).

——. 1941a. *J. Aust. Inst. agric. Sci.* **7**: 74 (grain infection in maize with *Gibberella fujikuroi*; **21**: 72).

——. 1941b. *Proc. Linn. Soc. N.S.W.* **66**: 425 (effects of temp. & soil moisture on maize seedling blight, with *G. fujikuroi*).

Ganguly, A. 1964. In *Sugarcane diseases of the world* Vol. II: 131 (general, *Cephalosporium sacchari*).

Hean, A. F. 1947. *Sci. Bull. Dep. Agric. S. Afr.* 255, 15 pp. (on sunn hemp; **28**: 216).

Ochoa, S. et al. 1974. *Rev. Inst. Colomb. agropec.* **9**: 425 (spear rot of oil palm; **56**, 4153).

Raja, K. T. S. et al. 1972. *Sugarcane Pathol. Newsl.* 8: 21 (brief review, *C. sacchari* & *G. fujikuroi* on sugarcane; **51**, 4326).

Singh, K. et al. 1974. *Ibid* 11–12: 24 (*Acremonium* spp. & sugarcane wilt; **54**, 1425).

Ullstrup, A. J. 1936. *Phytopathology* **26**: 685 (on maize; **15**: 794).

Gibberella stilboides Gordon ex Booth, *The genus* Fusarium: 119, 1971.

stat. conid. *Fusarium stilboides* Wollenw., 1924
F. lateritium var. *longum* Wollenw., 1931
F. stilboides var. *minus* Wollenw., 1931
F. lateritium var. *longum* formae A1 & A2 Storey, 1932
F. lateritium var. *stilboides* (Wr.) Bilay, 1955.

Matings of certain strains resulted in the formation of perithecia in groups of 2–3 on a small cushion-like stroma in cortex of organic culture substrate (wheat stem). Stroma burst overlying epidermis to form a small bluish black cushion. PERITHECIA ovate to globose when fresh; 100–140 μ high, 80–130 μ diam. In LS wall 19–23 μ wide with 3–4 outer layers of dark walled, irregularly oval cells 7–8 $\times$ 4–5 μ; and 3–4 inner layers of delicate, compressed, elongate, hyaline cells. ASCI clavate, 8 spored, 60–70 $\times$ 9–11 μ. ASCOSPORES smooth walled, elliptical; many remain 1 septate, others with 1–2 further septa, 12–18 $\times$ 4–5.5 μ, may become light brown later. MICROCONIDIA absent but a few 0–3 septate conidia may develop sparsely from aerial mycelium. SPORODOCHIA usually present. MACROCONIDIA oblong, cylindrical to fusiform, straight to falcate, apex often beaked, thin walled, pedicillate, pinkish to orange or red in mass; aseptate 7–14 $\times$ 2–2.5 μ, 1 septate 9–23 $\times$ 2.3–3.5 μ, 3 septate 15–35 $\times$ 2.5–3.5 μ, 6–9 septate 35–85 $\times$ 3.5–5 μ. CHLAMYDOSPORES absent in mycelium but present in conidia. On potato sucrose agar (pH 6.5) an initially flat white mycelial growth turning pink; colony becoming tufted, deep carmine to port-wine coloured; sterile, blue to olive green SCLEROTIA develop (perfect state after C. Booth l.c.; stat. conid. C. Booth & J. M. Waterston, *CMI Descr.* 30, 1964).

The background to Storey's bark disease of coffee was described by Siddiqi et al (1963); and see Storey; Wallace et al. The disease was described from E. Africa. The fungus has been reported from the W. Indies and W. Africa; in India it was found to be associated with a dieback of citrus (mandarin). As judged by the symptoms the disease appears in three forms on coffee: on suckers (Storey's bark), on the main stem (scaly bark) and at or near soil level (collar rot). On suckers a slightly sunken lesion forms, it has a watersoaked margin and becomes dark with an orange halo. A constriction in the stem appears, the stem is girdled, leaves wilt but do not fall, and the shoots die in 1 to >8 months. On the main stem sunken, cankerous lesions are formed; the bark becomes rough and death of the parts of the plant above the lesion may occur. Spread of the fungus from such stem lesions to the suckers can take place.

At the collar slowly developing lesions are caused (*c.* 15 weeks after infection) and death may occur after 3 more weeks. Leaf and berry infection is of little importance.

The fungus penetrates via wounds, particularly those caused through pruning; it does not enter through the roots and is restricted to parts of the non-vascular tissue. In the collar rot there is a purplish brown to pale violet discolouration in the wood. The opt. temp. for conidial germination is 30° and for growth in vitro 25°. Viability in infected bark is retained for 12 months and pathogenicity for 6; viability in field soil is lost after 1 month (Siddiqi et al. 1968). Pruning cuts should be protected with a fungicide paint (captan has been recommended), weeding, mulching and fertiliser application should be correctly carried out, and suckers may be sprayed with captan.

Baker, C. J. 1970. *Kenya Coff.* **35**: 226 (symptoms & control; **50**, 686).

Banerjee, S. K. et al. 1967. *Z. PflKrankh. PflPath. PflSchutz* **74**: 350 (on citrus with *Glomerella cingulata*; **47**, 818).

Siddiqi, M. A. et al. 1963. *Trans. Br. mycol. Soc.* **46**: 91 (symptoms, pathogenicity & identity of pathogen; **42**: 548).

——. 1965. *E. Afr. agric. For. J.* **31**: 11 (report from Malawi, control; **45**, 1079).

——. 1968. *Trans. Br. mycol. Soc.* **51**: 129 (life history of pathogen & conditions favouring disease; **47**, 1891).

Storey, H. H. 1932. *Ann. appl. Biol.* **19**: 173 (symptoms & aetiology; **11**: 711).

Wallace, G. B. et al. 1955. *E. Afr. agric. J.* **21**: 25 (bark diseases of coffee; **35**: 449).

Gibberella xylarioides Heim & Saccas, *Rev. Mycol. Suppl. Colon.* **15**: 97, 1950.

stat. conid. *Fusarium xylarioides* Steyaert, 1948

F. oxysporum forma *xylarioides* (Stey.) Delassus, 1954

F. lateritium f. sp. *xylarioides* (Stey.) Gordon, 1965.

PERITHECIA violaceous, embedded, singly or in groups, in dark purple stromata; globose with flattened base, 200–400 × 180–300 μ. ASCI cylindrical, thin walled, shortly pedicillate, 90–110 × 7–9.5 μ with 8 monostichous ascospores. ASCOSPORES hyaline to straw coloured, fusoid, 1–3 septate, finely roughened, 12–14.5 × 4.5–6 μ. *Fusarium* state on potato sucrose agar (pH 6.5) pale to beige with sparse white mycelium; purple discolouration develops later accompanied by dark bluish black, discrete stromata, some of these represent perithecial initials. MICROCONIDIA aseptate, allantoid, curved, 5–10 × 2.5–3 μ. MACROCONIDIA fusoid, falcate, 2–3 septate, 20–25 × 4–5 μ. CHLAMYDOSPORES oval to globose, smooth or roughened, 10–15 × 8–10 μ (C. Booth & J. M. Waterston, *CMI Descr.* 24, 1964; and see Booth, *The genus* Fusarium: 115).

G. xylarioides causes a tracheomycosis of coffee. It is probably a minor disease but attracted attention through causing serious losses in the Ivory Coast and Zaire (1948–49). The crop in central and W. Africa is largely of *Coffea canephora* (robusta type) which occurs wild in African equatorial forests. *C. liberica*, a lesser crop, *C. dewevrei* and *C. excelsa* are also susceptible. There are *c.* 60 *Coffea* spp. and since over half of these are indigenous in tropical Africa, where the disease has been reported on wild coffee, many other spp. may be susceptible. The distribution is: Central African Republic, E. Africa, Guinea, Ivory Coast and Zaire; a record from Salvador is considered doubtful (Map 464). Collapse (chlorosis and wilt) can occur at any age. A necrosis is usually found near the collar and blue black areas are seen in the wood. The stroma is found in fissures of the bark. Penetration appears to occur through wounds either above or below ground. The incubation period varies with age of the host, from 6–10 days on plants 6 months to 2 years old to 4–6 months on those 7–10 years old. Opt. temp. for germination for conidia and ascospores is similar (25–27°). The pathogen is probably a soil inhabitant and the rate of spread to uninfected areas is slow. Sanitary measures are effective (host destruction and fungicide treatment of the excavated area), but inspection rounds at monthly intervals may be necessary. Resistance, which may be associated with rapid suberisation in wounds and the conc. of caffeine and chlorogenic acid, occurs in *C. canephora* and *C. liberica*. Resistant rootstocks have been used.

Blittersdorff, R. Von et al. 1976. *Z. PflKrankh. PflSchutz* **83**: 529 (cultural comparison of isolates; **56**, 2515).

Delassus, M. 1954. *Bull. scient. Minist. Colon. Sect. tech. Agric. trop.* 5: 345 (review; **35**: 182).

Fraselle, J. V. et al. 1953. *Bull. inf. I.N.E.A.C.* 2: 373 (control; **34**: 451).

Graaff, N. A. Van der 1978. *Neth. J. Pl. Pathol.* **84**: 117 (resistance & disease distribution pattern).
Heim, R. et al. 1950. *C. r. Acad. Sci. Paris* **231**: 536 (general & morphology in vitro; **30**: 228).
Jacques-Felix, H. 1954. *Bull. scient. Minist. Colon. Sect. tech. Agric. trop.* 5: 296 (general; **35**: 182).
Meiffren, M. 1955. *Phytiat. Phytopharm.* **4**: 131 (inoculation; **35**: 449).
Moreau, C. et al. 1954. *Bull. scient. Minist. Colon. Sect. tech. Agric. trop.* 5: 349 (morphology; **35**: 182).
Porteres, R. 1959. *Café–Cacao–Thé* **3**: 3 (resistance; **38**: 598).
Saccas, A. M. 1951. *Agron. trop. Nogent* **6**: 453 (general; **31**: 63).
——. 1956. *Ibid* **11**: 7 (general; **35**: 522).

Gibberella zeae (Schw.) Petch, *Annls Mycol.* **34**: 260, 1936.
Sphaeria zeae Schweinitz, 1822
stat. conid. *Fusarium graminearum* Schwabe, 1838.

The conidial state is most commonly found in nature or isolated from infected material. PERITHECIA occur on many gramineous hosts and are superficial on a thin stroma, forming in clusters around the lower nodes or the bases of infected stems. Ovoid with a rough tuberculate outer wall, 140–250 μ diam.; showing varying degrees of lateral collapse when dry. Wall of 2 layers, an outer stromatic layer, 17–31 μ wide, of globose cells, 5–12 × 1.3–3.5 μ, and a thin inner layer of compressed thin walled cells. ASCI 60–85 × 8–11 μ, clavate with a short stipe and 8, or occasionally 4–6, distichous or obliquely monostichous ascospores. ASCOSPORES hyaline to light brown, curved fusoid with rounded ends, 0–1 and finally 3 septate, 19–24 × 3–4 μ. MICROCONIDIA absent. MACROCONIDIA formed from doliiform phialides, 10–14 × 3.5–4.5 μ, formed laterally or on short multibranched conidiophores; sporodochia may form in older cultures. Conidia often formed sparsely, falcate, sickle shaped or markedly dorsiventral, 3–7 septate, 25–50 × 3–4 μ, with a well developed, often pedicellate foot cell. CHLAMYDOSPORES intercalary, single in chains or clumps, globose, thick walled, hyaline to pale brown with smooth or slightly roughened outer wall; 10–12 μ diam. Many strs do not form chlamydospores on standard media. Opt. temp. growth 24–26° (from C. Booth, *CMI Descr.* 384, 1973; and see Booth, *The genus* Fusarium: 179).

The name *G. saubinettii* has been frequently used for the perithecia of *F. graminearum*; it is based on *Gibbera saubinettii* Mont. which is a synonym of *Gibberella cyanogena* (Desm.) Sacc. (Booth 1971, l.c.). *G. zeae* is worldwide on temperate cereals, maize, rice, sorghum and on some dicotyledons (causing leaf and flower rots of some ornamentals). The diseases (which can be spread through seed) caused are variously called: seedling blight, pre- and post-emergence blight, root and foot rot, culm decay, head or kernel blight (scab or ear scab) and stalk rot. On rice the disease (on the glumes and nodes) is called scab (see Ou, *Rice diseases*). The fungus has been widely studied as a pathogen in a disease complex on maize (stalk rot). With *Diplodia maydis* (q.v.) it is one of the more important pathogens of the stalk rot syndrome (others include: *G. fujikuroi* and its var. *subglutinans, Khuskia oryzae, Macrophomina phaseolina* and *Pythium* spp.). The fungus is homothallic but heterothallic forms exist. In Australia (E.) Francis et al. distinguished 2 populations of the fungus depending on whether cultures formed perithecia or not. Tschanz et al. (1976) gave an opt. temp. of *c.* 29° for perithecial formation. But ascospore discharge was favoured by lower temps, there being none at >26°. Ikeya, working with isolates from rice and wheat, gave an opt. temp. of 28° for in vitro growth. *G. zeae* caused more damage to rice seedlings at 20° and 24° than at 28° and 32°. The literature on maize stalk rot has been reviewed (up to mid 1964) by Christensen et al.

The fungus is a lower temp. organism than *D. maydis*. In USA *G. zeae* is relatively more prevalent in the N. and E. and stalk rot caused by it varies considerably in intensity in the cooler regions of the USA maize belt. Both these fungi (and *M. phaseolina*) have a similar type of parasitism. Maize is attacked as the plants approach maturity. There is a sudden onset of a greyish green colour in the leaves, softening and discolouration of the outer parts of the lower internodes, shredded appearance of the pith and brown to black lesions near the lower nodes. When stems are split the diseased area has a reddish discolouration. After silking the plants may wilt suddenly. *G. zeae* invades the roots and then the stem as plants mature (Gates; and see Pearson). The symptoms are not particularly characteristic and isolations may well be needed. Forms of the fungus differ both in pathogenicity and in the effects of temp. Purss in Australia found that isolates from gramineous plants other than maize caused crown rot in wheat but they caused no symptoms typical of field infection in maize. Maize seed is not generally affected; a standard chemical treatment is to use captan (e.g. Kommedahl et al.). The fungus carries over on field debris (e.g., Wearing et al.); perithecia form on such material (and stems) and ascospores may infect ears and stalks. In Canada Mortimore et al. (1972) found that a reduction in competition for light and water decreased stalk rot by 63 and 20% in resistant and susceptible hybrids, respectively. In 1969 they stated that the deleterious effects of stress from interplant competition and ear development had their max. effect on stalk rot susceptibility and kernel weight during the month before physiological maturity.

Resistance to stalk rot is largely dependent on cell condition; living cells, in good condition, presumably limit spread of the fungus (Gates). Stalk rot ratings are highly correlated with pith condition ratings based on areas of dead cells (Pappelis 1965, 1970b). Injuries to maize roots and leaves increased susceptibility to stalk rot and the rate of cell death; most such increases were caused by root injury. Cell death took place in the nodes as the change from resistance to susceptibility occurred (Pappelis 1970a). Incidence of stalk rot was negatively correlated with stalk strength (Abney et al.). Damage by the fungus increased where the maize borer (*Ostrinia nubilalis*) was present (Chez et al.). Certain resistant hybrids had greater growth rates, leaf areas and stalk densities than susceptible crosses. The density of pith tissue from the lower stalk at physiological maturity was negatively correlated with subsequent stalk rot incidence (Wall et al.). Control is through resistant hybrids and balanced soil fertility.

Abney, T. S. et al. 1971. *Phytopathology* **61**: 1125 (effect of host nutrition; **51**, 1433).

Bonn, W. G. et al. 1970. *Can. J. Bot.* **48**: 1335 (effects of C & N nutrition on growth & macroconidial formation; **50**, 1086).

Cappellini, R. A. et al. 1971. *Mycologia* **63**: 641 (pH, nutrients & macroconidial germination; **51**, 91).

Chez, D. et al. 1977. *Phytoprotection* **58**: 5 (effect of *Ostrinia nubilalis*; **56**, 4989).

Christensen, J. J. et al. 1966. *Monogr. Am. phytopathol. Soc.* 3, 59 pp. (monograph, stalk rot of maize, 409 ref.; **47**, 1528).

Francis, R. G. et al. 1977. *Trans. Br. mycol. Soc.* **68**: 421 (characteristics of 2 *Gibberella zeae* populations in E. Australia; **56**, 5542).

Gates, L. F. 1970. *Can. J. Pl. Sci.* **50**: 679 (pith cell condition & disease incidence; **50**: 2906).

—— et al. 1972. *Ibid* **52**: 929 (effects of leaf removal on yield & disease; **53**, 137).

Ikeya, J. 1933. *Forschn Geb. PflKrankh. Kyoto* **2**: 292 (on rice; **13**: 263).

Kommedahl, T. et al. 1975. *Phytopathology* **65**: 296 (seed treatment with antagonists; **54**, 3880).

Mortimore, C. G. et al. 1969. *Can. J. Pl. Sci.* **49**: 723 (effects of reducing interplant competition; **49**, 1358).

—— ——. 1972. *Can. Pl. Dis. Surv.* **52**: 93 (as above; **52**, 1117).

Naik, D. M. et al. 1978. *Can. J. Bot.* **56**: 1113 (stimulation by maize pollen; **57**, 5465).

Pappelis, A. J. 1965. *Phytopathology* **55**: 623 (pith condition & disease; **44**, 3035).

—— et al. 1966. *Ibid* **56**: 850 (effects of soil fertility on host cell death in stalk tissue; **45**, 3526).

—— ——. 1969. *Ibid* **59**: 129 (ear removal & cell death rate in stalk tissue; **48**, 2353).

——. 1970a. *Ibid* **60**: 355 (effects of root & leaf injury on host cell death & disease; **49**, 2464).

——. 1970b. *Forschn Geb. PflKrankh. Kyoto* **7**: 85 (inoculations with *Diplodia maydis*; **50**, 3696).

Pearson, N. L. 1931. *J. agric. Res.* **43**: 569 (parasitism on seedlings; **11**: 170).

Purss, G. S. 1969. *Aust. J. agric. Res.* **20**: 257 (strs causing crown rot in gramineous hosts & maize stalk rot; **48**, 2352).

Szécsi, A. 1970. *Acta phytopathol. Acad. Sci. hung.* **5**: 35 (cell degrading enzymes; **51**, 1432).

Tschanz, A. T. et al. 1975. *Phytopathology* **65**: 597 (ascospore discharge; **54**, 4802).

—— ——. 1976. *Mycologia* **68**: 327 (effects of temp. & light on perfect state; **56**, 562).

Wall, R. E. et al. 1965. *Can. J. Bot.* **43**: 1277 (host growth pattern & resistance to root & stalk rot; **45**, 794).

Wearing, A. H. et al. 1978. *Trans. Br. mycol. Soc.* **70**: 480 (distribution & survival in maize soils of E. Australia; **57**, 4925).

Wolf, J. C. et al. 1977. *Appl. Environ. Microbiol.* **33**: 546 (zearalenone & perfect state; **56**, 4851).

GLOEOCERCOSPORA Bain & Edgerton ex Deighton, *Trans. Br. mycol. Soc.* 57: 358, 1971.

Gloeocercospora Bain & Edgerton, 1943.
(Tuberculariaceae)

MYCELIUM internal, of septate branched hyphae. STROMA small or none. SPORODOCHIA suprastomatal, originating from hyphae which emerge through the stomata, pulvinate, composed of more or less colourless, repeatedly branched hyphae with short cells, the ultimate cells functioning as CONIDIOPHORES. Conidial scars terminal, minute, unthickened. CONIDIA (blastospores) colourless, filiform, straight or curved, pluriseptate, smooth, slimy (F. C. Deighton l.c. who validated the generic name and the names *G. sorghi* q.v. and *G. inconspicua* Demaree & Wilcox; he doubted whether the latter was correctly assigned to *Gloeocercospora*).

Gloeocercospora sorghi Bain & Edgerton ex Deighton, *Trans. Br. mycol. Soc.* **57**: 359, 1971.
Gloeocercospora sorghi Bain & Edgerton, 1943.

MYCELIUM hyaline, septate, branching. SPRORODOCHIUM situated between guard cells and above stomatal opening. CONIDIOPHORES hyaline, septate, short, 5–10 μ in length, simple or branched. CONIDIA hyaline, septate, elongate to filiform, variable length, 20–195 × 1.4–3.2 μ (82.5 × 2.4 μ); generally borne in a slimy matrix, salmon in mass. SCLEROTIA 0.1–0.2 mm diam., black lenticular to spherical, in necrotic host tissue (from J. L. Mulder & P. Holliday, *CMI Descr.* 300, 1971. The synonymy given in this description is incorrect, *Ramulispora sorghi* q.v.; see also a description by Rawla from Indian material, *Ramulispora* q.v.).

Zonate leaf spot was first reported from sorghum in 1943 from USA (Louisiana); on turf grasses the disease is called copper spot. Besides the main host the fungus has been reported on *Agrostis* spp., *Cynodon dactylon* (Bermuda grass), *Pennisetum* spp., *Vetiveria zizanioides*, maize and sugarcane. *G. sorghi* has a widespread distribution but has not been reported from Europe, W. or Central Asia, Canada or New Zealand (Map 339). On sorghum the initial lesions are red brown, watersoaked and sometimes have a narrow, pale green halo. They enlarge, become dark red (in some cvs dark brown) and elongate, parallel to the veins. Possibly by coalescence, semi-circular, irregular lesions (several cm diam.) are formed. Smaller spots have a light brown centre surrounded by a reddish border; but larger ones may have alternate light and dark zones, and the whole leaf can be covered. Younger, red lesions are often so numerous as to form red blotches. The pinkish, gelatinous, conidial fructifications (over the stomata) are easily visible. Spherical sclerotia (0.1–0.2 mm diam.) form within the tissue, in a somewhat linear fashion. On bent grass (*A. canina*) irregular, copper-tinted spots become coalescent. The leaf spots on sorghum can be confused with those caused by *R. sorghi* and *R. sorghicola*. In the former case the spots are elliptical and lack the dark red markings caused by *G. sorghi*. The clearest morphological distinction between these 3 fungi is given by the sclerotia but these are not always present on the leaf spots. Sclerotia of *G. sorghi* are immersed in the necrotic tissue, whilst those of the other 2 fungi are superficial (Harris, *Ramulispora* q.v.).

Growth in vitro is opt. at 28–30° and sclerotia are formed in culture; up to 6 germ tubes may form from each conidium. Penetration (without appressoria) is stomatal (as quoted by Tarr, *Diseases of sorghum, Sudan grass and broom corn*). But Myers et al. reported penetration through epidermal cells and leaf trichomes. The cultural characteristics are better maintained by storing dried sclerotia. Sclerotia germinate by each producing a single sporodochium and this is not affected by storage at −4° and 4° for up to 32 days. The fungus is probably soilborne in crop residue and also seedborne. Sclerotia overwintered in USA (Mississippi), above or on the surface of soil, were still viable; but those placed 8 cm below soil level did not form cultures. Sclerotia were less abundant on pearl millet and an isolate from sorghum caused only slight infection on the former host; no races have, however, been described. *G. sorghi* became serious in Alabama (1964) on *S. vulgare* × *S. sudanense* hybrids which all seemed equally susceptible. Glasshouse screening of 1509 lines revealed no resistance on inoculation, although some field resistance apparently occurs. An organic cadmium fungicide gave good control of copper spot on *A. canina*. Seed treatment for control is advisable.

Bain, D. C. et al. 1943. *Phytopathology* **33**: 220 (morphology & general; **22**: 302).

——. 1950. *Ibid* **40**: 521 (seed infection; **29**: 556).

Dean, J. L. 1968. *Ibid* **58**: 113 (sclerotial germination & overwintering; **47**, 1536).

Deighton, F. C. 1971. *Trans. Br. mycol. Soc.* **57**: 358 (validation of *Gloeocercospora, G. sorghi* & *G. inconspicua*; **51**, 1194).

Fry, W. E. et al. 1975. *Physiol. Pl. Pathol.* **7**: 23 (hydrogen cyanide detoxification; **55**, 1777).

Howard, F. L. 1947. *Greenk. Reptr* **15**: 10 (fungicidal control on turf; **26**: 492).

Keil, H. L. 1946. *Phytopathology* **36**: 403 (as above).

Myers, D. F. et al. 1978. *Ibid* **68**: 1147 (development in sorghum).

Puranik, S. B. et al. 1966. *Bull. Indian phytopathol. Soc.* 3: 50 (notes on; **47**, 1113k).

GLOMERELLA Schrenk & Spauld., *Bull. Bur. Pl. Ind. U.S. Dep. Agric.* 44: 29, 1903.
Gnomoniopsis Stonem. non Berlese, 1898.
(for further synonymy and description see Arx, J. A. von et al., *Beitr. Kryptogflora Schweiz* **11** (1): 185, 1954)
(Polystigmataceae)

PERITHECIA membranous, dark brown, spherical to flask shaped, sometimes evidently hairy, embedded in plant tissue, scattered or in small clusters united by a rudimentary stroma or with slight development of a clypeus round the mouth, ostiole well developed, sometimes papillate or even beaked, wall composed of a few layers of angular brown cells. ASCI unitunicate, clavate, very short stalked, 8 spored as a rule. ASCOSPORES hyaline, oblong, obtuse, usually slightly curved, aseptate; paraphyses slender, hyaline (from Schrenk et al. l.c. and R. W. G. Dennis, *British ascomycetes*: 258, 1968).

Very many *Glomerella* spp. (including 7 of those mentioned below) were reduced to synonymy with the ubiquitous *G. cingulata* (q.v.) by Arx et al. (l.c.). Some spp. have conidial states in *Colletotrichum*. One other sp. is described, *G. tucumanensis*. The following are mentioned elsewhere, under *G. cingulata: G. folliicola* (citrus), *G. glycines* (and *C. dematium* f. sp. *truncata* q.v.), *G. gossypii* (cotton), *G. lagenarium* (Cucurbitaceae) and *G. psidii*; also *G. lindemuthianum* (*C. lindemuthianum* q.v.) and *G. phomoides* (*C. coccodes* q.v.). *G. major* Tunstall, from *Camellia sinensis* (tea), was distinguished from *G. cingulata* by the larger ascospores and conidia (7 × 25 μ compared with 4 × 14 μ and 8 × 25 μ compared with 5 × 15 μ, respectively). *G. magna* S. F. Jenkins & N. N. Winstead was described as causing an anthracnose of cucurbits. The symptoms on leaves, stems and cotyledons were similar to the common anthracnose on this crop (*G. cingulata*, Cucurbitaceae q.v.). But they differed considerably on fruits; the dark lesions spread rapidly to cover the entire fruit, the older lesions showed concentric zonation. The most important susceptible crops were: watermelon (*Citrullus lanatus*), squash and pumpkin (*Cucurbita maxima*) and melon (*Cucumis melo*). *G. magna* was distinguished from *G. cingulata* by its larger asci and ascospores, 72–125 × 10–20 μ and 19–48 (av. 35) × 4–7 (av. 5.3) μ, respectively; the fungus is heterothallic.

Jenkins, S. R. et al. 1964. *Phytopathology* **54**: 452 (*Glomerella magna* sp. nov.; **43**, 2780).
Pady, S. M. et al. 1971. *Trans. Br. mycol. Soc.* **56**: 81 (spore discharge in *Glomerella*; **50**, 2100).
Shear, C. L. et al. 1913. *Bull. Bur. Pl. Ind. U.S. Dep. Agric.* 252, 110 pp. (the genus *Glomerella*, as pathogens).
Tunstall, A. C. 1935. *Trans. Br. mycol. Soc.* **19**: 331 (*G. major* sp. nov.; **14**: 720).
Winstead, N. N. et al. 1966. *Phytopathology* **56**: 134 (effect of light on perithecial formation in *G. magna*; **45**, 3014).
Yamamoto, W. 1961. *Sci. Rep. Hyogo Univ. Agric.* Ser. agric. Biol. **5**: 1 (stat. conid. of *Glomerella, Guignardia* & *Physalospora* in Japan, Korea & Taiwan; **45**, 748).

Glomerella cingulata (Stonem.) Spaulding & Schrenk, *Science* Ser. 2 **17**: 751, 1903.
Gnomoniopsis cingulata Stonem., 1898
stat. conid. *Colletotrichum gloeosporioides* (Penz.) Sacc., 1882
Vermicularia gloeosporioides Penz., 1880.

PERITHECIA on various parts of the host but usually on dead leaves, twigs, etc., solitary or aggregated, globose to obpyriform, dark brown to black, 85–300 μ diam.: wall up to 8 cells thick, sclerotial on outside, pseudoparenchymatous within, ostiole slightly papillate, circular, canal lined with periphyses. ASCI 8 spored, clavate to cylindrical, thickened at apex, 35–80 × 8–14 μ, interspersed with paraphyses forming group at base of perithecium. ASCOSPORES narrowly oval to cylindrical to fusiform, sometimes slightly curved, aseptate, hyaline, *c.* 12 μ, occasionally becoming faintly brown and 1 septate before germination. ACERVULI on necrotic areas or clearly defined lesions on any part of host, usually setose, sometimes sparsely setose or glabrous, rounded, elongated or irregular in shape, may be 500 μ diam. SETAE variable in length rarely > 200 μ long, 4–8 μ wide, 1–4 septate, brown, slightly swollen at base and tapered to the apex on which conidia are occasionally borne. CONIDIA cylindrical with obtuse ends, sometimes slightly ellipsoid with rounded apex and narrow truncate base, hyaline, aseptate, uninucleate, 9–24 × 3–6 μ, formed on aseptate, hyaline or faintly brown cylindrical phialidic conidiophores.

COLONIES on PDA greyish white to dark grey. Aerial mycelium variable, some isolates develop an even, fretted mat, others show diurnal zonation of dense and sparse development, whilst a few have little aerial mycelium or aerial mycelium only in tufts associated with fructifications. CONIDIA more variable in size and shape than on the host, formed in setose or glabrous acervuli or on solitary phialides on mycelium, usually pale salmon in mass. PERITHECIA occasionally formed in cultures, more common in older ones, often associated with stromatic structures and darker or more tufted mycelium than are acervuli. Reverse of colony unevenly white to grey, darker with age, sometimes very dark where large numbers of appressoria formed. APPRESSORIA formed on mycelium in old cultures or slide cultures, sepia brown, ovate to obovate, sometimes lobed, 6–20 × 4–12 μ, borne on hyaline or light brown undifferentiated or sigmoid supporting hyphae. Arx et al., *Beitr. Kryptogflora Schweiz* **11**(1), 1954 gave 120 synonyms for the perfect state and Arx (1957a, *Colletotrichum* q.v.) gave >600 for the imperfect. Many of the taxa in these suggested synonymies have not been fully investigated and detailed studies may reveal characters which enable them to be distinguished (J. E. M. Mordue, *CMI Descr.* 315, 1971).

Some distinguishing features of the other *Colletotrichum* spp., which are fully described, are briefly:

C. capsici: exceptionally complex development of appressoria and their supporting hyphae in slide cultures.

C. coccodes: distinguished from *C. gloeosporioides* chiefly by size and shape of conidia and frequency of sclerotium formation; mostly on tomato and Irish potato.

C. dematium: conidia are falcate, fusoid, 18–26 × 3–3.5 μ; plurivorous.

C. falcatum: only strongly pathogenic on sugarcane (*Glomerella tucumanensis* q.v.).

C. graminicola: the irregular or lobed appressoria with their thickened supporting hyphae, and conidia generally slightly narrower in proportion to their length, are distinctive compared with *C. falcatum*; pathogenic on Gramineae other than sugarcane.

C. lindemuthianum: distinguished from *C. gloeosporioides* by its slow growth and dark pigmentation in culture; mostly pathogenic on *Phaseolus* spp.

C. musae: distinguished from *C. gloeosporioides* by slightly longer and broader conidia, and a generally higher temp. for growth; mostly pathogenic on *Musa*.

The synonymy of Arx (l.c.) has been followed (but not necessarily in other *Colletotrichum* spp.) and the names given by authors in *Colletotrichum* and *Gloeosporium*, where they differ from *C. gloeosporioides*, have been mostly omitted. Simmonds (1965, *Colletotrichum* q.v.) in Australia designated *C. gloeosporioides* var. *minor* which he considered to be the same as *Glomerella cingulata* var. *minor* Wollenw. (Wollenweber et al., *Colletotrichum* q.v.). This var. is commoner in Queensland than the other form; it is an important cause of fruit rot in avocado, mango and papaw. Its conidia av. 3.7 × 14 μ and ascospores av. 4.2 × 15.6 μ, compared with av. 4.8 × 13.8 μ and av. 5.4 × 18.3 μ, respectively, for the other form or var. (Simmonds, l.c.). Arx (1957c, *Colletotrichum* q.v.) considered that 288 spp. previously included in *Gloeosporium* were the stat. conid. of *G. cingulata*. Shear et al. made a full, early study of the pathogen.

G. cingulata is plurivorous and an important pathogen (and a saprophyte), particularly in the tropics and subtropics. Edgerton, in an early account of differences in temp. requirements for growth in vitro amongst isolates from several plant hosts, divided them into several groups. The upper group had an opt. of 27–29°. Simmonds (l.c.) agreed in general, the opt. for *G. cingulata* was 26.5–28.5°. The fungus is extremely variable in pathogenicity, cultural characteristics and the formation of both spore stages. Both homothallic and heterothallic forms occur. Isolates fall into one of 3 culture types: with fertile perithecia and conidia, with fertile perithecia but few or no conidia, or with conidia only (Wheeler 1956). The genetics of spore formation has been intensively studied, see Wheeler (1954) for a review. Some account of heterocaryosis has been given by Stephan. Both UV and blue light increased formation of perithecia (Grand-Pernot). Conidia were more abundant on autoclaved or surface sterilised seed or seedlings, and neither far nor near UV light was required for sporulation (Crosier et al.).

The fungus causes diseases (usually called anthracnoses) on very many crops (on fruits, leaves and stems); stem dieback, fruit and blossom rots, and seedling diseases. One of its most important and common forms of infection is the cause of latent infection in tropical fruits. Numerous, macroscopically invisible, restricted lesions are formed at all stages of fruit growth by cuticular penetration via the appressoria. Under certain conditions, which occur during ripening (on or off the plant), transport or storage, a few of these incipient infections may develop to cause necroses. These are the typical sunken, black, anthracnose lesions (occurring on many parts of the plant) with their acervuli and mucilagenous, pinkish, pin head like masses of conidia. Such latent infection and its development has been most studied in *C. musae* (q.v., see Simmonds 1941) on banana, and Wood discussed the phenomenon in *Physiological plant pathology*. Baker and others did some of the early investigations of latent infection by *G. cingulata* in the tropics (and see Nolla for references to earlier work). Besides cuticular penetration of young tissue, host entry also takes place through wounds or when the host is weakened. The diseases caused on the following crops (or crop groups) are described separately: citrus, coffee, cotton, Cucurbitaceae, kenaf, Leguminosae, mango and rubber. Some lesser diseases are mentioned briefly below.

On *Acacia dealbata* (silver wattle) Hashimoto described a seedling blight with seed infection; seed treatment (water, 70° for 5 min.), dithane or Bordeaux+ethyl mercury phosphate gave control (and see Terashita). Anthracnose of *Agave sisalana* (sisal) was controlled with oxycarboxin. Fruit rots, seedling collapse, leaf fall and stem dieback have been described in *Annona* (Alvarez García; Chowdhury). Aragaki et al. described a necrosis of the spadix in *Anthurium*; this could lead to a complete rot; maneb gave control. For infection on banana, fruits of avocado and *Bixa orellana* (annatto) see Binyamini et al.; Kaiser et al.; Oste et al.; Peregrine. Both ferbam and copper have been used with effect against avocado pear anthracnose; and Spalding et al. described controlled atmosphere storage for control in this crop. An inflorescence dieback, button shedding and nut rot in *Areca catechu* was reported from India (Saraswathy et al.). Control of cashew (*Anacardium occidentale*) anthracnose was described from Brazil (Menezes et al.). For *Allium* (onion), seven curls disease, see Remiro et al.

Brown blight of tea has been studied largely in Japan; there is a general account by Kono (with many ref. and a discussion of latency). Young and wounded leaves are most susceptible, conidial sporulation is most abundant at 24–28° and infection is increased by cold injury, high temps and flooding. Anthracnose of papaw can be severe on the fruit in Hawaii; the small watersoaked fruit spots develop into sunken lesions with maturity. Stanghellini et al. considered that the fungus was primarily a wound pathogen and could not infect attached, unwounded fruit (mature or immature). Leaf spots which become light tan to ash grey develop shotholes and cause premature defoliation on papaw seedlings; the condition increased with higher soil pH levels (Trujillo et al.); see Dastur for infection of flower buds and young fruit. Amongst fungicides used for control dithane and mancozeb are effective, but copper can russet papaw fruit (Raabe et al.; Tsai). On *Derris elliptica* Ling described circular to irregular, yellow brown to deep brown leaf lesions, up to 1.7 cm diam.; lesions also occurred on petioles, and on stems there were brown to grey white cankers which caused stem breakage. Both Ling and Hoof found variation in susceptibility amongst *Derris* spp.; and see Kumamoto. Ferbam, zineb and benomyl+propineb gave control of anthracnose of yam (Fournet et al.; Goto; Singh et al.). On cassava the disease has been described by Chevaugeon.

Anthracnose of tobacco is largely a disease of the seedbed, although the disease can occur and spread in the field. Lucas gave an account in *Diseases of tobacco*; see Cole and Riley for the disease in E. Africa. The first symptoms on seedlings are small, dull or light green, watersoaked leaf spots; these enlarge up to 1 cm diam., become pale brown or grey white with a darker border; small plants are killed. On larger plants severe spotting causes leaf tear and distortion. Elongate, brown lesions or cankers on the leaf midrib, petiole and stem can cause damage in the field. Dithiocarbamate fungicides (and seedbed sterilisation) have been used in control. Hartill found that frequent spraying of seedlings with mancozeb was better than thiram. Anthracnoses on *Piper nigrum* and *P. betle* (black pepper and betel), and guava, are fairly often reported from S.E. Asia. On black pepper the leaf lesions darken with age, sometimes have a yellow or light brown border, spread to form large irregular necrotic areas and leaf fall results. The spikes are also attacked and the fungus has been associated with a hollow and light condition in the berries called pollu (Thomas et al.; Vimuktanandana et al.). On guava (*G. psidii* (Del.) Sheld. is probably not distinct from *G. cingulata*) infection seems most serious on the fruit on which corky, scab like or canker symptoms occur; copper gives control in India (Tandon et al.; Venkatakrishniah).

A leaf spot and stem canker of the pasture legumes *Stylosanthes* spp. was described by Sonoda et al. Leaflet lesions are 1–2 mm diam., light in the centre with a dark margin, with some shotholing and abscission. The elliptical stem lesions are 2–4 mm in length. Irwin et al. recognised 2 diseases on *Stylosanthes* in N. Australia. They were distinguished by their symptoms and cultural characteristics. In one (type A) 2 races were reported. In Ghana Dakwa et al. described a severe defoliation of cacao. On the leaves there were watersoaked areas, yellow green patches and older lesions became necrotic. Anthracnose also occurs on cacao pods; small, almost black, sunken lesions, with the characteristic pink pustules of the conidia.

For the relatively unimportant and apparently unspecialised pathogenic variants of *G. cingulata* control measures are often a matter of good husbandry. Since the fungus is also a common saprophyte and frequently on plant debris, infection of crops is always possible where: humidity conditions are high, crop debris accumulates in the field or builds up in soil, and the growing crop is physiologically weakened, attacked by other fungi or physically damaged by tools or animals. The most serious forms of the diseases (latent infections) caused are probably those on fruits, particularly when these are transported long distances. To prevent such infections developing into disfiguring and damaging rots it may be necessary to spray regularly in the field and/or to use fungicide dip treatments after harvest. Control of this form of disease, caused by several *Colletotrichum* spp., has been most fully investigated in *C. musae*.

Alvarez García, L. A. 1949. *J. Agric. Univ. P. Rico* **33**: 27 (on *Annona*; **30**: 573).

Aragaki, M. et al. 1960. *Pl. Dis. Reptr* **44**: 865 (on *Anthurium*; **40**: 364).

Baker, R. E. D. et al. 1937. *Ann. Bot.* **1**: 59 (storage & latent infections of tropical fruits; **16**: 395).

——. 1938. *Ibid* **2**: 919 (as above; **18**: 193).

—— et al. 1940. *Trop. Agric. Trin.* **17**: 128 (review, as above; **19**: 663).

Binyamini, N. et al. 1972. *Phytopathology* **62**: 592 (latent infection in avocado; **52**, 456).

Cheng, Y. H. et al. 1968. *Pl. Prot. Bull. Taiwan* **10**: 63 (on sisal; **49**, 1046).

Chevaugeon, J. 1957. *C. r. hebd. Séanc. Acad. Sci. Paris* **244**: 2549 (on cassava; **36**: 713).

Chowdhury, S. 1947. *Curr. Sci.* **16**: 384 (on *Annona reticulata*; **27**: 245).

Cole, J. S. 1959. *Ann. appl. Biol.* **47**: 698 (on tobacco; **39**: 440).

Crosier, W. F. et al. 1969. *Proc. Assoc. off. Seed Anal. N. Am.* **59**: 82 (conidial sporulation on sterilised plant material & UV light; **51**, 101).

Dakwa, J. T. et al. 1978. *Pl. Dis. Reptr* **62**: 369 (on cacao).

Dastur, J. F. 1920. *Ann. appl. Biol.* **6**: 245 (on papaw).

Driver, C. H. et al. 1955. *Mycologia* **47**: 311 (a sexual hormone; **35**: 311).

Edgerton, C. W. 1915. *Phytopathology* **5**: 247 (effect of temp.).

Fournet, J. et al. 1975. *Nouv. Agron. Antilles Guyane* **1**: 115 (on yam; **55**, 1587).

Goto, K. 1929. *J. Soc. trop. Agric. Taiwan* **1**: 301 (on yam; **9**: 429).

Grand-Pernot, F. Le 1972. *Fruits* **27**: 339 (effect of light on perithecial formation; **52**, 2157).

Hartill, W. F. T. 1967. *Rhod. Zamb. Malawi J. agric. Res.* **5**: 63 (fungicide control on tobacco; **46**, 2825).

Hashimoto, H. 1968. *Bull. Fukuoka-ken For. Exp. Stn* 20, 29 pp. (on *Acacia dealbata* & control; **48**, 2581).

Hoof, H. A. Van 1950. *Bull. bot. Gdns Buitenz.* Ser. 3 **18**: 473 (on *Derris*; **29**: 476).

Irwin, J. A. G. et al. 1978. *Aust. J. agric. Res.* **29**: 305 (on *Stylosanthes*).

Kaiser, W. J. et al. 1966. *Mycologia* **58**: 397 (on banana; **45**, 3178).

Kono, M. 1965. *Spec. Bull. Res. Inst. Fed Sci. Kinki Univ.* 1, 66 pp. (on tea; **45**, 2947).

Kumamoto, Y. 1949. *Ann. phytopathol. Soc. Japan* **13**: 7 (on *Derris*).

Lin, C. K. 1945. *Am. J. Bot.* **32**: 296 (nutrients & conidial germination; **24**: 428).

Ling, L. 1951. *Pl. Dis. Reptr* **35**: 13 (on *Derris*; **30**: 389).

Lingappa, B. T. et al. 1965, 1966 & 1969. *J. gen. Microbiol.* **41**: 67; **43**: 91; **56**: 35 (self-inhibition of conidial germination, growth & dimorphism; **45**, 985, 3063; **48**, 2166).

Menezes, M. et al. 1975. *Fitossanidade* **1**: 70, 77 (fungicides on cashew; **57**, 1429, 1430).

Nolla, J. A. B. 1926. *J. Dep. Agric. P. Rico* **10**(2): 25 (on citrus, mango & avocado; **6**: 288).

Oste, C. A. et al. 1974. *Revta ind. agric. Tucumán* **51**(1): 37 (on avocado; **55**, 814).

Peregrine, W. T. H. 1970. *PANS* **16**: 331 (on *Bixa orellana*; **49**, 3422).

Raabe, R. D. et al. 1964. *Hawaii Fm Sci.* **13**: 1 (fungicides on papaw; **45**, 841).

Remiro, D. et al. 1975. *Summa Phytopathol.* **1**: 51 (on onion; **54**, 4285).

Riley, E. A. 1954 & 1955. *Trop. Agric. Trin.* **31**: 307; **32**: 150 (on tobacco; **34**: 188; **35**: 241).

Saraswathy, N. et al. 1975. *J. Pl. Crops* **3**: 65 (on areca palm; **55**, 3252).

—— ——. 1977. *Pl. Dis. Reptr* **61**: 172 (on areca palm; **56**, 4636).

Shear, C. L. et al. 1913. *Bull. U.S. Dep. Agric.* 252, 110 pp. (general).

Singh, R. D. et al. 1966. *Pl. Dis. Reptr* **50**: 385 (fungicide control on yam; **45**, 3020).

Sonoda, R. M. et al. 1974. *Trop. Agric. Trin.* **51**: 75 (on *Stylosanthes* spp.; **53**, 2245).
Spalding, D. H. et al. 1975. *Phytopathology* **65**: 458 (avocado, controlling atmosphere storage; **55**, 336).
Stanghellini, M. E. et al. 1966. *Ibid* **56**: 444 (host response to infection in papaw; **45**, 2563).
Stephan, B. R. 1967. *Zentbl. Bakt. ParasitKde* Abt. 2 **121**: 41, 58, 73 (morphological variability & heterokaryosis; **47**, 456).
Tandon, I. N. et al. 1969. *Indian Phytopathol.* **22**: 322 (on guava; **49**, 2940).
Tandon, R. N. et al. 1954. *Proc. Indian Acad. Sci.* Sect. B **40**: 102 (as above; **34**: 735).
Terashita, T. 1963. *Bull. Govt For. Exp. Stn Meguro* 155: 1 (on *A. dealbata*; **43**, 2075).
Thomas, K. M. et al. 1939. *Madras agric. J.* **27**: 348 (on black pepper; **19**: 494).
Trujillo, E. E. et al. 1969. *Pl. Dis. Reptr* **53**: 323 (on papaw; **48**, 2501).
Tsai, W. H. 1969. *Jnl Taiwan agric. Res.* **18**: 51 (on papaw; **50**, 3063).
Venkatakrishniah, N. S. 1952. *Proc. Indian Acad. Sci.* Sect. B **36**: 129 (on guava; **33**: 243).
Vimuktanandana, Y. Y. et al. 1940. *Philipp. Agric.* **29**: 124 (on black pepper; **19**: 728).
Wheeler, H. E. 1954. *Phytopathology* **44**: 342 (genetics & evolution of heterothallism; **34**: 165).
——. 1956. *Mycologia* **48**: 349 (sexual versus asexual reproduction; **36**: 45).

CITRUS

Two forms of *G. cingulata* occur on citrus: one which infects *Citrus* spp. and cvs generally (citrus anthracnose); and one which only infects the most commonly grown cv. of *C. aurantifolia* (the Mexican, West Indian or Keys lime) and the closely related cv. Dominica Thornless (withertip or lime anthracnose). The form on lime is frequently referred to as *Gloeosporium limetticola* Claus. Withertip has in the past caused heavy damage in the wetter areas of several islands in the West Indies (see Fawcett, *Citrus diseases and their control*) and was more recently investigated in Tanzania (Zanzibar; Wheeler). The most conspicuous symptom is the withering of young shoots which are killed. When leaf infection is less severe leaf spots and distortion of the lamina occur. Flower clusters and young fruit (perhaps not >*c.* 12 mm diam.) are infected and drop. Infected fruit which is older and escapes very severe infection develops scabby and cork-like areas, becomes split, cracked and misshapen; these symptoms do not occur on leaves or young stems. Young (but not old) leaves are directly penetrated, the appressoria being formed in 7–8 hours. Withertip is only likely to be serious in wet regions with seasons of intense rainfall. Fulton showed that other forms of citrus (including 6 other lime cvs) were not affected by the str. of the pathogen which caused withertip. Bordeaux or its copper equivalent are generally recommended for control and phenylmercuric fungicides were effective in Zanzibar. Timing of the spray applications is important, not only with respect to rainfall distribution but also to the flushing behaviour of the trees since the new foliage and blossoms are the susceptible tissues. It is probably advisable to grow Mexican lime only in relatively dry areas.

Any *Citrus* sp. (at any age) may be infected by the unspecialised, pathogenic str. of the fungus. *G. foliicola* Nishida (a form of *Glomerella cingulata*) was described from *C. reticulata* (see Takimota). Symptoms of citrus anthracnose occur as leaf spots, dieback of small branches, spots and tear-stain markings on fruit. Dieback occurs on mature or senescent twigs. If spread is slow the leaves on the affected stems turn yellow, wither and fall, but if it is fast the leaves may dry up before abscission. Gumming may occur along the invaded stem or at the sharp juncture between healthy and diseased tissues. The stem condition may lead to fruit drop; blossom drop can also occur. On the fruits the anthracnose spots may reach 1 cm diam.; they are at first in shades of brown before blackening, more or less circular, sunken and become dry and hard. The lesion may extend through the rind. Another type of fruit infection leads to a soft decay which spreads fairly extensively and causes drop. Various necrotic patterns can be caused on fruits due to rain wash of the conidia. Adam et al. investigated latent infection in orange (and see Tokunaga et al.). Citrus anthracnose is considered to be a secondary condition; it rarely develops on vigorous and healthy trees. Trees that are damaged or weakened, overcropped or under fertilised, or subject to drought and severe cold are liable to infection under high RH conditions. Fungicide treatments are sometimes recommended (e.g., Agarwala et al.; Tanaka et al.).

Fruit anthracnose of Robinson tangerine can be economically important (Barmore et al.; Brown; Brown et al.). Mature, green fruit is very susceptible when degreened with ethylene. This treatment evidently stimulates appressoria to form infection hyphae, thus increasing the incidence of anthracnose. But fruits with a good orange break (25% external orange colour) resist infection and do not develop the disease. Washing or drenching with benomyl before degreening reduces anthracnose; as did applications of ethephon 5–7 days before harvest for 3 seasons.

Adam, D. B. et al. 1949. *Aust. J. scient. Res.* Ser. B **2**: 1 (latent infection in orange; **29**: 22).

Agarwala, R. K. et al. 1957. *Indian J. agric. Sci.* **27**: 205 (on lime; **37**: 408).

Alippi, H. E. 1971. *Revta Fac. Agron. Univ. nac. La Plata* **47**: 19 (atypical anthracnose; **51**, 2486).

Barmore, C. R. et al. 1978. *Pl. Dis. Reptr* **62**: 541 (preharvest ethephon application).

Brown, G. E. 1975. *Phytopathology* **65**: 404 (factors affecting postharvest development of pathogen; **54**, 49).

—— et al. 1976. *Proc. Fla St. hort. Soc.* **89**: 198 (effects of ethylene, fruit colour & fungicides on susceptibility; **57**, 2941).

—— ——. 1977. *Phytopathology* **67**: 120 (effect of ethylene on susceptibility; **56**, 4055).

——. 1977. *Ibid* **67**: 315 (fine structure of penetration; **56**, 5059).

——. 1978. *Ibid* **68**: 700 (hypersensitive host response after ethylene treatment).

Dey, P. K. 1933. *Ann. Bot.* **47**: 305 (cuticular penetration; **12**: 566).

Fulton, H. R. 1925. *J. agric. Res.* **30**: 629 (relative susceptibility of citrus to withertip form; **4**: 666).

Martin, J. T. et al. 1966. *Ann. appl. Biol.* **57**: 491, 501 (withertip form, fungitoxicites of cuticular & cellular components of leaves & of plant furocoumarins; **45**, 3049).

Roberts, M. F. et al. 1963. *Ibid* **51**: 411 (withertip form, leaf cuticle in relation to infection; **42**: 761).

Takimoto, S. et al. 1936. *Studia citrol.* **7**: 176 (*Gloeosporium folliicola*; **17**: 26).

Tanaka, H. et al. 1968. *Bull. hort. Res. Stn Japan* Ser B **8**: 99, 111 (anthracnose tear stain of fruit, factors affecting & fungicides; **50**, 2957).

Tokunaga, Y. et al. 1973. *Rep. Tottori mycol. Inst.* **10**: 693 (latent infection; **53**, 3024).

Wheeler, B. E. J. 1963. *Trans. Br. mycol. Soc.* **46**: 193 (withertip form, general; **42**: 761).

——. 1963. *Ann. appl. Biol.* **51**: 237 (withertip form, fungicides; **42**: 680).

—— et al. 1963. *Ibid* **51**: 403 (withertip form, Hg residues in host; **42**: 760).

Yamada, S. et al. 1965. *Bull. hort. Res. Stn Okitsu* Ser. B **4**: 107, 119 (anthracnose tear stain of fruit, the pathogen, host range & morphology, **45**, 1379).

COFFEA

The specific, virulent str. of *G. cingulata* which causes the important coffee berry disease (CBD) occurs only in parts of Africa. Muthappa and Feitosa et al. considered that it did not occur in S. India and Brazil, respectively. The disease has been very intensively studied in Kenya, see the monograph by Firman et al. Boisson and Rayner reviewed the early work; and for general accounts see also Griffiths (1969, Kenya), Hindorf (1975), Muller (Cameroon), Nutman (1970) and Saccas (Central African Republic). Rossetti et al. provided a bibliography. Infection occurs on the green berries. It causes a dark brown rot which destroys the beans and the berry dries out. First lesions may occur on the lateral surface of the green berry; these are rapidly enlarging, roughly circular, slightly sunken, dark brown spots. There may be several, coalescing lesions which form irregular necrotic areas; and the base of the berry may be attacked initially. The characteristic *Colletotrichum* acervuli (pinkish conidial masses) develop on the lesions. Abscission of the berries may occur at the top or bottom of the fruit stalk. Lesions that become ash grey (except for a dark brown periphery) are associated with the onset of dry conditions (scab symptoms). Scab lesions are relatively inactive and sparsely sporulating.

The forms of *Colletotrichum* on coffee have often been referred to as *C. coffeanum* Noack. Hocking (1971) refers the fungus to this name *sensu* Small. The main cultural characteristics of the different strs or spp. of the genus from coffee were tabulated by Firman et al. Hocking (1966) mentioned the earlier work on these forms and defined the anthracnoses of coffee, known in E. Africa as brown blight, in contrast to berry disease. Several forms that occur on branches (causing dieback), leaves, blossoms and berries were largely saprophytic with no consistent differences in any slight pathogenicity shown. The term brown blight was used, in reference to berries, only for infections of ripe or ripening berries caused by any str. which does not seriously damage green berries like the virulent one (CBD str.). Latent infections on green berries can become active on fruit ripening. Hocking (1971) inoculated fruits of 19 plants with the CBD str. and 1 form avirulent to green berries. The former infected fruits of 3 plant spp. and the latter infected those of 14 spp. Gibbs' (1969) *Colletotrichum* isolates from berries fell into 4 groups or strs distinguishable on cultural grounds; one of these was the CBD str. and the others were not pathogenic. The CBD str. made up only a small proportion of the *Colletotrichum* population complex on bark but its inoculum greatly increased when green berries became infected and formed sporulating lesions. Hindorf distinguished (on morphological grounds in vitro) isolates as *C. coffeanum* (CBD str.), *C. gloeosporioides* (*G. cingulata*) and *C. acutatum* (*Colletotrichum* q.v.). Hocking et al. infected unwounded green berries with ascospores from *G. cingulata*. Fertile perithecia were more frequent on branch prunings and fallen leaves than on green berries; but see Vermeulen who had no evidence for ascospore infection of such berries. The CBD str. did not form the perfect state in vitro. The subsequent description refers to this str. only except where stated.

Germination of, and infection by, the conidia can take place at 15–28°, free water is required and there is direct penetration of the green berry cuticle. Most infection occurs 6–10 weeks after flowering; at 4 and 14 weeks there is less infection. The opt. temp. for lesion formation is *c*. 22°. Conidia which have been dried for 24 hours can cause some infection. The interplay of climate, weather and the ecological conditions in the coffee tree have been studied in detail in Kenya (e.g., Bock, Gibbs, Griffiths, Hocking, Mulinge, Nutman and Waller). The CBD str. and the largely saprophytic forms of *Colletotrichum* occur as an interacting complex with other micro-organisms on the maturing bark of the tree. The CBD str. is much less abundant on the bark than the avirulent strs. At the beginning of the rains when a diseased crop is absent initial infection arises from the fungus in the bark. As diseased berries build up foci of increasing abundant sporulation the severe secondary spread of the disease follows and epidemics spread through water and splashborne inoculum. This inoculum becomes abundant in the upper canopy and is washed down through it and to neighbouring trees (Waller 1971, 1972). The disease is more severe at higher altitudes. In the Cameroons Muller (1970) described a higher incidence at 1700 m compared with 1200 m. In Kenya coffee is grown at altitudes of 1300–2000 m. The higher disease incidence at upper altitudes is apparently due to an indirect effect of temp. and not to rainfall. The temp. acts on the *Colletotrichum* population of the tree. The sporulation of one of the avirulent strs increases with temp. from 20° to 30° whilst 2 others and the CBD str. show no such increase. Thus temp. conditions at the lower altitudes favour the higher temp. str. Since, presumably, the 4 strs are competing for the same ecological niche in the tree this str. has a competitive advantage at lower altitudes. Therefore it is supposed that at these altitudes the CBD str. is under a more severe competitive disadvantage than it would be at the lower temps of the higher altitudes. It was found in Kenya that the population of the CBD str. in the trees did increase with altitude (and see Hindorf 1973c). Infection is positively correlated with the number of rain days, particularly with those having at least 1 mm of rain and with periods having at least 5 hours wetness (Cook 1975).

Control is with fungicides and some work on host resistance has been described. Much of the early work on spraying for control in Kenya was quoted by Griffiths et al. (1971a). The use of fungicides against berry disease has led to unforseen, undesirable side effects. These and others (the negative effects) were described by Griffiths (1972). Early on it had been found in Kenya that applying fungicides to coffee, in the absence of any apparently significant disease, could increase yields (the tonic effect). Control procedure was initially based on a short programme of spraying at a specific period early in the season. But some years later it was found that this was giving unsatisfactory control (see Gibbs 1971; Griffiths et al. 1971). Spraying was then extended to operate through the long rains season and this was more effective. Cuprous oxide, captafol, benomyl and chlorothalonil are amongst the recommended fungicides (Griffiths et al. 1971a; Pereira; Pereira et al.; Vine et al.).

Early season fungicide sprays not only gave inadequate control but resulted in loss of crop. Sprayed trees had large numbers of sporulating berries (active lesions) compared with unsprayed ones. This gave high inoculum levels when the crop was ripening. Where fungicide was inadequate it could eventually lead to an increase in the sporulation of the CBD str. on the bark. Such sporulation is the source of the important primary inoculum at the beginning of the rainy season (Furtado; Gibbs 1972; Griffiths 1972; Hindorf 1973c; Nutman et al. 1969b). Mulinge et al. (1974a) referred to the possible mechanisms, still not clear, whereby fungicide application increases the CBD str. population. A further complicating factor in Kenya is the need to control *Hemileia vastatrix* (q.v.), for both rust and berry disease may occur together at epidemic levels.

The former is generally severe at lower altitudes (*c.* 1700 m and below); it is at such altitudes that this disease situation becomes very complex with the variable effects on yield that rust has. High altitude fungicide assessments are also needed (Vine et al. 1973b). Cook et al. and Okioga described strs of the pathogen resistant to systemic fungicides.

Bock, K. R. 1956. *E. Afr. agric. J.* **22**: 97 (symptoms, infection & culture inter alia; **36**: 402).
Boisson, C. 1960. *Rev. Mycol.* **25**: 263 (review, 127 ref.; **40**: 681).
Cook, R. T. A. 1975. *Kenya Coff.* **40**: 190 (effect of weather on infection; **55**, 1804).
—— et al. 1976. *Ann. appl. Biol.* **83**: 365 (strs resistant to benzimidazoles; see **56**, 5647).
Feitosa, M. E. et al. 1977. *Arq. Inst. biol. S. Paulo* **44**: 33 (*Colletotrichum* on coffee in Brazil, São Paulo; **57**, 2947).
Firman, I. D. et al. 1977. *Phytopathol. Pap.* 20, 53 pp. (monograph, 181 ref.; **56**, 4538).
Furtado, I. 1969. *Trans. Br. mycol. Soc.* **53**: 325 (effects of Cu fungicides; **49**, 484).
Gassert, W. L. 1978. *Z. PflKrankh. PflSchutz* **85**: 30, 84, 98 (epidemiology, berry susceptibility & inoculum sources in Ethiopia; **57**, 3985, 4974, 4975).
Gibbs, J. N. 1969. *Ann. appl. Biol.* **64**: 515 (forms of *Colletotrichum* from coffee inoculum sources; **49**, 1378a).
——. 1971. *Ibid* **67**: 343 (factors affecting field fungicide programmes; **50**, 3773).
——. 1972. *Ibid* **70**: 35 (effects of fungicides on fungal populations in coffee bark; **51**, 3372).
Griffiths, E. 1969. *Span* **12**: 92 (Kenya, review; **48**, 2983).
—— et al. 1971a. *Ann. appl. Biol.* **67**: 45 (fungicide control; **50**, 2283).
—— ——. 1971b. *Ibid* **67**: 75 (disease rainfall & cropping patterns; **50**, **2284**).
——. **1972**. *Trop. Sci.* **14**: 79 (negative effects of fungicides; **52**, 126).
—— ——. 1972. *Trans. Br. mycol. Soc.* **58**: 313 (assessment of CBD str.; **51**, 4031).
Hindorf, H. 1970. *Z. PflKrankh. PflSchutz* **77**: 328 (differentiation of *Colletotrichum* from coffee; **50**, 1241).
——. 1973a. *Phytopathol. Z.* **77**: 97 (as above; **53**, 2197).
——. 1973b. *Ibid* **77**: 216 (as above, qualitative differences in population; **53**, 2198).
——. 1973c. *Ibid* **77**: 324 (distribution of *Colletotrichum* from different parts of coffee plant; **53**, 2199).
——. 1975. *J. Coff. Res.* **5**: 43 (review, 36 ref.; **56**, 277).
Hocking, D. 1966. *Ann. appl. Biol.* **58**: 409 (weakly pathogenic strs and/or saprophytes; **46**, 1265).
—— et al. 1967. *Nature Lond.* **214**: 1144 (ascospore formation, discharge & infection; **46**, 2727).
——. 1967. *E. Afr. agric. For. J.* **32**: 365, 367, 371; resistance; **46**, 3105).
——. 1971. *Turrialba* **21**: 234 (inoculation of plants with 2 strs from coffee; **51**, 2495).
Mulinge, S. K. 1970. *E. Afric. agric. For. J.* **36**: 227 (levels of strs in bark of coffee cvs; **50**, 3771).
——. 1970. *Ann. appl. Biol.* **65**: 269 (disease in relation to berry growth & altitude; **49**, 2864).
——. 1971. *Ibid.* **67**: 93 (effect of altitude on distribution of strs; **50**, 2285).
——. 1971. *Trans. Br. mycol. Soc.* **56**: 478 (distribution of strs; in coffee trees; **50**, 3770).
—— et al. 1974a. *Ibid* **62**: 495 (effects of fungicides on CBD str., *Hemileia vastatrix*, foliation & yield; **54**, 460).
—— ——. 1974b. *Ibid* **62**: 610 (effects of temp. on population of strs on coffee bark).
Muller, R. A. 1970. *Café–Cacao–Thé* **14**: 114 (disease development in Cameroon; **50**, 685).
——. 1973. *Ibid* **17**: 281 (susceptibility of berries during development & irrigation as a control method; **53**, 3028).
Muthappa, B. N. 1972. *J. Coff. Res.* **2**: 16 (*Colletotrichum* on coffee in India; **52**, 1543).
Nutman, F. J. et al. 1960. *Trans. Br. mycol. Soc.* **43**: 489, 643 (factors affecting conidial germination, infection & relation to disease distribution; **40**: 170, 361).
—— ——. 1961. *Ibid* **44**: 511 (infection of bearing wood & disease incidence; **41**: 388).
—— ——. 1969a. *Ann. appl. Biol.* **64**: 85, 101 (seasonal variations in, & effect of fungicides on, sporulation; **49**, 152a, c).
—— ——. 1969b. *Ibid* **64**: 335 (stimulating effect of fungicides on *G. cingulata*; **49**, 1027).
—— ——. 1969c. *E. Afr. agric. For. J.* **35**: 118 (climate & disease; **49**, 1653).
——. 1970. *PANS* **16**: 277 (review, 29 ref.; **49**, 3287).
Okioga, D. M. 1975. *Kenya Coff.* **40**: 170 (strs resistant to systemic fungicides; **55**, 1260).
——. 1976. *Ann. appl. Biol.* **84**: 21 (strs resistant to carbendazim & chemically similar compounds; **56**, 720).
Pereira, J. L. 1972. *Int. Pest Control* **14**: 6 (control with *H. vastatrix* by multi-row application; **52**, 3699).
—— et al. 1973. *Expl Agric.* **9**: 209 (redistribution of fungicides in coffee trees; **52**, 3326).
Rayner, R. W. 1952. *E. Afr. agric. J.* **17**: 130 (survey of work to 1950; **32**: 77).
Rossetti, V. et al. 1975. *Arq. Inst. Biol. S. Paulo* **42**: 265 (bibliography, 192 ref.; **57**, 174).
Saccas, A. M. et al. 1969. *Café–Cacao–Thé* **13**: 131 (general account for Central African Republic; **48**, 2982).

Vermeulen, H. 1970. *Neth. J. Pl. Pathol.* **76**: 277, 285 (bark colonisation & role of perfect state; **50**, 684a & b).
Vine, B. H. et al. 1973a. *Ann. appl. Biol.* **75**: 359 (fungicides; **53**, 3027).
———. 1973b. *Ibid* **75**: 377 (problems in fungicide evaluation; **53**, 2195).
Waller, J. M. 1971. *Expl Agric.* **7**: 303 (weather, climate & disease; **51**, 1509).
——. 1971. *Kenya Coff.* **36**: 119 (as above; **51**, 371).
——. 1972. *Ann. appl. Biol.* **71**: 1 (waterborne dispersal & fungicide control; **52**, 129).

CUCURBITACEAE

The fungus causing cucurbit anthracnose is usually called *Colletotrichum lagenarium* (Pass.) Ell. & Halst. or *C. orbiculare* (Berk. & Mont.) Arx. Arx (1975a, *Colletotrichum* q.v.) considered the pathogen to be another specialised form of *Glomerella cingulata* and gave a synonymy. Perfect states in *Glomerella* were described by Stevens and Watanabe et al. (no Latin diagnosis). *G. magna* was described as causing an anthracnose of cucurbits (*Glomerella* q.v.). The disease is most severe on *Citrullus lanatus* (watermelon), *Cucumis melo* (melon) and *C. sativus* (cucumber). *Cucurbita* spp. are not seriously affected but Rodigin reported infection of *C. pepo* (marrow) by a str. that differed in pathogenicity from the one causing diseases in cucumber, melon and watermelon (and see Layton).

Symptoms are seen on leaves, stems and fruits, and damping off in cucumber occurs. Yellowish, watersoaked lesions form on the leaves; they are angular or roughly circular, 1 cm or more in diam., becoming brown (or black in watermelon); the spots coalesce and the whole leaf dries up. Linear or oval, slightly sunken lesions which dry out occur on the stems and petioles. Fruit pedicels may be infected in which case the young fruits become shrivelled, malformed and do not develop. Maturing fruits show conspicuous, circular spots, sometimes 5–6 cm diam; these are at first raised but become sunken, often quite markedly so; the rot is usually confined to the rind and develops the characteristic sporulating acervuli. The fruit spots can be very numerous. The characteristic direct penetration from appressoria occurs during a single leaf wetness period. In inoculations Leben et al. did not find that several subsequent moist periods increased the number of lesions in cucumber. Ishida et al. (1969) found a high percentage of germination of conidia at 20–30° but the opt. temp. range for appressorial formation was less (20–26°). At 24° 80–90% of the germination was followed by the formation of appressoria, the remaining conidia germinated by a germ tube only and did not form appressoria; with a rise in temp. (28–32°) an increasing percentage germinated without forming appressoria. If conidia were pre-treated at 32° for 2–12 hours they only formed germ tubes. Therefore high temps suppress appressorial development. In inoculations Layton obtained severe infection at 20–30°. The temp. range over which cucurbit anthracnose occurs is probably very wide. Rankin was not able to isolate the fungus from watermelon seed but seed inoculation resulted in infected seedlings. Horn et al. detected a low percentage of seed infection in cucumber. Sowing seed immediately after soil infestation resulted in infected seedlings but not when sown 10 months after infestation.

In field trials (Caruso et al. 1977b) cucumber plants were partially protected against challenge inoculation with *C. lagenarium* by prior inoculation with the same fungus. Protected plants had fewer and smaller lesions (and see Caruso et al. 1977a; Kuć et al.). Localised infection with tobacco necrosis virus also protected against *C. lagenarium* (Jenns et al.). Work on other aspects (e.g., cell degrading enzymes) will be found in *Rev. appl. Mycol.* **43**, 2167; **46**, 3344; **47**, 2946; *Rev. Pl. Pathol.* **49**, 361; **52**, 3113; **53**, 3237, 3806, 4907; **54**, 1494; **55**, 2016, 5998; **56**, 2735, 5458; **57**, 414, 3692.

Seven races were differentiated on 12 cvs by Jenkins et al. who described 3 additional races (5, 6 and 7) to the previously reported 4 (Dutta et al.; Goode 1958; Winstead et al. 1959). Race 5 differed from the previously described races in being weakly virulent on cucumber and highly virulent on watermelon. Race 6 was very virulent on watermelon; and race 7 weakly virulent to the cucumber Pixie, this distinguished it from race 3. Resistance to some races is controlled by single dominant genes and more complex resistance in cucumber occurs (Busch et al.). Resistant watermelon cvs have been developed in USA (e.g., Henderson et al.); one (Smokylee) was recently described by Crall and others are given by Chupp and Sherf in *Vegetable diseases and their control.* Akai et al. described resistance in cucumber and *Cucurbita*. Seed treatment does not appear to be very important; thiram and mercurials have been used. On melons in the field dichlone, maneb, and zineb were effective in Australia, France and

Turkey. Dithicarbamates have been generally used in USA. Cucumbers may be attacked in glasshouses; and in routine sterilisation and disinfestation measures it should be borne in mind that the fungus can survive saprophytically; for example, decaying woodwork should be treated. Plants in glasshouses may have to be sprayed and sanitation measures taken.

Akai, S. et al. 1957. *Ann. phytopathol. Soc. Japan* **22**: 113 (resistance; **37**: 752).
—— ——. 1958. *Pl. Dis. Reptr* **42**: 1074 (as above; **38**: 115).
—— ——. 1959. *Forschn Geb. PflKrankh. Kyoto* **6**: 97 (as above; **39**: 207).
Anderson, J. L. et al. 1962. *Phytopathology* **52**: 650 (histology; **42**: 171).
Arx, J. A. von et al. 1961. *Phytopathol. Z.* **41**: 228 (strs on cucumber & other plants; **41**: 197).
Busch, L. V. et al. 1958. *Phytopathology* **48**: 302 (inheritance of resistance & spread in cucumber; **37**: 751).
Caruso, F. L. et al. 1977a & b. *Ibid* **67**: 1285, 1295 (protection by prior inoculation; **57**, 4192, 4193).
Crall, J. M. 1971. *Circ. Fla Univ. agric. Exp. Stn* S211, 10 pp. (Smokylee, a resistant watermelon cv.; **51**, 983).
Dutta, S. K. et al. 1960. *Bot. Gaz.* **121**: 163, 166 (races 1–4, pathogenicity of biochemical mutants; **40**: 139).
Goode, M. J. 1958. *Phytopathology* **48**: 79 (races 1–3; **37**: 624).
——. 1967. *Ibid* **57**: 1028 (lesion development in resistant & susceptible cucumber; **47**, 718).
Henderson, W. R. et al. 1977. *J. Am. Soc. hort. Sci.* **102**: 693 (resistance in watermelon; **57**, 3168).
Henry, C. E. et al. 1967. *Can. J. Microbiol.* **13**: 618 (virulent & avirulent mutants from race 3; **46**, 2576).
Horn, N. L. et al. 1957. *Pl. Dis. Reptr* **41**: 69 (seed & insect transmission; **36**: 677).
Ishida, N. et al. 1968. *Forschn Geb. PflKrankh. Kyoto* **7**: 73 (lipid degradation, conidial germination & temp.; **49**, 1930).
—— ——. 1968. *Mycopathol. Mycol. appl.* **35**: 68 (fine structure in formation of appressoria; **47**, 3657).
—— ——. 1969. *Mycologia* **61**: 382 (temp., conidial germination & formation of appressoria; **49**, 34).
Jenkins, S. F. et al. 1964. *Pl. Dis. Reptr* **48**: 619 (races 1–7; **44**, 310).
Jenns, A. E. et al. 1977. *Physiol. Pl. Pathol.* **11**: 207 (protection by tobacco necrosis virus; **57**, 1477).
Kuć, J. et al. 1975. *Ibid* **7**: 195 (protection by prior inoculation; **55**, 2954).
—— ——. 1977. *Phytopathology* **67**: 533 (as above; **57**, 336).
Layton, D. V. 1937. *Res. Bull. Iowa agric. Exp. Stn* B223: 37 (general on cucurbits; **17**: 429).
Leben, C. et al. 1968. *Phytopathology* **58**: 264 (effect of leaf wetness on lesion number; **47**, 2024).
McLean, D. M. 1967. *Pl. Dis. Reptr* **51**: 885 (races 1 & 2, interaction; **47**, 717).
Rankin, H. W. 1954. *Phytopathology* **44**: 675 (in seed; **34**: 511).
Rodigin, M. N. 1935. *Trans. Bykovskaya regional exp. Stn Cult. Cucurbits* 3: 59 (on marrow; **15**: 698).
Senyürek, M. et al. 1977. *Bitki Koruma Bült.* **17**: 150 (in Turkey).
Stevens, F. L. 1931. *Mycologia* **23**: 134 (perfect state; **10**: 771).
Watanabe, T. et al. 1952. *Ann. phytopathol. Soc. Japan* **16**: 137 (perfect state; **32**: 465).
Winstead, N. N. et al. 1959. *Pl. Dis. Reptr* **43**: 570 (resistance in watermelon to races 1–3; **38**: 647).
—— ——. 1963. *Phytopathology* **53**: 961 (formation of cell degrading enzymes with *Glomerella magna*; **43**, 643).
Yasumori, H. 1957. *Ann. phytopathol. Soc. Japan* **22**: 119 (host penetration; **37**: 752).

GOSSYPIUM

Glomerella gossypii (Southw.) Edg. (stat. conid. *Colletotrichum gossypii* Southw.) is the name usually given in the literature for the pathogen causing cotton anthracnose; but whether the fungus is morphologically distinct from *G. cingulata* is doubtful. There is doubt, too, as to whether *G. gossypii* is the same fungus as *C. gossypii* (J. E. M. Mordue, personal communication, March 1978). Arx et al. (*G. cingulata* q.v.) reduced *G. gossypii* to synonymy. Leakey et al., who briefly discussed the taxonomy, considered that there was nothing to distinguish these *Glomerella* spp. *C. indicum* (*Colletotrichum* q.v.) also causes an anthracnose of cotton and was reduced to synonymy with *C. dematium*. *C. gossypii* var. *cephalosporioides* was described (without a Latin diagnosis) by Costa et al. from Brazil. This var. differed in its greater virulence to certain cvs of *Gossypium hirsutum*. The disease in Brazil was called ramulosis or excess budding; Malaguti preferred the name little broom for the condition which is characterised by excessive sprouting of axillary and terminal buds.

Cotton anthracnose is very widespread but its degree of severity depends very much on local conditions; the weather is probably the most important. Smith stated that in USA it is widespread from Virginia to Texas and Oklahoma but is delimited by the 102 cm isohyet which runs N. and S. through the eastern part of the last 2 states. W. of it low rainfall and RH greatly limit disease incidence which is max. along the Atlantic and Gulf of Mexico coasts (and see Miller et al.; Simpson et al.). The main damage is done to seedlings and bolls (boll rot, frequently in association with other organisms); symptoms are found on cotyledons, young stems, leaves and bolls. Reddish or light-coloured spots form on the cotyledons with similar lesions on the stems, above or below soil level; seedling damping off is frequent. Necrotic lesions may be caused on leaves often in association with damage or infection by *Xanthomonas malvacearum*. As the crop develops no further symptoms may be seen until boll formation. Infection of the flower or young boll may result in an internal boll rot. On the boll small, round, water-soaked spots become red brown, sunken, and often coalesce to form large sporulating lesions over much of the boll. In severe infections penetration of the capsule wall takes place with infection of seed and lint. The latter becomes discoloured (pinkish brown, grey or grey black) and compacted (the tight lock), instead of opening into the healthy fluffy mass. There is both internal and external infection of the seed; see Halfon-Meiri et al. for detection in seed.

The boll phase of cotton anthracnose is closely connected with the presence and behaviour of other fungi, insects and bacteria. The result is a boll rot disease syndrome which is complex and has been extensively studied, especially in USA (Bagga; Bagga et al.; Cauquil et al. 1969; Crawford; Edgerton 1912; Kuo et al.; Leakey et al.; McCarter et al.; Marsh et al.; Ray; Roncadori; Weindling et al.; Wiles). Several interacting factors affect the amount of boll rot that may occur: amount of wet weather, inoculum level, infection by other fungi and by bacteria, insect attack and disease escape by the host or its resistance. *Glomerella cingulata* is both a primary and secondary invader, and there is evidence for latent infection of cotton bolls. Direct penetration of the boll occurs through the intact wall provided there is a long, continuous period of high RH. Penetration is probably more frequent through natural openings, apex, peduncle and along the edges of the carpels. Under short periods of high RH boll infection only follows through a cracked suture or some damage to the boll, brought about through insect puncture or infection by fungi and bacteria; many spp. of fungi have been recorded as occurring in both the internal and external boll tissue, whether or not a rot occurs.

Inoculum builds up on host debris in the field, it occurs on debris brought to the processing plants and infests the seeds on which it is fairly persistent (Arndt). Under lower inoculum levels attack by *Meloidogyne incognita* may increase the severity of the seedling disease phase of cotton anthracnose (Cauquil et al. 1970). Some variation in pathogenicity between isolates has been described. Bollenbacher et al. (1967), working with a moderately and a highly pathogenic str., found that seedlings of *Gossypium hirsutum* (24 cvs) were extremely susceptible to the highly pathogenic str. Eastern USA cvs of *G. hirsutum* were also mostly susceptible to the moderately pathogenic str. but nearly all Acala cvs were partly or very resistant to this str. *G. arboreum* cv. Nanking was very resistant to both isolates.

The seedling blight can be largely controlled by fungicidal treatment of the seed. Where boll rot is likely to be severe other control measures are: rotation, ploughing, defoliation (to allow rapid drying of the bolls), using cvs with smaller leaves and an open type of growth, limiting vegetative growth by reducing nitrogen fertiliser, controlling boll damaging insects and using cvs that are somewhat resistant.

Arndt, C. H. 1944. *Phytopathology* **34**: 861 (temp. & seedling infection; **24**: 100).

——. 1946. *Ibid* **36**: 24 (seed storage & fungus survival; **25**: 261).

——. 1953. *Ibid* **43**: 220 (as above; **33**: 81).

Atkinson, G. F. 1890. *J. Mycol.* **6**: 173 (general).

Bagga, H. S. et al. 1969. *Phytopathology* **59**: 255 (boll rot organisms & cvs; **48**: 2412).

——. 1970. *Ibid* **60**: 158 (pathogenicity of boll rot organisms; **49**, 2077).

Bollenbacher, K. et al. 1967. *Pl. Dis. Reptr* **51**: 632 (susceptibility of cvs to seedling anthracnose; **47**: 215).

—— ——. 1971. *Ibid* **55**: 879 (susceptibility of 8 *Gossypium* spp. & cvs to seedling anthracnose; **51**: 1537).

—— ——. 1971. *Phytopathology* **61**: 1394 (effects of N compounds on seedling resistance; **51**, 2534).

Cauquil, J. et al. 1969. *Coton Fibr. trop.* **24**: 193 (boll rot & genetical selection; **48**, 2995).

—— ——. 1970. *Phytopathology* **60**: 448 (root knot nematode & fungi infecting seedlings; **49**, 2874).

Costa, A. S. et al. 1939. *Revta Agric. S. Paulo* **2**: 151 (var. *cephalosporioides*; **18**: 798).

Crawford, R. F. 1923. *Phytopathology* **13**: 501 (isolation from within seed; **3**: 271).

Edgerton, C. W. 1909. *Mycologia* **1**: 115 (*Glomerella gossypii* sp. nov.).

——. 1912. *Phytopathology* **2**: 23 (flower infection).

——. 1912. *Bull. La agric. Exp. Stn* 137: 20 (rots of boll).

Follin, J. C. 1969. *Coton Fibr. trop.* **24**: 337, 345 (variation in pathogen; **44**, 768a & b).
——. 1970. *Ibid* **25**: 387 (races; **50**, 1253).
Halfon-Meiri, A. et al. 1977. *Seed Sci. Technol.* **5**: 129 (detection in seed with *Xanthomonas malvacearum*; **56**, 3046).
Kuo, C. C. et al. 1963. *Acta Phytophylac. sin.* **2**: 409 (factors affecting boll rot; **43**, 1651).
Leakey, C. L. A. et al. 1966. *Ann. appl. Biol.* **57**: 337 (insect damage & boll rot in Uganda; **45**, 2504).
Ludwig, C. A. 1925. *Bull. S. Carol. agric. Exp. Stn* 222, 52 pp. (on seed; **5**: 737).
Luke, W. J. et al. 1970. *Cotton Grow. Rev.* **47**: 20 (role of bract in boll rot; **49, 1663**).
McCarter, S. M. et al. 1970. *Pl. Dis. Reptr* **54**: 586 (micro-organisms associated with boll rot; **50**, 102).
Malaguti, G. 1955. *Agron. trop.* **5**: 73 (differing symptom expression; **36**: 405).
Marsh, P. B. et al. 1965. *Pl. Dis. Reptr* **49**: 138 (infection of bolls; **44**: 1881).
Miller, P. R. et al. 1943. *Ibid* Suppl. 141: 53 (surveys, seedling blight & boll rots; **22**: 479).
Ray, W. W. et al. 1942. *Phytopathology* **32**: 233 (pathogenicity of fungi to cotton seedlings; **21**: 331).
——. 1946. *Bull. Okla agric. Exp. Stn* B 300, 26 pp. (boll rot survey; **26**: 152).
Roncadori, R. W. 1969. *Phytopathology* **59**: 1356 (fungal infection of developing bolls; **49**: 1030).
Simpson, M. E. et al. 1973. *Pl. Dis. Reptr* **57**: 828 (climate & boll rot distribution; **53**, 2207).
Smith, A. L. 1953. *Plant diseases U.S. Dep. Agric. Yearbk*: 303.
Southworth, E. A. 1890. *J. Mycol.* **6**: 100 (*Colletotrichum gossypii* sp. nov.).
Ullstrup, A. J. 1938. *Phytopathology* **28**: 787 (variability in vitro & in vivo; **18**: 175).
Weindling, R. et al. 1941. *Ibid* **31**: 158 (fungi associated with boll & seedling diseases; **20**: 201).
Wiles, A. B. 1963. *Ibid* **53**: 984 (seedborne fungi & boll rots; **43**, 468).

HEVEA BRASILIENSIS

Glomerella cingulata on rubber can attack any green parts of the tree but is primarily a leaf infecting fungus. It has recently been intensively studied (particularly by Wastie) in Malaysia (W.) where, as one of the causes of the diseased condition called secondary leaf fall (SLF), *G. cingulata* is economically important. Several other *Colletotrichum* spp. (or forms of other spp.) which have been described from this crop appear to be of no practical importance (*C. dematium* q.v.; John; Wastie 1968). The other agents that cause SLF, alone or combined, are: *Oidium heveae* (q.v.), the yellow tea mite (*Polyphagotarsonemus latus*) and a thrips (*Scirtothrips dorsalis*). In Malaysia the commonest cause of SLF is *O. heveae* with which the mite may be associated; *P. latus* may also cause defoliation on its own. *G. cingulata* is mainly important in the later stages of refoliation. *S. dorsalis* is the least important of the 4 organisms and, when present, is almost invariably associated with 1 or more of the other 3. This disease complex arises since rubber is deciduous. In Malaysia the tree characteristically sheds its leaves during the drier weather (wintering) in the first 3 months of the year; wintering can be less regular in other climatic regions. It affects trees mostly more than *c.* 5 years old. Refoliation begins almost immediately afterwards and then the young leaves are susceptible to attack by these organisms. A severe attack produces a pale green carpet of fallen leaflets which are seen on approaching the trees. The weather at the time of refoliation is an important factor determining the intensity of attack. In S. and Central America SLF can be caused by *Microcyclus ulei* (q.v.) which is still confined to that region. Severe leaf infection at refoliation can lead to the repeated loss of leaves and branch dieback. Fallen leaflets are shrivelled and have necrotic tips; in wet weather the conidia of *G. cingulata* form all over the leaf surface. When older leaves are attacked they are less likely to fall; numerous small spots with a narrow brown margin and a chlorotic halo form, they become characteristically raised. The margin of the leaflet becomes shrivelled and distorted, particularly at the tip. Infection of young buddings can lead to dieback and death of the scion (Anon.).

Opt. temp. for growth and sporulation is 26–32°, whereas conidial germination can exceed 90% at 21.5–30.5°. Germination at RH 99% is appreciably less (*c.* 50%) than at RH 100%, and is negligible below RH 97%. Germination decreases by up to 30% after 3 hours storage at RH 80% and is reduced by sunlight. The lower leaf surface is more susceptible than the upper. Lesion development after 72 hours is quickest at 21° and slower at 26.5°. Leaflets 15 days old are normally resistant. There is a well defined increase in resistance between the eleventh and thirteenth day of leaf development; this period coincides with the formation of the cuticle. Isolates differ in their degree of pathogenicity. The conidia have a daily max. conc. at 2300 hours with a fall in numbers during daytime and rain. The study of SLF epidemiology in Malaysia (W.) over many years (see Wastie 1972b & c) has led to the following general conclusions. The March rainfall and the number of rain days appear to be the most important factors but there are also smaller effects of mean RH, temp. and hours of sunshine. In the clone PB86 there was moderate to severe defoliation when the RH was 97–100% for 13.5 hours/day. *G. cingulata* is favoured by higher total amount and frequency of rainfall whereas *O. heveae* is more severe in drier weather. The geographic distribution of the 2 pathogens is correlated with the rainfall pattern. Clonal susceptibility, wintering characteristics and stage of refoliation are also important in determining the nature and severity of an outbreak of SLF.

Studies in Malaysia and Sri Lanka on control have demonstrated the various factors and approaches that are involved. Where the disease is largely caused by *O. heveae* the long used sulphur dust applications are effective (several treatments throughout the refoliation period). There is no standard fungicide treatment for *G. cingulata* but captafol and chlorothalonil have shown promise. Development of an oilborne prophylactic spray which is effective against mature leaf fall caused by *Phytophthora* spp. is a possibility. It is important to determine in which climatic regions *G. cingulata* leaf fall is likely to be severe. Another factor is clonal susceptibility. Wastie (1973) assessed the susceptibility of nearly 200 clones. He found a satisfactory agreement between the assessment of *G. cingulata* leaf disease in the nursery and field susceptibility; although the latter is modified for some clones by their wintering pattern. A less usual control method which has been investigated is to induce (or hasten) the onset of wintering with a chemical defoliant. The object is to control the time when refoliation takes place, i.e. this should be in drier weather and the developing leaves, therefore, escape infection. Cacodylic acid has been used in Malaysia (Rao).

Anon. 1969. *Plr's Bull. Rubb. Res. inst. Malaya* **97**: 100 (advisory).
Anon. 1970. *Ibid* **106**: 7 (advisory).
Carpenter, J. B. et al. 1954. *Pl. Dis. Reptr* **38**: 494 (in Costa Rica; **34**: 177).
John, K. P. 1952. *J. Rubb. Res. Inst. Malaya* **14**: 11 (a form of *Colletotrichum*; **32**: 338).
Rao, B. S. 1972. *Ibid* **23**: 248 (chemical defoliation for control; **53**, 253).
Saccas, A. M. 1959. *Agron. trop.* **14**: 409 (in the Central African Republic; **39**: 191).
Wastie, R. L. 1967. *Planter Kuala Lumpur* **43**: 553 (fungicide control; **47**, 1958).
——. 1968. *Pl. Prot. Bull. F.A.O.* **16**: 11 (*Colletotrichum* spp. on rubber; **47**, 3059b).
——. 1970. *Proc. Crop Prot. Conf. Kuala Lumpur Malaysia*: 197 (control of secondary leaf fall; **51**, 3698a).
—— et al. 1970. *Trans. Br. mycol. Soc.* **54**: 117 (variability in pathogen; **49**, 2616a).
——. 1972a. *Ann. appl. Biol.* **72**: 273 (factors affecting formation, germination & viability of conidia; **52**, 2359).
——. 1972b. *Ibid* **72**: 283 (meteorological & other factors affecting infection; **52**, 2360).
——. 1972c. *J. Rubb. Res. Inst. Malaya* **23**: 232 (weather & other factors affecting secondary leaf fall; **53**, 252).
——. 1973. *Ibid* **23**: 339 (nursery screening for resistance & field susceptibility; **53**, 2296).
Wimalajeewa, D. L. S. 1967. *Q. Jl Rubb. Res. Inst. Ceylon* **43**: 4 (factors affecting conidial germination; **47**, 299).

HIBISCUS CANNABINUS

Kenaf anthracnose is sometimes referred to as *Colletotrichum hibisci* Poll. which may be morphologically the same as *C. gloeosporioides*. Pulsifer compared the 2 fungi (using an isolate of the latter from cotton) and found that symptoms produced by them on kenaf were indistinguishable. *C. hibisci* also caused damping off in cotton seedlings. Seed transmission is very important and, therefore, infection may first be seen on seedlings: necrotic spots on the cotyledons, rapid spread to cause stem lesions, tip blight and death of young plants. The death of the apical part of the growing plant may be the most obvious field symptom. The capsules are attacked and the seeds, which can be light and non-viable, are penetrated (More et al.). Pulsifer (1957a) isolated the fungus from seed of okra (*H. esculentus*). Summers differentiated 3 races and found *Gossypium hirsutum* to be highly resistant. Summers et al. found an effect of temp. on infection using race 2 on 9 kenaf lines. At 20° there was no difference between lines, the entire plant being killed in most cases; at 25° some lines survived and at 30° some were highly resistant. Resistant lines have been used for many years in the American tropics. Crandall et al. (1972) reported severe infection in Zambia of cvs which were resistant to the 3 known races. Some resistance was found to this apparently new race which also attacked selections of roselle (*H. sabdariffa*). Various hot water treatments for control of the fungus in seed have been described: presoaking at 20° for 12 hours, then 15 minutes at 50° or 10 minutes at 52° (Kuo et al.); presoaking at 20° for 24 hours, then 15–20 minutes at 50° (Siang et al.); presoaking at 15° for 8 hours, then 10 minutes at 52° or 5 minutes at 55° (Wu et al.). Sy et al. reported that the fungus remained viable in the seed for 31 months and was not fully controlled by hot water.

Crandall, B. S. et al. 1954. *Pl. Dis. Reptr* **38**: 311 (resistance in cvs, hybrids and other *Hibiscus* spp.: **33**: 674).

—— ——. 1972. *Ibid* **56**: 1049 (probable new race; **52**, 3335).

Follin, J. C. et al. 1974. *Coton Fibr. trop.* **29**: 331 (inheritance of resistance; **54**, 3339).

——. 1975. *Ibid* **30**: 465 (expression of resistance, **55**, 5219).

Goebel, S. 1963. *Ibid* **18**: 120 (in Mali, reaction of cvs & *H. sabdariffa*; **42**: 765).

Kuo, S. G. et al. 1959. *Acta phytopathol. sin.* **5**: 45 (seed disinfection; **39**: 108).

More, W. D. et al. 1969. *Indian J. agric. Sci.* **39**: 432 (general; **50**, 2978).

Pate, J. B. et al. 1955. *Pl. Dis. Reptr* **39**: 776 (reactions to 2 races; **35**: 454).

Pulsifer, H. G. 1957a. *Iowa St. Coll. J. Sci.* **31**: 504 (general; **36**: 698).

——. 1957b. *Ibid* **32**: 57 (causing damping off on cotton; **37**: 169).

Siang, W. N. et al. 1956. *Acta phytopathol. sin.* **2**: 141 (control in seed & crop survival; **37**: 169).

Summers, T. E. 1954. *Pl. Dis. Reptr* **38**: 483 (3 races; **34**: 94).

—— et al. 1955. *Ibid* **39**: 650 (effect of temp. on susceptibility; **35**: 453).

Sy, C. M. et al. 1958. *Acta phytopathol. sin.* **4**: 25 (survival in seed & soil, control in seed; **38**: 6).

Wilson, F. D. et al. 1965. *Circ. agric. Exp. Stns Univ. Fla* S168, 12 pp. (2 resistant cvs; **45**, 3352).

Wu, S. L. et al. 1956. *Acta phytopathol. sin.* **2**: 127 (survival in field & control in seed; **37**: 169).

MANGIFERA INDICA

Mango anthracnose (*Glomerella cingulata*) is one of the most serious diseases of the crop. All floral and young vegetative tissue is infected in trees of any age; infection may be severe in the nursery. The small necrotic spots on leaves enlarge to *c.* 0.5 cm diam., they are circular or irregular, usually surrounded by a zone which is paler than the healthy parts of the leaf; shotholes form and leaves fall. Infection of the inflorescence may cause complete necrosis of the blossoms and attacked young fruits fall. The typical black anthracnose lesions form on older fruit; these may be so abundant as to cause necrosis of almost the whole surface. Rain wash of conidia results, on infection, in the characteristic, black, tear drop and streak patterns. Branch dieback also occurs. Latent infection of fruit is common and, although when picked fruit may appear healthy, anthracnose lesions form on ripening. The close relation between anthracnose in mango fruit and the physiology of ripening, and consequently the need to prevent infection of the fruit in the field, was early recognised (Wardlaw et al.). After picking and as the fruit softens the anthracnose fungus appears before the other common storage saprophytes.

Although cvs differ in susceptibility to infection, many desirable ones require fungicide treatment in the field. Spraying is often obligatory and it needs to be frequent from the onset of flowering. Copper, including Bordeaux, is still used in some areas; captan, maneb and zineb can also be effective (Aragaki et al.; Tandon et al.). In USA (Florida) McMillan (1973) reported that benomyl sprays increased the numbers of total and marketable fruit over those obtained with maneb which was followed in effectiveness by tribasic copper sulphate and captafol. Whether postharvest treatment of fruit and field spraying is needed may depend either on the effectiveness of the latter or on what other disease control measures need to be taken in the field. In South Africa Jacobs et al. stated that postharvest treatment is necessary there, since copper (not so effective against anthracnose) must be used to control *Erwinia mangiferae*. They found that a fruit dip in benomyl at 55° for 5 minutes gave satisfactory control in cold-stored mangoes. Wax treatment immediately afterwards overcame the problem of reduced fruit lustre caused by this dip. In several countries some control of latent infections has been shown by hot water treatment of harvested fruit. Conditions will vary with the cv.: temps are 50–55° for 5–15 minutes, the higher temps at the shorter times. This treatment is said to be effective for 3–6 weeks depending on the storage temp. Spalding et al. found that benomyl or thiabendazole in hot water gave better control than the latter alone, but that heat injury symptoms appeared in a week when fruits were held afterwards at 13°, a normal shipping temp.; there was no such injury at a post-treatment temp. of 21° (and see Muirhead).

Aragaki, M. et al. 1960. *Pl. Dis. Reptr* **44**: 318 (fungicides; **39**: 727).
Jacobs, C. J. et al. 1973. *Ibid* **57**: 173 (fungicides, postharvest treatment; **52**, 3390).
McMillan, R. T. 1972. *Proc. Fla St. hort. Soc.* **85**: 268 (fungicides; **53**, 2266).
——. 1973. *Trop. Agric. Trin.* **50**: 245 (as above with *Oidium* sp.; **53**, 1029).
Muirhead, I. 1976. *Aust. J. exp. Agric. Anim. Husb.* **16**: 600 (benomyl & hot water postharvest treatment; **56**, 1215).
Ocfemia, G. O. et al. 1925. *Philipp. Agric.* **14**: 199 (general, with avocado; **5**: 109).
Pennock, W. et al. 1962. *J. Agric. Univ. P. Rico* **46**: 272 (hot water treatment; **42**, 478).
Quimo, T. H. et al. 1975. *Philipp. Agric.* **58**: 322 (pathogenicity; **56**, 2157).
Rorer, J. B. 1915. *Bull. Dep. Agric. Trin.* 14: 164 (general).
Sattar, A. et al. 1939. *Indian J. agric. Sci.* **9**: 511 (general; **18**: 750).
Smoot, J. J. et al. 1963. *Pl. Dis. Reptr* **47**: 739 (hot water treatment; **43**, 174).
Spalding, D. H. et al. 1972. *Ibid* **56**: 751 (fungicides & hot water postharvest treatment; **52**, 2696).
Tandon, I. N. et al. 1968. *Indian Phytopathol.* **21**: 212, 331 (field fungicides & hot water postharvest treatment; **48**, 1255; **50**, 197).
Wardlaw, C. W. et al. 1936. *Mem. Low Temp. Res. Stn Trin.* 2, 47 pp. (storage; **15**: 592).

Glomerella tucumanensis (Speg.) Arx & Müller in *Beitr. Kryptogflora Schweiz* **11**: 196, 1954.
stat. conid. *Colletotrichum falcatum* Went, 1893
Physalospora tucumanensis Speg., 1896.

PERITHECIA on various parts of the host but abundant on leaf sheaths and blades, solitary or aggregated, often forming short lines between vascular bundles, globose, immersed, dark brown to black, 65–250 μ diam.; wall up to 8 cells thick, sclerotial on outside, pseudoparenchymatous within, ostiole slightly papillate, circular. ASCI 8 spored, clavate, thickened at apex, 50–118 × 7.4–19.2 μ. ASCOSPORES aseptate, hyaline, straight or slightly fusoid, ellipsoid or ovoid when mature, 12–30 × 5–11.1 μ. Paraphyses numerous, septate, unbranched, with granular contents extending almost to ostiole. ACERVULI on leaf and stem lesions, setose, rounded or elongated, 70–300 μ across. Setae *c.* 150 μ long, 4–6 μ wide, 1–4 septate, brown, slightly swollen at base, tapered to rounded tip on which conidia are occasionally borne. CONIDIA sickle shaped, hyaline, aseptate, 19.9–27.2 (23.5) × 3.9–4.9 (4.5)μ, formed from aseptate, hyaline, cylindrical, phialidic conidiophores *c.* 8–20 μ long and 4–8 μ wide.

COLONIES on PDA greyish white with sparse aerial mycelium on young portions and small dense felty patches elsewhere; there are no sclerotia or dendroid mycelial strands. Acervuli on surface of agar, with salmon pink spore masses sometimes forming pionnotes. Reverse of colony greyish white, darker with age. Some cultures, known as the dark race, show abundant greyish white aerial mycelium, a faint suggestion of dendroid growth, no distinct acervuli and very poor sporulation; variation from light to dark occurs. Appressoria formed in slide cultures are cinnamon buff to ochraceous tawny, broadly clavate, rarely irregular in shape, 6–20 (13.5) × 6–17 (10.5) μ, borne on hyphae which are undifferentiated or only faintly thickened and coloured.

The relationship between this sp. and *Colletotrichum graminicola* (q.v.) is controversial. Arx (1975a, *Colletotrichum* q.v.) regarded *C. falcatum* as a synonym of *C. graminicola* and used the latter name for the conidial state of *Glomerella tucumanensis*. Although occasionally recorded on other hosts *G. tucumanensis* is strongly pathogenic only on *Saccharum*; on a group basis the two spp. can be distinguished by the morphology of appressoria and conidia (J. E. M. Mordue, *CMI Descr.* 133, 1967. Sutton 1968 (*C. graminicola* q.v.) compared the appressoria of these 2 *Colletotrichum* spp. and considered them to be distinct).

Red rot of *Saccharum* spp. is one of the oldest (first described from Java in 1893) and most important sugarcane fungus diseases. It has caused epidemics in recent years in India, Hawaii and Nigeria. Butler et al. gave an early account, Abbott (1938) a monograph, and there are reviews by Abbott (1953 & 1961), Chona (1961) and Mian. The disease is widespread in sugarcane areas (Map 186) and has been investigated mainly in Hawaii, India and USA. Losses occur largely through reduction of germination in setts and in lowering the sucrose content. *Erianthus* and *Sorghum* spp. can be attacked; see Le Beau for host specificity. Red rot is a disease of setts, standing cane and leaves. The stalks must be split to observe the first symptoms. These are a sour odour and a reddening of the internal tissues, at first most intense in the vascular tissue but then spreading to the pith. The red colour in the internodes has a characteristic, blotched appearance due to white areas which generally run transversely. The pith dries, becomes necrotic, the rind falls in and cane may be easily broken. Fructification occurs on the plant surface. Setts may be completely destroyed, with dark mycelium in the pith and whole stools can be destroyed after wilting. On the leaves, reddish lesions (later with pale centres) occur, and on the midribs these are considerably elongated. Severity of the symptoms varies with the cv.

Invasion probably takes place mainly via the leaves (though it can occur in the stem directly) in young tissue. Leaf sheaths can be directly penetrated, symptoms occurring after 48 hours, but wounds are often sites of invasion including those caused by insects, for example, the moth borer larvae (*Diatraea saccharalis*). Besides mycelial spread in the host this also takes place with spores moving in the vessels. Spread from the leaf to stem may lead to active stem invasion or to a dormant condition; the latter then causing infection on germination of infected setts. Direct infection of planted setts has also been reported (e.g., Chona 1950; Edgerton et al.; Manser; Steib et al.). Spread of red rot appears to depend in part on whether cane is grown continuously or not. In the latter case spread via setts is more important, whilst in the former later spread through water splash in growing cane causes more infection. Where temps are higher and lead to more rapid cane germination and growth, sett infection and injury become less important. Infested soil and irrigation water are to be avoided; rain and dew increases the disease on the leaves.

G. tucumanensis is homothallic. There is considerable variation in vitro (opt. growth *c.* 30°) presumably through heterokaryosis. Isolates have been grouped morphologically into dark and light mycelial forms. These differ also in virulence, amount of sporulation, growth rate, spore germination, and sucrose and pectinase activity. Isolates of *Colletotrichum* which may be referred to either *C. falcatum* or *C. graminicola* (q.v.) differ in virulence depending on host origin, those from sugarcane not being generally pathogenic to *Sorghum* and *Erianthus* spp. which both give isolates not pathogenic to cane. Survival in soil extends over a few months (Agnihotri et al.; Kar et al.; Manocha; Manocha et al.; Singh, G. P. et al.; Sarma). Singh, P. (1973) gave the standard conditions for opt. conidial germination.

Control has concentrated on resistance, particularly from *S. spontaneum*; see the reviews for earlier work and *Rev. Pl. Pathol.* **51**, 625, 626, 627; **53**, 1532; **56**, 377, 3695; **57**, 3593. The so-called physiological resistance is little understood. Morphological resistance is associated with thickness of epidermis, cuticle and rind; and is also found in forms where there are relatively few vascular strands which are continuous through the nodes. This anatomical feature of 'closed' nodes, where the vessels have septa, prevents the spores spreading throughout the cane stem. It has been suggested that resistance is governed by 1 or a few genes from *S. spontaneum* plus a dominant inhibitor gene from *S. officinarum* and masking the expression of resistance (Atkinson; Azab et al.; Varma et al.). Some cvs may have morphological but not physiological resistance and vice versa. Control through resistance remains not entirely effective. Changes in virulence and low correlation between testing and general field performance affect it. Treatment of setts with benomyl, carboxin and methoxyethylmercury chloride is said to be effective (Chand et al.; Lewin et al.). Singh, K. found that heat treatment (54° for 48 hours) of setts was promising.

Abbott, E. V. 1938. *Tech. Bull. U.S. Dep. Agric* 641, 96 pp. (monograph, 75 ref.; **18**: 344).

——. 1953. *Plant diseases U.S. Dep. Agric. Yearbk*: 536 (review).

——. 1961. In *Sugarcane diseases of the world* Vol. 1: 263 (review, 46 ref.).

Agnihotri, V. P. et al. 1974. *Sugcane Pathol. Newsl.* 11–12: 19 (strs & virulence; **54**, 1424).

Atkinson, R. E. 1939. *Proc. int. Soc. Sugcane Technol.* 6th Congr. 1938; 684 (nature of resistance; **18**: 619).

Azab, Y. E. et al. 1952. *Phytopathology* **42**: 282 (inheritance of resistance; **32**: 213).

Butler, E. J. et al. 1913. *Mem. Dep. Agric. India bot. Ser.* **6**: 151 (general).

Carvajal, F. et al. 1944. *Phytopathology* **34**: 206 (morphology of perfect state; **23**: 358).

Chand, J. N. et al. 1974. *Sci. Cult.* **40**: 69 (systemic fungicides; **54**, 4131).

Chona, B. L. et al. 1942. *Indian Fmg* **3**: 70 (epidemic; **21**: 347).

——. 1950. *Indian J. agric. Sci.* **20**: 363 (source & mode of infection; **31**, 87).

—— ——. 1950. *Indian Phytopathol.* **3**: 196 (effect of temp. & pH in vitro; **31**: 400).

——. 1961. *Proc. Indian Sci. Congr.*: 197 (review; **41**: 407).

Dutta, A. K. et al. 1965. *Indian Phytopathol.* **18**: 274 (perithecia, effect of hexose phosphates & UV; **46**, 729).

Edgerton, C. W. et al. 1944. *Phytopathology* **34**: 827 (penetration & spread in host; **24**: 120).
Kar, K. et al. 1965. *Indian Sug.* **15**: 161 (a virulent str.; **45**, 207).
Le Beau, F. J. 1950. *Phytopathology* **40**: 430 (in vivo behaviour of isolates of *Colletotrichum* from *Sorghum*, *Saccharum* & *Erianthus*; **29**: 535).
Lewin, H. D. et al. 1976. *Sugcane Pathol. Newsl.* 17: 17 (fungicides; **56**, 3699).
Manocha, M. S. 1963. *Indian Jnl Microbiol.* **3**: 109 (production of sucrase & pectinase; **45**, 1498).
—— et al. 1964. *Indian J. agric. Sci.* **34**: 261 (strs & nutrition; **44**, 2869).
Manser, P. D. 1960. *Proc. Jamaica Assoc. Sug. Technol.* 1959: 17 (penetration via damage by *D. saccharalis*; **41**: 59).
Mian, A. L. 1969. *PANS* **15**: 482 (review, 57 ref.; **49**, 1471).
Padwick, G. W. 1940. *Indian Fmg* **1**: 263 (epidemic; **20**: 228).
Rao, K. C. et al. 1968. *Curr. Sci.* **37**: 532 (phenols & resistance; **48**, 1295).
Sarma, M. N. 1970. *Sci. Cult.* **36**: 52 (new str.; **50**, 1952).
Singh, G. P. et al. 1964. *Phytopathology* **54**: 1100 (hydrolytic enzyme activity & virulence; **44**, 827).
Singh, K. 1973. *Pl. Dis. Reptr* **57**: 220 (hot air treatment; **52**, 3815).
Singh, P. 1965. *Phytopathol. Z.* **54**: 79 (production of sucrase; **45**, 1174).
——. 1973. *Acta phytopathol. Acad. Sci. hung.* **8**: 19 (opt. conditions for conidial germination; **53**, 2429).
Sreeramulu, T. et al. 1970. *Pl. Dis. Reptr* **54**: 226 (spread on leaves & weather; **49**, 2622).
Steib, R. J. et al. 1951. *Phytopathology* **41**: 522 (mode of infection; **30**: 544).
Varma, S. C. et al. 1949. *Indian J. agric. Sci.* **19**: 383 (resistance & xylem structure at nodes).

GNOMONIA Ces. & de Not., *Comment. Soc. Crittogam. Ital.* **1**: 231, 1863.
(Diaporthaceae)

ASCOMATA (PERITHECIA) without a stroma, scattered, in tissue of leaves or herbaceous stems, long beaked. ASCI unitunicate, fusiform, clavate or nearly cylindrical, loosening from ascogenous cells at an early stage and lying free in the ascoma. ASCOSPORES 1 septate, septum in or near the middle, mostly fusiform, pyriform or ellipsoid, hyaline (from R. W. G. Dennis, *British Ascomycetes*: 324, 1968 and key by E. Müller & J. A. von Arx in *The fungi* (edited by G. C. Ainsworth et al.) Vol. IVA: 115, 1973).

G. iliau Lyon, causing iliau disease of sugarcane, was originally described as serious in Hawaii. But, although the fungus was reported from several countries (Map 158), it is now rarely seen and is apparently of no economic importance. The PERITHECIA are 325–480 × 240–340 μ with beaks 350–550 μ long, ASCI 8–14 × 60–80 μ, ASCOSPORES 5–7 × 22–30 μ (Edgerton). Infection of young stems takes place near soil level, the mycelium binds the leaf sheaths tightly together and to the stem which is penetrated. The outer leaves die and young plants may be killed or so weakened that breakage occurs. The perithecial necks protrude above the surface which thereby becomes characteristically rough to the touch. The illustrations after Lyon are given by J. P. Martin in *Sugarcane diseases of the world* Vol. 2: 115, 1964. A recently and fully described member of the genus is *G. manihotis* Punith. from stored cassava chips (Punithalingam, *Guignardia* q.v.).

Edgerton, C. W. 1913. *Phytopathology* **3**: 93 (morphology & symptoms).
Lyon, H. L. 1912. *Bull. Pathol. Physiol. Exp. Stn Hawaii. Sug. Plrs' Assoc.* **11**, 32 pp. (general).

GUIGNARDIA Viala & Ravaz, *Bull. Soc. mycol. Fr.* **8**: 63, 1892.
(Dothidiaceae)

STROMA prosenchymatous or plectenchymatous, often reduced or absent. ASCOMATA globose, subglobose or tympaniform, seldom papillate, ostiolate at maturity; wall of an outer layer of pigmented, mostly thick-walled pseudoparenchyma, somewhat darker around the outside, and an inner layer of subhyaline isodiametric pseudoparenchymatous cells. ASCI clavate, broadly rounded above, attenuated at base to a shorter or longer stalk, bitunicate, with membrane thickened near top. ASCOSPORES aseptate, hyaline, faintly greenish or brownish, ovoidal, ellipsoidal or rhomboidal, at times with distinct hyaline, gelatinous caps at one or both ends, mostly < 20 μ (seldom > 25 μ) long; filamentous paraphyses absent but some remains of the former pseudoparenchyma may occur (from H. A. Van der Aa 1973, *Phyllosticta* q.v.).

Reference should be made to Van der Aa (l.c.) and Punithalingam for taxonomic discussions. The latter dealt with the connections between the ascogenous and conidial states. The former gave a description of a spermogonial state with which *Guignardia* spp. are often associated. This state is redescribed and classified in *Leptodothiorella* Höhnel sensu Sydow ex Van der Aa, gen. nov. The spermatia (in ostiolate and papillate pycnidial like structures) are formed in basipetal succession as blastoconidia and separated from the spermatogenous cells by a septum; cylindrical or dumb-bell shaped, straight or slightly curved, bi-guttulate at the swollen ends, 5–15 × 0.5–3 μ. The imperfect states should be in *Phyllosticta* (q.v.). Among the 9 *Guignardia* spp. fully described by Punithalingam are: *G. arachidis* Punith. (on *Arachis hypogaea*), *G. capsici* Punith. (on *Capsicum frutescens*), *G. cocoicola* Punith. (on *Cocos nucifera*), *G. heveae* Syd. (synonyms: *G. heveae* Fragoso & Cif., *G. heveicola* Cif., on *Hevea brasiliensis*), *G. mangiferae* Roy (on *Mangifera indica*), *G. musae* Racid. q.v. (on *Musa*) and *G. perseae* Punith. (on *Persea americana*).

Punithalingam, E. 1974. *Mycol. Pap.* 136, 63 pp. (Sphaeropsidales in culture II; **54**, 1154).

Guignardia citricarpa Kiely, *Proc. Linn. Soc. N.S.W.* **73**: 295, 1949.

stat. conid. *Phyllosticta citricarpa* (McAlp.) Van der Aa, 1973

Phyllostictina citricarpa (McAlp.) Petrak, 1953

Phoma citricarpa McAlp., 1899.

ASCOCARPS amphigenous on dead leaves, not formed in fruit or leaf lesions, solitary or aggregated, globose, immersed, dark brown to black, 95–125 μ diam., wall up to 5 cells thick, sclerotioid on the outside, pseudoparenchymatic and thin walled within, ostiole nonpapillate, circular, 10–17.5 μ diam. ASCI clavate cylindrical, shortly stipitate, 8 spored, 40–65 × 12–15 μ, wall thick bitunicate. ASCOSPORES aseptate, hyaline, multi-guttulate, cylindrical but swollen in the middle, ends obtuse, each with a colourless appendage, 12.5–16 × 4.5–6.5 μ. Pseudoparaphyses absent. PYCNIDIA in fruit and leaf lesions, amphigenous on dead leaves, solitary, sometimes aggregated, globose, immersed, mid to dark brown, 115–190 μ diam., wall up to 4 cells thick, sclerotioid on the outside, pseudoparenchymatous within; ostiole darker, slightly papillate, circular, 12–14.5 μ diam. CONIDIA obovate to elliptical, hyaline, aseptate, multi-guttulate, apex slightly flattened with a colourless subulate appendage, base truncate, 8–10.5 × 5.5–7 μ, formed as blastospores from hyaline, aseptate, cylindrical conidiophores up to 9 μ long (after B. C. Sutton & J. M. Waterston, *CMI Descr.* 85, 1966).

The fungus is widespread on many hosts (Map 53). McOnie (1064a & e) in South Africa showed that it existed in at least 2 physiologic forms. This was long after *G. citricarpa*, as the cause of black spot of citrus, had been the subject of extensive investigations in Australia (Kiely) and South Africa (Wager). One form is pathogenic and causes black spot, the other does not cause disease in citrus (although it occurs on this crop) and is a widespread saprophyte on many other hosts. The pathogenic form is considered to be confined to citrus. Its distribution is: Australia (NSW, Queensland, Victoria), China (S.E.), Hong Kong, Indonesia (Java), Kenya, Mozambique, New Hebrides, Philippines, Zimbabwe, South Africa (Natal, Transvaal), Taiwan and Zambia. Only the more widespread, non-pathogenic form has been reported from N., Central and S. America. A full, illustrated description of this form from *Saccharum* was given by Hudson (*Leptosphaeria michotii* q.v.). Black spot has also been studied in Taiwan (Huang et al.; Lee; Lee et al.; Liu) and reviews were given by Schüepp and Whiteside (1965).

Kiely (1948, 1960; and see McOnie 1964d) gave several names to the types of spotting on the fruits. Factors which affect these differing symptom patterns include temp. and the physiological state in the host. The characteristic limited spot begins as a very small brown or reddish depression in the rind; it enlarges to 3–5 mm diam., has a greyish centre with pycnidia, an almost black ring and a green or yellow halo (hard spot). The spots can be very numerous. Freckle spots were described as deep orange to brick red, numerous, becoming brown, 2–3 mm diam. with a sunken centre but no green halo. Virulent spot was a rapidly spreading lesion arising later as the fruits mature; irregular, 2 cm or more in diam., a central area (dark with numerous pycnidia), a brown periphery and an outermost, narrow, brick red area; fruits fall and the entire thickness of the rind is affected. Speckled blotch consists of numerous, 1–2 mm diam. spots becoming dark brown, either scattered over the fruit surface or grouped in clusters or streaks; around the speckled blotch lesions are often very small dark speckles or dots which may be raised. Leaf lesions are circular up to *c.* 4 mm diam., centres grey or light brown with a darker border and usually a chlorotic halo; the spots may be few or numerous. Both pycnidia and ascocarps develop abundantly on fallen leaves.

The air dispersed ascospores (after wetting of the ascocarps) and water dispersed conidia infect only young fruits and leaves. McOnie (1967) demonstrated direct penetration from the ascospores, 66% of which had formed appressoria after 47–48 hours. Moist periods are required for infection after which there is a very long latent period up to *c.* 3 years for leaves and up to *c.* 15 months for fruit. Leaves rarely show symptoms until they have fallen. Severe disease occurs only at relatively high temps, 24° or more. Infected fruit, showing no symptoms, will develop them in store at >21°. Symptoms on Valencia oranges increased with an increase in temp. from 20° to 27° (and upon exposure to light, Broderick et al.). Ascocarps in dead leaves matured in 4–5 weeks at 21–28° (Lee et al.). Under normal good management conditions most inoculum is ascospores from fallen leaves. But conidial inoculum can be significant from fruits left on the tree, old fruit stalks and dead twigs (Whiteside 1967). The annual infection cycle (see diagram by Kiely 1948) is affected by the amount and distribution of rainfall when there is susceptible tissue on the trees (such tissue may escape infection in dry periods), by temp. and by the characteristics of the spore inoculum. In culture the pathogenic form is slower growing, has a less luxurious growth and only develops pycnidia; compared with the non-pathogenic or saprophytic form which readily develops ascocarps. Both types have been detected in the symptomless rind of fruits. Frean, on the basis of in vitro sporulation, reported a third type which formed both fructifications.

Control is with fungicides. Kiely (1970) discussed spray programmes in Australia (including control of *Diaporthe citri*) on Valencia orange; Bordeaux + nonionic surfactant at petal fall followed 9–15 weeks later by zineb or mezineb + white spraying oil. Kiely (1976) recommended Bordeaux at petal fall followed 16 weeks later by benomyl. In Taiwan captafol gave control (Liu) and Tsia found programmes of benlate + oil and mancozeb effective. In South Africa McOnie et al. reported that benomyl was better than mancozeb. Seberry et al. found that a 15 second dip in a water wax emulsion reduced development of *G. citricarpa* on orange fruit stored at 13–26.5°. In South Africa the discontinuation of fumigation with hydrocyanic acid (replaced by parathion) coincided with an increase in black spot incidence. Fumigation for 30 minutes killed the mycelium of the pathogen in fruit (Smith).

Brodrick, H. T. 1971. *Phytophylactica* **3**: 69, 89 (toxicity of limonene & citrus peel extracts to the fungus; osmotic pressure in orange peel & symptoms of black spot; **51**, 2487, 2488).

—— et al. 1970. *Ibid* **2**: 157 (light & temp. effects on symptoms & sporulation; **51**, 1492).

Frean, R. T. 1966. *S. Afr. J. agric. Sci.* **9**: 777 (nutrition & behavioural patterns in culture; **46**, 3436).

Huang, C. S. et al. 1972. *Jnl Taiwan agric. Res.* **21**: 256 (leaf infection & ascospore discharge; **53**, 951).

Kiely, T. B. 1948. *Proc. Linn. Soc. N.S.W.* **73**, 249 (general; **29**: 208).

——. 1950. *Sci. Bull. Dep. Agric. N.S.W.* 71, 88 pp. (fungicides & epidemiology; **30**: 566).

——. 1960. *Agric. Gaz. N.S.W.* **71**(9): 474 (speckled blotch symptoms; **40**: 306).

——. 1970. *Sci. Bull. Dep. Agric. N.S.W.* 80, 11 pp. (fungicides, with *Diaporthe citri*; **51**, 362).

——. 1976. *Rural Newsl.* 59: 35 (fungicides; *Hort. Abstr.* **48**, 4951).

Lee, Y. S. 1969. *Jnl Taiwan agric. Res.* **18**(2): 45 (pathogenicity of isolates; **50**, 2953).

—— et al. 1973. *Ibid* **22**(2): 135 (climate & ascospore discharge; **53**, 1384).

Liu, K. C. 1966. *Ibid* **15**(2): 20 (fruit infection & fungicides; **47**, 1126).

McOnie, K. C. 1964a, b, c, d. *Phytopathology* **54**: 40, 64, 1448, 1488 (pathogenic & saprophytic forms, source of inoculum, ascospore discharge & infection, speckled blotch symptoms; **43**, 1922a & b, **44**, 1566a & b).

——. 1964e *S. Afr. J. agric. Sci.* **7**: 347 (pathogenic & saprophytic forms; **44**, 701).

——. 1967. *Phytopathology* **57**: 743 (germination of & infection by ascospores; **46**, 3437).

—— et al. 1969. *S. Afr. Citrus J.* 423: 7 (fungicides; **49**, 2075).

Schüepp, H. 1961. *Phytopathol. Z.* **40**: 258 (short review, 21 ref.; **40**: 532).

Seberry, J. A. et al. 1967. *Aust. J. exp. Agric. Anim. Husb.* **7**: 593 (control; **47**, 1127).

Smith, J. H. 1965. *Phytopathology* **55**: 486 (effects of hydrocyanic acid; **44**, 2495).

Tsia, Y. P. et al. 1977. *Pl. Prot. Bull. Taiwan* **19**: 140 (fungicides; **57**, 2940).

Wager, V. A. 1952. *Sci. Bull. Dep. Agric. S. Afr.* 303, 52 pp. (general; **32**: 373).

Whiteside, J. O. 1965. *Rhod. agric. J.* **62**: 87 (review; **45**, 1070).

——. 1967. *Rhod. Zamb. Malawi J. agric. Res.* **5**: 171 (inoculum sources; **47**, 1125).

Guignardia musae Raciborski, *Bull. int. Acad. Sci. Cracovie* No. 3: 388, 1909.
Guignardia musae Stevens, 1925
Sphaeropsis musarum Cooke, 1880
Phoma musae (Cooke) Sacc., 1884
Macrophoma musae (Cooke) Berl. & Vogl., 1886
stat. conid. *Phyllosticta musarum* (Cooke) Van der Aa, 1973
Phoma musae C. W. Carpenter, 1919.

COLONIES on oat agar slow growing, greenish grey becoming black with abundant submerged mycelium, margin smooth to lobed, reverse fuscous black with agar almost of same colour. ASCOCARPS abundant, amongst pycnidia, solitary or as extensive stromatic structures in dendroid pattern, subglobose at base with long cylindrical necks; wall stromatic of dark brown cells on the outside. ASCI subclavate to cylindrical, stipitate, 55–85 × 8–10 μ, 8 spored, wall thick, bitunicate. PESUDOPARAPHYSES absent. ASCOSPORES 13–16 × 4–7 μ, distichous, hyaline, aseptate, fusiform to ellipsoidal, wider around mid region, the obtuse ends with a gelatinous plug. PYCNIDIA amongst ascocarps, brown to black, solitary or grouped; wall of many layers of cells, outer cells sclerotised. CONIDIOGENOUS CELLS holoblastic, simple, hyaline, obpyriform to cylindrical. CONIDIA (blastospores) hyaline, obovoid 8–15 × 4–8 μ with gelatinous envelope and hyaline apical appendage (5–) 8–12 μ long. SPERMOGONIA sparse, globose. SPERMATIAL CELLS hyaline, cylindrical to dumb-bell shaped, aseptate, 4–7 × 0.5–1 μ (from E. Punithalingam, *Mycol. Pap.* 136: 33, 1974).

The synonymy is largely that given (with ref.) by Punithalingam (l.c.) and which differs from that of Van der Aa (*Phyllosticta* q.v.). The former gave a fully illustrated description of the fungus, with comments on the taxonomy, and the first unequivocal demonstration of the connection between the perfect and imperfect states. This minor disease (freckle) of banana caused by *G. musae* was most recently investigated by Meredith. The fungus is widespread but the infections it causes have not apparently been reported from Central and S. America. Symptoms on the fingers may appear 2–4 weeks after the bunch has shot but are most noticeable at harvest. The fruit spotting occurs as red brown, superficial lesions, usually circular, up to 2 mm diam. and with dark green watersoaked haloes. Coalescence of the spots causes large, irregular diseased areas to form, and dense aggregations of spots may develop as streaks or circular areas. At harvest freckle may be so intense that the fingers are completely covered; development of the pycnidia renders the surface rough to the touch. The proximal hand is often more severely affected than the distal hands. On leaves the dark brown or black spots may be small (< 1 mm diam.) or larger (up to 4 mm diam.). In the former case the freckled areas have a sooty appearance, and spots may spread as streaks or circular areas. In the latter case the spots may have fawn or greyish centres with yellowish green haloes. Coalescence of spots occurs and freckle is usually restricted to the older leaves.

The water dispersed conidia germinate to form the characteristic irregularly lobed appressoria, usually > 15 μ diam.; cuticular penetration (after 24–72 hours at 24°) takes place. The necrosis caused rarely extends beyond the fifth cell layer beneath the epidermis. Cvs apparently differ in susceptibility, for example, Gros Michel is not affected but freckle can be severe on Cavendish cvs. Some AAB genotypes (e.g. Horn plantain, pome and silk) get severe freckle whilst others appear resistant, as are some ABB genotypes. Only fruit infection has been described as being of economic importance but leaves with freckle may be significant sources of inoculum. In Hawaii maneb sprays to the leaves and fruit reduced the disease. In Taiwan folpet and captan gave some control.

Huang, C. C. et al. 1967. *Pl. Prot. Bull. Taiwan* **9**: 31 (in Taiwan; **49**, 1097).

Kumabe, B. et al. 1964. *Hawaii Fm Sci.* **13**: 7 (control with maneb).

Meredith, D. S. 1968. *Ann. appl. Biol.* **62**: 329 (general; **48**, 538).

Punithalingam, E. et al. 1975. *C.M.I. Descr. pathog. Fungi Bact.* 467 (*Guignardia musae*).

HAPLOBASIDION Eriksson, *Bot. Zbl.* **38**: 786, 1889.

(Dematiaceae)

COLONIES compact or effuse, brown or olivaceous brown, velvety or hairy. MYCELIUM immersed. STROMA none. SETAE and HYPHOPODIA absent. CONIDIOPHORES macronematous, mononematous, unbranched, straight or flexuous, subhyaline to brown, smooth. Each conidiophore terminates in a vesicle or ampulla, the lower part of which is dark brown; the upper part, which frequently becomes flattened or invaginated, is much paler. Conidiophores often proliferate percurrently forming further ampulae at higher levels. CONIDIOGENOUS CELLS polyblastic, usually discrete, determinate, spherical, subspherical, ellipsoidal or clavate, covering the upper surface of the ampulla but sometimes integrated, terminal, the ampulla itself then being the conidiogenous cell. When conidia and conidiogenous cells become detached from the ampulla they sometimes leave annular scars on its surface. CONIDIA catenate, chains simple or branched, dry, borne on the rounded upper part of the conidiogenous cell, simple, spherical or subspherical, subhyaline to brown, verruculose, aseptate (M. B. Ellis, *Demat. Hyphom.*: 353, 1971).

H. musae M. B. Ellis causes diamond leaf spot (also called Malayan leaf spot) of banana. It was described in 1957 but, as Firman has pointed out, the disease was first reported by C. H. Knowles from Fiji in 1916. COLONIES hypophyllous, ellipsoidal or round, olivaceous brown, velvety. CONIDIOPHORES 50–110 × 4–6 μ, ampullae 9–12 μ diam. CONIDIOGENOUS CELLS spherical, 4–8 μ diam. CONIDIA brown, verruculose, 4–6 μ diam. (Ellis, l.c.; Ellis et al.). The pathogen occurs in Fiji, Malaysia (W.), Samoa and Tonga (Map 474). On the upper leaf surface the spots are greyish white with straight edges, diamond shaped with a black border, 4–5.5 × 3–4 mm, longer axis parallel to the veins, border 0.5 mm wide. These lesions can be surrounded by a watersoaked area often several times their size. Since 1964 the disease has been recognised as potentially serious in Fiji but little is known of the biology of *H. musae*. Infection probably occurs on the young leaves soon after emergence. There is severe infection in high rainfall areas (> 250 cm a year), and cool season temps favour diamond leaf spot more than black leaf streak (*Mycosphaerella fijiensis* q.v.). In Fiji oil sprays in the cool season for the control of the latter disease have apparently resulted in it being replaced by the former one.

Ellis, M. B. 1957. *Mycol. Pap* 67, 15 pp. (*Haplobasidion, Lacellinopsis* & *Lacellinia*, *H. musae* sp. nov.).

—— et al. 1976. *C.M.I. Descr. pathog. Fungi Bact.* 496 (*H. musae*).

Firman, I. D. 1971. *PANS* **17**: 315 (general in Fiji; **51**, 1688).

HELICOBASIDIUM Pat., *Bull. Soc. bot. Fr.* 32: 171, 1885.

Helicobasis Clem. & Sh., 1931.
(Auriculariaceae)

BASIDIOCARP resupinate, effused or occasionally pustulate, arid, floccose or fleshy fibrous, composed of more or less loosely interwoven hyphae. PROBASIDIA basal, persistent or not; if persistent thin walled or slightly thick walled and irregular in shape. METABASIDIA not readily detached, emergent, typically circinnately coiled apically. Parasitic on subterranean parts of plants (from key by R. F. R. McNabb in *The fungi* (edited by G. C. Ainsworth et al.) Vol. IVB: 311, 1973; see notes by M. A. Donk, *Persoonia* **4**: 145, 1966; and for further descriptions see Boedijn et al.).

The diseases caused by the fungi which have been placed in the genus have been commonly called violet root rot; the mycelial mats are often purplish. The fungi are generally referred to: *H. compactum* (Boedijn) Boedijn, *H. mompa* Tanaka or *H. purpureum* Pat. (imperfect state *Rhizoctonia crocorum* DC. ex Fr.). They are soil inhabitants and the behaviour of *H. purpureum* in this environment has been studied in some detail (see the discussions by Garrett, *Biology of root-infecting fungi* and *Pathogenic root-infecting fungi*; Valder). This sp. spreads through the soil as mycelial strands from a food base. It shows an affinity for continuous surfaces in soil. Sclerotia are formed along the mycelial strands or around small lateral roots along which they may be strung like beads on a thread (resembling mouse dung). Valder found sclerotial survival to be poor in acid soil but it was unaffected by a wide range of conditions in slightly alkaline soil. The poor survival in acid soil was due to spontaneous germination. *H. purpureum* occurs on temperate plants (see Hering) and an example of a disease caused by it is the one on carrot described by Whitney; deep lesions are formed on the fleshy root.

H. compactum was fully described by Reid who also discussed the 2 other spp. This sp. is apparently more common in the tropics but few diseases have been recently described in any detail. Examples are violet root rot of teak in Tanzania and an infection of coffee (collar canker) in Guatemala (Hocking et al.; Schieber et al.). An exhaustive description of the biology of *H. mompa* in Japan was given by Ito (1949) who showed it to be plurivorous. Later papers on this sp. from Japan are by Araki (soil conditions), Ito (on Gramineae), Takai (toxin) and several workers on the infection of sweet potato (*Bull. natn. Inst. agric. Sci. Tokyo* Ser. C 8, 173 pp., 1957; 37: 557).

Araki, T. 1967. *Bull. natn. Inst. agric. Sci. Tokyo* Ser. C. 21: 1 (soil conditions, *Helicobasidium mompa* & *Rosellinia necatrix* on fruit trees; **46**, 3127).

Boedijn, K. B. et al. 1930. *Archf. Theecult. Ned.-Indië* 1: 3–59 (*Helicobasidium* & *Septobasidium* spp. on cultivated plants in Indonesia; **9**, 562).

——. 1931. *Bull. Jard. bot. Buitenz.* Ser. 3 **11**: 165 (spp. as above in Indonesia).

Hering, T. F. 1962. *Trans. Br. mycol. Soc.* **45**: 488 (hosts of *H. purpureum*; **42**, 304).

Hocking, D. et al. 1967. *Pl. Prot. Bull. F.A.O.* **15**: 10 (*H. compactum* on teak; **46**, 3222).

Ito, K. 1949. *Bull. Govt For. Exp. Stn Meguro* 43, 126 pp. (biology of *H. mompa*, 136 ref.; **30**: 337).

——. 1969. *Ibid* 224: 111 (immunity of Gramineae to *H. mompa*; **49**, 1404).

Reid, D. A. 1975. *Trans. Br. mycol. Soc.* **64**: 159 (morphology of *H. compactum* & discussion on *H. mompa* & *H. purpureum*; **54**, 2705).

Schieber, E. et al. 1967. *Pl. Dis. Reptr* **51**: 267 (*H. compactum* on coffee; **46**, 2234).

Takai, S. 1966. *Bull. Govt For. Exp. Stn Meguro* 195, 55 pp. (on toxin helicobasidin; **46**, 1437).

Valder, P. G. 1958. *Trans. Br. mycol. Soc.* **41**: 283 (biology of *H. purpureum*; **38**: 66).

Whitney, N. J. 1954. *Can. J. Bot.* **32**: 679 (*Rhizoctonia crocorum* on carrot; **34**: 203).

HELMINTHOSPORIUM Link ex Fr.; Link, *Magazin Ges. naturf. Freunde Berl.* 3: 10, 1809; Fries, *Syst. Mycol.* 1: XLVI, 1821.
(Dematiaceae)

COLONIES effuse, dark hairy. MYCELIUM immersed, STROMA usually present, dark, often large. SETAE and HYPHODIA absent. CONIDIOPHORES macronematous, mononematous, unbranched, often caespitose, straight or flexuous, cylindrical or subulate, mid to very dark brown, smooth or occasionally verruculose with small pores at the apex and laterally beneath the septa. CONIDIOGENOUS CELLS polytretic, integrated, terminal and intercalary, determinate, cylindrical. CONIDIA solitary, catenate in one sp., acropleurogenous, developing laterally often in verticils through very small pores beneath the septa whilst the tip of the conidiophore is actively growing, growth of the conidiophore ceasing with the formation of terminal conidia, simple, usually obclavate, sometimes rostrate, subhyaline to brown, smooth, pseudoseptate, frequently with a prominent dark brown or black scar at the base (M. B. Ellis, *Demat. Hyphom.*: 388, 1971).

Spp. are often incorrectly placed under *Helminthosporium* and should be in *Drechslera* (q.v. for the distinction between the 2 genera). Ellis (1961) gave a key and descriptions of 10 spp.; and see Ellis, *More Demat. Hyphom.* These include *H. velutinum* Link ex Ficinus & Schubert and *H. solani* Dur. & Mont. A synonymy and description of the latter, which causes the widespread (Map 233) silver scurf of Irish potato, was given by Ellis (1968). The former sp. is common on dead stems of herbaceous plants and trees. A rot of tomato fruits caused by *H. carposaprum* Pollack, considered to be a weak pathogen, showed cream buff lesions, a dark area around the infection point and an outer zone of bay, burnt sienna or mahogany red. The CONIDIOPHORES are dark, erect, multiseptate, 6–10 × 140–500 μ, tapering to 4–6 μ at the paler apex; thick-walled bulbous basal cell, 10–12 × 12–16 μ. CONIDIA usually in long chains, cylindrical with obtuse to rounded base and apex, straight or curved, subhyaline to dilute olivaceous, 1–5 septate, 6–12 × 28–220 (mostly 8–10 × 120–160) μ.

Ellis, M. B. 1961. *Mycol. Pap.* 82, 55 pp. (*Helminthosporium* inter alia; **41**: 373).

——. 1968. *C.M.I. Descr. pathog. Fungi Bact.* 166 (*H. solani*).

McColloch, L. et al. 1946. *Phytopathology* **36**: 988 (*H. carposaprum* sp. nov.; **26**: 175).

Talbot, P. H. B. 1973. *APPS Newsl.* **2**: 3 (genus sensu lato; keys to perfect & imperfect states, & to common Australian spp. on cereals; **54**, 2098).

HEMILEIA Berk. & Br., *Gdnrs' Chron.*: 1157, 1869.
Hemileiopsis Racib., 1900.
(Pucciniaceae)

PYCNIA and AECIDIA unknown. UREDIA minute, superstomal, rarely subepidermal; UREDOSPORES borne in clusters on short pedicels or sterigmata at the apex of sporogenous sporophores, emerging through the stomata; sporophores few to many, distinctly separate in early stages, later showing lateral coalescence; uredospores bifacially ovate to reniform, smooth on the lower flat side and covered with dense aculeate processes on the convex side. TELIA like uredia; TELIOSPORES borne in clusters at apices of sporogenous sporophores on short stalks, ovate, napiform, crescentic or tridentate, orange yellow when fresh; epispore thin, hyaline, without germ pores; germinating immediately at maturity by the prolongation of spore apex; promycelium external and 4 celled (from M. J. Thirumalachar & B. B. Mundkur, *Indian Phytopathol.* **2**: 213, 1949; q.v.).

Members of the genus occur mostly on plants in the Apocynaceae, Asclepiadaceae and Rubiaceae; see Gopalkrishnan who gave notes on the morphology of 32 spp. The 2 pathogens both occur on coffee.

Gopalkrishnan, K. S. 1951. *Mycologia* **43**: 271 (morphology; **30**: 583).
Rajendren, R. B. 1972. *Mycopathol. Mycol. appl.* **47**: 81 (development & parasitism in 17 spp.; **52**, 45).
Thirumalachar, M. J. et al. 1947. *Ann. Bot.* **11**: 77 (on Rubiaceae in India, Mysore; **26**: 199).

Hemileia coffeicola Maublanc & Roger, *Bull. Soc. mycol. Fr.* **50**: 195, 1934.
Uredo coffeicola Maubl. & Roger, 1934.

SORI hypophyllous, scattered over the entire leaf surface and giving a powdery appearance; consisting of 1–3 large, inflated feeder hyphae and swollen rounded cells below the stomata and whose tips bear numerous pedicels on which the spores are borne. UREDOSPORES more or less reniform, 34–40 × 20–28 μ, wall hyaline, strongly warted on the convex face, smooth on the straight or concave face, 1 μ thick. TELIOSPORES more or less spherical, 20–26 μ diam., wall hyaline, smooth, 1 μ thick. Distinguished from *H. vastatrix* (q.v.) by the sori being scattered over the entire leaf surface instead of being confined to distinct spots, by the few feeder hyphae which swell to 20–30 μ diam. as they pass from the mesophyll to the substomatal chamber; and by the uredospores which have larger but less numerous spines. (G. F. Laundon & J. M. Waterston, *CMI Descr.* 2, 1964.)

This more recently described coffee rust, see review by Saccas, has not so far become a serious problem although infection can cause complete defoliation. It has been reported only from: Cameroon, Central African Republic, Fernando Po, N. Nigeria, São Thomé and Principe, Uganda and Zaire (Map 470). The powdery, orange uredospores completely cover the abaxial leaf surface and the adaxial surface remains green for some time. *H. coffeicola* infects the 3 commercial spp. of coffee (*C. arabica*, *C. canephora*, *C. liberica*) and occurs on other spp. *C. racemosa* was infected by collections from Cameroon, Central African Republic and Nigeria but not by those from São Thomé. All cvs of *C. arabica*, including the whole range of genotypes resistant to races of *H. vastatrix*, are susceptible. But other spp. and interspecific hybrids with *C. arabica* have shown resistance. In São Thomé serious disease is said to occur only at over 500 m. The economic importance of *H. coffeicola* may increase.

Maublanc, A. et al. 1934. *C. r. hebd. Séanc. Acad. Sci. Paris* **198**: 1069 (in Cameroon; **13**: 507).
——. 1934. *Bull. Soc. mycol. Fr.* **50**: 193 (in Cameroon, *H. coffeicola* sp. nov.; **14**: 303).
Rodrigues, C. J. 1957. *Revta Café port.* **3**: 48 (resistance; **37**: 536).
Roger, L. 1937. *Annls agric. Afr. occid fr.* **1**: 92 (in Cameroon; **16**: 379).
Saccas, A. M. 1972. *Bull. Inst. Fr. Café Cacao* **11**: 5 (review, 18 ref.; **52**, 3700).

Hemileia vastatrix Berk. & Br., *Gdnrs' Chron.*: 1157, 1869.

SORI hypophyllous, densely scattered, powdery appearance on yellowish orange, rounded blotches *c.* 3–25 mm diam.; consisting of numerous narrow interwoven feeder hyphae and somewhat rounded cells below the stomata, bearing clavate filaments emerging through the stomata, their tips bear numerous pedicels on which the spores are borne. UREDOSPORES more or less reniform, 28–36 × 18–28 μ; wall hyaline strongly warted on the convex face, smooth on the straight or concave face, 1 μ thick. TELIOSPORES more or less spherical, 20–28 μ diam., wall hyaline, smooth, 1 μ thick. Distinguished from *H. coffeicola* (q.v.), which also occurs on *Coffea* spp., by being restricted to yellow orange coloured spots on the leaf; by the formation of numerous, narrow feeder hyphae forming an interwoven mass in the substomal cavity and by uredospores which have smaller and more numerous spines (G. F. Laundon & J. M. Waterston, *CMI Descr.* 1, 1964).

There is no adequate monograph in English on this coffee rust which causes one of the classical plant diseases. Because of it Sri Lanka became a tea rather than a coffee producing country. A good historical account was given by Large (*The advance of the fungi*); and see *Famine on the wind* by Carefoot and Sprott; Mayne 1969. For a full coverage of the extensive literature the reviews and bibliographies should be consulted. The latter: Carmargo; Ramos; Stevenson et al.; the former: Nutman et al. 1970; Rayner 1960 & 1972; Razafindramamba; Saccas et al. The early account of this very important disease by Butler (*Fungi and disease in plants*) should be read for history, fungal morphology and host parasite histology. For further data on cytology, teliospore germination, basidiospores, uredia, haustoria and fine structure see Chinnapa et al.; Rajendren; Rajendren et al.; Ragunathan; Rijkenberg et al.; Rijo et al.; Vishveshwara et al.

Symptoms are generally confined to the leaves and first appear as small, yellowish spots on the under surface. The colour deepens to orange with growth and at *c.* 3 mm diam. sporulation begins; the leaf blotches may enlarge up to 1.5 cm. On the upper surface livid or brownish patches eventually form in places corresponding to the lesions below. Later the centre of the spot on the lower surface becomes grey or brown and dies. The spots are generally rounded but growth checks at veins or coalescence may lead to irregularity in their outlines. Infection can cause extensive, premature defoliation. *H. vastatrix* (Map 5) is widespread in Africa and S.E. Asia; it also occurs in Argentina, Brazil, Nicaragua and Paraguay, and parts of Oceania (Fiji, New Caledonia, New Hebrides and Samoa). Unlike some major fungal diseases on tropical crops the first authenticated outbreak of this rust (Sri Lanka 1868) did not occur close to the centre of origin of the crop. The main economic sp. (*C. arabica*) is indigenous in Ethiopia where the rust occurs (e.g. Sylvain). The disease soon spread to other parts of Asia. In Africa the fungus apparently took *c.* 70 years to spread from E. to W. A comparison with the much more rapid spread across Africa in the other direction by *Puccinia polysora* is an obvious one to make. In 1970 *H. vastatrix* was reported from Brazil where it is now established. The fact that it took some years to discover this latest spread by such a notorious pathogen is yet one more example of the inadequacy of plant disease coverage in large areas of the tropics. Assessments of coffee rust in S. America have been given by Firman (1972); Schieber; Waller.

Before its spread to Brazil *H. vastatrix* had been most studied in E. Africa and S. India. The basidiospores do not infect coffee (no alternate host is known); infection and spread are through the uredospores. Penetration is through the stomata which only occur on the lower leaf surface. Most uredospore germination occurs at 22–24° (liquid water is required) in the dark or at low light intensities, and on av. in 2.6–4.7 hours; appressoria are formed in 6.5–8.5 hours. Light more intense than 2.5-foot candles greatly reduces germination and lesion development; for max. infection 9 hours in darkness is required, for max. germination 4 hours. Prior exposure to strong light for up to 2 hours has a progressively inhibitory effect on germination. Relatively high concs of uredospores are needed for infection. The mean incubation time is 4–7 weeks (increasing with low temps and dry conditions); and 5–7 weeks after infection there can be 50% sporulation. Susceptibility is not affected by leaf age or crop size. Leaf lesions can develop after 3.5 hours' exposure to conditions favouring infection but lesion numbers increase with up to at least 12 hours' exposure. Stored dry uredospores lose viability by 50% in <2 days. At room temp. infectivity was lost after 20 days and viability after 60 days. At 10–15° viability was lost after 80 days (Hocking; Montoya et al.; Nutman et al. 1963; Rayner 1961; Zambolim et al.).

There has been controversy (see Firman 1965a) over whether the uredospores are very largely splash dispersed or whether they can be also truly windborne. Some experimental evidence appears to favour the predominance of the former mechanism. But diurnal periodicity and wind dispersal were recently described (Becker; Becker et al.). Uredospores adhere strongly to one another, to leaves and to any surfaces however smooth. They can be removed from leaves fluttering at 9–10 kph but only in clusters. Relatively small numbers of uredospores appear to be caught in volumetric traps. Rain splash disperses uredospores in quantities adequate for infection, and their numbers increase with the amount and intensity of rain. Showers of >8 mm are necessary for effective dispersal (Becker et al.; Bock 1962a; Burdekin; Nutman; Nutman et al. 1960 & 1962). Insects have also been implicated in spread (Ananth et al.; Crowe). Outbreaks of coffee rust are closely related both to the rainfall pattern and to temp. In S. India the disease begins to increase after the beginning of the S.W. monsoon, with a max. intensity and heavy leaf fall occurring before the rains end. In E. Africa the seasonal periodicity differs E. and W. of the rift valley. In the E. 2 wet seasons give 2 disease peaks; in the W. with a more or less continuous rainy season there is one extended epidemic. More disease occurs at 1500–1650 m than at 1740–1800 m. The course and severity of outbreaks depends on 3 interacting factors: distribution and intensity of rainfall; degree of tree leafiness; amount of residual inoculum at the end of the dry season (Bock 1962b; Mayne 1930 & 1931).

A discussion of the earlier work on control, with particular reference to Kenya, was given by Firman (1965a). In an important exercise Shaw demonstrated the eradication of *H. vastatrix* from Papua. Mayne (1971) reviewed control in S. India over the last 100 years. Copper spraying against this leaf rust is an established practice. Work in E. Africa and India has shown that the timing of fungicide applications is critical, particularly the first one before the beginning of the rains, but also for those in the early part of the wet season (see Mayne 1937, for India). Bock (1959) found that captan (3 applications) gave control although this fungicide was often unsatisfactory against coffee berry disease (*Glomerella cingulata*). Burdekin (1964) found that zineb was effective, although copper fungicides gave the max. yield increase and were more effective in preventing leaf fall. Low volume spraying in E. Africa was described by Firman et al.; Park et al. (1964a); and the ageing of copper fungicides by Park et al. (1964b). In this region fungicide control is complicated by the need to spray against the virulent coffee str. of *G. cingulata*. In South Africa captafol and copper gave satisfactory control (Broderick et al.). Spraying in Brazil was described by Chaves et al.; Andrade and others.

Considerable work has been done on resistance of the vertical pattern (race specific resistance). Anon.; Bettencourt et al.; D'Oliveira; Goujon; Rodrigues et al. (1975b); and Sreenivasan have given general accounts. D'Oliveira and his co-workers designated 30 races on 17 differentials. The work in Portugal has demonstrated the existence of 6 dominant genes controlling resistance to the known races (Goujon). The following brief comment is mostly taken from Rodrigues et al. (1975b) who stated that 5 genes for resistance have been identified. Resistance in *C. arabica* (tetraploid; 65% of the cultivated area) is less (and more critical) than that in *C. canephora* (diploid; 33% of the cultivated area). The coffees have been divided into groups depending on their reactions to the races. No *C. arabica* selection resistant to all races is known but *C. canephora* gives some selections that are resistant to all known races (group A). The selection Hybrid de Timor (*arabica* × *canephora*) appeared spontaneously in Timor (Indonesia). It is a self-fertile tetraploid which gives seedlings that have good agronomic characteristics and 95% of which are resistant to all known races. The commonest race is II (from 30 countries) followed by race I (15 countries); the remaining races have a much more limited distribution. In India 12 races (Sreenivasan) are known, 6 of which have not been reported from elsewhere. In Kenya 6 (Thitai et al.), in Timor 9 (4 not reported from elsewhere) and in Brazil 4 races (Ribeiro et al.; Schieber) are known. The problem of control by resistance is mostly confined to *C. arabica*; this problem is much less important in *C. canephora* (and other diploid coffees). Resistance has only played a small role in control in Africa and Asia, because of the incomplete resistance in cvs and the early success of chemical control (Waller).

Ananth, K. C. et al. 1961. *Indian Coff.* **25**: 37 (spread by thrips; **40**, 467).

Andrade, I. P. R. and others. 1972. Coffee Inst. Brazil, 55 pp. (fungicides; **52**, 737–741).

Anon. 1971. Coffee Rust Res. Centre, Oeiras, Portugal, 18 and 29 pp. (resistance; **51**, 4027).

Bakala, J. et al. 1975. *Café–Cacao–Thé* **19**: 291 (in vitro culture; **55**, 4120).

Becker, S. et al. 1975. *Phytopathol. Z.* **82**: 359 (wind dispersal; **55**, 748).

—— ——. 1977. *Z. PflKrankh. PflSchutz* **84**: 526 (rain splash & wind dispersal; **57**, 2135).

——. 1977. *Ibid* **84**: 577, 691 (seasonal & diurnal periodicity; **57**, 2136, 2946).

Bettencourt, A. J. et al. 1968. *Bragantia* **27**: 35 (breeding for resistance; **51**, 1507).

Bock, K. R. 1959. *Kenya Coff.* **24**: 286 (use of captan; **39**: 412).

——. 1962a. *Trans. Br. mycol. Soc.* **45**: 63 (field dispersal; **41**: 600).

——. 1962b. *Ibid* **45**: 289 (seasonal periodicity in Kenya; **42**: 198).

Bowden, J. et al. 1971. *Nature Lond.* **229**: 500 (wind spread across Atlantic; **50**, 2286).

Broderick, H. T. et al. 1975. *Citrus Subtrop. Fr. J.* 502: 5 (fungicides in South Africa; **56**: 719).

Burdekin, D. A. 1960. *Kenya Coff.* **25**: 212 (dispersal; **40**: 169).

——. 1964. *E. Afr. agric. For. J.* **30**: 101 (fungicides; **44**, 1561).

Carmargo, M. H. 1971. *Circ. Inst. Agron. S. Paulo* 8, 176 pp. (bibliography; **52**, 127).

Hemileia vastatrix

Castillo-Zapata, J. et al. 1976. *Cenicafé* **27**: 3 (resistance; **56**, 4059).
Chaves, G. M. et al. 1971. *Seiva* **31**: 120 (fungicides in Brazil; **51**, 4029).
Chinnapa, C. C. et al. 1968. *Caryologia* **21**: 75 (cytology; **47**, 3115).
Crowe, T. J. 1963. *Trans. Br. mycol. Soc.* **46**: 24 (spread by insects; **42**, 684).
D'Oliveira, B. 1965. Coffee Rust Res. Centre, Oeiras, Portugal, 144 pp. (report 1960–65; **45**, 1076).
Figueiredo, P. et al. 1974. *Arq. Inst. biol. S. Paulo* **41**: 47 (biology of some races; **54**, 1283).
—— ——. 1977. *Biológico* **43**: 32 (incubation period & seasonal periodicity; **57**, 4475).
Firman, I .D. 1965a. *Trop. Agric. Trin.* **42**: 111 (review of control, with *Glomerella cingulata*, in Kenya; **44**, 2796).
—— et al. 1965b. *Ann. appl. Biol.* **55**: 123 (fungicides, low volume, in Kenya; **44**, 2145).
——. 1972. *Pl. Prot. Bull. F.A.O.* **20**: 121 (disease in Brazil; **52**, 3701).
Goujon, M. 1971. *Café–Cacao–Thé* **15**: 308 (distribution & biology, review of resistance, 53 ref.; **51**, 4030).
Hocking, D. 1968. *Trans. Br. mycol. Soc.* **51**: 89 (effect of light on uredospore germination & infection; **47**, 1890).
Mayne, W. W. 1930. *Bull. Mysore Coff. Exp. Stn* 4, 16 pp. (seasonal periodicity in India; **10**: 239).
——. 1931. *Ibid* 6, 22 pp. (as above; **11**: 368).
——. 1937. *Ibid* 15, 46 pp. (fungicides; **17**: 31).
——. 1969. *Biologist* **16**: 58 (historical essay, 1869–1969; **48**, 2981).
——. 1971. *Wld Crops* **23**: 206 (review, fungicides in India over last 100 years; **51**, 373).
Monaco, L. C. 1977. *Ann. N.Y. Acad. Sci.* **287**: 57 (consequences of disease in Brazil; **56**, 533).
Montoya, R. H. et al. 1974. *Experientia* **18**: 239 (effect of light & temp. on uredospore germination & infection; **55**, 3178).
Mulinge, S. K. et al. 1974. *Phytopathology* **64**: 147 (effects of pyracarbolid; **53**, 4817).
Musmeci, M. R. et al. 1974. *Ibid* **64**: 71 (uredospore self inhibitor; **53**, 2575).
Nutman, F. J. 1959. *Kenya Coff.* **24**: 451 (dispersal; **39**: 311).
—— et al. 1960. *Trans. Br. mycol. Soc.* **43**: 509 (dispersal; **40**, 106).
—— ——. 1962. *Nature Lond.* **194**: 1296 (as above; **42**, 22).
—— ——. 1963. *Trans. Br. mycol. Soc.* **46**: 27 (biology; **42**, 684).
—— ——. 1970. *PANS* **16**: 606 (review, 32 ref. also in *Kenya Coff.* **36**: 139; **50**, 1799; **52**, 411).
—— ——. 1971. *Ibid* **17**: 385 (on spread to Brazil; **51**, 372).
Orillo, F. T. et al. 1961. *Philipp. Agric.* **45**: 223 (resistance selection; **41**: 600).
Park, P. O. et al. 1964a. *Ann. appl. Biol.* **53**: 133 (fungicide, high & low vol. with Cu; **43**, 2279).
—— ——. 1964b. *Ibid* **54**: 335 (ageing of Cu fungicides; **44**, 1562).
Ragunathan, C. 1924. *Ann. R. bot. Gdns Peradeniya* **8**: 109 (teliospores; 3: 647).
Rajendren, R. B. et al. 1965. *Trans. Br. mycol. Soc.* **48**: 265 (abnormal development of uredia; **44**, 3053).
——. 1967. *Nature Lond.* **213**: 105 (new type of nuclear life cycle; **46**, 1581).
——. 1967. *Mycologia* **59**: 918 (atypical & typical germination of uredinoid teliospores; **47**, 526).
——. 1968. *Mycopathol. Mycol. appl.* **36**: 107 (histology & morphology of haustoria; **49**, 150).
Ramos, H. C. 1970. *Bibliog. 1, Spec. Suppl. Inst. interam. Ciencas Agric.* 108 pp. (bibliography, 1210 items; **50**, 1242).
Rayner, R. W. 1960. *Wld Crops* **12**: 187, 222, 261, 309 (reviews; **40**: 169).
——. 1961. *Ann. appl. Biol.* **49**: 497 (uredospore germination & penetration; **41**: 152).
——. 1967. *Nature Lond.* **215**: 90 (infection; **46**, 3446).
——. 1972. *Publ. Misc. Inst. interam. Ciencas Agricolas O.E.A.* 94, 68 pp. (review, 92 ref.; **52**, 2617).
Razafindramamba, R. 1958. *Rev. Mycol.* **23**: 177 (review, 100 ref.; **38**: 144).
Ribeiro, I. J. A. et al. 1975. *Summa Phytopathol.* **1**: 19 (races in São Paulo, Brazil; **54**, 3940).
Rijkenberg, F. H. J. et al. 1973. *Phytopathology* **63**: 281 (haustoria & intracellular hyphae, with *Puccinia sorghi*; **52**, 3522).
Rijo, L. 1972. *Agron. lusit.* **33**: 427 (histology of hypersensitive reaction; **52**, 4080).
—— et al. 1974. *Can. J. Bot.* **52**, 1363 (fine structure; **54**, 859).
Rodrigues, C. J. et al. 1975a. *Physiol. Pl. Pathol.* **6**: 35 (phytoalexin-like response; **55**, 249).
—— ——. 1975b. *A. Rev. Phytopathol.* **13**: 49 (races & resistance, 78 ref.).
Saccas, A. M. et al. 1971. *Bull. Inst. Fr. Café Cacao* 10: 123 pp. (review, 310 ref.; **51**, 1506).
Schieber, E. 1972. *A. Rev. Phytopathol.* **10**: 491 (economic impact in America, 83 ref.).
——. 1975. *Ibid* **13**: 375 (status in S. America, control).
Shaw, D. E. 1968. *Res. Bull. Dep. Agric. Port Moresby* 2: 20 (disease outbreaks 1892–1965 and 1965 eradication; **47**, 2721).
——. 1970. *Papua New Guin. agric. J.* **22**: 59 (eradication; **50**, 3767).
——. 1975. *Res. Bull. Dep. Agric. Stock Fish. Port Moresby* 13: 33 (as above; **54**, 3939).
Sreenivasan, M. S. 1971. *Indian Coff.* **35**: 421 (races; **52**, 735).
Stevenson, J. A. et al. 1953. *Spec. Publ. Pl. Dis. Surv.* 4, 80 pp. (bibliography; **34**: 367).

Sylvain, P. G. 1955. *Turrialba* **5**: 37 (*Coffea arabica* in Ethiopia).
Thitai, G. N. W. et al. 1977. *Kenya Coff.* **42**: 241 (races; **57**, 1734).
Vishveshwara, S. et al. 1960. *Indian Coff.* **24**: 118 (basidiospore cytology; **39**: 707).
——— ———. 1962. *Phyton* **18**: 75 (abnormalities in teliospore germination; **42**: 123).
Waller, J. M. 1972. *PANS* **18**: 402 (disease in America; **52**: 1148).
Zambolim, L. et al. 1974. *Experientia* **17**: 151 (effects of temp. & RH on uredospore viability; **54**, 5354).

HENDERSONULA Speg., *An. Soc. Cient. Arg.* 10: 160, 1880.

(Sphaerioidaceae)

PYCNIDIA dark brown, spheroid, ampulliform or ovate, unilocular, ostiolate, immersed, irregularly aggregated, outer layer forming a distinct specialised wall; phialides short cylindrical, hyaline. CONIDIA at maturity 2 septate, median cell dark brown, end cells hyaline. CONIDIOGENOUS CELLS enteroblastic, phialidic (from key by B. C. Sutton in *The fungi* (edited by G. C. Ainsworth et al.) Vol. IV A: 560, 1973).

H. toruloidea Nattrass (synonyms: *Torula dimidiata* Penzig; *Exosporina fawcettii* Wilson) is a wound (or weak) pathogen of many tropical and temperate, frequently woody, crops; causing diebacks and cankers. It can also cause fruit and storage rots. The fungus was recently described by Punithalingam et al. who gave a host list: STROMATA black, solitary, up to 1.5 mm, raised, immersed at first, eventually bursting through the surface of the bark; uni- or multilocular, locules globose to laterally compressed or sometimes irregular; locular wall dark brown and composed of many layers of thick-walled cells. CONIDIOPHORES (phialophores) hyaline, simple, arising from the innermost layer of cells lining the locules. CONIDIA ellipsoid to ovoid, initially aseptate, hyaline, becoming brown and 3 celled at maturity, usually $10–14 \times 4–6\ \mu$ with the central cell darker than the end cells. MYCELIUM composed of branched brown hyphae which eventually break up into 1 or 2 celled THALLOSPORES (toruloid state); temp. range for growth in culture 18–39°, opt. 33°.

Calavan, E. C. et al. 1954. *Phytopathology* **44**: 635 (on citrus; **34**: 365).
Giha, O. H. 1975. *Pl. Dis. Reptr* **59**: 899 (on *Ficus benghalensis*; **55**, 4301).
Meredith, D. S. 1963. *Trans. Br. Mycol. Soc.* **46**: 473 (on banana; **43**, 1716).
Nattrass, R. M. 1933. *Ibid* **18**: 189 (*Hendersonula toruloidea* sp. nov.; **13**: 382).
Paxton, J. D. et al. 1965. *Phytopathology* **55**: 21 (on *Juglans regia*; **44**, 1975).
Punithalingam, E. et al. 1970. *C.M.I. Descr. pathog. Fungi Bact.* 274 (*H. toruloidea*).

HYPOXYLON Bull. ex Fr., *Summ. veg. Scand.*: 383, 1849.

(synonymy in Miller)
(Xylariaceae)

STROMA convex, applanate, pulvinate or semiglobose, determinate or widely effused, erumpent from bark or wood, with very little or well-developed entostroma but not zoned, either fleshy leathery, woody or carbonaceous, with a variously coloured or black ectostroma. PERITHECIA embedded in periphery of stroma, semiglobose or angular from mutual pressure, with centrum composed of a basal and wall layer of asci interspersed with filiform paraphyses, and with very fine periphyses in ostiolar canal. ASCI cylindrical, unitunicate, with slight thickening at tip and a wide pore surmounted by a crown. ASCOSPORES 8, aseptate, semiglobose to ellipsoid with smooth wall, light brown to opaque and a longitudinal germ pore. Conidia formed first on exposed ectostroma, later with some spp. developing on any part of old stromata in favourable weather. CONIDIOPHORES branched, hyphomycetous, hyaline to greenish brown. CONIDIA minute, continuous, hyaline, borne apically, 1 to many, becoming lateral by the sympodial growth of the hyphae (from J. H. Miller).

A full taxonomic treatment was given by Miller; Whalley et al. have studied the taxonomy of the British spp.; see Martin on the Xylariaceae and Jong et al. on the conidial states. Hawksworth has recently described 4 *Hypoxylon* spp.: *H. mammatum* (Wahl.) J. H. Miller (synonym: *Sphaeria mammata* Wahl.); *H. mediterraneum* (de Not.) Ces. & de Not. (synonym: *S. mediterranea* de Not.); *H. rubiginosum* (Pers. ex Fr.) Fr. (synonym: *S. rubiginosa* Pers.); and *H. serpens* (Pers. ex Fr.) Kickx (synonym: *S. serpens* Pers.); see Miller and Hawksworth for further synonyms. *Hypoxylon* spp. are wood inhabitants and generally weak pathogens. Most occur on dicotyledons, few on monocotyledons and still fewer on conifers. In general the host ranges are not limited.

Agnihothrudu described *H. rubiginosum* var. *tropica* J. H. Miller on rubber (*Hevea brasiliensis*) and *H. serpens* (q.v.) var. *effusum* (Nits.) J. H. Miller on tea. *H. mediterraneum* causes charcoal disease of *Quercus suber*; it has reached epidemic levels in Algeria, Morocco and Tunisia. This spp. also attacks the bark of several *Eucalyptus* spp. in N. Africa and Portugal. In India it was found to be associated with a dying condition of sal (*Shorea robusta*), the var. *microspora* J. H. Miller. The disease occurred particularly on poor sites and the fungus became pathogenic following loss of tree vigour, hastening death or preventing tree recovery (Bakshi; Boyce et al.). *H. mammatum* (Map 465) occurs on trunks and branches of temperate trees causing cankers, particularly *Populus* spp

Agnihothrudu, V. 1967. *Indian Phytopathol.* **20**: 196 (morphology, *Hypoxylon serpens, H. rubiginosum* & *Tunstallia aculeata*; **47**, 2655).

Bakshi, B. K. 1963. *Indian For.* **89**: 265 (*H. mediterraneum* on *Shorea robusta*; **42**: 578).

Boyce, J. S. et al. 1959. *Ibid* **85**: 575 (as above; **39**: 197).

Greenhalgh, G. N. et al. 1967. *Trans. Br. mycol. Soc.* **50**: 183 (structure of ascus apex in *H. fragiforme*).

—— ——. 1968. *Ibid* **51**: 57 (conidial states in British Xylariaceae).

—— ——. 1970. *Ibid* **55**: 89 (British spp., stromal pigments, substrate & taxonomy).

Hawksworth, D. L. 1972. *C.M.I. Descr. pathog. Fungi Bact.* 356, 357, 358, 359 (*H. mammatum, H. rubiginosum*; *H. serpens*; *H. mediterraneum*).

Jong, S. C. et al. 1972. *Tech. Bull. agric. Exp. Stn Wash. St.* 71, 51 pp. (conidial states; **55**, 5584).

Martin, P. 1967. *Jl S. Afr. Bot.* **33**: 205, 315 (studies in Xylariaceae 1 & 2).

——. 1968. *Ibid* **34**: 153, 303 (as above 3 & 4).

——. 1969. *Ibid* **35**: 149, 267, 393 (as above 5, 6 & 7).

——. 1970. *Ibid* **36**: 73 (as above 8; **49**, 3141).

——. 1976. *Ibid* **42**: 71 (as above).

Miller, J. H. 1961. *A monograph of the world spp. of* Hypoxylon, 158 pp. Univ. Georgia Press (**41**: 584).

Rajagopalan, C. 1965. *Trans. Kans. Acad. Sci.* **68**: 541 (4 spp. wood decay fungi including *H. rubiginosum*; **45**, 2638).

Rogers, J. D. 1977. *Can. J. Bot.* **55**: 2394 (surface of light coloured ascospores; **57**, 1633).

Taligoola, H. K. et al. 1976. *Trans. Br. mycol. Soc.* **67**: 517 (genus in Uganda forests).

Whalley, A. J. S. et al. 1971. *Ibid* **57**: 161 (chemical races of *H. rubiginosum*).

—— ——. 1973. *Ibid* **61**: 435, 455 (numerical taxonomy, cultural & perfect states; key for British spp.; **53**, 2890, 2891).

—— ——. 1975. *Ibid* **64**: 229, 369 (numerical taxonomy; *Hypoxylon* cultural states & *Nodulisporum* spp.; cultural states applanate spp. & other members of the genus; **54**, 3747; **55**, 2564).

——. 1976. *Mycopathologia* **59**: 155 (numerical taxonomy of some spp.; **56**, 1492).

——. 1977. *Ibid* **61**: 99 (stromal pigments & taxonomy; **57**, 1001).

——. 1977. *Bull. Br. mycol. Soc.* **11**: 45 (key to British spp.; **56**, 4867).

Hypoxylon serpens (Pers. ex Fr.) Kickx, *Flore cryptogamique des environs de Louvain*: 115, 1835.

stat. conid. *Geniculosporium serpens* Chesters & Greenhalgh, 1964.

(for synonyms see Miller, *Hypoxylon* q.v., and Hawksworth)

STROMATA superficial or erumpent through the bark, discrete to effuse, often forming extensive patches, 1–1.5 mm thick, surface whitish grey at first, becoming bronze and finally dark purplish brown to black when mature, smooth but with conspicuous projecting perithecia, matt to slightly nitid, carbonaceous. PERITHECIA projecting from the stroma, usually numerous, globose to ovate, 0.5–1 mm diam., rarely to 1.5 mm tall, ostioles papillate. ASCI subcylindrical to cylindrical, stalked, 8 spored, $65–130 \times 5–10$ μ, stalk 30–70 μ tall. PARAPHYSES numerous but gelatinising at maturity, simple, filiform, sparsely septate, 2–4 μ wide. ASCOSPORES uniseriate, inequilaterally ellipsoid, brown to dark brown, with a longitudinal furrow which may be indistinct, $10–17 \times 5–8$ μ. STAT. CONID. forming greyish patches up to 0.3 mm tall on decorticate wood. CULTURES on malt agar growing moderately slowly at 25°, whitish grey to brownish above, partly submerged and aerially weakly flocculose. CONIDIOPHORES pale brownish below but becoming hyaline above, irregularly sparsely branched, to 320 μ tall, 2–4.5 μ diam. CONIDIA arising on geniculations from the conidiogenous cells, hyaline to very pale fuscous, simple, obovoid to ellipsoid with a minute basal frill, $2.5–6 \times 1.5–3$ μ. On wood of many genera of dicotyledonous trees, including decorticate logs and stumps. In Europe it is particularly common on *Fagus sylvatica* (D. L. Hawksworth, *CMI Descr.* 358, 1972).

H. serpens causes a wood rot of tea in S. India.

This disease can be serious and removal of infected tissue by pruning has been recommended (Venkata Ram 1968 & 1971). Pruning must be severe and properly timed (Apr.–May) in India. If all traces of the fungus are not removed in this way it will continue to spread in the bush (Venkata Ram 1967). Sarmah in *Diseases of tea and associated crops in north-east India* described tarry root rot as one of the primary fungal diseases of tea. Agnihothrudu later identified the cause of this disease as *Hypoxylon nummularium* Bull. ex Fr. The fungus and its vars were described by Miller (*Hypoxylon* q.v.). Tarry root rot occurs on mature bushes and infection causes them to die suddenly. A black, smooth, hard, effused stroma forms on the stem at ground level, spreading upwards; abnormal callus growth may form. No external symptoms occur on the roots but in the wood there are brown reticulations and thin black lines. The control measures are the general ones applied to primary root diseases of tea in N.E. India, see Sarmah (l.c.).

Agnihothrudu, V. 1964. *Rep. Tocklai Exp. Stn Indian Tea Assoc.* for 1963: 64 (*Hypoxylon nummularium* inter alia; **44**, 3142).

Kenerley, C. M. et al. 1976. *Mycologia* **68**: 688 (in culture; **56**, 1443).

Stiers, D. L. 1977. *Cytologia* **42**: 697 (fine structure of ascus apex inter alia; **57**, 1009).

Venkata Ram, C. S. 1967. *Rep. Tea scient. Dep. United Plrs' Assoc. S. India* for 1966–67: 23 (*H. serpens*; **47**, 621b).

——. 1968. *Ibid* for 1967–68: 35 (*H. serpens*; **48**, 587a).

——. 1969. *Ibid* for Apr.–Dec. 1968: 21 (*H. serpens*; **49**, 2170c).

——. 1971. *Bull. United Plrs' Assoc. S. India* 28: 6 (*H. serpens* & Zn deficiency; **51**, 1868a).

KABATIELLA Bubák, *Hedwigia* 46: 297, 1907. (Melanconiaceae)

ACERVULI subcuticular, subepidermal or subperidermal, with or without subicle or setae; stromatic tissue restricted to the base of the fructification, forming conidia on the upper surface; dehiscing by regular or irregular splitting of the overlying host tissues; acervuli and phialides hyaline. CONIDIOGENOUS CELLS consistently polyphialidic, enteroblastic. CONIDIA hyaline, aseptate, oval to pyriform, formed more or less synchronously from 2–8 apertures (from key by B. C. Sutton in *The fungi* (edited by G. C. Ainsworth et al.) Vol. IVA: 555, 1973).

Hermanides-Nijhof listed *Kabatiella* as a synonym of *Aureobasidium* Viala & Boyer which was described. *K. zeae* Narita & Y. Hiratsuka causes eyespot of maize; CONIDIOPHORES 10–15 × 4–6 μ, CONIDIA 16.2–47.5 (av. 32.5) × 2–3.5 (av. 2.6) μ. Dingley transferred the fungus to *Aureobasidium* and gave a description of its morphology. The disease has been reported from: Argentina, Austria, Canada, China, France, Germany (W.), Japan (Hokkaido), New Zealand, USA (N. & central) and Yugoslavia (Map 506). Symptoms occur mostly on the leaves: round to oval 1–4 mm diam. spots with a translucent, pale yellow halo, becoming pale cream to tan in the centre and surrounded by a brown or brownish purple ring; they can be numerous enough to become coalescent and form large necrotic areas. Less well defined, larger spots may occur on the leaf sheaths and cob husks. The fungus shows both good growth and conidial formation on PDA at 20–26°. Little is known of the biology of *K. zeae*. Although it has caused damage in USA (reducing yield and quality in maize) it appears to be a relatively minor disease. It is probably one of cool conditions.

K. caulivora (Kirchn.) Karak. causes scorch of clover (*Trifolium* spp.; frequently called northern anthracnose) and infects lucerne (*Medicago sativa*). Hermanides-Nijhof placed this name as a synonym of *A. caulivorum* (Kirchn.) W. B. Cooke (synonym: *Gloeosporium caulivorum* Kirchner; see Cooke). The fungus was compared with *Colletotrichum trifolii* causing southern anthracnose (*Colletotrichum* q.v., Massenot et al.; Sampson). *C. trifolii* has straight conidia; those of *K. caulivora* are sickle shaped and smaller (12–22 × 3.5–5 μ) than those of *K. zeae*. Witkowska and Beale gave general accounts of clover scorch. Necrotic lesions are caused on stems, petioles and flowering shoots; the stems break and the leaves and flowers wither. Penetration via the conidia is epidermal (min. 3 hour moisture required for infection); the disease is most severe at 20–24° and symptoms are absent at 28°. The seed (testa) is penetrated and infection spreads to the cotyledons to cause damping off which is rapid at 18.4° and slower at 29.5°. *K. caulivora* was recovered from diseased plant debris after *c.* 13 months and was still viable in clover seed, stored at room temp., after 2.25 years (Cole et al.; Darunday et al.; Leach). Helms described the development of infection in *T. subterraneum* plants of differing age; and see Chatel et al. for var. variation in resistance in the same crop.

Arny, D. C. et al. 1971. *Phytopathology* **61**: 54 (*Kabatiella zeae* in USA; **50**, 3697).
Beale, P. 1976. *J. agric. S. Aust.* **79**: 49 (*K. caulivora*, progress report; **56**, 1644).
Berkenkamp, B. 1969. *Can. J. Bot.* **47**: 453 (viability of *K. caulivora* **48**, 1786).
Chatel, D. L. et al. 1973. *Tech. Bull. Dep. Agric. W. Aust.* 17, 11 pp. (*K. caulivora*, resistance in *Trifolium subterraneum*).
Cole, H. et al. 1958. *Phytopathology* **48**: 326 (*K. caulivora*, general; **37**: 731).
Cooke, W. B. 1962. *Mycopathol. Mycol. appl.* **17**: 1 (taxonomy in the 'black yeasts', *Aureobasidium* inter alia; **42**: 3).
Darunday, Z. D. et al. 1967. *Crop Sci.* **7**: 613 (*K. caulivora*, factors affecting disease; **48**, 3025).
Dingley, J. M. 1973. *N.Z. Jl agric. Res.* **16**: 325 (*K. zeae*, general; **53**, 513).
Helms, K. 1975. *Phytopathology* **65**: 197 (*K. caulivora* infection & host age; **54**, 5466).
Hermanides-Nijhof, E. J. 1977. *Stud. Mycol.* 15: 141 (*Aureobasidium* & allied genera).
Leach, C. M. 1962. *Phytopathology* **52**: 1184 (*K. caulivora*, seedborne, **42**: 324).
Narita, T. et al. 1959. *Ann. phytopathol. Soc. Japan* **24**: 147 (*K. zeae* sp. nov.; **39**, 304).
——— ———. 1959. *Bull. Hokkaido natn. agric. Exp. Stn* 4: 71 (*K. zeae*, general).
Sakuma, T. et al. 1970. *Res. Bull. Hokkaido natn. agric. Exp. Stn* 96: 96 (*K. caulivora*, factors affecting conidial germination; **50**, 3007).
——— ———. 1970. *Ann. phytopathol. Soc. Japan* **36**: 250 (*K. caulivora*, histology; **50**, 1855).
Schneider, R. et al. 1972. *Phytopathol. Z.* **74**: 238 (*K. zeae*, general; **52**, 709).
Witkowska, A. 1974. *Biul. Inst. Hodowl. Aklim. Rośl.* 5–6: 45 (*K. caulivora*, review, 52 ref.: **56**, 2113).

KHUSKIA Hudson, *Trans. Br. mycol. Soc.* **46**: 358, 1963.
(Polystigmataceae)

PERITHECIA clustered, black, globose or ovoid, ostiole papillate, erumpent. ASCI clavate, short stalked, unitunicate, 8 spored. PARAPHYSES septate, thin walled, longer than the asci. ASCOSPORES hyaline, granular, curved, inequilateral, tapering towards base, irregularly biseriate, aseptate; on germination becoming 1 septate, the 2 cells unequal (from H. J. Hudson l.c.).

Khuskia oryzae Hudson, *Trans. Br. mycol. Soc.* **46**: 358, 1963.
stat. conid. *Nigrospora oryzae* (Berk. & Br.) Petch, 1924
Monotospora oryzae Berk. & Br., 1873
Basisporium gallarum Moll., 1902.

PERITHECIA formed in clusters of 1–7, in uniseriate or irregular rows, up to 2 mm long, subepidermal, erumpent, globose or ovoid, up to 250 μ diam. with papillate ostioles. Around each perithecial group is a blackened area of host tissue. ASCI short stalked, unitunicate, clavate, 55–75 × 8.5–12 μ, with 8 biseriate spores; thin walled, septate paraphyses present. ASCOSPORES hyaline, granular, curved, inequilateral, 16–21 × 5–7 μ tapering to the base, with rounded ends, initially aseptate but on discharge from the ascus and germination develop a single transverse septum dividing the spore unequally into 2 cells. CONIDIA smoky brown or jet black, globose or subglobose, 10–16 × 10–13 μ, commonly 12–14 μ, produced singly and terminally on short, pale brown, inflated, simple conidiophores (A. Sivanesan & P. Holliday, *CMI Descr.* 311, 1971).

A saprophyte, or at most a weak parasite, widely distributed in warmer regions and found particularly on debris of monocotyledons. It may be considered plurivorous (especially on monocotyledons) but has been most reported as attacking maize, *Sorghum* spp. and rice. In USA it is one of the lesser pathogens concerned in the stem rot complex in maize. On this host symptoms develop towards maturity, mostly on the shanks, husks and ears but also on the stems and stalks where blackish, shallow lesions can occur. Ears may snap off at harvest; the cob becomes shredded and rotten through disintegration of the parenchyma; sparse mycelium and dark sporulation develop in the furrows between the kernels and on the seed itself.

In vitro growth and conidial germination are opt. at 25° or less; good growth takes place without vitamins. Previous insect attack can facilitate infection. In Jamaican banana fields *N. oryzae* and *N. sphaerica* (q.v.) showed a diurnal periodicity with a peak at 0800–1000 hours. Mason (1927, *Nigrospora* q.v.) and Hudson considered that these 2 spp. could be separated on the basis of conidial size but Standen (1943), studying isolates from maize, did not consider this to be justified. The fungus causes discolouration and damage to seed but is not strictly seedborne. The rot caused by *K. oryzae* is not likely to be of importance unless the host is damaged or debilitated in any way, and a favourable growing environment with good seed and culture should prevent attacks on cereals.

Alfaro, A. 1946. *Bol. Pat. Veg. y Eut. Agr. Madrid* **16**: 321 (associated with damage by *Siteroptes graminum* on wheat; **26**: 388).

Durrell, L. W. 1925. *Res. Bull. Iowa agric. Exp. Stn* 84: 139 (general & conidial germination, maize; **5**: 86).

Focke, I. 1964. *NachrBl. dt. PflSchutzdienst Berl.* **18**: 4 (symptoms on maize; **43**, 3205).

Hudson, H. J. 1963. *Trans. Br. mycol. Soc.* **46**: 355 (*Khuskia* & *K. oryzae*, gen. & sp. nov.; **43**, 945).

Kulik, T. A. 1963. *Zashch. Rast. Mosk.* **8**: 16 (on sorghum; **43**, 727).

Laemmlen, F. F. et al. 1973. *Phytopathology* **63**: 308 (association with *S. reniformis* in cotton; **52**, 3707).

Lilly, V. G. et al. 1949. *Proc. W. Va Acad. Sci.* **19**: 27 (vitamins & growth in vitro; **28**: 478).

Meredith, D. S. 1961. *J. gen. Microbiol.* **26**: 343 (diurnal periodicity of conidia; **41**: 242).

Mohamed, H. A. et al. 1965. *Pl. Dis. Reptr* **49**: 244 (on maize in Egypt; **44**: 2123).

Molliard, M. 1902. *Bull. Soc. mycol. Fr.* **18**: 167 (morphology of stat. conid.).

Penčić, V. 1969. *Arh. poljopr. Nanke Teh.* **22**: 118 (on maize; **50**: 2904).

Petch, T. 1924. *J. Indian bot. Soc.* **4**: 21 (morphology of stat. conid.; **3**: 610).

Săvulescu, T. et al. 1932. *Phytopathol. Z.* **5**: 153 (on maize in Rumania; **12**: 20).

——— ———. 1933. *Annls Inst. Rech. Agron. Roumanie* **5**: 3 (as above; **13**: 571).

Standen, J. H. 1939, 1944 & 1945. *Phytopathology* **29**: 656; **34**: 315, **35**: 552 (on maize in USA; **18**: 734; **23**: 339; **24**: 499).

——. 1943. *Iowa St. Coll. J. Sci.* **17**: 263 (morphology; **23**: 383).

LEPTOSPHAERIA Ces. & de Not., *Comment. Soc. Crittogam. Ital.* 1: 234, 1863.

(see L. Holm on nomenclature, *Taxon* **24**: 480, 1975)

(Pleosporaceae)

PSEUDOTHECIA scattered, immersed or becoming superficial by shedding of host tissue, black, globose to conical with well developed ostioles; wall of a scleroplectenchyma of isodiametric, thick walled cells, stout, often strongly thickened laterally at the base; pseudothecial beak short or lacking. ASCI thick walled, bitunicate, 8 spored. ASCOSPORES usually somewhat fusiform, 2 to many transverse septa, hyaline at first, later yellowish to yellowish brown or brown. Mostly on stems of herbaceous dicotyledons (after R. W. G. Dennis, *British ascomycetes*: 389, 1968; and key by E. S. Luttrell in *The fungi* (edited by G. C. Ainsworth et al.) Vol. IVA: 189, 1973; q.v. both for taxonomic notes and ref.).

L. taiwanensis Yen & Chi (sugarcane leaf blight) is one of several *Leptosphaeria* spp. described from *Saccharum* (see Sivanesan). The ascospores are 3 (or rarely 4) septate, distinctly constricted at the septa dark brown when mature, 39–46 × 6.6–12.5 μ. *L. michotii* (q.v.) and *L. sacchari* (q.v.) on this host have 2 septate ascospores, 14–23 × 3.5–6 μ; and 3 septate ascospores, 18–25 × 3.5–6 μ, respectively. *L. taiwanensis* has also been reported to have an imperfect state, *Cercospora taiwanensis* Matsumoto & Yamamoto. The first symptoms, on the immature leaves in the spindle, are small, narrow, elliptical or elongate, yellowish spots, dotted red, and seen on both leaf surfaces. The spots became elongate, reddish brown, streaks which often coalesce into bands; the leaves die. The pseudothecia develop in the margins of old lesions. The disease has only been studied in Taiwan. Good growth occurs on PDA on maize meal media at 25–30°. Some isolates develop the perfect state in culture. The best temps for ascospore germination are 26–30°; at 26° there was no germination at 90% RH or less. Sugarcane cvs differ widely in their reaction to infection (Leu et al.); in Taiwan spraying with mineral oil reduced the number of leaf lesions and sanitary measures (removal of infected leaves from fields used as sources of planting material) have been recommended.

Several *Leptosphaeria* spp. have been described from banana, coffee, maize, rice and sorghum but none appear to be of any economic importance. *L. pratensis* (*Stagonospora* q.v.) causes a disease on lucerne and other legumes. *L. narmari* J. Walker & A. M. Smith and *L. korrae* J. Walker & A. M. Smith were described as root and stolon parasites of grasses in Australia, including Bermuda grass (*Cynodon dactylon*). *L. elaeidis* Booth & Robertson from oil palm has an imperfect state in *Pestalotiopsis*.

Booth, C. et al. 1961. *Trans. Br. mycol. Soc.* **44**: 24 (*Leptosphaeria elaeidis* sp. nov.; **40**: 554).

Hsieh, W. H. et al. 1970 & 1971. *Pl. Prot. Bull. Taiwan* **12**: 152; **13**: 127 (ascospore formation & germination in *L. taiwanensis*; **51**, 2815, 2816).

Leu, L. S. et al. 1970. *Ibid* **12**: 59 (control; **51**, 2814).
——. 1971. *Sugcane Pathol. Newsl.* 7: 4, 25 (resistance to *L. taiwanensis*; **51**, 2817, 2818).
——. 1974. *Rep. Taiwan Sug. Exp. Stn* 63: 59 (as above; **54**, 3462).
——. 1976. *Ibid* 71: 45 (as above; **56**, 2650).
Lucas, M. T. et al. 1967. *Trans. Br. mycol. Soc.* **50**: 85 (stat. conid. of British spp. of *Leptosphaeria*; **46**, 2192).
Matsumoto, T. et al. 1934. *J. Soc. trop. Agric. Taiwan* **6**: 584 (*Cercospora taiwanensis* sp. nov. inter alia; **14**: 395).
Müller, E. 1950. *Sydowia* **4**: 185 (genus in Switzerland; **33**: 321).
Sivanesan, A. 1976. *C.M.I. Descr. pathog. Fungi Bact.* 506 (*L. taiwanensis*).
Walker, J. et al. 1972. *Trans. Br. mycol. Soc.* **58**: 459 (*L. narmari* & *L. korrae* spp. nov.).
Yen, W. Y. et al. 1954. *Proc. int. Soc. Sug. Cane Technol. 1953* 8: 939 (*L. taiwanensis* sp. nov. & life history; **35**: 634).

Leptosphaeria maculans (Desm.) Ces. & de Not., *Comment. Soc. Crittogam. Ital.* **1**: 235, 1863.
Sphaeria maculans Desm., 1846
Pleospora maculans (Desm.) Tul., 1863
stat. conid. *Phoma lingam* (Tode ex Fr.) Desm., 1849
S. lingam Tode, 1791
S. lingam Tode ex Fr., 1823
Phyllosticta brassicae Westend., 1851
Phoma brassicae (Thüm.) Sacc., 1884
P. lingam var. *napobrassicae* (Rostrup) Grove, 1935.
(for full nomenclatural synonyms see Boerema et al., *Persoonia* **3**: 20, 1964)

ASCOCARPS on stems and leaves, immersed, becoming erumpent, globose, black with protruding ostioles, 300–500 μ diam. ASCI cylindrical to clavate, sessile or short stipitate, 8 spored, 80–125 × 15–22 μ; bitunicate. ASCOSPORES biseriate, cylindrical to ellipsoidal, ends mostly rounded yellow brown, slightly or not constricted at the central septum, guttulate, 35–70 × 5–8 μ. Pseudoparaphyses filiform, hyaline and septate. PYCNIDIA on stems and leaves of 2 types; type I (sclerotioid form) immersed, becoming erumpent, gregarious, variable in shape, convex, soon becoming sunken and concave, sometimes without any definite shape, with narrow ostioles (or pores), 200–500 μ across; wall composed of several layers of thick-walled cells (sclerenchymatous). Type II globose, black, 200–600 μ diam.; wall composed of several layers of cells, thick walled on the outermost layer. CONIDIA hyaline, shortly cylindrical, mostly straight, some curved, guttulate, with one guttule at each end of the conidium, aseptate, 3–5 × 1.5–2 μ. On *Brassica* spp. (E. Punithalingam & P. Holliday, *CMI Descr.* 331, 1972).

The fungus causes canker, dry rot or black leg of *Brassica* spp. Several cruciferous genera are attacked and the pathogen is very widespread, mostly in temperate and subtropical regions (Map 73). There is a monograph by Ndimande. The early symptoms are seen on seedlings as pale lesions on the stem, cotyledons and first true leaves. These become greyish with the pycnidia developing in their centres. On older plants in the field the lesions on above ground parts often have purplish margins. The stem, bulb and roots are attacked causing necrotic, girdling cankers and transverse splits; severe infection of this type leads to wilt or the plant toppling over. Pycnidia develop abundantly on all infected parts but the perfect state is rarer and does not appear to have been reported from the tropics. The disease is a major one in Australia and France (McGee et al. 1978).

Black leg usually develops under cool conditions and is, therefore, only likely to be severe in the tropics at high altitudes. A storage rot can be caused (Hoof). *L. maculans* invades the seeds forming a dormant mycelium beneath the seed coat. Neergaard reported a longevity of 3 years and 8 months (and see Hoof). Petrie et al. (1974) described seed infestation on *Cheiranthus cheiri* (wallflower), *Raphanus sativus* var. *oleifera, Sisymbrium altissimum* and *Thlaspi arvense.* Barbetti (1978) considered that *R. raphanistrum* (wild radish) could be a reservoir for the fungus in W. Australia. In the field 19% of plants developed cankers when seed with a 5.9% infection rate was planted (Wood et al. 1977a; see also Allen et al.; Lloyd). Airborne ascospores also introduce the disease and a persistence of 3 years in soil organic matter occurs (Snyder et al.). Wood et al. (1977b) found more infection of rape seedlings with ascospores compared with conidia. McGee and McGee et al. (1977) found that ascospores from crop residues were a major source of infection. Crops grown on, or near to, fields which had grown rapeseed in the previous year could be failures through infection by *L. maculans.* Barbetti (1976) found little spread of infection when plants which were 8.5 weeks old were inoculated, compared with those inoculated at earlier ages.

No races have been differentiated. Petrie et al. (1974) referred to the 3 major strs of the fungus. Thurling et al. investigated the responses of 53 cvs to infection by 3 different populations of the pathogen. There was variation in disease development both between cvs and populations. McGee et al. (1978) isolated virulent and avirulent strs from rapeseed plants. A different str. occurred in *T. arvense*. Petrie (1978) described a highly virulent str. in Canada.

The production of disease free seed, careful handling in the seedbed, and a rotation (3–4 years) are recommended. The standard hot water treatment for seed (25–30 minutes at 50°) is sometimes unreliable. A thiram soak has been found effective and complete control in seed was given by germisan; also 24 hrs in 0.2% benomyl and thiabendazole is effective (Jacobson et al.; Neergaard). Brunin (1972b) found dusting with benomyl to be effective but Brown et al. reported poor control with this fungicide (see also Gabrielson et al.).

Alabouvette, C. et al. 1970. *Annls Phytopathol.* **2**: 463 (plant residues & ascospore dispersal; **50**, 2027a).

Allen, J. D. et al. 1961. *N.Z. Jl agric. Res.* **4**: 676 (effect of seed infection; **42**: 62).

Bakel, J. M. M. Van. 1968. *Meded. Proefstn Groenteteelt volle Grand* 41 (general; **47**, 1996).

Barbetti, M. J. 1975. *Aust. J. exp. Agric. Anim. Husb.* **15**: 705 (effects of temp. on disease development: **55**, 2943).

——. 1976. *Ibid* **16**: 911 (role of conidia in spread; **56**, 3760).

——. 1978. *APPS Newsl.* **7**: 3 (infection of rape & cruciferous weeds; **57**, 4673).

Bousquet, J. F. et al. 1977. *C. r. hebd. Séanc. Acad. Sci. Paris* Ser. D **284**: 927 (toxin; **57**, 1468).

Brown, A. G. P. et al. 1976. *Aust. J. exp. Agric. Anim. Husb.* **16**: 276 (use of benomyl; **55**, 5390).

Brunin, B. et al. 1970. *Annls Phytopathol.* **2**: 477 (infection by ascospores; **50**, 2027b).

——. 1972a. *Ibid* **4**: 87 (histology; **51**, 4451).

——. 1972b. *Phytiat. Phytopharm.* **21**: 143 (use of benomyl; **52**, 3107).

Calvert, O. H. et al. 1949. *Phytopathology* **39**: 848 (mycelial strs & stimulation of pycnidial production; **29**: 282).

—— ——. 1949. *J. agric. Res.* **78**: 571 (variation in vitro; **29**: 190).

Clayton, E. E. 1927. *Bull. N.Y. agric. Exp. Stn* 550 (general; **7**: 610).

Cunningham, G. H. 1927. *Bull. N.Z. Dep. Agric.* 133 (general; **7**: 70).

Gabrielson, R. L. et al. 1977. *Pl. Dis. Reptr* **61**: 118 (control in seed; **56**, 5251).

Henderson, M. P. 1918. *Phytopathology* **8**: 379 (historical & general).

Hoof, H. A. Van. 1959. *Meded. Dir. Tuinb.* **22**: 256 (general & disease in storage; **39**: 358).

Jacobson, B. J. et al. 1971. *Pl. Dis. Reptr* **55**: 934 (control in seed; **51**, 811).

Lloyd, A. B. 1959. *N.Z. Jl agric. Res.* **2**: 649 (transmission in seed; **39**: 200).

McGee, D. C. et al. 1977. *Aust. J. agric. Res.* **28**: 47 (crop loss & factors affecting disease; **56**, 5252).

——. 1977. *Ibid* **28**: 53 (infection sources, inoculum, environment & disease; **56**, 5253).

—— ——. 1978. *Phytopathology* **68**: 625 (variation in pathogenicity).

Ndimande, B. 1976. Agric. Coll. Sweden, Uppsala. 130 pp. (monograph, 48 ref.; **56**, 2262).

Neergaard, P. 1969. *Friesia* **9**: 167 (survey of occurrence in seed & control; **48**, 2641).

Petrie, G. A. et al. 1968. *Can. J. Bot.* **46**: 869 (perfect state on *Thlaspi arvense*; **48**, 51).

——. 1973. *Can. Pl. Dis. Surv.* **53**: 26 (herbicide damage & infection; **53**, 1601).

—— ——. 1974. *Ibid* **54**: 119 (infestation of crucifer seed; **54**, 5609).

——. 1978. *Ibid* **58**: 21 (a highly virulent str.).

Pound, G. S. 1947. *J. agric. Res.* **75**: 113 (variability, mainly in vitro; **27**: 305).

Smith, H. C. et al. 1964. *Trans. Br. mycol. Soc.* **47**: 159 (on the perfect state; **44**, 65).

Snyder, W. C. et al. 1950. *Pl. Dis. Reptr* **34**: 21 (long persistence in soil & host debris; **29**: 448).

Thurling, N. et al. 1977. *Aust. J. exp. Agric. Anim. Husb.* **17**: 445 (infection of cvs by pathogen populations; **57**, 832).

Wood, P. McR. et al. 1977a. *Ibid* **17**: 1040 (seed infection & spread; **57**, 4184).

—— ——. 1977b. *J. Aust. Inst. agric. Sci.* **43**: 79 (seedling infection with ascospores & conidia).

Leptosphaeria michotii (West.) Sacc., *Fung. ital.* tab. 279, 1878.

Sphaeria michotii West., 1863
Sphaerella michotii (West.) Auersw., 1869
Sclerópleella michotii (West.) Hohnel, 1920
Leptosphaeria trimera Sacc., 1875
Didymosphaeria taiwanensis Yen & Chi, 1954.

ASCOCARPS immersed, globose, dark brown, with a papillate ostiole, 100–200 μ diam. ASCI cylindrical to clavate, sessile, 8 spored, 55–100 × 12–15 μ; ascus wall thick, bitunicate. ASCOSPORES biseriate, cylindrical, light brown to brown, 2 septate (very occasionally 3), constricted at the lower septum giving the cell below an obovate shape, minutely verruculose when mature, 14–23 × 3.5–6 μ. Pseudoparaphyses filiform, hyaline, septate. PYCNIDIA immersed, globose, pale brown, epiphyllous, up to 200 μ diam., wall composed of thin walled, pseudoparenchymatic cells. CONIDIA hyaline at first then becoming brown, aseptate, oval to elliptical, 5–13.5 × 3–5 μ.

Hudson noted the occurrence of 1 septate conidia referable to the form genus *Microdiplodia* and the aseptate *Coniothyrium* type. Webster matched the stat. conid. with the material on which the name *C. scirpi* Trail is based. Easily distinguished from *Leptosphaeria sacchari* (q.v.) and *L. eustomoides*, which also occur on *Saccharum*, by its brown, 2 septate ascospores (G. Morgan-Jones, *CMI Descr.* 144, 1967).

This minor disease (leaf blast) of sugarcane was reported as new from Taiwan in 1954. It is, however, widespread; the fungus occurs on many genera of monocotyledons of tropical and temperate plants (see *CMI Descr.* 144). It has been recorded on rice in Japan, Korea and Uganda. Yellowish, elongate, narrow lesions are formed on the leaves parallel to the veins; generally 3–25 × 0.5–1 mm, becoming purple red. Coalescence may occur and may result in the whole leaf becoming reddish and withering. In vitro growth is opt. at 24–26°. No control measures appear to have been recommended.

Booth, C. 1960. *Commonw. phytopathol. News* **6**: 44 (on sugarcane in Kenya and *Coniothyrium* state).

Hudson, H. J. 1962. *Trans. Br. mycol. Soc.* **45**: 395 (morphology & in fungal succession on sugarcane; **42**: 146).

Webster, J. 1955. *Ibid* **38**: 347 (morphology, taxonomy & *Coniothyrium* state).

Yen, W. Y. et al. 1954. *J. Sug. Cane Res. Taiwan* **8**: 83 (on sugarcane; **36**: 729).

———. 1964. In *Sugarcane diseases of the world* Vol. 2: 29 (general).

Leptosphaeria sacchari van Breda de Haan, *Meded. Proefstat. Suikerr. W. Java* 3: 25, 1892.

ASCOCARPS immersed, globose pale brown, with a protruding papillate ostiole, 100–140 μ diam. ASCI cylindrical to clavate, short stipitate or sessile, 8 spored, 60–72 × 9–13 μ; ascus wall thick, bitunicate. ASCOSPORES irregularly biseriate, ellipsoid, ends broadly obtuse, hyaline or subhyaline, 3 septate, guttulate, second cell slightly swollen, 18–25 × 3.5–6 μ. Pseudoparaphyses filiform, hyaline, septate. PYCNIDIA immersed globose, pale brown, epiphyllous, up to 100 μ diam. wall composed of thin walled pseudoparenchymatic cells, slightly thicker walled around the ostiole. CONIDIA hyaline, pale brown in mass, ellipsoid to sub-fusoid, guttulate, aseptate, 9–14 × 3–5 μ.

The name *Phyllosticta saccharicola* P. Henn. probably refers to this fungus although the dimensions given for the conidial length by that author are larger. *P. sorghina* Sacc. (syn. *P. sacchari* Speg.) has smaller conidia. *Leptosphaeria eustomoides* Sacc., which also occurs on sugarcane, can be distinguished from *L. sacchari* by its narrower, yellowish brown, non-guttulate ascospores, the end cells of which are more pointed. *L. spegazzinia* Sacc. & Syd. (synonym: *L. sacchari* Speg.) also occurs on the same host and has 3 septate ascospores, constricted at the centre, pale olivaceous, 25 × 5–6 μ (Hudson 1960; G. Morgan-Jones, *CMI Descr.* 145, 1967).

This fungus (widespread, Map 330) is associated with ringspot of sugarcane. Although described a considerable time ago, and now generally considered to be the primary cause of ringspot, adequate proof is lacking. Besides the fungi already mentioned *Drechslera sacchari* (q.v.) also seems to be associated with *L. sacchari*. Losses from this disease appear to be slight but it should not be overlooked in any general breeding programme for resistance. The leaf lesions are fairly characteristic; at first small and purplish, though they have been described as dark green. They are usually elongate, but may be spherical, with a chlorotic border. The spots may coalesce and their centres usually become pale as they dry out from a reddish brown. The margin with uninfected tissue is well defined. The lesions may be up to 1.5 cm long. Fructifications will be found on the older spots. Little is known of the life history.

Abbott, E. V. 1964. In *Sugarcane diseases of the world* Vol. 2: 53 (general).

Bourne, B. A. 1934. *Tech. Bull. Fla Exp. Stn* 267, 76 pp. (general; **14**: 57).

Hudson, H. J. 1960. *Trans. Br. mycol. Soc.* **43**: 607 (*L. sacchari* & other *Leptosphaeria* spp. on sugarcane, morphology & taxonomy; **40**: 429).

Luc, M. 1953. *Rev. Mycol.* **18**, Suppl. Colon. 2(10): 1 (morphology; **34**: 64).

LEPTOSPHAERULINA McAlpine, *Fungus diseases of stone-fruit trees in Australia and their treatment*: 103, 1902.

Pseudoplea Höhnel, 1918.

(Pseudosphaeriaceae)

ASCOCARPS small, spherical, parenchymatous, membranous, pale brown, immersed ascostromata erumpent at apex and opening by a broad pore in a more or less raised collar or short neck of darker brown cells, completely filled with a few large saccate asci. ASCI thick walled, bitunicate, arising individually and successively within the centrum parenchyma, remaining more or less separated by this parenchymatous tissue at maturity. ASCOSPORES irregularly clustered in the ascus, extremely variable in size, shape, septation and colour; oblong, ellipsoid or short cylindrical, pharagmosporous or muriform, 3 or more transverse septa and 0 to several longitudinal septa, surrounded by a thin gelatinous sheath, typically hyaline but often becoming brown. Parasites or saprophytes on leaves and herbaceous stems (J. H. Graham & E. S. Luttrell, *Phytopathology* **51**: 681, 1961).

Graham et al. recognised and described 6 spp. on legumes. Booth et al. (*L. trifolii* q.v.) placed 4 of these spp. in synonymy with *L. trifolii*, i.e. *L. australis, L. arachidicola, L. argentinensis* and *L. briosiana*. These workers should be consulted for the extensive synonymy.

Graham, J. H. et al. 1961. *Phytopathology* **51**: 680 (on legumes; **41**: 310).

Karan, D. 1964. *Mycopathol. Mycol. appl.* **24**: 85 (new plant hosts from India; **44**, 1813).

—— et al. 1968. *Ibid* **35**: 193 (as above; **47**, 3642).

Wehmyer, L. E. 1958. *Sydowia* **12**: 490 (*Leptosphaerulina* McAlp. antedates *Pseudoplea* Höhn.).

Leptosphaerulina trifolii (Rostrup) Petrak, *Sydowia* **13**: 76, 1959.

Sphaerulina trifolii Rost., 1899

Pseudoplea trifolii (Rost.) Petr., 1921

Pleospora trifolii (Rost.) Petr., 1927

Pseudosphaeria trifolii (Rost.) Höhn., 1929

Leptosphaerulina briosiana (Poll.) Graham & Luttrell, 1961

L. arachidicola Yen, Cheng & Huang, 1956

L. australis McAlp., 1902

L. vignae Tehon & Stout, 1928

L. argentinensis (Speg.) Graham & Luttrell, 1961.

(see *CMI Descr.* 146 for further synonymy)

PSEUDOTHECIA immersed in leaf tissues, globose, erumpent at the apex and opening by a broad pore, 120–200 μ diam., wall membranous, pale brown, composed of 1–2 layers of cells. Pseudothecia filled with a few large, saccate, thick walled, bitunicate ASCI, 50–90 × 40–60 μ, embedded in and separated by thin walled, hyaline parenchyma. ASCOSPORES 8, irregularly arranged in the ascus, oval, clavate or ellipsoid, 3–4 transversely and 0–2 vertically septate, hyaline, when mature often slightly coloured, 25–50 × 10–20 μ. MYCELIUM in culture grey, floccose to felted, soon covered with black crust of pseudothecia (often compressed) and usually beaked. Pseudothecia, asci and ascospores tend to be larger in culture. No stat. conid. observed (C. Booth & K. A. Pirozynski, *CMI Descr.* 146, 1967).

Some confusion has occurred between this fungus and *Pleospora herbarum* (q.v.). The latter belongs to the Pleosporaceae rather than the Pseudosphaeriaceae, has darker ascospores and more numerous septation. In culture *P. herbarum* forms an imperfect state (*Stemphylium botryosum*) and both ascospores and conidia usually penetrate the host via the stomata. *L. trifolii* does not form an imperfect state and penetration is generally through the cuticle. It appears distinct from *L. americana* (Ell. & Ev.) Graham & Luttrell, as the latter has larger and generally 6 septate ascospores (Booth et al., l.c.; McDonald).

L. trifolii is widespread and plurivorous, especially on economic plants in the Cruciferae, Euphorbiaceae, Gramineae, Leguminosae and Solanaceae. Barrière et al. gave a review. The disease caused is a leaf spot, sometimes called pepper spot (or burn or leaf scorch). It has been mostly studied on lucerne and ref. to the pathogen (frequently called *L. briosiana*) on other hosts have been largely omitted. The wide variation in ascospore size which occurs on the host has resulted in the large synonymy. But on standardised PDA isolates from the Gramineae and Leguminosae form ascospores that are uniform in size and septation. *L. trifolii* was described by Leath et al. (1974) as causing considerable loss of lucerne in E. USA, particularly in the northern part of the region. Numerous very small black spots form on the leaves; under favourable field conditions for development of the disease the spots enlarge, become brown with darker, often irregular, margins and chlorotic haloes. Areas of the leaves are killed off and leaflets become curled and shrivelled. Pandey et al. described 3 symptom patterns on lucerne: resistant (black spots with no chlorosis) under light of 450-foot candles and an 8-hour day; moderately susceptible (black or brown spots with a trace of chlorosis) under 1100 foot candles and an 8–12-hour day; susceptible (brown spots with tan centres, dark margins and chlorotic haloes) under 2000-foot candles or more and a day length of 12 hours or more. On groundnut, as *L. crassiasca* (Sechet) Jackson & Bell (probably a synonym of *L. trifolii*), tests on susceptibility showed that cvs and breeding lines that had a notable amount of leaf necrotic scorch showed few pepper spots; whilst those with many pepper spots generally had little leaf scorch (Porter et al.).

Pseudothecial development is stimulated by certain UV wavelengths and has an opt. temp. of 24–26.6°; that for linear growth is 21–24.5° (Leach). The airborne ascospores were caught in greater numbers between 1600 and 2000 hours than at any other 4-hour period (Sundheim et al. but see Mallaiah et al.). They germinate, mostly after 1–2 hours, at 16–28°; germination is most rapid at 24° and nearly as rapid at 20°. A moisture period of at least 36–48 hours is needed for infection; penetration is cuticular. Ascospores alone can survive for 4 days on lucerne leaves and they overwinter in the pseudothecia on host debris. Older leaves of lucerne are more resistant than younger ones. Lesions on the youngest leaves are the largest and they decrease in size progressively on the older leaves. Pandey et al. described 3 races; and see Raynal. In 1974 it was stated that none of the planted cvs in USA is moderately resistant (Leath et al.).

Barrière, Y. et al. 1974. *Annls Phytopathol.* **6**: 341 (review, 53 ref.; **55**, 296).
Elliot, A. M. et al. 1964. *Phytopathology* **54**: 1443 (effect of temp. & moisture on formation & ejection of ascospores & on survival).
Jackson, C. R. et al. 1968. *Oléagineux* **23**: 387 (on groundnut; **47**, 2928).
Leach, C. M. 1972. *Mycologia* **64**: 475 (an action spectrum for light induced sexual reproduction; **52**, 336).
Leath, K. T. 1971. *Phytopathology* **61**: 70 (quality of light required for sporulation; **50**, 2729).
—— et al. 1974. *Ibid* **64**: 243 (relation of lesion size to leaf age & light intensity; **53**, 4019).
McDonald, W. C. 1958. *Ibid* **48**: 365 (with *Pleospora herbarum*; **38**: 11).
Mallaiah, K. V. et al. 1976. *Indian J. agric. Sci.* **46**: 372 (ascospore dispersal; **56**, 3339).
Martinez, E. S. et al. 1963. *Phytopathology* **53**: 938 (factors affecting growth, sporulation, pathogenicity & dissemination; **43**, 495).
Miller, J. H. 1925. *Am. J. Bot.* **12**: 224 (general; **4**: 541).
Moyer, B. G. et al. 1976. *Can. J. Bot.* **54**: 1839 (sporogens, sterols & sporulation; **56**, 563).
Pandey, M. C. et al. 1970. *Phytopathology* **60**: 1456 (effect of light, races & selection for resistance; **50**, 1291).
Porter, D. M. et al. 1971. *Pl. Dis. Reptr* **55**: 530 (susceptibility of groundnut; **51**, 928).
Raynal, G. 1978. *Rev. Zool. agric. Path. Vég.* **77**: 1 (evidence for races).
Sundheim, L. et al. 1965. *Phytopathology* **55**: 546 (ascospore germination & spread, infection & disease development; **44**, 2831).
Wilcoxson, R. D. et al. 1967. *Indian Phytopathol.* **20**: 199 (effect of temp. on ejection of ascospores; **47**, 3012).

LEVEILLULA Arnaud, *Annls Epiphyt.* 7: 94, 1919–20 (1921).
(Erysiphaceae)

ASCOCARPS (cleistothecia, perithecia) like those of *Erysiphe*, i.e. with simple, hyphoid appendages and several ASCI, but CONIDIA borne singly and sometimes the CONIDIOPHORES emerge through the stomata unlike other genera in the Erysiphaceae. Like *Phyllactinia* the abundant MYCELIUM is partly internal, permeating the host tissues (partly from C. E. Yarwood in *The fungi* (edited by G. C. Ainsworth et al.) Vol. IV A: 71, 1973).

The genus is generally considered to be monotypic. The imperfect state is the form genus *Oidiopsis* Scalia; see *Erysiphe* for some general comments and ref. on the powdery mildews. Golovin divided the genus into 6 sections on the basis of conidial morphology.

Eliade, E. 1972/73. *Lucr. Grăd. bot. Buc.*: 533 (monograph; **53**, 2089)
Golovin, P. N. 1956. *Trudy̆ bot. Inst. Akad. Nauk SSSR II* **10**: 195 (monograph; **37**: 400).

Leveillula taurica (Lév.) Arnaud, *Annls Epiphyt.* 7: 94, 1921.

Erysiphe taurica Lév., 1851
E. taurica var. *andina* Speg., 1902
E. taurica var. *zygophyllii* Maire, 1905
stat. conid. *Oidiopsis taurica* (Lév.) Salmon, 1906
Ovularia indica Rao, 1968.

MYCELIUM permeating the host tissue, amphigenous, often covering the whole plant, persistent, effused, densely compacted, tomentose to membranaceous or crustaceous, usually white, often pale buff in places, sometimes completely evanescent. CLEISTOTHECIA when formed generally scattered, rarely gregarious, often somewhat embedded in a dense superficial mycelium, 135–250 μ diam., globose or becoming concave at maturity. Peridium of polygonal cells, up to 10 μ diam. APPENDAGES numerous, like hyphae, simple, densely interwoven, short, indistinctly branched, colourless to olivaceous brown. ASCI usually *c.* 20 but sometimes less or up to 35 in each ascocarp, 2 spored, ovate, distinctly stipitate, 70–110 × 25–40 μ. ASCOSPORES large, cylindrical to pyriform sometimes slightly curved, variable in size, 25–40 × 12–22 μ. CONIDIA borne singly on short hyphal branches, predominantly of 2 distinct shapes, cylindrical and navicular, varying in size on different hosts, 25–95 (usually 50–79) × 14–20 μ (K. G. Mukerji, *CMI Descr.* 182, 1968).

L. taurica has an endoparasitic habit. Its distribution and host range are very wide (Hirata; Map 217); but attempts to divide the sp. on the grounds of host distribution or morphology have not been generally accepted. For early ref. to this powdery mildew see the review by Palti (1971) from whom this account is largely taken. Although the distribution of the fungus is worldwide the number of plant hosts on which it is found varies with the geographical region. The greatest host ranges occur in the Mediterranean region, in central and W. Asia. It is most widespread on red pepper, and it also occurs widely on eggplant and tomato (Palti 1974). The diseases on these 3 crops, and on globe artichoke (*Cynara scolymus*), are of the most economic importance. The xerophytic characteristics of *L. taurica* make it of particular importance as a pathogen in arid regions with crops grown under irrigation.

The symptoms vary somewhat according to host or plant organ. But they are commonest on the leaves as fairly bright, chlorotic spots on the upper leaf surface, whilst on the lower surface these spots have a powdery, white appearance. The spots may be diffuse or angular (vein delimited). In a few hosts leaf shedding may occur but on many the affected leaves wither and remain attached. Penetration can be stomatal or cuticular. Crops become more susceptible with maturity and individual leaves are rarely affected in early growth. Conidial germination is highest at RH 100% but occurs freely at 40–95% and 10–30%. In growth chambers more powdery mildew developed on red pepper at high than at low day humidities; shedding of infected leaves was more pronounced at low day humidities. On tomato there was more disease at low humidities but the leaves were not shed; a 25° and 15° temp. regime resulted in more disease than did one of 20° and 10°. The opt. temps for the disease on red pepper and tomato were 20° and 25°, respectively (Reuveni et al.). Cultural control measures should aim at planting when crops can escape infection; at reducing the amount of inoculum that may arise from nearby older, infected crops; and by the use of sprinkler rather than furrow irrigation, where *L. taurica* is favoured by low humidity. The general powdery mildew fungicide, sulphur, is effective. Benomyl, dinocap and tridemorph have also been used.

Hirata, K. 1958. *Bull. Fac. Agric. Niigata Univ.* **10**: 146 (host range & geographical distribution).
Ondieki, J. J. 1973. *Acta Hortic.* 33: 137 (fungicide control on red pepper **54**, 635).
Palti, J. 1971. *Phytopathol. Mediterranea* **10**: 139 (review, 45 ref.; **51**, 193).
——. 1974. *Ibid* **13**: 17 (divergency in distribution on major crops; **55**, 3113).
Reuveni, R. et al. 1973. *Phytopathol. Z.* **76**: 153 (effects of temp. & RH on disease on red pepper & tomato; **53**, 97).
—— ——. 1974. *Ibid* **80**: 79 (temp. & disease development; effect of infection on leaf abscission in red pepper **54**, 1981).

MACROPHOMA (Sacc.) Berl. & Vogl., *Atti Soc. Venet.-Trent. Sci. Nat.* **10**: 4, 1886.
(synonymy in B. C. Sutton, *Mycol. Pap.* 141: 117, 1977, no satisfactory description is available, Sutton, personal communication, February 1978)
(Sphaerioidaceae)

Three *Macrophoma* spp. are discussed briefly but their taxonomic position may need reinvestigation.

M. allahabadensis Kapoor & Tandon causes a fruit rot of guava (*Psidium guajava*). The black PYCNIDIA are an av. 132.6 μ diam., ostiolate with a broad papilla (av. 33.3 μ diam.) with dark brown hairs; CONIDIOPHORES short, hyaline; CONIDIA aseptate, hyaline, ellipsoidal to fusiform, av. 16.7 × 4.3 μ. Diseased fruit show a brownish and watersoaked discolouration of the skin. The mycelium eventually becomes dark brown to black, covering the whole fruit; pycnidia form all over the surface. Infection of fruit of banana, citrus, eggplant, mango and pea inter alia was reported by Kapoor et al. (1971), but pycnidia were seen on guava fruit only.

A blight of mango (*Mangifera indica*) is caused by *Macrophoma mangifera* Hingorani & Sharma. PYCNIDIA mostly hypophyllus, subepidermal, separate, globose or subglobose, 77–231 μ diam., ostiole 7–17.5 μ diam.; CONIDIOPHORES slender, hyaline, 8–11 × 1.5–2 μ; CONIDIA aseptate, hyaline, av. 19.8 × 6.5 μ. Symptoms appear as small spots on leaves and twigs; the leaf spots enlarge, are at first round becoming irregular, brown to dark brown with slightly raised and broad, dark purplish margins. Most infection occurs on the leaves; half a leaf may be affected. Stem lesions are elliptical and girdling. On the fruit watersoaked, circular lesions are formed; these result in a rot, particularly in storage. Resistance in mango was reported by Desai et al.

M. theicola Petch causes branch canker of tea (*Camellia sinensis*) in India and Sri Lanka. PYCNIDIA immersed, cracking outer cortex when mature, 0.25 mm diam., black, thin walled, ostiole *c.* 25 μ diam., not projecting; CONIDIA narrow oval or fusoid, aseptate, hyaline, ends obtuse or subtruncate, 27–32 × 5–7 μ (T. Petch, *Ann. R. bot. Gdns Peradeniya* 6: 234, 1917). The disease, which is apparently secondary following drought and sun scorch, is described by Hainsworth in *Tea pests and diseases and their control.* Growth of the fungus occurs on stems between the bark and wood; the former disintegrates leaving the exposed white wood. The elongated lesions are surrounded by a ring of callus growth. Sometimes this callus is attacked and a second (and third) callus is formed beyond the first. Affected bushes are killed slowly, branch by branch, until the infection reaches the collar. Control aims at preventing the predisposing conditions; shade, ground covers and mulches should be adopted. Diseased branches should be cut back to healthy wood. Dzhalagoniya reported inoculations and only plants in Theaceae were infected. But other hosts, including *Albizia* spp., have been reported.

Desai, M. V. et al. 1963. *Indian Phytopathol.* **16**: 239 (host susceptibility to *Macrophoma mangiferae*; **43**, 2365).

Dzhalagoniya, K. T. 1975. *Mikol. i Fitopatol.* **9**: 133 (*M. theicola* in USSR, Georgia; **54**, 5542).

Hingorani, M. K. et al. 1956. *Indian Phytopathol.* **9**: 195 (*M. mangiferae* sp. nov.).

—— ——. 1960. *Ibid* **13**: 137 (*M. mangiferae*; **41**: 98).

Kapoor, I. J. et al. 1970. *Ibid* **23**: 122 (*M. allahabadensis* sp. nov. **49**: 3407).

—— ——. 1971. *Phytopathol. Z.* **70**: 137 (*M. allahabadensis*; **51**, 520).

MACROPHOMINA Petrak, *Annls Mycol.* 21: 314, 1919.

(Sphaerioidaceae)

See *M. phaseolina* for a description.

Goidànich, G. 1947. *Annali Sper. agr.* N.S. **1**: 449 (revision of genus; **27**: 387).

Macrophomina phaseolina (Tassi) Goid, *Annali Sper. agr.* N.S., 1: 449, 1947.

Macrophoma phaseolina Tassi, 1901
M. phaseoli Maubl., 1905
Rhizoctonia bataticola (Taub.) Briton-Jones, 1925.

(further synonymy in *CMI Descr.* 275)

SCLEROTIA within roots, stems, leaves and fruits, black, smooth, hard, 100 μ to 1 mm diam. (in culture 50–300 μ). PYCNIDIA dark brown, solitary or gregarious on leaves and stems, immersed becoming erumpent, 100–200 μ diam., opening by apical ostioles; pycnidial wall multicellular with heavily pigmented thick-walled cells on the outermost side. CONIDIOPHORES (phialides) hyaline, short obpyriform to cylindrical, 5–13 × 4–6 μ. CONIDIA aseptate; hyaline, ellipsoid to obovoid, 14–30 × 5–10 μ.

Pycnidia production in culture is obtained on propylene oxide sterilised tissues, on groundnut meal irradiated with UV light, filter paper treated with vegetable oil, peptone or asparagine agar. There has been considerable controversy over the use of the name *Rhizoctonia lamellifera* Small. The type (IMI 35132), designated by Small (1924) on *Grevillea robusta*, appears indistinguishable from *Macrophomina phaseolina* and for this reason Small

(1926) withdrew *R. lamellifera* in favour of *R. bataticola*. Another epithet should be chosen for the fungus which plant pathologists regard as distinct from *M. phaseolina* but refer to as *R. lamellifera*. (P. Holliday & E. Punithalingam, *CMI Descr.* 275, 1970. A perfect state has been described: *Orbilia obscura* Ghosh, Mukherji & Basak, *Jute Bull.* **27**: 135, 1964; **44**, 3374.)

M. phaseolina causes diseases generally called charcoal rot or ashy stem blight. It is plurivorous (Young lists 284 hosts) and widespread in the tropics and subtropics. The biology of the fungus was recently discussed by Dhingra et al. and a bibliography was also given in another paper. Besides these sources of the numerous ref. further general treatments were given by: Holliday et al. (l.c.); Reichert et al.; Young. The main, common symptoms are: a dry or wet, dark rot of the lower stem; drying of the leaves; lodging and poor fruiting; canker; damping off; fruit and leaf lesions.

The fungus is a common root inhabitant sensu Garrett who, in 1956 (*Biology of root infecting fungi*), referred to its imposter role since it had been implicated in more diseases than was justified. Although it usually invades immature, unthrifty, wounded or senescent plant tissue (especially roots) severe losses can be caused. The opt. temps for growth in vitro and for the disease are 30–37°. In culture at 35° the linear growth rate can be 26 cm in 8 days. In sterile soil growth can be 21 cm in 10 days. There is little growth in unsterilised soil. *M. phaseolina* colonises dead plant material in the soil but is a poor competitor. The survival units are the sclerotia. Survival times are up to *c.* 4 years in roots and 75% survival has been reported in soil for 1 year (26° and 50–55% moisture holding capacity). Such times in soil are longer under dry conditions.

Plants are attacked as seedlings; when approaching maturity and at flowering; when temps are high and soil moisture levels are low; when water stress occurs in the host; and when the plant has been infected by other organisms. Under a susceptible crop build up of inoculum takes place. The fungus is seedborne, sometimes internally. Infection occurs from sclerotia lying close to the host tissue. A superficial mycelium develops on the plant surface and penetration occurs from appressoria. Mycelium is inter- and intracellular. The cortex is destroyed. Crops on which extensive work has been done include: castor, cotton, Irish potato, jute, maize, *Phaseolus* spp., pine, sorghum, soybean and sunflower. There appears to be no significant host specificity. Dhingra et al. (1978) described the symptoms.

Control of charcoal rot in brief can only be given in very general terms. The local situation will need to be taken into account. Good growing conditions will prevent most of the diseases. Conditions which favour the pathogen, for example, extreme moisture levels, should be avoided. Correct time of planting, rotation and seed treatment with fungicides are important. Mixed cropping and organic amendments to the soil may reduce infection. Host resistance may be an important factor when considering different crops or cvs within a crop.

Ashby, S. F. 1927. *Trans. Br. mycol. Soc.* **12**: 141 (taxonomy & morphology; **6**: 757).

Dhingra, O. D. et al. 1977. Universidade Federal de Vicosa, Minas Gerais, Brazil, 244 pp. (annotated bibliography 1905–75, 904 ref.; **57**: 906).

——. 1978. As above, 166 pp. (general, 279 ref.; **57**, 4287).

Kulkarni, N. B. et al. 1966. *Mycopathol. Mycol. appl.* **28**: 257 (taxonomy & nomenclature; **45**, 2417).

Maiello, J. M. 1978. *Mycologia* **70**: 176 (origin of the pycnidium; **57**, 4356).

Michail, S. H. et al. 1977. *Acta phytopathol. Acad. Sci. hung.* **12**: 311 (pycnidial induction in culture; **57**, 2809).

Moreau, C. 1956. *Rev. Mycol.* **21** Suppl. Colon. 2 (general; **37**: 458).

Papavizas, G. C. 1977. *Soil Biol. Biochem.* **9**: 337, 343 (factors affecting survival of sclerotia; **57**, 1622, 1623).

Reichert, I. et al. 1947. *Palest. J. Bot. Rehovot Ser.* **6**: 107 (general, 165 ref.; **28**: 84).

Shokes, F. M. et al. 1977. *Phytopathology* **67**: 239 (effect of water potential on growth & survival; **56**, 4857).

Short, G. E. et al. 1978. *Ibid* **68**: 736, 742 (quantitiative enumeration in soybean & inoculum potential).

Small, W. 1924. *Trans. Br. mycol. Soc.* **9**: 152 (morphology & taxonomy; **3**: 748).

——. 1926. *Ibid* **10**: 287 (as above; **5**: 451).

Young, P. A. 1949. *Bull. Texas agric. Exp. Stn* 712, 33 pp. (general; 91 ref.).

MAGNAPORTHE Krause & Webster, *Mycologia* 64: 110, 1972.

(Diaporthaceae)

Nonstromatic. PERITHECIA dark, globose; neck long, cylindrical, slightly or not protruding beyond the surface of the leaf sheath. ASCI unitunicate, thin walled, short stipitate, floating freely in the perithecium at maturity and deliquescing. ASCOSPORES long, fusiform, curved, 3 celled, slightly constricted at the septa, hyaline or yellowish brown at maturity; reliquiae wide and band like, partially gelatinising at maturity (R. A. Krause & R. K. Webster, l.c.).

The genus was established to accommodate the fungus which causes stem rot of rice, formerly known as *Leptosphaeria salvinii*. This fungus has unitunicate asci and, therefore, does not belong in *Leptosphaeria* which has bitunicate asci (Loculascomycetes). The authors of the genus discussed one of the characteristics of the Diaporthales, the nature of the 'paraphyses'; their new term 'reliquiae' describes the tissue that is left behind as the asci grow up through the pseudoparenchymatous centrum.

Krause, R. A. et al. 1972. *Mycologia* **64**: 103 (morphology, taxonomy & sexuality of rice stem rot pathogen, *Magnaporthe salvinii*; **51**, 3999).

Barr, M. E. 1977. *Ibid* **69**: 952 (*Magnaporthe, Telimenella* & *Hyponectria*, Physosporellaceae; **57**, 3821).

Magnaporthe salvinii (Catt.) Krause & Webster, *Mycologia* **64**: 110, 1972.
Leptosphaeria salvinii Catt., 1879
stat. conid. *Nakataea sigmoidea* Hara, 1939
Helminthosporium sigmoideum Cav., 1889
Vakrabeeja sigmoidea (Cav.) Subram., 1956
Curvularia sigmoidea (Cav.) Hara, 1959.

Colonies effuse, black. MYCELIUM partly immersed, partly superficial. Spherical or subspherical black SCLEROTIA mostly 200–300 μ diam. are formed on natural substrata and in culture; these are often referred to as *Sclerotium oryzae*. PERITHECIA dark, globose, embedded in outer leaf sheaths; 250–650 μ diam. (av. 420 μ), length with neck 500–1100 μ (av. 820 μ, wall 5–12 cells thick, cells dark, elongated; neck slightly or not protruding beyond surface of leaf sheath. ASCI long cylindrical, thin walled, short stalked, 104–165 × 8.7–17.4 (av. 135 × 9.7) μ, deliquescing as ascospores mature. ASCOSPORES biseriate, slightly twisted, 8/ascus, 3 septate, slightly constricted at septa, 35–65 × 8.7 (av. 52–8.7) μ, fusiform, curved; in culture all cells equally granular and hyaline; in nature distal cells hyaline and less granular than middle cells which may be hyaline, yellow or yellowish brown. CONIDIOPHORES unbranched or rarely branched, brown, smooth, septate, bearing towards and at the apex a number of thin walled, cylindrical or broadly conical denticles each cut off by a septum to form a separating cell. CONIDIA borne on the denticles, simple, usually falcate, often sigmoid, smooth, almost always 3 septate, 40–83 (mostly 45–55) μ long, 11–14 μ thick in the broadest part, tapering rather abruptly at the ends; cells unequally coloured, the cell at each end hyaline or very pale, intermediate cells pale to mid-pale brown (perfect state from R. A. Krause & R. K. Webster l.c.; stat. conid. from M. B. Ellis & P. Holliday, *CMI Descr*. 344, 1972).

A var. *irregulare* of *Helminthosporium sigmoideum* has been distinguished by its more irregular sclerotia which are dull, rough and embedded in the substratum compared with the smooth, spherical and superficial ones of the other form (Cralley et al.). A form on *Zizania latifolia*, not pathogenic to rice, was considered to be a new var. (Chang; Hsieh et al.). *M. salvinii* is heterothallic.

Stem rot of rice is widespread with its host (Map 448). Symptoms are usually first seen in the later growth stages. Necrotic lesions begin on the outer leaf sheath near the water line; these spread gradually to the inner sheaths and the stem base. At about maturity lodging may occur and sclerotia will be found within the tissues. Partial sterility is caused and the grain becomes light and chalky. Late infection can lead to small tillers but whether the pathogen causes excessive tillering, as earlier reports stated, is in doubt. Sclerotia form conidia when floating free or detached on moist substrates. Symptoms seem to vary with the growing conditions. Krause et al. (1973) described a disease index and reported up to 22% loss in the field in USA (California).

The pathogen may be abundant but little disease occurs; predisposing factors (tendency to lodge or become damaged) for infection are important. Initial infection, through appressoria and infection cushions, takes place from sclerotia and secondary infection by conidia seems less important. Penetration proceeds from outer leaf sheaths to inner stem. High nitrogen and leaf removal increase the disease, and sodium silicate, added to the soil, decreases it. In Japan early cvs tend to get more severe infection than late ones. Spread is largely through the sclerotia but the roles of the conidium and ascospore seem to have been little studied. Suzuki found that conidia were at a max. at 1000–1400 hours and a min. at 1700–0600 hours. Both sclerotia and perithecia can survive high summer temps and the former may remain viable for 6 years. The viability of sclerotia was 21–26% after differing rice residue management treatments. Alternate wetting and drying of sclerotia reduced their viability. Sclerotia from wet soils, after incubation at 24°, were significantly less viable than those from soil which had dried or those subjected to alternate wetting and drying (Keim et al.; Krause et al. 1972; Tullis et al. 1941). The fungus grows well in vitro at 25–30° and forms of it vary in virulence (e.g., Ferreira et al. 1975; Jain).

In USA Webster et al. examined the effects of various tillage methods on stem rot. Open field burning of residues was the most effective method for minimising sclerotial inoculum. Ploughing buried much inoculum which, therefore, did not infect water sown rice. Spraying fungicides on debris and stubble after harvest has been recommended in Taiwan (Chen et al.). Jackson et al. found that triphenyltin hydroxide was effective in field trials. Rice cvs differ in their degree of susceptibility (Ferreira et al. 1975a, 1976; Goto et al.; Krause et al. 1973; Srivastava et al.).

Chang, H. S. 1977. *Bot. Bull. Acad. sin. Taipei* **18**: 39 (isolate from *Zizania latifolia*: **57**, 592).

Chen, C. C. et al. 1972. *Mem. Coll. Agric. Taiwan Univ.* **13**: 24 (chemical control; **53**, 3482).

Corbetta, G. 1953. *Phytopathol. Z.* **20**: 260 (general in Italy; **33**: 445).

Cralley, E. M. et al. 1935. *J. agric. Res.* **51**: 341 (var. *irregulare*; **15**: 47).

Ferreira, S. A. et al. 1975a & b. *Phytopathology* **65**: 672, 968 (sporulation, growth & virulence; genetics of resistance & virulence; **55**, 1793, 1794).

——— ———. 1976. *Ibid* **66**: 1151 (evaluation of virulence; **56**, 2062).

Goto, K. et al. 1954. *Bull. Div. Pl. Breed. Cult. Tokai-Kinki natn. agric. Exp. Stn* 1: 27 (cv. resistance & seasonal development; **35**: 229).

Hsieh, S. P. Y. et al. 1975. *Pl. Prot. Bull. Taiwan* **17**: 372 (on *Z. latifolia*; **55**, 5729).

Hsu, H. T. 1968. *Ibid* **10**: 7 (pectolytic activity in infected rice plants; **51**, 4824).

Inoue, Y. 1961. *Spec. Bull. 1st Agron. Div. Tokai-Kinki natn. agric. Exp. Stn* 3, 118 pp. (infection process; **41**: 654).

Jackson, L. F. et al. 1977. *Phytopathology* **67**: 1155 (fungicide; **57**, 2518).

Jain, S. S. 1973. *Oryza* **10**(1): 59 (virulence; **54**, 4491).

Keim, R. et al. 1974. *Phytopathology* **64**: 1499 (soil moisture, temp. & sclerotial viability; **54**, 3310).

——— ———. 1975. *Ibid* **65**: 283 (fungistasis of sclerotia; **54**, 4489).

Krause, R. A. et al. 1972. *Mycologia* **64**: 1333 (recovery of sclerotia from soil & their viability; **52**, 4069).

——— ———. 1973. *Phytopathology* **63**: 518 (disease index, losses & cv. susceptibility; **53**, 162).

Lo, T. C. et al. 1964. *Pl. Prot. Bull. Taiwan* **6**: 121 (distribution & economic significance).

Luthra, J. C. et al. 1936. *Indian J. agric. Sci.* **6**: 973 (symptoms & infection; **16**: 123).

Misawa, T. et al. 1955. *Ann. phytopathol. Soc. Japan* **19**: 125; **20**: 65 (effect of N in vitro; biotin & thiamine requirements; **35**: 790, 843).

——— ———. 1962. *Ibid* **27**: 102 (cause of lodging, pectolytic & cellulytic enzymes in vitro; **42**: 195).

Misra, A. P. et al. 1966. *Indian Phytopathol.* **19**: 14 (sclerotial viability; **47**, 182).

Mundkur, B. B. 1935. *Indian J. agric. Sci.* **5**: 939 (sclerotial, morphological differences; **15**: 313).

Nonaka, F. et al. 1956. *Sci. Bull. Fac. Agric. Kyushu Univ.* **15**: 431, 435 (effect of leaf removal; dissemination of conidia; **36**: 496).

——— ———. 1958. *Ibid* **16**: 439, 447, 459, 473 (infection of late & early cvs; effect of silicic acid; N & carbohydrate in disused culms; **38**: 142).

Ono, K. et al. 1960. *Spec. Rep. Forecast. Dis. Insect Pest.* 4: 94 (infection & ecology).

Park, M. et al. 1932. *Ceylon J. Sci.* Sect. A, **11**: 342 (morphology, symptoms, inoculations & sclerotial viability; **11**: 599).

——— ———. 1934. *Ibid* **12**: 1, 11, 25 (inoculations & infection, strs; **15**: 113, 114).

Singh, R. A. et al. 1966. *Phytopathol. Z.* **57**: 24 (report from India of perfect state, carry over; **46**, 1574).

Srivastava, M. P. et al. 1971. *Indian J. agric. Sci.* **41**: 93 (resistance of cvs; **51**, 2448).

Suzuki, H. et al. 1971. *Proc. Assoc. Pl. Prot. Hokuriku* **19**: 1 (infection by *Helminthosporium sigmoideum* var. *irregulare*; **53**, 533).

———. 1974. *Bull. Hokuriku natn. agric. Exp. Stn* 16: 43 (conidial dispersal with *Cochliobolus miyabeanus*; **55**, 1242).

Togashi, K. 1955. *Bull. Fac. Agric. Niigata Univ.* 7: 34 (conidial formation & germination).

Tullis, E. C. 1933. *J. agric. Res.* **47**: 675 (perfect state, growth in vitro & inoculation; **13**: 322).

——— et al. 1933. *Bull. Ark. agric. Exp. Stn* 295, 23 pp. (temp. & growth, histology of infection; **13**: 395).

——— ———. 1941. *Phytopathology* **31**: 279 (viability of sclerotia in uncultivated soil; **20**: 381).

Watanabe, B. 1952. *Bull. Kyushu agric. Exp. Stn* **1**: 271 (disease & stem physiology).

Webster, R. K. et al. 1976. *Phytopathology* **66**: 97 (distribution & survival under different tillage methods; **55**, 3174).

MARASMIELLUS Murr., *N. American Flora* 9(4): 243, 1915.
(Tricholomataceae)

Habit marasmioid–collybioid or mycenoid, also frequently pleurotoid, revivescent or putrescent; some spp. forming carpophoroids and/or protocarpic tubers. PILEUS glabrous, subtomentose, tomentose flocculose, not glutinous. HYMENOPHORE lamellate, variously attached to the stipe, often intervenose but not favoloid-anastomosing, mostly intermixed-inserted but lamellulae sometimes very few or none, rarely in some specimens of a population hymenophore strongly reduced STIPE central, eccentric, lateral or absent, unshining or at the utmost with a dull shine underneath a pruinate, pubescent, flocculose or tomentose covering, but eventually often glabrescent, not black and seta like, insititious or subinsititious, rarely with a fibrillose–tomentose basal mycelium but then the epicutis with a distinct Rameales structure and/or stipe eccentric to lateral. CONTEXT either not at all gelatinised or with gelatinised (i.e. thin walled hyphae imbedded in a gelatinous mass) zones and/or pockets but then always with non-anastomosing lamellae and fusoid basidioles (in pleurotoid spp.) and with a distinct Rameales structure (if collybioid). Odour usually not characteristic, rarely strong (of sauerkraut or garlic). Epicutis typically a Rameales structure, i.e. of lacerate elements with knobs, diverticulate and short ramifications, as characteristic of sect. Rameales, but this structure often poor or weak and in a minority of spp. not any more recognisable as a Rameales structure but replaced by mostly thin, mostly densely interlaced hyphae or a cutis-like structure not gelatinised even if the underlying trama is gelatinised; hyphae inamyloid, with or more rarely without clamp connections, not forming chains of very voluminous cells. BASIDIOLES most or all fusiform or subfusiform, rarely in otherwise typical forms with a majority of clavate cylindrical basidioles; cystidia always either like the cheilocystidia or similar, or else pseudoparaphysoid, or absent; cheilocystidia usually numerous, rarely rare or scattered, gloeo-, chryso- and macrocystidia never present; covering of the stipe like that of the pileus or containing dermatocystidia or hairs, often hairs like those of *Crinipellis* (but inamyloid). Spore print pure white to pale cream. SPORES small to large, narrow to broad, with thin, inamyloid, homogenous wall, acyanophilus, smooth, sometimes dimorphic, sometimes (sect. Nigripedes) tetrahedric or cross shaped or merely with an eccentric bulge or spur on the outer side (much like in some Campanellae) (R. Singer, *The Agaricales in modern taxonomy*: 315, 1975).

Marasmiellus differs from *Marasmius* (q.v.) in the structure of the epicutis of the pileus and the inamyloid hyphae (even in the stipe; Singer l.c.). Several spp. of the latter have been transferred to the former and other genera. *Marasmiellus albus-corticis* (Secr.) Singer (*Lilloa* 22: 300, 1951) was described causing a sheath blight and wilt of *Eleusine coracana* (finger millet); synonyms: *Agaricus candidus* Bolt ex Secr., *A. albus-corticis* Secr. and *Marasmius candidus* (Bolt ex Secr.) Fr. Plants show circular to elliptical, necrotic blotches *c.* 5–15 cm from ground level. The sheaths become bound together by the mycelium and death results (Parambaramani et al.).

Marasmiellus cocophilus Pegler (1969; Bock et al.; and see D. N. Pegler, *A preliminary agaric flora of East Africa, Kew Bull. addit. Ser.* 6: 120, 1977) causes lethal bole rot of coconut in Kenya and Tanzania (Map 515). The disease most probably also occurs on the E. African coast S. of Tanzania and possibly on the W. coast of the Malagasy Republic. Pegler compared the fungus with *Marasmius palmivorus* (*Marasmius* q.v.) and *Marasmiellus dealbatus* (Berk. & Curt.) Singer (*Sydowia* 9: 397, 1955). The former grows on leaves of coconut, oil palm and rubber. It can be distinguished from *M. cocophilus* (basidiospores $14.5 \times 3.8\ \mu$) by the larger and more robust carpophores and the broadly ellipsoid spores, $8\text{–}9 \times 4.5\text{–}5\ \mu$. *M. dealbatus* most closely approaches *M. cocophilus* but it is known only as a saprophyte on forest litter and has smaller basidiospores, $8\text{–}9 \times 2.5\text{–}3.5\ \mu$.

The diagnostic symptom of lethal bole rot is a dry reddish brown rot at the base of the bole. An LS shows a distinctive yellowish margin (0.5–1 cm wide) around the decay; this is most pronounced in 2–4-year-old palms. In the area of rot there are cavities lined with a profuse white mycelium; these are rare in 4–8-year-old trees and absent in mature ones. Root systems of diseased palms are collapsed. The first visible symptom is a general wilt; the leaves become yellow and then bronze. The spike of unfurled leaves dies first; the fronds die back from the tip and may show midrib necrosis. In older palms the outer whorl of dead or dying fronds, dead spathes and inflorescences remain attached and droop, leaving an erect tuft of dying leaves. A soft rot, with a nauseating odour, is confined to the inner unfurled leaves and, in older palms, to the developing axillary spathes. Later the rot spreads to the bases of older leaves. The innermost, youngest leaves of the bud and stem below the bud are healthy.

The disease is typically one of young palms, and seedlings are highly susceptible on transplanting from the nursery to the field. Most deaths occur in seedlings and palms up to 8 years old. Sporophores occur on exposed roots, on leaf bases of seedlings at ground level, and on the soil surface at holes where diseased palms have been removed. They have not been produced in culture. The fungus penetrates via wounded roots and it appears to be a persistent coloniser of coconut debris in the soil. Invasion of older tissue is slow. Spread may be randomly scattered, infection originating from diseased seedlings; or a very slow spread from palm to palm may take place. The latter type is more typical of 15–20-year-old plantings. Control measures are: early selection and transplanting of seedlings; damaged roots pruned and disinfected; take seedlings from healthy areas; soil sterilisation in nurseries; and no cultivation between palms, especially where the disease is present.

M. inoderma (Berk.) Singer (*Sydowia* **9**: 385, 1955) has been reported on maize, *Musa* (sheath rot) and rice; synonym: *Marasmius inoderma* Berk.; for the distinction from *Marasmiellus semiustus* (Berk. & Curt.) Singer (*Pap. Mich. Acad. Sci.* **32**: 129, 1968) see Singer in *Beih. nova Hedwigia* **44**: 235, 1973. Sabet et al. described a root rot and wilt of maize in Egypt caused by *M. inoderma*. They described the fungus and its sporophores were formed in culture. In Australia Allen called it a common saprophyte which might attack banana already weakened in some way. In *Musa* (see Ramos for *M. textilis*) the outer leaf sheaths and leaves dry up, growth is stunted and leaves emerge slowly or not at all. Mycelium can be seen on and between the dead leaf sheaths. On the inner sheaths there are watersoaked, brown oval lesions, several cm diam., and these may penetrate the pseudostem. Roots can also be attacked and killed. Control is cultural and any measure to reduce RH and strengthen the plants should be taken. Sheath rot of abaca has also been reported (*Rev. appl. Mycol.* **33**: 604) to be caused by *Marasmiellus stenophyllus* (Mont.) Singer (*Sydowia* **15**: 58, 1962); synonyms: *Marasmius stenophyllus* Mont. and *M. subsynodicus* Murr. Ordosgoitty et al. described a dry rot of plantain in Venezuela caused by *Marasmiellus troyanus* (Murr.) Dennis (*Kew Bull. addit. Ser.* 3: 31, 1970); synonyms: *Marasmius troyanus* Murr., *Collybia troyana* (Murr.) Dennis. *Marasmiellus scandens* (Massee) Dennis & Reid causes a thread blight in tropical crops (Guillaumin; and *Marasmius* q.v., Dennis et al.; Petch 1924); synonyms: *M. scandens* Massee, *M. byssicola* Petch.

Allen, R. N. 1970. *Agric. Gaz. N.S.W.* **81**(9): 524 (*Marasmiellus inoderma* on *Musa*; **50**, 1312).

Bock, K. R. et al. 1970. *Ann. appl. Biol.* **66**: 453 (*M. cocophilus* on coconut, aetiology; **50**, 2416).

Guillaumin, J. J. 1971. *Annls Phytopathol.* **3**: 143 (*M. scandens* in Ivory Coast; **51**, 2317).

Ordosgoitty, F. A. et al. 1974. *Agron. trop.* **24**: 33 (*M. troyanus* on *Musa*; **54**, 2367).

Parambaramani, C. et al. 1975. *Curr. Sci.* **44**: 358 (*M. albus-corticis* on finger millet; **55**, 210).

Pegler, D. N. 1969. *Kew Bull.* **23**: 523 (*M. cocophilus* sp. nov.; **49**, 1110).

Ramos, M. M. 1941. *Philipp, J. Agric.* **12**: 31 (*M. semiustus* on abaca; **20**: 579).

Sabet, K. A. et al. 1970. *Trans. Br. mycol. Soc.* **54**: 123 (*M. inoderma* on maize; **49**, 2466).

Singer, R. 1973. *Beih. nova Hedwigia* 44, 517 pp. (monograph, *Marasmiellus* inter alia).

Singh, R. A. 1972. *Riso* **21**: 373 (*M. inoderma* on rice; **53**, 4438).

MARASMIUS Fr., *Gen. Hymen.*: 9, 1836.

(synonymy in R. Singer, *The Agricales in modern taxonomy*: 350, 1975)

(Tricholomataceae)

Habit collybioid (to almost mycenoid) or pleurotoid, mostly somewhat toughish and reviving after drying out when remoistened. HYMENOPHORE lamellate, rarely smooth or venose for a long time but eventually mostly becoming lamellate, lamellae sometimes intervenose or anastomosing; epicutis either hymeniform or not, if not, consisting of broom cells or at any rate strongly diverticulate nodulose hyphal elements (and then either the whole trama or only that of the STIPE more or less distinctly pseudoamyloid or at least the stipe seta-like and shining blackish and insititious, or absent to small and eccentric to lateral); trama of the PILEUS monomitic; pseudoamyloid or inamyloid; no gelatinous layers; clamp connections present or more rarely absent; if absent, epicutis with broom cells. Black rhizomorphs often present, sometimes stipe of carpophore rising from it; telepods often formed; if there is a basal mycelium, the epicutis is always hymeniform; some spp. form endotrophic mycorrhiza, none ectotrophs or lichens.

The stiff stipe, like horsehair, often shining and glabrous, or itself beset by seta-like bodies, often rising from or accompanied by black rhizomorphs, the occasional garlic odour, the complete absence of mycelium at the base of the fully insititious stipe, the most distinct pseudoamyloid hyphae (at least in the rind of the stipe) and the precise structure of the cheilocystidia and elements of the epicutis, difficult to describe in a generalised way for comparison with *Marasmiellus*, and the general aspect are usually sufficient indications for the generic identity of a sp. (Singer l.c.).

Spp. of the genus are frequently mentioned in the old literature on plant pathology in the tropics. Some have been transferred (or await transfer) to *Marasmiellus* (q.v.), *Crinipellis* (q.v.) and other genera. The fungi concerned were usually described as causing thread (frequently whitish) or horsehair (brownish) blights, and sheath rots of stem bases. See Petch and Briton-Jones et al. for some account of their habits. Petch (1924) discussed the characteristics of the external mycelia in thread blights of *Corticium* spp. (sensu lato) and those of the marasmioid type; these adhere, particularly, to green stems and leaves. Whilst the *Marasmius* (or related) spp. involved may cause death of the foliage they are of no economic importance. As common saprophytic inhabitants of tropical forests they frequently spread to perennial crops (e.g., tea, cacao, citrus, coffee, banana and rubber) under the then somewhat primitive systems of agriculture. Horsehair blights form thin rhizomorphic mycelium which spreads freely over bushes, trees and leaf litter; they are saprophytic (Petch 1915).

The following are thread blights: *Amyloflagellula pulchra* (Berk. & Br.) Singer (*Darwiniana* **14**: 14, 1966, synonyms: *Cyphella pulcher* Berk. & Br. and *M. pulcher* (Berk. & Br.) Petch); *M. cyphella* Dennis & Reid. The following are horsehair blights: *Crinipellis actinophorus* (Berk. & Br.) Singer (*Sydowia* **9**: 397, 1955, synonyms: *M. actinophorus* Berk. & Br. and *M. coronatus* Petch); *M. crinisequi* F. Muell. ex Kalch. (synonyms: *M. equicrinis* F. Muell. ex Berk., *M. repens* P. Henn. and *M. graminum* (Lib.) Berk. var. *equicrinis* Dennis); *M. rigidichorda* Petch (*Trans. Br. mycol. Soc.* **31**: 34, 1948), this is a *Crinipellis* but the combination has not been made (all D. N. Pegler, personal communication, January 1978).

A bunch rot of oil palm is caused by *M. palmivorus* Sharples in Malaysia; it also causes white fan blight of rubber (*Hevea brasiliensis*) and occurs on other plants. The disease has been noted in Africa (Turner 1967b). The fungus is a *Marasmiellus* but the combination has not been made (D. N. Pegler, personal communication, January 1978). Proof of pathogenicity of *Marasmius palmivorus* on oil palm may still be lacking (Turner & Bull, *Diseases and disorders of the oil palm in Malaysia*). The fungus, normally saprophytic, builds up on the first fruit bunches (not harvested) of young palms and on bunches inadequately pollinated. At high inoculum levels ripe fruit bunches are attacked and white mycelium extends over the bunch surface. The fruit pericarp is penetrated and a soft, brown, wet rot develops; it is sharply defined from healthy tissue. Mycelium may also occur on leaf bases, especially on the frond immediately behind the affected bunch. In wet weather the sporophores (pileus white, 2–8 cm diam.) are abundant. In dry weather the mycelium and pileus are pinkish, and the sporophore much smaller. The disease may be particularly severe on palms growing under sub-optimal conditions, such as acid sulphate soils.

Control is largely preventive through sanitation measures. Rotting bunches, male flowers and abortive or very poorly pollinated bunches should be removed. The removal of immature inflorescences, to avoid the formation of large numbers of small bunches, in young palms may assist control. A high incidence of bunch rot, however, may indicate that the crop is being grown under poor conditions and these should be corrected. It is important to ensure that pollination is adequate. Organo-mercurial fungicides may be effective if applied as soon as possible after infection has occurred. The chemical is applied to the top of each affected bunch.

Pont (Australia) described root and stalk rots of maize caused by *M. sacchari* Wakker var. *hawaiiensis* Cobb and *M. graminum* (Lib.) Berk. & Br. var. *brevispora* Dennis. The former is a *Marasmiellus* but the combination has not been made (D. N. Pegler, personal communication, January 1978); it is not the same as *M. sacchari* Singer (=*Collybia sacchari* (Sing.) Singer (*Sydowia* **15**: 55, 1962)). *Marasmius sacchari* var. *hawaiiensis*: PILEUS up to 2.6 cm, white campanulate at first then expanding; STIPE up to 2.6

cm, white, bulbous and villose at point of attachment to substrate, central, sometimes excentric; GILLS simple or bifurcate, adnate; BASIDIOSPORES hyaline, clavate, papillate at point of attachment, 10–15 × 3.4–4.5 μ, print white; opt. growth on PDA at 30–32°; on maize, *Euchlaena mexicana* and *Cyperus rotundus*. Symptoms on maize can appear at any growth stage. Infected seedlings wilt and there is a conspicuous white rot of the roots and crown. On older plants there is at first a mild chlorosis of the foliage and a faint striped mottle of apical leaves. Stunting, firing of the lower leaves; rolling, bleaching and wilt of the apical leaves all occur. A white mycelium cements the lower leaf sheaths, particularly at and below soil level, to the stem. Beneath such leaf sheaths necrotic streaking is seen on the stem, and there is parenchyma and vascular breakdown from discoloured areas in the sclerenchyma. The necrosis may extend to >6 internodes. The internal tissue is eventually decomposed and white mycelium is seen in the cavity.

M. graminum var. *brevispora*: PILEUS up to 4–5 mm diam., convex with a shallow umbilicus and a dark central papilla, striate-sulcate to the umbilicus, orange; STIPE slender, brown and smooth, up to 2.5 cm or more in length; GILLS and undersurface of pileus white, up to 12, adnate; BASIDIOSPORES hyaline, oval or elliptical, 9–10 × 4 μ; opt. growth on PDA at 30–32°; on maize. The roots, mesocotyl and basal internodes are discoloured and rotted. Lower leaf sheaths are cemented as in *M. sacchari* var. *hawaiiensis* but the extensive internal and external necrosis does not occur. Leaf sheaths higher up the stem become infected. The straw coloured leaf sheaths have a fine white mycelium on their inner surfaces. Both these maize pathogens penetrate via the roots and possibly wounded stem tissue. The seed or mesocotyl may be initially infected. Root lesions are caused on other crop plants. Incidence of the diseases is aggravated by hot, dry conditions. Pont found that one hybrid maize grown widely in N. Queensland had a reasonable level of field resistance. Philipp reported a maize root and stalk rot, probably *M. graminum*, in Germany; the fungus had a wide host range.

M. sacchari, *M. plicatus* Wakker and *Marasmiellus stenophyllus* (*Marasmiellus* q.v) have been implicated in the aetiology of the very minor diseases (basal stem, root and sheath rots) of sugarcane (see Divinagracia). The condition is recognised by the white mycelial growth over the lower leaf sheaths, and between the sheaths and stems. The underground parts of the stem and young shoots are invaded. Sporophores form at the base of the plant. Viégas described a root rot of weakened coffee plants caused by *Marasmius viegasii* Singer (apud Viégas).

Abe, T. et al. 1957. *Scient. Rep. Fac. Agric. Saikyo Univ.* 9: 41 (*Marasmius crinisequi* on tea; **37**: 507).

Briton-Jones, H. R. et al. 1934. *Trop. Agric. Trin.* **11**: 55 (thread blights; **13**: 539).

Dennis, R. W. G. et al. 1957. *Kew Bull.* (2): 287 (tropical marasmioid fungi, taxonomy; **37**: 458).

Divinagracia, G. G. 1957. *Philipp. Agric.* **40**: 469 (*M. sacchari* on sugarcane; **36**: 617).

Petch, T. 1915. *Ann. R. bot. Gdns Peradeniya* **6**: 43 (horsehair blights).

——. 1924. *Ibid* **9**: 1 (thread blights; **4**: 66).

Philipp, A. 1959. *Kühn-Arch.* **73**: 42 (*M. graminum* on maize; **40**: 44).

Pont, W. 1973. *Qd J. agric. Anim. Sci.* **30**: 225 (*M. sacchari* var. *hawaiiensis* & *M. graminum* var. *brevispora* on maize; **54**, 142).

Turner, P. D. 1965. *Planter Kuala Lumpur* **41**: 387 (*M. palmivorus* on oil palm; **45**, 1128).

——. 1967a. *Ibid* **43**: 240 (as above, occurrence & control; **46**, 3535).

——. 1967b. *Expl Agric.* **3**: 129 (as above, fungicides; **46**, 2297).

Viégas, A. P. 1957. *Bolm Suptdcia Serv. Café S. Paulo* 32 (368): 7 (*M. viegasii* on coffee; **37**: 536).

MELAMPSORA Cast., *Observations Uredinees* 2: 18, 1843.

(see G. Laundon, *Mycol. Pap.* 99: 11, 1965)
(Melampsoraceae)

SPERMOGONIA subcuticular or subepidermal, conical or hemispherical, without paraphyses but sometimes with flexuous hyphae. AECIDIA caeomoid usually foliicolous, without peridium or sometimes with peripheral hyphae like paraphyses which may unite to form a rudimentary peridium. AECIDIOSPORES catenulate, globoid or ellipsoid, with verrucose walls. UREDIA subepidermal, pulverulent, with a thin, evanescent peridium. UREDOSPORES borne singly on pedicels, globoid or ellipsoid, with indistinct pores; with capitate or clavate paraphyses. TELIA subcuticular or subepidermal, forming crusts consisting of a single layer of spores. TELIOSPORES aseptate, adhering laterally, with coloured walls, with one indistinct apical pore, germinating in spring; BASIDIA typically 4 celled, producing globoid, colourless or yellowish BASIDIOSPORES (from

M. Wilson & D. M. Henderson, *British rust fungi*: 64, 1966).

M. ricini Noronha (synonyms: *Uredo ricini* Biv.-Bern., *M. ricini* Pass., *Caeoma ricini* Schlect., *Melampsorella ricini* de Toni) causes castor (*Ricinus communis*) rust. It was described by Ajrekar and Punithalingam, the latter's description is: PYCNIA and AECIDIA unknown. UREDIA chiefly hypophyllous up to 1.5 mm, seldom epiphyllous, in groups, circinate and often coalescing. UREDOSPORES ovoid to ellipsoidal, 20–30 × 12–20 μ, with orange contents when fresh, walls 1.5–2 μ, pale yellow and verruculose with 3–4 subequatorial germ pores. PARAPHYSES intermixed with uredospores, capitate, hyaline, 50–85 μ long and 15–25 μ wide at the apex. TELIA amphigenous; subepidermal, light brown or in colourless spots *c*. 1–2 mm and sometimes with a chlorotic halo on the dorsal side. TELIOSPORES prismatic, often rounded at the apex, 25–60 × 6–12 μ. No other rust has been recorded on castor (Punithalingam). Noronha successfully inoculated several *Euphorbia* spp. *Melampsora ricini* occurs widely in Africa but has not been reported from the W.; it is also present in Burma, China, Cyprus, Greece, India, Israel, Italy (including Sardinia), Portugal, Sri Lanka, Taiwan and Yemen (Map 467). The pathology of castor rust has not apparently been studied; although there are early Indian reports of its destructiveness.

The very widespread (Map 68) flax rust (*M. lini* (Ehrenb.) Desm.) has been studied in depth. Several *Melampsora* spp. occur on *Populus* spp., 2 of which were recently described by Walker; see also Browne (*Pests and diseases of forest plantation trees*) and Arthur (*Manual of the rusts in United States and Canada*) for other *Melampsora* spp.

Ajrekar, S. L. 1912. *J. Bombay nat. Hist. Soc.* July, misc. notes 32: 1092 (*Melampsora ricini*, morphology).

Laundon, G. F. et al. 1965. *C.M.I. Descr. pathog. Fungi Bact.* 51 (*M. lini*).

Noronha, E. de A. 1952. *Agron. lusit.* 14: 229 (*M. ricini*, inoculations: 32: 591).

Punithalingam, E. 1968. *C.M.I. Descr. pathog. Fungi Bact.* 171 (*M. ricini*).

Walker, J. 1975. *Ibid* 479, 480 (*M. larici-populina* & *M. medusae*).

MELANOPSICHIUM Beck., *Annln naturh. Mus. Wien* 9: 122, 1894.
(Ustilaginaceae)

SORI formed in cavities in various parts of the host, enclosed by a more or less compact and permanent gelatinous envelope, containing compact, hard, conspicuous, dark, gall-like spore masses; blackish galls consisting of a mixture of plant tissue, mycelium and spore-bearing hyphae. SPORES firmly held together in a slimy matrix, developing singly on mycelial branches and germinating by the formation of a septate promycelium with terminal and lateral sporidia (from B. B. Mundkur & M. J. Thirumalachar, *Ustilaginales of India*: 45, 1952).

M. eleusinis (Kulk.) Mundkur & Thirum. (synonym: *Ustilago eleusinis* Kulkarni) causes smut of finger millet (*Eleusine coracana*). SORI ovariicolous, scattered in the spike, in lysigenous cavities in the ovaries; each cavity surrounded by host tissue, at first small and filled with mycelium forming a dense mass, later enlarging by gelatinisation of host tissue, irregular in shape and size; no columella. SPORES set free in a gelatinous matrix, later pulverulent, globose or subglobose, auburn, 7–11 μ diam. (mean 9.5 μ); epispore densely pitted with a rather uneven margin. Germination by a septate promycelium producing both lateral and terminal sporidia (from B. B. Mundkur & M. J. Thirumalachar, *Mycol. Pap.* 16: 1, 1946). Khanna et al. found that the spores are echinulate not pitted.

The infected grains are converted into galls (sori or sacs) which become 6–7 times the size of the grain and up to 15–16 mm diam. The outer coat is at first greenish, occasionally pinkish, then becoming brown and black. Although this smut has been reported to be seedborne it would seem that such transmission does not occur. Infection is airborne and floral infection is likely. The disease, mostly studied in India, seems of little economic importance. *M. missouriense* Whitehead & Thirum. was described from soybean (SPORES reddish brown, densely echinulate, 6.1–11.3 × 7.5–12 μ). Infected pods are transformed into hard, charcoal-like galls. *M. esculentum* (P. Henn.) Mundkur & Thirum. (l.c. 1946) is found on *Zizania latifolia*.

Khanna, A. et al. 1971. *Curr. Sci.* 40: 529 (on finger millet, electron microscopy of spores; 51, 2252).

Kulkarni, G. S. 1922. *Ann. appl. Biol.* 9: 184 (on finger millet, morphology & symptoms; 2: 308).

Thirumalachar, M. J. et al. 1947. *Phytopathology* **37**: 481 (on finger millet, morphology & transmission; **26**, 542).

Whitehead, M. D. et al. 1960. *Mycologia* **52**: 189 (on soybean; **41**: 115).

MICROCYCLUS Sacc., *Annls Mycol.* **2**: 165, 1904.

Melanopsammopsis Stahel, 1917

Dothidella sensu Thiess & Syd., 1915 (non Speg., 1881).

(further synonymy in Müller & Arx, *Beitr. Kryptogflora Schweiz* **11**: 368, 1962)

(Dothidiaceae)

ASCOSTROMATA superficial on leaves or other parts of the plant and developing from a hypostroma which penetrates the substrate. HYPOSTROMA forming a dark cushion or crust-like stroma on the surface, often broadening upwards. One or more (frequently several) locules within the stroma, spherical or flat sided when crowded, when mature with a papillate ostiole, when immature locules contain a pseudoparenchyma of colourless roundish to angular cells. ASCI developing from a basal stroma, cylindrical, claviform, bitunicate, thickening towards the top, 8 spored. ASCOSPORES hyaline or less frequently yellowish, 1 septate (after E. Müller & J. A. von Arx l.c.).

Microcyclus ulei (P. Henn.) Arx in Müller & Arx, *Beitr. Kryptogflora Schweiz*, **11**: 373, 1962.

Dothidella ulei P. Henn., 1904

Melanopsammopsis ulei (P. Henn.) Stahel, 1917

Aposphaeria ulei P. Henn., 1904

stat. conid. *Fusicladium macrosporum* Kuyper, 1911.

STROMATA globose, adaxial, carbonaceous, superficial, frequently crowded and sometimes in rings round the edge of a shothole, 200–450 μ diam., but frequently fused laterally. Cell walls of pseudoparenchyma uniformly thick and dark. ASCOCARPS internal diam. 100–200 μ, separate, with papillate ostiole. ASCI bitunicate, clavate, 50–80 × 12–16 μ, with 8 sub-distichous spores. ASCOSPORES hyaline, 1 septate, constricted, ellipsoidal; cells unequal, the longer cell (with a more acute apex) lies towards the base of the ascus, 12–20 × 2–5 μ. PYCNIDIA ostiolate, 120–160 μ diam. PYCNOSPORES hyaline, dumbbell shape 6–10 μ long, borne on hyphal elements, 12–20 × 2–3 μ. CONIDIOPHORES simple, arising from a subepidermal stroma, at first aseptate with a subglobose base, later multicellular, hyaline then greyish, up to 140 × 4–7 μ. CONIDIA borne singly, acrogenous, usually 1 septate but occasionally aseptate, hyaline then greyish, ellipsoidal; proximal cell broader with a thickened truncate end and usually a very characteristic single twist which is absent in the aseptate conidia; 2 celled, 23–63 × 5–10 μ; 1 celled, 15–43 × 5–9 μ. (P. Holliday, *CMI Descr.* 225, 1970.)

South American leaf blight (one of the classical plant diseases) of rubber (*Hevea brasiliensis*) is the most serious leaf disease of this crop whose main area of production is S.E. Asia, especially W. Malaysia. *M. ulei* occurs on wild *Hevea* spp. in the Amazon and Orinoco river basins of S. America where the genus is indigenous. The fungus is restricted to tropical America and has been reported from: Bolivia, Brazil, Colombia, Costa Rica, French Guiana, Guatemala, Guyana, Haiti, Honduras, Nicaragua, Panama, Peru, Surinam, Trinidad and Venezuela (Map 27); and see Hilton for an account of the spread of the disease, partly brought about by man. Besides the economic host the pathogen has been reported on *H. benthamiana*, *H. guianensis* and *H. spruceana*. It is not apparently known whether the remaining *Hevea* spp. have susceptible genotypes.

Rubber cultivations in Asia are largely derived from a single introduction of seed of *H. brasiliensis* collected in Brazil by Henry Wickham in 1876 from an area west of the lower Tapajoz river. A few plants from this collection reached Sri Lanka and Singapore via the Royal Botanic Gardens, Kew. Up to 1942 there were attempts, based on seed and clones imported from Asia, to establish the crop in S. and Central America. But these descendants of Wickam's trees proved extremely susceptible as *M. ulei* spread to the cultivations from wild *Hevea* spp. In 1933 infection destroyed 2000 acres of a planting begun in 1927 on the river Tapajoz. A planting of 16 000 acres (1934–42) in the same area was only partly saved from complete destruction by crown budding with partially resistant material. In Costa Rica a report published in 1941 stated that all but 10% of a stand of 36 000 trees was decimated. These early attempts to establish rubber in tropical America have been popularly described by Carefoot & Sprott (*Famine on the wind*). They originated in the increasing exploitation of rubber after the discovery of vulcanisation (see V. Baum, 1945, *The weeping wood*, Michael Joseph; and R. Collier, 1968, *The river that God forgot*, Collins).

The sexual and asexual states were described separately by Hennings and Kuyper and it was not until 1917 that Stahel showed unequivocally that they were forms of the same fungus, then causing serious disease in the young rubber plantings of Guyana and Surinam. After the initial studies on the biology of the disease by Stahel and Langford (1945) there was little work (apart from breeding and field selection for resistance) until that by Chee and Holliday (from 1964) under the aegis of the Rubber Research Institute of Malaysia. The symptoms are most conspicuous on the leaves. On young, brown red pigmented ones the conidial lesions form as very dark, powdery irregular areas on the lower surface. The leaf becomes distorted and usually falls. Older, pale green, soft leaves show conspicuous olive green areas, covered with masses of easily detached conidia, up to 2 cm diam. Such leaves do not usually fall and as they become hardened the black pycnidia form on the upper surface, increasing in number and size and often arranged in rings. The stromata become more massive as the ascocarps mature (mainly on the upper surface) and the centre of the original conidial lesion decays away to leave a shothole. Severe infection leads to dieback and death in both young and mature trees. Symptoms of the 3 spore stages can also occur on green stems, petioles and inflorescences which can become hypertrophied, and young green fruits are destroyed. Internally, pallisade, cortical and pith tissue become enlarged. Conidial germination is complete in 3–4 hours at an opt. of 24°. Growth in culture (best at 23°) is slow. Ascospores germinate readily at the same opt. temp. Penetration is cuticular (the pycnospores do not appear to cause infection) and it probably occurs within 5–6 hours of spore deposition, under moist conditions. Leaves up to 10 days old are most susceptible, becoming thereafter increasingly resistant. Conidia begin to form within a week of infection and the perfect state is mature about 8–9 weeks later. It is not known if heterothallism occurs.

The conidia are dry spores, passively abscinded before take off through air currents. In a dry state one or both cells are collapsed but they become turgid on adhering to the surface of a water drop. The diurnal periodicity, most marked on dry days, shows a max. conc. of conidia at 1000 hours with much less dispersal in the evening, night and early morning. Rain causes large, transient increases in the numbers of conidia, especially in the forenoon and early afternoon. It can, therefore, modify the periodicity typical of dry days. In dry weather the ascospore peak is at 0600 hours but on rainy days there can be heavy ascospore discharge later in the day. Ascospores are rapidly killed by high light intensities (e.g. 4 minutes' exposure to UV). Conidia can survive for up to 16 weeks on glass slides under desiccation and for 4 weeks RH 65%. The opt. temp. for infection is *c*. 24°. Four races have been reported but the experimental evidence is still inadequate. Since the pathogen infects the young leaves it can cause secondary leaf fall after the normal wintering of mature rubber. But it can also be serious on young trees that are not wintering and the complete destruction of nurseries and young plantings has been reported. Spread is both through the conidia and the ascospores. Where rainfall is <3–4 mm/day conidial sporulation is very low or absent, at twice this amount it is very high. A dry season of 4 consecutive months (<7–8 cm/month) limits annual incidence appreciably on mature trees, but its effect on juvenile trees would be less marked because of the difference in the manner of leaf renewal. When no such dry season exists, and annual rainfall is at least 250 cm, incidence can be severe.

Although in certain areas of Central and S. America (where there is a long dry season) susceptible clones can be grown, in most cases resistant material must be used. High resistance occurs in the upper Amazon region in wild populations of *H. brasiliensis*, and also in *H. benthamiana* and *H. pauciflora*. The early breeding work used the highest yielding (but susceptible) Asian clones available in the 1930s and 1940s. The second and third generations (back crosses to susceptible parent) in pure lines are less resistant than in hybrid ones (*brasiliensis* × *benthamiana*). Only first generation clones appear to have been adequately tested for yield and other characters. Nothing is known of the inheritance of resistance; but that from at least one clone of *H. benthamiana* (Ford 4542) may be controlled by dominant genes. Selection and breeding is continuing in Brazil, Guatemala, Liberia, Malaysia and Sri Lanka. An unusual control method (used with some success) is to bud a resistant crown clone on to a high yielding (oriental), but susceptible, panel or trunk clone. Although the method has some disadvantages in terms of cost, it could reduce the time required to produce a resistant tree with a yield comparable to present-day commercial clones. Early attempts at chemical control of South American leaf blight in this tree crop were confined to nurseries. Dithiocarbamates were effective. Aerial spraying is now being tried in Brazil and benomyl, mancozeb, propineb and oxycarboxin have shown some effectiveness (Alencar et al.; Mainstone et al.; Rocha et al.; Rogers et al.). Chee (1978b) found benomyl and thiophanate methyl to be the most effective in small-scale trials in Trinidad.

This disease poses a definite threat to the whole natural rubber producing industry of Asia, primarily, and also Africa; in particular to Malaysia, the largest single producer. It is known that all the Asian cultivations consist of selections which are extremely susceptible, although the temps in W. Malaysia might limit *M. ulei*. If the pathogen first appeared in this area, i.e. before being known in any other part of Asia, an elaborately planned attempt to eradicate it would be made. The infected area and a guard belt 1/4 mile wide would be defoliated with 5% n-butyl 2,4,5-T in 3 gal/gas oil/acre followed by 1% pentachlorophenol in the same solvent and the same rate, both applied by fixed wing aircraft (Altson; Hutchison).

Alencar, M. H. et al. 1975. *Revta Theobroma* **5**(3): 12 (aerial spraying; **57**, 268).

Altson, R. A. 1955. In Hilton *J. Rubb. Res. Inst. Malaya* **14**: 338 (quarantine, preventive & eradicant measures; **35**: 716).

Bos, H. et al. 1965. *Ibid* **19**: 98 (breeding for resistance; **45**, 342).

Camargo, A. P. de et al. 1967. *Bragantia* **26**: 18 pp. (climate; **48**: 1937).

Chee, K. H. 1976. *Ann. appl. Biol.* **84**: 135, 147 (resistance & spore dispersal; **56**, 1255, 1256).

——. 1976. *Trans. Br. mycol. Soc.* **66**: 499 (discharge, germination & viability of spores; **56**, 363).

——. 1978a. *Ibid* **70**: 341 (culture; **57**, 5101).

——. 1978b. *Ann. appl. Biol.* **90**: 51 (fungicides).

——. 1978c. *South American leaf blight of* Hevea: *A bibliography*. 41 pp., Unit Rubb. Res. Inst. Malaysia, Univ. W. Indies, Trinidad (annotated, 212 ref.; **57**, 5100).

Figari, A. 1965. *Turrialba* **15**: 103 (resistance; **45**, 19).

Goncalves, J. R. C. 1968. *Trop. Agric. Trin.* **45**: 331 (resistance; **48**, 271).

Hilton, R. N. 1955. *J. Rubb. Res. Inst. Malaya* **14**: 287 (general history in America, effect on Malaysian rubber; **35**: 716).

Hoedt, T. G. E. 1953. *Arch. Rubb. cult.* **30**: 1 (resistance, survey; **33**: 113).

Holliday, P. 1969. *Ann. appl. Biol.* **63**: 435 (conidial dispersal, climate; **48**, 3641).

——. 1970. *Phytopathol. Pap.* 13, 31 pp. (monograph, 138 ref.; **49**, 2617).

Honig, P. et al. 1947. *Chron. Natur.* **103**: 63 (resistance; **26**: 416).

Hutchison, F. W. 1958. *J. Rubb. Res. Inst. Malaya* **15**: 241 (eradicant aerial spraying; **38**: 623).

Langdon, K. R. 1965. *Pl. Dis. Reptr* **49**: 12 (races; **44**, 1943).

Langford, M. H. 1945. *Tech. Bull. U.S. Dept Agric.* 882, 31 pp. (general).

—— et al. 1953. *Turrialba* **3**: 102 (chemical control; **33**: 177).

—— ——. 1954. *Pl. Dis. Reptr* Suppl. 225: 42 (control; **34**: 176).

——. 1957. *Turrialba* **7**: 104 (resistant material; **37**: 506).

Mainstone, B. J. et al. 1977. *Plrs' Bull. Rubb. Res. Inst. Malaysia* 148: 15 (aerial spraying; **57**, 1835).

Miller, J. W. 1966. *Pl. Dis. Reptr* **50**: 187 (races; **45**, 1893).

——. 1966. *Phytopathology* **56**: 718 (pathogen toxin; **45**, 2942).

Rands, R. D. 1924. *Bull. U.S. Dept Agric.* 1286 (general; **4**: 309).

—— et al. 1955. *Circ. U.S. Dep. Agric.* 976, 79 pp. (American rubber development).

Rocha, H. M. et al. 1975. *Revta Theobroma* **5**(3): 3 (aerial spraying; **57**, 267).

—— ——. 1978. *Fitopat. Brasiliera* **3**: 163 (fungicides in a nursery).

Rogers, T. H. et al. 1976. *Proc. Int. Rubb. Conf.* Vol. 3, Kuala Lumpur: 266 (fungicides; **57**, 263).

Sorensen, H. G. 1942. *Agric. Am.* **2** (Oct.): 191 (crown budding; **22**: 495).

Stahel, G. 1917. *Meded. Dep. Landb. Surinam* 34: 111 pp. (monograph).

Subramaniam, S. 1969. *J. Rubb. Res. Inst. Malaya* **21**: 11 (selection & resistance).

——. 1970. *Ibid* **23**: 39 (performance of resistant clones in Malaysia; **50**, 1947).

Townsend, C. H. T. 1960. *Econ. Bot.* **14**: 189 (breeding for resistance; **40**: 486).

Tysdal, H. M. et al. 1953. *Agron. J.* **45**: 234 (general, breeding for resistance; **33**: 178).

Wycherley, P. R. 1968. *Planter Kuala Lumpur* **44**: 1 (*Hevea* introduction to Asia).

Yoon, P. K. 1967. *Plr's Bull. Rubb. Res. Inst. Malaya* 92: 240 (crown budding).

MICROSPHAERA Lév., *Annls Sci. nat. Bot.* III, **15**: 154, 1851.

(Erysiphaceae)

ASCOCARPS (cleistothecia, perithecia) with repeatedly, dichotomously branched appendages, with the branches not hooked or coiled like those of *Uncinula* (not described) and forming a compact head on a long stalk; with several ASCI. CONIDIOPHORES usually abundant. MYCELIUM superficial (partly from C. E. Yarwood, *The fungi* (edited by G. C. Ainsworth et al.) Vol. IVA: 71, 1973).

The imperfect state is *Oidium* Sacc. = *Acrosporium* Nees ex Gray; *Erysiphe* (q.v.) for some general comments and ref. on the powdery mildews. The powdery mildew on soybean has recently been described as *M. diffusa* Cooke & Peck by Paxton et al.; and see McLaughlin et al. It occurs on other legumes (Mignucci et al. 1978). CLEISTOTHECIA 86–120 μ diam., mature appendages mostly 180–280 μ in length; tips up to ×6 dichotomously branched, up to 60 μ from the first fork; the first branches at wide angles but the last at so narrow an angle as to be subparallel; neither thickened nor recurved (in this differing from most spp. of the genus). The fungus is rarely reported on soybean but an epidemic was reported from local areas in Wisconsin, USA in 1974. There was more disease on seedlings at 18° than at 24° and 30° (Mignucci et al. 1977). Cvs vary considerably in their reactions to infection and some show no mildew. Grau et al. reported a resistance form as being inherited as a single dominant. Two *Microsphaera* spp. recently described are *M. grossulariae* (Wallr.) Lév. (European gooseberry mildew) and *M. penicillata* (Wallr. ex Fr.) Lév. (alder and lilac mildew; and on many other hosts). Ref. are to *M. diffusa* unless stated otherwise.

Arny, D. C. et al. 1975. *Pl. Dis. Reptr* **59**: 288 (local epidemics in USA & cv. reactions; **54**, 5157).

Demski, J. W. et al. 1974. *Ibid* **58**: 723 (cv. reaction; **54**, 621).

Dunleavy, J. M. 1977. *Ibid* **61**: 32 (mildew reaction in 50 cvs; **56**, 3333).

——. 1978. *Crop Sci.* **18**: 337 (losses).

Grau, C. R. et al. 1975. *Pl. Dis. Reptr* **59**: 458 (inheritance of resistance; **55**, 982).

Lehman, S. G. 1931. *J. Elisha Mitchell scient. Soc.* **46**: 190 (on soybean; **11**: 19).

McLaughlin, M. R. et al. 1977. *Phytopathology* **67**: 726 (morphology; **57**, 868).

Mignucci, J. S. et al. 1977. *Pl. Dis. Reptr* **61**: 122 (effects of temp.; **56**, 4786).

—— ——. 1978. *Phytopathology* **68**: 160 (on legumes; **57**, 5213).

Mukerji, K. G. 1968. *C.M.I. Descr. pathog. Fungi Bact.* 183 (*Microsphaera penicillata*).

Paxton, J. D. et al. 1974. *Mycologia* **66**: 894 (identification of soybean powdery mildew; **54**, 1957).

Sivanesan, A. 1970. *C.M.I. Descr. pathog. Fungi Bact.* 252 (*M. grossulariae*).

MONILIOPHTHORA H.C. Evans, Stalpers, Samson & Benny, *Can. J. Bot.* **56**: 2530, 1978.

HYPHAE hyaline, septate, septa without clamp connections but with dolipores. CONIDIA in chains, globose to ellipsoidal, formed in basipetal succession. Teleomorph unknown (from H. C. Evans, J. A. Stalpers, R. A. Samson & G. L. Benny l.c.).

The genus was erected to accommodate the important pathogen of cacao which was formerly known as *Monilia roreri* Cif., and which was found to be a basidiomycete. The following description is after Evans et al. (1978): Growth (opt. 25–26°) on malt agar slow, colony diam. 8–15 mm after 2 weeks; woolly to felty; first pale ochraceous salmon or pinkish buff, becoming cinnamon buff to clay coloured or wood brown to apricot buff. HYPHAE in advancing zone hyaline, thin walled, septate, sometimes with swellings, 1.5–5 μ wide. Aerial hyphae: (a) as in advancing zone but with slightly thickened walls; (b) hyaline to pale brownish, thick walled, aseptate, (1–) 1.5–2 (–3) μ wide, skeletoid, rarely branched. CONIDIOPHORES branched, forming a basipetally maturing chain of conidia. CONIDIA easily separable, thick walled, pale yellow, brown in mass, typically globose to subglobose, (6.5–) 8–15 (–25) μ diam.; sometimes ellipsoid, 8–20 × 5–14 μ; wall up to 2 μ thick; chlamydospores or chlamydospore-like swellings may occur. Crystals (tetragonal) present. Perfect state not known. Evans et al. (1978) also described conidiogenesis and septal structure, and discussed the taxonomic position.

Moniliophthora roreri causes an extremely destructive pod rot of cacao to which the name frosty pod rot was recently given (Evans et al. 1978). Accounts of the disease have been given by Ampuero; Evans et al. (1977); Holliday (1970); Jorgensen. The fungus, which occurs on other *Theobroma* spp. and *Herrania*, is restricted to parts of N.W. South America (Colombia, Costa Rica, Ecuador, Panama, Peru and Venezuela; Map 13). It appears to be absent from the low-lying areas east of the N. Andean cordilleras (Amazon and Orinoco river basins; H. C. Evans, personal communication, December 1978) where the other important, and also restricted, cacao pathogen (*Crinipellis perniciosa* q.v.) is common.

Penetration through the pod cuticle occurs from the conidia, apparently at any stage of pod development. Young pods form pronounced swellings but symptoms may be absent until lesions form 45–90 days after penetration. The pods show signs of premature and irregular ripening; then dark brown lesions form and these coalesce to cover the whole pod surface. Pods infected at late growth stages may show a restricted necrosis. A white to cream pseudostroma grows over the pod surface 3–8 days after lesion initiation. This is soon covered with a dense, cream, powdery conidial mass which becomes grey or brown. The colour change may be associated with a thickening of the conidial wall. The internal pod rot may be watery, moist or even dry; the beans are stuck together and to the pod wall. If the infected tissue is exposed by cutting the pod across a thick, cream, sporulating pseudostroma soon covers the cut surfaces (Evans et al. 1978). Reports from Colombia that the fungus penetrates via insect wounds and punctures have not been confirmed in Ecuador. There is evidence for a systemic infection phase in the pod cushions and the stems. The conidia are air dispersed. No firm recommendations for control can be given (see Evans et al. 1977 and Jorgensen for discussions). The present approach in Colombia and Ecuador (apart from cultural attempts to reduce inoculum) is to investigate methods whereby cacao trees would set their crop at a time (drier seasons) when conditions for infection are unfavourable. Such a control method should also reduce losses from *C. perniciosa*.

Ampuero, C. E. 1967. *Cocoa Grow. Bull.* 9: 15 (review; **47**, 1496).

Baker, R. E. D. et al. 1954. *Rep. Cacao Res. Trin. 1953*: 8 (distribution, cacao in Colombia).

Brown, D. A. Ll. 1974. Federation Nacional de Cafeteros, Colombia, 13 pp. (bibliography, 157 ref.; **54**, 1212).

Evans, H. C. et al. 1977. *PANS* **23**: 68 (cacao diseases in Ecuador; **56**, 3951).

———. 1978. *Can. J. Bot.* **56**: 2528 (*Moniliophthora* gen. nov., taxonomy & morphology).

Holliday, P. 1953. *J. agric. Soc. Trin.* **53**: 397 (distribution, cacao in Colombia).

——. 1970. *C.M.I. Descr. pathog. Fungi Bact.* 226 (as *Monilia roreri*).

Jorgensen, H. 1970. *Cacao Turrialba* **15**(4): 4 (review, 39 ref.; **51**, 3915).

MONILOCHAETES *infuscans* Halst. ex Harter, *J. agric. Res.* 5: 791, 1916. (Dematiaceae)

Monilochaetes: Hyphae dark, erect, rigid, septate, not in definite fascicles; conidia distinctly different from the sporophores and hyphae, hyaline, slightly brown with age, continuous, not in chains, acrogenous.

M. infuscans: On host definite vegetative hyphae are lacking. CONIDIOPHORES septate, erect, unbranched, dark; attached to host, singly or in pairs, by a bulb-like enlargement, 40–175 μ long, 4–6 μ wide, bearing rarely a hyaline, aseptate, oblong spore. In cooked rice the hyphae are much branched, septate, brown; conidiophores brown except at terminal cell, which is frequently hyaline to slightly brown, septate, branched, stout, 30–225 × 4–6 μ. CONIDIA abundant, aseptate, ovoid to oblong, 12–20 × 4–7 μ, solitary, terminal. Parasitic on the underground parts of *Ipomoea batatas* (from L. L. Harter, l.c., and see Taubenhaus).

Scurf of the tubers of sweet potato has been reported from: Argentina, Australia, Azores, Brazil, China, Hawaii, Israel, Japan, New Zealand, Zimbabwe, Sierra Leone, Taiwan, USA, US Trust Territories (Pacific) (Map 246). The disease (largely investigated in USA) is a minor one but market losses can be important because of the disfiguring effects of infection. Also the somewhat superficial damage to the outer tuber tissue layers can lead to moisture loss and shrinking in storage. The apparent restriction of scurf to the more temperate sweet potato growing areas is presumably partly a temp. effect. It may also be partly due to the differing methods of propagation. In the tropics this is almost always by direct field planting of stem cuttings. In temperate regions propagation is by sprouts or slips obtained by planting smallish root tubers in nursery beds (or hotbeds); when the sprouts are some 25 cm long they are transplanted to the field. The fungus is thus spread from infested nursery beds to the field where the developing tubers are infected. Symptoms begin as small brown spots on the tuber surface; these enlarge and coalesce to give discoloured, superficial, necrotic areas of varying size and shape, and without a definite outline. There is no general rupture of the epidermis. But Daines (1972) found that the fungus invaded tissue in greater depth in the cvs Julian and Nemagold compared with Yellow Jersey. Infection may cover most of the tuber surface. Above-ground parts of the plant are not attacked but stems and petioles can be infected if covered with infested soil.

The fungus is soilborne and survives on crop

debris and probably other organic matter in the soil. Its growth in culture is very slow. The disease is more severe in heavy soils. Control is through the use of clean tuber stock grown in non-infested soil, fungicidal dip treatments of planting material (stems and tubers) and rotation (2–3 years on lighter soils and 3–4 years on heavier ones). Daines has described recent work on control with fungicides which are more effective when planting is in nursery beds at 32° compared with those at 24°. Complete control on harvested tubers was given by a 0.5 minute ferbam dip at 55° or 5 minutes at 49°. A water dip at these times and temps also gives good control. Thiabendazole and benomyl as dips at these temps are also effective, and gave better control than when used at 18°. Although cvs differ in reaction to infection there is apparently no high host resistance.

Daines, R. H. 1955. *Pl. Dis. Reptr* **39**: 617, 739 (disease development in storage and fungicides; **35**: 546).
——. 1970. *Phytopathology* **60**: 1474 (effects of plant bed, prebedding air & fungicide dip temp. in control; **50**, 1549).
——. 1971. *Pl. Dis. Reptr* **55**: 746 (effects of water & ferbam dips & temps; **51**: 1031).
——. 1972. *Ibid* **56**: 122 (effects of fungicides & temps; **51**: 3688).
Kantzes, J. G. et al. 1958. *Bull. Md agric. Exp. Stn* A-95, 28 pp. (nutrition of *M. infuscans*, pathogenicity & control; see **37**: 558).
Martin, W. J. et al. 1966. *Phytopathology* **56**: 1257 (disease & length of root attachment of tuber; **46**, 1087).
Nielsen, L. W. et al. 1966. *N.Z. Jl agric. Res.* **9**: 1032 (resistance, with *Ceratocystis fimbriata*; **46**, 726).
Taubenhaus, J. J. 1916. *J. agric. Res.* **5**: 995 (general).

MYCENA (Pers. ex Fr.) S. F. Gray, *Nat. Arr. Br. Pl.* **1**: 619, 1821.
(Tricholomataceae)
(A full description and synonymy of the genus, mostly saprophytic, was given by R. Singer in *The Agaricales in modern taxonomy*: 384, 1975)

Mycena citricolor (Berk. & Curt.) Sacc., *Sylloge Fung.* **5**: 263, 1887.
Agaricus citricolor Berk. & Curt., 1868
Stilbum flavidum Cooke, 1881
Omphalia flavida, Maubl. & Rangel, 1914.
(see R. W. G. Dennis, *Kew Bull.* **1950**: 434)

SPOROPHORE very minute, yellowish. PILEUS thin, membranaceous, hemispheric–campanulate, sunken or subumbilicate in the centre, then more or less flattened, glabrous, radially striate, 1.5–2.5 mm diam., with acute margin. STIPE setiform, straight thin, of the same colour, very minutely velvety, *c.* 1–1.5 cm long, 0.25 mm thick, base not swollen. GILLS few, rather distant, yellowish, somewhat waxy, triangular, attenuated at each end, more or less decurrent. BASIDIA clavate 14–17.4 × 5 μ. BASIDIOSPORES minute, ellipsoid or ovoid, apiculate below, hyaline, eguttulate or with 1 guttule, 4–5 × 2.5 μ. GEMMIFER consisting of a stalk (PEDICEL) and a head (GEMMA). Stalk yellow, solid, cylindrical, curved at maturity, slender, *c.* 2 mm long, 0.12 mm diam. at base, 0.05 mm diam. below head; gemma av. diam. *c.* 0.36 mm, an oblate spheroid, solid, pseudoparenchymatous, tough, coriaceous, attached by a basal collar (apophysis) which encloses stalk *c.* 0.1 mm from its extreme end; head covered with many aerial, radiating filaments (after A. H. R. Buller, *Researches on fungi* Vol. VI: 397, 1934).

Buller (l.c.), who described the history, morphology and behaviour of *M. citricolor* in great detail, considered that the pedicel and gemma are homologous with the stipe and pileus, respectively. The gemmifer produces no spores and the gemma is the infective body. The fungus causes cock's eye of coffee or American leaf spot. It has been largely investigated on this crop but *M. citricolor* is plurivorous and Carvajal gave 100 plant hosts in 50 families, and see Sequeira. Ashby showed the connection between the 2 forms in pure culture. The distribution is: Central America, West Indies, S. America (tropics), Mexico, USA (Florida; Map 9).

Leaves, stems and berries of coffee are attacked. Leaf spots are usually circular, 6 mm diam. (up to 13 mm), dark then light coloured; the central tissue may disintegrate giving a shothole. The hair-like projections (filaments) of the gemmifer stage (1–4 mm long) are seen on the spots on the upper surface, but to some extent on the lower surface, in wet weather. Infection can cause almost complete defoliation of a coffee bush. Gemma formation is light induced and the mycelium is luminous. The fungus spreads in wet conditions to crops from wild hosts in upland forested areas with relatively cool temps. The perfect state seems rare on coffee and basidiospores are not considered to be important in the epidemiology. The normally slow rate of spread indicates that the unusually large airborne propagating unit (the gemma) is of primary importance. After deposition of the gemmae on leaves they infect through the growth of numerous hyphae which penetrate the cuticle.

Cultural control measures, i.e. prompt destruction of infective material and barriers of hedge plants, other crops or non-agricultural land have been successful (Wellman). This disease appears to be an unusual example of one spread by a fungus aerially but relatively slowly, so that it is most likely to be controlled by sanitation. The effects of *M. citricolor*, should it spread outside the Americas, seem uncertain. The severity of the disease would presumably depend on how effective sanitary measures could be. The apparent difficulty in applying such measures under the conditions of smallholder cultivations is often reflected by the damage which can be caused (e.g., annual losses of 20% in Costa Rica and Guatemala). Perez used copper fungicides.

Ashby, S. F. 1925. *Kew Bull.* **1925**, 8: 325 (perfect state in pure culture; **5**: 160).
Carvajal, B. F. 1939. *Revita Inst. Def. Café C. Rica*, Suppl. **7** (52), 40 pp. (general; **18**: 589).
Salas, J. A. et al. 1972. *Hilgardia* **41**: 213 (perfect state in culture; **52**; 1149).
Sequeira, L. et al. 1954. *Pl. Physiol.* **29**: 11 (auxin inactivation & relation to leaf drop; **33**: 601).
——. 1959. *Turrialba* 8: 136 (hosts; **38**: 745).
Uribe Arango, H. 1947. *Revta Fac. nac. Agron. Medellin* 7(26): 249 (general; **27**: 132).
Wellman, F. L. 1950. *Turrialba* 1:12 (spread & control; **31**: 15).
Perez S., V. M. 1954. *Suelo tico* 7(30): 177 (fungicides in Costa Rica; **34**: 34).

MYCOCENTROSPORA Deighton, *Taxon* 21: 716, 1972.

Centrospora Neergaard, 1942
Ansatospora Newhall, 1944.
(Dematiaceae)

COLONIES effuse, at first hyaline, later variously coloured, finally almost black. MYCELIUM superficial and immersed; with torulose groups of swollen, mid to dark brown cells. STROMA none. SETAE and HYPHOPODIA absent. CONIDIOPHORES macronematous, mononematous, straight or flexuous, geniculate in the upper part, usually branched, colourless, smooth. CONIDIOGENOUS CELLS integrated, terminal, polyblastic, sympodial, clavate or cylindrical with broad flat scars. CONIDIA solitary, acropleurogenous, often appendiculate with a septate lateral appendage from the basal cell; or sometimes simple, obclavate, rostrate, frequently strongly curved, truncate at the base, colourless or with the broader cells rather pale brown, smooth, multiseptate (M. B. Ellis, *Demat. Hyphom.*: 264, 1971 as *Centrospora*; the genus and spp. also described by Deighton).

Deighton, F. C. 1971. *Mycol. Pap.* 124 13 pp. (*Centrospora*; **50**, 2169).

Mycocentrospora acerina (Hartig) Deighton, *Taxon* 21: 716, 1972.

Cercospora acerina Hartig, 1880
Ansatospora macrospora (Osterw.) Newhall, 1944 (nomen non rite publicatum)
Centrospora acerina (Hartig) Newhall, 1946.
(more synonymy in Deighton, *Mycol. Pap.* 124, 1971; Newhall 1946)

COLONIES at first hyaline, later green, grey or reddish purple, finally almost black. HYPHAE mostly cylindrical, colourless, 4–8 μ thick, but torulose groups of mid to dark brown cells swollen to 17–30 μ are often formed, especially near the centre of colonies. CONIDIOPHORES usually *c.* 50 μ long, 5–7 μ thick, markedly geniculate towards the apex with broad, flat scars. CONIDIA 60–250 (usually 150–200) μ long, 8–15 μ thick in the broadest part, tapering to 1–2 μ at the apex, 4–5 μ wide at the base, with 4–24 (mostly 8–11) septa; septate appendage from basal cell when present usually directed downwards, 30–150 μ long, 2–3 μ thick (M. B. Ellis, *Demat. Hyphom.*: 265, 1971 as *Centrospora acerina*; a description is also given by Deighton 1971 l.c.).

M. acerina has been reported from many hosts; Channon found that the leaves of a wide range of seedling hosts were susceptible to artificial infection. Among the diseases it causes are licorice rot of carrot, black crown rot of celery, a canker of parsnip and a leaf spot of lettuce. On carrot the symptoms are large, blackish, rather deeply sunken lesions with an indefinite, brownish, watersoaked margin. These are often soft, sometimes cover the entire circumference of the roots, and are 4–6 cm long (Srivastava). The pathogen is widespread in Europe; it occurs in N. America and New Zealand but has not apparently been recorded from the tropics. It has a low temp. requirement (opt. *c.* 17° on PDA) and more in vitro growth occurred at 6° than at 27° (Neergaard et al.; Newhall 1944; Truscott). Spread is through soil or water (Iqbal et al.) and the disease often becomes apparent in storage on celery after 7–8 weeks. In carrots it was found, on a crop overwintered in the soil, when the roots were being washed clean before packing. The fungus can carry over in the soil through the conidia, some of whose cells form chlamydospores (Wall et al.). In parsnip wounding roots before inoculation did not increase lesion development. In celery Newhall (1944) found that 3 year rotations to eliminate the fungus were inadequate. A benomyl dip at harvest gave control in celery kept in a cool store for 12 weeks and in hand lifted carrots for 7 months (Derbyshire et al.).

Channon, A. G. 1965. *Ann. appl. Biol.* **56**: 119 (general on parsnip; **45**, 681).

Derbyshire, D. M. et al. 1971. *Proc. 6th Br. Insectic. & Fungic. Conf.*: 167 (vegetable storage diseases; **51**, 2157p).

Griffin, M. J. et al. 1977. *Pl. Pathol.* **26**: 147 (on lettuce; **57**, 3202).

Iqbal, S. H. et al. 1969. *Trans. Br. mycol. Soc.* **53**: 486 (pathogenicity of aquatic isolates; **49**, 1526).

Neergaard, P. et al. 1951. *Phytopathology* **41**: 1021 (general; **31**: 269).

Newhall, A. G. 1944. *Ibid* **34**: 92 (general on celery; **23**: 324).

——. 1946. *Ibid* **36**: 893 (taxonomy; **26**, 133).

Srivastava, S. N. S. 1958. *Trans. Br. mycol. Soc.* **41**: 223 (general on carrot; **37**: 692).

Truscott, J. H. L. 1944. *Can. J. Res.* C **22**: 290 (general on celery; **24**: 174).

Wall, C. J. et al. 1978. *Trans. Br. mycol. Soc.* **70**: 157 (survival of chlamydospore forming conidia; **57**, 3682).

———. 1978. *Ibid* **71**: 143 (quantitative survival of conidia in soil).

MYCOSPHAERELLA Johanson, *Svenska Vetensk.-Akad. Ofvers.* 9: 163, 1884.

(Dothideaceae)

(for nomenclature see L. Holm, *Taxon* **24**: 482, 1975)

PSEUDOTHECIA small, usually black, separate, more or less globose, immersed in host tissue and usually on dead leaves. ASCI relatively large, bitunicate, often ventricose, 8 spored, without pseudoparaphyses. ASCOSPORES 1 septate (near middle), hyaline or nearly so or finally brownish (from R. W. G. Dennis, *British Ascomycetes*: 362, 1968; and key by E. S. Luttrell in *The fungi* (edited by G. C. Ainsworth et al.) Vol. IVA: 173, 1973).

For the taxonomy of the genus see Arx and Müller et al. *M. aleuritis* Ou (as *M. aleuritidis*; stat. conid. *Cercospora aleuritidis* Miyake; synonym: *Cercosporina aleuritidis* Sacc.) causes angular leaf spot of tung (*Aleurites*). The PERITHECIA are gregarious in small angular patches limited by small veins, amphigenous but mostly hypophyllous, innate, globose, black, 60–100 μ diam., papillate at maturity; ASCI fasciculate, cylindric–clavate, aparaphysate, 8 spored, $35–45 \times 6–7$ μ; ASCOSPORES hyaline, ellipsoid, biseriate, 2 celled, the upper cell somewhat larger, $9–15 \times 2.5–3.2$ μ. SPERMOGONIA similar to, but less prominent than, perithecia, $50–70 \times 45–60$ μ; spermatia (?) hyaline, rod shaped, $3.5–4.5 \times 0.5$ μ. CONIDIOPHORES amphigenous, arising from stromata, 5 to many fasciculate, usually simple, somewhat denticulate, dilute olivaceous, 1–5 septate, $22–65 \times 4–5.5$ μ; CONIDIA acicular to obclavate, straight or slightly curved, hyaline, finally olivaceous tinged, 2–12 septate, 35–135 (mostly 50–80) $\times$ 3–4.5 μ (after Ou). *M. websteri* Wiehe causes a leaf spot of *A. montana*.

M. aleuritis is a weak pathogen of tung which has been largely studied in USA on *A. fordii*; it is widespread (Map 278). The reddish brown, more or less circular leaf spots are 4–16 mm diam., usually sharply delimited especially on the upper surface, spots may be angular, vein limited; coalescence leads to extensive dead areas and a ragged appearance; infected leaves fall prematurely. Dark brown to black, ovate elliptic, slightly sunken lesions are caused on petioles, young branches and seedling stems, up to 2 cm long × 1.5 cm wide. The leaf spots caused by *M. websteri* are brown with ashy grey centres and a faint halo, up to 2 × 3 cm diam., distinctly zoned (ringspot), shotholing occurs and infected leaves have a shredded appearance. In USA much primary infection occurs through ascospores released from fallen leaves in the spring. The conidia infect the lower leaf surface more readily than the upper one. Selections of *A. fordii* vary in resistance and fungicides give some control, although this is apparently uneconomic (Zwet et al.).

The 2 weakly parasitic or saprophytic *Mycosphaerella* spp. on banana are referred to under *M. musicola*. Many spp. of the genus which are generally, at most, weak parasites and do not cause significant diseases, have been described from tropical crops. Examples are: *M. caricae* Syd. on *Carica papaya* (Stevens et al.), *M. holci* Tehon on *Sorghum dochna*, *M. manihotis* Syd. on *Manihot esculenta*

(Ghesquière), *M. molleriana* (Thüm.) Lindau and *M. nubilosa* (Cooke) Hansf. on *Eucalyptus*, *M. pinicola* (Fautr.) Naom. on *Pinus* (Leather) and *M. pueraricola* Weimer & Luttrell on *Pueraria thunbergiana*.

Arx, J. A. von 1949. *Sydowia* **3**: 28 (taxonomy; **29**: 122).

Bain, D. C. 1960. *Pl. Dis. Reptr* **44**: 190 (*Mycosphaerella aleuritis*, symptoms; **39**: 505).

Draper, A. D. et al. 1965. *Phytopathology* **55**: 926 (*M. aleuritis*, leaf infection; **45**, 528).

Ghesquière, J. 1932. *Bull. Inst. r. Colon Belge* **3**: 160 (*M. manihotis*; **12**: 137).

Leather, R. I. 1968. *Agric. Sci. Hong Kong* **1**: 17 (fungicide control of *M. pinicola*; **48**: 1351).

Merrill, S. et al. 1959. *Proc. Am. Soc. hort. Sci.* **74**: 232 (resistance to *M. aleuritis*; **41**: 734).

Müller, E. et al. 1962. *Beitr. Kryptogflora Schweiz* **11**: 353 (in genera of didymosporous Pyrenomycetes; **42**: 603).

Ou, S. H. 1940. *Sinensia Shanghai* **11**: 175 (*M. aleuritidis* sp. nov., general.; **21**: 55).

Stevens, F. L. et al. 1932. *Philipp. Agric.* **21**: 9 (*M. caricae*, morphology & synonymy; **11**: 662).

Tehon, L. R. 1937. *Mycologia* **29**: 434 (*M. holci* sp. nov. inter alia; **17**: 69).

Weimer, J. L. et al. 1948. *Phytopathology* **38**: 348 (*M. puerari cola* sp. nov., general; **27**: 478).

Wiehe, P. O. 1953. *Mycol. Pap.* 55, 4 pp. (*M. websteri* sp. nov., comparison with *M. aleuritis*; **34**: 557).

Zwet, T. van der 1961. *Proc. La Acad. Sci.* **24**: 81 (*M. aleuritis*, survey of distribution; **41**: 473).

—— et al. 1963. *Phytopathology* **53**: 734 (sp. as above, technique & ascospore dispersal; **43**, 246).

————. 1965. *Bull. Miss. agric. Exp. Stn* 705, 23 pp. (sp. as above, general).

————. 1966. *Pl. Dis. Reptr* **50**: 54 (sp. as above, effect on yield & oil content of fruit; **45**, 1462).

————. 1966. *Ibid* **50**: 717 (sp. as above, fungicides; **46**, 701).

Mycosphaerella arachidis Deighton, *Trans. Br. mycol. Soc.* **50**: 328, 1967.

Mycosphaerella arachidicola W. A. Jenkins, 1938 stat. conid. *Cercospora arachidicola* Hori, 1917.

PERITHECIA scattered, mostly along margins of lesions, amphigenous, partly embedded in host tissue, erumpent, ovate to nearly globose, 47.6–84 × 44.4–74 μ, black ostiole slightly papillate. ASCI cylindrical, club shaped, short stipitate, fasciculate, 27–37.8 × 7–8.4 μ, aparaphysate, bitunicate, 8 spored. ASCOSPORES uniseriate to imperfectly biseriate in the ascus, 2 celled, the upper cell somewhat larger, slightly curved, hyaline, 7–15.4 × 3–4, av. 11.2 × 3.6 μ. SPERMOGONIA scattered in and along margins of lesions produced by the conidial state, ovate to globose, black, amphigenous but perhaps more often epiphyllous, embedded in leaf tissue but later erumpent, ostiolate, 45–75 × 30–75, av. 62.3 × 48.5 μ. SPERMATIA small, rod shaped, hyaline, 1.5–3 × 0.5 μ, arising endogenously, usually in fours within spermatiferous cells and liberated through sterigma-like processes. STAT. CONID. amphigenous but confined mostly to the upper surface. STROMA present, slight, 25–100 μ diam., dark brown. CONIDIOPHORES in dense fascicles, 5 to many, pale olivaceous or yellowish brown, darker at the base, geniculate, mostly once, unbranched, septate, 15–45 × 3–6 μ. CONIDIA subhyaline, slightly olivaceous, obclavate to clavate, mildly to much curved, up to 12 septa, base round, truncate, tip subacute, 35–110 × 3–6 μ (perfect state W. A. Jenkins, *J. agric. Res.* **56**: 324, 1938; stat. conid. J. L. Mulder & P. Holliday, *CMI Descr.* 411, 1974).

The change in name of the perfect state was necessary because of the earlier described *M. arachidicola* Khokhryakov, also on groundnut. Later this fungus was described as *Didymosphaeria arachidicola* (Khokhr.) Alcorn, Punith. & McCarthy (the other synonym is *M. argentinensis* Frezzi). Its stat. conid. is *Ascochyta adzamethica* Schoschiaschvili (synonyms: ? *A. arachidis* Woronich. and *Phoma arachidicola* Marasas, Pauer & Boerema). *D. arachidicola* causes net blotch of groundnut.

The morphological differences between the conidial states of *M. arachidis* and *M. berkeleyi* have been given briefly under the latter. Early leaf spot of groundnut is widespread with its economic host (Map 166). The disease is also called tikka, without distinction from late leaf spot caused by *M. berkeleyi* on the same host. Accounts of the biology of *M. arachidis* often include the other sp. also, no distinction being made between the 2 pathogens. All ref., where both fungi are being discussed, have been placed under *M. berkeleyi* only. The leaf lesions caused by *M. arachidis* are circular, 1–10 mm diam., reddish brown to black on the upper leaf surface and lighter shades of brown on the lower one. Distinct chlorotic haloes form early on the upper surface. The lesions tend to be larger than those caused by *M. berkeleyi* and the characteristic dark stroma of the latter is absent. The conidia form on both leaf surfaces, the conidiophores being somewhat diffuse.

Severe attacks cause defoliation. Both fungi can often be found together on the same plant (see under *M. berkeleyi* for information on their disease behaviour patterns in the field). These patterns differ mainly in the rate of seasonal build up of infection (early and late leaf spots), the differential damage caused which is affected by climate, types of cvs grown and other agronomic factors such as plant spacing. Field temp. and RH requirements appear to be similar for opt. development of the 2 diseases. In Malawi lesion intensity caused by *M. arachidis* infection was greatest in the Dec. plantings compared with those in Jan.–Feb. (Farrell et al.). Smith et al. found a diurnal periodicity for the conidia with a max. conc. at 1100–1500 hours. In India (both pathogens) the max. conc. was in the forenoon. In sand nutrient culture Bledsoe et al. found that more leaf spots formed on plants without magnesium compared with those with it. Landers and Smith described in vitro growth.

Gibbons et al. exposed 8 *Arachis* spp. and a groundnut cv. to natural infection. *A. glabrata*, *A. hagenbeckii* and *A. repens* developed no lesions; the rest showed a gradation in susceptibility. Resistance appeared to be associated with small stomatal apertures on the leaf. Control is largely through fungicides and cultural measures should be taken (*M. berkeleyi* q.v.); and see Littrell et al. on resistance to benomyl; Parvin et al. on forecasting.

Alabi, R. O. et al. 1977. *Trans. Br. mycol. Soc.* **68**: 295, 296 (effect on leaf content & cell degrading enzymes; **56**, 4794, 4795).

Alcorn, J. L. et al. 1976. *Ibid* **66**: 351 (*Didymosphaeria arachidicola* comb. nov.; **55**, 3822).

Bledsoe, R. W. et al. 1946. *Pl. Physiol.* **21**: 237 (effect of Mg deficiency on disease; **26**: 181).

Farrell, J. A. K. et al. 1968. *Rhod. Zamb. Malawi J. agric. Res.* **5**: 241 (effects of time of planting, spacing & fungicides in Malawi; **47**, 2013).

Frezzi, M. 1969. *Revta Investnes agropec.* Ser. 5 **6**: 147 (*Mycosphaerella argentinensis* sp. nov.; **49**, 1228).

Gibbons, R. W. et al. 1967. *Rhod. Zamb. Malawi J. agric. Res.* **5**: 57 (resistance in *Arachis* spp.; **46**, 2904).

Landers, K. E. 1964. *Phytopathology* **54**: 1236 (growth in vitro; **44**, 935).

Littrell, R. H. 1974. *Ibid* **64**: 1377 (resistance to benomyl & related fungicides; **54**, 3069).

Marasas, W. F. O. et al. 1974. *Phytophylactica* **6**: 195 (*Phoma arachidicola* sp. nov.; **54**, 3573).

Oso, B. A. 1972. *Trans. Br. Mycol. Soc.* **59**: 169 (conidial germination; **52**, 939).

Parvin, D. W. et al. 1974. *Phytopathology* **64**: 385 (forecasting; **53**, 4209).

Smith, D. H. 1971. *Ibid* **61**: 1414 (production of conidial inoculum; **51**, 2260).

—— et al. 1973. *Ibid* **63**: 703 (spore air dispersal, with *Leptosphaerulina crassiasca*; **53**, 1148).

Mycosphaerella berkeleyi W. A. Jenkins, *J. agric. Res.* **56**: 330, 1938.

stat. conid. *Cercosporidium personatum* (Berk. & Curt.) Deighton, 1967

Cercospora personata (B. & C.) Ell. & Ev., 1855

Cladosporium personata B. & C., 1875

Passalora personata (B. & C.) S. A. Khan & K. Kamal, 1961

Septogloeum arachidis Racib., 1898

Cercospora arachidis P. Henn., 1902.

PERITHECIA scattered, mostly along margins of lesions produced by the conidial stage, amphigenous, partly embedded in host tissue, erumpent, broadly ovate to globose, 84–140 × 70–112 μ, black, ostiole slightly papillate. ASCI cylindrical club shaped, short stipulate, fasciculate, 30–40 × 4–6 μ, aparaphysate, bitunicate, 8 spored. ASCOSPORES uniseriate to imperfectly biseriate in the ascus, 2 celled, the upper cell somewhat larger, slightly constricted at the septum, hyaline, 10.9–19.6 × 2.9–3.8 μ, av. 14.9 × 3.4 μ. SPERMOGONIA scattered in and along borders of lesions produced by the conidial stage, ovate to globose, mostly heavy walls, black, amphigenous but perhaps more often epiphyllous, embedded in leaf tissue but later erumpent, ostiolate, 75–90 × 70–90 μ. SPERMATIA small, rod shaped, hyaline, 1–3 × 0.5–1 μ, arising endogenously, usually in fours within spermatiferous cells and liberated through sterigma like processes. STAT. CONID. amphigenous but more often on the lower surface. STROMA dense, pseudoparenchymatous, up to 130 μ diam. CONIDIOPHORES numerous, sometimes in concentric circles on the spot, in dense fascicles, pale to olivaceous brown, smooth, geniculate, continuous or sparingly septate, 10–100 × 3–6.5 μ; conidial scars conspicuous, prominent, thickened, 2–3 μ wide. CONIDIA medium olivaceous, mostly concolorous with the conidiophores, cylindric, obclavate, usually straight or slightly curved, wall usually finely roughened, rounded at the apex, base shortly tapered with a conspicuous hilum, 1–9 septa usually not constricted, mostly 3–4 septate, 20–70 × 4–9 μ (perfect state W. A. Jenkins, l.c.; stat.

conid. J. L. Mulder & P. Holliday, *CMI Descr.* 412, 1974).

M. berkeleyi differs from *M. arachidis* (q.v.), on the same host (groundnut), in that the conidia have a finely roughened wall, usually fewer septa, and are both shorter and broader. The conidiophores also differ in that they occur almost entirely on the lower leaf surface; they form as dark brown, stromatic raised areas, usually concentrically arranged, in contrast to the evenly distributed, olivaceous and more diffuse conidiophores of *M. arachidis*. Jenkins (l.c.) described lobed haustoria (and see Roldan et al.) for *M. berkeleyi* but they were not found in *M. arachidis*. Jenkins (1939) described the development of the spermogonia and perithecia of *M. berkeleyi*.

Late leaf spot (also with *M. arachidis* called tikka) of groundnut is widespread with its economic host (Map 152). Accounts of the biology of *M. berkeleyi* also often include *M. arachidis*, no distinction being made between the 2 pathogens (Gibbons; Tarjot); see Elston et al. for effects on growth of the host. Ref. concerning both fungi are given here. The circular leaf lesions (up to *c.* 8 mm diam.) become very dark brown or black with the chlorotic haloes on the upper surface being less distinct and developing later than those of *M. arachidis*. The symptoms of *M. berkeleyi* are most clearly identified by the distinct dark stroma of the imperfect state which forms on the under surface. *M. arachidis* forms this state on both leaf surfaces and has no such stroma. Severe attacks cause defoliation. Both pathogens can often be found in the same area and on the same plant; but the severe losses that can be caused by each varies with the geographical region and time during crop growth. In S.E. USA *M. arachidis* was said to be most frequent (Smith et al. 1972). In Malawi, in the areas where 70% of the groundnut crop was grown, this sp. was far more important than *M. berkeleyi* (Mercer 1973). In Tanzania Hemingway (1955) compared both spp. The percentage of leaf lesions caused by *M. berkeleyi* became increasingly preponderant as the season advanced. On one cv. a tenfold increase in disease took 7 days whilst for *M. arachidis* it took 23 days. Some 3 months after planting 93% of the leaf lesions were those of *M. berkeleyi* which showed greater spore formation and caused more rapid defoliation. In India differences in disease caused by the 2 pathogens and the effects of different sowing dates have been described (Nath et al.; Ramakrishna et al.). If early maturing cvs are being grown the crop may be nearly mature before *M. berkeleyi* has built up.

In USA (both pathogens) rapid disease build up occurred with temps of 23.5–26.5° and 6–8 hours at an RH of at least 95%. At lower temps longer periods of this RH were needed (Jensen et al.). In Israel considerable damage occurred in sprinkler-irrigated fields (Frank). In India (both pathogens) the air-dispersed conidia showed a diurnal periodicity with a max. conc. in the forenoon (Sreeramulu). *M. berkeleyi* and *M. arachidis* are both carried over in host debris and in neither case does seed transmission seem of much importance. There is some evidence for variation in pathogenicity and symptoms but races have not been clearly characterised (Miller; Sulaiman et al.). Growth of the imperfect state in vitro has been described (Purkayastha; Shanta 1956, 1961; Sulaiman et al. 1969).

Crop rotation, destruction of volunteers, and deep burial of host debris gives some control (rotation may need to be >2–3 years, Mazzani et al.). Kucharek found that a 1 year rotation with maize or soybean reduced the disease considerably and allowed a delay in the first fungicide application. Fungicide treatment is generally needed and there has been extensive work on this aspect, for example, from: India (Chahal et al. 1972b; Lewin et al.; Tandon et al.; Vidhyasekaran et al.); Israel (Frank); Malaysia (Chee); Malawi (Corbett et al.; Mercer); Mauritius (Felix et al.); Nigeria (McDonald et al.); Zimbabwe (Whiteside et al.); Thailand (Prasartsee et al.); Uganda (Mukiibi) and USA (Porter; Smith et al.). In India Rajput et al. also found that seed dressings were effective. These fungicides have given control: Bordeaux, cuprous oxide, oil or water-based copper oxychloride, sulphur (as lime, dust or wettable), maneb, zineb, mancozeb, chlorothalonil, fentin acetate (or as hydroxide), thiophanate, thiabendazole and benomyl (where Smith et al. 1972 found that an oil water emulsion was better than water). Hemingway (1957) found resistance associated with thicker palisade tissue and stomatal size. In India field tests with bunch and semi-spreading cvs showed that only 2.4% of the former were highly resistant, whilst of the latter 43.2% were placed in this category. None of the semi-spreading cvs was highly susceptible (Muhammad et al.; and see Abdou et al.; Aulakh et al.; Chahal et al., 1972a; Sulaiman et al. 1971). Clark et al. described resistance to benomyl and related fungicides, and Smith et al. (1974) forecasting.

Abdou, Y. A. M. et al. 1974. *Peanut Sci.* **1**: 6 (resistance in wild *Arachis* spp.; **54**, 3070).

Aulakh, K. S. et al. 1972. *Indian J. agric. Sci.* **42**: 952 (screening for resistance; **52**, 4288).

Chahal, A. S. et al. 1972a. *Pl. Dis. Reptr* **56**: 601 (reaction of cvs; **52**, 937).

—— ——. 1972b. *Ibid* **56**: 1099 (control with benomyl; **52**, 2789).

Chee, K. H. 1972. *Planter Kuala Lumpur* **48**: 146 (fungicides; **52**, 536).

Clark, E. M. et al. 1974. *Phytopathology* **64**: 1476 (resistance to benomyl & related fungicides; **54**, 1965).

Corbett, D. C. M. et al. 1966. *Rhod. Zamb. Malawi J. agric. Res.* **4**: 13 (fungicides; **45**, 1964).

Elston, J. et al. 1976. *Ann. appl. Biol.* **83**: 39 (effects of infection on crop growth; **55**, 3378).

Felix, S. et al. 1974. *Rev. agric. sucr. Ile Maurice* **53**: 185 (fungicides; **54**, 5649).

Frank, Z. R. 1967. *Phytopathol. Mediterranea* **6**: 48 (general in Israel; **47**, 3285).

Gibbons, R. W. 1966. *Pl. Prot. Bull. F.A.O.* **14**: 25 (brief review, 32 ref.; **45**, 3010).

Hemingway, J. S. 1954. *E. Afr. agric. For. J.* **19**: 263 (general in Tanzania including control; **34**: 74).

——. 1955. *Trans. Br. mycol. Soc.* **38**: 243 (epidemiology, seasonal disease build up in Tanzania; **35**: 342).

——. 1957. *Emp. J. exp. Agric.* **25**: 60 (resistance; **36**: 448).

Hoof, H. A. Van 1950. *Contr. gen. agric. Res. Stn Bogor* 114, 17 pp. (review & control; **31**: 101).

Jenkins, W. A. 1938. *J. agric. Res.* **56**: 317 (perfect states of *Mycosphaerella arachidis* & *M. berkeleyi*, morphology, symptoms & culture; **17**: 651).

——. 1939. *Ibid* **58**: 617 (structure & development of *M. berkeleyi* perfect state; **18**: 571).

Jensen, R. E. et al. 1965. *Pl. Dis. Reptr* **49**: 975 (effects of temp., RH & rain; **45**, 1591).

—— ——. 1966. *Ibid* **50**: 810 (as above, forecasting; **46**, 807).

Kucharek, T. A. 1975. *Pl. Dis. Reptr* **59**: 822 (control by rotation; **55**, 2042).

Lewin, H. D. et al. 1971. *Madras agric. J.* **58**: 480 (fungicides; **51**, 933).

McDonald, D. et al. 1976. *Niger. J. Pl. Prot.* **2**: 43 (fungicides; **56**, 5894).

Mazzani, B. et al. 1971. *Agron. trop.* **21**: 329 (effect of rotation, Venezuela; **51**, 4552).

Mercer, P. C. 1973. *PANS* **19**: 201 (chemical control & effects on parameters of yield; **52**: 4290).

——. 1976. *Ibid* **22**: 57 (ultra low volume fungicide spraying; **55**, 4387).

Miller, L. I. 1954. *Va J. Sci.* **5**: 239 (variation in pathogenicity, temp. & reaction of cvs; see **33**: 463).

Muhammad, S. V. et al. 1968. *Indian J. agric. Sci.* **38**: 941 (reaction of cvs; **48**, 3764).

Mukiibi, J. 1975. *E. Afr. agric. For. J.* **41**: 95 (fungicides; **57**: 360).

Nath, V. R. et al. 1967. *Indian J. agric. Sci.* **37**: 362 (effect of differing sowing dates on disease; **47**, 1730).

Porter, D. M. 1970. *Pl. Dis. Reptr* **54**: 955 (control with benomyl; **50**, 2601).

Prasartsee, C. et al. 1971. *Thai J. agric. Sci.* **4**: 79 (control with benomyl & thiabendazole; **51**, 4554).

Purkayastha, R. P. 1968. *Indian J. mycol. Res.* **6**: 69 (growth & conidial sporulation in vitro; **51**, 934).

Rajput, N. H. et al. 1970. *J. agric. Res. Punjab* **8**: 385 (fungicide spraying & seed dressings; **51**, 4553).

Ramakrishna, V. et al. 1968. *Indian Phytopathol.* **21**: 31 (symptoms & disease development in the field; **48**, 999).

Reyes, G. M. et al. 1940. *Philipp. J. Agric.* **11**: 371 (reactions of cvs; **20**: 443).

Roldan, E. F. et al. 1939. *Philipp. Agric.* **27**: 669 (general; **18**: 433).

Shanta, P. 1956. *Proc. Indian Acad. Sci.* Sect. B **44**: 271 (growth & conidial sporulation in vitro; **36**: 441).

——. 1960. *J. Madras Univ.* Sect. B **30**: 167 (environment, disease incidence & survival; **41**: 76).

——. 1961. *Phytopathol. Z.* **41**: 59 (growth requirements in vitro).

Smith, D. H. et al. 1972. *Phytopathology* **62**: 1029 (control with benomyl oil water emulsion; **52**, 1738).

—— ——. 1974. *Pl. Dis. Reptr* **58**: 666 (forecasting; **54**, 282).

Sreeramulu, T. 1970. *Indian J. agric. Sci.* **40**: 173 (conidial dispersal; **50**, 1524).

Sulaiman, M. et al. 1969. *Beitr. trop. subrop. Landwirt. Tropenvet-Med.* **6**: 103 (temp. & culture in vitro, symptoms & variation in pathogenicity; **48**, 3767).

—— ——. 1971. *Ibid* **9**: 61 (reactions of Indian cvs; **51**, 2085).

Tandon, I. N. et al. 1968. *Indian Phytopathol.* **21**: 281 (fungicides; **50**, 414).

Tarjot, M. 1959. *Rev. Mycol.* **24**: 13 (brief review, 22 ref.; **38**: 643).

Vidhyasekaran, P. et al. 1968. *Indian J. agric. Sci.* **38**: 373 (fungicides; **47**, 3284).

Whiteside, J. O. et al. 1967. *Rhod. agric J.* **64**: 128 (fungicides; **47**, 1356).

Woodroof, N. C. 1933. *Phytopathology* **23**: 627 (early taxonomy, morphology & symptoms; **13**: 74).

Mycosphaerella brassicicola (Duby) Lindau,

Die Naturlichen Pflanzenfamilien Leipzig **1**: 424 (Feb.) 1897.

Sphaeria brassicicola Duby (as *brassicaecola*), 1830

Sphaerella brassicicola Ces. & de Not. (as *brassicaecola*), 1863

Mycosphaerella brassicicola (Duby) Oudem (as *brassicaecola*), (Mar.) 1897
stat. conid. *Asteromella brassicae* (Chev.) Boerema & van Kest., 1964
Asteroma brassicae Chev., 1826.
(further synonymy in *CMI Descr.* 468; Boerema et al.; Vanterpool reported a *Cercospora* stage)

PSEUDOTHECIA mostly on leaves but sometimes on stems and seed pods, globose, dark brown with apical papillate ostioles, 100 × 130 μ diam. ASCI bitunicate, 8 spored, 30–45 × 12–18 μ. ASCOSPORES irregularly biseriate, hyaline, cylindrical, 18–23 × 3–5 μ, rounded at the ends and not constricted at the septum. 'PYCNIDIA' (SPERMOGONIA) solitary or gregarious, globose, dark brown, with papillate ostioles, 100–200 μ; wall pseudoparenchymatic, composed of several layers of cells. 'CONIDIA' (spermatial cells) hyaline, cylindrical, aseptate, 3–5 × 1 μ (E. Punithalingam & P. Holliday, *CMI Descr.* 468, 1975).

Butler & Jones (*Plant pathology*) gave an account of ringspot (also called black ringspot) of *Brassica* spp. (and see Dring). The fungus is widespread (Map 189) but generally restricted to cool, moist regions. The symptoms are most conspicuous on the outer leaves. The small necrotic spots enlarge to an av. 0.5–1 (up to 2) cm diam.; spots are generally circular with light brown or grey centres, a definite edge is bounded by a narrow watersoaked area and a chlorotic zone. Where the lesions are numerous the whole leaf becomes yellowish and shows curled, cracked and ragged edges. Within the lesion the pseudothecia (hypophyllous at first, later amphigenous) form in a typically zonate manner. They may be extremely numerous and the lesion appears black. Rectangular or oval lesions are caused on the midribs, stems and siliqua.

Host penetration is probably stomatal. More infection occurred at 16° and 20° than at 12° and 24°; the most rapid development of lesions was at 20° but the most lesions formed at 16°. The opt. temp. for ascospore formation was 16–20°. Mature perithecia did not develop on lesions unless the RH was at saturation for at least 4 days. Mycelial growth in vitro was best at 20° (Nelson et al.). The fungus is homothallic (Snyder). Spread is through the air dispersed ascospores. Huber et al. reported seed infection, but it is not clear how important seed-borne spread is in the epidemiology of ringspot. The spermatia are not infective.

In control cultural measures should be taken: clean, healthy seedbeds, destruction of infected plant remains, particularly where crops are being grown for seed, and isolation of seedbeds. Seed treatment with hot water (45° for 20 minutes) followed by thiram has been recommended. High volume sprays of maneb, mancozeb and dichlone were the best of 19 fungicides tested in Australia. In USA benomyl gave almost complete control and chlorothalonil gave 90% control. Benomyl was also effective in New Zealand where June–July planting resulted in severe infection in contrast to earlier planting. In the Netherlands (Quak) no resistance was found in 40 cvs of Brussels sprout (*Brassica oleracea* var. *gemmifera*).

Boerema, G. H. et al. 1964. *Persoonia* **3**: 17 (nomenclature of *Mycosphaerella brassicicola* & *Leptosphaeria maculans*; **43**, 2120).
Dring, D. M. 1961. *Trans. Br. mycol. Soc.* **44**: 253 (life history, morphology & nomenclature; **41**: 112).
Huber, G. A. et al. 1949. *Phytopathology* **39**: 869 (seed treatment; **29**: 341).
Nelson, M. R. et al. 1959. *Ibid* **49**: 633 (effects of temp. & RH; **39**: 358).
Quak, F. 1957. *Meded. Dir. Tuinb.* **20**: 317 (on *Brassica oleracea* var. *gemmifera*; **37**: 689).
Rogers, I. S. et al. 1970. *Exp. Rec.* (5): 12 (fungicides in Australia; **50**, 340).
Snyder, W. C. 1946. *Phytopathology* **36**: 481 (homothallism, spermatia non-infective; **25**: 591).
Vanterpool, T. C. 1968. *Proc. Can. phytopathol. Soc.* **35**: 20 (overwintering as dark, thick walled mycelium, a *Cercospora* state; **48**, 4f).
Weimer, J. L. 1926. *J. agric. Res.* **32**: 97 (general; **5**: 459).
Welch, N. C. et al. 1969. *Calif. Agric.* **23**(12): 17 (fungicides in USA; **49**, 2218).
Wilson, G. J. 1971. *N.Z. comml Grow.* **26**: 11 (control in New Zealand; **52**, 2406).

Mycosphaerella citri Whiteside, *Phytopathology* **62**: 263, 1972.
stat. conid. *Stenella* sp.

PSEUDOTHECIA immersed in decomposing leaves, densely grouped, subepidermal, amphigenous, up to 90 μ diam. with a papillate ostiole. ASCI obclavate, bitunicate, 8 spored, 25–35 × 8–10 μ. ASCOSPORES hyaline, fusiform, 1 septate, straight or slightly curved, 6–12 × 2–3 μ. SPERMOGONIA containing rod-shaped spermatia, 2–3·5 × 1 μ, are often formed prior to pseudothecial production. CONIDIOPHORES sparsely produced, arising from extramatricular hyphae, simple, erect, deep olivaceous, paler towards the apex, septate, lightly rough walled, 12–40 × 2–3.5 μ. In culture the conidiophores are formed singly on verrucose hyphae, simple or branched, smooth or slightly rough walled, septate, olivaceous brown, up to 150 μ long. CONIDIOGENOUS CELLS polyblastic, integrated, sympodial, apex more or less rounded to slightly tapered, with prominent apical or lateral, thick, flat but not raised scars, the older scar lying flat against the side of the conidiogenous cell which often elongates after bearing one group of conidia to form the next group a little distance beyond. CONIDIA pale olivaceous, verrucose, cylindric, straight or slightly bent, mostly simple, rarely catenulate or branched, rounded at the apex, tapering to a truncate base with a thick hilum, 10–70 × 2–3.5 μ (A. Sivanesan & P. Holliday, *CMI Descr*. 510, 1976; and see Whiteside l.c.).

Greasy spot of citrus (black melanose or greasy melanose) was known for many years in Florida, USA before its aetiology was determined. Earlier in Japan (Tanaka et al.) a similar (or identical) disease had been attributed to *M. horii* Hara, 1917 (see Saccardo, *Sylloge Fung*. **24** Sect. 2: 884, 1928). This fungus, which was later described by Yamada, was considered by Whiteside (1970a, 1972b) to be different from *M. citri*, although all the precise differences are not clear. Some characteristics of *M. horii* taken from Yamada are: PERITHECIA 60–100 μ diam.; ASCI with 8 biseriate ascospores, 29.3–58.5 × 2.7–10.6 (av. 48.3 × 5.8) μ; ASCOSPORES 1 septate, 7.9–15.9 × 2.3–3.7 (av. 12 × 2.9) μ; CONIDIA (*Cercospora* type) 0–8 septate, 5.3–58.5 (mostly 16.5–24.9) × 1.3–4 (mostly 2.7–2.8) μ. Both fungi form pycnidia with aseptate spores 3–4 μ long. *M. horii* has 4 clearly defined oil globules in the ascospores, this was not found in *M. citri* (Whiteside 1970a). The latter also differs in that the ascospores are not constricted at the septum and they are smaller than those of *M. horii*. Fisher, who erroneously attributed greasy spot to a new sp. of *Cercospora* (*C. citri-grisea* Fisher), illustrated the symptoms and reviewed the literature. The disease has also been reported from Argentina, Belize and Surinam (Map 524).

Symptoms are found on leaves and fruit. On the former small blister-like areas appear, mostly on the under surface; on the upper surface each spot or blister is marked by a slightly larger chlorotic spot. The spots are at first a light yellow orange (translucent) and then become brown and black; in the field they resemble irregular, diffusely margined flecks of dirty grease. The same appearance is eventually seen on the upper surface. The necrotic areas may become confluent and affect most of the leaf; there may be a continuous necrotic band along the leaf edge, showing a yellow area when viewed on the upper surface. The infected leaves fall. A leaf symptom known as small brown spot has also been described in Japan and USA. On fruit very small black spots are formed. Collectively they look like an unsightly blemish, often further accentuated by delayed colouring of the rind due to the retention of chlorophyll in the adjacent living cells. The black specks consist of dead guard cells and frequently a few dead cells immediately beneath the substomatal chamber. The fruit condition previously known as pink pitting is a more severe form of this black spot (or rind stippling) condition and has been given the name greasy spot rind blotch, to cover all the symptoms caused by *M. citri* on the fruit (Whiteside 1972c).

In Florida symptoms appear 2–8 months after infection and the subsequent interval to defoliation is very variable. Severe leaf fall frequently occurs by the end of winter. Penetration of the host from conidia (and presumably from ascospores) occurs through the stomata. Mycelium from germinating spores ramifies over the leaf surface. Perithecia are most abundant on the lower surface and airborne ascospores from fallen leaves are thought to be the most important inoculum source. These spores were trapped after rain or nights with heavy dew. Ascospore germination on glass is low at RH 96% but increases rapidly above this; in sucrose it increases rapidly above RH 92%. Germ tube growth is most on cornmeal agar at 25–30°, and at RH 100% on glass. On agar more rapid germination occurs at 30° (50% after 4 hours) than at 25 or 20°. Ascospores remain viable for at least 6 days. More infection was recorded on leaves of rough lemon under conditions of daily wetting and drying than under those of continuous wetness; and on leaves sprayed with sucrose (Whiteside 1974).

In Surinam Brussel investigated the relationships between *M. citri*, a rust mite (*Phyllocoptruta oleivora*) and a fungus on the mite. More greasy spot occurred on trees infested with the mite. Copper sprays tended to increase populations of the mite and scale insects by eliminating the fungi which attacked them. Control of the disease should, therefore, aim at preventing build up of the mite. In Florida one fungicide spray application in June or July is effective; satisfactory materials are benomyl, tribasic copper sulphate and oil. Whiteside (1973b) examined the action of oil. Zineb has also been used. Using ground sprays of benomyl ascospore inoculum was reduced for a few weeks (Whiteside 1973c).

Brussel, E. W. van 1975. *Agric. Res. Rep. Wageningen* 842, 66 pp. (interrelationships between the pathogen, citrus rust mite & *Hirsutella thompsonii*; **57**, 2537).

Fisher, F. E. 1961. *Phytopathology* **51**: 297 (citrus greasy & tar spots; **40**: 752).

Tanaka, S. et al. 1952. *Bull. Tokai-Kinki natn. agric. Exp. Stn* (Hort. Div.) 1, 15 pp. (general, *Mycosphaerella horii*).

Whiteside, J. O. 1970a. *Phytopathology* **60**: 1409 (aetiology & epidemiology; **50**, 1231).

——. 1970b. *Ibid* **60**: 1859 (symptoms on fruit; **50**, 2280).

——. 1972a. *Proc. Fla St. hort. Soc.* **85**: 24 (benomyl spray coverage requirements; **53**, 2192).

——. 1972b. *Phytopathology* **62**: 260 (*M. citri* sp. nov., histology; **51**, 3364).

——. 1972c. *Pl. Dis. Reptr* **56**: 671 (symptoms on fruit; **52**, 1145).

——. 1973a. *Ibid* **57**: 691 (control with benomyl, Cu, captafol & other fungicides; **53**, 952).

——. 1973b. *Phytopathology* **63**: 262 (action of oil; **52**, 3697).

——. 1973c. *Proc. Fla St. hort. Soc.* **86**: 19 (use of ground sprays against ascospores; **54**, 4933).

——. 1974. *Phytopathology* **64**: 115 (factors affecting infection; **53**, 3490).

Yamada, S. 1956. *Bull. Tokai-Kinki natn. agric. Exp. Stn* (Hort. Div.) 3: 49 (morphology of *M. horii*).

Mycosphaerella cruenta Latham, *Mycologia* **26**: 525, 1934.

Cercospora cruenta Sacc., 1880

stat. conid. *Pseudocercospora cruenta* (Sacc.) Deighton, 1976.

PERITHECIA scattered or slightly aggregated, mostly hypophyllous, innate but erumpent at maturity, globose, black, ostiole slightly papillate, 52–70 × 63–87 μ. ASCI fasciculate, cylindric–clavate, aparaphysate, bitunicate, excentrically papillate at apex of inner membrane, 8 spored, 35–52 × 7–11 μ. ASCOSPORES unequally 2 celled, upper cell larger, very slightly curved, hyaline, 11–19.2 × 3.5 (mostly 14–17.5 × 3.5) μ. SPERMOGONIA globular to flask shaped, black, subepidermal at first, later erumpent, ostiolate, 31–77 × 24–70 μ. SPERMATIA rod shaped, hyaline, 2–2.5 × 0.8 μ. CONIDIOPHORES amphigenous, arising usually from substomatal stromata, loosely fasciculate, usually simple but may be forked, somewhat subdenticulate, dilutely olive. CONIDIA acicular to obclavate, slightly curved, 35–154 × 3.5–4.5 μ, 1–8 septate, hyaline becoming olivaceous (after D. H. Latham l.c.; the stat. conid. was described by J. L. Mulder & P. Holliday, *CMI Descr.* 463, 1975, as *C. cruenta*; and see Da Ponte).

This leaf spot of cowpea (*Vigna unguiculata*) is widespread in warmer regions. *M. cruenta* is also found on *Calopogonium, Lablab niger, Phaseolus* spp. and *Stizolobium deeringianum*. On *P. aureus* (green gram) the leaf spots are more or less circular, purplish red and mostly with light grey centres, usually 3–4 mm diam.; large, irregular necrotic patches, limited by the veins, are formed; stem lesions are purplish up to 4 cm long and pods can be destroyed. On cowpea the reddish brown leaf spots (becoming necrotic) can be >1 cm diam., outline irregular, coalescence may occur and leaflets wither. This apparently minor disease has been little studied. Most mycelial growth is at *c.* 26° and most conidial germination at 25°. The fungus survived in infected tissue for up to 21 days at 35°. In India a high proportion of cowpea lines were resistant. In this host control of resistance by both dominant and recessive single genes has been described (Fery et al. 1975, 1976). Of several fungicides tested benomyl was the most effective.

Da Ponte, J. J. 1976. *Bolm Soc. cearense Agron.* 17: 1 (*Cercospora* spp., identity on common bean & cowpea).

Ekpo, E. J. A. et al. 1977. *Phytopathol. Z.* **89**: 249 (conidial germination; **57**, 876).

Fery, R. L. et al. 1975. *HortScience* **10**: 627 (inheritance of resistance in cowpea; **55**, 4392).

——. 1976. *J. Am. Soc. hort. Sci.* **101**: 148 (as above; **55**, 4954).

——. 1977. *Pl. Dis. Reptr* **61**: 741 (loss, resistance & fungicides; **57**, 1917).

Kannaiyan, S. et al. 1974. *Labdev B* **12**: 150 (fungicides; **56**, 922).

Latham, D. H. 1934. *Mycologia* **26**: 516 (hosts, symptoms, morphology & taxonomy; **14**: 280).

Verma, P. R. et al. 1969. *Indian Phytopathol.* **22**: 61 (hosts & resistance; **49**, 606).

——. 1972. *Phytopathol. Mediterranea* **11**: 25 (factors affecting in vitro growth; **52**, 1424).

Verma, P.R. et al. 1975. *Ibid* **14**: 23 (survival with *C. beticola*; **55**, 4893).
Singh, K. et al. 1975. *Indian J. Mycol. Pl. Pathol.* **5**: 108 (fungicides; **56**, 923).
Welles, C. G. 1924. *Phytopathology* **14**: 351 (general on green gram; **4**: 75).

Mycosphaerella fijiensis Morelet, *Annls Soc. Sci. nat. Archeol. Toulon Var* **21**: 105, 1969.
stat. conid. *Pseudocercospora fijiensis* (Morelet) Deighton, 1976
Cercospora fijiensis Morelet, 1969

PERITHECIA amphigenous but mostly epiphyllous, scattered, immersed, ostiole narrow or thick, papillate, globose, 47–85 μ diam., wall dark brown, 3 or more layers, cells polygonal. ASCI bitunicate, obclavate, numerous; paraphyses absent. ASCOSPORES hyaline, biseriate, 1 septate, slightly constricted at the septum, fusiform, larger cell uppermost in the ascus, 12.5–16.5 $\times$ 2.5–3.8 μ. STROMA absent. CONIDIOPHORES amphigenous, predominantly on the lower leaf surface, appearing at the initial speck or streak stage, singly or in small groups (2–8); pale olivaceous brown, paler towards the apex, straight or curved, geniculate, rarely branched, 0–5 septate, 16–62(32) $\times$ 4–7(5.5) μ, basal swelling up to 8 μ, conidial scars conspicuous. CONIDIA subhyaline to pale olivaceous, obclavate or cylindro–obclavate, straight or curved, septate 1–10 (mostly 5), apex obtuse, truncate or round base with a visible thickened hilum, 20–132(72) $\times$ 2.5–5 μ, tapering towards the apex. Colonies on PDA slow growing, compact, prominently raised, grey to pale buff or olive green, reverse black, surface velvety (J. L. Mulder & P. Holliday, *CMI Descr.* 413, 1974).

The perfect state is almost identical to that of *M. musicola* (q.v. for further comment). But the conidiophores and conidia have slightly thickened scars, and a stroma is absent. *M. fijiensis* var. *difformis* Mulder & Stover shows, like *M. fijiensis* (and unlike *M. musicola*), thickened scars. But in this var. the conidiophores are formed singly, in small groups and on a stroma. These pathogens appear to be very closely related but are considered to be distinct (Meredith et al. 1969, 1970a; Mulder et al. 1976; Stover 1976, 1978).

M. fijiensis causes black leaf streak and var. *difformis* causes black Sigatoka of banana. The first disease was described from Fiji in 1963 (Leach; Rhodes) but as Stover (1978) described there are earlier records. A centre of origin in the New Guinea and Solomon Islands region was postulated; black leaf streak occurs in most of the islands of the S. Pacific and in Hawaii, Malaysia (W.), Philippines, Taiwan and Thailand (Map 500). Experience in Fiji and Honduras shows that *M. fijiensis* tends to replace *M. musicola* (Stover et al.). Black Sigatoka has been reported from Honduras, Taiwan and some S. Pacific islands (Stover 1976).

There is little difference in the symptoms caused by these 2 diseases. Initially, reddish brown specks appear on the lower leaf surface; they elongate to become streaks up to 20 $\times$ 2 mm with the long axis parallel to the leaf veins, and at this stage are more clearly seen on the lower surface. The streaks can be extremely numerous; they darken, become almost black, and develop into fusiform or elliptical spots with light brown, watersoaked margins and dark centres. The centres dry out, becoming light grey or buff, sunken, surrounded by narrow, dark brown or black borders and often by chlorotic zones beyond. Necrosis of the whole leaf can occur in *c.* 3 weeks. There are some differences in the macroscopic symptoms between black leaf streak and Sigatoka (*M. musicola*). In the latter disease the early stage streaks are clearly seen on the upper surface and are yellowish; in the former one the streaks are darker at similar stages in development. But there are no clear macroscopic differences between the mature spots of black leaf streak and Sigatoka.

Work on *M. fijiensis* so far has been largely concerned with comparisons with *M. musicola* and control with fungicides. Black leaf streak reaches epidemic proportions in Fiji where it has apparently replaced Sigatoka as the important disease (see Firman 1976 for an advisory account). It prevents fruit of export quality forming and bunches maturing. Perithecia are formed in abundance and, therefore, ascospores are considered important in spread, perhaps more so than conidia. In Fiji and Hawaii, in dry weather, ascospore conc. reached a max near 0600 hours as dew formed. On rainy days peak conc. was reached shortly after rain began. Seasonal increases in ascospores were associated with those of rainfall and RH (Meredith et al. 1973). In early work the oil water dithiocarbamate emulsions used against Sigatoka were employed in control. But maneb as a water based spray has given good disease control. The possible side effects of oil sprays (when used to control Sigatoka in Fiji), in creating conditions which may have led to its replacement by black leaf streak as the more serious disease, have been discussed (Firman 1970; Firman et al.). In comparisons of the aerial applications of these oil water emulsions Stover found that, to control black leaf streak in the Philippines, both more oil and fungicide, and a greater frequency of application were needed compared with Sigatoka control in Central America. Oil water emulsions with benomyl have given better control than those containing maneb (Firman 1972; Long). In Hawaii all the important commercial banana cvs are highly susceptible. The epidemic spread of *M. fijiensis* from the Pacific region to the major banana-producing areas of the world (as happened in the case of Sigatoka) appears to be a real possibility.

Firman, I. D. 1970. *Nature Lond.* **225**: 1161 (possible side effects of fungicides on banana & coffee diseases; **49**, 2400).
—— et al. 1970. *Ann. appl. Biol.* **66**: 293 (fungicides; **50**, 1311).
——. 1972. *Ibid* **70**: 19 (as above; **51**, 2703).
——. 1976. *Advis. Leafl. S. Pacific Comm.* 1, 4 pp. (advisory; **56**, 308).
Hapitan, J. C. et al. 1970. *Philipp. Agric.* **54**: 46 (general; **51**, 1687).
Leach, R. 1964. *Coun. Pap. Fiji* 38, 20 pp. (general; **45**, 1867).
——. 1964. *Wld Crops* **16**: 60 (general; **44**, 1180).
Long, P. G. 1971. *Pl. Dis. Reptr* **55**: 50 (control with benomyl; **50**, 3051).
——. 1973. *Trop. Agric. Trin.* **50**: 75 (fungicides; **52**, 2993).
Meredith, D. S. et al. 1969. *Trans. Br. mycol. Soc.* **52**: 459 (symptoms & comparison of stat. conid. with that of *Mycosphaerella musicola*; **48**, 3071).
—— ——. 1970a. *Trop. Agric. Trin.* **47**: 127 (history, spread & distribution; **49**, 2937).
—— ——. 1970b. *Ibid* **47**: 275 (susceptibility of cvs; **50**, 817).
—— ——. 1973. *Trans. Br. mycol. Soc.* **60**: 547 (ascospore release & dispersal; **53**, 4160).
Mulder, J. L. et al. 1976. *Ibid* **67**: 77 (validation of *M. musicola*, diagnosis of *M. fijiensis* & *M. fijiensis* var. *difformis* var. nov.; **56**, 2154).
Rhodes, P. L. 1964. *Commonw. phytopathol. News* **10**: 38 (symptoms; **44**, 191).
Stover, R. H. 1971. *Pl. Dis. Reptr* **55**: 437 (contrast of fungicide control with *M. musicola*; **51**, 504).
——. 1976. *Trop. Agric. Trin.* **53**: 111 (distribution & cultural characteristics of leaf spot pathogens on bananas; **55**, 4210).
—— et al. 1976. *Pl. Prot. Bull. F.A.O.* **24**: 36 (*M. musicola* & *M. fijiensis* var. *difformis*, first central American epidemics; **56**, 2591).
——. 1977. *Trans. Br. mycol. Soc.* **68**: 122 (extranuclear inherited tolerance of benomyl in *M. fijiensis* var. *difformis*; **56**, 3639).
——. 1977. *Ibid* **69**: 500 (a non-virulent, benomyl tolerant *Cercospora* from leaf spots; **57**, 3037).
——. 1978. *Trop. Agric. Trin.* **55**: 65 (distribution & origin of *M. fijiensis* in S.E. Asia; **57**, 3036).

Mycosphaerella musicola Mulder, *Trans. Br. mycol. Soc.* **67**: 77, 1976.
stat. conid. *Pseudocercospora musae* (Zimm.) Deighton, 1976
Cercospora musae Zimmerman, 1902
C. musae Massee, 1914.

On leaves. PERITHECIA dark brown or black, amphigenous, scattered on mature leaf spots, erumpent, short protruding ostiole, dark wall well defined, 46.8–72 (av. 61.8) μ diam. ASCI oblong, clavate, 28.8–36 × 8–10.8 μ. ASCOSPORES 1 septate, hyaline, obtuse ellipsoid, upper cell slightly broader, 14.4–18 (av. 16.7) × 3–4 μ; constriction at the septum not marked, oil globules irregular and often inconspicuous. PARAPHYSES none. SPERMOGONIA mostly on the lower surface, wall of 2–3 layers of pale to medium brown, rectangular or polygonal cells, 46–77 × 37–63 (av. 58 × 48) μ. SPERMATIA hyaline, more or less rod shaped, 2–5 × 0.8–1.4 (av. 3.5 × 1) μ (Philippines, Meredith & Lawrence 1970b). CONIDIAL STAGE amphigenous, more abundant on the upper surface. STROMA dark brown to black, 15–35 μ diam. erumpent through stomata. CONIDIOPHORES in dense fascicles, borne terminally on stromatal hyphae, pale to very pale olivaceous brown, straight or slightly curved, rarely branched, nonseptate, non-geniculate, narrow towards apex, rounded mostly ampuliform, conidial scars absent, 5–25 × 2–6 μ. CONIDIA pale to olivaceous, cylindric to obclavate cylindric, straight, curved or undulate, apex rounded or obtuse, unthickened basal hilum, 3–5 or more septate, 10–80 (up to 110) × 2–6 μ (perfect state from R. Leach, *Trop. Agric. Trin.* **18**: 91, 1941; for a Latin diagnosis see J. L. Mulder & R. H. Stover, *Trans. Br. Mycol. Soc.* **67**: 77, 1976; stat. conid. from J. L. Mulder & P. Holliday, *CMI Descr.* 414, 1974).

M. musicola, causing Sigatoka of banana, can be distinguished (morphologically) from *M. fijiensis* in that the conidia and conidiophores do not have slightly thickened scars as is the case in *M. fijiensis* (Mulder et al. 1974 l.c.; Mulder et al. 1976, *M. fijiensis* q.v.). In 1976 Deighton (*Cercospora* q.v.) transferred the imperfect states of both these important banana pathogens to *Pseudocercospora* (q.v.) from *Cercospora*. In Deighton's description of the former genus the conidial scars are described as unthickened, i.e. of the same thickness as the wall of the conidiogenous cell. There is a stroma in *M. musicola*; this is absent in *M. fijiensis*. Leach's (l.c.) name was validated by Mulder et al. 1976 (l.c.). Meredith et al. gave a comparative description of the stat. conid. of the fungus causing Sigatoka. Further details of the perfect state were given in Meredith's monograph (1970c) and by Stover (*Banana, plantain and abaca diseases*).

Stover (1963a, 1969a) compared *M. musae* (Speg.) Syd. and *M. minima* Stahel (both associated with banana leaf spots) with the pathogens *M. fijiensis* and *M. musicola*. *M. musae* and *M. minima* are weakly parasitic or saprophytic and are probably widespread on banana; no imperfect states are known and of the 2 only *M. musae* has a spermogonia stage. Morphologically, *M. musae* differs from *M. musicola* in having smaller ascospores which have faster-growing germ tubes, and colonies on agar media also develop quicker. *M. minima* is distinguished by its larger ascospores which are more constricted at the septum (Stover 1963a). On agar the colonies are also distinct, slower growing than *M. musae* and distinctively black.

The name of this banana leaf spot is from that of the valley in Fiji where the disease first attracted serious attention. It is widespread (Map 7) but has not been reported from the producing areas of the Canary Islands, Egypt and Israel. Stover (1962) described the chronology and effects of its spread from Fiji and Australia where serious losses were caused in 1912 and 1924, respectively. The centre of origin of the disease is thought to have been in the Indo-Malayan region. By the 1930s Sigatoka had been found in Africa, and the Central America and West Indies region where major epidemics occurred. Subsequently the disease spread to both the E. and W. of tropical S. America. Stover (1978, *M. fijiensis* q.v.) tabulated the approximate dates of identification of *M. musicola* and *M. fijiensis* in S.E. Asia. The former is now recorded in Australia (E. and N.), Brunei, S. China, Indonesia (Java, Sumatra and Timor), Malaysia (W., Sabah & Sarawak) and Philippines. *M. musicola* is considered not to be present in most of the islands of the S. Pacific. Wherever *M. fijiensis* has invaded areas occupied by *M. musicola* it has replaced the latter as the dominant banana leaf spot. (See *M. fijiensis* for further comment on Sigatoka.)

Symptoms are first seen as a yellowish green speck (1 mm long) on the leaves; this becomes a streak (3–4 × 1 mm) which broadens and lengthens. The streak turns brown or rusty and has an ill-defined margin. The outline becomes more definite and the brown centre has a yellow to light brown halo, sometimes with a watersoaked border. In the fully developed leaf spot (12–15 × 2–5 mm) the sunken central area is at first dark, becoming grey with a dark brown or black border and sometimes a chlorotic halo. Coalescence of the spots can lead to the death of large areas of the lamina. Generally spotting is more frequent towards the leaf apex and edges. Streak and tip spotting is caused by ascospore infection, mostly on the apical third of the leaf. Line spotting (mostly basal) is caused by conidial infection. Besides the obviously damaging effects of leaf destruction, spotting detrimentally affects the physiology of fruit ripening (at lower levels of infection), reduces yield and increases the time from shooting to bunch maturity.

Besides the 2 extensive accounts of Sigatoka that have been referred to there are also those given from Australia (Pont 1960; Simmonds), Jamaica (Leach), Taiwan (Wang) and Surinam (Stahel). Most infection is through the furled heart leaf and the first unfurled leaf. Penetration is stomatal for both spore types. Conidial germination is opt. at 25–29°; most growth of the germ tube on the leaf surface is at 22–28°. Conidia germinate on the leaf surface in a few hours and the germ tubes grow, in a water film, for up to 6 days before penetration of the host begins. The germ tubes of ascospores may have a shorter period of epiphytic growth. Most infection is through the lower leaf surface and the time to the appearance of the first macroscopic symptoms is very variable. In Central America this so-called incubation period is 20–25 days in drier weather and 30–60 days when wetter. The structure of the mature leaf restricts lateral spread of the pathogen and linear (rather than elliptical) lesions are produced. Most sporodochia form at the brown spot stage before the centre begins to pale. Ascospores can form 4 weeks after the yellow streak stage. Conidia occur on both leaf surfaces, abundantly at

25–28° and declining sharply below 22°. There are also less perithecia at this lower temp.

The conidia and ascospores are water and air dispersed, respectively. Sporodochia are probably more abundant than perithecia. Ascospores are released largely through rain over a wide temp. range. In Honduras the peak ascospore dispersal occurred at *c.* 0200 hours (lower rainfall season) and at *c.* 1000 hours (higher rainfall season). Conidia are the major source of inoculum. The absolute and relative numbers are affected by climate and weather. Sigatoka is severe at 23–28° under humid conditions. Seasonal variations are not so marked for conidial formation as they are for that of ascospores. Ascospores are most abundant during rainy weather whilst conidia can be formed in large numbers during drier, lower rainfall periods, provided that leaf surface water films are present. In the large planted areas of Central America (several hundred hectares) sites where disease incidence is higher (so called hot spots) are characterised by a higher RH in daytime and more hours of leaf wetness. Detailed scales for determining and comparing disease intensity have been described (Stover et al. 1970b; Stover 1971). *M. musicola* is probably heterothallic but only the imperfect state has been produced in agar culture (Calpouzos 1954; Stover 1963b).

Control is with fungicide sprays. Until the late 1950s Bordeaux was used. The elaborate use of this fungicide in the huge acreages of banana cultivations of Central America provides a classical example of this particular control method, and one that was greater in scale than any other with Bordeaux. The discovery of the effectiveness of petroleum oil (alone or with added chemical compounds) was made in the French West Indies (see Cuille) in control studies on Sigatoka. The action of oil in plant disease control (mostly arising from work on this disease) was reviewed by Calpouzos (1966). The properties of the oils that are needed have also been more recently discussed by Walker. Oil reduces: spore germination, germ tube growth, appressorial formation and number of spots whose development is also retarded. Development of disease after host penetration is thus inhibited. Control is given by oil alone or now (much more usually) by oil in water emulsions with mancozeb or maneb (e.g., Pont 1970); benomyl in such emulsions is effective (Ganry et al.; Stover et al. 1968b, Stover 1969b). Spraying is done during cool, still periods at low volume, using aircraft or mist blowers from the ground. Rates of application vary with local climate and disease intensity. Fungicide spraying against Sigatoka was described by Meredith (1970a) in a review of the status of major banana diseases. No races of *M. musicola* are known. Although there are some differences in reaction all the edible triploid (genome AAA) banana cvs are highly susceptible. When the B genome is introduced (AAB and ABB) resistance in the triploids occurs (Vakili; Shillingford). Breeding for resistance to *M. musicola* was described by Shepherd (*Fusarium oxysporum* f. sp. *cubense* q.v.).

Brun, J. 1964. *Fruits* **19**: 125 (perfect state; **43**, 2685).

Calpouzos, L. 1954. *Nature Lond* **173**: 1084 (sporulation in vitro; **33**: 490).

—— et al. 1962. *Pl. Dis. Reptr* **46**: 758 (relation between climate & disease incidence; **42**: 334).

——. 1966. *A. Rev. Phytopathol.* **4**: 369 (action of oil in control of plant disease, 78 ref.).

Cuillé, J. 1965. *PANS* **11**: 281 (aerial application of mineral oil; **45**, 517).

Dantas, B. 1948. *Bolm téc. Inst. agron. N.* 14, 45 pp. (occurrence in Brazil; **28**: 529).

Ganry, J. et al. 1973. *Fruits* **28**, 671 (methods with systemic oil fungicide; **53**, 3109).

Goos, R. D. et al. 1963. *Trans. Br. mycol. Soc.* **46**: 321 (infection & disease development in the glasshouse; **43**, 1098).

Kranz, J. 1968. *Z. PflKrankh. PflSchutz* **75**: 327, 518 (conidial formation & development, effect of weather factors on disease; **48**, 537, 1248).

Leach, R. 1946. Govt Printer, Kingston, Jamaica. 118 pp. (monograph; **26**: 250).

McGahan, M. W. et al. 1965. *Phytopathology* **55**: 1179 (effect of leaf anatomy on lesion morphology; **45**, 1111).

Meredith, D. S. 1970a. *Rev. Pl. Pathol.* **49**: 539 (status of major banana diseases).

—— et al. 1970b. *Trans. Br. mycol. Soc.* **54**: 265 (conidial state morphology in Pacific region; **47**, 2938).

——. 1970c. *Phytopathol. Pap.* 11, 147 pp. (monograph, 13 pp. ref.; **49**, 2562).

Merle, P. et al. 1958. *Fruits* **13**: 143 (ground & aerial spraying with Cu in oil in Cameroon; **37**: 545).

Pont, W. 1960. *Qd J. agric. Sci.* **17**: 211 (epidemiology & control in Australia; **41**: 240).

——. 1970. *Qd agric. J.* **96**: 709 (fungicides against banana leaf diseases; **50**, 1901).

Price, D. 1960. *Span* **3**: 122 (climate & control in Cameroon; **39**: 726).

Shillingford, C. A. 1975. *Trop. Agric. Trin.* **52**: 157 (cv. susceptibility in sprayed & unsprayed plants; **54**, 4053).

Simmonds, J. H. 1939. *Qd J. agric.* **52**: 633 (effect of climate & control in Australia).

Mycosphaerella pinodes

Stahel, G. 1937. *Trop. Agric. Trin.* **14**: 257 (general in Surinam; **17**: 190).

Stover, R. H. 1962. *Ibid* **39**: 327 (world spread; **42**: 272).

——. 1963a. *Can. J. Bot.* **41**: 1481 (*Mycosphaerella musicola* & associated ascomycetes; **43**, 1093).

——. 1963b. *Ibid* **41**: 1531 (heterothallism; **43**, 1094).

——. 1964. *Phytopathology* **54**: 1320 (perithecial formation & discharge; **44**, 1179).

——. 1965. *Trop. Agric. Trin.* **42**: 351 (effect of temp. on spore germination, germ tube growth & on conidial formation; **45**, 838).

—— et al. 1966. *Ibid* **43**: 117 (relation of infection sites to leaf development & spore type; **45**, 2189).

——. 1968a. *Ibid* **45**: 1 (climate & perithecial & sporodochial formation; **47**, 1212).

—— ——. 1968b. *Can. J. Bot.* **46**: 1495 (action of oil on life cycle of pathogen; **48**, 1249).

——. 1969a. *Trop. Agric. Trin.* **46**: 325 (*Mycosphaerella* spp. associated with banana leaf spots; **49**, 521).

——. 1969b. *Pl. Dis. Reptr* **53**: 830 (use of benomyl in aerial spraying; **49**, 804).

——. 1970a. *Phytopathology* **60**: 856 (role of conidia in epidemiology; **49**, 3401).

—— ——. 1970b. *Trop. Agric. Trin.* **47**: 289 (methods of measuring spotting prevalence & severity; **50**, 818).

——. 1971. *Ibid* **48**: 185 (proposed international scale for estimating disease intensity; **51**, 1685).

——. 1971. *Phytopathology* **61**: 139 (ascospore survival; **50**, 3052).

——. 1974. *Trop. Agric. Trin.* **51**: 531 (effect of measured levels of disease on fruit quality & leaf senescence; **54**, 1818).

Vakili, N. G. 1968. *Trop. Agric. Trin.* **45**: 13 (response of *Musa acuminata* & edible cvs to infection).

Walker, L. A. 1972. *PANS* **18**: 34 (properties of spray oils & phytoxicity on banana; **51**, 4215).

Wang, H. C. 1963. *Pl. Prot. Bull. Taiwan* **5**: 49 (symptoms, inoculation, morphology & physiology of pathogen).

——. 1966. *Ibid* **8**: 485 (disease incidence & distribution of perfect state).

——. 1971. *J. Agric. For. Taiwan* **20**: 147 (cultural types, ascospore discharge, climate & disease; **52**, 3389).

Mycosphaerella pinodes (Berk. & Blox.) Vestergr., *Bih. K. svenska Vetensk Akad. Handl.* 22: 15, 1896.

Sphaeria pinodes Berk. & Blox., 1861
Sphaerella pinodes (Berk. & Blox.) Niessl, 1875
Mycosphaerella pinodes (Berk. & Blox.) Stone, 1912
Didymellina pinodes (Berk. & Blox.) Höhnel, 1918
Didymella pinodes (Berk. & Blox.) Petrak, 1924
stat. conid. *Ascochyta pinodes* Jones, 1927 (as *A. pinodes* (Berk. & Blox.) Jones).

PSEUDOTHECIA on stems and pods, globose, dark brown, with apical papillate ostioles, 90 × 180 μ diam. ASCI cylindrical to subclavate, bitunicate, short stipitate or sessile, 8 spored, 50–80 × 10–15 μ. ASCOSPORES irregularly biseriate, hyaline, ellipsoid, guttulate, constricted at the septum, rounded at the ends, 12–18 × 4–8 μ. PYCNIDIA on stems, leaves, pods and seeds, solitary or gregarious, initially immersed becoming erumpent, dark brown to black, globose, 100 × 200 μ diam., opening by papillate ostioles; wall composed of pseudoparenchymatous cells slightly thicker walled on the outermost layer. CONIDIA hyaline, slightly constricted at the septa, ellipsoid, guttulate, 1 septate (sometimes 2 or even 3 septate), 8–16(–18) × 3–4.5(–5)μ (from E. Punithalingam & P. Holliday, *CMI Descr.* 340, 1972).

The stat. conid. should not be confused with *Phoma medicaginis* var. *pinodella* (q.v. formerly *Ascochyta pinodella*) which has mostly aseptate conidia. These 2 fungi (and *A. pisi* q.v.) have frequently been studied together (e.g. Clauss; Hare et al.; Jones; Sattar; Sorgel; Sorgel et al.; Wallen; Wallen et al.). The conidia of *A. pinodes* are larger than those of *P. medicaginis* var. *pinodella* (4.5–8(–10) × 2–3 μ). *A. pisi* on oat agar after 8–12 days at 18° forms an exudate of carrot-red conidial masses. *A. pinodes* forms such masses as a light buff to flesh-coloured exudate (Jones). *M. pinodes* occurs on *Pisum sativum*, *Lathyrus*, *Phaseolus* and *Vicia*.

Leaf, stem and pod spot, and foot rot of pea is widespread, especially in temperate and subtropical regions (Map 316). The main host can only be grown in cool parts of the tropics where the disease may occur. Lesions on the above ground parts begin as very small purplish spots which enlarge, becoming more or less zonate and dark brown, without a definite margin; they may be circular or irregular in shape, with a darker centre. Infection spreads via the petiole to the stem causing girdling lesions; the tap root and proximal parts of the lateral roots can be attacked. Flowers become spotted and pods poorly filled. Infection leads to post- and pre-emergence damping off, dwarfing or death of older plants, and discolouration and shrinkage of seed. Because of the severe foot rot syndrome, and more wide ranging dispersal mechanisms, *M. pinodes* causes a more damaging disease than *A. pisi*.

Penetration through the cuticle and stomata has been described but whether both processes occur with both spore forms is not clear. Opt. temp. for linear growth and sporulation in vitro, and for conidial germination, is *c.* 24°, and infection is high at this temp. or below; from Canada an opt. of 15–18° is reported. Spread is through waterborne conidia, air-dispersed ascospores, in soil, host debris (in which the pathogen survives between crops) and seed. Chlamydospores and sclerotia occur, and in the soil conidia can be transformed into the former; there is some saprophytic ability. Five races were recently reported (Ondřej). *M. pinodes* is homothallic. Light stimulates pycnidial production and most perithecia in vitro occurred at 16°. The effect of light, enzyme production, lesion development on detached leaflets and fungal growth inhibitors have all been the subject of comparative investigations with *A. pisi.* Disease free seed must be used. A 24 hour soak in captan or thiram at 30° and other thiram and orthocide treatments were effective. Cultural control measures should be taken and there is some resistance. For further ref. on this pathogen see *A. pisi* with which it may occur in the field.

Baumann, G. 1953. *Kuhn-Arch* **67**: 305 (homothallism, temp. requirements & control; **35**: 339).
Carter, M. V. et al. 1961. *Aust. J. agric. Res.* **12**: 879 (spread & survival in perfect state; **41**: 345).
——. 1963. *Aust. J. biol. Sci.* **16**: 800 (ascospore dispersal; **43**, 1782).
*Clauss, E. 1961. *Naturvissenschaften* **61**: 106 (phenolic content & resistance with *Ascochyta pisi*; **40**: 502).
*——. 1964. *Zuchter* **34**: 260 (pectinase & cellulase formation with *A. pisi*).
Dickinson, C. H. et al. 1968. *Ann. appl. Biol.* **62**: 473 (soil survival with *A. pisi*; **48**, 1389).
*Hare, W. W. et al. 1944. *Res. Bull. Wisc. agric. Exp. Stn* 150, 31 pp. (general with *A. pisi*; **26**: 91).
*Jones, L. K. 1927. *Bull. N.Y. agric. Exp. Stn* 547, 46 pp. (comparative studies with *A. pisi*; **7**: 611).
Kerling, L. C. P. 1949. *Tijdschr. PlZiekt.* **55**: 41 (penetration, histology, spread by ascospores; **28**: 433).
Ondřej, M. 1973. *Ochr. Rost.* **9**: 49 (5 races; **52**, 4275).
*Sattar, A. 1934. *Trans. Br. mycol. Soc.* **18**: 276 (comparative studies with *A. pisi* & *A. rabiei*; **13**: 611).
Sheridan, J. J. 1973. *Ann. appl. Biol.* **75**: 195 (survival on pea haulm; **53**, 1620).
*Sorgel, G. 1952. *Zuchter* **22**: 4 (nature of differential resistance with *A. pisi*; **32**: 228).
——. 1953. *Arch. Mikrobiol.* **19**: 247 (development in vitro; **34**: 424).
*—— et al. 1954. *Zuchter* **24**: 56 (course of infection & temp. with *A. pisi* **34**: 122).
*——. 1956. *Phytopathol. Z.* **28**: 187 (germination & temp. with *A. pisi*; **36**: 444).
Stubbs, L. L. 1942. *J. Agric. Vict. Dep. Agric.* **40**: 260 (natural seed infection & control; **21**: 438).
Van Warmelo, K. T. 1966. *Bothalia* **9**: 183 (perfect state & cytology; **46**, 1144).
*Wallen, V. R. 1965. *Can. J. Pl. Sci.* **45**: 27 (comparative significance in epidemiology with *A. pisi*; **44**, 2658).
—— et al. 1967. *Ibid* **47**: 395 (epidemiology & control; **46**, 3596).
*——. 1967. *Can. J. Bot.* **45**: 2243 (isolation from & incidence in soil, changes in virulence, **47**, 1721).

* These papers also concern the fungus now known as *Phoma medicaginis* var. *pinodella* q.v.

Mycosphaerella zeae-maydis Mukunya & Boothroyd, *Phytopathology* **63**: 530, 1973.
stat. conid. *Phyllosticta maydis* Arny & Nelson, 1971.

PSEUDOTHECIA within leaf tissue with tips exposed at maturity; dark brown, globose, 86.5–192 (av. 141) μ diam.; wall of several layers of irregular, isodiametric pseudoparenchymatous cells, 8–25 (av. 16) μ diam. Mature ascocarps ostiolate, papillate, 14.5–24 (av. 18.5) μ diam.; ostioles apparently open wide at spore discharge. ASCI cylindrical or clavate, straight or curved, short stipitate, 41–65 × 9.5–12 (av. 49.5 × 11.5) μ. ASCOSPORES 8, biseriate, hyaline, straight or curved, 13.5–20 × 5–6 (av. 16 × 5) μ, 2 celled, cells with rounded ends the upper larger than the lower, guttulate and markedly constricted at the septum. PYCNIDIA within leaf tissue, ostioles usually towards upper surface, globose, reddish brown, 60–150 μ diam. CONIDIA aseptate, hyaline, shape variable but mostly oblong ellipsoidal to subcylindrical, typically biguttulate; 8–20 × 3–7.5 (mostly 12–15 × 4–6)μ, length : breadth ratio commonly *c.* 3 : 1 (perfect state D. M. Mukunya & C. W. Boothroyd l.c., stat. conid. after D. C. Arny & R. R. Nelson, 1971).

Yellow leaf blight of maize was first noticed in 1967 in Canada and USA. It occurs in Canada (Ontario), the N.E. states of USA, France and Kenya (Map 505). The authors of the sp. compared it with 3 other *Mycosphaerella* spp., including *M. maydis* (Pass.) Lindau and *M. zeicola* Stout, described from maize. *M. zeicola* has smaller pseudothecia (80–110 μ) and ascospores with median septa that have only a slight constriction. *M. maydis* has smaller ascospores. The conidia of *Phyllosticta zeae*

Stout from maize are smaller, 4.5–7 × 2–3.5 μ; yellow leaf blight had been attributed to this fungus. Infected leaves have rectangular to oblong, elliptical (usually 7–10 × 15–20 mm), necrotic lesions with a chlorosis of the surrounding tissue and running parallel to, but not delimited by, the veins; the centre is buff and the border brown. Leaves of any age are attacked and severe infection causes death, beginning with the older leaves. Lesions may occur on leaf sheaths and ear husks, and seedlings can be stunted or killed. The leaf lesions resemble those caused by *Cochliobolus heterostrophus* (q.v.).

The fungus was called a weak pathogen by Sutton et al. Most conidial germination occurs at RH 100% and 12–18° (after 4–5 hours). Colonisation, as determined by lesion size, was greatest at 21–27° after 14–16 days. Opt. temp. for the formation of pycnidia was 18° (after 7–11 days; Castor et al.). Conidia form appressoria and penetrate the cuticle. Carry over on maize debris occurs, presumably as the perfect state. More disease has been found in field plots with high rather than low plant densities, although air temps and dew duration were similar. Crops with min. tillage were affected more than those with conventional tillage. Disease severity was correlated with the amount of infected maize residues from previous crops (Sutton et al.). Primary inoculum early in the growing season is assumed to be mostly of ascospores. Resistance in maize lines with normal cytoplasm has been demonstrated. Cvs with Texas male sterile cytoplasm are susceptible in varying degrees. There is evidence for a cytoplasmic control of susceptibility. *M. zeae-maydis* forms a host specific toxin which causes plant responses similar to those caused by *C. heterostrophus* toxin. Measures which should reduce inoculum in the field such as rotation and deep ploughing have been suggested.

Arny, D. C. et al. 1970. *Pl. Dis. Reptr* **54**: 281 (history, reactions of inbreds & crosses; **49**, 3248).
——. 1971. *Phytopathology* **61**: 1170 (*Phyllosticta maydis* sp. nov.; **51**, 2430).
Ayers, J. E. et al. 1970. *Pl. Dis. Reptr* **54**: 277 (reactions of inbreds & crosses; **49**, 3247).
Bootsma, A. et al. 1973. *Phytopathology* **63**: 1157 (RH & conidial germination; **53**, 1355).
Castor, L. L. et al. 1977. *Ibid* **67**: 85 (effect of environment on infection; **56**, 3548).
Comstock, J. C. et al. 1973. *Ibid* **63**: 1357 (host specificity of toxin to Texas male sterile maize; **53**, 1798).
Jimenez-Diaz, R. M. et al. 1976. *Ibid* **66**: 1169 (penetration; **56**, 1585).
Mukunya, D. M. et al. 1973. *Ibid* **63**: 529 (*Mycosphaerella maydis* sp. nov.; **53**, 510).
——. 1975. *Crop Sci.* **15**: 495 (genetics of resistance; **55**, 2696).
Scheifele, G. L. et al. 1969. *Pl. Dis. Reptr* **53**: 656 (reactions of inbred lines; **49**, 135).
Sutton, J. C. et al. 1972. *Can. Pl. Dis. Surv.* **52**: 89 (effects of cultural practices; **52**, 1523).
Yoder, O. C. 1973. *Phytopathology* **63**: 1361 (selective pathotoxin; **53**, 1799).

MYCOVELLOSIELLA Rangel, *Archos Jard. bot. Rio de J.* **2**: 71, 1917.

Vellosiella Rangel, 1915
Ragnhildiana Solheim, 1931
Cercodeuterospora Curzi, 1932.
(Dematiaceae)

Foliicolous, sometimes causing a leaf spot but often only an indeterminate chlorotic area on the upper leaf surface. Caespituli amphigenous but commonly almost entirely hypophyllous, effuse, whitish or yellowish or, more usually, pale to deep olivaceous brown; in several spp. not easily visible but in most velvety or floccose. PRIMARY MYCELIUM immersed. STROMA usually absent or poorly developed, but sometimes more or less well developed. SECONDARY MYCELIUM external, usually arising from a few hyphae which penetrate a stoma, extensive, assurgent or repent, usually much branched, often climbing the leaf hairs of the host plant or forming ropes. CONIDIOPHORES mostly arising terminally and as lateral branches from the secondary mycelial hyphae; sometimes also in small fascicles direct from the hyphae which penetrate the stoma and distinct fascicles of conidiophores often occur in *M. concors*. CONIDIOGENOUS CELLS integrated, polyblastic, sympodial, usually not geniculate but sometimes slightly so, cicatricised, the conidial scars conspicuously thickened. CONIDIA more or less colourless, pale yellowish brown or deeper olivaceous brown, narrowly ellipsoid, fusiform, subcylindric or occasionally slightly obclavate, usually smooth (but distinctly rough walled in *M. maclurae*), usually markedly catenate but in some spp. very variable in degree of catenation, in others frequently catenate and in some spp. never catenate, the hilum and terminal and lateral scars conspicuously thickened (F. C. Deighton, *Mycol. Pap.* 137: 2, 1974; see also M. B. Ellis, *Demat. Hyphom.*: 303, 1971).

The distinctive characters of *Mycovellosiella* (conidia 0–3 or more septate) are the thickened conidial scars and the formation of an assurgent or repent secondary external mycelium, on the hyphae of which the conidiophores are borne terminally and as lateral branches (Deighton, who has fully described the genus and spp.). *M. cajani* (P. Henn.) Rangel ex Trotter (synonyms: *Cercospora cajani* P. Henn.; *Vellosiella cajani* (P. Henn.) Rangel) occurs on pigeon pea (*Cajanus cajan*). Deighton named 3 vars of this sp.: *cajani, indica* (synonym: *Cercospora indica* U. B. Singh) and *trichophila* (synonyms: *Cercodeuterospora trichophila* Curzi; *Dendryphium cajani* Sawada). *M. concors* (Casp.) Deighton (synonyms: *Fusisporium concors* Caspary; *Cercospora concors* (Casp.) Sacc.; *C. heterosperma* Bresadola) occurs on Irish potato (*Solanum tuberosum*). *M. puerariae* Shaw & Deighton (=*Ramularia puerariae* Sawada, not validly published) causes yellow leaf mould of *Pueraria lobata*. This fungus was not found on *P. phaseoloides* in Papua New Guinea. *M. tarrii* Deighton was described from eggplant (*Solanum melongena*).

Deighton, F. C. 1974. *Mycol. Pap.* 137, 75 pp. (on *Cercospora* & allied genera V. *Mycovellosiella* and a new sp. of *Ramulariopsis*; **54**, 3746).
Onim, J. F. M. 1976. *SABRAO J.* 8: 121 (resistance to *M. cajani* in pigeon pea; **57**, 1923).
Shaw, D. E. et al. 1970. *Trans. Br. mycol. Soc.* **54**: 326 (*M. puerariae* sp. nov.; **49**, 2901).

Mycovellosiella phaseoli (Drummond) Deighton, *Mycol. Pap.* 137: 70, 1974.
Ovularia phaseoli Drummond, 1945
Ramularia phaseoli (Drummond) Deighton, 1967
R. phaseolin Petrak, 1960.

Leaf spots usually yellowish areas on upper leaf surface up to 1 cm wide, indefinite margin; sometimes becoming brown with a yellow margin. Caespituli hypophyllous, typically as white floury aggregations, evenly and closely distributed over the spots. PRIMARY MYCELIUM forming a loose hyphal aggregation, up to 40 μ wide and 25 μ high, filling the substomatal cavities. Developing through the stomata to form the SECONDARY MYCELIUM which bears the conidiophores terminally and as lateral branches. CONIDIOPHORES colourless, smooth, septate, branched, branches up to 85 μ long, 3–4 μ wide diminishing near the apex, flexuous, slightly or strongly geniculate; conidial scars small, easily visible, thickened; old scars usually prominent and on a small shoulder or occasionally at the end of a short lateral rounded conical peg (the old conidiophore apex). CONIDIA colourless, catenulate in branched chains, smooth, ellipsoid, the terminal ones broadly obtuse at apex, mostly continuous but sometimes 1 septate, mostly 4–21 × 3–5.5 μ. Growth on PDA very slow, cushion like, with a white cottony covering over dark mycelial growth immersed in the agar; sporulates in culture well as small colonies 1 mm diam. (after F. C. Deighton, *Trans. Br. mycol. Soc.* **50**: 125, 1967; as *R. phaseoli*).

Floury leaf spot of common bean is widespread (Map 436), mostly at high altitudes in the tropics. In India extensive defoliation was reported. In Colombia the disease was considered to be potentially serious but in Brazil (Minas Gerais) it was not thought to be of economic importance. No work on the pathology of *M. phaseoli* appears to have been done. Benomyl and thiophanate were effective fungicides in India.

Cardona-Alvarez, C. et al. 1958. *Pl. Dis. Reptr* **42**: 778 (in Colombia **37**: 749).
Deighton, F. C. 1967. *Trans. Br. mycol. Soc.* **50**: 123 (morphology, taxonomy & distribution; **46**, 2136).
Singh, B. M. et al. 1976. *Indian J. Mycol. Pl. Pathol.* **6**: 148 (fungicides with *Phaeoisariopsis griseola*; **57**, 1903).
Sohi, H. S. et al. 1965. *Indian Phytopathol.* **18**: 384 (in India; **45**, 2299).
Vieira, C. et al. 1965. *Revta Ceres* **12**: 311 (in Brazil; **46**, 1149).

MYROTHECIUM Tode ex Fr., *Syst. Mycol.* **1**: XLV, 1821.
Myxormia Berk & Br., 1850
Myrotheciella Speg., 1910
Exotrichum Syd., 1914
Starkeyomyces Agnihothrudu, 1956.
(Tuberculariaceae)

FRUCTIFICATION cupulate, sporodochial or synnematous, formed from closely compacted conidiophores arising from a more or less developed stroma and bearing a mass of slimy green to black spores which become hard on drying. Fructification surrounded by differentiated marginal hyphae which may be free or laterally compacted into a plectenchymatous wall. Hyaline or darkened SETAE sometimes present, arising from the basal stroma. CONIDIOPHORES hyaline, olivaceous or slightly darkened, macronematous, irregularly and repeatedly branched forming several branches at each node, the ultimate branches bearing the conidiogenous cells (phialides) in whorls. PHIALIDES hyaline or darkened at the apex, occasionally percurrent, sometimes with a flared collarette, compacted into a dense parallel layer. CONIDIA aseptate, hyaline or dilute olivaceous, black in mass, slimy (M. Tulloch, *Mycol. Pap.* 130: 8, 1972).

The genus was revised by Tulloch who described 13 spp. with a key; see also Ellis, *Demat. Hyphom.*, 1971 for a key and description. *M. roridum* is the only plant pathogen. *M. cinctum* (Corda) Sacc. (striate conidia) and *M. verrucaria* (Alb. & Schw.) Ditm. ex Fr. are both plurivorous and widespread, the former particularly on grasses, sedges and soil, the latter particularly in soil. Some of the spp. produce phytotoxins.

Nguyen, T. H. et al. 1973. *Trans. Br. mycol. Soc.* **61**: 347 (seedborne spp. & pathogenicity; **53**, 2927).

Pidoplichko, N. M. et al. 1971. In *Metabolites of soil micromycetes* (editor N. M. Pidoplichko): 157 (taxonomy; **51**, 3794).

Tulloch, M. 1972. *Mycol. Pap.* 130, 42 pp. (monograph, 42 ref.; **52**, 1040).

Myrothecium roridum Tode ex Fr. *Syst. Mycol.* **3**: 217, 1828.

SPORODOCHIA (on leaf or stem) sessile or rarely slightly stipitate, polymorphic in surface view; size very variable, diam. 16–750(258) μ, depth 50–200 (70) μ. Spore mass wet when young, drying hard, shiny black, convex, surrounded by a fringe of entangled white hyphae. Marginal hyphae contorted, hyaline, usually tapering towards a blunt apex, branched septate, cells 20–30(28.6) × 1–2.2 (1.7) μ. CONIDIOPHORES arising directly from the mycelium in culture or from epidermal cells. In the host the basal cells are short 6.3–13.5 (8.6) × 1.8–2.3(2) μ and closely compacted, forming a stromatic layer within the epidermis. Hyaline, cylindrical, branched below and then into 2–3 branches bearing the phialides, closely compacted to form a sub-hymenial layer; septate, cells 6.2–23(14.7) × 1.4–3.6(1.9) μ. PHIALIDES in whorls of 2–5 at apex of conidiophore branches, mostly cylindrical, rarely slightly clavate, hyaline, some dilute olivaceous, occasionally the apex darkened, 9–27(14.2) × 0.9–1.8(1.4) μ, closely compacted into parallel rows forming a dense hymenial layer. CONIDIA rod shaped, ends rounded, one end rarely slightly truncate, smooth walled, hyaline, to dilute olivaceous black in mass, guttulate, 4.5–10.8(7.2) × 1.3–2.7(1.8) μ.

CULTURES on PDA 4–6 cm diam. at 25° after 14 days; white, floccose, wrinkled and often raised in the centre; reverse pinkish buff to light pinkish cinnamon. Sporulation throughout, often in concentric zones, consisting of small groups of conidiophores forming rudimentary sporodochia. These may enlarge into wet black expanses surrounded by fringing entangled white hyphae. Sporodochia surface white at first, then olivaceous black, shiny and wet. Hyphae smooth walled, hyaline, rarely branched, septate, with cells 13.5–36(25.3) × 1.8–3.1 (2.4) μ. *Myrothecium verrucaria* (Alb. & Schwein.) Ditmar ex. Fr., growing on a wide range of plant material, differs from *M. roridum* in its ovate conidia (M. Fitton & P. Holliday, *CMI Descr.* 253, 1970).

M. roridum is plurivorous and widespread (Map 458). Serious losses have been reported on *Gossypium* spp. (India), coffee (Guatemala) and temperate (and ornamental) annuals in USA, Canada and Europe. It has been reported also on *Brassica*, *Cucumis melo* (melon); *C. sativus* (cucumber), *Cyamopsis tetragonoloba* (cluster bean), *Eichhornia crassipes* (water hyacinth), soybean, *Hibiscus esculentus* (okra), tomato, sesame, eggplant, *Trapa bispinosa* (water chestnut) and *Vigna unguiculata* (cowpea). *M. advena* Sacc. has caused a similar disease on coffee in India.

The necrotic lesions caused are commonest on the leaves but can occur on petioles, stems and fruits. Small, sometimes watersoaked lesions increase to *c.* 2.5 cm diam., these may be zonate and coalesce, defoliating the plant. The viscous, circular, conidial masses are first greenish darkening to almost black, mostly abaxial and sometimes arranged as rings on the leaf lesions. Stem cankers have been reported on cotton and sunken lesions on the fruit of melon. On tomatoes a firm black rot develops with a sharply delimited border between healthy and diseased tissue; the rotted part can be removed readily in one piece. On many hosts infection is most serious under nursery or glasshouse conditions. Infection leads to stem lesions, dieback of the crown, and decay at soil level, resembling damping off. Shotholing can occur. The spore masses develop a characteristic white halo of mycelium.

The diseases caused are possibly more serious in warmer climates. Penetration of intact tissue may not occur without the aid of a toxin which creates favourable conditions for infection. The toxin, similar to necrocitin, is not directly linked to pathogenicity since both pathogenic and non-pathogenic isolates produce it (Cunfer et al. 1970; Pawar et al. 1966). Inoculations have shown that wounded tissue is more readily attacked. Spread is largely through soil, in which the fungus is a common saprophyte with the capacity to become actively parasite under conditions that are not clearly definable. Seed transmission may occur and infection of cotton bolls and carpel walls is reported. Inoculated seed of *C. tetragonoloba* gave 16% germination compared with 93% for healthy seed.

Growth in vitro is better in light. Opt. temp. for conidial germination is 25–28°, 95% germination occurred at 25° and 30°; that for linear growth in vitro is similar. On *Trifolium pratense* disease was most severe at 28–32°, the min. temp. for development being 12–15°. Tropical isolates have slightly higher temp. opt. for growth compared with temperate ones. No races have been reported, and the literature conflicts as to whether isolates from one host differ in pathogenicity to another, widely different one. Control is probably through general cultural measures: avoidance of contaminated soil and destruction of infected plant debris, both of which can lead to inoculum build up. Foliage sprays have been successful in some cases.

Arya, H. C. 1956. *Indian Phytopathol.* **9**: 174 (on cluster bean).

——. 1959. *Ibid* **12**: 164 (physiological strs; **40**: 455).

Brooks, F. T. 1945. *Trans. Br. mycol. Soc.* **27**: 155 (differential pathogenicity; **24**: 372).

Chauhan, M. S. et al. 1970. *Cotton Grow. Rev.* **47**: 29 (physiology; **49**, 1661).

Cognee, M. et al. 1964. *Coton Fibr. trop.* **20**: 343 (on cotton; **44**, 3367).

Cunfer, B. M. et al. 1969. *Phytopathology* **59**: 1306 (with *Myrothecium verrucaria* on clover; **49**, 503).

——. 1970. *Ibid* **60**: 341 (toxin; **49**, 2519).

Fergus, C. L. 1957. *Mycologia* **49**: 124 (on *Gardenia*; **36**: 528).

Lakshminarayana, C. S. et al. 1978. *Pl. Dis. Reptr* **62**: 231 (on soybean; **57**, 5215).

Littrell, R. H. 1965. *Ibid* **49**: 78 (on *Gloxinia*; **44**, 1586).

McClean, D. M. et al. 1961. *Ibid* **45**: 728 (on melon; **41**: 199).

Munjal, R. L. 1960. *Indian Phytopathol.* **13**: 150 (general on cotton **41**: 134).

Nag Raj, T. R. et al. 1958. *Ibid* **11**: 153 (*M. advena*, leaf infection of coffee seedlings; **38**: 598).

Nath, R. et al. 1966. *Ibid* **19**: 224 (on eggplant; **46**, 2927).

Nespiak, A. et al. 1961. *Nature Lond.* **192**: 138 (antibiotic properties; **41**: 210).

Padmanabhan, S. Y. 1948. *Curr. Sci.* **17**: 56 (on cowpea; **27**: 309).

Pawar, V. H. et al. 1966. *Hindustan Antibiot. Bull.* **8**: 126 (toxin necrocitin).

———. 1970. *Proc. 1st int. Symp. Pl. Pathol. 1966–67*, New Delhi: 173 (parasitism).

Ponnappa, K. M. 1970. *Hyacinth Control Jnl* **8**: 18 (on water hyacinth; **50**, 2654).

——. 1971. *Indian J. Mycol. Pl. Pathol.* **1**: 90 (on water chestnut; **51**, 4763).

Preston, N. C. 1936. *Trans. Br. mycol. Soc.* **20**: 242 (on *Viola*; **16**: 319).

——. 1943. *Ibid* **26**: 242 (*Myrothecium*, morphology & taxonomy; **23**: 191).

Rao, V. G. et al. 1974. *Riv. Patol. veg.* Ser. IV **10**: 411 (on cucumber; **55**: 3791).

Schieber, E. et al. 1968. *Pl. Dis. Reptr* **52**: 115 (on coffee; **47**, 3446).

Schneider, R. et al. 1967. *NachrBl. dt. PflSchutzdienst. Stuttg.* **19**: 75 (on *Hypocyrta*; **47**, 227).

Singh, D. B. et al. 1967. *Indian J. Microbiol.* **7**: 39 (on sesame; **47**, 868).

Singh, D. V. et al. 1967. *Sci. Cult.* **33**: 70 (on *Vinca*; **46**, 2449).

Stevenson, J. A. et al. 1947. *Pl. Dis. Reptr* **31**: 147 (on tomato fruit; **26**: 473).

Tandon, R. N. et al. 1963. *Curr. Sci.* **32**: 426 (as above; **43**, 1151).

Tewari, J. P. et al. 1977. *Can. Pl. Dis. Surv.* **57**: 37 (on *Brassica*; **57**, 5737).

Urhan, O. 1951. *Biol. Inf. Colombia* **2**: 33 (in coffee nurseries; **32**: 251).

Wilhelm, S. et al. 1945. *Pl. Dis. Reptr* *29*: 700 (on *Antirrhinum*; **25**: 345).

NECTRIA Fr., *Summ. veg. Scand.*: 387, 1849.

Dialonectria Cke., 1884
Lasionectria Cke., 1884
Stilbocrea Pat., 1900
Creonectria Seaver, 1909
Nectriopsis Maire, 1911.
(synonymy after Dingley, 1951)
(Hypocreaceae)

PERITHECIA superficial, scattered freely on surface of host, often yellow, orange or red, caespitose or gregarious on a pulvinate, often erumpent, pseudoparenchymatous stroma or a byssoid subiculum. ASCI unitunicate, 8 spored, spores liberated by rupturing of apex of ascus. ASCOSPORES one septate, elliptical, oval or fusiform, hyaline or lightly coloured. Pseudoparaphyses usually present (after J. M. Dingley, 1951).

Dingley described the genus and the New Zealand *Nectria* spp. Booth (1959, 1960) dealt with the British spp. and the nomenclature of some fusaria in relation to their nectrioid perithecial states (and see Booth, *The genus* Fusarium; also Perrin; Samuels; Samuels, *Acremonium* q.v.). *Nectria* spp. are frequently associated with cankers and diebacks of timber and ornamental trees (mostly temperate); they behave in general as wound pathogens (see Browne, *Pests and diseases of forest plantation trees*; Hepting, *Diseases of forest and shade trees of the United States*). Hainsworth (*Tea pests and diseases and their control*) discussed the infection of damaged tea bark without identifying the *Nectria* spp. concerned. *N. haematocca* Berk. & Br. (perfect state of *Fusarium solani* q.v.) is frequently recorded in the tropics on woody hosts, for example, *Dalbergia sisso*, *Maesopsis eminii* (Brown), *Shorea robusta* and *Tectona grandis*. *N. flavo-lanata* Berk. & Br., *N. dealbata* Berk. & Br. and *N. ochroleuca* (Schw.) Berk. also occur on tropical plants, but none of them appear to be of any economic importance. *N. radicicola* Gerlach & Nilsson causes storage and root rots of several crops; it is widespread but mostly in temperate regions. Pathogenic and saprophytic strs are found. Infection is generally from soil where thick walled chlamydospores can be formed. The stat. conid. is *Cylindrocarpon destructans* (Zins.) Scholten. Chung described it causing a rot of *Panax schinseng*. *N. galligena* Bresadola occurs on many hosts but it is only important as the cause of diseases of apple and pear (European canker; Swinburne).

Booth, C. 1959. *Mycol. Pap.* 73, 115 pp. (*Nectria*, British spp.; **39**: 557).
——. 1960. *Ibid* 74, 16 pp. (nomenclature of some fusaria & their nectrioid states; **39**: 557).
——. 1967. *C.M.I. Descr. pathog. Fungi Bact.* 147, 148 (*N. galligena* & *N. radicicola*).
Brown, K. W. 1964. *E. Afr. agric. For. J.* **30**: 54 (*N. haematococca* on *Maesopsis eminii*; **44**, 254).
Chung, H. S. 1975. *Rep. Tottori mycol. Inst.* **12**: 127 (*N. radicicola* on ginseng).
Dingley, J. M. 1951. *Trans. R. Soc. N.Z.* **79**: 177 (genus *Nectria*).
——. 1957. *Ibid* **84**: 467 (life histories of New Zealand *Nectria* spp.; **36**: 729).
Doyle, A. F. 1978. *Mycologia* **70**: 355 (metabolites & pigments in classification).
Emechebe, A. M. et al. 1976. *Pl. Dis. Reptr* **60**: 227 (*N. haematococca* on passion fruit; **55**, 5852).
Hanlin, R. T. 1971. *Am. J. Bot.* **58**: 105 (morphology of *N. haematococca*; **50**, 2769).
Perrin, R. 1976. *Bull. Soc. mycol. Fr.* **92**: 335 (key for European *Nectria*; **56**, 2390).
Qureshi, A. A. et al. 1972. *Can J. Bot.* **50**: 2443 (morphology & nutrition in perithecial formation by *N. haematococca*; **52**, 2174).
Samuels, G. J. 1976. *Mem. N.Y. bot. Gdn* **26**(3), 126 pp. (revision of fungi formerly classified as *Nectria* subgenus *Hyphonectria*; **55**, 4540).
——. 1978. *N.Z. Jl Bot.* **16**: 73 (some *Nectria* spp. with *Cylindrocarpon* states).
Swinburne, T. R. 1975. *Rev. Pl. Pathol.* **54**: 787 (*N. galligena*, review, 133 ref.).

NEMATOSPORA Peglion, *Atti Accad. naz Lineci* Ser. 5, 6: 278, 1897.
(Spermophthoraceae)

COLONIES yeast like or mycelial; MYCELIUM septate, pseudomycelium often present, cells hyaline, uninucleate or multinucleate, single cells oval, round or cylindrical. Distinct CONIDIOPHORES lacking; asexual reproduction by blastosporic buds (sprout cells), thick walled cells, like chlamydospores, also present. ASCI usually 8 spored, cylindrical, with rounded ends, often compressed in the middle, ascus wall deliquescent. ASCOSPORES act as gametospores, homothallic or (?) heterothallic; spindle or needle shaped with a flagellum-like appendage at one end, the other end bluntly pointed with a refractive cytoplasmic zone in the middle; zygote germinating by forming a sprout cell (proascus), a germ tube or an ascus (from L. R. Batra, *Tech. Bull. U.S. Dep. Agric.* 1469: 22, 1973).

In *Nematospora* the ascospores are needle shaped whilst in *Eremothecium* (q.v.) they are sickle shaped or bent. The genus and spp. (and their biology) were fully described by Batra (q.v. for a bibliography). *N. lycopersici* Schneider differs from *N. coryli* (q.v.) in having sprout cells that are usually spherical rather than ellipsoid or cylindrical; and in using dulcitol and mannitol. *N. phaseoli* Wingard may be a distinct sp. or synonymous with *N. coryli* or *N. lycopersici* (Batra). The spp. are saprophytes and plant parasites, usually associated with insects, and are the most important pathogens in the family. The diseases that are caused are collectively called stigmatomycosis; they are described briefly under *N. gossypii* (and see Mukerji).

Ashby, S. F. et al. 1926. *Ann. Bot.* **40**: 69 (morphology; **5**: 389).
Batra, L. R. 1973. *Tech. Bull. U.S. Dep. Agric.* 1469, 71 pp. (monograph on family, 200 ref.; **53**, 3833).
Mukerji, K. G. 1968. *C.M.I. Descr. pathog. Fungi Bact.* 184, 185 (*Nematospora coryli* & *N. gossypii*).
Schneider, A. 1917. *Phytopathology* **7**: 52 (*N. lycopersici*).
Wingard, S. A. 1925. *Bull. Torrey bot. Club* **52**: 249 (*N. phaseoli*; **5**: 202).

Nematospora coryli Peglion, *Zentbl. Bakt. ParasitKde* Abt. 2 7: 754, 1901.

Vegetative element yeast like, hyphae few, septate at maturity, vegetative reproduction by budding. ASCI (? sporiferous sacs or sporangia) abundant, cylindrical to naviculate, 60–70 × 6–8 μ with 2–8 or more ascospores arranged lengthwise. ASCOSPORES acicular to fusiform, 30–40 × 2–3 μ, with a distinct septum at or near the centre, the upper cell slightly broader at the septum, held together in a mass after liberation by appendages 35–50 μ long. Distinguished from *N. gossypii* (q.v.) in having a yeast-like vegetative phase and septate ascospores, broader at the septum (K. G. Mukerji, *CMI Descr.* 184, 1968).

Batra gave additional morphology and other characters (*Nematospora* q.v.). *N. coryli* is widespread in the tropics (Map 163) and recorded on many hosts including: cashew (*Anacardium occidentale*), citrus, coffee, cotton and many leguminous plants. It occurs on seed and fruit causing the conditions known as yeast spot (of beans and soybean) or stigmatomycosis. *N. coryli* is frequently found with *N. gossypii* and, since the diseased conditions caused by the 2 spp. are comparable (as is the host range), some account of the pathology is given under *N. gossypii* only. Batra (l.c.) gave a list of hosts, associated insects and geographic localities.

Daugherty, D. M. et al. 1970. *Trans. Mo. Acad. Sci.* **4**: 24 (culture; **52**, 1332).
Koopmans, A. 1977. *Genetica* **47**: 187 (cytology; **57**, 1008).

Nematospora gossypii Ashby & Nowell, *Ann. Bot.* **40**: 69, 1926.
Ashbya gossypii (Ashby & Nowell) Guilliermond, 1927
Ashbia gossypii Cif. & Fragoso, 1928.

HYPHAE hyaline, often vacuolated or containing a granulated material and numerous hyaline droplets, at first non-septate, at maturity septate, branching dichotomously. Vegetative reproduction by lateral buds or transverse fission. ASCI abundant, developing from hyphal segments, clavate cylindrical or most frequently sigmoid, single, in groups or chains, 100–200 × 10–20 μ, vacuolated or containing a granular protoplasm; sac breaks or autolyses to release fascicles of mature ascospores. ASCOSPORES grouped parallel into 2 or more fascicles of 2–6 arranged lengthwise in the sac; 4–32/sac; 25–37 × 2–5 μ, acicular to fusiform, often with a thin, central septum; held together after liberation by intertwined appendages 50–100 μ long. In some strs hyphal elements become swollen and thick walled, termed 'buld forms', distinguished from *N. coryli* (q.v.) in not having yeast-like vegetative phase and septate ascospores (K. G. Mukerji, *CMI Descr.* 185, 1968).

Batra, who retained the sp. in *Ashbya* Guillier., gave additional morphology and other characters (*Nematospora* q.v.). *N. gossypii* is widespread in Africa and also occurs in America and Asia (Map 153). It is found particularly on cotton causing stigmatomycosis (internal boll rot or staining) with *N. coryli*; and see Mukerji (l.c.) for other hosts. Both fungi infect other crop plants, especially *N. coryli* which occurs on seeds of *Phaseolus* spp. Plants in the Malvaceae and Leguminosae are most commonly attacked. Neither fungus directly penetrates fruits or seeds nor do they normally enter through the feeding punctures of uninfested insects. They are mechanically transmitted from plant to plant on the mouth parts of hemipterous insects. In cotton these are chiefly *Dysdercus* spp. known as stainer bugs. The fungi are not soil inhabitants but persist on fruits, seeds and fallen cotton bolls; they occur within certain insects which may remain infective for some days.

Cotton bolls attacked at *c.* 1–2 weeks show arrested growth, are shed or abort and seeds are killed. Infection at *c.* 3–4 weeks causes a reduction in boll size; dirty, yellowish brown lint; the carpels open, are contorted and partially reflexed; centre of lock is hard, not expanding and appearing webbed. Infection at later growth stages has little or no effect on boll size but the lint, although intact, may be stained from dark brown to yellow. Infection of seeds of Leguminosae by *N. coryli* causes them to show sunken cotyledon tissue and wrinkled seed coats which may rupture. Some seeds (e.g., lima bean and soybean) show a brownish discolouration of testas and cotyledons. Infections of fruits of many other crops have been described, for example, cashew, citrus, coffee, pigeon pea and tomato. Control measures in cotton are directed at the insect vectors.

Frazer, H. L. 1944. *Ann. appl. Biol.* **31**: 271 (transmission in cotton; **24**: 190).

Marsh, R. W. 1926. *Ann. Bot.* **40**: 883 (inoculation of cotton; **6**: 163).

Moore, E. S. 1930. *Sci. Bull. Dep. Agric. S.Afr.* 94: 12 (internal boll disease of cotton; **10**: 519).

Pearson, E. O. 1947. *Ann. appl. Biol.* **34**: 527 (as above, development in relation to time of infection: **27**: 361).

Pridham, T. G. et al. 1950. *Mycologia* **41**: 603 (general, 81 ref.; **30**: 124).

Ullyett, G. C. 1930. *Sci. Bull. Dep. Agric. S.Afr.* 94: 3 (general on cotton).

Wallace, G. B. 1932. *Trop. Agric. Trin.* **9**: 127 (on coffee; **11**: 572).

NEOCOSMOSPORA E. F. Smith, *Bull. U.S. Dep. Agric.* 17: 45, 1899.

(Hypocreaceae)

ASCOMATA (PERITHECIA) not immersed in a stroma, usually developing singly, superficial, globose, not rostrate. ASCI 8 spored, clavate to cylindric, unitunicate. ASCOSPORES aseptate, monostichous, globose, brown, with a distinct wrinkled (sculptured) epispore (wall). PARAPHYSES present, inconspicuous, broad, loosely jointed, unbranched, of *c.* 5 cells.

Smith (l.c.) gave a full description of the genus and *N. vasinfecta* E. F. Smith (stat. conid. a *Cephalosporium*), and see Udagawa. The PERITHECIA of this sp. are mostly 250–350 × 200–300 μ, orange vermillion; ASCI 8 spored; ASCOSPORES thick walled, colourless until mature and then light brown, exospore wrinkled, usually 10–12 μ diam. *N. vasinfecta* is a widespread soil inhabitant which has been described as causing minor diseases by attacking the roots of young plants in nurseries. Nisikado et al. gave hosts and described a seedling wilt of *Albizia julibrissin* (silk tree); the opt. temps for mycelial growth, conidial and perithecial formation were *c.* 30°, 24–29° and *c.* 27°, respectively. Most ascospore germination occurred at 30°. Ref. to other crops on which the fungus has caused disease are given for: *Cajanus cajan* (pigeon pea), *Cyamopsis tetragonoloba* (cluster bean), *Crotalaria* spp., soybean and *Pisum sativum* (pea). *N. africana* Arx is distinct in that its ascospores are smaller and smooth at maturity. *N. vasinfecta* does not, apparently, cause serious diseases and good husbandry in producing young plants will presumably prevent damaging infection.

Doguet, G. 1956. *Annls Sci. nat. Bot.* **17**: 353 (perithecial development in *Neocosmospora vasinfecta* & *N. africana*; **37**: 27).

Kern, H. et al. 1971. *Phytopathol. Z.* **72**: 327 (toxin formation, pathogenicity on pea; **51**, 3781).

Nisikado, Y. et al. 1937. *Ber. Ohara Inst. landw. Forsch.* 7: 594, 557 (on *Albizia julibrissin*; effects of temp.; **17**: 146, 147).

Phillips, D. V. 1972. *Phytopathology* **62**: 612 (on soybean; **52**, 535).

Roos, A. 1977. *Phytopathol. Z.* **88**: 238 (physiology & pathology; **56**, 4766).

Sarojini, T. S. 1954. *J. Madras Univ.* B **24**: 137 (on pigeon pea; **34**: 727).

Singh, R. S. 1951. *Sci. Cult.* **17**: 131 (on cluster bean; **31**: 386).

Thakur, R. N. et al. 1971. *Indian J. Mycol. Pl. Pathol.* **1**: 84 (on *Crotalaria*, including *C. mucronata*; **51**, 3386).

Udagawa, S. I. 1963. *Trans. mycol. Soc. Japan* **4**: 121 (morphology of *N. vasinfecta* & *N. africana*; **43**, 2836).

Van Warmelo, K. T. 1976. *Mycologia* **68**: 1181 (fine structure of ascospores).

NIGROSPORA Zimm., *Zentbl. Bakt. ParasitKde* Abt 2 8: 220, 1902.

Basiporium Molliard, 1902
Dichotomella Sacc., 1914.
(Dematiaceae)

COLONIES at first white with small, shining black conidia easily visible under a low-power dissecting microscope, later brown or black when sporulation is abundant. MYCELIUM all immersed or partly superficial. STROMA none. SETAE hyphopodia absent. CONIDIOPHORES micronematous or semi-macronematous, branched, flexuous, colourless to brown, smooth. CONIDIOGENOUS CELLS monoblastic, discrete, solitary, determinate, ampulliform or subspherical, colourless. CONIDIA solitary, with a violent discharge mechanism, acrogenous, simple, spherical or broadly ellipsoidal, compressed dorsiventrally, black, shining, smooth, aseptate (M. B. Ellis, *Demat. Hyphom.*: 319, 1971).

Mason considered that these 3 spp. could be separated on conidial size: *N. oryzae* (stat. conid. of *Khuskia oryzae* q.v. and Hudson; conidia 10–16, mostly 12–14 μ diam.), *N. sphaerica* (Sacc.) Mason (conidia 14–20, mostly 16–18 μ diam.) and *N. sacchari* (Speg.) Mason (conidia 17–24, mostly 20–22 μ diam.). Another sp. *N. panici* Zimm. has conidia 25–30 μ diam. Measurements are from Ellis (l.c.). The sp. described by Bremer from *Saccharum* and attributed to *N. panici* appears to be *N. sphaerica*. Another sp. was described from rice, *N. padwickii* Prasad., Agni. & Agar. and distinguished by its large conidia av. 35×38 μ. *N. vietnamensis* Jechová from rotting orange has very small conidia, 4.4×5.9 μ. McLennan et al. described *N. musae* (no Latin diagnosis) when investigating squirter disease of banana fruit; this disease was generally attributed to *N. sphaerica* (Simmonds). *N. sphaerica* is a cosmopolitan sp. especially widespread in the tropics on many plants, and sometimes isolated from food and soil. Jechová attributed squirter to *N. maydis* (Garovaglio) Jechová (basionym *Sporotrichum maydis* Garovaglio). Potlaichuk described 10 *Nigrospora* spp. and Webster described the violent discharge (rare in Hyphomycetes) of the conidium of *N. sphaerica*. Meredith (*K. oryzae* q.v.) investigated dispersal of *Nigrospora* spores in Jamaican banana fields. Bur-Ravault et al. associated *N. sphaerica* with a relatively dry rot of pineapple.

Squirter is rare outside Australia. There may be no external symptoms or there may be a blue black skin discolouration of the peel over the area that is worst affected internally; this symptom occurs as the fruit begins to turn yellow, no symptoms are seen on green fruit. The internal rot characteristic of squirter develops soon after removal from the ripening room. The pulp shows a dark soft rot which becomes liquid; this is readily seen when pressure is applied to the fruit. The internal rot is usually connected with a rot of the stalk. In the early stages of development there is a characteristic zone of healthy pulp directly beneath the stalk. This separates the rotting central pulp from the stalk which is connected to the rot by a central core (or isthmus) of necrotic tissue. Later this healthy zone becomes rotted like the rest of the pulp. The disease is found most commonly in singles when these are the form in which bananas are packed and transported. The fungus (isolates of which differ in producing a rot) enters the cut ends of stalks of green fruit. Squirter occurs only in the winter and early spring. Packing as hands or clusters reduces squirter, as do post-harvest measures in cultural hygiene and careful handling. Fruit dipping (at the correct time of year) with salicylanilide gives control; more recently thiabendazole was found to be equally effective (Rippon et al.). Control with dips of thiabendazole or benomyl were preferred to salicylanilide since the latter only gave control of *Colletotrichum musae* at concs which are phytotoxic (Allen).

Allen, R. N. 1970. *Aust. J. exp. Agric. Anim. Husb.* **10**: 490 (control on banana with thiabendazole, benomyl & salicylanilide; **51**, 1691).

Bremer, G. 1926. *Meded. Proefstn Java–Suiklnd.* 22: 885 (on sugarcane; **6**: 319).

Burden, O. J. 1969. *Qd agric. J.* **95**: 621 (control of squirter; **49**, 803).

Bur-Ravault, L. et al. 1964. *Fruits* **19**: 325 (on pineapple; **43**, 3277).

Jechová, V. 1963. *Ceská Mycol.* **17**: 12 (*Nigrospora vietnamensis* sp. nov., *N. sphaerica* synonym of *N. maydis* comb. nov., taxonomy; **42**: 476).

McLennan, E. I. et al. 1933. *Res. Bull. Aust. Inst. Sci. Ind.* 75, 36 pp. (*N. musae* sp. nov. & squirter; **13**: 43).

Magee, C. J. 1934. *Agric. Gaz. N.S.W.* **45**: 262 (squirter; **13**: 643).

—— et al. 1939. *Ibid* **50**: 22 (control on banana with salicylanilide; **18**: 327).

Mason, E. W. 1927. *Trans. Br. mycol. Soc.* **12**: 152 (taxonomy of *Nigrospora* on monocotyledons; **6**: 757).

Potlaichuk, V. I. 1952. *Mikrobiologiya* **21**: 219 (taxonomy of *Nigrospora*; **33**: 322).

Prasad, N. et al. 1960. *Curr. Sci.* **29**: 352 (*N. padwickii* sp. nov.; **40**: 303).

Rippon, L. E. et al. 1970. *Agric. Gaz. N.S.W.* **81**: 416 (control on banana with thiabendazole; **50**, 816).

Simmonds, J. H. 1933. *Qd agric. J.* **40**: 98 (*N. sphaerica* & squirter; **13**: 42).

—— et al. 1937. *Ibid* **47**: 542 (control of squirter; **16**: 763).

Webster, J. 1952. *New Phytol.* **51**: 229 (discharge of conidia in *N. sphaerica*; **31**: 567).

OIDIUM sensu Sacc., *Michelia* **2**: 15, 1880.
Acrosporium Nees ex S. F. Gray, 1821.
(Moniliaceae)

Parasitic on higher plants. Vegetative HYPHAE

superficial, abundant, septate, branched, decumbent or creeping, developing lobulate haustoria within epidermal cells of host. CONIDIOPHORES erect, mostly simple, producing basipetal chains of conidia and with a generative or meristematic region toward the apex, the conidiophore merging imperceptibly with the chain of conidium initials which exhibit a gradual maturation toward the distal part of the chain. CONIDIA acrogenous, large, ovate or elliptical, hyaline or bright coloured, aseptate (C. V. Subramanian, *Hyphomycetes*: 832, 1971, as *Acrosporium*).

The fungi belonging to this genus are known as the powdery mildews. Where known the perfect states are members of the Erysiphaceae (see C. E. Yarwood in *The fungi* (edited by G. C. Ainsworth et al.) Vol. IV A: 71, 1973; and *Erysiphe* q.v. for ref.). In the tropics the perfect state, if known, is rarely found; for some forms (or spp.) no perfect state has been described. Some general characteristics of the powdery mildews are described briefly under *Erysiphe*.

O. mangiferae Berthet causes powdery mildew of mango. The conidia are mostly 33–43 × 18–22 μ (Uppal et al. 1941); 33–43 × 21.7–28 μ (Palti et al.). They germinate best at 20–25°. The fungus attacks young leaves, blossoms and fruit; leaves become distorted and infected tissue is killed. Fungicide control may be required. In Israel benomyl was less effective than sulphur, and 10 cvs were placed in 3 groups differing in susceptibility (Palti et al.; and see Rodriguez-Landaeta et al.). In Egypt dinocap was effective (Gafar et al.).

Three *Oidium* spp. causing powdery mildew of papaw have been described and their differences have been set out by Yen. *O. caricae* Noack (conidia elliptical, 24–30 × 17–19 μ, Chona et al.); *O. indicum* Kamat (conidia barrel shaped, 31.2–46.8 × 12.7–33.4 μ, Chiddarwar); *O. caricae-papayae* Yen (conidia 36–44.4 × 15.6–21.6 μ). Another powdery mildew *Ovulariopsis papayae* Van der Bijl (conidia 60–90 × 14–23 μ) was described from this fruit crop. Obrero et al. described *Oidium caricae* as being common in Hawaii and causing premature leaf fall and yellowing in high rainfall areas; mixtures of dinocap and zinc+maneb gave control. In India Chiddarwar described *O. indicum* as causing leaf fall and dieback mainly in nurseries.

O. tingitaninum Carter attacks young shoots causing leaf fall and twig dieback of citrus. Carter gave conidial measurements of 20–28 × 10–15 μ; Petch considered the fungus on citrus in Sri Lanka to be different with conidial measurements of 36–42 × 15–18 μ. No recent control measures for citrus powdery mildew appear to have been reported. Some other *Oidium* spp., causing diseases which may need fungicide control, are: *O. arachidis* Chorin on groundnut (conidia 31–44 × 13–15.4 μ in an amended diagnosis); *O. bixae* Viegas on *Bixa orellana* (Capretti); *O. erysiphoides* Fr. (conidia on eggplant, 28–40 × 14–20 μ, on sesame, 23–36 × 14–20 μ; Roy); and *O. piperis* Uppal, Kamat & Patal (on *Piper betle*, conidia mostly 34–47.5 × 13.7–20.4 μ; see Jhamaria et al.; Nema et al. for control).

Bijl, P. A. van der 1921. *Trans. R. Soc. S. Afr.* **9**: 187 (*Ovulariopsis papayae* sp. nov.; **1**: 106).

Bisby, G. R. 1952. *Trans. Br. mycol. Soc.* **35**: 236 (nomenclature, conservation of *Oidium* Sacc.; **32**: 342; and see *Taxon* **2**: 101, 1953).

Capretti, C. 1961. *Riv. Agric. subtrop. trop.* **55**: 13 (*O. bixae*; **40**: 701).

Carter, C. N. 1915. *Phytopathology* **5**: 193 (*O. tingitaninum* sp. nov.).

Chiddarwar, P. P. 1955. *Curr. Sci* **24**: 239 (*O. indicum* sp. nov.; **35**: 307).

Chona, B. L. et al. 1959. *Indian Phytopathol.* **12**: 186 (*O. caricae*; **40**: 372).

Chorin, M. 1961. *Bull. Res. Coun. Israel* D **10**: 148 (*O. arachidis* sp. nov.; **41**: 193).

—— et al. 1966. *Israel Jnl Bot.* **15**: 133 (*O. arachidis*; **46**, 3607).

Gafar, K. et al. 1970. *Agric. Res. Rev. Cairo* **48**(3): 10 (*O. mangiferae*, fungicides; **51**, 1696).

Holubová-Jechova, V. 1975. *Folia geobot. phytotax.* **10**: 433 (usage of *Oidium*; **55**, 3473).

Jhamaria, S. L. et al. 1970. *Hindustan Antibiot. Bull.* **12**: 71 (*O. piperis*, fungicides; **50**, 212).

Khurana, S. M. P. 1971. *Phytopathol. Z.* **70**: 181 (interactions, *O. caricae* & papaw viruses; **51**, 514).

Linder, D. H. 1942. *Lloydia* **5**: 165 (taxonomy; **22**: 178).

Nema, A. G. et al. 1965. *Pesticides* **9**: 34 (*O. piperis*, fungicides).

Obrero, F. P. et al. 1968. *Pl. Dis. Reptr* **52**: 814 (*O. caricae*, fungicides; **48**, 543).

Overeem, C. van et al. 1926. *Icones Fungorum Malayensium* **16**(4), 6 pp. (*O. caricae*; **6**: 42).

Palti, J. et al. 1974. *Pl. Dis. Reptr* **58**: 45 (*O. mangiferae*, general including control; **53**, 2638).

Petch, T. 1915. *Phytopathology* **5**: 350 (*O. tingitaninum*?).

Raabe, R. D. 1966. *Pl. Dis. Reptr* **50**: 519 (*O. caricae*, susceptibility of senescent papaw leaves; 45, 3380).

Rodriguez-Landaeta, A. et al. 1963. *Revta Fac. Agron. Maracay* 3: 40 (*O. mangiferae*; **43**, 2009).
Roy, A. K. 1965. *Pl. Prot. Bull. F.A.O.* **13**: 42 (*O. erysiphoides*; **45**: 80).
Uppal, B. N. et al. 1941. *J. Univ. Bombay* A **9**: 12 (*O. mangiferae*; **20**: 413).
——. 1946. *Proc. Indian Acad. Sci.* B **24**: 255 (*O. piperis* sp. nov.; **27**: 43).
Yen, W. Y. 1966. *Rev. Mycol.* **31**: 311 (*O. caricae-papayae* sp. nov.).

Oidium heveae Steinmann, De ziektenen en plagen von *Hevea brasiliensis* in Nederlandsch-Indië: 91, 1925.
Acrosporium heveae (Steinm.) Subram. 1971.

MYCELIUM white, hyaline, branched, septate. CONIDIA formed in basipetalous chains, ellipsoidal or barrel shaped, very variable in size, 25–42 × 12–17 μ (A. Sivanesan & P. Holliday, *CMI Descr.* 508, 1976).

No perfect state of *O. heveae* has been described and the fungus appears to be confined to rubber for all practical purposes (powdery mildew on *Hevea brasiliensis*), although other hosts have been reported (e.g. Young 1954). *O. heveae* occurs in: Brazil (São Paulo), Brunei, Burma, Cambodia, Congo, India (S.), Indonesia, Malawi, Malaysia (W., Sabah, Sarawak), Papua New Guinea, Sri Lanka, Tanzania, Thailand, Uganda, Vietnam and Zaire (Map 4). The fungus is the commonest cause of the diseased condition in S.E. Asia known as secondary leaf fall (SLF). Another cause is *Glomerella cingulata* (q.v. under *H. brasiliensis* for some account of the characteristics and control of SLF, and deciduousness, i.e. 'wintering' in rubber). Rubber leaflets infected by *O. heveae* at the shining brown to yellowish stages become shrivelled, necrotic and fall. At the light green (cuticle hardening) stage, *c.* 10 days after bud burst, the infected leaflets do not absciss. The mildew colonies are most conspicuous on the lower surface, and leaflets which do not fall show characteristic chlorotic, translucent blotches. Repeated defoliation leads to branch dieback. Only young leaves are attacked.

The conidia are presumably air dispersed. The fungus survives on self-sown seedlings, the shaded lower canopy and in nurseries. Predisposing factors to infection are the degree of clonal resistance and the weather; the clonal 'wintering' pattern is also important. The expression of resistance may be obscured by disease avoidance. A susceptible clone which 'winters' early may avoid SLF by refoliating in drier weather before the disease has become widespread. In Malaysia (W.) the most important weather parameters are the March rain and number of rain days; RH, temp. and sunshine also have some effect. *O. heveae* infection and spread is favoured by relatively cool weather with humid, overcast and misty mornings. Compared with *G. cingulata* it is more severe in drier conditions (Wastie 1972c, *G. cingulata* q.v. under *H. brasiliensis*).

Prophylactic control with sulphur dust is a long-established control method; but is now generally only done on very susceptible and high yielding clones (Leitch; Wastie; Wastie et al.). Some 4–6 applications are probably required, at 5–7-day intervals, beginning as the leaves unfold. A forecasting system for use in dusting has been described (Lim). Disease avoidance, by accelerating refoliation, may be induced by additional applications of nitrogen (Anon.). Indirect control with a defoliant (sodium cacodylate) is effective and economically feasible in Malaysia where severe SLF occurs (Azaldin et al.; Rao). Its aim is to advance natural defoliation, i.e. the normal 'wintering' time, so that refoliation is brought forward to occur in drier weather. The contact defoliant is applied *c.* 1 month ahead of natural 'wintering'. Another approach to control, on which work has been done, is in the selection of resistant canopy clones for crown budding on susceptible but high yielding trunk or panel clones.

Anon. 1976. *Plrs' Bull. Rubb. Res. Inst. Malaysia* 146: 120 (increased N to avoid disease; **56**, 5771).
Azaldin, M. Y. et al. 1974. *Proc. Rubb. Res. Inst. Malaysia Plrs' Conf.*: 161 (use of chemical defoliant; **54**, 2948).
Bolle-Jones, E. W. et al. 1956. *Nature Lond.* **177**: 619 (Zn deficiency predisposing to infection; **35**: 547).
Fernando, T. M. 1971. *Q. Jl Rubb. Res. Inst. Ceylon* **48**: 100 (epidemiology & control; **51**, 3529c).
Leitch, T. A. T. 1971. *Planter Kuala Lumpur* **47**: 134 (control with S; **51**, 613).
Lim, T. M. 1971. *Proc. Rubb. Res. Inst. Malaysia Plrs' Conf.*: 169 (forecasting for chemical control; **54**, 1417).
Populer, C. 1972. *Publ. Inst. natn. Etude agron. Congo (INEAC), Scientifique* 115, 368 pp. (epidemiology, 358 ref.; **52**, 845).
Rao, B. S. et al. 1973. *Proc. Rubb. Res. Inst. Malaysia Plrs' Conf.*: 267 (use of chemical defoliant; **54**, 2947).
Wastie, R. L. et al. 1969. *J. Rubb. Res. Inst. Malaya* **21**: 64 (economics of control; **49**, 2165b).

Wastie, R. L. 1969. *Planter Kuala Lumpur* **45**: 587 (control with S; **49**, 2615).
Young, H. E. 1952. *Q. Circ. Rubb. Res. Inst. Ceylon* **27**: 3 (review; **31**: 630).
——. 1954. *Ibid* **30**: 51 (control; **34**: 544).

OLIVEA Arthur, *Mycologia* **9**: 60, 1917.
(Pucciniaceae)

PYCNIA mammilliform, subcuticular, without ostiolar filaments. AECIDIA deep seated, subepidermal, protected by host tissues; peridium apparently wanting; AECIDIOSPORES catenulate with intercalary cells obovate, strongly echinulate–verrucose, with rod-like warts, walls coloured; in appearance simulating uredospores. UREDIA from a minute, subcuticular hymenium, expanding into a globose mass of strongly incurved paraphyses, having their bases united; UREDOSPORES borne singly on pedicels, obovate, stellately angular, echinulate, the wall coloured, pores *c*. equatorial, at the angles. TELIA replacing the uredia in the basket of paraphyses; TELIOSPORES numerous, free, aseptate, sessile, colourless, cylindraceous, wall thin, smooth (from J. C. Arthur, l.c.; and see Thirumalachar et al., *Indian Phytopathol.* 2: 236, 1949).

Mulder et al. described leaf rust of teak (*Tectona grandis*), *O. tectonae* (T. S. & K. Ramakrishnan) Mulder (synonyms: *Uredo tectonae* Racib., *Chaconia tectonae* T. S. & K. Ramakrishnan, *Olivea tectonae* Thirum., nomen nudum). The life cycle is not known. UREDIA hypophyllous, pulverulent, subepidermal, later erumpent, orange coloured. Uredial PARAPHYSES cylindric, swollen at apex, incurved, wall 2.5 μ thick, peripheral. UREDOSPORES ovoid to ellipsoid, orange yellow, echinulate, 18.4–25.5 × 16–21 μ (18–28 × 14–22 μ); germ pores indistinct, wall 2–2.5 μ thick. TELIA subepidermal sometimes mixed with the uredia. Paraphyses as in the uredia, slightly less swollen at apex. TELIOSPORES clavate, sessile, in clusters on basal cells, 22–51 × 4–11 μ; wall hyaline, thin.

The telial stage of *U. tectonae* was described by Ramakrishnan et al. and Thirumalachar independently. It is proposed that this sp. be placed in *Olivea* owing to the presence of telial paraphyses, a character absent from *Chaconia*. Because Thirumalachar had not validly published *O. tectonae* a new combination is made, based on *C. tectonae* (J. L. Mulder & I. A. S. Gibson, *CMI Descr*. 365, 1973).

Teak rust occurs in Bangladesh, Burma, India, Indonesia, Sri Lanka, Taiwan and Thailand (Map 499). The infected leaves have a grey flecked appearance on the upper surface and there are masses of orange uredia on the underside, in corresponding positions. Nursery plants and young plantings are most susceptible. There may be severe checks to growth from premature defoliation. Infection is said to be favoured by a hot, relatively dry environment and crowded growing conditions. Some control may be obtained by opening up dense stands and fungicides may be practicable in the nursery. *O. tectonae* is a potentially important teak pathogen in the tropics.

Khan, A. H. 1951. *Pakist. J. For.* **1**: 209 (control; **31**: 214).
Mulder, J. L. et al. 1973. *CMI. Descr. pathog. Fungi Bact*. 365 (*Olivea tectonae*).
Ramakrishnan, T. S. et al. 1949. *Indian Phytopathol.* **2**: 17 (morphology & taxonomy; **29**: 545).
Thirumalachar, M. J. 1949. *Curr. Sci.* **18**: 175 (as above; **28**: 553).

ONCOBASIDIUM Talbot & Keane, *Aust. J. Bot*. **19**: 203, 1971.
(Ceratobasidiaceae)

SPOROPHORES white, as effuse, adherent patches, membranous to subhypochnoid. Hyphal system monomitic. Basal hyphae thin walled but firm, hyaline to yellowish, smooth, not encrusted, without clamp connections, with prominent dolipores, more or less horizontal and up to 10 μ wide. BASIDIA holobasidiate, obovate, later elongating becoming capitate clavate. STERIGMATA 4, conical, straight or curved, stout. BASIDIOSPORES repetitive, smooth, hyaline, thin walled, not amyloid, often multi-guttulate. Conidial and sclerotial states unknown (from P. H. B. Talbot & P. J. Keane l.c.).

The genus was erected to accommodate an obligate pathogen of cacao causing vascular streak dieback in Papua New Guinea and Malaysia (W.; Map 507). *O. theobromae* Talbot & Keane: basal HYPHAE up to 10 μ wide, long celled, up to 200 μ, branching at a wide angle; ascending hyphae narrower, up to 5–6 μ diam., shorter celled hyaline, binucleate. Cystidiate structures absent. BASIDIA arising from clustered monilioid hyphae, subcyclindrical base

6–8 μ wide and a more or less abruptly inflated apex (10–)12–16 μ wide, the whole metabasidium (18–23–)26–36 μ long. STERIGMATA 6–12 μ long, up to 4 μ wide at the base. BASIDIOSPORES broad ellipsoid with one side flattened, (12–) 15–25 × (5–)6.5–8.5 μ (from Talbot et al. l.c.). The new genus was differentiated largely on basidial morphology and the authors compared it with *Ceratobasidium* and *Thanatephorus*; they also commented on *Koleroga*.

This disease is a distinctive dieback and is not to be confused with other forms of cacao dieback which may or may not be associated with pathogens and/or insects. Keane (1974) stated that vascular streak dieback is the most damaging disease of cacao in the wetter areas of Malaysia (W.) and has caused serious losses in Papua New Guinea. It is likely that the pathogen occurs elsewhere on the crop. The first symptoms on a tree branch, or the main shoot of a seedling, is a chlorosis of 1–2 leaves on the second or third growth flush behind the tip. These leaves may, therefore, be up to 1 m behind the tip; not at the tip, as is usually the case in environmentally induced dieback. Diseased leaves have small, sharply defined green spots scattered on a yellow background; they fall within a few days of turning yellow. Leaves above and below the first diseased one begin to show the yellow and green pattern, and also fall. Lenticels become enlarged, giving the bark a rough appearance in the affected parts. Axillary buds begin to grow and the disease spreads to lateral branches, particularly those formed from these buds on an infected stem. Growth of the diseased part is slow. The growing tip eventually dies, the disease spreads to other branches and a mature tree can be killed. The white, closely adherent sporophores may cover the leaf scars. The vascular traces in the leaf scars are discoloured. A diseased stem which is split longitudinally shows the diagnostic brown streaking in the xylem, extending beyond the leaf fall region.

O. theobromae grows specifically in the xylem vessels of diseased stems and leaves before symptom development. The sporophores develop from hyphae which grow from the vessels exposed on leaf scars when the leaves fall. Basidiospores shed on to 3-month-old seedlings (expanding leaves) led to disease symptoms after 3 months. Penetration probably occurs readily through undamaged and unhardened leaves. In moist weather sporulation takes place on the leaf scars and surrounding bark; the sporophores can survive for up to 1 month. Spore release is mainly at 2400–0600 hours. Wet conditions are essential for the spread of vascular streak dieback and spores probably do not survive >1 day in the field. Spore germination occurs immediately in free water on agar but there is no further growth. Cacao cvs differ in their susceptibility to the pathogen.

Keane, P. J. et al. 1972. *Aust. J. biol. Soc.* **25**: 553 (aetiology; **52**, 669).

——. 1974. In Phytophthora *disease of cocoa* (editor P. H. Gregory): 283, Longman (general).

Prior, C. 1977. *J. gen. Microbiol.* **99**: 219 (growth on cacao callus; **56**: 3955).

——. 1978. *Ann. appl. Biol.* **88**: 357 (inoculation method; **57**, 4382).

Talbot, P. H. B. et al. 1971. *Aust. J. Bot.* **19**: 203 (*Oncobasidium* gen. nov. & *O. theobromae* sp. nov.; **51**, 1285).

OPERCULELLA Kheswalla, *Indian J. agric. Sci.* 11: 317, 1941.

(B. C. Sutton reduced the genus to synonymy with *Phacidiopycnis*, *Mycol. Pap.* 141: 140, 145, 1977) (Sphaerioidaceae)

PYCNIDIA unilocular, discoid to subglobose, immersed at first, later erumpent, opening with an apical pore or by a hinged lid. CONIDIOPHORES of 2 kinds; the shorter ones simple, lining the walls of the pynidium and bearing spores terminally; the longer ones branched, sometimes septate and bearing spores laterally and terminally. CONIDIA irregular in shape, aseptate, hyaline; they emerge through a minute apical pore or force open the top (circumscissile lid) of the spherical or discoid pycnidium (from K. F. Kheswalla l.c.).

The genus is monotypic. *O. padwickii* Kheswalla has been reported only from India (Map 514) where it was described on gram or chick pea (*Cicer arietinum*). The PYCNIDIA are finally carbonaceous, 270–810 μ diam.; the shorter CONIDIOPHORES av. 83 μ in length and the longer ones bear spores laterally, on minute sterigmata, and terminally. CONIDIA hyaline, yellowish white in mass, 7.4–16.6 × 5.5–11.1 μ. The fungus causes a foot rot of chick pea. The affected plants begin to dry from the tip downwards. The leaves become pale green, later yellowish and finally falling. The collar becomes dark brown and sometimes the roots are also attacked. The foliage is not directly infected. Good growth and sporulation in culture occur at 25°.

Carry over is on crop debris. In the Punjab low pH and soil moisture, and a sandy soil, increase the disease (Singh et al.).

Bahl, N. et al. 1973, 1974, 1975. *Indian Phytopathol.* **26**: 622; **27**: 413; **28**: 429 (nutrition in culture; **54**, 5661; **55**, 2537; **56**, 2783).
Kheswalla, K. F. 1941. *Indian J. agric. Sci.* **11**: 316 (*Operculella* gen. nov. & *O. padwickii* sp. nov.; **21**: 120).
Singh, G. et al. 1974. *J. Res. Punjab agric. Univ.* **11**: 400 (reactions of cvs; **55**, 2440).
——. 1975. *Indian Phytopathol.* **28**: 546 (carry over; **57**, 3715).
——. 1976. *Indian J. Ecol.* **3**: 81 (soil factors; **57**, 880).

PENICILLIUM Link ex Gray, *Nat. Arr. Br. Pl.* Vol. 1: 554, 1821.

(see Hawksworth et al. & Jørgensen et al. for typification & nomenclature, respectively)
(Moniliaceae)

Vegetative MYCELIUM colourless, or pale or brightly coloured, never dematiaceous, septate, either predominantly submerged or partly submerged and partly aerial, with aerial portion closely matted, loosely floccose, or partially as ropes of hypae. CONIDIOPHORES arising from, and more or less perpendicular to, submerged or aerial hyphae, either detached from one another or to some degree aggregated into fascicles or compacted into definite coremia, septate, smooth or rough, terminating in a broom-like whorl of branches (the penicillus); the latter consisting of a single whorl of spore bearing organs (PHIALIDES), or twice to several times verticillately branched, with the branching system symmetrical or asymmetrical, the final branches being the phialides. CONIDIA produced by abscission, forming unbranched chains, globose, ovoid, elliptical or pyriform, smooth or rough, in most cases green during growth but sometimes colourless or in other pale colours. PERITHECIA produced by some spp., either sclerotium like, ripening tardily from the centre outwards; or soft, ripening quickly. SCLEROTIA produced by several spp. In penicilli with >1 stage of branching the branches bearing the phialides are (as in *Aspergillus*) called METULAE. The branches supporting the metulae, if comparatively short and obviously part of the penicillus, are known as RAMI (from G. Smith, *An introduction to industrial mycology* 6th edn: 173, 1969).

An earlier description is given in the monograph by Raper et al. who gave accounts of other genera to which these fungi have been referred. Smith (l.c.) discussed the related genera *Gliocladium* Corda and *Paecilomyces* Bainier, and gave a key to the series. A series consists of 1 to several closely related spp. 'The taxonomy of the genus presents much greater difficulties than that of the related genus *Aspergillus*' (Smith l.c. p. 172). The genus and some spp. are also described by Subramanian, *Hyphomycetes*. Kulik compiled the spp. described since 1949 or not included in Raper et al. For nomenclature in *Penicillium* and *Aspergillus* see Raper. The taxonomy of the perfect states (*Eupenicillium* Ludwig) was monographed by Scott (and see Stolk et al.). A synoptic key to *Eupenicillium* and to sclerotigenic *Penicillium* spp. was given by Pitt* (1974) who also described taxonomic criteria based on temp. and water relations (1973). Dart et al., investigating the long chain fatty acids, obtained groupings which did not correspond to those of Pitt (1973).

Like *Aspergillus* spp. the members of the genus are common moulds, frequent laboratory contaminants, often components of soil and rhizosphere fungal populations. *Penicillium* spp. are associated, as wound pathogens, with many rots of agricultural produce; they have a widespread use in industrial fermentations. Barkai-Golan described these fungi causing decay of stored fruits and vegetables in Israel; and see Mislivec for those on maize. *P. expansum* Link ex F. S. Gray (blue mould of apple; see Onions) has been recorded on tropical fruits. *P. oxalicum* Currie & Thom is a major causal factor in storage diseases of cush cush yams (*Dioscorea trifida*) in Guadeloupe. Ricci et al. gave control measures for seed and edible tubers. In the former case immersion in malathion + benomyl, and in the latter 10 min. in thiabendazole followed by storage in the dark at 18° and RH 60–80%. Yamamoto et al., on rots of Chinese yam (*D. batatas*), described *P. sclerotigenum* Yamamoto as one of the fungi involved. *P. corymbiferum* Westl. causes a decay of garlic (*Allium sativum*) clones before emergence; wounds are a predisposing factor to disease; pre-plant fungicides are effective (Smalley et al.). The role of allicin in the resistance of garlic to *Penicillium* spp. was investigated by Durbin et al. *P. funiculosum* Thom was shown by Rohrbach et al. to cause the diseased conditions in pineapple known as interfruitlet corking, leathery pocket and fruitlet core rot.

* J. I. Pitt 1979. *The genus* Penicillium *and its teleomorphic states* Eupenicillium *and* Talaromyces, Academic Press.

Hepton et al. described the fungus as invading the non-cutinised floral parts *c.* 3 months after floral differentiation. Diseased fruit (interfruitlet corking) are characterised by conspicuously corked areas between the fruitlets. Some internal corking may occur, this is associated with poorly developed fruitlets and fruit deformity.

Abe, S. 1956. *J. gen. appl. Microbiol. Tokyo* 2(1–2): 1; 2(3): 195 (monograph, taxonomy; **36**: 214, 500).

Barkai-Golan, R. 1974. *Mycopathol. Mycol. appl.* **54**: 141 (*Penicillium* & decay of stored fruits & vegetables in Israel; **54**, 1204).

Dart, R. K. et al. 1976. *Trans. Br. mycol. Soc.* **66**: 525 (relationships based on long chain fatty acids; **56**, 83).

Durbin, R. D. et al. 1971. *Phytopathol. Mediterranea* **10**: 227 (allicin & resistance in garlic; **52**, 3143).

Hawksworth, D. L. et al. 1976. *Taxon* **25**: 665 (typification; **56**, 1963).

Hepton, A. et al. 1968. *Phytopathology* **58**: 74 (*P. funiculosum* on pineapple; **47**, 1631).

Hess, W. M. et al. 1968. *Mycologia* **60**: 290 (surface characteristics of conidia).

Jørgensen, M. et al. 1977. *Taxon* **26**: 581 (nomenclature; **57**, 2028).

Kulik, M. M. 1968. *Agric. Handb. U.S. Dep. Agric.* 351, 80 pp. (new *Penicillium* spp.; **48**, 383).

Mislivec, P. B. et al. 1970. *Mycologia* **62**: 67, 75 (*Penicillium* in harvested & stored maize, temp. & RH; **49**, 2842a & b).

Moura, R. M. et al. 1976. *Fitopat. Brasileira* **1**: 67 (*P. sclerotigenum* on yam; **56**, 3378).

Onions, A. H. S. 1966. *CMI. Descr. pathog. Fungi Bact.* 97 (*P. expansum*).

Pidoplichko, N. M. 1972. *Penicillium* spp., keys for the identification of spp., 150 pp. (Naukova Dumka, Kiev, USSR; **52**, 2168).

Pitt, J. I. 1973. *Mycologia* **65**: 1135 (identification, taxonomic criteria based on temp. & water relations; **53**, 2096).

——. 1974. *Can. J. Bot.* **52**: 2231 (synoptic key to *Eupenicillium* & to sclerotigenic *Penicillium* spp.; **54**, 2702).

Raper, K. B. et al. 1949. *A manual of the Penicillia*, 875 pp. Baillière, Tindall & Cox (monograph, **29**: 122).

——. 1957. *Mycologia* **49**: 644 (nomenclature in *Aspergillus* & *Penicillium*; **37**: 216).

Ricci, P. et al. 1975. *Nouvelles Agronomiques Antilles Guyane* **1**: 153 (storage diseases in yam; **55**, 1588).

———. 1979. *Trop. Agric. Trin.* **56**: 41 (as above).

Rohrbach, K. G. et al. 1976. *Phytopathology* **66**: 392 (*P. funiculosum* on pineapple; **56**, 312).

Sampson, R. A. et al. 1976. *Stud. Mycol.* 11, 47 pp. (revision of subsection *Fasciculata* & allied spp.; **55**, 2154).

Scott, De B. 1968. *Res. Rep. S. Afr. Coun. scient. ind. Res.* 272, 150 pp. (*Eupenicillium*, monograph; **47**,3737).

Smalley, E. B. et al. 1962. *Phytopathology* **52**: 666 (*P. corymbiferum* on garlic; **42**: 170).

Stolk, A. C. et al. 1967. *Persoonia* **4**: 391 (taxonomy & nomenclature in relation to sclerotioid ascocarpic states).

Yamamoto, W. et al. 1955. *Sci. Rep. Hyogu Univ. Agric.* 2(1): 69 (*Penicillium* & *Fusarium* rots of yam; **37**: 130).

———. 1956. *Ibid* 2(2): 23 (*P. expansum* on sweet potato inter alia; **37**: 170).

Penicillium digitatum Sacc. *Sylloge Fung.* **4**: 78, 1886.

P. olivaceum Wehmer, 1895.

(there are probably many more synonyms in the literature)

COLONIES on Czapek agar, and all similar synthetic media, grow very restrictedly and thinly. Growth on malt agar rapid, plane velvety, dull yellow green to greyish olive with age; odour strong suggesting decaying citrus fruit; reverse colourless to pale dull brown. CONIDIAL APPARATUS very fragile and tending to break up into many cellular elements, but when seen complete often resembling the skeleton of a hand, asymetric, very irregular in the number and dimension of the parts, but usually consisting of a few branches and/or metulae-bearing phialides, producing long divergent chains of elliptical conidia. CONIDIOPHORES short, 30–100 × 4–5 μ, smooth walled, arising from the basal felt or submerged mycelium. METULAE variable, 15–30 × 4–6 μ. PHIALIDES few in number, 15–28 × 3.5 μ. CONIDIA variable, smooth, subglobose to cylindrical but usually elliptical, 3.4–12 × 3–8 (usually 6–8 × 4–6) μ (A. H. S. Onions, *CMI Descr.* 96, 1966).

Green mould of citrus fruits is worldwide in orchards and a common cause of decay after harvest. *P. digitatum* is commonly considered with blue mould (*P. italicum* q.v.) of the same crop. Fawcett (*Citrus diseases and their control*) described and contrasted the symptoms of the 2 moulds in detail with coloured illustrations. Green mould has a pasty, wrinkled mycelium on the fruits and which shows much in advance of the conidia. The decay margin is an indefinite band with the margin not well defined in advance of the mycelium on the surface. In wrapped fruits the wrapper adheres closely to the rotting fruit (Fawcett l.c.). Onions (l.c.) stated that an infected fruit softens and begins to shrink, and if exposed to the air becomes a hollow, mummified shell. Both moulds are considered under *P. digitatum*. In both cases they are mainly postharvest, wound pathogens but can affect fruits on the tree. The symptoms first appear as soft watersoaked areas *c.* 1 cm diam.; these enlarge to lesions several cm in diam. and sporulation follows over the watersoaked areas. Green mould has less tendency than blue mould to spread from infected fruits to adjacent sound fruits by contact.

Both fungi have an opt. temp. for decay of fruit near 24°; decay is rapid at 18.5–26.5°. Both the growth and decay rates tend to be faster for *P. digitatum*. Green showed that inoculations were successful when wounding (needle puncture) went beyond the flavedo into the albedo. Bates found that needle prick inoculations into the rind did not cause infection unless the wounds penetrated the rind to the inner pulp or one or more oil vesicles were ruptured. Resistance to infection through shallow wounds was broken down by a spore inoculum in orange juice or a sucrose solution. Nadel-Schiffman et al. found that *P. digitatum* gave higher infection rates than *P. italicum*. Conidial inoculum (with or without additives) did not cause infection through unwounded surfaces. Conidia in water applied to wounds made between oil vesicles caused infection only if such wounds were deep into the albedo. The flavedo was resistant to infection even when damaged. When wounds damaged oil vesicles lesions developed more rapidly (Kavanagh et al.). Brown found that degreening oranges at RH 90–96% at 30° reduced the incidence of *P. digitatum*; 27° was less effective. These conditions induced lignification in injured flavedo and so reduced infection. But injuries with severe peel oil damage were usually invaded even when fruit was degreened at RH 90–96%; oil prevented lignification. For cell degrading enzymes see Cole et al.; Garber; Garber et al.; Miyakawa. In grapefruit the rootstock affected the amount of fruit decay (mostly *P. digitatum*; McDonald et al. 1974a).

Cultural control measures should aim at keeping down conidial inoculum and careful handling of fruits to minimise rind injury. There is an extensive literature on the postharvest application of chemicals for control of both green and blue moulds. Most recently used fungicides are benomyl, biphenyl, guazatine, sodium O-phenylphenate and thiabendazole; some ref. to their use are given for: Australia, Cyprus, Egypt, India, Israel, Morocco, New Zealand and USA. Resistance in the 2 fungi to the benzimidiazole fungicides has been reported from several countries (Gutter 1975; Harding 1972; Kuramoto; Muirhead; Smoot et al.; Wild et al. 1975b). Fungicides against such resistance strs have been reported by Harding (1976); Hartill et al.; Rippon et al. (1976).

Bates, G. R. 1933. *Nature Lond.* **132**: 751 (infection through oil glands; **13**: 228).
——. 1936. *Publs Br. S. Afr. Co.* (Mazoe Citrus Exp. Stn) 4b: 87 (infection; **15**: 715).
Brown, G. E. 1973. *Phytopathology* **63**: 1104 (green mould in degreened oranges; **53**, 1818).
Cole, A. L. J. 1970. *Ann. Bot.* **34**: 211 (pectolytic enzymes; **49**, 2072).
—— et al. 1970. *Phytochemistry* **9**: 695 (formation of hemicellulases; **49**, 2758).
—— ——. 1970. *Ann. appl. Biol.* **66**: 75 (infection with *Trichoderma viride*; **50**, 681).
Davis, P. L. et al. 1965. *Phytopathology* **55**: 1216 (effects of orange rind components & pH on conidial germination; **45**, 1074).
—— ——. 1972. *Ibid* **62**: 488 (effects of citrus fruit volatile components on conidial germination; **51**, 4019).
Fawcett, H. S. et al. 1927. *J. agric. Res.* **35**: 925 (temp. & growth; **7**: 316).
Fergus, C. L. 1952. *Mycologia* **44**: 183 (nutrition in vitro).
French, R. C. et al. 1978. *Phytopathology* **68**: 877 (effect of nonanal, citral & citrus oils on conidial germination).
Garber, E. D. et al. 1965. *Bot. Gaz.* **126**: 36 (pectolytic & cellulytic enzymes; **44**, 2370).
—— ——. 1966. *Can. J. Bot.* **44**: 1645 (pectolytic enzymes; **46**, 1213).
——. 1967. *Phytopathol Z.* **59**: 147 (electrophoresis of endopolygalacturonases, *P. italicum*; **46**, 3669).
Green, F. M. 1932. *J. Pomol.* **10**: 184 (infection; **12**: 167).
Gutter, Y. 1969. *Pl. Dis. Reptr* **53**: 474, 479 (fungicides in Israel; **48**, 3492a & b).
——. 1970. *Israel Jnl agric. Res.* **20**: 91, 135 (as above; **50**, 97, 1796).
—— et al. 1971. *Ibid* **21**: 105 (as above; **51**, 2489).
——. 1974. *Phytopathology* **64**: 1477 (as above; **54**, 3324).
——. 1975. *Ibid* **65**: 498 (resistance to fungicides; **55**, 247).
Harding, P. R. et al. 1967. *Pl. Dis. Reptr* **51**: 51 (fungicides in USA; **46**, 1003).
——. 1968. *Ibid* **52**: 623 (as above; **47**, 3442).
——. 1972. *Ibid* **56**: 256 (resistance to fungicides; **51**, 4018).

——. 1976. *Ibid* **60**: 643 (fungicides in USA; **56**, 1612).
Hartill, W. F. T. et al. 1977. *N.Z. Jl exp. Agric.* **5**: 291 (fungicides in New Zealand; **57**, 3473).
Isshak, Y. M. et al. 1974. *Agric. Res. Rev. Cairo* **52**(3): 85 (fungicides in Egypt; **54**, 4504).
Kavanagh, J. A. et al. 1967. *Ann. appl. Biol.* **60**: 375 (infection; **47**, 1129).
—— ——. 1971. *Ibid* **67**: 35 (effect of additives on conidial germination & infection; **50**, 2276).
Kokkalos, T. I. 1973. *Tech. Pap. Minist. Agric. nat. Resources* 3, 9 pp. (fungicides in Cyprus; **54**, 3933).
Kuramoto, T. 1976. *Pl. Dis. Reptr* **60**: 168 (resistance to fungicides; **55**, 4722).
McDonald, R. E. et al. 1974a. *HortScience* **9**: 455 (effect of rootstocks on postharvest fruit decay; **54**, 2251).
——. 1974b. *Pl. Dis. Reptr* **58**: 1143 (fungicides in USA; **54**, 2817).
Miyakawa, T. 1962. *Ann. phytopathol. Soc. Japan* **27**: 129 (protopectinase activity; **42**: 197).
——. 1963. *Ibid* **28**: 17 (decomposition of pectin in, & action of d-galacturonic acid on, the peel; **42**: 547).
Muirhead, I. F. 1974. *Aust. J. exp. Agric. Anim. Husb.* **14**: 698 (resistance to fungicides; **54**, 3325).
Nadel-Schiffman, M. et al. 1956. *Ktavim* **6**: 61 (infection; **36**: 24).
Rippon, L. E. et al. 1973. *Aust. J. exp. Agric. Anim. Husb.* **13**: 724 (fungicides; **53**, 2572).
—— ——. 1973. *Agric. Gaz. N.S.W.* **84**(3): 133 (as above; **53**, 554).
—— ——. 1974. *Ibid* **85**(3): 38 (as above; **54**, 3930).
—— ——. 1976. *APPS Newsl.* **5**: 45 (control of benomyl resistant strs; **56**, 4057).
Roth, G. 1967. *Phytopathol. Z.* **58**: 383 (insect punctures & infection; **46**, 3442).
Smoot, J. J. et al. 1974. *Pl. Dis. Reptr* **58**: 933 (resistance to fungicides; **54**: 855).
Subramanian, T. M. et al. *Madras agric. J.* **60**: 90 (fungicides; **53**, 173).
—— ——. 1973. *Indian J. agric. Sci.* **43**: 284 (as above; **53**, 3025).
Vanderplank, J. E. 1934. *Sci. Bull. Dep. Agric. S. Afr.* 127, 20 pp. (infection; **13**: 763).
Vanderweyen, A. et al. 1966. *Al Awamia* **18**: 1 (fungicides in Morocco; **46**, 3102).
Wild, B. L. et al. 1975a. *Aust. J. exp. Agric. Anim. Husb.* **15**: 108 (fungicides; **54**, 3931).
—— ——. 1975b. *Phytopathology* **65**: 1176 (resistance to fungicides, **55**, 2244).

Penicillium italicum Wehmer, *Hedwigia* **33**: 211, 1894.
(see Raper & Thom 1949, for related spp., *Penicillium* q.v.)

COLONIES on Czapek agar restricted, central area raised, few radial furrows, margins thinner; some strs plane, though usually showing some fasciculation, others strongly fasciculate, with prostrate coremia borne at edges of older colonies; sporing irregularly, pale grey green to graphalium green, sometimes more yellowish to pea green; odour aromatic; exudate lacking or limited; reverse zonate, grey to yellowish brown. CONIDIAL APPARATUS asymmetric penicilli bearing tangled chains of conidia. CONIDIOPHORES smooth, variable in length up to $250 \times 4–5\ \mu$, originating from the substrate or occasionally from superficial hyphae, singly or often in bundles or coremia up to 1 mm or more in length, which in some strs extend several mm beyond the edge of the colony. Penicilli comparatively long 50–70 μ, usually consisting of a main axis and 1–3 branches, $12–25 \times 2.8–4.4\ \mu$, occasionally much longer and often rebranched. METULAE few in number, single or in groups of 2–4, $15–20 \times 3.5–4\ \mu$, frequently larger. PHIALIDES few in number, $8–12 \times 3\ \mu$ but may be larger. CONIDIA cylindrical at first, becoming elliptical or even subglobose, smooth, $4–5 \times 2.5–3.5\ \mu$, up to $9 \times 5\ \mu$. When first produced cylindrical, merging imperceptibly into the phialides. SCLEROTIA occasional in fresh isolates (from A. H. S. Onions, *CMI Descr.* 99, 1966).

Blue mould of citrus fruits is worldwide in orchards and a common cause of decay after harvest. *P. italicum* commonly occurs with green mould of citrus (*P. digitatum* q.v.). In the literature on citrus the 2 fungi are frequently considered together; and, on the basis of individual sp., most work refers to *P. digitatum* where all ref. are placed and where both diseases are discussed. Fawcett (*Citrus diseases and their control*) described and contrasted the symptoms of the 2 moulds in detail, with coloured illustrations. He referred to the fruit rot caused by *P. italicum* as blue contact mould since 'it has a greater tendency than *P. digitatum* to spread from one fruit to the next by contact'. Blue mould shows a powdery mycelium on the fruit surface and which does not extend much beyond the blue conidial mass. The decay margin is a definite, well defined, soft, water-soaked band just in advance of the mycelium on the surface. In wrapped fruit the wrapper does not readily adhere to the rotting fruit (Fawcett l.c.). Onions (l.c.) stated that *P. italicum* reduces the fruit to a soft and often a slimy, shapeless mass.

Periconia circinata

PERICONIA Tode ex Fr.; Tode, *Fungi Mecklenb.*, **2**: 2, 1791; Fries, *Syst. Mycol.* **1**: XLVII, 1821.
Sporocybe Fr., 1825
Sporodum Corda, 1836
Trichocephalum Costantin, 1887
Harpocephalum Atkinson, 1897
Berkeleyna O. Kuntze, 1898.
(Dematiaceae)

COLONIES effuse or, in a few spp., small and compact, grey, brown, olivaceous brown or black, hairy. MYCELIUM mostly immersed but sometimes partly superficial. STROMA frequently present, mid to dark brown, pseudoparenchymatous. Separate SETAE absent but in a few spp. the apex of the conidiophore is sterile and setiform. HYPHOPODIA absent. CONIDIOPHORES macronematous and sometimes also micronematous, mononematous. Macronematous conidiophores mostly with a stipe and spherical head, looking like round headed pins, branches present or absent, stipe straight or flexuous, in one sp. torsive, pale to dark brown often appearing black and shining by reflected light, smooth or rarely verrucose; sometimes the apex is sterile and setiform. CONIDIOGENOUS CELLS monoblastic or polyblastic, discrete on stipe and branches, determinate, ellipsoidal, spherical or subspherical. CONIDIA catenate, chains often branched, arising at one or more points on the curved surface of the conidiogenous cell, simple, usually spherical or subspherical, pale to dark brown, verruculose or echinulate, aseptate (M. B. Ellis, *Demat. Hyphom.*: 344, 1971, with key to 17 spp.).

The spp. in UK and India have been described. *P. manihoticola* (Vincens) Viégas (synonyms: *Haplographium manihoticola* Vincens, *P. heveae* Stevenson & Imle) has been described on *Hevea* spp. and *Manihot*. CONIDIOPHORES up to 500 μ long, 25–40 μ thick at the base, 18–26 μ immediately below the head; apical cell 18–40 × 15–26 μ. CONIDIA spherical, brown, verrucose, 25–45 μ diam. (Ellis l.c.). The fungus was reported as causing minor damage to *H. brasiliensis* (rubber) seedlings in Central America. The leaf spots are circular to oval, sometimes irregular or elongate along the veins but not vein limited and similar on both surfaces, 2–10 mm diam., frequently coalescing, particularly found on young leaves which may fall; spots brown becoming ashen in the centre with a brown border, on mature leaves there is often a chlorotic halo; the necrotic areas may show splitting and fall away; spread to petioles and young branches can occur. *P. shyamala* A. K. Roy (*Indian Phytopathol.* **18**: 332, 1965) causes large, pale tan spots on leaves of *Manihot*. *P. sacchari* Johnston occurs on *Saccharum*. CONIDIOPHORES up to 450 μ long, 10–20 μ thick at the base, 6–10 μ immediately below the head; branches up to 40 μ long, 5–10 μ thick. CONIDIA oblong or cylindrical, brown, verrucose, 15–30 × 8–14 μ (Ellis l.c.).

Chen, C. T. et al. 1976. *Pl. Dis. Reptr* **60**: 1083 (*Periconia manihoticola* on cassava; **56**, 3798).
Mason, E. W. et al. 1953. *Mycol. Pap.* 56, 127 pp. (British spp., key).
Rao, P. R. et al. 1964. *Mycopathol. Mycol. appl* **22**: 285 (Indian spp., key).
Roy, A. K. 1964. *Indian Phytopathol.* **17**: 75 (*P. sacchari* on sugarcane; **44**, 228).
Stevenson, J. A. et al. 1945. *Mycologia* **37**: 576 (*P. manihoticola* on *Hevea* spp.; **25**: 9).
Subramanian, C. V. 1955. *J. Indian bot. Soc.* **34**: 339 (Indian spp., key).

Periconia circinata (Mangin) Sacc., *Sylloge Fung.* **18**: 569, 1906.
Aspergillus circinatus Mangin, 1899.

COLONIES on PDA buff to grey, central sporing region becoming black, pale sepia in reverse. MYCELIUM composed of branched, septate, hyaline to pale brown, smooth or verruculose, 2–5 μ thick hyphae, often swollen to 7 μ, darker and more coarsely warted at the point of origin of the conidiophores. Conidia formed on the hyphae as well as on the macronematous conidiophores. CONIDIOPHORES arising singly or in small groups. Stipe characteristically circinate at the tip, dark brown, septate, up to 200 μ long, 6–11 μ thick at the base, 7–9 μ thick immediately below the head. A few short fat branches with smooth or sometimes warted walls are formed just behind the apex, on these and over the end of the conidiophore are borne a small number of ovoid, pale brown, smooth SPOROGENOUS CELLS 5–7 μ diam. CONIDIA formed singly or in short chains on sporogenous cells, spherical, dark brown, echinulate, 15–22 μ diam. (M. B. Ellis, *CMI Descr.* 167, 1968).

This sp., which causes milo disease of *Sorghum*, may be confused with *P. macrospinosa* Lefebvre & A. G. Johnson (Ellis) which was first described from

Sorghum. *P. circinata* is distinguished by its circinate conidiophores, the smaller conidia (18–35 μ in *P. macrospinosa*) and the smaller spines on them. Milo disease, now readily controlled with resistant cvs, occurs in the milo sorghums (*S. subglabrescens*), most other *Sorghum* spp. being resistant in the field. The disease is apparently restricted to the southern sorghum growing areas of USA, although *P. circinata* has been reported from Australia (Queensland), France, Tanzania and UK (Map 282). Milo was long attributed to *Pythium arrhenomanes* (q.v.; Elliot et al.) and selection of effectively resistant cvs had taken place before its aetiology was known. Leukel, who demonstrated the true pathogen, gave a review of the earlier work.

Milo disease is a root and crown rot. In heavily infested soil the roots of young plants show water-soaked lesions, reddish discolouration of the cortex and vascular tissue, and destruction of the smaller roots. The larger roots become, particularly in the vascular region, dark red or brown; the colour spreads to the crown and sometimes the stalk, a first diagnostic symptom. Above ground symptoms, as a wilt, differ in severity depending on the degree of soil infestation by *P. circinata*. These symptoms (leaf droop, yellowing, drying out and stunting) may begin a few weeks after sowing and no heads form. In less heavily infested soil above ground signs of disease appear later; they show retarded growth and poorly filled heads. The fungus persists in the soil for at least 4 years and control by rotation has not been practical. The disease develops equally well at 18–21° and 24–27°; it is soilborne and seed transmission is probably of little importance.

Some isolates produce toxins (2 distinct ones have been demonstrated) which inhibit the growth of roots from susceptible cvs and produce the symptoms normally seen (Gardiner et al.; Pringle et al.; Scheffer et al.). Conidia from isolates which did or did not form toxin could cause cortical lesions on roots but only the toxin forming ones caused extensive vascular infection and death (Odvody et al.). Early breeding and selection for resistance had shown that susceptibility was dominant (or partly so) and controlled by a single major gene (Bowman et al.; Heyne et al.). This conclusion was supported by Schertz et al. who treated seedlings with free toxin. The host pathogen interaction was controlled at a single locus. Susceptible cvs (*PcPc*) died, resistant ones (*pcpc*) grew normally, and the heterozygotes showed an intermediate response. A mutation frequency from susceptibility to resistance was determined. A selection of resistant cvs was given by Tarr, *Diseases of sorghum, sudan grass and broom corn* (and see Karper; Melchers et al.; Wagner).

Bowman, D. H. et al. 1937. *J. agric. Res.* **55**: 105 (inheritance of resistance; **16**: 807).

Bronson, R. C. et al. 1977. *Phytopathology* **67**: 1232 (heat & ageing induced tolerance of host to toxins from *Cochliobolus victoriae* & *Periconia circinata*; **57**, 4383).

Dunkle, L. D. et al. 1975. *Ibid* **65**: 1321 (heat activation of conidial germination; **55**, 2529).

Elliot, C. et al. 1937. *J. agric. Res.* **54**: 797 (general; **16**: 740).

Ellis, M. B. 1968. *C.M.I. Descr. pathog. Fungi Bact.* 168 (*P. macrospinosa*).

Gardiner, J. M. et al. 1972. *Physiol. Pl. Pathol.* **2**: 197 (effects of host specific toxin on host plasma membranes; **51**, 4819).

Heyne, E. G. et al. 1944. *J. Am. Soc. Agron.* **36**: 628 (reaction of the F_1; **24**: 147).

Karper, R. E. 1949. *Agron. J.* **41**: 536 (resistant cvs; **29**: 207).

Leukel, R. W. 1948. *J. agric. Res.* **77**: 201 (review & aetiology; **28**: 170).

Melchers, L. E. et al. 1940. *Pl. Dis. Reptr* Suppl. 126: 165 (resistant spp., cvs & hybrids; **20**: 298).

——. 1943. *Tech. Bull. Kans agric. Exp. Stn* 55, 24 pp. (resistance in *Sorghum* spp. & cvs; **23**: 101).

Odvody, G. N. et al. 1977. *Phytopathology* **67**: 1485 (toxin formation & pathogenicity; **57**, 4451).

Pringle, R. B. et al. 1963. *Ibid* **53**: 785 (purification of toxin; **43**, 83).

——. 1966. *Ibid* **56**: 1149 (amino acid composition of toxin; **46**, 311).

——. 1967. *Ibid* **57**: 530 (multiple host specific toxins; **46**, 3090).

Quinby, J. R. et al. 1949. *Agron. J.* **41**: 118 (effect of milo on yields; **28**: 311).

Scheffer, R. P. et al. 1961. *Nature Lond.* **191**: 912 (toxin & its assay; **41**: 225).

Schertz, K. F. et al. 1969. *Crop Sci.* **9**: 621 (inheritance of host reaction to toxin; **51**, 322).

Wagner, F. A. 1936. *J. Am. Soc. Agron.* **28**: 643 (resistance of sorghum spp., cvs & selections; **16**: 33).

PERONOSPORA Corda, *Icon. Fung.* 1: 20, 1837.

(Peronosporaceae)

MYCELIUM intercellular; haustoria in a few spp. short, like knobs, but in most filamentous and more or less branched. CONIDIOPHORES (SPORANGIOPHORES) consisting of erect trunks *c.* × 2–10 dichotomously branched, the branches more or less reflexed and the terminal ones sharp pointed; habit more graceful than in *Plasmopora*. CONIDIA typically larger than in that genus, aseptate, coloured, lacking an apical papilla, germinating by a germ tube. OOSPORES more or less globose, smooth or variously marked, germinating by germ tubes (after H. M. Fitzpatrick, *The lower fungi Phycomycetes*: 221, 1930).

Peronospora is distinguished from other genera in the Peronosporaceae by the dichotomous branching, acute tips of the conidiophore and the non-papillate conidia. The genus was last monographed in 1923 by Gäumann whose erection of many new spp. has not been generally accepted. Gustavsson gave a taxonomic revision and a general account of the Nordic spp.; Rao described spp. from India. The haustoria and manner of host penetration have also been described. Two spp. which cause important diseases but are not discussed are: *P. farinosa* (Fr.) Fr. on *Beta vulgaris, Spinacia oleracea* and other Chenopodiaceae (see Yerkes et al., *P. parasitica* q.v.); *P. arborescens* (Berk.) Casp. on *Papaver somniferum* (see Scharif for a review).

Fraymouth, J. 1956. *Trans. Br. mycol. Soc.* **39**: 79 (haustoria of the Peronosporales; **36**: 114).
Gäumann, E. 1923. *Beitr. Kryptogflora Schweiz* **5**(4), 360 pp. (monograph; **3**: 241).
Gustavsson, A. 1959. *Op. bot. Soc. bot. Lund.* **3**(1), 271 pp., (2), 61 pp. (taxonomic revision & general account of Nordic spp.; **39**: 159).
Ikata, S. et al. 1941. *Ann. phytopathol. Soc. Japan* **10**: 326 (haustoria; **22**: 226).
Ikata, Y. 1942–43. *Ibid* **12**: 97 (host penetration; **30**: 339).
Rao, V. G. 1968. *Nova Hedwigia* **16**: 269 (in India; **49**, 1983).
Scharif, G. 1970. *Iran. Jnl Pl. Pathol.* **6**: 1 (review of *Peronospora arborescens*, 40 ref.; **50**, 1323).
Thind, K. S. 1942. *J. Indian bot. Soc.* **21**: 197 (in Punjab; **22**: 112).

Peronospora destructor (Berk.) Casp. ex Berk., *Notices of British fungi*: 341, 1860.
Botrytis destructor Berk., 1841
Peronospora schleideni Unger, 1947.

SPORANGIOPHORES non-septate, various shades of violet, emerging from stomata, 122–150 μ long, 7–18 μ wide at the base, tapering to acute sterigmata at tips, ×2–6 monopodially branched, bearing 3–63 sporangia. SPORANGIA (CONIDIA) pyriform to fusiform, attached to sporangiophore by pointed end, 18–29 × 22–40 μ, thin walled, slightly papillate at proximal end, germinating by 1 or 2 germ tubes. Mycelium non-septate, intercellular, 4–13 μ. HAUSTORIA filamentous, coiled within cells, 1.3–5 μ diam. OOGONIA 43–54 μ. OOSPORES globular, 30–44 μ (J. Palti, correct author, *CMI Descr.* 456, 1975).

Gustavsson (*Peronospora* q.v.) pointed out that *P. destructor* has relatively large conidia (sporangia) *c.* 30–50 μ compared with that (20–30 μ) for other members of the genus. Downy mildew (see general accounts by Cook and Yarwood), mainly of onion and shallot (*Allium cepa*) but also recorded on Welsh onion (*A. fistulosum*), garlic (*A. sativum*), chives (*A. schoenoprasum*) and leek (*A. ampeloprasum* var. *porrum*). The disease is very widespread (Map 76) but only serious in cooler climates in damp conditions. The first symptoms may arise from direct infection of young plants by conidia or from diseased bulbs when systemic infection arises; young plants can be killed. In direct infection discrete, oval, chlorotic lesions appear on the leaves which can be dwarfed, distorted and show a tip dieback. Inflorescence stems are similarly attacked; seed is shrivelled or does not mature. The dull, whitish growth of conidial sporulation on the infected areas may have a violet tinge. In systemically infected plants the externally visible symptoms do not appear for some time. The plants appear abnormally turgid and eventually develop a glazed yellow look before sporulation begins on the surface. Infection of young plants reduces bulb growth considerably. Infected bulbs may show browning and wrinkling of the outer fleshy scale; in storage they soften, rot and sprout prematurely.

Penetration by the conidia is stomatal. Infection may be initiated in the seedling by oospores. The air-dispersed conidia are formed at 10–13° and the opt. temps for conidial germination are similar. Oospores are apparently erratic in formation but in some years they can be very abundant; they occur in all above ground parts of the plant including the surface of the seed, although seed transmission has not apparently been clearly shown. Oospores, with several years survival, are considered to be a source of primary infection. McKay (1937) obtained 60–85% germination and found (1957) that some were still viable after 25 years. Jovićević reported most oospore germination at 16–21°. Virányi (1974a, 1975b) considered that carry over was as mycelium in the bulb and that oospores were not important in this respect. In Israel Palti et al. found that onion downy mildew was affected by RH as conditioned by stand density, row direction and plant age. Sprinkling irrigation resulted in more infection compared with furrow trickle.

Some resistance has been described (e.g., Berry; Ershov et al.; Jones et al.; Vitanov) but commercial cvs generally require other control methods. Excessively damp growing conditions, infected bulbs and nearby crops, and infested soil should be avoided. Only firm bulbs should be used for planting, and attacked young plants rogued. Recommendations for heat treatment of bulbs (40° for 24 hours or 45° for 8 hours) need further trial. Fungicides should be applied when symptoms first appear; zineb has been most frequently used but maneb, dithane and difolatan are also effective (and see Weille for forecasting).

Anon. 1973. *Advis. Leafl. Minist. Agric. Fish. Fd* 85, 4 pp. (advisory; **54**, 3587).

Berry, S. Z. 1959. *Phytopathology* **49**: 486 (resistance; **39**: 140).

Boelema, B. H. 1967. *Fmg S. Afr.* **43**: 24 (fungicides; **47**, 1357).

Cook, H. T. 1932. *Mem. Cornell Univ. agric. Exp. Stn* 143, 40 pp. (general, 56 ref.; **12**: 484).

Ershov, I. I. et al. 1968. *Trudy̆ prikl. Bot. Genet. Selek.* **40**: 131 (resistance; **48**, 2087).

Haltebourg, M. 1966. *Al Awamia* **21**: 53 (fungicides; **48**, 1415).

Hiura, M. 1930. *Byochu-gai Zasshi April* 10 pp. (distribution in host, carry over; **10**: 7).

——. 1930. *Agric. Hort. Tokyo* **5**: 1008 (conidial germination & temp.; **10**: 153).

Jones, H. A. et al. 1939. *Hilgardia* **12**: 531 (resistance; **19**: 324).

Jovićević, B. M. 1964. *Zašt. Bilja* **15**: 117 (general; **45**, 308).

Jung, B. J. 1964. *Jnl Pl. Prot. Korea* **3**: 11 (fungicides; **45**, 1598).

McKay, R. 1935. *Nature Lond.* **135**: 306 (oospore germination; **14**: 488).

——. 1937. *Ibid* **139**: 758 (as above; **16**: 651).

——. 1939. *Jl R. hort. Soc.* **64**: 272 (general; **18**: 778).

——. 1957. *Scient. Proc. R. Dubl. Soc.* **27**: 295 (oospore longevity; **36**: 508).

Murphy, P. A. et al. 1926. *Ibid* **18**: 237 (carry over; **6**: 138).

—— ——. 1932. *J. Dep. Agric. Repub. Ire.* **31**: 60 (spread & control; **11**: 689).

Newhall, A. G. 1938. *Phytopathology* **28**: 257 (air dispersal; **17**: 646).

Palti, J. et al. 1972. *Phytopathol. Mediterranea* **11**: 30 (disease & crop environment; **52**, 1743).

Takahashi, M. et al. 1958. *Ann. phytopathol. Soc. Japan* **23**: 117 (formation & germination of oospore; **39**: 206).

Van Doorn, A. M. 1959. *Tijdschr. PlZiekt.* **65**: 193 (occurrence & control; **39**: 524).

Virányi, F. 1974a & b. *Acta phytopathol. Acad. Sci. hung.* **9**: 311, 315 (carry over, conidial sporulation & germination; **54**, 4282, 4283).

——. 1975a. *Ibid* **10**: 321 (epidemiology in Hungary; **55**, 5444).

——. 1975b. *Növényved. Kut. Intéz. Evk.* **13**: 243 (biology in Hungary; **56**, 1367).

Vitanov, M. 1970. *Grad. loz. Nauka* **7**: 123 (resistance; **50**, 3325).

Wakefield, E. M. et al. 1936. *Trans. Br. mycol. Soc.* **20**: 97 (nomenclature inter alia; **15**: 467).

Weille, G. A. De 1975. *Meded. Verh. K. ned. met. Inst.* 97, 83 pp. (forecasting; **55**, 2055).

Whetzel, H. H. 1904. *Bull. Cornell Univ. agric. Exp. Stn* 218: 139 (general).

Yarwood, C. E. 1943. *Hilgardia* **14**: 595 (general, 87 ref.; **22**: 336).

Peronospora manshurica (Naumov) Sydow ex Gäuman, *Beitr. Kryptogflora Schweiz.* **5**: 221, 1923.

Peronospora trifolium de Bary var. *manshurica* Naum., 1914

P. sojae Lehman & Wolf, 1924.

(synonymy from Gustavsson, *Peronospora* q.v.)

CONIDIOPHORES single or several, emerging from the stomata, 240–900 μ high, stalk over 0.5–0.75 of its length 7–9 μ thick; 3–5 branched, dichotomous, curved, erect; ends bifurcate, short, rectangular, almost straight. CONIDIA brownish, broadly ellipsoid to almost globose, 14–30 mostly 20–25 (av. 22.7) μ long, 14–29 mostly 18–24 (av. 21.2) μ wide. OOSPORES numerous in dry leaves, 24–28 μ diam., epispore yellowish, smooth or irregularly reticulate. On living leaves of *Glycine max* (from Gäumann, *Peronospora* q.v. Measurements given by Lehman et al. are: CONIDIOPHORES 300–500 × 5–8 μ, CONIDIA 20 × 34 μ, OOSPORES 20–23 μ).

Soybean downy mildew is fairly widespread in cooler regions and has been reported from: Australia, Bermuda, Brazil, Canada (E. and British Columbia), Colombia, and E. and W. Asia, Iran,

Mexico, New Zealand, Zimbabwe, South Africa and USA (S. and E.) and parts of Europe (Map 268). The small, chlorotic leaf spots are irregular with an indefinite margin; they become necrotic, 5–10 mm diam., with a conspicuous dark brown border and a chlorotic halo; the downy masses of greyish conidiophores form on the lower surface. Diseased areas crack, with portions of the leaf falling away to give a tattered appearance. Oospores develop with the onset of chlorosis. Infected pods show internal mycelial growth and seeds become covered with oospores (Johnson et al.). If such seeds are sown systemically infected plants arise. Pathak et al. found that an 8-year-old seed sample gave *c.* 20% viable oospores. A 1-year-old sample, which had been treated with captan, gave a 6–11% viability count.

Day temps of 20–30° after inoculation have no differential effect on lesion development. Temps before inoculation also have no effect when leaves of similar physiological age are used. The number of lesions increases and their size decreases with increasing leaf age at inoculation. A self-inhibitor of conidial germination has been reported and germinating oospores placed on water agar become surrounded by a zone which inhibits the germination of conidia. As with other members of the genus the conidia are presumably air dispersed in the early forenoon. Oospores kept in contact with seed can cause infection for up to one year. The effect of light and temp. on conidial germination has been briefly described (Pederson), and the fine structure of the haustorial interface examined (Peyton).

The fungus is pathogenically very variable. In 1971 15 new races were described (Dunleavy; Geeseman; Grabe; Lehman and others). In 1977 9 more races were found (Dunleavy). Single dominant genes control resistance to some races but Bernard et al. found a single gene controlling resistance to several races in 2 cvs; the phenotype of the heterozygote gave a varying reaction (intermediate to resistant) depending on the host and pathogen genotypes, and environment. In addition to the use of resistant cvs, seed treatment may be required; fermate, spergon and thiram have been used. Little chemical control in the growing plant has been described; in Colombia of 8 fungicides tested (applied 30–60 days after germination) Bordeaux and copper oxychloride were the best (Varela et al.).

Bernard, R. L. et al. 1971. *J. Hered.* **62**: 359 (a gene for general resistance; **51**, 3646).

Buitrago, G. L. A et al. 1970. *Acta agron. Palmira* **20**: 1 (races in Colombia; **50**, 2049).

Dunleavy, J. M. 1959. *Phytopathology* **49**: 537 (8 races in USA; **39**: 204).

—— et al. 1962. *Proc. Iowa Acad. Sci.* **69**: 118 (inhibition of spore germination; **43**, 308).

—— ——. 1970. *Pl. Dis. Reptr* **54**: 901 (resistance sources for 9 races; **50**, 1517).

——. 1970. *Crop Sci.* **10**: 507 (resistance tests with 14 races; **51**, 3645).

——. 1971. *Am. J. Bot.* **58**: 209 (new races; **50**, 2594).

——. 1977. *Pl. Dis. Reptr* **61**: 661 (9 new races in USA; **57**, 1911).

Geeseman, G. E. 1950. *Agron. J.* **42**: 257, 608 (3 races & inheritance of resistance; **30**: 94; **31**: 45).

Grabe, D. F. et al. 1959. *Phytopathology* **49**: 791 (2 new races; **39**: 364).

Hildebrand, A. A. et al. 1951. *Sci. Agric.* **31**: 505 (systemic infection; **31**: 529).

Johnson, H. W. et al. 1942. *Pl. Dis. Reptr* **26**: 49 (on seed; **21**: 360).

Jones, F. R. et al. 1946. *Phytopathology* **36**: 1057 (systemic infection; **26**: 156).

Koretskiĭ, P. M. 1970. *Mikol. i Fitopatol.* **4**: 3 (germination & viability of oospores; **49**, 2238).

Lehman, S. G. et al. 1924. *J. Elisha Mitchell scient. Soc.* **39**: 164 (general, as *P. sojae* sp. nov.; **3**: 627).

——. 1953. *Phytopathology* **43**: 460 (race 4; **33**: 275).

——. 1958. *Ibid.* **48**: 83 (4 new races; **37**: 625).

Millikan, D. F. et al. 1965. *Ibid* **55**: 932 (nucleic acid synthesis in infected host; **45**, 297).

Murakami, S. et al. 1977. *Bull. Tohoku natn. Exp. Stn* **55**: 229 (resistance; **57**, 353).

Pathak, V. K. et al. 1978. *EPPO Bull.* **8**: 21 (on seed; **57**, 5214).

Pederson, V. D. 1960. *Proc. Iowa Acad. Sci.* **67**: 103 (self-inhibition of conidial germination; **41**: 627).

——. 1964. *Phytopathology* **54**: 903 (light, germination & storage of conidia).

——. 1965. *Ibid* **55**: 1071 (temp., germination & storage of conidia; **45**, 7031).

Peyton, G. A. et al. 1963. *Am. J. Bot.* **50**: 787 (fine structure of host & parasite interface; **43**, 624).

Riggle, J. H. 1974. *Phytopathology* **64**: 522 (histology of leaf infection in resistant & susceptible cvs; **53**, 4664).

——. 1977. *Can. J. Bot.* **55**: 153 (histology on a nonhost and cvs of 3 reaction categories; **56**: 4785).

Varela, G. R. et al. 1969. *Acta agron. Palmira* **19**: 7 (fungicides; **49**, 1521).

Wyllie, T. D. et al. 1965. *Phytopathology* **55**: 166 (effects of temp. & leaf age on disease development; **44**, 2005).

Peronospora parasitica (Pers. ex Fr.) Fr., *Sum. Veg. Scand.*: 493, 1849.
Botrytis parasitica Pers. ex Fr., 1832.
(extensive synonymy in Yerkes et al.; sometimes referred to as *P. brassicae*)

MYCELIUM intercellular, hyaline coenocytic; haustoria large, elongate, club shaped, often branched. CONIDIOPHORES arising from the stomata, usually in groups, 200–300 μ long, dichotomously branched $\times$ 6–8, tips bifurcate, branching acute and branches often thickened a little above each fork. CONIDIA hyaline, broadly ovoid, 24–27 $\times$ 15–20 μ, germinating by a germ tube. OOGONIA irregularly rounded, swollen into crest-like folds, pale yellow. ANTHERIDIA tendril like, on separate hyphae. OOSPORES globose, 30–40 μ diam. (largely from E. J. Butler, *Fungi and diseases in plants*: 298, 1918).

Yerkes et al. rejected the splitting of *P. parasitica* into many spp. by Gäumann (*Peronospora* q.v.), and recognised only one on the Cruciferae; and see Dickinson et al. The fungus may, therefore, be considered plurivorous within this plant family and economically most important on cultivated *Brassica*. It is widespread in cooler regions. The symptoms are variable depending on the part of the plant which is attacked. Downy mildew is often first seen as abundant conidial sporulation on the lower surface of cotyledons, the upper surface showing pale spotting; necrosis of the seedling results. Spotting can also occur (less frequently) on leaves, stems and inflorescences. In mature cabbage infection causes severe grey to black lesions which spread to large areas of the lamina and successively from the outer to inner leaves. A rot, as with other brassicas, can develop in storage. Infection of cauliflower curds leads to dark purple discolourations, often as broad streaking with dark grey, necrotic, internal areas; normal inflorescences do not develop. On radish cracking occurs in the roots and infection has an adverse effect in storage.

Penetration, which has been examined ultrastructurally (Chou), of the cotyledons is cuticular; the appressoria forming at the junctions between the epidermal walls (adaxial surface). Systemic infection both above and below ground can occur; but in seedlings it may be restricted to the hypocotyls and cotyledons. The most favourable temps for infection and sporulation are 15–20%. Systemic infection at the cotyledon stage occurred with <16 hours of light but there was no infection with a higher amount of light (Polyakov). Oospores are formed abundantly in the host and are assumed to be the main means of survival between crops; but infection by the oospores does not seem to have been demonstrated. In an examination of 11 single conidial isolates De Bruyn found, on inoculation, that 3 formed oospores but that 7 required an opposite str. and were, therefore, heterothallic. Although there is carry over by mycelium, and oospores occur in the capsule, the evidence for seed transmission seems in doubt. Davison described the cytochemistry and fine structure of hyphae and haustoria, and development of the conidiophores. Shiraishi et al. also described the fine structure of infection. Ingram investigated pathogenicity in tissue culture where the results obtained with callus were frequently at variance with those obtained when using intact brassica plant parts. Asada et al. described cell wall lignification following infection; and see Matsumoto et al.

The data on pathogenic forms and/or races appears inadequate. More than one worker has designated 3 forms, from *Brassica*, *Capsella* and *Raphanus*. Isolates from *R. sativus* may or may not attack *Brassica*. Jafar, using isolates from *Matthiola* spp., found no infection in *Alysum*, *Brassica*, *Capsella*, *Cheiranthus*, *Iberis*, *Malcholmia*, *Raphanus* or *Sinapis*. McMeekin (1969) considered that *P. parasitica* may not be so host specific as to exclude the possibility that weeds may be hosts for the forms found on commercial cvs. Natti examining forms of *B. oleracea* described 2 races, resistance to which was controlled by 2 independent, dominant genes. In 1971 it was said that the development of resistant cvs was at an early stage. Greenhalgh et al. (1976) described the role of flavour volatiles in resistance. Control is through cultural measures (destruction of plant material in which carry over between crops may take place) and fungicides; those recommended in recent years are: chloranil, copper, copper–zinc, dichlofluanid, maneb and zineb.

Asada, Y. et al. 1969. *Ann. phytopathol. Soc. Japan* **35**: 160 (formation of lignin after infection; **49**, 592).
——. 1971. *Ibid* **37**: 311 (as above; **51**, 2957).
——. 1971. *Physiol. Pl. Pathol.* **1**: 377 (as above; **51**, 2023).
——. 1972. *Phytopathol. Z.* **73**: 208 (nature of lignin from infected radish root; **51**, 4452).
Chang, I. H. et al. 1963. *Acta phytopathol. sin.* **6**: 153 (oospores as infection source & systemic infection; **43**, 1184).

Chang I. H. et al. 1964. *Ibid* **7**: 33 (form on & within *Brassica* and on *Capsella* & *Raphanus*; **44**, 903).
Channon, A. G. et al. 1968. *Ann. appl. Biol.* **62**: 23 (tests with fungicides on detached cotyledons; **47**, 3616).
——— ———. 1970. *Pl. Pathol.* **19**: 151 (field tests with dichlofluanid; **50**, 2025).
Chou, C. K. 1970. *Ann. Bot.* **34**: 189 (fine structure of penetration & early haustorial formation; **49**, 2216).
Chu, H. T. 1935. *Ann. phytopathol. Soc. Japan* **5**: 150 (penetration & haustorial formation; **15**: 334).
Davison, E. M. 1968. *Ann. Bot.* **32**: 613, 623, 633 (cytochemistry & fine structure of hyphae & haustoria; development of, & distribution of substances within, conidiophores; **47**, 3375).
De Bruyn, H. L. G. 1937. *Genetica* **19**: 553 (heterothallism; **16**: 793).
Dickinson, C. H. et al. 1977. *Trans. Br. mycol. Soc.* **69**: 111 (host range & taxonomy; **57**, 833).
Felton, M. W. et al. 1946. *J. agric. Res.* **72**: 69 (effect of temp. & other factors; **25**: 243).
Gäumann, E. 1926. *Landw. Jb. Schweiz* **40**: 463 (pathogenic forms & hosts; **5**: 711).
Greenhalgh, J. R. et al. 1975. *Phytopathol. Z.* **84**: 131 (differential reactions of 3 crucifers; **55**, 2414).
——— ———. 1976. *New Phytol.* **77**: 391 flavour volatiles & resistance in *B. oleracea*; **56**, 453).
——— ———. 1976. *Ann. appl. Biol.* **84**: 278 (crucifer seedling resistance; **56**, 978).
Hiura, M. et al. 1934. *Trans. Sapporo nat. Hist. Soc.* **13**: 125 (pathogenic forms & hosts; **14**: 1).
Ingram, D. S. 1969. *J. gen. Microbiol.* **58**: 391 (tissue culture & pathogenicity; **49**, 1821).
Jafar, H. 1963. *N.Z. Jl agric. Res.* **6**: 70 (infectivity of isolates from *Matthiola*; **42**: 688).
Jones, W. 1944. *Sci. Agric.* **24**: 282 (infection of cauliflower; **23**: 247).
Kiermayer, O. 1958. *Öst. bot. Z.* **105**: 515 (effect of host growth substances after infection; **38**: 670).
Kupryanova, V. K. 1957. *Trudȳ leningr. Obshch. Estest.* **42**: 760 (carry over between crops; **37**: 428).
Le Beau, F. J. 1945. *J. agric. Res.* **71**: 453 (systemic invasion of cabbage seedlings; **25**: 196).
McMeekin, D. 1960. *Phytopathology* **50**: 93 (role of the oospores; **39**: 514).
——. 1969. *Ibid* **59**: 693 (infection of *B. oleracea* vars & radish; **48**, 3698).
Matsumoto, I. et al. 1978. *Ann. phytopathol. Soc. Japan* **44**: 22 (lignin induction in roots).
Natti, J. J. 1958. *Pl. Dis. Reptr* **42**: 656 (resistance in *Brassica*; **37**: 611).
—— et al. 1967. *Phytopathology* **57**: 144 (races; **46**, 2126).
Polyakov, I. M. et al. 1964. *Trudȳ vses. Inst. Zasheh. Rast.* **21**: 18 (light & infection; **45**, 905).
Preece, T. F. et al. 1967. *Pl. Pathol.* **16**: 117 (host penetration; **46**, 3638).
Ramsey, G. B. 1935. *Phytopathology* **25**: 955 (infection in storage; **15**: 188).
——. 1954. *Ibid* **44**: 384 (infection of radish; **34**: 199).
Sansome, E. et al. 1974. *Trans. Br. mycol. Soc.* **62**: 323 (cytology & life history with *Albugo candida*; **53**, 3827).
Shiraishi, M. et al. 1973 & 1974. *Mem. Coll. Agric. Ehime Univ.* **18**: 157; **19**: 137 (fine structure of infection; **55**, 3231, 5391).
——— ———. 1975. *Ann. phytopathol. Soc. Japan* **41**: 24 (as above; **55**, 955).
Thung, T. H. 1926. *Tijdschr. PlZiekt.* **32**: 161 (infection; **5**: 643).
Tsu, H. T. 1936. *J. agric. Assoc. China* **148**: 17 (infection).
Wager, H. 1900. *Ann. Bot.* **14**: 263 (fertilisation).
Wang, T. M. 1944. *Chin. J. scient. Agric.* **1**: 249 (pathogenic forms & hosts; **25**: 379).
——. 1949. *Phytopathology* **39**: 541 (host penetration & histology; **29**: 71).
Yerkes, W. D. et al. 1959. *Ibid* **49**: 499 (taxonomy of *Peronospora* spp. in Cruciferae & Chenopodiaceae; **39**: 159).

Peronospora tabacina Adam, *J. Dep. Agric. Vict.* **31**: 412, 1933.

MYCELIUM intercellular with branched haustoria. CONIDIOPHORES emerging, one or more, from a stoma; up to 820 μ, 5–8 times dichotomously branched, curvature of branches increasing to ultimate branches which diverge obtusely and are slightly curved or recurved, 8 (or less)–14 μ long, ending bluntly. CONIDIA ovoid to ellipsoid, thin walled, hyaline to dilute violet colour, 13–19 × 16–29 (av. 17–22) μ; germinating by a germ tube. OOSPORES 35–60 (av. 46) μ, epispore smooth or slightly but irregularly roughened. ANTHERIDIA paragynons (from D. B. Adam). On *Nicotiana* spp.; also *Capsicum, Lycopersicon esculentum, Solanum melongena* and other Solanaceae.

The name given by Adam, since it is the one that is commonly used, is adopted here. But the later proposals of Shepherd (1970) and Skalický may be a preferable interpretation of the pathogen's nomenclatural position. Cooke in 1891 had referred the fungus that causes blue mould of tobacco to the earlier described *P. hyoscyami* de Bary. Adam rejected this and considered that the tobacco pathogen was distinct on the grounds of oospore characteristics and the restriction of *P. hyoscyami* to *Hyoscyamus*. He also considered that another member of the genus, *P. nicotianae* described from *N. longiflora* by Spegazzini, was distinct; this name has since been rejected by later workers, including Skalický, as a nomen confusum. Skalický and others (e.g., Clayton et al. 1943; Kröber et al. 1964) considered that *P. tabacina* and *P. hyoscyami* could not be distinguished on morphological grounds. Skalický proposed that they should be combined as *P. hyoscyami* with 2 ff. sp.: *hyoscyami* (on *H. bohemicus* and *H. niger*) and *tabacina* on *Nicotiana* and other Solanaceae). His synonymy for *P. hyoscyami* is: *P. effusa* Rabenh. var. *hyoscyami* Rabenh., *P. nicotianae* Speg. (nomen confusum), *P. dubia* Berlese and *P. tabacina* Adam. Shepherd (1970), who has recently discussed the taxonomy, placed the Australian (APT) str. 1 in f. sp. *tabacina*. He proposed also ff. sp. *hybrida* (APT str. 2) and *velutina* (APT str. 3); the oospore size in *hybrida* (33.7 μ) is significantly different from that in *velutina* (21.6 μ). The separation of these 4 ff. spp. was given by Shepherd (1970):

Occurrence of sporulation on:

ff. sp	*Hyoscyamus niger*	*Nicotiana langsdorfii*	SO 1 hybrid (*N. tabacum* × *N. debneyi* breeding line)
hyoscyami	+	0	0
tabacina	0	+	0
hybrida	0	+	+
velutina	0	0	+

Full host lists are given by Skalický and Shepherd. Smith investigated the effect of various factors on the size of conidia.

There is no monograph on blue mould, a major disease of tobacco; for the earlier work reference should be made to the reviews by Hill (1957), McGrath et al. and Rayner et al. Many of the more recent studies have been done by workers in Australia: Angell, Cruickshank, Dean, Hill, Mandryk, Paddick, Pont, Shepherd, Wark, Wuttke and others. There are also general accounts by Lucas (*Diseases of tobacco*) and Wolf (*Tobacco diseases and decays*). Blue mould is an old disease in, and was long restricted to, parts of Australia and USA. It has been considered to be endemic on wild *Nicotiana* spp. in these regions but its precise origin may never be known. Since Australian *Nicotiana* spp. have generally more resistance than those indigenous to USA, and because of greater variability in pathogenicity in Australia, this latter country may be where the disease originally arose. In 1958 it was detected in Eurasia (southern England) for the first time. Some 45 months later *P. tabacina* had swept through the tobacco growing areas of Europe, N. Africa and W. Asia (causing enormous loss), reaching E. Iran *c.* 4500 km away. Blue mould is still absent from many countries, including much of the tropics; these are: China, India, S.E. Asia and its archipelago, Japan, Oceania (including New Zealand), Africa (except Algeria, Egypt, Libya, Morocco and Tunisia), S. America (within the tropics) and the West Indies (except Cuba and Dominican Republic; Map 23). Accounts of the epidemic in Europe and its consequences have been given by: Berger (for E. Germany), Corbaz (Switzerland), Klinkowski and Peyrot (Europe), Soydan (Turkey) and Żanardi for Italy.

The first symptoms may occur in the seedbed on small plants. Isolated groups of seedlings show leaves which have become pale, somewhat erect and slightly cupped; on older plants yellowish leaf blotches may be the first signs of disease. The chlorosis spreads and the seedlings develop the grey to bluish downy coating of conidia and conidiophores, mostly on the lower leaf surface. Leaves of older seedlings become deformed, younger plants are killed and infection can spread with extreme rapidity, to result in almost total loss of the seedlings. In the field the leaves show pale green, circular areas becoming chlorotic, necrotic and coalescing; large areas of the lamina are killed. Lesions can occur on the inflorescence. Infection, initiated either in the seedbed or the field, enters the stem and becomes systemic. On younger field plants systemic infection can be sufficiently severe to cause lodging. Older plants become stunted and may show wilt symptoms. Damage to leaves (that appeared healthy) during curing has also been described.

Initial penetration, probably mostly through the leaves by a conidial germ tube, can be cuticular or stomatal; the relative importance of these patterns of entry has not apparently been described. An infection period of *c.* 4 hours is required. Sporulation can occur in 4 days from infection reaching a max. 3–4 days later. It is inhibited by a water film on the host surface. The mycelium spreads from the leaf parenchyma into the vascular system of the stem. As the xylem ages it forms a barrier to spread. If the fungus becomes established in the cambium at an early stage, xylem development is inhibited and the stem becomes brittle and weak. Internal spread also takes place from the stem to axillary buds and thence to shoots and leaves where sporulation occurs. Direct infection of the young plant stem can also take place.

Blue mould is a disease of moderate temps those familiar with it have commented: long periods of high temp. and low RH are unfavourable for conidial germination; warm days and cool nights are best for conidial sporulation; min. temps seem the best guide to disease spread; a favourable temp. range of 10–20° in Europe. Germ tube growth of the conidium is quickest at 24–27° but after 12 hours there is most growth at 15–18°. The shortest incubation periods are at 16–24° and most conidial germination and sporulation occurs over 15–23 (–25)°; similar temps are most favourable for spread in the host. Day temps that are >25° for relatively long periods reduce disease incidence appreciably. A dark period is required for conidial formation; under continuous light the response is proportional to the length of dark exposure over 1.5–7 hours. High RH (close to saturation) is required for sporulation but the conidia are released in conditions of decreasing RH. Hygroscopically induced distortion of the conidiophores precedes or coincides with the release of the conidia which are air dispersed and have a diurnal periodicity with a forenoon max. (1000–1100 hours, Australia); this max. is earlier on clear days (0100–0500) than on cloudy ones (0800–1000, USA). As temps increase the opt. longevity occurs at a lower RH. After 30 days at 15° there is *c.* 10% conidial germination at 50 and 80% RH; and 19% at 0% RH. But lower survival rates for conidia have been reported. At the same temps shading has no effect on the disease. With conidia in water suspension the number of leaf lesions produced drops from 73% to 1% after storage in water for 24 hours, and in water at 35° few conidia survive for 30–60 minutes.

The epidemic pattern of blue mould differs in different geographical areas; the reasons for this appear to be temp., rainfall (amount and distribution) and pathogenic variation. In USA the disease is primarily one of the seedbed; in Australia and Europe it is also of great importance in the field. In the former country, where rainfall conditions are favourable when field plants are young, the disease is worst at this stage but in other areas, where adequate rain only falls later, older plants are more severely attacked. Soil conditions which extend the juvenile phase and increase leaf succulence will lead to more infection. The biology of the oospores has been little studied. They occur in the leaf mesophyll, other parts of the host and have been reported from the seed pod. Although oospores may be a source of primary inoculum in Europe and USA they are considered of little importance, in this respect, in Australia. Carry over of the pathogen occurs in host debris, volunteers and wild *Nicotiana* spp. Seed transmission seems in doubt. In Australia the introduced *N. glauca* is an important carry over host and *N. repanda* is in Texas.

Strs of *P. tabacina* have been described from Australia (Wark 1960). There is also some evidence that Australian and European isolates differ from American ones in that the latter infect red pepper and eggplant; further work seems needed here. The evidence for the existence of different pathogenic strs on *Nicotiana* spp. in Europe seems to be accumulating (e.g. Egerer; Govi; Janowski; Mikhaĭlova et al.; Ramson et al.; Ternovskiĭ et al. 1973). The ff. sp. and their characterisation, to which the 3 Australian strs have been assigned, were mentioned earlier. Str. APT 1 is the normal form on the susceptible, primary host, *N. tabacum*. Certain tobacco lines resistant to APT 1 are attacked by APT 2 and 3. Str. APT 2 does not compete successfully with str. APT 1 on susceptible tobacco and it has a somewhat different temp. range. More recent work on sources of resistance to str. APT 2 has been described by Gillham et al.; O'Brien 1973; Wark et al. 1976; Wuttke 1972).

Control is through cultural measures, soil sterilants and fungicides, and resistance. It begins in the seedbed which should have the usual, cultural, sanitary precautions; particular attention being applied to carry over from plant debris, volunteer or perennial hosts. Steam or the usual chemical sterilants may be required and healthy seed should be used. Benzol vapour has long been used, particularly in Australia, for control of the disease in the seedbed. But this has been replaced by the dithiocarbamates (e.g., mancozeb, maneb, ferbam and zineb). These fungicides are widely applied in the field where 7–8 weekly applications may be required. Maneb has been associated with a fall in leaf quality under conditions of nutrient stress. Dusting has been used on seedbeds but low or high volume sprays are more satisfactory for field use. Forecasting has been investigated for field spraying but at present rainfall is the only guide; warning services are in general use.

Resistance in *Nicotiana* spp. (high resistance is

absent in *N. tabacum*) has long been known and investigated. Clayton in 1962 summarised the previous 20 years work and described further progress in 1967 and 1968. Mandryk (1971) reported one of the more recent series of tests when he compared the reaction of 34 *Nicotiana* spp. in whole plant and disk tests, using 4 categories; 5 spp. had the highest resistance. Australian spp. are generally more resistant than American ones. *N. debneyi*, from Australia, has been used most in breeding and F_1 hybrids (with *N. tabacum*) are used on a commercial scale in Australia and Europe, although not satisfactory in all respects. *N. goodspeedii* has also been used and a single dominant from both this sp. and *N. debneyi* confers resistance, but minor resistance genes are also involved (see Dean et al.; Marini et al.; Wuttke 1969). Other *Nicotiana* spp. have been investigated and Ternovskiĭ et al. (1965) stated that *N. exigua* and *N. ingulba* are also important in Europe. In Australia sources of resistance to str. APT 2, which attacks tobacco cvs resistant to str. APT 1, occur in *N. debneyi, N. excelsior* and *N. velutina* (Wark et al. 1976; Wuttke 1972).

Adam, D. B. 1933. *J. Dep. Agric. Vict.* **31**: 412 (morphology & nomenclature; **13**: 132).

Angell, H. R. 1965. *Aust. Tob. J. Mareeba* **6**: 6 (association of disease with subsoil moisture; **46**, 2514).

Bailov, D. et al. 1964. *Rasteniev. Nauki* **1**: 43 (biochemical factors in host resistance, **44**, 1673).

Berger, P. et al. 1961. *Ber. Inst. Tabakforsch. Dresden* **8**: 31 (review of distribution in Europe, weather & control; **41**: 170).

——. 1965. *Ibid* **12**: 138 (occurrence in central Europe; oospores as primary inocula; **45**, 1186).

Clayton, E. E. et al. 1943. *Phytopathology* **33**: 101 (comparative morphology; **22**: 279).

——. 1945. *J. agric. Res.* **70**: 79 (resistance testing in American & Australian *Nicotiana* spp.; **24**: 251).

——. 1962. *Bull. Inf. CORESTA* **2**: 25 (summary of resistance work, resistance from *N. debneyi* in hybrids with *N. tabacum*; **42**: 489).

——. 1967. *Tob. Sci.* **11**: 91 (progress in breeding for resistance 1937–54).

——. 1968. *Ibid* **12**: 120 (as above 1957–67; **48**, 2545).

Cohen, Y. 1976. *Aust. J. biol. Sci.* **29**: 281 (effects of light & temp. on sporulation; **56**, 1720).

Collins, B. G. 1964. *Aust. J. exp. Agric. Anim. Husb.* **4**: 178 (effect of atmospheric conditions on sporulation; **43**, 3316).

Corbaz, R. 1961. *Phytopathol. Z.* **42**: 39 (review of blue mould in Europe, resistant cvs; **41**: 252).

——. 1970. *Rev. suisse Agric.* **2**: 90 (10 years of control in Switzerland; **51**, 4346).

Cruickshank, I. A. M. 1958. *Aust. J. biol. Soc.* **11**: 162 (moisture & the production & discharge of conidia; **37**: 679).

—— et al. 1960. *J. Aust. Inst. agric. Sci.* **26**: 269 (effect of stem infestation on foliage reaction; **40**: 431).

—— ——. 1961. *Aust. J. biol. Soc.* **14**: 45 (infection & host transpiration & growth; **40**: 630).

——. 1961. *Ibid* **14**: 58, 198 (effect of temp. on conidial sporulation & germination; **40**: 630; **41**: 252).

——. 1963. *Ibid* **16**: 88 (effect of light on conidial formation; **42**: 490).

Dean, C. E. et al. 1968. *Crop Sci.* **8**: 93 (transfer of resistance to F_1 hybrids; **50**, 3158).

Dean, J. C. 1970. *Qd J. agric. Anim. Sci.* **27**: 269 (agronomic aspects of fungicide control; **50**, 3163).

Egerer, A. 1972. *Ber. Inst. Tabakforsch. Dresden* **19**: 5 (resistance in *Nicotiana* spp.; **52**, 1257).

Fantechi, F. et al. 1964. *Tobacco Roma* 713: 241 (resistant cvs; **45**, 867).

Gillham, F. E. M. et al. 1977. *Aust. J. exp. Agric. Anim. Husb.* **17**: 652 (resistant breeding lines; **57**, 1399).

Golenia, A. 1970. *Pr. nauk. Inst. Ochr. Rośl.* **12**: 11, 57 (factors affecting field spread, resistance in *Nicotiana* spp. & tobacco cvs; **51**, 662a & b).

Govi, G. 1971. *Tobacco Roma* **75**(738): 1 (strs in Italy; **52**, 488).

Hill, A. V. et al. 1933. *J. Coun. scient. ind. Res. Aust.* **6**: 260 (carry over in tobacco & wild *Nicotiana* spp. & chemical control; **13**: 331).

——. 1957. *Tech. Pap. Div. Pl. Ind. C.S.I.R.O. Aust.* 9, 16 pp. (review, 67 ref.; **37**: 510).

——. 1960. *Nature Lond.* **185**: 940 (conidial release; **39**: 624).

——. 1961. *Aust. J. biol. Soc.* **14**: 208 (conidial diurnal periodicity; **41**: 252).

——. 1962. *Aust. J. agric. Res.* **13**: 650 (effect of soil organic matter on incidence; **42**: 342).

—— ——. 1962. *Aust. J. exp. Agric. Anim. Husb.* **2**: 12 (resistance tests in American & Australian *Nicotiana* spp.; **41**: 546).

——. 1962 & 1963. *Nature Lond.* **195**: 827; **199**: 396 (conidial longevity, pathogenic strs; **42**: 152, 704).

—— ——. 1965. *Aust. J. agric. Res.* **16**: 597 (temp. & development of blue mould; **45**, 222a).

——. 1965 & 1966. *Ibid* **16**: 609; **17**: 133 (effect of temp. & pathogenic str. on host growth, effect of inoculum potential on leaf infection; **45**, 222b, 2603).

——. 1966. *Bull. Inf. CORESTA* 1: 7 (pathogenic strs; **46**, 430).

—— ——. 1967 & 1968. *Aust. J. agric. Res.* **18**: 575; **19**: 759 (epidemiology, effect of temp. & shade on plant growth & disease development; **46**, 3563a; **48**, 916).

——. 1969. *Aust. J. biol. Sci.* **22**: 393, 399 (factors affecting conidial viability & host lesion production; **48**, 3151a & b).

Peronospora tabacina

Jadot, R. 1966. *Parasitica* **22**: 55, 208 (effects of light on conidial sporulation & of temp. on incubation; **46**, 424, 1731).

Janowski, F. 1972. *Biul. Inst. Ochr. Rośl.* 52: 107 (new virulent str.; **53**: 673).

Klinkowski, M. et al. 1960. *NachrBl. dt. PflSchutzdienst Berl.* **14**: 61 (review, 216 ref.; **40**: 63).

——. 1962. *Biol. Zbl.* **81**: 75 (epidemic in Europe & N. Africa; **42**: 152).

Kröber, H. et al. 1961. *NachrBl. dt. PflSchutzdienst Stuttg.* **13**: 81 (host range; **41**: 63).

—— ——. 1962. *Ibid* **14**: 82 (resistant lines; **42**: 152).

—— ——. 1964. *Phytopathol. Z.* **51**: 79, 241 (infection by oospores, morphology & taxonomy; **43**, 3315; **44**, 1243).

——. 1967. *Ibid* **58**: 46 (moisture period & infection; **46**, 1732).

Leppick, R. A. et al. 1972. *Phytochemistry* **11**: 2055 (an inhibitor of conidial germination; **51**, 4350).

McGrath, H. et al. 1958. *Pl. Dis. Reptr* Suppl. 250, 35 pp. (review, 200 ref.; **37**: 679).

Mandryk, M. 1957. *J. Aust. Inst. agric. Sci.* **23**: 319 (control with benzol; **37**: 510).

——. 1960. *Aust. J. agric. Res.* **11**: 16 (histology of stem infection & spread in host; **39**: 500).

——. 1962. *Ibid* **13**: 10 (acquired host resistance & soil N; **41**: 482).

——. 1966. *Ibid* **17**: 39 (stem infection by 3 pathogenic strs; **45**, 2243).

——. 1971. *Aust. J. exp. Agric. Anim. Husb.* **11**: 94 (host & non-host necrotic reactions, *Nicotiana* spp. & resistance; **50**, 3157).

Marani, A. et al. 1972. *Euphytica* **21**: 97 (inheritance of resistance; **51**, 4347).

Mikhaĭlova, P. et al. 1977. *Rasteniev. Nauki* **14**: 123 (new virulent str.; **57**, 288).

O'Brien, R. G. 1970. *Qd J. agric. Anim. Sci.* **27**: 137 (fungicide control: **50**, 1964).

——. 1973. *APPS Newsl.* **2**: 2 (f. sp. *hybrida*; **54**, 1869).

Paddick, R. G. et al. 1967. *Aust. J. agric. Res.* **18**: 589 (epidemiological patterns; **46**, 3563b).

——. 1971. *Aust. J. exp. Agric. Anim. Husb.* **10**: 506 (spray warning systems; **51**, 1888).

Pawlik, A. 1961. *Z. PflKrankh. PflPath. PflSchutz* **68**: 193 (overwintering & oospore germination; **40**: 766).

Peyrot, J. 1962. *Pl. Prot. Bull. F.A.O.* **10**: 73 (in Europe, a review; **42**: 703).

Pinckard, J. A. 1942. *Phytopathology* **32**: 505 (mechanisms of conidial dispersal; **21**: 498).

Pont, W. et al. 1961. *Qd J. agric. Sci.* **18**: 1 (epidemiology; **41**: 253).

—— ——. 1965. *Qd agric. J.* **91**: 680 (fungicide control; **45**, 1192).

—— ——. 1967. *Qd J. agric. Anim. Sci.* **24**: 187 (as above; **47**, 311).

Ramson, A. et al. 1973. *NachrBl. dt. PflSchutzdienst DDR* **27**: 112 (new virulent form; **53**, 1546).

Rayner, R. W. et al. 1962. *Misc. Publs Commonw. mycol. Inst.* 16, 16 pp. (review, 262 ref. to abstracts in *Rev. appl. Mycol.*; **41**, 739).

Rider, N. E. et al. 1961. *Aust. J. agric. Res.* **12**: 1119 (field environment, conidial sporulation & forecasting; **41**: 482).

Rotem, J. et al. 1968. *Pl. Dis. Reptr* **52**: 310 (effect of soil moisture on infection; **47**, 2832).

—— ——. 1970. *Phytopathology* **60**: 54 (effect of temp. on the disease; **49**, 2175).

Shaw, C. G. 1949. *Ibid* **39**: 675 (*P. tabacina* & *P. nicotianae* in Washington; **29**: 124).

Shepherd, C. J. et al. 1962. *Trans. Br. mycol. Soc.* **45**: 233 (inhibitors of conidial germination; **42**: 51).

——. 1962. *Aust. J. biol. Sci.* **15**: 483 (conidial germination in vitro; **42**: 218).

—— ——. 1963. *Ibid* **16**: 77 (as above; **42**: 489).

—— ——. 1964. *Ibid* **17**: 878 (effect of metabolites & antimetabolites on conidial sporulation; **44**, 1245).

—— ——. 1967. *Ibid* **20**: 87, 1161 (a necrotrophic reaction in *Nicotiana* spp.; **46**, 2515; **47**, 1276).

——. 1970. *Trans. Br. mycol. Soc.* **55**: 253 (nomenclature & ff. sp.; **50**, 964).

—— ——. 1971. *Ibid* **56**: 443 (dynamics of development of watersoaking & necrosis on leaf disks; **51**, 660).

—— ——. 1971. *Aust. J. biol. Sci.* **24**: 219 (factors affecting conidial viability; **51**, 661).

Skalický, V. 1964. *Acta Univ. Carol. Ser. Biol. Suppl.* 2: 38 (*P. hyoscyami* f. sp. *tabacina* comb. nov.).

Smith, A. 1970. *Trans. Br. mycol. Soc.* **55**: 59 (factors affecting conidial size; **50**, 963).

Soydan, A. 1971. *Tekel Enstitüleri Yayinlari* 13, 134 pp. (general & spread in Turkey; **52**, 1259).

Stover, R. H. et al. 1951. *Sci. Agric.* **31**: 225 (epidemiology in Canada; **31**: 150).

Ternovskiĭ, M. F. et al. 1965. *Infekts. Zabol. Kul'tur. Rast. Mold.* (4): 15 (tests on resistance in *Nicotiana* spp. & cvs; **45**, 602).

—— ——. 1973. *Mikol. i Fitopatol.* **7**: 40 (new race; **52**, 2742).

Todd, F. A. 1961. *Pl. Dis. Reptr* **45**: 319 (account of European outbreak & control; **40**: 708).

Waggoner, P. E. et al. 1958. *Phytopathology* **48**: 46 (discharge & diurnal periodicity of conidia; **37**: 376).

Wark, D. C. et al. 1960. *Nature Lond.* **187**: 710 (pathogenic strs; **40**: 247).

—— ——. 1965. *Ibid* **207**: 214 (resistance in *N. tabacum* to stem infection; **44**, 3154).

—— ——. 1976. *Tob. Int.* **178**: 127 (resistance to str. APT 2; **56**, 3211).

Wolf, F. A. et al. 1934. *Phytopathology* **24**: 337 (general in USA; **13**: 602).

—— ——. 1936. *Ibid* **26**: 760 (morphology & general; **16**: 65).

——. 1947. *Ibid* **37**: 721 (origin & history of early collections of *P. tabacina*; **27**: 163).
Wuttke, H. H. 1969. *Aust. J. exp. Agric. Anim. Husb.* **9**: 545 (inheritance of resistance; **49**, 847).
——. 1972. *Bull. Aust. Tob. Grow.* **20**: 6 (resistance to APT strs 1 & 2; **53**, 4872).
Zanardi, D. 1960. *Italia agric.* **97**: 1075 (review in Italy; **40**: 432).
—— et al. 1961. *Tabacco Roma* **65** (698) 106 pp. (6 papers in a general account of blue mould in Italy; **40**: 709).

Peronospora trifoliorum de Bary, *Annls Sci. nat. Ser. 4 Bot.* **20**: 117, 1863.
(extensive synonymy in Chilton et al.)

MYCELIUM intercellular, hyaline, coenocytic; haustoria in *Trifolium repens* sturdy usually with a thickened base. CONIDIOPHORES dichotomously branched at acute angles, arising in tufts, branches with bifurcate tips; length very variable depending on host, 150–500 μ long, 4–10 μ wide. CONIDIA broadly ellipsoidal or ovoid, may be tinged violet in mass; germinating by a germ tube, 19–37 × 9–32 μ, size very variable depending on host. OOSPORES globose, smooth, light or yellow brown; diam. varying from 20 to 45 μ, on *Medicago sativa* only.

Clover and lucerne downy mildew is widespread (Map 343), largely in temperate areas; it has been mostly recorded on *Medicago* and *Trifolium*. A minor disease although it can be locally serious. Infection is either as primary lesions (chlorotic spotting or banding with violet grey mycelial wefts on the lower surface) or as systemic infection which can be traced from the leaflet bases to the developing shoots. Germination of the conidia is best at 18°; 8–12% germination occurs at 5–15° and none takes place at 30°. On lucerne penetration is stomatal and the fungus develops profusely at 15–22°. Light inhibits germ tube growth but does not decrease germination. In lucerne the tendency to systemic infection is inherited and differences in resistance amongst cvs have been reported from N. America. *M. lupulina* is very resistant (Melhus et al.).

Berkenkamp, B. et al. 1978. *Can. J. Pl. Sci.* **58**: 893 (resistance in lucerne).
Chilton, S. J. P. et al. 1943. *Misc. Publs U.S. Dep. Agric.* 499, 152 pp. (fungi reported on spp. of *Medicago*, *Melilotus* & *Trifolium*; **22**: 483).
Faizieva, F. 1968 & 1969. *Uzbek. biol. Zh.* **12**: 13; **13**: 37 (morphology, as *Peronospora aestivalis*, on 6 wild *Medicago* spp. & lucerne; **48**, 839, 1792).
Fried, P. M. et al. 1977. *Phytopathology* **67**: 890 (conidial development & effects of RH & light on discharge & germination; **57**, 1251).
Gopal, S. et al. 1971. *Biol. Pl.* **13**: 396 (biochemical changes in infected leaves of *Trigonella foenum-graecum*; **51**, 1774).
Hanson, E. W. et al. 1964. *Crop Sci.* **4**: 229 (resistance in lucerne; **44**, 740).
Jones, F. R. et al. 1946. *Phytopathology* **36**: 1057 (systemic infection in lucerne and by *P. manshurica* in soybean; **26**: 156).
Martin, T. J. et al. 1975. *Ibid* **65**: 638 (cell wall ingrowths of nonhaustorial hyphae; **54**, 5468).
Melhus, I. E. et al. 1929. *Proc. Iowa Acad. Sci.* **36**: 113 (general, effects of temp.; **10**: 316).
Waite, S. B. 1971. *Utah Sci.* **32**: 98 (effect of temp. & light; **51**, 4109).
——. 1974. *Phytopathol. Z.* **79**: 368 (penetration; **54**, 1323).

Peronospora viciae (Berk.) Casp., *Monatsber. K. Preuss. Akad. Wiss. Berl.*: 330, 1855.
Botrytis viciae Berk., 1846.

MYCELIUM intercellular, haustoria filiform to coiled, branched. SPORANGIOPHORES (conidiophores) emerge in clusters of 5–7 from each stoma, long, stiff, straight, unbranched for two thirds or more of their height, 160–750 × 8–13 μ (main axis 100–450 μ) μ; × 5–8 dichotomously branched, primary branches straight to slightly curved, upper ones curved and spreading, ultimate branchlets diverging at obtuse or right angles, pointed, unequal, short, 18–22 × 2–3 μ, bearing a single sporangium at each end. SPORANGIA (conidia) oval to elliptical, narrowed a little towards attached end, 15–30 × 15–20 μ, pale violet to pale greyish in mass aı maturity. OOSPORES spherical, light brown to deep yellowish pink, 25–37 μ (42 μ with persistent oogonial wall), epispore thick and marked by large, raised reticulations (K. G. Mukerji, *CMI Descr.* 455, 1975).

Descriptions of *P. viciae* were given by Butler (*Fungi and disease in plants*) and Campbell. The latter described the oospores, which are abundant in systemically infected plants, as greenish yellow, reticulate, 26–28 μ. *P. pisi* (de Bary) Syd. is probably the same as *P. viciae* but Campbell preferred the former name for the American fungus. *P. viciae-sativae* Gäum. is a synonym.

The fungus causes pea (*Pisum sativum*) downy mildew and it occurs on some other cultivated legumes: sweet pea (*Lathyrus odoratus*), grass pea (*L. sativus*), common vetch (*Vicia sativa*) and broad bean (*V. faba*). The pathogen is widespread and amongst the regions that it has been reported from are: Argentina, Australia, Canada, Ethiopia, Europe (general), India, Israel, Kenya, Malawi, New Zealand, Pakistan, Zimbabwe, South Africa, Tanzania, Turkey, Uganda and USA. On pea several symptom patterns have been described. Infection of the hypocotyl and epicotyl (but not the roots) of the seedling leads to a primary systemic invasion and death within 3 weeks. Inoculum from such plants directly infects the leaves, stipules and pods of the growing crop. On the leaf the light green lesions, becoming chlorotic and then necrotic, show the slightly violaceous conidiophore masses on the lower surface. Pods are infected directly by conidia and not by mycelial growth through the pedicel and peduncle. Mycelial growth occurs on the inner surface and affects the seed, causing reduction in size and necrotic lesions. Severely infected seeds do not germinate or develop. Systemically infected seedlings may die before emergence above ground. Secondary systemic infection arises from invasion of meristematic tissue, for example, apical buds; it can lead to plant distortion, short internodes and reduction in size. Oospores are found in infected tissue including the seed testa and they can be abundant when conditions ($<90\%$ RH and $>20°$) do not allow conidial sporulation.

Mence and Pegg have recently studied the disease in detail in UK. Penetration of leaf disks is cuticular. Resistance of pea plants and individual leaves to infection increases with age but decreases at senescence. After an incubation of 6–10 days most conidia on leaves are produced at 12–20°; exposure to 20–24° for 10 days reduces subsequent sporulation. No conidia are produced in continuous light or at $<90\%$ RH; they are produced for up to 6 weeks after infection. Conidia washed off leaves give 20–60% germination but shaken off dry only 11–19%. Most lose viability 3 days after shedding. The opt. temps for conidial germination are 4–8° and conidia produced at 12° or 16° show reduced germination. Opt. temps for infection are 12–20° and a min. leaf wetness period of 4 hours is necessary. Conidia show a diurnal periodicity with a peak at *c.* 0700 after 1–2 hours of insolation. Temps $>18°$ slow development of the disease.

Infested soil and host debris should be avoided since this provides primary inoculum. The effectiveness of seed treatment appears unproven although granosan and thiram have been recommended. Spraying with maneb or zinc+maneb at the onset of flowering reduced the number of diseased pods by *c.* 50% (Olofsson). In Sweden, however, spraying was only economic in 2 years out of 4 and effective disease forecasting is needed. Cvs are being screened for resistance in UK.

Allard, C. 1970. *Annls Phytopathol.* **2**: 87 (general; **50**, 374).

——. 1971. *Phytiat. Phytopharm.* **20**: 23 (control with mancozeb; **51**, 2968).

Assaul, B. D. 1961. *Zashch. Rast. Mosk.* **6**: 53 (control including seed treatment; **41**: 495).

Campbell, L. 1935. *Tech. Bull. agric. Exp. Stn Wash. St.* 318, 42 pp. (general; **15**: 194).

Glassock, H. H. 1963. *Pl. Pathol.* **12**: 92 (on broad bean; **43**, 616).

Hickey, E. L. et al. 1977. *Can. J. Bot.* **55**: 2845 (fine structure of pathogen in pea; **57**, 2698).

————. 1978. *Protoplasma* **97**: 201 (cytochemistry of host pathogen interface).

Mence, M. J. et al. 1971. *Ann. appl. Biol.* **67**: 297 (factors affecting local infection & systemic colonisation; **50**, 3280).

Olofsson, J. 1966. *Pl. Dis. Reptr* **50**: 257 (control with maneb or Zn+maneb; **45**, 2294).

Pegg, G. F. et al. 1970. *Ann. appl. Biol.* **66**: 417 (effects of temp., RH & light on production, germination & infectivity of conidia; **50**, 2573).

————. 1972. *Ibid* **71**: 19 (effect on yield; **51**, 4473).

Ryan, E. W. 1971. *Irish J. agric. Res.* **10**: 315 (infection methods in resistance screening; **51**, 2969).

Snyder, W. C. 1934. *Phytopathology* **24**: 1358 (infection of pods; **14**: 340).

White, N. H. et al. 1944. *Tasm. J. Agric.* **15**: 92 (resistance in pea; **24**: 132).

PESTALOTIOPSIS Stey, *Bull. Jard. bot. Etat. Brux.* 19: 300, 1949.

(Melanconiaceae)

ACERVULI subepidermal, irregularly erumpent through the epidermis, or as a longitudinal crack; black. CONIDIOPHORES of uniform length, short. CONIDIA 4 septate; 3 median cells brown or dark or greenish; end cells hyaline; apical cell with simple or branched bristles (cystidia); basal cell ending in filiform, simple (occasionally branched) pedicel (from R. L. Steyaert, Latin diagnosis, l.c.).

Guba did not accept *Pestalotiopsis* Stey., preferring *Pestalotia* de Not. Mordue (1971) described the type sp. of the former genus as *Pestalotiopsis guepini* (Desm.) Stey. (synonym: *Pestalotia guepini* Desmazières). She noted that many workers do not accept the segregation of *Pestalotiopsis* from *Pestalotia*, therefore the latter name is still in use for *Pestalotiopsis guepini* and many other spp. which fall within the concept of *Pestalotiopsis*. *Pestalozzia* is a later spelling of *Pestalotia*. Sutton described developmental studies in the genus and 5 spp. including *Pestalotiopsis guepini*. Griffiths et al. studied conidial structure. Members of the genus are frequently reported in the literature of tropical (and other) plants as causing leaf spots, and sometimes spots or cankerous lesions on fruits. None of the diseases caused appears to be of much significance and nearly all ref. have been omitted. *Pestalotiopsis* spp. can probably at most be regarded as weak and/or wound pathogens. Two are described: those causing a leaf spot on palms and grey blight of tea. Others that are fairly commonly reported in the tropics are *P. mangiferae* (P. Henn.) Stey. on mango; *P. psidii* (Pat.) Mordue on guava (fruit canker and grey leaf spot); and *P. versicolor* (Speg.) Stey. on sapodilla (*Mankilkara achras*). The guava canker can be severe after harvest. *P. funerea* (Desm.) Stey. (see Mordue 1976) can be damaging in conifer nurseries. Reference should be made to Guba and Steyaert (1949) for identification and descriptions of the many spp. There appears to be little host specificity that might be important in a disease situation. If the disease caused becomes locally important control will need to be devised for the specific condition. It may be found that the primary cause of infection will be other than the *Pestalotiopsis* sp. concerned.

Bilgrami, K. S. et al. 1968. *Proc. natn. Acad. Sci. India* Sect. B **38**: 181 (variations in morphology of acervulus; **49**, 1990).
Doyer, C. M. 1925. *Meded. phytopathol. Lab. Willie Commelin Scholten* 9, 72 pp. (taxonomy, morphology & pathology; **5**: 391).
Dube, H. C. et al. 1966. *Mycopathol. Mycol. appl.* **28**: 305 (acervulus morphology; **46**, 65).
Guba, E. F. 1961. *Monograph of* Monochaeta *and* Pestalotia, 342 pp., Harvard Univ. Press (**41**: 85).
Griffiths, D. A. et al. 1974. *Trans. Br. mycol. Soc.* **62**: 295 (conidial structure; **53**, 3824).
Kaushik, C. D. et al. 1972. *Indian Phytopathol.* **25**: 61 (*Pestalotiopsis psidii* on guava; **52**, 2695).
Mordue, J. E. M. 1971. *C.M.I. Descr. pathog. Fungi Bact.* 320 (*P. guepini*).
——. 1976. *Ibid* 514, 515 (*P. funerea* & *P. psidii*).
Patel, M. K. et al. 1950. *Indian Phytopathol.* **3**: 165 (*P. disseminata* on guava; **31**: 390).
Sarkar, A. 1960. *Lloydia* **23**: 1 (*P. mangiferae* on mango; **40**: 421).
Steyaert, R. L. 1949. *Bull. Jard. bot. Etat. Brux.* **19**: 285 (monographic taxonomic study; **28**: 489).
——. 1953. *Trans. Br. mycol Soc.* **36**: 235 (spp. from Ghana & Togo; **33**: 184).
Sutton, B. C. 1961. *Mycol Pap.* 80, 16 pp. (development & descriptions of 5 spp.; **41**: 391).
Wilson, K. I. et al. 1970. *Sci. Cult.* **36**: 109 (*P. versicolor* on sapodilla; **50**, 1313).

Pestalotiopsis palmarum (Cooke) Steyaert in *Bull. Jard. Bot. Etat. Brux.* 19: 322, 1949.

Pestalotia palmarum Cooke, 1875.

ACERVULI associated with lesions or on discoloured or dead areas; globose to lenticular or ellipsoidal, rupturing the epidermis by a pore which becomes wide and irregular. Conidia emerge in black cirrhi which are diffuse and spreading at maturity. CONIDIOPHORES (annellophores) formed from upper surface stroma, hyaline, cylindrical to obovoid, 1–4 μ diam. and 5–18 μ long, with 1–2 successive proliferations. CONIDIA fusiform, straight, rarely curved, 5 celled, slightly constricted at septa, 17–25 (20) μ long $\times$ 4.5–7.5 (6) μ wide; 3 median cells 11.5–16.5 (13) μ long, olivaceous with the 2 superior median cells or single median cell darker than the inferior one, or with dark band at septum separating 2 superior median cells. Apical and basal cells hyaline. Apical appendages 3, rarely 2 or 4, hyaline, cylindrical to the obtuse apices, 5–25 (16) μ long, basal appendages hyaline, straight, 2–6 μ long.

Colonies on PDA almost white, with aerial mycelium diffuse towards the somewhat irregular advancing edge and denser on older parts of the colony. Acervuli develop from small yellowish clumps of hyphae and give rise to conspicuous greenish black spore masses. Colonies usually show diurnal zonation in mycelial growth and acervulus formation. Reverse shows little pigmentation of colony or discolouration of medium. On potato carrot agar growth sparse with little aerial mycelium and scattered small acervuli.

Some other related spp. occur on palms, for example, *Pestalotia phoenicis* Vize which has small conidia; *Pestalotiopsis papposa* Steyaert and the stat. conid. of *Leptosphaeria elaeidis* Booth & Robertson which have larger conidia. All 3 have spathulate apical appendages. *P. palmarum* on its hosts shows a relatively small range of variation and is normally distinguished from other species by the shape, dimensions and pigmentation of its conidia (J. E. M. Mordue & P. Holliday, *CMI Descr*. 319, 1972).

A minor leaf spot largely restricted to the Palmae and reported as causing disease in *Areca catechu, Borassus flabellifer, Cocos nucifera, Chamaerops humulis* and *Elaeis guineensis*. *P. palmarum* also occurs on several other palm genera and has been observed on *Capsicum* sp., *Hevea brasiliensis, Manilkara hexandra* and *Musa* spp. Small yellow brown spots, becoming white to grey with a dark brown margin, oval, > 1 cm long, and elongating parallel to the veins, sometimes coalescing and very abundant, with the dark acervuli on the upper surface in the centre. The disease, widespread in the tropics, has been described mostly from coconut. Very severe spotting (especially on young plants) can occur where growth conditions are poor, following insect attack and under wet conditions with dense planting where shade is excessive. Linear growth in vitro has been reported as *c*. 25° and the cellulytic properties have been investigated. In *Musa* direct penetration of the cuticle, with the formation of appressoria, occurs. From India a *Rhynchosphaeria* state has been reported. Unless conditions become extremely favourable for the pathogen in fields of older plants, control measures are only likely to be needed in nurseries. Opt. growth of seedlings is necessary; prevent insect attack, avoid overcrowding and adopt sanitary precautions. Bordeaux and zineb sprays have been recommended.

Agnihothrudu, V. et al. 1965. *J. Indian bot. Soc*. **44**: 290 (*Rhynchosphaeria* state; **45**, 1501).
Briolle, C. E. 1968. *Oléagineux* **23**: 519 (control in coconut nurseries; **47**: 3528).
Chowdhury, S. 1946. *J. Indian bot. Soc*. **25**: 131 (general on *Borassus flabellifer*; **26**: 104).
Cortez, F. 1928. *Philipp. Agric*. **17**: 223 (on coconut; **8**: 170).
Leininger, H. 1911. *Zentbl. Bakt. ParasitKde* Abt. 2 **29**: 3 (morphology & development in vitro).
Perišić, M. et al. 1969. *Zašt. Bilja* **19**: 325 (on *Chamaerops humulis* in glasshouse; **49**, 3434).
Roepke, W. 1935. *Meded. LandbHoogesch. Wageningen* **39**: 3 (on coconut following damage by *Chalcocelis albiguttata*).
Sasikala, M. et al. 1971. *Pl. Dis. Reptr* **55**: 185 (on *Manilkara hexandra* used as a rootstock for *M. achras*; **50**, 2395).
Stevens, F. L. 1932. *Philipp. Agric*. **21**: 80 (on coconut following damage by *Promecotheca cumingii*; **11**: 780).
Vakili, N. G. 1963. *Pl. Dis. Reptr* **47**: 644 (on *Musa*, inoculum from *Orbignya cohune*; **43**, 170).

Pestalotiopsis theae (Sawada) Steyaert in *Bull. Jard. Bot. Etat. Brux*. **19**: 327, 1949.
Pestalotia theae Sawada, 1915.

ACERVULI associated with lesions and on dead discoloured tissue, irregularly distributed on both surfaces of leaves but most frequently epiphyllous; globose to lenticular and rupture the epidermis by a pore which becomes wide and irregular. Conidia emerge in black masses which are diffuse and spreading at maturity. CONIDIOPHORES (annellophores) formed from upper surface of stroma, hyaline, cylindrical or obovoid to obpyriform, 1–5 μ diam., 10–15 μ long with 1–5 successive proliferations. CONIDIA fusiform, straight, rarely curved, 5 celled, hardly constricted at septa, 23–35 (27) μ long, 5.5–8 (7.2) μ wide; 3 median cells 15–22 (18) μ long, equally dark olivaceous. Apical and basal cells hyaline. Apical appendages 3, rarely 2 or 4, hyaline, cylindrical to the spathulate apices, 15–50 (30) μ long. Basal appendage hyaline, straight, 4–10 μ long.

On PDA mycelium hyaline, with whitish aerial mycelium sparse towards the somewhat uneven advancing edge but denser on older parts of the colony. Acervuli develop from small yellowish hyphal aggregations and from conspicuous greenish black spore masses. Colonies usually show a clear diurnal zonation in mycelial growth and acervulus formation. Reverse shows little pigmentation or discolouration of the medium. On potato carrot agar growth sparse with little aerial mycelium and scattered small acervuli.

Pestalotiopsis guepini (Desm.) Steyaert has slightly smaller conidia with shorter, non-spathulate appendages (*CMI Descr*. 320). *Monochaetia karstenii* (Sacc. & Syd.) Sutton and *M. karstenii* var. *gallica* (Stey.) Sutton have conspicuously branched apical appendages. The long apical appendages of *P. theae* with their slightly swollen tips distinguish the species from others of similar conidium dimensions which may be found occasionally on *Camellia sinensis* (J. E. M. Mordue & P. Holliday, *CMI Descr*. 318, 1972).

P. theae, causing grey blight of tea, occurs on some unrelated hosts. Symptoms begin as small brown leaf spots, later enlarging to 1 cm diam. or more and showing (on the upper surface) a greyish centre with light to dark brown margins. Lesions are usually circular or oval with concentric zonations marked out on the upper surface by the dark acervuli; on young leaves these zonations are often absent. Coalescence of the spots may occur. A stalk rot of tea cuttings has been described from India. The fungus is widespread in Africa and Asia, and also occurs in Australia and S. America. The disease it causes is considered to be secondary. Old leaves about to fall are infected and it is found on leaves of any age which have already been weakened or damaged by insects, excess inorganic fertiliser, potash or nitrogen deficiency, hard plucking, drought, hail, sun scorch or waterlogging. Light is required for sporulation. Work on the utilisation of cellulose, reaction to hormones, alkaloids, carbon and nitrogen sources and trace elements in vitro has been reported and the toxicity of cycloheximide and dodine described. The conditions predisposing the plant to attack should be determined and corrected.

Bertus, L. S. 1927. *Ann. R. bot. Gdns Peradeniya* **10**: 197 (general & comparison with *Pestalotiopsis palmarum*; **6**: 698).
Grover, R. K. et al. 1963. *Can. J. Bot.* **41**: 569 (effect of light in vitro; **42**: 630).
——. 1968. *Indian Jnl Microbiol.* **8**: 207 (toxicity of cycloheximide & dodine; **49**, 1133).
Mandahar, C. L. 1970. *Sci. Cult.* **36**: 467 (effect of light in vitro; **50**, 3139).
—— et al. 1971. *Sydowia* **24**: 209 (as above & C:N ratio; **51**, 2831).
Nojima, T. 1929. *Bull. Kagosima Imp. Coll. agric. For.* 7, 34 pp. (*P. theae* on *Diospyros kaki*; **9**: 536).
Roy, A. K. 1964. *Sci. Cult.* **30**: 242 (effect of hormones on growth in vitro with *Sphaerostilbe repens*; **44**, 831).
——. 1965. *Ibid* **31**: 205 (effect of alkaloids on growth in vitro with *S. repens* & *Phellinus noxius*; **45**, 213).
Venkata Ram, C. S. 1960. *Rep. Tea scient. Dep. United Plrs' Assoc. S. India* 1959–60: 59 (associated with *Glomerella cingulata* in stalk rot of tea cuttings; **40**: 184).

PHAEOCYTOSTROMA Petrak, *Annls Mycol.* **19**: 45, 1921.

Pleocyta Petrak & H. Syd., 1927
Phaeocytosporella Stout, 1930.
(Sphaerioidaceae)

MYCELIUM immersed, branched, septate, light or dark brown. CONIDIOMATA eustromatic, separate, rarely confluent, immersed, unilocular, multilocular or convoluted, dark brown, thick walled; wall of dark brown, thick walled textura angularis. OSTIOLE single, circular, central, slightly papillate or dehiscence by breakdown of upper wall of the conidioma. CONIDIOPHORES hyaline, septate and branched at the base, less so above, cylindrical to filiform, formed from the inner cells of the locular walls. CONIDIOGENOUS CELLS enteroblastic, phialidic, determinate, discrete or integrated, cylindrical, hyaline, with apical, periclinal thickening, little or no collarette, channel narrow or wide. CONIDIA aseptate, brown, eguttulate, smooth, thin or thick walled, base truncate, apex obtuse, cylindrical or ellipsoid. PARAPHYSES hyaline, septate, branched, filiform obtuse at the apices (B. C. Sutton, personal communication, August 1978; Sutton 1964).

Sutton discussed the taxonomy of the genus, described the spp. and gave a key. *P. ambiguum* (Mont.) Petrak (as *ambigua*) has pyriform conidia which are larger (14–19 × 7–10.5 μ) than those of *P. sacchari* (q.v.). Synonyms of *P. ambiguum* are: *Sphaeropsis ambigua* Mont., *P. istrica* Petrak, and *Phaeocytosporella zeae* Stout. *P. ambiguum* occurs on maize. Krüger isolated it from stems showing straw coloured blotches with a thin dark border. Roots were infected and infection of seedlings increased with soil temp. from 20 to 35°. Pycnidia were abundant on oatmeal agar at 27°. The fungus has been reported from maize seed and may be associated with the maize stalk rot syndrome.

Criag, J. et al. 1958. *Pl. Dis. Reptr* **42**: 622 (*Phaeocytostroma ambiguum* on maize; **37**: 657).
Krüger, W. 1965. *S. Afr. J. agric. Sci.* **8**: 587 (as above; **45**, 439).
——. 1968. *Phytopathol. Z.* **62**: 174 (fungi on maize seed & toxic effects of seed fungicides; **48**, 143).
Sutton, B. C. 1964. *Mycol. Pap.* 97, 42 pp. (Coelomycetes III, *Phaeocytostroma* inter alia; **44**, 3277).

Phaeocytostroma sacchari (Ell. & Ev.) Sutton, *Mycol. Pap.* 97: 26, 1964.

Trullula sacchari Ell. & Ev., 1892
Coniothyrium sacchari (Massee) Prill. & Delacr., 1897
Pleocyta sacchari (Massee) Petrak & Sydow, 1927.

(full synonymy in *CMI Descr.* 87)

PYCNIDIA immersed in culms, indeterminate in

shape and size, mostly elongated, multilocular, dark brown to black, 650 μ long × 350 μ wide, wall 5–10 cells thick composed of dark brown, rectangular, thick walled cells longitudinally arranged; black spore masses are exuded through an irregular, longitudinal slit in the epidermis. CONIDIA ellipsoidal to ovate, smooth walled, aseptate, pale brown, apex obtuse, base distinctly truncate, 11–14.5 × 3.5–5 μ, formed as phialospores from hyaline, erect, cylindrical, rarely branched or septate conidiophores which are 5–20 × 1.5–2 μ. CONIDIOPHORES formed all round the pycnidial cavity and mixed with hyaline, aseptate, occasionally branched, flexuous paraphyses, 15–35 × 2 μ long. On *Saccharum* spp.

Differs from *P. ambiguum* (Mont.) Petrak with pyriform conidia measuring 14–19 × 7–10.5 μ wide and which has been recorded from sugarcane field soil by Sutton (l.c.); and from *P. sacchari* var. *penniseti* Sutton which has slightly larger conidia and paraphyses (B. C. Sutton & J. M. Waterston, *CMI Descr.* 87, 1966).

Hudson (1962, *Leptosphaeria michotii* q.v.) commented on *P. sacchari* ecologically. Abbott et al. (1964) stated that rind disease occupies a position of historical interest among sugarcane diseases considerably out of proportion with its actual economic importance. Johnston reviewed the early history of *P. sacchari* and its confusion with other fungi, especially *Glomerella tucumanensis* (q.v.). The disease has also been called sour rot. The most obvious symptoms are the numerous, black, coiled, hairy, conidial threads exuded from the pycnidia from the rind surface under moist conditions. The disease is severe in cane weakened in any way and infection occurs through wounds, spread being largely through water. The distribution is widespread (Map 255). In severe cases the fungus may spread to the underground parts and kill the stool. Germination in setts may be affected, not directly but by increasing damage which has been begun by more serious pathogens. Early harvesting should reduce incidence.

Abbott, E. V. et al. 1964. In *Sugar-cane diseases of the world* Vol. 2: 125 (review, 14 ref.).

Johnston, J. R. 1917. *J. Dept Agric. P. Rico* **1**: 16 (history & aetiology).

PHAEOISARIOPSIS Ferraris, *Annls Mycol.* **7**: 280, 1909.
(Dematiaceae)

COLONIES effuse, olivaceous brown, cottony or hairy. MYCELIUM immersed. STROMA present, prosenchymatous, brown or olivaceous brown. SETAE and HYPHOPODIA absent. CONIDIOPHORES macronematous, mononematous, and caespitose or synnematous, individual threads unbranched, straight or flexuous, pale to mid brown or olivaceous brown, smooth. CONIDIOGENOUS CELLS polyblastic, integrated, terminal, sympodial, cylindrical or clavate, cicatrised; scars thin but visible, flattened against the side of the conidiogenous cell. CONIDIA solitary, dry, acropleurogenous, simple, obclavate or cylindrical, pale olive, olivaceous brown or brown, smooth or verruculose, mostly with 3 or more transverse septa (M. B. Ellis, *Demat. Hyphom.*: 268, 1971).

P. bataticola (Cif. & Bruner) M. B. Ellis (synonym: *Cercospora bataticola* Cif. & Bruner) was described on sweet potato leaves in America. Ellis (*More Demat. Hyphom.*) described 10 spp.

Ciferri, R. et al. 1931. *Phytopathology* **21**: 93 (on sweet potato; **10**: 404).

Phaeoisariopsis griseola (Sacc.) Ferraris, *Annls Mycol.* **7**: 280, 1909.
Isariopsis griseola Sacc. 1878.

CONIDIOPHORES caespitose or forming synnemata, up to 500 μ long, threads 2–4 μ thick near the base, swelling to 5–6 μ near the apex. CONIDIA mostly obclavate, conico-truncate at the base, very pale olive or olivaceous brown, smooth, 3–6 septate, 30–70 μ long, 5–8 μ thick in the broadest part, 1.5–2 μ wide at the base (M. B. Ellis, *Demat. Hyphom.*: 269, 1971. *Cercospora solimani* Speg. is also a synonym, F. C. Deighton, *Trans. Br. mycol. Soc.* **68**: 282, 1977).

Angular leaf spot of common bean and Lima bean (*Phaseolus lunatus*) is a minor, widespread (Map 328) disease. The fungus has been reported from cowpea (*Vigna*) and soybean but apparently only causes disease on *Phaseolus*. The greyish lesions on the leaves (becoming light brown with age) are characteristically limited by the veins and premature defoliation can be common. Roughly circular spots, red brown with dark brown borders, form on the pods; stem lesions are dark brown and elongate. Infected seed will result in primary infections on the cotyledons. Conidial germination is best at 20–28° and in vitro growth at *c.* 24°, penetration is stomatal. The disease developed more slowly at 16° than at 20° and 28°, and there was no infection at 32° (Cardona-Alvarez et al.). Carry over is on host debris and the seed where the fungus is borne internally. Seed infection declined to 10% in 9 months and there was none after one year (Orozco-Sarria et al.). *P. griseola* occasionally causes epidemics; for example, such an outbreak in Colombia (yield losses of up to 60%) resulted from inadequate rotation, close cropping and irrigation (Barros et al. 1958a).

Díaz et al. suggested that races may exist and Hocking described a new virulent form which arose in Tanzania. In this country angular leaf spot (as seems to be usual) causes little damage except in exceptionally wet seasons. The new form caused large leaf lesions, circular, symmetrical and which (atypically) spread across the veins. It produced 1 cm diam. lesions in 6 days compared with the minute spots after 8 days produced by the common form. The synnemata of the new form were longer (148 μ compared with 117 μ) and they were found on both leaf surfaces, whereas those of the angular spot form were restricted almost entirely to the lower surface. Healthy seed, proper cultural measures and avoidance of the more susceptible cvs should give adequate control. Dithiocarbamate fungicides are effective. Fortugno used benomyl and triforine. Control of resistance by a single recessive was described by Santos Filho et al.

Barros, O. et al. 1958a. *Pl. Dis. Reptr* **42**: 420 (epidemic in Colombia; **37**: 618).

——. 1958b. *Pl. Prot. Bull. F.A.O.* **6**: 97 (control with zineb; **37**: 692).

Brock, R. D. 1951. *J. Aust. Inst. agric. Sci.* **17**: 25 (resistance in common bean cvs; **31**: 3).

Cardona-Alvarez, C. et al. 1956. *Phytopathology* **46**: 610 (general; **36**: 369).

Díaz, P. C. et al. 1965. *Agron. trop.* **14**: 261 (general, seed transmission; **47**. 2007).

Fortugno, C. 1974. *IDIA* 317–320: 26 (general; **57**, 2704).

Hocking, D. 1967. *Pl. Dis. Reptr* **51**: 276 (new virulent form in Tanzania; **46**, 2343).

Milatović, I. 1958. *Zašt. Bilja* 45: 39 (fungicides; **38**: 437).

Olave L., C. A. 1958. *Acta agron. Palmira* **8**: 197 (resistance in common bean; **39**: 257).

Orozco-Sarria, S. H. et al. 1959. *Phytopathology* **49**: 159 (seed transmission; **38**: 436).

Santos Filho, H. P. et al. 1976. *Revta Ceres* **23**: 226 (resistance; **57**, 854).

PHAEORAMULARIA Muntañola, *Lilloa* 30: 182, 209, 1960.

(Dematiaceae)

COLONIES effuse, olivaceous or brown, often velvety. MYCELIUM mostly immersed. STROMA present but small. SETAE and HYPHOPODIA absent. CONIDIOPHORES macronematous, mononematous, caespitose, emerging through stomata, unbranched or loosely branched, straight or flexuous, rather pale olivaceous or brown, smooth. CONIDIOGENOUS CELLS polyblastic, integrated, terminal, becoming intercalary or occasionally discrete, sympodial, cylindrical, cicatrised, RAMOCONIDIA sometimes present. CONIDIA dry, in branched or unbranched chains, acropleurogenous, simple, cylindrical with rounded ends, ellipsoidal or broadly fusiform, hyaline or olivaceous, smooth with 0, 1 or several transverse septa (M. B. Ellis, *Demat. Hyphom.*: 307, 1971).

Ellis (*More Demat. Hyphom.*) described 22 spp. briefly. Deighton fully described the confused nomenclature and morphology of *P. capsicicola* (Vassilj.) Deighton. The synonymy is: *Cercospora capsicicola* Vassiljevskiy, *C. capsici* Marchal & Steyaert (non *C. capsici* Heald & Wolf), *C. unamunoi* E. Castellani, *C. capsici* Unamuno (non *C. capsici* Kovachevsky) and *Phaeoramularia unamunoi* (E. Castell.) Muntañola (not validly published). The fungus causes brown leaf mould (also called velvet spot) of *Capsicum* and is widespread in warm temperate, subtropical and tropical regions. It can cause serious defoliation. The leaf spot usually shows only indefinite yellowish areas on the upper surface, on the corresponding lower surface there are effuse, densely velutinous, olivaceous, suborbicular patches of conidiophores and conidia up to 1.5 cm wide, sometimes with a yellowish halo. In some severe infections an irregularly suborbicular brown spot is formed, visible on both leaf surfaces, often with a narrow deeper brown margin and with irregular line zonations. CONIDIOPHORES pale to moderately deep olivaceous, smooth, straight, not geniculate, continuous or 1–2 septate above a basal septum, up to 40 μ long, mostly 4–4.5 μ wide (Muntañola gave

22–62 × 4–6 μ). CONIDIAL SCARS conspicuously thickened, CONIDIA very pale to moderately deep olivacous, usually subcylindric, often in branched chains, mostly straight, smooth, mostly 0–1 septate, not uncommonly 2–4 septate, 12–92 × 3–7 (mostly 3–4) μ. In luxuriant specimens the conidia may bear shorter or longer lateral branches (after Deighton). The spp. described by Ellis (l.c.) include: *P. dioscorea* (Ellis & Martin) Deighton (synonym: *Cercospora dioscoreae* Ellis & Martin) on *Dioscorea*, and *P. manihotis* (*Cercosporidium* q.v.) on *Manihot*.

Castellani, E. 1948. *Riv. Agric. subtrop. trop.* **42**: 20 (general on *Capsicum*; **27**: 406).

Deighton, F. C. 1976. *Trans. Br. mycol. Soc.* **67**: 140 (synonymy & morphology of *Phaeoramularia capsicicola*; **56**, 2318).

Kovachevsky, I. C. 1938. *Z. PflKrankh. PflPath. PflSchutz* **48**: 321 (morphology & symptoms, on *Capsicum*; **17**: 790).

Muntañola, M. 1954. *Phytopathology* **44**: 233 (general on *Capsicum*; **34**: 8).

PHAKOPSORA Dietel, *Ber. dtsch. Bot. Ges.* **13**: 333, 1895.
(Melampsoraceae)

PYCNIA and AECIDIA unknown. UREDIA subepidermal, erumpent, surrounded by encircling incurved PARAPHYSES developing from hyphoid peridium and opening out by a narrow pore; UREDOSPORES developing singly, obovate–globoid or ellipsoid, pale yellow, with obscure germ pores. TELIA subepidermal, non-erumpent, lenticular, black; TELIOSPORES aseptate, chestnut brown or golden brown; formed in succession from the basal hymenium, younger spores wedging in between older ones and forming a compact crust, germinating after a rest period (from M. J. Thirumalachar & B. B. Mundkur, *Indian Phytopathol.* **2**: 237, 1949, who placed *Physopella* Arth. in synonymy).

P. apoda (Har. & Pat.) Mains (*Physopella zeae* q.v.) occurs on *Pennisetum*, *Phakopsora setariae* Cummins on *Setaria* and *P. elettariae* Cumm. on cardamom (*Elettaria cardamomum*). *P. setariae* differs from *P. apoda* in having shorter and narrower uredospores, and a thinner apical wall in the outer teliospores.

Cummins, G. B. 1956. *Bull. Torrey bot. Club* **83**: 221 (tropical rusts, *Phakopsora setariae* inter alia; **35**: 722).

Thirumalachar, M. J. et al. 1949. *Mycologia* **41**: 283 (some spp. of *Phakopsora* & *Angiospora*).

Phakopsora gossypii (Arthur.) Hirat. f., *Ured. Stud.*: 266, Oct. 1955.
Kuehneola gossypii Arth., 1912 (as *K. gossypii* (Lagerh.) Arth.)
Cerotelium gossypii (Arth.) Arth., 1917 (as *C. gossypii* (Lagerh.) Arth.)
Phakopsora gossypii (Arth.) Dale, Dec. 1955.
(further synonymy in *CMI Descr.* 172)

UREDIA primary ('AECIDIAL UREDIA') epiphyllous, secondary ('UREDIAL UREDIA') hypophyllous, scattered in groups, slightly pulverulent, yellowish brown, up to 0.5 mm, in purplish spots 1–5 mm across. PARAPHYSES peripheral, clavate, usually incurved, 7–15 × 45–65 μ, wall smooth, colourless, 1 μ thick. UREDOSPORES ellipsoidal or obovoidal, 16–19 × 19–27 μ, wall light yellow to hyaline, 1–1.5 μ, echinulate, pores inconspicuous, 2 equatorial. TELIA rare, hypophyllous, scattered, inconspicuous naked or somewhat pulverulent, light cinnamon brown. TELIOSPORES in crusts of laterally adherent spores, 5–8 spores in depth, 10–13 × 70–110 μ, angular to irregularly oblong, 10–14 × 24–32 μ long, wall smooth, pale brown, 1–1.5 μ and 3–6 μ at the apex (E. Punithalingam, *CMI Descr.* 172, 1968).

One of the cotton (*Gossypium* spp.) rusts occuring largely on the leaves and causing defoliation; the others are *Puccinia schedonnardi* and *P. cacabata* (q.v.), both of which are heteroecious with their aecidia on *Gossypium* and uredia and telia on grasses. Although *Phakopsora gossypii* has been reported to have caused appreciable losses (up to 25%) in various countries little or nothing appears to be known of its pathology. It is fairly widespread in cotton-growing regions but not reported from Egypt (Map 258). Apart from sanitation there are no control measures; it has been stated that weak plants are more severely attacked.

Phakopsora pachyrhizi Sydow, *Annls Mycol.* **12**: 108, 1914.
(synonymy in Sathe)

UREDIA hypophyllous, densely scattered, often over whole leaf surface, minute, round, *c.* 200 μ diam., ochraceo–ferruginous, with numerous PARAPHYSES often curved inwards, hyaline or subhyaline, 25–45 μ long, apex thickened, 8–13 μ wide, and surrounded by a very thick membrane; peridium distinct, consisting of angular, thinly tunicate cells (membrane *c.* 2 μ thick), 10–15 × 8–12 μ. UREDOSPORES globose, subglobose, ovoid or ellipsoidal, briefly or more thickly echinulate, pale yellowish brown, 20–28 × 17–23 μ, epispore 1.5 μ thick; germ pores scarcely conspicuous. TELIA hypophyllous, scattered or aggregated, irregular, erumpent, minute, 0.15–0.25 mm diam., dark blood red brown. TELIOSPORES 4–6 superposed, variable, mostly clavate, oblong or angulate, yellow to brownish, 20–35 × 8–15 μ; epispore 1.5 μ thick, slightly thicker at the apex, particularly in the upper spores (up to 5 μ thick), and darker (from H. & P. Sydow l.c.; see *CMI Descr.* 589).

This soybean rust (see review by Yang) infects many leguminous plants (see Koegh; Vakili et al.) including: *Cajanus*, *Canavalia*, *Crotalaria*, *Dolichos*, *Lupinus*, *Pachyrhizus*, *Phaseolus*, *Pueraria* and *Vigna*. Serious losses have been reported from Asia (see McLean). *Phakopsora pachyrhizi* is fairly widespread in E. and S. Asia, and also occurs in Australia and Africa; in the Americas there are records for Barbados, Brazil, Costa Rica, Cuba, Dominican Republic, Guatemala, Puerto Rico, Venezuela and Virgin Islands (Map 504). Marchetti et al. reported a broad opt. (15–25°) for uredospore germination (and see Singh et al.). Max. infection of soybean is at 20–25° with 10–12 hours of dew, and at 15–17.5° with 16–18 hours. The min. dew period for infection is 6 hours at 20–25° and 8–10 hours at 15–17°. Infection does not occur at > 27.5°. Uredial primordia are evident 5–7 days after inoculation and sporulation begins 2–4 days later. Deverall et al. described infection. Variation in pathogenicity has been described from Australia, Puerto Rico and Taiwan. Control is through resistance and fungicides (e.g., Australia: benomyl, mancozeb; China: dichlone, dinocap, zineb, Bordeaux; Philippines: mancozeb; Taiwan: benomyl, maneb, zineb; Thailand: benomyl, mancozeb, maneb, oxycarboxin).

Cheng, Y. W. et al. 1968. *Jnl Taiwan agric. Res.* **17**: 30 (resistance; **48**, 2667).

Deverall, B. J. et al. 1977. *Trans. Br. mycol. Soc.* **69**: 411 (infection; **57**, 3706).

Hung, C. H. et al. 1961. *Agric. Res. Taiwan* **10**: 35 (fungicides; **43**, 1204).

Jan, C. R. et al. 1971. *Mem. Coll. Agric. Taiwan Univ.* **12**: 173 (fungicides; **53**, 1139).

Koegh, R. 1976. *APPS Newsl.* **5**, 51 (hosts & distribution in Australia, NSW; **56**, 4283).

Lin, S. Y. 1966. *Jnl Taiwan agric. Res.* **15**: 24 (races; **47**, 1355).

McLean, R. et al. 1976. *APPS Newsl.* **5**: 34 (resistance; **56**, 2764).

Marchetti, M. A. et al. 1975. *Phytopathology* **65**: 822 (uredial development; **55**, 980).

—— ——. 1976. *Ibid* **66**: 461 (temp., dew period & infection; **56**, 485).

Sangawongse, P. 1973. *Thai J. Agric. Sci.* **6**: 165 (resistance & fungicides; **53**, 4663).

—— et al. 1977. *Ibid* **10**: 1 (fungicides; **57**, 2708).

Sathe, A. V. 1972. *Curr. Sci.* **41**: 264 (nomenclature & identity; **51**, 4524).

Singh, K. P. et al. 1977. *Indian J. Mycol. Pl. Pathol.* **7**: 27 (inoculation & effect of temp. on uredospore germination; **57**, 4230).

Torres, C. Q. et al. 1976. *Trop. Grain Legume Bull.* 6: 20 (fungicides in Philippines; **56**, 5889).

Vakili, N. G. et al. 1976. *Pl. Dis. Reptr* **60**: 995 (hosts & strs in Puerto Rico; **56**, 4267).

Wang, C. S. 1961. *J. agric. Assoc. China* **35**: 51 (fungicides; **42**: 506).

Yang, C. Y. 1977. *Bull. Inst. trop. Agric. Kyushu Univ.* **2**: 78 (review, 74 ref.; **57**, 5219).

PHELLINUS Quél., *Enchir. Fung.*: 172, 1886. (Hymenochaetaceae)

Pegler et al. fully described *P. ignarius* (L. ex Fr.) Quél. (synonyms: *Polyporus ignarius* L. ex Fr., *Fomes ignarius* (L. ex Fr.) Kickx) which has been accepted as the type sp.; but see Donk (*Persoonia* **1**: 253, 1960). *Phellinus ignarius* causes white heart rot of temperate, deciduous trees. *P. pomaceus* (Pers.) Maire causes heart rot of plum. *P. robustus* (P. Karst.) Bourd. & Galz. (synonym: *F. robustus* P. Karst.) occurs mainly on *Quercus* (yellow trunk rot of oak) but has been recorded on many other temperate tree genera. The fungus is readily distinguished from *P. ignarius* by the bright yellowish context, scanty setae and larger spores.

Pegler, D. N. et al. 1968. *C.M.I. Descr. pathog. Fungi Bact.* 194, 196, 197 (*Phellinus ignarius*, *P. pomaceus* & *P. robustus*).

Phellinus noxius (Corner) G. H. Cunn., *Bull. N.Z. Dep. scient. ind. Res.*, 164: 221, 1965.

Fomes noxius Corner, 1932.

CARPOPHORE perennial, solitary or imbricate, sessile with a broad basal attachment, commonly resupinate. PILEUS 5–13 × 6–25 × 2–4 cm., applanate, dimidiate or appressed-reflexed; upper surface deep reddish brown to umbrinous, soon blackening; at first tomentose, glabrescent, sometimes with narrow concentric zonation, developing a thick crust; margin white then concolorous, obtuse. CONTEXT up to 1 cm thick, golden brown, blackening with KOH, silky-zonate fibrous woody. PORE SURFACE greyish brown to umbrinous; PORES irregular, polygonal, 6–8 per mm, 75–175 μ diam., dissepiments 25–100 μ thick, brittle and lacerate; TUBES stratified, developing 2–5 layers, 1–4 mm to each layer, darker than context, carbonaceous. BASIDIOSPORES 3.5–5 × 3–4 (4.2 × 3.2) μ, ovoid to broadly ellipsoid, hyaline, with a smooth, slightly thickened wall, and irregular guttulate contents. BASIDIA 12–16 × 4–5 μ, short clavate, 4 spored. SETAE absent. SETAL HYPHAE present both in context and the dissepiment trama. Context setal hyphae radially arranged, up to 600 × 4–13 μ, unbranched or rarely branching, with a thick dark chestnut brown wall and capillary lumen; apex acute to obtuse, occasionally nodulose. Tramal setal hyphae diverging to project into the tube cavity, 55–100 × 9–18 μ, with a thick dark chestnut brown wall (2.5–7.5 μ) and a broad obtuse apex. HYPHAL SYSTEM dimitic with generative and skeletal hyphae, non-agglutinated in the context, but strongly agglutinated in the dissepiments. Generative hyphae 1–6.5 μ diam., hyaline or brownish, wall thin to somewhat thickening, freely branching, simple septate. SKELETAL HYPHAE 5–9 μ diam., unbranched, of unlimited growth, with a thick reddish brown wall (–2.5 μ) and continuous lumen, non-septate. Cultural characteristics unknown.

Phellinus lamaensis (Murr.) Heim, a closely related sp. with comparable geographical distribution, is readily separated from *P. noxius* by the presence of hymenial setae and narrow (up to 7 μ diam. only) setal hyphae of the dissepiment trama similar to those in the context (D. N. Pegler & J. M. Waterston, *CMI Descr*. 195, 1968).

P. noxius causes brown root rot, particularly of rubber (*Hevea brasiliensis*); it causes upper stem rot of oil palm, and attacks cacao and tea. It has been recorded on > 50 genera of tropical trees. It occurs in E. and W. Africa, S.E. Asia and Australia (Map 104). The characteristic symptoms on the roots distinguish it from those of the other root pathogens (*Rigidoporus lignosus* and *Ganoderma philippii* q.v.) which may occur on the same hosts, particularly rubber. The rhizomorphs on the roots form a continuous skin, brown internally but becoming darker on the outside with age; soil particles adhere to the rhizomorphs giving a very characteristic rough, dark appearance. The wood rot is at first pale brown and brown zone lines appear as a network beneath the bark. In softer woods the wood becomes lighter and friable, developing a honeycomb structure. In the root disease complex on rubber in W. Malaysia *P. noxius* is less common and less important than *R. lignosus* and *G. philippii*. It has both a similar persistence in the soil and a rate of advance along roots as *R. lignosus*, but the distance between the tip of the rhizomorph and the site of host penetration is much less. Also infection by air dispersed basidiospores is apparently more important; infection of rubber tree stumps and damaged branches (leading to dieback and possibly trunk snap) takes place. The underlying principles which guide control measures have been discussed under *R. lignosus*.

The disease on oil palm, known as upper stem rot to distinguish it from basal stem rot, has been studied in W. Malaysia. It is a relatively minor disease, usually affecting palms > 10 years old, but young palms can be attacked and it might become more prominent where this crop replaces a previous one of rubber. Airborne spores cause infection in a leaf base. A dark rot with dark zonations develops, giving a honeycomb appearance; this can extend to > half the cross-sectional area of the trunk and leads to frond wilt and trunk snap which is sometimes the first indication of a diseased palm. Sporophores may not appear for 1–3 years after the initial infection. *G. applanatum* (q.v.) is also a factor in oil palm stem rot, giving a light brown diseased area with irregular dark zonation, a more honeycombed appearance and a yellow advancing edge. It appears that this pathogen is a secondary invader. Incidence of the disease will presumably depend in part on how efficient control measures before planting have been, i.e. in general those that apply to rubber. Diseased leaf bases can be detected in the field since they produce a dull sound when struck with a wooden pole. In the early stages of rot, surgery and fungicidal paint treatment is effective. Thrower described *P. noxius* as causing the most important root disease of cacao in Papua New Guinea. The rhizomorph crust can spread up the trunk to 1 m above soil level; the cortex is dead 7–10 cm behind its advancing margin. Satyanarayana described control in tea.

Fidalgo, O. 1968. *Mem. N.Y. bot. Gdn* **17**: 109 (morphology & taxonomy, *Phellinus noxius* & allied spp.).
Corner, E. J. H. 1932. *Gdns' Bull. Straits Settl.* **5**: 317 (comparison as *Fomes noxius*, *F. lamoensis* & *F. pachyphloeus*; **12**: 55).
Lim, T. M. 1970. *Malaysian Crop Prot. Conf.* Nov.: 221 (on rubber; **51**, 3698).
Navaratnam, S. J. 1965. *Malaysian agric. J.* **45**: 175 (on oil palm, general; **45**, 542).
Ramakrishnan, T. S. et al. 1962 & 1964. *Rubb. Bd Bull.* **6**: 8; **7**: 67 (symptoms & control; **42**: 571; **44**: 224).
Riggenbach, A. 1958. *Nature Lond.* **182**: 1390 (thiamine requirement; **38**: 128).
Satyanarayana, G. 1973. *Two Bud* **20**: 22 (control in tea; **54**, 984).
Sharples, A. et al. 1930. *Malayan agric. J.* **18**: 184 (on oil palm; **9**: 648).
Thompson, A. 1931 & 1937. *Bull. Straits Settl. Fed. Mal. St. Dept agric.* Sci. Ser. 6, 23 pp.; 21, 28 pp. (on oil palm; **11**: 105; **17**: 313).
Thrower, L. B. 1965. *Trop. Agric. Trin.* **42**: 63 (on cacao; **44**, 1483).

PHIALOPHORA Medlar, *Mycologia* **7**: 200, 1915.
Cadophora Lagerberg & Melin, 1928.
(see Schol-Schwarz for synonymy)
(Dematiaceae)

COLONIES effuse, brown or olivaceous brown to black. MYCELIUM partly superficial, partly immersed. STROMA none. SETAE and HYPHOPODIA absent. CONIDIOPHORES semi-macronematous, mononematous, branched, straight or flexuous, pale to mid brown or olivaceous brown, smooth. CONIDIOGENOUS CELLS monophialidic, integrated and terminal or discrete, determinate, ampulliform, lageniform or subulate, with well-defined collarettes. CONIDIA aggregated in slimy heads, semi-endogenous, simple, straight or curved, ellipsoidal or oblong rounded at the ends, colourless to rather pale brown or olivaceous brown, smooth, aseptate (M. B. Ellis: *Demat. Hyphom.*: 524, 1971).

The genus was revised by Schol-Schwarz (*P. radicicola* Cain is not described) who summarised its characteristics: flask shaped phialides with a collarette and aseptate, slimy conidia. Walker (1975, *Gaeumannomyces* q.v.) discussed the fungi in the literature under *P. radicicola*. It was suggested that the name be reserved for Cain's fungus. McKeen described its behaviour (runner hyphae with finer hyphae penetrating the outer root tissue) on maize. The fungus was not considered to be very pathogenic although possibly a factor in root degeneration. Hoes described a disease (yellows) of sunflower caused by a *Phialophora* sp. in Canada (Manitoba). The first symptoms in the field are seen towards flowering. The leaves become a dull light green and then large areas turn dull yellow, usually beginning at the apex and margin. Necrotic, angular patches (5–10 mm long) form interveinally; leaf margins are necrotic, plants stunted, flower heads are small and they may be sterile. The symptoms differ from those caused by *Verticillium dahliae* (q.v.) where the initial sign is the bright yellow interveinal chlorotic mottling. The leaf necrosis (with the characteristic chlorotic edge) is much more pronounced with *V. dahliae* infection. Yellows might be confused with nutrient deficiency or water excess. Plants with yellows have brown roots with a disintegrating cortex.

The pathogen was isolated from petioles and stems (at various heights) and from roots, it occurred in the vascular tissue which becomes systemically invaded. Inoculated sunflower plants were 48 cm high with a mean head diam. of 3.6 cm (after 87 days) compared with 60 and 7.3 cm, respectively, for the controls. Symptoms caused by *V. dahliae* appeared earlier, developed faster and caused more necrosis than those caused by *Phialophora* sp. Incidence and severity of yellows coincided with poorly drained spots. Healthy sunflower cvs were observed next to the severely diseased inbred CM144 on which yellows had been first observed. Tirilly et al. found their *Phialophora* isolates from sunflower to be the same as *P. asteris* (Dowson) Burge & Isaac (on aster). They proposed *P. asteris* f. sp. *helianthi* on the grounds of specific pathogenicity to sunflower. *P. parasitica* Ajello, Georg & Wang is associated with wilt and dieback of mature trees.

Balis, C. 1970. *Ann. appl. Biol.* **66**: 59 (*Phialophora radicicola* on grasses & temperate cereals; **50**, 1148).
Cain, R. F. 1952. *Can. J. Bot.* **30**: 338 (the genus & *P. radicicola* sp. nov.; **32**: 280).
Cole, G. T. et al. 1969. *Ibid* **47**: 779 (phialides of *Phialophora*, *Penicillium* & *Ceratocystis*).
—— ——. 1973. *Mycologia* **65**: 661 (taxonomy, generic description, common wood inhabiting *Phialophora* spp.).
Hawksworth, D. L. et al. 1976. *CMI. Descr. pathog. Fungi Bact.* 503, 504, 505 (*P. asteris*, *P. cinerescens* & *P. parasitica*).

Phialophora gregata

Hawksworth, D. L. et al. 1976. *Trans. Br. mycol. Soc.* **66**: 427 (*P. parasitica* & diseased conditions; **56**, 626).

Hoes, J. A. 1972. *Phytopathology* **62**: 1088 (on sunflower; **52**, 2343).

McKeen, W. E. 1952. *Can. J. Bot.* **30**: 344 (*P. radicicola* on maize; **32**: 249).

Schol-Schwarz, M. B. 1970. *Persoonia* **6**: 59 (revision of genus; **49**, 3676).

Scott, P. R. 1970. *Trans. Br. mycol. Soc.* **55**: 163 (*P. radicicola* on grasses & wheat; **50**, 626).

Tirilly, Y. et al. 1976. *Bull. Soc. mycol. Fr.* **92**: 349 (comparative studies as vascular plant parasites; **56**, 2620).

Phialophora gregata (Allington & Chamberlain) W. Gams., Cephalosporium-*artige Schimmelpilze* (*Hyphomycetes*): 199, 1971.

Cephalosporium gregatum Allington & Chamberlain, 1948.

COLONIES slow growing, 14–16 mm diam. in 10 days, thin transparent, whitish. Sporulation sparse, better on oatmeal and cornmeal agars, and more typical. PHIALIDES (when typical) in irregular clusters on ramified sporophores; very short, bottle shaped, occasionally irregularly curved, with distinct collarette, 5–9 (up to >15) μ long, at the base 2–3 μ, reducing to 1.2–1.5 μ. On other media the phialides are very thick walled (*c.* 0.5 μ), compact, 3–4 μ broad, without collarette. CONIDIA in small mucilagenous heads; ovoid, occasionally slightly compressed at the base, $3.5–4.8 \times 2.3–2.5$ μ, length : breadth ratio 1.2–2. The conidia germinate by budding or forming secondary conidia. CHLAMYDOSPORES absent (after W. Gams l.c.; and see Allington et al.).

Brown stem rot of soybean was recently reviewed by Abel. It has been fully investigated in USA where yield losses by the more virulent str. can be nearly 40% (Gray 1972a). This systemic disease also occurs in Canada, Egypt, Japan and Mexico. In Japan the fungus affected fields of adzuki bean (*Phaseolus angularis*). Gray (1974) found that only the virulent (or defoliating) str. caused the characteristic syndrome in the field. The first symptoms are internal browning of the pith and vascular elements; this can occur within 8 weeks of planting in infested soil (Gray et al. 1973). Towards maturity (3 weeks before in inoculated plants; Gray 1972b) the leaves wither, wilt and become necrotic (the interveinal areas showing the initial necrosis). The non-defoliating str. only causes vascular browning (Gray 1971, 1972b) which in itself is not diagnostic of brown stem rot. Tachibana compared the pathogenicity of *P. gregata* with that of *Verticillium dahliae*. Phillips (1973) reported the existence of 3 virulence types. Gray et al. (1975) considered that the differences in symptom pattern were due to the formation of a toxin by the defoliating str.

The fungus is soilborne and survives between crops on soybean residues (Gray 1972a). Sporulation occurs on all parts of these residues and Lai et al. (1969c) found the opt. temp. for conidial germination to be 23° in a silt clay loam at 20% moisture content (and see Lai et al. 1969a). The opt. temp. for growth on agar is similar. The results of earlier work on the effects of temp. on brown stem rot were not clear cut. But Gray (1974) found that with an isolate of the defoliating str. neither temp. nor plant age were limiting. Defoliation took place at 22° and 28°. Penetration occurs through the roots and conidia probably spread in the plant (Schneider et al.). Inoculum builds up in the soil with successive crops (Dunleavy), and for control there should be a 3-year rotation with maize or some other non-host crop; for the effects of crop rotations see Dunleavy et al. Although certain soybean lines have resistance this is not high in commercial cvs.

Abel, G. H. 1977. *Rev. Pl. Pathol.* **56**: 1065 (review, 65 ref.).

Allington, W. B. et al. 1948. *Phytopathology* **38**: 793 (aetiology; **28**: 204).

Chamberlain, D. W. et al. 1954. *Ibid* **44**: 4 (factors affecting disease development; **33**: 574).

——. 1961. *Ibid* **51**: 863 (effects on stem water flow; **41**: 497).

Dunleavy J. 1966. *Ibid* **56**: 298 (disease build up; **45**, 2303).

Dunleavy, J. M. et al. 1967. *Ibid* **57**: 114 (control by rotations; **46**, 2139).

Gray, L. E. 1971. *Ibid* **61**: 1410 (variation in pathogenicity; **51**, 4519).

——. 1972a. *Ibid* **62**: 1362 (recovery from overwintered host debris; **52**, 2785).

——. 1972b. *Pl. Dis. Reptr* **56**: 580 (effect on yield; **52**, 534).

—— et al. 1973. *Ibid* **57**: 853 (disease development & yield; **53**, 2007).

——. 1974. *Phytopathology* **64**: 94 (effects of temp., host age & fungal isolate; **53**, 3700).

—— et al. 1975. *Ibid* **65**: 89 (toxin; **54**, 4265).

Kobayashi, K. et al. 1977. *Physiol. Pl. Pathol.* **11**: 55 (wilt inducing compounds from culture filtrates; **57**, 863).

Lai, P. Y. et al. 1969a, b, c, d. *Phytopathology* **59**: 343, 986, 1646, 1950 (sporulation in vivo; temp. & conidial germination; effect of host & soil moisture on sporulation & conidial germination; growth in vivo & in vitro; **48**, 2668, 3752; **49**, 1222, 1850).
Nicholson, J. F. et al. 1973. *Pl. Dis. Reptr* **57**: 269 (effects of crop spacing & history; **53**, 334).
Phillips, D. V. 1971. *Phytopathology* **61**: 1205 (effect of air temp. on disease; **51**, 2061).
——. 1972. *Ibid* **62**: 1334 (effects of photoperiod, host age & development stage; **52**, 2786).
——. 1973. *Pl. Dis. Reptr* **57**: 1063 (variation in pathogenicity; **53**, 2754).
Schneider, R. W. et al. 1972. *Phytopathology* **62**: 345 (disease development in host; **51**, 4520).
Tachibana, H. 1971. *Ibid* **61**: 565 (comparison with *Verticillium dahliae*; **51**, 918).
—— et al. 1972. *Ibid* **62**: 1314 (resistance & its modification by soybean mosaic virus; **52**, 2787).

PHOMA Sacc., nom. cons., *Michelia* **2**: 4, 1880.
Plenodomus Preuss, 1851
Peyronellaea Goid. ex Goid. apud Togliani, 1952.
(see B. C. Sutton, *Mycol. Pap.* 141: 150, 1977 for further synonymy which includes *Deuterophoma* Petri q.v.)
(Sphaerioidaceae)

PYCNIDIA mostly glabrous but sometimes hairy or setose especially towards the ostiole, usually globose–subglobose or globose–ampulliform to obpyriform but also more irregular in shape, separated or in small groups, usually subepidermal then erumpent with mostly one, but sometimes more, distinct, impressed but more often papillate openings (ostiole or porus); wall pseudoparenchymatous or prosenchymatous, sometimes pseudosclerenchymatous, the outer cells mostly dark and thick walled, the inner cells hyaline and more or less isodiametric, giving rise to CONIDIOGENOUS CELLS which are usually indistinguishable from the inner cells of the pycnidial wall but for a single aperture.

Under light microscopy the conidiogenesis may be characterised as monopolar repetitive 'budding'. The 'bud' of the first conidium arises from a papillate extension; subsequently, conidia arise as 'buds' in basipetal succession from the apex of the conidiogenous cells surrounded by a distinct collarette. Under electron microscopy the CONIDIOGENOUS CELLS appear to be phialides producing, from a fixed conidiogenous locus, enteroblastic conidia which secede by a 3-layered septum. The first conidial initial is produced within the inner layer of the papillate thickening of the wall at the apex of the conidiogenous cell. The upper part of the papilla wall sooner or later dissolves, but its basal part remains as a conspicuous collarette. The walls of successively produced conidia arise from the fixed meristem as outgrowths of the basal layer of the 3-layered septum remaining after secession of the previous conidium. Differentiation of the conidial wall is associated with abundant production of mucilage.

CONIDIA hyaline or sometimes slightly coloured (yellow to pale brown), globose, obovoidal, ellipsoidal or clavate, mostly once or twice as long as wide, generally (2)2.5–10(12) × (0.5)1–3.5(5) μ; aseptate but secondary septation may occur resulting in 2 (or even more) celled conidia; the percentage of septate conidia depends on the environment and may vary between 0–95% (in vivo) (from G. H. Boerema & G. J. Bollen, *Persoonia* 8: 134, 1975).

Phoma spp. can be confused with those of *Ascochyta* (q.v. for a comment and Boerema et al.; Brewer et al.). Conidia in *Phoma* spp. are essentially aseptate; those in *Ascochyta* spp. are essentially 2 celled.

The taxonomy, nomenclature, morphology, culture, hosts and other substrates of *Phoma* have been extensively studied in recent years by Boerema and associates. A bibliography of > 150 ref. was given by Boerema et al. (1973a). The difficulties in the classification of fungi like *Phoma* were discussed by Boerema (1969) who stated that in order to have a practical and realistic classification it is essential that the taxa should be based on a combination of morphological and cultural characteristics. This group of fungi has few workable morphological criteria. Boerema (1965a) described development of the conidia; and see Brewer et al., *Ascochyta* q.v.

Phoma spp. occur as fairly specialised pathogens but other spp. are general, weak and plurivorous parasites, saprophytes and soil fungi. Some have perfect states, for example, in *Didymella, Leptosphaeria* and *Pyrenochaeta* (q.v.). Five of the commoner of these generalised spp. were recently and fully described, they are widely distributed: *Phoma exigua* Desm., *P. glomerata* (Corda) Wollenw. & Hochapf., *P. herbarum* Westend., *P. macrostoma* Mont. and *P. pomorum* Thüm. *P. herbarum* (the type sp.) is a ubiquitous saprophyte of great variability.

Phoma

Its taxonomy, synonymy and morphology were described by Boerema (1964, 1970a). This sp. and *P. exigua* are soil fungi (see Domsch & Gams, *Fungi in agricultural soils*). Dorenbosch compiled a key to 9 ubiquitous soilborne *Phoma* type fungi. *P. exigua, P. glomerata, P. macrostoma* and *P. pomorum* were described (with synonymy and hosts) by Boerema et al. (1973a). *P. exigua* occurs as a widespread, weak, wound parasite and as specialised, pathogenic vars. Gangrene of Irish potato is caused by the vars *foveata* (pigment forming and more pathogenic) and *exigua* (non-pigment forming and weakly pathogenic, plurivorous), see Tichelaar; var. *linicola* occurs on flax (see Boerema 1967, 1969; Boerema et al. 1967; Maas). *Ascochyta phaseolorum* (q.v.) was considered synonymous with *P. exigua* (Boerema 1972). Morgan-Jones described *P. glomerata* and *P. prunicola* (Opiz) Wollenw. & Hochapf., the latter has been reported from lemon. The differences between *P. macrostoma* (Boerema et al. 1970b), *P. exigua* and *P. viburni* (Roum. ex Sacc.) Boerema & Griffin were described by Boerema et al. (1974).

P. insidiosa Tassi (described by Punithalingam et al.) causes a minor, probably widespread, leaf spot of *Sorghum* and *Setaria* spp., and other Gramineae. The macroscopic symptoms are variable and not particularly distinctive. Leaf lesions have an irregular outline, sometimes beginning at the tip or edge, and are brown to grey with narrow red purple margins. The scattered pycnidia occur sometimes in clusters or lines, interveinally. Spotting and pycnidia form on grain and glumes. Transmission is probably from seed and infection reduces both germination and subsequent growth. *P. insidiosa* may remain viable on seed for a year. Good growth in vitro occurs at 25–30° and conidia give a high germination at 25°. Cultures often cause the agar medium to become reddish. In India the pathogen has been reported as mainly infecting the grain. In USA, roots, stems and leaves of sorghum were successfully inoculated, and also maize and sugarcane. Apart from the use of healthy or chemically treated seed little control is required (Koch et al.; Rumbold et al.).

P. destructiva Plowr. (not *P. destructiva* (Plowr.) Jamieson; synonyms: *Diplodina destructiva* (Plowr.) Petr., misapplied; *Phyllosticta lycopersici* Peck) causes a fruit rot of tomato; also a leaf spot of tomato and Irish potato. The fungus and the diseases it causes were fully described by Jamieson. *Phoma destructiva* is often confused with *Didymella lycopersici* (q.v., stat. conid. *Diplodina lycopersici*) which causes a stem and fruit rot of tomato (Boerema et al. 1973a). Laundon gave the cultural characteristics of these 2 fungi on tomato and the common *P. exigua* (pyc. = pycnidia, pycnidial):

D. lycopersici	*P. exigua*	*P. destructiva*
(Dark; potato yeast extract dextrose agar; *c.* 24°		
Few or no pyc.	Few or no pyc.	Plentiful pyc.
Sterile stilboid bodies in 6 weeks	No such bodies	No such bodies
(Light; 12 hours near UV and white fluorescent and 12 hours dark)		
Clean zonate aerial hyphal mat	Irregular discoloured patchy hyphal mat	Few or no aerial hyphae
Regular margin	Scalloped margin	Regular margin
Moderate pyc. formation	Moderate pyc. formation	Profuse pyc. formation
Not forming pyc. clusters	Forming pyc. clusters	Not forming pyc. clusters

In the field *P. destructiva* causes small, slightly sunken, brown spots near the stem end of fruit. In ripening fruit and on fruit in transit the spots enlarge, becoming black, definite in outline, sunken with a surrounding area of watersoaked tissue and up to 4 cm diam. Frequently, the tissue breaks and a greyish white mycelium develops in the cracks. The lesions become leathery on the surface with prominent pycnidia. On the leaves the spots are brown to almost black, definite in outline, coalescing to form irregular blotches and showing concentric zonations not unlike those caused by *Alternaria solani* (q.v.). Obrero et al. described sunken elongate stem lesions. *P. destructiva* (widespread) in USA is said to be most severe as the fruit rot phase in transit; leaf infection being less important than other tomato leaf pathogens. In Guadeloupe Fournet described it as a serious leaf pathogen. Jamieson infected leaves without wounding but not fruit. Fruit decay is rapid at 21° and in vitro growth has been reported to be best at *c.* 21° and *c.* 28°. The decay spreads more rapidly in ripe fruit. The fungus is carried over in infested crop debris. Control measures are general in nature and fit in with those directed at some other tomato pathogens (Jones et al.). Seedbed precautions to prevent damping off and early leaf infection should be taken. Spraying as for *A. solani* will also give control. Care in harvesting and handling (avoid wet weather and damage to fruit) is needed. Borax fruit dips in the packing process reduce infection (Tisdale et al.).

Other examples of diseases on tropical crops caused by fungi like *Phoma* were given by Echandi (coffee), Hopkins (papaw), McDonald (sunflower), Olembo (acacia), Ponte (coriander), Prakash et al. (mango), Quiniones et al. (kenaf) and Zainun et al. (*Stylosanthes*). *Plenodomus destruens* Harter occurs on sweet potato and causes a foot rot. This fungus appears to be a *Phoma* (see comment by B. C. Sutton, *Mycol. Pap.* 141: 157, 1977). The disease is of minor importance in the tropics. In the field necrotic lesions develop slowly on the stems at around soil level; they may extend upwards for *c.* 12 cm and become girdling; leaf chlorosis and drop, and vine death follow. Early infection of field plants may cause a complete loss of crop. Infection can take place at the rooting nodes some distance from the main stem. Pycnidia develop in the diseased areas. If root tubers have formed on the attacked plant these can be affected and a brown firm rot is caused. The rot may be dormant in storage or the fungus may spread through wounds to hitherto healthy tubers. Control is through seed tuber selection and treatment, clean plant beds and crop rotation. If transplants from tubers are used for field plantings they should be obtained from tubers which have been grown in non-infested soil. A rotation of at least 2 years may be advisable (survival is probably only on host debris if it is for more than one season). Seed tubers can be treated with a fungicide and general sanitation precautions should be taken, both in the field and in storage.

Aulakh, K. S. et al. 1969. *Pl. Dis. Reptr* **53**: 219 (on tomato; **48**, 1978).

Boerema, G. H. 1964. *Persoonia* **3**: 9 (*Phoma herbarum*; **43**, 2241).

——. 1965a. *Ibid* **3**: 413 (conidial development in *Phoma*; **45**, 394).

—— et al. 1965b. *Ibid* **4**: 47 (*Phoma* referred to *Peyronellaea*; **45**, 1697).

———. 1967. *Ibid* **5**: 15 (*Phoma exigua* & vars; **47**, 1442).

——. 1967. *Neth. J. Pl. Pathol.* **73**: 190 (*Phoma* causing gangrene of Irish potato; **47**, 1245).

——. 1969. *Trans. Br. mycol. Soc.* **52**: 509 (use of term forma specialis for *Phoma*; **48**, 3366).

——. 1970a. *Persoonia* **6**: 15 (*P. herbarum*; **49**, 3674).

———. 1970b. *Ibid* **6**: 49 (*P. macrostoma*).

——. 1972. *Neth. J. Pl. Pathol.* **78**: 113 (*Ascochyta phaseolorum* synonymous with *P. exigua*; **52**, 290).

———. 1973a. *Studies in mycology* 3, 50 pp. (*Phoma* & *Ascochyta* spp. described by Wollenweber & Hochapfel; **53**, 433).

———. 1973b. *Persoonia* **7**: 131 (*Phoma* spp. referred to *Peyronellaea* 4; **53**, 75).

———. 1974. *Trans. Br. mycol. Soc.* **63**: 109 (*Phoma* spp. from *Viburnum*; **54**, 1314).

——. 1976. *Ibid* **67**: 289 (the *Phoma* spp. studied in culture by R. W. G. Dennis; **56**, 2394).

———. 1977. *Kew Bull.* **31**: 533 (*Phoma* referred to *Peyronellaea* 5; **56**, 3900).

Dennis, R. W. G. 1946. *Trans. Br. mycol. Soc.* **29**: 11 (British fungi ascribed to *Phoma* & related fungi; **25**: 526).

Dorenbosch, M. M. J. 1970. *Persoonia* **6**: 1 (key to 9 soilborne *Phoma* type fungi; **49**, 3673).

Echandi, E. 1957. *Revta Biol. trop.* **5**: 81 (on coffee in Costa Rica; **37**: 661).

Fournet, J. 1971. *Annls Phytopathol.* **3**: 215 (on tomato; **51**, 2864).

Harter, L. L. 1913. *J. agric. Res.* **1**: 251 (*Plenodomus destruens*, general).

Hopkins, J. C. F. 1938. *Proc. Trans. Rhod. scient. Assoc.* **35**: 128 (on papaw; **17**: 611).

Jamieson, C. O. 1915. *J. agric. Res.* **4**: 1 (on tomato).

Jones, J. P. et al. 1966. *Phytopathology* **56**: 929 (control of soilborne pathogens of tomato; **46**, 156).

Koch, E. et al. 1921. *Ibid* **11**: 253 (on sorghum; **1**: 170).

Laundon, G. F. 1971. *N.Z. Jl Bot.* **9**: 610 (records of fungal plant diseases in New Zealand 2; **51**, 3903).

Maas, P. W. T. 1965. *Neth. J. Pl. Pathol.* **71**: 113 (on flax; **45**, 462).

McDonald, W. C. 1964. *Phytopathology* **54**: 492 (on sunflower; **43**, 2700).

Morgan-Jones, G. 1967. *C.M.I. Descr. pathog. Fungi Bact.* 134, 135 (*Phoma glomerata* & *P. prunicola*).

Obrero, F. P. et al. 1968. *Pl. Dis. Reptr* **52**: 946 (on tomato; **48**, 1136).

Olembo, T. W. 1972. *E. Afr. agric. For. J.* **38**: 201 (on *Acacia mearnsii*; **53**, 285).

Ponte, J. J. Da 1967. *Bolm Soc. cearense Agron.* **8**: 37 (on coriander; **48**, 556).

Prakash, O. et al. 1977. *Pl. Dis. Reptr* **61**: 419 (on mango; **57**, 711).

Punithalingam, E. et al. 1972. *C.M.I. Descr. pathog. Fungi Bact.* 333 (*P. insidiosa*).

Quiniones, S. S. et al. 1952. *Philipp. Agric.* **36**: 235 (on kenaf; **33**: 481).

Rumbold, C. et al. 1921. *Phytopathology* **11**: 345 (on sorghum; **1**: 170).

Sharma, K. R. et al. 1973. *Mycologia* **65**: 709 (asexual complementation in *P. exigua*; **53**, 817).

Tichelaar, G. M. 1974. *Neth. J. Pl. Pathol.* **80**: 169 (use of thiophanate methyl to distinguish the 2 *Phoma* vars. causing gangrene of Irish potato; **54**, 2426).

Tisdale, W. B. et al. 1937. *Bull. Fla agric. Exp. Stn* 308, 28 pp. (on tomato; **17**: 139).

Phoma medicaginis

Zainun, W. et al. 1974. *APPS Newsl.* **3**: 70 (on *Macroptilium* & *Stylosanthes*; **54**, 1779).

Phoma medicaginis Malbr. & Roum. var. **pinodella** (Jones) Boerema, *Neth. J. Pl. Pathol.* 71: 88, 1965.

Ascochyta pinodella L. K. Jones, 1927

Phoma trifolii E. M. Johnson & Valleau, 1933.

COLONIES on oat and malt agars somewhat felty, greyish brown, turning black, sometimes sectoring with abundant crystals arranged like a fan. PYCNIDIA subglobose to variable, 200–300 μ diam. CONIDIOGENOUS CELLS hyaline, short, obpyriform, enteroblastic, phialidic, arising from the hyaline cells lining the pycnidial cavity. CONIDIA hyaline, mostly aseptate, sometimes 2 celled, 4.5–8 (–10) $\times$ 2–3 μ; in old cultures sometimes yellow brown. CHLAMYDOSPORES dark brown, spherical to irregular, smooth to rough, terminal or intercalary and formed singly or in chains. Pycnidia on *Pisum sativum* occur on upper tap roots, stems, leaves and pods (E. Punithalingam & I. A. S. Gibson, *CMI Descr.* 518, 1976).

The work on the fungi associated with the conditions usually known as foot rot, black stem or spring black stem of Leguminosae, particularly pea (*P. sativum*), lucerne and red clover (*Trifolium pratense*), was re-examined and extended by Boerema et al. The isolates could be grouped into 2 cultural types; one was very uniform and the other very variable. Both groups can apparently infect these 3 hosts; but the uniform group is generally associated with lucerne and the other with clover and pea. The former was designated *Phoma medicaginis* Malbr. & Roum. (var. *medicaginis*; synonyms: *P. herbarum* West. f. *medicaginum* West., *Diplodina medicaginis* Oud. and *Ascochyta imperfecta* Pk.). The latter was designated *P. medicaginis* var *pinodella*. The fungus known as *A. pinodella* has often been studied with the more serious pea pathogens *A. pisi* and *Mycosphaerella pinodes* (q.v.). Reference should also be made to the earlier comparisons (taxonomy, culture and pathogenicity) between isolates from clover and lucerne, largely in USA (Edmunds et al.; Ellingboe 1959a; Johnson et al.; Schenk et al.). Spring black stem of lucerne has also been extensively studied in Canada (survey by Mead 1964b). Isolates from both main hosts (lucerne and red clover) have wide host ranges in legumes. Those from lucerne caused symptoms on 34 spp. (Edmunds et al.). Mead (l.c.) included *Cajanus cajan* and *Arachis hypogaea* as hosts.

P. medicaginis is widespread, particularly in northern temperate areas (Map 263). The symptoms on lucerne occur on all above ground parts. On stems and petioles the elongate (up to 8 cm) lesions are dark brown to black, and when severe extend deeply, killing the shoots and causing severe defoliation. On leaves and pods there are irregular brown to purplish spots, variable in size. Blossoms, buds, crowns and (in some cases) the upper parts of the roots are infected; blossoms drop. Infected pods frequently contain shrivelled seeds. Seeds from infected plants germinate poorly and the seedlings that result may be killed; the pathogen is present as mycelium on the surface of the seed coat (Mead 1953, 1964a). In the field Kernkamp et al. found correlations between the numbers of light seeds, severe infection and infected seeds, also between the numbers of infected seeds and poor germination; yield, quality and viability of seeds were reduced.

The conidia germinate on the host to form appressoria and penetration is mostly direct, although it may be stomatal. Germination of conidia and growth in vitro (mycelium and pycnidia) are best at 19–22° or a little higher, but in the field disease develops most at *c*. 15° and under wet conditions. Renfro et al. found the disease to be about twice as severe on plants kept at 18° and 24° compared with those at 30°; a 24–48 hour moisture period was required for severe disease. Pycnidia form on the dead stems in the growing season and carry over occurs through these fructifications on crop debris. Infection of the new shoots by the waterborne conidia takes place as they grow up through the stubble or crop debris. The fungus persists in the soil of lucerne fields but disappears 1–2 years after ploughing and planting with a non-legume crop. It will persist in dry stem material for up to 10 years and on seeds for 3 years. Banttari et al. (1964) reported that feeding injury by, and honey dew of, the aphid *Acyrthosiphon pisum*, increased disease severity. Variation in pathogenicity occurs but there are no distinct races of the fungus; but see Ali et al. for pathotypes.

Control is through healthy seed (organo mercurials are effective eradicants), rotation (3 years) with non-legume crops and the use of cvs with some resistance which has been obtained by hybridisation in *Medicago*. Early cutting reduces inoculum for later growth. No fungicidal control that is likely to be economic appears to have been reported.

Ali, S. M. et al. 1978. *Aust. J. agric. Res.* **29**: 841 (pathotypes including *Aschochyta pisi* & *A. pinodes*).
Banttari, E. E. et al. 1963. *Phytopathology* **53**: 1233 (behaviour on resistant & susceptible hosts; **43**, 1058).
——. 1964. *Ibid* **54**: 1415 (effect of pea aphid on disease; **44**, 1133).
Bean, G. A. et al. 1961. *Crop Sci.* **1**: 233 (factors affecting infection; **41**: 393).
Boerema, G. H. et al. 1965. *Neth. J. Pl. Pathol.* **71**: 79 (comparative study, including taxonomy; **44**, 3395).
Cormack, M. W. 1945. *Phytopathology* **35**: 838 (general; **25**: 118).
Edmunds, L. K. et al. 1960. *Ibid* **50**: 105 (host range, pathogenicity & taxonomy; **39**: 473).
Ellingboe, A. H. 1959a. *Ibid* **49**: 764 (comparative study, including taxonomy; **39**: 322).
——. 1959b. *Ibid* **49**: 773 (growth in culture; **39**: 322).
Hancock, J. G. et al. 1965. *Ibid* **55**: 346, 356 (cell degrading enzymes with *Colletotrichum trifolii* & *Pleospora herbarum*; **44**, 2182a, b).
——. 1965. *Phytopathol. Z.* **54**: 53 (effect of infection on conc. of oxidative enzymes; **45**, 1419).
Johnson, E. M. et al. 1933. *Res. Bull. Ky agric. Exp. Stn* 339: 57 (general; **13**: 32).
Kernkamp, M. F. et al. 1953. *Phytopathology* **43**: 378 (effects of seed infection; **33**: 234).
Mead, H. W. 1953. *Can. J. agric. Sci.* **33**: 500 (seed & seedling infection; **33**: 358).
——. 1962. *Can. J. Bot.* **40**: 263 (factors affecting sporulation; **41**, 606).
——. 1963. *Ibid* **41**: 312 (effects of temp. on fungal growth & disease; **42**: 467).
——. 1964a. *Ibid* **42**: 1101 (seed infection; **44**, 159).
——. 1964b. *Can. Pl. Dis. Surv.* **44**: 134 (review, 57 ref.; **44**, 476).
Peterson, M. L. et al. 1942. *Phytopathology* **32**: 590 (general; **21**: 528).
Renfro, B. L. et al. 1963. *Ibid* **53**: 1340 (factors affecting infection, including temp.; **43**, 1059).
Rumbaugh, M. D. et al. 1962. *Crop Sci.* **2**: 13 (inheritance of resistance; **41**: 527).
Schenk, N. C. et al. 1956. *Phytopathology* **46**: 194 (comparative study, including taxonomy; **35**: 828).
Toovey, F. W. et al. 1936. *Ann. appl. Biol.* **23**: 705 (general in UK; **16**: 258).

PHOMOPSIS (Sacc.) Sacc., *Annls Mycol.* 3: 166, 1905.
(synonymy in B. C. Sutton, *Mycol. Pap.* 141: 152, 1977).
(Sphaerioidaceae)

MYCELIUM immersed, branched, septate, hyaline to pale brown. CONIDIOMATA eustromatic, immersed, brown to dark brown, separate or aggregated and confluent, globose, ampulliform or applanate, unilocular, multilocular or convoluted, thick walled; walls of brown, thin or thick-walled textura angularis, often somewhat darker in the upper region, lined by a layer of smaller celled tissue. Ostiole single, or several in complex conidiomata, circular, often papillate. CONIDIOPHORES branched and septate at the base and above, occasionally short and only 1–2 septate, more frequently multiseptate and filiform, hyaline, formed from the minor cells of the locular walls. CONIDIOGENOUS CELLS enteroblastic, phialidic, determinate, integrated, rarely discrete, hyaline, cylindrical, apertures apical on long or short lateral and main branches of the conidiophores; channel and periclinal thickening minute. CONIDIA of 2 basic types, but in some spp. with intermediates between the two: A conidia hyaline, fusiform, straight, usually biguttulate (1 guttule at each end) but sometimes with more guttules, aseptate; B conidia hyaline, filiform, straight or more often hamate, eguttulate, aseptate (B. C. Sutton, personal communication, November 1978).

Some recently and fully described *Phomopsis* spp. occur on tropical crops, for example, cacao, cashew, clove, oil palm, *Oryza*, sorghum and sugarcane (Early et al. 1972; Punithalingam 1974, 1975a). They do not appear to cause significant disease. Purss described a disease on *Annona squamosa* (sugar or custard apple) caused by *P. anonacearum* Bondarseva-Monteverde. Infection of the fruit resulted in purple lesions (1.5 cm diam., up to blotches over half the surface) near or at the apical end. The lesions became hard and cracked; the internal necrosis was <0.7 cm in depth. The fungus was also associated with a marginal leaf scorch. Dhingra et al. described a fruit rot of papaw caused by *P. carica-papayae* Petrak & Cif. Soetardi described a dieback of young tissue of 4-month-old seedlings of *Hevea brasiliensis* (rubber) caused by *P. heveae* (Petch) Boedijn which was considered to be a pathogen of unthrifty plants. Several *Phomopsis* spp. are associated with diebacks and cankers of conifers and other trees (*P. juniperivora* q.v.). *P. cucurbitae* McKeen and *P. sclerotioides* van Kesteren (recently described) occur on cucurbits. The latter causes a serious disease (black rot) of cucumber under glass

Boedijn, K. B. 1929. *Recl. Trav. bot. néerl.* **26**: 396 (*Phomopsis heveae* inter alia; **9**: 560).
Dhingra, O. D. et al. 1971. *Curr. Sci.* **40**: 612 (*P. carica-papayae*; **51**, 2715).
Early, M. P. et al. 1972. *Trans. Br. Mycol. Soc.* **59**: 345 (*P. anacardii*; **52**, 1692).
Petrak, F. et al. 1930. *Annls Mycol.* **28**: 377 (*P. carica-papayae* inter alia; **10**: 340).
Punithalingam, E. 1974. *Trans. Br. mycol. Soc.* **63**: 229 (*P. elaeidis, P. eugeniae, P. folliculicola* inter alia; **54**, 1164).
——. 1975a. *Ibid* **64**: 427 (*P. oryzae, P. sacchari, P. sorghicola* inter alia; **54**, 4831).
—— et al. 1975b. *C.M.I. Descr. pathog. Fungi Bact.* 469, 470 (*P. cucurbitae* & *P. sclerotioides*).
Purss, G. S. 1953. *Qd J. agric. Sci.* **10**: 247 (*P. anonacearum* inter alia; **34**: 42).
Soetardi, R. G. 1949. *Archf. Rubbercult.* **26**: 279 (*P. heveae*; **28**: 589).

Phomopsis juniperivora Hahn, *Phytopathology* 10: 248, 1920 (as *juniperovora*).

PYCNIDIA on stems and leaves chiefly solitary, initially immersed becoming erumpent, dark brown, lenticular to conical, sometimes subspherical up to 400 μ wide and opening by apical ostioles; wall several cells thick, composed of outer sclerotised and heavily pigmented cells and inner pseudoparenchyma lining the cavity. Pycnidial cavity usually unilocular but sometimes divided. CONIDIOPHORES hyaline, simple or branched, septate and completely lining the inside of the cavity. CONIDIOGENOUS CELLS enteroblastic, phialidic, hyaline, simple cylindrical to obclavate, borne on septate conidiophores. A CONIDIA (phialospores) hyaline, aseptate, fusiform to ellipsoid, 2 guttulate, with 1 guttule at each end, 8–10 (–12)×2–3 μ. B CONIDIA (phialospores) hyaline, elongated filiform, curved often strongly hooked, 20–30×0.5–1 μ. COLONIES on PDA show white to olivaceous, floccose aerial mycelium with numerous pycnidia distributed on the agar (E. Punithalingam & I. A. S. Gibson, *CMI Descr.* 370, 1973).

P. juniperivora occurs on a wide range of conifers, see *CMI Descr.* 370 and Hahn (1940, 1943) for hosts. But the pathogen has been largely studied in USA as causing a blight of *Juniperus*, particularly *J. virginiana* (eastern red cedar). Hahn (1928, 1930, 1943) described the saprophytes or weak parasites *P. conorum* (Sacc.) Died. and *P. occulta* Trav. on conifers and compared the latter with *P. juniperivora*. *P. lokoyae* Hahn causes a canker of *Pseudotsuga menziesii* (Douglas fir; Thomas; perfect state *Diaporthe lokoyae* Funk). *Phomopsis salmalica* Khan was described as causing a dieback of *Bombax malabaricum* (semul). *P. juniperivora* has been reported from: Canada (E.), Germany, Kenya, Mozambique, Netherlands, New Zealand, Portugal, South Africa, UK, Uruguay and USA (central, S. and E.; Map 502).

Juniper blight is largely a disease of seedlings, transplants and saplings. Although trees >3–4 years old are not severely damaged they can show infections of smaller branches which lead to death and the consequent disfigurement of whole trees. The fungus kills the cambium and the underlying wood becomes dark brown. Lesions in young plants spread from the laterals to the main stem and so to further laterals; sunken cankers form and the main stem can be girdled. Extremely serious losses can occur in nurseries. Peterson, in a recent study, infected seedlings of *J. virginiana* and *J. scopulorum* (Rocky mountain juniper) with conidia in suspensions. Lesions formed on the foliage after 3–5 days but only on newly formed, light green leaves; infection then spread to older leaves and stems. Stems can be directly attacked through wounds and there was generally more damage on younger seedlings. Growth in vitro, germination of A conidia and germ tube growth have an opt. temp. near 24°. Germination began after 4 hours and both this and germ tube growth were similar in the light or dark. Most infection occurred at 24–28° and at least 7 hours at RH 100% was required; there was no infection at RH 82 or 91%. Temps under 24° tended to reduce the rate of lesion development. The fungus persisted in inoculated trees and after 24 months could still be isolated. Pycnidia formed 3 weeks after inoculation. *J. virginiana* is somewhat more susceptible than *J. scopulorum*. Schoeneweiss described the relative damage due to *P. juniperivora* for 188 spp., vars and cvs in 4 coniferous genera.

Control is largely with fungicides. Phenyl mercury compounds are effective, but benomyl has given good results, for example, Gill found that it gave better control than other fungicides. It is not clear why the disease is much more important in USA than in Africa and Europe.

Funk, A. 1968. *Can. J. Bot.* **46**: 601 (*Diaporthe lokoyae* sp. nov.; **47**, 2571).
Gill, D. L. 1974. *Pl. Dis. Reptr* **58**: 1012 (fungicides; **54**, 3010).
Hahn, G. G. 1917. *J. agric. Res.* **10**: 533 (general).
——. 1926. *Phytopathology* **16**: 899 (general & hosts; **6**: 327).
——. 1928. *Trans. Br. mycol. Soc.* **13**: 278 (*Phomopsis conorum*; **8**: 279).
——. 1930. *Ibid* **15**: 32 (*Phomopsis* on conifers; **10**: 278).
——. 1940. *Pl. Dis. Reptr* **24**: 52 (hosts; **19**: 444).
——. 1943. *Mycologia* **35**: 112 (general taxonomy with *P. occulta*; **22**: 281).
Khan, A. H. 1961. *Mycopathol. Mycol. appl.* **14**: 241 (*P. salmalica* sp. nov. inter alia; **41**: 256).
Otta, J. D. 1974. *Pl. Dis. Reptr* **58**: 476 (fungicides; **53**, 4155).
——. 1978. *Can. J. Bot.* **56**: 727 (occurrence of A & B conidia in vivo & in vitro; **57**, 4656).
Pero, R. W. et al. 1970. *Phytopathology* **60**: 491 (activity of juniper diffusates on conidia; **49**, 2672).
Peterson, G. W. 1973. *Ibid* **63**: 246 (factors affecting infection; **52**, 3851).
Scheld, H. W. et al. 1963. *Pl. Dis. Reptr* **47**: 932 (effects of temp., survival; **43**, 598).
Schoeneweiss, D. F. 1969. *J. Am. Soc. hort. Sci.* **94**: 609 (susceptibility of conifers; **49**, 2673).
Smyly, W. B. et al. 1973. *Pl. Dis. Reptr* **57**: 59 (control with benomyl; **62**, 3099).
Thomas, G. P. 1950. *Can. J. Res.* Ser. C. **28**: 477 (*P. lokoyae*; **30**: 253).

Phomopsis theae Petch, *Ann. R. bot. Gdns Peradeniya* **9**: 324, 1925.

PYCNIDIA on bark, shoots and leaves immersed, black, conical or more or less globose up to 400 μ diam., with protruding apical ostioles; cavity rarely divided; pycnidial wall multicellular, with heavily pigmented and sclerotised cells around the ostiolar region. CONIDIOPHORES (phialides) hyaline, simple or branched, septate, 5–20 μ long, arising from the innermost layer of cells lining the cavity. A CONIDIA hyaline, aseptate, fusiform to ellipsoidal, 2–3 guttulate, usually with 1 guttule at each end of conidium, 5–8 (–9) $\times$ 2–2.5 μ. B CONIDIA hyaline, elongated, filiform, curved, sometimes strongly hooked, 20–30 $\times$ 0.5–1 μ. COLONIES on PDA and oat agar show floccose white aerial mycelium with pycnidia scattered on the agar (E. Punithalingam & I. A. S. Gibson, *CMI Descr.* 330, 1972).

Collar and branch canker of tea has been reported from India, Kenya, Malawi, Sri Lanka, Tanzania and Uganda (Map 493). The fungus is a facultative parasite and the disease it causes has been investigated in recent years in Sri Lanka. Here it began to attract attention again about 1957 when some 300 tea bushes were affected at one site. The disease had been attributed in the past to *Leptothyrium theae* Petch; see Shanmuganathan (1965) who compared the pathogenicity of *P. theae*, *L. theae* and 2 other fungi on tea. The first obvious symptoms are the yellow or brown leaves on a branch or a whole bush. A canker will be found at the collar or at the base of a branch. Old cankers are recognised by the raised margin due to development of a callus. The diseased areas are regular or irregular in shape, often sunken, grey to black and the underlying tissues are dead. The bark, cambium and sapwood are invaded. Where the cankers are girdling either a branch or a whole plant is killed. Cankers can arise at the crotch, base of a twig, a cut or a leaf scar. Woody tissue is invaded through wounds; but young green shoots can be penetrated without previous wounding and the resulting necrotic lesions cause death down to brown wood. This shoot dieback is not considered to be an important cause of damage but may be of significance in increasing the amount of conidial inoculum.

In Sri Lanka this canker has occurred mostly in plants 2–8 years old and at the higher elevations (1200 m or more). The following conditions tend to increase disease incidence (determined by size or number of cankers): low rainfall, low soil and bark moisture, relatively shallow root penetration and where susceptible clones are brought into bearing by bending, compared with cutting across. Pruning and mulching with polythene sheet tend to reduce disease incidence. Tea clones have been grouped into those with high, intermediate and low resistance. Where soil and climatic conditions favour canker the more resistant clones should be used. Measures should also be taken to avoid poor shallow soils, induce deeper root penetration and more rapid growth of young plants, and minimise evaporation of water from the soil. Benomyl is effective.

De Silva, R. L. et al. 1968. *Tea Q.* **39**: 87 (effect of shallow topsoil; **48**, 2535).
Hester, D. N. 1973. *Tea E. Afr.* **2**: 9 (clonal resistance & use of benomyl; **54**, 2955).
Shanmuganathan, N. 1965. *Tea Q.* **36**: 14 (aetiology; **44**, 3145).

Shanmuganathan, N. et al. 1966. *Ibid* **37**: 221 (infection of green shoots; **46**, 1719).
——. 1967. *Ibid* **38**: 320 (effect of soil moisture; **47**, 1961).
——. 1969. *Ibid* **40**: 164 (clonal resistance; **50**, 262).
——. 1971. *Pl. Dis. Reptr* **55**: 1021 (uptake of benomyl; **51**, 1870).
——. 1972. *Tea Q.* **43**: 36 (effects of cultural treatments; **52**, 1250).
Venkata Ram, C. S. 1973. *Bull. United Plrs' Assoc. S. India* 30: 5 (effects of genetic factors & cultural practices; **54**, 3464).

Phomopsis vexans (Sacc. & Syd.) Harter, *J. agric. Res.* **2**: 338, 1914.
Phoma vexans Sacc. & Syd., 1899
P. solani Halst., 1884.

PYCNIDIA on stems, leaves and fruit, solitary or gregarious, initially immersed, becoming erumpent, black, globose to irregular up to 350 μ wide and opening by 20–50 μ wide ostioles; wall composed of heavily pigmented cells on the outer side and the ostiolar region; cavity undivided and lined with hyaline cells. CONIDIOPHORES (phialides) hyaline, simple or branched, sometimes septate, 10–16 μ long arising from the innermost layer of cells lining the cavity. A CONIDIA fusoid to ellipsoidal, 2 guttulate, rarely 3, with 1 guttule at each end of conidium, 5–8 (–9) $\times$ 2–2.5 μ. B CONIDIA hyaline, filiform, curved, rarely erect, 20–30 $\times$ 0.5–1 μ.

Colonies on PDA and oat agar show white floccose aerial mycelium with numerous pycnidia scattered irregularly. Gratz found the ascigerous state of *Phomopsis vexans* in culture and proposed the name *Diaporthe vexans* Gratz, as *D. vexans* (Sacc. & Syd.) Gratz, but gave no Latin description. On *Solanum melongena* and *S. wendlandii* (E. Punithalingam & P. Holliday, *CMI Descr.* 338, 1972).

The fungus causes a disease on eggplant, and is variously known as tip over, stem blight or canker, leaf blight or spot and fruit rot; damping off can also occur. It is widespread (Map 329) and has caused serious losses in India, Mauritius, Philippines, Puerto Rico and USA. The leaf spots are conspicuous, up to 3 cm diam., irregular in outline and may coalesce; lower leaves tend to be infected first. In stem lesions the cortex dries and cracks, plants become stunted and girdling cankers cause death. Fruit spots are also conspicuous, pale, sunken and may affect the whole fruit which drops or remains attached, becoming mummified after a soft decay. Pycnidia are abundant.

Penetration is direct or stomatal. The opt. temp. for in vitro growth is 28–30° and light is necessary for sporulation. There is no physiologic specialisation. Primary sources of inoculum are from seed and host debris. Naturally infected seed germinates less well and more slowly. Fruit from healthy plants should be treated with a sterilant before seed extraction and a hot water (50° for 30 min.) seed treatment given. Seedbeds should be sterilised and a 3-year rotation practised. A fermate dust has controlled the leaf spot form of the disease. Several *Solanum* spp. have shown immunity or high resistance (e.g., *S. indicum*, *S. mammosum*, *S. pyracanthum* and others, see Kalda et al.). Partially resistant vars are in general use in USA.

Decker, P. 1951. *Phytopathology* **41**: 9 (resistant cvs Florida Market & Florida Beauty).
Divinagracia, G. G. 1969. *Philipp. Agric.* **53**: 173, 185 (host penetration; requirements for in vitro growth; **51**, 1014, 1015).
Edgerton, C. W. et al. 1921. *Bull. La. agric. Exp. Stn* 178, 44 pp. (general; **1**: 197).
Felix, S. et al. 1965. *Rev. agric. sucr. Ile Maurice* **44**: 182 (epidemic in Mauritius, control; **45**, 1273).
Gratz, L. O. 1942. *Phytopathology* **32**: 540 (perfect state; **21**: 514).
Harter, L. L. 1914. *J. agric. Res.* **2**: 331 (general).
Howard, F. L. et al. 1941. *Proc. Am. Soc. hort. Sci.* **39**: 337 (resistance; **22**: 194).
Kalda, T. S. et al. 1976. *Vegetable Sci.* **3**(1): 65 (resistance in *Solanum* spp.; **57**, 3214).
Lapis, D. B. et al. 1967. *Philipp. Agric.* **50**: 276 (growth & sporulation in vitro; **47**, 388).
Nolla, J. A. B. 1929. *J. Dept Agric. P. Rico* **13**: 35 (general; **8**: 698).
Pawar, V. H. et al. 1958. *Indian Phytopathol.* **10**: 115 (epidemic, general; **37**: 569).
Porter, R. P. 1943. *Pl. Dis. Reptr* **27**: 167 (infection of & transmission by seed; **22**: 511).
Toole, E. H. et al. 1941. *Proc. Am. Soc. hort. Sci.* **38**: 496 (effect on seed germination; **20**: 621).

PHYLLACHORA Nits. ex Fuckel, *Symb. Mycol.*: 216, 1869.
Diachora, J. Müll., 1894
Pseudomelasmia P. Henn., 1902
Metachora Syd. & Butler, 1911
Endophyllachora Rehm, 1913
(synonymy in C. R. Orton, *Mycologia* **36**: 20, 1944; other reported synonyms are: *Catacauma* Thiess.

& Syd., *Geminospora* Pat., *Halstedia* Stev., *Trabutiella* Thiess. & Syd.; and see L. Holm, *Taxon* **24**: 483, 1975)

(Polystigmataceae)

FRUCTIFICATIONS parasitic, foliicolous, simple or usually compound at maturity, made up of few to numerous ascocarps, generally crowded in the mesophyll so that their lateral walls form a dark brown palisade-like tissue when viewed in TS; apical and basal regions of the ascocarps usually extend radially or in the direction of the leaf axis to form a blackish clypeus more or less conspicuous in the epidermal region of one or both leaf surfaces; ascocarps ostiolate through the overarching clypeus. PARAPHYSES filiform. ASCI cylindrical to broadly ellipsoid, unitunicate, operculum not usually conspicuous. ASCOSPORES aseptate, hyaline, variously arranged. CONIDIA of uncertain phyllogeny and borne in similar fructifications are rather constantly present in some spp., variously shaped. SPERMATIA (?) short filiform, rather commonly present (from C. R. Orton l.c.).

Members of the genus cause the condition in plants known as leaf tar spots. Parbery (1967), who monographed the graminicolous spp. of *Phyllachora*, stated that *P. maydis* Maubl. is one of the few sp. that can cause severe leaf damage. There are no important pathogens in the genus but some may occasionally (and locally) be serious. Orton described the graminicolous spp. of N. America and Kamat et al. the Indian spp. *P. cynodontis* (Sacc.) Niessl. can cause wilt and defoliation on *Cynodon dactylon* (Bermuda grass). *P. lespedezae* (Schw.) Sacc. has caused losses on the cover crop *Lespedeza stipulacea* in S.E. USA; *L. striata* was largely immune in the field. Most infection occurred on young leaves. Germinating ascospores penetrated the cuticle via an appressorium (Lopez-Rosa et al.). *P. musicola* Booth & Shaw was described from banana leaves and the condition called black cross. On a leaf, from a point on a main lateral vein, the fungus grows out as a 4-pointed star. The 2 dominant arms develop along the veins for *c*. 3 cm on either side; simultaneously 2 shorter arms grow out at right angles and parallel to the midrib for *c*. 0.75 cm on either side of a lateral vein. Maize tar spot (*P. maydis*) is occasionally reported on (e.g., Liu; Malaguti et al.). The leaf lesions are elliptical, at first yellowish, with irregular borders, amphigenous, the stroma forms a dark crust towards the centre. Infection seems most serious on older plants. Parbery (1967) placed *P. sorghi* Höhnel (on sorghum) as a synonym of *P. sacchari* P. Henn. On rubber (*Hevea brasiliensis*) *P. huberi* P. Henn is apparently debilitative on nursery plants in parts of tropical S. America.

Anahosur, K. H. et al. 1974. *Curr. Sci.* **43**: 451 (spermogonia & ascocarps in *Phyllachora sorghi*; **54**, 3290).

——. 1977. *Mysore Jnl agric. Sci.* **11**: 91 (reaction of sorghum cvs to *P. sorghi*; **58**, 195).

Barrochina, J. J. T. 1969. *Boln Patol. veg. Ent. agric.* **31**: 15 (*P. cynodontis* ?; **50**, 3002).

Booth, C. et al. 1961. *Papua New Guin. agric. J.* **13**: 157 (*P. musicola* sp. nov.; **41**: 163).

Doidge, E. M. 1942. *Bothalia* **4**: 421 (revised descriptions of S. African spp.).

Hanson, C. H. et al. 1956. *Agron. J.* **48**: 369 (*P. lespedezae*; **36**: 34).

Kamat, M. N. et al. 1978. *Monogr. Ser. Univ. agric. Sci. Hebbal* 4, 100 pp. (monograph on Indian spp.).

Liu, L. J. 1973. *J. Agric. Univ. P. Rico* **57**: 211 (*P. maydis*; **54**, 434).

Lopez-Rosa, J. H. et al. 1966. *Phytopathology* **56**: 1136 (disease caused by *P. lespedezae*; **46**, 368).

Malaguti, G. et al. 1972. *Agron. trop.* **22**: 443 (*P. maydis*; **52**, 3308).

Meredith, D. S. 1969. *Trans. Br. mycol. Soc.* **53**: 324 (*P. musicola*; **49**, 524).

Miller, J. H. 1954. *Am. J. Bot.* **41**: 825 (ascocarps of *P. lespedezae*; **34**, 155).

Orton, C. R. 1944. *Mycologia* **36**: 18 (graminicolous spp. in N. America; **23**: 301).

Parbery, D. G. 1967. *Aust. J. Bot.* **15**: 271 (monograph on graminicolous spp.; **47**, 452).

——. 1971. *Ibid* **19**: 207 (additions & corrections to graminicolous spp.; **51**, 1195).

PHYLLACTINIA Lév., *Annls Sci. nat. Bot.* III **15**: 144, 1851.

(Erysiphaceae)

ASCOCARPS (cleistothecia, perithecia) with straight, unbranched appendages that have a basal swelling, subglobose; several ASCI. CONIDIOPHORES on superficial hyphae, base usually straight, sometimes twisted. CONIDIA pointed. Like *Leveillula* the MYCELIUM is partly internal (partly from C. E. Yarwood, *The fungi* (edited by G. C. Ainsworth et al.) Vol. IV A: 71, 1973).

The imperfect state is *Ovulariopsis* Pat. & Har.; *Erysiphe* (q.v.) for some general comments and ref. on the powdery mildews. *P. dalbergiae* Pirozynski occurs on *Dalbergia sissoo* (powdery mildew of sisam, or sissoo) and other *Dalbergia* spp. Synonyms: *P. yarwoodii* Patwardhan, *P. subspiralis* (Salm.) Blumer and *P. corylea* (Pers.) Karst. var. *subspiralis* Salm. The fungus (occurring in Asia) is readily distinguished from other *Phyllactinia* spp. by the spirally coiled conidiophores and the reticulated surface of the conidia. *P. guttuta* (Wallr. ex Fr.) Lév. is the widespread powdery mildew of temperate, dicotyledonous trees. Synonyms: *Erysiphe betulae* DC., *E. alni* DC., *E. guttata* Fr., *Alphitomorpha guttata* Wallr., *P. suffulta* (Rabenh.) Sacc. and *P. corylea* (Pers.) Karst. emend. Salmon. *P. guttata* is best known on *Corylus* on which it can cause defoliation; it is distinguished from other members of the genus by the size of the cleistothecia, 160–230 (mostly <200) μ.

Cullum, F. J. et al. 1977. *Trans. Br. mycol. Soc.* **68**: 316 (cleistocarp development & dehiscence in *Phyllactinia guttata*; **56**, 4362).
Kapoor, J. N. 1967. *C.M.I. Descr. pathog. Fungi Bact.* 157 (*P. guttata*).
Mukerji, K. G. 1968. *Ibid* 186 (*P. dalbergiae*).
Yu, Y. N. et al. 1978. *Acta microbiol. sin.* **18**: 102 (*Phyllactinia* in China; **58**, 100).

PHYLLOSTICTA Desm., non cons, *Annls Sci. nat.* Ser. 3 **8**: 28, 1847.
Phyllostictina H. & P. Syd., 1916.
(*Guignardia* q.v. for validation of *Phyllostictina* and a description of the type material *Phyllosticta convallariae* Pers. by Punithalingam, 1974; and see B. C. Sutton, *Mycol. Pap.* 141: 154, 1977 for further synonymy)
(Sphaerioidaceae)

STROMA prosenchymatous or plectenchymatous, well developed in pure culture, often reduced or absent on leaves. PYCNIDIA globose, pyriform or tympaniform, separate or in small groups, embedded in a subepidermal stroma, uni- or multilocular, ostiolate; wall prosenchymatous or pseudoparenchymatous, variable in thickness, not sharply delimited from the stroma, outer cells mostly dark and thick walled, inner cells hyaline, isodiametric, giving rise to conidiogenous cells. CONIDIOGENOUS CELLS short cylindrical or conical, forming blastoconidia in basipetal succession and abstricting them from a fixed locus with a broad base. CONIDIA aseptate, hyaline, gobose, obovoidal, ellipsoidal or clavate; young ones often more strongly tapering towards the base; broadly rounded apically, flattened or indented, surrounded by a slime layer and an apical appendage, usually containing characteristic greenish guttules; generally 8–20 × 5–10 μ, often forming appressoria on germination (from H. A. Van der Aa).

The genus was monographed by Van der Aa. It should not be confused with *Phoma* which has aseptate conidia without a slime layer or an apical appendage, and a different development of the conidium. Many *Phyllosticta* spp. have been described from tropical crops but few appear to cause any significant diseases. Two are described under their *Guignardia* perfect states, on banana and citrus. There are 2 recent reports of chemical control of leaf spots caused by *P. elattariae* Chowdhury (on cardamom) and *P. zingiberi* Ramakrishnan (on ginger; there is also *P. zingiberis* Stevens & Ryan). *Mycosphaerella zea-maydis* (q.v.) also has a *Phyllosticta* state.

Aa, H. A. Van der 1973. *Studies in mycology* 5, 110 pp. (monograph; **53**, 1246).
Mailum, N. P. et al. 1969. *Philipp. Agric.* **53**: 202 (on ginger; **51**, 1755).
Ramakrishnan, T. S. 1942. *Proc. Indian Acad. Sci.* Sect. B **15**: 167 (on ginger; **21**: 471).
Rao, D. G. et al. 1974. *J. Pl. Crops* **2**(1): 14 (on cardamom; **54**, 2923).
Sohi, H. S. et al. 1973. *Pesticides* **7**(5): 21 (on ginger; **54**, 527).

PHYSALOSPORA Niessl, *Verh. naturf. Ver. Brünn* **14**: 170, 1876.
(Polystigmataceae)

STROMA OR CLYPEUS absent. PERITHECIA embedded singly in the substrate; saprophytes or parasites which mature their perithecia in dead tissue only; perithecial wall fleshy, ostiole usually darker, often with bristles. ASCI large, unitunicate, usually broadest about the middle, rather thick walled. ASCOSPORES aseptate, >20 μ long. The genus is close to *Glomerella* (q.v.) which has ascospores <20 μ long (from R. W. G. Dennis, *British ascomycetes*: 257, 1968; see also Arx & Müller, *Beitr. Kryptogflora Schweiz* 11(1), 1954).

P. perseae Doidge (on avocado) was reported to be of economic importance in South Africa; the disease has been called blossom blight and brown rot. Infection spreads from the flowers to the main inflorescence stalk which may show characteristic longitudinal cracks. Fruits that are nearly set may drop and their rinds become brown and leathery. Girdling cankers on branches 5–8 cm diam. are formed. PERITHECIA 170 × 100 μ; ASCI 80–100 × 18–23 μ; ASCOSPORES are more or less distichous, hyaline, ellipsoid or subfusoid, tapering abruptly at each end to a blunt apex, 20–21 × 8–10 μ. Doidge also described a pycnidial state. Marlroth referred to fungicide control.

P. psidii Stevens & Peirce causes a disease on guava (*Psidium guajava*); Uppal described it as serious in India (Maharashtra). Infection shows in the bark which cracks; death of the branches can lead to the tree being killed. PERITHECIA (occurring on the dead bark) 120–165 × 270–345 μ; ASCI 72–100 × 26–33 μ; ASCOSPORES elliptical, hyaline 30–37 × 13–16 μ. *P. rhodina* is the perfect state described for *Botryodiplodia theobromae* (q.v.). Tarr (*Diseases of sorghum, sudan grass and broom corn*) gave the morphology of, and symptoms caused by, the 3 *Physalospora* spp. on sorghum which appear to be of no economic importance. Several spp. of the genus have been described from tropical crops (e.g., cacao, castor, cinnamon, cotton, quinine and rubber) but their economic importance, and possibly their taxonomic position, is uncertain.

On maize 2 spp. have recently been referred to by Ullstrup (1973): *P. zeae* Stout and *P. zeicola* Ell. & Ev. The former was transferred to *Botryosphaeria* by Arx et al. (l.c. p. 40). *P. zeae* causes grey ear rot. A whitish mycelium forms on and between the husks and kernels at the base of the ear. It becomes deep slate grey and covers the entire ear. Ears infected at an early age are light in weight and, therefore, at maturity they are upright (heavy, healthy ears droop). The fungus also infects the uppermost parts of the culm just beneath the tassel. In the field this is shown by straw coloured flag leaves and tassels; the latter droop. Beneath the sheath of infected flag leaves a dense mass of whitish mycelium is frequently found. Small, black, irregularly shaped sclerotia occur beneath the pericarp of the kernels; they also are found in the cob pith. Perithecia and pycnidia (*Macrophoma zeae* Tehon & Daniels) on large straw coloured lesions (25–50 × 100–450 mm) on the leaves. PERITHECIA 75–235 μ diam.; ASCI 85–150 × 13–22 μ; ASCOSPORES hyaline to very dilute olivaceous, aseptate, 19–25 × 6.5–8 μ; PYCNIDIA 65–120 μ becoming carbonaceous with age, ostiole protruding, nonrostrate, 25–35 × 15–17 μ; CONIDIA hyaline to greenish, aseptate, 17–31 × 6.5–8.5 μ (taken from Ullstrup (1946) who later (1973) commented on the apparent disappearance of grey ear rot; based on maize ears examined in Indiana, USA).

P. zeicola causes an ear and stalk rot of maize, largely confined to the gulf states in USA. A dark brown, felt-like mould forms on the kernels and husks. The interior of the kernel is black with pycnidia. The whole of the ear, or only a few scattered kernels, can be attacked. The pycnidia, characteristically, emerge through longitudinal cracks in infected stems; the pith is blackened. PERITHECIA 250–330 μ diam.; ASCI 75–80 × 12–15 μ; ASCOSPORES aseptate, hyaline, 18–20 × 8–10 μ; PYCNIDIA (*Diplodia frumenti* Ell. & Ev.) black 170–400 μ diam.; CONIDIA when mature dark, 1 septate, with longitudinal striations, 19–31(25.3) × 11–15(13.3) μ μ (from Eddins; Eddins et al.; the latter compare the morphology of *P. zeicola* and *P. rhodina* and describe the imperfect state of the former).

Doidge, E. M. 1922. *Bothalia* **1**: 179 (*Physalospora perseae* sp. nov., symptoms & culture; **3**: 92).
Eddins, A. H. 1930. *Phytopathology* **20**: 733 (*Diplodia frumenti*; **10**: 96).
—— et al. 1933. *Ibid* **23**: 63 (*P. zeicola*; **12**: 366).
Marlroth, R. H. 1947. *Fmg S. Afr.* **22**: 615 (*P. perseae*; **27**: 30).
Stevens, F. L. et al. 1933. *Indian J. agric. Sci.* **3**: 912 (*P. psidii* sp. nov.; **13**: 269).
Ullstrup, A. J. 1946. *Phytopathology* **36**: 201 (*P. zeae*: **25**: 391).
——. 1973. *PANS* **19**: 545 (*P. zeae*; **53**, 2997).
Uppal, B. N. 1936. *Int. Bull. Pl. Prot.* **10**: 99 (*P. psidii*; **15**: 664).

PHYSODERMA Wallroth, *Flora Crypt. Germ.* 2: 192, 1833.
Urophlyctis Schroet., 1889.
(Physodermataceae)

RHIZOMYCELIUM endobiotic, tenuous portions filamentous, delicate, fine or comparatively coarse; intercalary enlargements or '*Sammelzellen*' oval, elliptical, subspherical, broadly or narrowly spindle shaped, turbinate and occasionally irregular, usually septate, bi- or multicellular. RHIZOIDS arising from most portions of the rhizomycelium, often reduced, digitate and like haustoria. EPHEMERAL ZOOSPORANGIA, when present, usually developing on monocentric, eucarpic thalli like *Phlyctochytrium* or *Rhizophydium*; usually epibiotic, gregarious, sessile, pyriform, oval elongate, slipper shaped, irregular, sunken and deeply lobed, sometimes star shaped and angular; dehiscing by delinquescence of apical papilla, proliferating; subtended by a small endobiotic apophysis, richly branched and bushy, or reduced and digitate rhizoids. ZOOSPORES from ephemeral sporangia usually smaller than those from resting sporangia, ellipsoidal, oval, almost spherical, usually tapering posteriorally, with a conspicuous, eccentric, refractive globule; movement similar to most rhizidaceous chytrids but somewhat slower and intermittently amoeboid. RESTING SPORANGIA formed on endobiotic rhizomycelium from 1 of the cells of the spindle shaped or turbinate enlargements or from a short outgrowth of 1 of these cells; terminal or intercalary, ellipsoidal, oval, truncate, sub-hemispherical and flattened on one surface, sometimes slightly irregular, smooth or grooved; amber, light to dark brown; usually with a thick epispore and thin endospore; contents coarsely granular or with 1 to several, large, refractive globules; with or without an encircling ring of blunt, digitate haustoria; germinating by an ENDOSPORANGIUM which pushes up an oval or a circular, saucer shaped lid, or irregularly cracks the epispore; endosporangium emerging partly, to emit zoospores by deliquescence of an apical papilla. SEXUAL ORGANS unknown, doubtful or lacking (from J. S. Karling).

The genus was fully described by Karling. Sparrow presented evidence in support of the distinction between *Physoderma* and *Urophlyctis*. There are few plant pathogens. Besides the 2 spp. described, the following *Physoderma* spp. are among those reported on economic hosts: *citri* Childs, Kopp & Johnson (*Citrus*), *corchori* Lingappa on jute (*Corchorus olitorius* and *C. angulatum*), *cynodontis* Pavgi & Thirum. on Bermuda grass (*Cynodon dactylon*), *echinochloae* Thirum. & Whitehead on barnyard grass (*Echinochloa crusgalli*), *lathyri* (Palm) Karl. on *Lathyrus* spp., *leproides* (Trab.) Karl. on beet (*Beta vulgaris*) and *trifolii* (Pass.) Karl. on clover (*Trifolium pratense*).

Childs, J. F. L. et al. 1965. *Phytopathology* **55**: 681 (*Physoderma citri* sp. nov.; **44**, 3050).

Brown, G. E. et al. 1969. *Ibid* **59**: 241 (*P. citri* in citrus albedo & callus tissue; **48**, 2388).

Das, C. R. et al. 1964. *Indian Phytopathol.* **17**: 180 (*P. corchori*; **44**, 1104).

Gopalkrishnan, K. S. 1951. *Phytopathology* **41**: 1065 (*P. graminis* on *Agropyron repens*; **31**: 285).

Karling, J. S. 1950. *Lloydia* **13**: 29 (monograph, 282 ref.; **30**: 125).

Lingappa, B. T. 1955. *Mycologia* **47**: 109 (*P. corchori* sp. nov.; **34**: 786).

Pavgi, M. S. et al. 1954. *Sydowia* **8**: 90 (*P. cynodontis* & *P. setaricola* spp. nov.; **34**: 400).

Prakash, G. et al. 1964. *Jute Bull.* **26**: 237 (*P. corchori* on jute; **44**, 147).

Sparrow, F. K. 1962. *Trans. mycol. Soc. Japan* **3**: 16 (distinction between *Physoderma* & *Urophlyctis*; **42**: 106).

——. 1973. *Trans. Br. mycol. Soc.* **60**: 339 (location of types including those for genera *Physoderma* & *Urophlyctis*).

Thirumalachar, M. J. et al. 1953. *Science N.Y.* **118**: 693 (*P. echinochloae* sp. nov.; **33**: 235).

—— ——. 1954. *Bull. Torrey bot. Club* **18**: 149 (*P. graminis* on Bermuda grass inter alia; **33**: 430).

Walker, J. 1957. *Aust. J. Sci.* **19**: 207 (*P. trifolii*; **36**: 702).

Physoderma alfalfae (Pat. & Lagerh.) Karling, *Lloydia* **13**: 44, 1950.

Cladochytrium alfalfae Pat. & Lagerh., 1895
Physoderma leproides (Trab.) Lagerh., 1898
Urophlyctis alfalfae (Lagerh.) Mag., 1902.

RHIZOMYCELIUM intracellular, tenuous portions branched, fine 0.5–1 μ diam. when young, becoming coarse, 3–5 μ diam. and very thick walled when old; terminal and intercalary swellings predominantly turbinate, 10–15 × 17–19 μ diam., consisting of 2–6, somewhat disk shaped cells and bearing groups of branched, digitate and blunt haustoria or rhizoids at the apex. RESTING SPORANGIA numerous, 6–40 in a host cell, almost hemispherical, flattened on one surface, 40–59 μ max. diam., with a sub-apical ring of digitate, blunt haustoria or rhizoids; wall smooth, golden brown, epispore 1.5–2 μ thick, endospore thin and hyaline; contents coarsely granular, refractive; wall cracking irregularly in germination to emit swarming zoospores. Parasitic in the leaf scales, buds and stems of *Medicago sativa* (from J. S. Karling l.c.).

Crown wart (or sometimes called marbled gall) of lucerne is widespread in Europe and N. America (especially W.); also reported from Argentina, Australia, Chile, Ecuador, India, Iran, Israel, Mexico, New Zealand, Pakistan, Peru and South Africa (Map 130). Wilson covered the early literature of this little known disease. The buds arising from the crown of the plant below soil are infected, and become rounded and glistening white. The characteristic galls of diseased tissue are mostly formed 2–3 cm below the soil surface. They arise from axillary buds which would not normally develop. Some galls reach the soil surface and become green. Their exterior develops corky tissue and brown sporangial masses form inside; many of the galls eventually disintegrate. In cases of severe disease a crust of infected tissue may form around the base of the healthy stems. Local infections on leaves give rise to small galls like blisters. Crown wart is of minor economic importance being generally very restricted, but Leach found it to be a problem in low, wet land of Oregon (USA). He tested 24 lucerne cvs for resistance; all were susceptible but in varying degrees. Control measures are not usually required.

Arnaud, G. 1923. *C. r. hebd. Séanc. Acad. Agric. Fr.* **9**: 494 (general observations; **2**: 493).

Heim, P. 1961. *Rev. Mycol.* **26**: 3 (morphology, histology & cytology; **40**: 691).

Jones, F. R. et al. 1920. *J. agric. Res.* **20**: 295 (general account).

Leach, C. M. et al. 1959. *Pl. Dis. Reptr* **43**: 619 (susceptibility in cvs; **38**: 753).

Rao, C. G. P. 1973. *Indian Phytopathol.* **26**: 205 (anatomy of gall; **53**, 4254).

Scott, C. E. 1920. *Science N.Y.* **52**: 225 (germination of resting sporangia).

Wilson, O. T. 1920. *Bot. Gaz.* **70**: 51 (early literature, morphology & life cycle).

Physoderma maydis (Miyabe) Miyabe in A. Ideta, *Handbook of the plant diseases of Japan* edition 4: 114, 1909–11.
Cladochytrium maydis Miyabe, 1903
Physoderma zeae-maydis Shaw, 1912.

RHIZOMYCELIUM intracellular, tenuous portions delicate, 1–2.3 μ diam., filamentous and frequently branched; intercalary enlargements variously shaped, elongate, oval, broadly and narrowly fusiform, usually septate, bi- or multicellular. EPHEMERAL ZOOSPORANGIA epibiotic, elongate, asymmetrical and somewhat slipper shaped, 13–66 μ long, 10–15 μ max. diam. with long axis parallel to surface of the host; proliferating once or more, exit papilla apical, broad and blunt; subtended by a small, spherical, endobiotic apophysis and rudimentary digitate, once-branched rhizoids which are confined to 1 host cell. ZOOSPORES from ephemeral sporangia small, ellipsoidal, 3 × 5 μ with a small, eccentric, hyaline globule; flagellum 15–20 μ long. RESTING SPORANGIA numerous and usually filling the host cell; terminal or intercalary on rhizomycelium, oval to nearly circular in median plane but markedly flattened on one surface, 18–24 × 20–30 μ, with a double, 3 μ thick, amber, smooth wall; contents coarsely granular with 1 to several large, refractive globules; forming a conical or pyriform, 8–13 μ long, thin walled endosporangium (in germination) which pushes up a shallow, saucer shaped, 13–19 μ diam. lid and emits zoospores after deliquescence of its apical papilla. ZOOSPORES from the endosporangium ellipsoidal, 5 × 7 μ with a large, eccentric, hyaline, refractive globule; flagellum 20–25 μ long. Parasitic on *Zea mays* and *Euchlaena mexicana* (from J. S. Karling, *Physoderma* q.v.). The ephemeral zoosporangia were described by Couch and Sparrow (1947).

Brown spot of maize occurs in central and W. Africa, Central America, Argentina, Australia, S. and E. Asia, Brazil, Colombia, Cuba, Hawaii, USA and Venezuela (Map 106). A relatively minor disease which has been studied almost entirely in USA where it is only locally serious in south-eastern, wetter areas. Walker (*Plant pathology*) stated that *P. maydis* has never become destructive in the maize belt of USA. The symptoms are virtually restricted to the leaves and stems, mostly on the lower part of the plant. Numerous, small, watersoaked spots appear, becoming chlorotic and reddish brown (like rust); they are seldom > 1 mm diam. on the lamina but may be up to 5 mm elsewhere. The whole leaf may be affected; leaves dry up, the plant is stunted and weakening of the stem can cause severe lodging. Penetration by the zoospore occurs only during a short development period in the meristematic tissue. Most infection takes place through tissue formed during the day and, since tissue cannot be infected the day after formation, the plant shows alternating, lateral bands of infected and non-infected tissue; one band becoming visible each day and indicating that all infections take place at the same site in the whorl. Lesions appear simultaneously 20 days after tissue formation, i.e. the incubation period (Broyles 1962).

The resting sporangia, which can be air dispersed, have an opt. temp. for germination (by zoospores) of 28° at an opt. pH of 8.5–9. They survive in soil or host debris for 3–4 years. No germination takes place in the dark, it increases with light intensity up to 12-foot candles and is highest under blue light. When resting sporangia are exposed to 12 hours of light and then placed in the dark for 60 hours germination is poor. When they are given 12 hours of light after 24 hours of darkness germination is good. It is possible that zoospores from the resting sporangia are the primary inocula and that those from the ephemeral ones are a secondary source of infection. In the inheritance of resistance, through 6 generations of 15 populations, pooled additive effects were significant in all 15; parents for each population were a susceptible inbred and an isogenic resistant line. Pooled effects were significant for 5 populations and epistasis for 4 (Thompson 1969). In USA (Illinois) there was little or no brown spot where infested debris was ploughed in (Burns et al.; and see Lal et al. 1977 for systemic fungicides).

Broyles, J. W. 1959. *P. Dis. Reptr* **43**: 18 (disease incidence & loss in USA, Mississipi; **38**: 592).

——. 1962. *Phytopathology* **52**: 1013 (penetration of meristematic tissue; **42**: 318).

Burns, E. E. et al. 1973. *Pl. Dis. Reptr* **57**: 630 (effects of tillage; **53**, 929).

Couch, J. M. 1953. *J. Elisha Mitchell scient. Soc.* **69**: 182 (ephemeral sporangia; **33**: 533).

Eddins, A. H. 1933. *J. agric. Res.* **46**: 241 (general; **12**: 431).

Harvey, P. H. et al. 1955. *Pl. Dis. Reptr* **39**: 973 (reaction of inbred lines; **35**: 763).

Hebert, T. T. et al. 1958. *Phytopathology* **48**: 102 (effect of temp. & light on germination of resting sporangia; **37**: 407).

Lal, B. B. et al. 1976. *Indian J. Mycol. Pl. Pathol.* **5**: 174 (occurrence & survival; **56**, 3000).

——. 1977. *Pl. Dis. Reptr* **61**: 334 (inoculation & systemic fungicides; **56**, 5616).

Moll, R. H. et al. 1963. *Crop Sci.* **3**: 389 (quantitative study on inheritance of resistance; **43**, 1311).

Ojerholm, E. 1934. *Bull. Torrey bot. Club* **61**: 13 (multiflagellate zoospores; **13**: 628).

Olson, L. W. 1978. *Protoplasma* **97**: 275 (meiospore).

Sparrow, F. K. 1947. *Am. J. Bot.* **34**: 94 (ephemeral sporangia; **26**: 357).

——. 1974. *Mycologia* **66**: 693 (endobiotic stage; **54**, 1247).

Thompson, D. L. et al. 1963. *Crop Sci.* **3**: 511 (inheritance of, & breeding for, resistance; **44**, 1077).

——. 1969. *Ibid* **9**: 246 (quantitative genetic estimates for resistance; **51**, 314).

Tisdale, W. H. 1919. *J. agric. Res.* **16**: 137 (general).

Voorhees, R. K. 1933. *Ibid* **47**: 609 (factors affecting germination of resting sporangia; **13**: 225).

PHYSOPELLA Arthur, *Result. Sci. Congr. int. Bot. Wien* 1905: 338, 1906.
Angiopsora Mains, 1934.
(Pucciniaceae)

UREDIA round, subepidermal with peripheral, incurved paraphyses more or less united at their bases, often forming a pseudoperidium with the tips of the paraphyses projecting into the sorus; UREDOSPORES sessile, ellipsoid or obovate–globoid, with echinulate walls. TELIA forming lenticular sori, tardily dehiscent. TELIOSPORES in chains of 2–7 at centre of sorus, the cells oblong or prismatic, with walls somewhat coloured and smooth. The limits of the genus have not been well defined. Similar spp. are sometimes referred to *Phakopsora* (q.v.), uredia with paraphyses intermixed with the uredospores, or *Cerotelium* (q.v.), uredia with membraneous peridium or hyphoid paraphyses (from J. C. Arthur, *Manual of the rusts in United States and Canada*: 60, 1934).

Cummins (*The rust fungi of cereals, grasses and bamboos*) described and gave a key to 13 *Physopella* spp. *P. ampelopsidis* (Diet. & P. Syd.) Cumm. & Ramachar causes leaf rust of grapevine and occurs on other *Vitis* spp. Its distribution is S. and E. Asia, and America (Map 87). The fungus was described by Punithalingam (synonyms: *Phakopsora ampelopsidis* Diet. & P. Syd., *Angiopsora ampelopsidis* (Diet. & P. Syd.) Thirum. & Kern, *Uredo vitis* Thum., *U. vialae* Lagerh. and *Physopella vitis* Arth.). PYCNIA and AECIDIA unknown. UREDIA hypophyllous, scattered, pulverulent, yellowish, *c.* 0.1–0.15 mm diam., UREDOSPORES broadly ellipsoid to obovate, 18–28 × 12–17 μ, wall almost colourless to pale yellow, 1–1.5 μ thick, minutely but closely echinulate, pores indistinct; PARAPHYSES cylindrical, numerous, curved and irregular, 30–60 × 6–10 μ, wall 1–1.5 μ thick, yellowish. TELIA hypophyllous scattered, roundish, 0.1–0.2 mm across, 3–4 cells thick; TELIOSPORES in catenulate chains, ovoid, 20–30 × 12–15 μ, wall smooth, nearly colourless, 1 μ thick. The other rust fungi on *Vitis* were mentioned by Punithalingam.

Cummins, G. B. et al. 1958. *Mycologia* **50**: 741 (*Physopella* replaces *Angiopsora*: **38**: 354).

Punithalingam, E. 1968. *C.M.I. Descr. pathog. Fungi Bact.* 173 (*P. ampelopsidis*).

Physopella zeae (Mains) Cummins & Ramachar, *Mycologia* **50**: 743, 1958.
Angiopsora zeae Mains, 1938.

UREDIA amphigenous, in elongated groups up to *c.* 5 mm long, of *c.* 5 sori, pale luteous, up to 1 mm diam., long covered, finally naked. UREDOSPORES ellipsoidal to obovoidal, 22–28 × 16–20 μ, wall hyaline, echinulate, 1.5–2 μ thick; pores difficult to see. TELIA hypophyllous, scattered or in somewhat concentric groups around the uredia, dark brown, subepidermal. TELIOSPORES aseptate, in short chains, usually 2 spores, cylindrical, somewhat truncate above and below, 22–34 × 12–18 μ, wall luteous to sienna, smooth, 1–1.5 μ thick at the sides, 3–4 μ thick above, sessile.

Originally believed to be synonymous with *P. pallescens* on *Tripsacum* but is now considered distinct. The 2 other maize rusts: *Puccinia polysora* (q.v.) and *P. sorghi* (q.v.) differ in their coloured uredospores and pedicillate teliospores (G. F. Laundon & J. M. Waterston, *CMI Descr.* 5, 1964).

Physopella zeae, the lesser of the 3 maize rusts, nevertheless causes an important leaf disease which has caused considerable damage in the Pacific coast lowlands of Guatemala and in Venezuela. The rust is restricted to central America, some West Indian islands (Cuba, Dominican Republic, Grenada, Jamaica, Puerto Rico, St Vincent and Trinidad) and also occurs in Colombia, Peru, Venezuela and possibly Ecuador (Map 469). Virtually nothing is known of the biology of *P. zeae*. Host resistance and 2 races have been reported.

Mains, E. B. 1938. *Mycologia* **30**: 42 (*Angiopsora zeae* n. sp.; **17**: 452).
Malaguti, G. 1962. *Agron. trop. Maracay* **12**: 103 (maize rusts in Venezuela; **43**, 1904).
Muller, A. S. 1949. *Res. Bull. Iowa agric. Exp. Stn* 371: 597 (maize diseases in Guatemala; **29**: 504).
Robert, A. L. 1962. *Phytopathology* **52**: 1010 (host ranges & races of maize rusts; **42**: 318).
Schieber, E. et al. 1963. *Ibid* **53**: 517 (comparative pathology of maize rusts; **42**: 676).

PHYTOPHTHORA de Bary, *Jl R. agric. Soc.* **12** Ser. 2(1): 240, 1876.
(see Waterhouse, 1970a, for the 8 genera which have been placed in synonymy)
(Pythiaceae)

MYCELIUM white in mass, in host often with haustoria. HYPHAE 3–8 μ up to 12 μ, irregularly swollen undulate or gnarled; sometimes with characteristic swellings; initial branching at right angles to parent hypha and often swollen for a short distance. CHLAMYDOSPORES usually spherical, intercalary, sometimes terminal, wall smooth, up to 2 μ thick, hyaline at first. SPORANGIOPHORES usually undifferentiated apart from a few spp. where branching is reminiscent of *Peronospora* but with nodal swellings; branching sympodial or irregular and from below the sporangium or from within an empty one. SPORANGIA usually terminal, single on long hyphae in sympodia or within (or just beyond) an evacuated sporangium; ellipsoid, ovoid, obpyriform or limoniform (when shed); apex differentiated by an internal hyaline thickening of the inner wall of a depth (up to 6 μ) constant for the sp. and sometimes protruding to form a papilla; wall smooth up to 2 μ thick; non-caducous or shed with a pedicel. Germination by zoospores emerging individually through apex (free at once or held momentarily in an evanescent vesicle) or by a germ tube(s). ZOOSPORES hyaline, ovoid to phaseoliform, biflagellate, anteriorly directed tinsel (shorter) and posteriorly directed whiplash (longer); when motility ceases the spherical cyst may show repetitional emergence but not diplanetism. OOGONIUM usually terminal, spherical or tapering to the stalk, delimited by a thick septum; wall hyaline, thin, becoming thicker and often yellow to brown, mostly smooth occasionally tuberculate or reticulate. ANTHERIDIUM usually single, monoclinous or diclinous, spherical oval, clavate or short cylindrical, often angular; amphigynous or paragynous (sometimes both), if latter usually applied to the oogonium close to the stalk. OOSPORE single more or less filling oogonium, spherical, smooth, hyaline (sometimes faintly yellow), outer wall very thin, inner wall 0.5–6 μ thick, when mature with large central globule (after G. M. Waterhouse 1963).

Phytophthora, one of the most important plant pathogenic genera, is long overdue for monographic revision; last done by Tucker (1931). There is a sourcebook by Ribeiro. The most recent key to the spp. is by Newhook et al. An earlier key by Waterhouse (1963) included descriptions of the spp.; Waterhouse (1970a) reproduced the original descriptions and see Blackwell for morphology. There are reviews on aspects of the genus: hosts and distribution (Tucker 1933); general and zoospores (Hickman); zoospore emergence and discharge (Drechsler; MacDonald et al.); chlamydospore germination (Mircetich); development and reproduction (Zentmyer et al. 1970a); sterols, growth and reproduction (Brasier; Hendrix; Ko); effects of light (Harnish, Lilly; Merz et al. 1968; Ribeiro et al.); variation (Erwin et al.); mating types, cytology and genetics (Brasier et al. 1978a; Elliott et al.; Gallegly; Sansome; Savage et al.); serology (Gill et al.; Hall et al.; Halsall 1976a; Merz et al. 1969); zoospore taxis (Halsall 1976b; Khew et al.; Zentmyer 1970b); nutrition and physiology (Cameron et al.; Halsall 1977; Leal et al.; Lin et al.; Lopatecki et al.; Mehrotra; Mitchell et al.; Robbins; Roncadori), and isolation (Masago et al.; Tsao et al.). Kouyeas et al., Frezzi, Lee et al. and Rao gave accounts of the genus in Greece, Argentina, Malaysia and India, respectively. Brasier (1971, 1975, 1978b) described the stimulation of sex organ formation in *Phytophthora* by antagonistic *Trichoderma* spp. and its ecological implications. Al-Hedaithy et al. investigated caducity.

Members of the genus frequently cause root and seedling diseases like those caused by *Pythium*, but *Phytophthora* spp. are often more specialised and always much more destructive plant pathogens. Zoospore emission should be checked to ensure that the fungus being identified is not a *Pythium* (q.v.). All parts of the plant (both woody and herbaceous) can be attacked and under suitable weather conditions serious epidemics occur frequently. Survival (often in the soil) under unfavourable conditions can be long, but precisely how this occurs is not always clear. Spread can take place in natural and man-made water channels. Many spp. are plurivorous but others have very restricted host ranges. Many spp. are heterothallic; 2 compatibility types which are bisexual. Interspecific crossing occurs. The somatic nuclei are probably diploid with meiosis occurring in the oogonia and antheridia.

Apart from the spp. described (19, including 2 vars) the following may be found in the tropics: *P. cactorum* (Lebert & Cohn) Schroeter; *P. cajani* Amin, Baldev & Williams (on *Cajanus cajan*); *P. cambivora* (Petri) Buisman; *P. erythroseptica* Pethybridge (see Stamps); *P. melonis* Katsura (on cucumber, *P. capsici* q.v.); *P. oryzae* (Ito & Nagai) Waterhouse (on rice); *P. mexicana* Hotson & Hartge (on tomato; see *Mycol. Pap.* 122: 37, 1970); *P. quininea* Crandall (on *Cinchona*) and *P. verrucosa* Alcock & Foister (see Waterhouse 1970a).

Al-Hedaithy, S.S.A. et al. 1979. *Trans. Br. mycol. Soc.* **72**: 1 (sporangium pedicel length & caducity).

Amin, K. S. et al. 1978. *Mycologia* **70**: 171 (*Phytophthora cajani* sp. nov.; **57**, 4722).

Blackwell, E. M. et al. 1930. *Trans. Br. mycol. Soc.* **15**: 294 (sporangia, conidia, chlamydospores & sphaeroconidia & their germination; **10**: 755).

Blackwell, E. 1949. *Mycol. Pap.* 30, 24 pp. (terminology; **28**: 632).

Blackwell, E. M. 1953. *Trans. Br. mycol. Soc.* **36**: 138 (haustoria in *P. infestans* & other spp.; **33**: 46).

Brasier, C. M. 1971. *Nature New Biol.* **231**: 283 (induction of sexual reproduction in A_2 isolates by *Trichoderma viride*; **51**, 1143).

——. 1975. *New Phytol.* **74**: 183, 195 (stimulation of sex organs by *Trichoderma* sp., effect in vitro & ecological implications; **54**, 3708, 3709).

—— et al. 1978a. *Trans. Br. mycol. Soc.* **70**: 297 (gametic & zygotic nuclei in oospores; **57**, 4831).

——. 1978b. *Ann. appl. Biol.* **89**: 135 (stimulation of oospore formation by *Trichoderma* spp. & ecological implications).

Cameron, H. R. et al. 1965. *Phytopathology* **55**: 653 (effect of N & pH on growth; **44**, 2989).

Crandall, B. S. 1947. *Mycologia* **39**: 218 (*P. quininea* sp. nov.; **26**: 467).

Drechsler, C. 1930. *J. agric. Res.* **40**: 557 (repetitional emergence; **9**: 611).

Elliott, C. G. et al. 1973. *Trans. Br. mycol. Soc.* **60**: 311 (genetics & life history in *P. cactorum*; **52**, 3571).

Erwin, D. C. et al. 1963. *A. Rev. Phytopathol.* **1**: 375 (variation, 208 ref.).

Frezzi, M. J. 1950. *Revta Invest. agric. B. Aires* **4**: 47 (in Argentina; **30**: 433).

Gallegly, M. E. 1970. In *Root diseases and soil-borne pathogens* (editors T. A. Toussoun et al.): 50 (genetical aspects of behaviour, 29 ref.).

——. 1970. *Phytopathology* **60**: 1135 (genetics, 35 ref.; **50**, 538c).

Gill, H. S. et al. 1978. *Phytopathology* **68**: 163 (identification by disk electrophoresis; **57**, 4834).

Hall, R. et al. 1969. *Ibid* **59**: 770 (acrylamide gel-electrophoresis in taxonomy; **48**, 3363).

Halsall, D. M. 1976a. *J. gen. Microbiol.* **94**: 149 (specificity of cytoplasmic & cell wall antigens from 4 spp.; **55**, 4520).

——. 1976b. *Can. J. Microbiol.* **22**: 409 (zoospore chemotaxis; **55**, 4519).

—— et al. 1977. *Ibid* **23**: 994 (effects of cations on formation & infectivity of zoospores; **57**, 960).

——. 1977. *Ibid* **23**: 1002 (as above; **57**, 961).

Harnish, W. N. 1965. *Mycologia* **57**: 85 (effect of light on formation of oospores & sporangia; **44**, 2047).

Hendrix, J. W. 1970. *A. Rev. Phytopathol.* **8**: 111 (sterols, growth & reproduction of fungi, 152 ref.).

Hickman, C. J. 1958. *Trans. Br. mycol. Soc.* **41**: 1 (review, 112 ref.; **37**: 519).

——. 1970. *Phytopathology* **60**: 1128 (zoospores, 57 ref.; **50**, 5386).
Katsura, K. 1976. *Trans. mycol. Soc. Japan* **17**: 238 (*P. castaneae* & *P. melonis* spp. nov.; **57**, 2024).
Khew, K. L. et al. 1973. *Phytopathology* **63**: 1511 (zoospore chemotaxis in 5 spp.; **53**, 2851).
—— ——. 1974. *Ibid* **64**: 500 (zoospore electrotaxis in 7 spp.; **53**, 4310).
Ko, W. H. 1978. *J. gen. Microbiol.* **107**: 15 (hormonal regulation of sexual reproduction).
Kouyeas, H. et al. 1968. *Annls Inst. Phytopathol. Benaki* **8**: 175 (in Greece; **48**, 3361).
Leal, J. A. et al. 1967. *Mycologia* **59**: 953 (relation of C:N ratio to sexual reproduction; **47**, 1777).
Lee, B. S. et al. 1974. *Malaysian agric. Res.* **3**: 13, 137 (in Malaysia; **54**, 790, 4867).
Lilly, V. G. 1966. *The fungus spore: Proc. 18th Symp. Colston Res. Soc.*: 259 (effects of sterols & light on spore formation & germination, 39 ref.).
Lin, C. K. et al. 1965. *Acta microbiol. sin.* **11**: 470 (N, Ca & organic acid requirements & pH in vitro; **45**, 1308).
Lopatecki, L. E. et al. 1956. *Can. J. Bot.* **34**: 751 (nutrition; **36**: 205).
MacDonald, J. D. et al. 1978. *Phytopathology* **68**: 751 (matric & osmotic components of water potential & zoospore discharge).
Masago, H. et al. 1977. *Ibid* **67**: 425 (selective inhibition of *Pythium* on medium for isolation of *Phytophthora*; **56**, 4882).
Mehrotra, B. S. 1949. *J. Indian bot. Soc.* **28**: 108 (presence of enzymes; **28**: 636).
Merz, W. G. et al. 1968. *Proc. W. Va Acad. Sci.* **40**: 135 (relationship of light & sterol in oospore formation; **48**, 2153).
—— ——. 1969. *Phytopathology* **59**: 367 (serological comparison of 6 spp.; **48**, 2213).
Mircetich, S. M. et al. 1970. In *Root diseases and soil-borne pathogens* (editors T. A. Toussoun et al.): 112 (germination of chlamydospores, 22 ref.).
Mitchell, D. J. et al. 1971. *Phytopathology* **61**: 787, 807 (effects of O_2 & CO_2 tensions on growth, sporangium & oospore formation; **51**, 110a,b).
Newhook, F. J. et al. 1978. *Mycol. Pap.* 143, 20 pp. (tabular key).
Rao, V. G. 1970. *Mycopathol. Mycol. appl.* **42**: 241 (in India, review, 123 ref.; **50**, 2171).
Ribeiro, O. K. et al. 1976. *Mycologia* **68**: 1162 (effects of qualitative & quantitative radiation on reproduction & spore germination in 4 spp.; **56**, 3866).
——. 1978. *A source book of the genus* Phytophthora, J. Cramer.
Robbins, W. J. 1938. *Bull. Torrey bot. Club* **65**: 267 (requirement for thiamine).
Roncadori, R. W. 1965. *Phytopathology* **55**: 595 (nutrition; **44**, 2728).
Sansome, E. 1965. *Cytologia* **30**: 103 (meiosis).
Savage, E. J. et al. 1968. *Phytopathology* **58**: 1004 (homo- and heterothallism & interspecific hybridisation; **47**, 3742).
Stamps, D. J. 1978. *C.M.I. Descr. pathog. Fungi Bact.* 593 (*P. erythroseptica*).
Tsao, P. H. et al. 1977. *Phytopathology* **67**: 796 (inhibition of *Mortierella* & *Pythium* on a medium with hymexazol; **57**, 421).
Tucker, C. M. 1931. *Res. Bull. Mo. agric. Exp. Stn* 153, 208 pp. (monograph, 240 ref.; **10**: 754).
——. 1933. *Ibid* 184, 80 pp. (host range & geographical distribution, 524 ref.; **12**: 594).
Waterhouse, G. M. 1930. *Trans. Br. mycol. Soc.* **15**: 311 (formation of conidia; **10**: 756).
——. 1963. *Mycol. Pap.* 92, 22 pp. (key to & descriptions of spp., & rejected names; **42**: 748).
—— et al. 1966. *C.M.I. Descr. pathog. Fungi Bact.* 111 (*P. cactorum*).
——. 1970a. *Mycol. Pap.* 122, 59 pp. (diagnoses or descriptions & figures from original papers; **50**, 1633).
——. 1970b. *Phytopathology* **60**: 1141 (taxonomy, 23 ref.; **50**, 538d).
Williams, F. J. et al. 1975. *Ibid* **65**: 1029 (on pigeon pea; **55**, 2047).
Zentmyer, G. A. et al. 1970a. *Ibid* **60**: 1120 (development & reproduction, 90 ref.; **50**, 538a).
——. 1970b. In *Root diseases and soil-borne pathogens* (editors T. A. Toussoun et al.): 109 (tactic responses of zoospores, 15 ref.).

Phytophthora arecae (Coleman) Pethybridge, *Scient. Prog. R. Dubl. Soc.* N.S. **13**: 555, 1913.
P. omnivora var. *arecae* Colem., 1910
P. cactorum var. *arecae* (Colem.) Sacc. & Trotter, 1912.

HYPHAE somewhat variable in width, up to 9 μ with occasional hyphal swellings. SPORANGIOPHORES narrow, 2.5 μ wide, without swellings, irregularly branched, each branch ending in short sympodia of 1–3 sporangia. SPORANGIA very broad ellipsoid, obturbinate or nearly spherical, not distorted and with no double apexes, 20–48 (33) × 30–71 (45) μ, length : breadth ratio 1.1–1.6 (usually 1.3–1.4); deciduous, pedicel 1–6 μ long; papilla and apical thickening hemispherical or slightly less so. CHLAMYDOSPORES absent or rare, occasionally abundant, usually 35–40 μ diam., wall 1 μ thick. SEX ORGANS absent or rare in single str. culture (have been found on the host); form when grown with opposite mating type, with *P. palmivora* (both mating types) or with *P. meadii* or *P. nicotianae* (non-oospore producing strs of last 2 spp.). OOGONIA spherical,

23–44 (av. 30)μ diam., wall thin, colourless, may become yellow or brown. ANTHERIDIA amphigynous frequently broader than long, av. 14×15 μ. OOSPORES spherical, 23–36 (av. 29) μ diam., wall 3 μ thick, colourless or slightly brownish. Opt. temp. growth 27–30°. Mycelium in culture fairly copious; few sporangia on dry agar medium but abundant in a water film (after G. M. Waterhouse, 1974a; *P. palmivora* q.v.).

This sp. is very close to *P. palmivora* and has not been extensively studied. It has been most frequently reported from India where it causes mahali or koleroga disease of *Areca catechu* and the disease has also been reported on coconut. The pathogen occurs in Sri Lanka; Baker reported a fungus of the arecae-meadii group from cacao in Trinidad and *P. arecae* was isolated in the Netherlands from tomato showing foot rot symptoms. The typical symptoms occur on the nuts of the areca palm; infection causes watersoaked areas on the green nuts which fall. The necrosis spreads from the nuts to the inflorescence stalks and leaf bases; mycelial growth may cover the fruit. Nut fall in coconut has also occurred in India. The fungus is dormant during the dry season possibly as mycelium in leaf sheaths or as oospores since these have been reported from the host. Infection should be reduced by destroying infected palms and copper sprays have been used.

Ashby, S. F. 1929. *Trans. Br. mycol. Soc.* **14**: 254 (oospore production in vitro with *Phytophthora meadii*; **9**: 272).

Baker, R. E. D. 1936. *Trop. Agric. Trin.* **13**: 330 (arecae-meadii group isolate from cacao; **16**: 312).

Coleman, L. C. 1910. *Bull. Dep. Agric. Mysore Mycol. Ser.* 2, 1: 92 pp. (general).

——. 1910. *Annls Mycol.* **8**: 591 (general).

Narasimhan, M. J. 1926 & 1927. *Mysore agric. Cal. Yb.*: 25; 36 (control, oospores in other plants; **5**: 361; **6**: 289).

Ramakrishnan, T. S. et al. 1956. *Proc. Indian Acad. Sci.* Sect. B **43**: 308 (formation of oospores, production in host as *P. palmivora*; **36**: 402).

Sundararaman, S. et al. 1924. *Mem. Dep. Agric. India bot. Ser.* **13**: 87 (on coconut; **4**: 165).

Uppal, B. N. et al. 1939. *Curr. Sci.* **8**: 122 (oospores & compatible types; **18**: 518).

Venkatarayan, S. 1932. *Phytopathology* **22**: 217 (on areca, oospores with *P. palmivora*; **11**: 509).

Verhoeff, K. et al. 1966. *Neth. J. Pl. Pathol.* **72**: 317 (on tomato; **46**, 1325).

Phytophthora botryosa Chee, *Trans. Br. mycol. Soc.* **52**: 428, 1969.

COLONIES slightly radiate on Difco lima bean agar; at 26° slow growing (reaching 48–60 mm in 4 days) with little aerial mycelium. HYPHAE fairly uniform 5 μ wide. CHLAMYDOSPORES infrequent 18.7 μ, mostly 14–30 μ, thin walled (2 μ), terminal. SPORANGIOPHORES thin walled, 1–2 μ diam. with simple sympodial branching, without nodal swellings. SPORANGIA produced abundantly in clumps, deciduous, oval, occasionally ovoid, pedicel short, 15×28 (max. 16×31) μ, length : breadth ratio over 1.65, papilla inconspicuous. Sex organs rare in single culture but readily produced when complementary strains grown together in intraspecific mating or with *P. palmivora* in interspecific mating. OOGONIA spherical, 25 (max. 30) μ. ANTHERIDIA amphigynous, 13×14 μ. Opt. temp. growth 26°. On petioles of *Hevea brasiliensis* (from K. H. Chee l.c.).

P. botryosa differs from *P. palmivora* in the clumping of the sporangia and their smaller size, the lower opt. temp., and in the few chlamydospores. Waterhouse (1974b, *P. palmivora* q.v.) gave further details on morphology and isolates. This sp. was described following a severe outbreak of pod rot and abnormal leaf fall in rubber in the extreme northern part of Malaysia (W.). The disease and the pathogen were subsequently found in Thailand where they were considered to have been present for many years. Leaf shed frequently occurs following an attack of the petiole where a necrotic lesion is caused and a latex drop is formed. Pods, leaves and green stems are also infected, becoming extensively necrotic. Abnormal leaf fall (caused also by *P. meadii* and *P. palmivora* q.v.) must be distinguished from secondary leaf fall in Asia caused by fungi (*Glomerella cingulata* and *Oidium heveae* q.v.). In the latter disease infection only occurs on the immature leaves developed during the annual refoliation; in the former one infection can occur on leaves of any age, and outbreaks coincide with monsoon rain conditions. Wastie described the effects of weather. Defoliation was closely correlated with the duration of leaf wetness and 100% RH 7 days earlier.

In Malaysia (W.) *P. botryosa* is confined to leaf fall areas (but it can cause black stripe on the same host, *P. palmivora* q.v.). It has been isolated from soil and was recovered from infected petioles placed in soil for 3–4 months. Chee (1971), in a comparison with

P. palmivora, found 23 plant spp. susceptible to *P. botryosa* and some differential host susceptibility between these 2 spp.; 68 plant spp. were resistant to both fungi. Oospores have been found in rubber fruit on the tree. Control of abnormal leaf fall by copper spraying is possible and obligatory where the particular monsoon rain conditions are very favourable for infection from year to year over large areas of the rubber crop. Such areas occur in S.W. India; outbreaks of the disease in Sri Lanka and Malaysia seem more sporadic. *P. botryosa* has not been reported from India or Sri Lanka. Some clones in the 1966–67 outbreak in N. Malaysia (W.) were much less attacked than others. Inoculations have shown that there are significant differences in clonal susceptibility.

Chee, K. H. et al. 1967. *Pl. Dis. Reptr* **51**: 443 (leaf fall & pod rot caused by *Phytophthora palmivora* in Malaysia, clonal differences; **46**, 2815).
—— ——. 1968. *Pl. Prot. Bull. F.A.O.* **16**: 1 (leaf fall & pod rot caused by *P. palmivora* & *P. botryosa* in Thailand; **47**, 2822).
——. 1969. *Trans. Br. mycol. Soc.* **52**: 425 (description, variability & distribution of *Phytophthora* spp. on rubber in Malaysia; **48**, 3135).
——. 1969. *J. Rubb. Res. Inst. Malaya* **21**: 79 (*Phytophthora* spp. on rubber, behaviour in soil & clonal susceptiblity; **49**, 2165d).
——. 1969. *Plrs' Bull. Rubb. Res. Inst. Mayaya* 104: 190 (review, advisory; **49**, 2166).
——. 1971. *Malaysian agric. J.* **48**: 54 (pathogenicity of *P. palmivora* & *P. botryosa* to cultivated plants; **51**, 1276).
Pillay, P. N. R. et al. 1968. *Pl. Prot. Bull. F.A.O.* **16**: 49 (susceptibility in 10 clones; **48**, 272).
Wastie, R. L. 1973. *J. Rubb. Res. Inst. Malaya* **23**: 381 (effect of weather; **53**, 2297).

Phytophthora capsici, Leonian, *Phytopathology* **12**: 403, 1922.

P. hydrophila Curzi, 1927

P. parasitica var. *capsici* (Leonian) Sarejanni, 1936.

HYPHAE fairly coarse (5–7 μ). SPORANGIOPHORES rather narrow (1.5–2 μ) but sometimes widening slightly at the base of the sporangium, irregularly branched. SPORANGIA very variable in size and shape, from nearly spherical through broadly ovoid or obturbinate to elongated, usually 30–60 × 25–35 μ, but may be over 100 μ (greatest diam.), markedly papillate, often distorted and developing asymmetrically on the sporangiophore, many with 2 papillae; not proliferating internally; probably not deciduous; fairly profuse on agar media. CHLAMYDOSPORES absent or rare. OOGONIA frequent in culture, av. 30 (max. 39) μ diam., wall yellowish, becoming increasingly yellowish brown but not much thicker with age. ANTHERIDIA very variable in shape and size (av. 15 × 17 μ), always amphigynous. OOSPORES nearly filling the oogonium, wall thin. CULTURES finely radiate, fluffy. Opt. temp. growth *c.* 28° (partly after G. M. Waterhouse 1963, *Phytophthora* q.v.). Satour et al. and Polach et al. described 2 compatibility groups, oospores being abundant only when these occur together in culture (but see Noon et al.).

P. capsici has been reported on a fairly wide range of hosts but as a pathogen occurs mostly on red pepper, Cucurbitaceae, eggplant and tomato. It is widespread in S. and central USA; it also occurs in parts of S. and Central America, West Indies, S. Europe, W. and E. Asia, and Canada (British Columbia; Map 277). Some isolates attributed to *P. capsici* are not pathogenic to red pepper. No other *Phytophthora* sp. apparently causes such severe disease in this host. In the field symptoms on red pepper occur on the stems, leaves and fruit, roots can also be attacked; the disease may be called blight. Infected areas on above ground green tissue usually begin as dull, watersoaked lesions which spread quite rapidly, becoming necrotic and causing stem girdling. On the fruit the lesion may become dry, straw coloured, papery, cracked and sunken; the diseased and healthy margin on the stem is well defined. Severe infection leads to death of mature plants. Infection of tomato and cucurbit fruit causes a soft rot. Fruit losses during transport and storage can occur, spread by contact taking place. Spread is through soil and water splash; there appears to be no evidence for air dispersal other than in water droplets. Direct penetration of the cuticle has been described for several hosts. Opt. temps for sporangial production and disease development are near to those for growth.

P. capsici has been intensively studied as a fungus and a pathogen; see: *Rev. appl. Mycol.* **22**: 380; **37**: 144, 453; **39**: 8, 671; **40**: 449; **42**: 452; **43**, 2825; **47**, 1776 (culture, sporangial formation and germination, chemotaxis, behaviour and flagella structure of zoospores); *Rev. Pl. Pathol.* **50**, 2777; **52**, 340; **53**, 2834; **56**, 5450, 5451 (auxotrophic mutants, culture, zoospore formation); **50**, 1634; **54**, 2056, 2096; **56**,

597, 3485 (genetics and cytology); **49**, 2352; **53**, 2854; **54**, 1085, 5315; **55**, 3834, 3835 (fine structure and hyphal anastamosis); **55**, 4414; **57**, 1545, 1546, 1547 (physiology of resistance).

There is variation in pathogenicity between isolates. Polach et al. reported 14 distinct pathogenic strs from 23 isolates (15 from *Capsicum annuum*). For example 1 isolate was pathogenic to tomato, eggplant, squash, watermelon and red pepper (including 4 resistant lines); 7 isolates attacked all hosts except these resistant lines of red pepper; 4 isolates attacked only susceptible red pepper lines and some were avirulent on all hosts. Fungal lines from oospores showed recombination for pathogenicity factors and compatibility type. Resistance in red pepper is controlled by both dominant and recessive genes (e.g., Saini et al.; Smith et al.). When weather conditions are favourable for the pathogen extremely destructive attacks can occur on susceptible *Capsicum* cvs. The usual cultural control precautions against a *Phytophthora* sp. should be taken; and fungicides (e.g., Clerjeau; Sohi et al.) may be effective against fruit infection if applied before there is too much inoculum build up.

Citropoulos, P. D. 1955. *Bull. Torrey bot. Club.* **82**: 168 (on tomato, soil survival; **34**: 555).
Clerjeau, M. 1974. *Rev. Zool. agric. Path. vég.* **73**: 83 (control on red pepper; **54**, 2584).
Divinagracia, G. G. 1969. *Philipp. Agric.* **52**: 148, 166 (formation of sporangia & oospores, pathogenicity; **51**, 2190).
Godoy, E. G. 1950. *Revta Fac. Agron. Univ. nac. La Plata* **24**: 235 (general on red pepper; **20**: 224).
Katsura, K. et al. 1954. *Scient. Rep. Fac. Agric. Saikyo Univ.* **6**: 38 (on watermelon; **35**: 269).
——. 1961. *Tech. Bull. Fac. Agric. Kyoto Univ.* 1, 70 pp. (types of sporangial germination).
Kimble, K. A. et al. 1960. *Pl. Dis. Reptr* **44**: 872 (resistance in red pepper; **40**: 328).
Kodama, T. et al. 1974. *Proc. Kansai Pl. Prot. Soc.* **16**: 1 (on cucumber & control; **54**, 3038).
Kreutzer, W. A. et al. 1940. *Phytopathology* **30**: 951, 972 (compatibility groups, on Cucurbitaceae & tomato; **20**; 317, 335).
—— ——. 1946. *Ibid* **36**: 329 (epidemiology & control of tomato fruit rot; **25**: 584).
Leonian, L. H. 1922. *Ibid* **12**: 401 (general on red pepper; **2**: 101).
Morita, H. 1974. *Proc. Kansai Pl. Prot. Soc.* **16**: 7 (control on cucumber; **54**, 3039).
Noon, J. P. et al. 1974. *Can. J. Bot.* **52**: 1591 (oospore formation in presence of chloroneb; **54**, 46).
Polach, F. J. et al. 1972. *Phytopathology* **62**: 20 (pathogenic strs & inheritance of pathogenicity; **51**, 3684).
Saini, S. S. et al. 1971. *Himachal J. agric. Res.* **1**: 1 (inheritance of resistance in red pepper; **51**, 1011).
—— ——. 1978. *Euphytica* **27**: 721 (as above).
Satour, M. M. et al. 1968. *Phytopathology* **58**: 183 (compatibility types, oospore progeny & pathogenicity; **47**, 2083).
Simonds, A. O. et al. 1944. *Ibid* **34**: 813 (penetration & histology in tomato fruit; **24**: 123).
Smith, P. G. et al. 1967. *Ibid* **57**: 377 (inheritance of resistance in red pepper; **46**, 2584).
Sohi, H. S. et al. 1973. *Pesticides* **7**(5): 30 (control on red pepper; **54**, 636).
Stephenson, L. W. et al. 1972. *Can. J. Bot.* **50**: 2439 (encirclement of oogonial stalk by antheridium; **52**, 2173).
Teramoto, M. 1974. *Proc. Kansai Pl. Prot. Soc.* **16**: 12 (pathogenicity on cucumber; **54**, 3040).
Tompkins, C. M. et al. 1937 & 1941. *J. agric. Res.* **54**: 933; **63**: 417 (on melon fruit, root rot of red pepper & *Cucurbita*; **16**: 793; **21**: 118).
Weber, G. F. 1932. *Phytopathology* **22**: 775 (general on red pepper; **12**: 112).
Wiant, J. S. et al. 1940. *J. agric. Res.* **60**: 73 (rot in transport of watermelon; **19**: 513).

Phytophthora cinnamomi Rands, *Med. Inst. PlZiekt. Buitenz.* **54**: 41, 1922.

HYPHAE (on malt agar) coralloid (i.e. with frequent rounded nodules), becoming broad (8 μ) and very tough; hyphal swellings typically spherical (av. 42, max. 60 μ diam.) and in clusters; wall not much thickened. SPORANGIOPHORES thin (3 μ), occasionally branched, more often proliferating through the empty sporangium. SPORANGIA (formed only in aqueous solutions) broadly ellipsoidal or ovoid, 33×57 (up to 40×110) μ, no papilla, apical thickening slight; not shed. SEX ORGANS rarely produced on agar media in single strain culture but abundant when grown with *P. cryptogea* or the opposite str. of *P. cinnamomi*. OOGONIA av. 40 (max. 58) μ diam., walls smooth, becoming yellow or golden with age. ANTHERIDIA always amphigynous, long, $21–23 \times 17$ μ. OOSPORE nearly filling the oogonium, wall colourless, 2 μ thick. Culture with profuse, tough, aerial mycelium, sometimes in a rosette pattern; opt. temp. growth 24–28°. The coralloid hyphae readily distinguish this sp. in culture on malt agar from *P. cambivora* (*CMI Descr.* 112; G. M. Waterhouse &

J. M. Waterston, *CMI Descr.* 113, 1966; Zentmyer et al. 1976 gave 20–32.5° as the opt. temp. range).

P. cinnamomi is plurivorous and worldwide (Map 302). The general impression from the very large literature seems to support the conclusions that the many diseases caused by it occur usually under moderate temp. conditions; and that crops growing in the lowland tropics are much less prone to attack. This brief account was written before the publication of Zentmyer's monograph (1980). But the summary of this work was seen (Zentmyer personal communication, January 1979) and what follows is partly from this summary. Only a few selected ref. are given. Some of these give information on general aspects of the pathogen and its biology: formation of sporangia (Ayers; Ayers et al.; Chee et al.; Chen et al.; Marx et al.) and of the perfect state (Ashby; Mircetich et al. 1966; Pratt et al. 1972a; Reeves et al.); mating types (Chang et al.; Galindo et al.; Shepherd et al.); diploidy and gametangial meiosis (Brasier et al.); fine structure (Hemmes et al.; Ho et al.); behaviour in, and recovery from, soil (Broadbent et al.; Hendrix et al.; Hwang et al.; Marks et al.; Pratt et al. 1972b; Weste et al.; Zentmyer et al. 1966); chlamydospore germination (Mircetich et al. 1968); zoospore chemotaxis (Allen et al.); recovery from native vegetation (Pratt et al. 1973); effect of light on sporangial formation (Zentmyer et al. 1977); role of ectomycorrhizae (Marx) and dosage response to methyl bromide (Munnecke et al.).

P. cinnamomi is largely a pathogen of woody plants. The main hosts include avocado, pineapple, eucalyptus, pine and other conifers, woody ornamentals and native forest plants in Australasia; few monocotyledons are attacked. An origin in the East Indian Archipelago (Malaysia to N.E. Australia) has been postulated (see Ko et al.). It mostly attacks and destroys the small absorbing roots, but can also rot larger roots and cause stem cankers. It has long been known that high soil moisture promotes infection by increasing sporangial formation and favouring zoospore release and spread. Disease development is more severe on soil with restricted drainage and the effect of temp. in vivo is similar to that in vitro. The fungus infects the small roots causing visible lesions in 24 hours. It has considerable competitive saprophytic ability and meets some of the criteria for a soil inhabiting fungus. It can grow through non-sterile soil to a limited extent and survives for long periods in the absence of a living host.

The study of the biology of *P. cinnamomi* became intensive in Australia with the demonstration about 1970 that it was the cause of dieback in the jarrah (*Eucalyptus marginata*) forests (and other plant communities) of W. Australia (see the reviews by Newhook et al.; Podger; Zentmyer 1980). This dieback was first recorded in 1921, before Rand's description of the fungus had been published. An estimated area of *c.* 100 000 hectares is now affected. An epidemic of this dieback began about 1969 in Victoria (Brisbane Ranges). The disease spread at rates of up to 175 m a year (flat terrain) and 400 m a year (down drainage lines). This destruction of natural forest vegetation is one of the most dramatic, probably unique, disease case histories in epidemiological plant pathology. Plant communities have been destroyed and their floristic compositions altered. Other diebacks occur in other parts of Australia but the precise role of *P. cinnamomi* in these was still a matter of inference in 1973 (Podger). The distribution of the fungus in coastal Australia is widespread, discontinuous and extending. In east coast forests N. of latitude 37°S. little damage has been caused. Podger discussed the disease, the environment and the origin of the pathogen which has caused grave concern over the future of much of the indigenous vegetation of southern Australia (Newhook et al.).

Allen, R. N. et al. 1973. *Trans. Br. mycol. Soc.* **61**: 287 (zoospore chemotaxis; **53**, 2845).

——. 1974. *J. gen. Microbiol.* **84**: 28 (as above; **54**, 706).

Ashby, S. F. 1929. *Trans. Br. mycol. Soc.* **14**: 260 (perfect state in vitro; **9**: 272).

Ayers, W. A. 1971. *Can. J. Microbiol.* **17**: 1517 (induction of sporangia; **51**, 3149).

—— et al. 1971. *Phytopathology* **61**: 1188 (soil solution, pseudomonads & sporangium formation; **51**, 2191).

Brasier, C. M. et al. 1975. *Trans. Br. mycol. Soc.* **65**: 49 (diploidy & gametangial meiosis with other *Phytophthora* spp.; **55**, 89).

Broadbent, P. et al. 1974. *Aust. J. agric. Res.* **25**: 121 (behaviour in root rot conducive & suppressive soils; **53**, 3809).

——. 1975. In *Biology and control of soil-borne plant pathogens*: 152, American Phytopathol. Soc. (suppressive soils in E. Australia).

Chang, S. T. et al. 1974. *Aust. J. Bot.* **22**: 669 (mating types; **54**, 3707).

Chee, K. H. et al. 1965. *N.Z. Jl agric. Res.* **8**: 104 (steroid factor, growth & sporulation; **44**, 2048).

Chee, K. H. et al. 1966. *Ibid* **9**: 32 (micro-organisms & sporulation; **45**, 1311).
Chen, D. W. et al. 1970. *Mycologia* **62**: 397 (sporangial formation; **49**, 3623).
Galindo, A. J. et al. 1964. *Phytopathology* **54**: 238 (mating types; **43**, 2364).
Hemmes, D. E. et al. 1975. *Can. J. Bot.* **53**: 2945 (fine structure of chlamydospores; **55**, 3946).
Hendrix, F. F. et al. 1965. *Phytopathology* **55**: 1183 (recovery from soil; **45**, 1685).
Ho, H. H. et al. 1977. *Mycologia* **69**: 701 (light & electron microscopy; **57**, 1002).
Hwang, S. C. et al. 1978. *Phytopathology* **68**: 726 (chlamydospores, sporangia & zoospores in soil).
Ko, W. H. et al. 1978. *Trans. Br. mycol. Soc.* **71**: 496 (isolates from Taiwan as evidence for an Asian origin).
Marks, G. C. et al. 1974. *Aust. For.* **36**: 198 (detection in soil; **54**, 1148).
Marx, D. H. et al. 1965. *Nature Lond* **206**: 673 (induction of sporangia by diffusates of soil micro-organisms; **44**, 2729).
——. 1975. In *Biology and control of soil-borne plant pathogens*: 112, American Phytopathol. Soc. (ectomycorrhizae in protection of pine).
Mircetich, S. M. et al. 1966. *Phytopathology* **56**: 1076 (oospores in roots & soil; **46**, 390).
—— ——. 1968. *Ibid* **58**: 666 (physiology of chlamydospore germination; **47**, 3004).
Munnecke, D. E. et al. 1974. *Ibid* **64**: 1007 (dosage response to methyl bromide; **54**, 686).
Newhook, F. J. 1970. In *Root diseases and soil-borne pathogens*: 173, Univ. California Press (in New Zealand).
—— et al. 1972. *A. Rev. Phytopathol.* **10**: 299 (role in Australian & New Zealand forests, 168 ref.).
Podger, F. D. 1975. In *Biology and control of soil-borne plant pathogens*: 27, American Phytopathol. Soc. (role in dieback diseases of Australian eucalyptus forests, 63 ref.).
Pratt, B. H. et al. 1972a. *Aust. J. biol. Soc.* **25**: 861 (oospore formation in the presence of *Trichoderma*; **52**, 1020).
—— ——. 1972b. *Trans. Br. mycol. Soc.* **59**: 87 (differentiation from other *Phytophthora* spp. in soil baiting; **52**, 1013).
—— ——. 1973. *Ibid* **60**: 197 (recovery from native vegetation in a remote area of NSW, Australia; **52**, 3564).
Reeves, R. J. et al. 1972. *Ibid* **59**: 156 (induction of oospores in soil by *Trichoderma*; **52**, 609).
Shepherd, C. J. et al. 1974. *Aust. J. Bot.* **22**: 231, 461 (mating types; temp., genetic diversity, morphology & behaviour; **54**, 1444, 1603).
—— ——. 1978. *Ibid* **26**: 123, 139 (mating behaviour of *Phytophthora* spp. & *P. cinnamomi*).
Weste, G. et al. 1977. *Ibid* **25**: 461 (population densities in forest soils; **57**, 3112).
Zentmyer, G. A. et al. 1966. *Phytopathology* **56**: 710 (saprophytism & persistence in soil; **45**, 3300).
—— ——. 1976. *Ibid* **66**: 982 (growth & temp.; **56**, 1400).
—— ——. 1977. *Ibid* **67**: 91 (visible & near visible radiation & sporangial formation; **56**, 3428).
——. 1980. *Monogr. Am. phytopathol. Soc.* **10**, in press (monograph).

PERSEA AMERICANA

Phytophthora cinnamomi causes a serious root rot of avocado which has been studied in Australia, South Africa and USA. It can also cause a stem canker (see Brun; Crandall; Cross), but this form of the disease seems less important. In root rot an infected tree shows leaves that are smaller than normal, pale or yellow green and may wilt. The leaves tend to fall and new growth is poor or absent. Branch dieback occurs and trees may set an abnormal, heavy crop of small fruit. The small roots show necrosis and in the advanced stages of root rot feeder roots are rare. Roots of (or over) *c.* 0.75 cm diam. are seldom infected.

The disease is a serious one in California. Zentmyer et al. found that in infested soil there was more damage to trees that were irrigated weekly than those which were irrigated every 2 weeks. Borst reported that trees which escaped infection were concentrated on calcareous soils; root rot was widespread on clays. In Queensland Pegg stated that infection occurred on a wide scale in the S.E. Nursery or older trees were attacked, with death taking place within a few months to 2–3 years. Control measures consist of: careful irrigation, soil fumigation of small infested areas, resistant rootstocks, soil fungicides and rotation. High soil moisture levels should be avoided. Dazomet and metham-sodium are effective fumigants. The Duke cv. gives a stock which is somewhat resistant. Repeated applications of fenaminosulf at and after planting are effective. Many crops show resistance to attack by *P. cinnamomi*; these include citrus, some deciduous fruits and most vegetables (Anon.; Boyce; Milne et al.; Pegg; Zentmyer).

Anon. 1977. *Citrograph* **62**: 235 (progress in control; **57**, 708).
Borst, G. 1971. *Ibid* **56**(4): 109 (soil & disease escape; **50**, 3058).

Boyce, A. M. 1957. *Ibid* **43**: 3 (including control; **37**: 295).
Brun, J. 1975. *Fruits* **30**: 339 (canker; **55**, 337).
Crandall, B. S. 1948. *Phytopathology* **38**: 123 (in Peru; **27**: 327).
Cross, G. F. 1953. *Fmg S. Afr.* **28**: 210 (trunk rot; **33**: 99).
Gilpatrick, J. D. 1969. *Phytopathology* **59**: 973, 979 (effects of soil amendments; **49**, 218a & b).
Ho, H. H. et al. 1977. *Ibid* **67**: 1085 (infection of avocado & other *Persea* spp.; **57**, 2596).
Labanauskas, C. K. et al. 1976. *Calif. Avocado Soc. Yearbk* 59: 10 (effect of infection & soil O_2 on plant nutrients; **57**, 709).
Milne, D. L. et al. 1975. *Citrus Subtrop. Fr. J.* 502: 22 (control on replanting; **56**, 795).
Pegg, K. G. 1970. *Qd agric. J.* **96**: 412 (in Australia & control; **50**: 824).
Roth, G. 1964. *Tech. Commun. Dep. Agric. tech. Serv. S. Afr.* 18, 10 pp. (tree decline; **44**, 775).
Sterne, R. E. et al. 1978. *Phytopathology* **68**: 595 (effect of root rot on host water relations; **58**, 307).
Zentmyer, G. A. et al. 1952. *Ibid* **42**: 35 (effect of irrigation; **31**, 338).
——. 1973. *Ibid* **63**: 267 (control with fenaminosulf; **52**, 3778).

OTHER CROPS

P. cinnamomi was originally described by Rands from Indonesia (Sumatra) where it caused a bark (stripe) canker of cinnamon (*Cinnamomum zeylanicum*). The plants show irregular, vertical stripes of dead bark, 1–5 cm wide, beginning at or below soil level and several metres long: the canker is sunken and at the advancing upper edge drops of an amber or wine coloured exudate appear; a zonation may develop and a narrow, black line divides the healthy and diseased tissue. On pineapple the fungus causes a wilt and a top rot (Blake et al.; Lewcock). Wilting plants show root destruction, stunting, poor growth and fruit malformation. In top rot symptoms usually appear before fruiting. The central leaves dry out, become necrotic and fall. The outer ones may remain apparently normal until a late stage in the disease. A slight pull detaches the leaf crown before the foliar symptoms are clear. There is an internal rot with a distinct brown margin. Top rot appears to be a minor disease which can be controlled through conventional, cultural measures. Recently Pegg (1977) described the soil application of sulphur in Australia as a control measure. A pH of <3.8, which limits *P. cinnamomi*, did not adversely affect the growth of pineapple.

Pegg (1973) described control measures for the stem canker of macadamia nut (*Macadamia integrifolia*). The disease is widespread in Queensland, Australia, and has also been reported from USA (California and Hawaii). In Queensland the measures recommended included avoidance of plant wounding, surgery and Bordeaux application, hygiene, soil disinfestation and possibly high grafts on less susceptible stocks. Root rots caused by *P. cinnamomi* on other tropical crops include *Cinchona* (quinine, Crandall), *Passiflora edulis* (passion fruit, Young) and *Eugenia caryophyllus* (clove, Lee). For the fungus in forests of Hawaii see Hwang et al.; Kliejunas et al.

Blake, C. D. et al. 1959. *Agric. Gaz. N.S.W.* **70**: 638 (pineapple wilt; **39**: **605**).
Crandall, B. S. 1947. *Phytopathology* **37**: 928 (on quinine; **27**: 255).
Hwang, S. C. et al. 1978. *Trans. Br. mycol. Soc.* **70**: 312 (in Hawaiian ohia forests; **57**, 5147).
Kliejunas, J. T. et al. 1976. *Phytopathology* **66**: 116, 457 (as above & dispersal; **55**, 4302; **56**, 114).
—— ——. 1977. *Pl. Dis. Reptr* **61**: 290 (as above, site & edaphic factors; **56**, 5828).
Lee, B. S. 1974. *MARDI Res. Bull.* 2(2): 26 (on clove; **55**: 4227).
Lewcock, H. K. 1935. *Qd agric. J.* **43**: 145 (pineapple top rot; **14**: 458).
Pegg, K. G. 1973. *Ibid* **99**: 595 (macadamia canker; **54**, 576).
——. 1977. *Aust. J. exp. Agric. Anim. Husb.* **17**: 859 (control on pineapple; **57**, 2205).
Rands, R. D. 1922. *Meded. Inst. PlZiekt. Buitenz.* 54, 53 pp. (on cinnamon, *Phytophthora cinnamomi* sp. nov.; **2**: 246).
Young, B. R. 1970. *N.Z. Jl agric. Res.* **13**: 119 (inter alia on passion fruit; **49**, 2943).

Phytophthora citricola Sawada, *Rep. Govt Res. Inst. Dept Agric. Formosa* 27: 22, 1927.
P. cactorum (Leb. & Cohn) Schroet, var. *applanata* Chester, 1932.

HYPHAE 6 μ wide, without characteristic or frequent hyphal swellings. Sporangia absent or rare on the usual solid media, readily formed in water solution. SPORANGIOPHORES long and slender (up to 3 μ wide) often with an occasional swelling, particularly at the point of branching; branching irregular and lax, not regularly sympodial, and not often occurring immediately beneath the sporangium. SPORANGIUM obpyriform, obovoid or occasionally ellipsoid, often distorted in shape or skew on the sporangiophore or with 2 widely divergent apexes; apex broadly papillate with a shallow apical thickening up to 3 μ deep; not shed; cross wall at base of sporangium flush with the base and without a septal plug. SEX ORGANS abundant on most media. OOGONIA 17–44 (mostly 27–32) μ diam., spherical with a smooth, thin wall, pale yellow to yellowish brown according to the medium. ANTHERIDIUM always paragynous, often diclinous, and applied to the oogonium at any point on the circumference; antheridial cell small, 10–12 × 7.5 μ. OOSPORE nearly filling the oogonium, slightly eccentrically placed, wall 2–2.5 μ thick, colourless. Cultures radiate with narrow, lanceolate-shaped sectors; opt. temp. growth 25–28° (G. M. Waterhouse & J. M. Waterston, *CMI Descr.* 114, 1966).

P. citricola has also been fully described by Waterhouse who stated that the sporangia can be up to 47 × 78 μ; Sawada's means are 25 × 43 μ. It can be distinguished from *P. cactorum* (*CMI Descr.* 111), with which it has been compared, by the nature of the apex, size and shape of the sporangium, the fact that the sporangium is not shed and serologically. *P. citricola* has a widespread distribution (particularly in non-tropical regions) and it may be present more often than the records suggest (Map 437). It has been recorded on many crops including avocado, citrus and temperate crops, for example, black root rot of hop (*Humulus lupulus*) and a dieback of *Rhododendron*. It occurs in the soil and is splash dispersed. It survives as mycelium in host tissue and as oospores which have been germinated in as little as 2 weeks after their formation. Biotypes of differing virulence have been described from hops. *P. citricola* is a cause of brown rot and other citrus diseases whose control is described under *P. citrophthora*.

Burrell, R. G. et al. 1966. *Phytopathology* **56**: 422 (serological distinction from *Phytophthora cactorum*; **45**, 2729).

Chester, K. S. 1932. *J. Arnold Arbor.* **13**: 232 (on lilac & comparisons with *P. cactorum* & *P. syringae*; **11**: 579).

Doepel, R. F. 1966. *Pl. Dis. Reptr* **50**: 494 (distribution of *Phytophthora* spp. on citrus in W. Australia; **45**, 3533).

Henry, A. W. et al. 1968. *Can. J. Bot.* **46**: 1419 (oospore germination; **48**, 1087).

Krober, H. 1959. *Phytopathol. Z.* **36**: 381 (on *Rhododendron*; **39**: 318).

Royle, D. J. 1966. *Rep. Dep. Hop Res. Wye Coll.* 1965: 39 (review on hop; **45**, 3381c).

Salerno, M. et al. 1960. *Atti Ist. bot. Univ. Lab. crittogam Pavia* **18**: 222 (on tomato; **40**: 433).

Schwinn, F. J. 1962. *Phytopathol. Z.* **45**: 217 (morphological comparisons with *P. cactorum*; **42**: 309).

Waterhouse, G. M. 1957. *Trans. Br. mycol. Soc.* **40**: 349 (description, separation of *P. cactorum* & *P. citricola*: **37**: 148).

Zentmyer, G. A. et al. 1974. *Mycologia* **66**: 830 (on avocado; **54**, 2364).

Phytophthora citrophthora (Smith & Smith) Leon., *Am. J. Bot.* 12: 445, 1925.

Pythiacystis citrophthora Smith & Smith, 1906.

HYPHAE up to 6–7 μ. SPORANGIOPHORES in water delicate 1–2 μ wide, scarcely widening at base of sporangium, irregularly branched with a swelling at the point of branching. SPORANGIA rather scanty on some agar media, in water very variable in size and shape, often with two widely divergent apexes, av. 40–45 × 27 μ but often 50–55 × 30 μ, may be up to 60 × 90 μ; papilla scarcely hemispherical, apical thickening up to 4 μ; deciduous with a pedicel 10–12 μ long. CHLAMYDOSPORES absent or few, or frequent in old cultures, av. 28 μ diam., wall 1.5–2 μ thick. SEX ORGANS unknown. Cultures on cornmeal agar have a finely radiate appearance, sometimes with flame-like sectors; opt. temp. growth 24–28°. May be distinguished from *P. nicotianae* var. *parasitica* (q.v.), which sometimes produces indistinguishable symptoms on similar hosts, by the 4–5° lower opt. for growth and by growth rates on differential media (Dimitman et al.; G. M. Waterhouse & J. M. Waterston, *CMI Descr.* 33, 1964).

Fawcett (*Citrus diseases and their control*) should be consulted for a detailed discussion of the history, symptoms and early investigations of the various diseases on this crop group caused by *Phytophthora* spp. They are: foot rot, collar rot or gummosis of the main stem and crown roots, rot of the smaller roots, leaf and twig blight, and brown rot of the fruit. They are largely described here under *P. citrophthora* (widespread on *Citrus* spp., Map 35). The pathogen has been reported from many other hosts (e.g., temperate Rosaceae) but has been studied almost entirely on citrus. The other spp. associated with some or all of these diseases are: *P. citricola, P. hibernalis,*

P. nicotianae var. *parasitica* and *P. palmivora* (q.v.). The general differences in the various symptoms are due to host and growing conditions, and not to the pathogen. Temp. conditions affect the geographical distribution of these fungi.

Symptoms of foot rot often begin near the soil line; dark, watersoaked areas form in the bark and a sour smell may occur in wet conditions. Infection may spread to the crown roots and upwards 50 cm or more. The bark is killed through to the cambium and the wood which becomes necrotic. Gum exudes (there are other causes of this exudation) often in large quantities and it is more noticeable in dry weather. The bark remains firm and as it dries out large, vertical, and smaller horizontal, cracks appear in it; it breaks away leaving bare, dead areas. The extent of these stem lesions will partly depend on the degree of host susceptibility; for example, if a somewhat resistant stock has been used symptoms will begin above ground on the scion. The chlorotic effect on the canopy may be seen only weeks or months later. Attacks on the smaller permanent roots and fibrous feeder roots can lead to unthriftiness in the whole tree. Leaf infection (generally more serious in the lower part of the canopy) begins as translucent spots followed by necrosis, shoots are attacked and clumps of dead leaves appear. Defoliation can be severe. On the fruit nearer the ground small, greyish spots appear, they spread becoming dull, pale brown and a soft rot eventually develops. There is a characteristic pungent smell. White mycelial growth can occur on the surface of fruit in storage.

The fungus occurs in the soil and water used for irrigation; it can also infect the seed. In favourable conditions, i.e. long periods of wet weather, it can cause very extensive damage to citrus. It occurs frequently with the other *Phytophthora* spp. on this crop but is rather less common than *P. nicotianae* var. *parasitica* in warmer areas. Thus in Australia it tends to be more prominent in New South Wales than in Queensland; in USA it may be less abundant in Florida than in California; and is more destructive on the fruit in Israel than either *P. hibernalis* or *P. nicotianae* var. *parasitica*. Leaves, fruit and small branches are directly infected through splash dispersal, and the lower stem through infested soil and water. *P. citrophthora* is more virulent to fruit than *P. nicotianae* var. *parasitica* (Oxenham et al.; Whiteside, *P. nicotianae* var. *parasitica* q.v.). Burnett et al. suggested that races might occur.

An account of control measures was given by Knorr (*Citrus diseases and disorders*). A resistant stock should be used; its choice largely depends on cultural factors and also on virus diseases; selection for resistance to *Phytophthora* spp. may be a secondary consideration. Susceptible hosts are lemon (*C. limon*), sweet lime (*C. aurantifolia*), pummelo (*C. grandis*), sweet orange (*C. sinensis*) and grapefruit (*C. paradisi*). Some cvs of these have effective resistance, for example, rough lemon; also some mandarins (*C. reticulata*) and tangelo. Stocks with resistance are sour orange (*C. aurantium*), Cleopatra mandarin, kumquat, Carrezo and Troyer citrange. Trifoliate orange (*Poncirus trifoliata*) has high resistance and Broadbent et al. found Australian wild lime (*Microcitrus australis*) and box orange (*Severina buxifolia*) to be highly resistant. Variation in resistance within *Citrus* spp. occurs but it would appear that resistance does not vary much between different geographic areas. Klotz et al. (1965 & 1968) found that bark was a factor in resistance and, therefore, the need to avoid wounding resistant stocks remained important; also found was an effect of the scion on the susceptibility of the stock.

Other measures against foot rot are: avoid heavy soils and alkalinity, bud high (25–40 cm), plant high, soon after planting paint a susceptible stock with a neutral copper, clean weed near main stem, keep irrigation channels from trees, avoid excess nitrogen fertiliser. Copper or chlorine in water, when this is being applied, has been recommended. Young trees should be inspected for infection, more frequently in the first 2 years. If a stem lesion is found remove the diseased bark and a 1 cm strip of healthy bark all round (the exposed cambium should not be scraped) and paint with a suitable protectant. The canopy should be kept at least 1 m above soil level. Copper and captan sprays are used against leaf and fruit infection. Seed can be treated for 10 min at 51–52° and planted in disinfested beds. Control in fruit in the packing house requires immersion in water at 46–49° for 4 minutes or more.

Arentsen, S. 1942. *Boln Sanid. veg. Santiago* **2**: 54 (differential host susceptibility; **22**: 477).

Broadbent, P. et al. 1971. *Proc. Linn. Soc. N.S.W.* **96**: 119 (reaction of *Citrus* & related genera to stem inoculations; **51**, 2482).

Burnett, H. C. et al. 1974. *Pl. Dis. Reptr* **58**: 355 (pathogenicity of isolates from citrus & weed *Morrenia odorata*; **53**, 3972).

Cohen, E. et al. 1972. *Phytopathology* **62**: 932, 1361 (respiration of infected fruit & post infection changes in compounds; **52**, 1144, 2615).
——— ———. 1978. *Pl. Dis. Reptr* **62**: 386 (prevention of spread by contact in packed fruit).
Dimitman, J. E. et al. 1960. *Phytopathology* **50**: 83 (separating *Phytophthora* spp. from citrus in vitro; **39**: 410).
Fawcett, H. S. et al. 1940. *Arq. Inst. biol. S. Paulo* **11**: 107 (pathogenicity & temp. relations in Brazil; **20**: 400).
Fraser, L. 1942. *J. Aust. Inst. agric. Sci.* **8**: 101 (general in New South Wales; **22**: 133).
——. 1949. *Proc. Linn. Soc. N.S.W.* **74**: 5 (general & host reaction; **28**: 571).
Frezzi, M. J. 1940 & 1942. *Revta argent. agron.* **7**: 165; **9**: 216 (general in Argentina; **21**: 195; **23**: 294).
Klotz, L. J. et al. 1930. *J. agric. Res.* **41**: 415 (relative resistance of *Citrus* cvs & spp.; **10**: 98).
——. 1943 & 1950. *Calif. Citrogr.* **28**: 200, 220; **36**: 48, 67 (general survey including control; **22**: 354; **30**: 464).
——— ———. 1958. *Phytopathology* **48**: 616 (fibrous root decay, control; **38**: 258).
——— ———. 1960. *Calif. Citrogr.* **46**: 63 (control in citrus seed; **40**: 306).
——— ———. 1961. *Pl. Dis. Reptr* **45**: 264, 268 (hot water control in lemon fruit & infection through contact; **40**: 608).
——— ———. 1965. *Calif. Citrogr.* **50**: 143 (effect of rootstock on Lisbon lemon & Valencia orange in infested soil; **44**, 1875).
——— ———. 1968. *Pl. Dis. Reptr* **52**: 952 (factors in resistance; **48**, 1181).
——— ———. 1969. *Citrograph* **54**: 228, 279 (general control; **48**, 2389).
——— ———. 1971. *Phytopathology* **61**: 1342 (effect of *Phytophthora* spp. & aeration on root growth in orange seedlings; **51**, 2481).
——— ———. 1972. *Citrograph* **57**: 267 (control with Cu high & low volume; **51**, 4020).
Nadel-Schiffman, M. 1951. *Riv. Patol. veg. Pavia* **30**: 232 (effect of climate in Israel on *Phytophthora* spp. on citrus; **31**: 604).
——. 1956. *Ktavim* **6**: 111 (control on citrus with preharvest Cu sprays; **36**: 24).
—— et al. 1968. *Israel Jnl agric. Res.* **18**: 209 (control of brown rot in grapefruit with Cu; **48**, 1707).
——— ———. 1969. *Phytopathology* **59**: 237 (incubation time & temp. in brown rot of citrus; **48**, 2390).
Raphaela Musumeci, M. et al. 1975. *Summa Phytopathol.* **1**: 275 (phenols & phytoalexins in inoculated tissue; **55**, 2724).
Rossetti, V. 1948. *Arq. Inst. biol. S. Paulo* **18**: 97 (differential host susceptibility; **28**: 213).
Smith, R. E. 1907. *Bull. Calif. agric. Exp. Stn* 190, 72 pp. (general, historical).
Stolzy, L. H. et al. 1960. *Calif. Citrogr.* **45**: 66, 76 (effect of irrigation; **39**: 704).
——— ———. 1965. *Phytopathology* **55**: 270 (water & aeration in root decay, **44**, 2142).
Yamamoto, S. 1968. *Shokubutsu Bôeki* **22**: 249 (control with Cl in sprinkler irrigation; **50**, 3748).

Phytophthora colocasiae Raciborski, *Parasit. Algen u. Pilze, Javas I, Batavia*: 9, 1900.

HYPHAE 3–4 μ. SPORANGIOPHORES slender (2–4 μ), on the leaf rather short, sometimes scarcely longer than the sporangia and unbranched; in culture branching is irregular to sympodial with a swelling at the point of branching. SPORANGIA elongated ellipsoidal, sometimes almost fusiform, mostly 45–60 × 23 μ, but up to 70 × 28 μ, length : breadth ratio 1.6; papilla projecting but not markedly so; deciduous, pedicel slender, 3.5–10 μ long. CHLAMYDOSPORES usually produced in culture, sometimes abundantly, usually 26–30 (up to 39) μ diam., wall up to 3 μ thick, yellow. OOGONIA usually develop in culture, av. 29 (up to 35) μ diam., becoming yellow. ANTHERIDIA 11–13 × 11 μ, predominantly or all amphigynous. OOSPORE loose in oogonium, wall up to 2.5 μ thick. Culture uniformly fluffy. Opt. temp. growth 27–30° (from G. M. Waterhouse 1963; *Phytophthora* q.v.).

This sp. is one of the more distinctive of the tropical members of the genus with its narrowly elongate sporangia and apparent restriction as a pathogen to *Colocasia* spp., and probably related genera. It is widespread in S. and E. Asia and parts of Oceania (where it can be severe); records from the Dominican Republic, Ethiopia and Fernando Po may need confirmation (Map 466). Severe disease has been reported in *C. esculenta* and *Xanthosoma sagittifolium*; foliage attack (the more obvious symptom) can lead to reduced rhizome yield. The small, dark necrotic leaf spots enlarge, becoming water-soaked, more or less circular, purplish then brown and sometimes with concentric markings. In wet weather the lesions may show a yellowish exudate. They are often most extensive at the leaf margins; their centres may fall out and coalescence leads to death of the leaf. Spread to the petiole and more rarely to the rhizome may occur. Penetration is cuticular and indirect sporangial germination is opt. at *c*. 27°. *Choanephora cucurbitarum* and *C. trispora* have been described as secondary invaders of the leaf spots on *Colocasia*.

Control with copper fungicides can be effective and in Hawaii mancozeb gave control. In the Solomon Islands Gollifer et al. found poor control with copper although the fungicide treatment gave increases in yield. Some, apparently slight, resistance has been found in *Colocasia* cvs and was reported in *X. sagittifolium* by Gomez.

Bergquist, R. R. 1974. *Ann. Bot.* **38**: 213 (fungicides in Hawaii; **53**, 4936).
Butler, E. J. et al. 1913. *Mem. Dep. Agric. India bot. Ser.* **5**: 233 (general).
Deshmukh, M. J. et al. 1960. *Curr. Sci.* **29**: 320 (resistance in *Colocasia*; **40**: 262).
Gollifer, D. E. et al. 1974. *Papua New Guin. agric. J.* **25**: 6 (in Solomon Islands & control; **55**, 2452).
Gomez, E. T. 1925. *Philipp. Agric.* **14**: 429 (general; **5**: 341).
Hicks, P. G. 1967. *Papua New Guin. agric. J.* **19**: 1 (testing for resistance in *Colocasia*; **47**, 2028).
Paharia, K. D. et al. 1964. *Sci. Cult.* **30**: 44 (as above; **43**, 2479).
Sinha, S. 1940. *Proc. Indian Acad. Sci.* Sect. B **11**: 167 (*Choanephora* as secondary invader; **19**: 514).
Trujillo, E. E. et al. 1964. *Hawaii Fm Sci.* **13**: 3 (control with Cu fungicide; **45**, 1274).
——. 1965. *Phytopathology* **55**: 183 (effects of RH & temp. on sporangial formation & germination; **44**, 2014).

Phytophthora cryptogea Pethybridge & Lafferty, *Scient. Proc. R. Dubl. Soc.* **15**: 498, 1919.

HYPHAE of very uneven width (up to 8 μ); MYCELIUM in water culture with groups of angular swellings (av. 11 μ, max. 20 μ), with 2 or 3 branch hyphae which soon swell in their turn and branch giving a net like appearance. SPORANGIOPHORES 2–3.5 μ, proliferating through the empty sporangium, occasionally sympodially from below. SPORANGIA not formed in agar culture, only in liquid media; regularly ovoid or obpyriform, non-papillate, without obvious apical thickening, apex flattening on mounting; often with a conspicuous central vacuole; av. 23 × 37–40 (max. 30 × 55) μ, length : breadth ratio 1.7. SEX ORGANS infrequent and only after a few weeks in single str. culture but develop promptly and abundantly when grown with a complementary str. or with *P. cinnamomi*. OOGONIA av. 30–32 (max. 38) μ diam., wall becoming yellow but not much thicker with age. ANTHERIDIA spherical, 10 μ diam. (max. 16 μ), or occasionally oval or short cylindrical, mostly amphigynous. OOSPORE nearly filling the oogonium wall. In culture uniform and fairly fluffy. Opt. temp. growth 22–25° (from G. M. Waterhouse 1963, *Phytophthora* q.v.; and see Stamps).

Waterhouse (l.c.) kept this sp. separate from *P. drechsleri* (q.v.). *P. cryptogea* has smaller sporangia and oogonia, and a lower opt. temp. for growth in vitro, compared with *P. drechsleri*. Bumbieris (1974) considered the 2 fungi to be 1 sp. *P. cryptogea* is plurivorous, occurring on temperate ornamental crops and other plants (see Moore, *British parasitic fungi*, Middleton et al. and Tomkins for hosts). The distribution is largely temperate: Argentina, Australia, Canada (British Colombia, Ontario), Egypt, Europe, Iran, New Zealand, Zimbabwe and USA (Map 99).

The diseases caused are root rots and foot rots; *P. cryptogea* may be associated with *P. erythroseptica* Pethybr. in pink rot of tubers of Irish potato; in tulip the disease is called shanking. Infection hardly ever appears to be serious if normal soil sterilisation is carried out in plant beds. Upstone et al. found that tomato grafted on stocks which had resistance to other pathogens out-yielded ungrafted plants grown in soil heavily contaminated with *P. cryptogea*. These stocks developed a resistance to this pathogen 2 months after planting. He found that some control was given by methyl bromide although its effect was not lasting. The highest yields were given by the grafted plants in soil treated with methyl bromide. This sterilant has been used against a foot rot of *Gerbera* and control by watering with fenaminosulf has also been reported. Bumbieris (1976) described the role of the fungus in a decline of *Pinus radiata*. It is associated with a root rot of safflower (*Carthamus tinctorius*); this disease has been largely attributed to *P. drechsleri* (Duniway; Klisiewicz). Csinos et al. described a phytoxin from *P. cryptogea* in work on a non-parasitic stunting of tobacco.

Brien, R. M. 1940. *N.Z. Jl Sci. Technol.* Sect. A **22**: 232 (foot rot of tomato; **20**: 501).
Buddin, W. 1938. *Ann. appl. Biol.* **25**: 705 (shanking of tulip with *Phytophthora erythroseptica*; **18**: 183).
Bumbieris, M. 1974. *Aust. J. Bot.* **22**: 655 (characteristics of *P. cryptogea* & *P. drechsleri*; **54**, 3755).
——. 1976. *Ibid* **24**: 703 (on *Pinus radiata*; **56**, 4232).
Csinos, A. et al. 1977. *Can. J. Bot.* **55**: 26, 1156 (non-parasitic stunting of tobacco & phytotoxin; **56**, 3213; **57**, 289).

Phytophthora drechsleri

Duniway, J. M. 1977. *Phytopathology* **67**, 331, 884 (on safflower, resistance to water transport & effect of water stress; **56**, 5134; **57**, 732).

Jacks, H. 1951. *N.Z. Jl Sci. Technol.* Sect. A **33**: 71 (control with cuprox on tomato; **31**: 522).

Klisiewicz, J. M. 1977. *Phytopathology* **67**: 1174 (on safflower, identity & virulence of *Phytophthora* spp.; **57**, 4077).

Middleton, J. T. et al. 1944. *J. agric. Res.* **68**: 405 (on *Sinningia* & other hosts; **23**: 390).

Scholten, G. 1970. *Neth. J. Pl. Pathol.* **76**: 212 (on *Gerbera*; **49**, 3338).

Stamps, D. J. 1978. *C.M.I. Descr. pathog. Fungi Bact.* 592 (*P. cryptogea*).

Tompkins, C. M. 1937. *J. agric. Res.* **55**: 563 (on ornamentals & other hosts; **17**: 181).

Upstone, M. E. et al. 1966. *Pl. Pathol.* **15**: 15 (control in tomato by grafting on resistant stocks; **45**, 2246).

——. 1968. *Ibid* **17**: 103 (control as above & with methyl bromide; **48**, 283).

Phytophthora drechsleri Tucker, *Res. Bull. Mo. agric. Exp. Stn* 153: 188, 1931.

P. erythroseptica var. *drechsleri* (Tucker) Sarejanni, 1936

Pythium teratosporon Sideris, 1931.

HYPHAE fairly uniform, not often > 5 μ wide; MYCELIUM in liquid cultures sometimes forming groups of swellings similar to those of *P. cryptogea*. SPORANGIOPHORES narrow (1–2 μ), sometimes widening slightly below the sporangium; usually simple with a single terminal sporangium, resuming growth through the base of the evacuated sporangium and producing new sporangia within or beyond the walls of the empty one. SPORANGIA very variable in shape from broadly obpyriform to elongated obpyriform, sometimes asymmetrical and sometimes tapering at the base, non-papillate, apical thickening barely detectable or a narrow crescent, apex often flattening on mounting; collapsing somewhat after dehiscence; 15–24 × 24–38 μ (Tucker l.c.); 26–30 × 36–50 (max. 40 × 70) μ (Waterhouse). SEX ORGANS sometimes formed promptly and abundantly in single culture, but if not, will produce them if grown with an opposite str., but not with *P. cinnamomi*. OOGONIA 36 (max. 53) μ diam., broadly clavate to subspherical, smooth, hyaline to light amber, becoming darker with age. ANTHERIDIA oval or cylindrical, amphigynous, 13 × 14–15 μ. OOSPORE spherical, smooth, nearly filling the oogonium, wall 3 μ thick, 17–45 (av. 25–26) μ. Opt. temp. growth 30–32.5° (Tucker), 28–31° (Waterhouse; from C. M. Tucker l.c., and G. M. Waterhouse, 1963, the latter under *Phytophthora*. See Cother et al. 1973b, 1974, for chlamydospores).

P. drechsleri belongs to the group in the genus with non-papillate sporangia and usually amphigynous antheridia. It is distinguished further by the rather small antheridia and by its high opt. temp.; there is growth of 5 mm or more in 24 hours at 34–36°. It has larger sporangia and oogonia than *P. cryptogea* (q.v.). Originally described from Irish potato *P. drechsleri* causes diseases on a variety of hosts. The pathogen's distribution is wide although reports of its occurrence in different countries are moderate: Argentina, Australia, Brazil, Canada, Colombia, Egypt, Greece, Hawaii, Iran, Japan, Lebanon, Malagasy Republic, Malaysia (Sarawak), Mexico, New Zealand, Papua New Guinea, Zimbabwe and USA (Map 281). Tompkins described it causing a tap root rot of sugar beet. From Japan and Iran diseases of Cucurbitaceae have been described (Katsura; Ershad); these result in damping off, watersoaked fruit lesions, stem, root and leaf necrosis. In Malagasy Republic Bouriquet et al. described a stem canker of a shade tree (*Albizia chinensis*) used in coffee. Brown, irregular areas form beneath the bark at or above ground level; there is a wet, cracked appearance and gum (red turning brown) is formed. The canker should be cut out and a fungicidal paint used. A stem and leaf infection of velvet bean (*Stizolobium* spp.) described from Australia (Queensland, Sturgess et al.) was considered to have arisen from the import of infected seed from Africa where *P. drechsleri* had been reported on this host. Other hosts include cassava (root rot, Oliveros et al.); eucalyptus (Weste); and *Pinus radiata* (Heather et al.). In Australia Cother et al. (1973a) found chlamydospores in common weeds and concluded that rotations would be of little value as a control measure. Shepherd et al. divided isolates from Australian forests into 2 ecotypes.

Most work has been done on the root rot caused in safflower (*Carthamus tinctorius*); in 1951 Thomas referred to extensive losses on this crop in irrigated areas of California, USA. Infection of the lower stem leads to the collapse of young plants and plants near flowering can also be killed. Infection of the roots may reduce yield without causing shoot symptoms. More disease occurred at 25° and 30° compared with 20° and 35° (Erwin; and see Thomas et al. 1977). Irrigation frequency and intensity also differentially affected root infection of cvs (Zimmer et al.). *P. drechsleri* seems generally heterothallic with 2 compatibility types although homothallic isolates were obtained in segregation work (Mortimer et al. 1977). Sansome discussed the ratios obtained in genetical work with oospores by Galindo et al., and auxotrophic mutants (tryptophan dependant) were described by Castro et al. Abundant oospores are produced at 15–24° but they develop more rapidly and germinate best at 24°; those produced in the host germinate better than those produced in vitro. More oospores occur in continuous dark compared with continuous light. Light or dark does not affect germination of oospores previously exposed to light in vitro but light is required for good germination of those produced in the dark (Klisiewicz). Some isolates of *P. drechsleri* from safflower and sugar beet are not host specific and can cause root rot in both hosts.

Allen, Thomas and others have reported studies on the effects of anti-fungal substances produced by both resistant and susceptible safflower cvs when these are infected. In the resistant cv. Biggs these substances reach higher levels than in a susceptible cv. and in 48–96 hours are in sufficient conc. to inhibit growth of the pathogen. Resistance in Biggs is controlled by a single recessive gene. There are indications that pathogenicity varies between isolates but no races seem to have been demonstrated. In USA, where safflower is grown under irrigation, cvs with some resistance are available (e.g., Thomas 1976; Thomas et al. 1960, 1976, 1977).

Allen, E. H. et al. 1971. *Phytochemistry* **10**: 1579 (identification of an anti-fungal polycetylene).
——. 1971. *Physiol. Pl. Pathol.* **1**: 235 (time & accumulation of anti-fungal substances; **51**, 564).
——. 1971 & 1972. *Phytopathology* **61**: 1107; **62**: 471 (anti-fungal substances, levels in resistant & susceptible cvs; **51**, 1766, 4259).
Barash, I. et al. 1965. *Ibid* **55**: 1257 (use & effect of C compounds on zoospores; **45**, 1121).
Bouriquet, G. et al. 1959. *Agron. trop. Nogent* **14**: 711 (bark canker of *Albizia chinensis*; **39**: 442).
Castro, F. J. et al. 1971. *Phytopathology* **61**: 283 (induction of auxotrophic mutants; **50**, 2777).
Cother, E. J. et al. 1973a. *Aust. J. biol. Sci.* **26**: 1109 (role of alternative hosts in survival; **53**, 1907).
——. 1973b. *Trans. Br. mycol. Soc.* **61**: 379 (formation of chlamydospores; **53**, 2428).
——. 1974. *Ibid* **63**: 273 (germination of chlamydospores; **54**, 1390).
——. 1975. *Aust. J. Bot.* **23**: 87 (host range & pathogenicity; **54**, 5355).
Duniway, J. M. 1975. *Phytopathology* **65**: 886, 1089 (water relations in safflower; effect of low water potential on sporangia in soil; **55**, 829, 2135).
——. 1975. *Can. J. Bot.* **53**: 1270 (formation of sporangia in soil at high matric potentials; **55**, 1095).
Ershad, D. 1969. *Iran, Jnl Pl. Pathol.* 5: 38 (root rot in cucurbits; **49**, 1530).
Erwin, D. C. 1952. *Phytopathology* **42**: 32 (root rot of safflower; **31**: 351).
Galindo, J. A. et al. 1967. *Ibid* **57**: 1300 (segregation from oospores & cytology of *P. drechsleri*, tentative identification; **47**, 1221).
Heather, W. A. et al. 1975. *Aust. J. Bot.* **23**: 285 (on *Pinus radiata*; **55**, 919).
Johnson, L. B. et al. 1969. *Phytopathology* **59**: 469 (effect of inoculum conc., light & temp. on infection of safflower; **48**, 3099).
——. 1970. *Ibid* **60**: 534, 1000 (symptom development & resistance, anti-fungal substances in safflower; **49**, 2958; **50**, 1331).
Katsura, K. 1958. *Sci. Rep. Fac. Saikyo Univ.* 10: 77 (on *Citrullus lanatus*; **39**: 146).
Klisiewicz, J. M. et al. 1968. *Phytopathology* **58**: 1022 (histology in resistant & susceptible safflower; **47**, 3527).
——. 1970. *Ibid* **60**: 1738 (factors affecting production & germination of oospores; **50**, 2776).
Mortimer, A. M. et al. 1975. *Genet. Res.* **25**: 201 (evidence for meiosis in gametangia; **55**, 3052).
——. 1977. *Arch. Microbiol.* **111**: 255 (genetics of secondary homothallism; **56**, 5481).
Oliveros, B. et al. 1974. *Pl. Dis. Reptr* **58**, 703 (root rot of cassava; **54**, 1088).
Pratt, B. H. et al. 1974. *Aust. J. Bot.* **22**: 9 (pathogenicity of isolates from forests to 3 crops; **53**, 3875).
Sansome, E. 1970. *Trans. Br. mycol. Soc.* **54**: 101 (selfing, a possible cause of disturbed ratios in *Phytophthora*; **49**, 2351).
Schneider, C. L. et al. 1965. *Pl. Dis. Reptr* **49**: 293 (comparison of isolates from sugar beet & safflower; **44**, 2431).
Shaw, D. S. et al. 1971. *Genet. Res.* **17**: 165 (genetical evidence for diploidy; **52**, 48).
Shepherd, C. J. et al. 1973. *Aust. J. biol. Sci.* **26**: 1095 (ecotypes from forests; **53**, 1962).
Stovold, G. 1973. *Aust. J. exp. Agric. anim. Husb.* **13**: 455 (with *Pythium* spp. on safflower; **53**, 643).
Sturgess, O. W. et al. 1960. *Tech. Commun. Bur. Sug. Exp. Stns Qd* 2: 9 (wilt of *Stizolobium*; **39**: 717).
Thomas, C. A. 1951. *Pl. Dis. Reptr* **35**: 454 (*Phytophthora* spp. on safflower; **31**: 206).
—— et al. 1960. *Phytopathology* **50**: 129 (resistant cvs in safflower; **39**: 483).

Thomas, C. A. et al. 1963. *Ibid* **53**: 368 (variation in pathogenicity; **42**: 625).
——. 1966. *Ibid* **56**: 985 (Ca, pectic substances & resistance in safflower; **45**, 3586).
—— ——. 1970. *Ibid* **60**: 63, 261, 1153 (resistance & anti-fungal substances; **49**, 2152, 2580; **50**, 1332).
—— ——. 1971. *Ibid* **61**: 1459 (light & anti-fungal substances; **51**, 2746).
——. 1976. *Pl. Dis. Reptr* **60**: 123 (inheritance of resistance in safflower; **55**, 4807).
—— ——. 1977. *Phytopathology* **67**: 698 (reaction of safflower cotyledons, effects of temp. & inheritance; **57**, 240).
Timmer, L. W. et al. 1970. *Mycologia* **62**: 967 (pigment production & inheritance; **50**, 2700).
Tompkins, C. M. et al. 1936. *J. agric. Res.* **52**: 205 (on sugar beet; **15**: 550).
Weste, G. 1975. *APPS Newsl.* **4**: 38 (on eucalyptus; **55**, 2910).
Yermanos, D. M. et al. 1960. *Agron. J.* **52**: 596 (effects of gibberellic acid on safflower; **40**: 375).
Zimmer, D. E. et al. 1967. *Phytopathology* **57**: 1056 (effect of irrigation & cvs of safflower on disease incidence; **47**, 592).

Phytophthora heveae Thompson, *Malay agric. J.* 17: 77, 1929.

HYPHAE mostly av. 4.5 μ (up to 7 μ); hyphal swellings occasional, the larger look like abortive reproductive organs. SPORANGIOPHORES 3.5 μ wide, irregularly branched; swellings may occur some distance behind the sporangium. SPORANGIA profuse on agar, often irregular in shape, asymmetric or bent; ovoid, ellipsoid or broadly obpyriform, 20–48 (29) × 27–66 (46) μ, length: breadth ratio 1.1–2.9 (mostly 1.4–1.5); deciduous with pedicels up to 10 μ long; papilla conspicuous with apical thickening 5 μ deep. CHLAMYDOSPORES absent. SEX ORGANS develop promptly on agar. OOGONIA often arise in clusters, spherical above but usually tapering to the stalk, 25 × 28 (up to 32 × 35) μ diam.; wall thin, colourless or slightly yellow. ANTHERIDIA small, spherical (9 μ diam.) or ellipsoid (9 × 1.5 μ), amphigynous. OOSPORES loose in the oogonium, 21 μ diam., wall colourless, 3 μ thick. In agar culture the numerous sporangia give a frosted look. Opt. temp. growth 25°. Characterised by: the abundant sex organs in single strain culture (also abundant in host), oogonia tapering to the stalk and rather small (as are the antheridia), rather gross irregular sporangia, lack of chlamydospores and low opt. temp. for growth (after G. M. Waterhouse, 1974: 71, *P. palmivora* q.v.; and see Stamps).

A little known sp. in the field and which has a marked homothallism. It has been reported only from Australia (New South Wales), Brazil, Guatemala, Malaysia (W. and Sabah), New Zealand and USA (Map 428). When originally described *P. heveae* was thought to be one of the causes of black stripe in rubber (*Hevea brasiliensis, P. palmivora* q.v.); see Chee for pathogenicity to this crop. It has also been recorded on cacao, Brazil nut (*Bertholletia excelsa*), eucalyptus and avocado (canker). Since the spore forms are readily produced in single str. culture *P. heveae* has been used to study the effects of light, temp., sterols and nitrogen on sporulation. Light stimulates asexual reproduction but inhibits the formation of oogonia. Dark is required for the maturation of oogonia but their germination is stimulated by light. The opt. temp. for oogonia production is 25° but sporangia are only formed above 27.5°. Brasier suggested that oospores may be active in dispersal, a possibly rare phenomenon in the *Phytophthora* spp. with relatively high temp. requirements.

Albuquerque, F. C. de et al. 1974. *Pesqui. Agropecu. Bras. Agron.* **9**: 101 (on Brazil nut; **55**, 5369).
Berg, L. A. et al. 1966. *Phytopathology* **56**: 583 (light & oospore germination in *Phytophthora* spp.; **45**, 3263a).
Brasier, C. M. 1969. *Trans. Br. mycol. Soc.* **52**: 105 (effect of light & temp. on reproduction with *P. palmivora*; **48**, 1533).
Chee, K. H. 1970. *J. Rubb. Res. Inst. Malaya* **23**: 13 (on rubber; **50**, 1946).
Gerrettson-Cornell, L. 1976. *APPS Newsl.* **5**: 8 (on eucalyptus with *P. boehmeriae*; **55**, 4876).
Gomez-Miranda, B. et al. 1965. *Microbiol. esp.* **18**: 235 (effect of light, sterol & N on asexual reproduction with *Phytophthora cactorum* & *Pythium* spp.; **45**, 1309).
Leal, J. A. et al. 1965. *Trans. Br. mycol. Soc.* **48**: 491 (effect of light on germination of oospores with *Phytophthora cactorum* & *P. erythroseptica*; **45**, 372).
—— ——. 1968. *Proc. W. Va Acad. Sci.* **40**: 47, 73 (L-glutamine, urea, ethanol & growth; **48**, 2150).
Stamps, D. J. 1978. *C.M.I. Descr. pathog. Fungi Bact.* 594 (*P. heveae*).
Thompson, A. 1929. *Malay. agric. J.* **17**: 53 (*Phytophthora* spp. in Malaysia, W.); 8: 674).
Turner, P. D. 1968. *Pl. Prot. Bull. F.A.O.* **16**: 33 (record on cacao; **47**, 3060b).
Zentmyer, G. A. et al. 1978. *Pl. Dis. Reptr* **62**: 918 (in Guatemala on avocado).

Phytophthora hibernalis Carne, *J. R. Soc. W. Aust.* **12**: 36, 1925.

HYPHAE av. 4 (up to 7) μ, rather undulate but without hyphal swellings or chlamydospores. SPORANGIOPHORES 1–1.5 μ wide, unbranched or producing long branches in an irregular sympodium. SPORANGIA elongated, ellipsoid or ovoid, often with the broadest part nearer the apex, av. 19 × 36 (max 28 × 56) μ, deciduous, with a long pedicel up to 55 μ long; apex scarcely papillate, but rounded, never flattened, with evident but shallow thickening 1–2 μ deep; after dehiscence a translucent globule is often left in the sporangium. OOGONIA readily produced on agar, av. 35 (max. 56) μ diam., usually spherical but sometimes slightly elongated towards the stalk, wall thin, colourless or faintly yellow with age. ANTHERIDIA usually amphigynous, sometimes paragynous, 12 (–14) × 10 μ. OOSPORE nearly fills the oogonium, wall 3 μ thick.

A low temp. fungus with an opt. near 20°. Cultures form rose pattern growth zones. Distinguished from *P. syringae* (*CMI Descr.* 32, with which it has sometimes been confused) by the absence of hyphal swellings, by the elongated shape of the sporangia which are deciduous, and by the predominantly amphigynous antheridia. On *Citrus limon, C. medica, C. paradisi, C. reticulata* and *C. sinensis* (G. M. Waterhouse & J. M. Waterston, *CMI Descr.* 31, 1964).

P. hibernalis causes brown rot of fruit, and leaf and twig blight of citrus, whilst *P. syringae* causes disease on several temperate hosts as well as infecting citrus. The reported distribution of *P. hibernalis* (Map 47) is: Argentina, Australia, France, Greece, Israel, Italy, New Zealand, Portugal, South Africa, Spain, Turkey, UK, USA (California) and Venezuela. Its low temp. requirements presumably limit its distribution and possible records from Fiji and Jamaica are doubtful. Isolates from lemon caused rots in eggplant, tomato and other fruit. Spread is by splash dispersal but since the sporangia are reported to be deciduous they may be air dispersed as well, although no specific studies on this aspect appear to have been reported. *P. hibernalis* is a factor in the *Phytophthora* canopy diseases of citrus when temps are low but frequently *P. nicotianae* var. *parasitica* (q.v.) and *P. citrophthora* (q.v.) are more important. Symptoms and control are described under the last sp. Although infection of the main stem in citrus can occur any role that *P. hibernalis* may play in the foot rot or stem gummosis syndrome seems slight.

Carne, W. M. 1925. *Jl R. Soc. West. Aust.* **12**: 13 (general; **5**: 295).
Pittman, H. A. 1932. *J. Dep. Agric. West. Aust.* **9**: 286 (citrus brown rot with *P. citrophthora*; **11**: 779).
Schiffman-Nadel, M. 1947. *Palest. J. Bot. Rehovot Ser.* **6**: 148 (morphology, temp. & growth in vitro, inoculations; **28**: 122).
Verneau, R. 1954. *Annali Sper. agr.* **9**: 133 (on citrus in Italy with other *Phytophthora* spp.; **34**: 717).

Phytophthora infestans (Mont.) de Bary, *Jl R. agric. Soc.* **12** Ser. 2(1): 240, 1876.
Botrytis infestans Montagne, 1845
Peronospora infestans (Mont.) Caspary, 1854.

SPORANGIOPHORES differentiated from the mycelium (in the host) by being broader and having a small swelling at the point of formation of each sporangium. SPORANGIA abundant on the host and on solid media, ellipsoid, ovoid or (when shed) limoniform, with a tendency to taper to the base, 19 × 29 (max. 31 × 59) μ, deciduous, pedicel short (up to 3 μ); papilla not very protuberant, apical thickening less than hemispherical, usually *c.* 3–3.5 μ. SEX ORGANS rare in host and in culture unless opposite strs are grown together. OOGONIA 38 (max. 50) μ diam., spherical to obpyriform, wall smooth, 1–2 μ thick, becoming reddish brown and thicker by an accretion from the medium. ANTHERIDIA not always present but always amphigynous when they are, 17 (max. 22) × 16 μ. OOSPORES 30 μ diam., loose in oogonium, wall colourless, up to 3–4 μ thick. Slow growing (4 mm)/day at opt. temp. in culture; opt. temp. growth 20°.

This description is taken from G. M. Waterhouse (1963, *Phytophthora* q.v.) who divided *P. infestans* into 2 f. sp., *infestans* (on Solanaceae) and *thalictri* (on *Thalictrium*). *P. infestans*, which is worldwide (Map 109), causes the classical and extensively investigated late blight of Irish potato and tomato. There is no monograph on this extremely important disease. Short accounts (mainly on Irish potato) can be found in: Butler and Jones, *Plant pathology*, Walker, *Plant pathology*, and D. H. Lapwood et al. in *Diseases of crop plants* (Western, editor). There are reviews on some aspects of late blight: social history (Carefoot & Sprott, *Famine on the wind*; Large, *The advance of the fungi*); epidemiology (Cox et al.; Van de Zaag); disease forecasting (Miller et al. 1957); host and parasite interaction, and genetic control (Gallegly et al. 1959; Gallegly 1968); tomato late blight (Miller et al. 1955; Turkensteen). Access to the very large literature should be made through these general accounts. The disease on Irish potato is not treated here, and late blight, although it may be present, has not so far been severe in the lowland tropics. The disease probably reaches its highest incidence at 13–19°, and above 25° it is much less common. Climatic seasons favourable for late blight are most common in maritime, temperate areas and in upland regions of the subtropics and tropics where much cloud cover and mists occur. But see Palti et al. for the special factors in semi-arid conditions where irrigation is used.

Penetration of the leaf by the zoospore is cuticular; in tomato fruit it probably occurs through the stalk or calyx end. Symptoms are usually first seen in the field as ill-defined, rapidly spreading, blackish lesions on the leaves; on the under surface sporangia form in damp conditions towards the periphery of the lesions. The pathogen spreads to petioles and stems causing complete above ground collapse. On tomato fruit symptoms can occur at any age; grey green, watersoaked spots enlarge to give ill-defined lesions; on green fruit the lesions become dark brown and somewhat wrinkled; in transit a soft, wet rot may develop. In vitro most sporangia form at 21°; above 20° they lose viability in 1–3 hours in dry air and in 5–15 hours in moist air. Opt. temp. for indirect germination is *c.* 13° and for direct is *c.* 24°. Zoospore motility is 15 minutes at 24° and *c.* 24 hours at 1–2°; 70% of zoospores germinate between 6° and 24°; germ tube elongation is most rapid at 21–24°. At 20–23° the first appearance of visible lesions occurs in 66–82 hours (Crosier). Host penetration requires a water film for 2–5 hours depending on temp. and at least RH 90% for sporangial formation. The sporangia are strongly deciduous with a diurnal periodicity (max. dispersal in the forenoon; Hirst). In Irish potato the surface soil remains infective for at least 32 days after the haulm has been killed with acid. Sporangia mixed with unsterilised soil (moisture content 20%) are infective for 11 weeks. Survival is largely through infected tubers, although where tomatoes are grown (in the summer gap between Irish potato crops) they may carry the fungus over. Abundant oospores occur naturally only in Mexico where the 2 compatibility types are found and where *P. infestans* may have its origin. Elsewhere only one type (A_1) apparently occurs. In Mexico the A_1 and A_2 types are equally distributed between isolates. The oospore is, therefore, of chief interest as a mechanism for variation (Gallegly 1968; Romero et al.) and not as a means of survival where late blight occurs.

Resistance, of the specific type, has a long history of investigation in Irish potato; less has been done in tomato (see Turkensteen). This resistance is controlled by single dominant genes. Four different R (resistance) genes, and the 4 races which could attack these genotypes, were first studied in *Solanum* (see Black). This gave 16 genotypes in addition to race 0 which will only attack hosts without any R genes. Ten R genes are known. *P. infestans* is now known to be so variable (mostly, it is considered, by mutation) that hope for control in Irish potato through specific resistance only has been abandoned.

The fungus passes naturally between the 2 main hosts and virulence varies both within a single race and according to the race genotype; see Graham et al. for an example of race selection induced by serial passage through foliage of resistant hosts. Kishi (who also did serial passage experiments) divided Irish potato and tomato isolates into 3 groups of differing pathogenicity to tomato. In the tomato isolates 68% were highly pathogenic to this host but only 4% of Irish potato isolates were so. Wilson et al. divided 29 Irish potato isolates (representing 10 races) into 4 pathogenicity groups with respect to tomato and depending on the host genotype. The races on Irish potato may be non-pathogenic to tomato or pathogenic to cvs with and without the gene Ph_1 which confers resistance to tomato race 0. Isolates of different races from either host may behave as races of the other host. There is, apparently, no close relationship between the particular races from each host and, therefore, the race characteristics of any given isolate should be considered independently with respect to host. Gallegly (1968) referred to 2 tomato races and Palti et al. (1963) to 4. Commercial tomato cvs differ in reaction to infection (Andrus; Bonde et al.; Dowley et al.; Eggert; Laterrot; Richards et al.); but in many areas adequately resistant cvs are not available. Non-specific resistance is known in both hosts (Grümmer, Gunther and others). Eggert (1970) pointed out the need to test for both tomato leaf and fruit resistance. Sources of resistance other than in *Lycopersicon esculentum* have been investigated (e.g., Glushchenko; Walter et al.) and found in the var. *cerasiforme*.

In many areas the weather pattern which precedes outbreaks of late blight is well established and warning systems are in operation; methods used for prediction vary (Bourke; De Weille; Large; Smith et al.; Wallin et al. 1960). In UK prediction is based on the occurrence, frequency and distribution of Beaumont periods (48 hours when min. temp. is not $<10^\circ$ and RH remains at $>75\%$); copper, dithiocarbamates and organo-tin compounds are effective fungicides which should be applied at 10-day intervals. In some areas protection of tomato fruit is more important than that of the foliage.

Andrus, C. F. 1946. *Pl. Dis. Reptr* **30**: 269 (screening of 46 cvs of tomato after the 1946 epidemic; **26**: 136.

Anon. 1967. *Advis. Leafl. Minist. Agric. Fish. Fd* 271, 8 pp. (advisory; **46**, 3542).

Black, W. 1952. *Proc. R. Soc. Edinb.* Sect B **64**: 312; **65**: 36 (inheritance of resistance & race classification; **33**: 47).

—— et al. 1953. *Euphytica* **2**: 173 (nomenclature of *Solanum tuberosum* R genes & races; **33**: 250).

Bonde, R. et al. 1952. *Bull. Maine agric. Exp. Stn* 497, 15 pp. (resistance in tomato; **32**: 154).

Bourke, P. M. A. 1957. *Tech. Note Rep. Ind. Comm. Meteor. Serv. Dublin* 23, 35 pp. (forecasting; **37**: 553).

Cox, A. E. et al. 1960. *Agric. Handb. U.S. Dep. Agric.* 174, 230 pp. (world review of epidemics on Irish potato, 325 ref.; **39**: 612).

Crosier, W. 1934. *Mem. Cornell Univ. agric. Exp. Stn* 155, 40 pp. (biology; **13**: 724).

De Weille, G. A. 1963. *Eur. Potato J.* **6**: 121 (weather & epidemiology; **43**, 542).

Dowley, L J. et al. 1975. *Phytopathology* **65**: 1422 (ontogenetic predisposition of tomato foliage to infection by race 0; **55**, 3291).

Eggert, D. 1970. *Phytopathol. Z.* **67**: 112 (infection, temp. & resistance in tomato; **49**, 2649).

——. 1970. *Acta phytopathol. Acad. Sci. hung.* **5**: 47 (resistance in tomato to race T_0; **51**, 1928).

——. 1972. *Arch. Züchtungforschung* **2**: 213 (resistance in tomato cvs; **53**, 1079).

Galindo, J. et al. 1960. *Phytopathology* **50**: 123 (compatibility types, heterothallism; **39**: 491).

Gallegly, M. E. 1952. *Ibid* **42**: 461 (races on tomato; **32**: 403).

—— et al. 1955. *Ibid* **45**: 103 (inheritance of resistance to race T_0; **34**: 554).

—— ——. 1958. *Ibid* **48**: 274 (compatibility types in Mexico; **37**: 676).

—— ——. 1959. In *Plant pathology – problems and progress, 1908–1958*: 168 (review, genetic control of host & parasite interaction, 108 ref.).

——. 1968. *A. Rev. Phytopathol.* **6**: 375 (review, genetics of pathogenicity, 64 ref.).

Glushchenko, E. Ya. 1971. *Trudȳ prikl. Bot. Genet. Selek.* **43**: 172 (resistance in *Lycopersicon*; **51**, 1930).

Graham, K. M. 1955. *Am. Potato J.* **32**: 277 (races in Canada; **35**: 217).

—— et al. 1961. *Phytopathology* **51**: 264 (mutability & serial passage through resistant hosts; **40**: 621).

Grümmer, G. et al. 1968. *Arch. PflSchutz* **4**: 143 (testing for resistance in tomato; **48**, 1979).

—— ——. 1969. *Theor. appl. Genet.* **39**: 232 (as above; **49**, 853).

Gunther, E. et al. 1970. *Ann. appl. Biol.* **65**: 255 (as above; **49**, 3001).

Hirst, J. M. 1953. *Trans. Br. mycol. Soc.* **36**: 375 (spore dispersal & diurnal periodicity in *P. infestans* inter alia; **33**: 615).

—— et al. 1960. *Ann. appl. Biol.* **48**: 471, 489 (epidemiology, climate & inoculum source in Irish potato; **40**: 380).

Hori, M. 1935. *Ann. phytopathol. Soc. Japan* **5**: 225 (penetration of solanaceous & non-solanaceous hosts; **15**: 311).

King, J. E. et al. 1968. *Trans. Br. mycol. Soc.* **51**: 269 (fine structure of germinating sporangia; **47**, 2810).

Kishi, K. 1962. *Ann. phytopathol. Soc. Japan* **27**: 172, 180 (physiologic specialisation on main hosts; **42**: 220).

——. 1962. *Bull. hort. Res. Stn Okitsu* Ser. B **1**: 142 (infection process in resistant & susceptible tomato; **42**: 279).

Kubicka, H. 1969. *Acta agrobot.* **22**: 281 (str. pathogenicity on tomato; **49**, 1486).

Lacey, J. 1965. *Ann. appl. Biol.* **56**: 363 (infectivity & survival in soil; **45**, 1147).

Large, E. 1953. *Pl. Pathol.* **2**: 1 (forecasting; **32**, 585).

Laterrot, H. 1975. *Annls Amél. Pl.* **25**: 129 (selection for resistance in tomato; **56**, 2214).

Marks, G. E. 1965. *Chromosoma* **16**: 681 (cytology of vegetative & asexual stages; **45**, 183).

Miller, P. R. et al. 1955. *Pl. Dis. Reptr* Suppl. 231, 89 pp. (world survey on tomato; **34**: 824).

—— ——. 1957. *A. Rev. Microbiol.* **11**: 77 (review, prediction of epidemics in *P. infestans* inter alia, 224 ref.).

Palti, J. et al. 1963. *Phytopathol. Mediterranea* **2**: 265 (disease development & control in Israel; **43**, 2392).

—— ——. 1973. *Phytoparasitica* **1**: 119 (limitations to forecasting with Peronosporaceae in Israel; **53**, 4765).

Pristou, R. et al. 1954. *Phytopathology* **44**: 81 (penetration in resistant & susceptible hosts; **33**: 685).

Richards, M. C. et al. 1946. *Pl. Dis. Reptr* **30**: 16 (resistance in *Lycopersicon*; **25**: 584).

Röder, K. 1935. *Phytopathol. Z.* **8**: 589 (strs on tomato; **15**: 405).

Romero, S. et al. 1969. *Phytopathology* **59**: 1310 (pathogenicity of single oospore cultures; **49**, 538).

Sansome, E. et al. 1973. *Nature Lond.* **241**: 344 (diploidy & chromosomal structural hybridity; **52**, 2177).

Sansome, E. et al. 1977. *J. gen. Microbiol.* **99**: 311 (polyploidy & induced gametangial formation; **56**, 4391).

Smith, L. P. et al. 1966. *Pl. Pathol.* **15**: 113 (forecasting; **46**, 412).

Smoot, J. J. et al. 1958. *Phytopathology* **48**: 165 (oospores; **37**: 503).

Turkensteen, L. J. 1973. *Meded. Inst. plziektenk, Onderz.* 633, 88 pp. (resistance in tomato, 103 ref.; **53**, 3625).

Van der Zaag, D. E. 1956. *Tijdschr. PlZiekt.* **62**: 89 (review, epidemiology; 117 ref.).

Wallin, J. R. 1953. *Phytopathology* **43**: 505 (production & survival of sporangia; **33**: 374).

—— et al. 1960. *Pl. Dis. Reptr* **44**: 227 (forecasting; **39**: 612).

Walter, J. M. et al. 1952. *Phytopathology* **42**: 197 (source of resistance in *Lycopersicon*; **31**: 636).

Warren, R. C. et al. 1975. *Trans. Br. mycol. Soc.* **64**: 73 (sporangial viability in relation to drying; **54**, 4614).

Warren, R. S. et al. 1970. *HortScience* **95**: 266 (use of tomato tissue culture in single gene resistance work; **50**, 279).

Wilson, J. B. et al. 1955. *Phytopathology* **45**: 473 (interrelationship of races in both main hosts; **35**: 245).

Phytophthora meadii McRae, *J. Bombay nat. Hist. Soc.* 25: 760, 1918.

HYPHAE variable, 2.5–6 μ diam., irregularly swollen on some agar media but no distinct hyphal swellings. SPORANGIOPHORES slender (3 μ), up to 6 μ and with nodal swellings from which short branches (1–3) arise. SPORANGIA variable in shape from almost spherical to elongated ellipsoid, often distorted or lobed or laterally inserted on the sporangiophore, base rounded, papilla nearly hemispherical (4 μ deep), 25–72 (48) × 14–40 (24) μ; length : breadth ratio 1.3–2; deciduous with a slender pedicel, 10–20 μ long. CHLAMYDOSPORES absent or rare, 16–36 (mostly 25–29) μ diam. SEX ORGANS occur in fruit of *Hevea brasiliensis* and in culture of single strs but appear more promptly and abundantly in mixed culture with *P. palmivora* or *P. arecae*. OOGONIA spherical to somewhat pyriform, 21–49 (33) μ diam., wall usually colourless and thin but may become yellowish and thicker (wrinkled or rough) with age. ANTHERIDIA amphigynous, rounded 12 × 13 (up to 16 × 16) μ. OOSPORES loose in the oogonium 15–33 (mostly 21–26) μ diam., wall yellowish, 4 μ thick. Opt. temp. growth 28–30°. Sporangia may be more abundant on transference to a water film. Differs from *P. palmivora* sensu Butler in the abundant aerial mycelium on some organic agars, irregular sporangiophores with swellings, irregular sporangia with larger stalks, few or no chlamydospores which are smaller when produced, and the larger oogonia (after G. M. Waterhouse, 1974a; *P. palmivora* q.v.).

Literature on the canopy diseases (abnormal leaf fall) of rubber (*H. brasiliensis*) attributed to *P. meadii* and *P. palmivora* is placed here. The former fungus has been recorded principally on rubber and, since it was originally described from India, it has only been identified with certainty (on rubber only) from Sri Lanka. Sideris recorded it on pineapple in Hawaii. Rubber diseases caused by *Phytophthora* in both S. India and Sri Lanka were long ascribed to *P. palmivora*, partly because McRae's sp. was not considered to be distinct. Recent work, however, in the latter country by Peries and others has confirmed that *P. meadii* is a good sp., and that it probably causes many of these diseases on rubber in the island. In India (Thankamma) both spp. may cause abnormal leaf fall and pod rot, and presumably also the diseases of the trunk bark.

The symptoms on the pods and leaves are similar to those caused by *P. botryosa* (q.v.). Necrosis of the leaves, pods and inflorescence leads to stem infection and dieback. Heavy sporulation on the green fruit has been thought to be a major source of inoculum. Oospores are formed at temps lower than the opt. for growth and sporangial formation. Dispersal (like that of *P. botryosa*) is probably through water. Control of the disease of the panel (black stripe) has been given under *P. palmivora*. The rubber growing areas of Sri Lanka and India lie in the path of the S.W. monsoon which brings extremely heavy rain in June, July and August. In the latter country a pre-monsoon spray of a copper fungicide is obligatory if high yields are to be maintained. High volume Bordeaux has long been used but low volume copper in oil applied from the ground or the air is effective. In Sri Lanka abnormal leaf fall appears less serious in that severe outbreaks (or detrimental effects on yield) occur less often and fungicide treatment has been questioned.

Kershaw, K. L. 1962. *Plrs' Chron.* **57**: 190, 211, 229 (fungicides in India; **42**: 145).

Lloyd, J. H. 1963. *Bull. Rubb. Res. Inst. Ceylon* 57, 67 pp. (fungicides in Sri Lanka; **44**, 1225).

McRae, W. 1918. *Mem. Dep. Agric. India bot. Ser.* **9**: 219 (general).

Peries, O. S. et al. 1965. *Trans. Br. mycol. Soc.* **48**: 631 (compatibility & variation in *Phytophthora* spp. from rubber; **45**, 1162).
——. 1966. *Ibid* **49**: 311 (general; **45**, 2944).
——. 1966. *Q. Jl Rubb. Res. Inst. Ceylon* **42**: 1 (nomenclature of *Phytophthora* spp. on rubber; **46**, 1091).
——. 1970. *Ibid* **46**: 1 (epidemiology; **50**, 3135).
——. 1975. *Pl. Dis. Reptr* **59**: 252 (races on rubber; **54**, 5064).
Ramakrishnan, T. S. 1961. In *Proc. Nat. Rubb. Res. Conf. Kuala Lumpur* 1960: 454 (symptoms & fungicide control in India; **40**: 624).
Satchuthananthavale, V. 1963. *Phytopathology* **53**: 729 (comparison of *Phytophthora* isolates from rubber; **43**, 216).
Sideris, C. P. et al. 1930. *Ibid* **20**: 951 (heart rot of pineapple; **10**: 325).
Thankamma, L. et al. 1968 & 1970. *Rubb. Bd Bull.* **10**: 43; **11**: 9 (isolation of, & compatibility between, strs of *P. meadii* & *P. palmivora*; **48**, 1934; **50**, 1360).

Phytophthora megasperma Drechsler, *J. Wash. Acad. Sci.* **21**: 525, 1931.

HYPHAE 3–8 μ wide; hyphal swellings rare on solid media but quite common in aqueous solutions as spheroidal to elongated ellipsoidal inflations strung out at greater or smaller intervals along the hyphae emergent into the water. SPORANGIOPHORES thin (2–2.5 μ wide), sometimes widening to up to 5 μ near the base of the sporangium; rarely with subsporangial branching, usually proliferating inside the empty sporangium. SPORANGIA not produced on solid media, regularly ovoid, usually 35–50 × 25–35 (max. 45 × 63) μ, no papilla, apical thickening slight, not shed. SEX ORGANS abundant on most agar media. OOGONIA spherical, 16–61 (mostly 40–50) μ diam., wall smooth, up to 1.5 μ thick, colourless or yellowish. ANTHERIDIUM irregularly spherical or ellipsoid, sometimes with a distal prolongation of the hypha, 14–20 × 10–18 μ, more often paragynous, sometimes amphigynous (up to 35%), monoclinous or diclinous. OOSPORE aplerotic, 11–54 (mostly 37–44) μ diam., wall smooth, up to 5 μ thick (unstained) or 7 μ (stained). CULTURES slightly radiate with a medium amount of aerial mycelium; opt. temp. growth 25°. The var. *sojae* Hildebrand differs in having oogonia av. 40 μ diam., rarely over 45 μ (G. M. Waterhouse & J. M. Waterston, *CMI Descr.* 115, 1966).

P. sojae Kaufmann & Gerdeman was described as a pathogen on soybean, and a root rot of this crop had also been attributed to *P. cactorum* (Leb. & Cohn) Schroet. *P. sojae* was not considered sufficiently distinct from *P. megasperma* and Hildebrand erected the var. nov. *sojae* in the *P. megasperma* complex. This var. has a strong specificity to soybean. It has smaller oogonia (av. 34.7–34.9 μ), *c.* 10–15 chromosomes and a slower growth rate compared with var. *megasperma* (oogonia, av. 43.8–47.7 μ; *c.* 22–27 (–30) chromosomes; Sansome et al.). *P. megasperma* has both a wide host range (G. M. Waterhouse et al. l.c.) and a wide, if limited distribution: Australia, Argentina, Canada, France, Germany, Greece, India, Irish Republic, Italy, Japan, New Zealand, Papua New Guinea, Philippines, Poland, UK, USA and Venezuela (Map 157). The fungus was originally described as causing a crown rot of hollyhock (*Althaea rosea*). It can cause a sett rot of sugarcane (Van der Zwet et al.), root rot of cauliflower (Tompkins et al.), a soft rot of carrot (White) and a disease of asparagus (Boesewinkel). The 2 most important diseases are the stem and root rots of soybean and lucerne. The pathogen on the former crop is usually referred to as the var. *sojae*. Although Irwin referred to this var. as causing a root rot of lucerne.

Soybean seedlings and adult plants can be killed. Symptoms of wilt and collapse result from stem lesions at soil level; these cause necrosis in the cortex and vascular tissue. The leaves are retained for a long time after death. Leaf infection can lead to the spread of lesions to the stem. Root infection by zoospores in lucerne was described by Irwin (1976) and Marks et al. (1971b). Extensive studies have been done on zoospore behaviour in the var. *sojae* (see review by Ho 1971). Zoospores were motile longest at 15° whilst cyst germination and germ tube formation were best at 25°. Erwin et al. (1971a) found more oospore germination from cultures incubated at 24° and 27° than those at 15°, 18° and 30°. Slusher et al. (1973) described oospores in soybean roots. Other aspects of the disease on this crop include: infection (Hildebrand; Kaufmann et al.; Morgan); zoospore variability (Hilty et al. 1962); effects of infection with *Heterodera glycines* and *Meloidogyne hapla* (Adeniji et al.; Wyllie et al.); effects of mycorrhiza (Chou et al.; Ross); yield loss and root reduction (Caviness et al.; Meyer et al.); fine structure (Hemmes et al.; Slusher et al. 1974); evidence for diploidy and heterokaryosis (Long et al.); and behaviour in soil (Ho 1969).

Phytophthora megasperma

Races occur in var. *sojae* and 9 have been designated so far (Athow et al.; Buzzell et al.; Haas et al.; Laviolette et al.; Morgan et al.; Schmitthenner 1972; Schwenk et al.). Control of resistance in soybean can be through dominant genes (e.g., Kilen et al. 1974; Mueller et al.; Smith et al.). Hartwig et al. found an allelomorphic series controlling resistance and susceptibility to 2 races. The resistance in the cv. Harasoy 63 and other cvs carrying the *Rps* gene for hypocotyl, hypersensitive resistance is expressed by the production of a phytoalexin which has been studied in depth and its structure determined. See *Rev. appl. Mycol.* **43**, 311, 1205; **45**, 661, 1585; **46**, 2569; **47**, 2335; **48**, 3751. *Rev. Pl. Pathol.* **49**, 2239, 2240; **51**, 912, 913, 914, 915, 4517; **52**, 3119, 3131; **53**, 1636; **55**, 978, 979, 6027, 6028, 6029, 6030; **56**, 4285; **57**, 1516, 4229, 4709. The phytoalexin is produced in similar amounts by roots, cotyledons, pods and tissue culture callus of cvs which do or do not possess the *Rps* gene. But in the resistant cv. (with *Rps*) there is a quicker and greater production of the phytoalexin. Its accumulation is noted 10–12 hours after inoculation, fungal invasion is halted in 24 hours and death of the pathogen follows. This resistance is expressed only in the hypocotyl when young (up to *c.* 2 weeks); resistance in older plants is due to other factors. The hypocotyl resistance type was the one observed in the original resistant selections. The resistance can be destroyed by heat, the critical change occurring at 45–47°; after 50° for 1 minute resistance can be regained but at this temp. for 2 minutes the change to susceptibility becomes irreversible. Long et al. (1975) reported a variable pathogenicity amongst single oospore and zoospore progeny but repeated selfing reduced this variability. Resistant soybean cvs should be grown where the disease occurs (Bernard et al.; Kilen et al. 1977; Schmitthenner 1963). Papavizas et al. reported on control by infusion of pyroxychlor into the seed.

Root rot of lucerne was at first ascribed to *P. cryptogea* (q.v.). The roots of variously aged plants are attacked in heavy, poorly drained soils; the soft rot produced causes a reduced growth rate, chlorosis, premature defoliation and wilt. Isolates from this host are slightly pathogenic to other crops and those from other hosts do not infect lucerne severely. Irwin (1976) described penetration of lucerne roots from zoospores but ascribed the pathogen to *P. megasperma* var. *sojae*. In tests on seedlings Pratt et al. (1976) found more severe disease at 20° and 24° than at 16° and 28°. In flooded soil the opt. temps for sporangial formation were 12–16° and the opt. temp. for indirect germination 16° (95% complete after 72 hours; Pfender et al.). Ribeiro et al. described a high temp. str. which was pathogenic to lucerne roots. Its opt. growth was at 33–36° and, experimentally, severe root rot took place at 29–32°. This isolate did not form oospores. Pratt et al. (1975b) found that the fungus remained infective for 2–3.5 years in some soils stored at 25°; but it is not a strong saprophytic competitor. No races in the form attacking lucerne have been described. For resistance in cvs and testing techniques see Bray et al.; Frosheiser et al.; Gray et al. 1973; Hine et al.; Irwin 1974; Leuschen et al.; Lu et al.; Pratt et al. 1975a; Rogers et al.

Adeniji, M. O. et al. 1975. *Phytopathology* **65**: 722 (infection with *Heterodera glycines* in soybean; **55**, 475).

Athow, K. L. et al. 1974. *Pl. Dis. Reptr* **58**: 789 (reaction of soybean strs to 4 races; **54**, 1067).

Bernard, R. L. et al. 1957. *Agron. J.* **49**: 391 (inheritance of resistance in soybean).

—— ——. 1964. *Soybean Dig.* **24**(4): 10 (resistant cvs; **43**, 2774).

Boesewinkel, H. J. 1974. *Pl. Dis. Reptr* **58**: 525 (on asparagus; **54**, 633).

Bray, R. A. et al. 1978. *Aust. J. exp. Agric. Anim. Husb.* **18**: 708 (selection for resistance in lucerne).

Bushong, J. W. et al. 1959. *Pl. Dis. Reptr* **43**: 1178 (on lucerne as *Phytophthora cryptogea*, symptoms, inoculation & temp.; **39**: 324).

Buzzell, R. I. et al. 1977. *Can. Pl. Dis. Surv.* **57**: 68 (races; **57**, 5780).

Caviness, C. E. et al. 1971. *Crop Sci.* **11**: 83 (effect on yield & chemical composition of soybean seed; **52**, 4286).

Chi, C. C. et al. 1978. *Can. J. Bot.* **56**: 795 (on lucerne, chemotaxis of zoospores; **57**, 5031).

Chou, L. G. et al. 1974. *Pl. Dis. Reptr* **58**: 221 (on soybean, effects of *Rhizobium japonicum* & *Endogone mosseae* with *Pythium ultimum*; **53**, 3257).

Erwin, D. C. 1954. *Phytopathology* **44**: 700 (morphology as *P. cryptogea*; **34**: 458).

——. 1965. *Ibid* **55**: 1139 (specificity to lucerne as *P. megasperma*; **45**, 819).

——. 1966. *Ibid* **56**: 653 (temp., variation in virulence & lucerne var. reaction; **45**, 3162).

——. 1968. *Mycologia* **60**: 1112 (effect of Ca on mycelial growth with *P. cinnamomi*; **48**, 2148).

—— et al. 1971a. *Ibid* **63**: 972 (oospore germination; **51**, 4518).

—— ——. 1971b. *Can. J. Microbiol.* **7**: 15, 945 (nutrient requirements in vitro, effect of thiamine in growth with bacteria; **40**: 476; **41**: 368).

Frosheiser, F. I. et al. 1973. *Crop Sci.* **13**: 735 (selection for resistance in lucerne; 55, 5254).
Gray, F. A. et al. 1973. *Phytopathology* **63**: 1185 (resistance in lucerne, technique; **53**, 1433).
—— ——. 1976. *Ibid* **66**: 1413 (disease in field & association with *Rhizobium*; **56**, 3605).
Haas, J. H. et al. 1976. *Ibid* **66**: 1361 (races; **56**, 2766).
Hartwig, E. E. et al. 1968. *Crop Sci.* **8**: 634 (inheritance of resistance in soybean; **50**, 3315).
Hemmes, D. E. et al. 1977. *Can. J. Bot.* **55**: 436 (fine structure of early gametangial interaction; **56**, 4874).
Hildebrand, A. A. 1959. *Ibid* **37**: 927 (full account, var. nov. *sojae* **39**: 451).
Hilty, J. W. et al. 1962. *Phytopathology* **52**: 859 (in soybean, pathogenic & cultural variability in single zoospore isolates; **42**, 288).
—— ——. 1966. *Ibid* **56**: 287 (comparisons of proteins from resistant & susceptible soybean cvs; **45**, 2304).
Hine, R. B. et al. 1975. *Ibid* **65**: 840 (resistance in lucerne; **55**, 787).
Ho, H. H. 1969. *Mycologia* **61**: 835 (behaviour in soil; **49**, 1221).
——. 1971. *Nova Hedwigia* **19**: 601 (zoospore behaviour to plant roots & review, 76 ref.; **51**, 2067).
Irwin, J. A. G. 1974. *Aust. J. exp. Agric. Anim. Husb.* **14**: 561 (disease reaction of lucerne cvs; **54**, 2312).
——. 1976. *Aust. J. Bot.* **24**: 447 (penetration of lucerne roots; **56**, 2559).
Johnson, H. W. et al. 1965. *Pl. Dis. Reptr* **49**: 753 (pathogenicity of *P. cryptogea*; **45**, 484).
Kaufmann, M. J. et al. 1958. *Phytopathology* **48**: 201 (general as *P. sojae*; **37**: 626).
Kennedy, B. W. et al. 1961. *Trans. Br. mycol. Soc.* **44**: 291 (asexual sporulation in vitro; **41**, 42).
Kilen, T. C. et al. 1974. *Crop Sci.* **14**: 260 (inheritance of resistance in soybean; **55**, 2038).
—— ——. 1977. *Ibid* **17**: 185 (screening for resistance in soybean to 3 races; **56**, 5291).
Klarman, W. L. et al. 1974. *Phytopathology* **64**: 971 (history of infected soybean hypocotyls; **54**, 1955).
Laviolette, F. A. et al. 1977. *Ibid* **67**: 267 (races; **56**, 5289).
Leuschen, W. E. et al. 1976. *Agron. J.* **68**: 281 (field performance of resistant & susceptible lucerne cvs; **56**, 4587).
Long, M. et al. 1975. *Phytopathology* **65**: 592 (wild type strs of var. *sojae*; **54**, 5644).
—— ——. 1977. *Ibid* **67**: 670, 675 (heterokaryosis & evidence for diploidy in var. *sojae*; **57**, 82, 83).
Lu, N. S. J. et al. 1973. *Crop Sci.* **13**: 714 (inheritance of resistance in lucerne; **55**, 5253).
Marks, G. C. et al. 1970. *Phytopathology* **60**: 1687 (detection & inoculum level in soil; **50**, 1859).
—— ——. 1971a. *Ibid* **61**: 510 (histology, resistance & susceptibility in lucerne; **50**, 3901).
—— ——. 1971b. *Can. J. Bot.* **49**: 63 (penetration, infection & histology in lucerne; **50**, 3900).
Meyer, W. A. et al. 1972. *Phytopathology* **62**: 1414 (root reduction & stem lesions in soybean; **52**, 3130).
Morgan, F. L. 1963. *Pl. Dis. Reptr* **47**: 880 (soybean leaf & stem infection; **43**, 310).
—— et al. 1965. *Phytopathology* **55**: 1277 (races; **45**, 917).
Mueller, E. H. et al. 1978. *Ibid* **68**: 1318 (inheritance of resistance in soybean to 4 races).
Papavizas, G. C. et al. 1976. *Pl. Dis. Reptr* **60**: 484 (control by pyroxychlor in soybean; **56**, 927).
Pfender, W. F. et al. 1977. *Phytopathology* **67**: 657 (sporangial formation & zoospore release in soil; **57**, 199).
Pratt, R. G. et al. 1973. *Ibid* **63**: 1374 (detection in soil; **53**, 2243).
—— ——. 1975a & b. *Ibid* **65**: 365, 1267 (resistance & susceptibility in lucerne cotyledons, activity in soil; **55**, 299, 3213).
—— ——. 1976. *Ibid* **66**: 81 (disease in lucerne seedlings, inter-relations of factors; **55**, 3629).
Ribeiro, O. K. et al. 1978. *Ibid* **68**: 155 (high temp. str. pathogenic to lucerne; **57**, 5030).
Rogers, V. E. et al. 1978. *Aust. J. exp. Agric. Anim. Husb.* **18**: 434 (lucerne resistance).
Ross, J. P. 1972. *Phytopathology* **62**: 896 (effect of mycorrhiza in soybean; **52**, 1331).
Salvatore, M. A. et al. 1973. *Ibid* **63**: 1083 (enzymically induced oospore germination; **53**, 2065).
Sansome, E. et al. 1974. *Trans. Br. mycol. Soc.* **63**: 461 (polyploidy associated with var. differentiation; **54**, 2694).
Schmitthenner, A. F. 1963. *Soybean Dig.* **23**(10): 20 (advisory, resistant soybean cvs; **43**, 309).
——. 1972. *Pl. Dis. Reptr* **56**: 536 (race 3; **52**, 293).
Schwenk, F. W. et al. 1974. *Ibid* **58**: 352 (race 4; **53**, 4201).
Slusher, R. L. et al. 1973. *Phytopathology* **63**: 1168 (oospores in roots; **53**, 1641).
—— ——. 1974. *Ibid* **64**: 835 (fine structure of host pathogen interface; **54**, 279).
Smith, P. E. et al. 1959. *Agron. J.* **51**: 321 (inheritance of resistance in soybean; **38**: 642).
Tompkins, C. M. et al. 1936. *J. agric. Res.* **53**: 685 (general on cauliflower; **16**: 292).
Tu, J. C. 1978. *Physiol. Pl. Pathol.* **12**: 233 (protection of soybean by *Rhizobium*; **57**, 4705).
Van der Zwet, T. et al. 1960. *Pl. Dis. Reptr* **44**: 519 (symptoms, inoculation & sugarcane resistance; **40**, 135).
—— ——. 1961. *Phytopathology* **51**: 634 (morphology & comparison with *P. erythroseptica* on sugarcane; **41**: 250).
White, N. H. 1945. *Tasm. J. Agric.* **16**: 59 (soft rot of carrot; **24**: 487).

Phytophthora nicotianae

Wyllie, T. D. et al. 1960. *Pl. Dis. Reptr* **44**: 543 (soil temp. & infection with *Meloidogyne hapla*; **40**: 138).

Phytophthora nicotianae Breda de Hann, *Meded. PlTuin Batavia* **15**: 57, 1896, var. **nicotianae.**

Phytophthora melongenae Sawada, 1915

P. terrestris Sherbakoff, 1917

P. parasitica var. *rhei* Godfrey, 1923

(these 3 fungi were considered to be the same as *P. nicotianae* var. *nicotianae*, G. M. Waterhouse, *Mycol. Pap.* 92, 1963)

HYPHAE fairly uniform in diam., up to 5 μ; hyphal swellings frequent, especially in water cultures and on the sporangiophores, spherical, oval or angular, all sizes up to 20 μ: as soon as the swellings develop 4–5 hyphae may grow out from various points giving the swelling a spiky appearance. SPORANGIOPHORES thin (2 μ wide), closely or sparsely branched, depending on the conditions, the branches usually arising from a hyphal swelling. SPORANGIA terminal or, less frequently, intercalary, broadly turbinate, the basal part nearly spherical and the apical 1/3 or 1/4 narrowed or prolonged into a beak, sometimes skewly orientated to the pedicel, av. 36–45 μ, often up to 60–70 μ long, not usually deciduous; papilla conspicuous, often 2 present, apical thickening hemispherical. CHLAMYDOSPORES frequent, forming early on the new mycelium, 20–40 μ diam., wall 1.5 μ thick, not usually turning brown. OOGONIA formed readily in single culture by some isolates but if they are sparse or absent after some weeks they may be induced by culturing with an opposite str.; mostly 28–30 (–32) μ diam., wall becoming slightly thicker with age and acquiring a golden brown accretion. ANTHERIDIA always amphigynous, but the oogonial stalk is sometimes quite eccentric so that the antheridium appears to be not amphigynous at first, 10–16 × 10 μ, rounded rather than angular. OOSPORE not filling the oogonium, 24–26 μ diam., wall 1.5–2 (–3) μ thick, usually colourless or faintly yellow. Cultures are diaphanously fluffy, usually without zones or striations; opt. temp. growth (25–) 30° (G. M. Waterhouse & J. M. Waterston, *CMI Descr.* 34, 1964).

Further details on the morphology and physiology of this var. were given by Waterhouse (1974a, *P. palmivora* q.v.). She stated that *P. nicotianae* differs from all strs of *P. palmivora* in its wide temp. range (particularly in its good growth at 37–37.5°) and that it also grows more slowly over the opt. temp. range. Var. *nicotianae* is distinguished by its rounded, beaked sporangia, spiky hyphal swellings and larger oogonia; var. *parasitica* (q.v.) differs in its cultural characters, smaller sporangia and smaller sex organs. Hendrix and others have made detailed studies on sterol physiology. Diseases caused on crops other than tobacco are discussed under *P. nicotianae* var. *parasitica*.

Black shank of tobacco is widely distributed in the warmer regions of cultivation. The disease has caused severe losses especially in USA where it has been largely studied. A recent, full account was given by Lucas (*Diseases of tobacco*). Provided temps are sufficiently high symptoms of damping off may occur in seedbeds but the classical symptoms are those which appear on field plants. The lower leaves become flaccid and wilt; the entire plant can wilt without recovery in a few days. The root system of an infected plant, not showing above ground symptoms, will have 1 or 2 laterals attacked and necrotic. The necrosis spreads to the stem where the cortex, pith and other tissues are infected and the lesion may spread 30 cm above soil level. If the stem is split longitudinally the dry, brown to black pith is found to be laminated, giving the discoid (like plates) appearance which is a characteristic symptom. Splash dispersal of inoculum causes above ground symptoms on plants which show no root infection. Leaf lesions, up to 8 cm diam., have a necrotic centre with marginal, concentric, chlorotic bands; these continue to develop in curing and thereby detract from leaf quality. Zoospores penetrate uninjured roots to which they are attracted. Wilt is probably due to vessel blocking by products of the host and pathogen interaction and not to any specific toxin.

The opt. temp. for growth in vitro corresponds with that for development in the field; the zoospores are motile for the longest periods at 20°. Spread is through soil and water. Soil survival is lengthy and, although inoculum potential is considerably reduced after a 4–5 year rotation with non-host crops, appreciable damage to highly susceptible tobacco can still occur after such a period. In the soil infective propagules increase rapidly in the rhizospheres of susceptible cvs and they are high in the upper 7–8 cm of soil where they overwinter. The development of black shank is related to soil calcium and where the disease is an important factor in production liming should be kept as low as possible. Isolates from different geographical areas probably differ in virulence. There is adaptive selection in the field. Apple and others found that: isolates from fields where resistant cvs were being grown were generally more virulent than those with susceptible cvs; highly virulent isolates produced abundant sporangia whose zoospores remained motile longer than did those produced by weakly virulent isolates; serial passage through tobacco increased virulence which tended to be lost after long culture in vitro. Three races have been differentiated (0, 1 and 3). Race 3 differs from races 0 and 1, apart from pathogenicity and metabolism, in tolerating lower soil temps (Apple 1962; Lamprecht et al.; Litton et al.; McIntyre et al. 1977, 1978; Stokes et al.; Taylor et al. 1978).

In the presence of root knot nematodes (*Meloidogyne* spp.) black shank becomes more severe. Nematode attack apparently predisposes the plant to more extensive attack by the fungus. Cvs resistant to the latter but not to nematodes can show severe black shank symptoms. The disease is less severe on resistant cvs where fumigation is done, although this treatment has little effect on disease incidence where black shank susceptible cvs are grown. More severe disease has also been found following inoculation with *Pratylenchus brachyurus*. Control is mainly through resistant cvs. Single dominant genes appear to control resistance to each race but other genes also affect it. Other control measures include rotation (e.g., Flowers et al. 1974; Mathews et al.; Miller et al. 1970), control of nematodes, and soil fumigation of plant beds (e.g., Miller et al. 1976; Noveroske; Prinsloo et al. 1977; Taylor et al. 1975).

Apple, J. L. 1957. *Phytopathology* **47**: 733 (pathogenic, cultural & physiological variation **37**: 313).
——. 1959. *Ibid* **49**: 37 (compatibility types; **38**: 423).
——. 1961. *Ibid* **51**: 386 (effect of host nutrition, N, P & K; **40**: 766).
——. 1961. *Pl. Dis. Reptr* **45**: 968 (effect of resistant cvs on virulence level in natural populations; **41**: 409).
——. 1962. *Phytopathology* **52**: 351 (races; **42**: 51).
——. 1963. *Pl. Dis. Reptr* **47**: 632 (persistence in soil; **43**, 266).
——. 1967. *Tob. Sci.* **11**: 79 (race 1 & implication for breeding; **47**, 889).
Collins, G. B. et al. 1971. *Can. J. Genet. Cytol.* **13**: 422 (inheritance of resistance to race 0; **51**, 2843).
Dukes, P. D. et al. 1961. *Pl. Dis. Reptr* **45**: 362 (effect of host passage on virulence; **40**: 710).
—— ——. 1961. *Phytopathology* **51**: 195 (chemotaxis of zoospores; **40**: 630).
—— ——. 1962. *Ibid* **52**: 191 (relationship of virulence to zoospore production & motility; **41**: 741).
—— ——. 1965. *Ibid* **55**: 666 (effect of O_2 & CO_2 tensions on growth; **44**, 3155).
——. 1968. *Tob. Sci.* **12**: 200 (inoculum potential & soil factors; **48**, 3658).
Flowers, R. A. et al. 1972. *Phytopathology* **62**: 474 (population density & distribution in soil; **51**, 4351).
—— ——. 1974. *Ibid* **64**: 718 (host & non-host effects on soil populations; **53**, 4582).
Gooding, G. V. et al. 1959. *Ibid* **49**: 274, 277 (inoculum potential & effect of temp. on sporangial formation & zoospore activity; **39**: 44).
Hanchey, P. et al. 1971. *Ibid* **61**: 33 (fine structure of infection process; **50**, 3164).
Helgeson, J. P. et al. 1972. *Ibid* **62**: 1439 (tissue culture & resistance; **52**, 3420).
Hendrix, J. W. et al. 1964. *Ibid* **54**: 987 (effect of fats & fatty acid derivatives on growth; **44**, 541).
—— ——. 1966. *Physiol. Pl.* **19**: 159 (mycelial growth on vegetable oils; **45**, 3291).
—— ——. 1967. *Tob. Sci.* **11**: 148 (stem resistance in tobacco).
—— ——. 1969. *Phytopathology* **59**: 1620 (cation & sterol effects on growth; **49**, 1284).
—— ——. 1970. *Mycologia* **62**: 195 (sterol or Ca requirement for growth on nitrate N; **49**, 2768).
Husain, A. et al. 1956. *Pl. Dis. Reptr* **40**: 629 (formation of polygalacturonase; **36**: 358).
Imagaki, H. et al. 1969. *Phytopathology* **59**: 1350 (effect of infection by *Pratylenchus brachyurus*; **49**, 1137).
Johnson, E. M. et al. 1954. *Ibid* **44**: 312 (compatibility types; **34**: 109).
Kincaid, R. R. et al. 1970 & 1972. *Ibid* **60**: 1513; **62**: 302 (effect of soil Ca & other factors; **50**, 970; **51**, 3543).
Lamprecht, M. P. et al. 1974. *Agroplantae* **6**(4): 73 (inheritance of resistance to race 2; **54**, 3473).
Lautz, W. 1957. *Pl. Dis. Reptr* **41**: 95 (resistance in *Nicotiana* spp.; **36**: 667).
Litton, C. C. et al. 1970. *Tobacco N.Y.* **171**(11): 45 (reaction of tobacco & *Nicotiana* spp. to races 0 & 1; **51**, 3545).
McCarter, S. M. 1967. *Phytopathology* **57**: 691 (effect of soil moisture & temp. on disease incidence; **46**, 3566).
McIntyre, J. L. et al. 1976. *Ibid* **66**: 70 (screening seedlings for resistance; **55**, 3740).
—— ——. 1977. *Mycologia* **69**: 756 (chemical differentiation of race 3; **57**, 1838).
—— ——. 1978. *Phytopathology* **68**: 35 (race 3; **57**, 5117).
Mathews, E. M. et al. 1959. *Agron. J.* **51**: 513 (effect of crop rotation; **39**: 126).
Miller, C. R. et al. 1970. *Proc. Soil Crop Sci. Soc. Fla* **30**: 376 (effects of rotation, chemicals & cvs on disease; **52**, 2368).

Miller, P. M. et al. 1976. *Phytopathology* **66**: 221 (soil fumigation; **55**, 5346).

Moore, E. L. et al. 1956. *Ibid* **46**: 545 (effect of nematode infection on host resistance; **36**: 358).

Moore, L. D. et al. 1969. *Ibid* **59**: 1974 (heat-induced susceptibility; **49**, 1775).

Noveroske, R. L. 1975. *Ibid* **65**: 22 (systemic fungicide; **54**, 3474).

Powell, N. T. et al. 1960. *Ibid* **50**: 899 (effect of *Meloidogyne* spp. on infection; **40**: 562).

Powers, H. R. 1954. *Ibid* **44**: 513 (mechanism of wilt; **34**: 326).

Prinsloo, G. C. et al. 1974. *Phytophylactica* **6**: 217 (races in South Africa; **54**, 5080).

—— ——. 1977. *Ibid* **9**: 25 (soil fumigation; **56**, 5183).

Reichle, R. E. 1969. *Mycologia* **61**: 30 (fine structure of zoospores; **48**, 2212).

Sasser, J. N. et al. 1955. *Phytopathology* **45**: 459 (effect of *Meloidogyne* spp. on infection; **35**: 47).

Schramm, R. J. et al. 1954. *J. Elisha Mitchell scient. Soc.* **70**: 255 (transpiration in infected tobacco; **34**: 823).

Stokes, G. W. et al. 1966. *Phytopathology* **56**: 678 (races & host resistance; **45**, 3220).

Taylor, G. S. et al. 1975. *Pl. Dis. Reptr* **59**: 434 (soil fumigation; **55**, 882).

—— ——. 1978. *Bull. Conn. agric. Exp. Stn* 773, 8 pp. (race 3).

Thung, T. H. 1938. *Meded. Proefst. vorstenl. Tabak Klaten* 86, 55 pp. (epidemiology in Java, Indonesia; **18**: 419).

Tisdale, W. B. et al. 1926. *Bull. Fla agric. Exp. Stn* 179, 61 pp. (general; **6**: 441).

Troutman, J. C. 1962. *Tech. Bull. Va agric. Exp. Stn* 158, 12 pp. (effect of pH).

—— ——. 1964. *Phytopathology* **54**: 225 (electrotaxis of zoospores; **43**, 2058).

Uozumi, T. 1964. *Bull. Morioka Tob. Exp. Stn* 1, 55 pp. (general; **47**, 890).

Valleau, W. D. et al. 1960. *Tob. Sci.* **4**: 92 (host resistance; **40**: 186).

Veech, J. A. 1969. *Phytopathology* **59**: 566 (comparison of isozyme patterns in resistant & susceptible infected tobacco cvs; **48**, 3152).

Wills, W. H. 1954. *J. Elisha Mitchell scient. Soc.* **70**: 231, 235 (effect of C, N and pH on growth in vitro; **34**: 822, 823).

——. 1964. *Phytopathology* **54**: 1133 (use of amino acids & pathogenicity; **44**, 837).

——. 1969. *Ibid* **59**: 346 (Ca nutrition in host & infection; **48**, 2556).

——. 1971. *Tob. Sci.* **15**: 47, 51 (inoculation & expression of resistance; **51**, 1898, 1899).

Wolf, F. T. et al. 1954. *J. Elisha Mitchell scient. Soc.* **70**, 235, 244, 255, 261 (sporangium & toxin formation, host transpiration & respiration; **34**: 823, 824).

Phytophthora nicotianae (Breda de Haan) var. **parasitica** (Dastur) Waterhouse, *Mycol. Pap.* 92: 14, 1963.

P. parasitica Dastur, 1913.

HYPHAE tough, up to 9 μ wide, irregular in width but without typical hyphal swellings. SPORANGIOPHORES more slender than the mycelial hyphae, irregularly or sympodially branched, the sympodia being close in moist air. SPORANGIA broadly ovoid or obpyriform to spherical, not noticeably narrowed at the apex, av. 30–38 max. 40–50 μ, sometimes intercalary, deciduous with a very short (2 μ) pedicel, apical thickening hemispherical. CHLAMYDOSPORES up to 60 μ, forming rather tardily (1–2 weeks), wall 3–4 μ, becoming yellowish brown with age. OOGONIA usually produced in single culture though often very sparsely or not until after some weeks, but readily produced when grown with an opposite str., av. 24–26 (–31) μ diam., wall becoming thick and yellow brown with age. ANTHERIDIA spherical or oval 10 × 12 (–16) μ. OOSPORE aplerotic, 18–20 μ, wall thick (2 μ) compared with the small size. CULTURES on oat- or cornmeal agar variable but commonly irregularly fluffy in an irregular rosette pattern; opt. temp. growth 30–32°. May be distinguished from *P. citrophthora* (q.v.), which sometimes produces indistinguishable symptoms on similar hosts, by the 4–5° higher opt. for growth and by growth rates on differential media (G. M. Waterhouse & J. M. Waterston, *CMI Descr.* 35, 1964).

Further details on the morphology and physiology of this var. and its differences from var. *nicotianae* (q.v.) and *P. palmivora* were given by Waterhouse (1974a, *P. palmivora* q.v.). Var. *parasitica* has smaller sporangia, oogonia and oospores. It is plurivorous, widespread, one of the most frequently encountered members of the genus, and causes many diseases which tend to be more serious at *c.* 28° and above. It tends to be less confined to warmer areas than the equally ubiquitous *P. palmivora* sensu lato. This comment does not appear to square with the opt. temp. for growth; this is said to be higher for *P. parasitica* which, however, has a wider temp. range. Kouyeas (76 isolates from 6 hosts) found that most oospores were produced when certain strs were paired, 18% of the isolates produced oospores in single str. culture. Various workers have investigated the effects of nutrients in vitro; and methods for obtaining zoospores, chlamydospores and oospores have been described by Menyonga et al., Tsao and Honour et al., respectively. The fine structures of zoospore formation, encystment and germination have also been described. Inoculum is soil or water dispersed and any part of the plant can be attacked. Isolates from one host may be either virulent only to that host or equally so to others. But no clear picture seems to have emerged yet with regard to physiological specialisation. The fungus carries over readily as survival structures in host debris. The literature to work on *P. nicotianae* var. *nicotianae* on hosts other than tobacco is referred to here and some individual diseases (omitting those on temperate crops) are discussed below.

Ashby, S. F. 1928. *Trans. Br. mycol. Soc.* **13**: 86 (morphology, oospore formation & comment on other papillate *Phytophthora* spp.; **7**: 601).

Bonnett, P. et al. 1978. *Annls Phytopathol.* **10**: 15 (variability & host specificity).

Fothergill, P. G. et al. 1964. *J. gen. Microbiol.* **36**: 67 (mineral nutrition with *P. infestans* & *P. erythroseptica*; **44**, 346).

Gooday, G. W. et al. 1971. *Trans. Br. mycol. Soc.* **57**: 178 (cellulose wall ingrowths; **51**, 1198).

Hemmes, D. E. et al. 1969. *Am. J. Bot.* **56**: 300 (fine structure in sporangia; **48**, 2211).

——. 1971. *Jnl Cell Sci.* **9**: 175 (fine structure of encystment & germination; **51**, 2241).

Hohl, H. R. et al. 1967. *Am. J. Bot.* **54**: 1131 (fine structure of zoospore formation; **47**, 770).

Honour, R. C. et al. 1974. *Mycologia* **66**: 1030 (oospore formation; **54**, 4355).

Kouyeas, V. 1953. *Annls Inst. Phytopathol. Benaki* **7**: 40 (sexuality; **34**: 738).

Menyonga, J. M. et al. 1966. *Phytopathology* **56**: 359 (zoospore formation; **45**, 2422).

Newton, W. 1956 & 1957. *Can. J. Bot.* **34**: 759; **35**: 445 (effects of optical isomers of alanine & leucine, & other N compounds; **36**: 205, **37**: 143).

Rao, V. G. et al. 1966. *Mycopathol. Mycol. appl.* **30**: 121 (temp. & sporangial germination, & effects of dyes & vitamins; **46**, 1459).

Thomson, S. V. et al. 1972. *Mycologia* **64**: 457 (atypical structures like sporangia; **51**, 3821).

Tsao, P. H. 1971. *Phytopathology* **61**: 1412 (chlamydospore formation; **51**, 3148).

AGAVE

A newly released hybrid (11648), derived from *A. amaniensis* and *A. angustifolia*, was attacked in Kenya by a disease variously called zebra leaf spot (from the characteristic dark purple and green, concentric markings), bole rot and spike rot. Several *Phytophthora* spp. have been referred to as the cause (*P. arecae, P. nicotianae* var. *nicotianae, P. n.* var. *parasitica* and *P. palmivora*). Inoculum is splash dispersed and infection usually begins on younger plants as a leaf attack; the following bole rot can lead to collapse in 3–4 weeks but in larger field plants this may take 4–6 months. No direct infection of the roots appears to occur. Mancozeb has given some control in addition to the customary cultural measures. *A. sisalana*, said to be unsuitable for breeding, has some resistance but this was also found in the progeny from crosses between the hybrid and *A. lespinassei.*

Allen, D. J. 1970. *E. Afr. agric. For. J.* **36**: 119 (infection of other hosts; **50**, 1255).

——. 1971. *PANS* **17**: 42 (breeding for resistance; **50**, 2979).

——. 1972. *Pl. Dis. Reptr* **56**: 678 (pathogenic variability & resistance; **52**, 1157).

Clinton, P. K. S. et al. 1963. *E. Afr. agric. For. J.* **29**: 110 (symptoms & isolations; **43**, 2288).

Peregrine, W. T. H. 1969. *Ann. appl. Biol.* **63**: 45 (fungicides; **48**, 1743).

——. 1969. *PANS* **15**: 558 (symptoms, isolation & inoculation; **49**, 1385).

Wienk, J. F. 1968 & 1969. *Ibid* **14**: 142; **15**: 562 (symptoms & infection; **47**, 1901).

——. 1968. *E. Afr. agric. For. J.* **33**: 261 (consideration of several *Phytophthora* spp.; **47**, 1902).

CITRUS

The citrus diseases caused by *Phytophthora* spp., including symptoms and control, are discussed more fully under *P. citrophthora*. Reference should also be made to *P. citricola, P. hibernalis* and *P. palmivora*, all of which cause diseases on this host group. On the lower stem the disease has been called collar rot, foot rot or gummosis, and on the fruit brown rot; a decay of feeder and other roots, and a branch blight or dieback can also be caused. The most generally important spp. on citrus are *P. nicotianae* var. *parasitica* and *P. citrophthora*. *P. hibernalis* has a temp. opt. 10–12° lower than the former and 4–8° lower than the latter; it does not appear to cause much foot rot and has a limited distribution. *P. citricola* is probably rare in the tropics, it causes brown rot of citrus and other diseases on temperate hosts.

There is no invasion of the stem where cork tissue

is fully formed except through wounds or other openings, and penetration seldom occurs where wounds are > 10 days old. *P. nicotianae* var. *parasitica* and *P. citrophthora* are the principal spp. attacking citrus in USA (California) and Klotz (1953) stated that they were isolated from foot rot lesions with about equal frequency. In a survey of irrigation water in citrus in California *P. citrophthora* was most commonly isolated, and throughout the year. *P. nicotianae* var. *parasitica* was the next most frequent, occurring in warmer periods. *P. syringae* was found least frequently and in cool periods. In Florida *P. nicotianae* var. *parasitica* is very common and *P. citrophthora* is restricted, although the latter sp. is predominant in brown rot epidemics and its zoospores infect fruit more readily than those of the former sp. Brown rot in Florida only occurs where there are long periods of fruit wetness. Thomson et al. considered that zoospores (or structures formed by them) in irrigation water could be significant in disease spread.

In Australia (Queensland) it was found that, except on Ellendale mandarin, local strs of the pathogen did not infect uninjured fruit and leaves, or cause tip blight. *P. citrophthora* readily infected both types of tissue but following injury both spp. caused brown rot at the same rate. The restricted occurrence of *P. citrophthora* in Queensland, and the fact that it is not a serious trunk and root pathogen, does not appear to be fully understood; on the other hand in New South Wales this sp. is the predominant one (Broadbent et al., *P. citrophthora* q.v.). Baker described gummosis on grapefruit caused by *P. palmivora* and *P. nicotianae* var. *parasitica* in Trinidad. Klotz et al. (1958) found differences in the amount of bark invasion between isolates; and larger lesions occurred on stocks with Lisbon lemon scions than on those with Washington navel orange. Differences in reaction to the fungus within clones have been more recently described by Carpenter et al. (1975) and Hutchison et al. Curative trunk treatments were described by Timmer and methods for applying methyl bromide by Grim et al. (1971).

Baker, R. E. D. 1934 & 1935. *Trop. Agric. Trin.* **11**: 236; **12**: 36 (symptoms, incidence & control on grapefruit; **14**: 505).
Carpenter, J. B. et al. 1962. *Phytopathology* **52**: 1277 (resistance in seedlings; **42**: 382).
———. 1975. *Pl. Dis. Reptr* **59**: 54 (tolerance in stocks; **54**, 3320).
Grimm, G. R. et al. 1962. *Proc. Fla St. hort. Soc.* **75**: 73 (variations in pathogenicity to sweet orange; **43**, 882a).
———. 1971. *Ibid* **84**: 52 (methods of applying methyl bromide in small plots; **52**, 2272).
Hutchison, D. J. et al. 1972. *Ibid* **85**: 38, 39 (resistance in clones & stocks; **53**, 1816, 3023).
———. 1973. *Ibid* **86**: 88 (resistance in clones; **54**, 4932).
Klotz, L. J. 1953. *Plant Diseases, Yearbook U.S. Dep. Agric.*: 734 (foot rot of citrus trees).
—— et al. 1958. *Phytopathology* **48**: 520 (effect of scions on pathogenicity; **38**: 143).
———. 1959. *Pl. Dis. Reptr* **43**: 830 (*Phytophthora* spp. in irrigation water; **39**: 24).
Knorr, L. C. 1956. *Ibid* **40**: 772 (brown rot & period of fruit wetness; **36**: 316).
Oxenham, B. L. et al. 1969. *Qd J. agric. Anim. Sci.* **26**: 173 (comparison of infection by *P. n.* var. *parasitica* & *P. citrophthora*; **49**, 480).
Stolzy, L. H. et al. 1975. *Soil Sci.* **119**: 136 (nutrition responses & root rot under high & low soil O_2; **55**, 2723).
Thomson, S. V. et al. 1976. *Phytopathology* **66**: 1198 (zoospore survival in irrigation water; **56**, 1607).
Timmer, L. W. 1977. *Ibid* **67**: 1149 (fungicidal trunk paints; **57**, 2533).
Toxopeus, H. J. 1934. *Landbouw Buitenz.* **9**: 385 (effect of temp. & RH on asexual sporulation; **14**: 301).
Tsao, P. H. 1969. In *1st Int. Citrus Symp. Proc. Univ. Calif.* Vol. 3: 1221 (saprophytic behaviour in soil).
Whiteside, J. O. 1970. *Pl. Dis. Reptr* **54**: 608 (factors affecting brown rot incidence in Florida; **50**, 96).
—— et al. 1971. *Phytopathology* **61**: 1233 (penetration of stems; **51**, 2480).

LYCOPERSICON ESCULENTUM

The disease on tomato has been reported mostly on the fruit (buckeye), but damping off and a cortical rot of roots and stems may also occur. Other *Phytophthora* spp. have been implicated but *P. nicotianae* var. *parasitica* is probably the commonest cause. Weststeijn gavé a detailed account of the disease in the Netherlands, including fungicide control and comparisons with *P. infestans*. Fruit infections are caused by *P. nicotianae* and *P. infestans*. The latter does not (like the former) invade the roots or the stem bases. Root infection developed most at 17–27°. On the fruit the lesions begin as watersoaked spots and the subsequent rot is firm at first, becoming soft. Fruit near the ground may be affected first. Sharma et al. (1974) found seed infection. Fruit penetration is cuticular; severe fruit rot occurred at 24–31° (Obrero et al.) and damping off is max. at 22°. Critopoulos found symptom differences after inoculating unwounded, immature tomato fruit with *P. capsici, P. nicotianae* var. *parasitica* and *P. drechsleri*. The first 2 spp. caused infection more readily than the last. Outbreaks of buckeye rot may be related to the predisposing conditions typical of many diseases caused by the genus, i.e. exceptionally favourable weather for infection and unsatisfactory agronomic methods. Soil sterilisation may be necessary in severe outbreaks. Copper, captan and maneb sprays may give control. Most common vars are susceptible although variations in reaction to infection have been found.

Critopoulos, P. D. 1954. *Phytopathology* **44**: 551 (symptoms caused by 3 *Phytophthora* spp.; **34**: 326).

Fulton, J. P. et al. 1951. *Ibid* **41**: 99 (root rot syndrome; **30**: 392).

Obrero, F. C. et al. 1965. *Pl. Dis. Reptr* **49**: 327 (variation in pathogenicity, & host & var. reaction; **44**, 2616).

Richardson, L. T. 1941. *Can. J. Res.* Ser. C **19**: 446 (general syndrome, effect of temp., oospores in roots, tests for resistance; **21**: 170).

Sharma, S. L. et al. 1974. *Indian J. agric. Sci.* **44**: 323 (var. reaction; **56**, 847).

——. 1975. *Indian Phytopathol.* **28**: 130 (seed infection; **55**, 4857).

Tompkins, C. M. et al. 1941. *J. agric. Res.* **62**: 467 (general, inoculations with 3 *Phytophthora* spp.; **20**: 501).

Wager, V. A. 1935. *S. Afr. J. Sci.* **32**: 235 (general; **15**: 265).

Weststeijn, G. 1973. *Neth. J. Pl. Pathol.* **79** Suppl. 1, 86 pp. (general in Netherlands & parts of Europe, 170 ref.; **53**, 279).

Wilson, J. D. 1956. *Phytopathology* **46**: 511 (fungicides & cv. reaction; **36**: 284).

PIPER BETLE

Foot rot of betel is generally attributed to this fungus although *P. palmivora* may also cause similar symptoms on this crop which is widely cultivated in India and S.E. Asia. The disease is serious in India where most of the investigations have been done (see the recent review by Saksena). Turner (1969, *P. palmivora* q.v.) considered that all *Phytophthora* isolates from this host should be placed under *P. palmivora*. The symptoms are generally a typical foot rot with infection of the cortical tissues around soil level and the leaves are also attacked. More recently the pathogen has been reported to cause a vascular wilt. The frequent, heavy losses in India are in part due to the highly traditional form of cultivation in small, very intensively cultivated, often irrigated plots. These are usually screened with a variety of materials to keep the RH high. Apart from attempting to modify cultural methods, for example, irrigation (which doubtless spreads and increases infection) copper fungicide drenches have given some control. Vars differ in their reaction to infection. Dastur invalidly described the pathogen as *P. parasitica* var. *piperina* which is synonymous with *P. nicotianae* var. *parasitica*.

Dastur, J. F. 1935. *Proc. Indian Acad. Sci.* Sect. *B*1: 778 (general; **14**: 717).

Saksena, S. B. 1977. *Indian Phytopathol.* **30**: 1 (review, 75 ref.)

OTHER CROPS

Infection of *Cinchona* spp. (quinine) has been described from S. and Central America. The leaves and young branches of nursery plants, and of suckers from older field plants, are killed. Spread to the main stem results in girdling cankers or elongated cankers which spread a considerable distance above ground level. Dieback and death can occur. *C. pitayensis, C. pubescens* and *C. officinalis* are attacked (Crandall et al.; Darley et al.; Szkolnik). A foot rot of roselle and kenaf (*Hibiscus sabdariffa* and *H. cannabinus*), and a root and collar rot of *H. rosa-sinensis* have been described (Alconero et al.; Follin; Mukerjee; Navaratnam). Heart, fruit, stem or root rot of pineapple causes a leaf wilt; the leaves turn yellowish to pink brown and are readily detached. A firm, white to yellow rot in the stem and base of the leaves has a brown margin. Young fruit may be rotted. A similar condition is also caused by *P. cinnamomi* (q.v.) (Chowdhury; Mehrlich). Frossard (1978) recommended captafol sprays.

The original description of the pathogen as *P. parasitica* by Dastur was based on isolates from *Ricinis communis* (castor) and *Sesamum indicum* (sesame). He described a leaf spot (concentrically marked necrosis with an ill-defined, wavy edge and causing premature leaf fall) on the former crop. The fungus also caused damping off and infection of the flowers and fruit. The leaves, stems and pods of sesame are attacked, and severe losses have been reported from Rajasthan. Copper and zineb give some control. The watersoaked, dark lesions on the fruit of eggplant (*Solanum melongena*) have been attributed to infection by *P. nicotianae* var. *nicotianae* in USA, Japan and the Philippines; to *P. nicotianae* var. *parasitica* in Greece and the Philippines; and to *P. n.* var. *parasitica* and *P. palmivora* in Indonesia (Java). Other parts of this crop are also attacked (Cabaccang et al.; Kendrick; Ocfemia; Reitsma et al.; Sarejanni). Hemmi et al. stated that oospores occurred in the plant.

Amongst the many other diseases described are: general infection of *Atropa belladona*; foliar blight of *Bougainvillea*; fruit rots of *Psidium guajava, Annona squamosa, Manilkara achras* and *Citrullus lanatus; Gossypium* boll rot and damping off; stem rots of *Acacia mearnsii, Passiflora edulis* and *Tephrosia vogelii*; a blight on racemes and nuts of *Macademia integrifolia*, and a bud rot of *Washingtonia* spp.; capsule rot of *Ellettaria cardamomum* and stem canker of *Cajanus cajan*.

Alconero, R. et al. 1969. *Pl. Dis. Reptr* **53**: 702 (on roselle & kenaf, differential pathogenicity of isolates; **49**, 165).

Alfieri, S. A. 1970. *Phytopathology* **60**: 1806 (on *Bougainvillea* & resistance; **50**, 2299).

Cabaccang, F. R. et al. 1965. *Philipp. Agric.* **49**: 222 (on eggplant, differential pathogenicity of isolates; **46**, 218).

Chowdhury, S. 1946. *Curr. Sci.* **15**: 82 (on pineapple; **25**: 305).

Crandall, B. S. et al. 1945. *Phytopathology* **35**: 138 (on *Cinchona* spp.; **24**: 289).

Darley, E. F. et al. 1951. *Ibid* **41**: 641 (on *Cinchona* spp. with *P. cinnamomi*; **30**: 628).

Dastur, J. F. 1913. *Mem. Dep. Agric. India bot. Ser.* **5**: 177 (on castor).

Follin, J. C. 1977. *Coton Fibr. trop.* **32**: 241 (on roselle, resistance; **57**, 4494).

Frossard, P. 1976. *Fruits* **31**: 617 (on pineapple, pH, Ca & fungicides; **56**, 2161).

——. 1978. *Ibid* **33**: 183 (on pineapple, fungicides; **58**, 311).

Gemawat, P. D. et al. 1964. *Indian Phytopathol.* **17**: 273 (on sesame, effect of temp. & control; **44**, 2856).

—— ——. 1965. *Ibid* **18**: 128 (on sesame, resistance; **45**, 1122).

Hemmi, T. et al. 1939. *Ann. phytopathol. Soc. Japan* **9**: 157 (on eggplant, effect of temp.; **19**: 383).

Hunter, J. E. et al. 1971. *Phytopathology* **61**: 1130 (on *Macademia integrifolia*; **51**, 1981).

—— ——. 1972. *Pl. Dis. Reptr* **56**: 689 (on *Macademia*, fungicides; **52**, 1286).

Kaiser, W. J. et al. 1978. *Ibid* **62**: 240 (on pigeon pea; **57**, 5792).

Kale, G. B. et al. 1957. *Indian Phytopathol.* **10**: 38 (on sesame, inoculation of other hosts; **38**: 708).

Kendrick, J. B. 1923. *Proc. Indiana Acad. Sci.* 1922: 299 (on eggplant, tomato & red pepper; **3**: 247).

Leu, L. S. et al. 1976. *Pl. Prot. Bull. Taiwan* **18**: 286 (on passion fruit; **56**, 3142).

Mehrlich, F. P. 1936. *Phytopathology* **26**: 23 (on pineapple with *P. cinnamomi* & *P. palmivora*; **15**: 378).

Middleton, J. T. 1943. *Bull. Torrey bot. Club* **70**: 244 (on *Atropa belladona*; **23**: 41).

Milne, D. L. et al. 1975. *Citrus Suptrop. Fr. J.* 502: 11 (on passion fruit, resistance; **56**, 799).

Mitra, M. 1929. *Trans. Br. mycol. Soc.* **14**: 249 (on cotton & guava; **9**: 240).

Mukerjee, N. 1966 & 1967. *Indian J. mycol. Res.* **4**: 61; **5**: 9 (on roselle; **51**, 1549, 1550).

Nambiar, K. K. N. et al. 1974. *J. Pl. Crops* **2**(1): 30 (on cardamom, fungicides; **54**, 2922).

Navaratnam, S. J. 1970. *Planter Kuala Lumpur* **46**: 220 (on *Hibiscus rosa-sinensis*; **50**, 1264).

Norton, D. C. et al. 1954. *Pl. Dis. Reptr* **38**: 854 (on watermelon).

Ocfemia, G. O. 1925. *Philipp. Agric.* **14**: 317 (on eggplant; **5**: 205).

Pegg, K. G. 1973. *Qd agric. J.* **99**: 655 (on passion fruit; **53**, 4533).

Pinckard, J. A. et al. 1973. *Phytopathology* **63**: 896 (on cotton bolls; **53**, 965).

Rao, V. G. et al. 1968 & 1970. *Mycopathol. Mycol. appl.* **34**: 346; **42**: 39 (on sugar apple, sapodilla & temp.; **47**, 2656; **50**, 1908).

Reitsma, J. et al. 1947. *Chronica Nat.* **103**: 60 (on eggplant in Java; **26**: 438).

Ruppel, E. G. et al. 1965. *Pl. Dis. Reptr* **49**: 676 (on *Tephrosia vogelii*; **45**, 535).

——. 1966. *Phytopathology* **56**: 489 (on *T. vogelii*, host differential susceptibility & histology; **45**, 2920).

Sarejanni, J. A. 1952. *Annls Inst. Phytopathol. Benaki* **6**: 14 (on eggplant; **33**: 275).

Sideris, C. P. et al. 1930. *Phytopathology* **20**: 951 (on pineapple; **10**: 325).

Szkolnik, M. 1951. *Pl. Dis. Reptr* **35**: 16 (on *Cinchona*; **30**: 388).

Turner, G. J. 1974. *Trans. Br. mycol. Soc.* **62**: 59 (on passion fruit; **53**, 3115).

Van den Boom, T. et al. 1970. *Phytophylactica* **2**: 71 (on passion fruit; **51**, 522).

Wheeler, J. E. et al. 1971. *Phytopathology* **61**: 1293 (on *Washingtonia* spp. inter alia; **51**, 2240).

Zeijlemaker, F. C. J. 1971. *Ibid* **61**: 144 (on *Acacia mearnsii*; **50**, 3192).

Phytophthora palmivora (Butler) Butler, *Scient. Rep. agric. Res. Inst. Pusa 1918–19*: 82, 1919.
Phytophthora faberi Maubl., 1909
P. theobromae Coleman, 1910.

Omnivorous tropical sp. of wide distribution; existing as morphologically and pathologically distinguishable strs. Growth rapid (opt. 27–28°) on standard organic media, forming abundant, large, terminal, prominently papillate sporangia in well developed sympodia. SPORANGIA when mature fall away from the sporangiophores by natural abscission with a short, often stout and more or less occluded pedicel; av. length 40–60 μ, av. breadth 25–35 μ, av. length: breadth ratio 1.4–2, but frequently appreciably smaller on natural substrata; discharging 10–40 (or more) zoospores. CHLAMYDOSPORES spherical and subspherical, terminal or intercalary, not separating by natural abscission, walls thin or thick, hyaline or brown, germinating by hyphae, 32–42 μ. Perfect state amphigynous, OOSPORES av. diam. 22–24 μ when suitable strs are grown together (after S. F. Ashby, *Trans. Br. mycol. Soc.* **14**: 34, 1929).

P. palmivora is treated sensu lato in these brief accounts of some of the more important diseases caused by this ubiquitous (and economically very important) plant pathogen of the wetter and warmer regions of the world. Newhook et al. (*Phytophthora* q.v.) remarked that the taxonomy of this sp., and those similar to it, is in a marked state of flux. These *Phytophthora* spp. have strongly papillate sporangia which do not proliferate internally, amphigynous antheridia and are nearly all heterothallic. Besides *palmivora* they include: *arecae, boehmeriae, botryosa, capsici, castaneae, citrophthora, heveae, meadii, mexicana* and *nicotianae* (Newhook et al.; Waterhouse 1963, *Phytophthora* q.v.). Many earlier workers (see for ref. Brasier et al. 1979; Waterhouse 1974) recognised forms in *P. palmivora* that varied in both morphology and pathogenicity. The emended description from Ashby, and based on Butler's fungus, is taken by Brasier et al. (1979) to be of the typical form of the fungus as fully described by Waterhouse (1974a). Brasier et al. (1979) refer to this type (or morphological form (MF) 1) as the S type. Waterhouse (1974a) also described atypical *palmivora* (MF 2) and further atypical strs; these all differ from the fungus described by Butler, Ashby and others. The main criterion for the typical form (MF 1) is that the sporangium bears a broad, short, occluded pedicel, 2–10 (mostly 5) μ in length. The sporangium is caducous but under natural conditions probably shed in water only. This form of the fungus may be considered *P. palmivora* sensu stricto (Waterhouse 1974b).

Griffin described 4 forms (MF 1–4) in the *palmivora* complex. These were separated on the basis of sporangial stalk, cultural and chromosome types, and their characteristics were described. Zentmyer et al. (1977; and see Idosu et al.) divided their isolates of this sp. from cacao into 3 or 4 groups, largely based on pedicel length. Groups I, II and III corresponded to MF 1, 3 and 4, respectively. Brasier et al. (1979), in one of the most important and recent taxonomic papers on *P. palmivora*, described their conclusions after examining *c.* 950 isolates from cacao (worldwide) and isolates from other hosts. Amongst the latter were some used by C. H. Gadd and S. F. Ashby in the 1920s. Their work is summarised below.

Most of the isolates could be placed in one of 3 distinct forms: S, L and MF 4. The first (as already stated) was taken to be typical *P. palmivora* (MF 1). The other 2 forms could not be identified with any known sp. of the genus, and both were considered to be distinct sp. The L form was described as a new sp., *P. megakarya* Brasier & Griffin; this corresponds to MF 3 (see Griffin). MF 4 was characterised but not formally described; this form from cacao was similar to isolates from black pepper (*Piper nigrum* q.v.). Other isolates from coconut and rubber (*Hevea brasiliensis*) were attributed to *palmivora* but were different from the S form. MF 2 was rejected as a valid taxon. The main distinguishing characteristics of the 3 forms are:

S (MF 1): n=9–12, small chromosomes; agar colony stellate, sharp edge, pedicel broad, short, occluded, mostly 5 μ long; sporangium with rounded base; mostly A2 compatibility type; oogonia spherical, meeting stalk fairly abruptly, stalk narrow, tubular or broadening towards oogonium, cardinal temps 10–11°, 28–30°, 34°; worldwide on cacao.

P. megakarya (L or MF 3): n=5–6, large chromosomes; agar colony faintly lobed, floral, diffuse edge; pedicel narrow, medium length, not occluded, 10–30 μ long; sporangium with rounded base; mostly A1 compatibility type; oogonia pyriform, tapering downwards to a funnel shaped base;

cardinal temps 10–11°, 24–26°, 29°; West Africa on cacao.

MF 4: n = 9–12, small chromosomes; agar colony petalloid, diffuse edge; pedicel narrow, long, not occluded, 20–150 μ long (up to 250 μ); sporangium with tapered base; A1 and A2 compatibility types; oogonia spherical to slightly oval, meeting stalk fairly abruptly, stalk narrow, tubular or broadening towards oogonium; cardinal temps 8°, 28–30°, 33°; Central and South America, West Indies on cacao.

Some conclusions of probable significance in the pathology of the *P. palmivora* complex may be drawn. Firstly, that some diseases attributed to this sp. may continue to be shown to be caused by hitherto undescribed spp. Secondly, that some of the spp. in the *palmivora* group (as given) may not be distinct. Thirdly, like the situation in citrus and rubber, several spp. in the group may well be found causing similar and/or different diseases in other single crops, as apparently happens in cacao. Besides taxonomy other general treatments were given by Chee (hosts and diseases caused), Tarjot (physiology) and Zentmyer (1974, mating types). (See also for fine structure: *Rev. Pl. Pathol.* **49**, 68; **51**, 2242; **52**, 2183, 2522, 3574; **54**, 2666; **55**, 579; **57**, 999.)

Brasier, C. M. 1972. *Trans. Br. mycol. Soc.* **58**: 237 (sexual mechanism; **51**, 3820).
—— et al. 1978. *Ibid* **70**: 295 (karyotype of MF 4; **57**, 4830).
—— et al. 1979. *Ibid* **72**: 111 (taxonomy of cacao isolates).
Chee, K. H. 1969. *Rev. appl. Mycol.* **48**: 337 (hosts, 82 ref.).
——. 1974. In Phytophthora *disease of cocoa* (editor, P. H. Gregory): 81, Longman (hosts & diseases).
—— et al. 1976. *Pl. Dis. Reptr* **60**: 866 (mating types in Malaysia; **56**, 3484).
Griffin, M. J. 1977. *PANS* **23**: 107 (taxonomy; **56**, 3954).
Huguenin, B. et al. 1971. *Annls Phytopathol.* **3**: 353 (formation & germination of oospores; **51**, 3774).
Idosu, G. O. et al. 1978. *Mycologia* **70**: 1101 (typical & atypical isolates from cacao).
Tarjot, M. 1974. In Phytophthora *disease of cocoa* (editor P. H. Gregory): 103, Longman (physiology).
Waterhouse, G. M. 1974a & b. *Ibid* 51, 71 (taxonomy & *Phytophthora* spp. on cacao).
Zentmyer, G. A. 1974. *Ibid* 89 (mating types).
—— et al. 1977. *Trans. Br. mycol. Soc.* **69**: 329 (taxonomy, also in *Can. J. Bot.* **56**: 1730; **57**, 1668).

CARICA PAPAYA

A papaw root rot, caused by *P. palmivora*, was described from Australia by Teakle. A similar and serious disease also occurs in Hawaii (at first attributed to *P. nicotianae* var. *parasitica*). Watersoaked lesions are also caused on the fruit and green stem. The fruits become mummified and fall. On the upper stem cankers develop and these may be girdling. Inoculum builds up in the soil which becomes seriously infested. A replant problem, where *Pythium aphanidermatum* (q.v.) may be also involved, therefore arises. In Hawaii epidemics of the above ground form of the disease occur in wet and windy weather. Sporangia are trapped in splash droplets and not during dry weather (Hunter et al.). New outbreaks can arise from seeds of infected fruit or from infected seedlings. Ko described a method of planting seedlings in non-infested soil; this protected the young plant until mature plant resistance arose; see also Kliejunas et al.; Ko et al.; Ramirez et al. for infection. Above ground control is obtained with fungicides: captan, captafol, mancozeb, maneb and basic copper sulphate.

Aragaki, M. et al. 1967. *Mycologia* **59**: 93 (temp. & sporangial germination; **46**, 2288).
Hine, R. B. et al. 1965. *Phytopathology* **55**: 1223 (inactivation by papain & proteolytic enzymes; **45**, 1444).
Huang, T. H. et al. 1976. *Pl. Prot. Bull. Taiwan* **18**: 293 (fungicides; **56**, 3139).
Hunter, J. E. et al. 1969. *Ann. appl. Biol.* **63**: 53 (symptoms, spread & fungicides; **48**, 1868).
—— ——. 1974. *Phytopathology* **64**: 202 (dispersal; **53**, 3567).
Kliejunas, J. T. et al. 1974. *Ibid* **64**: 426 (effect of motility on seedling disease & substrate colonisation in soil; **53**, 4070).
Ko, W. H. 1971. *Ibid* **61**: 780 (root rot control; **51**, 515).
—— et al. 1974. *Ibid* **64**: 1307 (infection & colonisation in soil; **54**, 1821).
Nakasone, H. Y. et al. 1973. *Proc. trop. Region Am. Soc. hort. Sci.* **17**: 176 (tolerance in host; **55**, 1375).
Parris, G. K. 1942. *Phytopathology* **32**: 314 (symptoms & pathogenicity; **21**: 382).
Ramirez, B. N. et al. 1975. *Ibid* **65**: 780 (inoculum density & infection in soil; **55**, 817).
Teakle, D. S. 1957. *Qd J. agric. Sci.* **14**: 81 (root rot in Australia; **37**: 415).

HEVEA BRASILIENSIS

The diseases described here are patch canker and

black stripe. The latter is also caused by *P. botryosa* and *P. meadii* (q.v.). Under these two spp. the rubber diseases of the canopy (green fruit and green branch decay, and abnormal leaf fall) are described (and see Agnihothrudu). *P. palmivora* can also cause canopy diseases although its importance in this respect is less than that of *P. botryosa* in Malaysia (W.) and *P. meadii* in Sri Lanka. But in India (and Costa Rica) *P. palmivora* may be the main cause of abnormal leaf fall with *P. meadii* also playing a role in India. *P. heveae* (q.v.), originally recorded from rubber, appears to be of less economic importance than the other 3 spp. Further work is needed on the interplay and role of these fungi (whose taxonomy is in a fluid state) in the diseases that they cause on rubber. Chee (1975) found a continuous range of variation in pathogenicity of *P. palmivora* isolates (in Malaysia) to tapping panels and detached petioles.

The symptoms (occurring in wet weather) of black stripe (also called black thread, black line or stripe canker, and bark or cambium rot) first appear as small areas of sunken, discoloured bark, just above the tapping cut. On paring away the bark narrow, vertical, black lines (invaded vessels) appear. These lines coalesce and may extend 15 cm below the cut and 5 cm above it; splits and cankers with latex exudation occur. The cambium is destroyed and bark regeneration prevented. Atypical, patch like lesions can also form on the renewing bark; these are not the result of coalescence of stripes. Patch canker is a condition of the untapped bark. Discrete, irregular patches are seen 0.5–2.5 m above soil level. Latex oozes as the bark and wood rots, and latex pads cause bulging and splitting; the callus formed on healing is difficult to tap. Infection of the tapping cut probably occurs through wash from above and tapping knives do not spread the disease. Clones vary in susceptibility. Infections are often worse near ground level, at high planting densities and through any factors that hinder the drying of the tapping cut after rain. Where the disease is severe a fungicide needs to be applied after every tapping (brush or spray) above and below the cut. Stickers have been used with antimucin, captafol, captan and cycloheximide in efforts to reduce the number of applications (e.g., Anon.; Chee 1972; Schreurs).

Agnihothrudu, V. 1975. In *Advances in mycology and plant pathology* (editors S. P. Raychaudhuri et al.; abnormal leaf fall of rubber, 27 refs.).

Anon. 1970. *Plr's Bull. Rubb. Res. Inst. Malaya* 109: 105 (black stripe, advisory; **50**, 1359).

Carpenter, J. B. 1954. *Phytopathology* **44**: 597 (leaf fall in Costa Rica; **34**: 318).

Chee, K. H. 1968. *Pl. Dis. Reptr* **52**: 132 (patch canker in Malaysia; **47**, 3550).

——. 1972. In *Proc. Rubb. Res. Inst. Malaysia Plr's Conf.*: 155 (fungicides; **54**, 1416).

——. 1973. *Mycopathol. Mycol. appl.* **50**: 275 (phenotypic differences among single oospore cultures of *Phytophthora palmivora* & *P. botryosa*; **53**, 664).

——. 1973. *Trans. Br. mycol. Soc.* **61**: 21 (formation, germination & survival of chlamydospores; **52**, 4206).

——. 1975. *Ibid* **65**: 153 (pathogenicity; **55**, 860).

Darley, E. F. et al. 1952. *Phytopathology* **42**: 547 (black stripe in Liberia; **32**: 508).

Dastur, J. F. 1916. *Mem. Dep. Agric. India bot. Ser.* **8**: 217 (history).

Deconinck, G. O. 1969. *J. Rubb. Res. Inst. Malaya* **21**: 88 (fungicides in Cambodia; see **49**, 2165e).

Manis, W. E. 1954. *Pl. Dis. Reptr* Suppl. 225: 49 (leaf fall & dieback in Costa Rica; **34**: 1777).

Schreurs, J. 1969 & 1971. *Neth. J. Pl. Pathol.* **75**: 113; **77**: 113 (fungicides in Liberia; **48**, 1935; **51**, 1852).

——. 1972. *Meded. LandbHoogesch. Wageningen* 72–63, 130 pp. (general in Liberia; **52**, 1655).

PALMAE

Several palm spp. can be infected (Reinking 1923) and bud rot, as the disease is usually called, has been described on areca palm (*Areca catechu*), palmyra palm (*Borassus flabellifer*, Butler; McRae), coconut (Shaw et al.; Tucker 1926; and others) and *Sabal causiarum* (Tucker 1927). Joseph et al., in an unusually recent study, successfully infected oil palm. *P. palmivora* caused a dry rot before the development of symptoms which are associated with other micro-organisms. Although severe damage has been described in the past from widely separated areas (e.g., on coconut in Jamaica and the Philippines, and on palmyra in India and Puerto Rico) bud rot is probably only locally serious when conditions for infection are favourable.

There is some doubt as to whether mature palms can be killed without damage through high winds. Infection occurs at the base of the youngest leaves and in the growing point. The first symptoms are usually the collapse of the youngest leaves, infection spreads outwards until all the central fronds are dead. In coconut this gives the characteristic appearance of a dead centre and a drooping fringe of green, healthy leaves. The bud becomes completely necrotic and spread downwards into harder tissue occurs, this invasion is delimited by a red brown margin.

The whole tree eventually dies. Nut fall is not particularly characteristic but young inflorescences with small nuts can be attacked and these drop. Sometimes palms will recover from infection, and in coconut this gives an appearance which has been called little leaf or bitten leaf. The new leaves which appear centrally often remain small but may eventually become normal. When infection occurs higher up, in the folded lamina of a young leaf, it spreads through the leaf and when this expands rows of necrotic spots appear; these rows are particularly distinctive in the digitate or palmate leaves. In the heavy rainfall regions, where bud rot usually occurs, there is little preventive control apart from ensuring good growing conditions. Infected palms should be destroyed. Bordeaux spraying has often been used.

Butler, E. J. 1910. *Mem. Dep. Agric. India bot. Ser.* **3**: 221 (general, particularly on palmyra).
Corbet, M. K. 1959. *Principes* **3**: 117 (on coconut).
Gadd, C. H. 1927. *Ann. Bot.* **41**: 253 (comparison, *Phytophthora palmivora* & *P. arecae*; **6**: 608).
Joseph, T. et al. 1975. *Pl. Dis. Reptr* **59**: 1014 (in coconut, symptoms & inoculations; **55**, 3688).
McRae, W. 1912. *Agric. J. India* **7**: 272 (leaf symptoms on palmyra).
——. 1923. *Mem. Dep. Agric. India bot. Ser.* **12**: 21, 57 (history & inoculations; **3**: 270).
Reinking, O. A. 1919. *Philipp. J. Sci.* **14**: 131 (general on coconut).
——. 1923. *J. agric. Res.* **25**: 267 (on coconut, morphology of *P. palmivora*; **3**: 396).
Seal, J. L. 1928. *Tech. Bull. Fla agric. Exp. Stn* 199, 87 pp. (on coconut).
Shaw, F. J. F. et al. 1914. *Annls Mycol.* **12**: 251 (on coconut in India).
Tucker, C. M. 1926 & 1927. *J. agric. Res.* **32**: 471; **34**: 879 (on coconut & *Sabal causiarum* in Puerto Rico; **5**: 488; **6**: 665).

PIPER NIGRUM

Foot rot of black pepper has a long record of destruction in the East Indies. Muller determined its aetiology and gave a full account of the disease in Java and Sumatra. A later, detailed study was made in Malaysia (Sarawak) by Holliday et al. Foot rot occurs in India; see Nambiar et al. for a general account. The earlier literature is fully covered by these three papers. This brief description of foot rot is placed under *P. palmivora* but the taxonomic position of the pathogen is not clear. The possibility that more than one *Phytophthora* sp. may cause disease(s) in this crop needs to be borne in mind. Holliday et al. considered their isolates to be atypical *P. palmivora*. Turner (1969a) also thought that foot rots of betel (*Piper betle*) and black pepper were due to such atypical forms. The disease on betel (a serious disease in India) had, by others, been attributed to *Phytophthora nicotianae* var. *parasitica* (q.v.). A wilt of *Piper methysticum* (yangona, kava) in Fiji may also be caused by *Phytophthora palmivora*. Tsao et al. examined isolates of *Phytophthora* from black pepper in Thailand. These were unlike any isolates of the genus previously reported from this crop. They differed from *P. palmivora* primarily in the long pedicels (av. 69 μ and 75 μ in 2 isolates) and the umbellate arrangement of the sporangia. Isolates from Africa, Central America and Malaysia were similar. Brasier et al. (1978, *P. palmivora* q.v.) considered that the fungus of Tsao et al. was the same as their morphological form 4 and probably a new sp. Their isolates of this form from cacao and black pepper were similar, and also resembled the form described by Holliday et al. Apart from Asia this foot rot has been found in Brazil, Jamaica and Puerto Rico. Turner (1971) found that the isolates from *Piper* did not infect some other crops and weeds (*Peperomia* spp. were not attacked). *Piper* spp. from S.E. Asia were susceptible but those from elsewhere varied in susceptibility.

If infection begins in the root system the first above ground symptoms are a yellowing, leaf droop and wilt followed within a few days by an almost complete leaf fall. At or below soil level on the stem there is a sharply defined edge between a dark, necrotic cortical lesion and the non-infected yellowish tissue. Similar diseased and healthy margins occur in the root system before the main stem becomes infected via the roots. In suitable weather conditions initial infection may be found on the leaves up to 3 m above ground; infection can then spread to the branches, inflorescences and fruit. The actively spreading leaf necrosis, under wet conditions, has a characteristic fimbriate edge. In alternating wet and dry conditions the leaf lesion becomes concentrically zoned. The leaf usually dehisces before the lamina becomes entirely necrotic. Young roots and leaves are readily infected with zoospores, in the latter case via the undersurface of immature leaves. Infection is spread in water and water droplets, and probably in soil and host debris. Under the clean weeding, traditional growing methods of Chinese farmers in S.E. Asia, and in high rainfall areas, spread of infection can be extremely rapid. A holding of > 1000 vines can be destroyed in 3–4 months, and spread from holding to holding (where contiguous) is equally rapid. The giant African snail (*Achatina fulica*) can also spread infection in its faeces. Young black pepper vines, in newly planted areas, generally escape infection. The pathogen clearly has some saprophytic and/or dormant existence. The former may occur on wild *Piper* spp. and Brasier has shown experimentally that oospores can be formed in the main host.

Apart from general cultural and preventive measures little can be recommended for control. Under certain conditions fungicide drenches and spraying might be effective. Several *Piper* spp. (*P. aduncum, P. columbrinum* and *P. oblianum*) have been found to be resistant and stocks from such material is one possibility. *P. nigrum* cvs in India, Indonesia and Malaysia have shown indications of differential susceptibility but whether this is sufficient to be of practical use remains in doubt.

Albuquerque, F. C. de 1968. *Pesq. agropec. bras.* **3**: 141 (*Piper colubrinum* as a resistant stock; **49**, 2955).

Alconero, R. et al. 1972. *Phytopathology* **62**: 144 (in Brazil & Puerto Rico; **51**, 3483).

Brasier, C. M. 1969. *Trans. Br. mycol. Soc.* **52**: 273 (formation of oospores in vivo; **48**, 2154).

Holliday, P. et al. 1963. *Phytopathol. Pap.* 5, 62 pp. (monograph, 95 ref.; **43**, 1104).

Leather, R. I. 1967. *Pl. Prot. Bull. F.A.O.* **15**: 15 (in Jamaica; **46**, 3158).

Muller, H. R. A. 1936. *Meded. Inst. PlZiekt. Buitenz.* 88, 73 pp. (monograph; **16**: 559).

Nambiar, K. K. N. et al. 1977. *J. Pl. Crops* **5**: 92 (general, particularly India, 85 ref.).

Turner, G. J. 1967. *Trans. Br. mycol. Soc.* **50**: 251 (spread by *Achatina fulica* in Sarawak; **46**, 3159).

——. 1969a & b. *Ibid* **52**: 411; **53**: 407 (on *Piper betle* in Sarawak, leaf symptoms; **48**, 3100a; **49**, 1447).

——. 1971. *Ibid* **57**: 61 (inoculation of *Piper* & *Peperomia* spp. & other plants; **51**, 558).

——. 1973. *Ibid* **60**: 583 (variation in pathogenicity; **52**, 4174).

Tsao, P. H. et al. 1977. *Mycologia* **69**: 631 (identity of *Phytophthora* sp. from black pepper in Thailand; **57**, 729).

THEOBROMA CACAO

Like the situation in citrus and rubber (*Hevea brasiliensis*) it is likely that the diseases of cacao attributed to *P. palmivora* will be found to have a more complex aetiology and epidemiology with more than one sp. of the genus involved. Such a situation may well be found in other tropical crops. The disease on cacao is usually called black pod. This is an old, established name and moreover this form of the disease results in the most crop loss, which is considerable in many regions where cacao is grown. The literature on black pod is very large but was well brought together by Gregory. Various authors (some are given here) in this book fully considered all aspects of the disease and its control, taxonomy of the pathogen (and related spp.), and diseases caused by other cacao pathogens with which black pod can be confused. See also the recent account by Thorold in *Diseases of cocoa*, and Thorold's review.

Fruits (called pods), unwounded, are attacked at all ages. Infection of young pods (cherelles) should not be confused with the physiological condition called cherelle wilt. On older pods the first sign of infection is a deep brown lesion on the shell; this is clearly circular unless it begins high up on the shoulder. The necrosis spreads concentrically, rapidly and over the whole surface; its margin with the still apparently healthy tissue is fairly sharp. Under wet conditions a white bloom of sporangia forms on the surface. The necrosis eventually spreads to the internal tissues but in a mature pod the seed may be unaffected. Pod infection should not be confused with an attack by *Botryodiplodia theobromae* (q.v.) which may be associated with a wound or which may follow the primary attack by *P. palmivora*. The rots caused by other fungal pathogens on cacao pods may also be confused with black pod infection, although they (*Crinipellis perniciosa, Moniliophthora roreri* and *Trachysphaera fructigena* q.v.) are of limited distribution. In some parts of the tropics and on some cvs (criollo groups) a canker form of the disease can be serious (see Firman for some recent work). Both the canker and cushion infection aspects have been somewhat neglected but were usefully discussed by I. D. Firman and A. L. Wharton, respectively (see Gregory). The fungus probably enters the bark via wounds. But the origin of most cankers and cushion infection is from infected pods. Conversely, pods can be invaded by mycelium from cushions and cankered bark. In the early stages canker is difficult to detect. Later there may be an exudation of reddish fluid through cracks in the bark. The tissue beneath the bark is at first a watersoaked grey green, usually becoming reddish brown (claret) and then brown. The fungus spreads faster in the cambium and inner cortex than in the outer bark. Canker plays a part in the cacao dieback syndrome. *P. palmivora* can also infect young vegetative shoots causing a general collapse and death of shoots and leaves in both old and young plants. It also occurs in cacao soils (see Newhook et al.) and can infect roots.

Few useful generalisations can be made about the epidemiology and control of cacao black pod. The factors concerned vary between geographical areas. The three dominant factors affecting disease build up are climate, strs (or spp.) of the pathogen(s) and agronomic practices. The fungus is endemic in all cacao growing areas and like all zoosporic fungi is favoured by wet conditions. A dry season of some months will probably have a significant effect on reducing disease incidence. With the long cropping periods in cacao it is important to pin point the main sources of inoculum, which is spread mostly by rain splash. Undoubtedly, the main one is the infected pod on the tree. Hence the often demonstrated fact that frequent harvesting of the pods (i.e. every 1–3 weeks) reduces losses. This simple cultural practice not only reduces overall inoculum but saves beans in infected pods, and reduces the risks of cushion attack. Improvements in the standards of cultivation will probably do more than any other single factor to make control measures more effective.

Although there is evidence for the existence of strs which differ in pathogenicity the precision with which they can be defined is inadequate. G. A. Zentmyer (see Gregory) gave a significant example of differential pathogenicity between cacao isolates from 5 countries. The cankered areas caused by them on seedlings of one clone differed from 78.6 cm^2 to 5.4 cm^2. Physiologic races in the generally accepted sense of the term have not been demonstrated. Given good husbandry control can be strengthened by applying fungicides and planting material with effective resistance. There is a long history of copper spraying against black pod (see A. M. Gorenz; A. E. H. Higgins & J. E. Clayphon in Gregory). Although pod resistance (in varying degrees) has been shown, its practical importance is not clear. Selections found resistant in one geographical area may not be so in another. The control of black pod is made more difficult not only by the extremely variable standards of cultivation and the differing effects of climate, but by the inherent variability of the pathogen and its extensive host range.

Dakwa, J. T. 1973. *Ghana J. agric. Sci.* **6**: 93 (relationship between disease incidence & weather; **53**, 1752).

Evans, H. C. 1973. *Ann. appl. Biol.* **75**: 331 (spread by invertebrates in Ghana; **53**, 1753).

Firman, I. D. 1978. *Trop. Agric. Trin.* **55**: 269 (incidence of canker & inoculation; **57**, 5373).

Gregory, P. H. (editor) 1974. Phytophthora *disease of cocoa*, 348 pp., 776 ref., Longman.

Hicks, P. G. 1975. *Papua New Guin. agric. J.* **26**: 10 (investigations 1962–71; **55**, 2180).

Jackson, G. V. H. et al. 1978. *Trans. Br. mycol. Soc.* **71**: 239 (sources of inoculum).

Lockwood, G. et al. 1978. *J. hort. Sci.* **53**: 105 (self incompatibility & harvesting in relation to losses from black pod; **57**, 4380).

Medeiros, A. G. 1977. *Ceplac Publ. Especial* 1, 220 pp. (sporulation, epidemiology & chemical control in Brazil, 182 ref.; **57**, 2057).

Newhook, F. J. et al. 1977. *Trans. Br. mycol. Soc.* **69**: 31 (*P. palmivora* in cacao plantation soils in Solomon Islands; **57**, 497).

Thorold, C. 1967. *Rev. appl. Mycol.* **46**: 225 (review, 128 ref.).

OTHER CROPS

These few examples of lesser diseases caused by *P. palmivora* show that, as for the major diseases, few parts of the plant are free from infection. But collar rots of young plants, cankers in both young and mature woody hosts, and fruit rots appear to predominate. Grasses and conifers are not apparently attacked. On avocado Conover described a leaf infection causing lamina distortion, and there was invasion of the stem. Examples of collar rots are on safflower (*Carthamus tinctorius*, Malaguti) and *Cinchona* (Celino; Kheswalla). On citrus Frezzi et al. found a brown fruit rot and Uppal et al. a gummosis (both more commonly caused by other *Phytophthora* spp.). Patch canker of durian (*Durio zibethinus*) is an old and significant disease of this fruit in Malaysia. Budded trees can be frequently attacked. Brownish red gum exudations appear on the stem whose bark turns dark brown or black. The diseased cortex becomes pinkish with darker specks compared with the creamy coloured healthy tissue. The plant is girdled and dieback results. Infection occurs through pruning wounds. Durian clones differ in their susceptibility (Navaratnam; Tai; Thompson 1934). A canker of mango was described by Lourd et al. Lesions on the trunk and branch showed a gum flow. Several workers have reported on infection of orchids, including vanilla (*Vanilla fragrans*; Hine; Maublanc et al.; Thompson 1959). Both a collar rot and fruit infection of okra (*Hibiscus esculentus*) were described by Balakrishnan. Other examples of fruit rots are in pineapple (heart rot; Boher), sapodilla (*Manilkara achras*; Rao et al.) and tomato (Ramakrishnan et al.).

Balakrishnan, M. S. 1947. *Proc. Indian Acad. Sci.* Sect. B. **26**: 142 (on okra; **27**: 172).

Boher, B. 1974. *Fruits* **29**: 721 (on pineapple, penetration; **54**, 2910).

Celino, M. S. 1934. *Philipp. Agric.* **23**: 111 (on *Cinchona*; **13**: 812).

Conover, R. A. 1948. *Phytopathology* **38**: 1032 (on avocado; **28**: 343).

Frezzi, M. J. et al. 1943. *Revta argent. Agron.* **10**: 227 (on citrus fruit; **23**: 15).

Hine, R. B. 1962. *Pl. Dis. Reptr* **46**: 643 (on orchids; **42**: 126).

Kheswalla, K. F. 1935. *Indian J. agric. Sci.* **5**: 485 (on *Cinchona*; **15**: 410).

Lourd, M. et al. 1975. *Fruits* **30**: 541 (on mango; **55**, 3674).

Malaguti, G. 1950. *Phytopathology* **40**: 1154 (on safflower; **30**: 344).

Maublanc, A. et al. 1928. *Agron. colon.* **17**(123): 77 (on vanilla; **7**: 599).

Navaratnam, S. J. 1966. *Malayan agric. J.* **45**: 291 (on durian; **45**, 3181).

Ramakrishnan, T. S. et al. 1947. *Proc. Indian Acad. Sci.* Sect. B. **25**: 39 (on tomato fruit; **26**: 318).

Rao, V. G. et al. 1962. *Pl. Dis. Reptr* **46**: 381 (on sapodilla; **41**: 732).

Tai, L. H. 1971. *Malaysian agric. J.* **48**(1): 1 (on durian; **51**, 1709).

——. 1973. *MARDI Res. Bull.* **1**(2): 5 (durian clonal susceptibility; **54**, 939).

Thompson, A. 1934. *Malayan agric. J.* **22**: 369 (on durian; **14**: 46).

——. 1959. *Ibid* **42**: 83 (on *Vanda*; **39**: 416).

Uppal, B. N. et al. 1936. *Indian J. Agric.* **6**: 803 (on citrus; **15**: 795).

Phytophthora phaseoli Thaxter, *Bot. Gaz.* **14**: 27, 1889.
(figures in *Rep. Conn. agric. Exp. Stn* 1889 Pl. III, 1890)

P. infestans var. *phaseoli* (Thaxter) Leonian, 1925.

This sp. is, morphologically, very near *P. infestans*, q.v. for a description. It differs in host range (being restricted to *Phaseolus*) and has been largely studied on *P. lunatus* (lima bean), in the longer SPORANGIOPHORES, larger SPORANGIA (22 × 32 μ, max. 27 × 50 μ), smaller OOGONIA (30 μ, max. 38 μ diam.), OOSPORES and ANTHERIDIA (10 × 15 μ, max. 11.5 × 17 μ). Slow growing in culture with an opt. temp. of 15–20°. Oospores occur in the host and are abundant in pods.

Lima bean downy mildew has been reported from only a few countries: Italy, Mexico, Philippines, Puerto Rico, Sri Lanka, USA (E. and central), USSR (Caucasas) and Zaire (Map 201). The disease has been studied almost entirely in USA, particularly by Hyre and others. There is a brief review of the earlier literature by several authors in *Pl. Dis. Reptr* Suppl. 257, 1959. The disease is most conspicuous on the pods on which a thick, white, felt-like growth develops; pods that are attacked when young remain undeveloped. Seedlings are most susceptible for *c.* 7 days after emergence. After 14 days the mature tissues are somewhat resistant, with only the apical areas of stems, leaves and flowers, and petioles of young leaves susceptible. Leaves of maturing plants in the field are not severely attacked. Penetration by the zoospore is cuticular but mature tissue is highly resistant. At 20° and near RH 100% *c.* 40% of the sporangia are viable after 27 hours, viability decreases rapidly at a lower RH. There is most indirect germination at 10–15° and the zoospore motility period decreases typically with each rise in temp. The opt. temp. for zoospore germination is *c.* 20° which is also about opt. for disease development. On lima bean seedlings most sporangia are formed at 18–22°. Sporulation is higher with an 8 hour day (27°, RH 70%) and 16 hour night (22°, RH 100%) regime than with one of 20 and 15°, respectively. Host colonisation decreases with rise in temp., and at 30–32° or more significant reductions in fungal growth in vivo occur. Most oospores are produced in lima bean seedlings which are transferred after infection to 15.5–18° compared with higher and lower temps. The pathogen survives between crops as oospores.

Resistance (inherited as a single dominants to each race) is known and has been bred for successively as each new race appeared. Wester (1970) gave 6 differential lima bean cvs on which 3 (A, B, C) may be distinguished. Thomas et al. (1976) designated race D. A chemical factor in resistance to race A has been suggested (Valenta et al.). Maneb, zineb and tribasic copper sulphate are effective fungicides. In

USA (N.E.) forecasting is done. In New Jersey Scarpa et al. reported that warnings were given when the dewpoint was equal to or greater than 20.5° for 12 hours and forecast to persist for at least another consecutive 28 hours.

Clinton, G. P. 1906. *Rep. Conn. agric. Exp. Stn* 1905: 278 (general).
Cox, R. S. et al. 1951. *Pl. Dis. Reptr* **35**: 354 (overwintering; **31**: 44).
——. 1954. *Phytopathology* **44**: 325 (effect of temp. on disease development; **34**: 124).
Crossan, D. F. et al. 1957. *Pl. Dis. Reptr* **41**: 156 (fungicides & resistance; **37**: 62).
Cunningham, H. S. 1947. *Bull. N.Y. St. agric. Exp. Stn* 723, 19 pp. (symptoms & control; **27**: 214).
Drechsler, C. et al. 1962. *Phytopathology* **52**: 164 (temp. & oospore development in vivo; **41**: 764).
Goth, R. W. et al. 1963. *Ibid* **53**: 233 (culture in vivo & in vitro; **42**: 600).
Hyre, R. A. 1953. *Ibid* **43**: 419 (temp. & other factors affecting viability & growth; **33**: 274).
——. 1957. *Pl. Dis. Reptr* **41**: 7 (forecasting; **36**: 630).
—— et al. 1962. *Ibid* **46**: 393 (forecasting).
——. 1964. *Phytopathology* **54**: 181 (effect of high temp. following infection; **43**, 2143).
——. 1966. *Ibid* **56**: 1274 (effect of high post-inoculation temp. on colonisation; **46**, 800).
——. 1966. *Pl. Dis. Reptr* **50**: 734 (effect of low post-inoculation temp. on colonisation; **46**, 801).
—— ——. 1967. *Ibid* **51**: 253 (effect of temp. on sporulation in vivo; **46**, 2344).
—— ——. 1968. *Ibid* **52**: 507 (low temp. storage; **47**, 3274).
—— ——. 1969. *Phytopathology* **59**: 514 (effect of day & night temps on host colonisation; **48**, 2665).
Saksena, R. K. et al. 1943. *Proc. Indian Acad. Sci.* Sect. B **18**: 45 (effect of N & vitamins in vitro; **23**: 352).
Scarpa, M. J. et al. 1964. *Pl. Dis. Reptr* **48**: 77 (forecasting; **43**, 1787).
Thomas, H. R. et al. 1952. *Phytopathology* **42**: 43 (race A, resistance & its inheritance; **31**: 366).
—— ——. 1976. *Pl. Dis. Reptr* **60**: 308 (race D; **56**, 1348).
Valenta, J. R. et al. 1962. *Phytopathology* **52**: 1030 (phenolic acid as resistance factor; **42**: 355).
Wester, R. E. et al. 1959. *Pl. Dis. Reptr* **43**: 184 (race B; **38**: 437).
—— ——. 1966. *Phytopathology* **56**: 95 (survival as oospores; **45**, 1583).
——. 1968. *Pl. Dis. Reptr* **52**: 563 (resistance to races A & B; **47**, 3273).
——. 1970. *Phytopathology* **60**: 1856 (host differentiation of 3 races; **50**, 2589).
—— ——. 1972. *Pl. Dis. Reptr* **56**: 65 (resistance to races A, B & C; **51**, 2981).
Zaumeyer, W. J. et al. 1969. *Ibid* **53**: 25 (spread of race B; **48**, 1402).

Phytophthora porri Foister, *Trans. Proc. bot. Soc. Edinb.* **30**: 277, 1931.

MYCELIUM branched, non-septate when young, septate and empty when old, coiled in spirals in region of sexual activity. SPORANGIA inversely pyriform or oval, 51 × 35 (82–31 × 52–23) μ, formed at the end of SPORANGIOPHORES or intercalary, usually without a papilla, occasionally with a broad papilla 5 × 10–12 μ, sometimes with a beaked papilla 12 × 10–12 μ; apical thickening 2.5–5 μ and hyaline, mouth of discharge broad, germination by a tube or by zoospores, 10–15 μ, which may germinate in turn by zoospores (repetitional diplanetism) or by a germ tube. ANTHERIDIA oval or flattened spheres, terminal or intercalary, not on the same hypha as oogonium; when amphigynous, 12.5 (19–7.3 μ), and when paragynous, 10–7.3 μ. Mature OOGONIUM spherical, with an unevenly thickened wall, 38–39 (46–29) μ, fertilised mostly by one paragynous antheridium, sometimes by several paragynous antheridia, sometimes by an amphigynous antheridium, and sometimes by both types at once. OOSPORES spherical, honey yellow when old, 32–33 (39–22) μ, with a thick wall, 3–4.5 μ. Conidia produced abundantly in water and in the host, less often in solid media; SEXUAL ORGANS formed abundantly in solid and liquid media, and in the host. Parasitic on the leaves and stems of *Allium porrum*, causing watersoaked areas followed by a whitening of the tips of leaves and other affected parts. (C. E. Foister l.c.; see also G. M. Waterhouse, *Mycol. Pap.* 92: 16, 1963; Stamps.)

The fungus, causing white tip disease, was first described from leek (*Allium ampeloprasum* var. *porrum*). It infects other *Allium* spp. according to Yokoyama who found that it was pathogenic on other plant genera only after wounding. Legge reported *P. porri* from *Campanula*; but isolates from this host did not infect leek and vice versa. Other hosts include tulip (Katsura et al. 1969) and cabbage; on the latter Geeson described a storage rot. The fungus has been reported from few countries (Map 204): Irish Republic, Japan, Netherlands, Norway, Switzerland and UK.

On *Allium* the leaf tips die back for several cm and this area becomes white. Leaves may become twisted and watersoaked areas develop in the middle or at the base of the leaf. The plants collapse. An infected field is a much lighter green than a healthy one. Rots in storage and damping off can occur. In the cabbage storage rot the heads develop extensive brown rots. The lower parts of outer, wrapper leaves are desiccated and yellow or brown. A dark brown discolouration spreads from the base of the stem to the leaves. The opt. temp. for growth of the fungus was given as 25° by Foister, and 15–20° by Geeson and Katsura et al. (1969). For control continuous cropping with *Allium* spp. should probably be avoided. Outbreaks in Japan can be forecast with a 75% probability (Yokoyama). Captafol is used in S.E England on leeks (see Griffin) and captan was effective in the Netherlands (Van Bakel).

Foister, C. E. 1931. *Trans. Proc. bot. Soc. Edinb.* **30**: 257 (aetiology, *Phytophthora porri* sp. nov.; **11**: 151).

Geeson, J. D. 1976. *Pl. Pathol.* **25**: 115 (storage rot of cabbage; **56**, 899).

Griffin, M. J. et al. 1977. *Ibid* **26**: 149 (on onion; **57**, 3205).

Katsura, K. et al. 1969. *Ann. phytopathol. Soc. Japan* **35**: 55 (symptoms, hosts, temp.; **48**, 2088).

——— ———. 1971. *Sci. Rep. Kyoto prefect. Univ. Agric.* **23**: 34 (chlamydospores; **53**, 807).

Legge, B. J. 1951. *Trans. Br. mycol. Soc.* **34**: 293 (on *Campanula*; **31**: 65).

Stamps, D. J. 1978. *C.M.I. Descr. pathog. Fungi Bact.* 595 (*P. porri*).

Van Bakel, J. M. M. 1964. *Meded. Dir. Tuinb.* **27**: 198 (fungicides; **43**, 2464).

Yokoyama, S. 1976. *Bull. Fukuoka agric. Exp. Stn* 22, 55 pp. (hosts, pathogenicity, loss, epidemiology & forecasting; 31 ref.; **56**, 4799).

Phytophthora vignae Purss, *Qd J. agric. Sci.* **14**: 141, 1957 (also as *Bull. Div. Pl. Ind. Qd* 107).

MYCELIUM hyaline, non-septate when young, occasional septa developing when old; within host tissue mycelium mostly intercellular but some intracellular. CHLAMYDOSPORES variable, mainly spherical, 12–21 (av. 17) μ, terminal and intercalary. Hyphal swellings in both solid and liquid media, irregular, both terminal and intercalary, single. SPORANGIA on host and in liquid nutrient media developed frequently on simple unbranched sporangiophores, frequently on a simple, monochasial, sympodial sporangiophore, rarely on a compound, sympodial SPORANGIOPHORE; sporangia variable in shape, mostly ovoid to obpyriform, $24–72 \times 15–54$ (av. 48×27) μ, non-papillate or inconspicuously papillate, germinating by zoospores. ZOOSPORES differentiated within sporangium, liberated singly, reniform when motile, spherical when encysted, av. diam. of encysted zoospores 10 μ. Homothallic; sexual organs abundant in artificial media and host tissue. OOGONIA colourless, smooth, almost spherical, $27–42 \times 24–46$ (av. 33×31) μ. ANTHERIDIA amphigynous, variable in shape from almost spherical to ovate, $12–27 \times 9–18$ (av. 16×15) μ. OOSPORES colourless to light brown, 18–32 (av. 26) μ in diam., oogonial cavity not filled. (G. S. Purss l.c.; see also G. M. Waterhouse, *Mycol. Pap.* 92: 17, 1963.)

Stem rot of cowpea (*Vigna unguiculata*) was described from Australia (Queensland) and was later reported from New South Wales. Elsewhere it has only been found in India (Uttar Pradesh, on cowpea and *Cajanus cajan*, pigeon pea) and Japan (on *Phaseolus angularis*, adzuki bean); see Map 512. Affected plants turn yellow and wilt rapidly under dry conditions; in moist weather they may survive for a considerable time. Plants show a brown, sunken, girdling lesion on the stem, beginning at soil level. The margin is usually watersoaked and when wet is covered by growth of the fungus with sporangia. The lesion may reach the top of the stem and a brown discolouration spreads a little internally. Lesions high up on an otherwise healthy stem may be common. The fungus builds up in the soil and the disease spreads rapidly in wet weather. The opt. temp. for in vitro growth is 28–30° and severe infection can occur at 19–28°. Four races were described from Australia; and 3 genes for resistance found in a collection of 500 cowpea strs and cvs. The cv. Red Caloona (Purss 1975) was reported to be resistant so far to the Queensland races of the fungus.

Kitazawa, K. et al. 1978. *Ann. phytopathol. Soc. Japan* **44**: 528 (in Japan).

Partridge, J. E. et al. 1976. *Phytopathology* **66**: 426 (association of phytoalexin kievitone with single gene resistance; **56**, 494).

Purss, G. S. 1957. *Qd J. agric. Sci.* **14**: 125 (aetiology, *Phytophthora vignae* sp. nov.; **37**: 433).

——. 1958. *Qd J. agric. Sci.* **15**: 1 (resistance; **38**: 112).

——. 1972. *Aust. J. agric. Res.* **23**: 453 (races; **51**, 4543).

——. 1975. *Qd agric. J.* **101**: 551 (resistant cv. Red Caloona; **55**, 6035).

PLASMOPORA Schroet., *Kryptogflora von Schlesien* 3(1): 236, 1889.
Rhysotheca G. W. Wilson, 1907
Pseudoplasmopora Sawada, 1922.
(Peronosporaceae)

MYCELIUM intercellular; haustoria unbranched and like knobs. SPORANGIOPHORES emerging from the stomata, erect, solitary to densely fasciculate, monopodially branched, the branches arising more or less definitely at right angles to the main axis; secondary branches also at right angles; the terminal branches apically obtuse. SPORANGIA small, hyaline, papillate, germination sometimes by a germ tube but in most cases by zoospores. OOSPORES yellowish brown; epispore wrinkled and sometimes somewhat reticulate; oogonial wall persistent but not fused with the epispore as in *Sclerospora* (from H. M. Fitzpatrick, *The lower fungi Phycomycetes*: 215, 1930).

The genus has not been monographed. It is chiefly distinguished from other members of the family by the slender sporangiophores which branch monopodially, usually at right angles, and the obtuse tips of the branches. Savulescu et al. (1952) gave a key to 24 spp. There are 2 important plant pathogens: *P. viticola* (Berk. & Curt. ex de Bary) Berl. & de Toni on grapevine (*Vitis*) and *P. halstedii* (described); *P. nivea* (Ung.) Schroet. is occasionally found attacking members of the Umbelliferae. *P. penniseti* Kenneth & Kranz was described from bulrush millet (*Pennisetum typhoides*). The small, water-soaked leaf stripes or spots expand into irregular elongate lesions (1–2 × 60 mm or longer) which may coalesce.

Kenneth, R. et al. 1973. *Trans. Br. mycol. Soc.* **60**: 590 (*Plasmopora penniseti* sp. nov. & taxonomic discussion; **52**, 4057).
Novotel'nova, N. S. 1963. In *Proc. 2nd Symp. on problems in the fungus and lichen flora of the Baltic Republics*, Vilnius, Lietuvos TSR Mokslu Akad. Bot. Inst.: 111–18 (*Plasmopora* spp. on Compositae, survey; **43**, 2493h).
Savulescu, T. 1941. *Bull. Sect. scient. Akad. roum.* **24**: 23 pp. (3 *Plasmopora* spp. on Compositae; **23**: 120).
—— et al. 1952 (?). Academy Rumanian People's Republic: 327–457 (Peronosporaceae of Rumania; key to 24 *Plasmopora* spp.; **34**: 323).

Plasmopora halstedii (Farl.) Berl. & de Toni in Sacc., *Sylloge Fung.* 7: 242, 1888.
Peronospora halstedii Farl., 1879.

Hypophyllous on cotyledons and leaves, the affected area small, 1–3 mm or extending over the entire leaf. SPORANGIOPHORES fasciculate, slender, 300–750 μ, 3–5 branched, ultimate branches 8–15 μ long, verticillate below the apex of the branching axis which is frequently swollen and like a ganglion; very variable, especially in the laxity of their branches and the development of the swellings from which the ultimate branches arise. SPORANGIA (zoosporangia) oval or elliptic, 18–30 × 14–25 μ. OOSPORES 30–32 μ; epispore yellowish brown, somewhat wrinkled (from G. W. Wilson, *Bull. Torrey bot. Club* **34**: 403, 1907, as *Rhysotheca halstedii*, and see Novikova; *P. helianthi* Novotel'nova is possibly a synonym, see *Index of Fungi* **3**: 364).

Downy or false mildew of sunflower is a serious disease which quite often appears in epidemic form; it has been extensively studied in USSR and N. America. *P. halstedii* is widespread in Europe, N. America, central and S.W. Asia; it also occurs in: Argentina, Brazil, Chile, Dominican Republic, Ethiopia, Japan, Kenya, Zimbabwe, Taiwan, Uganda and Uruguay (Map 286). Novotel'nova (1962, 1966) reported on the host range (>80 hosts in 35 genera of Compositae), described 3 ff. sp. (1 on annual and 2 on perennial sunflowers) within the fungus on *Helianthus*, and split *P. halstedii* with respect to other host genera into more ff. sp. and several spp. Orellana gave the hosts within N. American *Helianthus* spp. Leppick discussed origin and distribution; he referred the forms within *P. halstedii* to 2 groups (N. American and Eurasian), there being host specificity within each. There are reviews by Novotel'nova (1966) and Scharif.

The symptoms may be first seen in the seedlings within a week of emergence. These become stunted, abnormally thickened and the thick, downward curling cotyledons and leaves have a prominent, chlorotic, epiphyllous mottling, like a mosaic. On older leaves there is a very characteristic symmetrical chlorotic pattern which follows the main veins. Severely infected plants which do not succumb earlier may survive only a few weeks. This infection is systemic and results from penetration of the host at an early stage of growth up to the 6–8 true leaf stage. Wehtje et al. described the infection of seedlings as primarily through (or near to) the zone of elongation of the radicle. In infections of apical buds after this stage only the middle and upper leaves will be infected. Infection may occur through the stomata. Primary infection is through the zoospores formed by both reproductive stages underground. Zimmer (1971) described secondary infection by windborne zoosporangia formed on the leaves of systemically infected plants. Seedlings become increasingly resistant with age. In the field (USA, mean air temps during germination and emergence 13.2°) plants remained susceptible for at least 15 days, but for only 5 days at 22–25°. Systemic invasion occurs through the hypocotyls, apical buds (where it may be more frequent), leaves and roots. In 3-day-old seedlings the opt. temp. for infection was 15° and for incubation 20–25°. Zoosporangia germinate mostly at 9–22°. Oospores are formed in seedlings in 1–2 weeks after inoculation at 14–23°, mainly in the collar and lateral roots (Cohen et al. 1973; Goosen et al.; Nishimura; Novotel'nova 1960; Raicu; Young et al.; Zimmer 1975).

Infection of the seed occurs (e.g., Novotel'nova 1963). Cohen et al. (1974a) reported a latent infection; they found oospores in the seeds of inoculated and naturally infected field plants. Infected seed gave higher numbers of plants with latent infections. This infection can also arise from invasion of apical buds, from below ground contact with systemically invaded plants and from infected, newly formed flower heads. Thus dissemination of the disease can be widespread before typical symptoms appear. Plants with latent infection may form infected seed.

Zimmer (1972b, 1974) found evidence for the existence of races. Lines that were resistant to the pathogen in USA were also so in Europe. But many resistant lines from Europe were susceptible in USA. Fernández et al. reported that a southern Spanish form of *P. halstedii* might be a different race from the French form. A form more virulent than those already known in Europe was described by Iliescu et al. Dominant genes control resistance (see also Belkov et al.; Leclercq et al.; Vear et al.; Vranceanu et al.; Zimmer et al. 1972a). The approach to control seems to be largely through resistance. No treatment for the internal infection of seed appears to have been recommended. The production of seed in disease free areas might be advisable. A rotation would need to be lengthy because of the long soil survival through the oospores. Infection can occur in a sunflower crop planted 4 years after a previous one. Measures which support rapid seedling emergence and reduce free soil moisture in the early growth stages should be beneficial (Zimmer 1975).

Allard, C. 1978. *Annls Phytopathol.* **10**: 197 (histology of systemic infection).

Belkov, V. et al. 1975. *Rasteniev. Nauki* **12**: 103 (inheritance of resistance; **55**, 2328).

Cohen, Y. et al. 1973. *Can. J. Bot.* **51**: 15 (factors affecting infection; **52**, 3020).

—— ——. 1974a & b. *Ibid* **52**: 231, 861 (seed & latent infection, disappearance of IAA in infected tissues; **53**, 4083, 4545).

—— ——. 1975. *Ibid* **53**: 2625 (phenolic compounds; **55**, 4235).

——. 1975. *Physiol. Pl. Pathol.* **7**: 9 (fluorescence microscopy & enzyme activity; **55**, 1902).

Fernández, M. J. et al. 1978. *An. hist. nac. Invest. agrarias* Ser. production Vegetal **8**: 107 (inheritance of resistance).

Goosen, P. G. et al. 1968. *Can. J. Bot.* **46**: 5 (symptoms, sporulation of & infection by sporangia; **47**, 1941).

Iliescu, H. et al. 1975. *Problème Prot. Pl.* **3**: 217 (differences in virulence; **56**, 4147).

Leclercq, P. et al. 1970. *Annls Amél. Pl.* **20**: 363 (selection for resistance; **52**, 210).

Leppick, E. E. 1962. *Pl. Prot. Bull. F.A.O.* **10**: 126 (distribution; **43**, 1106).

——. 1966. *Ibid* **14**: 72 (origin & host specialisation **46**, 708).

Marte, M. et al. 1976. *Caryologia* **29**: 405 (fine structure & cytochemistry of haustoria; **56**, 5135).

Nikolić, V. 1952. *Zăst. Bilja* **9**: 42 (general; **31**: 609).

Nishimura, M. 1922 & 1926. *J. Coll. Agric. Hokkaido Imp. Univ.* **11**: 185; **17**: 1 (general; **2**: 315; **5**: 670).

Novikova, N. D. 1958. *Nauch. Dokl. vȳssh. Shk.* 2: 83 (morphology & host distribution of oospores & sporangia; **39**: 320).

Novotel'nova, N. S. 1960. *Bot. Zh. SSSR* **45**: 1283 (general, temp. & sporangial germination; **40**: 366).

——. 1962. *Sborn. Dokl. nauch. Konf. po Zashch. Rash. Tartu*: 129 (ff. sp. on *Helianthus*; **43**, 2011).

——. 1962. *Bot. Zh. SSSR* **47**: 970 (*Plasmopora halstedii* as an aggregate sp., host range; **42**: 4).

——. 1963. *Ibid* **48**: 845 (spread & distribution in host, seed infection; **42**: 768).

——. 1965. *Ibid* **50**: 301 (differences in sporangial morphology; **44**, 2220).

——. 1966. Moscow–Leningrad, Izdat. 'Nauka'. 150 pp. (review, 8 pp. ref.; **46**, 2085).

Orellana, R. G. 1970. *Bull. Torrey bot. Club* **97**: 91 (host reaction in N. American *Helianthus* spp.; **50**, 218).

Raicu, C. 1971. *An. Inst. Cerc. Prot. Pl.* **9**: 153 (infection; **53**, 3577).

Scharif, G. 1971. *Ent. Phytopathol. appl.* **31**: 1 (in Iran & review, 55 ref.; **51**, 1769).

Vear, F. et al. 1971. *Annls Amél. Pl.* **21**: 251 (2 genes for resistance; **52**, 818).
Virányi, F. 1977. *Acta phytopathol. Acad. Sci. hung.* **12**: 263 (detection of systemic infection; **57**, 3056).
Vranceanu, V. et al. 1970 & 1971. *Anal. Inst. cere. cereal Plante tehn.* Ser. C **36**: 299; **37**: 209 (immune inbred line; control of resistance by single dominant; **50**, 1335; **51**, 2754).
Wehtje, G. et al. 1978. *Phytopathology* **68**: 1568 (seedling infection).
Young, P. A. et al. 1927. *Am. J. Bot.* **14**: 551 (symptoms, infection & histology; **7**: 243).
—— ——. 1929. *Science N.Y.* **69**: 1783 (infection of seedlings; **8**: 579).
Zhyvylo, V. I. et al. 1970. *Ukr. bot. Zh.* **27**: 785 (oospore formation & temp.; **50**, 1931).
Zimmer, D. E. et al. 1971. *Pl. Dis. Reptr* **55**: 11 (severe outbreak in USA; **50**, 2412).
—— ——. 1972a. *Ibid* **56**: 428 (field testing for resistance; **51**, 4262).
—— ——. 1972b. *Crop Sci.* **12**: 749 (inheritance of resistance & suspected races; **52**, 4176).
——. 1974. *Phytopathology* **64**: 1464 (physiologic specialisation between forms in America & Europe; **54**, 1836).
——. 1975. *Ibid* **65**: 751 (factors affecting incidence; **55**, 831).

PLEIOCHAETA (Sacc.) Hughes, *Mycol. Pap.* 36: 39, 1951.
(Dematiaceae)

COLONIES effuse, grey, olivaceous brown or black. MYCELIUM mostly immersed. STROMA none. SETAE and HYPHOPODIA absent. CONIDIOPHORES macronematous, mononematous, usually branched, flexuous, frequently geniculate, hyaline or pale olivaceous, smooth. CONIDIOGENOUS CELLS polyblastic, integrated, terminal becoming intercalary, sympodial, cylindrical, cicatrised; scars broad, flat, thin. CONIDIA solitary, dry, acropleurogenous, appendiculate, roughly cylindrical, narrowed at the apex, truncated at the base, colourless or with the cell at each end hyaline or subhyaline and intermediate cells straw coloured to golden brown, smooth, multiseptate; the apical cell bears several long hyaline, subulate appendages which are sometimes branched (M. B. Ellis, *Demat. Hyphom.*: 263, 1971).

P. setosa (Kirchn.) Hughes (synonym: *Ceratophorum setosum* Kirchn.) causes a disease (brown spot) on several hosts (CONIDIOPHORES up to 150 μ long, 8–13 μ thick; CONIDIA 4–8 (mostly 5) septate, 60–90 × 14–22 μ scar 7–11 μ wide, appendages up to 100 μ long, 3–4 μ thick at the base, tapering to about 1 μ; Ellis l.c.; Ellis et al.). Germar made a full study of brown spot on lupins (*Lupinus* spp.; and see Harvey). Shanmuganathan recently described the severity of the disease on *Crotalaria* in Sri Lanka. Pegg reported *P. setosa* on common bean in Australia as a wound pathogen. The disease followed sand abrasion and was confined to light sandy soils; the fungus was seedborne on *Crotalaria* weed hosts and cowpea (*Vigna*) was also affected. *P. albizziae* (Petch) Hughes (synonym: *Ceratophorum albizziae* Petch) occurs on *Albizia* shade (Haan).

Ellis, M. B. et al. 1976. *C.M.I. Descr. pathog. Fungi Bact.* 495 (*Pleiochaeta setosa*).
Germar, B. 1939. *Z. PflKrankh. PflPath. PflSchutz* **49**: 482 (general on lupin; **19**: 99).
Green, D. E. 1933. *Jl R. hort. Soc.* **58**: 144 (on lupin; **13**: 166).
—— et al. 1949. *Ibid* **74**: 310 (on *Cytisus*; **28**: 457).
Haan, de I. 1938. *Archf. Theecult. Ned.-Indië* **12**: 303 (on *Albizia*; **18**: 281).
Harvey, I. C. 1974. *Protoplasma* **82**: 203 (light & electron microscope observations of conidiogenesis in *P. setosa*).
——. 1975. *Trans. Br. mycol. Soc.* **64**: 489 (development & germination of chlamydospores in *P. setosa*; **54**, 5286).
——. 1977. *Can. J. Bot.* **55**: 1261 (infection of lupin; **57**, 200).
Hogetop, C. 1937. *Revta agron. Porto Alegre* **1**: 346 (on lupin; **17**: 824).
Hughes, S. J. 1951. *Mycol. Pap.* 36, 43 pp. (*Mastigosporium, Camposporium* & *Ceratophorum*; **30**: 584).
Pegg, K. G. 1968. *Qd. J. agric. Anim. Sci.* **25**: 219 (on common bean; **48**, 310).
Pirozynski, K. A. 1974. *Fungi Canadenses* 12 (*P. setosa*).
Pulselli, A. 1928. *Boll. Staz. Patol. veg. Roma* **8**: 50 (on *Cytisus* & lupin; **7**: 582).
Shanmuganathan, N. 1970. *Tea Q.* **41**: 64 (on *Crotalaria*; **50**, 1863).
Wells, H. D. et al. 1969. *Pl. Dis. Reptr* **53**: 774 (on lupin seed with other fungi; **49**, 784).

PLEOSPORA Rabenh. in Klotzsch, *Herb. Viv. mycol.* ed. II, No. 547, 1857.
(Pleosporaceae)

ASCOSTROMATA immersed or erumpent, globose, ellipsoid, flattened or pyriform, glabrous or variously tomentose and/or setose, thin or thick walled, pseudosphaeriaceous, with pseudoparaphyses more or less persistent; ostiole usually small, papillate to conical, sometimes penicillate or setose. ASCI bitunicate, narrowly clavate to broadly clavate, more or less thick walled with a claw-like base. ASCOSPORES muriform, pale yellow brown to dark red brown, variable in size, form and septation (after L. E. Wehmeyer, *A world monograph of the genus* Pleospora *and its segregates*: 38, 1961).

The genus was monographed by Wehmeyer (and see Simmons, *Stemphylium* q.v.). *P. betae* (Berl.) Nevodovsky (synonym: *P. bjöerlingii* Byford) causes the seedborne disease of sugar beet called black leg (*CMI Descr.* 149, as *P. björlingii*). *P. papaveracea* (De Not.) Sacc. causes a destructive disease of poppy, including *Papaver somniferum* (subsp. *hortense*, seed and seed oil; subsp. *somniferum*, opium). The stat. conid. is a *Dendryphion* (see Ellis, *Demat. Hyphom.*: 504, 1971). Shoemaker gave the stat. conid. as *D. penicillatum* (Corda) Fr. and a full synonymy which included *D. papaveris* (Saw.) Saw., *Brachycladium penicillatum* Corda and *Helminthosporium papaveris* Saw. *Pleospora infectoria* Fuckel, common on temperate cereals, has an *Alternaria* state (Ellis l.c.: 468).

Björling, K. 1945. *Meddn. St. Växtsk Anst* **44**, 96 pp. (*Pleospora betae*, monograph; **25**: 54).

Booth, C. 1967. *C.M.I. Descr. pathog. Fungi Bact.* 149 (*P. betae*).

Gates, L. F. 1959. *Ann. appl. Biol.* **47**: 502 (*P. betae*, seed treatment with ethyl mercury phosphate & other compounds; **39**: 201).

Meffert, M. E. 1950. *Z. ParasitKde* **14**: 442 (*P. papaveracea*, general; **31**: 85).

Rădulescu, E. et al. 1964. *Rev. roum. Biol.* Ser. bot. **9**: 19 (as above; **44**, 1195).

Shoemaker, R. A. 1968. *Can. J. Bot.* **46**: 1143 (type studies of *P. calvescens*, *P. papaveracea* & some allied spp.; **48**, 50).

Wehmeyer, L. E. 1961. *A world monograph of the genus* Pleospora *and its segregates*, Univ. Michigan Press (**41**: 513).

Zambettakis, C. 1952. *Annls Epiphyt.* Ser. C **3**: 11 (*P. papaveracea*, general; **32**: 99).

Pleospora herbarum (Fr.) Rabenhorst, *Herb. Viv. mycol.* ed. II, No. 547, 1856.

Sphaeria herbarum Fr., non Pers., 1823
Pleospora grossulariae (Fr.) Fuckel, 1870
P. allii (Rabenh.) Ces. & de Not., 1861
P. armeriae (Corda) Ces., 1863
P. dianthi de Not., 1863
P. evonymi Fuckel, 1870
P. australis Cooke, 1879
stat. conid. *Stemphylium botryosum* Wallr., 1833.

ASCOSTROMATA scattered, immersed to erumpent in host tissue, globose or somewhat flattened, 100–500 μ diam. ASCI bitunicate, cylindrical to clavate, 90–250 × 20–50 μ with 8 irregularly distichous ascospores. ASCOSPORES light to dark yellow brown, ellipsoid to clavate, 7 septate, slightly constricted at the 3 primary transverse septa, finally muriform, 26–50 × 10–20 μ. In culture aerial MYCELIUM filamentous, sparse, hyaline to brown, branched, 5–8 μ wide, thicker hyphae develop later on the agar surface and are darker in colour; from below the culture appears brown. CONIDIOPHORES erect, flexuous 1–7 septate, 20–72 × 4–6 μ, pale brown to brown, with a swollen apical sporogenous cell, 7–12 μ diam. and slightly roughened towards the apex. They possess a single apical pore, 5–8 μ diam. Several successive sporogenous cells may form by proliferation through the apical pore. CONIDIA oblong, olive to brown, ovoid to subdoliiform, occasionally constricted at 1–3 transverse septa and at 1–3 longitudinal septa if these are complete, 19.5 × 28.5 μ; with a single basal pore, 8 μ diam. and a roughened outer wall (C. Booth & K. A. Pirozynski, *CMI Descr.* 150, 1967; and see Ellis, *Demat. Hyphom.*: 167, 1971 for stat. conid.; Wiltshire for taxonomy, *Stemphylium* q.v.).

The fungus is plurivorous, worldwide and very common in temperate and subtropical regions. It has been confused with *Leptosphaerulina trifolii* q.v. for differences between the 2 fungi. Some crops on which it has been recorded are: broad bean, cabbage, carrot, cauliflower, chick pea, red pepper, citrus, cucumber, garlic, lettuce, lucerne, marrow, melon, onion, leek, pea, pine, sugarcane, sweet potato and tomato. *P. herbarum* is a saprophyte and a weak pathogen; it occurs on the above ground parts (including the seed) of herbaceous hosts and may be associated with (or follow) other fungal pathogens. It can be a factor in storage rots of fruits and vegetables. Some forms may be host specific. Opt. temps for growth in vitro vary from 22 to 28°, mostly *c.* 25°. Those for decay or the disease in the field are rather lower and, therefore, the fungus is most active under relatively cool conditions.

Both temp. and light affect sporulation, and the interaction of these factors and their effects have been studied in detail by Leach (and *Alternaria dauci* q.v.). Conidiophores (inductive phase) form under the stimulation of near UV and relatively high temps; whilst conidia (terminal phase) form at lower temps and this phase is inhibited by near UV and blue light. The terminal phase is completely inhibited by light at high temps but only partly so at low ones. Protoperithecium initiation can be triggered by UV (photo-induction) or low temp. (cold induction). In the first case maturation requires relatively long, continuous exposure to low temp. (e.g., 21 days at 5–10°). Terminal incubation temps of 10–15° are most favourable for maturation of protoperithecia; maturation can occur in the dark. In the second case induction occurs in the dark at 5–10°; max. formation of protoperithecia needs *c.* 8 days exposure to low temps. The cold treatment trigger can be negated by subsequent exposure to a high temp. (e.g., 30°). Maturation requires a fairly long terminal incubation at low temps (e.g., 35 days at 12.5°). Further ref. are given under *Stemphylium*.

Bashi, E. et al. 1975. *Phytoparasitica* **3**: 63 (effect of light on sporulation in vivo with *Alternaria solani*; **55**, 3108).
Carrol, F. E. 1972. *J. Cell Sci.* **11**: 33 (fine structure of conidial initiation; **52**, 362).
Corlett, M. 1973. *Can. J. Microbiol.* **19**: 392 (surface structure of conidium & conidiophore; **52**, 2859).
Ellis, M. 1931. *Trans. Br. mycol. Soc.* **16**: 102 (culture).
Leach, C. M. 1963. *Mycologia* **55**: 151 (quantitative & qualitative relationship of monochromatic radiation to sexual & asexual reproduction; **42**: 601).
——. 1968. *Ibid* **60**: 532 (action spectrum for light inhibition of terminal phase of photosporogenesis; **47**, 3723).
——. 1971. *Trans. Br. mycol. Soc.* **57**: 295 (regulation of perithecium development by light & temp.; **51**, 1139).
Meredith, D. S. 1965. *Pl. Dis. Reptr* **49**: 1006 (release of conidia under decreasing vapour pressure; **45**: 1597).
Van Warmelo, K. T. 1971. *Bothalia* **10**: 329 (somatic nuclear division & conidial nuclei; **50**, 3524).

LYCOPERSICON ESCULENTUM

Ramsey described a rot of tomato fruit in USA and attributed the pathogen to *Pleospora lycopersici* E. & E. Marchal. An imperfect state (now *Stemphylium sarciniforme; Stemphylium*, Leguminosae q.v.) was incorrectly connected with it. Middleton found that *S. sarciniforme* infected legumes but not tomato and that *P. lycopersici* only attacked tomato (1–3 mm diam. leaf spots, somewhat irregular in outline and with a diffuse yellow halo). *S. sarciniforme* has no reported perfect state. Wehmeyer (*Pleospora* q.v.) considered that *P. lycopersici* is probably a form of *P. herbarum*. Rotem et al. (1966a) described a new tomato disease (foliage blight) in Israel caused by *P. herbarum*. Because of its host specificity it was proposed as a new form (f. sp. *lycopersici*). Two other *Stemphylium* spp. occur on tomato, *S. lycopersici* and *S. solani* (q.v.). The conidia of the former differ from those of *P. herbarum* in that they are pointed, conical at the apex and have a different length : breadth ratio. Those of the latter are also pointed at the apex but differ from *S. lycopersici* in that they are constricted at the median septum. The conidia of *S. lycopersici* are usually constricted at the 3 major transverse septa. *S. floridanum* on tomato was considered to be a synonym of *S. lycopersici* by Yamamoto (*S. solani* q.v.). Schneider et al. considered that in W. Germany a tomato leaf spot disease was caused by a specialised form of *P. herbarum* different from that reported from Israel.

On fruits small brown V shaped to oval spots appear at the edge of the stem scar or at wounds elsewhere. The lesions enlarge, they are brown with grey to grey brown mycelium on the surface, with the perithecia developing in their centres. A moderately firm, brown decay spreads 1–2 cm into the fruit. Symptoms can begin to develop slowly in transit in 3–4 days; they spread more rapidly as ripening occurs. On mature, green, inoculated fruit most decay develops at 18.5–21°; there is little decay at $<7.2°$ or $>26.7°$. Decay develops more rapidly in ripe tomatoes but the temp. effects remain similar.

The f. sp. *lycopersici* attacks tomato leaves and has been described from Canada, Israel and UK. In Israel the lesions (mostly on the lamina) were described as mainly irregular but also circular, 1–10 mm diam., light to dark brown or greyish, slightly sunken and occasionally with a chlorotic halo. In transmitted light a net-like pattern is often discernible within the lesion; old lesions may crack in the centre. In UK the leaf spots were described as irregularly shaped, light brown with dark margins; on older leaves coalescence of the lesions occurred and shotholes formed (Dickens et al.). In Canada (Nova Scotia) the disease was not considered to be important. Lesions (2–10 mm diam.) occurred first on the older leaves as small chlorotic spots whose centres became necrotic, they developed concentric markings, coalesced and the necrotic tissue cracked (Gourley). The fruit is apparently not attacked by this f. sp.

In Israel the disease has been investigated with

the other pathogens of tomato foliage (see Rotem et al. 1977 for a review). The opt. temp. for leaf colonisation is 25° but infection is relatively insensitive to temps of 15–25°. The duration of wetting has a greater effect on the disease than temp. Long periods of wetness are required for sporulation. Several short wet periods interrupted by dry ones can replace long wet periods but the sporulation is lower. More conidia were formed when the wet period was extended from 8 to 16 hours/night and the night and day temps were raised from 10 and 20° to 15 and 25° or 20 and 30° (Bashi et al. 1973a, 1975). More disease occurred under overhead sprinkling irrigation compared with that in the furrow. In Israel the disease occurs in autumn and winter. Fungicide control was given by anilazine, chlorothalonil and propineb (Rotem) but not by benomyl (Dickens et al.). Tomato cvs resistant to *S. solani* carried a high degree of resistance to *P. herbarum* (f. sp. *lycopersici*) and *S. lycopersici*; the indications being that resistance is dominant. Some plants of other *Lycopersicon* spp. showed a higher level of resistance than tomato cvs.

Bashi, E. et al. 1973a. *Phytoparasitica* **1**: 87 (effects of wetness, temp. & other factors; **53**, 3184).
—— ——. 1973b. *Phytopathology* **63**: 1542 (resistance; **53**, 2683).
—— ——. 1975. *Ibid* **65**: 532 (sporulation under wet & dry regimes; **55**, 2611).
Dickens, J. S. W. et al. 1973. *Pl. Pathol.* **22**: 70 (in UK; **53**, 1559).
Gourley, C. O. 1971. *Can. Pl. Dis. Surv.* **51**: 135 (in Canada; **51**, 3562).
Middleton, J. T. 1939. *Phytopathology* **29**: 541 (with *Stemphylium sarciniforme*; **18**: 716).
Ramsey, G. B. 1935. *J. agric. Res.* **51**: 35 (on tomato fruit; **14**: 799).
Rotem, J. et al. 1966a. *Can. J. Pl. Sci.* **46**: 265 (*S. botryosum* f. sp. *lycopersici*; **45**, 2961).
—— ——. 1966b. *Pl. Dis. Reptr* **50**: 635 (effect of irrigation on tomato foliage diseases in Israel; **46**, 742).
——. 1968. *Israel Jnl agric. Res.* **18**: 87 (fungicides; **47**, 3217).
—— ——. 1977. *Phytoparasitica* **5**: 45 (review of *Stemphylium* complex in Israel, 48 ref.; **57**, 785).
Schneider, R. et al. 1976. *Phytopathol. Z.* **87**: 264 (in W. Germany; **56**, 3724).

OTHER CROPS

P. herbarum attacks clovers, trefoils, sainfoin and lucerne (Focke; Hughes; Nelson; Smith and *Stemphylium* q.v.). A leaf spot (sometimes called ringspot) is caused. The spots are circular or irregular, light to very dark brown, and with a straw coloured halo; the lesions can increase in size, coalesce and kill much of the leaf. Sporulation results in a sooty appearance. Small black linear lesions may be caused on stems and petioles. The necrotic leaves may remain attached but defoliation has been reported. Infection of lucerne reduces yield and quality of seed; fewer pods, with fewer and lighter seeds, are formed. In temperate regions infection in early spring is caused by ascospores from crop debris; later infections are caused by ascospores and conidia, both are air dispersed and on germination penetrate the stomata. The fungus can occur in the seed. There is little evidence for differential pathogenicity in the forms which infect *Medicago, Melilotus* and *Trifolium.* Lucas et al. described a new resistant lucerne cv. from USA (North Carolina). On lucerne Borges et al. recently described the host and pathogen interaction, fungal biotypes (7) and host resistance (including other *Medicago* spp.).

On onion (*Allium cepa* var. *cepa*) infection through the leaves and stems causes small white sunken spots which enlarge to sunken linear areas that are brown to black with a velvety appearance. Stems and leaves are girdled and break (Teodoro). Infection probably mostly follows that by *Alternaria porri* and *Peronospora destructor*. On *Capsicum* spp. the leaf spots become circular to subcircular, tan to brown, with a light centre and dark margin, 1–2 cm diam. and sometimes concentrically ringed (Braverman, as f. sp. *capsicum*). Zachos found infection of chick pea (*Cicer arietinum*) seeds; it prevented germination. Infection of citrus fruits, through wounds, causes a firm, dark brown or almost black rot, often at the stem end and several cm in diam. (Cocchi). On melon (*Cucumis melo*) small watersoaked lesions form, sometimes profusely, on the lower leaves; they enlarge, becoming yellow, light then dark brown, up to 1.5 cm diam. or more and with a chlorotic halo 1–2 mm wide. The older leaves fall and young ones may be attacked (Petzer). Du Toit described *P. herbarum* on sweet potato. On leaves the chocolate brown spots can spread to kill most of the lamina. Decay of the tubers (1–3 weeks after storage) usually begins at a previously damaged area; the rot is at first soft, brown and moist, but as decay spreads the tissues dry out and they become woody. On lettuce the small watersoaked lesions become more or less circular (sometimes they are vein limited), zonate and 2–3 cm diam., shotholes develop and both young and old leaves show symptoms. The lettuce form was called f. sp. *lactucum* (Padhi et al.) and captan or maneb were recommended for control in New Zealand (Slade; see also Barash et al.; Sivan et al.). In Egypt the fungus caused circular to irregular, dark brown leaf spots, slightly sunken 2–5 mm diam. but eventually affecting much of the lamina of broad bean (Abdou et al.).

Abdou, Y. A. et al. 1969. *Pl. Dis. Reptr* **53**: 157 (on broad bean; **48**, 2080).
Barash, I. et al. 1975. *Pl. Physiol.* **55**: 646 (toxin stemphylin; **54**, 4707).
——— ———. 1978. *Phytoparasitica* **6**: 95 (effect of toxin on lettuce cvs).
Borges, O. L. et al. 1976. *Phytopathology* **66**: 715, 749 (biotypes & host pathogen interaction in lucerne; **56**, 749, 750).
——— ———. 1976. *Crop Sci.* **16**: 456, 458 (selection & inheritance of resistance in lucerne; **56**, 1174, 1175).
Braverman, S. W. 1968. *Phytopathology* **58**: 1164 (on red pepper; **47**: 3659).
Cocchi, F. 1931. *Boll. Staz. Patol. veg. Roma* **11**: 179 (on lemon; **11**: 449).
De Sarasola, M. A. R. et al. 1957. *Revta Fac. Agron. Univ. nac. La Plata* **33**: 83 (on lettuce; **37**: 622).
Du Toit, J. J. 1951. *Sci. Bull. Dep. Agric. S. Afr.* 301, 74 pp. (on sweet potato; **30**: 486).
Focke, I. 1966. *Zentbl. Bakt. ParasitKde* Abt 2, **120**: 288 (on lucerne; **46**, 1027).
Heath, M. C. et al. 1973. *Physiol. Pl. Pathol.* **3**: 107 (in vitro & in vivo conversion of phaseollin & pisatin; **52**, 3365).
Higgins, V. J. et al. 1969. *Phytopathology* **59**: 1493, 1500 (induction & degradation of a phytoalexin, with *Setosphaeria turcica*; **49**, 1065a, b).
——— ———. 1974. *Ibid* **64**: 105 (conversion of phaseollin to phaseollinisoflavan; **53**, 3350).
———. 1975. *Physiol. Pl. Pathol.* **6**: 5 (induced conversion of phytoalexin maackiain to dihydromaackiain; **55**, 301).
Hughes, S. J. 1945. *Trans. Br. mycol. Soc.* **28**: 86 (on *Onobrychis viciifolia*; **25**: 166).
Lucas, L. T. et al. 1973. *Pl. Dis. Reptr* **57**: 946 (resistance in lucerne; **53**, 2242).
Nelson, R. R. 1955. *Phytopathology* **45**: 352 (on lucerne; **35**: 21).
Padhi, B. et al. 1954. *Ibid* **44**: 175 (on lettuce; **34**: 125).
Petzer, C. F. 1958. *S. Afr. J. agric. Sci.* **1**: 3 (on melon; **37**: 625).
Sivan, J. et al. 1976. *Ann. appl. Biol.* **82**: 425 (on lettuce; **55**, 2984).
Slade, D. A. 1963. *N.Z. Jl Agric.* **107**: 73 (on lettuce; **43**, 638).
Smith, O. F. 1940. *J. agric. Res.* **61**: 831 (on lucerne & *Trifolium*; **20**: 306).
Teodoro, N. G. 1923. *Philipp. agric. Rev.* **16**: 233 (on onion; **3**: 499).
Zachos, D. G. 1952. *Annls Inst. Phytopathol. Benaki* **6**: 60 (on chick pea; **34**: 126).

PORIA Pers. ex S. F. Gray, *Nat. Arr. Br. Pl.* Vol. 1: 639, 1821.
(Polyporaceae)

SPOROPHORES resupinate, annual or perennial, consisting of context attached to the substrate that supports the tubes, or in perennial forms the new tube layers formed below the old ones; tubes opening by pores that vary from > 1 mm wide to 10 per mm or rarely even smaller, lined with a hymenium that in most spp. does not continue around the end of the dissepiments, and consisting of sterile elements such as hyphal pegs, cystidia and setae, and non-septate basidia bearing smooth or echinulate, hyaline to brown basidiospores. Young sporophores of pileate spp. of polypores may be resupinate for some time and can be confused very easily with spp. of *Poria* (J. L. Lowe, *Tech. Publs N.Y. St. Coll. For.* 90: 12, 1966, q.v. for synonymy).

Donk discussed nomenclature (*Persoonia* **1**: 266, 1960) and Wright gave a description. Members of the genus are largely saprophytic on wood.

Donk, M. A. 1967. *Persoonia* **5**: 47 (European polypores II, *Poria*).
Lowe, J. L. 1963. *Mycologia* **55**: 453 (*Poria* & similar fungi from tropics; **43**, 47).
———. 1966. *Tech. Publs N.Y. St. Coll. For.* 90, 183 pp. (Polyporaceae of N. America, genus *Poria*; **46**, 554).
Wright, J. E. 1964. *Mycologia* **56**: 692 (pseudoamyloid in pore fungi).

Poria hypobrunnea Petch, *Ann. R. bot. Gdns Peradeniya* 6: 137, 1916.

CARPOPHORE annual, more rarely persistent, entirely resupinate, widely effused, firmly attached to the substrate, 3–6 mm thick; arising from dark red mycelial strands over the substrate surface. Pore surface at first yellowish white to ochraceous then reddish pink, becoming brownish red on ageing, finally dark purplish grey; margin whitish to cream, narrow, tomentose. PORES 8–11 per mm, round to angular, often variable; edge thin to thick, entire, sometimes with a white prunina. TUBES up to 3 mm long, tough to cartilaginous, concolorous with the subiculum. SUBICULUM dark purplish brown, firm, felty, up to 3 mm thick. HYPHAL SYSTEM dimitic, with generative and skeletal hyphae. GENERATIVE HYPHAE 1.5–3.5 μ diam., hyaline, thin walled, frequently branching, loosely interwoven, septate but lacking clamp connections. SKELETAL HYPHAE abundant, 2.5–6 μ diam., deep yellowish brown to brown with a very thick wall which at times almost obliterates the lumen, non-septate, sinuous, very occasionally branching. BASIDIOSPORES 4–5 (–6) × 3.5–5 (4.8 × 4.5) μ, subglobose to ovoid, although adaxially flattened and sometimes appearing triangular, hyaline, inamyloid, smooth, thin walled. BASIDIA 9–10.5 × 4.5–5 μ, broadly clavate. CYSTIDIA 15–45 × 5–10 μ abundant to rare, thin walled, hyaline, either covered by a crystalline deposit or non-encrusted with a mucronate apex, often more or less embedded.

This sp. would be more correctly placed in *Chaetoporus* P. Karst., differing from *C. vinctus* (Berk.) Wright in the darker pore surface and subiculum. Petch (*Diseases of the tea bush*: 154, 1923) records *P. hypolateritia* as a tree pathogen but examination of all his material in Herb. Kew shows it to be young material of *P. hypobrunnea*. The type specimen of *P. hypolateritia* is somewhat fragmentary but shows all the characteristics of *Schizopora paradoxa* (*P. versipora*) and the two are, therefore, regarded as synonymous. On cacao, rubber, tea, dadap (*Erythrina* sp.), *Tephrosia candida, Crotalaria anagyroides, Spathodea campanulata, Crotalaria incrassata* (D. N. Pegler & I. A. S. Gibson, *CMI Descr.* 322, 1972).

Setliff, in a study of *P. vincta* (Berk.) Cooke, re-examined the basidiocarps associated with root rots and stem canker of rubber and tea which have been referred to either *P. hypobrunnea* or *P. hypolateritia* (Berk.) Cooke. *P. vincta* was divided into 2 vars: var. *vincta* Setliff and var. *cinerea* (Bres.) Setliff. It was considered that basidiocarps named *P. hypolateritia* by Petch are the same as *P. vincta* var. *vincta*, but that the type of *P. hypolateritia* is microscopically distinct from this var. *P. hypobrunnea* was reduced to synonymy by Setliff under *P. vincta* var. *cinerea*. The *Poria* diseases of rubber and tea were, therefore, attributed to the 2 vars of *P. vincta*. Var. *vincta* is buff to cinnamon orange in pore surface colouration and the context is buff. Var. *cinerea* has a reddish brown to grey pore surface and a dark brown to almost black context. The two vars may also differ in distribution; var. *cinerea* is unknown in tropical America and var. *vincta* is widespread in the tropics (except Africa). Petch (*Diseases of the tea bush*) stated that *P. hypolateritia* is, perhaps, the commonest cause of root disease in the higher tea districts of Sri Lanka, whilst *P. hypobrunnea* is common on rubber but appears to be rare on tea. Agnihothrudu considered that the *Poria* associated with branch canker of tea in N.E. India was *P. punctata* (Fr.) Karst. and not *P. hypobrunnea* to which the condition had been attributed. The sporophore on tea in this region of India has a reddish brown context with basidiospores ovoid to globose, 5–7 × 4–6 μ; *P. hypobrunnea* has a dark coloured context.

The symptoms on tea roots are the rhizomorphs. These are at first white, smooth and soft becoming compact, tough and red to dark red. They form a network initially but later fuse into sheets. The interior of the strands is white. When old the rhizomorphs become black and soil particles may adhere to the mycelium. Care should be taken not to confuse the syndrome with that of *Ganoderma philippii* (q.v.) in the earlier stages and that of *Phellinus noxius* (q.v.) in the later ones. The flat plates of the sporophores (a tube layer on a basal one) on the collar or stem differ in colour (as already pointed out) depending on the var. or sp. The symptoms on rubber are much the same. In N.E. India the disease (branch canker) on tea is mainly one of the stems. It begins in the upper parts of the bush, spreading downwards and eventually (in several years) killing the plant. The affected wood turns yellowish and soft, thin, irregular brown lines occur inside; the sporophores form on the underside of the larger branches or at the collar.

The pathogen(s) is a common saprophyte on woody plant material. In Malaysia (a minor disease on rubber) and Sri Lanka infection spreads to the roots of healthy plants by contact from stumps or other massive inocula. In N.E. India infection apparently spreads through airborne basidiospores. Here the fungus behaves as a wound parasite penetrating the frame of the tea bush where this has been damaged (Sarmah, *Diseases of tea and associated crops in north-east India*). Control consists of removing all dead wood at each pruning, protecting medium pruned tea from sun scorch, and smoothing off all cuts and wounds which should be treated with a bitumen formulation. In the case of root infection the general measures for root diseases (outlined under *Rigidoporous lignosus*) broadly apply. In Sri

Lanka good control on tea has been achieved with methyl bromide after removal of diseased bushes. When tea is young cultural measures should be taken to prevent *Poria* root rot becoming extensive.

A root disease of exotic forest trees in E. Africa is caused by *Poria vincta* var. *cinerea*. The first symptom is a chlorosis of the foliage at the base of the lower branches; the foliage dies, becomes a conspicuous fulvous colour and falls. The whole tree is affected. Roots are covered with white mycelial fans which develop into a black fungal sheath (diagnostic) scattered with white flecks. The bark and wood beneath the sheath is permeated by the fungus; a white spongy rot with many black zone lines results. Hosts are given by Ivory who found that the fungus survived after 6 months burial in soil.

Agnihothrudu, V. 1965. *Curr. Sci.* **34**: 538 (identity of *Poria* on tea in N.E. India; **45**, 596).
Ivory, M. H. 1973. *E. Afr. agric. For. J.* **39**: 180 (on exotic forest trees in E. Africa; **53**, 4126).
Setliff, E. C. et al. 1971. *Pl. Dis. Reptr* **55**: 257 (as above; **50**, 3189).
——. 1972. *Mycologia* **64**: 689 (taxonomy & morphology of *P. vincta*; **52**, 361).
Shanmuganathan, N. 1964. *Tea Q.* **35**: 22 (control on tea in Sri Lanka; **44**, 3146).
—— et al. 1965. *Ibid* **36**: 144 (as above; **45**, 1500).
—— ——. 1967. *Ibid* **38**: 311 (as above & general effects of methyl bromide; **47**, 1962).

PROTOMYCES Unger, *Die Exantheme der Pflanzen*: 341, 1833.

(Protomycetaceae)

In this family the spore sac, variously interpreted as an ascus or a synascus, representing a number of fused asci, is produced singly by germination of a thick walled chlamydospore. Each spore sac contains numerous small, ellipsoid spores which congregate in a ball at the tip of the sac and are shot out in a mass. After ejection they fuse in pairs and germinate to give a mycelium which reinfects the host. The CHLAMYDOSPORES are scattered irregularly through a somewhat hypertrophied portion of host tissue. In *Protomyces* these spores are smooth walled and intercalary in the mycelium (*Protomycopsis* q.v.; from R. W. G. Dennis, *British Ascomycetes*: 80, 1968).

The members of the genus are parasitic and cause galls on the stems, leaves or fruits of Compositae or Umbelliferae. Tubaki and Valadon et al. gave descriptions of some spp., covering life history (including germination of chlamydospores), culture and taxonomy; and see Reddy et al.

Reddy, M. S. et al. 1975. *Mycotaxon* 3: 1 (taxonomic revision of *Protomycetales*; **55**, 2566).
Tubaki, K. 1957. *Mycologia* **49**: 44 (biology of 3 *Protomyces* spp.).
Valadon, L. R. G. et al. 1962. *Trans. Br. mycol. Soc.* **45**: 573 (life history & taxonomy of *P. inundatus*).

Protomyces macrosporus Unger, *Die Exantheme der Pflanzen*: 344, 1833.

(Reddy et al., *Protomyces* q.v., gave *Physoderma gibbosum* Wallr. and *Protomyces cari* Blytt as synonyms)

CHLAMYDOSPORES more or less spherical, 50–70 μ diam. with wall 5 μ thick. ASCUS subglobose, similar to the chlamydospores but thin walled, produced from the resting spores; endospores (blastospores) broadly elliptical, 3 × 4.5 μ (partly from R. W. G. Dennis, *British Ascomycetes*: 81, 1968).

The fungus is widely distributed on Umbelliferae (leaves and stems) and causes an important disease (stem gall) on coriander (*Coriandrum sativum*) in India. Butler (*Fungi and disease in plants*) described and illustrated *P. macrosporus*. Gupta (1954) in India calculated a mean loss/plant of *c.* 15% (mean disease intensity); a 10% increase in disease intensity resulted in an 8.9% loss in expected yield. Infection occurs on all green parts of the plant above ground but not in the seed itself, although the inflorescence and pericarp are attacked. Galls (tumour-like swellings) form on infected plants; the fungus causes hypertrophy and hypoplasia. The galls are usually elongate (they may be up to 1.5 cm long), at first soft becoming woody with age. They develop a corky surface which cracks and are irregularly distributed. An infected bud which develops becomes a completely diseased shoot with presumably systemic infection. Infected leaves become chlorotic and minute galls develop on the mid and lateral veins. The inflorescence may show outgrowths on the surface but here the mature galls are very small; only some parts of a flower may be invaded. Uniform invasion of the fruit makes it abnormally large but partial invasion may lead to distortion (Gupta 1962; Rao). Seed germination from infected plants can be reduced.

The chlamydospores are released into the soil from infected crop debris and from fruit galls in uncleaned seed lots. They are the survival structures but heat resistant, dried endospores have been reported (Pavgi et al. 1969a). In India no other members of the Umbelliferae, known to be susceptible, appear to have any role in the life history of the pathogen. Chlamydospores from galls stored indoors at a low temp. gave <25% germination at 22–24° after 4 months (Mukhopadhyay et al. 1964). Gupta (1973b) found 37% chlamydospore germination after 1 year and 27% after 2, with material stored at room temp. Immersion in water assists germination. Penetration of the young plant, which is susceptible up to *c*. 40 days old, is through the epidermis by the germinating endospores. But no details of the process seem to have been described. Seedlings can show stem infection which has arisen via the integuments (Pavgi et al. 1969b). Mathur (in pot experiments) found most plants to be infected at pH 7.5, when most chlamydospore germination occurred; the least infection was at pH 4.6. There is some evidence for the existence of strs which differ in pathogenicity. The cvs in India also show differences in reaction to infection; Narula et al. found 1 cv. to be highly resistant (and see Gupta et al. 1973a). Partial field control has been given by seed treatment with thiram, but combined seed and soil treatments (same fungicide) gave better results (Nene et al.).

Gupta, J. S. 1954. *Indian Phytopathol.* 7: 53 (losses; **34**: 485).
——. 1962. *Agra Univ. J. Res. Sci.* 11: 307 (anatomy of infected flowers & fruits; **42**: 697).
—— et al. 1963. *Proc. natn. Acad. Sci. India* Sect. B **33**: 507 (morphologic forms; **43**, 2692).
—— ——. 1964. *Ibid* **34**: 243 (variations in pathogenicity; **44**, 1635).
Gupta, R. N. et al. 1973a. *Indian Phytopathol.* **26**: 337 (resistance in cvs; **53**, 4540).
——. 1973b. *Ibid* **26**: 581 (longevity of chlamydospores; **54**, 949).
Heim, P. 1967. *Botaniste* Sér. 50, 1–6: 259 (morphology & taxonomy; **46**, 3360).
Mathur, S. B. 1962. *Indian Phytopathol.* **15**: 75 (infection & soil pH; **42**: 41).
Mukhopadhyay, A. N. et al. 1964. *Experientia* **20**: 619 (chlamydospore germination & culture; **44**, 1190).
—— ——. 1971. *Ann. phytopathol. Soc. Japan* **37**: 215 (infection & environment; **51**, 1753).
Narula, P. N. et al. 1963. *Sci. Cult.* **29**: 206 (tests on cvs; **43**, 178).
Nene, Y. L. et al. 1966. *Mycopathol. Mycol. appl.* **29**: 142 (fungicides; **45**, 3186).
Pavgi, M. S. et al. 1969a & b. *Ann. phytopathol. Soc. Japan* **35**: 265, 271 (dispersal & survival; seed viability in infected plants; **49**, 1106a & b).
—— ——. 1970. *Cytologia* **35**: 359 (cytology of chlamydospore germination; **52**, 811).
—— ——. 1972. *Ibid* **37**: 619 (development of infected fruit; **52**, 2706).
Rao, C. G. P. 1972. *Indian Phytopathol.* **25**: 483 (anatomy of abnormal growth in infected plants; **53**, 1043).

PROTOMYCOPSIS Magnus, *Pilzflora un Tyrol*: 322, 1905.
(Protomycetaceae)

Protomyces q.v. for a note on the characteristics of this family. The CHLAMYDOSPORES are scattered irregularly through a somewhat hypertrophied portion of host tissue; they differ from those of *Protomyces* in being formed singly at tips of hyphal branches and having the wall finely punctate (rough) at maturity (from R. W. G. Dennis, *British Ascomycetes*: 82, 1968). This author points out that the small, round, yellowish leaf spots induced by *Protomycopsis* spp. must be distinguished from those caused by *Entyloma*, in which the host tissue is filled with pale spherical chlamydospores. The critical difference between these genera is, of course, the different product of chlamydospore germination. Also it would be unexpected to find an *Entyloma* with chlamydospores of >25 μ diam. Resting sporangia of *Physoderma* and *Synchytrium* may also be mistaken for chlamydospores of Protomycetaceae but they lie within the host cell and are not produced on a septate mycelium. They are commonly embedded in wart-like galls which are often brightly coloured.

In discussing the *Protomycopsis* spp. described from India Prasad et al. divided them into 2 spp. *P. phaseoli* Ramakrishnan & C. V. Subramanian (synonyms: *P. patelii* Pavgi & Thirum., *P. smithiae* Thirum., Bhatt, Patel & Dhande, *P. crotalariae* Joshi and *Synchytrium phaseoli* Patel, Kulk. & Dhande) causes dark, ash grey angular spots (angular black leaf spot) on *Phaseolus mungo* (black gram), *P. radiatus, Vigna unguiculata* (cowpea) and *Crotalaria triquetra*. Its chlamydospores (16–34 μ diam.) have a convolute to warty exospore. The other sp., *Protomycopsis ajmeriensis* J. S. Gupta, causes circular dark brown leaf galls on *Sesbania aculeata* and *S. aegyptica* (chlamydospores of a comparable diam.

have an areolate to reticulate exospore). Haware et al. (1971, 1972) gave some description of *P. thirumalacharii* Pavgi causing purple leaf spot of *S. grandiflora*. In black gram and cowpea infection by *P. phaseoli* first shows as pale yellow specks, mostly on the lower leaf surface; the final dark, angular spots are larger (2–3 cm diam.) in cowpea. (But Reddy et al., *Protomyces* q.v., considered these spp. on Leguminosae to be doubtful.)

Haware, M. P. et al. 1971. *Ann. phytopathol. Soc. Japan* **37**: 242 (survival in *Protomycopsis phaseoli* & *P. thirumalacharii*; **51**, 4511).
——. 1972. *Friesia* **10**: 43 (physiology of chlamydospore germination in spp. as above; **52**, 3546).
——. 1976. *Cytologia* **41**: 459 (cytology of chlamydospore development; **56**, 1439).
——. 1976. *Indian J. agric. Sci.* **46**: 280 (host reaction to *P. phaseoli*; **56**, 3325).
——. 1976. *Mycopathologia* **59**: 105 (chlamydospore germination; **56**, 1920).
Pavgi, M. S. et al. 1970. *Pathol. Microbiol.* **35**: 297 (conidial discharge in *P. thirumalacharii*; **53**, 1908).
Prasad, N. et al. 1962. *Indian Phytopathol.* **15**: 24 (genus in India; **42**: 105).
Rao, C. G. P. 1972. *Ibid* **25**: 449 (anatomy of *Sesbania aculeata* infected by *P. ajmeriensis*; **53**, 183).

PSEUDOCERCOSPORA Speg., *An. Mus. nac. Hist. nat. B. Aires* **20**: 438, 1910.
Helicomina L. S. Olive, 1948
Ancylospora Sawada, 1944 (no Latin diagnosis)
Cercocladospora G. P. Agarwal & S. M. Singh, 1974?
(Dematiaceae)

MYCELIUM internal or both internal and external. STROMA present or absent. CONIDIOPHORES well developed, short or long, simple or branched, continuous or septate, aggregated in fascicles emerging through the stomata and/or produced terminally and laterally on external mycelial hyphae. CONIDIOGENOUS CELLS integrated, sympodial and polyblastic, denticulate, subgeniculate or more or less sinuous, leaving the old conidial scars situated at the end of longer or shorter denticles or on distinct or less conspicuous or scarcely visible shoulders, sometimes pseudo-percurrent on one and the same conidiophore that is also denticulate, proliferating through the apex and either displacing the old apical scar into a lateral position or, more usually, completely rupturing the old apical scar, to produce pseudo-annellations, smooth or sometimes slightly verrucose towards the apex. CONIDIAL SCARS unthickened, i.e. of the same thickness as the wall of the conidiogenous cell. CONIDIA coloured, pale or deeper brown, usually obclavate to cylindric, sometimes subcylindric, slightly or more obviously tapered at the base towards the unthickened hilum, straight or curved, sometimes strongly curved, smooth or finely verrucose or rugulose (sometimes slightly wrinkled rough), pluriseptate, usually not catenate (from F. C. Deighton, *Mycol. Pap.* 140: 8, 1976; see also Ellis, *Demat. Hyphom.*: 208).

The genus and many spp. are fully described by Deighton (l.c.). *P. purpurea* (Cooke) Deighton (synonym: *Cercospora purpurea* Cooke) causes significant damage on avocado (brown spot) in Brazil and Cameroon. The leaf spots are at first indistinct, becoming large, irregular blotches, up to 1.5 cm diam., pale brown, a slightly darker line margin, sometimes no distinct spots and sporulation appears semi-effuse on lower leaf surface, leaf dies; fruit is also attacked. STROMA dark, 15–125 μ diam.; CONIDIOPHORES $20–200 \times 3–4.5$ μ; CONIDIA 1–9 septate, $20–100 \times 2–4.5$ μ (after Chupp; *Cercospora* q.v.). In Cameroon, where brown spot requires control in the coastal zone, Bordeaux, benomyl and copper in oil are effective. Varietal resistance in avocado was reported from Brazil. *P. psophocarpi* (Yen) Deighton (synonym: *C. psophocarpi* Yen) was described from winged bean (*Psophocarpus tetragonolobus*). This plant is said to be of potential economic importance and the fungus can cause complete destruction of the foliage (see Price et al. who described some aspects of its biology). *C. aranetae* Borlaza & Roldan, also described from winged bean (but not validly published), was considered by Deighton (l.c.) to be different from *P. psophocarpi*. Amongst the *Pseudocercospora* spp. described fully by Deighton (l.c.) are: *P. colocasiae* Deighton (on *Colocasia*); *P. puerariae* (H. & P. Syd.) Deighton (on *Pueraria*; synonym: *Cercospora puerariae* H. & P. Syd.); *P. stahlii* (F. L. Stev.) Deighton (on *Passiflora*; synonyms: *Helminthosporium stahlii* F. L. Stev., *Cercospora stahlii* (F. L. Stev.) Subramanian, *C. passiflorae-foetidae* Yen, *C. passiflorae-longipedis* Yen). Three spp. are described from *Dioscorea*: *P. contraria* (H. & P. Syd.) Deighton, *P. hiratsukana* (Togashi & Katsuki) Deighton and *P. ubi* (Racib.) Deighton.

Albuquerque, F. C. 1962. *Rev. Soc. Agron. Vet. Pará* **8**: 35 (*Pseudocercospora purpurea*, note; **53**, 1363).
Borlaza, P. B. et al. 1964. *Araneta J. Agric.* **11**: 181 (*Cercospora aranetae* sp. nov.; **46**, 3216).
Gaillard, J. P. 1971. *Fruits* **26**: 225 (*P. purpurea*, fungicides; **51**, 527).
Hino, T. et al. 1976. *Summa Phytopathol.* **2**: 127 (as above, resistance: **56**, 305).
Price, T. V. et al. 1978. *Trans. Br. mycol. Soc.* **70**: 47 (*P. psophocarpi* in Papua New Guinea; **57**, 4250).

Pseudocercospora fuligena (Roldan) Deighton, *Mycol. Pap.* 140: 144, 1976.
Cercospora fuligena Roldan, 1938.

Leaf spots indistinct, at times only effuse yellow patches are visible. Fruiting hypophyllous, scattered or in angular areas delimited by leaf veins, brown to dark brown. STROMA absent or only slight. CONIDIOPHORES in fascicles, 2–7 or very dense, pale to medium brown, darker towards the basal region, simple, unbranched, slightly geniculate, finely echinulate or smooth, septate, conidial scars thickened, 16–70 × 3.5–5 μ. CONIDIA subhyaline, obclavate–cylindrical, base truncate with thickened hilum, apex round, mostly straight but occasionally curved, distinctly multiseptate, 15–120 × 3.5–5 μ (J. L. Mulder & P. Holliday, *CMI Descr.* 465, 1975, as *C. fuligena*).

P. fuligena causes a leaf spot or leaf mould of tomato; the latter common name is best kept for the more important pathogen (*Fulvia fulva* q.v.) on the same crop. Early symptoms are faint, sunken, chlorotic areas with indefinite margins; the infected areas become necrotic and have an indefinite discoloured halo on both leaf surfaces; defoliation can occur. The fungus has been reported from W. Africa, E. Asia, Oceania, Uganda and USA (Florida; Map 382). The disease needs further investigation. Opt. mycelial growth occurs at 26–28°, and stomatal and cuticular penetration have been described. Severe outbreaks have occurred in India where infection spread from local cvs to the cvs Marglobe and Red Ball. In USA (Florida) the disease may become economically important since the susceptible cvs Florida MH-1 and Walter are increasingly grown. The cv. Homestead was mildly susceptible and cv. Floradel appeared to be resistant (Blazquez et al.; and see Yamada).

Blazquez, C. H. et al. 1974. *Phytopathology* **64**: 443 (symptoms, host reaction; **53**, 4592).
Magda, T. R. et al. 1970. *Philipp. Phytopathol.* **6**: 75 (penetration & sporulation; **53**, 1949).
Mohanty, U. N. et al. 1955. *Sci. Cult.* **21**: 269 (in India; **35**: 335).
Yamada, S. 1951. *Ann. phytopathol. Soc. Japan* **15**: 61 (general, including resistance; **31**: 151).

PSEUDOCERCOSPORELLA Deighton, *Mycol. Pap.* 133: 38, 1973.
(Moniliaceae)

MYCELIUM internal. STROMATA none or well developed. CONIDIOPHORES emerging through the stomata or erumpent through the cuticle of neighbouring cells; short, thin walled, smooth, subcylindric, usually simple, continuous or with few septa, colourless or almost so (sometimes faintly greenish). CONIDIOGENOUS CELLS integrated, sympodial, polyblastic; conidial scars inconspicuous, truncate, not thickened. CONIDIA subcylindric or very slightly obclavate cylindric, colourless or almost so, smooth, thin walled, mostly pluriseptate, with a truncate unthickened hilum, not catenulate or in a few spp. catenulate (from F. C. Deighton l.c.).

Deighton described the spp. of this genus and also *Cercosporella* Sacc. Several spp. were transferred from *Cercosporella* to *Pseudocercosporella*. These include *P. capsellae* (q.v.), *P. herpotrichoides* (Fron) Deighton (eyespot of temperate cereals and grasses; see Booth et al.) and *P. ipomoea* (Sawada) ex Deighton (on leaves of water spinach, *Ipomoea aquatica*). *C. carthami* Mourashkinsky on safflower (*Carthamus tinctorius*) is not referred to by Deighton. He described the characteristic method of liberation of the conidium in *Cercosporella*; it is quite different from that of *Pseudocercosporella*. The distinction between *Cercospora* and *Cercosporella* was also demonstrated.

Booth, C. et al. 1973. *C.M.I. Descr. pathog. Fungi Bact.* 386 (*Pseudocercosporella herpotrichoides*).
Castellani, E. 1940. *Agric. colon.* **34**(8), 10 pp. (*Cercosporella carthami* & *Puccinia carthami*; **25**: 319).
Deighton, F. C. 1973. *Mycol. Pap.* 133, 62 pp. (*Cercosporella, Pseudocercosporella* gen. nov. & *Pseudocercosporidium* gen. nov.; **53**, 73).
Kanchaveli, L. A. et al. 1972. *Trudy̆ Inst. Zasheh. Rast. Gruz. SSR* **23**: 312 (*Cercosporella* in USSR, Georgia; **53**, 2879).
Mourashkinsky, K. E. 1926. *Bull. W. Siberian Sect. Russian geogr. Soc.* 5, 6 pp. (*C. carthami* sp. nov. & diseases of safflower; **6**: 354).

Pseudocercosporella capsellae (Ell. & Ev.) Deighton, *Mycol. Pap.* 133: 42, 1973.
Cylindrosporium capsellae Ell. & Ev., 1887
Cercosporella brassicae (Fautr. & Roum.) Höhn., 1924
Cercosporella albomaculans (Ell. & Ev.) Sacc., 1895
Ramularia rapae Pim, 1897.
(see Deighton l.c. for further synonymy)

Sporulation amphigenous, whitish, densely distributed over the leaf spot. MYCELIUM internal; hyphae colourless, 2.5–4 μ wide. STROMA variable, sometimes small and of swollen, colourless, loosely aggregated hyphae below a stoma and in surrounding epidermal cells; sometimes *c.* 65 μ wide and 50 μ high, of more densely aggregated hyphae which, in parts of the stroma, may be slightly brown. Stromata often contiguous and confluent. CONIDIOPHORES numerous, emerging through the stomata and erumpent through the cuticle, colourless, smooth, ampulliform, with a cylindric portion 2–4 μ wide, 5–15 μ long, protruding above the stoma or leaf surface, continuous, straight or somewhat flexuous. Conidial scars truncate, unthickened, *c.* 1 μ diam. The terminal scar can be seen clearly but old displaced scars cannot be seen and it is not clear whether more than a single conidium is produced from each conidiogenous cell. CONIDIA colourless, cylindric or very slightly narrower towards the apex (slightly obclavate cylindric), smooth, straight or slightly curved, obtuse at the apex, slightly narrowed at the extreme base to the truncate, unthickened hilum, 1–5 septate, mostly 3 septate, not constricted, 30–90 μ long, mostly up to 60 μ long and usually 2–3 μ wide but occasionally up to 4 μ wide (from F. C. Deighton l.c.).

P. capsellae causes white spot of *Brassica* spp. and other crucifers including radish; see Petrie for a recent description. It is widespread (Map 197). On some hosts the leaf lesions are conspicuous, round, semi-transparent, whitish with brownish grey centres and well defined brown or tan margins, on turnip and wild mustard the fully formed spots are *c.* 3 cm diam. On cabbage the young lesions av. 2 mm diam. and are dark grey or black, dendritic but becoming more or less rectangular or rounded with well defined margins; mature lesions are 1–2 cm diam. and ashy black. Growth in vitro is best at 20–24° and these temps are most favourable for conidial germination and germ tube growth. The fungus may persist in leaf debris for at least 9 months and may be seedborne. Crucifer white spot appears to be of little importance. For control captafol, maneb and metiram have been used in USA.

Chandler, W. A. 1965. *Pl. Dis. Reptr* **49**: 419 (fungicides, with *Colletotrichum higginsianum*; **44**, 2657).
Crossan, D. F. 1954. *Tech. Bull. N. Carol. agric. Exp. Stn* 109, 23 pp. (general, 33 ref.; **35**: 736).
Miller, P. W. et al. 1948. *Phytopathology* **38**: 893 (general; **28**: 263).
Petrie, G. A. et al. 1978. *Can. Pl. Dis. Surv.* **58**: 69 (morphology, symptoms, hosts & toxin).

PSEUDOPERONOSPORA Rostovtsev, *Izv. mosk. Sel'.-khoz. Inst.* **9**: 47, 1903.
Peronoplasmopora (Berl.) Clinton, 1905.
(Peronosporaceae)

MYCELIUM intercellular, branching; haustoria small, usually simple. SPORANGIOPHORES pseudomonopodially branched, primary branches usually arising at acute angles, the ultimate branches acute. SPORANGIA typically coloured, rarely hyaline, elliptical, conspicuously papillate both apically and basally; germinating by zoospores. OOSPORES thin walled, smooth or roughened. OOGONIA thin walled (partly from G. W. Wilson, *Bull. Torrey bot. Club* **9**: 47, 1903; see Waterhouse in *The fungi* Vol. IVB: 175, 1973).

Fitzpatrick (*The lower fungi Phycomycetes*) briefly discussed the taxonomic history of this genus which is close to *Peronospora* (q.v.) and *Plasmopora* (q.v.). *Pseudoperonospora* is generally accepted as distinct but a recent, detailed, comparative study of the 3 genera does not apparently exist. Hoerner upheld the validity of *Pseudoperonospora* and gave a list of 8 spp. with their hosts and early literature. *P. humuli* (Miyabe & Takah.) G. W. Wilson causes the important downy mildew of hop (*Humulus lupulus*); see Western (editor), *Diseases of crop plants*. *P. cannabina* (Otth.) Curzi is a minor pathogen on hemp (*Cannabis sativa*).

Hoerner, G. R. 1940. *J. Wash. Acad. Sci.* **30**: 133 (on nomenclature; **19**: 435).
——. 1940. *Pl. Dis. Reptr* **24**: 170 (spp. list & hosts; **19**: 567).

Pseudoperonospora cubensis (Berk. & Curt.) Rostov., *Annls Inst. Agron. Moscow* 9: 47, 1903.
Peronospora cubensis Berk. & Curt., 1868
Plasmopora cubensis (Berk. & Curt.) Humphrey, 1891
Peronoplasmopora cubensis (Berk. & Curt.) Clinton, 1904 (1905).
(synonymy in G. W. Wilson, *Bull. Torrey bot. Club* **34**: 413, 1907)

Hypophyllous or rarely amphigenous. MYCELIUM intercellular with short ovate haustoria. SPORANGIOPHORES usually arising through a stoma, mostly 1–2 (up to 5–6), 180–400 (usually 200–300) μ long, 5–9 μ wide; ×3–5 branched, roughly dichotomous, branches at an acute angle, 5–20 μ long, slightly curved, tips subacute; resembling some *Peronospora* spp. rather than *Plasmopora*. SPORANGIA hyaline when immature, becoming smoky grey or purplish, grey black in mass, ovoid or ellipsoid, thin walled, 14–25 × 20–40 μ, papillate distally, deciduous; usually germinating by zoospores, 10–13 μ diam. at encystment. OOGONIA obovoid to ellipsoid or irregularly pyriform; 33–41 × 33–40 μ (Chen et al.); 28–56 × 24–44 μ (Hiura et al.). ANTHERIDIA clavate to globose, 14–22 × 10–16 μ. OOSPORES spherical, rarely ellipsoid to obovoid, hyaline or pale yellow, wall smooth, 1.5–3.5 μ thick; 25–33 × 25–31 μ (Chen et al.); 22–42 μ (Hiura et al; see also Palti 1975).

Cucurbit downy mildew is a worldwide and important leaf disease; it has not been reported from parts of N. and W. Europe (Map 285). The most important host is cucumber (*Cucumis sativus*); others on which it is most frequently found are: melon (*C. melo*), squash and pumpkin (*Cucurbita* spp.) and watermelon (*Citrullus lanatus*; see Palti 1974). It is less frequent on other Cucurbitaceae and hosts outside this family. Characteristic, angular, chlorotic areas form on the upper leaf surface; they become necrotic, coalesce and kill the leaf. The leaves tend to be destroyed from the base upwards. The fruit is not directly attacked but the few that form remain small or do not ripen normally if they reach full size.

Penetration by the zoospores is through the stomata when fully differentiated. Opt. temps for sporangial germination, infection and disease development are 15–20°. The shortest dew period for infection is 2 hours at an opt. of 20°. Increasing the dew periods gave infection at lower, less favourable temps (Cohen 1977a). Under opt. moist conditions sporulation can occur *c.* 6 hours after infection; it is most abundant on the chlorotic lesions. A diurnal periodicity for the airborne sporangia with a peak at *c.* 0800 hours was described from Israel and USA; under natural conditions probably most do not survive a day (Cohen et al. 1971b; Schenck 1968). Sporangial infectivity decreases with rising temp. and RH but a wet period is required for sporulation. The effect of light and dark on sporulation appears to be an indirect one, acting through the host. There is a decrease in lesion growth when leaves are kept in the dark for 5–8 days after inoculation but not when kept thus 2 days before. This growth is checked in proportion to the length of the dark period, especially when this is 3 or more days. Lesion area decreases with a fall in light intensity; increases in light and daily photoperiods are associated with more sporulation. Since oospores are rare or absent carry over is through mycelium in the host. On new crops in cooler areas infection spreads from warmer ones or from glasshouses. Differences in pathogenicity between forms of *P. cubensis* on several cucurbits have been described by Hughes and Iwata (1941, 1942 & 1951).

Both polygenic and oligogenic resistance occur and breeding programmes have produced resistant cucumber, melon and watermelon cvs (see Barnes, Blazquez, Boelema, Cohen 1976, Ezuka, Ivanoff, Jenkins, Thomas, Vliet, Winstead and others). Chemical control is given by daconil, dithane, maneb and maneb + zineb. Young crops should not be grown too near old ones. Measures which reduce the length of leaf wetness periods should be taken.

Barnes, W. C. et al. 1954. *Pl. Dis. Reptr* **38**: 620 (lesion patterns on resistant cucumber; **34**: 274).

——. 1955. *Proc. Am. Soc. hort. Sci.* **65**: 409 (breeding for resistance in cucumber with *Colletotrichum lagenarium*; **35**: 71).

Blazquez, C. H. 1970. *Pl. Dis. Reptr* **54**: 52 (resistant cucumber cvs; **49**, 2264).

Boelema, B. H. 1967. *Fmg S. Afr.* **43**: 3 (control with dithane & resistant cucumber cv.; **47**, 969).

Chen, C. P. et al. 1959. *Zhibing Zhishi* **3**: 144 (oospores; **40**: 393).

Clinton, G. P. 1905. *Rep. Conn. agric. Exp. Stn* 1904: 329 (general).

Cohen, Y. et al. 1969. *Israel Jnl Bot.* **18**: 135 (lesion development, temp., moisture & sporulation; **49**, 1532).

——. 1971a. *Phytopathology* **61**: 265, 594, 736 (lesion development, light, dark, moisture & epidemiology; **50**, 3361; **51**, 995, 996).

Cohen, Y. et al. 1971b. *Trans. Br. mycol. Soc.* **57**: 67 (sporangial dispersal & viability; **51**, 994).
———. 1974. *Can. J. Bot.* **52**: 447 (fine structural & physiological change in sporangia under water stress, with *Phytophthora infestans*; **53**, 3340).
——. 1976. *Phytoparasitica* **4**: 25 (resistance in *Cucumis*; **56**, 3300).
——. 1977a. *Can. J. Bot.* **55**: 1478 (effects of temp., leaf wetness & inoculum conc. on infection; **57**, 842).
———. 1977b. *Physiol. Pl. Pathol.* **10**: 93 (effects of light & temp. on growth of sporangia & sporangiophores; **56**, 5264).
Doran, W. L. 1932. *Bull. Mass. agric. Exp. Stn* 283, 22 pp. (general; **12**: 5).
Duvdevani, S. et al. 1946. *Palest. J. Bot. Rehovot Ser.* **5**: 127 (temp. & moisture in annual cycle; **26**: 477).
Ezuka, A. et al. 1974. *Bull. Tokai-Kinki natn. agric. Exp. Stn* 27: 42 (resistance in cucumber; **54**, 2534).
Hiura, M. 1929. *Res. Bull. Gifu Imp. Coll. Agric.* 6, 58 pp. (effect of temp. & other conditions; **8**: 698).
—— et al. 1933. *Jap. J. Bot.* **6**: 507 (oospores; **13**: 349).
Hughes, M. B. et al. 1952. *Pl. Dis. Reptr* **36**: 365 (pathogenic forms; **32**: 299).
Inaba, T. et al. 1971. *Ann. phytopathol. Soc. Japan* **37**: 340 (effect of light on lesion development; **51**, 4571).
———. 1975. *Bull. natn. Inst. agric. Sci. Tokyo* Ser. C. **29**: 65 (lesion development, sporulation & photosynthesis; **54**, 4681).
Ivanoff, S. S. 1944. *J. Hered.* **35**: 35 (resistance in melon; **23**: 373).
Iwata, Y. 1938. *Ann. phytopathol. Soc. Japan* **8**: 124 (host penetration; **18**: 86).
——. 1941 & 1942. *Ibid* **11**: 101, 172 (forms on cucumber & squash; **30**: 356).
——. 1951. *Forschn Geb. PflKrankh. Kyoto* **4**: 124 (form on *Benincasa hispida*; **33**: 136).
——. 1951. *Bull. Fac. Agric. Mie Univ.* **2**: 34 (leaf maturity & infection; **31**: 102).
Jenkins, J. M. 1942. *J. Hered.* **33**: 35 (resistance in cucumber; **21**: 317).
——. 1946. *Ibid* **37**: 267 (inheritance of resistance; **26**: 140).
Kajiwara, T. et al. 1957. *Ann. phytopathol. Soc. Japan* **22**: 201 (development of sporangiophore & sporangia; **37**: 752).
———. 1959. *Ibid* **24**: 109 (effect of light on sporulation; **39**: 367).
———. 1968. *Ibid* **34**: 85 (light & sporangial longevity; **47**, 2346).
——. 1969. *Bull. natn. Inst. agric. Sci. Tokyo* Ser. C. **23**: 63 (fine structure of haustorial interface; **50**, 1027).
Kurosawa, E. 1927. *Trans. nat. Hist. Soc. Formosa* 17, 18 pp. (temp. & zoospore emergence; **7**: 295).
Palti, J. 1974. *Phytoparasitica* **2**: 109 (divergencies in distribution; **54**, 4685).
——. 1975. *C.M.I. Descr. pathog. Fungi Bact.* 457 (*Pseudoperonospora cubensis*).
Perl, M. et al. 1972. *Physiol. Pl. Pathol.* **2**: 113 (transfer of assimilates from leaves to sporangia; **51**, 3680).
Schenck, N. C. 1968. *Phytopathology* **58**: 91 (sporangial dispersal; **47**, 1365).
——. 1969. *Proc. Fla St. hort. Soc.* **82**: 151 (epidemiological comparison with *Didymella bryoniae*; **50**, 2621).
Thomas, C. E. 1970. *Pl. Dis. Reptr* **54**: 108 (differential sporulation on watermelon; **49**, 2270).
——. 1978. *Ibid* **62**: 221 (resistance in melon; **57**, 5183).
Vliet, G. J. A. Van et al. 1974. *Euphytica* **23**: 251 (inheritance of resistance in cucumber; **54**, 603).
Winstead, N. N. et al. 1957. *Pl. Dis. Reptr* **41**: 620 (resistance in lines & cvs of watermelon; **37**: 65).

PUCCINIA Persoon, ex Persoon, *Syn. meth. Fung.*: 228, 1801.
(Pucciniaceae)

SPERMOGONIA subepidermal. AECIDIA subepidermal in origin, aecidioid with a membranous peridium and catenulate spores or uredinoid with aecidiospores borne singly or endophylloid with aecidioid sori but the spores germinating to form basidia. UREDIA subepidermal, paraphysate or araphysate; UREDOSPORES borne singly, wall usually echinulate. TELIOSORI subepidermal, persistently so in some spp., TELIOSPORES typically 2 celled, less commonly 1, 3 or 4 celled, one pore in each cell, pedicels various, long or short, persistent or deciduous, wall coloured, basidia external, typically 4 celled. The largest genus of the rust fungi with some 3–4 thousand spp. (from M. Wilson & D. M. Henderson, *British rust fungi*: 122, 1966).

The plant pathogens in *Puccinia* are particularly inportant on members of the Gramineae, see Cummins (*The rust fungi of cereals, grasses and bamboos*) for their systematics. The pathogens on the following crops are described separately: *Allium*, Bermuda grass, cotton, *Digitaria*, groundnut, guava, maize, millet, safflower, sorghum, sugarcane and sunflower. *P. pitteriana* P. Henn. occurs on Irish potato and tomato, the only rust on the latter host, and on which it apparently causes no appreciable damage. It is restricted to Brazil, Colombia, Costa Rica, Ecuador, Mexico, Paraguay, Peru and Venezuela (Map 113; Laundon et al. 1971c). Briton-Jones et al.

described the canker of cypre (*Cordia alliodora*) in Trinidad and caused by *P. cordiae* (P. Henn.) Arth. (synonym: *P. corticola* Arth. & Rorer). *P. paulensis* Rangel occurs on *Capsicum* (Davidson). In Pakistan (Rizvi et al.) *P. citrulli* Syd. & Butler decreased yields of watermelon (*Citrullus lanatus*) near Karachi. They considered that their fungus differed from the form on *C. colocynthis*. *P. chondrillina* Bub. & Syd. has been fairly extensively studied in its use for the control of skeleton weed (*Chondrilla juncea*) in Australia (Cullen et al.; Groves et al.; Hasan; Hasan et al.).

In 1968 Nagy reported a >50% loss on tarragon (*Artemisia dracunculus*) by the rust called *P. dracunculina* Fahr. *P. tanaceti* DC. also occurs on this crop (Tanev) and Cummins proposed the combination *P. tanaceti* DC. var. *dracunculina* (Fahr.) Cumm. (*Mycotaxon* **5**: 406, 1977). The var. has uredospores 17–20 μ diam. whilst in *P. tanaceti* they are 22–28 μ. *P. absinthii* DC. is presumably a synonym of *P. tanaceti*. Zineb was used in the control of tarragon rust. *P. apii* Desm. (celery rust) is of little economic importance. *P. asparagi* DC. (Map 216, widespread) causes asparagus rust which can be serious and justify chemical control. It can infect onion. *P. menthae* Pers. (Map 211, widespread with crop) causes mint rust (Laundon et al. 1964; Laundon et al. 1971a; Waterston 1965).

Puccinia spp. cause diseases on several ornamental crops, for example, antirrhinum (*P. antirrhini* Diet. & Holw.); canna (*P. thaliae* Diet.); chrysanthemum (*P. chrysanthemi* Roze, black rust; *P. horiana* P. Henn., white rust); geranium (*P. leveillei* Mont.); hollyhock (*P. malvacearum* Mont.); iris (*P. iridis* Rabenh.); pelargonium (*P. granularis* Kalch. & Cooke, *P. pelargonii-zonalis* Doidge (see Laundon et al. 1971b; Punithalingam; Sivanesan).

Briton-Jones, H. R. et al. 1930. *Mem. Imp. Coll. trop. Agric. Trin.* Mycol. Ser. 3, 8 pp. (*Puccinia cordiae* on cypre; **10**: 215).
Cullen, J. M. et al. 1973. *Nature Lond.* **244**: 462 (*P. chondrillina*, biological control; **53**, 35).
Davidson, R. W. 1932. *Mycologia* **24**: 221 (*P. paulensis* inter alia; **11**: 606).
Groves, R. H. et al. 1975. *Aust. J. agric. Res.* **26**: 975 (*P. chondrillina*, biological control; **55**, 2614).
Hasan, S. et al. 1972. *Pl. Dis. Reptr* **56**: 858 (as above; **52**, 2466).
——. 1972. *Ann. appl. Biol.* **72**: 257 (as above; **52**, 1379).
—— ——. 1973. *Ibid* **74**: 325 (as above; **53**, 390).
Hooker, A. L. 1967. *A. Rev. Phytopathol.* **5**: 163 (genetics & expression of plant resistance to *Puccinia*, 102 ref.).
Johnson, T. et al. 1967. *Ibid* **5**: 183 (world situation in cereal rusts, 74 ref.).
Laundon, G. F. et al. 1964. *C.M.I. Descr. pathog. Fungi Bact.* 7 (*P. menthae*).
—— ——. 1971a. *Ibid* 284 (*P. apii*).
—— ——. 1971b. *Ibid* 285 (*P. iridis*).
—— ——. 1971c. *Ibid* 286 (*P. pittieriana*).
Nagy, F. 1968. *Acta phytopathol. Acad. Sci. hung.* **3**: 331 (*P. tanaceti* var. *dracunculina* on tarragon; **48**, 1264).
Punithalingam, E. 1968. *C.M.I. Descr. pathog. Fungi Bact.* 175, 176 (*P. chrysanthemi* & *P. horiana*).
Rizvi, S. R. H. et al. 1960. *Pakist. J. scient. ind. Res.* **3**: 163 (*P. citrulli* on watermelon; **41**: 118).
Rowell, J. B. 1968. *A. Rev. Phytopathol.* **6**: 243 (chemical control of cereal rusts, 102 ref.).
Sivanesan, A. 1970. *C.M.I. Descr. pathog. Fungi Bact.* 262–7 (*P. antirrhini, P. granularis, P. leveillei, P. malvacearum, P. pelargonii-zonalis* & *P. thaliae*).
Tanev, I. 1969. *Rasteniev. Nauki* **6**: 99 (*P. tanaceti* on tarragon; **48**, 2512).
Waterston, J. M. 1965. *C.M.I. Descr. pathog. Fungi Bact.* 54 (*P. asparagi*).

Puccinia allii Rud., *Linnaea* **4**: 392, 1829.
Puccinia mixta Fuckel, 1870
P. porri Wint., 1881
P. blasdalei Diet. & Holw., 1893
Uromyces ambiguus (DC.) Lev., 1847
U. durus Diet., 1907.

PYCNIA amphigenous in small groups. AECIDIA amphigenous, surrounding the pycnia, cupulate. AECIDIOSPORES spheroidal to ellipsoidal 16–28 μ, wall very pale, finely verrucose, 1–2 μ thick. UREDIA amphigenous, irregularly scattered, luteous, elongated, mostly 1–3 mm long, powdery. UREDOSPORES spheroidal to ellipsoidal, 23–32 × 20–26 μ; wall hyaline to luteous, echinulate, 1–2.5 μ thick. TELIA amphigenous, irregularly scattered, sometimes surrounding the uredia, usually stromatic and covered by the epidermis, sometimes open, blackish, very variable in size. TELIOSPORES usually mostly 2 celled, sometimes mesospores numerous or all mesospores, clavate to cylindrical, often angular and irregular, 36–65 × 18–28 μ; MESOSPORES ellipsoidal, 20–32 × 17–24 μ; wall sienna to chestnut, usually paler below, smooth, 1–2 μ thick at sides, 2–8 μ thick above, pedicels very pale, short and fragile.

P. allii-japonici Diet., *P. granulispora* Ellis & Gall., *P. mutabilis* Ellis & Gall., *Uromyces aemulus*

Arth. and *U. bicolor* Ellis (= *U. aterrimus* Diet. & Holw.) are very similar and would probably better be considered synonymous (G. F. Laundon & J. M. Waterston, *CMI Descr.* 52, 1965).

Laundon et al. (l.c.) also refer to 7 other rusts on cultivated *Allium* spp. and 1 on wild *Allium* spp. *P. asparagi* DC. (*CMI Descr.* 54, 1965) on *Asparagus officinalis* has been recorded on perennial onion as aecia, and is reported to have been inoculated to *Allium cepa* producing aecidia, uredia and telia. *P. asparagi* differs from *P. alli* chiefly in having uredospores with 4 equatorial pores and in the shorter, more rounded teliospores and absence of paraphyses.

Rust of *Allium* spp. is widespread (Map 400) especially in temperate and subtropical areas where it can probably be more serious locally than in the tropics. The symptoms appear as elongate lesions on the leaves which split longitudinally to reveal the sporodochia. The lower leaves may yellow and wilt and bulb size can be reduced. Economic damage is frequently localised and sporadic but serious losses have been reported from Brazil, Israel and S. Africa (on chive), UK (leek), Japan (Welsh onion), Hong Kong (chive), Tanzania (onions and garlic) and USA (California).

Little appears to be known of the details of the life history. The fungus appears to overwinter readily and in USA it is not uncommon on wild *Allium* spp. Uredospores and teliospores are found but not always aecidia. Seed transmission is said to be unimportant. Physiologic specialisation appears to exist but there has been no adequate study on this. Uredospores from garlic in USA were used to infect garlic, Australian brown and Italian red onions and asparagus. Heavy infection occurred on both garlic and onion but the pustules were smaller and fewer on the latter; no infection was found on asparagus. The rust has been reported to appear on garlic whilst onion remains uninfected (USA); Welsh onion was attacked but not other *Allium* spp. (Japan); only Chinese chive (*A. tuberosum*) was attacked (Hong Kong); and in UK it can be common on leek and chive but rare on onion. There are no data on the effect of weather or temp., high nitrogen applications may increase disease incidence. There are no specific control measures. Spraying has been successful with several common fungicides; it needs to be applied early and probably frequently. Cultural precautions against carry over and other hosts, i.e. *Allium* spp. and asparagus should be taken. Leek cvs show different levels of chlorosis.

Dixon, G. R. 1976. *J. natn. Inst. agric. Bot.* **14**: 100 (incidence on leek in UK; **56**, 4296).

Doherty, M. A. et al. 1978. *Physiol. Pl. Pathol.* **12**: 123 (antagonism by *Bacillus cereus*; **57**, 4260).

Goto, K. 1933–5. *J. Soc. trop. Agric.* **5**: 167; **6**: 44; **7**: 38 (morphology & inoculation; **13**: 73; **14**: 75, 735).

Leather, R. I. 1968. *Agric. Sci. Hong Kong* **1**: 13 (fungicides; **48**, 1419).

Sawada, K., 1928–9. *Rep. nat. Hist. Soc. Formosa* **18**: 148; **19**: 180 (taxonomy; **8**: 479; **9**: 82).

Tavel, C. von 1932. *Ber. schweiz. bot. Ges.* **41**: 123 (taxonomy; **11**: 619).

Walker, J. C. 1921. *Phytopathology* **11**: 87 (*Puccinia asparagi* on onion; *Botrytis* sp. secondary parasite).

Yarwood, C. E. et al. 1941. *Pl. Dis. Reptr* **25**: 202 (inoculation from garlic).

Puccinia arachidis Speg., *Anal. Soc. cient. Argent.* **17**: 90, 1884.
Uredo arachidis Lagerh., 1895
Uromyces arachidis P. Henn., 1896

PYCNIA & AECIDIA unknown. UREDIA amphigenous but mostly hypophyllous, scattered, umber brown, up to *c.* 1 mm diam. UREDOSPORES ellipsoidal 22–30 × 18–22 μ; wall sienna, finely echinulate, 1.5–2 μ thick; pores mostly 2, occasionally 3 or 4, equatorial. TELIA like the uredia, but almost black. TELIOSPORES ellipsoidal sometimes with 3 or 4 cells, slightly constricted at septa, 38–42 × 14–16 μ; wall chestnut, smooth, thickened above; pedicel colourless, *c.* 55 μ long (G. F. Laundon & J. M. Waterston, *CMI Descr.* 53, 1965).

The only rust recorded on groundnut and possibly a disease of rather local importance. It occurs also on *Arachis marginata, A. nambyquarae* and *A. prostrata*. *P. arachidis* appears, for a long time, to have been present only in the tropical Americas and parts of central and E. Asia. It has caused severe losses in the West Indies and parts of southern USA. The rust has now caused further concern by its recent spread to other parts of Asia (including India), Australia and Africa (Map 160).

The first symptoms are evident 8–10 days after inoculation as whitish flecks on the lower leaf surface; about a day later yellowish flecks are seen on the upper surface and pustules begin to form on the lower. These rupture after *c.* 2 days; spores may occur on the upper surface. Although penetration takes place on both sides of the leaf the symptoms appear earlier if infection is via the lower surface. The sori are 0.3–0.6 mm diam., circular, darkening at maturity; coalescence is common and leaflets may be shed. In USA infection patterns in the field can be erratic. The rust is thought not to overwinter in the S.E. of the country; but inoculum is blown in from southerly countries (Hammons). Uredospores germinate best at *c*. 22° and most infection occurs at around this temp. (Fang; Kono). Soybean, lucerne, common bean and pea are said to be immune. Some resistance has been described. Fungicide control by chlorothalonil, mancozeb and oxycarboxin has been reported from USA; and see Siddaramaiah et al. for India. Sprays will probably also have to control *Mycosphaerella arachidis* and *M. berkeleyi* simultaneously.

Bromfield, K. R. et al. 1970. *Pl. Dis. Reptr* **54**: 381 (resistance; **49**, 3062).
Castellani, E. 1959. *Olearia* **13**: 261 (general; **39**: 524).
Cook, M. 1972. *Pl. Dis. Reptr* **56**: 382 (resistance; **51**, 4549).
Fang, H. C. 1977. *Pl. Prot. Bull. Taiwan* **19**: 218 (general; **57**, 5225).
Foudin, A. S. et al. 1974. *Phytopathology* **64**: 990 (uredospore characteristics & self inhibitor; **54**, 2560).
Hammons, R. O. 1977. *PANS* **23**: 300 (in USA & Caribbean; **57**, 1915).
Harrison, A. L. 1971. *J. Am. Peanut Res. Educ. Assoc.* **3**: 96 (fungicides; **53**, 749).
Kono, A. 1977. *Bull. Fac. Agric. Miyazaki Univ.* **24**: 217, 225 (susceptibility of cvs & uredospore germination; **57**, 4714, 4715).
Siddaramaiah, A. L. et al. 1977. *Pesticides* **11**(12): 38 (fungicides; **57**, 5788).

Puccinia cacabata Arth. & Holw. in Arth., *Proc. Am. Phil. Soc.* **64**: 179, 1925.
Aecidium gossypii Ell. & Ev., 1896
Uredo chloridis-berroi Speg., 1925
Puccinia stakmanii Presley, 1943
U. chloridis-polydactylidis Viagas, 1945.

Heteroecious, macrocyclic. PYCNIA amphigenous, numerous in raised circular groups; yellow to orange becoming brown, 90–120 μ diam. AECIDIA mainly hypophyllous, in groups surrounding pycnia, 2–5 mm diam.; peridium orange becoming yellow or colourless; peridial cells oblong, strongly echinulate 17–30 × 11–20 μ. AECIDIOSPORES globose to oblong, 13–19 × 18–25 μ; wall finely echinulate, yellow to hyaline, 2–5 μ thick. UREDIA mainly epiphyllous, rarely amphigenous, becoming confluent; early naked, cinnamon brown. UREDOSPORES globose to broadly ellipsoid, 19–24 × 23–31 μ; wall pale to cinnamon brown, 1.5–2.5 μ, moderately echinulate; pores 3–4 (usually 3) distinct, equatorial. TELIA amphigenous, crowded, round, elliptical or linear, sometimes confluent; naked early with ruptured epidermis conspicuous. TELIOSPORES variable in shape and colour, rounded at both ends or slightly pointed at the apex, 19–26 × 27–39 μ; sometimes constricted at the septum; wall chestnut brown or slightly coloured, 2–4 μ at the base, 4–9 μ at the apex; pedicel greater than spore length up to 90 μ or more, hyaline or slightly coloured. AECIDIAL stage on *Gossypium* spp. Uredial and telial stages on *Bouteloua* spp. and *Chloris* spp. (J. L. Mulder & P. Holliday, *CMI Descr*. 294, 1971).

Other rusts on cotton are *Phakopsora gossypii* (q.v.), only the uredial and telial stages being known; and *Puccinia schedonnardi* Kellerm. & Sw. (*CMI Descr*. 293, 1971). The latter differs from *P. cacabata* in that the uredia are amphigenous and the uredospores have scattered pores (6–8) and a thinner spore wall. The aecidial stage of *P. schedonnardi* occurs not only on *Gossypium* spp. but on spp. of *Abutilon, Callirhoe* and *Hibiscus*; it appears to be of less importance than *P. cacabata*.

The aecidial stage appears to have been reported from USA and Mexico only. The uredial and telial stages seem more widespread and also occur in Argentina, Bahamas, Bolivia, Brazil and Dominican Republic (Map 486). The pycnial lesions on the upper surface of cotton leaves are surrounded by a maroon zone. The aecidia are commonest on the leaves but may occur on stems and severe infection causes defoliation. Carpel infection leads to dwarfing of the locules and attacks on the peduncle contribute to breakage and boll loss. Infection of cotton seedlings may cause death.

This rust causes disease in the irrigated cotton of S.W. USA and N. Mexico where losses of 40% or more can occur under favourable climatic conditions. Teliospores, which probably require a long maturation period, germinate at high RH at an opt. of 25°. This temp. is also opt. for sporidial germination which, however, falls off more rapidly above 25° than does that of teliospores. In the field pycnia appear 4 days after infection and aecidia in 14 days.

Puccinia carthami

The natural occurrence of the disease is erratic and the degree of severity depends largely on weather and closeness of the telial hosts to cotton. Sporidia from overwintering teliospores on the wild grass hosts spread to cotton in the late spring and summer. Most vars, both American and Egyptian, appear susceptible although resistance has been reported. In USA 10 wild *Gossypium* spp. were also susceptible but 5 other spp. in the Malvaceae were resistant. Zineb is effective as a foliage spray before infection.

Blank, L. M. et al. 1963. *Phytopathology* **53**: 921 (life history; **43**, 471).

Brown, J. G. 1938. *Pl. Dis. Reptr* **22**: 380 (*Puccinia schedonnardi*; **18**: 392).

Duffield, P. C. 1958. *Diss. Abstr.* **19**: 419 (inoculation, temp. & resistance; **38**: 520).

Hensen, J. F. et al. 1956. *Mycologia* **48**: 126 (uredial & telial stage; **36**: 32).

Presley, J. T. 1942. *Phytopathology* **32**: 97 (connection of telial & aecidial stages; **21**: 253).

—— et al. 1943. *Ibid* **33**: 382 (morphology, symptoms, inoculation of cotton, resistance; **22**: 385).

Smith, T. E. 1960. *Pl. Dis. Reptr* **44**: 77 (severe outbreak in New Mexico; telial hosts; **39**: 470).

Puccinia carthami Corda, *Icon. Fung.* **4**: 15, 1840.

Dicaeoma carthami (Corda) Kuntze, 1898
Bullaria carthami (Corda) Arth. & Mains, 1922
Puccinia kentrophylli Syd., 1902.

PYCNIA amphigenous, flask shaped or spherical 80–100 μ diam. and in small groups. AECIDIA UREDINOID ('aecidial uredia') amphigenous, chestnut brown up to *c.* 0.4 mm, scattered. UREDOSPORES globose, $21–27 \times 21–24$ μ; wall light chestnut brown, echinulate, 1.5–2 μ thick; pores 2–3 rarely 4, equatorial. TELIA like UREDIA. TELIOSPORES ellipsoid, $36–44 \times 24–30$ μ slightly or not constricted at septa, rounded or somewhat obtuse at both ends, wall chestnut brown, minutely verrucose, 2.5–3.5 μ thick at the side, pores of the upper and lower cell almost sunken, pedicels colourless, short, fragile, up to 30 μ long.

Safflower (*Carthamus tinctorius*) rust also occurs on *C. glaucus, C. lanatus, C. oxycantha, C. syriacus* and *C. tenuis*. Three other rusts are recorded on *Carthamus*. *Puccinia carduncelli* Syd. (from Sicily) can be distinguished from *P. carthami* in having uredospores with 2 super-equatorial pores and teliospores of variable size and shape. *P. verruca* Thuem. is a microcyclic sp. with compact telia in large groups and teliospores with thickened apex. *Melampsora ricini* Noronha (as *Melampsorella ricini*) (*CMI Descr.* 171) has been reported on safflower from Ethiopia, and can be distinguished by the paraphysate uredia and hyaline to pale yellow uredospores (E. Punithalingam, *CMI Descr.* 174, 1968).

One of the most important diseases of safflower and one which has become serious in the younger industry of western USA. *P. carthami* occurs in S. Europe, N.E. Africa, parts of Asia, W. and central USA, Canada and Mexico (Map 424). Symptoms first occur at the seedling stage when lesions will be found on the hypocotyl, the tap and lateral roots; these may coalesce to form one large lesion. The hypocotyl may become abnormally elongated, hypertrophied, discoloured, rough and split longitudinally. Affected plants are pulled up easily and stems may be girdled. Infection spreads from the surviving seedlings to the leaves of maturing plants causing the second or foliar phase of the disease. At the 6–8 leaf stage wilt can occur.

The disease cycle usually begins through direct cuticular penetration by the sporidia from germinating teliospores carried over on the seed, soil or crop debris. The pycnia precede dikaryotisation and the uredospores set up secondary infection cycles in the growing crop. Whether the uredinoid aecidia also cause infection is not clear. Seedling infection is greater at 5–15° than at 20–24°. Soil temps of 15–25° favour infection and little takes place at 30–35°; soil moisture levels have little effect. In the field in Nebraska teliospores survive for 12 months but not for 21. On safflower straw they survive for at least 45 months at 5°. Teliospores survive higher temps than uredospores. The latter have an opt. temp. for germination of 18–20°, remain viable for >1 year at 8–10° and for 3 weeks on infected plants at 30–31°. Above 40° teliospores tend to be formed directly; they germinate best at 18–24° (Calvert et al.; Klisiewicz 1972, 1973; Prasada et al.; Schuster; Zimmer 1963). Four races have been differentiated and more may occur. In some virulence is inherited as a recessive in 1 or 2 gene pairs. *C. arborescens, C. oxycantha* and *C. palaestinus* are susceptible to races 1, 2 and 3.

Chemical treatment of the seed reduces incidence effectively and its efficiency is increased if the storage period is extended. In USA considerable selection for resistance has been done (see Zimmer and others). Lines which show resistance as seedlings generally do so also at the later foliar infection stage. Hypocotyl elongation is greater in susceptible lines and the degree of elongation is correlated with that of sporulation. Progeny tests using a highly resistant strain of *C. oxycantha* show resistance to be dominant in a single gene pair. Resistance appears to be mostly of a vertical nature. But one selection from Nebraska showed resistance at the seedling stage (and, exceptionally, not at the foliar stage), this being due to a restriction of the pathogen to the cortex, probably by a well defined peri-vascular ring. This type of resistance is also governed by a single dominant factor. *Carthamus* spp. which differ from cultivated safflower in chromosome number do not readily hybridise but they are resistant to *P. carthami*.

Bernaux, P. 1953. *Bull. Soc. mycol. Fr.* **68**: 327 (general; **33**:50).
Binder, R. G. et al. 1977. *Phytopathology* **67**: 472 (stimulation of teliospore germination by host polyacetylenes; **56**, 5749).
Calvert, O. H. et al. 1954. *Ibid* **44**: **609** (temp. & seedling infection; **34**: 321).
Daly, J. M. et al. 1957. *Ibid* **47**: 163 (growth, respiration & infection; **36**: 616).
——. ——. 1958. *Ibid* **48**: 91 (infection & changes in auxin levels; **37**: 499).
Darpoux, H. 1946. *Annls Epiphyt.* **12** *Sér. Pathol. Vég. Mém.* 4: 91 (morphology; **26**: 214).
Deshpande, B. G. et al. 1963. *Indian Phytopathol.* **16**: 347 (cytology & heterothallism; **43**, 3285).
Klisiewicz, J. M. 1972. *Phytopathology* **62**: 436 (teliospore germination; **51**, 4258).
——. 1973. *Ibid* **63**: 795 (as above; **53**, 1491).
—— et al. 1977. *Ibid* **67**: 787 (effects of flooding & temp.; **57**, 731).
McCain, A. H. 1963. *Ibid* **53**: 184 (inheritance of pathogenicity; **42**: 566).
Prasada, R. et al. 1950. *Ibid* **40**: 363 (effect of temp.; **29**: 534).
Sackston, W. E. 1953. *Pl. Dis. Reptr* **37**: 522 (infection in Canada; **33**: 260).
Schuster, M. L. et al. 1952. *Phytopathology* **42**: 211 (infection; **31**: 631).
——. 1956. *Ibid* **46**: 591 (infection, survival, temp. & soil moisture; **36**: 275).
Siddiqui, M. R. et al. 1959. *Indian Phytopathol.* **12**: 59 (heterothallism; **39**: 332).
Thomas, C. A. 1952. *Phytopathology* **42**: 108 (seed transmission & treatment; **31**: 459).
——. 1958. *Pl. Dis. Reptr* **42**: 1089 (races; **38**: 94).
——. 1971. *Ibid* **55**: 1046 (seed treatment with nonmercurials; **51**, 2747).
Zimmer, D. E. 1962. *Phytopathology* **52**: 1177 (resistance & hypocotyl infection; **42**: 335).
——. 1963. *Pl. Dis. Reptr* **47**: 486 (races).
——. 1963. *Phytopathology* **53**: 316 (life cycle; **42**: 625).
——. 1965. *Ibid* **55**: 296 (histology; **44**, 2219).
——. 1965. *Pl. Dis. Reptr* **49**: 623 (seed treatment; **45**, 171).
——. 1967. *Ibid* **51**: 586, 589 (resistance tests & seed treatment; **46**, 3529).
——. 1967. *Phytopathology* **57**: 772 (seedling & foliar resistance; **46**, 3528).
—— et al. 1968. *Ibid* **58**: 1340, 1451 (inheritance of resistance; **48**, 887, 888).
——. ——. 1968. *Pl. Dis. Reptr* **52**: 876 (effect on yield; **48**, 557).
——. ——. 1969. *Crop Sci.* **9**: 491 (inheritance of resistance; **51**, 563).
——. 1970. *Phytopathology* **60**: 1157 (fine structure & resistance; **50**, 1329).
——. ——. 1970. *Pl. Dis. Reptr* **54**: 364 (effect on yield; **49**, 2957).

Puccinia cynodontis Lacroix in Desm., *Pl. Crypt.* II. 655, 1859.
Aecidium plantaginus Ces., 1859
Dicaeoma cynodontis Ktze., 1898.

Heteroecious, macrocyclic. PYCNIA epiphyllous. AECIDIA hypophyllous. AECIDIOSPORES oblong, 19–24 × 23–29 μ; wall 1.5–2 μ thick. UREDIA mainly hypophyllous, cinnamon brown, 0.5–1 × 0.2–0.5 mm. UREDOSPORES globose, brown, 20–26 × 19–23 μ, wall finely verrucose, 1.5–3 μ thick; pores 2–3, equatorial. TELIA mainly hypophyllous, brown to black, 0.5–1.5 × 0.2–0.5 mm. TELIOSPORES ellipsoid, oblong 15–25 × 28–60 μ, obtuse or attentuated at each end, constricted slightly at the septa; wall chestnut brown, paler at one end, 1.5–2.5 μ thick at the sides, 6–12 μ above; pedicels colourless or subhyaline, 1.5 times the length of the spores. Aecidial stages on *Plantago* spp. Uredial and telial stages on *Cynodon* spp., particularly *C. dactylon* (J. L. Mulder & P. Holliday, *CMI Descr.* 292, 1971).

Leaf rust of Bermuda grass is widespread in tropical and temperate regions. The host seldom appears to be killed but severe damage has been reported from Nigaragua. In USA commercially grown strs of *C. dactylon* are not attacked, as they are elsewhere, but a var. of *C. magennisii* grown there is very susceptible. Uredospore germination is max. at 20° but there is little difference between 20° and 30°. Better and more rapid germination occurs in the dark, within 1 hour at 20–30°. The higher the light intensity the lower the germination. High infection requires at least 10 hours of leaf wetness, and a high light intensity restricted infection which appeared to be affected more by light than temp. Since the aecidial stage has not apparently been found in USA the uredospores presumably survive there during the dormancy of the telial host. Collections from Oklahoma yielded 4 races which were differentiated on 3 clones of *C. dactylon*. No specific control measures have been reported. Sources of resistance seem more likely in *C. dactylon* than in *C. magennisii* or *C. transvaalensis*.

Vargas, J. M. et al. 1967. *Phytopathology* **57**: 405 (general; **46**, 2450).

Puccinia erianthi Padwick & Khan, *Mycol. Pap.* 10: 10, 1944.

UREDIA mostly hypophyllous, irregularly scattered on purplish spots, dark brown, linear up to 2 mm long. Paraphyses 35–70 μ long, clavate to capitate, head 12–20 μ diam.; wall pale luteous to bay, 3–8 μ thick above. UREDOSPORES ellipsoidal or obovoidal, $28–36 \times 20–26$ μ; wall bay, finely echinulate, 1.5–2 μ thick; pores 4–5 equatorial. TELIA like the uredia but black and compact. TELIOSPORES clavate, obtuse above, slightly constricted at the septum, $36–55 \times 18–24$ μ; wall bay, paler below, smooth, 1–2 μ thick at the sides, 4–6 μ thick above. Pedicels bay (darker than the base of the spore) about 10 μ long.

This pathogen was confused with *P. kuehnii* (q.v.) but may be distinguished from this sp. by the smaller uredospores, the absence of uredospores with a thickened apical cell, the darker spore wall, and the conspicuous and abundant capitate paraphyses. Comparisons between the 2 spp. have been tabulated by Kandasami et al. (G. F. Laundon & J. M. Waterston, *CMI Descr.* 9, 1964; see Sathe, *P. melanocephala*, priority).

This rust has become of more importance recently in India on sugarcane than the earlier described *P. kuehnii* on the same crop. Cummins discussed the rusts on *Saccharum* spp. and related hosts. He referred to *P. miscanthi* Kiura which is similar to *P. erianthi* but may be distinguished by the slightly shorter uredospores, and teliospores which are considerably shorter with a slightly thinner apical wall. It does not seem to be clear what, if any, differences there are between *P. erianthi* and *P. kuehnii* with respect to their host ranges on sugarcane and wild *Saccharum* and *Erianthus* spp. in India.

Small yellowish spots are first seen on both leaf surfaces. These darken and enlarge, becoming usually not >1 cm in length and much narrower in width. The uredospores are followed by necrosis of the lesions which may coalesce and cause the premature death of young leaves. It is the leaves that are mostly affected. The distribution is China, India, Malagasy Republic, Nepal, Reunion, South Africa and E. Africa (Map 462). Severe outbreaks of this rust have occurred in recent years in India, so that if it became more widely distributed it may be of greater importance than *P. kuehnii*. The opt. germination for uredospores is *c*. 20° and the incubation period 10–11 days. These spores are viable for 5 weeks at 18–28°. Uredospores from *S. spontaneum* and *E. fulvus* were pathogenic to sugarcane but teliospores did not infect their respective hosts. Races (6) have been reported and resistance is known. (Outbreaks of this rust have occurred very recently in Australia (Queensland) and the West Indies.)

Chona, B. L. et al. 1950. *Curr. Sci.* **19**: 151 (morphology; **30**: 1).

Cummins, G. B. 1950. In Uredineana IV (editor A. L. Guyot) *Encycl. Mycol.* **24**: 5–90 (*Puccinia* on Andropogoneae; **33**: 383).

Egan, B. T. 1964. In *Sugarcane diseases of the world* Vol. 2: 61 Elsevier (general, with *P. kuehnii*, 22 ref.).

Kandasami, P. A. et al. 1959. *Curr. Sci.* **28**: 209 (comparison, morphology of *P. erianthi* & *P. kuehnii*; **39**: 39).

Muthaiyan, M. C. et al. 1966. *Indian Phytopathol.* **19**: 317 (race 6; **46**, 3557).

Payak, M. M. 1956. *Sci. Cult.* **21**: 688 (identity; **35**: 791).

Sahni, M. L. et al. 1965. *Indian Phytopathol.* **18**: 191 (general; **45**, 1176).

Sathe, A. V. 1971. *Curr. Sci.* **40**: 42 (nomenclatural revision; **51**, 622).

Sharma, S. L. 1952. *Ibid* **21**: 288 (resistance; **32**: 400).

Sreeramulu, T. et al. 1971. *J. Indian bot. Soc.* **50**: 39 (spore periodicity of *P. erianthi* & *P. kuehnii*; **51**, 2822).

Vaheeduddin, S. et al. 1955. *Sci. Cult.* **21**: 329 (Hyderabad outbreak; **35**: 396).

Puccinia helianthi Schw., *Schr. nuturf. Ges. Leipzig* **1**: 73, 1822.
Aecidium helianthi-mollis Schw., 1822
Puccinia heliopsidis Schw., 1822
Caeoma helianthi Schw., 1832
P. helianthorum Schw., 1832
P. xanthifoliae Ell. & Ev., 1891.

PYCNIA amphigenous or epiphyllous, *c.* 3–6 in groups up to 1 mm diam. AECIDIA hypophyllous, surrounding the pycnia in ±crowded groups *c.* 2–5 mm diam., cupulate, 0.2–0.3 mm diam. AECIDIOSPORES ellipsoidal, 16–26 μ diam.; wall hyaline, irregularly verrucose, 1 μ thick. UREDIA mostly hypophyllous, irregularly scattered cinnamon, 0.5–1 mm diam. UREDOSPORES ellipsoidal obovoidal, often ±cylindrical, 25–32 × 19–25 μ; wall sienna, very finely echinulate, 1–2 μ thick, often slightly thickened at apex and base; pores 2± equatorial and opposite. TELIA like the uredia but dark umber. TELIOSPORES cylindrical to clavate, slightly constricted at septum, rounded above, rounded or attenuated below, 40–60 × 18–30 μ; wall sienna to chestnut, smooth, 1–2.5 μ at the sides, 8–12 μ thick above; pores apical; pedicels pale luteous, fragile, *c.* 60–150 μ long.

Several other rusts have been described on *Helianthus. Coleosporium helianthi* (USA) has hyaline uredospores and sessile teliospores. *P. minuscula* (S. America) has uredospores with *c.* 6 scattered pores and teliospores with unthickened apical wall. *P. massalis* (USA) has similar uredospores to those of *P. helianthi* but has teliospores with a thick side wall (3–5 μ). All these spp. produce similar aecidia on *Helianthus*, though *P. minuscula* has more coarsely ornamented aecidiospores. *Uromyces junci* is a widespread heteroecious sp. producing pycnia and aecidia on *Helianthus* and which has smaller aecidiospores. On *Helianthus, Heliopsis, Iva* and *Xanthium* spp. (G. F. Laundon & J. M. Waterston, *CMI Descr.* 55, 1965).

Sunflower rust was also described by Parmelee (q.v. for hosts) whose synonymy includes *P. viguierae* Pk. The rust also attacks Jerusalem artichoke (*Helianthus tuberosus*). The fungus is worldwide (Map 195), occurring wherever sunflower is grown, and often causes severe losses in temperate and subtropical regions; it appears to be of less importance in the lowland tropics. Fick et al. and Siddiqui et al. described losses. Bailey gave an account of the disease. In early work in USA and Yugoslavia the opt. temp. for uredospore germination was given as *c.* 18° (for aecidiospores *c.* 16°). Later Sood et al. (1972) found that uredospores germinated equally well at 10–25° but less germination occurred at 30°. Appressorium formation was highest at 10–25° and it fell off at 30° (for races 1 and 3), but none occurred at 30° for races 2 and 4. For races 1 and 3 the opt. temp. for penetration was 20°, for race 2, 15–20° and for race 4, 20–25°. Germination was reduced by increasing the light from 220 to 17 600 lux. On detached sunflower leaves uredia developed in 9–10 days after uredospore inoculation at 20–22° and in 14–20 days at 10°. Telia developed in 14–20 days at 20–22° and in 22 days at 10° (Hennessy et al. 1970). In a susceptible cv. Sood et al. (1970) found that uredospore formation began 6 days after inoculation. Sackston (1960) found that this rust was particularly tolerant of low temps compared with other rusts. Bailey reported a viability of *c.* 6 months for teliospores (about the same time for uredospores) at 8–23° and RH 20–40%, thus indicating that the latter can overwinter. Carry over is mostly on crop debris and seed is probably unimportant in spread of the disease (but see Perišić).

Several races have been differentiated and Zimmer et al. (1976), in an examination of *Helianthus* spp. in N. central USA, found plants resistant to all known races (and see Hennessy et al. 1972). Resistance is controlled by dominant factors (Jabbhar Miah et al.; Putt et al. 1955, 1963; Sackston 1962). Fungicide control has been described from Australia (maneb), Hungary (zineb, zineb+sulphur), India (fentin hydroxide, mancozeb), Kenya (oxycarboxin, zineb) and Yugoslavia (Bordeaux). Control recommendations have included a 3-year rotation; ploughing under or removal and destruction of crop debris.

Antonelli, E. F. 1968. *Boln genét. Inst. Fitotéc Castelar 1968*(5): 17 (genotypic determination of pathogenicity; **49**, 224).
Bailey, D. L. 1923. *Tech. Bull. Minn. agric. Exp. Stn* 16, 31 pp. (general; **3**: 400).
Brown, A. M. 1936. *Can. J. Res.* Sect. C **14**: 361 (infertility of 4 strs; **16**: 184).
Coffey, M. D. et al. 1972. *Can. J. Bot.* **50**: 231, 1485 (fine structure with *Melamspora lini*; **51**, 3203; **52**, 665).
——. 1973. *Trans. Br. mycol. Soc.* **60**: 245 (nutrition in culture; **52**, 3557).

Craigie, J. H. 1959. *Can. J. Bot.* **37**: 843 (nuclear behaviour in diploidisation of haploid infections).
Fick, G. N. et al. 1975. *Pl. Dis. Reptr* **59**: 737 (effects on yield; **55**, 1904).
Hennessy, C. M. R. et al. 1970. *Can. J. Bot.* **48**: 1811 (life cycle on detached leaves; **50**, 1334).
——— ———. 1972. *Ibid* **50**: 1871 (specialisation on wild sunflowers in USA, Texas; **53**, 4543).
Jabbhar Miah, M. A. et al. 1970. *Phytoprotection* **51**: 1, 17 (races, genetics of resistance & pathogenicity; **49**, 3430).
Kurnik, E. et al. 1962. *Pflanzenschutzberichte* **28**: 47 (fungicides in Hungary; **42**: 27).
Middleton, K. J. et al. 1972. *APPS Newsl.* **1**: 18 (fungicide in Australia; **54**, 1835).
Nozzolillo, C. et al. 1960. *Can. J. Bot.* **38**: 227 (growth on host tissue cultures; **39**: 714).
Parmelee, J. A. 1977. *Fungi Canadenses* 95, 2 pp. (synonymy, morphology, hosts & Canadian distribution).
Pawar, K. B. et al. 1976. *Pesticides* **10**(10): 37 (fungicides in India; **56**, 2173).
Perišić, M. M. 1957. Spec. edn Inst. Pl. Prot., Beograd 8, 52 pp. (general, in Yugoslavia; **37**: 497).
Putt, E. D. et al. 1955. *Nature Lond.* **176**: 77 (resistance; **34**: 652).
——— ———. 1955. *Can. J. agric. Sci.* **35**: 557 (inheritance of resistance; **35**: 301).
——— ———. 1957. *Can. J. Pl. Sci.* **37**: 43 (resistance sources; **37**: 104).
——— ———. 1963. *Ibid* **43**: 490 (genes for resistance; **43**, 1107).
Ramasamy, R. et al. 1973. *Madras agric. J.* **60**: 594 (fungicides; **54**, 5031).
Sackston, W. E. 1960. *Can. J. Bot.* **38**: 883 (uredospore longevity; **40**: 311).
———. 1962. *Ibid* **40**: 1449 (occurrence, distribution & significance of races; **42**: 397).
Siddiqui, M. Q. et al. 1977. *Aust. J. agric. Res.* **28**: 389 (effects on growth & yield; **56**, 5139).
Singh, J. P. 1975. *Pl. Dis. Reptr* **59**: 200 (fungicides in Kenya; **54**, 5032).
Sood, P. N. et al. 1970. *Can. J. Bot.* **48**: 2179 (penetration & infection; **50**, 3093a).
——— ———. 1971. *Ibid* **49**: 21 (effect of light & temp. during spore formation on germination of uredospores; **50**, 3093b).
——— ———. 1972. *Ibid* **50**: 1879 (effect of light & temp. on germination & infection; 53, 4544).
Zimmer, D. E. et al. 1973. *Pl. Dis. Reptr* **57**: 524 (evaluation for resistance, with *Verticillium albo-atrum*; **53**, 646).
——— ———. 1976. *Phytopathology* **66**: 208 (resistance in wild *Helianthus* spp. of N. central USA; **55**, 4808).

Puccinia kuehnii Butler, *Ann. Mycol. Berl.* **12**: 82, 1914.
Uromyces kuehnii Krueger, 1890
Uredo kuehnii Wakk. & Went, 1898.

UREDIA amphigenous, irregularly scattered, mostly on brownish or purplish spots, yellow brown, linear, up to 1.5 mm long. PARAPHYSES inconspicuous cylindrical to capitate, up to 45 μ long, wall 1–2 μ thick. UREDOSPORES ellipsoidal or obovoidal, 32–45 × 25–28 μ; wall luteous, echinulate, 1–2 μ thick, 1.5–5 (up to 8) μ above, pores 4–6, equatorial. TELIA (after Butler l.c.) hypophyllous, minute, linear, blackish. TELIOSPORES oblong–clavate, apex rounded or truncate, not or slightly constricted, base attenuated, 10–18 × 25–40 μ; wall pale yellowish (? immature), smooth, not thickened; pedicels hyaline, short. Can be distinguished from *P. erianthi* (q.v.) by the uredospores with apical thickening, and which are larger and paler. (G. F. Laundon & J. M. Waterston, *CMI Descr.* 10, 1964.)

This rust occurs on wild *Saccharum* spp. and sugarcane, and was described originally from Java in 1890. Symptoms in the field are similar to those of *P. erianthi*. Although in the past the rust has eliminated sugarcane cvs in Fiji and Queensland, Australia, it is now of little importance. Its distribution is fairly wide in S. and E. Asia, Australasia and Oceania, and parts of E. Africa. It also occurs in Cuba, Dominican Republic and Mexico (Map 215). For ref. see *P. erianthi.*

Puccinia oahuensis Ell. & Ev., *Bull. Torrey bot. Club* **22**: 435, 1895.
Uredo digitariaecola Thüm.
U. digitariae-ciliaris Mayor
U. duplicata Rangel
U. syntherismae Speg.
P. digitariae Pole Evans
P. digitariae-velutinae Viennot-Bourgin.
(dates for synonyms not given in *CMI Descr.* 516).

UREDIA amphigenous, mostly hypophyllous, irregularly scattered, light yellowish brown, circular or elongated with the veins, up to 2 mm long, sparsely paraphysate. PARAPHYSES mostly clavate, usually incurved, thin walled, hyaline, 35–65 × 10–16 μ. UREDOSPORES oval or obovoid, sometimes subglobose, 24–35 (–40) × 20–26 (–30) μ; wall golden or cinnamon brown, echinulate *c.* 1.5 μ thick; pores 4 or

5 (occasionally 3 or 6), equatorial or in some specimens tending to be scattered. TELIA similarly arranged but dark brown and long covered; hyaline to brownish peripheral paraphyses occasionally occur. TELIOSPORES clavate, obovoid–clavate or oblong, 27–46 (–52) × 16–22 (–26) μ; wall chestnut brown, smooth, 1–1.5 (–2) μ thick at sides, 2–5 (–7) μ above; pedicels hyaline to golden brown, thin walled, up to 20 μ long, persistent though sometimes collapsing.

Of other rusts common on *Digitaria*, *Puccinia digitariae-vestitae* Ramachar & Cummins and *P. substriata* Ell. & Barth. have slightly larger teliospores; in the former the uredospores have 5–8 scattered pores, in the latter they differ little from *P. oahuensis* but the sori lack paraphyses. *P. esclavensis* Diet. & Holw. and its vars have comparatively short, broad teliospores and verrucose uredospores. *P. levis* (Sacc. & Bizz.) Magn. teliospores are usually strongly diorchidioid and its uredospores have only 2 germ pores. The uredospores of *Uromyces pegleriae* Pole Evans have 7–9 scattered pores (J. E. M. Mordue & C. Critchett, *CMI Descr.* 516, 1976; see also P. Ramachar, *Mycopathol. Mycol. appl.* **25**: 15, 1965).

Digitaria rust occurs on several *Digitaria* spp. including *D. decumbens* (pangola grass). The fungus is widespread, particularly in warmer regions. The affected plants show minute, irregular, yellowish brown to orange brown spots; they increase in size and a chlorotic halo forms; usually 0.5 mm diam. (up to 1 mm). The pustules coalesce to form large, irregular, necrotic areas; affected leaves die prematurely. This rust has recently been reported as being serious on pangola grass in the West Indies. The uredospores have an opt. temp. for germination of *c.* 18° and the incubation time is 6–8 days. These spores survived 20–50 days at 25° and RH 55–100%. Alternate wetting and drying induced teliospore germination. The frequency of cutting and the use of nitrogenous fertilisers affect disease levels. Resistance has been found and this may be more effective in control than fungicide treatment.

Jailloux, F. 1970. *Annls Phytopathol.* **2**: 55 (general; **50**: 722).

—— et al. 1970. *Agron. trop.* **25**: 765 (fungi on *Digitaria*; **50**, 1288).

Liu, L. J. 1969. *J. Agric. Univ. P. Rico* **53**: 132 (general; **48**, 3023).

Puccinia polysora Underw., *Bull. Torrey bot. Club* **24**: 86, 1897.

Dicaeoma polysorum (Underw.) Arth., 1906.

PYCNIA and AECIDIA unknown. UREDIA on both sides of leaves and sheaths, densely scattered, more or less circular, mostly 0.2–1 mm diam. UREDOSPORES somewhat ellipsoidal, often subangular with more or less straight and parallel sides, 28–38 × 22–30 μ; wall pale luteous to buff, minutely and sparsely echinulate, 1–2 μ thick; pores 4–5, equatorial. TELIA dark, remaining covered by epidermis, circular, 0.2–0.5 mm diam. TELIOSPORES more or less obovate or clavate, usually irregular and subangular, obtuse or truncate above, slightly constricted at septum, 35–50 × 16–26 μ. MESOSPORES numerous; wall sienna, luteous or pale luteous, smooth, 1–1.5 μ thick at sides, 1.5–2.5 μ thick above; pedicels mostly basal, pale luteous, up to 30 μ long.

Cammack (1958b) found 2 main size groups of uredospores in an examination of 65 worldwide collections; the smaller spore form occurred in S.E. Asia and neighbouring islands (but not Borneo where the spores were larger than either group) and the larger in the W. Indies, Africa and the S. Indian ocean; USA material was intermediate. Two other rusts are recorded on maize. *Puccinia sorghi* (q.v.) differs in the slightly smaller, darker coloured and more regular (always well rounded, never angular) uredospores, and in the teliospores being well rounded, with thicker apical walls and long pedicels. Macroscopically *P. sorghi* usually has much larger sori with a more sparse distribution. *Physopella zeae* (q.v.) has smaller, hyaline uredospores and covered telia containing sessile teliospores. On *Erianthus divaricatus, E. alopecuroides, Euchlaena mexicana, Tripsacum dactyloides, T. laneolatum, T. latifolium, T. laxum, T. pilosum* and *Zea mays*. (G. F. Laundon & J. M. Waterston, *CMI Descr.* 4, 1964; see Pavgi for morphology and taxonomy of the *Puccinia* spp. on maize and sorghum; Cummins).

P. polysora (sometimes called southern maize rust) is one of the rusts which are presumably indigenous on this crop in America where it had been confused with *P. sorghi*, and does not seem to have been considered serious. It was confined to this region until 1948–9 when outbreaks on maize were reported from W. Africa and the Philippines; these were apparently independent and originated from spread (eastwards and westwards) from the Caribbean. The rapid spread W. to E. across Africa in 3–4 years is an excellent example of the severe losses (up to 70%) which can result when a virulent pathogen first appears in a highly susceptible crop. A situation paralleled 10 years later through the introduction of *Peronospora tabacina* (q.v.) to Europe. There are now few tropical areas from which *Puccinia polysora* has not been reported and it is present over a large area of USA (Map 237). There is a review by Huguenin to 1959, a bibliography on maize rusts by Wood et al. (1956) and a general account by Cammack (1961).

The rust is a tropical one and is unlikely to be serious in maize in the temperate zone; in the tropics only lowland, wet areas are seriously affected. Spread is through the uredospores which have a diurnal periodicity tending to a forenoon peak. Teliospores are rare in W. Africa and they have not been observed to germinate. In Nigeria a decrease in incidence was noted from the wetter to the drier areas. The opt. germination for uredospores is 27–28° and penetration is stomatal; at this temp. pustule formation occurs after 9 days (at 24° after 14 days and at 32° lesions are abortive). In severe infections pustule density may be 50/cm². At 25° and RH 65% the uredospores are viable for 2 months, and they survive between crops throughout the dry months especially on irrigated plots. There is no evidence for seedborne infection.

Considerable breeding in Africa has led to the production of resistant, local forms of maize using resistant material from the American continent. Resistance to the races of *P. polysora* is controlled by dominant or incompletely dominant genes. Three E. African races have been reported (Storey et al. 1958, 1967) although EA2, which developed in a glasshouse (Ryland et al. 1955), was later said to have been discarded (Schieber 1971). In USA 7 further races have been described (*Physopella zeae* q.v.) and Ullstrup used 8 differentials to distinguish the 10 races. In Nigeria 2 more races were characterised using 12 differentials. It has been considered that there are adequate resistance sources in American material against any new races which might arise.

Anon. 1958. W. African maize research unit, review 1955–57. *Ibadan, Federal Dept agric. Res.* 54 pp. (general; **38**: 514).

Cammack, R. H. 1955. *Nature Lond.* **176**: 1270 (diurnal periodicity of uredospores; **35**: 383).

——. 1958a. *Ann. appl. Biol.* **46**: 186 (infection gradients; **37**: 657).

——. 1958b. *Trans. Br. mycol. Soc.* **41**: 89 (world distribution & uredospore size; **37**: 656).

——. 1959. *Ibid* **42**: 27 (introduction to Africa; **38**: 591).

——. 1959. *Ibid* **42**: 55 (life cycle in Africa; **38**: 591).

——. 1961. In *6th Commonw. mycol. Conf. Kew* 1960: 134 (general; **40**: 724).

Cummins, G. B. 1941. *Phytopathology* **31**: 856 (identity, *Puccinia polysora, P. sorghi* & *Physopella zeae*; **21**: 73).

Futrell, M. C. et al. 1975. *Crop Sci.* **15**: 597 (resistance controlled by a dominant gene; **55**, 2682).

Hemingway, J. S. 1955. *E. Afr. agric. J.* **20**: 191 (effect on maize yield in Tanzania; **34**: 716).

Huguenin, B. 1959. *Rev. Mycol.* **24**: 289 (review, 73 ref.; **39**: 698).

Lallmahomed, G. M. et al. 1968. *Pl. Dis. Reptr* **52**: 136 (differentiation of 2 Nigerian races; **48**, 137).

Le Conte, J. 1959. *Riz et Rizic.* **5**: 98 (selection for resistance; **38**: 689).

Nattrass, R. M. 1953. *Nature Lond.* **171**: 527 (altitude & incidence in Kenya; **32**: 309).

——. 1954. *E. Afr. agric. J.* **19**: 260 (symptom patterns; **33**: 719).

Orian, G. 1954. *Nature Lond.* **173**: 505 (distribution & occurrence in Indian ocean; **33**: 346).

Pavgi, M. S. 1972. *Mycopathol. Mycol. appl.* **47**: 207 (morphology & taxonomy of *Puccinia* spp. on maize & sorghum; **52**, 2223).

Reyes, G. M. 1953. *Philipp. J. Agric.* **18**: 115 (epidemic in Philippines).

——. 1957. *Pl. Prot. Bull. F.A.O.* **6**: 39 (spread to Philippines; **37**: 350).

Rhind, D. et al. 1952. *Nature Lond.* **169**: 631 (epidemics & losses in W. Africa; **31**: 325).

Ryland, A. K. et al. 1955. *Ibid* **176**: 655 (E. African race 2; **35**: 12).

Saccas, A. M. 1955. *Agron. trop.* **10**: 499 (general in W. Africa; **35**: 11).

Schieber, E. et al. 1964. *Pl. Dis. Reptr* **48**: 425 (distribution, *P. polysora* & *P. sorghi* in Mexico; **43**, 2893).

——. 1971. *Pl. Prot. Bull. F.A.O.* **19**: 25 (distribution, as above, in Africa & pathogenicity on American maize; **51**, 1412).

Stanton, W. R. et al. 1953. *Nature Lond.* **172**: 505 (testing for resistance in Nigeria; **32**: 676).

—— ——. 1956. *Robigo* **2**: 19 (testing for resistance & chronological list of world spread; **36**: 523).

Storey, H. H. et al. 1954. *Nature Lond.* **173**: 778 (genes for resistance; **33**: 420).

—— ——. 1957. *Heredity Lond.* **11**: 289 (breeding for resistance, E. African races 1 & 2; **37**: 533).

—— ——. 1958. *Emp. J. exp. Agric.* **26**: 1 (as above; **37**: 406).

——. 1959. *Heredity Lond.* **13**: 61 (linkage of resistance genes; **38**: 405).
——. 1967. *Ann. appl. Biol.* **60**: 297 (resistance & genetics in 3 E. African races; **47**: 505).
Turner, G. J. 1974. *Trans. Br. mycol. Soc.* **62**: 205 (possible spread by bees; **53**, 2987).
Ullstrup, A. J. 1965. *Phytopathology* **55**: 425 (American race 7 & differentiation of 10 races; **44**, 2275).
Von Meyer, W. C. 1964. *Proc. Indiana Acad. Sci.* **73**: 89 (infection histology in resistant & susceptible maize; **44**, 1538).
Wood, J. I. et al. 1956. *Spec. Publ. USDA* 9, 37 pp. (bibliography & spread in maize rusts; **37**: 160).

Puccinia psidii Wint., *Hedwigia* **23**: 171, 1884.

UREDIA amphigenous, in groups on brownish or blackish spots up to 5 mm diam., pale yellowish, 0.1–0.5 mm diam. UREDOSPORES ellipsoidal to obovoidal, 21–26 × 16–19 μ; wall hyaline, finely echinulate, 1.5–2.5 μ; no pores seen. TELIOSPORES appearing in the uredia, ellipsoidal to cylindrical, rounded above, slightly constricted at the septum, 30–48 × 19–22 μ; wall buff, smooth, 1–1.5 μ thick at the side, 2–4 μ thick above; pedicels fragile, often deciduous (G. F. Laundon & J. M. Waterston, *CMI Descr.* 56, 1965).

Rust of guava (*Psidium guajava*) and pimento (*Pimenta dioica*) has also been recorded on other members of the Myrtaceae: *Eucalyptus citriodora, Eugenia jambos* (rose apple), *E. malaccensis* (pomerac or Malay apple) and *Pimenta racemosa* (bay tree). The pathogen is confined to parts of central and S. America, USA (Florida) and some of the West Indian islands (Map 181), and where both pimento and guava are indigenous. Although it was first described from the latter crop little work was done on the disease until it attracted attention by causing an epidemic and extremely serious losses in the Jamaican pimento industry *c.* 1934. It infects the young fruit, inflorescence, immature leaves and shoots; severe attacks cause defoliation and a destructive dieback.

In Jamaica the sudden appearance of the rust on pimento suggested that a mutant had arisen in the form that was common on pomerac and rose apple. Uredospores from the latter hosts did not infect pimento whilst those from pimento only caused small non-sporulating lesions on rose apple. No natural infection of guava was observed and guava seedlings were immune to infection with inoculum from pimento and pomerac. There, therefore, appear to be 3 forms of this rust, of which the one on pimento may be the most virulent. The uredospores play the main role in dissemination and the teliospores (whose biology has not been investigated) are less frequent. Penetration of pimento by the uredospore is through the cuticle of both leaf surfaces. At 17° germination occurs within 6 hours and a short, unbranched germ tube forms an appressorium within 12 hours. The detailed description of this process by Hunt (1968) constitutes a first record of cuticular penetration by this spore type, and it may be more common than was thought, at least for dicotyledons. High uredospore germination (opt. *c.* 16°) in vivo occurs only on young tissue (leaves) and was <0.5% on leaves 30–35 days old, those *c.* 40 days old being highly resistant. Rain releases large numbers of uredospores which begin to be formed *c.* 13 days from infection.

The disease on pimento in Jamaica has only proved serious at *c.* 300 m and over. This appears to be a temp. effect; at low altitudes serious losses can be avoided and above 21° infection is inhibited. Some chemical control by spraying has been reported but commercial application is difficult. The sweet guava cvs are more susceptible.

Andrade, A. C. 1951. *Arq. Inst. biol. S. Paulo* **20**: 127 (control on guava with Bordeaux; **31**: 391).
Anon. 1963 & 1964. *Rep. Sci. Res. Coun. Jamaica* 3: 39; 4: 38 (dissemination & uredospore germination; **45**, 2204).
Hunt, P. 1966. *Proc. Caribb. Reg. Am. Soc. hort. Sci. 1965* **9**: 46 (as above; **46**, 2599c).
——. 1968. *Trans. Br. mycol. Soc.* **51**: 103 (host penetration; **47**, 2037).
Joffily, J. 1944. *Bragantia* **4**: 475 (on *Eucalyptus*; **24**: 480).
Maclachan, J. D. 1938. *Phytopathology* **28**: 157 (general on pimento in Jamaica; **17**: 554).
Smith, F. E. V. 1934. *J. Jamaica agric. Soc.* **38**: 276 (outbreak of disease; **13**: 653).
——. 1935. *Ibid* **39**: 408 (general; **14**: 792).

Puccinia purpurea Cooke, *Grevillea* **5**: 15, 1876.
Uredo sorghi Pass., 1867
Puccinia sanguinea Deitel, 1897
Dicaeoma purpureum (Cooke) Kentze, 1898
P. prunicolor Syd. & Butl., 1906.

UREDIA amphigenous, irregularly scattered on dark

reddish brown stained spots, rust coloured, circular or elongated with the veins, up to 2 mm long, paraphysate. PARAPHYSES mostly clavate, a few capitate, usually incurved, 40–90 × 12–18 μ; wall ochraceous, 2–8 μ thick above. UREDOSPORES ellipsoidal or obovoidal, 28–42 × 24–30 μ; wall fulvous or rust colour often paler at the equator, finely echinulate, 1.5–2 μ thick. Pores 4–8, scattered or more or less equatorial. TELIA like the uredia but black. TELIOSPORES ellipsoidal, obovoidal, or occasionally somewhat clavate, usually slightly constricted, obtuse, 40–60 × 25–32 μ; wall bay, smooth, 2–3 μ thick at the side, 4–6 μ above; pedicels basal, hyaline or yellowish, up to 120 μ long.

Other spp. of rusts recorded on sorghums include *P. jaagii* Boedijn, which lacks paraphyses and has diorchidioid teliospores; *P. cesatii* Schroet., which has verrucose uredospores; and *Uredo geniculata* Cummins, with smaller uredospores. On *Cymbopogon citratus, Sorghum almum, S. arundinaceum, S. cullorum, S. halepense, S. nitens, S. nitidum, S. virtatum* and *S. bicolor*. In USA Le Roux et al. (*P. sorghi* q.v.) produced infection on *Oxalis corriculata* using teliospores from sorghum. Race 13 produced from one of the resulting aecidia was pathogenic to 4 maize lines whilst other clonal lines from the same aecidiospore population were pathogenic only to sorghum, Sudan grass, or hybrids of these (from G. F. Laundon & J. M. Waterston, *CMI Descr.* 8, 1964).

A full account of sorghum leaf rust was given by Tarr, *Diseases of sorghum, Sudan grass and broom corn*. It probably occurs wherever sorghums are cultivated (Map 212). Although it can cause considerable damage on very susceptible cvs, losses are often slight since the disease does not become noticeably serious until the crop begins to mature and it occurs mostly on the older leaves. Purple red or tan flecks appear at first; this characteristic purpling is generally present but can be absent where the cv. does not contain pigmentation factors. Spots, on which the sori develop, occur over much of the leaf surface. The rust appears 2–3 months after sowing. Germinating uredospores penetrate via the stomata, always with appressorial formation; these spores lose viability after 14 days at 26–29°. Teliospores germinate at 10–20° and require no resting period. They remain viable for *c.* 9 months (if kept dry) at 5–7°. Resistance occurs in most sorghum groups but although cvs differ considerably in their reaction to infection there is little or no evidence for races. In one cross data indicated that resistance was controlled by a single dominant gene.

Agrawal, S. C. et al. *Sci. Cult.* **39**: 235 (seed treatment & spraying with oxycarboxin; **53**, 933).
Coleman, O. H. et al. 1961. *Crop Sci.* **1**: 152 (inheritance of resistance; **41**: 89).
Dalmacio, S. C. 1969. *Philipp. Agric.* **53**: 53 (infection; **50**: 656).
Johnston, C. O. et al. 1931. *Phytopathology* **21**: 525 (reaction of sorghum; **10**: 653).
Prasada, R. 1948. *Indian Phytopathol.* **1**: 119 (biology of teliospores; **29**: 148).
Soumini, C. K. 1950. *Ibid* **2**: 35 (uredospore viability, reaction of sorghum; **29**: 505).

Puccinia sorghi Schwein., *Trans. Am. phil. Soc.* Ser. 2, **4**: 295, 1832.
Puccinia maydis Bérenger, 1845
P. zeae Bérenger, 1851
Aecidium oxalidis Thuem., 1876
Tilletia epiphylla Berk. & Br., 1882
Dicaeoma sorghi (Schwein.) Kuntze, 1898.

PYCNIA up to *c.* 6, on each side of the leaf, grouped on an area up to 0.5 mm diam. in the centre of the spot. AECIDIA on underside of leaf only, surrounding the pycnia in a zone up to 2 mm wide, cupulate, 0.15–0.2 mm diam. AECIDIOSPORES spherical or ellipsoidal, 12–24 μ diam., wall hyaline, verrucose, 1–2 μ thick. UREDIA on both sides of leaves and leaf sheaths, dense or sparse, irregularly scattered, circular and *c.* 1 mm diam. or elongated and up to *c.* 10 mm long. UREDOSPORES spherical or ellipsoidal, 24–29 × 22–29 μ; wall amber, minutely echinulate, 1.5–2 μ thick; pores 3–4, equatorial. TELIA like the uredia but black. TELIOSPORES ellipsoid, cylindrical or clavate, obtuse or subacute, usually slightly constricted at septum, 35–50 × 16–23 μ; wall umber or sienna, slightly paler below, smooth, 1–1.5 μ thick at the sides, 3–6 μ above; pedicels basal, pale luteous, up to 80 μ long.

With regard to the other rust fungi occurring on maize, *P. sorghi* can be distinguished from *P. polysora* (q.v.) by the larger, usually more sparse and elongated uredia; the regular, well rounded, darker uredospores; the black open telia and the well-rounded teliospores with thicker apical walls and long pedicels; and from *Physopella zeae* (q.v.) which has hyaline uredospores and covered telia containing sessile teliospores. Pycnia and aecidia on some 30 spp. of *Oxalis* including *O. corniculata* (some of these by artificial inoculation only). Uredia and telia on *Euchlaena mexicana, E. perennis* and *Zea mays* (G. F. Laundon & J. M. Waterston *CMI Descr.* 3, 1964).

The uredial and telial stages of this heterothallic (Allen; Cummins) maize rust are practically co-extensive with the distribution of the main host (Map 279). In the tropics it may be less common than *P. polysora* (being a lower temp. rust) but can be abundant in highlands above 1300 m. The aecidial stage is limited in distribution to temperate regions, Europe, Mexico, S. Africa and USA; there is also a record for Nepal (*Commonw. phytopathol. News* 7: 59, 1961). Kim et al. (1976b) described the effects on yield. Penetration by the germinating uredospore is stomatal, usually with an appressorium but in rare cases without one. The temp. range for germination of uredospores (8–26°) is broader than that for aecidiospores. Opt. germination in the former occurred with 2 hours in the dark at 17° followed by 4 hours in the light at 24°. In 8 inbred maize lines inoculation with uredial clonal lines gave incubation periods of 9.2, 7.1, 6 and 5.3 days at 28, 24, 20 and 16°, respectively. Inoculation at 24° and 16° (day and night) gave a 6-day incubation period. Uredospore pustule formation is most rapid at 15–20°. These spores have a peak conc. at noon and early afternoon (Mahindapala; Pavgi et al. 1961; Symananda et al.; Weber; Zerekidze et al.). Both host stages have been cultured on detached leaves, and self-inhibitors of spore germination described. In temperate areas the uredospores probably do not overwinter and in Europe initiation of the disease may occur through aecidiospores.

Although fungicide treatment (zineb+copper oxychloride, fermate and dithane) has given some control this is largely through resistance. In USA, where most of the work on resistance has been done, the disease has varied in incidence partly due to the introduction of hybrid lines with limited resistance factors. The reaction of inbred maize lines used as differential hosts (14 races have been described) varies markedly with light and temp. The inheritance of resistance is complex; in some lines it is dominant (e.g., single dominants in each of 11 maize inbred lines from S. and C. America), in others recessive (multiple genes) and incomplete dominance has been reported. Using 6 inbred maize lines, representing 2 gene groups conditioning resistance and susceptibility, 96 uredospore cultures inbred on the aecidial host segregated into 8 pathogenic types. Mapping the loci of the resistant genes, in allelic series, has been done (Hagan; Hooker; Kim; Lee; Le Roux; Malm; Mains; Rhoades; Roy; Russell; Saxena; Stakman; Wellensiek; Wilkinson and others.)

Allen, R. F. 1934. *J. agric. Res.* **49**: 1047 (heterothallism; **14**: 438).
Arthur, J. C. 1904. *Bot. Gaz.* **38**: 64 (aecidial stage).
Cummins, G. B. 1931. *Phytopathology* **21**: 751 (heterothallism; **10**: 784).
Dickson, J. G. et al. 1959. *Am. J. Bot.* **46**: 614 (effect of environment on host & pathogen interaction; **39**: 227).
Flangas, A. L. et al. 1961. *Ibid* **48**: 275 (genetic control of pathogenicity; **41**: 302).
Hagan, W. L. et al. 1965. *Phytopathology* **55**: 193 (genes for resistance in 11 maize lines; **44**, 1857).
Hecke, L. 1906. *Annls Mycol.* **4**: 418 (aecidial stage).
Hilu, H. M. 1965. *Phytopathology* **55**: 563 (histology of infection in susceptible & resistant maize; **44**: 2780).
Hooker, A. L. et al. 1957. *Ibid* **47**: 187 (resistance factors in maize; **36**: 583).
—— ——. 1962. *Ibid* **52**: 122 (single, major dominant genes conditioning resistance in 6 maize lines; **41**: 595).
——. 1963. *Ibid* **53**: 221 (second major gene locus for resistance in 3 Australian maize lines; **42**: 545).
——. 1966. *Ibid* **56**: 536 (detached leaf culture of uredial & aecidial stages; **45**, 2817).
Kellerman, W. A. 1906. *J. Mycol.* **12**: 9 (aecidial stage).
Kim, S. K. et al. 1976a. *Pl. Dis. Reptr* **60**: 551 (resistance sources in Hawaii; **56**, 1088).
—— ——. 1976b. *Crop Sci.* **16**: 874 (effects on yield & agronomic traits in maize; **56**, 4488).
—— ——. 1977. *Ibid* **17**: 456 (inheritance of general resistance; **57**, 150).
Lee, B. H. et al. 1963. *Ibid* **3**: 24 (relationships of alleles for resistance on chromosome 10; **44**, 1856).
Le Roux, P. M. et al. 1957. *Phytopathology* **47**: 101 (races & genetics of *Puccinia sorghi* on maize & *P. purpurea* on sorghum; **36**: 466).
——. 1934. *Ibid* **24**: 405 (aecidial stage; **13**: 572).
Mahindapala, R. 1978. *Ann. appl. Biol.* **89**: 411, 417 (infection).
——. 1978. *Ibid* **90**: 155 (epidemiology).
——. 1978. *Trans. Br. mycol. Soc.* **70**: 393 (occurrence in England; **57**, 4918).
Mains, E. B. 1931. *J. agric. Res.* **43**: 419 (inheritance of resistance; **11**: 170).
Malm, N. R. et al. 1962. *Crop Sci.* **2**: 145 (resistance due to recessive genes in 2 maize lines; **41**: 652).
Pavgi, M. S. et al. 1961. *Phytopathology* **51**: 224 (effect of temp. & light on infection structures; **40**: 677).
—— ——. 1961. *Mycologia* **52**: 608 (cytology of teliospores; **41**: 150).

Puccinia substriata

Pavgi, M. S. 1969. *Mycopathol. Mycol. appl.* **38**: 175 (morphology & variation in 2 uredospore lines; **48**, 3469).

Pryor, A. 1976. *Physiol. Pl. Pathol.* **8**: 307 (catechol oxidase not involved in resistance; **55**, 5712).

Rhoades, M. M. et al. 1939. *Genetics Princeton* **24**: 302 (gene for resistance to race 3 on chromosome 10; **18**: 670).

Rhoades, V. H. 1935. *Proc. natn. Acad. Sci. U.S.A.* **21**: 243 (location of a gene for resistance; **14**: 626).

Rijkenberg, F. H. J. et al. 1974. *Mycologia* **66**: 319 (fine structure of sporogenesis in pycnia; **53**, 4411).

———. 1974. *Protoplasma* **81**: 231 (fine structure of aecidia; **53**, 4348).

———. 1975. *Ibid* **83**: 233 (cell fusion in aecidia; **54**, 2709).

Roy, M. K. et al. 1966. *Indian Phytopathol.* **19**: 305 (races on 6 differentials; **47**, 153).

Russell, W. A. et al. 1959. *Agron. J.* **51**: 21 (inheritance of dominant genes for resistance; **38**: 365).

———. 1962. *Crop Sci.* **2**: 477 (location of genes for resistance; **42**: 380).

Saxena, K. M. S. et al. 1974. *Can. J. Genet. Cytol.* **16**: 857 (structure of gene *Rp3* for resistance; **54**, 3867).

Stakman, E. C. et al. 1928. *Phytopathology* **18**: 345 (7 races distinguished on 8 differentials; **7**: 575).

Syamananda, R. et al. 1959. *Ibid* **49**: 102 (effect of temp. & light on host & pathogen interaction; **38**: 512).

Van Dyke, C. G. et al. 1969. *Ibid* **59**: 33, 1934 (unusual symptom type, fine structure of infection; **48**, 1659; **49**, 1627).

Wallin, J. R. 1951. *Pl. Dis. Reptr* **35**: 207 (epiphytotic in USA; **30**: 563).

Weber, G. F. 1922. *Phytopathology* **12**: 89 (factors affecting uredospore germination; **1**: 380).

Wellensiek, S. J. 1927. *Ibid* **17**: 815 (histology, 2 races; **7**: 371).

Wilkinson, D. R. et al. 1968. *Ibid* **58**: 605 (resistance genes in maize lines from Africa & Europe; **47**, 2699).

Zerekidze, R. I. et al. 1970. *Mikol. i Fitopatol.* **4**: 28 (effect of temp. & light on uredospore germination & infection of maize; **49**, 2053).

Zogg, H. 1949. *Phytopathol. Z.* **15**: 143 (epidemiology in Switzerland; **29**: 20).

Puccinia substriata Ell. & Barth. vars **indica** and **penicillariae** (Speg.) Ramachar & Cummins, *Mycopathol. Mycol. appl.* **25**: 26, 1965.

Puccinia penniseti Zimm., 1891
P. penicillariae Speg., 1914
P. penniseti-spicati Petrak, 1959.

PYCNIA mostly epiphyllous, about 5–50 in the centre of the spot, subepidermal, flask shaped. AECIDIA mostly hypophyllous, often grouped in a zone around the pycnia or occasionally irregularly scattered over large areas of the leaf surface, cupulate, 0.25–0.35 mm diam. AECIDIOSPORES ellipsoidal, 16–24 μ diam., wall hyaline, finely verrucose, 1 μ thick. UREDIA amphigenous, irregularly scattered, often elongated along the veins, mostly *c.* 1–2 mm long, cinnamon, often with inconspicuous pale clavate or cylindrical paraphyses. UREDOSPORES ellipsoidal, obovoidal or clavate, 33–48 × 23–31 μ; wall luteous, very finely and sparsely echinulate, 1.5–2 μ thick. Pores 3–5 more or less equatorial and equidistant. TELIA like the uredia but black. TELIOSPORES ellipsoidal, cylindrical or clavate, constricted at the septum, obtuse or truncate above, 40–80 × 22–27 μ; wall sienna, smooth, 1–2 μ thick at the sides, 3–8 μ thick at the apex; pedicels sienna, *c.* 20 μ long. The only other frequent rust on *Pennisetum*, *Phakopsora apoda* (Har. & Pat.) Mains, differs in its pale, thin walled uredospores and sessile teliospores.

Pycnia and aecidia on *Solanum anoplocadium*, *S. melongena* (eggplant), *S. panduriforme*, *S. pubescens*, *S. torvum* and *S. xanthocarpum*. Uredospores and teliospores on *Beckeropsis*, *uniseta*, *Cenchrus pennisetiformis*, *Chloridion cameronii*, *Pennisetum clandestinum*, *P. purpureum*, *P. typhoides* (pearl or bulrush millet), other *Pennisetum* spp. and *Setaria glauca* (from G. F. Laundon & J. M. Waterston, *CMI Descr.* 6, 1964 as *Puccinia penicillariae* Speg., *An. Mus. Nac. Buenos Aires* **26**: 119, 1914).

Ramachar et al. (l.c.) divided the common rust of millet into *P. substriata* var. *penicillariae* and *P. substriata* var. *indica*. The former is the common African form which also occurs in Malagasy Republic, India and Sri Lanka; its aecidial state has not been demonstrated. The latter is the common rust in India and no specimens from outside this country were seen; it differs from var. *penicillariae* in having small and very tardily dehiscent telia, and narrower (19–24 μ), usually paler, teliospores; the aecidial state is on eggplant. Aecidia have been collected on eggplant in Africa and the Philippines but their relationship to *P. substriata* has not been demonstrated.

Millet rust is widespread in W., E. and S. Africa. As var. *indica* it was recently reported from USA (S.E.). It occurs in Venezuela and the aecidial state was described from Brazil (São Paulo; *Rev. Pl. Pathol.* **51**, 3029; Map 225). Severe attacks can lead to extensive leaf necrosis. Survival of the uredospores (max. germination at 20°) under natural conditions is probably < a month and at 41–42° only remained viable for 8 days. The teliospores have no resting period and germination takes place in 48 hours. The fungus is homothallic, the basidium is constantly 2 celled and the basidiospores usually quadrinucleate; in other respects *P. substriata* has a nuclear cycle similar to that of other macrocyclic, homothallic rusts. The evidence for the existence of races appears to conflict. In India the length : breadth ratios of uredospores from monospore cultures originating from 22 localities indicated 3 groups; and 2 morphological types of uredia have been noted. But when 10 isolates were tested on 42 cvs no distinct races could be distinguished (Dalela 1964). Resistance, the only probable control at present, needs further investigation

Basu-Chaudhary, K. C. 1955. *Agra Univ. J. Res.* **4**: 31 (effect of temp. on uredospore survival; **34**: 373).

—— et al. 1955. *Ibid* **4**: 575 (20 cvs bulrush millet all susceptible as seedlings; **36**: 22).

Dalela, G. G. et al. 1957. *Proc. Indian Acad. Sci.* Sect. B **27**: 245 (forms based on uredospore dimensions; **38**: 742).

—— ——. 1962. *Agra Univ. J. Res.* **11**: 113 (inoculation on Gramineae; **42**: 679).

——. 1962. *Indian Phytopathol.* **15**: 156 (homothallism; **43**, 1317).

——. 1964. *Ibid* **17**: 63 (no races distinguished; **44**, 693).

Kapooria, R. G. 1968. *Neth. J. Pl. Pathol.* **74**: 2 (cytology of teliospore & basidiospore; **47**, 1532).

—— et al. 1969. *Trans. Br. mycol. Soc.* **53**: 153 (effect of millet leaf mycoflora on development; **49**, 136).

——. 1970. *Pl. Dis. Reptr* **54**: 646 (production of telia by uredospore strs; **50**, 660).

——. 1973. *J. gen. Microbiol.* **77**: 443 (morphological variability in teliospores; **53**, 521).

——. 1973. *Ghana J. Sci.* **13**: 153 (inoculation of *Pennisetum* spp.; **54**, 438).

Kulkarni, U. K. 1956. *Trans. Br. mycol. Soc.* **39**: 48 (initiation of dicaryon; **36**: 97).

——. 1958. *Ibid* **41**: 65 (development & cytology; **37**: 658).

Misra, A. et al. 1971. *Indian J. Mycol. Pl. Pathol.* **1**: 103 (uredospore germination & host incubation; **51**, 4820).

Ramakrishnan, T. S. et al. 1948. *Indian Phytopathol.* **1**: 97 (morphology & aecidial host; **29**: 148).

—— ——. 1956. *Proc. Indian Acad. Sci.* Sect. B **43**: 190 (general, hosts & 2 types uredia; **35**: 673).

Wells, H. D. 1978. *Pl. Dis. Reptr* **62**: 469 (on eggplant in USA; **58**, 198).

PYRENOCHAETA de Not., *Mem. R. Accad. Torina* Ser. 2 **10**: 347, 1849.
(synonymy in B. C. Sutton, *Mycol. Pap.* 141: 172, 1977)
(Sphaerioidaceae)

MYCELIUM immersed or superficial, pale brown to hyaline, branched, septate. CONIDIOMATA pycnidial, separate, rarely aggregated, globose, immersed or superficial, brown, unilocular, thin to thick walled; wall of pale to medium brown textura angularis. OSTIOLE single, central, circular. Setae abundant around the ostiole, less so over the rest of the pycnidium, erect or lax, dark brown, thick walled, smooth, septate, usually tapered at the apices. CONIDIOPHORES long, filiform, branched at the base, multiseptate, hyaline, formed from the inner cells of the conidiomatal wall. CONIDIOGENOUS CELLS enteroblastic, phialidic, determinate, integrated, arising as very short lateral branches immediately below transverse septa, hyaline, smooth; channel, colarette and periclinal thickening minute. CONIDIA hyaline, aseptate, smooth, more or less guttulate, straight, cylindrical to ellipsoid (B. C. Sutton, personal communication, November 1978; see Schneider for taxonomy).

P. oryzae Shirai ex Miyake, causing sheath brown spot (also called yellow sheath spot or sheath blotch) of rice, was illustrated by E. Punithalingam in Ou, *Rice diseases*, and described briefly by Haskioka who considered it of no great importance. The blackish brown pycnidia are 69–176 × 69–137 μ; ostioles are surrounded by > 10 multiseptate, black setae, 60–140 × 4–5 μ; conidia 2.5–5.4 (av. 4.3) × 1.2–2.7 (av. 2.1) μ, guttulate at both ends. (*P. nipponica* Hara, also from rice, differs in the smaller pycnidia 100–150 μ, setae 45–75 × 2.5–3 μ and conidia 3–4 × 1–1.5 μ.) Large brown lesions are caused on the leaf sheaths; they are elongate, several cm long and becoming paler in the centre. Two *Pyrenochaeta* spp. have been described from sugarcane (*Saccharum*): *P. sacchari* Bitancourt (pycnidia 50–100 μ; setae 1–20, 5–40 × 2.5–5 μ; conidia 6–12 × 3 μ); and *P. indica* Viswanathan (pycnidia 30–190 μ; setae 1–50, 5–71 × 1.7–3.4 μ; conidia 3–6 × 2.5 μ). Other spp. described from tropical crops are *P. cajani* R. A. Singh & Pavgi on *Cajanus cajan* (pigeon pea), *P. dolichi* Mohanty on *Dolichos uniflorus* (horse gram) and *P. glycines* Stewart on *Glycine max* (soybean).

Pyrenochaeta lycopersici

Bitancourt, A. A. 1938. *Arq. Inst. biol. S. Paulo* **9**: 299 (*Pyrenochaeta sachari* sp. nov.; **18**: 346).

Hashioka, Y. 1972. *Riso* **21**: 11 (rice diseases due to Sphaeropsidales; **53**, 527).

Mohanty, A. K. et al. 1969. *Indian Phytopathol.* **22**: 309 (growth of *P. dolichi* in vitro; **49**, 3058).

Mohanty, N. N. 1958. *Ibid* **11**: 85 (*P. dolichi* sp. nov.; **38**: 702).

Pavgi, M. S. et al. 1965. *Mycopathol. Mycol. appl.* **27**: 97 (*P. cajani* sp. nov. inter alia; **45**, 1588).

Schneider, R. 1976. *Ber. dt. bot. Ges.* **89**: 507 (taxonomy; **56**, 2890).

Stewart, R. B. 1957. *Mycologia* **49**: 115 (*P. glycines* sp. nov.; **36**: 569).

Viswanathan, T. S. 1957. *Curr. Sci.* **26**: 117 (*P. indica* sp. nov.; **36**: 729).

Pyrenochaeta lycopersici Schneider & Gerlach, *Phytopathol. Z.* **56**: 117, 1966.

PYCNIDIA solitary, globose to subglobose, 150–300 μ diam. brown to black, darker around the neck region, with papillate ostioles beset with 3–12 light brown, septate setae up to 120 × 7 μ. Pycnidial wall many cells thick, composed of outer sclerotised and heavily pigmented cells, and inner hyaline pseudoparenchymatic cells lining the pycnidial cavity. CONIDIOPHORES simple, mostly branched, septate, arising from the hyaline cells lining the inside of the pycnidial cavity. CONIDIOGENOUS CELLS enteroblastic, phialidic, hyaline, simple, cylindrical, borne on septate conidiophores. CONIDIA (phialospores) hyaline, aseptate, acropleurogenous, formed from the apex and short lateral branches immediately below the septa, cylindrical to allantoid, 4.5–8 × 1.5–2 μ. Occurs in a sterile mycelial state within corky root of tomatoes, and pycnidia have not been found in the natural state. The description is based on type material kindly provided by R. Schneider (E. Punithalingam & P. Holliday, *CMI Descr.* 398, 1973. White et al. described the formation of MICROSCLEROTIA (a viability of >2 years), 65.1 (38–112) × 32.7 (9–45) μ, in culture, isolated tomato root segments and in roots of intact tomato plants. In such roots they formed in single cells of the outer cortex.)

Corky root (also called brown root rot) of tomato has not been described from the tropics. It has been reported in: Belgium, Canada, France, Germany, Italy, Netherlands, New Zealand, Rumania, Scandinavia and UK. In New Zealand *P. lycopersici* causes a disease on tobacco and lesions on this host were caused by isolates from tobacco, tomato and strawberry (Taylor et al.). Tomato seedlings have also been infected with isolates from red pepper. The pathogen was first isolated *c.* 1929; it was long known as the grey sterile fungus and was not identified for nearly 37 years. It was, however, shown to be pathogenic to tomato in 1944. The disease had been previously thought to be partly due to other fungi, notably *Colletotrichum coccodes* (q.v.) which is common in soils infested with *P. lycopersici* but is only a secondary invader of tomato roots. Reviews of the early work have been given by Ebben, Last et al. (1966a) and Termohlen (1962). *P. lycopersici* has been compared with *P. terrestris* (q.v.) from onion but the latter was only weakly pathogenic to tomato and the former does not produce the characteristic red pigment of the onion pathogen.

The fungus is soilborne and in Europe it seems to occur only in crops grown in close succession under glass. The plants lack vigour, become stunted and give poor yields. There is a cortical rot of the fine and medium-sized roots; larger roots become corky with swollen, cracked and furrowed bark, almost cankered; finally, the stem base may rot. *P. lycopersici* increases sooner and more extensively in the second year of cropping, decreases with soil depth and can increase even when tolerant rootstocks are used; it can be isolated from such stocks. Corkiness symptoms tend to decrease from one season to the next; an effect probably associated with increasing populations of the fungus which prevent the formation of the large lateral roots that form cork. In UK it has been shown that disease build up is progressive through 5 seasons of successive tomato crops. Five weeks after planting in the first year there was 4% disease, rising to 12% in 3 months. In later years (and where early season infection was 8%) incidence subsequently increased fivefold. Monthly increments of fruit were inversely proportional to disease incidence. (See *Rev. Pl. Pathol.* **53**, 3185; **54**, 2472, 3165; **55**, 70, 1471, 5928; **56**, 3228 for cell degrading enzymes.)

Predicting when economically damaging soil infestations might build up should help to determine the best timing for soil sterilants and fungicides. Methyl bromide and chloropicrin have been used successfully (Cirulli; Taylor et al.; recent Reports of the Glasshouse Crops Research Institute, UK); benomyl and thiabendazole drenches have been investigated. In the Netherlands a major gene for resistance (derived from *L. glandulosum*) was found to have a low degree of dominance when heterozygous and its expression to be affected by the environment (Hogenboom; and see Volin et al.).

Cirulli, M. 1968. *Annali. Fac. Agr. Univ. Bari* **22**: 453 (soil fumigation; **49**, 2655).
Clerjeau, M. 1976. *Annls Phytopathol.* 8: 9 (temp., growth & pathogenicity; **56**, 2878).
Delon, R. et al. 1973. *Ibid* **5**: 151 (histology; **53**, 4594).
Ebben, M. H. et al. 1956. *Ann. appl. Biol.* **44**: 425 (associated fungal flora; **36**: 359).
—. 1974. *A. Rep. Glasshouse Crops Res. Inst.* for 1973: 127 (review, 46 ref.; **54**, 3489).
— —. 1978. *Pl. Pathol.* **27**: 91 (comparison of a tolerant line & susceptible cv. in infested soils).
Gerlach, W. et al. 1964. *Phytopathol. Z.* **50**: 262 (the *Pyrenochaeta* state; **44**, 244).
— —. 1966. *Ibid* **56**: 19 (inoculation & comparison with *P. terrestris*; **46**, 155).
Hogenboom, N. G. 1970. *Euphytica* **19**: 413 (inheritance of resistance; **50**, 2504).
Last, F. T. et al. 1963. *N.A.A.S. q. Rev.* **62**: 68 (aetiology, symptoms, soil fumigation).
— —. 1966a. *Scient. Hort.* **28**: 36 (review; **45**, 3415).
— —. 1966b. *Ann. appl. Biol.* **57**: 95 (epidemiology; **45**, 1520).
— —. 1968. *Ibid* **62**: 55 (effect of cultural treatments on disease incidence & damage done; **47**, 3571).
— —. 1969. *Ibid* **64**: 449 (build up; **49**, 1489).
Noordam, D. et al. 1957. *Tijdschr. PlZiekt.* **63**: 145 (symptoms, microsclerotia; **37**: 250).
Preece, T. F. 1964. *Trans. Br. mycol. Soc.* **47**: 375 (inoculation, symptoms & microsclerotia; **44**, 245).
Richardson, J. K. et al. 1944. *Phytopathology* **34**: 615 (proof of pathogenicity; **24**: 78).
Schneider, R. et al. 1966. *Phytopathol. Z.* **56**: 117 (morphology; **46**, 437).
Taylor, J. B. et al. 1971. *N.Z. Jl Sci.* **14**: 276 (on tobacco & other hosts, soil fumigation; **51**, 1900).
Termohlen, G. P. 1958. *Tijdschr. PlZiekt.* **63**: 369 (symptoms & pathogenicity; **37**: 739).
—. 1962. *Ibid* **68**: 295 (review; **42**: 492).
Volin, R. B. et al. 1978. *Euphytica* **27**: 75 (inheritance of resistance; **57**, 5687).
White, J. G. et al. 1973. *Ann. appl. Biol.* **73**: 163 (microsclerotia; **52**, 1674).
—. 1976. *Trans. Br. mycol. Soc.* **67**: 497 (effects of temp., light, & aeration on microsclerotial formation; **56**, 4203).

Pyrenochaeta terrestris (Hansen) Gorenz, Walker & Larson, *Phytopathology* 38: 838, 1948.
Phoma terrestris Hansen, 1929.

PYCNIDIA on onion roots, usually solitary, immersed, later becoming erumpent, globose, up to 400 μ diam., brown to black, heavily pigmented around the ostiole, sometimes slightly beaked and beset with brown septate setae, 60–180 μ long. Pycnidial wall several cells thick, composed of outer pigmented, thick walled cells and inner thin walled, pseudoparenchymatic cells lining the entire pycnidial cavity. CONIDIOGENOUS CELLS enteroblastic, phialidic, hyaline, simple, obpyriform, arising from the innermost layer of pseudoparenchymatic cells lining the pycnidial cavity. CONIDIA (phialospores) aseptate, ovoid to allantoid, biguttulate with a guttule at each end, ends rounded, $4–7 \times 1.5–2$ μ (E. Punithalingam & P. Holliday, *CMI Descr.* 397, 1973).

Pink root of onion (*Allium cepa* var. *cepa*) and other *Allium* spp.; the fungus is widespread as a soil inhabitant and is frequently isolated from the roots of many crops on which no disease is caused (Kreutzer 1941; Sprague). Isolates from such crops can cause disease symptoms in onion (Schneider). Pink root has been mostly investigated in USA. The general, uncharacteristic, above ground symptoms of stunting, chlorosis, and tip burn and dieback of the leaves can occur at any growth stage. In seedlings there may be complete collapse whilst in older plants the upper leaves droop. Infected roots become characteristically pale pink, reddish and sometimes darken to red or purple; few or many roots may be rotted and the diseased plants are pulled easily. Bulbs are not attacked although the outer scales may be penetrated. New roots formed throughout the growing season may be infected and killed successively. The pink pigment produced on the host forms in culture and is a means of identification (Watson). Siemer et al. and Pfleger et al. (1971) described methods of determining the amount of soil inoculum. Gorenz et al. (1949) showed that most disease occurred at 28° compared with 24, 20 and 16°; the best in vitro growth is at 28° or a little lower, and isolates may differ in this respect. UV light induces abundant sporulation in vitro (Hess et al.). No races have been described although isolates may differ in pathogenicity. *P. terrestris* infection and host resistance in association with *Fusarium oxysporum* f. sp.

cepae (q.v.) were investigated by Kehr et al. and Woolliams. The fine structure of root penetration has been described and the cell wall degrading enzymes extensively studied (Horton and Keen).

Onion cvs with some resistance have been developed in N. America. Nichols et al. reported control of resistance by a single recessive gene but there was evidence that more than one gene was involved (and see Jones et al.). Struckmeyer et al. described the reaction of a susceptible and 2 resistant onion cvs to infection by a moderately virulent isolate and a highly virulent one. Resistance was expressed in the inner cortical cells. The highly virulent isolates destroyed the roots of the susceptible cv. in 15 days and those of the resistant cvs in 20 days. The moderately virulent one destroyed the roots of the susceptible cv. in 25 days. Cvs of leek, chives and Welsh onion have adequate resistance but those of onion, shallot and garlic are generally susceptible. Long rotations prevent damaging build up in the soil. Planting at temps below the opt. for the pathogen and ensuring vigorous growth will reduce damage. Soil disinfestation has been used.

Davis, G. N. et al. 1937. *Phytopathology* **27**: 763 (pathogenicity with *Fusarium oxysporum*; **17**: 7).
Gorenz, A. M. et al. 1948. *Ibid* **38**: 831 (morphology, pycnidia in culture; **28**: 202).
—— ——. 1949. *J. agric. Res.* **78**: 1 (temp. & other factors affecting pathogenicity; **29**: 134).
Gunasekaran, M. et al. 1973. *Can. J. Microbiol.* **19**: 491 (fine structure & lipid changes during ageing; **52**: 2510).
Hansen, H. N. 1929. *Phytopathology* **19**: 691 (general; **9**: 154).
——. 1930. *Science N.Y.* **71**: 424 (variation in vitro; **9**: 621).
Hess, W. M. et al. 1964. *Phytopathology* **54**: 113 (UV induced sporulation; **43**: 2150).
——. 1968. *Can. J. Microbiol.* **14**: 205 (fine structure of hyphal cells; **47**, 2632).
——. 1969. *Am. J. Bot.* **56**: 832 (fine structure of root penetration; **49**: 1230).
Horton, J. C. 1964. *Proc. Iowa Acad. Sci.* **71**: 84 (pectolytic & cellulytic enzymes; **45**, 1599a).
—— et al. 1964. *Ibid* **71**: 88 (glucose inhibition of cellulase synthesis; **45**, 1599b).
—— ——. 1966. *Can. J. Microbiol.* **12**: 209 (regulation of induced cellulase synthesis by utilisable C compounds; **45**, 2725).
—— ——. 1966. *Phytopathology* **56**: 908 (high root sugar content as a resistance mechanism; **46**, 209).
Jones, H. A. et al. 1956. *J. Hered.* **47**: 33 (inheritance of resistance).
Keen, N. T. et al. 1966. *Can. J. Microbiol.* **12**: 443 (induction & repression of endopolygalacturonase synthesis; **45**, 3060).
—— ——. 1966. *Phytopathology* **56**: 603 (polygalacturonase; **45**, 3294).
Kreutzer, W. A. 1939. *Ibid* **29**: 629 (pigment; **19**: 4).
——. 1941. *Ibid* **31**: 907 (penetration, histology, infection of hosts other than *Allium*; **21**: 117).
Kulik, M. M. et al. 1960. *Pl. Dis. Reptr* **44**: 54 (differences in pathogenicity & pycnidial formation; **40**: 328).
Nichols, C. G. et al. 1965. *Phytopathology* **55**: 752 (expression & inheritance of resistance; **45**, 312).
Perry, B. A. et al. 1955. *Proc. Am. Soc. hort. Sci.* **66**: 350 (resistant cvs; **35**: 743).
Pfleger, F. L. et al. 1971. *Phytopathology* **61**: 1299 (soil assay; **51**, 2101).
—— ——. 1974. *Can. J. Bot.* **52**: 43 (fungitoxic compound from onion; **53**, 3715).
Porter, D. R. et al. 1933. *Phytopathology* **23**: 290 (resistance in *Allium* spp.; **12**: 547).
Schneider, R. 1965. *Phytopathol. Z.* **53**: 249 (isolation from crops other than *Allium* & pathogenicity; **45**, 311).
Siemer, S. R. et al. 1971. *Phytopathology* **61**: 146 (soil assay; **50**: 3326).
Sneh, B. et al. 1974. *Ibid* **64**: 275 (isolation from soil; **53**, 3716).
Sprague, R. 1944. *Ibid* **34**: 129 (on Gramineae; **23**: 261).
Struckmeyer, B. E. et al. 1962. *Ibid* **52**: 1163 (histology in susceptible & resistant onion cvs; **42**: 355).
Taubenhaus, J. J. et al. 1921. *Bull. Texas agric. Exp. Stn* 273, 42 pp. (general; **1**: 405).
Tims, E. C. 1953. *Pl. Dis. Reptr* **37**: 533 (on shallot; **33**: 332).
Watson, R. D. 1961. *Ibid* **45**: 289 (identification on an agar medium; **40**: 644).
Wright, J. R. et al. 1965. *Physiol. Pl.* **18**: 1044 (use & formation of carbohydrates).
—— ——. 1966. *Ibid* **19**: 702 (mannitol formation; **46**, 210).

PYRICULARIA Sacc., *Michelia* 2: 20, 1880.
(often as *Piricularia*)
(Dematiaceae)

COLONIES effuse, thinly hairy, grey, greyish brown or olivaceous brown. MYCELIUM immersed, CHLAMYDOSPORES sometimes found in culture. STROMA none. SETAE and HYPHOPODIA absent. CONIDIOPHORES macronematous, mononematous, slender, thin walled, usually emerging singly or in small groups through stomata, mostly unbranched, straight or flexuous, geniculate towards the apex, pale brown smooth. CONIDIOGENOUS CELLS polyblastic, integrated, terminal, sympodial, cylindrical, geniculate, denticulate; each denticle cylindrical, thin walled, cut off usually by a septum to form a separating cell. CONIDIA solitary, dry, acropleurogenous, simple, obpyriform, obturbinate or obclavate, hyaline to pale olivaceous brown, smooth, 1–3 (mostly 2) septate; hilum often protuberant (M. B. Ellis, *Demat. Hyphom.*: 218, 1971; see *P. grisea* and *P. oryzae* for perfect state).

Asuyama (*P. oryzae* q.v.) in 1963 discussed the *Pyricularia* spp. on rice, other cereals and grasses. The position is still confused and a re-examination of these fungi is needed (e.g., Kato et al.). Although host specificities in these 2 and other spp. have often been demonstrated, the 2 commonest spp. (*P. grisea* and *P. oryzae*) are morphologically very similar. Asuyama, therefore, considered that the similar spp. on Gramineae might be better classified on the basis of host or hosts. Other plant families from which members of the genus have been described are: Cannaceae, Commelinaceae, Cyperaceae, Musaceae and Zingiberaceae (Hashioka). Webster described a perfect state (*Massarina aquatica* Webster) of *P.* (as *Piricularia*) *aquatica* Ingold (synonym: *Dactylella aquatica* (Ing.) Ranzoni). The spp. on tropical cereals (other than the 2 mentioned) that are more often encountered in the literature are: *P. penniseti* Prasada & Goyale (Hashioka 1973 gives some synonymy) on *Pennisetum typhoides*; and *Pyricularia setariae* Misikado on *Setaria italica*. The disease caused by *P. penniseti* on bulrush millet is probably the one described by Mehta et al. *P. setariae* has been reported on seed of foxtail millet and may be serious enough to warrant control measures (Goel et al.; Liang et al.). *P. setariae* can infect *Eleusine coracana* (finger millet) and bulrush millet (Palaniswami et al.; Sivaprakasam et al.). These 2 fungi can be spread in seed (Kato et al.; Singh et al.). Ellis (*More Demat Hyphom.*) described *P. didyma* M. B. Ellis from *Pennisetum purpureum*.

Goel, L. B. et al. 1967. *Pl. Dis. Reptr* **51**: 138 (*Pyricularia setariae*, seedborne; **46**, 2010).

Hashioka, Y. 1971. *Trans. mycol. Soc. Japan* **12**: 126 (spp. on Cannaceae, Musaceae & Zingiberaceae; **51**, 3834).

——. 1973. *Ibid* **14**: 256 (spp. on Commelinaceae, Cyperaceae & Gramineae; **53**, 3373).

Kato, H. et al. 1977. *Ann. phytopathol. Soc. Japan* **43**: 392 (*Pyricularia* spp., spread, pathogenicity & control; **57**, 3453).

Liang, P. Y. et al. 1959. *Acta phytopathol. sin.* **5**: 89 (*P. setariae* on foxtail millet; **39**: 576).

Matsuyama, N. et al. 1977. *Ann. phytopathol. Soc. Japan* **43**: 419 (isozyme patterns in *Pyricularia*; **57**, 3335).

Mehta, P. R. et al. 1952. *Indian Phytopathol.* **5**: 140 (*Pyricularia* sp. on bulrush millet; **33**: 600).

Palaniswami, A. et al. 1970. *Madras agric. J.* **57**: 686 (*P. setariae* on foxtail millet; **50**, 2918).

Prasada, R. et al. 1970. *Curr. Sci.* **39**: 287 (*P. penniseti* sp. nov.; **50**, 661).

Singh, D. S. et al. 1977. *Indian Phytopathol.* **30**: 242 (*P. penniseti*).

Sivaprakasam, K. et al. 1974. *Indian J. agric. Sci.* **44**: 249 (fungicides against *P. setariae* on finger millet; **55**, 1782).

Suryanarayanan, S. 1958. *Phytopathol. Z.* **33**: 341 (thiamine & growth; **38**: 184).

Webster, J. 1965. *Trans. Br. mycol. Soc.* **48**: 449 (*Massarina aquatica* sp. nov.).

Pyricularia grisea Sacc., *Michelia* **2**: 20, 1880.

CONIDIOPHORES up to 150 μ long, 2.5–4.5 (usually 3–4) μ thick. CONIDIA mostly 17–28 (20.9) × 6–9 (7.6) μ (M. B. Ellis, *Demat. Hyphom.*: 219, 1971).

A full description is given under the very similar *P. oryzae*. Ellis (l.c.) gave the conidial measurements for *P. oryzae*: 17–23 (21.2) × 8–11 (9.6) μ, i.e. broader than those for *P. grisea*. He (like Sprague, *Diseases of cereals and grasses in North America*) retained *P. grisea* for the forms on Gramineae (other than rice) and some other hosts. Other workers have considered that the 2 spp. are morphologically indistinguishable. H. Asuyama (in Ou, *P. oryzae* q.v.), in discussing the morphology, taxonomy, host range and life cycle of *P. oryzae*, gave an account of *Pyricularia* on cereals, other than rice, and grasses. It was suggested that these 2 fungi on Gramineae be considered as one sp. and distinguished on the basis of hosts; 11 host groups were given. Other spp. described from this plant family might also be included (*Pyricularia* q.v.).

A perfect state (*Ceratosphaeria grisea* Hebert) developed in the culture of 2 *P. grisea* isolates from crab grass (*Digitaria sanguinalis*). The PERITHECIA are long beaked with cylindric to subclavate, nonpersistent ASCI with fusiform, hyaline, 4 celled ASCOSPORES, 5 (4–7) × 21 (17–24) μ; see Yaegashi et al. and *P. oryzae* for ref. to a perfect state. The state was transferred to *Magnaporthe* (q.v.) independently by Barr and Yaegashi et al. (1978). Its name is

Pyricularia oryzae

M. grisea (Hebert) Barr (synonyms: *C. grisea* Hebert; *M. grisea* (Hebert) Yaegashi & Udagawa); and see Kato.

Besides being often reported (e.g., as grey leaf spot) from cereals and grasses (causing relatively minor diseases) a form of *P. grisea* causes an important disease (pitting, also Johnston fruit spot) of preharvest banana fruit. The fungus is common on leaf trash of this crop. Meredith (1962b) gave a description of *P. musae* Hughes, a common invader of banana leaves. Ellis (*Demat. Hyphom.*) transferred this saprophyte to a new genus, *Pyriculariopsis parasitica* (Sacc. & Berl.) M. B. Ellis (*Helminthosporium parasiticum* Sacc. & Berl. is another synonym). Hashioka (1971, *Pyricularia* q.v.) discussed the *Pyricularia* spp. on banana and described another sp. on this plant, *P. angulata* Hashioka. It was distinguished by the thin epispore and angular appressoria. Halmos (1969) could not infect rice with *P. grisea* from banana or banana with *P. oryzae* from rice.

Pitting disease was described by Meredith (1963a). Round sunken pits (av. 4–6 mm diam.) appear on the fruit as it matures or after harvest. A reddish brown zone surrounds the sunken centre and beyond this is a narrow, greenish, watersoaked halo. The centre of the pit may split. Pits do not extend to the pulp. Pitting seldom occurs on fruit until 70 days after the flower has emerged. The number of pits can increase fivefold or more when fruit is in transit and during ripening. Fingers on the side of the bunch facing away from the pseudostem are more severely pitted than those on the opposite side. The large proximal hands have more pits than the smaller distal ones. Smaller pits form on the finger stalk and crown cushion. Spots resembling those on fruit, but larger and less sunken, may occur on the transition leaf (bract) and on young water suckers during abundant rain. As tissue decays sporulation occurs and Halmos (1970) found that most conidia arose from hanging, dead, trash leaves of banana plants. Meredith (1962a) found a peak in conidial dispersal at *c.* 0400 hours. Germinating conidia form an appressorium in 4–8 hours in a saturated atmosphere; opt. temp. *c.* 24–26°. The fungus may then enter a latent period (as subcuticular hyphae), and renew activity when the fruit approaches maturity (Meredith 1963b). Young green fruit that are infected in situ rarely develop pits before fruit is 70 days old. But pits can form in 3 weeks when fruit approaching harvesting grade is infected. The disease is at a max. at high rainfall periods and is generally only serious at such times. It reduces the quality of fruit. Control involves removal of leaf and bract trash, weekly spraying with dithiocarbamates and bagging with a perforated polyethylene tube when all fingers have emerged.

Bailey, A. G. et al. 1961. *Phytopathology* **51**: 197 (on maize; **40**: 605).

Guyon, M. 1970. *Fruits* **25**: 685 (fungicides; **50**, 2392).

Halmos, S. 1969. *Pl. Dis. Reptr* **53**: 878 (*Pyricularia oryzae* in Honduras & host specificity, with *P. grisea*; **49**, 1643).

——. 1979. *Phytopathology* **60**: 183 (inoculum sources; **49**, 2135).

Hebert, T. T. 1971. *Ibid* **61**: 83 (perfect state; **50**, 2770).

Hoette, S. 1936. *Proc. R. Soc. Vict.* **48**: 90 (general; **16**: 195).

Jailloux, F. et al. 1976. *Agron. trop.* **31**: 58 (on *Digitaria decumbens*; **55**, 4161).

Kato, H. 1977. *Rev. Pl. Prot. Res.* **10**: 20 (review, perfect state of *Pyricularia*, 56 ref.).

Malca, M. I. et al. 1957. *Pl. Dis. Reptr* **41**: 871 (on *Stenotaphrum secundatum*; **37**: 296).

Meredith, D. S. 1962a. *Nature Lond.* **195**: 92 (conidial dispersal; **41**: 732).

——. 1962b. *Trans. Br. mycol. Soc.* **45**: 137 (*P. musae*, morphology; **41**: 611).

——. 1963a. *Ann. appl. Biol.* **52**: 453 (general; **43**, 1715).

——. 1963b. *Pl. Dis. Reptr* **47**: 766 (latent infection; **43**, 171).

Wardlaw, C. W. 1934. *Trop. Agric. Trin.* **11**: 8 (general & other banana fruit spots; **13**: 454).

Yaegashi, H. et al. 1976. *Phytopathology* **66**: 122 (development & cytology of perfect state; **55**, 3349).

—— ——. 1976. *Ann. phytopathol. Soc. Japan* **42**: 556 (conditions for obtaining perithecia; **56**, 3874).

—— ——. 1978. *Can. J. Bot.* **56**, 180, 2184 (*Magnaporthe grisea* comb. nov. & correction; **57**, 3822).

Pyricularia oryzae Cav., *Fungi Longob exsicc.* No. 49 cum diagnose, 1891. Published again as sp. nov. in Briosi & Cavara *I Funghi Parass.* No. 188, 1892, and as sp. nov. Cavara, *Atti ist. bot. Univ. Pavia* Ser. 2. **2**: 280, 1892.

Dactylaria oryzae (Cav.) Saw., 1916.

CULTURES greyish in colour. CONIDIOPHORES single or in fascicles, simple, rarely branched, showing a sympodial growth. CONIDIA formed singly at the tip of the conidiophore at points arising sympodially and in succession, pyriform to obclavate, narrowed towards tip, rounded at the base, 2 septate, rarely 1 or 3 septate, hyaline to pale olive, 14–40 × 6–13 (mostly 19–23 × 7–9)μ with a distinct protruding basal hilum. CHLAMYDOSPORES often produced in culture, thick walled, 5–12 μ diam. (C. V. Subramanian, *CMI Descr.* 169, 1968).

Kato et al. (1976) described a perfect state which was obtained in matings of *P. oryzae* (from rice) and *Pyricularia* sp. (from *Eleusine coracana*). Yaegashi et al. also obtained a perfect state in some matings between *Pyricularia* isolates from 20 gramineous hosts. In both cases these states resembled *Ceratosphaeria grisea* Hebert (now *Magnaporthe grisea, P. grisea* q.v.). Ueyama et al. (1975) also reported a similar state using an isolate from rice and a *Pyricularia* sp. from *E. coracana*.

P. oryzae causes blast of rice. The fungus is not readily distinguished from the earlier described *P. grisea*. The taxonomy and nomenclature (see Padwick, *Manual of rice diseases*, for a discussion) of these 2 spp. and other *Pyricularia* spp. on Gramineae and other families is confused. H. Asuyama (in Ou) surveyed these fungi and the grass and cereal hosts that they infect. Host specificity has often been reported but the spp. on grasses may be morphologically very similar. Asuyama suggested a classification on the basis of host(s) group, and gave 11 such groups. Full accounts of rice blast were given by Ou (*Rice diseases*, >350 ref.) and Ramakrishnan (*Diseases of rice*). Useful accounts of all aspects of the disease were given at a symposium in 1963 (in Ou); and see *Bull. Tohoku natn. agric. Exp. Stn* 39, 1970 (*Rev. Pl. Pathol.* **50**, 1777–81). Reviews of the epidemiology and weather factors were given by Kato (1974b) and Suzuki (1975), respectively. In 1971 a seminar on the horizontal resistance of rice to *P. oryzae* was held in Colombia (proceedings published in 1975 by Centro Internacional de Agricultura Tropical (CIAT) Cali, Colombia); and see Ezuka. The bibliographies of the works quoted should be consulted, particularly since they contain ref. to work in Japan which is not available in English. Only some more recent papers (and the reviews) are noted here.

Blast, which has been most intensively studied in Japan, is the major fungus disease of rice and it occurs in nearly all cultivated areas (Map 51). The symptoms consist of lesions on leaves, nodes, panicles and grain; but seldom on the leaf sheaths. The leaf spots are elliptical with more or less pointed ends, grey or whitish centres and brown or reddish brown margins. They begin as watersoaked dots and when fully developed are 1–1.5 × 0.3–0.5 cm. Spots on susceptible cvs growing under moist shaded conditions show very little brown margin and may have a chlorotic halo. Small necrotic lesions, which may indicate a resistant host (e.g., Sridhar et al.) or a condition(s) unfavourable for development, have been called chronic, whilst the large, greyish lesions are acute. Heavy spotting kills the leaf and seedlings, or plants at the tillering stage can be killed. Infection at the nodes causes necrosis, stem break and death of those parts above the node. Brown lesions occur on the panicles; infection at the base of the panicle causes neck rot and the panicles often fall over. The pattern of symptoms will vary with the climate of the rice growing region; the amount and distribution of moisture being the most important factors.

Penetration by the germinating conidium is mostly direct through the cuticle. The upper leaf surface and younger leaves are more readily penetrated. The opt. temp. for conidial germination, mycelial growth and sporulation is 28° or a little below. Free water is required for conidial germination and near saturation for infection. Appressoria form in 11 hours at an opt. of 24–25°. The most rapid host penetration (*c.* 16 hours) occurs at this temp. The min. dew period for infection is 5–8 hours at 25–26° (*c.* 12 hours at 15.6° and 9–10 hours at 21.1°). The incubation period is 4–5 days at 26–28° (5–6 days at 24–25° and 7–9 days at 17–18°). Infection is mostly in the dark; light tends to suppress conidial germination and germ tube growth. Sporulation is absent at <RH 93%; it is opt. at *c.* 28° (at >RH 93%). Conidia can be formed *c.* 6 days after inoculation. Av. daily temps of 19–28° are favourable for blast development, see Kato (1974a) and Kato et al. (1974c) for recent work on the effects of temp. on lesion growth and sporulation. The dry conidia are wind and splash dispersed with a peak between 0001 and 0400 hours (and see Ou et al. 1974); dispersal is at a min. between 1400 and 1700 hours. Under natural conditions conidial survival is short, probably <1 day under sunny conditions (e.g., Panzer et al.).

Blast incidence and infection patterns may differ considerably between different geographical regions. In Japan the disease occurs at the 3 plant growth stages: seedling blast in the nursery, leaf blast at tillering and booting, and panicle blast after heading. In the tropics the disease is most destructive at the seedling stage and as panicle blast (neck rot). In cooler areas temp. is an important (critical) controlling factor. In the tropics temps (night temps are important) are generally favourable, rain and RH being the more critical factors. Blast will be more severe where the rainy season(s) is relatively long. In the tropics conidia are present throughout the year; in the temperate zones survival occurs on straw or seed (e.g., Lamey). Under dry conditions (at room temps) conidia can survive for a year and mycelium for almost 3 years. Upland rice in the tropics can get more blast compared with lowland rice, this may be due to the longer dew periods in the former crop. Large amounts of nitrogenous fertiliser tend to increase the disease. Rice plants with a high silica content are more resistant. For recent work on: toxins see the *Rev. Pl. Pathol.* **54**, 1668, 2230; **55**, 3453; **56**, 203; biochemistry, **51**, 2451, 3349; **53**, 1368, 1371, 1372; **54**, 838, 839, 1262; **55**, 221, 3168; and fine structure, **48**, 2371b; **49**, 3265; **50**, 3721; **51**, 1465, 1466; **52**, 3313; **54**, 3301, 3909; **55**, 5180, 5181; **56**, 699, 1114.

Isolates of *P. oryzae* differ considerably in cultural characteristics, nutritional requirements and pathogenicity (Ou, *Rice diseases*, q.v. for an account). Many races have been distinguished in Asia (e.g., India, Japan, Korea, Philippines and Taiwan). The position is made more complex by the many gramineous hosts that have been reported for the fungus (H. Asuyama in Ou). In Asia *P. oryzae* appears to be pathogenically very unstable whilst in USA (where 20 races have been recognised) the position is less so (Marchetti et al.). Manibhushanrao et al. also described variations in resistance patterns due to temp. Kozaka et al. tested the reactions of 691 foreign rice cvs to 21 races (representing the major ones in Asia). Most cvs in each country were generally susceptible to the local races but resistant to those prevalent in distant countries. Japonica cvs differed greatly from indica cvs in reaction to various races; and the indica group could be divided into 2 sections. Veeraraghavan and Yamada et al. described methods for classifying races. Because of this extreme proneness for forming new races increasing attention will no doubt be given to work on horizontal resistance (see Ezuka for a review; Sakai et al.).

Control of rice blast is through cultural methods, resistance and fungicides; see Neergaard et al. for detection on seed. Excessive nitrogenous fertiliser should be avoided and time of planting may affect disease incidence. In the nursery planting measures to prevent a too high RH may be beneficial. The use of resistant cvs will depend on local conditions. The genes controlling resistance to races are generally dominant (see Ou, *Rice diseases*; and Kiyosawa et al. for an account of current issues in breeding for resistance). For the more recent main ref. to work on resistance (breeding, inheritance and races) see *Rev. Pl. Pathol.* **50**, 667, 668, 1776, 2936, 2937, 2938, 3723, 3725; **51**, 348, 1463a & b, 1464, 2454; **52**, 721, 2270, 2934, 3686; **53**, 1370, 3474, 3475; **54**, 2809, 2810, 3302, 4481, 4484; **55**, 731, 1789, 3577; **56**, 1112, 1115, 1594, 4037, 5006; **57**, 547, 588, 589, 3462, 3957. The work on fungicide control is also very extensive and cannot be covered here. A review of early work is given by Kozaka. A brief account, with rates of, and intervals between, applications, is given in *PANS Manual* No. 3, 1976. This includes reference to the antibiotics blasticidin-S and kasugamycin; also edifenphos, pentachlorobenzyl alcohol and benomyl+thiram (seed treatment). Other fungicides include fuji one and tricyclazole (new systemics; Froyd et al.; Tsumura); and the organophosphorous compound, hinosan (Umeda). Resistance in *P. oryzae* to both blasticidin-S and kasugamycin has been described (Hwang et al.; Miura et al; Taga et al.). For predicting outbreaks of rice blast see Kiyosawa; Sasaki et al.; Suzuki 1974.

Ahn, S. W. et al. 1974. *Ann. phytopathol. Soc. Japan* **40**: 337 (effect of near UV light on sporulation in vitro; **54**, 3901).

Chakrabarti, N. K. et al. 1970. *Phytopathology* **60**: 171 (effect of light on sporulation; **49**, 2059).

Chiba, S. et al. 1972. *Ann. phytopathol. Soc. Japan* **38**: 15 (infection rate in the field; **52**, 1135).

Eruotor, P. G. et al. 1978. *Crop Sci.* **18**: 505 (inheritance of resistance to 2 races).

Ezuka, A. 1972. *Rev. Pl. Prot. Res.* **5**: 1 (horizontal resistance, review, 93 ref.; **53**, 160).

Froyd, J. D. et al. 1976. *Phytopathology* **66**: 1135 (systemic fungicide tricyclazole; **56**, 2058).

—— ——. 1978. *Ibid* **68**: 818 (methods for applying tricyclazole).

Fukutomi, M. et al. 1976. *Trans. mycol. Soc. Japan* **17**: 74 (fine structure *Pyricularia* sp., ascus & ascospore walls; **56**, 598).

Hwang, B. K. et al. 1977. *Phytopathology* **67**: 421 (resistance to blasticidin S; **56**, 5624).

Kato, H. 1974a. *Jap. agric. Res. Q.* **8**: 19 (epidemiological aspect of sporulation; **54**, 444).

——. 1974b. *Rev. Pl. Prot. Res.* **7**: 1 (epidemiology, review, 96 ref.; **55**, 732).

—— et al. 1974c. *Phytopathology* **64**: 828 (effect of temp. on lesion enlargement & sporulation; **54**, 445).

—— ——. 1976. *Ann. phytopathol. Soc. Japan* **42**: 507 (perfect state; **56**, 2894).

Kiyosawa, S. 1972. *Ibid* **38**: 30 (technique for forecasting epidemics; **52**, 402).
—— et al. 1975. *Agric. Hort. Tokyo* **50**: 25, 258, 629 (current issues in breeding for resistance, many ref.; **56**, 2497–99).
—— ——. 1977. *Ann. phytopathol. Soc. Japan* **43**: 508, 517, 549 (tests for field resistance & ethylene evolution in infected, stunted plants; **57**, 5481, 5482, 5483).
Kozaka, T. 1969. *Rev. Pl. Prot. Res.* **2**: 53 (chemical control, review, 44 ref.; **49**, 1369).
—— et al. 1970. *Bull. natn. Inst. agric. Sci. Tokyo* Ser. C. 24: 113 (reactions of foreign rice cvs to major races in Asia; **51**, 2455).
Lamey, H. A. 1970. *Pl. Dis. Reptr* **54**: 931 (on seed; **50**, 2941).
Manibhushanrao, K. et al. 1972. *Phytopathology* **62**: 1005 (low night temp. & disease development; **52**, 1528).
Marchetti, M. A. et al. 1976. *Pl. Dis. Reptr* **60**: 721 (races in USA; **56**, 2495).
Miura, H. et al. 1975. *Ann. phytopathol. Soc. Japan* **41**: 415 (resistance to kasugamycin; **55**, 4084).
—— ——. 1976. *Ibid* **42**: 117 (as above; **55**, 5734).
Namai, T. et al. 1977. *Ibid* **43**: 175 (light & sporulation; **57**, 1175).
Neergaard, P. et al. 1970. *Proc. int. Seed Test. Assoc.* **35**: 157 (detection on seed; **50**, 1041).
Ohmori, K. et al. 1970. *Ibid* **36**: 319 (effect of light on sporulation; **50**, 3724).
Oran, Y. K. 1975. *Bitki Koruma Bült.* Suppl. 1, 49 pp. (general in S.E. Anatolia; **55**, 4091).
Ou, S. H. (technical editor) 1965. *The rice blast disease* (international symposium, Philippines, 1963), Johns Hopkins Press, 507 pp. (general reviews on all aspects; **44**, 2131).
—— et al. 1970. *Pl. Dis. Reptr* **54**: 1045 (races derived from monoconidial cultures; **51**, 2456).
—— ——. 1974. *Ibid* **58**: 544 (periodicity of conidial release; **54**, 1261).
Panzer, J. D. et al. 1976. *Riso* **25**: 257 (epidemiology in USA; **56**, 1116).
Sakai, Y. et al. 1974. *Bull. Hiroshima prefect. agric. Exp. Stn* 35: 21 (horizontal resistance; **55**, 731).
Sasaki, T. et al. 1971. *Rep. Soc. Pl. Prot. N. Japan* 22: 38 (climate, weather & first infection; **52**, 4065).
—— ——. 1972. *Phytopathology* **62**: 1126 (technique for forecasting epidemics; **52**, 1895).
Sekiguchi, Y. et al. 1970. *Bull. Chugoku agric. Exp. Stn* Ser. E 6: 81 (environment & primary infection; **51**, 346).
Sridhar, R. et al. 1972. *Pl. Dis. Reptr* **56**: 960 (lesions formed by different races on a cv.; **52**, 3686).
Suzuki, H. 1974. *Jap. agric. Res. Q.* 8: 78 (forecasting epidemics; **54**, 3303).
——. 1975. *A. Rev. Phytopathol.* **13**: 239 (weather & epidemiology, review, 51 ref.).
—— et al. 1977. *Trans. mycol. Soc. Japan* **18**: 385 (inoculum potential; **57**, 4947).
Tago, M. et al. 1978. *Phytopathology* **68**: 815 (ascospore analysis of kasugamycin resistance).
Tsumura, T. 1976. *Jap. Pestic. Inf.* 27: 20 (systemic fungicide fuji I; **56**, 2496).
Ueyama, A. et al. 1975. *Trans. mycol. Soc. Japan* **16**: 420 (perfect state, *Pyricularia* sp.; **55**, 4549).
—— ——. 1977. *Ibid* **18**: 312 (mating types; **57**, 2931).
Umeda, Y. 1973. *Jap. Pestic. Inf.* 17: 25 (fungicide hinosan; **53**, 2182).
Veeraraghavan, J. 1975. *F.A.O. int. Rice Comm. Newsl.* 24: 128 (race classification; **55**, 2226).
Yaegashi, H. et al. 1976. *Ann. phytopathol. Soc. Japan* **42**: 511 (perfect state, *Pyricularia* sp.; **56**, 2895).
Yamada, M. et al. 1976. *Ibid* **42**: 216 (race classification; **55**, 5736).
Yamanaka, S. et al. 1976. *Trans. mycol. Soc. Japan* **17**: 483 (light & sporulation; **57**, 1722).
Yoshino, R. et al. 1974. *Bull. Hokuriku natn. agric. Exp. Stn* **16**: 61 (shade, sunshine & disease; **55**, 1244).

PYTHIUM Pringsheim, *Jahrb. wiss. Bot.* 1: 304, 1858.

(synonymy and conservation in Waterhouse 1968 and *Taxon* **17**: 88, 1968)
(Pythiaceae)

MYCELIUM well developed, consisting of much branched hyphae, occasionally bearing appressoria, sometimes forming tangled complexes, irregular toruloid elements and chlamydospores. ZOOSPORANGIUM either entirely filamentous and undifferentiated from the vegetative hyphae, simple or branched, acrogenous or intercalary; or consisting of a series of basal, complex lobulations and a filamentous discharge tube or a well defined sphaeroidal structure sharply distinct from its supporting hypha, and acrogenous, intercalary or laterally sessile, with an emission tube of variable length, sometimes internally proliferous. ZOOSPORES somewhat reniform, each containing a single vacuole and with 2 oppositely directed flagella of *c.* equal length, emerging from a shallow, longitudinal groove, expelled from the sporangium as an undifferentiated mass into a delicate vesicle produced by the tip of the discharge tube and where cleavage and maturation take place; capable of repeated emergence before finally encysting and germinating; probably always monoecious. OOGONIA terminal or intercalary, spherical or subspherical when terminal, ellipsoidal to limoniform when intercalary, smooth walled or variously echinulated, for the most part forming a single oospore with or without conspicuous periplasm. ANTHERIDIA none or 1 to several, hypogynous, monoclinous or diclinous, allantoid, clavate, globose, suborbicular or trumpet shaped, terminal or intercalary, borne on a short or long stalk, or sessile, usually 1 to 4 (may be lacking or if present up to 25) to an oogonium, forming a distinct fertilisation tube. OOSPORES usually borne singly within the oogonium, plerotic or aplerotic, wall smooth or reticulate, thin or inspissate, the granular protoplasm usually bearing a conspicuous reserve globule and a lateral refringement body; upon germination forming 1 or several germ tubes, or zoospores. Saprophytic and parasitic on plant and animal material in water and soil (F. K. Sparrow, *Aquatic Phycomycetes*: 1031, 1960).

The genus was last monographed by Middleton in 1943. Waterhouse gave the diagnoses (or descriptions) and figures from the original papers (1968), a key to 89 spp. and a list of spp. rejected from the key (1967). Literature surveys were given by Rangaswami and Tomkins (*Thanatephorus cucumeris* q.v.). There are several regional accounts of *Pythium* spp.: Argentina (Frezzi), Australia (Teakle, Vaartaja et al.), India (Rao), Japan (Takahashi), Netherlands (Plaats-Niterink), New Zealand (Robertson), South Africa (Wager) and Taiwan (Hsieh). Nine spp. are described but many others cause relatively minor diseases. Some account of general aspects of the genus has been given fairly recently: pathogenicity (Hendrix, F. F. et al. 1973), taxonomy and genetics (Hendrix, F. F. et al. 1974), physiology and biochemistry of growth and reproduction (Hendrix, J. W. 1974), infection (Endo et al.), germination, growth and survival in soil (Stanghellini), isolation from soil (Lumsden et al. 1975) and effect of temp. on oospore formation (Hsu et al.). In recent years much work (with *Phytophthora*) has been done on the effects of sterols; both on vegetative growth and reproduction (e.g., Al-Hassan et al.; Child et al.; Defago et al.; Haskins et al. 1964; Hendrix, J. W. 1964). Only a few selected ref. can be given here.

Pythium spp. are very common soil inhabitants with long survival rates. They have the competitive advantage of a rapid growth rate which allows early colonisation of the more ephemeral nutrient substrates in the soil. They are general parasites with no narrow host specificity, and in any 1 disease syndrome several spp. may often be encountered. It may, therefore, be necessary to determine their relative importance in the aetiology of any given disease. They most commonly occur as factors in, and causes of, both pre- and post-emergent damping off; in general root rots (cortex or whole root) of young plants; and, by invading roots of more established plants, may be factors in replant problems. They occur also as invaders of fruit borne close to soil level, and may cause rotting diseases both in transit and in storage. Less frequently *Pythium* spp. attack growing shoots, leaves and young fruit of annual crops. In a recent study of damping off Burdon et al. found that varying the plant density in a host population had a very similar effect on the frequency of primary infection to that caused by varying the inoculum potential. A simple negative relationship was found between the mean distance separating adjacent plants and both the rate of disease advance and the rate of disease multiplication in a randomly inoculated seedling stand. *Pythium* spp. are often isolated with spp. of *Phytophthora*; and care is needed to distinguish the 2 genera (zoospore emission differs). Examples of disease complexes are: carrot (Barr et al.; Howard et al.; Kalu et al.; Sutton; Wisbey et al.); common bean (Lumsden et al. 1976; Pieczarka et al.) and eucalyptus (Marks et al.).

Control measures are largely preventive. In the seedling and young plant bed vigorous growth, avoidance of overcrowding and prevention of excessive soil moisture are necessary; sterilisation of beds is often desirable. Fungicidal seed treatment and soil fumigation are frequently effective, though in the latter case fumigants may increase damping off owing to selective action against competitors (Gibson et al.). Older plants which lack inherent vigour and balanced nutrients may be severely set back by root infection, and there are several examples of synergism with fungi, viruses and nematodes. Host resistance is not generally high but can be sufficient to be of economic importance.

Agnihotri, V. P. et al. 1967. *Can. J. Microbiol.* **13**: 1509 (effect of N compounds on growth).

Al-Hassan, K. K. et al. 1968. *Mycologia* **60**: 1243 (effect of oil & sterols on sexual reproduction; **48**, 2151).

Ayers, W. A. et al. 1975. *Phytopathology* **65**: 1094 (factors affecting formation & germination of oospores).

Barr, D. J. S. et al. 1975. *Can. Pl. Dis. Surv.* **55**: 77 (in rusty root of carrot; **56**, 455).

Burdon, J. J. et al. 1975. *Aust. J. Bot.* **23**: 899 (effects of host & inoculum densities on frequency of infection foci in damping off; **55**, 4337).

——— ———. 1975. *Ann. appl. Biol.* **81**: 135 (epidemiology of damping off & host density; **55**, 2008).
Child, J. J. et al. 1969. *Mycologia* **61**: 1096 (effect of C & N nutrition on sterol induced sexuality; **49**, 2767).
——— ———. 1969. *Can. J. Microbiol.* **15**: 599 (effect of cholesterol & polyene antibiotics on the permeability of the protoplasmic membrane; **48**, 3323b).
——— ———. 1971. *Can. J. Bot.* **49**: 329 (induction of sexuality by cholesterol; **50**, 3463).
Defago, G. et al. 1969. *Can. J. Microbiol.* **15**: 509 (induction of oogenesis by polyenic antibiotics & interaction with cholesterol; **48**, 3323a).
Endo, R. M. et al. 1974. *Proc. Am. phytopathol. Soc.* **1**: 215 (anatomy, cytology & physiology of infection, 55 ref.).
Fothergill, P. G. et al. 1962. *J. gen. Microbiol.* **29**: 325 (nutrition; **42**: 246).
Frezzi, M. J. 1956. *Rev. Invest. agric. B. Aires* **10**: 113 (in Argentina; **37**: 337).
Gibson, I. A. S. et al. 1961. *Phytopathology* **51**: 531 (anomalous effect of quintozene on damping off; **41**: 181).
Haskins, R. H. et al. 1964. *Can. J. Microbiol.* **10**: 187 (stimulation of sexual reproduction by sterols; **43**, 2533).
——— ———. 1976. *Can. J. Bot.* **54**: 2193 (fine structure of sexual reproduction in *Pythium acanthicum*: **56**, 1489).
Hendrix, F. F. et al. 1970. *Can. J. Bot.* **48**: 377 (distribution in soil, with *Phytophthora*; **49**, 2334).
——— ———. 1973. *A. Rev. Phytopathol.* **11**: 77 (as plant pathogens, 165 ref.).
——— ———. 1974. *Proc. Am. phytopathol. Soc.* **1**: 200 (taxonomy & genetics, 30 ref.).
——— ———. 1974. *Mycologia* **66**: 681 (taxonomic value of reproductive cell size; **54**, 1163).
Hendrix, J. W. 1964. *Science N.Y.* **144**: 1028 (sterol induction of reproduction & growth stimulation, with *Phytophthora*; **44**, 47).
———. 1974. *Proc. Am. phytopathol. Soc.* **1**: 207 (physiology & biochemistry of growth & reproduction, 35 ref.).
Howard, R. J. et al. 1978. *Phytopathology* **68**: 1293 (on carrot).
Hsieh, H. J. 1978. *Bot. Bull. Acad. sin.* **19**: 199 (in Taiwan; **58**, 92).
Hsu, D. S. et al. 1972. *Mycologia* **64**: 447 (effect of temp. on oospore formation).
Kalu, N. N. et al. 1976. *Can. J. Pl. Sci.* **56**: 555 (on carrot; **56**, 1332).
Klemmer, H. W. et al. 1965. *Phytopathology* **55**: 320 (lipid stimulation of sexual reproduction & growth, with *Phytophthora*; **44**, 2049).
Lumsden, R. D. et al. 1975. *Can. J. Microbiol.* **21**: 606 (isolation from soil; **55**, 594).
——— ———. 1976. *Phytopathology* **66**: 1203 (on common bean; **56**, 1812).
McKeen, W. E. 1975. *Can. J. Bot.* **53**: 2354 (fine structure of oogonium; **55**, 2149).
Marks, G. C. et al. 1974. *Am. J. Bot.* **22**: 661 (on eucalyptus; **54**, 5094).
Middleton, J. T. 1943. *Mem. Torrey bot. Club* **20**: 1–171 (monograph; **22**: 373).
Pieczarka, D. J. et al. 1978. *Phytopathology* **68**: 403, 409, 766 (on common bean).
Plaats-Niterink, A. J. Van der 1975. *Neth. J. Pl. Pathol.* **81**: 22 (in Netherlands; **54**, 3734).
Rangaswami, G. 1961. *Pythiaceous fungi* ICAR, New Delhi (review, with *Phytophthora*, bibliog. 112 pp.).
Rao, V. G. 1963. *Mycopathol. Mycol. appl.* **21**: 45 (in India; **43**, 1271).
Ridings, W. H. et al. 1969. *Phytopathology* **59**: 732 (thiamine requirements; **48**, 3322).
Robertson, G. I. 1973. *N.Z. Jl. agric. Res.* **16**: 357, 367 (occurrence & pathogenicity; **53**, 426, 463).
———. 1976. *Ibid* **19**: 97 (pathogenicity; **55**, 4892).
Saksena, R. K. 1941. *Proc. Indian Acad. Sci.* Sect. B **14**: 141 (thiamine & growth; **21**: 91).
——— et al. 1952. *J. Indian bot. Soc.* **31**: 281 (S & N requirements; **32**: 584).
Schlösser, E. et al. 1968. *Arch. Mikrobiol.* **61**: 246 (effect of sterols on metabolism; **48**, 378).
Stanghellini, M. E. 1974. *Proc. Am. phytopathol. Soc.* **1**: 211 (spore germination, growth & survival in soil, 46 ref.).
Sutton, J. C. 1975. *Can. J. Pl. Sci.* **55**: 139 (on carrot; **54**, 4234).
Takahashi, M. 1973. *Rev. Pl. Prot. Res.* **6**: 132 (in Japan, 42 ref.; **55**, 1111).
Teakle, D. S. 1960. *Qd J. agric. Sci.* **17**: 15 (in Australia; **40**: 288).
Vaartaja, O. et al. 1964. *Aust. J. biol. Sci.* **17**: 436 (frequency in soil; **44**, 52).
Wager, V. A. 1931. *Sci. Bull. Dep. Agric. For. Un. S. Afr.* 105, 43 pp. (diseases caused by Pythiaceae; **11**: 330).
Waterhouse, G. M. 1967. *Mycol. Pap.* 109, 15 pp. (key; **47**, 51).
———. 1968. *Ibid* 110, 71 pp. (diagnoses or descriptions & figs. from original papers; **47**, 3741).
Wisbey, B. D. et al. 1977. *Can. J. Pl. Sci.* **57**: 235 (on carrot; **56**, 5260).

Pythium aphanidermatum (Edson) Fitzpatrick, *Mycologia* **15**: 168, 1923.
Rheosporangium aphanidermatum Edson, 1915
Nematosporangium aphanidermatum (Edson) Fitzp., 1923.
(other synonymy in Middleton 1923; *Pythium* q.v.)

MYCELIUM white, well developed and woolly but not as copious or robust as *P. butleri* (q.v.). Hyphae up to 7.5 μ diam. SPORANGIA produced only in liquid culture, often smaller and less profusely branched than those of *P. butleri*, swellings sometimes only forming outgrowths (like buds) from the main sporangial element; encysted zoospores 9 μ diam. OOGONIA usually abundant in culture and in the host, terminal on lateral hyphae, 19–29 (23) μ, wall smooth. ANTHERIDIA terminal or intercalary, barrel shaped or broadly clavate, 9–11 × 10–14 μ, more usually produced adjacent to the oogonium. OOSPORE very loose in the oogonium (17.5 μ diam.), wall 2 μ thick, smooth. Growth on agar media rapid, covering a 9 cm plate in 24 hours at 36°, opt. 34–36° (G. M. Waterhouse & J. M. Waterston, *CMI Descr.* 36, 1964).

Widespread (Map 309) and plurivorous, including many crops: common bean, cotton, cucurbits (cottony leak), eggplant, ginger, maize, papaw, pineapple, safflower, sugarcane, soybean, tobacco (black stem rot), tomato and turf grasses. Plants in 25 families (*CMI Descr.* 36) are attacked and the host range is similar to that of *P. butleri*. In decay of vegetables a luxuriant, cottony, mycelial weft develops and causes an extensive soft rot. This can occur in the field, in transit or in storage; in the first case infection is found more commonly in fruit touching (or close to) the soil. Damping off and root rots of several crops can occur (and see Kim et al. for infection of above ground parts). Temps favourable for infection correspond with those opt. for in vitro growth. On rye grass (*Lolium*) infection was negligible below 20° but severe damage resulted when plants were pre-conditioned for 8–24 hours at 25°, 4–8 hours at 30° or 2–4 hours at 35° (Freeman). Papaw seedlings were killed in 6–8 days at 36° and 11–13 days at 30°. Infection was slight at 25°. *P. aphanidermatum* is one factor (*Phytophthora palmivora* q.v.) in the papaw replant problem in Hawaii (see Trujillo et al.). In soybean most infection was at 24–36°. This was contrasted with infection of the same host by *Pythium debaryanum* and *P. ultimum* (q.v.) where most disease occurred at 15–20° and also at higher temps provided that there was a preconditioning cold treatment for > 4 days (Thompson et al.). Tomato fruit on the soil surface were infected in 3–4 hours at 24° (Pearson et al.).

Direct penetration of the roots has been shown, for example, in bent grass (*Agrostis palustris*) and cotton (Kraft et al.; Spencer et al.). Zoospores are attracted to the root region of elongation, to wounds in the epidermis and to the exposed stele in cut roots; there is a preferential attraction to wounds. The germination tubes point towards the source of stimulation. Glutamic acid, and mixtures of sugars combined with mixtures of amino acids induced the same responses obtained with roots. Exudates from mung bean (*Phaseolus aureus*) seed stimulated mycelial growth from zoospores and the exudates increased virulence towards this plant. Mycelial growth in the soil is opt. at 28–31° and occurs only from a foodbase. Oospores survive 12 months in saturated and air-dried field soil at 4°, and in the latter at 40° for the same time, but survival was only 1 month in saturated soil at 40°. In China a survival in soil of 4 years has been reported. Oospores germinated best at 35° (Stanghellini et al.). Burr et al. described detection and behaviour in soil. The production of pectolytic and cellulytic enzymes, the effect of N sources on growth and the fine structure in *P. aphanidematum* have been described.

Seed treatment with dithiocarbamates may increase plant stands as has been indicated for pea (*Pisum sativum*). Organic mercury fungicides and captan are effective against infection of temperate grasses (cottony blight) and conventional procedures against damping off should be used. Root attack by nematodes may increase infection by *Pythium aphanidematum* as has been indicated in okra attacked by *Meloidogyne* sp. Some resistance was found in common bean (*P. debaryanum* q.v.). Williams et al. found an increased incidence of the fungus when benomyl and related fungicides were used. This was attributed to suppression of soil competitors and antagonists.

Burr, T. J. et al. 1973. *Phytopathology* **63**: 1499 (propagules & density in field soil; **53**, 2835).

Cardoso, E. J. B. N. et al. 1977. *Summa Phytopathol.* **3**: 31 (effect of light & other factors on oospore germination; **57**, 1603).

Chakravarty, D. K. et al. 1967. *Phytopathol. Z.* **59**: 49 (pectolytic enzymes; **46**, 3342).

Chiba, S. et al. 1977. *Bull. Morioka Tob. Exp. Stn* 12: 93 (on tobacco; **57**, 773).

Dennett, C. W. et al. 1977. *Phytopathology* **67**: 1134 (evidence for a diploid life cycle; **57**, 2835).

Drechsler, C. 1925. *J. agric. Res.* **30**: 1035 (morphology & symptoms on cucumber; **5**: 71).

——. 1926. *Phytopathology* **16**: 47 (on eggplant; **5**: 465).

Freeman, T. E. 1960. *Ibid* **50**: 575 (effect of temps on infection of *Lolium* sp.; **40**: 52).

—— et al. 1966. *Pl. Dis. Reptr* **50**: 292 (pathogenicity on cereals, no specialisation; **45**, 2802).

Harter, L. L. et al. 1927. *J. agric. Res.* **34**: 443 (transit disease of common bean; **7**: 2).
Hickman, C. J. 1944. *Trans. Br. mycol. Soc.* **27**: 63 (identity, morphology & host range; **23**: 359).
Hine, R. B. 1965. *Mycologia* **57**: 36 (effect of age & nutrition on endogenous respiration; **44**, 2050).
Kaiser, W. J. et al. 1971. *Pl. Dis. Reptr* **55**: 244 (control in pea; **50**, 3284).
Kim, S. H. et al. 1974. *Phytopathology* **64**: 373 (infection of common bean above ground; **53**, 4186).
Klisiewicz, J. M. 1968. *Ibid* **58**: 1384 (damping off in safflower with other *Pythium* spp.; **48**, 886).
Kobayashi, N. et al. 1974. *Trans. mycol. Soc. Japan* **15**: 358, 365 (encystment of zoospores, fine structure; **55**, 87, 88).
Kraft, J. M. et al. 1967. *Phytopathology* **57**: 86, 374, 866 (infection of roots of *Agrostis palustris*; effect of N on growth with *P. ultimum*; stimulation by green gram exudate; **46**, 2050; 2621; 3599).
Luna, L. V. et al. 1964. *Ibid* **54**: 955 (growth in soil; **44**, 355).
Mitra, M. et al. 1928. *Mem. Dep. Agric. India bot. Ser.* **15**: 79 (on Cucurbitaceae; **7**: 487).
Payak, M. M. et al. 1973. *Phytopathol. Z.* **77**: 65 (maize infection; **53**, 928).
Pearson, R. C. et al. 1973. *Pl. Dis. Reptr* **57**: 1066 (tomato infection; **53**, 2319).
Royle, D. J. et al. 1964. *Can. J. Microbiol.* **10**: 151, 201 (factors affecting behaviour of zoospores on roots; **43**, 2532).
Ruben, D. M. et al. 1978. *Am. J. Bot.* **65**: 491 (fine structure of oospore germination; **58**, 103).
Seshadri, K. et al. 1964. *Indian Jnl expl Biol.* **2**: 178 (site of origin & structure of the flagella; **44**, 1443).
——— ———. 1970. *Mycopathol. Mycol. appl.* **40**: 145 (nuclear structure & behaviour in vegetative hyphae; **50**, 40).
Singh, R. S. 1963. *Indian Phytopathol.* **16**: 48 (on okra & association with *Meloidogyne* sp.; **43**, 1213).
Spencer, J. A. et al. 1967. *Phytopathology* **57**: 1332 (infection of & attraction by cotton roots, with *P. debaryanum, P. ultimum* & *P. paroecandrum*; **47**, 1572).
Stanghellini, M. E. et al. 1973. *Ibid* **63**: 133, 1493, 1496 (germination of oospores & effect of soil water on this & disease incidence; **52**, 3547; **53**, 2836; **54**, 42).
Suzuki, K. et al. 1978. *Ann. phytopathol. Soc. Japan* **44**: 241 (zoospore chemotaxis).
Tasugi, H. et al. 1935–6. *Ibid* **5**: 245 (morphology, symptoms, effects of temp., on common bean; **15**: 339).
Thompson, T. B. et al. 1971. *Phytopathology* **61**: 933 (effect of temp. on infection of soybean with *P. debaryanum* & *P. ultimum*; **51**, 910).
Trujillo, E. E. et al. 1965. *Ibid* **55**: 1293 (effect of papaw residue on papaw root rot & of temp. on infection, with *Phytophthora nicotianae* var. *parasitica*; **45**, 1450).
Turner, M. T. et al. 1968. *Ibid* **58**: 1509 (pectolytic enzymes; **48**, 1068).
Williams, R. J. et al. 1975. *Ibid* **65**: 217 (increased disease in cowpea after benomyl treatment; **54**, 4271).
Winstead, N. N. et al. 1961. *Ibid* **51**: 270 (production of cell wall degrading enzymes in vivo & in vitro; **40**: 662).
Yu, T. F. et al. 1945. *Lingnan Sci. J.* **21**: 45 (on white gourd, *Benincase hispida*, effect of temp. & survival in soil; **24**: 488).

Pythium arrhenomanes Drechsler, *Phytopathology* 18: 874, 1928.

(synonymy in Middleton 1943, *Pythium* q.v.)

HYPHAE rarely >5.5 μ diam.; aerial MYCELIUM strong to profuse, on some agar media lobulate, hyphal complexes are produced. SPORANGIA consist of lobulate complexes of spherical, ovoid or irregular lateral swellings, up to 20 μ diam., connected by shorter or longer elements. Zoospore production in most strs meagre on transference of pieces of agar media to water but more prolific if infected roots are used. Evacuation tube of variable length, usually 50–75 μ long. Zoospores 20–50 or more, 12 μ diam. when rounded up. OOGONIA terminal or more rarely, intercalary, (18–) 24–36 (–56) (av. 30) μ, wall smooth, 0.5 μ thick. ANTHERIDIA numerous, usually 5–15, but may be up to 25/oogonium, radially arranged, crook necked, the rounded apical end 6–9 μ diam. making narrow apical contact and tapering back to the antheridial hyphae (each of which may produce as many as 4 antheridia) all distinct from the oogonial hyphae. After fertilisation the branches become increasingly difficult to see and the antheridia degenerate. OOSPORES not formed in 50% of the oogonia on cornmeal agar, plerotic or slightly aplerotic, (15–) 22–33 (–54) (av. 28) μ, reserve globule 12–19 μ, wall yellowish, 1.2–2 μ. Grows rapidly on agar, 14–16 mm in 24 hours (room temp.); opt. 28–30°. Often found with *P. graminicola* (q.v.) with which it has been compared by Drechsler (1936). (G. M. Waterhouse & J. M. Waterston, *CMI Descr.* 39, 1964.)

P. arrhenomanes is plurivorous but largely restricted to Gramineae (recorded on 30 genera), and in the tropics may particularly be found causing a root rot of maize, sorghum and sugarcane. It is widespread in N. America and has been recorded in S.E. Asia, Africa (S.), Australia, Brazil, Britain, Fiji,

Pythium butleri

Hawaii, Italy, Papua New Guinea and Venezuela (Map 420). It probably occurs in more regions than the records suggest. The disease on sugarcane was reviewed by Rands (1961) and the pathogen (sometimes associated with others) was one of the most important factors in the history of the failures of this crop through root rot. The root tips become necrotic and soft, laterals are killed back, a more general necrosis develops and the root system consists of short stubby roots, deficient in laterals and finer roots. The fungus occurs in both cortex and stele. Root attack, as expected, leads to the non-specific above ground symptoms of poor, unthrifty growth. Infection is more severe at cool temps (*c.* 19°) and becomes progressively less serious at temps above this. Consequently, root rot is a more important factor in the cooler, cane growing regions; especially in the winter months when root growth is slow, i.e. where forced dormancy can occur. Antagonism by other micro-organisms has been investigated. Sugarcane isolates were more tolerant of higher temps than those from maize. Although str. differences in pathogenicity to the former host have been shown, physiologic specialisation sensu stricto probably does not occur. The disease can still be severe on heavy soils and resistant vars are commonly used; this resistance is indirect, i.e. host adaptation to particular field conditions. Cultural measures may be required in control, especially at planting.

On maize and sorghum the fungus causes pre- and post-emergent damping off, and a root rot of older plants which can reduce size, vigour and yield. It had been considered to be implicated in the aetiology of milo disease of sorghum now shown to be due to *Periconia circinata* (q.v.). As in the case of sugarcane root rot this pathogen causes more serious disease at temps appreciably lower than the opt. for growth in vitro would suggest. Resistant vars have been developed in maize and sorghum and standard seed treatment and cultural measures may be necessary. Five *Pythium* spp. were pathogenic to pineapple in Hawaii (Klemmer), *P. arrhenomanes* being the most virulent, and field fumigation has been recently reported against this sp. in that country.

Anderson, E. J. 1966. *Down to Earth* **22**: 23 (use of fumigant 1-3 dichloropropene, 1-2 dichloropropane; **46**, 3351).

Bowman, D. H. et al. 1937. *J. agric. Res.* **55**: 105 (inheritance of resistance in sorghum; **16**: 807).

Cooper, W. E. et al. 1950. *Phytopathology* **40**: 544 (antagonism of actinomycetes; **29**: 583).

Elliot, C. 1942. *J. agric. Res.* **64**: 711 (relative susceptibility in maize inbreds; **21**: 449).

Hsu, J. C. 1963. *Rep. Taiwan Sug. Exp. Stn* 32: 143 (effect of time of year & soil types in inoculations on sugarcane; **44**, 826).

Johan, H. et al. 1928. *J. agric. Res.* **37**: 443 (symptoms, effect on maize & of temp.; **8**: 375).

Johnson, L. F. 1954. *Phytopathology* **44**: 69 (antagonism of micro-organisms; **33**: 667).

Klemmer, H. W. et al. 1964. *Pl. Dis. Reptr* **48**: 848 (pathogenicity to pineapple of 8 *Pythium* spp.; **44**, 1184).

Luke, H. H. et al. 1954. *Phytopathology* **44**: 377 (antagonism of micro-organisms; **34**: 183).

Napi-Acedo, G. et al. 1965. *Philipp. Agric.* **49**: 279 (penetration & infection of maize roots; **46**, 118).

Rands, R. D. et al. 1934. *J. agric. Res.* **49**: 189 (variability between isolates & effect of temp.; **14**: 94).

—— ——. 1938. *Ibid* **56**: 53 (in sugarcane, effect of soil constituents; **17**: 487).

—— ——. 1938. *Tech. Bull. U.S. Dep. Agric.* 666, 96 pp. (general on sugarcane, 102 ref.; **18**: 343).

——. 1961. In *Sugarcane diseases of the world* Vol. 1: 289 (review, 34 ref.).

Roldan, E. F. 1932. *Philipp. Agric.* **21**: 165 (general on maize; **11**: 777).

Van der Zwet, T. 1958. *Phytopathology* **48**: 345 (effect of flooding in unsterile soil; **37**: 718).

Wagner, F. A. 1936. *J. Am. Soc. Agron.* **28**: 643 (resistance in sorghum; **16**: 33).

Pythium butleri Subramaniam, *Mem. Dep. Agric. India bot. Ser.* **10**: 193, 1919.

Aerial MYCELIUM white, copious and robust, especially on the host; hyphae up to 9 μ diam., appressoria abundant, clustered with spherical ends; in the host inter- and intracellular without haustoria. SPORANGIA produced only in liquid media, composed of lateral hyphal elements 40–150 μ long, with basal part (about one half) crowded with up to 6 lobulate branches, the apical region tapering in width (15–6 μ) at the tip, where the contents pass into an apical vesicle before the zoospores are delimited; encysted zoospores 11 μ diam., germinating by hyphae or by a single zoospore (repeated emergence). OOGONIA usually abundant in the host and produced readily in agar media on short (4 μ) or long (up to 70 μ) lateral hyphae, 15–33 (26.5) μ diam., wall colourless thin and smooth. ANTHERIDIA usually single, intercalary, commonly diclinous (a few monoclinous), the hyphal part 9–22 μ long, the inflated part 10–17 μ deep, barrel shaped or irregularly lobed, broadly applied to the oogonium with little or no tapering, persistent. OOSPORES very loose in the oogonium, av. 22.5 μ diam., wall smooth, colourless or slightly yellowish, 2 μ thick; when mature has one central, hyaline globule (from C. Drechsler 1955; Subramaniam, 1919 (l.c.); and see Waterhouse et al.).

The fungus grows rapidly on agar (opt. 34°) covering a 10 cm plate in 24 hours; good zoospore production at 18–30°. Distinguished from *P. aphanidermatum* (q.v.) in the more copious, robust aerial mycelium; the larger, more swollen and more branched sporangia; larger zoospores, oogonia and oospores; larger, usually diclinous, antheridia (Waterhouse et al.).

Sprague (*Diseases of cereals and grasses in North America*) stated that *P. butleri* is probably widely distributed in the tropics and subtropics. Various diseases are caused: foot rot of papaw, soft rot of ginger rhizome, root rot of grasses, damping off of tobacco, tomato, and red pepper, cottony rot of cucurbit fruit, maize stalk rot and stem rot of common bean. The pathogen causes most severe disease at the relatively high temp. which is opt. for growth in vitro; this has been shown for *Agrostis, Poa* and common bean. On bean a characteristic soft cortical rot develops on the stems, the roots are also attacked but in maize only stem infection seems to occur. Some maize inbreds are more resistant than others and varietal differences in bean have been noted. There are no special control measures.

Desai, B. G. et al. 1969. *Phytopathol. Mediterranea* **8**: 238 (on fruit of *Luffa cylindrica*; **49**, 1862).
Drechsler, C. 1955. *Sydowia* **9**: 451 (detailed description & distinction from *Pythium aphanidermatum*; **35**: 849).
Elliot, C. 1943. *J. agric. Res.* **66**: 21 (general on maize; **22**: 202).
Harter, L. L. et al. 1931. *Phytopathology* **21**: 991 (general on common bean, effect of temp. **11**: 218).
Monteith, J. 1933. *Ibid* **23**: 23 (on *Agrostis*, effect of temp.; **12**: 450).
Oshima, N. et al. 1969. *Pl. Dis. Reptr* **53**: 766 (outbreak on common bean, effect of temp.; **11**, 218).
Robbins, W. J. et al. 1938. *Bull. Torrey, bot. Club* **65**: 453 (thiamine & growth; **18**: 268).
Waterhouse, G. M. et al. 1964. *C.M.I. Descr. pathog. Fungi Bact.* 37 (*P. butleri*).

Pythium debaryanum (as '*de Baryanum*') Hesse, *Maug. Diss. Halle*: 14–34 (after Oct.) 1874. (synonymy in Middleton 1943, *Pythium* q.v.)

SPORANGIA always spherical and terminal on side branches cut off by a cross wall at the base, with a short, blunt ended process 15–26 (19) μ diam. CONIDIA colourless, terminal or intercalary, terminal ones catenulate and deciduous, with a moderately thick wall. Ellipsoidal, intercalary cells (or conidia) do not form zoospores but germinate by a tube after being shed. OOGONIA spherical, terminal on short side branches or intercalary, with a smooth thin, entire wall, 15–28 (21) μ diam. ANTHERIDIA 1–6 per oogonium, mono- and diclinous, when monoclinous arising some distance below the oogonium, not adjacent to it; antheridial cell 20 μ long, not much swollen, straight, point of contact subterminal. OOSPORE aplerotic with a small eccentrically placed vacuole, wall clear, thin, smooth, 12–20 (17) μ diam. (Middleton l.c., Waterhouse 1968, *Pythium* q.v., and see Butler).

Middleton (l.c.) stated that *P. debaryanum* and *P. ultimum* are the spp. most often encountered. Also that the former is distinguished from the latter by its plurality of antheridia which are mono- or diclinous; when monoclinous they are stalked and originate some distance from the oogonium. *P. ultimum* usually has only a single antheridium which characteristically arises adjacent to the oogonium; it is sessile and peculiarly shaped. Drechsler described the formation of zoospores from oospores and compared development with *P. ultimum* (q.v.) and *P. irregulare*. Sansome described meiosis.

The fungus is plurivorous (> 130 hosts in *Rev. appl. Mycol. Host–Pathogen index* Vols 1–40, 1922–61) and widespread (Map 208). The best temps for in vitro growth are 28–32° but those reported to be most favourable for infection are variable and temp. adaptation in different geographic regions may occur, although this aspect has not been fully investigated. In temperate zones 10–20° is generally favourable for damping off but higher temps (24–28°) have been found best for post-emergent damping off. One report shows a higher disease rating at 28° compared with 24°, 20°, and 16° on common bean (Adegbola 1969; and see Buchholtz; Chi et al.; Greeves et al.; Halpin et al.). Also, on soybean, *P. debaryanum* and *P. ultimum* were more virulent at 15–20°, but by pre-conditioning at 4–12° for 4 days they became virulent at 24–36° which were the most favourable temps for pathogenicity of *P. aphanidermatum* (q.v.).

P. debaryanum is most commonly found as a cause of damping off but will also attack a plant beyond the seedling stages, both above and below ground. In tests on clover seedlings this sp. was found to be more pathogenic than *P. ultimum, P. splendens* (q.v.) and 4 other *Pythium* spp. (Halpin). In cotton, with *P. aphanidermatum*, zoospores were attracted to the elongation and root hair zones of unwounded radicles but wounding greatly increased attraction. The pectolytic enzymes, antagonism of soil micro-organisms and fine structure have been studied. Variations in pathogenicity between isolates have often been reported and more recently in those from cotton, forest nurseries and common bean. Variations in host resistance also occur, for example, in cvs of bean and soybean. Treatment with methyl bromide or fenaminosulf improved root and shoot growth of young cotton plants with a root rot mainly caused by *P. debaryanum* and *P. irregulare* (Roncadori et al.). This beneficial effect was rather more at a lower temp. regime than at a higher one.

Adegbola, M. O. K. 1969. *Phytopathology* **59**: 1113, 1484 (temp. & infection of common bean, compared with *Pythium aphanidermatum, P. rostratum, P. ultimum*; penetration; **49**, 297, 1215).

——. 1970. *Ibid* **60**: 1477 (5 *Pythium* spp. and resistance in common bean, variation in virulence; **50**, 1505).

Ashour, W. E. 1954. *Trans. Br. mycol. Soc.* **37**: 343 (pectolytic enzymes with *Botrytis cinerea*; **34**: 666).

Buchholtz, W. F. 1938. *Phytopathology* **28**: 448 (temp. & pathogenicity to beet; **18**: 5).

Butler, E. J. 1913. *Mem. Dep. Agric. India bot. Ser.* **5**: 262 (morphology).

Chi, C. C. et al. 1962. *Phytopathology* **52**: 985 (effect of leguminous seedling age, temp. & soil moisture on infection; **42**: 324).

Cooper, B. A. et al. 1967. *Mycologia* **59**: 658 (cell wall structure; **46**, 3677).

Drechsler, C. 1952. *Bull. Torrey bot. Club* **79**: 431 (development of germinating oospores with *P. ultimum*; **32**: 161).

——. 1953. *J. Wash. Acad. Sci.* **43**: 213 (morphological comparison with *P. irregulare* & *P. ultimum*; **32**: 638).

Greeves, T. N. et al. 1936. *Ann. appl. Biol.* **23**: 264 (temp. & damping off in *Brassica*; **15**: 769).

Gupta, S. C. 1956. *Ann. Bot.* **20**: 179 (pectolytic enzymes; **35**: 385).

——. 1958. *Proc. Indian Acad. Sci.* Sect. B **47**: 43 (as above; **37**: 452).

Halpin, J. E. et al. 1952, 1954, 1958. *Phytopathology* **42**: 245; **44**: 572; **48**: 481 (effect of host age & temp. on pathogenicity of 7 *Pythium* spp., including *P. arrhenomanes, P. debaryanum, P. splendens* & *P. ultimum*, to clover & lucerne; **32**: 22; **34**: 302; **38**: 149).

Hawker, L. E. 1963. *Nature Lond.* **197**: 618 *J. gen. Microbiol.* **31**: 491 (fine structure).

Kerling, L. C. P. 1947. *Trans. R. Soc. S. Aust.* **71**: 253 (on pea, morphology with 3 *Pythium* spp. **28**: 203).

Lehman, S. G. et al. 1926. *J. agric. Res.* **33**: 375 (morphology, on soybean; **6**: 75).

Likais, R. 1952. *Arch. Mikrobiol.* **18**: 49 (effect of soil & antagonism; **33**: 627).

McElroy, F. D. et al. 1971. *Phytopathology* **61**: 586 (on carrot; **51**, 955).

Miyake, K. 1901. *Ann. Bot.* **15**: 653 (cytology & fertilisation).

Roncadori, R. W. et al. 1972. *Phytopathology* **62**: 373 (effect of soil fumigants, temp. & host age in cotton root rot, with *P. irregulare*; **51**, 4045).

Sansome, E. 1963. *Trans. Br. mycol. Soc.* **46**: 63 (meiosis; **42**: 531).

Strissel, J. F. et al. 1970. *Phytopathology* **60**: 961 (stunting of soybean; **50**, 406).

Vaartaja, O. et al. 1961. *Ibid* **51**: 505 (pathogenicity of isolates from forest nurseries, with *P. ultimum*; **41**: 181).

Van Luijk, A. 1934. *Meded. phytopathol. Lab. Willie Commelin Scholten* **13**: 23 (taxonomy & nomenclature with *P. ultimum*; **14**: 259).

Wood, R. K. S. et al. 1958. *Ann. Bot.* **22**: 309 (pectolytic enzymes).

Pythium graminicola Subramaniam (as *P. graminicolum*), *Bull. agric. Res. Inst. Pusa* 177: 5, 1928.

HYPHAE up to 7 μ diam., aerial MYCELIUM strong to profuse; no swellings on agar media. SPORANGIA formed only in water and consisting of lobulate complexes aggregated or strung out on the hypha. Zoospores readily produced, 15–48/sporangium, 8–11 μ diam. when encysted, with repeated emergence. OOGONIA terminal or intercalary 17–36 (25) μ diam., wall smooth, persistent. ANTHERIDIA 1–3 (–6) usually monoclinous, occasionally diclinous, crook necked, making flattened apical contact in the equatorial region of the oogonium. Persisting after fertilisation. OOSPORES in all the oogonia, usually plerotic, sometimes aplerotic, 15–35 (23) μ, wall may become slightly brownish, 2 μ thick.

Grows rapidly on agar, 10–15 mm in 24 hours (room temp.); opt. 28–30°. Often found with *P.*

arrhenomanes (q.v.) from which it differs in the fewer, usually monoclinous, persistent antheridia, smaller oogonia, all fertile, and the ready formation of zoospores (G. M. Waterhouse & J. M. Waterston, *CMI Descr*. 38, 1964).

P. graminicola, like *P. arrhenomanes*, is very largely restricted to Gramineae, and both are frequently associated in the same diseased conditions of damping off, seedling root blight and progressive deterioration of the root system of older plants. Drechsler compared their morphology. *P. graminicola* is widespread (Map 296) in N. America, E. and S. Asia; it is also recorded from Argentina, Australia, Belize, Brazil, Guyana, Hawaii, Jamaica, Mauritius, Puerto Rico, Salvador, South Africa, Sudan and Europe. It may be found on sugarcane (where *P. arrhenomanes* is more important), maize, millet, rice and sorghum. Host lists on grasses were given by Sprague (*P. butleri* q.v.) and Couch (*Diseases of turf grasses*). The pathogen has been studied in detail on cereals in N. America. Some non-grass crops on which it has been reported are: arrowroot, cotton, ginger, lucerne, onion, pea, pineapple, soybean and turmeric.

Considerable damage can be done to grasses and cereals where these are grown continuously. On *Bromus inermis* there was less root rot at 21.3° than at 27.2° (Hawk et al.). In soil the pathogen was most frequent at a depth of 8–15 cm, very infrequent at 68–76 cm and more abundant under maize than from lucerne and Kentucky blue grass (*Poa pratensis*). Isolates show host specialisation (Hampton et al. 1962) to maize and *Setaria* as well as to temperate cereals, and cvs differ in their reaction to infection. The effects on growth of sugarcane have been compared with those caused by *Pythium tardicerescens* Vanterpool (Koike). *P. graminicola* caused more shoot height reduction alone than when present with *P. tardicrescens*. Also on this crop either additive and/or synergistic effects have been noted with nematode infection by *Meloidogyne incognita*, *Helicotylenchus dihystera* and *Pratylenchus zeae*. The pathogenicity of *P. brachyurus* with *Pythium graminicola* has also been reported. Both additive and synergistic effects, detrimental to growth, have been found in infection with sugarcane mosaic virus.

Apt, W. J. et al. 1962. *Phytopathology* **52**: 798, 1180 (pathogenicity with nematodes; **42**: 278, 341).

Drechsler, C. 1936. *Ibid* **26**: 676 (morphological comparison with *Pythium arrhenomanes*; **15**: 825).

Erwin, D. C. et al. 1957. *Pl. Dis. Reptr* **41**: 988 (susceptibility in 5 maize vars; **37**: 407).

Hampton, R. O. et al. 1959. *Iowa St. Coll. J. Sci.* **33**: 489 (seasonal occurrence on maize; **38**: 593).

—— ——. 1962. *Iowa St. J. Sci.* **37**: 43 (host specialisation on cereals; **42**: 313).

Hawk, V. B. et al. 1949. *J. Am. Soc. Agron.* **40**: 721 (temp. & infection of *Bromus inermis*; **28**: 17).

Knaphus, G. et al. 1958. *Iowa St. Coll. J. Sci.* **33**: 201 (distribution in soil; **38**: 396).

Koike, H. et al. 1970. *Phytopathology* **60**: 1562 (pathogenicity with *Pratylenchus brachyurus*; **50**, 1953).

—— ——. 1971. *Ibid* **61**: 1090 (pathogenicity with sugarcane mosaic virus, str. H; **51**, 1855).

——. 1971. *Pl. Dis. Reptr* **55**: 766 (pathogenicity with *Pythium tardicrescens*; **51**, 1865).

——. 1971. *Sugcane Pathol. Newsl.* 7: 29 (infection by *P. tardicrescens*; **51**, 2827).

Lenney, J. F. et al. 1966. *Nature Lond.* **209**: 1365 (nutrition, sexual reproduction & vegetative growth; **45**, 2398).

Ramakrishnan, T. S. et al. 1955. *Indian Phytopathol.* **7**: 111 (on turneric; **35**: 396).

Santo, G. S. et al. 1970. *Phytopathology* **60**: 1537 (pathogenicity with *Pratylenchus zeae*).

Srinivasan, K. V. 1958. *Madras agric. J.* **45**: 89 (on sugarcane; **38**: 422).

Pythium mamillatum Meurs, Proefschr. Verkrijg. Graad Doct. Rijks-univ. Utrecht: 44, 1928.
(further details in *Phytopathol. Z.* **1**: 111, 1929)

MYCELIUM in culture moderately dense, uniform or growing in rosette fashion in some media. HYPHAE up to 9 μ diam. SPORANGIA spherical, broadly ellipsoidal or broadly ovoid, terminal or intercalary, non-papillate, 17–21 μ diam. collapsing after evacuation, discharge tube very short. Also hyphal bodies of all shapes and sizes (spherical up to 27 μ diam., ellipsoidal up to 42 μ long). OOGONIA abundant, spherical, terminal or occasionally intercalary, rarely ellipsoidal, 13–20 (17) μ diam., wall with projections which usually cover the oogonium but are occasionally few, of irregular length 2–7 (4.5) μ, very slightly tapered and rounded at the tip often bent or curved. ANTHERIDIA usually single, monoclinous or diclinous, stalked, end slightly inflated (to 6 μ), spherical or oval, sometimes crook necked, applied by the apex to near the equator of the oogonium. OOSPORE nearly plerotic or slightly aplerotic, wall smooth, 1 μ thick. Growth on agar fairly rapid (25 mm in 24 hours at 25°); opt. 28–30° (G. M. Waterhouse & J. M. Waterston, *CMI Descr.* 117, 1966).

The fungus is plurivorous (see *CMI Descr.* 117 for hosts) with a wide distribution. Damping off and root rot have been caused in cauliflower, cotton, cucumber, lucerne, onion, pea, pineapple and sugarcane. The fungus has been investigated in Australia as a cause of a seedling blight of pea (with *P. ultimum* q.v.). It was found that isolates from different soils differed in pathogenicity. *P. mamillatum* is one of the organisms involved in damping off in S. Australian nurseries with, inter alia, *P. irregulare* and *P. ultimum.* It is a pioneer coloniser of organic matter in the soil; high soil moisture, low temp., pH 5.5–7 and a fine soil texture are opt. for saprophytic colonisation. Exudates from turnip seedlings stimulate oospores to germinate.

Angell, H. R. 1951. *Aust. J. agric. Res.* **2**: 286 (effect of source of isolates on pea & poppy; **31**: 459).
——. 1952. *J. Aust. Inst. agric. Sci.* **18**: 99 (effect of watering methods on hosts above; **32**: 601).
——. 1952. *Aust. J. agric. Res.* **3**: 128 (effect of soil on hosts above; **32**: 601).
Barton, R. 1957. *Nature Lond.* **180**: 613 (oospore stimulation by turnip; **36**: 802).
——. 1960. *Trans. Br. mycol. Soc.* **43**: 529 (effect of substrate composition in soil; **40**: 156).
——. 1961. *Ibid* **44**: 105 (restriction to pioneer colonisation of substrates; **40**: 592).
Hickman, C. J. 1944. *Ibid* **27**: 49 (morphology; **23**: 359).
Vaartaja, O. 1967. *Phytopathology* **57**: 765 (damping off in nurseries in S. Australia; **46**, 3226).

Pythium myriotylum Drechsler, *J. Wash. Acad. Sci.* **20**: 404, 1930.

MYCELIUM copious on agar or in water. HYPHAE straight or sinuous, not much branched except at the terminations, up to 8.5 μ diam., forming numerous clavate or knob-like appressoria, often in branching clusters. SPORANGIA terminal or intercalary, consisting of long, simple or branched portions of unswollen hyphae or (often) with lateral, swollen, lobulate or digitate branches up to 17 μ diam. Evacuation tube long and slender (up to 3.5 μ), widening slightly at the apex (6 μ). OOGONIA abundant, terminal or (less often) intercalary, 15–44 μ, mostly 23–30 (26.5)μ diam., wall smooth. ANTHERIDIA up to 10, usually 3–6/oogonium, usually diclinous, stalks slender (1–2 μ), often once or twice branched, investing the oogonium, the end slightly swollen, usually crook necked, making apical and sometimes basal contact, eventually disappearing. OOSPORE aplerotic, 12–37 (21)μ diam., with a single reserve globule, wall up to 2 μ thick, smooth, contents often pale golden. Growth very rapid, particularly at high temp. (40 mm in 24 hours at 35–40°). Opt. *c.* 37°. Differs from *P arrhenomanes* (q.v.) and *P. graminicola* (q.v.) in growth at cardinal temps (G. M. Waterhouse & J. M. Waterston, *CMI Descr.* 118, 1966).

The fungus is plurivorous and probably fairly widespread in warm climates. It causes seedling root rots in common bean, grasses, lucerne, papaw, soybean, tobacco and tomato; ginger rhizome rot, and fruit rots in eggplant, cucumber and watermelon. *P. myriotylum* is also a cause of breakdown of groundnut pods; amongst other soil inhabitants it was considered the most important and can also cause a wilt of the same host (Frank; Garcia et al.; Garren; Porter). On common bean damping off was max. at 30–35° (Gay; and see Dow et al.; Drechsler 1952). In comparisons with *P. aphanidermatum* (q.v.) in 12 crops, isolates differed in pathogenicity and the 2 pathogens could differ in the degree of attack on a single host. *P. myriotylum* was highly virulent to tomato seedlings at 35° but was less so at 23–31° than *P. aphanidermatum* which was most virulent at 27–35° (Littrell). The former survived longer in 2 soils to which it is not indigenous compared with a soil in which it is. Csinos et al. found evidence for a phytotoxin in work on tomato.

Csinos, A. et al. 1978. *Can. J. Bot.* **56**: 2334 (on tomato, evidence for toxin).
Dow, R. L. et al. 1975. *Ibid* **53**: 1786 (histology in common bean; **55**, 4367).
Drechsler, C. 1943. *Phytopathology* **33**: 261 (morphological comparison with other *Pythium* spp.: **22**: 373).
——. 1952. *Pl. Dis. Reptr* **36**: 13 (stem rot in common bean with *P. butleri* & *P. ultimum*; **31**: 366).
Frank, Z. R. 1972. *Pl. Dis. Reptr* **56**: 600 (on groundnut with *Fusarium solani*; **52**, 938).
Garcia, R. et al. 1975. *Phytopathology* **65**: 1375 (on groundnut, interactions with other fungi; **55**, 3821).
Garren, K. H. 1966. *Phytopathol. Z.* **55**: 359 (pod rot in groundnut & associated soil fungi: **46**, 204).
——. 1967. *Pl. Dis. Reptr* **51**: 601 (as above; **46**, 3287).
——. 1970. *Ibid* **54**: 840 (pod rot in groundnut with *Rhizoctonia solani*; **50**, 1526).

——. 1971. *Phytopathology* **61**: 596 (persistence in indigenous & non-indigenous soils; **50**, 3502).
Gay, J. D. 1969. *Pl. Dis. Reptr* **53**: 707 (effect of temp. & soil moisture on common bean damping off: variation in pathogenicity; **49**, 298).
Littrell, R. H. 1970. *Phytopathology* **60**: 704 (effect of temp. on tomato infection & comparison with *P. aphanidermatum*; **49**, 3129).
McCarter, S. M. et al. 1968. *Pl. Dis. Reptr* **52**: 179 (damping off in several hosts; **47**, 2434).
—— ——. 1970. *Phytopathology* **60**: 264 (comparative pathogenicity with *P. aphanidermatum*, variation in isolates; **49**, 2398).
Porter, D. M. 1970. *Ibid* **60**: 393 (wilt of groundnut; **49**, 2250).
Shahare, K. C. et al. 1962. *Indian Phytopathol.* **15**: 77 (on ginger; **42**: 41).
Southern, J. W. et al. 1976. *Phytopathology* **66**: 1380 (on soybean with *P. irregulare*; **56**, 4284).
Teakle, D. S. 1956. *Qd J. agric Sci.* **13**: 241 (on lucerne; **37**: 244).

Pythium splendens Braun, *J. agric. Res.* **30**: 1061, 1925.

MYCELIUM fairly copious and thick. HYPHAE straight and not greatly branched, up to 7.5 μ broad, bearing terminal, or occasionally intercalary, ovoid CHLAMYDOSPORES up to 55 μ diam., av. 36 μ, with a thin, smooth, colourless wall and sometimes a short (up to 4 μ) pedicel, with dark dense, granular contents. SPORANGIA rarely produced. SEX ORGANS also rarely produced. OOGONIA 25–38 (35)μ diam., wall smooth thin. ANTHERIDIA 1–8, usually monoclinous or diclinous, straight or crook necked, large, 16 × 12 (often up to 20 × 15)μ. OOSPORE very aplerotic, av. 26 μ diam., wall smooth, 1–2 μ thick. Moderate growth on agar media (20 mm in 24 hours at 27–32°), opt. *c.* 30° (G. M. Waterhouse & J. M. Waterston, *CMI Descr.* 120, 1966).

The fungus is plurivorous (see *CMI Descr.* 120 for hosts) and widespread, especially in warm temperate and tropical areas. It causes damping off in a large number of crops which include aroids, Easter lily, lucerne, maize, oil palm, papaw, pineapple and safflower. The disease (blast) on oil palm seedlings (particularly at 6–8 months) has caused severe losses in Africa and more recently in Malaysia. *P. splendens* invades the root system causing a cortical rot which stops at the bulb. This attack is followed by *Rhizoctonia lamellifera* (*Macrophomina phaseolina* q.v. and see *CMI Descr.* 275), both organisms being necessary for the complete syndrome. Above ground the leaves lose their colour, becoming pale yellow green and then necrotic. A toxin has been implicated in the aetiology. Cultural measures are used in control, the most important being the use of shade; where this is inadequate blast is much more severe. Besides lowering the soil temp. satisfactory soil moisture levels and host nutrition should be maintained (Aderungboye et al.; Bachy; Rajagopalan; Robertson; Turner).

P. splendens is one pathogen found in safflower root rot and temp. effects the relative susceptibility of safflower cvs. In studies on clover and lucerne with 7 *Pythium* spp. *P. splendens* was one of the most pathogenic (Halpin, *P. debaryanum* q.v.). In cucumber seedling infection a lignin barrier to spread has been described. Control by soil fumigation of the root rot on Chinese evergreen (*Aglaonema simplex*) is reported (Miller; Tisdale et al.) and root rots also occur in aroids and Easter lily (*Lilium longiflorum* var. *eximum*). *P. splendens* has been described as heterothallic which may be the exception in a genus hitherto considered to be largely homothallic; 3 other *Pythium* spp. are also heterothallic (Plaats-Niterink).

Aderungboye, F. O. et al. 1976. *Pl. Soil* **44**: 397 (ecology on oil palm; **56**, 1691).
—— ——. 1976. *Niger. J. Pl. Prot.* **2**: 8 (control on oil palm; **56**, 5755).
Bachy, A. 1958. *Oléagineux* **13**: 653 (effect of shade on incidence of blast of oil palm; **37**: 733).
McClure, T. T. et al. 1942. *Bot. Gaz.* **103**: 684 (resistance in seedlings of cucumber).
Miller, H. N. 1958. *Proc. Fla St. hort. Soc.* **71**: 416 (control by soil fumigation; **38**: 696).
Moreau, C. et al. 1958. *Rev. Mycol.* **23**: 201 (blast of oil palm; **38**: 157).
Plaats-Niterink, A. J. Van Der 1968. *Acta bot. neerl.* **17**: 320 (heterothallism in *Pythium heterothallicum*, *P. intermedium* & *P. sylvaticum*; **48**, 387).
——. 1969. *Ibid* **18**: 489 (heterothallism in *P. splendens*; **49**, 690).
Rajagopalan, K. 1974. *Jnl Niger. Inst. Oil Palm Res.* **5**: 23 (effect of irrigation & shade on oil palm; **56**, 334).
Robertson, J. S. 1959. *Jl W. Afr. Inst. Oil Palm Res.* **2**: 310 (full account of blast of oil palm; **39**: 184).
——. 1959. *Trans. Br. mycol. Soc.* **42**: 401 (co-infection of oil palm by *P. splendens* and *Rhizoctonia lamellifera*; **39**: 486).
Thomas, C. A. 1970. *Pl. Dis. Reptr* **54**: 300 (effect of temp. on infection of safflower; **49**, 2959).

Tisdale, W. B. et al. 1949. *Phytopathology* **39**: 167 (on aroids, *Aglaonema simplex* and *Lilium longiflorum* var. *exinum*; **28**: 397).

Turner, P. D. 1966. *Planter Kuala Lumpur* **42**: 103 (blast of oil palm; **45**, 2578).

Zimmer, D. E. et al. 1969. *Pl. Dis. Reptr* **53**: 473 (root rot of safflower; **48**, 3612).

Pythium ultimum Trow, *Ann. Bot.* **15**: 300, 1901.

SPORANGIA chiefly terminal and spherical, 12–28 (20) μ diam., occasionally intercalary and barrel shaped, 14–23 × 28 μ, germinating only by germ tubes. OOGONIA smooth, terminal, spherical, rarely intercalary, 20–23 (21) μ diam. ANTHERIDIA usually 1/oogonium, monoclinous from immediately below the oogonium, curved, sometines 2/oogonium, then often of diclinous origin and straight. OOSPORES aplerotic, single, spherical, 15–18 (16) μ diam., with a smooth, thick wall containing a single central reserve globule and refringent body. Opt. growth in vitro 28° or less (J. T. Middleton 1943, *Pythium* q.v.).

Drechsler described the formation of zoospores from oospores and compared this sp. with *P. irregulare* and *P. debaryanum* (q.v.) whose morphological differences (from *P. ultimum*) are mentioned under the latter sp. In germinating oospores the empty evacuation tube is generally longer in *P. ultimum* than in *P. debaryanum*.

P. ultimum is plurivorous (virtually on as many hosts as *P. debaryanum*) and widespread (Map 207). Leach in a study with pea (*Pisum sativum*), spinach, sugar beet and watermelon (*Citrullus lanatus*) fully demonstrated one of the general principles of damping off. This is that pre-emergence infection is most severe at temps that are relatively less favourable to the host than to the pathogen, as measured by the ratio of their growth rates. Therefore, crops with high temp. requirements would tend to be more subject to damping off at low temps. The characteristic lessening of disease with age of the seedling has more recently been demonstrated in soybean and *Antirrhinum*. In the latter restriction of the development of *P. ultimum* in the roots of 25-day-old plants (compared with that in 5-day-old ones or younger) could not be related to lignification. Lumsden et al. found that a 1–10 weeks' incubation in soil was needed for conversion from thick walled dormant oospore to the thin-walled condition before germination took place. Thin walled spores, but not thick walled ones, germinated in 2 hours on nutrient media. (See also Hancock for soil factors.)

Flentje and others fully investigated soil moisture, seed exudate and host factors in the pre-emergence rotting of pea in Australia. One of the effects of high soil moisture is that it facilitates the diffusion of plant exudates thereby increasing disease; spore germination near seed can occur within 3–4 hours from sowing and host penetration within 24 hours (Kerr; Stanghellini et al.). Griffin concluded that the occurrence of *P. ultimum* in wet environments is due as much to its tolerance of poor gaseous exchange as to its requirement for abundant water. Bainbridge found that sporulation was affected quantitatively by soil moisture. At 5, 10 and 15% moisture content oospores were 10.1, 72.7 and 81.4%, respectively, of all spores; at 20 and 25% moisture only oospores were found. Apparently oogonia were formed in water filled pores and sporangia only in air filled ones. Stanghellini et al. (1971a) showed that sporangia survived in soil for 11 months, that max. (*c.* 80%) germination occurred in soil 3 hours after the addition of nutrients, and that germ tube growth (*c.* 300 μ/hour) was independent of soil moisture between field capacity and near saturation. The effects of oxygen conc. on infection of soybean and nitrogen sources on growth in vitro (Kraft, *P. aphanidermatum* q.v.) have been studied, and also pectolytic activity (Mellano et al. 1970b). In pea Perry considered that faults in the seed coat enhanced susceptibility. (See Kraft et al. 1969b; McDonald et al.; Munnecke et al.; Ohh et al.; Richardson for control in pea.)

Harter et al. described mottle necrosis of sweet potato. The typical, soft, grey, cheesy type tuber symptoms develop most at <20° (opt. temp. for most decay being 12–16°). Above this temp. the amount of decay falls; it becomes dry, dark in colour and irregularly outlined (marble type). This disease has also been associated with *P. aphanidermatum, P. splendens* (q.v.) and *P. scleroteichum* (Drechsler). Examples of other diseases caused by *P. ultimum* are given for: *Antirrhinum* (Mellano et al.), common bean (Dickson et al.; Stanghellini et al. 1971b; York et al.), cotton (Arndt), cucumber (Gabrielson et al.; Haglund et al.), maize (Kisiel et al.; Singh), melon (McKeen et al.), safflower (Kochman et al.), soybean (Brown et al.; Hildebrand et al.; Laviolette et al.) and tobacco (Gayed et al.). In maize attack by *Tylenchus agricola* and *Tylenchorhynchus claytoni* did not increase disease caused by *P. ultimum*. Synergistic effects with cucumber mosaic virus have been described by Nitzany; with *Fusarium solani* f. sp. *pisi*, bean yellow mosaic and pea common mosaic viruses by Escobar et al. and also with *P. debaryanum*. In general, control is through conventional cultural treatments including fungicide treatment of seed and soil fumigation. Resistance has been found in pea and common bean.

Agnihotri, V. P. et al. 1967. *Can. J. Bot.* **45**: 1031 (effect of *Pinus resinosa* exudates on growth & sporangial germination; **46**, 3587).

——. 1967. *Phytopathology* **57**: 1116 (effect of soil factors & temp. on sporangial germination; **47**, 443).

Arndt, C. H. 1943. *Ibid* **33**: 607 (damping off of cotton; **22**: 478).

Bainbridge, A. 1970. *Trans. Br. mycol. Soc.* **55**: 485 (sporulation at different soil moisture tensions; **50**, 1620).

Brown, G. E. et al. 1966. *Phytopathology* **56**: 407 (effect of O_2 on seed rot of soybean; **45**, 3007).

Dickson, M. H. et al. 1974. *Pl. Dis. Reptr* **58**: 774 (resistance in common bean; **54**, 1499).

Drechsler, C. 1934. *J. agric. Res.* **49**: 881 (*Pythium scleroteichum* sp. nov., mottle necrosis of sweet potato; **14**: 467).

Escobar, C. et al. 1967. *Phytopathology* **57**: 1149 (testing pea cvs & synergism with other organisms; **47**, 952).

Flentje, N. J. 1964. *Aust. J. biol. Sci.* **17**: 643, 651 (pre-emergence rot in pea, seed & soil factors; **44**, 916a, b).

—— et al. 1964. *Ibid* **17**: 665 (host & pathogen interaction in pea; **44**, 916c).

Gabrielson, R. L. et al. 1968. *Pl. Dis. Reptr* **52**: 806 (chemical control of damping off in cucumber; **48**, 670).

Griffin, D. M. 1963. *Trans. Br. mycol. Soc.* **46**: 368 (behaviour at small soil water suctions; **43**, 925).

Gayed, S. K. et al. 1978. *Can. Pl. Dis. Surv.* **58**: 15 (on tobacco with *Rhizoctonia solani*; **57**, 5674).

Haglund, W. A. et al. 1972. *Phytopathology* **62**: 287 (differential action of soil fungicides in damping off of cucumber; **51**, 3022).

Hancock, J. G. 1977. *Hilgardia* **45**: 107 (factors affecting soil populations; **57**, 2819).

Hare, W. W. 1949. *J. agric. Res.* **78**: 311 (stem blight in pea, effect of temp.; **28**: 557).

Harter, L. L. et al. 1927. *Ibid* **34**: 893 (on sweet potato; **6**: 748).

Hildebrand, A. A. et al. 1952. *Sci. Agric.* **32**: 574 (general on soybean in Canada; **32**: 358).

Kerr, A. 1964. *Aust. J. biol. Sci.* **17**: 676 (effect of soil moisture & pea seed exudates; **44**, 917).

Kisiel, M. et al. 1969. *Phytopathology* **59**: 1387 (on maize with nematodes & *Fusarium* sp.; **49**, 1015).

Kochman, J. K. et al. 1969. *Aust. J. exp. Agric. Anim. Husb.* **9**: 644 (on safflower; **49**, 1752).

Kraft, J. M. et al. 1969a. *Phytopathology* **59**: 149 (additive effect with *F. solani* f. sp. *pisi*; **48**, 2066).

——. 1969b. *Pl. Dis. Reptr* **53**: 776 (control with soil fumigants in pea; **49**, 895).

Laviolette, F. A. et al. 1971. *Phytopathology* **61**: 439 (infection & age of soybean seedlings; **50**, 3316).

Leach, L. D. 1947. *J. agric. Res.* **75**: 161 (growth rates of host & pathogen as factors determining the severity of pre-emergence damping off; **27**: 344).

Lumsden, R. D. et al. 1975. *Phytopathology* **65**: 1101 (soil environment & oospore germination).

McDonald, W. C. et al. 1961. *Can. Pl. Dis. Surv.* **41**: 275 (resistance in pea; **41**: 495).

McKeen, C. D. et al. 1968. *Can. J. Bot.* **46**: 1165 (general on melon; **48**, 675).

Marchant, R. 1968. *New Phytol.* **67**: 167 (fine structure of sexual reproduction; **47**, 1443).

Mellano, H. M. et al. 1970a. *Phytopathology* **60**: 936 (infection & seedling age in *Antirrhinum majus*; **50**, 117).

——. 1970b. *Ibid* **60**: 943 (pectolytic activity & development in *A. majus*; **50**, 118).

Moore, L. D. et al. 1961. *Pl. Dis. Reptr* **45**: 616 (on Gramineae; **41**: 40).

Munnecke, D. E. et al. 1967. *Phytopathology* **57**: 969 (persistence of thiram in soil & control in pea; **47**, 41).

Nitzany, F. E. 1966. *Ibid* **56**: 1386 (synergism with cucumber mosaic virus; **46**, 1395).

Ohh, S. H. et al. 1978. *Pl. Dis. Reptr* **62**: 196 (captan seed treatment & resistance in pea; **57**, 5191).

Perry, D. A. 1973. *Trans. Br. mycol. Soc.* **61**: 135 (infection of pea seed; **53**, 1129).

Poole, R. F. 1934. *Phytopathology* **24**: 807 (ring rot in sweet potato; **13**: 799).

Richardson, L. T. 1973. *Pl. Dis. Reptr* **57**: 3 (synergism between chloroneb & thiram in pea; **52**, 3474).

Schlub, R. L. et al. 1978. *Phytopathology* **68**: 1186 (soybean seed coat cracks & soil infection).

Short, G. E. et al. 1976. *Ibid* **66**: 188 (factors affecting pea seed & seedling rot; **55**, 5413).

Singh, R. S. 1964. *Mycopathol. Mycol. appl.* **22**: 182 (on maize, effect of temp.; **43**, 3206).

Stanghellini, M. E. et al. 1971a. *Phytopathology* **61**: 157 (survival of sporangia in soil; **50**, 2744).

——. 1971b. *Ibid* **61**: 165 (behaviour in soil near seed of common bean; **50**, 2745).

Tomkins, C. M. et al. 1939. *J. agric. Res.* **58**: 461 (on Cucurbitaceae; **18**: 497).

Vaartaja, O. 1977. *Phytopathology* **67**: 67 (response to a soil extract containing an inhibitor; **56**, 3443).

Wilhelm, S. 1965. *Ibid* **55**: 1016 (control by fumigation with chloropicrin, several hosts; **45**, 382).

York, D. W. et al. 1977. *Pl. Dis. Reptr* **61**: 285 (inheritance of resistance in common bean; **56**, 5878).

RAMULARIA sensu Sacc., *Sylloge Fung*. **4**: 196, 1886.
(*Ramularia* Unger, *Die Exanthema der Pflanzen*: 169, 1833)
Acrotheca Fuckel, 1860.
(Moniliaceae)

Vegetative HYPHAE hyaline, septate, phytophilous, endoparasitic, sometimes forming spots on leaves. CONIDIOPHORES hyaline, mostly hypophyllous and emerging through stomata, simple or branched, non-septate or septate, sporiferous towards the tip. CONIDIA acrogenous, variable in shape, ovate cylindrical, hyaline, solitary, aseptate or septate, thin walled (after C. V. Subramanian, *Hyphomycetes*: 436, 1971 as *Ramularia* Unger; Hughes (1949) gave a taxonomic history of the genus).

Several spp. of *Ramularia* cause diseases on tropical crops. They are relatively minor ones and/or have not been extensively studied. Other spp. occur on temperate ornamentals (see Moore, *British parasitic fungi*). *R. bellunensis* Speg. causes a bud rot of pyrethrum (*Chrysanthemum cinerariaefolium*) and occurs on other Compositae. The fungus has been described from: Germany, Italy, Kenya, Tanzania, UK and Zaire (Map 292); see Delhaye; Nattras. *R. carthami* Zaprometoff causes brown leaf spot of safflower (*Carthamus tinctorius*). The fungus was described by Subramanian, *Hyphomycetes*: 438, and the disease it causes has been studied in USSR (see Egorova). Another sp. (*R. carthamicola* Darpoux) from the same host has been described. Raghunath reported on the *Ramularia* spp. that occur on Umbelliferae. *R. coriandri* Moesz can be serious on coriander (*Coriandrum sativum*) in USSR (Andreeva). *R. foeniculi* Sibilia is found on fennel (*Foeniculum vulgare*; Prasad et al.; Singh). On temperate legumes 2 spp. have been fully discussed: *R. deusta* (Fuckel) Baker, Snyder & Davis on *Lathyrus* spp. and *R. onobrychidis* Allescher on sainfoin (*Onobrychis viciifolia*; Hughes 1949).

R. oryzae Deighton & Shaw, causing white leaf streak of rice, was recently described from Papua New Guinea. The fungus is also found in: British Solomon Islands, Malaysia (Sabah), Nigeria and Sierra Leone. The leaf spots (visible on both surfaces) are oblong linear, 1–2.5 mm long, 0.5 mm wide, white or greyish white, surrounded by a brown margin which is usually very narrow, but sometimes up to 0.5 mm wide, sometimes with a rather more diffuse brown area outside the margin of the spot. Younger lesions may show a white streak only on the upper leaf surface, many of the streaks on the lower surface being brown all over. The lesions are usually numerous and on heavily infected leaves may be contiguous, although each remains clearly delimited by its own narrow brown margin. The spot is limited laterally by the leaf veins (Deighton et al.). These symptoms should not be confused with those caused by *Sphaerulina oryzina* (q.v.).

A shoot blight of *Eucalyptus* spp. in Australia (New South Wales) is caused by *R. pitereka* Walker & Bertus. The disease is found on seedlings >3 months old. It distorts and twists young shoots and causes leaf spots and stem lesions. Diseased shoots are shining white, due to a massive development of the fungus pushing up the waxy cuticle and rupturing it. The erumpent pustules are up to 100 μ diam. and closely packed. On leaves the spots are 1–2 mm diam. up to large irregular areas which often develop along one edge and result in distortion of the leaf. The spots are brown with a thin red purple margin. Sunken brown lesions up to 1.5 cm long occur on stems and petioles. *R. pitereka* is unusual amongst members of the genus in its occurrence on young growth of woody hosts; most spp. occur on herbaceous plants. Possibly closely related to *R. pitereka* is the fungus (*Sporotrichum destructor*, not validly published) that causes a canker of adult trees of *E. ficifolia* (Walker et al.).

Andreeva, L. T. 1966. *Sb. rabot po maslich. kul'turam* (3): 91 (*Ramularia coriandri* in USSR; **47**, 278).
——. 1969. *Mikol. i Fitopatol.* **3**: 331 (as above, culture & hosts; **49**, 225).
Baker, K. F. et al. 1950. *Mycologia* **42**: 403 (*R. deusta* on *Lathyrus* spp.; **30**: 41).
Darpoux, H. 1946. *Annls Epiphyt*. **12**: 297 (fungi on safflower, *R. carthamicola* sp. nov.; **27**: 383).
Deighton, F. C. et al. 1960. *Trans. Br. mycol. Soc.* **43**: 516 (*R. oryzae* sp. nov.; **40**: 104).
Delhaye, R. J. 1952. *Bull. Inf. I.N.E.A.C.* 1(4): 305 (*R. bellunensis*; **32**: 627).
Egorova, N. P. 1963. *Trud. Tashkent Sel'.-khoz. Inst.* **15**: 277 (*R. carthami*; **44**, 782).
Hughes, S. J. 1949. *Trans. Br. mycol. Soc.* **32**: 34 (*R. onobrychidis* & history of genus; **28**: 457).
——. 1951. *Mycol. Pap.* 38, 8 pp. (*Acrotheca*; **30**: 547).
Moesz, G. V. 1930. *Magy. bot. Lap.* **29**: 35 (*R. coriandri* sp. nov. inter alia; **10**: 343).

Nattras, R. M. 1947. *Nature Lond.* **160**: 120 (*R. bellunensis*; **26**: 454).
Prasad, N. et al. 1961. *Curr. Sci.* **30**: 65 (*R. foeniculi*; **40**: 620).
Raghunath, T. 1963. *Ibid* **32**: 324 (*Ramularia* spp. on Umbelliferae; **43**, 180).
Singh, R. D. 1974. *Indian J. Mycol. Pl. Pathol.* **4**: 166 (*R. foeniculi*; **55**, 4231).
Vimba, E. K. 1970. Izdatel'stvo 'Zinatne' Riga, Latvia, USSR (*Ramularia* spp. in Latvia; **52**, 1033).
Walker, J. et al. 1971. *Proc. Linn. Soc. N.S.W.* **96**: 108 (*R. pitereka* sp. nov.; **51**, 1971).
Zaprometoff, N. G. 1926. *Morbi Plantarum*, Leningrad **15**: 141 (*Bolez. Rast.*) (*R. carthami* sp. nov.; **7**: 165).

Ramularia gossypii (Speg.) Ciferri, *Atti Ist. Bot. Univ. Pavia* Ser. 5, **19**: 124, 1962.
Cercosporella gossypii Speg., 1886
Ramularia areola Atkinson, 1890
Symphyosira areola (Atk.) Sawada, 1959.

Leaf spots irregular, hypophyllous, occasionally amphigenous, angular, delimited by the leaf veins, pale becoming darker. Fruiting amphigenous but mostly hypophyllous. Substomatal stroma present, size variable (mostly 30 μ diam.). Primary mycelium internal. CONIDIOPHORES borne on the substomatal stroma, emerging through the stoma, forming a loosely arranged fascicle, hyaline or slightly coloured, frequently branched at the base, septate, often bifurcate at the broad apex, bearing conspicuous conidial scars, 27–75 × 3.5–7 μ (mostly 30–55 × 3.5–4 μ). CONIDIA hyaline, straight, cylindrical, usually pointed at both ends, or occasionally rounded, 1–3 septate, mostly 2 septate, catenulate, sometimes branched conidia occur, scars thickened, lying flat against the conidial wall, 14–37 × 2.5–5 μ. On *Gossypium* spp. (J. L. Mulder & P. Holliday, *CMI Descr.* 520, 1976).

The fungus causing grey mildew of cotton (widespread with the host, Map 260) has been described as having a perfect state (*Mycosphaerella areola* Ehrlich & F. A. Wolf) but without conclusive proof. The symptoms occur chiefly on the older leaves as the plants mature. The spots are hypophyllous, rarely amphigenous, pale at first, becoming darker, 1–10 (mostly 3–4) mm diam., angular, irregular in shape, limited by the leaf veins, conidia in profusion and give a frosted appearance to the spots. Rathaiah (1977) gave 25–30° as the best temps for conidial germination and germ tube growth. Differences in virulence were found (Rathaiah 1976). Grey mildew is unlikely to be severe in dry climates. But in rain fed cultivations it can become severe enough to warrant specific control measures, as has fairly recently been reported in the Malagasy Republic. In this country the pathogen occurs particularly in such cultivations when rainfall is high. Control was obtained by spraying with benomyl.

Atkinson, G. F. 1890. *Bot. Gaz.* **15**: 166 (morphology).
Cauquil, J. et al. 1973. *Coton Fibr. trop.* **28**: 279 (in Malagasy Republic, control; **53**, 569).
Ehrlich, J. et al. 1932. *Phytopathology* **22**: 229 (*Mycosphaerella areola* sp. nov.; **11**: 512).
Rathaiah, Y. 1973. *Coton Fibr. trop.* **28**: 287 (conidial sporulation in culture; **53**, 570).
——. 1974. *Ibid* **29**: 263 (factors affecting growth & sporulation in culture; **54**, 162).
——. 1976. *Phytopathology* **66**: 1007 (reaction of spp. & cvs; **56**, 1620).
——. 1977. *Ibid* **67**: 351 (conidial germination & infection; **56**, 5068).

RAMULISPORA Miura emend. Olive, Lefèbvre & Sherwin, *Phytopathology* **36**: 198, 1946. (Tuberculariaceae)

SPORODOCHIA amphigenous, produced through stomata of infected leaves, arising from substomatal stromata. CONIDIOPHORES hyaline, simple or branched, short. CONIDIA acrogenous, hyaline, filiform, with lateral branches, produced in gelatinous aggregates; superficial SCLEROTIA present (L. S. Olive, C. L. Lefèbvre & H. S. Sherwin l.c.).

The author of *R. sorghicola* E. Harris (causing a common leaf spot of *Sorghum bicolor* in Nigeria) compared this sp. with *R. sorghi* (q.v.) and *Gloeocercospora sorghi* (q.v.). The clearest morphological distinction between the 3 fungi is given by the sclerotia but these are not always present on the leaf. The sclerotia of *R. sorghicola* bear septate setae and are fewer than those of *R. sorghi* which are tuberculate and glabrous. Sclerotia of both these spp. are superficial while those of *G. sorghi* are immersed in the necrotic tissue. Anahosur tabulated the distinctions between these fungi and described *R. sorghicola*. This fungus causes leaf spots which first appear as small watersoaked areas with an indistinct brick red margin, later elliptical, up to 7 × 3 (mostly 3–4 × 1–5) mm, becoming irregular by confluence, limited by the veins, with a conspicuous dark red margin (tan in some sorghum vars), up to 1 mm wide and pinkish grey to straw coloured necrotic centre, sometimes bearing scattered black sclerotia. *R. sorghicola* has been reported from several countries in central Africa and also from: Haiti, India (see Nagarajan et al.), Malaysia (Sabah), Pakistan and Philippines. *R. sacchari* Rawla was described on *Saccharum* in India. It differs from *R. sorghi* in its narrower conidiophores, shorter conidia and smaller sclerotia; also in cultural and nutritional characteristics (Rawla, who also discussed other *Ramulispora* spp. on Gramineae). *R. alloteropsis* Thirum. & Narasimhan occurs on *Pennisetum* spp. and other Gramineae.

Anahosur, K. H. 1978. *C.M.I. Descr. pathog. Fungi Bact.* 586 (*Ramulispora sorghicola*).
Harris, E. 1960. *Trans. Br. mycol. Soc.* **43**: 80 (*R. sorghicola* sp. nov.; **39**: 705).
Nagarajan, K. et al. 1971. *Indian Phytopathol.* **24**: 644 (reaction of sorghum cvs to *R. sorghicola* & *R. sorghi*; **52**, 2263).
Rawla, G. S. 1973. *Trans. Br. mycol. Soc.* **60**: 283 (*Gloeocercospora* & *Ramulispora* in India; **52**, 3572).

Ramulispora sorghi (Ell. & Ev.) Olive & Lefèbvre, *Phytopathology* **36**: 198, 1946.
Septorella sorghi Ell. & Ev., 1903
Ramulispora andropogonis Miura, 1920
Titaeospora andropogonis Tai, 1932.

Spots elongate elliptical with straw coloured centres, surrounded by reddish purple to tan borders according to the host var. SCLEROTIA amphigenous, gregarious, superficial on the centres of lesions, subglobose, coarsely tuberculate, glabrous, subcarbonaceous, 53–170 μ. SPORODOCHIA amphigenous, developing from subepidermal stromata, becoming erumpent through stomata. CONIDIOPHORES fasciculate, 10–35 × 2–3 μ. CONIDIA filiform with 1–3 branches, 5–53 × 1.1–2.5 μ, hyaline, curved, tapering towards apex, 38–86.3 × 1.9–3 μ, 3–8 septate (from L. S. Olive, C. L. Lefèbvre & H. S. Sherwin l.c. and q.v. for further details on morphology and taxonomy; see Rawla (*Ramulispora* q.v.) for a full description of the fungus based on Indian material; Anahosur).

R. sorghi causes sooty stripe of *Sorghum* spp. *R. sorghicola* (*Ramulispora* q.v.) differs in that the sclerotia are fewer (they are very profuse in *R. sorghi*) and bear septate setae. Both these spp. have superficial sclerotia while those of *Gloeocercospora sorghi* (q.v.) are immersed. The straw coloured, elongate elliptical lesions caused by *R. sorghi* are distinct from the roughly circular or semicircular lesions with broad dark red zonations caused by *G. sorghi*. Sooty stripe is widespread in central and southern Africa, and it also occurs in USA, parts of S. America, southern and eastern Asia, and Australia. The disease appears to be a relatively minor one but assumes a local importance at times. Its name derives from the sooty appearance of the centre of the leaf lesions; this is due to the growth of the numerous, black, superficial sclerotia which are readily brushed off. The leaf spots begin as small, oblong, reddish purple lesions. These enlarge to become elongate elliptical, have straw coloured centres and have tan to purplish borders; they can be several cm long and 1–2 cm in width; large areas of necrotic tissue in the leaves may develop. The conidia aggregate in gelatinous masses (both on the host and in culture) and are presumably water dispersed. *R. sorghi* has been isolated from sorghum seed. Carry over of the fungus is through the sclerotia, but surface sporodochia are also survival structures where infected leaf debris remains on the soil. In Nigeria, under the conditions of a natural epidemic, 5% of the lines tested were resistant; 47% of those from Volta, 10% from Nigeria and 6% from Mali were resistant. There was more resistance in the Conspicuum race than in the Cafforum race; and the data showed that W. Africa is the point of origin of resistant sorghums. Apart from the use of resistance there has been no experimental work on control; indeed, there is little such work on the general biology of this pathogen.

Anahosur, K. H. 1978. *C.M.I. Descr. pathog. Fungi Bact.* 585 (*Ramulispora sorghi*).
Futrell, M. C. et al. 1966. *Pl. Dis. Reptr* **50**: 606 (resistance in races of sorghums; **46**, 312).
Lele, V. C. et al. 1966. *Indian Phytopathol.* **19**: 357 (seedborne infection; **47**, 807).
Odvody, G. N. et al. 1973. *Phytopathology* **63**: 1530 (method for extraction of sclerotia from soil & carry over; **53**, 3000).
Olive, L. S. et al. 1946. *Ibid* **36**: 190 (taxonomy, morphology, culture & pathogenicity; **25**: 392).
Rawla, G. S. et al. 1975. *Trans. Br. mycol. Soc.* **64**: 532 (trace element & other growth factors, with *R. sacchari*; **54**, 4814).

RHIZINA Fr., *Syst. Mycol.* **2**: 33, 1822.
(Helvellaceae)

APOTHECIA discoid, large, with dark brown to black hymenium and tough brown flesh, attached to the soil or wood by numerous, cylindrical, branched, root-like processes; distinct hairs on excipulum lacking; paraphyses elements (like setae) accompany normal paraphyses. ASCI not blue in iodine. ASCOSPORES fusiform, apiculate, reticulate (rough), with 3 or more large guttules or biguttulate (after key by R. P. Korf in *The fungi* (edited by G. C. Ainsworth et al.) Vol. IV A: 269, 1973; and R. W. G. Dennis, *British Ascomycetes*: 13, 1968).

Rhizina undulata Fries, *Syst. Mycol.* **2**: 33, 1822.
R. inflata (Schäff.) Quél., 1886.

HYMENOPHORE flat, convex or undulating, often irregularly lobed, dark brown or black with paler margins, the hymenophore often extends to several cm in diam. and is 1–2 mm thick; flesh reddish brown, tough, fibrous, under surface pale ochraceous bearing numerous cylindrical, branched, whitish root like structures. ASCI operculate, 350–450 × 10–14 μ with 8 monostichous or obliquely monostichous ascospores. ASCOSPORES aseptate, fusiform, 24–40 × 9–11 μ, with a hyaline apiculus at each end; some spores are slightly roughened but this does not appear to be a consistent feature. PARAPHYSES stiff (like hairs), brown, 4 μ diam.; they are interspersed between the asci and extend beyond, where the swollen tips form a protective surface covered with an amorphous brown crust. Germination of ascospores on PDA is greatly stimulated by a short period of heat treatment (35–45°) and then incubation at 22° (Jalaluddin 1967); growth is then rapid, producing abundant yellowish white MYCELIUM composed of HYPHAE 3–4 μ wide and presenting a woolly appearance as the culture gradually turns orange brown. No CONIDIA or CHLAMYDOSPORES were observed but swollen hyphae up to 14 μ diam. develop after a few days. (C. Booth & I. A. S. Gibson, *CMI Descr.* 324, 1972; and see Ginns.)

R. undulata causes group dying in conifers (*Abies, Larix, Picea, Pinus, Pseudotsuga* and *Tsuga* spp.). It occurs in southern Africa, N. America, Europe and Japan (Map 489). The disease has been described in UK (Murray et al.; Jalaluddin), Japan (Sato et al.), the Netherlands (Gremmen) and Sweden (Hagner). Trees 15–60 years old can be killed. Roots of all ages are attacked; the larger ones become covered with a white or yellowish mycelium which can be traced directly to the sporophores. Outbreaks of the disease occur around old fire sites (or where the soil has been heated, e.g. during asphalt road construction in the Netherlands). The disease spreads outwards for a few years and then ceases. Sato found a centrifugal spread of 3–5 m/year for 4–5 years. Typical attacks in UK are described by Jalaluddin (1976b). Extensive damage to the roots is done before above ground symptoms appear; these symptoms are not diagnostic and the apothecia should be found.

The discharged ascospores lie in the soil, becoming dormant. They are stimulated to germinate at 35–45° (Jalaluddin 1967a; 38–40°, Gremmen 1971) and cause infection. The former worker reported a survival rate for a few ascospores of 2 years on glass slides in soil; and that infective mycelia only developed from ascospores if the soil was acid and fresh conifer roots were present. Sato reported the occurrence of apothecia 3 months after fire when fresh pine or larch roots were present; their numbers increased in stands > 20 years old, and they were absent in 2–5-year-old stands. The root systems of freshly cut conifer stumps and those of standing trees can provide food bases. The first sporophores form around the original fire site; they subsequently develop in wider, irregular, incomplete and discontinuous rings. The mycelium may occur 2 m beyond the ring in the soil and litter. Sato gave 25° as opt. for mycelial growth.

The main control is to have no fires which heat the soil. Phillips et al. stated that widely spaced 'spot' fires were less harmful than those over large areas of ground. Other control measures are: avoidance of highly susceptible conifer spp., trenching, plant barriers, re-afforestation by sowing, and delay transplanting for 2 years in infested sites.

Ginns, J. 1974. *Fungi Canadenses* 16 (morphology).

Gremmen, J. 1971. *Eur. J. For. Pathol.* **1**: 1 (in Netherlands; **51**: 2905).

——. 1976. *Ned. Bosb. Tijdschr.* **48**: 181 (effects of fire; **56**, 421).

Hagner, M. 1962. *Norrlands SkogsvFörb. Tidskr.* 2: 245 (in Sweden; **42**: 638).

Jalaluddin, M. 1967a & b. *Trans. Br. mycol. Soc.* **50**: 449, 461 (mycelial growth & ascospore germination; observations in E. Anglia, UK; **47**, 350a & b).

Murray, J. S. et al. 1961. *For. Rec. Lond.* 46, 19 pp. (in UK; **40**: 713).
Norkrans, B. et al. 1963. *Physiol. Pl.* **16**: 1 (cellulytic & pectolytic enzymes).
Phillips, D. H. et al. 1976. *Leafl. For. Commn.* 65, 7 pp. (advisory in UK; **55**, 5958).
Sato, K. et al. 1974. *Bull. Govt For. Exp. Stn* 268: 13 (in Japan; **54**, 4660).

RHIZOCTONIA DC. ex Fr., in Sacc., *Sylloge Fung.* **14**: 1175, 1899.

See *Thanatephorus cucumeris* for a description of its sclerotial state *R. solani* and the most important sp. The basic characters of the genus are the formation of sclerotia of uniform texture with hyphal threads emanating from them and the association of the mycelium with roots of living plants. The sclerotial cells in *Rhizoctonia* are essentially similar and, while the outer cells may be darker and thicker walled, there is no obvious differentiation into a rind and a medulla. Mycelia with differentiated, *Sclerotium* type sclerotia can, therefore, be readily excluded from *Rhizoctonia* (J. R. Parmeter & H. S. Whitney who gave notes on some *Rhizoctonia* spp. in Parmenter 1970, *T. cucumeris* q.v.).

Mordue described *R. carotae* Rader and *R. tuliparum* Whetzel & Arthur (synonym: *Sclerotium tuliparum* Klebhan). The former causes crater rot of carrot in cold storage and the latter grey bulb rot in ornamentals.

Gladders, P. et al. 1977. *Trans. Br. mycol. Soc.* **68**: 115 (*Rhizoctonia tuliparum*, infection cushion; **56**, 3432).
———. 1978. *Ibid* **71**: 129 (*R. tuliparum*, a winter active pathogen).
Mordue, J. E. M. 1974. *C.M.I. Descr. pathog. Fungi Bact.* 407, 408 (*R. tuliparum* & *R. carotae*).
Tu, C. C. et al. 1975. *Can. J. Bot.* **53**: 2282 (morphology, development & cytochemistry of spp. in the *Rhizoctonia* complex; **55**, 3472).
Ui, T. 1973. *Rev. Pl. Prot. Res.* **6**: 115 (*Rhizoctonia* diseases in Japan, 52 ref.; **55**, 80).

Rhizoctonia oryzae-sativae (Sawada) Mordue, *C.M.I. Descr. pathog. Fungi Bact.* 409, 1974.
Sclerotium oryzae-sativae Sawada, 1922.

COLONIES on PDA at first colourless, gradually developing a pale brown shade. Aerial mycelium variable, usually a thin felt but sometimes uneven and tufted. SCLEROTIA either scattered over most of colony or aggregated at centre or edge of plate, 0.5–2 mm across and irregularly globose; often covered by loose mycelium, and appear rough and warted. Cells of HYPHAE at advancing edge of colony usually 5–6.5 μ wide and up to 300 μ long. Branches usually arise on the distal half of the cell, are constricted at point of origin and septate shortly above. Older mycelium shows greater variation in cell diam. (3.5–7 μ) and length, with monilioid cells $21–37 \times 6–11$ μ in some isolates. The homogenous, prosoplectenchymatous sclerotia are composed of characteristically near globose cells 13–30 μ across interspersed with undifferentiated hyphae. On the host sclerotia are found inside the hollow stem or in large cavities in the leaf sheath, are often oblong in shape and are larger than in culture. No perfect state is known. Structure of the sclerotia differs from that of *R. solani* (*Thanatephorus cucumeris* q.v.) and the hyphae are generally slightly narrower than those of that sp. On *Oryza sativa*; has been found on *O. cubensis*, *Juncellus serotinus* and *Zizania latifolia* (from J. E. M. Mordue l.c.).

Ou (*Rice diseases*) described the diseases on this crop caused by *Rhizoctonia* and *Sclerotium* spp. *R. solani* is described under *T. cucumeris*. *R. oryzae* Ryker & Gouch, causing sheath spot of rice, was considered to be distinct. *S. oryzae-sativae* has homogenous sclerotia like those of the type sp. of *Rhizoctonia*. It shows little affinity with *Sclerotium* in which the sclerotia are differentiated into cortex and medulla (Mordue l.c.). Other sclerotial fungi attacking rice include: *S. fumigatum* Nakata ex Hara, *S. hydrophilum* Sacc. and *S. oryzicola* Nakata & Kawamura. *R. oryzae-sativae* causes lesions on rice leaf sheaths, usually 0.5–1 cm, with pale centres and brown margins. Several may occur together. Infection of the culm may lead to lodging and death. The fungus occurs in China, Japan, Malaysia, Sri Lanka, Taiwan and Vietnam. It is carried over in crop residues and soil. The opt. temp. for mycelial growth and sclerotial formation is 32°.

Endo, S. 1940. *Ann. phytopathol. Soc. Japan* **10**: 7 (morphology & pathogenicity; **19**: 615).
Ryker, T. C. et al. 1938. *Phytopathology* **28**: 233 (*Rhizoctonia oryzae* sp. nov., no Latin diagnosis; **17**: 622).

RHIZOPUS Ehrenberg ex Corda, *Icon. fung.* **2**: 20, 1838.
(see C. W. Hesseltine, *Mycologia* **47**: 349, 1955)
(Mucoraceae)

MYCELIUM as aerial, arching STOLONS (often several cm long) arising from the points of contact with the substrate and where tufts of repeatedly branched RHIZOIDS form. SPORANGIOPHORES arising from the stolons opposite the point of origin of the rhizoids; usually unbranched (simple), fasciculate. SPORANGIA terminal, large, globose, many spored; columella prominent, more or less hemispherical; wall not cutinised and at maturity almost wholly disappearing. SPORANGIOSPORES globose to oval or angular, smooth or with longitudinal striations, rarely echinulate. ZYGOSPORES formed from the rhizoids or stolons; suspensors lack outgrowths (from H. M. Fitzpatrick, *The lower fungi Phycomycetes*: 245, 1930).

Rhizopus spp. are common moulds and laboratory contaminants. They can be readily distinguished from the ecologically similar *Mucor* spp. by the stolons and rhizoids. Dabinett et al.; Inui et al. have given recent accounts of the taxonomy. The 2 *Rhizopus* spp. described are those most commonly found associated with plant disease conditions; and they are often found in association. *R. arrhizus*, which is probably a form *R. oryzae* (J. A. Lunn, personal communication, October 1977), is discussed under the latter sp. Lunn has also recently described *R. microsporus* van Tieghem, *R. rhizopodiformis* (Cohn) Ziof and *R. sexualis* (Smith) Callen. The last causes a rot of soft fruit, particularly strawberry (*Fragaria*). The extensive, early work in USA on the *Rhizopus* rot of sweet potato is considered under the fungus which is the main cause of this rot, *R. stolonifer*.

Dabinett, P. E. et al. 1973. *Can. J. Bot.* **51**: 2053 (numerical taxonomy; **53**, 3370).
Inui, T. et al. 1965. *J. gen. appl. Microbiol. Tokyo* **11**, Suppl., 121 pp. (taxonomy; 93 ref.; **47**, 1793).
Lunn, J. A. 1977. *C.M.I. Descr. pathog. Fungi Bact.* 522, 523, 526 (*Rhizopus rhizopodiformis, R. microsporus, R. sexualis*).
Smith, G. 1969. *An introduction to industrial mycology* edn 6. Edward Arnold.

Rhizopus oryzae Went & Prinsen Geerligs, *Verh. Kon. Ned. Akad. Wetensch.* Sect. 2, **4**(2): 16, 1895.
Rhizopus nodosus Namyslowski, 1906
R. delemar (Boidin) Wehmer & Hanzawa, 1912
R. maydis Bruderlein, 1917.

COLONIES on PDA at 25° fast growing, 5–8 mm high, spreading by stolons which are attached to the substrate by rhizoids, white cottony at first becoming brownish grey to blackish grey depending on amount of sporulation. SPORANGIOPHORES up to 2.8 mm in length and 7–20 μ diam., smooth walled, aseptate, simple or branched, arising from stolons opposite rhizoids usually in groups of 3 or more, rarely single. Branches often arise from a swelling on a primary sporangiophore or stolon. SPORANGIA globose, or often with a flattened base, 30–210 μ diam., white at first, then black, many spored. Sporangial wall echinulate. COLUMELLAE globose, subglobose or oval, up to 90 × 120 μ, pale brown. Collar poorly defined or absent. When dehisced the columella usually inverts to an umbrella-like form with spores often adhering to the outer edge. Apophysis present which can usually only be observed in mature sporangia before they dehisce. Rhizoids and stolons hyaline to light brown. Rhizoids branched. SPORANGIOSPORES unequal, mainly angular subglobose, rhomboidal or limoniform, occasionally elongate (? where division not completed in the sporangium), (4) 5–8 (10) μ in length, striate. Zygospores not observed but reported by Mil'ko et al. *Mykrobiol. Zh.* **28**: 30, 1966. Some isolates have numerous CHLAMYDOSPORES (called gemmae by various authors). Grows well at 37° and some isolates grow weakly at 45°. In air, soil, compost, and pathogenic for plants, man and other warm blooded animals. (J. A. Lunn, *CMI Descr.* 525, 1977.)

R. oryzae is a common saprophyte which can cause soft rots of fruits and vegetables (frequently with other fungi). There was early work on the fungus in USA on the important soft rot of sweet potato which is almost invariably caused by *R. stolonifer* (q.v.). In work on this crop *R. oryzae* was found to have opt. temps for sporangiospore germination of 36–38°, for mycelial growth 31–35°, and for rotting 32–35°; i.e. rather higher temps compared with *R. stolonifer* (q.v. Harter et al. 1921b; Lauritzen et al. 1925; Weimer et al.). *R. arrhizus* Fisch. is the name mostly used in describing the fungus (presumably *R. oryzae; Rhizopus* q.v.)

causing barn rot (or pole rot) of tobacco. Hildebrand described the fungus as a wound parasite causing a root rot of beet and compared it with *R. stolonifer*. *R. oryzae* caused most rotting at 30–40°, whilst *R. stolonifer* showed most infection at 14–16° and had an opt. temp. for growth in culture of 24°. In sunflower an injury (possibly caused by birds) led to infection and a head rot in Israel, and see Middleton in Australia. Rogers et al. found a positive correlation between head rot and infestation by larvae of the sunflower moth (*Homoeosoma ellectellum*). Control was given by a single application of oxine copper. Control of a tuber rot of *Dioscorea floribunda*, with other fungi, was given by benomyl (Bammi et al.). *R. oryzae* may occur in diseased conditions of seedbeds. It is infected by an aggressive mycoparasite, *Syncephalis californica* Hunter & Butler, in soil.

Tobacco barn rot has been mainly studied in Australia, Canada, New Zealand and Zimbabwe. Barn rots (several causes, see Lucas, *Diseases of tobacco*) occur on the crop after it has been housed. One of these is caused by *R. oryzae*. Under certain conditions (temp. <41° and RH >65%; Gayed 1972a) the fungus grows from the basal part of the midrib of the sessile leaf and spreads on the lamina. This worker reported it (as *R. arrhizus*) to have an opt. temp. for growth on PDA close to 35°. It was recovered from diseased, cured leaves after kiln drying at 74° for 24–32 hours, and was not killed on cured leaves when exposed to dry heat in the oven at 80–82° for 72 hours. Stephen also described the conditions under which this barn rot spreads. Gayed (1972b) found that injury to the midrib increased the disease. Control is by avoiding the temp. and RH conditions most favourable for the fungus as far as possible; i.e. temps of 35–40° and a high RH. Also the kilns should have an even air circulation and not be overloaded; leaves used should be uniformly ripe. Dipping or spraying with dichloran or dichlofluanid reduces disease incidence (Cole; Hartill; Hartill et al.; Paddick et al.).

Arnan, M. et al. 1970. *Can. J. Pl. Sci.* **50**: 283 (on sunflower in Israel; **50**, 1339).

Bammi, R. K. et al. 1972. *Pl. Dis. Reptr* **56**: 990 (control of fungi in tuber rot of *Dioscorea floribunda*; **52**, 4179).

Cole, J. S. 1975. *Rhod. Jnl agric. Res.* **13**: 15 (on tobacco, fungicide control; **54**, 4630).

Ekundayo, J. A. et al. 1964. *J. gen. Microbiol.* **35**: 261 (spore swelling & germ tube emergence).

——. 1966. *Ibid* **42**: 283 (spore germination; **45**, 2022).

Gayed, S. K. 1972a. *Can. J. Pl. Sci.* **52**: 103 (on tobacco; **51**, 3546).

——. 1972b. *Lighter* **42**(3): 29 (as above, effect of midrib injury; **52**, 1262).

Hartill, W. et al. 1974. *N.Z. Jl exp. agric.* **2**: 189 (as above, fungicide control; **54**, 1430).

Hartill, W. F. T. 1963. *Tob. Forum Rhod.* Nov.: 23 (as above, general control; **43**, 835).

Hess, W. M. et al. 1973. *Protoplasma* **77**: 15 (fine structure of dormant & germinated spores; **52**, 2521).

Hildebrand, A. A. et al. 1943. *Can. J. Res.* Ser. C **21**: 235 (on beet; **22**: 507).

Hunter, W. E. et al. 1975. *Mycologia* **67**: 863 (*Syncephalis californica* sp. nov.).

—— ——. 1977. *Phytopathology* **67**: 664 (parasitism by *S. californica*).

Middleton, K. J. 1977. *Aust. J. exp. Agric. Anim. Husb.* **17**: 495 (on sunflower in Queensland, Australia).

Paddick, R. G. et al. 1973. *Ibid* **13**: 612 (on tobacco, fungicide control; **53**, 2314).

Rogers, C. E. et al. 1978. *Pl. Dis. Reptr* **62**: 769 (on sunflower in USA).

Stephen, R. C. 1958. *Emp. J. exp. Agric.* **26**: 247 (on tobacco, general; **37**: 680).

Weber, D. J. et al. 1965. *Phytopathology* **55**: 159, 262 (mode of action of 2,6-dichloro 4-nitroaniline; specificity of proline in spore germination; **44**, 2051a & b).

——. 1966. *Ibid* **56**: 118 (proline & glutamic acid metabolism in spore germination; **45**, 1674).

Rhizopus stolonifer (Ehrenb. ex Fr.) Lind, *Danish Fungi*: 72, 1913.

Mucor stolonifer Ehrenb. ex Fr., 1832
M. stolonifer Ehrenb., 1818 (as 'stonolifer')
Rhizopus nigricans Ehrenb., 1820
R. stolonifer var. *luxurians* Schroeter, 1886
R. niger Ciaglinksi & Hewelke, 1893
R. artocarpi Racib., 1900
M. niger Gedoelst, 1902.

COLONIES on PDA at 25° white cottony at first becoming heavily speckled by the presence of sporangia and then brownish black in age, spreading rapidly by means of stolons fixed at various points to the substrate by rhizoids. SPORANGIOPHORES up to 34 μ diam., and 1–3.5 mm in length, smooth walled, aseptate, light brown, simple, arising in groups of 3–5 from stolons opposite rhizoids. SPORANGIA 100–350 μ diam., globose or subglobose with somewhat flattened base, white at first then black, many spored. COLUMELLAE 63–224 × 70–140 μ, subglobose to dorsiventrally flattened, light brownish grey, umbrella shaped when dehisced. Collar poorly defined or absent. Apophysis present, visible below young columella. Rhizoids and stolons hyaline to dark brown. SPORANGIOSPORES (5) 8–20 (26) μ, irregular, round, oval, elongate, angular, brownish black singly, contents homogeneous, strongly striate. Heterothallic. ZYGOSPORES produced when compatible isolates are grown together, 103–180 (220) μ, globose, or compressed between suspensors, brownish black, thick walled, verrucose. Suspensors 62–118 μ wide, swollen, usually unequal and somewhat granular. CHLAMYDOSPORES not seen. No growth at 37°. On soil, fruit and vegetables and decaying plant material, and associated with disorders of man and animals. (J. A. Lunn, *CMI Descr.* 524, 1977.)

R. stolonifer is a common mould, saprophyte and wound pathogen; it causes soft rots of fruits and vegetables (Harter et al. 1922). The diseases are sometimes known as whiskers (from the profuse mycelial growth) or leak (from the soft, watery rot). The fungus can attack plants in the field but more frequently it appears as a pathogen after harvest, particularly in storage. Its cell degrading enzymes have been described; see *Rev. appl. Mycol.* **1**: 273; **2**: 464, 565; **38**: 462; **43**, 825, 920; **45**, 3485; **48**, 1508, 3242.

The fungus was studied early and intensively in USA as the cause of soft rot of sweet potato after harvest. Although other *Rhizopus* spp. were implicated in the disease, *R. stolonifer* is easily the commonest sp. The infection of the root tuber is through wounds; a watery rot spreads rapidly and the tuber may be completely rotted in 4–5 days. Liquid exudes when the skin is broken and plant tissue turns brownish. The greyish sporangiophores, mycelium and black sporangia develop abundantly and rapidly on the broken skin. The affected tissue dries out and the shrivelled roots become hard, shrunken and brittle. The tissue may rot in zones giving a ring rot effect. Infection does not occur directly from germinating spores; a certain amount of growth is required before invasion of the host takes place. Opt. temps for mycelial growth, sporulation and spore germination lie between 23–28°, and for tuber decay 15–23°. Srivastava et al. gave an opt. temp. for growth on PDA of 28°, and for development in sweet potato he gave 20°. The discrepancy was attributed to the fact that most pectolytic enzyme is produced by the fungus at the lower temp. At 23° decay is most rapid at RH 75–84%; it tends to decrease at lower or higher humidities (Daines; Harter et al. 1921, 1923; Lauritzen et al.; Weimer et al.).

Control requires careful handling of the tubers at all stages, timely harvesting in fair weather, surface drying before collecting in the field, clean storage facilities and maintenance of the correct temp. and RH during curing and storage. Curing is done at relatively high temps (*c.* 28°) and *c.* RH 90%. Storage is at *c.* 13° and RH 85–90%. Treatment with dicloran or sodium-o-phenylphenate after harvest (Welch et al.) or after curing but before packaging (Martin) can reduce soft rot.

R. stolonifer is one of the organisms associated with cotton boll rot. Briton-Jones and Kirkpatrick studied the fungus on this crop in Egypt. The soft rot is olive green to blackish (Shapovalov compared it with the pink colour produced by *Aspergillus* sp. on cotton); the boll dries up, hardens, and the sutures split prematurely; sporulation causes the fibre to appear dirty. The fungus is confined to the lock or locks. In Egypt during early boll maturation infection occurs via the insect *Creontiades pallidus* which carries the spores on the rostrum. This insect was more abundant than *Nezara viridula*. Of the 2 boll worms, which allow entry by the fungus, *Pectinophora gossypiella* was the most important, with infection occurring through the exit holes only. There was a lapse of 48 hours between infection and symptoms in bolls near maturity.

Some other examples of crops infected by *R. stolonifer* are as follows. It has been investigated as a cause of peach rot (ref. omitted), but note Pierson who found that, whilst the growth rate of the fungus dropped sharply at 32.2° on glucose yeast agar, it remained at a fairly high level on the fruit. The role of *Drosophila melanogaster*, from which *R. stolonifer* was isolated, in tomato fruit rots was described by Butler et al. The fungus has also been reported as a cause of seedbed losses in groundnuts, being pathogenic to seedlings up to 4 days old; on maize seed; causing a postharvest decay of red pepper fruit; and on papaw (Adams et al.; Gibson et al.; Kurata et al.; Patil et al.). Infection of jackfruit (*Artocarpus heterophyllus*) can, apparently, be severe enough to warrant specific control measures by fungicides (McMillan). The male flowers and young fruit can be severely attacked; infection without host injury was reported by Chowdhury and Crisanto. Thorne (1975), in the rot on carrots, found that conidia invaded root tissue only when suspended in a pectinase solution or in washings from their culture. But young mycelial inocula readily attacked the tissue.

Adams, J. F. et al. 1920. *Phytopathology* **10**: 535 (on germinating maize).
Bouwkamp, J. C. et al. 1971. *Pl. Dis. Reptr* **55**: 1097 (fungicide control on sweet potato root pieces; **51**, 3039).
Briton-Jones, H. R. 1923. *Bull. Minist. Agric. Egypt tech. scient. Serv.* 19, 8 pp. (on cotton; **2**: 449).
Buckley, P. M. et al. 1968. *J. Bact.* **95**: 2365 (fine structure of germinating spores with *Rhizopus arrhizus*).
Bussel, J. et al. 1969. *Phytopathology* **59**: 946 (effect of anaerobiosis on germination & survival of spores; **49**, 29).
Butler, E. E. et al. 1963. *Ibid* **53**: 1016 (on tomato, role of *Drosophila melanogaster* with other fungi; **43**, 576).
Chowdhury, S. 1949. *J. Indian bot. Soc.* **28**: 45 (on jackfruit; **28**: 530).
Crisanto, J. 1924. *Philipp. Agric.* **12**: 465 (as above; **4**: 41).
Daines, R. H. 1942. *Bull. New Jersey agric. Exp. Stn* 698, 14 pp. (on sweet potato **23**: 54).
Fothergill, P. G. et al. 1957. *J. gen. Microbiol.* **17**: 631 (mineral nutrition).
Gibson, I. A. S. et al. 1953. *Emp. J. exp. Agric.* **21**: 226 (on groundnut inter alia; **33**: 65).
Harter, L. L. et al. 1921a. *J. agric. Res.* **22**: 511 (susceptibility of sweet potato cvs).
—— ——. 1921b. *Phytopathology* **11**: 279 (on sweet potato with *R. oryzae*; **1**: 272).
—— ——. 1922. *Ibid* **12**: 205 (on fruits & vegetables; **1**: 433).
—— ——. 1923. *J. agric. Res.* **26**: 363 (physiologic variation).
Hawker, L. E. et al. 1963. *J. gen. Microbiol.* **32**: 295 (fine structure of spore maturation & germination with *R. sexualis*).
Kirkpatrick, T. W. 1925. *Bull. Minist. Agric. Egypt tech. scient. Serv.* 54, 28 pp. (on cotton; **4**: 540).
Kurata, M. et al. 1975. *Bull. Kochi prefect. Inst. agric. For. Sci.* 7: 15 (on red pepper; **56**, 3368).
Lauritzen, J. I. et al. 1923. *J. agric. Res.* **24**: 441 (on sweet potato, **3**: 546).
—— ——. 1925. *Ibid* **30**: 793 (as above, effect of temp.; **4**: 699).
—— ——. 1926. *Ibid* **33**: 527 (as above, effect of RH; **6**: 182).
McClure, T. T. 1959. *Phytopathology* **49**: 359 (as above, effects of chilling, recuring & hydrowarming after storage; **38**: 710).
McMillan, R. T. 1974. *Proc. Fla St. hort. Soc.* 87: 392 (on jackfruit, fungicides; **55**, 6055).
Marsh, P. B. et al. 1965. *Phytopathology* **55**: 52 (on cotton, effects on fibre; **44**, 1577).
Martin, W. J. 1964. *Pl. Dis. Reptr* **48**: 606 (on sweet potato, fungicides; **44**, 537).
Matsumoto, T. T. et al. 1967. *Phytopathology* **57**: 881 (sensitivity to chilling; **46**, 3505).
—— ——. 1969. *Ibid* **59**: 863 (chilling induced fine structural changes in spores; **48**, 3318).
Menke, G. H. et al. 1964. *Z. PflKrankh. PflPath. PflSchutz* **71**: 128 (on carrot, physiology; **43**, 3076).
Patil, S. S. et al. 1973. *Pl. Dis Reptr* **57**: 86 (on papaw inter alia, fungicides; **52**, 3000).
Pierson, C. F. 1966. *Phytopathology* **56**: 276 (on *Amygdalus persica*, temp.; **45**, 2543).
Shapovalov, M. 1927. *J. agric. Res.* **35**: 307 (on cotton with *Aspergillus*; **7**: 96).
Smith, W. L. et al. 1965. *Phytopathology* **55**: 604 (effects of temp. & RH on spore germination with *Sclerotina fructicola*; **44**, 3106).
Srivastava, D. N. et al. 1959. *Ibid* **49**: 400 (on sweet potato, infection; **39**: 37).
Stobo, K. M. et al. 1974. *Phytopathol. Mediterranea* **13**: 133 (on carrot; **55**, 5393).
Thorne, S. N. 1972. *Jnl Fd Technol.* 7: 139 (as above).
Thorne, S. 1975. *J. Sci. Fd Agric.* **26**: 933 (as above; **55**, 5988).
Weimer, J. L. et al. 1923. *J. agric. Res.* **24**: 1 (temp. & *Rhizopus* spp.; **2**: 564).
Welch, N. C. et al. 1966. *Calif. Agric.* **20**(11): 14 (on sweet potato, fungicides; **46**, 1324).

RHYNCHOSPORIUM Heinsen, in Sacc., *Sylloge Fung.* 18: 540, 1906.
(Moniliaceae)

Parasitic, causing spots on leaves, sterile MYCELIUM sparse in mesophyll of host, mycelium subcuticular at first, later developing into a superficial fertile stroma (not a true sporodochium) more or less covering the leaf spot. CONIDIOPHORES absent. CONIDIA one septate, hyaline, on cells of stroma (from R. M. Caldwell, *J. agric. Res.* **55**: 183, 1937).

R. oryzae Hashioka & Yokogi causes leaf scald of rice. The *conidia* are borne on superficial stromata arising on the lesions; they are fusiform or oblong fusoid, tapering toward both ends, curved, 10.8–13.2 × 3.7–4 μ, septate near the middle, but often the 2 cells are somewhat unequal, rarely 2 septate, not constricted at the septum, epispore very thin (S. H. Ou, *Rice diseases*: 226, 1972). Ou et al. described a perfect state in *Metasphaeria albescens* Thüm. *R. secalis* (Oudem.) J. J. Davis (causing leaf scald of barley, rye and other temperate Gramineae) has conidia which are characteristically obliquely beaked, 11–35 (19) × 3–5.5 (4) μ (Owen). *R. orthosporum* Caldwell has uniformly cylindrical conidia, 14.4–19.4 × 2.3–4.7 μ; described from *Dactylis glomerata*.

R. oryzae, which Peregrine et al. considered to be a potentially serious plant pathogen, has been reported from: Australia, Brunei, Costa Rica, Ghana, Guatemala, Honduras, India, Ivory Coast, Japan, Malaysia, Nicaragua, Nigeria, Panama, Salvador, Thailand, USA (Louisiana) and Vietnam (Map 492). The symptoms usually show on mature leaves, mostly near the tip but sometimes at the margin. The lesions are oblong to irregular, water-soaked, more or less restricted by the veins, developing into large ellipsoidal areas, 1–5 × 0.5–1 cm, bands of dark brown and light brown tissue give the lesions their striking zonate pattern. Lesion growth and coalescence may result in a large part of the leaf being affected. The leaves dry out, turn a bleached straw colour, with a brown margin and faint zonation. Browning is caused on the inflorescence and leaf sheath (Hashioka et al.). De Guttiérrez described other symptoms. Good in vitro growth occurs at 20–27° (opt. of 20° and 24° have been reported). No specific control measures appear to have been described and the biology of *R. oryzae* is relatively unknown.

Caldwell, R. M. 1937. *J. agric. Res.* **55**: 175 (*Rhynchosporium secalis* general & *R. orthosporum* sp. nov.; **17**: 32).

De Guttiérrez, L. C. 1960. *Pl. Dis. Reptr* **44**: 294 (*R. oryzae* in Costa Rica; **39**: 703).

Hashioka, Y. et al. 1955. *Contr. Lab. Pl. Dis. Sci. Fac. Agric. Gifu Univ.* 6: 46 (*R. oryzae*; **34**: 482).

Makino, M. et al. 1959. *Sci. Rep. Fac. Agric. Meijo Univ.* 3: 17 (*R. oryzae* conidial germination & fungicide tolerance; **38**: 594).

Ou, S. H. et al. 1978. *Pl. Dis. Reptr* **62**: 524 (perfect state of *R. oryzae*).

Owen, H. 1973. *C.M.I. Descr. pathog. Fungi Bact.* 387 (*R. secalis*).

Peregrine, W. T. H. et al. 1974. *PANS* **20**: 177 (*R. oryzae* in Brunei; **53**, 4436).

RIGIDOPORUS Murrill, *Bull. Torrey Bot. Club* 32: 478, 1905.
(Polyporaceae)

HYMENOPHORE annual, at times reviving, epixylous, sessile, dimidiate, conchate, simple or imbricate; surface pelliculose, multizonate, margin thin, incurved when dry; CONTEXT thin, white, woody, very rigid when dry, TUBES minute, regular, light brown, mouths pruinose when young; SPORES smooth, hyaline (W. A. Murrill l.c.; see D. N. Pegler for taxonomy in *The fungi* (edited by G. C. Ainsworth et al.) Vol. IV B: 407, 410, 1973).

R. ulmarius (Sow. ex Fr.) Imazeki occurs on *Ulmus* (elm butt rot) and other temperate trees. Donk (*Persoonia* **4**: 341, 1966) stated that up until now the genus has been used for more or less distinctly pileate spp. although some of these may form strictly resupinate sporophores. The generic limits will need to be extended also to include some so-called 'resupinate' spp. which for some time have been treated in a distinct genus under the misapplied names *Podopora* and *Physisporinus* (= *Poria*).

Pegler, D. N. et al. 1968. *C.M.I. Descr. pathog. Fungi Bact.* 199 (*Rigidoporus ulmarius*).

Rigidoporus lignosus (Klotzsch) Imazeki, *Bull. Govt For. Exp. Stn Meguro* 57: 118, 1952.
Polyporus lignosus Klotzsch, 1833
Fomes lignosus (Klotzsch) Bres., 1912
F. auberianus (Mont.) Murr., 1905.

CARPOPHORE annual, rarely perennial, sessile with broad basal attachment, often imbricate, occasionally resupinate. PILEUS 3–10 × 4–22 × 0.3–1.5 cm, applanate, dimidiate, thin; upper surface reddish brown, with a bright yellow margin later fading to a uniform wood colour, velutinate then glabrescent, concentrically zonate–sulcate; margin thin, decurrent. CONTEXT up to 1 cm thick, white to yellowish, fibrous to woody. PORE SURFACE bright reddish brown, fading on drying; pores irregular, round to angular, 5–9 per mm, 50–140 μ diam., dissepiments 15–40 μ thick; tubes occasionally stratified, up to 6 mm long, reddish brown. BASIDIOSPORES 3.5–4.5 × 3.5–4 (4.2 × 3.7) μ, subglobose, hyaline, smooth, thin walled, with few contents. BASIDIA short clavate, 4 spored. CYSTIDIA absent. HYPHAL SYSTEM monomitic, non-agglutinated. GENERATIVE HYPHAE 2–7 μ diam., hyaline or with a pale brownish tint, wall thin to slightly thickened, freely branching, simple septate. Radial growth in culture 5–5.5 cm in 7 days at 25° on malt agar; mat white becoming brownish with marginal rhizomorphic strands. Conidia reported. Distinguished from *R.*

zonalis (q.v.) by the absence of cystidia (D. N. Pegler & J. M. Waterston, *CMI Descr.* 198, 1968).

This fungus, causing white root rot of tropical crops, was first described as a pathogen of rubber (*Hevea brasiliensis*) by H. N. Ridley in Malaysia (W.) in 1904. It occurs on many plants but has been particularly studied on rubber where it caused huge losses during the development of cultivation since the turn of the century. Sharples (*Diseases and pests of the rubber tree*) refers to losses of 18% over areas of 700 acres in Malaysia and, quoting Steinmann, he mentions losses in 2 Sumatran estates (2750 acres) of 70 000 trees. Fox (1961a) stated that in a replanting neglected during the last war a density of 145 trees/acre had been reduced to 94/acre of which a further *c.* 19/acre were also infected. In 16 years nearly half the original stand had become infected; the life of a modern clone is 30–40 years. It is unfortunate that there is no exhaustive, recent, monographic treatment of *R. lignosus* since this evolution of the study of white root rot, the interplay of agriculture, biology and economics in a tropical tree crop root disease, is unique. Early accounts can be found in the books on rubber diseases by Petch and Sharples; Fox (1961a & b, 1965 and *Armillariella mellea* q.v. 1970) has contributed review treatments. Reference should also be made to the annual reports of the rubber research institutes of Malaysia and Sri Lanka, particularly the former.

R. lignosus is widespread in S.E. Asia, Sri Lanka, central, E. and W. Africa. Although it has been reported from parts of Central and S. America the identity of the fungus there with the one described in the Old World is uncertain (Map 176). The external white rhizomorphs are firmly attached to the roots and collar, they become yellowish and later reddish. The leading edge of mycelium, advancing vigorously, appears as an almost continuous sheath, like a fan on the bark surface; the branching, well defined rhizomorphs develop later from it and they seldom appear above soil level. The rot is undifferentiated. Trees which have had their tap roots destroyed may show a broad fluting of the lower stem due to a relatively greater girthing rate over the main laterals. Where only the tap root has been attacked foliar symptoms of chlorosis and leaf fall may be long delayed. Above ground symptoms usually occur, although in Africa they can be absent. The carpophores appear much later being commonly found on long dead laterals, tap roots, trunks and stumps.

Penetration is by rhizomorphs during root contact and by spore invasion of cut surfaces; the former process has been studied more intensively. On roots entry is through lenticels, scars of feeding roots and wounds (moribund roots may be more susceptible); entry at the collar is facilitated by fissuring of the bark. The rhizomorphs on the roots are epiphytic for considerable distances beyond the point of penetration (up to 2 m) and in soil can grow at 30 cm/month on roots. But where epiphytic growth has reached the collar this may be deeply penetrated whilst several feet of the lateral roots, along which the rhizomorphs grew, have only epiphytic growth. Growth inwards (along roots) is almost twice as fast as that outwards. Disturbance of the soil and host wounding tend to increase spread. Three-year-old trees can be killed in 6 months. Host defence mechanisms are the development of flaky bark by the cork cambium, wound barriers and growth of callus. The wood is penetrated through the medullary rays. Inoculum potential is a very important factor and infective sources are considered as potential rather than actual. Stumps are major sources of new infections which decrease in frequency as the size of the infective source becomes smaller. Slight growth may occur through soil from a substantial food base but ceases if this is cut off. Fox (1965) pointed out that Bancroft's observations on white root rot are amongst the first that demonstrate the phenomenon of inoculum potential. In one experiment woody inocula of 280 cm^3 killed 2.5% of 6-year-old trees whilst those of 850 cm^3 killed 17.5%. No 13- or 23-year-old trees were killed by the larger inocula. The extent of viability in the soil depends on the size of the infected source and the speed with which normal decay takes place. In stumps the fungus can persist for several years and up to 4 years in roots of 7.5 cm diam. in soil without a leguminous cover; with such a cover viability drops to 12–15 months. Isolates do not differ in culture though different opt. temps for growth have been reported. There is a requirement for thiamine. Fructifications have been produced on rubber wood in contact with agar cultures which have also been used to infect trees in situ in Sri Lanka and Malaysia.

In the early history of this disease and comparable ones (*Armillariella mellea, Ganoderma philippii* and *Phellinus noxius* q.v.) spread arose from felled trees in indigenous forest cleared for cultivation. Control problems now relate more to areas where an existing woody crop is being replanted or replaced by another one. These diseases are of no importance where planting is done on land with a long history of cultivation of non-woody crops. Rubber trees of all ages are susceptible and there are no clonal differences in susceptibility. In a new stand max. incidence of the disease occurs in the third to fourth year and control measures in the first 5 years of growth are the most vital. In replanting the objectives (e.g., see Newsam, Newsam et al.) in control are:

1. to reduce the chances of spore infection of cut surfaces by facilitiating wood decay, protecting and reducing the number of such surfaces;
2. to promote the most rapid, general decay of all woody tissue by a quick kill (mechanical clearing is very effective in reducing subsequent disease but has the disadvantage of cost and causing excessive loss of organic matter);
3. to accelerate decay of potential inocula through the early years of tree growth, create soil and root conditions which dissipate rhizomorphic inocula rendering it harmless and increase microbial antagonism (leguminous covers are superior to others);
4. to detect actual infective sources by tree inspection;
5. to prevent further epiphytic growth on root and collar by fungicide treatment.

The following recommendations for control also serve as a guide for root pathogens of tropical tree crops in general and which have similar ecological niches. They are based on Malaysian practice (Anon.; and see Peries and others for Sri Lanka).

a. All trees of the old planting are poisoned standing or felled; if the latter, poison stump and creosote cut surface. Mechanical clearing may be carried out but there are agricultural and economic disadvantages.
b. Establish mixed, non-woody leguminous cover between planting rows which are kept clean weeded (*c.* 2 m wide).
c. Inspect foliage quarterly when stand is a year old. Where canopy symptoms are found remove the diseased tree and inspect collars of neighbouring trees in row until a healthy tree is found. Treat diseased trees (collar inspected) using the recommended fungicide to prevent further epiphytic growth. Reinspect in later rounds.
d. Eradicate and burn all diseased sources and diseased roots of young trees in clean weeded plant row. Sources outside row are left alone and diseased roots traced only to edge of row.
e. If a stump outside planting row continually causes new infections isolate by a trench, sever all lateral roots and remove dead ones. Leave perched on tap root.
f. Before final thinning begins collar inspect a part of each field primarily to detect centres of the slower moving pathogens, *G. philippii* and *P. noxius*, and because large trees are slower to develop foliar symptoms. Extend area sampled if disease is found in parts of the field where the earlier foliar inspection had failed to detect it.

Anon. 1974. *Plrs' Bull. Rubb. Res. Inst. Malaysia* 133: 111; 134: 157 (root diseases of rubber; detection, recognition & control; **54**, 1860, 3459).

Bancroft, K. 1912. *Bull. Dep. Fed. Malay St.* 13 (inoculum potential).

Boisson, C. 1968. *C. r. hebd. séance. Acad. Sci. Paris* Sér. D. **267**: 1435 (in vitro growth; **49**, 675).

Bose, S. R. et al. 1957. *Trans. Br. mycol. Soc.* **40**: 456 (taxonomy & identity; **37**: 267).

Chevaugeon, J. 1959. *Rev. Mycol.* **24**: 39 (in Ivory Coast, review; **39**: 52).

Cronshey, J. F. H. et al. 1939. *Arch. Rubb. cult. Nederl.-Indie* **23**: 163 (cultural control in Sumatra; **19**: 44).

De Jong, W. H. 1933. *Ibid* **17**: 83 (inoculum potential; **12**: 720).

Fassi, B. 1964. *Publs Inst. natn. Etude agron. Congo belge* Ser. Sci. 105: 54 pp. (disease in Zaire; **44**, 223).

Fox, R. A. 1961a. *Rep. Commonw. mycol. Conf. 1960*: 41 (review, control in Malaysia; **40**: 723).

——. 1961b. *Ibid*: 97 (role of fungicides; **40**: 724).

——. 1961c. *Proc. Nat. Rubb. Res. Conf. Kuala Lumpur 1960*: 473 (culture, inoculation & identity of pathogen; **40**: 624).

——. 1965. In *Ecology of soil-borne plant pathogens* (editors K. F. Baker et al.): 348, John Murray (biological eradication in control, 58 ref.; **45**, 1679g).

——. 1966. *J. Rubb. Res. Inst. Malaya* **19**: 231 (fungicidal dressings; **46**, 1705).

Hanson, E. W. 1938. *Phytopathology* **28**: 8 (temp. & in vitro growth; **17**: 484).

Harrar, J. G. 1937. *Tech. Bull. Minn. agric. Exp. Stn* 123 (identity & in vitro growth; **17**: 553).

Hilton, R. N. 1961. *Proc. Nat. Rubb. Res. Conf. Kuala Lumpur 1960*: 496 (spore dispersal; **40**: 625).

Hutchison, F. W. 1953. *Arch. Rubb. cult. Nederl.-Indie* (extra number): 117 (cultural control; **33**: 257).

Rigidoporus zonalis

Hutchison, F. W. 1961. *Proc. Nat. Rubb. Res. Conf. Kuala Lumpur 1960*: 483 (factors affecting incidence & control; **40**: 624).
John, K. P. 1958. *J. Rubb. Res. Inst. Malaya* **15**: 223 (inoculation; **38**: 335).
——. 1960. *Ibid* **16**: 173 (viability in soil of *Rigidoporus lignosus, Ganoderma philippii* & *Phellinus noxius*; **40**: 427).
——. 1965. *Ibid* **19**: 17 (spore infection of stumps; **45**: 201).
——. 1966. *Ibid* **19**: 226 (inoculum potential & host age; **46**: 1706).
Martin, R. et al. 1969. *Ibid* **21**: 96 (general, from Ivory Coast; **49**, 2165f).
Momoh, Z. O. 1976. *PANS* **22**: 43 (on teak; **55**, 3763).
Napper, R. P. N. 1932. *J. Rubb. Res. Inst. Malaya* **4**: 5, 34 (cultural control; **12**: 52, 54).
Newsam, A. 1954. *Q. Circ. Rubb. Res. Inst. Ceylon* **29** (1953): 78 (general in replanted areas; **34**: 61).
——. 1963. *Plrs' Bull. Rubb. Res. Inst. Malaya* 68: 177 (covers & root disease; **43**, 2042).
—— et al. 1964. *Ibid* 75: 207, 225, 238 (conference on root disease, various aspects; **44**, 1222a,d & f).
——. 1967. *Ibid* 92: 175 (clearing methods & control; **47**, 881).
—— ——. 1967. *J. Rubb. Res. Inst. Malaya* **20**: 1 (decay of rubber wood; **47**, 1257).
Palmer, J. A. 1967. *Plrs' Bull. Rubb. Res. Inst. Malaya* 92: 183 (tree poisoning).
Peries, O. S. et al. 1965. *Q. Jl Rubb. Res. Inst. Ceylon* **41**: 81 (control; **45**, 1891).
——. 1970. In *Root diseases and soil-borne pathogens* (editors T. A. Toussoun et al.): 191, Univ. California Press (economics of control).
—— ——. 1973. *Ann. appl. Biol.* **73**: 1 (histology on rubber; **52**, 2010).
Petch, T. 1928. *Trans. Br. mycol. Soc.* **13**: 238 (nomenclature; **8**: 266).
——. 1928. *Tea Q.* **1**: 64 (identity; **8**: 72).
Riggenbach, A. 1960. *Phytopathol. Z.* **40**: 187 (in vitro growth, thiamine & enzymes; **40**: 486).
Van Hell, W. F. 1948. *Arch. Rubb. cult.* **26**: 221 (control in Sumatra; **28**: 240).
Wijewantha, R. T. 1964. *Trop. Agric.* **41**: 69 (environment & incidence in Sri Lanka; **43**, 2041).
Wong, P. W. 1964. *J. Rubb. Res. Inst. Malaya* **18**: 231 (inhibitory substances in *Mikania cordata*; **45**, 590).

Rigidoporus zonalis (Berk.) Imazeki, *Bull. Govt For. Exp. Stn Meguro* **57**: 119, 1952.
Polyporus zonalis Berk., 1843
Fomes rugulosus (Lév.) Cke., 1885
Polystictus rigidus (Lév.) Cke., 1886
Rigidoporus surinamensis (Miq.) Murr., 1907.

CARPOPHORE annual, often imbricate, substipitate to sessile, usually attached by a narrow base. PILEUS 0.5–7 × 1–9 × 0.1–0.5 cm, dimidiate to flabelliform; upper surface pinkish buff to reddish brown, drying to pale yellow brown, velutinate then glabrescent, concentrically zonate sulcate; margin concolorous, thin, decurved. CONTEXT up to 4 mm thick, white to yellowish, fibrous to woody. PORE SURFACE white to dingy livid grey; PORES irregular, round to angular, 6–9 per mm, 40–150 μ diam., dissepiments 25–50 μ thick; tubes 2–5 mm long, concolorous with the context or slightly darker. BASIDIOSPORES 4.5–6 × 4–5 (5 × 4.5) μ, subglobose, hyaline, smooth, thin walled, containing a large refractive oil guttule. BASIDIA 12–15 × 6–8.5, short clavate, 4 spored. CYSTIDIA present, rare to abundant, 25–43 × 6–12 μ, narrowly clavate, wall thickening distally, originating in the trama and becoming heavily encrusted with crystals towards the apex. HYPHAL SYSTEM monomitic, non-agglutinated. GENERATIVE HYPHAE 1.5–7 μ diam., hyaline or nearly so, wall thin to slightly thickened, freely branching, simple septate. In CULTURE radial growth 4–5 cm, 7 days at 25° on malt agar, mat white to cream with raised lines and ridges and marginal rhizomorphic strands. Test for extracellular oxidase positive, encrusted hyphal endings present, obverse side marbled.

Readily distinguished from *R. lignosus* (q.v.) by the constant presence of encrusted cystidia. On *Bauhinia purpurea, Ceiba pentandra, Ficus benghalensis, Hevea brasiliensis, Mangifera indica, Sapindus trifoliatus, Shorea robusta, Syzygium cuminii, Tectona grandis* and *Toona ciliata* (D. N. Pegler & J. M. Waterston, *CMI Descr.* 200, 1968).

R. zonalis causes a widespread white pocket rot, and occurs as a wound parasite on roots and butts of dicotyledons; it has been reported as causing a serious root rot of teak in India in association with *Peniophora* sp. It is present also as a saprophyte on timber in mines; wide differences in timber resistance to decay have been noted in Australia. The fungus is heterotrophic for thiamine and its growth was stimulated by methanolic and aqueous extracts of *Eucalyptus* wood. The resistance of 6 different timbers to fungal decay was related to the amount of methanolic extract present and to its toxicity as estimated by inhibition of *R. zonalis*. Spread is by airborne spores which can infect wood, and mycelium has been observed to spread over unsterilised wood and moist sand.

Bakshi, B. K. et al. 1963. *Trans. Br. mycol. Soc.* **46**: 426 (comparison with *Rigidoporus lignosus*; **43**, 1154).
Bose, S. R. et al. 1937. *Proc. R. Soc.* Ser. B **123**: 193 (enzymes, *R. zonalis* inter alia; **17**: 88).
——. 1938. *Mycologia* **30**: 683 (encrusted cystidia as diagnostic feature; **18**: 341).
Long, W. H. 1939. *Bull. Torrey Bot. Club* **66**: 625 (pocket rot symptoms; **19**: 244).
Osborne, L. D. et al. 1964. *Trans. Br. mycol. Soc.* **47**: 601 (thiamine requirement & timber durability; **44**, 1307).
—— ——. 1966. *Holzforschung* **20**: 160 (distribution & spread in mines; **46**, 2543).
Singh, S. et al. 1973. *Indian For.* **99**: 421 (heartwood rot in teak; **53**, 1580).
Thrower, L. B. et al. 1961. *Emp. For. Rev.* **40**: 242 (variability of Australian timbers to decay; **41**: 184).

ROSELLINIA de Not., *Giorn. Bot. Ital.* **1**: 334, 1844 (Sacc., *Sylloge Fung.* **1**: 252, 1882).
Pleosporopsis Oersted, 1865
Sphaeropyxis Bon., 1864.
(Xylariaceae)

PERITHECIA superficial, ostiolate, usually on bark, not embedded in a common stroma but often occurring in dense swarms upon a common blackish hyphal mat (subiculum), subglobose, smooth, black, ostiole papillate. ASCI unitunicate, with a well-developed apical ring, wall persistent. ASCOSPORES black, aseptate, often with minute colourless appendages (partly from R. W. G. Dennis, *British Ascomycetes*: 285, 1968).

R. aquila (Fr.) de Not. has ascospores that are uniseriate, elliptical, black, 17–24 × 7–8 μ, one side often somewhat flattened and bearing a longitudinal furrow, usually a minute round hyaline appendage at each end. The fungus is widely distributed in the N. hemisphere on dicotyledonous trees and conifers; it has been recorded on coffee.

Nowell, W. 1916. *W. Indian Bull.* **16**: 31 (*Rosellinia* root diseases in the lesser Antilles).
Rogers, J. D. et al. 1974. *Can. J. Bot.* **52**: 5 (cytology of *R. aquila* & *R. mammiformis*; **53**, 3375).
Saccas, A. M. 1956. *Agron. trop. Nogent* **11**: 551, 687 (*Rosellinia* spp. on coffee in Ubangui-Chari; **36**: 318).

Rosellinia arcuata Petch, *Ann. R. bot. Gdns Peradeniya* **6**: 175, 1916.

PERITHECIA gregarious, globose, brownish black or black, at first embedded in a subiculum of brown, septate hyphae, superficial, 1.5–2.5 mm diam., with a papillate ostiole. The perithecial wall is brittle and of 3 distinct layers. The outer layer is of several layers of irregularly rounded to polygonal, thick walled, dark brown cells, the thinner middle layer of flattened, elongated, thin walled cells and the innermost layer of polygonal cells which gradually become less thickened and less coloured as they progress towards the interior. ASCI cylindrical, long stalked, unitunicate, 8 spored, 250–325 × 5–7.5 μ, with an apical apparatus blued by iodine. ASCOSPORES monostichous, dark brown, aseptate, cymbiform, 30–48 × 3–6 μ, with the ends pointed and often abruptly narrowed, and with a longitudinal germ slit. PARAPHYSES numerous and filiform. Conidial state *Dematophora* type produced independently or in association with the perithecia. Synnemata up to 2–5 mm high; stipe 50–100 μ thick; branches 3–3.5 μ wide. CONIDIA ellipsoid or ovoid, hyaline to pale brown, 4–6 × 2–3 μ (A. Sivanesan & P. Holliday, *CMI Descr.* 353, 1972).

R. arcuata differs from *R. necatrix* (q.v.) in the absence of pyriform swellings on the hyphae and it has smaller ascospores than *R. bunodes* and *R. pepo* (q.v.). It causes black root rot, mainly of tropical and subtropical woody hosts; plurivorous but only serious on tea. The advancing edge of the mycelium is white shading to black. On the root surface the black network of strands gives a woolly appearance and beneath the bark fans (like stars) of white mycelium spread out on the wood. On tea the fungus may spread up the stem for a short distance; the bush often dies suddenly, the leaves remaining attached for some time. Records of this fungus are apparently infrequent. It has been reported from: the Central African Republic, Hong Kong, India, Indonesia (Java and Sumatra), Kenya, Papua New Guinea, Sri Lanka and Zaire.

This soil inhabitant is of some importance as a pathogen on tea where, in N.E. India, it is a primary root disease. In Sri Lanka it is probably less important, although a more frequent cause of loss than is *R. bunodes*. It spreads (as mycelium in organic debris) radially in tea plantations attacking vigorous bushes and can also cause severe damage in nurseries. When an outbreak occurs surrounding bushes should be destroyed, infective material removed and a cover crop such as *Crotalaria anagyroides* sown. Methyl bromide, which has been used primarily against *Poria hypolaterita* (? *P. hypobrunnea* q.v.), is an effective soil sterilant for *R. arcuata*.

Gadd, C. H. 1928. *Tea Q.* 1: 55 (inoculation on tea; **8**: 69).

Rosellinia bunodes (Berk. & Br.) Sacc., *Sylloge Fung.* **1**: 254, 1882.

Sphaeria bunodes Berk. & Br., 1873
Rosellinia echinata Massee, 1901
R. zingiberi Stevens & Atienza, 1931.

PERITHECIA densely aggregated, rarely free, 1.5–3 mm diam., partly immersed at first in a brown hyphal mat which forms a thin crust on the host substrate at the base of the perithecium, with the outer surface closely ornamented with warts (like scales) which are more or less pyramidal and concentrically arranged. Ostiole is papillate. The wall of the perithecium is 2 layered. The outer, thicker layer is composed of thick walled, dark brown, more or less rounded to polygonal cells and the thinner, inner layer is made up of polygonal, flattened, elongated, pale brown cells. The warts, 120–160 μ wide at the base and 125–200 μ high, are composed of cells similar to those on the outer wall. ASCI cylindrical, long stalked, unitunicate, 8 spored, 250–375 × 12–15 μ, with an apical apparatus blued by iodine. ASCOSPORES obliquely mono-bi-tristichous, dark brown, aseptate, cymbiform, 80–120 × 5–9 μ, with the ends prolonged into thread-like structures up to 25 μ long and with a longitudinal germinating slit. PARAPHYSES numerous and filiform. Conidial state *Dematophora* type as in *Rosellinia necatrix* (q.v.) produced independently or in association with the perithecia. Synnemata 1.7–3 mm high, stipe 30–100 μ thick, branches 2.5–3 μ wide. CONIDIA ellipsoid or obovoid, hyaline to pale brown, 4–10 × 2–5 μ (A. Sivanesan & P. Holliday, *CMI Descr.* 351, 1972).

R. bunodes causes black root rot mainly of tropical and subtropical woody hosts; plurivorous (*CMI Descr.* 351 lists 31 genera) but described mostly from cacao, quinine (*Cinchona* spp.), coffee, rubber (*Hevea brasiliensis*) and tea. Wilt and death of the whole plant or single branches may be the first signs of attack. At the collar the mycelial sheet is at first cream white shading to purplish black and may extend well above soil level in damp conditions. On the root surface the firm, black, branching strands are firmly applied and thicken into irregular knots. In the cortex the strands have a black periphery and a white core; in the wood they appear like black threads or sometimes as dots in TS. In culture the mycelium is white, later buff with black strands. *R. bunodes* has often been associated with *R. pepo* (q.v.). The latter has smaller ascospores and the characteristics of the mycelial fans and rhizomorphs differ; *R. pepo* has white, star like, rhizomorphic strands beneath the bark like *R. arcuata* (q.v.) which has smaller ascospores than *R. pepo*. *R. bunodes* is widespread in tropical America and also reported from the Central African Republic, India (Nilgris and Maharashtra), Indonesia (Java and Sumatra), Malaysia (W), Philippines, Sri Lanka and Zaire (Map 358).

R. bunodes appears to have a wider distribution than *R. pepo* and this has led to its causing damage to a greater range of crops. Their characteristics in the field were described by Nowell (and see his *Diseases of crop plants in the Lesser Antilles*) and Waterston (*R. pepo* q.v.). Both are inhabitants of soil and organic litter on the surface, through which they spread as mycelial fans. Losses have occurred mostly on woody crops on land recently cleared from forest. Spread through the soil is only slight and infection by spores probably unimportant. Mycelium spreads from the surface litter to the roots of the crop and thence to the collar, killing the plant tissues. Damage, particularly to coffee, has recently been reported from central and S. America but with the adoption of cultural measures of control the diseases caused should be of little significance. Where woody crops occur near forest, on land recently cleared from such habitats, removal of dead, plant debris on the surface may be necessary, especially near the stem. Leguminous covers on land cleared from forest or an earlier woody crop may be beneficial. Infected trees should be destroyed and the source of infection removed.

Castaño, J. J. 1953 & 1954. *Boln Inf. Colombia* 4 (37): 28; (38): 30; 5 (51): 32 (*Rosellinia* on coffee; **32**: 480; **33**: 480).

Fernández-Borrero, O. et al. 1964. *Cenicafé* 15: 126 (inoculations on coffee with *R. pepo*; **46**, 1582).

Garcia, L. A. A. 1945. *J. Agric. Univ. P. Rico* **29**: 1 (symptoms & pathogenicity, *R. bunodes* inter alia; **27**: 471).

Lopez-Duque, S. et al. 1966. *Cenicafé* **17**: 61 (effect of soil RH & pH on mycelial growth & infection, with *R. pepo*; **47**, 1135).

Nowell, W. 1916. *W. Indian Bull.* **16**: 31 (general with *R. pepo*).
Stevens, F. L. et al. 1931. *Philipp. Agric.* **20**: 171 (on ginger as *R. zingiberi*; **11**: 73).

Rosellinia necatrix Prill., *Bull. Soc. mycol. Fr.* **20**: 34, 1904.
stat. conid. *Dematophora necatrix* Hartig, 1883.

PERITHECIA densely aggregated, globose, black, shortly pedicellate at the base, 1–2 mm diam., embedded in ropy subiculum of brown, septate HYPHAE which form a thin crust on the host substrate at the base of the perithecium. The hyphae of the subiculum are of two types; some uniform in thickness, 5–8 μ wide and others exhibiting characteristic pyriform swellings, 2–3 times the diam. of the hyphae and formed immediately above a septum. The ostiole is papillate. The wall of the perithecium is 3 layered. The thicker outer layer is composed of dark brown, thick walled, rounded to polygonal cells and the innermost layer of cells similar to those of the outermost layer, these cells becoming gradually less coloured as they progress towards the interior. The pedicel is composed of fused, intertwined filaments with thickened walls. ASCI cylindrical, long stalked, unitunicate, 8 spored, 250–380 × 8–12 μ, with an apical apparatus blued by iodine. ASCOSPORES monostichous, cymbiform, straight or curved, dark brown, 30–50 × 5–8 μ, with a longitudinal germ slit running parallel to the long axis of the spore for about a third of its length. PARAPHYSES numerous and filiform.

CONIDIOPHORES produced independently or in association with perithecia on brown ropy synnemata which project straight outwards as rigid columns. Synnemata up to 1.5 mm high. STIPE 40–300 μ thick, composed of flexuous, intertwined, repeatedly branched threads, 2–3.5 μ thick, often branched dichotomously towards the apex and splaying out to form a pale to brown head. CONIDIOGENOUS CELLS polyblastic, integrated and terminal on branches or discrete, sympodial, geniculate, denticulate with short, thin walled, separating cells which break across the middle leaving a minute frill at each geniculation which corresponds to a frill at the base of each conidium. CONIDIA solitary, acropleurogenous, simple, ellipsoid or obovoid, hyaline to pale brown, aseptate, smooth, 3–4.5 × 2–2.5 μ (A. Sivanesan & P. Holliday, *CMI Descr.* 352, 1972).

The fungus causes white root rot, mainly on temperate fruit crops (see *CMI Descr.* 352 for hosts; and Mantell et al.). Crops on which *R. necatrix* has been reported include citrus, coffee, lucerne, ramie (*Boehmeria nivea*) and tea; distribution is wide (Map 306). It is distinguished from some other root rots of woody hosts by the absence of well defined rhizomorphs. The roots (young ones of tea being attacked first) are invested with a white mycelium which on older roots turns to brown and almost black; superficial black sclerotia may occur and the hyphae are swollen near the septa. Wilt and death of the tree may be slow or fairly rapid and infection is generally confined to the roots.

A low temp. pathogen with an opt. in soil of 20° (one of 24° has also been reported) or below and probably of little or no importance in the lowland tropics, although described from crops in these regions. Khan gave a review including the host range (170 spp.). Accounts of the disease on temperate tree crops are not given here. Most recent studies have been in Japan on fruit trees and tea (Abe and others). The fungus spreads from recently cleared forested areas. Woody mulches and grass covers increase spread whilst legume covers reduce it. On tea the disease is most severe in spring and summer. Infection of cereals occurs and survival in soil is over several years. Infective debris should be removed in replanted areas which have a history of infection in the previous crop. Chloropicrin injected into the soil gives control. Tilling before fumigation may alter its effectiveness in different soils which require different injection depths.

Abe, T. et al. 1953, 1954, 1955 & 1956. *Sci. Rep. Fac. Agric. Saikyo Univ.* 5: 93; 6: 153; 7: 49; 8: 74 (general on tea, nutrients in vitro & phytoxicity of culture filtrates; **34**: 822; **35**: 330; **36**: 65; **37**: 182).
Ieki, H. et al. 1969. *Ann. phytopathol. Soc. Japan* **35**: 76 (detection in soil; **48**, 2182).
Khan, A. H. 1959. *Biologia Lahore* **5**: 199 (review, 102 ref.; **40**: 90).
Mantell, S. H. et al. 1973. *Trans. Br. mycol. Soc.* **60**: 23 (on narcissus in the Scilly Isles; **52**, 2298).
Setsumi, I. 1970. *Bull. seric. Exp. Stn Japan* **24**: 287 (chloropicrin soil treatment; **50**, 2749).
Tanaka, H. et al. 1966. *Bull. hort. Res. Stn Japan* B, **5**: 119 (inoculation of citrus, effect of soil and plant covers; **45**, 3299).

Rosellinia pepo Pat., *Bull. Soc. mycol. Fr.* **24**: 9, 1908.

PERITHECIA solitary to gregarious, seated upon or partially immersed in a subiculum or byssus of reddish brown to black mycelium. They are globose to spherical, 1.5–3 mm diam., with a carbonaceous outer wall and conical, ostiolar papilla; the surface of the outer wall is ornamented and appears granulose. ASCI 8 spored, unitunicate, hyaline, cylindrical, 250–300 × 8–12 μ, with a prominent apical apparatus which gives an amyloid reaction, and a long, pedicellate stalk. PARAPHYSES numerous, hyaline, filiform. ASCOSPORES aseptate, inequilateral, fusiform to navicular with pointed ends, with a smooth outer wall which is hyaline at first, becoming pale brown and finally dark brown; 50–69 × 7–9 μ; a longitudinal germ slit 15–20 × 1 μ is present along one side of the spore. CONIDIAL state was referred to the genus *Graphium* by Saccas (*Rosellinia* q.v.) but it appears to be of the *Dematophora* type as in *R. necatrix* and *R. bunodes* (q.v.). SYNNEMATA 1–2 mm high and 30–70 μ diam., are composed of an aggregation of upright, parallel, brown, septate hyphae 3–4 μ diam. CONIDIA aseptate, hyaline, oblong to ovoid, 4–6 × 2 μ (C. Booth & P. Holliday, *CMI Descr.* 354, 1972).

Black root rot (mainly of tropical and subtropical woody hosts) caused by *R. pepo* is recorded less often than *R. bunodes* with which is has often been associated in the American tropics. It appears to be restricted to Central America, W. Indies and W. Africa; other records are doubtful (Map 298). At the collar the mycelial fan has a light grey margin, shading to brown or purplish black. On roots the greyish strands (like cobwebs) become black and coalesce into a woolly or felty mass. Beneath the bark, white fans (like stars) of mycelium occur on the surface of the wood. Thin plates of mycelium in the wood appear as zig-zag lines in section. In culture deep brown, later olive green; strands white becoming black. The behaviour of this soil inhabitant and its pathogenicity are similar to those described for *R. bunodes*. The best growth in vitro is at 25–27°. Waterston compared the characteristics, both in vivo and in vitro, of this pathogen with *R. bunodes* (q.v. for control).

Waterston, J. M. 1941. *Trop. Agric. Trin.* **18**: 174 (contrasted with *R. bunodes*; **21**: 162).

SAROCLADIUM W. Gams & D. Hawksw., *Kavaka* 3: 57, 1975 (publ. 1976).
(Moniliaceae)

RESEMBLES *Verticillium* and *Gliocladium*. COLONIES spreading with numerous CONIDIOPHORES arising from aerial or submerged hyphae, irregularly and densely branched with the branches and phialides arising in dense broom-like rows along the supporting cells. PHIALIDES narrowly cylindrical, hardly tapering towards the apices, lacking collarettes. CONIDIA aseptate, cylindrical, hyaline (from W. Gams & D. L. Hawksworth, l.c.).

Gams, W. et al. 1975. *Kavaka* 3: 57 (taxonomy, genus & spp.; **55**, 5189).

Sarocladium oryzae (Sawada) W. Gams & D. Hawksw., *Kavaka* 3: 58, 1975 (publ. 1976).
Acrocylindrium oryzae Sawada, 1922.

COLONIES on the natural host superficial, spreading pale pinkish or whitish; on malt agar reaching 13–15 mm diam. in 10 days at room temp., pale saffron, with a thin compacted cottony aerial mycelium, reverse darker orange ochre. VEGETATIVE HYPHAE 1–1.5 μ wide; in the aerial mycelium some wider hyphae up to 2.5 μ diam. with irregularly gnarled outlines. Sporulation abundant in the central parts of the colonies. Fertile hyphae giving a strong metachromophilic reaction with aniline blue. CONIDIOPHORES irregularly branched; sometimes with several branches arising in whorls at one level but more commonly with branches arising laterally in dense rows over a considerable length of the supporting cell; secondary branches often exceeding the primary ones; repeatedly branched conidiophores up to 60 μ or more in length and 2–2.5 μ wide at the base. CONIDIOGENOUS CELLS (phialides) arising from conidiophores or directly from undifferentiated vegetative hyphae, 30–40 μ long and 1.5–2 μ wide at the base when formed at the apex of slender conidiophores, 6–20 μ long and 1–1.5 μ wide when arising in dense broom-like fascicles; tips tapering to 0.6–1 μ in width, lacking any distinct collarette. CONIDIA formed in slimy masses, cylindrical with rounded ends, sometimes becoming slightly curved, hyaline, thin walled, smooth, aseptate, 3.5–7 × 0.8–1.5 μ. CHLAMYDOSPORES absent. Opt. growth at 27–33° reaching 27 mm diam. after 10 days (from W. Gams & D. L. Hawksworth l.c.).

S. oryzae causes sheath rot of rice. Ou (*Rice diseases*) stated that the disease is common in S.E. Asia and can result in severe damage in Taiwan (see Chakravarty et al. for loss). The fungus also occurs in Japan, Kenya, Nigeria and USA (Louisiana). In describing it Gams et al. (l.c.) also described *S. attenuatum* W. Gams & D. Hawksw. This fungus differs from *S. oryzae* in having more regularly verticillate conidiophores (although the branches of these are strongly appressed), somewhat less frequent solitary phialides, and the conidia tend to be longer, 4.5–8 (–14) × 0.6–1 μ, slightly narrower, taper gradually and have truncated ends. It is not clear whether both fungi are concerned in the disease or whether rice sheath rot should be attributed solely to *S. oryzae*. Agnihothrudu was apparently describing *S. attenuatum*.

The symptoms occur mostly on the upper leaf sheaths and are most conspicuous on the flag leaf sheath. The lesions are oblong to linear blotches with irregular margins up to 1.5 cm long; at first brown becoming grey in the centre with brown margins or greyish throughout. The spots coalesce to cover the whole sheath. The panicles of diseased plants emerge only partly or not at all; glumes are discoloured dark red or purple brown. The mechanism of infection needs further investigation. Amin et al., Chen et al. and Tasugi et al. all noted that infection was apparently increased by insect attack or other stem damage. Shahjahan et al. found that *S. oryzae* was seedborne and survived on straw and grain for > 1 year. Rice cvs vary in their reaction to infection (Amin; Chen et al.; Chung; Subramanian et al.).

Agnihothrudu, V. 1973. *Kavaka* **1**: 69 (in India, symptoms; **54**, 449).
Amin, K. S. et al. 1974. *Pl. Dis. Reptr* **58**: 358 (in India, general; **53**, 3964).
——. 1976. *Ibid* **60**: 72 (resistance; **55**, 4105).
Chakravarty, D. K. et al. 1978. *Ibid* **62**: 226 (loss; **57**, 4965).
Chen, C. C. et al. 1964. *Jnl Taiwan agric. Res.* **13**(2): 39 (general, resistance & effects of fertiliser; **45**, 2107).
Chen, M. J. 1957. *J. Agric. For. Taiwan* **6**: 84 (general).
Chung, H. S. 1975. *Korean J. Pl. Prot.* **14**: 23 (resistance; **55**, 236).
Hsieh, S. P. Y. et al. 1977. *Pl. Prot. Bull. Taiwan* **19**: 30 (host sterility & infection; **57**, 596).
Kawamura, E. 1940. *Ann. phytopathol. Soc. Japan* **10**: 55 (aetiology).
Shahjahan, A. K. M. et al. 1977. *Pl. Dis. Reptr* **61**: 307 (in USA; **56**, 5628).
Subramanian, C. L. et al. 1975. *Curr. Sci.* **44**: 405 (resistance; **55**, 237).
Tasugi, H. et al. 1956. *Bull. natn. Inst. agric. Sci. Tokyo* Ser. C. 6: 151 (general in Japan; **37**: 162).

SCIRRHIA Nits., in Fuckel, *Jahrb. Nass. Vereins f. Maturkunde* **23–24**: 220, 1870.
(Dothideaceae)

STROMA dark coloured, elongated, elliptical to linear, intra-epidermal or deeper; with partially erumpent, uniloculate PSEUDOTHECIA along the apex. ASCI thick walled, bitunicate. ASCOSPORES 1 septate near middle, hyaline to pale yellowish (from key by E. S. Luttrell in *The fungi* (edited by G. C. Ainsworth et al.) Vol. IV A: 173, 1973; R. W. G. Dennis, *British Ascomycetes*: 360, 1968).

Scirrhia acicola (Dearn.) Siggers, *Phytopathology* **29**: 1076, 1939.

Oligostroma acicola Dearn., 1926
Systremma acicola (Dearn.) Wolf & Barbour, 1941
Mycosphaerella dearnessii Barr, 1972
stat. conid. *Lecanosticta acicola* (Thüm.) Syd., 1924
Cryptosporium acicolum Thüm., 1878
Septoria acicola (Thüm.) Sacc., 1884
L. pini Syd., 1922.

(for a detailed discussion on taxonomy see Siggers; Wolf et al.)

ASCOSTROMATA (pseudothecia) on pine needles amphigenous, dark, linear, within dead tissue, becoming partially erumpent; uni- or multilocular, ostiolate, 300–2500 μ long; 300–500 μ wide; locules 40–80 μ diam. ASCI bitunicate, cylindrical to clavate, 8 spored, 35–50 × 7–9 μ. ASCOSPORES 2 celled, upper cell larger, dilute brown, 15–19 × 3.5–4.5 μ, each cell commonly with 2 prominent oil globules. SPERMOGONIA 40–80 μ diam. SPERMATIA 2–4 × 1 μ. ACERVULI on pine needles within brown spots, subepidermal becoming erumpent, stromatic with a multicellular dark brown basal stroma. CONIDIOPHORES simple or branched, septate, brown, wall smooth. CONIDIOGENOUS CELLS holoblastic, annellidic, pale brown, cylindrical, wall smooth. CONIDIA (blastospores) pale brown to dark brown, straight, or curved, mostly 1–3 septate, thick walled, wall rough to coarsely verruculose, 15–35 × 3–4 μ, base truncate, apex rounded (perfect state from F. A. Wolf & W. J. Barbour, *Phytopathology* **31**: 61, 1941; stat.

conid. from E. Punithalingam & I. A. S. Gibson, *CMI Descr.* 367, 1973).

Brown spot needle blight of *Pinus* spp. occurs in Canada, Colombia, Cuba, Greece, USA, USSR (Georgia) and Yugoslavia (Map 482). A list of pine spp. on which *S. acicola* occurs was given by Punithalingam et al. (l.c.). Siggers gave an account of the disease. Small, light green grey spots form on the needles; they become necrotic and form a narrow band killing the distal part of the needle. Spots may form throughout the length of the needle. *S. acicola* causes an important disease of *P. palustris* at the nursery stage and in the early years after planting out. Infection may kill young trees if severe defoliation takes place over at least 3 successive years. Less severe attacks may delay emergence of the young trees from the 'grass stage'. Germination of the conidia on the host was described by Setliff et al. Infection of immature needles was followed by an incubation period of 8–11 weeks in *P. palustris* (2-year-old plants) but *P. taeda* was not affected (Snow). On seedlings of *P. taeda* the incubation period was 4–7 months (Parris). Susceptibility of needles decreases with age. Seedlings inoculated at 8 weeks showed severe infection; those at 14 weeks were moderately resistant. It was suggested that light enhances infection by stimulating stomatal opening; without light the fungus behaves mainly as a wound pathogen. Exposure of seedlings to high RH before and after inoculation increased infection and shortened the incubation period. Infection occurs over a wide temp. range but most occurred under a 30° day and a 21° night; a regime of 35° and 27° was inhibitory (Kais 1976b). The fungus has a growth opt. just above 25°. Spread is by splash dispersed conidia. Kais (1971) reported dispersal of ascospores during wet periods. Nicholls et al. found no evidence of windborne ascospores and considered that conidia were the most important agents of spread. Endospores have been described but their function is unknown. Various fungicides are effective. Bordeaux gives good control; chlorothalonil is also satisfactory as a substitute for Bordeaux. Benomyl, captafol, cycloheximide and maneb have been used. Heritable resistance occurs in *P. palustris.*

Crosby, E. S. 1966. *Phytopathology* **56**: 720 (endospores; **45**, 3229).

Derr, H. J. et al. 1970. *For. Sci.* **16**: 204 (resistance in *Pinus palustris*; **50**, 314).

Kais, A. G. 1971. *Pl. Dis. Reptr* **55**: 309 (spore dispersal in S. USA).

——. 1975a. *Ibid* **59**: 686 (fungicides; **55**, 1975).

——. 1975b. *Phytopathology* **65**: 1389 (effects of light, RH & temp. on infection; **55**, 3317).

——. 1977. *Ibid* **67**: 686 (effects of needle age & inoculum potential on susceptibility).

McGrath, W. T. et al. 1972. *Pl. Dis. Reptr* **56**: 99 (sporulation peaks & Bordeaux; **51**, 3598).

Nicholls, T. H. et al. 1973. *Ibid* **57**: 55 (spread; **52**, 3445).

Parris, G. K. 1967. *Ibid* **51**: 552 (field infection; **46**, 3230).

Setliff, E. C. et al. 1974. *Phytopathology* **64**: 1462 (conidial germination on pine needles; **54**, 3004).

Siggers, P. V. 1944. *Tech. Bull. U.S. Dep. Agric.* 870, 36 pp. (general; **23**: 505).

Snow, G. A. 1961. *Phytopathology* **51**: 186 (inoculation; **40**: 636).

Wolf, F. A. et al. 1941. *Ibid* **31**: 61 (morphology & taxonomy; **20**: 185).

Scirrhia pini Funk & Parker, *Can. J. Bot.* **44**: 1171, 1966.

stat. conid. *Dothistroma pini* Hulbary, 1941
Cytosporina septospora Doroguine, 1911
D. septospora (Doroguine) Morelet, 1968.

ASCOSTROMATA (pseudothecia) on pine needles, subepidermal, becoming erumpent, black multilocular, up to 600 μ wide, wall pseudoparenchymatous and composed of dark brown highly sclerotised cells on the outside. ASCI cylindrical to subclavate, intermixed among chains of subglobose hyaline cells (pseudoparaphysoids), $35–50 \times 7–10\ \mu$, bitunicate, 8 spored. ASCOSPORES biseriate, hyaline, fusiform to ellipsoidal, 2 celled, slightly constricted at the septum, the 2 cells unequal, guttulate, apices rounded, $11–16 \times 3–4\ \mu$. PYCNIDIAL STROMA on needles, subepidermal becoming erumpent, dark brown to black, up to 1.5 mm, sometimes divided, basal stroma multicellular, composed of dark brown sclerotised cells. CONIDIOGENOUS CELLS holoblastic, hyaline, lining the inside of the pycnidial stroma. CONIDIA (blastospores) hyaline, straight or slightly curved, 1–5 septate (usually 3), $25–60 \times 2–3\ \mu$. MICROCONIDIA (spermatia) hyaline, $1–2 \times 0.5\ \mu$. Ascospore and conidial isolates are indistinguishable on culture media. COLONIES on malt agar floccose with purplish brown mycelium accompanied by reddening of the agar through diffusion of pigment(s) into the medium. Three vars of the stat. conid., on the basis of conidial dimensions, have been reported: *D. pini* Hulbary var. *pini*, *D. pini* var. *linearis* Thyr & Shaw and *D. pini* var. *keniensis* Ivory. On pines including *Pinus radiata* and its hybrids, *P. halepensis, P. canariensis, P. caribaea, P. ponderosa, P. nigra* and others, *Pseudotsuga menziesii* and *Larix decidua* (E. Punithalingam & I. A. S. Gibson, *CMI Descr.* 368, 1973).

Red band needle blight of pines has been described as probably the most important disease of tropical exotic pine crops by Gibson (1972) who fully reviewed the work on *S. pini*. The fungus is widespread (Map 419) and became serious in Africa, Chile and New Zealand after its spread to plantings of *P. radiata* which is susceptible and the most important economic sp. The disease has also become serious in N. America, damaging in particular *P. nigra, P. ponderosa* and *P. contorta* in amenity, shelterbelt and Christmas tree crops. Ito et al., who gave some more hosts, considered red band needle blight to be of minor importance in Japan. Infection causes a foliage blight and necrosis. The first symptom is a chlorotic spot on the needle; it spreads to become a necrotic band. The distal parts of the needle die and eventually bear conidial stromata; the asci are also formed at this stage. The dead needles of infected pines often show a reddish tint and banding. The most serious results of an attack are a reduction of growth rate brought about through defoliation; tree death can occur. Conidia, the primary inoculum, are splash dispersed.

Penetration by the germinating conidia is through the stomata and appressoria may or may not be formed. Cool wet conditions are required for severe outbreaks of the disease which has mostly occurred between 25° and 40° N. and S. of the equator and in tropical highlands. Both conidial germination and the infection process have an opt. temp. of *c.* 17°. There is a variable period between infection and the appearance of symptoms; in E. Africa on *P. radiata* this was 32 days in the warmest months and 67 days in the colder ones (Ivory 1972b). At 13–24° and RH 70–100% Parker found that infection by conidia was greater at the lower temps and at the higher RH values. On *P. nigra* and *P. ponderosa* Peterson reported an incubation time of 11–16 weeks. Gadgil (1974) found that infection increased greatly under continuous moisture with a day and night temp. regime of 20° and 12° and, to a lesser extent, with one of 24° and 16°. Infection was greater on foliage >1 year old than on that <1 year. Gadgil et al. found that the severity of infection (cuttings) decreased with decreasing light; a light effect on the host rather than on the fungus.

Some spp. (*P. radiata* and *P. canariensis*) are most susceptible when young and become more resistant with age. In *P. radiata* the first signs of resistance may show at 8 years, but the trees may be 15 years old before this whole tree characteristic shows. Higher levels of resistance are needed where rainfall is high. Apparently light has no effect on host penetration. Resistance was induced by shade 5–20 days after inoculation but shade applied later was progressively less effective (Ivory 1972a). Other pines (*P. ponderosa* and *P. nigra*) remain fully susceptible throughout their life span. *P. patula, P. elliottii, P. taeda* and other spp. are usually resistant. Inoculum in cast, infected foliage on the forest floor survives for 2–6 months in moist conditions.

Gibson discussed control measures, the gross impact of the disease and net loss on the crop. In 1954 it was found that Bordeaux spraying was effective. The practical difficulties of applying copper fungicides could not then be overcome in E. Africa but they have been used successfully in New Zealand (2–4 low volume, aerial applications). Copper can also be used in nurseries. Treatments with anilazine, benomyl and chlorothalonil give results similar to those of copper which has, apparently, both protectant and eradicant actions. The disease's long incubation period contributes to the slow re-establishment of infection after fungicide treatment. In E. Africa *P. radiata* selections which show early development of mature plant resistance have been made; no complete resistance in this sp. has been found. In Africa the substitution of the susceptible *P. radiata* by the resistant *P. patula* and other spp. has meant lower production, nevertheless. A promising approach to control has been through the vegetative propagation of *P. radiata* from older trees which are completely resistant. In Chile red band needle blight, unlike New Zealand and Africa, has not proved to be chronically serious (Gibson 1974).

Bassett, C. 1972. *Fm For.* **14**: 47 (distribution, host susceptibility & control in New Zealand; **53**, 1103).

Gadgil, P. D. 1974. *N.Z. Jl For. Sci.* **4**: 495 (effects of temp. & leaf wetness on infection; **54**, 5580).

—— et al. 1976. *Ibid* **6**: 67 (effect of light intensity on infection; **56**, 872).

——. 1977. *Ibid* **7**: 83 (leaf wetness & infection; **57**, 1868).

Gibson, I. A. S. 1972. *A. Rev. Phytopathol.* **10**: 51 (review, 91 ref.).
——. 1973. *PANS* **19**: 342 (review of fungicide control; **53**, 1972).
——. 1974. *Eur. J. For. Pathol.* **4**: 89 (impact & control; **54**, 1465).
Gilmour, J. W. et al. 1973. *N.Z. Jl For. Sci.* **3**: 120 (low-volume aerial application of Cu; **53**, 300).
Harvey, A. M. et al. 1976. *J. gen. Microbiol.* **95**: 268 (inhibition of RNA synthesis by toxin dothistromin; **56**, 429).
Ito, K. et al. 1975. *Bull. Govt For. Exp. Stn Tokyo* 272: 123 (in Japan; **54**, 4201).
Ivory, M. H. 1972a. *Trans. Br. mycol. Soc.* **59**: 205 (resistance induced by maturity & shade; **52**, 2385).
——. 1972b. *Ibid* **59**: 365 (penetration & development in host, effect of temp.; **52**, 2391).
Parker, A. K. 1972. *Phytopathology* **62**: 1160 (effects of temp. & RH on infection; **52**, 2050).
Peterson, G. W. 1973. *Ibid* **63**: 1060 (conidial dispersal, infection in E. Nebraska, USA; **53**, 1584).
—— et al. 1978. *Ibid* **68**: 1422 (development on & in needles in E. Nebraska, USA).
Shaw, C. G. et al. 1977. *Phytopathology* **67**: 1319 (impact on diam. growth with *Armillariella*; **57**, 4162).

SCLEROPHTHORA Thirumalachar, Shaw & Narasimhan, *Bull. Torrey bot. Club* 80: 304, 1953. (Pythiaceae)

MYCELIUM parasitic in higher plants, hyaline, coenocytic, sporangial stage like *Phytophthora*. SPORANGIOPHORES hyphoid, very little differentiated from the hyphae within the host (micronemous), simple or sympodially and successively branched. SPORANGIA large, limoniform or obpyriform, apically poroid, born singly at the apices of the sporangiophores, germinating in water by division of the cell contents into biciliate zoospores. OOGONIA like *Sclerospora*, wall thickened and confluent with the wall of the OOSPORE, oospore germination indirect by sporangial formation (M. J. Thirumalachar, C. G. Shaw & M. J. Narasimhan l.c.).

The type sp. is *S. macrospora* (q.v.) which was originally placed in *Sclerospora* and later in *Phytophthora*. Payak et al. summarised the differences between these 2 genera and *Sclerophthora*; they gave a key to the 4 *Sclerophthora* spp. The unbranched or sympodial sporangiophores with their citriform or obpyriform sporangia are like *Phytophthora*. The mature oospore has a thick epispore which fuses with the oogonial wall to form a thick brown covering; the oospore is plerotic. In *Phytophthora* the oospore is aplerotic and the oogonial wall is smooth, not usually ornamented as in *Sclerophthora* and never spiny (G. M. Waterhouse in *The fungi* (edited by G. C. Ainsworth et al.) Vol. IV B: 167, 181, 1973). Unlike *Phytophthora*, *Sclerospora* and *Sclerophthora* spp. are virtually restricted to Gramineae (on the leaves and inflorescences) and variously cause the distinctive symptoms of chlorotic streaking, stunting, shedding, proliferation and hypertrophy. The 2 spp. not described here are *S. cryophila* Jones and *S. lolii* Kenneth. Both have smaller oospores (20–37.5 and 25–29 μ diam., respectively) than *S. macrospora* and *S. rayssiae*, and smaller sporangia (12–23 × 23–46 and 25–35 × 41–55 μ, respectively) than the former sp.

Jones, W. 1955. *Can. J. Bot.* **33**: 350 (*Sclerophthora cryophila* on *Dactylis glomerata*; **35**: 193).
Kenneth, R. G. 1963. *Israel Jnl Bot.* **12**: 136 (*S. lolii* on *Lolium rigidum*).
Payak, M. M. et al. 1970. *Indian Phytopathol.* **23**: 183 (diseases caused by *Sclerophthora*, key to spp.).
Srinivasan, M. C. et al. 1962. *Bull. Torrey bot. Club* **89**: 91 (*S. cryophila* on new grass hosts in India; **42**: 28).

Sclerophthora macrospora (Sacc.) Thirumalachar, Shaw & Narasimhan, *Bull. Torrey bot. Club* 80: 304, 1953.

Sclerospora macrospora Sacc., 1890
S. kriegeriana Magnus, 1895
S. oryzae Brizi, 1919
Kawakamia macrospora (Sacc.) Hara, 1940
Phytophthora macrospora (Sacc.) Ito & Tanaka, 1940.

MYCELIUM intercellular, aggregating near vascular bundles. SPORANGIOPHORES emerging from stomata, external hyphae 8–28 (av. 14) μ long and 1–4 μ wide; undifferentiated from hyphae in the host, sympodial. SPORANGIA in clusters of 4–5, limoniform, obovate or ellipsoidal, hyaline to slightly purplish, moderately papillate; 58–98 × 30–65 μ (natural material) or 65–113 × 33–55 (av. 87 × 44) μ (in water). ZOOSPORES at first ovate or irregularly kidney shaped, somewhat globose when motile, spherical at rest, 13–16 × 10–13 (av. 11 × 14) μ; may produce zoosporangia 10–16 (av. 13) μ with germ tubes 1.6–2.5 μ wide. OOGONIA somewhat globose, light greenish to greenish brown, 50–95 × 55–100 (mostly 57–73 × 63–75) μ and av. 65 × 69 μ; wall 2.5–7.5 μ thick, commonly 3.8–5 (av. 4.3) μ. ANTHERIDIA laterally attached, hyaline to light yellow, obovate to ellipsoidal, wall slightly thickened; 13–23 × 23–41 (av. 15–28) μ, wall 1.8–3.8 (av. 2.5) μ thick. OOSPORES hyaline, somewhat globose, attached closely to the wall of the oogonium 43–70 × 43–73 (av. 57–60) μ; wall 3.8–10 (av. 6.5) μ thick (after I. Tanaka as *Phytophthora macrospora*).

S. macrospora was made the type sp. for the genus *Sclerophthora*. The host range is wide and includes both tropical and temperate cereals and grasses. Ou (*Rice diseases*, 1972) stated that > 43 genera in the Gramineae (and see Semeniuk et al.) are affected. The fungus has a widespread, scattered distribution (Map 287). Infection is often systemic. Peglion described the germination of the oospore and these spores from temperate hosts were examined in some detail by Miles et al. and Whitehead. Cytology was described by McDonough, morphological changes induced in several hosts by Akai et al. (1962) and fine structure by Fukutomi et al. Following some general ref. the diseases caused on rice, sugarcane and maize are described.

Akai, S. et al. 1962. *Ann. phytopathol. Soc. Japan* **27**: 239 (production of indoleacetic acid in vitro; **42**: 448).
—— ——. 1966. *Phytomorphology* **16**: 291 (changes induced in the external morphology of several hosts; **47**, 518).
Fukutomi, M. et al. 1966. *Trans. mycol. Soc. Japan* **7**: 199 (fine structure of zoospores & germ tubes; **47**, 2627).
—— ——. 1969. *Forschn Geb. Pflkrankh. Kyoto* **7**: 79 (fine structure of antheridium & oogonium; **49**, 1647).
—— ——. 1971. *Mycopathol. Mycol. appl.* **43**: 249 (as above; **50**, 3729).
McDonough, E. S. 1946. *Trans. Wis. Acad. Sci. Arts Lett.* **38**: 211 (cytology of sexual organs).
Miles, L. E. et al. 1942. *Phytopathology* **32**: 867 (morphology, on oat & wheat; **22**: 62).
Peglion, V. 1930. *Boll. Staz. Patol. veg. Roma* **10**: 153 (oospore germination; sporangia in wheat & other Gramineae; **10**: 174).
Peyronel, B. 1929. *Ibid* **9**: 353 (morphology of sporangia on wheat; **9**: 513).
Semeniuk, G. et al. 1964. *Phytopathology* **54**: 409 (life cycle & host list, in Dakota, USA; **44**, 398).
—— ——. 1966. *Ibid* **56**: 351 (further hosts; **45**, 2517).
Tanaka, I. 1940. *Ann. phytopathol. Soc. Japan* **10**: 127 (description as *Phytophthora macrospora*; **20**: 457).
Tokura, R. 1969. *Bull. Kyoto Univ. Educ.* Ser. B **36**: 35 (germination of zoospores on cellulose film; **50**, 3461).
Ullstrup, A. J. 1955. *Pl. Dis. Reptr* **39**: 839 (symptoms on further hosts; **35**: 528).
Whitehead, M. D. 1958. *Phytopathology* **48**: 485 (pathology & histology on 6 graminicolous hosts, especially maize; **38**: 136).

ORYZA SATIVA

All aspects of this rice downy mildew (called yellow wilt by Renfro) were reported on by several workers at a symposium (Akai). At early disease development new leaves show chlorotic or white spots or patches and they may become distorted and twisted. At flowering the symptoms are usually more pronounced; the panicles become distorted, small in size, and emergence is abnormal and partially suppressed; they usually remain green longer than is normal. The flowers are reduced or abortive. Direct infection from zoospores occurs through the plumule and the first leaf, symptoms appearing on the second leaf. The opt. temps for germination of the sporangia and oospores are virtually the same, 18–20° (an opt. range for sporangia of 18–23° has also been reported). These temps are also about the opt. for infection and development of the disease. Oospores can germinate within 48 hours in wet conditions; they can continue to germinate for 2–3 years under field conditions and remain viable for 5 years. New infections arise from oospores (which germinate to form a sporangium) both from rice and wild grasses. Oospores which have passed through the grasshoppers *Oxya velox* and *O. japonica* can germinate and cause infection. Zoospores are attracted towards germinating seed. The fungus has been cultured in vitro on a synthetic medium. Yellow wilt is not serious; seed treatment with fungicides gives control and the pathogen is suppressed at high temps. of at least 30°.

Akai, S. (editor) 1964. *Spec. Res. Rep. Dis. Insect Forecast.* 17, 183 pp. (general; **43**, 2904).
Katsube, T. 1977. *A. Rep. Soc. Pl. Prot.* 28: 44 (different nursery practice & disease incidence; **57**, 2521).
Katsura, K. et al. 1954. *Sci. Rep. Fac. Agric. Saikyo Univ.* 6: 49 (sources of primary infection & temp. effects; **35**: 230).
——. 1965. *Ann. phytopathol. Soc. Japan* **31**: 186 (infection & disease development).
Renfro, B. L. 1970. *Indian Phytopathol.* **23**: 177 (common names of diseases caused by *Sclerophthora* & *Sclerospora* inter alia).

Sclerophthora rayssiae

Takatsu, S. et al. 1957. *Ann. phytopathol. Soc. Japan* **22**: 123 (germination of, & infection by, oospores; **37**: 720).

Tasugi, H. 1953. *Bull. natn. Inst. Agric. Sci. Tokyo*, Ser. C **2**: 1 (general, effect of temp., oospore germination, viability and passage through *Oxya* spp.; **34**: 395).

SACCHARUM

A minor disease of sugarcane, for which there is no characteristic common name. Steindl gave a general account and there is a more recent description of symptoms and histology by Roth from South Africa. Infection causes extreme stunting of the whole or part of a stool and excessive tillering. Leaves are reduced; they show chlorotic to white streaking (up to several cm long) between the veins, or blotches; tips and edges dry up prematurely. Oospores, which can be fairly readily seen, are formed in large numbers within the mesophyll. The asexual stage is apparently not readily observed in the field; the sporangia occur mostly on the abaxial surface. Zoospores are released at 15.5–26.5°. Spread of the disease occurs mainly through infected setts but it can also originate from wild grasses. Areas in fields that are liable to be poorly drained should be avoided and healthy seed pieces used.

Farrar, L. L. et al. 1956. *Proc. int. Soc. Sugarcane Tech.* 9th Congr. Vol. 1: 1111 (general).

Orian, G. 1954. *Rev. agric. sucr. Ile Maurice* **33**: 64 (outbreak in Mauritius, control; **34**: 321).

Roth, G. 1967. *Z. PflKrankh. PflPath. PflSchutz* **74**: 83 (symptoms & histology; **46**, 2307).

Steindl, D. R. L. et al. 1961. In *Sugarcane diseases of the world* Vol. 1: 311 (general, 17 ref.).

ZEA MAYS

A general account of crazy top of maize was given by Ullstrup (1970). The disease is of minor importance when compared with others caused by the related fungi (*Sclerophthora rayssiae* var *zeae, Sclerospora philippinensis* and *S. sacchari* q.v.). Symptoms in part depend on whether infection becomes systemic or not. The striking field symptom, from which the common name is derived, is extensive phylloidy of the floral parts; a leafy tassel may become > 40 cm in diam. Excessive tillering is an early symptom. Leaves are distorted, do not unfurl or become narrow and leathery; they show a chlorotic striping. Both stunting and an increase in height of plants is found. Seedlings show a mild, general chlorosis or chlorotic striping. Pellucid dots on the leaves indicate the presence of oospores. Sporangia are formed on leaves fairly abundantly at 12–28°; opt. temp. for indirect (zoospore) germination is 12–16°. Infection probably occurs much as it does in rice, through the young above ground parts of the germinating seedling. Seed transmission can occur but is of little practical importance. Adequate soil drainage to prevent zoospore production and healthy seed should give good control.

Koehler, B. 1939. *Phytopathology* **29**: 817 (symptoms & other hosts; **19**: 12).

Sun, M. H. et al. 1970 & 1971. *Ibid* **60**: 1316; **61**: 913 (temp. & production & germination of sporangia; pathogenicity).

Ullstrup, A. J. 1952. *Ibid* **42**: 675 (general; **32**: 554).

—— et al. 1969. *Pl. Dis. Reptr* **53**: 246 (general).

——. 1970. *Indian Phytopathol.* **23**: 250 (general; 34 ref.).

Sclerophthora rayssiae Kenneth, Koltin & Wahl, *Bull. Torrey bot. Club* **91**: 189, 1964.

SPORANGIOPHORES arising from the stomata, singly or more, generally short stalked. SPORANGIA limoniform or ovate, thin walled, hyaline, with persistent, wedge shaped pedicel at the base; apex may protrude, poroid, 19–28 × 29–55 μ, germinating to form 6–10 reniform zoospores, 7.5 × 11 μ at below 20°; at 20° sporangia sometimes germinate directly. SEXUAL ORGANS abundant in mesophyll, solitary or clumped in groups; not tending to collect in any particular area of the lamina. OOGONIA unevenly thickened, usually sinuous, 44–59 (up to 61.4) μ diam. ANTHERIDIA paragynous, closely appressed to oogonium. OOSPORES globose, occasionally subglobose, always smooth and moderately thin walled, light golden amber with wall a deep golden brown, 30–44 (mostly 33.3) μ diam. (from R. Kenneth, Y. Koltin & I. Wahl l.c.).

This sp. was described from barley in Israel. It can be distinguished from *S. macrospora* by the considerably smaller oospores and smaller sporangia, the scattered sexual organs in the host which does not develop symptoms of hypertrophy and phylloidy of the inflorescence. *S. rayssiae* has been reported from India where it can cause a severe disease (brown stripe) of maize. Payak et al. (1967) separated the maize form as var. *zeae* on the grounds of slightly larger sporangia, lack of golden or amber brown colour in oogonia (33–45 μ diam.) and oospores, smaller oospores (30–37 μ diam) and a hyaline, glistening oospore wall. The only other host reported is *Digitaria sanguinalis*.

Symptoms on maize occur on the leaves as narrow, chlorotic streaks, variable in length and 3–7 mm wide, delimited by the veins; the streaks become red to purple and give a blotched effect when close together, seed development may be suppressed and the whole plant may dry out. Leaf shredding and malformations (common symptoms in downy mildews of grasses) have not been reported. Primary infection of seedlings occurs from oospores (presumably germinating to form sporangia) in the soil or in host debris; they survive for 5 years in the latter substrate. Internal infection of the seed has been described (Singh, R. S. et al. 1968). Secondary infection takes place from sporangia and zoospores from leaves. Inoculation of maize seedlings showed a decline of infection with age of the host (10–60 days old). The highest germination of both sporangia (indirect) and zoospores occurs at *c.* 22°, a similar temp. also being most favourable for the formation of sporangia on the host. At least 12 hours of leaf wetness is required for infection. In India plantings of maize made before or after the S.W. monsoon can escape infection (Singh, J. P. et al. 1970). Frequent spraying with dithane, captan or ziram soon after germination has given control; and see Lal for systemic fungicides. Resistance is also known; it appears to be controlled by several genes with minor and major effects and its inheritance is partially dominant.

Asnani, V. L. et al. 1970. *Indian Phytopathol.* **23**: 220 (inheritance of resistance).

Handoo, M. H. et al. 1970. *Ibid* **23**: 231 (as above).

Lal, S. et al. 1976. *Pesticides* **10**(7): 28 (systemic & nonsystemic fungicides).

Nene, Y. L. et al. 1970. *Indian Phytopathol.* **23**: 216 (fungicides).

Payak, M. M. et al. 1967. *Phytopathology* **57**: 394 (var. *zeae* on maize, symptoms & morphology; **46**, 2707).

——. 1970. *Proc. 1st int. Symp. Pl. Pathol.* 1966–67, New Delhi: 383 (downy mildews on maize, resistant lines & distribution in India).

Singh, I. S. et al. 1975. *Indian J. Genet. Pl. Breed.* **35**: 123, 128 (inheritance of resistance; **56**, 4026, 4027).

Singh, J. P. et al. 1970. *Indian Phytopathol.* **23**: 194 (infection & spread, effects of temp., control).

——. 1971. *Ibid* **24**: 777 (testing for resistance; **52**, 2257).

——. 1971. *Indian Jnl expl Biol.* **9**: 530 (infection by & survival of oospores; **51**, 3978).

Singh, R. S. et al. 1968. *Pl. Dis. Reptr* **52**: 446 (occurrence within seed & treatment with paratoluene sulphonamide; **47**, 2705).

SCLEROSPORA Schroet., *Hedwigia* 18: 86, 1879.

(Schroeter gave a formal diagnosis in *Kryptogflora Schlesien* 3 Pilze 1: 236, 1886)
(Peronosporaceae)

MYCELIUM intercellular, bearing small, usually knob-like, unbranched haustoria. SPORANGIOCONIDIOPHORES typically stout, composed of a main trunk and a compact group of rather short apical branches which are 1 to several times divided; branching dichotomous to indefinite. SPORANGIOCONIDIA germinate by 1 or more germ tubes or by zoospores. OOSPORES unlike those of other genera of the family in that the exospore is confluent with the wall of the oogonium (from H. M. Fitzpatrick, *The lower fungi Phycomycetes*: 212, 1930; Fitzpatrick distinguished the genus from others in the Peronosporaceae by the oospore wall being confluent with that of the oogonium and by the stout conidiophore with heavy branches clustered at the apex).

Germination of the oospore is direct by means of a germ tube. Ito divided the genus into 2 groups: those with a sporangium germinating by zoospores (Eusclerospora) and those with a conidium germinating by a germ tube (Peronosclerospora). Shaw (1978) erected the gen. nov. *Peronosclerospora* (Ito) Shaw and transferred these spp. to it from *Sclerospora: dichanthiicola, maydis, miscanthi, philippinensis, sacchari, spontanea, sorghi* and *westonii*. These transfers have not been adopted here. Waterhouse gave the descriptions of the spp. from the original papers and keys based on the asexual and sexual states. Narayanan described 10 spp. from India.

The genus has not been monographed and many of the spp. (on the leaves, stems and inflorescences of Gramineae) are important, obligate pathogens on tropical cereals; particularly in S. and E. Asia, and in Africa. There is a review of the *Sclerophthora* and *Sclerospora* spp. on maize by Fredericksen et al. Also see the general treatments of downy mildews in: *Indian Phytopathol.* **23**: 173–435, 1970; *Trop. Agric. Res. Ser.* Japan 8, 259 pp., 1975; Safeeulla. Six spp.

are described. The detailed, morphological accounts of Weston should be consulted and his spp. not described here are: *noblei* (on *Sorghum plumosum*, Australia); *northii* (on *Erianthus maximus* var. *seemanni*, Fiji); and *spontanea* (on *Zea mays* and other hosts, the Philippines). Other *Sclerospora* spp. are: *dichanthiicola* Thirumalachar & Narasimhan (on *Dichanthium annulatum*, India); *iseilematis* Thirumalachar & Narasimhan (on *Iseilematis laxis* and *Eleusine corocana*, India); *farlowii* Griffiths (on *Chloris elegans,* USA); *westonii* Srinivasan, Narasimhan & Thirumalachar (on *Iseilema laxum*, India).

Fredericksen, R. A. et al. 1977. *A. Rev. Phytopathol.* **15**: 249 (maize downy mildews, 175 ref.).
Griffiths, D. 1907. *Bull. Torrey bot. Club* **34**: 207 (*Sclerospora farlowii* sp. nov.).
Ito, S. 1913. *Bot. Mag. Tokyo* **27**: 218 (Eusclerospora & Peronosclerospora).
Narayanan, S. A. 1963. *Mycopathol. Mycol. appl.* **20**: 315 (*Sclerospora* in India; **43**: 2236).
Safeeulla, K. M. 1976. *Biology and control of the downy mildews of pearl millet, sorghum and finger millet*, 304 pp., Wesley Press, India.
Shaw, C. G. 1970. *Indian Phytopathol.* **23**: 364 (taxonomy, brief review, 16 ref.).
——. 1978. *Mycologia* **70**: 594 (*Peronosclerospora* gen nov.).
Srinivasan, M. C. et al. 1961. *Bull. Torrey bot. Club* **88**: 91 (*S. westonii* sp. nov.; **41**: 41).
Thirumalachar, M. I. et al. 1949. *Indian Phytopathol.* **2**: 46 (*S. iseilematis* sp. nov.; **29**: 511).
————. 1952. *Phytopathology* **42**: 596 (*S. dichanthiicola* sp. nov.; **32**: 487).
Waterhouse, G. M. 1964. *Misc. publs Commonw. mycol. Inst.* 17, 30 pp. (original descriptions, host list & keys; **44**, 648).
Weston, W. H. 1921. *J. agric. Res.* **20**: 669 (*S. spontanea* sp. nov.).
——. 1929. *Phytopathology* **19**: 961, 1107 (*S. northii* & *S. noblei* spp. nov.; **9**: 249, 320).
——. 1942. *Ibid* **32**: 206 (stat. conid. of *S. noblei*; **21**: 336).

Sclerospora graminicola (Sacc.) Schroet., *Kryptogflora Schlesien* 3 Pilze 1: 236, 1886.

Protomyces graminicola Sacc., 1876
Peronospora graminicola (Sacc.) Sacc., 1882
Sclerospora graminicola (Sacc.) Schroet. var. *setariae-italicae* Traverso, 1902.

Obligate parasite, forming haustoria. CONIDIOPHORES ephemeral, bloated, hyaline, determinate, of variable length, arising singly or in groups through stomata, hypophyllous or amphigenous; club shaped, aseptate, (10) 15.8–23.7 μ wide above (composite records); at apex, at irregular intervals, short stout branches of unequal length arise, one of which is usually an extension of trunk, branching again or not, forming1–6 relatively short sterigmata (*c.* 8 μ long). CONIDIA (SPORANGIA) hyaline, thin walled, broadly elliptical or subglobose, widest above mid spore, with poroid apical papilla of dehiscence, 14–31 (37) × 12–21 (25) μ (composite records), occasionally more elongated giant sporangia, 28–57 × 17–26 μ (composite records). ZOOSPORES (4–8 per sporangium), reniform, 9–12 μ. OOSPORES form in mesophyll of plants, between veins, are smooth, globose, 19–40 (45) μ diam. (composite records), with thick walls (1.9–2.9 μ), within light brown to reddish brown retentive oogonial wall which may have smooth or slightly uneven outline, mostly uniform thick but not ridged, convoluted or knobbed. Germination described by most observers as direct by hyphal germ tubes, and by one as indirect, by formation of sporangium. The fungus can be grown in axenic culture on special media, Arya et al. 1969 (R. Kenneth, *CMI Descr.* 452, 1975).

This sp. differs from the other 5 spp. described here in the germination of the conidio-sporangia by zoospores, in the oogonial wall and in the greater host range. Weston should be consulted for further information on morphology, and see a review by Nene et al. (*Claviceps* q.v.). Safeeulla et al. (1956) have compared the gametogenesis and oospore formation in this sp. with *S. sorghi* and *Sclerophthora macrospora* (q.v.).

S. graminicola has the widest host range of the *Sclerospora* spp.; it is widespread (Map 431) but not reported from Australasia, W. Indies and some countries in S.E. Asia. It has been recorded on these genera: *Agrostis, Echinochloa, Eleusine, Euchlaena, Panicum, Pennisetum, Saccharum, Setaria, Sorghum* and *Zea*. It causes green ear disease on bulrush millet (*Pennisetum typhoides*) and foxtail millet (*Setaria italica*); and diseases on common millet (*Panicum miliaceum*), maize and sorghum; although on the last 2 hosts the pathogen seems of little importance. Green ear disease is important in India. Melhus et al. gave an extensive review of the early literature and Butler (1907 and *Fungi and disease in plants*) gave a full description of the symptoms. On the 2 more important hosts (bulrush and foxtail millets) the most conspicuous symptom is the whole or partial transformation of the inflorescence into a loose, green head composed of structures like leaves, all parts of the flower become hypertrophied. Other symptoms in all hosts are: no flowering, dwarfing, excessive tillering, abnormally erect leaves, chlorotic leaf streaks which become necrotic, shredding of the leaves (due to breakdown of the intervascular tissue), a grey white growth of sporangiophores on the abaxial surface and oospores which are visible in the necrotic tissue. Where the crop is grown for fodder the symptoms other than those on the flower head will be more conspicuous.

Most infection probably occurs through oospores (germinating by a germ tube or like sporangia) in soil debris and less from those on the seed surface. Pande described the germination of oospores which require no resting period. It can take place within 24 hours of host germination; the cotyledon, meristem and successive leaves are then invaded systemically. Root invasion has been reported but probably only the above ground parts have true systemic infection. Stunting of the host can occur within a few days but infection may be latent. High infection at the 2-leaf stage has been reported at 24–32° or at lower temps. Host susceptibility declines with age and secondary, local infection of young tissue in older plants is less important than seedling infection. Water saturation at the leaf surface is needed for asexual reproduction which occurs abundantly over a wide temp. range (18–25°). Similar temps are favourable for zoospore release. The conidia are deciduous and actively dispersed. Abundant oospores are formed in the mesophyll. They can germinate within 48 hours but more germination occurs after a resting period of *c*. 45 days (but see Pande). Oospore germination is abundant at 20–25° but more rapid at 30–31°. Arya et al. (1962) described seed infection.

There is evidence for physiologic specialisation. Uppal et al. described 2 forms; 1 on several *Setaria* spp. and *Euchlaena mexicana*, and the other on bulrush millet, neither form caused infection of common millet. Tasugi (1934) proposed 4 strs based on different *Setaria* spp., including 1 on *S. viridis* and another on *S. italica*. Safeeulla et al. (1963) obtained infection on bulrush millet but not on foxtail millet or sorghum. King et al. (*Sclerospora sorghi* q.v.) found that in soil with oospores of *S. sorghi* and *S. graminicola* the latter only infected bulrush millet and the former only sorghum.

Control is most likely through host resistance; open pollinated lines of bulrush millet have been reported to be more resistant in India. Rotation may be ineffective unless combined with roguing because of the long survival of the oospores; this has been variously reported as 3–5 years in the soil and 8–30 months in the laboratory. Seed treatment (agrosan; hot water at 50° for 1 hour or 55° for 10 minutes) can be effective. The evidence for infection being carried within as well as outside the seed needs confirmation. In any case seed from infected plants should not be used for planting.

Appadurai, R. et al. 1975. *Indian J. agric. Sci.* **45**: 179 (inheritance of resistance; **56**, 3550).

Arya, H. C. et al. 1962. *Indian Phytopathol.* **15**: 166 (seed transmission; **43**, 1318).

—— ——. 1969. *Ibid* **22**: 446 (growth in tissue culture; **50**, 1208).

Butler, E. J. 1907. *Mem. Dep. Agric. India bot. Ser.* **2** (1), 24 pp. (history, symptoms & morphology).

Chaudhuri, H. 1932. *Phytopathology* **22**: 241 (general; **11**: 507).

Evans, M. M. et al. 1930. *Ibid* **20**: 993 (germination of oospores; **10**: 306).

Gill, K. S. et al. 1975. *Crop Improvement* **2**: 128 (inheritance of resistance; **57**, 159).

Hiura, M. 1935. *Res. Bull. Gifu Coll. Agric.* 35: 121 (general; **14**: 576).

Kenneth, R. 1966. *Scr. hierosol.* **18**: 143 (general in Israel with *Sclerospora sorghi*; **45**, 3567).

Kulkarni, G. S. 1913. *Mem. Dep. Agric. India bot Ser.* **5**: 268 (symptoms, proposes *S. graminicola* var. *andropogonis-sorghi*).

McDonough, E. S. 1937. *Mycologia* **29**: 151 (cytology).

——. 1938. *Phytopathology* **28**: 846 (penetration & histology in *Setaria* spp.; **18**: 174).

——. 1943. *Am. J. Bot.* **30**: 809 (cytology of cytoplasm & inclusions; **23**: 384).

Melhus, I. E. et al. 1928. *Res. Bull. Iowa agric. Exp. Stn.* 111: 297 (general on *Setaria viridis* & maize; **7**: 712).

Pande, A. 1972. *Mycologia* **64**: 426 (oospore germination; **51**, 3993).

Pu, M. H. et al. 1949. *Phytopathology* **39**: 512 (infection of foxtail millet & temp.; **29**: 22).

Safeeulla, K. M. et al. 1956. *Phytopathol. Z.* **26**: 41 (periodicity in formation of sporangia; **35**: 602).

—— ——. 1963. *Pl. Dis. Reptr* **47**: 679 (temp., sporangial germination & inoculation; **42**: 760).

Singh, H. et al. 1965. *Phytomorphology* **15**: 338 (morphological & histological changes induced in bulrush millet; **45**, 1786).

Suresh, S. 1969. *Madras agric. J.* **56**: 88 (resistance tests in bulrush millet; **51**: 2436).

Suryanarayana, D. 1952. *Indian Phytopathol.* **5**: 66 (infection of seedling).
——. 1962. *Ibid* **15**: 247 (infection by oospores; **43**, 1319).
——. 1966. *Bull. Indian phytopathol. Soc.* **3**: 72 (general, with *S. sorghi*; **47**, 1113p).
Takasugi, H. et al. 1933. *Res. Bull. agric. Exp. Stn S. Manchuria Rly Co.* 11: 1 (oospore germination; **13**: 436).
Tasugi, H. 1933. *J. Imp. agric. Exp. Stn Nishigahara* **2**: 225 (conditions affecting spore germination; **12**: 623).
——. 1934. *Ibid* **2**: 345 (pathogenicity & physiologic forms; **13**: 629).
——. 1935. *Ibid* **2**: 459 (conditions affecting infection by oospores; **14**: 577).
Tiwari, M. M. et al. 1967. *Indian Phytopathol.* **20**: 356 (growth in tissue culture; **47**, 3100).
Uppal, B. N. et al. 1931. *Phytopathology* **21**: 337 (physiologic forms; **10**: 517).
Weston, W. H. 1924. *J. agric. Res.* **27**: 771 (morphology & development of asexual stage; **3**: 718).
—— et al. 1928. *Ibid* **36**: 935 (general on *Setaria magna*; **8**: 98).
Yu, T. F. 1944. *Chin. J. scient. Agric.* **1**: 199 (resistance in foxtail millet; **24**: 312).

Sclerospora maydis (Racib.) Butler, *Mem. Dep. agric. India bot. Ser.* 5: 275, 1913.

Peronospora maydis Racib., 1897
Sclerospora javanica Palm, 1918.

MYCELIUM intercellular, with haustoria. CONIDIOPHORES arising from the stomata; length variable, 200–550 μ; basal cell 60–180 μ (Semangoen), 180–300 μ (Waterhouse, *Sclerospora* q.v.) long. CONIDIA 12–29 × 10–23 (Semangoen), 23–17.5 (Waterhouse); germination direct by germ tube. SEX ORGANS unknown.

Butler (l.c.) originally used this name, realising that the fungus was not a *Peronospora*, for his Pusa (India) downy mildew on maize. But Palm (*Meded. Lab. PlZiekt. Buitenz.* **32**: 18, 1918) showed that Raciborski's fungus from Java, on the same host, was different from Butler's; his name for the Java pathogen (*S. javanica*) is, however, illegitimate. Butler then erected a new name for his Indian downy mildew (*S. indica*) but this fungus had already been named *S. philippinensis* (q.v.) by Weston in the Philippines. The latter is, therefore, the correct name for the pathogen causing the disease generally called Philippine downy mildew of maize and first described by Butler as *S. maydis*.

S. maydis (smaller conidia than *S. philippinensis*) causes Java downy mildew or sleepy disease of maize. It has been insufficiently studied and compared with the other members of the genus *Sclerospora* on this host, especially those restricted to parts of Asia and Australasia. The pathogen appears to be restricted to parts of Indonesia and India but it is a more important disease in the former country. It has been reported from Africa but this probably requires confirmation. Symptoms on young maize are narrow, chlorotic leaves, stunting and death. Growth may be apparently normal but chlorotic stripes form on the leaves, sometimes only on those at the base. Conidial sporulation occurs under the usual conditions of high RH. Seed taken from infected plants and sown in sterile soil gave systemically infected, chlorotic seedlings (Purakusumah). Penetration by conidial germ tubes on young leaves is stomatal. Apart from the type host only *Euchlaena mexicana* has been infected. Cultural measures for control should be taken (Hirose et al. considered the effect of roguing) and resistance has been found in Indonesia.

Daran, D. V. 1953. *J. scient. Res. Djakarta* **2**: 1 (symptoms, viability of conidia; **34**: 448).
Hirose, S. et al. 1978. *Jap. J. Crop Sci.* **47**: 141 (effect of roguing on disease; **57**, 4443).
Purakusumah, H. 1965. *Nature Lond.* **207**: 1312 (seed transmission; **45**: 108).
Reitsma, J. et al. 1949. *Meded. alg. Proefstn. AVROS* 89, 20 pp. (symptoms & resistance; **29**: 256).
Semangoen, H. 1970. *Indian Phytopathol.* **23**: 307 (general, 21 ref.).
Sudjadi, M. et al. 1978. *Ann. phytopathol. Soc. Japan* **44**: 142 (histology of seedling systemic infection).
Van Hoof, H. A. 1954. *Tijdschr. PlZiekt.* **60**: 221 (*Sclerospora* in Indonesia; **34**: 223).

Sclerospora miscanthi T. Miyake, 1912.

(Latin diagnosis by S. Ito et al. 1935. *Trans. Sapporo nat. Hist. Soc.* **14**: 19).

CONIDIOPHORES 97–300 μ long (up to 438 μ), 12–37 μ wide, branched dichotomously at the tips and bearing 20 or more oval conidia. CONIDIA 37–49 × 14–23 (av. 42 × 18) μ, germinating directly. OOGONIA angular, brownish, 38–70 × 26–67 (av. 62 × 54) μ, wall 3–16 μ thick. OOSPORES 32–57 (av. diam. 47) μ, (mostly 44–47 μ, wall 2.3–2.4 μ thick); abundant in retted tissue (Chu 1953; Matsumoto et al. 1962 & 1964).

Leaf splitting is a minor disease of sugarcane characterised by conversion of the leaf into a bundle of fibres like a whip. It occurs in Taiwan (where it was first described), in the Philippines and possibly in Fiji and Papua New Guinea (Chu 1964). The early symptoms are similar to those of *S. sacchari* (q.v.). On the leaves, narrow, greenish yellow stripes appear at first, often extending to the full length of the leaf. These streaks or stripes (with age) become more yellow, mottled, reddish brown and then dark red. Finally, the tissue between the vascular bundles disintegrates and the lamina splits into the separated fibres. Affected plants are stunted and die; oospores are produced in large numbers in the host. Healthy setts planted in soil infested with oospores become infected, showing symptoms when the plants are 1.2–1.5 m tall. If plant material is taken from diseased stems the symptoms appear at an earlier stage. Matsumoto et al. compared *S. miscanthi* (originally described from *Miscanthus japonicus*) with *S. sacchari*. No other hosts apart from *Miscanthus* and *Saccharum* appear to have been reported for the former pathogen. Resistant cvs of sugarcane are known and Chu (1953) obtained control by placing diseased setts in water at 46° for 20 minutes and then at 52° for the same period.

Chu, H. T. 1953. *Rep. Taiwan Sug. Exp. Stn* 10: 113 (general description; **34**: 486).

——. 1964. In *Sugarcane diseases of the world* Vol. 2: 37 (general, 7 ref.).

Matsumoto, T. et al. 1962 & 1964. *Rep. Taiwan Sug. Exp. Stn* 28: 127; 33: 53 (comparison between *Sclerospora miscanthi* & *S. sacchari*, morphology & inoculations; **43**, 2718).

Miyake, T. 1912. *Spec. Bull. Sug. Exp. Stn Formosa 1911* 1, 61 pp. (original description).

Sclerospora philippinensis Weston, *J. agric. Sci.* **19**: 118, 1920.

Sclerospora maydis Reinking 1918

S. indica Butler, 1931.

MYCELIUM intercellular in all parts except the root, branched, slender, usually *c.* 8 μ diam. but irregularly constricted and inflated; haustoria simple, vesiculiform to subdigitate, small, *c.* 8 μ long and 2 μ diam. CONIDIOPHORES always produced in night dew and growing out of stomata, erect, 150–400 μ long, 15–26 μ thick, bearing a basal cell in the lower part, dichotomously branched ×2–4 above, branches robust, sterigmata conoid to subulate, 10 μ long, slightly curved. CONIDIA elongated ellipsoid, elongated ovoid, or rounded cylindrical, varying in size, usually 27–39 × 17–21 (av. 18 × 34) μ, hyaline, with thin episporium, minutely granular within, slightly rounded at the apex, with a minute apiculus at the base; always germinating by a tube. OOGONIA av. 22.9 μ, wall smooth with fragments of oogonial stalk or antheridial cell frequently adhering. OOSPORES regularly spherical, central to eccentric, 15.3–22.6 (19.2) μ, wall 2–3.9 μ thick, content homogeneous, finely granular with masses of oil reserves (stat. conid. from W. H. Weston l.c.; perfect state from Napi-Acedo & Exconde who pointed out that the oospores are relatively smaller than those known for other *Sclerospora* spp.; and see Holliday).

Philippine downy mildew of maize or sleepy disease; the latter common name (also used for Java downy mildew, *S. maydis* q.v.) was used by Butler for his Pusa (India) downy mildew of maize. A note on the early confusion between these two maize pathogens, *S. philippinensis* and *S. maydis*, is given under the latter. The former has larger conidia than those of *S. maydis* (usually 18 × 23 μ) whose oospores are unknown. Uppal et al. compared *S. indica* and *S. philippinensis* and concluded that they were the same. Elazegui et al. compared the latter with *S. sacchari* (q.v.) which is the important downy mildew of sugarcane. Exconde et al. (1968) described the host range which includes spp. in *Saccharum*, *Sorghum* and *Euchlaena*; infection of 8 (out of 76) grass spp. was obtained. Weston et al. gave a very detailed description of host morphological effects caused by infection of *E. luxurians*. The distribution is apparently restricted: India, Indonesia, Nepal, Pakistan, Philippines and Thailand (Map 497).

The symptoms of chlorotic streaking on leaves may be first seen from the seedling stage up to the time when the tassels and silks are developed. The chlorotic areas spread and dry out and the oospores are formed, scattered, in the necrotic leaf tissue. There is no shredding of the leaves. There are abnormalities in vegetative growth, abortive development of the ear and partial or complete sterility. *S. philippinensis* causes some of these symptoms when inoculated to sugarcane. The disease has been almost entirely studied in the Philippines where it has been described as the most serious one on maize. The outlines of the life cycle are apparently similar to other members of the genus. Infection in the seedling stage by oospores and later by conidia; plants up to 3–4 weeks old are highly susceptible. Germinating conidia penetrate the stomata. The conidia are deciduous and wind dispersed. The relative importance of the 2 spore stages in providing primary inoculum is not known. Invasion of the primary host is systemic when infected at an early stage. Advincula et al. described seedborne infection. In India (Chona et al.) the occurrence of *S. philippinensis* on *Saccharum spontaneum*, a perennial wild grass, is considered to be a source of inoculum (and see Dalmacio et al. 1972).

At present, breeding for resistance seems likely to be the main control and this is being done in the Philippines. But protection of young plants during the most susceptible period of growth was given by covering the seedlings after emergence for 2–3 days and spraying with chloroneb 3–4 times over the next 12 days (Schultz; and see Estrada et al.; Exconde et al. 1975/1976; Raymundo et al. for the use of fentin hydroxide, mancozeb and mineral oil). Roguing has also been recommended.

Advincula, B. A. et al. 1975/1976. *Philipp. Agric.* **59**: 214 (seedborne infection; **56**, 4982).
Carangal, V. et al. 1970. *Indian Phytopathol.* **23**: 285 (breeding for resistance).
Chona, B. L. et al. 1955. *Ibid* **8**: 209 (occurrence on *Saccharum spontaneum*; **36**: 241).
Dalmacio, S. C. et al. 1969/1970. *Philipp. Agric.* **53**: 35 (penetration & infection of young plants of maize; **50**, 652).
—— ——. 1972. *Philipp. Phytopathol.* **8**: 72 (factors affecting disease incidence; **53**, 2159).
Elazegui, F. A. et al. 1967/1968. *Philipp. Agric.* **51**: 767 (comparison of *Sclerospora philippinensis* & *S. sacchari*; **49**, 693).
Estrada, B. A. et al. 1975/1976. *Ibid* **59**: 256 (fungicides; **56**, 4983).
Exconde, O. R. et al. 1968/1969. *Ibid* **52**: 175, 189 (host range, disease incidence & weather in Philippines; **49**, 1359, 1360).
——. 1970. *Indian Phytopathol.* **23**: 275 (general).
—— ——. 1975/1976. *Philipp. Agric.* **59**: 237 (fungicides; **56**, 4981).
Holliday, P. 1975. *C.M.I. Descr. pathog. Fungi Bact.* 454 (*S. philippinensis*).
Mochizuki, N. et al. 1974. *Jap. agric. Res. Q.* **8**: 185 (resistance; **54**, 3868).
Napi-Acedo, G. et al. 1967/1968. *Philipp. Agric.* **51**: 279 (oospores; **48**, 1665).
Raymundo, A. D. et al. 1976/1977. *Ibid* **60**: 52 (fungicides; **57**, 569).
Schultz, O. E. 1972. *Phytopathology* **62**: 500 (control with chloroneb).
Titatarn, S. et al. 1974/1975. *Philipp. Agric.* **58**: 90 (virulence patterns & morphology; **56**, 686).
Uppal, B. N. et al. 1936. *Indian J. agric. Sci.* **6**: 715 (merging *S. indica* with *S. philippinensis*; **15**: 794).
Weston, W. H. 1920. *J. agric. Res.* **19**: 97 (*S. philippinensis* sp. nov., symptoms on maize).
——. 1923. *Ibid* **23**: 239 (production & dispersal of conidia with *S. spontanea*; **2**: 359).
—— et al. 1929. *Ibid* **39**: 817 (infection & morphological abnormalities of *Euchlaena luxurians*; **9**: 319).

Sclerospora sacchari Miyake, *Bull. Sug. Exp. Stn Daimokko* 1: 12, 1911 (1912).

(Latin diagnosis by S. Ito, *Bot. Mag. Tokyo* **27**: 217, 1913)

MYCELIUM intercellular with bulbous haustoria. CONIDIOPHORES fugaceous, erect, 1–4 projecting from each stoma, hyaline, 125–190 × 18–25 μ, base bulbous, 1–2 septate, narrowing above, often with a foot cell, middle part × 2–3 broader, dichotomously branched twice or thrice at the tip, each branch swollen in the middle, ultimate branchlets bearing 2–4 conical, pointed and slightly curved stalks (sterigmata), up to 12–30 μ long. CONIDIA hyaline, ellipsoid, elongate ovoid to oblong, apex rounded, base slightly pedunculate or rounded, 25–55 × 15–25 μ (generally 36 × 18 μ), wall thin and smooth. OOSPORE yellow to yellowish brown, globular, av. 50 μ diam. wall thick 3.5–5 μ (K. G. Mukerji & P. Holliday, *CMI Descr.* 453, 1975; see also Leece and Matsumoto et al., the latter compared this fungus with *S. miscanthi* (q.v.) which also occurs on sugarcane).

If *Sclerophthora macrospora* (q.v.), originally described as a *Sclerospora*, is excluded the oospores of *S. sacchari* are the largest in the genus. Several grasses (including cereals) are attacked (*Euchlaena*, *Saccharum*, *Sorghum* and others). The 2 main downy mildew diseases caused are on maize and sugarcane. On the latter the disease has also been called cane dew and leaf stripe. The limited distribution is: Fiji, India, Indonesia (W. Irian), Japan, Philippines and Thailand (Map 21). *S. sacchari* has been virtually eradicated from Australia (Queensland; see *Rev. appl. Mycol.* **42**: 485; **43**, 1129). Reports of its presence in Central America are doubtful.

SACCHARUM

A general account was given by Hughes et al. The syndrome differs in some characteristics depending on whether diseased setts have been used or whether infection takes place at a later stage. In the first case a pale mottling is seen on the young spindle and the very young plant may die. If not the leaves become abnormally narrow, discoloured and upright; the stalk is thinner and general stunting makes the plant conspicuous. The characteristic, almost white, downy, asexual sporulation (largely on the lower surface of the leaf) occurs in suitably damp conditions. Where infection takes place at a later stage, after growth of the sett, the first symptoms occur on the inner leaf of the spindle as pale green to yellow striping. These longitudinal stripes increase in length as each leaf unfolds until they may cover the whole leaf; they are rarely found on the leaf sheaths and never develop initially on mature leaves. Regularity in the stripes may be lost and large mottled areas formed. The chlorotic areas become necrotic, a deep red to brown, and oospores can be seen lying between the veins. Disintegration of interveinal leaf tissue and the consequent shredding occurs; this can also be found in plants arising from diseased setts. In plants infected later the development of the sexual state coincides with an abnormal stem elongation; these jump-up canes, as they have been appropriately called, stand out from normal cane. They are thin, weak, have more internodes, fewer and shorter leaves which may not unfold, and eventually wither and shred.

Infection seems to occur mostly from the conidia which have an opt. temp. for formation and germination of *c.* 25° or a little lower. Their viability is lost at this temp. within an hour if the RH is <95%. Stomatal penetration has been seen in detached leaf disks and a higher germination noted on such disks compared with that on water agar. Natural infection mostly occurs through buds and the young tissues of the spindle are not an important site of entry. Matsumoto et al. obtained infection with oospores. Infection of buds leads to systemic infection, both in cuttings and young standing cane. Tillers from such systemically infected plants become infected likewise. Cane infected at an older stage will show a few infected stems but no systemic infection. Most conidia are formed on the younger leaves and are 5–10 times more abundant on the abaxial surface. They are released mostly at *c.* 0100–0300 hours. Conidia from maize readily infect sugarcane.

Control is through resistance and cultural measures which have been successful in, and fully described from, Queensland (Hughes et al.). Such measures mainly involve inspection, roguing, use of disease free plant material and elimination of reinfection from other hosts, notably maize. Reddi et al. described an intensive method for testing host resistance. Wang et al. reported good control in infected setts with a hot water treatment of 1 hour at 45°, followed by drying (at room temp.) and a further treatment of 1 or 0.5 hour at 52 or 55°, respectively.

Chu, T. L. et al. 1963. *Rep. Taiwan Sug. Exp. Stn* 30: 1 (testing resistance in first year seedlings; **42**: 629).

Hughes, C. G. 1951. *Q. Bull. Bur. Sug. Exp. Stns Qd* **14**: 163 (testing cvs for resistance; **30**: 581).

—— et al. 1961. In *Sugarcane diseases of the world* Vol. 1: 141 (general; 23 ref.).

Leece, C. W. 1941. *Tech. Commun. Bur. Sug. Exp. Stns Qd* 5, 25 pp. (111–135) (general on sugarcane, maize & other hosts; **21**: 347).

Leu, L. S. et al. 1962 & 1963. *Rep. Taiwan Sug. Exp. Stn* 26: 17; 30: 11 (symptoms, infection of & spread in host; **42**: 47, 629).

—— ——. 1970. *Sugcane Pathol. Newsl.* 4: 40; 5: 16 (conidial sporulation & germination; **49**, 2623; **50**, 926).

Matsumoto, T. et al. 1961. *Rep. Taiwan Sug. Exp. Stn* 24: 1, 7, 95 (germination of oospores, formation of & infection by conidia & oospores; **41**: 544).

Reddi, K. et al. 1970. *Sugcane Pathol. Newsl.* 5: 38 (field method for testing resistance; **50**, 925).

Rivera, J. R. 1962. *Philipp. Sug. Inst. Q.* **8**: 83 (general account; **42**: 629).

Robinson, P. E. et al. 1956. *9th Meet. int. Soc. Sugcane Technol.*: 986 (testing for resistance).

Stevenson, M. D. 1970. *Sugcane Pathol. Newsl.* 5: 12 (leaf characters & susceptibility; **50**, 924).

Wang, C. S. et al. 1958. *J. agric. Assoc. China* **22**: 51 (hot water treatment of setts; **38**: 625).

Yang, S. M. et al. 1962. *Rep. Taiwan Sug. Exp. Stn* 27: 67 (effect of temp. & RH on conidial biology; **42**: 47).

ZEA MAYS

Much of the work on the maize form of the disease has been done in Taiwan and has recently been reviewed both by Chang (1970) and Sun who considered that it was a factor limiting maize production in the island. This crop and sugarcane can be equally susceptible and also the fungus passes readily from one host to the other. Infection at a very early stage in growth leads to extreme stunting and death. In other cases the typical chlorotic, leaf streaks are formed but there is no shredding. Systemic infection also results in small, poorly filled ears (formed in abnormally large numbers), elongated ear shanks, imperfect tassels with grain and sterility.

Penetration (after 2 hours at 25° and 28°) is through the stomata from conidia; the role of the oospore in the epidemiology is uncertain. The characteristics of the conidia are much as has already been described; they are abundant on both leaf surfaces. One-month-old plants are highly resistant and local lesions with little or no sporulation may be caused on older plants. In control the risk of infection from susceptible sugarcane must be eliminated and roguing carried out. No general field control by fungicides is yet possible but resistance, which appears to be controlled by a single dominant gene, is known.

Chang, S. C. 1970. *Indian Phytopathol.* **23**: 270 (review of work in Taiwan, 20 ref.).
—— et al. 1970. *Rep. Corn Res. C. Tainan Dais Taiwan* **8**: 1 (diurnal periodicity for conidia; **50**, 3701).
—— ——. 1970. *Ibid* **8**: 25 (resistance in Indonesian cvs; **50**, 3702).
—— ——. 1972. *Ibid* **9**: 16 (effect of light on sporulation; **51**, 4816).
——. 1972. *Ibid* **9**: 22, 28 (temp., infection & incubation; **51**, 4814, 4815).
Singh, R. S. et al. 1968. *Pl. Dis. Reptr* **52**: 446 (seed transmission with *Sclerophthora rayssiae* var. *zeae*; **47**, 2705).
—— ——. 1968. *Labdev Jnl Sci. Technol.* **6-B**: 197 (disease in India; **48**, 1166).
Sun, M. H. 1970. *Indian Phytopathol.* **23**: 262 (review with particular reference to Taiwan, 16 ref.).

Sclerospora sorghi Weston & Uppal,

Phytopathology **22**: 582, 1932.
(A new name for *Sclerospora graminicola* var. *andropogonis-sorghi* Kulk., *Mem. Dep. Agric. India bot. Ser.* **5**: 268, 1913)
S. andropogonis-sorghi (Kulk.) Mundkur 1951, nomen nudum
S. sorghi-vulgaris (Kulk.) Mundkur 1951, nomen nudum.

Obligate parasite, forming haustoria. CONIDIOPHORES ephemeral, bloated, hyaline, determinate, of variable length, arising singly or in groups through stomata, hypophyllous or amphigenous; club shaped, 1–2 septate, 15–31 μ wide above (composite records), usually with 3 or more thick short branches arising from one point at apex, branching again or not, forming 1 to many sterigmata (until 13 μ long). CONIDIA hyaline, thin walled, obovate, non-papillate, non-poroid, 15–19 (32) × (12) 15–23 (27) μ (composite records). OOSPORES produced in mesophyll of systemically infected plants between vascular bundles; smooth, globose, hyaline to yellow. 25–43 μ diam. (composite records), walls 1–3 μ thick, single and *c*. central with adhering, irregularly thick walled, retentive, rusty-brown oogonial wall with strongly wavy margin; germination by hyphal germ tube and possibly other means. The fungus can be grown in axenic culture on special media (R. Kenneth, *CMI Descr.* 451, 1975; and see Weston et al.).

Safeeulla et al. described gametogenesis and oospore formation, and compared it with *S. graminicola* and *Sclerophthora macrospora* (q.v.). Sansome considered that meiosis occurs in the antheridium and oogonium, the vegetative state being diploid. Weston et al. and Tarr (*Diseases of sorghum, Sudan grass and broom corn*) compared *Sclerospora sorghi* with *S. graminicola* which is much less important on *Sorghum* spp. *S. sorghi* has smaller conidia and oospores than *S. sacchari* (q.v.), and smaller conidia but larger oospores than *S. philippinensis* (q.v.). *S. sorghi* primarily causes downy mildew of *Sorghum bicolor* and other *Sorghum* spp. It also attacks maize and has been reported from *Euchlaena mexicana, Panicum* and *Pennisetum typhoides*. It is widespread in Africa and has also been reported from Argentina, Australia, Bangladesh, China, India, Israel, Philippines, Thailand and USA (Map 179). Frederiksen et al. (1973) gave a review.

Symptoms arise from systemic infection (probably brought about by oospores) or from local infection where conidia are presumably the main infective agents. Where infection in sorghum occurs at a very early stage of growth the young plants are stunted, the leaves narrow, somewhat stiff and upright with white to pale yellow streaks which become red to purple. Intravascular necrosis follows with the formation of numerous oospores which can be seen as dark lines parallel to the veins. The interveinal tissue disintegrates to give the characteristic shredding symptoms. No shredding apparently occurs in maize and it has been reported not to occur in *S. arundinaceum* var. *sudanense* and *S. halepense* in Israel. The plants are sterile. Symptoms from systemic infection may become apparent only at about floral differentiation. Such plants have heads which are partly or wholly sterile and the chlorotic pattern on the leaves differs. Non-systemic (local) infection results in white to yellow streaking on the leaves which may dry up completely; there is no shredding and oospores may be absent; the head may be stunted. Similar leaf symptoms occur in maize where the oospores are scattered and shredding is absent. Phylloidy is found in the tassels.

Where oospores are present in the soil (see Pratt; Pratt et al.) these may be the primary inoculum source but airborne infection of leaves by conidia occurs. Plants are most susceptible at the youngest stages but systemic infection can arise in plants inoculated at 2–3 weeks. Seedlings that are infected when very young often die at the fourth leaf stage. Systemic infection can appear in sprouting sorghum tillers after cutting down healthy shoots. Infection can probably occur over a wide temp. range. In maize Bonde et al. found 14–22° in a 2 hour dew period to be near the opt.; Cohen et al. obtained high infection in darkness at 20–25°. There are also local infections and penetration through stomata occurs. In mature maize seed the fungus is confined to the pericarp and pedicel. It appears that hosts other than maize and sorghum are rarely attacked naturally (Bonde et al.; Cohen et al.; Jones; Jones et al.; Kenneth; Kenneth et al.; King et al.; Puranik et al.; Uppal et al.).

Control is most likely through resistance and there is now a fairly extensive literature; see *Rev. appl. Mycol.* **34**: 638; **46**, 990, 2008; *Rev. Pl. Pathol.* **49**, 3251; **50**, 3700; **51**, 323; **52**, 3304; **53**, 144; **54**, 3893; **55**, 4687; **57**, 570, 571. Balasubramanian found mancozeb to be effective. Jones et al. (1972) prevented disease transmission in maize seed by reducing its moisture content to 9% and storing it for 40 days before planting.

Balasubramanian, K. A. 1976. *Curr. Sci.* **45**: 416 (fungicide; **56**, 190).
Bonde, M. R. et al. 1978. *Phytopathology* **68**: 219 (dew period temp., conidial germination & systemic infection; **57**, 4921).
Cohen, Y. et al. 1977. *Ibid* **67**: 515 (spread & airborne conidia; **56**, 5614).
Doggett, H. 1970. *Indian Phytopathol.* **23**: 350 (in E. Africa, resistance in sorghum heritable).
Frederiksen, R. A. et al. 1969. *Pl. Dis. Reptr* **53**: 566, 995 (symptoms, distribution & loss in maize in Texas).
—— ——. 1970. *Indian Phytopathol.* **23**: 321 (distribution in USA; resistance in sorghum & maize).
—— ——. 1973. *Res. Monogr. Texas agric. Exp. Stn* 2, 32 pp. (review; **54**, 1725).
Govinda, H. C. et al. 1970. *Indian Phytopathol.* **23**: 378 (downy mildew diseases of cereals in Mysore).
Jones, B. L. 1971. *Phytopathology* **61**: 406 (penetration of leaves of sorghum; **50**, 3711).
—— et al. 1972. *Ibid* **62**: 817 (in carpellate flowers & seed of maize; **52**, 1114).
——. 1978. *Ibid* **68**: 732 (systemic infection in sorghum & Sudan grass).
Kenneth, R. 1966. *Scr. hierosol.* **18**: 143 (general in Israel with *Sclerospora graminicola*; **45**, 3567).
——. 1970. *Indian Phytopathol.* **23**: 371 (general).
—— et al. 1973. *Phytoparasitica* **1**: 13 (systemic infection in sorghum & maize; **53**, 1802).
King, S. B. et al. 1970. *Indian Phytopathol.* **23**: 342 (general in Nigeria with *S. graminicola*).
Malaguti, G. et al. 1977. *Agron. trop.* **27**: 103 (on maize in Venezuela; **57**, 3930).
Nagarajan, K. et al. 1970. *Indian Phytopathol.* **23**: 356 (host reaction in sorghum).
Patel, M. K. 1949. *Curr. Sci.* **18**: 83 (oospores in maize; **28**: 392).
Pratt, R. G. et al. 1978. *Phytopathology* **68**: 1600 (in soil & relationship with disease incidence).
——. 1978. *Ibid* **68**: 1606 (oospore germination in the presence of host & non-host roots).
Puranik, S. B. et al. 1968. *Mysore J. agric. Sci.* **11**: 143 (oospores in floral parts of sorghum; **48**, 455).
Safeeulla, K. M. et al. 1955. *Mycologia* **47**: 177 (gametogenesis & oospore formation; **37**: 164).
Sansome, E. 1963. *Proc. Sci. Assoc. Niger.* **6**: 48 (cytology).
Uppal, B. N. et al. 1932. *Phytopathology* **22**: 587 (infection of maize & *Euchlaena mexicana*; **11**: 635).
Weston, W. H. et al. 1932. *Phytopathology* **22**: 573 (diagnosis of *S. sorghi* & comparison with *S. graminicola*; **11**: 634).

SCLEROTINIA Fuckel, *Symb. Mycol.*: 330, 1870.

(Sclerotiniaceae)

Apothecia arising from well defined tuberoid SCLEROTIA, freed from the substrate at maturity or only loosely enclosed within it, rind black, medulla usually white. MICROCONIDIA borne in sporodochia, superficial or in cavities in the host (subgenus *Myriosclerotinia*). MACROCONIDIA state a *Botrytis*, *Ovularia* or wanting. APOTHECIA brown, disk saucer shaped or flat, receptacle downy or scurfy. ASCI 8 spored (rarely 4 spored). ASCOSPORES elliptical, inequilateral or very slightly reniform, hyaline, aseptate, often biguttulate, usually uniseriate (R. W. G. Dennis, *Mycol. Pap.* 62: 144, 1956).

Dennis (1956) accepted the older, broader concept of the genus. Whetzel erected several generic segregates from *Sclerotinia*, one of which was *Botryotinia* for spp. which had a *Botrytis* conidial state. He also re-defined *Sclerotinia* more narrowly as forming a sclerotium which was not formed in the host tissues (i.e. was loosely associated with them) and did not incorporate such tissues, and further did not have an imperfect (macroconidial) state. The broad concept of *Sclerotinia* appears to be still accepted by most plant pathologists and some taxonomists. Dennis (1956) placed the *Sclerotinia* spp. that he considered under 4 subgenera; of these the economically important spp. came under *Botryotinia* or *Sclerotinia*. Noviello et al. discussed the significance to be attached in classification to the method of development of the sclerotium. Whetzel described the stroma of *Botryotinia* as firmly attached to the host. Buchwald (1949) and Dumont et al. accepted Whetzel's concepts regarding *Sclerotinia, Botryotinia* and *Monilinia* (the last genus was erected by Honey to accommodate those spp. previously in *Sclerotinia* but with a *Monilia* state; and see Hino). Some plant pathologists still treat *Monilinia* as part of *Sclerotinia*. Following Whetzel's work Korf et al. gave the nomenclatural arguments for a new genus (*Whetzelina*) for *Sclerotinia* sensu Whetzel. Buchwald et al. made a plea for the retention of the well known name *S. sclerotiorum* (q.v.), an important plant pathogen, as the type sp. for *Sclerotinia* sensu Whetzel. Dennis (1974) and Korf continued the nomenclatural discussion. Some *Botrytis* spp. of economic importance have no known perfect state.

Six spp. are described separately. Two of these (*S. porri* and *S. squamosa*) are restricted to *Allium* and both have *Botrytis* states. *B. allii* (q.v.) and *B. byssoidea* (*Botrytis* q.v.) also occur on *Allium* and cause diseases (on onion) but have no known perfect states. Also on *Allium* is *S. globosa* (Buchwald) Webster (stat. conid. *B. globosa* Raabe). Raabe described this sp. as causing pale, watersoaked spots on leaves of garlic. But it has been more recently described from *A. ursinum* in W. Europe (Buchwald 1953; Hennebert; Webster; Webster et al.). Like the conidial state of *S. squamosa*, that of *S. globosa* shows a well marked, concertina-like collapse of the conidiophores. The conidia of the latter are spherical or subspherical, mostly 12–18 μ diam.; in the former they are elliptical, mostly $15–21 \times 13–16$ μ (length : breadth ratio 1.25–1.45). *S. sphaerosperma* Gregory was described from *A. triquetrum*; its *Botrytis* state has conidia mostly 20–26 μ diam.

S. trifoliorum Eriksson has been recorded on many hosts but is particularly parasitic on Leguminosae and causes clover (*Trifolium*) rot (see A. J. H. Carr in *Diseases of crop plants*, editor Western). Keay described *S. trifoliorum* var. *fabae* from broad bean (*Vicia faba*; see Loveless). *S. trifoliorum* has been recorded on lucerne (*Medicago sativa*) and *S. sativa* Drayton & Groves was also described from this host amongst others (see Cormack). Turner et al. described the parasitism of *Coniothyrium minitans* Campbell on *Sclerotinia* spp. and preliminary work on the control of *S. trifoliorum* by *C. minitans*. *S. homeocarpa* F. T. Bennet causes dollar spot of grass turf, probably mostly in temperate regions (see Drew Smith, *Fungal diseases of turf grasses*). *S. intermedia* Ramsey was described as one of the *Sclerotinia* spp. causing rot during transport and storage of vegetables (Sugimoto et al.).

Buchwald, N. F. 1949. *K. Vet.-og Landbohøisk. Aarsskr.*: 75 (taxonomy of Sclerotiniaceae; **28**: 546).

——. 1953. *Phytopathol. Z.* **20**: 241 (perfect state of *Botrytis globosa* on *Allium ursinum*; **33**: 463).

—— et al. 1972. *Friesia* **10**: 96 (plea for retention of *Sclerotinia sclerotiorum* as type for *Sclerotinia*; **52**, 2861).

Cormack, M. W. 1946. *Scient. Agric* **26**: 448 (*S. sativa* & related spp.; **26**: 61).

Dennis, R. W. G. 1956. *Mycol. Pap.* 62, 216 pp. (revision of British Helotiales & notes on European spp.; **35**: 848).

——. 1974. *Kew Bull.* **29**: 89 (*Whetzelina* Korf & Dumont, a superfluous name; **54**, 735).

Drayton, F. L. et al. 1943. *Mycologia* **35**: 517 (*S. sativa* sp. nov.; **23**: 108).

Dumont, K. P. et al. 1971. *Ibid* **63**: 157 (Sclerotiniaceae, generic nomenclature).

Gregory, P. H. 1941. *Trans. Br. mycol. Soc.* **25**: 26 (on *Botrytis* & *Sclerotinia*; **20**: 466).

Hennebert, G. L. 1958. *Bull. Jard. bot. Etat Brux.* **28**: 193 (morphology of conidiophore of *B. globosa*; **38**: 174).

Hino, I. 1929. *Bull. Miyazaki Coll. Agric. For.* **1**: 67 (conidial forms in *Sclerotinia*; **8**: 607).

Honey, E. E. 1928. *Mycologia* **20**: 127 (*Monilia* states of *Sclerotinia*; **7**: 744).

Keay, M. A. 1939. *Ann. appl. Biol.* **26**: 227 (comparative study on *S. minor, S. sclerotiorum, S. serica* & *S. trifoliorum*; **18**: 628).

Korf, R. P. et al. 1972. *Mycologia* **64**: 248 (*Whetzelina* gen. nov.; **51**, 3822).
——. 1974. *Mycotaxon* **1**: 146 (typification of *Sclerotinia*).
Loveless, A. R. 1951. *Ann. appl. Biol.* **38**: 252 (*S. trifoliorum* var. *fabae* & clover rot; **31**: 187).
Noviello, C. et al. 1962. *Mycologia* **53**: 237 (technique for investigating stroma formation; **41**: 586).
Raabe, A. 1938. *Nova Hedwigia* **78**: 1 (*B. globosa* sp. nov.; **18**: 140).
Ramsey, G. B. 1924. *Phytopathology* **14**: 323 (*S. intermedia* sp. nov.; **4**: 12).
Sugimoto, T. et al. 1959. *Mem. Fac. Agric. Hokkaido Univ.* **3**: 121 (*S. intermedia* on carrot; **39**: 258).
Turner, G. J. et al. 1975. *Pl. Pathol.* **24**: 109 (control of *S. trifoliorum* by *Coniothyrium minitans*; **55**, 297).
—— ——. 1976. *Trans. Br. mycol. Soc.* **66**: 97 (parasitism by *C. minitans* on *S. trifoliorum*; **55**, 4614).
Webster, J. et al. 1951. *Ibid* **34**: 187 (*B. globosa* on *A. ursinum*).
——. 1954. *Ibid* **37**: 168 (*S. globosa* in UK).
Whetzel, H. H. 1945. *Mycologia* **37**: 648 (the genera & spp. of Sclerotiniaceae; **25**: 235).

Sclerotinia fuckeliana (de Bary) Fuckel, *Symb. Mycol.*: 330, 1870.
Peziza fuckeliana de Bary, 1866
stat. conid. *Botrytis cinerea* Pers. ex Pers., 1822.
(a synonymy in Hennebert 1973, *Botrytis* q.v.)

SCLEROTIA hard, black, plano convex, firmly attached, very variable in size and shape; round to elongate, more or less curved, sometimes convoluted, very irregular, 1–15 mm in length, sometimes in confluent masses; composed of hyaline, thick walled hyphae with a thin pseudoparenchymatous rind of dark brown to black, almost isodiametric cells, *c*. 5–10 μ diam. APOTHECIA arising from sclerotia singly or clustered, long stipitate, at first deeply concave, expanding to plane and finally convex, often umbilicate, 1.5–7 mm diam., 3–10 (–15) mm in height; stipe rather stout, 0.5–1.5 mm diam.; disk avellaneous to wood brown or verona brown to bistre with delicate, paler margin; stipe concolorous becoming darker below to blackish at the base, nearly glabrous or minutely pubescent; tissue prosenchymatous, in the stipe hyaline, more or less vertically parallel to slightly interwoven, septate hyphae *c*. 5–10 μ diam.; in hypothecium hyphae become more intricately woven; outside a pseudoparenchymatous excipulum is differentiated, of large, hyaline, elongated to nearly isodiametric or polygonal cells mostly 10–20 μ diam. but sometimes up to 60μ in length; subhymenium differentiated as a narrow, brownish zone of closely woven hyphae, *c*. 2.5–5 μ diam. ASCI cylindric with a long, tapering stalk, 8 spored, 120(105)–140 (–160) × 6.5–9.5 μ. ASCOSPORES hyaline, aseptate, ellipsoid to ovoid, uniseriate, 9–12 × 4.5–6 μ. PARAPHYSES hyaline, filiform, septate, unbranched or very occasionally so, 2–2.5 μ diam., tips sometimes slightly swollen.

CONIDIOPHORES arising from mycelium or sclerotia; erect, brown, septate, simple or usually branched towards the tip, tip rounded or slightly swollen, 5–22 μ diam.; 1–3 mm in height. CONIDIA greyish brown in mass, nearly hyaline microscopically, ellipsoid to ovoid, aseptate, on small stalks arising from tips of conidiophores, forming botryose clusters, 8–14 (–18) × 6–9 μ. SPERMATIA hyaline, globose, aseptate, 2–2.5 μ diam., formed endogenously from hyaline flask shaped phialides which may arise singly from the mycelium or more commonly are aggregated into clusters forming spermodochia (after J. W. Groves & C. A. Loveland, *Mycologia* **45**: 422, 1953 as *Botryotinia fuckeliana* (de Bary) Whetzel see R. W. G. Dennis (*Mycol. Pap.* 62, 1956) who placed *B. fuckeliana* as a synonym and see Buckley et al.; Gull et al.; Hawker et al.; Katumoto et al. for fine structure).

Gregory reported on de Bary's original description and on specimens of the perfect state. Groves et al. (1939, 1953) described heterothallism. Morgan examined 107 characters biometrically for 33 isolates of the complex taxon *B. cinerea*. His results suggested that the isolates could be placed in several groups (termed races) but that differences were not great enough to warrant placing any isolate in a new taxon.

This fungus, a general pathogen and saprophyte, causes the ubiquitous grey mould of the conidial state on herbaceous, aerial plant tissue. It is commoner in cooler regions and frequently occurs on temperate flowers and fruit. Tissues are rapidly rotted, dense, dark masses of conidiophores develop on their surfaces with sclerotia which may give rise to apothecia. Ellis et al. gave an account of the stat. conid. and the diseases caused. *S. fuckeliana* is of historical interest since it was the fungus used in the classical studies on host penetration and disintegration. Initial work by A. de Bary was followed by that of W. Brown and others; these results were discussed by Wood, *Physiological plant pathology*. Thin or immature cuticle can be directly penetrated by the germinating conidium. The process has generally been considered to be a mechanical one; more recently McKeen considered that penetration does not take place by either mechanical or chemical means alone. Infection can also occur through wounds or weakened plant organs. After penetration (Louis), spread in, and destruction of, the host is very rapid. Attacks may occur during transport and in storage, as well as in the field. For cell degrading enzymes see Ashour; Berg et al. 1969; Gäumann et al.; Tani et al.; Tribe; Verhoeff et al. Attacks can occur on any plant at moderate temps (probably appreciably below 20°) where these may be below opt. for the host, and under any conditions likely to cause high humidities. In vitro growth rate, conidial germination and appressorial formation are best at 20° or somewhat lower (Berg et al. 1968; Shiraishi et al.). Harada et al. gave *c.* 25° for max. in vitro growth rate, and best temps of 20–25° for sporulation (conidia) and of 10–20° for sclerotial formation. The conidia are air and splash dispersed (most carried on the surface of water drops). Jarvis (1962) found a diurnal rhythm with a peak conidial conc. at around noon and high numbers during rain; opt. temps for conidial formation and mycelial growth were 15 and 20°, respectively. The effects of light on conidial sporulation have been examined by Hite; Honda et al.; Jarvis 1972; Suzuki et al.; Tan; Tan et al. Krause et al. described the effects of ozone.

Fungicide control is frequently required; some aspects of the pathology of *S. fuckeliana* are referred to later. There appears to be an increasing literature on the resistance of the fungus towards the newer fungicides, for example, benomyl (Bollen et al. in cyclamen; Ehrenhardt et al. various sources; Jarvis et al. 1973; Pourtois et al. in strawberry; Miller et al. in lettuce and tomato); dicloran, tecnazene and quintozene (Bolton; Esuruoso et al. 1971; Webster et al.).

Ashour, W. E. 1954. *Trans. Br. mycol. Soc.* **37**: 343 (pectinase formation; **34**: 666).

Baker, K. F. 1946. *Pl. Dis. Reptr* **30**: 145 (outbreaks in USA, California; **26**: 3).

Berg, L. van den et al. 1968. *Can. J. Bot.* **46**: 1477 (effects of RH & temp. on growth & survival with *Sclerotinia sclerotiorum*; **48**, 1082).

——— ———. 1969. *Ibid* **47**: 1007 (effect of RH on formation of extracellular pectolytic enzymes with *S. sclerotiorum*; **48**, 3315).

Blakeman, J. P. 1975. *Trans. Br. mycol. Soc.* **65**: 239 (conidial germination & nutrient conditions on leaf surfaces; **55**, 1652).

Bollen, G. J. et al. 1971. *Neth. J. Pl. Pathol.* **77**: 83 (resistance to benomyl & other systemic fungicides; **51**, 410).

Bolton, A. T. 1976. *Can. J. Pl. Sci.* **56**: 861 (resistance of conidia to fungicides; **56**, 2525).

Brodie, I. D. S. et al. 1976. *Physiol. Pl. Pathol.* **9**: 227 (conidial germination & competition for exogenous substrates by leaf surface micro-organisms; **56**, 2865).

——— ———. 1977. *Trans. Br. mycol. Soc.* **68**: 445 (effects on conidia caused by leaching with water; **56**, 5462).

Buckley, P. M. et al. 1966. *J. Bact.* **91**: 2037 (fine structure of conidia).

Chu Chou, M. et al. 1968. *Ann. appl. Biol.* **62**: 11 (effect of pollen grains on infection; **47**, 3350).

Ehrenhardt, H. et al. 1973. *NachrBl. dt. PflSchutzdienst. Stuttg.* **25**: 49 (resistance to benomyl; **52**, 2836).

Ellis, M. B. et al. 1974. *C.M.I. Descr. pathog. Fungi Bact.* 431 (*Botrytis cinerea*).

Esuruoso, O. F. et al. 1968. *Trans. Br. mycol. Soc.* **51**: 405 (conidial germination in the presence of fungicides; **48**, 22).

——— ———. 1971. *Ann. appl. Biol.* **68**: 271 (resistance of conidia to fungicides; **51**, 1101).

Gäumann, E. et al. 1947. *Helv. chim. Acta* **30**: 24 (adaptive enzymes; **26**: 313).

——— ———. 1947. *Ber. schweiz. bot. Ges.* **57**: 258 (effect of temp. on enzymatic efficiency with *Aspergillus niger*; **27**: 490).

Greenaway, W. 1973. *Ann. appl. Biol.* **73**: 319 (assay of benomyl; **52**, 3535).

Gregory, P. H. 1949. *Trans. Br. mycol. Soc.* **32**: 1 (de Bary's description & specimens; **28**: 490).

Groves, J. W. et al. 1939. *Mycologia* **31**: 485 (perfect state; **18**: 820).

——— ———. 1953. *Ibid* **45**: 415 (connection of states, heterothallism & taxonomy; **33**: 385).

Gull, K. et al. 1971. *J. gen. Microbiol.* **68**: 207 (fine structure of conidial germination; **51**, 2132).

Harada, Y. et al. 1972. *Bull. Fac. Agric. Hirosaki Univ.* 19: 22 (culture; **53**, 821).

Hawker, L. E. et al. 1963. *J. gen. Microbiol.* **33**: 43 (fine structure of conidial germination; **43**, 1564).

Hite, R. E. 1973. *Pl. Dis. Reptr* **57**: 131, 760 (effect of light on growth & conidial sporulation; **52**, 2852; **53**, 1231).

Honda, Y. et al. 1978. *Pl. Physiol.* **61**: 711 (light spectrum & conidial sporulation; **57**, 5303).

Jarvis, W. R. 1962. *Nature Lond.* **193**: 599 (splash dispersal of conidia; **41**: 436).

———. 1962. *Trans. Br. mycol. Soc.* **45**: 549 (conidial dispersal in raspberry crop; **42**: 332).

———. 1972. *Ibid* **58**: 526 (phototropism; **51**, 4688).

——— et al. 1973. *Pl. Pathol.* **22**: 139 (resistance to benomyl with *Penicillium corymbiferum*; **53**, 3309).

Kamoen, O. et al. 1963. *Meded. landbHogesch. Ghent* **28**: 839 (effect of light on conidial sporulation & sclerotial formation; **43**, 1512o).

Katumoto, K. et al. 1974. *Bull. Fac. Agric. Yamaguti Univ.* 25: 965 (fine structure; **54**, 4579).
Krause, C. R. et al. 1978. *Phytopathology* **68**: 195 (effects of ozone; **57**, 4806).
Lauber, H. P. 1971. *Schweiz. landw. Forsch.* **10**: 1 (variability & cytology, heterokaryosis; **51**, 3837).
Louis, D. 1963. *Annls Epiphyt.* **14**: 57 (host penetration; **42**: 599).
McKeen, W. E. 1974. *Phytopathology* **64**: 461 (epidermal penetration of broad bean; **53**, 4198).
Miller, M. W. et al. 1974. *Trans. Br. mycol. Soc.* **62**: 99 (resistance to benomyl; **53**, 2814).
Morgan, D. J. 1971. *Ibid* **56**: 319 (taxonomy; **51**, 133a).
Nonaka, F. et al. 1967. *Agric. Bull. Saga Univ.* 24: 93 (temp., growth, conidial sporulation & sclerotial types; **47**, 3716).
Paul, W. R. C. 1929. *Trans. Br. mycol. Soc.* **14**: 118 (variation in vitro; **8**: 528).
Pourtois, A. et al. 1976. *Annls Phytopathol.* **8**: 1 (resistance to benomyl; **56**, 2857).
Shiraishi, M. et al. 1970. *Ann. phytopathol. Soc. Japan* **36**: 230, 234 (effects of temp.; **50**, 1604).
Suzuki, Y. et al. 1977. *J. gen. Microbiol.* **98**: 199 (blue & near UV reversible photoreaction in conidial development; **56**, 2864).
Tan, K. K. 1973, 1974 & 1975. *Trans. Br. mycol. Soc.* **61**: 145; **62**: 105; **63**: 203; **64**: 213, 223 (effects of light on growth & conidial sporulation, reversibility of sporulation, recovery from blue light inhibition of sporulation, interaction of light & sporulation; **53**, 416, 2838; **54**, 334, 4810, 4811).
——. 1974. *J. gen. Microbiol.* **82**: 191, 201 (blue light inhibition of conidial sporulation & red far red photoreaction in recovery from this inhibition; **54**, 317, 318).
—— et al. 1974. *Trans. Br. mycol. Soc.* **63**: 157 (UV absorbing compounds & conidial sporulation; **54**, 333).
——. 1976. *J. gen. Microbiol.* **93**: 278 (light induced synchronous conidiation; **55**, 5039).
Tani, T. et al. et al. 1969. *Ann. phytopathol. Soc. Japan* **35**: 1 (cell degrading enzymes; **48**, 2165).
Tribe, H. T. 1955. *Ann. Bot.* **19**: 351 (as above; **34**, 667).
Verhoeff, K. et al. 1978. *Phytopathol. Z.* **91**: 110 (endo-polygalacturonase in conidia; **57**, 4625).
Webster, R. K. et al. 1970. *Phytopathology* **60**: 1489 (resistance to dicloran in vitro; **50**, 1591).

LACTUCA SATIVA

Grey mould is a serious disease on temperate lettuce crops (frames, glasshouse or field). In UK the disease has been called red leg (Abdel-Salam) and an account was given by Butler & Jones, *Plant pathology*. In seedlings infection of the stem (near soil level) or leaves causes collapse; stem lesions are reddish brown. In older plants the stem rot is followed by a dull grey appearance in the plant which wilts suddenly. The stem becomes separated from the root. The leaf tips and margins are attacked and a slimy, soft rot of the whole plant results. The sclerotia occur in the stem tissue. *S. fuckeliana* attacks weakened plants more readily; it may follow marginal scorching or infection by *Bremia lactucae* (q.v.). Control measures were fully discussed by Ogilvie, *Diseases of vegetables* (and see Anon.; Ballantyne; Crüger; Schmidt; Way et al.). In the field well drained situations are essential, deep planting and excessive watering or RH should be avoided, and a balanced plant nutrition maintained. Fungicides may be applied to the seedbeds before transplanting, for example, dicloran, quintozene or tecnazene. Affected seedlings (red leg stem symptoms) should be discarded and destroyed. In frames and the field thiram has been frequently recommended. General hygiene measures should be maintained at all times. Organic material worked into the soil just before planting may lead to serious infection. Cvs vary in their susceptibility.

Abdel-Salam, M. M. 1934. *J. Pomol.* **12**: 15 (general; **13**: 559).
Anon. 1973. *Advis. Leafl. Minist. Agric. Fish.* 559, 3 pp. (advisory; **54**, 3580).
Ballantyne, B. 1964. *Agric. Gaz. N.S.W.* **75**: 1048 (fungicides with *Sclerotinia sclerotiorum*; **43**, 3083).
Crüger, G. 1962. *Z. PflKrankh. PflPath. PflSchutz* **69**: 513 (review & fungicides; **42**: 429).
Delon, R. et al. 1977. *Can. J. Bot.* **55**: 2463 (fine structure of host & pathogen interaction; **57**, 1924).
Inoue, Y. et al. 1967. *Res. Progr. Rep. Tokai-Kinki natn. agric. Exp. Stn* 4: 49 (epidemiology & control in Japan; **47**, 713).
Krauss, A. 1971. *Z. PflErnähr. Dung. Bodenk* **128**: 12 (effect of fertilising lettuce with major elements; **51**, 968).
Newhook, F. J. 1951. *Ann. appl. Biol.* **38**: 169, 185 (antagonism; **31**: 269).
Schmidt, T. 1964. *Pflanzenarzt* **17**: 18 (fungicides in Austria; **43**, 3084).
Watanabe, Y. et al. 1970. *Bull. Tokai-Kinki natn. agric. Exp. Stn* **20**: 167 (epidemiology & control in Japan; **50**, 1024).
Way, J. M. et al. 1959. *Ann. appl. Biol.* **47**: 685 (fungicides in UK; **39**: 453).
Wood, R. K. S. 1951. *Ibid* **38**: 203 (antagonism; **31**: 270).

LEGUMINOSAE

The aetiology of chocolate spot on leaves of broad bean (*Vicia faba*), whose diseases are not considered, involves *S. fuckeliana* and *Botrytis fabae* Sardiña. The latter is more virulent and causes the most severe symptoms in the aggressive form of the disease (see Butler & Jones, *Plant pathology*). *B. fabae* (no perfect state known), which has larger conidia (16–25 × 13–16 μ) than *B. cinerea*, is widespread in Asia and Europe, and also reported from: Angola, Argentina, Australia, Canada (Nova Scotia), Chile, Egypt, Ethiopia, Morocco, New Zealand, South Africa and Uruguay (Map 162).

The behaviour and pathogenicity of these 2 fungi on broad bean leaves have been very fully investigated. Deverall et al. (1961) demonstrated that *B. fabae* caused lesions much more rapidly than *B. cinerea* and was, therefore, the more effective pathogen. Mansfield et al. (1974) described 2 general patterns of infection by *B. cinerea*. In one, where few or no lesions developed, conidial germination and germ tube growth were inhibited on the leaf surface. In the other limited lesions formed with marked browning at the inoculation site. Inhibition occurred within the leaf at sites bearing limited lesions; the invading hyphae were restricted to brown epidermal cells. On the other hand *B. fabae* was not inhibited on the leaf surface and spread inter- and intracellularly beneath the inoculation drop and into the surrounding tissues. Wastie found that 3–4 conidia/drop of *B. fabae* caused lesions in 50% of inoculations whilst 500 conidia/drop were needed for *B. cinerea* to obtain comparable lesion formation.

The phytoalexin detected in broad bean after infection by these fungi was called wyerone acid (Fawcett et al.; Purkayastha et al. 1965). *B. fabae* metabolises the acid to a much less anti-fungal form at a faster rate than *B. cinerea* (Deverall; Deverall et al. 1969; Mansfield et al. 1973). The fungitoxicity of wyerone acid is affected by pH and by components of natural media, for example, pollen (Deverall et al. 1972; Mansfield et al. 1971). The differing abilities of these 2 fungi to spread from lesions after both have induced wyerone acid formation probably depends both on their differing sensitivities to the phytoalexin and their capacity to metabolise it to less toxic products (Mansfield et al. 1974b). But later Rossall et al. concluded that, although the comparative insensitivity of *B. fabae* to wyerone acid may play a role in its virulence to broad bean, it seems unlikely that the different pathogenicities of these 2 fungi can be explained solely on the basis of a difference in the phytoalexin conc. needed for inhibition. Purkayastha (1969) found that substances in culture filtrates caused wilt and necrosis of broad bean shoots, the filtrates from *B. fabae* being the more active in this respect. For more recent work on phytoalexins see *Rev. Pl. Pathol.* **55**, 1605, 4942, 6023, 6024; **56**, 1345; **57**, 1511, 2701, 3177, 5196, 5767.

Other legumes attacked include: chick pea, common bean, lima bean, lucerne and pea. Infection of the pods of the last crop results in seeds with a chalky appearance. The typical soft, watersoaked decays occur, often beginning near soil level or spreading from blossom infections. Control is largely a matter of cultural practice: maintaining vigorous growth, reducing humidity, avoiding plant damage, removing debris and infected parts of the growing crop, and ensuring that the harvested crop is free from grey mould infection since this will later spread rapidly in transport and storage, especially where ventilation is inadequate. No regular use of fungicides in this crop group has been employed. In the ref. the host is broad bean except where stated otherwise.

Deverall, B. J. et al. 1961. *Ann. appl. Biol.* **49**: 461 (interactions between phenolase of host & pectic enzymes of pathogen; **41**: 191).

——. 1967. *Ibid* **59**: 375 (biochemical changes in infection droplets; **46**, 3280).

—— ——. 1969. *Ibid* **63**: 449 (role of phytoalexin in controlling lesion development; **48**, 3742).

—— ——. 1972. *Ibid* **72**: 301 (effect of pH & composition of test solutions on inhibition by wyerone acid; **52**, 1410).

Fawcett, C. H. et al. 1969. *Neth. J. Pl. Pathol.* **75**: 72 (isolation & properties of wyerone acid; **48**, 1496).

—— ——. 1971. *Physiol. Pl. Pathol.* **1**: 163 (increase in wyerone acid after infection; **50**, 3301).

Joshi, M. M. et al. 1969. *Indian Phytopathol.* **22**: 125 (on chick pea; **49**, 610).

Kendrick, J. B. et al. 1950. *Phytopathology* **40**: 228 (on lima bean; **29**: 488).

Mansfield, J. W. et al. 1971. *Nature Lond* **232**: 339 (action of pollen in breaking host resistance; **51**, 887).

—— ——. 1973. *Physiol. Pl. Pathol.* **3**: 393 (metabolism of wyerone acid; **53**, 740).

—— ——. 1974a, b. *Ann. appl. Biol.* **76**: 77, 227 (rate of fungal & lesion development & changes in wyerone acid conc. in *Botrytis cinerea* & *B. fabae* infection; **53**, 4921, 4922).

Polach, F. J. et al. 1975. *Phytopathology* **65**: 657 (on common bean; **55**, 2968).
Purkayastha, R. P. et al. 1965. *Ann. appl. Biol.* **56**: 139, 269 (growth of *B. cinerea* & *B. fabae* in leaves & anti-fungal substances; **45**, 296, 915).
——. 1966. *Indian J. mycol. Res.* **4**: 51 (*B. cinerea* & *B. fabae* mixed inoculum & lesion development; **51**, 888).
——. 1969. *Proc. natn. Inst. Sci. India* Ser. B **35**: 385 (phytotoxicity of culture filtrates; **51**, 2983).
Rossall, S. et al. 1978. *Ann. appl. Biol.* **89**: 359 (activity of wyerone acid against *Botrytis*).
Sode, J. 1971. *Ugeskr. Agron.* **116**: 523 (on pea; **51**, 4476).
Wastie, R. L. 1962. *Trans. Br. mycol. Soc.* **45**: 465 (inoculum potential; **42**, 355).
Wijngaarden, T. P. et al. 1968. *Neth. J. Pl. Pathol.* **74**: 8 (pea infection, temp. & nutrients; **47**, 1348).
Zakopal, J. et al. 1966. *Ochr. Rost.* **2**: 243 (on lucerne; **46**, 2767).

LYCOPERSICON ESCULENTUM

Grey mould on tomato has been extensively studied by Verhoeff. Infection takes place through stem wounds and leaf scars; flowers can also be attacked. The characteristic powdery growth then develops on the rotting areas. A symptom which occurs on green, immature fruit is called ghost or water spot (Ainsworth et al.; Darby; Ferrer et al.; Verhoeff 1970). A minute, necrotic, raised spot forms in a circular zone of normal colour and bounded by a narrower zone or ring which is pale green or silvery. These spots are more frequent at the calyx end and remain conspicuous (turning yellow) as the fruit ripens. They are 2–5 mm diam. and typically form when the RH is high. The germ tubes from conidia penetrate the epidermis and the fungus secretes an enzyme which diffuses outwards to form the ring. If many conidia cause the symptoms these may become like scabs and blisters can form. If the RH remains high a rot may spread in the fruit. The fungus may cause fruit fall. High nitrogen reduces this and also the number of stem lesions, affected leaf stalks and clusters. Lesions develop more rapidly in young tissue near stem apices than in old tissue towards the base of the stem. In USA (Florida) tomatoes grown on alkaline soils have less grey mould. The calcium–phosphorous balance is important and Stall et al. found that there was least disease where leaf calcium was high and leaf phosphorous low.

Less infection occurs if leaves are cut off close to the stem when deleafing. Stem pieces similarly should be trimmed off level with the stem. Wounds and cuts can be treated with captan, dicloran, salicylanilide or tecnazene. Smith et al. (1971) found that benomyl soil drenches were effective against flower and leaf infection but not against ghost spot (see also Channon et al.). Smith et al. (1975), on unheated crops, reported good control if the whole plant was sprayed initially with dichlofluanid as soon as the second truss was flowering and subsequent sprays were restricted to the upper section of the stem, including the 4 or 5 youngest trusses of buds, flowers and fruit. When used as a post-infection spray it was *c.* 8 weeks before dichlofluanid markedly reduced disease incidence. Morgan et al. described thermal fogging with this fungicide; and see Hartill for the use of vinclozolin and glycophene. Honda et al., by the use of UV absorbing vinyl film, reduced sporulation and, therefore, the disease on cucumber and tomato fruit. In glasshouses good sanitation standards and any measures that reduce RH and water deposition are very necessary.

Ainsworth, G. C. et al. 1938. *Ann. appl. Biol.* **25**: 308 (ghost spot; **17**: 633).
Channon, A. G. et al. 1973. *Ibid* **75**: 31 (benomyl, with *Fusarium oxysporum* f. sp. *lycopersici*, **53**, 1554).
—— ——. 1977. *Ibid* **85**: 359 (benomyl & dichlofluanid inter alia; **56**, 4202).
—— ——. 1978. *Ibid* **90**: 345 (carbendazim, oil & sensitive & resistant strs).
Chastagner, G. A. et al. 1978. *Phytopathology* **68**: 1172 (conidial dispersal).
Darby, J. F. 1955. *Pl. Dis. Reptr* **39**: 91 (ghost spot & grey mould control in Florida; **35**: 50).
Davison, R. M. et al. 1956. *N.Z. Jl Sci. Technol.* Sect. A **38**: 177 (incorporation of fungicides in fruit setting sprays; **36**: 139).
Ferrer, J. B. et al. 1959. *Phytopathology* **49**: 411 (ghost spot; **39**: 50).
Fletcher, J. T. et al. 1976. *Ann. appl. Biol.* **82**: 529 (benomyl resistance; **55**, 2897).
Hartill, W. F. T. 1976. *N.Z. Comm. Grower* **31**: 20 (vinclozolin inter alia; **56**, 4698).
Honda, Y. et al. 1977. *Pl. Dis. Reptr* **61**: 1041 (control by inhibiting sporulation; **57**, 4173).
Morgan, W. M. et al. 1978. *Pl. Pathol.* **27**: 6 (dichlofluanid by thermal fogging).
Newhook, F. J. et al. 1956. *N.Z. Jl Sci. Technol.* Sect. A **38**: 166, 180 (incorporation of fungicides in fruit setting sprays; **36**: 139).
——. 1957. *Ibid* **38**: 473 (antagonism of saprophytes; **37**: 114).
Shishiyama, J. et al. 1970. *Pl. Cell Physiol. Tokyo* **11**: 937 (characteristics & activity of cutin esterase; **50**, 2732).

Smith, P. M. et al. 1971. *Pestic. Sci.* **2**: 201 (systemic fungicides; **51**, 3739).
—— ——. 1975. *Ann. appl. Biol.* **80**: 49 (control on unheated crops with non-systemic fungicides; **54**, 4639).
Stall, R. E. 1963. *Phytopathology* **53**: 149 (effects of lime on disease; **42**: 574).
—— et al. 1965. *Ibid* **55**: 447 (disease incidence & Ca & P balance; **44**, 2613).
Verhoeff, K. 1965. *Neth. J. Pl. Pathol.* **71**: 167 (mycelial development in plants, nutrient levels & internodes of differing age; **45**, 1523).
——. 1967. *Ibid* **73**: 117 (effect of deleafing on occurrence of stem lesions; **46**, 3571).
——. 1968. *Ibid* **74**: 184 (effect of soil N & deleafing on disease occurrence in commercial conditions; **48**, 933).
——. 1968. *Meded. Dir. Tuinb.* **31**: 250 (control & plant nutrient levels; **48**, 286).
——. 1970. *Neth. J. Pl. Pathol.* **76**: 219 (ghost spot; **49**, 3006).
—— et al. 1972. *Ibid* **78**: 179 (in vitro & in vivo formation of cell wall degrading enzymes; **52**, 861).
—— ——. 1975. *Phytopathol. Z.* **82**: 333 (toxicity of tomatine; **55**, 893).
——. 1978. *Annls Phytopathol.* **10**: 137 (cell degrading enzymes).
White, R. A. J. 1961. *N.Z. Jl Agric.* **103**: 55 (fungicides; **41**, 65).
Wilson, A. R. 1966. *Acta Hortic.* **4**: 135 (general note).

OTHER CROPS

Only a selection of grey mould disease on other crops is given; the 40-year index (1922–61) to the *Rev. appl. Mycol.* gives >200 plant hosts. In vegetables instances of rots in storage are in carrot and red pepper. Lockhart et al. described the control of these diseases in the former crop where *S. fuckeliana* was the predominant pathogen. Carrots which were washed and graded before storage for 15–16 weeks at 0° and RH 95–100% had less decay than those stored directly from the field. Treatment of washed carrots with sodium o-phenyl phenate or thiabendazole gave a further decrease in decay. Goodliffe et al. (1975) found that carrots may enter storage with an incipient infection which can then develop into a crown rot and spread to adjacent roots. Infection of this crop has been fully investigated recently (Goodliffe et al.; Harding et al.; Heale et al.; Sharman et al.).

The fungus attacks red pepper fruit causing circular to elliptical and sharply outlined necrotic lesions. Infection can take place through uninjured tissue if the crop is weakened by chilling injury. On injured red pepper the decay rate increased with temp. (0–21°). In natural infections there was more decay at 4.4° compared with 7.2, 10 and 12.8°. In unwounded fruit there was progressively more decay with the time (0–20 days) that they were held at 0° after inoculation (Lauritzen et al.; McColloch et al.). Winstead et al. gave an example of conditions likely to lead to an outbreak of grey mould when they described severe attacks in red pepper seedbeds under plastic sheeting.

Clark et al. described brown stains in onion, a superficial discolouration of the dry scales from the neck to the shoulder of the bulb. The necrosis has a dark brown to black margin with a lighter brown centre; there are also dark spots (5–15 mm). Under natural conditions the fleshy scales are apparently not attacked. Sclerotia are formed in the abaxial epidermis. In an examination of 13 cvs Southport red globe had less brown stain than 11 yellow cvs. A typical example of grey mould attack spreading from infected blossoms after inoculum build up on moribund tissue was given by McKeen for cucumber (*Cucumis sativus*). Infection of cabbage was described by Bochow et al.; Yoder et al.

On crops other than vegetables some recent work has been done on a raceme blight of the edible nut *Macadamia* in Hawaii (Holtzmann; Hunter et al.; Rohrbach et al.). The disease occurs only on the flowers, which are attacked after anthesis or at senescence; the infected inflorescences remain attached and form few nuts. Incidence is positively correlated with temps of 18–22°, RH 95–100% and the presence of moisture. Benomyl was shown to be effective. The apparent absence of the disease during heavy rain was attributed to scrubbing of the conidia and stripping of senescing blossoms.

Hartill et al. (on tobacco) described the effects of flower removal on infection by *S. fuckeliana* and *S. sclerotiorum*; the former was the dominant pathogen on leaves and stems. Infection was correlated with the density of corollas falling on the leaves and flower head removal caused a marked decrease in infection by both fungi. Nattrass reported infections on *Eucalyptus* (and see Magnani; Raggi) and Aramina fibre or Congo jute (*Urena lobata*). Cankers were caused on branches and leaders, where diseased leaf tissue joined the stems, of *E. globulus c.* 1 m tall (at 2700 m altitude). Many girdled leaders were broken at the cankered area. On Aramina fibre cankers were caused at any point on the main stem or laterals, and on their surfaces there was a profuse growth of conidiophores and conidia. Sclerotia were seen when cankers were buried in moist sand. There was occasional zoning and gum exudation on the cankers. Infection took place through the flowers. Pawsey gave an account of grey mould in conifer nurseries and its control with Bordeaux which has also been used in citrus (Klotz et al.; and see Moreau). Sackston obtained *S. fuckeliana* and *S. sclerotiorum* from safflower and sunflower. Both fungi cause flower head rots of these crops (Bakos et al.; Barash et al.); fenaminosulf or thiovit were recommended (with cultural measures) as sprays before harvest in sunflower (Alekseeva et al.; and see Courtillot et al.). Hendrickx described an infection of coffee berries (warty disease) and considered the pathogen to be a distinct form. A severe outbreak of infection on coffee was reported by Baker; and see Javed.

Alekseeva, S. P. et al. 1971. *Zashch. Rast. Mosk.* **16**: 30 (on sunflower; **51**, 1772).

Baker, C. J. 1972. *Kenya Coff.* **37**: 266 (on coffee; **52**, 1544).

Bakos, Z. et al. 1967. *Növenytermeles* **16**: 391 (on sunflower; **47**, 1940).

Barash, I. et al. 1964. *Phytopathology* **54**: 923 (on safflower; **44**, 516).

Bochow, H. et al. 1976. *Arch. Phytopathol. PflSchutz* **12**: 261 (susceptibility of cabbage in store; **56**, 2720).

Clark, C. A. et al. 1973. *Pl. Dis. Reptr* **57**: 210 (reaction of onion cvs; **53**, 355).

—— ——. 1973. *Phytopathology* **63**: 1231 (on onion; **53**, 2017).

—— ——. 1975. *Ibid* **65**: 338 (on onion, role of phenols with *Colletotrichum circinans*; **54**, 4710).

Courtillot, M. et al. 1973. *Phytiat. Phytopharm.* **22**: 189 (on sunflower; **53**, 4546).

Goodliffe, J. P. et al. 1975. *Ann. appl. Biol.* **80**: 243 (incipient infections of carrot; **54**, 4678).

—— ——. 1977. *Ibid* **87**: 17 (factors in resistance of cold stored carrots; **57**, 1474).

—— ——. 1978. *Physiol. Pl. Pathol.* **12**: 27 (6-methoxy mellein in resistance & susceptibility of carrot root tissue; **57**, 4187).

Harding, V. et al. 1978. *Ann. appl. Biol.* **89**: 348 (inhibitors in carrot root tissue treated with conidia).

Hartill, W. F. T. et al. 1974. *N.Z. Jl agric. Res.* **17**: 147 (on tobacco; **54**, 224).

Heale, J. B. et al. 1977. *Physiol. Pl. Pathol.* **10**: 51 (induced resistance in root & tissue cultures; **56**, 4745).

—— ——. 1977. *Ann. appl. Biol.* **85**: 453 (infection of carrot & length of storage; **56**, 4257).

—— ——. 1978. *Ibid* **89**: 310 (cytochemical changes in carrot root tissue treated with conidia).

Hendrickx, F. L. 1959. *Publs Inst. natn. Etude agron. Congo belge* Ser. Sci. 19, 12 pp. (on coffee; **20**: 59).

Holtzmann, O. V. 1963. *Pl. Dis. Reptr* **47**: 416 (on macadamia; **42**: 708).

Hunter, J. E. et al. 1972. *Phytopathology* **62**: 316 (as above, epidemiology; **51**, 4388).

—— ——. 1973. *Ibid* **63**: 939 (as above, reduction in nut set; **53**, 1100).

Javed, Z. U. R. 1977. *Kenya Coff.* **42**: 53 (coffee & survival of conidia & sclerotia; **56**, 5650).

——. 1978. *Ibid* **43**: 13 (fungicides on coffee; **57**, 4479).

Klotz, L. J. et al. 1946, *Calif. Citrogr.* **31**: 247 (on lemon; **25**: 558).

Lauritzen, J. I. et al. 1930. *J. agric. Res.* **41**: 295 (on red pepper in storage; **10**: 57).

Lockhart, C. L. et al. 1972. *Can. Pl. Dis. Surv.* **52**: 140 (on carrot in storage; **52**, 4267).

McColloch, L. P. et al. 1966. *Mkt Res. Rep. agric. Res. Serv. U.S.D.A.* 754, 9 pp. (on red pepper in storage; **48**, 1016).

McKeen, C. D. 1952. *Scient. Agric.* **32**: 670 (on cucumber; **32**: 535).

Magnani, G. 1963. *Pubbl. Cent. Sper. agric. for.* **6**: 209 (on eucalyptus; **43**, 2734).

Moreau, C. 1960. *Fruits* **15**: 69 (on sweet orange in transit; **39**: 705).

Nattrass, R. M. 1949. *Emp. For. Rev.* **28**: 60 (on eucalyptus; **28**: 316).

——. 1951. *E. Afr. agric. J.* **16**: 181 (on Aramina fibre; **31**: 16).

Pawsey, R. G. 1964. *Leafl. For. Comm.* 50, 7 pp. (in conifer nurseries; **44**, 869).

Raggi, C. A. 1947. *Publçóes misc. Minist. Agric B. Aires* Ser. A 3(29): 11 pp. (on eucalyptus; **26**: 516).

Rohrbach, K. G. et al. 1970. *Pl. Dis. Reptr* **54**: 694 (fungicides on macadamia; **50**, 1399).

Sackston, W. E. 1960. *Ibid* **44**: 664 (on seed of safflower & sunflower with *S. sclerotiorum*; **40**: 119).

Sharman, S. et al. 1977. *Physiol. Pl. Pathol.* **10**: 63 (penetration of carrot roots; **56**, 4746).

Winstead, N. N. et al. 1958. *Pl. Dis. Reptr* **42**: 981 (on red pepper; **37**: 754).

Yoder, O. C. et al. 1975. *Can. J. Bot.* **53**: 691, 1972 (infection & susceptibility of stored cabbage; **54**, 5132; **55**, 2411).

Sclerotinia minor Jagger, *J. agric. Res.* **20**: 333, 1920.

APOTHECIA arising singly from black irregular sclerotia, 0.5–2 mm across. DISK concave, becoming flattened, light buff, 1–9 mm across. RECEPTACLE smooth, brown, seated on a slender cylindrical stalk,

5–12 mm long. EXCIPULUM of nearly isodiametric cells almost 20 μ across. ASCI cylindrical, indistinctly stalked, broadly rounded above, the pore slightly blued in Melzer's reagent, 8 spored, 125–180 × 8–15 μ. ASCOSPORES uniseriate, elliptical or slightly inequilateral, 9–20 × 5–10 μ. PARAPHYSES cylindrical, 3–4 μ thick, not enlarged at apex (R. W. G. Dennis, *Mycol. Pap.* 62: 147, 1956). See J. C. Jagger (l.c.) who also gave: MICROCONIDIA globose, hyaline, 3–4.2 μ, borne apically on short obclavate conidiophores; appressoria abundant. Sclerotia black, irregular, often anastomosing to form irregular flattened bodies several mm in length.

S. minor is plurivorous and widespread. It causes diseases (frequently on vegetables) similar to those caused by the more frequently reported *S. sclerotiorum* (q.v.). The 2 fungi differ in that *S. minor* has smaller sclerotia, apothecia that differ in shape and colour, and larger asci and ascospores (Keay; *Sclerotinia* q.v.). Lettuce is the most frequently mentioned crop on which disease (drop or, less commonly, collar rot) is described. A rapidly advancing, watery brown rot spreads upwards from the stem base and the lower leaves; collapse of the plant follows. White mycelium with sclerotia form on the above ground parts and these may be readily seen on turning the diseased plant upside down. *S. minor*, like *S. sclerotiorum*, causes disease only under cool conditions (most in vitro growth at *c.* 22°). Sclerotia carry over the fungus between crops and planting seasons. Jarvis et al. analysed epidemic progress on lettuce in New Zealand. Their results indicated that plants infected early sometimes provide inocula for later secondary spread within a crop, and there is sometimes evidence of plant to plant spread. They suggested that primary infections are attributable to air-dispersed ascospores (from the fructifying, soilborne sclerotia), and that secondary plant to plant spread is associated with mycelial inocula. Seed transmission is of little or no importance.

Hawthorne (1974) discounted the importance of ascospores in spread; primary inoculum probably arose from mycelium from sclerotia on the soil surface. Lettuce cvs which had their lower leaves in continual contact with the soil as they senesced got more infection than those with an upright habit. Hawthorne (1976) described apothecial formation in the field in New Zealand. In the laboratory apothecial stipe formation was opt. at 15°.

Rotations should be used if possible; in any case removal of host residues should be done and possibly ploughing to bury infested material for 1–2 years. A black polyethylene film (as a mulch) reduced infection (Hawthorne 1975). The usual glasshouse soil sterilisation methods are effective. The use of fungicides has often been reported, sometimes in work to control *S. sclerotiorum* and *S. fuckeliana* (q.v.) as well. Quintozene and dicloran are effective, and thiram has been used. Benomyl is reported to be better than other fungicides (Rainbow; Waffelaert). Elia et al. found significant differences in resistance between lettuce cvs but Schmidt found no resistance.

Alghisi, P. 1961. *Riv. Patol. veg. Pavia* Ser. 3, **1**: 65, 83, 357 (pectic enzymes & growth in vitro; **40**: 342; **41**: 212).

Barkai-Golan, R. 1974. *Mycopathol. Mycol. appl.* **54**: 297 (formation of cellulase & polygalacturonase; **54**, 3077).

Beach, W. S. 1921. *Bull. Pa. agric. Exp. Stn* 165, 27 pp. (general).

Chivers, A. H. 1929. *Phytopathology* **19**: 301 (variation in culture with *Sclerotinia intermedia*; **8**: 607).

Corda, P. 1969. *Note fitopatol. Sardegna* 13, 6 pp. (on tomato in the glasshouse, fungicides; **51**, 690).

Elia, M. et al. 1964. *Phytopathol. Mediterranea* **3**: 37 (resistance in lettuce; **45**, 1977).

Flachs, K. 1931. *Gartenbauwissenschaft* **5**: 541 (general; **11**: 492).

Goidànich, G. 1939. *Boll. Staz. Patol. veg. Roma* **19**: 293 (general; **19**: 319).

Hartill, W. F. T. et al. 1976. *N.Z. J. Bot.* **14**: 355 (ascospore discharge with *S. sclerotiorum*; **56**, 4363).

Hawthorne, B. T. 1973. *N.Z. Jl agric. Res.* **16**: 559 (formation of apothecia; **53**, 2070).

——. 1974. *Ibid* **17**: 387 (effect of lettuce growth on infection; **54**, 629).

——. 1975. *N.Z. Jl exp. Agric.* **3**: 273 (effect of mulching on disease in lettuce; **55**, 1565).

——. 1976. *N.Z. Jl agric. Res.* **19**: 383 (formation of apothecia in field; **56**, 1365).

Jarvis, W. R. et al. 1972. *Ann. appl. Biol.* **70**: 207 (epidemiology on lettuce in New Zealand; **51**, 4567).

Louvet, J. et al. 1965. *Phytiat. Phytopharm.* **14**: 199 (fungicides; **46**, 58c).

Marras, F. 1961. *Studi sassar.* **9**: 13 pp. (general on vegetables; **43**, 619).

Rainbow, A. F. 1970. *N.Z. Jl Agric.* **121**(6): 58 (general on tomato; **50**, 3180).

——. 1972. *Ibid* **124**(3): 71 (fungicides; 51, 3664).

Schmidt, H. 1965. *Arch. PflSchutz* **1**: 179 (general on lettuce with *S. fuckeliana*; **45**, 2676).

Sereni, D. 1944. *Palest. J. Bot. Rehovot Ser.* **4**: 78 (general; **24**: 218).

Waffelaert, P. 1969. *Phytiat. Phytopharm.* **18**: 39 (fungicides; **49**, 1529).
Wasewitz, H. 1938. *Angew. Bot.* **20**: 70 (general; **17**: 433).

Sclerotinia porri van Beyma, *Meded. phytopathol. Lab. Willie Commelin Scholten* **10**: 46, 1927.
Botryotinia porri (van Beyma) Whetzel, 1945.

APOTHECIA borne singly or in clusters on black sclerotia which may be small and lenticular or large and irregular. DISK *c.* 3 mm across, light grey brown. RECEPTACLE trumpet shaped with a long slender stalk, downy and light grey below, smooth and orange brown above. Anatomy not described. ASCI 175–200 × 12 μ, 8 spored. ASCOSPORES elliptical to reniform, biguttulate, 15–24 × 8–10.5 μ. PARAPHYSES colourless, cylindrical, slightly enlarged upwards. MACROCONIDIA elliptical, 7–19 × 5–10.5 μ. MICROCONIDIA present but undescribed (from R. W. G. Dennis, *Mycol. Pap.* 62: 157, 1956, and said by this author to be inadequately characterised). The very large, irregular, often cerebriform SCLEROTIA formed in culture are characteristic. CONIDIA 9–18 × 6–13 (mostly 11–14 × 7–10) μ, length: breadth ratio 1.35: 1.5 (Ellis, *Demat. Hyphom.*: 181, 1971). Beyma (l.c.) gave the size of the SCLEROTIA in culture as 20 mm in height and 35 mm in width.

There is little information on *S. porri* (homothallic) originally described from leek (*Allium ampeloprasum* var. *porrum*) seed. It has been reported from Belgium, Brazil, Bulgaria, Netherlands, Norway and UK. A neck rot infection of onion was reported by Cronshey (and see Røed) when he inoculated with isolates from wild garlic. Kovachevski described a dry rot of garlic and pathogenicity to leek but rarely to onion.

Cronshey, J. F. H. 1947. *Nature Lond.* **160**: 798 (in UK; **27**: 170).
Elliot, M. E. 1964. *Can. J. Bot.* **42**: 1393 (self fertility; **44**, 945).
Kovachevski, I. C. 1958. *Izv. bot. Inst. Sof.* **4**: 331 (in Bulgaria; **38**: 113).
Røed, H. 1952. *Acta Agric. scand.* **2**: 232 (in Norway; **32**: 416).

Sclerotinia ricini Godfrey, *Phytopathology* **9**: 565, 1919.
stat. conid. *Botrytis ricini* Buchw., 1949.

APOTHECIA 1 to several from 1 sclerotium, 5–30 mm high (usually 6–15 mm), infundibuliform to cyathiform and discoid, long stipitate, cinnamon brown to chestnut brown. STALK concolorous, cylindrical, slender, smooth, flexuous, attenuated below, without rhizoids. DISK at first closed, expanding to saucer shaped, margin sometimes recurved, exterior roughened, 1–7 mm diam. (usually 1.5–4 mm). ASCI cylindrical to cylindro-clavate, apex slightly thickened with a pore, 50–110 × 6–10 (usually 80–100 × 8) μ. ASCOSPORES 8, ellipsoidal, often subfusoid, hyaline, continuous, biguttulate, 9–12 × 4–5 μ. PARAPHYSES abundant, filiform, septate, hyaline, 1.5–2 μ diam. Stat. conid. forming widespread cobweb-like or woolly mass, pale drab grey to drab. CONIDIOPHORES long, slender smooth, slightly constricted at base, olivaceous when mature, dichotomously branched, terminal branching compact, apices non-inflated, thin walled, collapsing when the conidia fall; proliferation sometimes occurs. MACROCONIDIA on sterigmata, globose, smooth, hyaline 6–12 (usually 7–10) μ, compactly grouped. MICROCONIDIA globose, hyaline, 2–3.5 μ, borne apically on short, obclavate, single or clustered conidiophores that develop on sides of hyphae or on tips of special branches; appressoria rare, 20–60 μ across base. SCLEROTIA black, rough, elongate, irregular, 1–25 (usually 3–9) mm in length; suberumpent to superficial, often anastomosing (after G. H. Godfrey, l.c.). Hennebert (1973, *Botrytis* q.v.) placed this sp. in his new genus *Amphobotrys*, and cited *B. bifurcata* J. H. Miller, Giddens & Foster, 1957 as a synonym.

Grey capsule mould of castor (*Ricinus communis*), first described from southern USA is widespread; Brazil, Bulgaria, Colombia, Ghana, India, Jamaica, Malawi, Mozambique, Nigeria, Zimbabwe, Sierra Leone, Solomon Islands and Sri Lanka. The pathogen has been recorded on other members of the Euphorbiaceae, including poinsettia (*Euphorbia pulcherrima*). Hansen et al. described it causing a rot of *Caladium* tubers. Infection of the panicles and capsules is the most serious but it also occurs on leaves, stems and seedlings which can be killed. Symptoms are first seen as small bluish spots which exude yellowish drops. The mycelium spreads and the inflorescence becomes covered with a grey, then olive drab mould; it darkens in colour, the stem droops and the immature capsules and blossoms are destroyed. Leaf lesions are almost circular, covered with mycelium in damp conditions and may show zonate markings. The sclerotia are sometimes seen on the surface of the stems and they also form on the inflorescence. Lines developed in USA differ in susceptibility (Orellana et al. 1962; Thomas et al. 1963, 1964). The disease is most severe on forms with compact panicles and dwarf internodes. Since the male flowers appear to be important as an inoculum source, more grey mould occurs in inflorescences where the staminate flowers are interspersed compared with those where they are absent or only occur at the base (Thomas et al. 1963a). Esuruoso found little resistance in cvs in Nigeria; introduced dwarf cvs were more susceptible than the local tall ones. Vanev recommended copper fungicides and captan.

Esuruoso, O. F. 1966. *Niger. agric. J.* **3**: 15 (susceptibility of 39 cvs; **51**, 1763).
——. 1969. *Ibid* **6**: 15 (reaction of capsules to a biochemical test; **51**, 560).
Godfrey, G. H. 1923. *J. agric. Res.* **23**: 679 (general; **3**: 377).
Hansen, H. N. et al. 1955. *Pl. Dis. Reptr* **39**: 283 (on *Caladium*; **35**: 19).
Orellana, R. G. 1959. *Ibid* **43**: 363 (on leaves of castor; **38**: 613).
—— et al. 1962. *Phytopathology* **52**: 533 (relation of leachable constituents of the capsule to susceptibility; **42**: 40).
—— ——. 1965. *Ibid* **55**: 468 (effect of gallic acid on germination, growth & sporulation; **44**, 2576).
Thomas, C. A. et al. 1963a. *Ibid* **53**: 249 (relation of inflorescence form to susceptibility; **42**: 624).
—— ——. 1963b. *Science N.Y.* **139**: 334 (cell degrading enzymes & resistance; **42**: 565).
—— ——. 1964. *Phytopathol. Z.* **50**: 359 (phenols & pectin in relation to capsule necrosis & maceration; **44**, 199).
Vanev, S. 1960. *Rastit. Zasht.* **8**: 17 (in Bulgaria, fungicides; **40**: 121).
——. 1962. *Izv. bot. Inst. Sof.* (10): 161 (general in Bulgaria; **43**, 185).

Sclerotinia sclerotiorum (Lib.) de Bary, *Vergh. Morph. Biol. Pilze, Mycet. Bact.*: 236, 1884.
Peziza sclerotiorum Lib. 1837
Sclerotinia libertiana Fuckel, 1870
Whetzelina sclerotiorum (Lib.) Korf & Dumont, 1972.

APOTHECIA arising singly or in small groups from cushion shaped or more or less cylindrical sclerotia, white at first, becoming black, formed inside decaying tissues of herbaceous plants or amongst dense white mycelium on their surface. SCLEROTIA very variable in size, usually *c.* 5–15 × 3–5 mm. DISK concave, yellowish brown, 1–10 mm across, usually 3–8 mm. RECEPTACLE cup shaped, concolorous or slightly paler, smooth or minutely downy, narrowed to a long, undulating light brown, smooth cylindrical stalk, 1–2 mm thick and up to 3 cm long. EXCIPULUM of subglobose, thin walled cells, 15–20 μ across. FLESH of hyaline interwoven hyphae *c.* 5–7 μ thick. ASCI cylindrical, indistinctly stalked, broadly rounded above, pore blue in Melzer's reagent, 110–130 × 6–10 μ, 8 spored. ASCOSPORES uniseriate, elliptical, 9–13 × 4–6.5 μ. PARAPHYSES cylindrical, 1–5 μ thick. No conidial state but microconidia occur in artificial culture. Parasitic on many herbaceous plants (after R. W. G. Dennis, *Mycol. Pap.* 62: 146, 1956, in subgenus *Sclerotinia*; see also Mordue et al.).

Korf et al. (*Sclerotinia* q.v.; see also Dennis; Korf) stated that more critical taxonomic studies in this sp. are required before conclusions can be reached on its relation to other *Sclerotinia* spp. Many spp. recognised by several workers (and including several plant pathogens) were placed as synonyms of *S. sclerotiorum* by Purdy (1955). *S. sclerotiorum* (homothallic) is plurivorous and very widespread. It is generally more important as a pathogen of vegetables in the field, during transit and in store (Ramsey). It also causes an important disease of sunflower. Woody plants, grasses and cereals are rarely attacked. The literature on the fungus is very large and there is no recent review. Some brief regional accounts exist and these give details on hosts, for example, Canada (Duczek et al.), Chile (Mujica), Greece (Démétriadès), Israel (Palti) and New Zealand (Brien). Morrall et al. studied variations in morphology, pathogenicity and pectolytic enzyme activity with isolates from Canada. Their conclusions agreed with Purdy's broad concept of the sp.; but see Wong et al. Willets et al. described the differences between the larger and smaller sclerotia in the *S. sclerotiorum* group.

The common disease name adopted here is cottony soft rot. It is descriptive of the profuse, external, white mycelial growth and the general watery rot that follows infection, both on plants in the field, after harvest and in store. Many other common names are used (e.g., white mould and watery soft rot) and a list of these was given by Chupp & Sherf in *Vegetable diseases and their control.* Most parts (above and at soil level) of herbaceous crops, at any age, can be attacked. The first symptoms are frequently the collapse of the plant due to stem infection near the soil. A soft rot develops; it is followed by the conspicuous mycelium and the sclerotia which are often formed in the pith. The rate of collapse of the plant is variable. Similar symptoms occur following infection at some height above soil level. Infections in this region frequently begin from withering or fallen petals which are directly infected by the ascospores (violently discharged and air dispersed). Mycelium arising from sclerotia in soil or host debris can also cause infection. Direct, cuticular penetration by ascospores occurs but only through a relatively thin cuticle and non-woody tissue becoming moribund. Infection also takes place through wounds. Symptoms on produce in transit and in store are similar to those found in the field. Seed is another infection source, either from contaminating sclerotia or as internal mycelium. Ascospores are probably the most important source of primary inoculum (Henderson).

The disease is one of relatively cool, moist conditions and temps in the lowland tropics are likely to be above the opt. for its development. In the Mediterranean region there is little or no disease in the summer (Démétriadès; Palti). Apothecia are formed from sclerotia in soil and plant debris mostly in winter and sping. Light is needed for growth of the hymenial bearing structure but not for that of the stipe (Henson et al.; and see Letham). Purdy (1956) reported an effect on formation of apothecial initials when sclerotia were kept at different temps. At 4° initials formed in 23–27 days, at 24° in 20–45 days, and at 12, 15, 18 and 21° in 7–10 days. Abrupt, fairly large temp. changes may increase sclerotial formation (Krasnokutskaya). The best temps for the formation of stromata and apothecia were 19–22°, and 26–28° retarded their formation. Bedi studied *S. sclerotiorum* mostly in vitro. Growth on PDA was most at 25° but most sclerotia formed at 15°; at lower temps the sclerotia were larger in size than at higher ones. When sclerotia, formed on PDA at 5, 10, 15, 20, 25 and 30°, were floated on water only those at the 2 intermediate temps developed normal apothecia. Bedi found that there was a short period of dormancy for the sclerotia. Those 9–10 weeks old formed apothecia after 9 days, younger sclerotia after 19–30 days (and see Krüger).

Survival of the sclerotia in soil may be for a long period but is very variable. Smith found that sclerotia dried for short periods and then re-moistened in soil are colonised by other micro-organisms and rot in 2–3 weeks. He considered that many conclusions from earlier work were invalid because dried sclerotia had been used without the realisation that drying is an additional treatment and can bias the results. Williams et al. (1965a) recovered sclerotia (95%) from soil after 2 years and soil moisture affected survival. Schmidt found that sclerotia decomposed more rapidly in uncultivated than in cultivated soil and at a depth of 3–4 cm than at 10 cm; and see Jones et al. for parasitism of the sclerotia by soil fungi. Adams reported a >15-month sclerotial survival at 2.5, 15 and 30.5 cm soil depths; at 61 cm survival was poor. In field plots inoculum density fell over 2 years to *c.* 30% of the original density. Apothecia form from sclerotia that are up to 5–6 cm below the soil surface and their formation diminishes with increasing depth (Chambers et al.; Williams et al. 1965b). Sclerotia tend to decay after forming apothecia but may form them in successive years. Control will depend partly on the crop and geographical area. Cultural treatments to increase the rate of sclerotial decomposition in the field have been tried (Moore; Stoner et al.) and fungicidal treatment of both the crop and the soil can be effective (Jones). Measures against the disease are given under individual crops (and see Anon.).

Most ref. on the physiological and biochemical aspects of *S. sclerotiorum* have been omitted as these are very extensive and not of immediate concern here; some have been noted in the *Rev. appl. Mycol.* and *Rev. Pl. Pathol.*:

Nutrition in vitro: **33**: 547; **34**: 740, 741; **48**, 375; **51**, 3779; **52**, 1416.

Cell degrading enzymes: **36**: 659; **51**, 4679; **55**, 5035; **56**, 2263; **57**, 965.

Physiology of sclerotial growth: **49**, 31; **50**, 27, 2137, 3480; **51**, 1148, 2194; **52**, 341, 3544, 3545; **54**, 3712; **55**, 2532; **56**, 2369, 2370, 2870, 3873, 5275, 5454, 5455.

Chemical content of sclerotia: **46**, 1894; **49**, 3644; **50**, 1640, 2716, 2717.

Effects of chemical compounds on sclerotial formation: **51**, 1149; **52**, 4273; **53**, 3788; **55**, 585, 5036.

Apothecia: **45**, 3064; **53**, 3339; **54**, 3146, 3192; **55**, 5033.

Fine structure: **47**, 2624; **50**, 28, 1641; **53**, 4345; **54**, 1156; **55**, 615; **56**, 2436.
Toxins: **31**: 30; **32**: 167; **57**, 4672.
Variability: **55**, 1117.

Adams, P. B. 1975. *Pl. Dis. Reptr* **59**: 599 (soil survival; **55**, 3460).

Anon. 1973. *Advis. Leafl. Minist. Agric. Fish. Fd* 265, 4 pp. (advisory; **54**, 388).

Bedi, K. S. 1956. *Indian Phytopathol.* **9**: 39 (method for obtaining apothecia; **36**: 610).

——. 1958. *Ibid* **11**: 29, 37, 40, 110 (effects of staling products, UV, other micro-organisms & refrigeration on development; **38**: 669).

——. 1961. *Indian J. agric. Sci.* **31**: 236 (viability of sclerotia; **42**: 244).

——. 1962. *Indian Phytopathol.* **15**: 55 (temp. & sclerotial formation; **42**: 100).

——. 1962. *Proc. Indian Acad. Sci.* Sect. B **55**: 213, 244 (light, air, moisture, temp. & formation of apothecia; **42**: 1).

——. 1963. *J. Indian bot. Soc.* **42**: 66, 204 (nutrient factors & apothecial formation; age & size of sclerotia; **43**, 918, 1831).

Brien, R. M. 1932. *N.Z. Jl Agric.* **44**: 127 (hosts; **11**: 547).

Chambers, S. C. et al. 1964. *J. Agric. West. Aust.* Ser. 4 5: 169 (survival in soil; **43**, 2756).

Coe, D. M. 1944. *Mycologia* **36**: 234 (variations in single ascospore isolates, homothallism; **24**: 36).

De Bary, A. 1886. *Bot. Ztg* **44**: 377 (general).

Démétriadès, S. D. 1951. *Annls Inst. Phytopathol. Benaki* 5: 40 (general in Greece, hosts; **31**: 250).

Duczek, L. J. et al. 1971. *Can. Pl. Dis. Surv.* **51**: 116 (hosts).

Held, V. M. et al. 1953. *Pl. Dis. Reptr* **37**: 515 (cross inoculations with *Sclerotinia minor* & *S. trifoliorum*; **33**: 232).

Henderson, R. M. 1962. *Jl R. Soc. West. Aust.* **45**: 133 (life cycle; **42**: 527).

Henson, L. et al. 1940. *Phytopathology* **30**: 869 (apothecial formation in vitro; **20**: 76).

Jones, D. et al. 1969. *Nature Lond.* **224**: 287 (parasitism & lysis of sclerotia by soil fungi; **49**, 682).

——. 1974. *Trans. Br. mycol. Soc.* **63**: 249 (fungicidal effects of dazomet; **54**, 1117).

Krasnokutskaya, O. N. 1967. *Byul. nauchno-tekhn. Inf. po maslich Kul'tur* **1**: 14 (effect of environment including temp. on apothecial formation; **47**, 2404).

Krüger, W. 1976. *NachrBl. dt. PflSchutzdienst.* **28**: 129 (apothecial development; **56**, 561).

Letham, D. B. 1975. *Trans. Br. mycol. Soc.* **65**: 333 (light stimulation of initial apothecial development; **55**, 1653).

Luc, L. H. et al. 1971. *Arch. PflSchutz* 7: 91 (resistance to systemic fungicides; **51**, 60).

Merriman, P. R. 1976. *Soil Biol. Biochem.* **8**: 385 (soil survival; **56**, 584).

Moore, W. D. 1949. *Phytopathology* **39**: 920 (flooding for sclerotial destruction; **29**: 341).

Mordue, J. E. M. et al. 1976. *C.M.I. Descr. pathog. Fungi Bact.* 513 (*S. sclerotiorum*).

Morrall, R. A. A. et al. 1972. *Can. J. Bot.* **50**: 767 (variation, morphology, pathogenicity & pectolytic enzymes; **51**, 4679).

Mujica, R. F. 1955. *Agric. téc.* **15**: 64 (hosts in Chile; **37**: 266).

Newton, H. C. et al. 1973. *Phytopathology* **63**: 424 (conductivity assay for measuring virulence; **52**, 3978).

Palti, J. 1963. *Phytopathol. Mediterranea* **2**: 60 (in Israel, hosts; **43**, 14).

Purdy, L. H. 1955. *Phytopathology* **45**: 421 (a broad concept of the sp.; **35**: 45).

——. 1956. *Ibid* **46**: 409 (sclerotial size, temp., light & apothecial formation; **36**: 128).

——. 1958. *Ibid* **48**: 605 (host penetration & infection; **38**: 306).

Ramsey, G. B. 1925. *J. agric. Res.* **31**: 597 (*Sclerotinia* spp. & vegetable rots; **5**: 269).

Schmidt, H. H. 1970. *Arch. PflSchutz* **6**: 321 (longevity of sclerotia in soil; **50**, 1621).

Smith, A. M. 1972. *Soil Biol. Biochem.* **4**: 131 as above; **52**, 3956).

Stoner, W. N. et al. 1953. *Pl. Dis. Reptr* **37**: 181 (control through cultivation of rice; **32**: 657).

Wakefield, E. M. 1924. *Phytopathology* **14**: 126 (synonymy & nomenclature; **3**: 557).

Willets, H. J. et al. 1971. *Trans. Br. mycol. Soc.* **57**: 515 (ontogenetic diversity of sclerotia; **51**, 2247).

Williams, G. H. et al. 1965a & b. *Ann. Appl. Biol.* **56**: 253, 261 (apothecial formation & sclerotial survival in the soil environment; **45**, 818).

Wong, A. L. et al. 1973. *Trans. Br. mycol. Soc.* **61**: 167 (electrophoresis of *Sclerotinia* spp.; **53**, 808).

—— ——. 1975. *J. gen. Microbiol.* **88**: 339 (taxonomy of *S. sclerotium* & related spp., mycelial interactions; **54**, 4828).

—— ——. 1975. *Ibid* **90**: 355 (electrophoresis of *S. sclerotiorum* & related spp. from different countries; **55**, 1076).

BRASSICA

On young plants (including cabbage, Chinese cabbage and rape) watersoaked areas appear on the winged areas of the petioles. These symptoms also occur on older plants, the lower leaves become necrotic, drop and the plant becomes a shapeless rotten mass. Stems and flower stalks are attacked, and there may be a break at the root stem junction. Sclerotia form under and between the leaves, and in stem hollows. Similar symptoms occur on other *Brassica* spp. and vars. McLean (1958) showed that healthy leaves could be infected through ascospore infection of flower parts which had fallen on the leaves. Thus the disease may occur seriously only on plants grown for seed. Seed infection by mycelium is found but provided that seed is carefully cleaned of sclerotia this may not be serious. Stelfox demonstrated the spread of ascospores on pollen and thereby a head blight of rape resulted. Cultural control (sanitary measures) should be carried out. In USA (Washington) aerial application of benomyl and ground application of calcium cyanamide gave control in cabbage seed fields, but the latter gave no protection against airborne inoculum. Calcium cyanamide can also be applied to the soil (Krüger).

Davis, W. H. 1925. *Phytopathology* **15**: 249 (general; **4**: 713).

Farmer, L. J. et al. 1971. *Pl. Dis. Reptr* **55**: 1136 (symptoms; **51**, 2958).

Gabrielson, R. L. et al. 1973. *Ibid* **57**: 164 (aerial & ground fungicide control; **52**, 3466).

Krüger, W. 1973. *Phytopathol. Z.* **77**: 125 (control; **53**, 2358).

McLean, D. M. 1958. *Pl. Dis. Reptr*. **42**: 663 (infection; **37**: 611).

——. 1959. *Circ. agric. Exp. Stn Wash.* 359, 4 pp. (soil application of fungicide; **39**, 515).

Neegaard, P. 1958. *Pl. Dis. Reptr* **42**: 1105 (mycelial seed infection; **38**: 169).

Schlösser, U. G. 1968. *NachrBl dt. PflSchutzdienst. Stuttg.* **20**: 8 (spread in seed; **47**, 1998).

Stelfox, D. et al. 1978. *Pl. Dis. Reptr* **62**: 576 (pollen spread of ascospores).

Yang, S. M. 1959. *Acta phytopathol. sin.* **5**: 111 (general, host range & ecology; **39**, 515).

HELIANTHUS ANNUUS

The disease on sunflower caused by *S. sclerotiorum* has been described mostly from Canada, Chile, USA and central and E. Europe. In the field the general symptoms of a wilt are seen at an advanced stage of growth; when at *c*. 2 m in height (Baribeau), or max. disease at the bud stage (Auger 1970a), with another max. at fertilisation which followed a decline as the flowers formed. The typical watersoaked soft rot is caused on the stem at soil level or just above. The parenchyma disintegrates leaving the vascular tissue to give a shredded appearance. The roots can be attacked and the plant may fall. Sclerotia form in the stem and in the flower heads, mostly beneath the seed in the parenchyma. At the latter site the sclerotia can be very large (crust sclerotia, up to 12 × 20 × 0.5 cm). Many more sclerotia were found in the rhizosphere compared with non-rhizosphere soil. These are frequently parasitised and killed by *Coniothyrium minitans* on or in the roots and in the stem base (Hoes et al.; Huang). Seed should be clean, i.e. without contaminating sclerotia. Tollenaar et al. reported internal infection of seed and control in seed treatment with N-(ethylmercury)-p-toluene sulphonanilide. Seedling infection has been described from USSR and this was reduced by adding calcium cyanamide to the planting hole; seed treatment with mercurials and thiram is also effective (Lukashevich). Auger (1970b) described the action of benomyl. Four year rotations have been recommended in Chile and Austria. No high resistance appears to have been found (but see Putt).

Auger, S. J. et al. 1970a. *Agric. téc.* **30**: 161 (effect of age on infection; **50**, 3095).

—— ——. 1970b. *Ibid* **30**: 188 (use of benomyl; **50**, 3096).

Baribeau, B. 1923. *Scient. Agric.* **3**: 397 (general; **3**: 213).

Bisby, G. R. 1924. *Ibid* **4**: 381 (symptoms, morphology; **4**: 95).

Crişan, A. 1962. *Studia Univ. Cluj* Ser. Biol. **7**: 45 (control in seed; **43**, 1376).

Fernando Mujica, R. 1950. *Agric. téc.* **10**: 74 (general control; **31**: 242).

Hancock, J. G. 1966. *Phytopathology* **56**: 975 (degradation of pectic substances; **46**: 42).

——. 1967. *Ibid* **57**: 203 (degradation of hemicellulose; **46**, 2086).

——. 1972. *Pl. Physiol.* **49**: 358 (infection & changes in host cell membrane permeability; **51**, 3486).

Hoes, J. A. et al. 1975. *Phytopathology* **65**: 1431 (viability & sclerotial separation from soil; **55**, 3686).

Huang, H. C. 1977. *Can. J. Bot.* **55**: 289 (*Coniothyrium minitans* & sclerotial survival; **56**, 4141).

—— et al. 1978. *Can. J. Pl. Sci.* **58**: 1107 (resistance screening with toxic metabolites).

Jones, E. S. 1923. *Phytopathology* **13**: 496 (aetiology & morphology; **3**: 274).

Krexner, R. 1969. *Pflanzenarzt* **22**: 20 (in Austria; **48**, 3611).

Lukashevich, A. I. 1964. *Zashch. Rast. Mosk.* **9**: 24 (soil treatment with calcium cyanamide; **43**, 3289).

Orellana, R. G. 1975. *Phytopathology* **65**: 1293 (influence of photoperiodism on susceptibility; **55**, 2832).

Pawlowski, S. H. et al. 1964. *Ibid* **54**: 33 (effects of infection; **43**, 2012).

Putt, E. D. 1958. *Can. J. Pl. Sci.* **38**: 380 (differences in host susceptibility; **38**: 9).

Tollenaar, H. et al. 1971. *Agric. téc.* **31**: 44 (internal seed infection & treatment; **51**, 3487).
Young, P. A. et al. 1927. *Bull. Mont. agric. Exp. Stn* 208, 32 pp. (general; **8**: 246).

PHASEOLUS VULGARIS

Common bean is one of the crops on which *S. sclerotiorum* has been most investigated; Kerr et al. described loss. There are recent papers on control with fungicides, both in the field standing crop and after harvest, from Australia, Canada and USA. Seedlings can be infected by ascospores when very young; penetration of the hypocotyl through infection cushions from mycelium has been described (Hungerford et al.; Lumsden et al.). In the field infection of pods, leaves and stems shows the typical watersoaked decay, developing rapidly, girdling stems and causing chlorosis and general plant collapse. Sclerotia form inside pods and stems or externally. Besides penetration by mycelium at or near soil level, this also occurs through infection of fallen petals by ascospores; lesions spread from the plant sites where the flower parts have lodged. Close planting, excessive fertiliser application and moisture levels, and continual cropping with the same crop tend to increase disease incidence; planting times may also affect this. The disease is less on crop lines with the bush or determinate growth habit in USA. The indeterminate cvs with a greater leaf area near the soil get more disease and this is associated with higher numbers of apothecia. The critical plant growth factor is denseness of the leaf canopy, particularly near the ground. Cvs with open canopies get less disease. The cultural manipulation of the crop may also affect incidence of infection (Blad et al.; Coyne et al.; Schwartz et al.; Starr et al.; Steadman et al.). Infected pods in containers for shipment can result in severe losses in transport causing the symptoms known as nesting. Seed transmission seems unimportant (Steadman).

In areas with a history of cottony soft rot regular spraying may be necessary in the field and probably postharvest treatment also. Field fungicides need to be applied from about first blossom; those used include dicloran (Beckman et al.; Nieldbalski et al.; Pegg, K. G. 1962; Pegg, K. N.), thiabenazole (Gabrielson et al.) and benomyl (Haas et al. 1973; Natti; Pegg K. G. 1972). Dicloran has also been used for postharvest treatments (McMillan; Pegg, K. G. 1962; Pegg, K. N.). Wells et al. found such treatment effective (in laboratory tests) using 30 second dips at 51–52° in water alone or with dicloran. Spalding et al. also found heated dips more fungicidal. Control during 10–14 days storage at 15.6° and RH 95% was given by 10 second dips in dicloran or thiabendazole.

Beckman, K. M. et al. 1965. *Pl. Dis. Reptr* **49**: 357 (fungicides; **44**, 2290).
Blad, B. L. et al. 1978. *Phytopathology* **68**: 1431 (effect of canopy structure & irrigation).
Blodgett, E. C. 1946. *Pl. Dis. Reptr* **30**: 137 (general; **25**: 592).
Boyle, C. 1921. *Ann. Bot.* **35**: 337 (initial infection; **1**: 116).
Chambers, S. C. et al. 1960. *J. Agric. West. Aust.* 4th Ser. **1**: 977 (general; **40**: 446).
Coyne, D. P. et al. 1974. *Pl. Dis. Reptr* **58**: 379 (effect of plant architecture on disease & yield; **53**, 4183).
—— ——. 1978. *Euphytica* **27**: 225 (effect of genotypic blends, cv. growth habits; **57**, 5768).
Gabrielson, R. L. et al. 1971. *Pl. Dis. Reptr* **55**: 234 (fungicides).
Haas, J. H. et al. 1972. *Can. J. Pl. Sci.* **52**: 525 (ecology & epidemiology; **52**, 531).
—— ——. 1973. *Can. Agric.* **18**: 28 (disease prediction & fungicides; **52**, 1730).
Hungerford, C. W. et al. 1953. *Phytopathology* **43**: 519 (general; **33**: 401).
Hunter, J. E. et al. 1978. *Pl. Dis. Reptr* **62**: 633 (effects of timing, coverage & spray oil on control).
Kerr, E. D. et al. 1978. *Crop Sci.* **18**: 275 (crop loss; **58**, 439).
Lumsden, R. D. 1969. *Phytopathology* **59**: 653 (infection & formation of cellulase; **48**, 3735).
——. 1970. *Ibid* **60**: 1106 (formation of phosphatidase; **50**, 391).
—— et al. 1973. *Ibid* **63**: 708 (infection of hypocotyl; **53**, 731).
Maxwell, D. P. et al. 1970. *Ibid* **60**: 1395 (formation of oxalic acid; **50**, 1009).
McMillan, R. T. 1969. *Proc. Fla St. hort. Soc.* **82**: 139 (postharvest control; **50**, 2578).
Natti, J. J. 1967. *Fm Res.* **33**: 10 (fungicides; **47**, 1352).
——. 1971. *Phytopathology* **61**: 669 (epidemiology & fungicides; **51**, 870).
Nieldbalski, J. F. et al. 1969. *Pl. Dis. Reptr* **53**: 573 (fungicides; **48**, 3734).
Pegg, K. G. 1962. *Qd J. agric. Sci.* **19**: 561 (fungicides in field & postharvest; **42**, 643).
——. 1972. *Aust. J. exp. Agric. Anim. Husb.* **12**: 81 (fungicides; **51**, 4491).
Pegg, K. N. 1965. *Qd agric. J.* **91**: 262 (fungicides in field & postharvest; **44**, 2934).

Schwartz, H. F. et al. 1978. *Phytopathology* **68**: 465 (effect of blossoming characteristics & canopy structure).
Spalding, D. H. et al. 1974. *Pl. Dis. Reptr* **58**: 59 (postharvest control; **53**, 3686).
Starr, G. H. et al. 1953. *Bull. Wyo. agric. Exp. Stn* 322, 11 pp. (general; **34**: 425).
Steadman, J. R. et al. 1973. *Pl. Dis. Reptr* **57**: 1070 (effects of plant spacing & growth habit; **53**, 2373).
——. 1975. *Phytopathology* **65**: 1323 (significance of spread by seed; **55**, 3807).
Wells, J. M. et al. 1973. *Pl. Dis. Reptr* **57**: 234 (postharvest control; **53**, 326).

OTHER CROPS

Infection of the tap root and lateral branches of groundnut near soil level can kill field plants and pods are attacked. Branch lesions turn dark brown and shredding of the tissues results (Porter et al.; Porter). Joshi described a cortical stem rot of safflower where the flower heads broke off as large sclerotia formed in the receptacle. Seed infection occurs (Sackston). Cass Smith reported on a twig blight of citrus where there had been frost damage. This type of damage, infection of blossoms and infection of citrus fruit has been described by Smith (and see Prota). In the stem infection of coriander (*Coriandrum sativum*) Gupta found (on plants at the 8–10 leaf stage) that deaths resulted at 19 and 24° but not at 29 and 34°. Soil drenches of benomyl gave control on melon (*Cucumis melo*; Netzer et al.) in a single application (and see Honda et al.). Temkin-Gorodeiski found that *S. sclerotiorum* predominated in isolations from a fruit and stylar end rot of squash (*Cucurbita pepo*). A combined dicloran wax treatment was effective (and to a lesser extent either alone) in preventing rot development and delaying the appearance of watery spots in storage (and see Young). Nicholson et al. found internal seedborne infection of soybean. Germination and field emergence in soybean were both reduced at $>25\%$ incidence in seed. Harvested seed from infected plants were flattened compared with non-infected seed.

Mundkur and Pätzold gave accounts of infection on roselle (*Hibiscus sabdariffa*) and Jerusalem artichoke (*Helianthus tuberosus*), respectively. On lettuce (Brown et al.) some heritable field resistance was found by Newton et al. and lines that had a raised growth habit indicated disease escape (and see Adams et al.). Chemical and cultural control of lettuce drop was discussed by Darby, and Elia found effective control with dicloran. The characteristic mode of host infection through ascospores infecting flower parts has also been described for tomato (Purdy et al.; see also Chamberlain; Letham et al.). Bardin gave an account of probable ascospore attack on a large field area of tomatoes in USA (California); this inoculum was considered to have originated in a frequently irrigated, permanent pasture *c.* 600 m away. Stem rots and losses in seedlings of tobacco occur from time to time (e.g., Boning; Hartill et al. 1973; Kheswalla; McLeod et al.; Ribaldi). The infection of tobacco by the *Sclerotinia* complex was investigated by Hartill et al. (1977). Infection of eggplant in the glasshouse was controlled with dichlofluanid or thiophanate (Marras et al.; and see Honda et al.).

Adams, P. B. et al. 1975. *Pl. Dis. Reptr* **59**: 140 (on lettuce; **54**, 4277).
Bardin, R. 1951. *Ibid* **35**: 246 (ascospore infection of tomato).
Boning, K. 1933. *Phytopathol. Z.* **6**: 113 (on tobacco; **12**: 729).
Brown, J. G. et al. 1936. *Tech. Bull. Ariz. agric. Exp. Stn* **63**: 475 (on lettuce; **16**: 13).
Cass Smith, W. P. 1945. *J. Dep. Agric. West. Aust.* Ser. 2 **22**: 77 (on citrus; **24**: 500).
Chamberlain, E. E. 1932. *N.Z. Jl Agric.* **45**: 260 (on tomato; **12**: 193).
Darby, J. F. 1961. *Pl. Dis. Reptr* **45**: 552 (control on lettuce; **41**: 195).
Elia, M. 1971. *Inftore fitopatol.* **21**(11): 3 (control on lettuce with *Sclerotinia fuckeliana* & *S. minor*; **51**, 3665).
Gupta, J. S. 1963. *Indian Phytopathol.* **16**: 210 (on coriander; **43**, 2373).
Hartill, W. F. T. et al. 1973. *Pl. Dis. Reptr* **57**: 932 (control in tobacco; **53**, 1943).
—— ——. 1977. *N.Z. Jl agric. Res.* **20**: 415 (*Sclerotinia* complex in tobacco; **57**, 1837).
Honda, Y. et al. 1977. *Pl. Dis. Reptr* **61**: 1036 (control in glasshouse eggplant & cucumber by apothecial inhibition; **57**, 4172).
Joshi, S. D. 1924. *Mem. Dep. Agric. India bot. Ser.* **13**: 39 (on safflower; **3**: 650).
Kheswalla, K. F. 1934. *Indian J. agric. Sci.* **4**: 663 (on tobacco; **14**: 126).
Letham, D. B. et al. 1976. *Pl. Dis. Reptr* **60**: 286 (on cauliflower & tomato; **56**, 1059).
McLeod, A. G. et al. 1958. *N.Z. Jl agric. Res.* **1**: 866 (control on tobacco; **38**: 279).
Marras, F. et al. 1971. *Studi sassar.* **18**: 126 (control on eggplant; **51**, 1013).
Mundkur, B. B. 1934. *Indian J. agric. Sci.* **4**: 758 (on roselle; **14**: 106).

Netzer, D. et al. 1970. *Pl. Dis. Reptr* **54**: 909 (control on melon; **50**, 1543).
—— ——. 1973. *Phytoparasitica* **1**: 33 (as above & persistence of systemic fungicides; **53**, 1109).
Newton, H. C. et al. 1972. *Pl. Dis. Reptr* **56**: 875 (resistance in lettuce; **52**, 2791).
Nicholson, J. F. et al. 1972. *Phytopathology* **62**: 1261 (*S. sclerotiorum* and *Diaporthe phaseolorum* var. *sojae* seedborne in soybean; **52**, 3485).
—— ——. 1973. *Mycopathol. Mycol. appl.* **50**: 179, 257 (effects of seed infection & temp. on seed germination & plant growth in soybean; **53**, 335, 336).
Pätzold, C. 1953. *Pflanzenschutz* **5**: 139 (on Jerusalem artichoke; **33**: 518).
Porter, D. M. et al. 1974. *Phytopathology* **64**: 263 (on groundnut; **53**, 3708).
——. 1977. *Pl. Dis. Reptr* **61**: 995 (control on groundnut; **57**, 4238).
Prota, U. 1964. *Riv. Patol. veg. Pavia* Ser. 3 **4**: 129 (on citrus; **43**, 2910).
Purdy, L. H. et al. 1953. *Pl. Dis. Reptr* **37**: 361 (ascospore infection of tomato; **33**: 56).
Ribaldi, M. 1959. *Tobacco Roma* **63**: 13 (on tobacco; **39**: 46).
Sackston, W. E. 1960. *Pl. Dis. Reptr* **44**: 664 (*S. sclerotiorum* & *S. fuckeliana* seedborne in safflower; **40**: 119).
Smith, C. O. 1916. *Phytopathology* **6**: 268 (on citrus).
Temkin-Gorodeiski, N. 1970. *Israel Jnl agric. Res.* **20**: 97 (storage rot in squash; **50**, 446).
Young, P. A. 1936. *Phytopathology* **26**: 184 (on *Cucurbita*; **15**: 477).

Sclerotinia squamosa (Viennot-Bourgin) Dennis, *Mycol. Pap.* 62: 157, 1956.

Botryotinia squamosa Viennot-Bourgin, 1953
stat. conid. *Botrytis squamosa* Walker, 1925.

APOTHECIA solitary or in pairs on globose or elliptical black sclerotia, 0.5–3 mm diam. DISK light brown, 3–5 mm across. RECEPTACLE concolorous, smooth with a slender stalk. FLESH composed of slender sub-parallel hyphae whose terminal cells become enlarged, pyriform or subglobose on the surface of the excipulum. ASCI 78–90 × 10–13 μ, 8 spored. ASCOSPORES ovate or citriform, 11–17 × 8–11 μ. PARAPHYSES enlarged upwards to 5 μ thick, with brown contents. CONIDIOPHORES comparatively rare at 20–22°, more abundant at cooler temps; seldom arising directly from the mycelium, more often in tufts from sclerotia; erect, becoming flattened and twisted with age; hyaline at first, turning dark with age, septate, slightly swollen at the base; branches common, constricted at the base, growing tips branch and re-branch before sporulation. CONIDIA on short, hyaline sterigmata arising from swollen apices of branches, becoming detached at maturity; side branches degenerate after fructification, the walls drawing back in characteristic concertina-like folds; side branches cut off by septa near base, leaving distinct scars on main stem. MACROCONIDIA obovoid to ellipsoid, smooth, aseptate, hyaline, ash grey in mass when young and somewhat darker with age, sterigmata seldom remaining attacked, 13–22 × 10–17 (mostly 15–20 × 12–15) μ. MICROCONIDIA globose, *c.* 3 μ diam., on short hyaline conidiophores. SCLEROTIA white becoming black with age; most common on dry outer scales of host, roughly circular, flat, like scales, 0.5–2 mm diam., rarely >0.13 mm thick, often forming large conglomerates (after R. W. G. Dennis, *Mycol. Pap.* 62: 157, 1956 and J. C. Walker, *Phytopathology* **15**: 710, 1925; and see Viennot-Bourgin 1953).

This fungus (heterothallic) is one of the 3 spp. of *Botrytis* which cause neck rot of the common onion bulb. The disease caused on the bulbs was called small sclerotial neck rot by Walker (1926, *B. allii* q.v.) who fully compared *B. squamosa* with the other 2 pathogens *B. allii* and *B. byssoidea* (*Botrytis* q.v.). *B. allii*, the main pathogen in the onion neck rot complex, does not have the characteristic folds (like a concertina) of *B. squamosa* in the conidiophores. The latter is widespread (Map 164), but less frequently recorded than the former, and has been reported from Australia, Belgium, Brazil, Bulgaria, Canada, France, Hong Kong, Italy, Japan, Korea, Mauritius, Netherlands, New Zealand, Poland, UK and USA. Hennebert gave a review. The symptoms on onion bulbs and leaves have been described under *B. allii*. Those on the bulbs are essentially similar; the rot caused by *S. squamosa* spreads rather more slowly and the characteristic, usually very thin sclerotia (like scales) adhere closely to the dry, bulb scales. The grey white, slightly sunken leaf spotting and pendant withering of the leaf tips can be quite severe, more so than leaf symptoms caused by other *Botrytis* spp. (Hancock et al. 1963; Hickman et al.; Page 1955). Hancock et al. (1964b) sprayed culture filtrates on onion leaves and induced quite similar symptoms. The relatively low temp. requirements are comparable for both *B. allii* and *S. squamosa*. Bergquist et al. (1972), who also described compatibility, reported that most conidial sporulation occurred with a 14 hour photoperiod of fluorescent and near UV light (and see Page 1956; Stinson et al.).

Penetration of the leaf from the conidia was described by Clark et al. (1976). The number of lesions increased with the periods of leaf wetness up to at least 18 hours. Dry leaves held at RH 92% developed no infection. Leaf blight increased with dew periods (12–60 hours) after inoculation. Both lesion numbers and blight were greater on older leaves. Conidial germination was max. at 15° and growth in culture max. at 24°. Late forenoon and early afternoon peaks of max. conidial dispersal have been reported for N. America and Japan, respectively. Dispersal peaks are triggered by rain. Sclerotia survived 21 months, at rates of 7 and 66%, when buried 3 and 15 cm below the soil surface, respectively. Conidia survived for <2 months in soil. *S. squamosa* survived for 17 months in association with seed but could not be recovered after 25 months. Seed infestation rates (USA) were low (max. 6%). Cull piles and seed production fields were important sources of inoculum. Sclerotia formed conidia at 3–27° (opt. 9°). They formed apothecia after being uncovered from soil at 10–15 cm depths (Ellerbrock et al.; Matsuo; Shoemaker et al. 1977b; Sutton et al.).

Control measures for the neck rot form of the disease are given under *B. allii*. Control of the leaf infection is given by dithiocarbamates (Page 1955; Shoemaker et al. 1968, 1971, 1977a). Lafon et al. found cufraneb better than benomyl and maneb. Bergquist et al. (1971) in glasshouse and field work examined host reaction to the leaf blight form of the disease. *Allium bouddhae* and *A. schoenoprasum* were immune and *A. fistulosum* highly resistant. In the F_1 (*A. cepa* var. *cepa* × *A. fistulosum*) reactions resistance appeared to be due to dominant factors. A higher degree of resistance was transferred to the interspecific hybrids from *A. fistulosum* than from 2 other *Allium* spp., of which 7 were examined. All common onion cvs (67) were extremely susceptible.

Bergquist, R. R. et al. 1971. *Pl. Dis. Reptr* **55**: 394 (leaf infection of *Allium* spp. & interspecific hybrids; **51**, 946).
—— ——. 1972. *Phytopathology* **62**: 889 (effect of light, carbohydrate & pH on conidial formation; **52**, 1342).
—— ——. 1972. *Mycologia* **64**: 1270 (mating type; **52**, 2858).
—— ——. 1973. *Ibid* **65**: 36 (genetics of variation).
Clark, C. A. et al. 1976. *Phytopathology* **66**: 1279 (host penetration with *Sclerotinia fuckeliana*; **56**, 2795).
—— ——. 1977. *Ibid* **67**: 96, 212 (pathogenicity & phyllosphere bacteria, nutrient dependency in conidial germination & pathogenicity with *S. fuckeliana*; **56**, 4296, 4297).
Cronshey, J. F. H. 1946. *Nature Lond.* **158**: 379 (perfect state).
Ellerbrock, L. A. et al. 1977. *Phytopathology* **67**: 219, 363 (survival of sclerotia & conidia, sources of primary inoculum; **56**, 5297, 5298).
Hancock, J. G. et al. 1963. *Ibid* **53**: 669 (leaf infection with *Botrytis allii* & *S. fuckeliana*; **43**, 318).
—— ——. 1964a. *Ibid* **54**: 928 (pectolytic and cellulytic enzymes in vivo & in vitro with *B. allii* & *S. fuckeliana*; **44**, 943a).
—— ——. 1964b. *Ibid* **54**: 932 (pectolytic & cellulytic enzymes; **44**, 943b).
Hennebert, G. L. 1964. *Parasitica* **20**: 138 (review, Belgium; **44**, 944).
Hickman, C. J. et al. 1943. *Trans. Br. mycol. Soc.* **26**: 153 (on leaves with other related spp.; **23**: 161).
Lafon, R. et al. 1969. *Phytiat. Phytopharm.* **18**: 23 (fungicides; **49**, 1524).
McLean, D. M. 1960. *Pl. Dis. Reptr* **44**: 585 (perfect state formation in vitro; **40**: 138).
Matsuo, A. 1975. *Proc. Kansai Pl. Prot. Soc.* 17: 1 (conidial dispersal with *S. fuckeliana* & *B. allii*; **55**, 2444).
Page, O. T. 1955. *Can. J. agric. Sci.* **35**: 358 (general on leaves & control; **35**: 341).
——. 1956. *Can. J. Bot.* **34**: 881 (effect of light & other factors on mycelial growth & sclerotial formation; **36**, 491).
Shoemaker, P. B. et al. 1968. *Pl. Dis. Reptr* **52**: 469 (fungicides from aircraft; **47**, 2934).
—— ——. 1971. *Ibid* **55**: 565 (fungicides; **51**, 947).
—— ——. 1977a & b. *Phytopathology* **67**: 409, 1267 (timing of initial fungicide application, dew & temp. in epidemiology; **56**, 5299; **57**, 4261).
Stinson, R. H. et al. 1958. *Can. J. Bot.* **36**: 927 (effect of light & temp. on growth & respiration; **38**: 461).
Sutton, J. C. et al. 1978. *Ibid* **56**: 2460 (host, weather & incidence of airborne spores).
Tanaka, K. et al. 1975. *Trans. mycol. Soc. Japan* **16**: 416 (fine structure of apothecium; **55**, 4548).
Viennot-Bourgin, G. 1952. *Rev. Pathol. vég.* **31**: 82 (general in France; **32**: 417).
——. 1953. *Annls Epiphyt.* **4**: 23 (perfect state; **33**: 199).

Sclerotium cepivorum Berk., *Ann. Mag. nat. Hist.* Ser. 1 **6**: 359, 1841.

Stromatinia cepivorum (Berk.) Whetz., 1945.

COLONIES on PDA white or faintly brownish grey,

usually with a fairly even sheet of aerial mycelium but sometimes showing tufts of longer hyphae. BLACK SCLEROTIA develop on colony surface, in some isolates in diurnal zones, in others more evenly dispersed; nearly spherical, *c.* 200–500 μ across, with smooth or shallowly pitted surface. Cells of primary hyphae at advancing edge of colony thin walled, with dense granular contents, usually 9–14 (–18) μ wide, the tip cell frequently 300–400 μ long, sometimes showing branching which may be dichotomous before the development of the first septum, cells behind tip *c.* 30–100 μ long; cells of secondary and subsequent branch hyphae are narrower than those of primary hyphae. In lactophenol, particularly with cotton blue, sufficient shrinkage of cell contents occurs to show the septal pore clearly. Sclerotium initials are derived from dichotomously branching hyphal tips, and usually only 1 initial is involved in the development of each sclerotium. The MATURE SCLEROTIA show a sharply differentiated rind with evenly thickened strongly pigmented walls, a narrow cortex (2–3 cells thick) of almost isodiametric, thin walled cells and a medulla of intertwined branched hyphae of smaller diam. (4–12 μ) than primary hyphae and with colourless, moderately thick walls (0.5–1.5 μ). Cortex and medulla show granular cell contents, rind cells do not. Interhyphal spaces of the medulla contain gelatinous material.

A phialidic spermatial state is usually present and occasional, small aggregates of cells (superficially similar to those of the sclerotium rind) occur. On host white hyphae at first spread over surfaces of roots, bulbs and underground parts of shoots, later penetrate and ramify inter- and intracellularly. Eventually the bulbs become desiccated and large numbers of sclerotia develop. No perfect state is known, the sclerotia germinating directly to mycelium. Whetzel's combination in *Stromatinia* was based on similarity of the sclerotia to those seen in *Stromatinia gladioli* (J. E. M. Mordue, *CMI Descr.* 512, 1976).

The fungus causes white rot of *Allium* spp. It is widely distributed (Map 331) but there are relatively few records of its occurrence in the tropics. The first above ground symptoms are yellowing and dieback. Young plants may wilt before bulb formation and collapse rapidly. The roots are decayed and the bulb easily pulled up. The roots and base of the bulb are covered with the conspicuous mycelium and numerous sclerotia which also develop within the scales. There is a soft decay of the bulb. Incipient decay in the field at harvest continues as a storage rot. The fungus is soilborne and survives in the soil for up to 10 years through the sclerotia; outbreaks on land which had not grown onions for 20 years have been recorded. There is no persistence in soil as growing mycelium. Only *Allium* spp. are attacked. Young et al. induced symptoms on other plants under sterile conditions only. Mycelium from germinating sclerotia penetrates the roots directly via appressoria or infection cushions (see Abd-El-Razik et al. 1973) and infection then spreads to the base of the bulb. The most rapid in vitro growth has been reported to occur near 20° (Asthana 1947a), at 15–20° (Papavizas) and at 20–24° (Walker 1926). The last worker found that there was most disease at soil temps of 10–20°; at 20–22° there was a marked reduction in disease which did not occur at 24° or more. Nattrass noted that in Egypt white rot became apparent when the av. soil temps (5 cm below the surface) were 18.7° and 22.5° in February and March, respectively. Persistence of the sclerotia in the soil is not affected by soil pH, nutrients or inorganic supplements; organic supplements may or may not reduce the numbers of sclerotia (Coley-Smith 1959). Adams et al. (1971) found that >50% of the sclerotia germinated at pH 4.5–7.8; at pH 5 or below <30% of plants were infected and at pH 6 or above 90% were infected. Penetration of the seed does not occur.

The germination of sclerotia has been studied in detail by Coley-Smith and his co-workers (and see King et al.). Sclerotia can germinate in soil, without abrasion of the rind, after 1 month. In natural soils no germination occurs except in the presence of *Allium* spp.; other plants are not stimulatory. This stimulation from *Allium* spp. diffusates is effective up to 1 cm from the root. In non-sterile soil the sclerotia are subject to mycostasis which is specifically reversed by *Allium* spp.; in sterile conditions there is no specific sclerotial response to these plants. Sclerotia maintained in a dormant condition in the soil will germinate when transferred to sterile water and without the addition of nutrients. The diffusates from onion, which stimulate sclerotial germination, have no apparent direct effect on antibiotics but enable *S. cepivorum* to tolerate them (Keyworth et al.). The stimulation (by distillates and condensates) appears to be entirely due to the alkyl sulphide content. Synthetic allylcysteine and n-propylcysteine (and their sulphoxides) stimulated germination of sclerotia when added to soil.

Calomel (mercurous chloride) was still being recommended in 1973 (Anon.) as a soil treatment used in drill application at sowing time. Control may be more effective as a seed treatment with calomel (Wiggell). Transplants can be dipped in calomel paste. Dicloran and quintozene are also effective when applied to the transplanting furrow (Locke). Benomyl and thiophanate-methyl gave control when dusted into the planting furrow or used as a root dip treatment (Maloy et al., see also Ryan et al.). Adams et al. (1977) also found these last 2 fungicides effective and potassium azide (granular form in soil) gave control. Entwistle et al. found complete control with iprodione. Merriman et al. attempted control by using a trap crop of onions; it was ineffective. The longer infested soil is used for other crops the better. Ahmed et al. demonstrated biological control using a pycnidial dust of *Coniothyrium minitans* Campbell. They recommended the method for practical development. Hughes found that a postharvest dicloran dip reduced the disease in storage. Utkhede et al. found some resistance in onion. Variations in pathogenicity are not important in the field.

Abd-El-Razik, A. A. et al. 1973. *Phytopathol. Z.* **76**: 108 (host penetration; **52**, 3898).

———. 1974. *Zentbl. Bakt. ParasitKde* Abt. 2. **129**: 253 (pectolytic enzyme formation & pathogenicity; **54**, 1082).

———. 1974. *Arch. Phytopathol. PflSchutz* **10**: 327 (cellulytic enzymes; **54**, 4284).

Adams, P. B. et al. 1971. *Phytopathology* **61**: 1253 (soil factors & inocolum density; **51**, 2096).

———. 1977. *Agric. Res. U.S.A.* **25**(7): 11 (fungicide treatment of soil; **57**, 1538).

Ahmed, A. H. M. et al. 1977. *Pl. Pathol.* **26**: 75 (biological control; **57**, 1537).

Allen, J. D. et al. 1968. *Pl. Soil* **29**: 479 (soil fungistasis; **48**, 3349).

Anon. 1973. *Advis. Leafl. Minist. Agric. Fish. Fd* 62 (advisory; **54**, 3082).

Asthana, R. P. 1947a, b, c. *Proc. Indian Acad. Sci. India* Sect. B. **26**: 93, 108, 117 (general; **27**: 134).

Coley-Smith, J. R. et al. 1957. *Nature Lond.* **180**: 445 (stimulation of sclerotial germination; **37**, 63).

——. 1959. *Ann. appl. Biol.* **47**: 511 (host range, persistence & viability of sclerotia; **39**, 205).

——. 1960. *Ibid* **48**: 8 (sclerotial germination; **39**, 647).

———. 1966. *Ibid* **58**: 273 (effect of *Allium* spp. on sclerotial germination in soil; **46**, 491).

———. 1967. *Ibid* **60**: 109 (sclerotial germination in aseptic conditions; **46**, 3609).

———. 1968. *Ibid* **62**: 103 (effect of *Allium* spp. on soil bacteria & sclerotial germination; **47**, 3648).

———. 1969. *Ibid* **64**: 289 (formation of alkyl sulphides by *Allium* spp. & effect on sclerotial germination; **49**, 1229).

———. 1970. *Ibid* **65**: 59 (testing sclerotial viability in *Sclerotium cepivorum* inter alia; **49**, 2330).

———. 1974. *Soil Biol. Biochem.* **6**: 307 (dry conditions & leakage & rotting of sclerotia; **54**, 2080).

Cotton, A. D. et al. 1920. *J. Minist. Agric. Fish.* **26**: 1093 (general).

Elnaghy, M. A. et al. 1971. *Pl. Soil* **34**: 109 (stimulation of sclerotial germination by host plant extract; **51**, 2095).

Entwistle, A. R. et al. 1978. *Agric. Res. Coun. Res. Rev.* **4**: 27 (seed & drench treatment with iprodione; **57**, 5796).

Harrison, D. E. 1954. *J. Dep. Agric. Vict.* **52**: 510 (general; **34**, 202).

Hughes, I. K. 1970. *Qd J. agric. Anim. Sci.* **27**: 393 (control in storage; **51**, 2093b).

Joshi, L. K. et al. 1966. *Indian Phytopathol.* **19**: 168 (deficient for thiamine; **46**, 2910).

Keyworth, W. G. et al. 1969. *Ann. appl. Biol.* **63**: 415 (induced tolerance of antibiotics in the presence of onion exudates; **48**, 3217).

King, J. E. et al. 1968. *Ibid* **61**: 407 (effect of volatile products of *Allium* spp. & extracts on sclerotial germination; **47**, 2932).

———. 1969. *Ibid* **64**: 303 (formation of volatile alkyl sulphides by microbial degradation of synthetic alliin & like compounds & sclerotial germination; **49**, 1229).

———. 1969. *Soil Biol. Biochem.* **1**: 83 (suppression of sclerotial germination by water expressed from soils; **48**, 3350).

Locke, S. B. 1968. *Pl. Dis. Reptr* **52**: 272 (experimental control with soil chemicals; **47**, 2933).

Maloy, O. C. et al. 1974. *Ibid* **58**: 6 (fungicides in furrow & root dip application; **53**, 2767).

Merriman, P. R. et al. 1978. *Soil Biol. Biochem.* **10**: 339 (use of onion as a trap crop).

Nattrass, R. M. 1931. *Bull. Minist. Agric. Egypt tech. scient. Serv.* 107, 9 pp. (general; **11**: 219).

Papavizas, G. C. 1970. *Mycologia* **62**: 1195 (C & N nutrition; **50**, 2719).

Ryan, E. W. et al. 1976. *Irish J. agric. Res.* **15**: 317, 325 (control by fungicidal pelleting of seed & dusting of setts; **56**, 3362, 3363).

Scott, M. R. 1956. *Ann. appl. Biol.* **44**: 576, 584 (growth of mycelium in soil & spread of disease from plant to plant; **36**: 567).

Tims, E. C. 1948. *Phytopathology* **38**: 378 (on shallot; **27**: 549).

Townsend, B. B. et al. 1954. *Trans. Br. mycol. Soc.* **37**: 213 (development of some fungal sclerotia; **34**: 262).

Utkhede, R. S. et al. 1978. *Can. J. Pl. Sci.* **58**: 819 (screening world onion collection for resistance).
———. 1978. *Phytopathology* **68**: 1080 (screening onion cvs for resistance).
Wakaida, M. et al. 1974. *Bull. Coll. Agric. Utsunomiya Univ.* 9: 21 (pectic enzymes; **55**, 2056).
Walker, J. C. 1924. *Phytopathology* **14**: 315 (general; **4**: 13).
——. 1926. *Ibid* **16**: 697 (effects of soil temp. & soil moisture; **6**: 204).
Wiggell, D. 1956. *Pl. Pathol.* **5**: 60 (control with calomel; **36**: 161).
Young, J. M. et al. 1969. *Pl. Dis. Reptr* **53**: 821 (in vitro infection & pathogenesis on non *Allium* plants; **49**, 720).

SEPTORIA Sacc., *Sylloge Fung.* 3: 474, 1884, nom. cons.

(see B. C. Sutton, *Mycol. Pap.* 141: 189, 1977)
(Sphaerioidaceae)

PYCNIDIA separate, immersed, glabrous, spheroid, ampulliform or ovate, sometimes papillate, ostiolate, unilocular, pigmented or hyaline; walls simple or several cells thick, outer layer forming a distinct specialised wall. CONIDIOGENOUS CELLS similar to the inner cells of the pycnidial wall, ampulliform to doliiform or elongated, occasionally proliferating, holoblastic sympodial or polyblastic. CONIDIA hyaline, filiform, multiseptate, lacking cellular or extracellular appendages and a cellular cap (from key by B. C. Sutton, *The fungi* (editors G. C. Ainsworth et al.) Vol. IV A: 560, 1973; and see *S. lycopersici* which is a typical *Septoria*).

The genus has not been monographed. The conidia are water dispersed. Jørstad gave keys and descriptions for 40 spp. on Gramineae in Norway. Rădulescu et al. described the spp. in Rumania. The spp. on *Chrysanthemum*, the development of their pycnidia and the cytology of *S. leucanthemi* Sacc. & Speg. were described by Punithalingam (1966 & 1974); Punithalingam et al. (1965). Two of the spp. on this host genus have been reported on pyrethrum (*C. cinerariaefolium*): *S. chrysanthemella* Sacc. (conidia 36–65 × 1.5–2.5 μ, 4–9 septate) and *S. obesa* Syd. (conidia 56–91 × 2.7–3.5 μ, 5–9 septate).

Sutton et al. described the worldwide common leaf blight of cultivated and wild celery, and celeriac, caused by *S. apiicola* Speg. (synonyms: *S. apii* Rostrup, *S. apii-graveolentis* Dorogin and *S. petroselini* Desm. var. *apii* Cavara; an expanded synonymy was suggested by Gabrielson et al.). MacMillan et al. made a detailed study of the conidia (22–56 (av. 35) × 2–2.5 μ, 1–5 septate) of *S. apiicola*. Abundant brown necrotic spots (up to 8 mm diam.), amphigenous, circular or vein limited, form on the leaves; lesions also occur on the edible petioles. Infected seed (there is penetration of the pericarp and testa) is the primary source of the disease (Sheridan). The diseased seedlings that result bear necrotic lesions and pycnidia. For control seed should be soaked in thiram for 24 hours at 30°.

S. bataticola Taub. (conidia 15–80 μ long) causes leaf spots with white centres and brown borders (up to 8 mm diam.) on sweet potato. *S. cannabis* (Lasch.) Sacc. (conidia 12–46 × 2–3 μ, usually 3 septate) causes ellipsoidal or polygonal, yellowish to greyish leaf spots and leaf fall of hemp (*Cannabis sativa*; Ferri: Watanabe et al.). Minor spots on citrus leaves and fruits are caused by several *Septoria* spp.; 3 of these are: *S. citri* Pass. (conidia 14–15 × 2–3 μ), *S. depressa* McAlp. (13–19 × 1.5–3.5 μ) and *S. limonum* Pass. (8–15 × 1.5–2 μ). On fruits infection causes 1–2 mm diam. spots with sunken centres, not deeper than the flavedo; the spots may become larger, up to 1 cm diam. They first appear when fruits are green but become more conspicuous with ripening as yellowing begins. The brown spots have a narrow, greenish halo at first, it later becomes reddish brown. The disease(s) caused is not important and can be controlled if necessary with copper and zinc fungicides (Anon.; Klotz et al.). *S. carthami* Murash. (conidia 67–93 × 3.7 μ) occurs on safflower. Amongst the spp. reported on rice is *S. oryzae* Catt. (conidia 21 × 3 μ) causing speckled blotch on leaves, leaf sheaths and glumes; Ou (*Rice diseases*) reproduced a key to 6 *Septoria* spp. and one var. on rice. The spp. on cucurbits include *S. cucurbitacearum* Sacc. leaf spot; and see Stout for *Septoria* on maize.

S. passiflorae Louw was fully described by the author of the sp. The most severe symptoms on passion fruit are on the leaves but blossoms, fruit and stems are also attacked. The pycnidia are 50–160 μ diam. and the conidia 14.5–30.5 (av. 23) × 1.5–3 (av. 2) μ, 0–3 septate. The leaf lesions are initially light green and indefinite, becoming light brown, circular to irregular, sharply demarcated, slightly sunken on lower surface, 5–10 mm diam. and surrounded by a yellowish zone; pycnidia are seen on the lower surface. Severe attacks cause premature leaf fall. Infection of the calyx may spread to the stalk and blossoms fall. When immature fruit is infected there develop small, more or less circular, sharply demarcated spots, at first green becoming brown and sometimes coalescing to give large, irregular, necrotic and somewhat sunken areas with a hard woody texture on the mature fruit. Lesions occurring on young stems may girdle and cause death of young shoots. Penetration is through the stomata and most infection occurs at 25–27.5° with a leaf moisture period of at least 48 hours, but there is some infection at these temps with a 24 hour moisture period. Mycelial growth in vitro and conidia germ tube length were both greater at 25–30°, and the least time for the initiation of conidial germination (9 hours) was at 25–27.5°. In Australia mancozeb gave control and resistance to benomyl was found (Peterson).

S. pisi Westend., causing a minor disease of pea (*Pisum sativum*), was described by Cruickshank who referred to earlier work. The pycnidia are 79–237 μ diam. and the conidia 24.5–59 (av. 33) × 3–3.5 (av. 3.2) μ, 1–3 septate. Synonyms given by Cruickshank are *S. leguminum* Desm. and *Rhabdospora hortensis* Sacc. The first symptoms on leaves are small yellowish areas which become light brown blotches, irregular, elongated, limited by the veins and bearing scattered pycnidia. The disease spreads over leaves, petioles, stipules and stems; young plants can be killed but on older plants the symptoms are not so severe. Host penetration is direct and opt. temp. for growth in vitro is 20–22°. The disease is of little importance.

Anon. 1963. *Pl. Dis. Leafl. N.S.W. Dep. Agric.* 126, 6 pp. (*Septoria depressa*; **44**, 2143).
Beach, W. S. 1919. *Am. J. Bot.* **6**: 1 (*Septoria* spp. & their hosts).
Cruickshank, I. A. M. 1949. *N.Z. Jl Sci. Technol.* A **31**(3): 17 (*S. pisi*; **30**: 400).
Fawcett, H. S. et al. 1940. *Calif. Citrogr.* **26**: 2 (*S. citri* & *S. limonum*; **22**: 131).
Ferri, F. 1959. *Annali Sper. agr.* **13**(6): Suppl.: 189 (*S. cannabis*; **39**: 471).
Gabrielson, R. L. et al. 1964. *Phytopathology* **54**: 1251 (*S. apiicola*; **44**, 951).
Garman, P. et al. 1920. *Trans. Ill. St. Acad. Sci.* **13**: 176 (tabulated characteristics of *Septoria* spp.).
Jørstad, I. 1967. *Skr. norske Vidensk-Akad.* Mat. Nat. N.S. 24, 63 pp. (40 *Septoria* spp. on Gramineae; **47**, 2081).
Klotz, L. J. et al. 1969. *Calif. Citrogr.* **54**: 530 (*S. citri* or *S. limonum*; **49**, 1024).
Louw, A. J. 1941. *Scient. Bull. Dep. Agric. S. Afr.* 229, 51 pp. (*S. passiflorea* sp. nov.; **22**: 393).
MacMillan, H. G. et al. 1942. *J. agric. Res.* **64**: 547 (*S. apiicola*, conidial structure & germination; **21**: 427).
Peterson, R. A. 1977. *APPS Newsl.* **6**: 3 (fungicide control of & benomyl resistance in *S. passiflorae*; **57**, 232).
Punithalingam, E. et al. 1965. *Trans. Br. mycol. Soc.* **48**: 423 (on *Chrysanthemum*; **45**, 467).
——. 1966. *Ibid* **49**: 19 (pycnidial development in *Septoria* spp. on *Chrysanthemum*).
——. 1974. *Ibid* **63**: 255 (cytology of *S. leucanthemi*; **54**, 885).
Rădulescu, E. et al. 1973. Editura Acad. Rep. Soc. România, Bucharest (*Septoria* spp. from Rumania; **52**, 3963).
Sheridan, J. E. 1966. *Ann. appl. Biol.* **57**: 75 (importance of seedborne inoculum in *S. apiicola*; **45**, 1973).
Stout, G. L. 1930. *Mycologia* **22**: 271 (on maize; **10**: 305).
Sutton, B. C. et al. 1966. *C.M.I. Descr. pathog. Fungi Bact.* 88 (*S. apiicola*).
Taubenhaus, J. J. 1914. *Phytopathology* **4**: 305 (*S. bataticola* sp. nov. inter alia).
Teterevnikova-Babayan, D. N. 1961. *Izv. Akad. Nauk armyan. SSR* Ser. Biol. **14**: 15 (taxonomy; **42**, 373).
Watanabe, T. et al. 1936. *Ann. phytopathol. Soc. Japan* **6**: 30 (*S. cannabis*; **15**: 805).
Zachos, D. G. 1957. *Pl. Prot. Bull. F.A.O.* **6**: 41 (*S. citri* & *S. depressa*; **37**: 354).

Septoria glycines Hemmi, *Trans. Sapporo nat. Hist. Soc.* **6**: 15, 1915.

PYCNIDIA amphigenous, mostly epiphyllous, immersed, becoming erumpent, yellow brown to dark brown, subglobose, 100–180 μ diam., with ostioles 40–70 μ wide; wall 2–4 layers of cells thick, composed of yellow brown cells thickened on the outside. CONIDIOPHORES hyaline, obclavate to obpyriform 6–10 × 3–4 μ, arising from the cells lining the inside of the pycnidium. CONIDIA hyaline, guttulate, straight or curved, basal and blunt, gradually tapering, rounded at apex, 30–50 × 1.5–2 μ and 2–4 septate.

Other spp. of *Septoria* recorded on *Glycine* are *S. sojina* Thüm. and *S. sojae* Syd. & Butler. Distinguished from *S. sojina* (10–18 × 4.5–5 μ) by the large narrow conidia and from *S. sojae* (40 × 0.5–1 μ) by the broad conidia. (E. Punithalingam & P. Holliday, *CMI Descr.* 339, 1972.)

Brown spot of soybean occurs in Asia (China, Japan, Nepal and USSR) and parts of Canada and

USA (E. and central); it has also been reported from Colombia and Rumania (Map 361). Early symptoms are usually on the cotyledons and the first true leaves; spots are red brown, somewhat angular, up to 5 mm diam. and very numerous. They spread acropetally, coalesce and cause defoliation. Spots form on the stems and pods which both bear pycnidia. Stomatal penetration occurs, and the best growth and conidial germination in vitro are at 24–28°. Pycnidia appear readily in culture and their formation is said not to be affected by light or dark (Hemmi). Transmission is frequently by seed which is penetrated mostly via the placenta and funicle; seed bears necrotic areas and pycnidia. In Canada infection by *S. glycines* was more severe in the presence of *Pseudomonas glycinea* which causes bacterial spot. Apart from the use of healthy seed, rotation may be advisable. In USA brown spot was three times more severe on 3 soybean vars in a crop following soybeans than in one after flax, maize, oats or wheat. Varietal differences in disease severity have been described from N. America. Ross investigated the use of benomyl with and without overhead irrigation.

Benedict, W. G. 1964. *Can. J. Bot.* **42**: 1135 (synergistic effect with *Pseudomonas glycinea*; **44**, 932).

Dutta, P. K. 1959. *Diss. Abstr.* **19**: 2707 (effect of rotation; **38**: 724).

Hemmi, T. 1940. *Mem. Coll. Agric. Kyoto* 47: 1 (general; **19**: 452).

MacNeill, B. H. et al. 1957. *Can. J. Bot.* **35**: 501 (host penetration & histology; **37**: 199).

Ross, J. P. 1975. *Pl. Dis. Reptr* **59**: 809 (irrigation & use of benomyl with *Phomopsis* sp.; **55**, 2036).

Wolf, F. A. et al. 1926. *J. agric. Res.* **33**: 365 (general; **6**: 74).

Septoria helianthi Ellis & Kellerman, *Am. Nat.* 17: 1165, 1883.

PYCNIDIA mostly epiphyllous, immersed, becoming erumpent, yellow brown to dark brown, subglobose, 100–150 μ diam. with ostioles 40–60 μ wide; pycnidial wall 2–4 layers of cells thick, composed of yellow brown cells thickened on the outside. CONIDIOPHORES hyaline, clavate to obpyriform, 4–7 × 3–4 μ, originating from the cells lining the inside of the pycnidium. CONIDIA hyaline, straight or slightly curved, basal end blunt, rounded at the apex, 50–85 × 2–3 μ and 3–5 septate.

Distinguished from *Septoria pauper* Ellis (45–55 × 1–1.5 μ, with 3–7 septa) and *S. helianthicola* Cooke & Harkn. (synonym: *Rhabdospora helianthicola* (Cooke & Harkn.) Sacc. (30–35 × 1 μ) by the large conidia. (P. Holliday & E. Punithalingam, *CMI Descr.* 276, 1970.)

Although little has been published on the biology of this sunflower leaf spot it has caused damage in Germany, Zimbabwe, USA, USSR and Zambia. The disease is fairly widespread: E. Europe and USSR in Asia, China, India, Japan and Australia, E. and S. Africa and N. America (Map 468). It also occurs on some other *Helianthus* spp. Yellowish spots up to 1.5 cm diam. develop over the whole lamina, gradually turning necrotic and becoming almost black. The numerous pycnidia are mostly on the upper surface. The lesions have a polygonal outline, being sharply delimited by the veins. Infection may begin on the cotyledons and young leaves, spreading to later developing leaves. Severe attacks cause defoliation and loss in yield. Spore germination takes 8–10 hours for the formation of germ tubes. At 18–21° infection develops on leaves in about 9 days. Overwintering occurs in host debris. Seed treatment is recommended although seed transmission does not appear to have been demonstrated. Routine measures such as destruction of infected host debris and rotation are advisable.

Frandsen, N.O.C. 1948. *Phytopathol. Z.* **15**: 88 (general; **28**: 401).

Josifovic, M. et al. 1957. *Bull. Acad. serbe Sci. Cl. Sci. math. nat.* **18**: 43 (in Yugoslavia; **38**: 751).

Septoria lactucae Pass., *Atti Soc. Crittog. Ital. (Milano)* 2: 34, 1879.

Septoria lactucae Peck, 1879
Ascochyta lactucae Rostrup, 1882
S. consimilis Ellis & Martin, 1885.

PYCNIDIA chiefly epiphyllous, immersed, becoming erumpent, subglobose, 100–200 μ diam., opening by very wide ostioles, 50–70 μ; wall 2–4 layers of cells thick, composed of yellow brown cells, heavily pigmented and thickened around the ostiolar region. CONIDIOPHORES hyaline, obpyriform, 5–10 × 3–5 μ, arising from the cells lining the inside of pycnidium. CONIDIA hyaline, straight or curved, blunt at base, rounded at apex, 25–40 (–50) × 1.5–2 μ with 1–2 (3) septa.

Six other *Septoria* spp. have been recorded

on *Lactuca* spp.; of these *S. ludoviciana* Ellis & Everh. (15–25 × 2 μ), *S. fernandezii* Unamuno (26–30.5 × 1.6–2 μ), *S. schembelii* Melnik (synonym: *S. lactucina* Petrak) (12–28 × 1–1.5 μ) and *S. unicolor* (26–32 × 1 μ) have shorter conidia; while *S. sikangensis* Petrak (20–52 × 2–3 μ with 3–6 septa) and *S. lactucina* Lobik (39.5–52 × 3–3.6 μ) have longer and wider conidia than *S. lactucae* (E. Punithalingam & P. Holliday, *CMI Descr.* 335, 1972; and see Neergaard).

A minor and little investigated but widespread (Map 485) leaf spot of lettuce. The early symptoms are small, numerous, irregularly outlined chlorotic spots beginning on the outer leaves. They enlarge, becoming brown or somewhat silvery, sometimes with a chlorotic halo and shotholes develop. Pycnidia, which may not be conspicuous, are found on both leaf surfaces. In severe attacks infection spreads to the younger leaves and flowers. Spread occurs through seed and pycnidia are embedded in the seed coat. This disease rarely seems important but under very wet conditions severe outbreaks can occur as in Australia (Victoria) in 1961 when up to 90% of the seed crop was destroyed. Crops are normally free from infection when grown from seed produced in a disease free area. Seed treatment (water at 47–48° for 30 min. or acid mercuric chloride) is effective although the former reduces germination (Smith). Fournet recommended benomyl in the nursery. Most cvs and other *Lactuca* spp. are susceptible.

Bond, T. E. T. 1941. *Trop. Agric.* **97**: 62 (symptoms & damage in Sri Lanka; **21**: 182).

Fournet, J. 1976. *Annls Phytopathol.* **8**: 41 (control; **56**, 3356).

Moore, W. C. 1940. *Trans. Br. mycol. Soc.* **24**: 345 (symptoms & distribution; **20**: 193).

Neergaard, P. 1938. *Bot. Tidsskr.* **44**: 359 (mycological notes I; **17**: 703).

Smith, P. R. 1961. *J. Agric. Vict. Dep. Agric.* **59**: 555 (general, seed infection & control; **41**: 267).

Septoria lycopersici Spegazzini, *An. Soc. cient. argent.* 13: 16, 1882.

PYCNIDIA immersed, amphigenous, globose, honey yellow to brown, 75–200 μ diam.; wall multicellular, composed of an outer layer of large but thin walled, honey yellow pseudoparenchymatic cells and an inner region of small celled hyaline pseudoparenchyma; ostiole circular or wide and irregular, up to 50 μ diam. CONIDIA hyaline, filiform, 2–6 septate, apex acute, base truncate to obtuse, 52–95 × 2 μ, formed as blastospores from hyaline, aseptate, cylindrical to clavate conidiophores, 7.5–10 × 4 μ (B. C. Sutton & J. M. Waterston, *CMI Descr.* 89, 1966; and see Harris).

An account of this leaf spot of tomato was given by MacNeill and a review of the disease for S. America by Viégas. *S. lycopersici* is worldwide (Map 108); it has been recorded on other *Lycopersicon* spp. and *Solanum* spp., including eggplant and *S. carolinense* (horse nettle; Pritchard et al. 1921). An infection of Irish potato in Peru was attributed to a different strain of the fungus by Jiménez et al. The disease, although plants of any age are susceptible, usually appears first on the lower leaves as the crop is beginning to set: stem and petiole infections are not important. The minute chlorotic spots become necrotic (the first leaf symptoms are sometimes described as watersoaked spots), increase to a few mm diam. and are roughly circular, numerous, with characteristic grey centres and dark borders. At very severe disease levels infection spreads to the younger leaves, lesions coalesce, leaves dry out and the resulting defoliation ends by leaving only a few leaves at the top of the stem.

Host penetration is through the stomata and some 48 hours at high humidities are necessary for lesions to develop. The opt. temps for growth in vitro, sporulation and conidial germination are all very similar, 25° or a little over. The disease can become epidemic where there are wet conditions and moderate temps of 20–26° over long periods. Benedict reported that light may be an important factor in the assessment of symptom expression; decreased light intensity tended to increase disease severity. The waterborne conidia are spread from leaf lesions in tomato, some weed hosts and crops debris on the soil surface. Although some contamination of the seed surface by conidia may occur, seed transmission in the epidemiology is not important, neither is survival in soil. Reports of differential pathogenicity between isolates of *S. lycopersici* (Cook; Kurozawa et al. 1975; MacNeill) have not led to the separation of distinct races.

Some resistance has been described in breeding work (Andrus et al.; Barksdale et al.; Lincoln et al.; Locke; Voskan'yan) but control is mainly through cultural measures and fungicides. Inheritance of resistance suggests control by a single dominant factor but selections in later generations do not have a resistance equal to that of the original parent. *L. hirsutum* and *L. peruvianum* have been used as sources of resistance (and see Kurozawa et al. 1977). In outdoor tomatoes sanitary precautions should be taken: clean seedbeds, removal of susceptible solanaceous weeds and infected crop debris, and deep ploughing in of the refuse from earlier crops. This leaf spot builds up in fields which have frequently grown tomatoes and long rotations (preferably with cereals or legumes) are advisable. Copper, maneb, mancozeb and zineb are all effective. Sarasola et al. compared 8 fungicides, applied through irrigation, and found that benomyl gave control although it was slightly phytotoxic. In the glasshouse infected plant material should be destroyed and fungicides applied.

Andrus, C. F. et al. 1945. *Phytopathology* **35**: 16 (inheritance of resistance; **24**: 252).

Arneson, P. A. et al. 1967. *Ibid* **57**: 1358 (hydrolysis of tomatine by the pathogen; **47**, 1300).

Barksdale, T. H. et al. 1978. *Pl. Dis. Reptr* **62**: 844 (inheritance of resistance).

Benedict, W. G. 1971. *Can. J. Bot.* **49**: 1721 (light & peroxidase activity; **51**, 1931).

——. 1972. *Ibid* **50**: 1931 (light, peroxidase activity & resistance; **52**, 1268).

Cook, A. A. 1954. *Phytopathology* **44**: 374 (variation in pathogenicity; **34**: 189).

Cummins, G. R. 1949. *Ibid* **39**: 509 (effects on defoliation; **29**: 62).

Edrinal, D. M. et al. 1940. *Philipp. Agric.* **29**: 593 (general; **20**: 182).

Harris, H. A. 1935. *Phytopathology* **25**: 790 (morphology; **15**: 65).

Jiménez, A. T. et al. 1972. *Fitopatologia* **5**: 15 (form on Irish potato; **52**, 482).

Kurozawa, C. et al. 1975. *Arg. Inst. biol. S. Paulo* **42**: 17 (varying pathogenicity; **57**, 299).

—— ——. 1977. *Summa Phytopathol.* **3**: 135 (testing for resistance in cvs & *Lycopersicon* spp.; **57**, 3623).

Lincoln, R. E. et al. 1949. *Phytopathology* **39**: 647 (inheritance of resistance; **29**: 180).

Locke, S. B. 1949. *Ibid* **39**: 829 (resistance; **29**: 233).

MacNeill, B. H. 1950. *Can. J. Res.* Sect. C. **28**: 645 (general; **30**: 436).

Pritchard, F. J. et al. 1921. *J. agric. Res.* **21**: 501 (on horse nettle).

—— ——. 1924. *Bull. U.S. Dep. Agric.* 1288, 18 pp. (control; **4**: 380).

—— ——. 1924. *Phytopathology* **14**: 156 (effects of temp. & RH; **3**: 614).

Rizinski, S. 1966. *Arb. poljopr. Nauke Teh.* **19**: 101 (general in Yugoslavia; **45**, 2962).

Sarasola, A. A. et al. 1971. *Jornadas Argentinas Biol.* 5, 11 pp. (soil application of benomyl; **53**, 4878).

Viégas, A. P. 1962. *Bragantia* **21**: 383 (review for S. America, 80 ref.; **43**, 578).

Voskan'yan, S. S. 1959. *Bull. appl. Bot. Pl. Breed.* **32**: 175 (resistant cvs; **43**, 2410).

SETOSPHAERIA Leonard & Suggs, *Mycologia* 66: 294, 1974.

(Pleosporaceae)

ASCOCARPS superficial on substrate, black, globose or ellipsoid, with or without a neck; ostiole surrounded by simple, short, rigid, brown hairs, similar hairs scattered over surface of upper part of ascocarp; peridium coriaceous–carbonaceous, pseudoparenchymatous. ASCI bitunicate, cylindrical to cylindrical–clavate, 1–8 spored, among filamentous pseudoparaphyses. ASCOSPORES phragmosporous, hyaline, fusoid, 2–6 septate (rarely more), enclosed in a thin, gelatinous sheath which projects well beyond the ends of the spore (K. J. Leonard & E. G. Suggs l.c.).

Leonard et al. (1974a) established the new imperfect genus *Exserohilum* to accommodate all the graminicolous spp. of '*Helminthosporium*' with protruberant hila. In a brief discussion Ellis (*More Demat. Hyphom.*: 396) recommended the continued use of *Drechslera* for all graminicolous spp. previously in *Helminthosporium*. The *Drechslera* spp. transferred to *Exserohilum* by Leonard & Suggs are: *frumentacei, halodes* (later placed in synonymy with *rostrata*, Leonard), *holmii, monoceras, pedicellata, rostrata* and *turcica*. The ascigerous states of *Exserohilum* were placed in the new genus *Setosphaeria* which is segregated from *Trichometasphaeria* on the basis of lack of a clypeus, lysigenous development of the ostiole, occurrence of setae on the perithecial wall, the absence of periphyses in the ostiole and the hyphomycetous states. *S. rostrata* Leonard causes disease in Gramineae (including *Eleusine*, maize and sugarcane) and other plants; stat. conid. *D. rostrata* (Drechsl.) Richardson & Fraser (synonyms: *H. rostratum* Drechsler, *Bipolaris rostrata* (Drechsl.) Shoemaker and *E. rostratum* (Drechs.) Leonard & Suggs).

Anahosur, K. H. et al. 1978. *C.M.I. Descr. pathog. Fungi Bact.* 587 (*Setosphaeria rostrata*).

Honda, Y. et al. 1978. *Mycologia* **70**: 343, 538, 547 (effect of light on conidial sporulation, stability of hilum protruberance).
Leonard, K. J. et al. 1974a. *Ibid* **66**: 281 (*Exserohilum* gen. nov. & *Setosphaeria* gen. nov., *S. prolata* & *E. prolatum* spp. nov.; **54**, 2095).
———. 1974b. *Pl. Dis. Reptr* **58**: 612 (reaction of maize to *E. prolatum*; **54**, 2216).
——. 1976. *Mycologia* **68**: 402 (synonymy of *E. halodes* with *E. rostratum* & *S. rostrata* sp. nov.; **56**, 599).

Setosphaeria turcica (Luttrell) Leonard & Suggs, *Mycologia* **66**: 295, 1974.
Trichometasphaeria turcica Luttrell, 1958
stat. conid. *Drechslera turcica* (Pass.) Subram. & Jain, 1966
Helminthosporium turcicum Pass., 1876
H. inconspicuum Cooke & Ellis, 1878.
(the nov. comb. *Exserohilum turcicum* (Pass.) Leonard & Suggs was made, *Mycologia* **66**: 291, 1974)

ASCOCARPS superficial (distinguished from sclerotia by their larger size and more regular contours), black, ellipsoidal to globose, 345–497 μ diam., 359–721 μ high, outer layers a pseudoparenchyma of small cells with thick dark brown walls. Over upper 1/3 and around ostiole numerous short, stiff, brown, hairs like spines. Inner wall of small, hyaline cells; ostiole not beaked. ASCI cylindrical or clavate–cylindrical, short stipitate, 176–249 (202) × 24–31 (27) μ, bitunicate. ASCOSPORES 1–6 (commonly 2–4), hyaline, fusoid, straight or slightly curved, 3 septate, constricted at septa (6 septa not uncommon); 42–78 (62) × 13–17 (15) μ (larger in asci with 1 or 2 spores); often surrounded by mucoid sheath extending beyond either end by as much as the spore length. CONIDIOPHORES emerging singly or in groups of 2–6 through stomata, straight or flexuous, brown up to 300 μ long, 7–11 (mostly 8–9) μ thick. CONIDIA straight or slightly curved, ellipsoidal to obclavate, pale to mid straw coloured, smooth, 4–9 pseudoseptate, 50–144 (115) μ long, 18–33 (commonly 20–24) μ thick in the broadest part, hilum conspicuously protuberant (perfect state from E. S. Luttrell, *Phytopathology* **48**: 282, 1958 as *T. turcica*; stat. conid. M. B. Ellis & P. Holliday *CMI Descr.* 304, 1971).

S. turcica on maize causes the disease known as northern leaf blight since, in USA, it is found mostly in the N., compared with the more southerly distribution of the maize blight caused by *Cochliobolus heterostrophus* (q.v.; see Hooker for a review). *S. turcica*, which is widely distributed (Map 257), attacks *Sorghum* spp., *Euchlaena mexicana* and other Gramineae. The leaf symptoms begin as small, dark, watersoaked areas, becoming irregular or elliptical, sometimes linear; greyish green, brown, then straw coloured or greyish, with red purple or tan borders, often 4 × 10 cm or larger, coalescing and leading to death of leaves. Tassel infection on maize is less conspicuous. Direct penetration of the leaf (with appressoria) by the germinating conidium occurs (Campi; Knox-Davies). Free water is required for infection which takes place within 6–18 hours at 18–27°. Lesions are seen after 7–12 days. Cox et al. described a high temp. str. Leach et al. (1977a) differentiated between types of conidial release: forcible discharge induced by a rapidly falling RH, wind and rain release. Most airborne conidia were trapped during the day and Meredith (and see Palmerley et al.) described a diurnal periodicity for the conidia with a forenoon max. Leach et al. (1977b) found that airborne conidia were abundant over a maize field on days after warm nights with long periods (10–12 hours) of high RH (*c.* >90%). Free moisture was not required for sporulation which was opt. at *c.* 20–26°. A dark period is required for sporulation (and see Benedict). The fungus survives on plant debris in the soil or on its surface and chlamydospores formed in conidia may be a factor in survival (Fullerton et al.). It has been reported to be seedborne (Chilton; Valleau) but Hooker stated that this seldom occurs. The role of the perfect stage in the life history appears to be virtually unknown.

Hooker has fully discussed genetics and variation (races) in the pathogen, types and the expression of resistance, inheritance of resistance and breeding. Only some of the more recent work on these aspects is quoted (Calub et al. 1973; Couture et al.; Drolsom; Dunn et al.; Frederiksen et al.; Gevers; Hilu et al.; Long et al.; Mace; Schaik et al.; Tarumoto et al.; Tuleen et al.; Turner et al.; Ullstrup). Isolates from maize and sorghum differ in pathogenicity; they can attack either host alone or both; ff. sp. *sorghi* and *turcica* have been described. Virulence in the pathogen is inherited independently; there was a 1 : 1 : 1 : 1 segregation for pathogenicity to maize alone, sorghum alone, maize and sorghum, and avirulence to both hosts (Hamid et al.; Lim et al.). In maize, resistance may be polygenic (number of lesions type) or monogenic (chlorotic lesion type) in inheritance.

Colless reported the breakdown of monogenic resistance by a new race (and see Bergquist et al. 1974). Chlorotic lesion resistance may be controlled by single dominant genes or, apparently, by recessives. Work on disease forecasting and aerial, fungicide spraying was described by Berger. Control is by the use of resistant hybrids but where economically feasible fungicide treatment has been used. Resistance in sorghum was studied by Drolsom; Frederiksen et al.; Tarumoto et al.; Tuleen et al.

Benedict, W. G. 1976. *Can. J. Bot.* **54**: 552 (light dependent morphogenesis of conidia in vitro; **55**, 5524).

Berger, R. D. 1972. *Proc. Fla St. hort. Soc.* **85**: 141 (timing of aerial fungicides; **53**, 1795).

——. 1973. *Phytopathology* **63**: 930 (lesion numbers related to numbers of trapped spores & fungicide; **53**, 1351).

Bergquist, R. R. et al. 1974. *Ibid* **64**: 645 (races; **54**, 139).

——. 1975. *Misc. Publ. Hawaii agric. Exp. Stn* 122: 6 (race 2; **56**, 4987).

Calub, A. G. et al. 1973. *Crop Sci.* **13**: 561 (effects of genetic background on monogenic resistance; **55**, 4683).

—— ——. 1974. *Ibid* **14**: 303, 359 (inhibitory compounds in inbreds; genetic & environment effects on formation of inhibitory compounds (**55**, 2693, 2694).

Campi, M. D. 1939. *Lilloa* **4**: 5 (general; **19**: 207).

Chilton, S. J. P. 1940. *Phytopathology* **30**: 533 (seed transmission in *Sorghum sudanense*; **19**: 602).

Colless, J. M. 1975. *Agric. Gaz. N.S.W.* **86**(4): 54 (races; **55**, 2216).

——. 1975. *APPS Newsl.* **4**: 23 (races; **55**, 1770).

Couture, R. M. et al. 1971. *Physiol. Pl. Pathol.* **1**: 515 (role of cyclic hydroxamic acids on monogenic resistance; **51**, 1431).

Cox, R. S. et al. 1957. *Pl. Dis. Reptr* **41**: 796 (high temp. str.; **37**: 280).

Drolsom, P. N. 1954. *Agron. J.* **46**: 329 (inheritance of resistance in *S. sudanense*; **33**: 728).

Dunn, G. M. et al. 1970. *Crop Sci.* **10**: 352 (gene dosage effects on monogenic resistance; **51**, 2425).

Frederiksen, R. A. et al. 1975. *Pl. Dis. Reptr* **59**: 547 (resistance in sorghum; **55**, 3563).

Fullerton, R. A. et al. 1974. *N.Z. Jl agric. Res.* **17**: 153 (survival; **53**, 3940).

Gevers, H. O. 1975. *Pl. Dis. Reptr* **59**: 296 (new major gene for resistance; **55**, 200).

Hamid, A. H. et al. 1975. *Phytopathology* **65**: 280 (inheritance of pathogenicity; **54**, 4419).

Hilu, H. M. et al. 1963. *Ibid* **53**: 909 (conidial sporulation & chlorotic lesion resistance; **43**, 459).

Hooker, A. L. 1975. *Rep. Tottori mycol. Inst.* 12: 115 (review, 40 ref.; **55**, 3002c).

Knox-Davies, P. S. et al. 1960. *Am. J. Bot.* **47**: 328 (cytology & perfect state).

——. 1974. *Phytopathology* **64**: 1468 (penetration of leaf; **54**, 1718).

Leach, C. M. 1976. *Mycologia* **68**: 63 (electrostatic theory for violent conidial discharge; **56**, 184).

——. 1976. *Phytopathology* **66**: 1265 (comment on spore release in *Drechslera* & other fungi; **56**, 2366).

—— et al. 1977a & b. *Ibid* **67**: 380, 629 (in New Zealand, release, dispersal & formation of conidia; **56**, 4986; **57**, 154).

Lim, S. M. et al. 1974. *Ibid* **64**: 1150 (inheritance of virulence; **54**, 1244).

Long, B. J. et al. 1975. *Crop Sci.* **15**: 333 (relationship of hydroxamic acid in maize & resistance; **55**, 1230).

Luttrell, E. S. 1964. *Am. J. Bot.* **51**: 213 (morphology of perfect state; **43**, 2240).

Mace, M. E. 1973. *Phytopathology* **63**: 243, 1393 (histochemistry of beta-glucosidase in susceptible & resistant isolines, inhibition of conidial germination by leaf diffusates; **52**, 3675; **53**, 2162).

Meredith, D. S. 1965 & 1966. *Ibid* **55**: 1099; **56**: 949 (conidial release & dispersal; **45**, 795; **46**, 114).

Nelson, R. R. 1959. *Ibid* **49**: 159 (heterothallism; **38**: 511).

Palmerley, R. A. et al. 1977. *Can. J. Bot.* **55**: 1991 (conidial release in controlled conditions; **57**, 1164).

Schaik, T. van et al. 1959. *S. Afr. J. agric. Sci.* **2**: 255 (genetics of resistance; **39**: 409).

Scheifele, G. L. et al. 1970. *Can. J. Bot.* **48**: 1603 (survival in races; **50**, 1759).

Tarumoto, I. et al. 1977. *Jap. J. Breed.* **27**: 216 (inheritance of resistance in hybrids, sorghum & Sudan grass; **57**, 4931).

Tuleen, D. M. et al. 1977. *Pl. Dis. Reptr* **61**: 657 (characteristics of resistance in sorghum; **57**, 1707).

Turner, M. T. et al. 1975. *Phytopathology* **65**: 735 (conidial formation on maize with & without monogenic resistance; **55**, 719).

Ullstrup, A. J. 1970. *Ibid* **60**: 1597 (comparison between monogenic & polygenic resistance; **50**, 1761).

Valleau, W. D. 1935. *Ibid* **25**: 1109 (seed transmission; **15**: 289).

Yoka, P. et al. 1975. *Bull. Soc. Hist. nat. Toulouse* **111**: 255 (enzymatic & toxic activities; **55**, 4062).

SOROSPORIUM Rudolphi, *Linnaea* 4: 116, 1829.

(Ustilaginaceae)

SORI in various parts of the host plant, forming dark coloured, pulverulent spore masses which are usually covered by a pseudomembrane and surrounding a simple or compound columella; SPORE BALLS of several spores loosely bound together, often separating into individual spores when old. SPORES olive or reddish brown, of medium size and germinating by the formation of a septate promycelium with lateral and terminal sporidia (from B. B. Mundkur & M. J. Thirumalachar, *Ustilaginales of India*: 49, 1952).

Sorosporium spp. with evanescent spore balls can be mistaken for *Ustilago* or *Sphacelotheca*; younger sori should be examined if possible as these are more likely to have intact spore balls. *Sorosporium* is sometimes confused with *Thecaphora* which has permanent spore balls, larger spores, fewer spores/ball and spores are characteristically verrucose on the free surface (Fischer, *Manual of North American smut fungi*).

S. paspali-thunbergii (P. Henn.) S. Ito (synonyms: *Ustilago paspali-thunbergii* P. Henn., *S. paspali* McAlp.) causes a smut disease of kodo millet (*Paspalum scrobiculatum*) in India. The spores are aggregated into dark coloured, loose balls 30–50 μ diam.; SPORES are globose angular or roughly pear shaped, dark brown to yellowish brown, with a thick smooth wall, 9–15 × 8–11 μ diam. The whole ear is transformed into a sorus up to 8 cm long by 0.5 cm wide; when young the sorus is covered by a cream membrane. Usually all the shoots of an affected stool form smutted heads. Diseased ears, like healthy ones, remain concealed in the flag leaf sheath and are not readily noticed until harvest. The spores have a resting stage of *c*. 4 months. The opt. temp. for germination is *c*. 30°. The smut is mainly seedborne and infection of the seedling takes place. Control is by fungicidal treatment of the seed (Sattar).

Among the spp. recorded on *Panicum* are: *S. cenchri* P. Henn. (synonym *S. syntherismae* (Peck) Farl., Fischer l.c.) and *S. digitariae* (Kunze) Padwick. *S. simii* Pole Evans, described on *Sorghum halepense* (Johnson grass), is discussed by Tarr (*Diseases of sorghum, Sudan grass and broom corn*).

Pole Evans, I. B. 1916. *S. Afr. J. Sci.* **12**: 543 (*Sorosporium simii* sp. nov.).
Fullerton, R. A. 1975. *Aust. J. Bot.* **23**: 45, 51 (*S. caledonicum* on *Heteropogon contortus*, seasonal development & histology; **55**, 780, 781).
Ito, S. 1935. *Trans. Sapporo nat. Hist. Soc.* **14**: 87 (notes from E. Asia, *Ustilago* & *Sorosporium* systematics; **15**: 829).
Johnson, H. W. et al. 1940. *J. agric. Res.* **61**: 865 (*S. cenchri*; **20**: 307).
Langdon, R. F. N. et al. 1975. *Aust. J. Bot.* **23**: 915 (ontogeny & sporogenesis in *Sorosporium* & *Ustilago*; **55**, 3947).
Martin, W. J. 1943. *Phytopathology* **33**: 569 (genetics, *S. cenchri* & *Sphacelotheca destruens*; **22**: 476).
Sattar, A. 1930. *Bull. agric. Res. Inst. Pusa* 201, 16 pp. (*S. paspali-thunbergii*; **9**: 774).

SPHACELOMA de Bary, *Ann. Oenologie* 4: 165, 1874.

Manginia Viala & Pacottet, 1904.
Melanobasidium Maubl., 1906
Melanobasis Clements & Shear, 1931
Melanophora Arx, 1957.
(synonymy from B. C. Sutton & F. G. Pollack, *Mycologia* **65**: 1125, 1973, whose description of the genus is given below with that by A. E. Jenkins and A. A. Bitancourt, *Ibid* **33**: 339, 1941; and see Sutton, *Mycol. Pap.* 141: 195, 1977)
(Melanconiaceae)

HYPHAE subcuticular or more often intra-epidermal, of more or less well developed hyaline or yellowish pseudoparenchyma with clusters of dense, short conidiophores, often in conical protruding bundles or a superficial layer, but sometimes forming an acervulus; sometimes as sporodochial masses, base hyaline darkening towards outer surface, scattered or becoming continuous, a pseudoparenchymatic stroma with a surface layer of usually dark, appressed conidiophores. CONIDIOPHORES usually short, cylindrical, often aseptate or 1 septate; sometimes longer, several septate, branched and geniculate; forming bundles or a velvety covering over the lesion; forming conidia acrogenously or pleurogenously, sometimes shortly catenulate. CONIDIA hyaline, aseptate, small, refringent, ovoid to oblong elliptical, often a guttule at each end, wall mucilagenous; occasionally cylindrical, septate, yellowish or dark; forming a slow, compact, often colourful, gummy growth on most agar media (from Jenkins et al. l.c.).

ACERVULI folliicolous, caulicolous; composed of hyaline to pale brown, isodiametric pseudoparenchyma. CONIDIOGENOUS CELLS formed from the upper cells of the pseudoparenchyma, monophialidic to polyphialidic, integrated or discrete, determinate, hyaline to pale brown, lacking a thickened region around the phialide channel. CONIDIA hyaline, aseptate, ellipsoidal (from Sutton et al. l.c.).

Sphaceloma fawcettii

Some plant diseases caused by *Sphaceloma* spp. have been called anthracnoses but this common general term is probably best reserved for diseases caused by *Colletotrichum* (q.v., Jenkins). The term scab is usually used for *Sphaceloma* spp., some of which have perfect states in *Elsinoë* (q.v.). Most scab diseases seem to be relatively minor and have not been intensively studied. Some examples are: *S. arachidis* Bitanc. & Jenkins on *Arachis hypogaea*; groundnut scab occurs only in Brazil (Map 231) and Soave et al. found that $<3\%$ of 639 lines tested were very resistant. *S. cardamomi* Muthappa on *Elettaria cardamomum* (cardamom scab from India); *S. glycines* Kurata & Kuribayashi on *Glycine max* (soybean scab from Japan; Tasugi et al. investigated resistance); *S. papayae* Jenkins & Bitanc. on *Carica papaya* (papaw scab from Tanzania, see Wallace et al.); *S. poinsettiae* Jenkins & Ruehle on *Euphorbia pulcherrima* (poinsettia scab has been reported from Brazil, Cuba, Fiji, Guatemala, Hawaii, Jamaica, Puerto Rico and USA (Arizona, Florida, Texas), Map 393); *S. psidii* Bitanc. & Jenkins on *Psidium guajava* (guava scab from Brazil); *S. ricini* Jenkins & Cheo on *Ricinis communis* (castor scab from China); and *S. santali* Thirum. on *Santalum album* (white sandalwood scab from India). Wani et al. discussed the *Sphaceloma* spp. on *Ficus* briefly, and described *S. anacardii* Wani & Thirum. on *Anacardium occidentale* (cashew scab) and *S. tectonae* Wani & Thirum. on *Tectona grandis* (teak scab). Superelongation, a disease of cassava, has recently been investigated in S. America (Kraus; Larios et al.). It may be caused by *S. manihoticola* Bitanc. & Jenkins.

Bitancourt, A. A. et al. 1940. *Arq. Inst. biol. S. Paulo* **11**: 45 (*Sphaceloma arachidis* sp. nov. inter alia; **20**: 427).

——. 1949–50. *Ibid* **19**: 93 (*S. psidii* sp. nov. inter alia; **30**: 434).

——. 1950–1. *Ibid* **20**: 1 (*S. manihoticola* sp. nov. inter alia; **30**: 434).

Jenkins, A. E. et al. 1941. *J. Wash. Acad. Sci.* **31**: 415 (*Elsinoë dolichi* & *S. ricini* spp. nov.; **21**: 45).

——. 1942. *Proc. biol. Soc. Wash.* **55**: 83 (*S. poinsettiae* sp. nov.; **21**: 455).

Kraus, J. P. 1978. *Proc. Cassava Prot. Workshop CIAT*, Cali, Colombia, 7–12/11/77.

Kurata, H. et al. 1954. *Ann. phytopathol. Soc. Japan* **18**: 119 (*S. glycines* sp. nov.; **37**: 568).

Larios, J. et al. 1977. *Fitopatologia* **12**: 1 (cassava scab; **57**, 898).

Muthappa, B. N. 1965. *Sydowia* **19**: 143 (*S. cardamomi* sp. nov.; **46**: 1061).

Soave, J. et al. 1973. *Revta Agric. S. Paulo* **48**: 129 (resistance of *S. arachidis*; **53**, 4670).

Tasugi, H. et al. 1958. *Ann. phytopathol. Soc. Japan* **23**: 159 (resistance to *S. glycines*; **39**: 364).

Thirumalachar, M. J. 1947. *Trans. Br. mycol. Soc.* **31**: 1 (*S. santali* sp. nov. inter alia; **27**: 100).

Wallace, G. B. et al. 1955. *Arq. Inst. biol. S. Paulo* **22**: 79 (*S. papayae*; **35**: 691).

Wani, D. D. et al. 1969. *Sydowia* **23**: 252, 261 (*S. anacardii* & *S. tectonae* spp. nov. inter alia; **50**, 2207b & d).

——. 1970. *Proc. 1st int. Symp. Pl. Pathol. New Delhi* 1966–7: 216 (*Sphaceloma* spp. on *Ficus*).

Sphaceloma fawcettii Jenkins var. scabiosa (McAlp. & Tryon) Jenkins, *Phytopathology* 26: 195, 1936.

Ramularia scabiosa McAlp. & Tryon, 1899.

LESIONS on fruits, leaves and twigs. Leaf lesions amphigenous, mostly hypophyllous, discoid, irregularly elongate along midrib and veins with margin entire becoming crenate, sinuate or dark surface smooth to warty with superficial cracks and a dark centre, cinnamon coloured, solitary or coalescent, confluent up to 1.25 mm diam. Fruit lesions similar to leaf lesions and twig lesions elongate. ACERVULI of hyaline to pale brown pseudoparenchyma, epidermal to subepidermal, often confluent. CONIDIOGENOUS CELLS formed directly from the upper cells of the pseudoparenchyma or from aseptate to septate conidiophores, monophialidic to polyphialidic, terminal, integrated, determinate, hyaline to pale brown, $8–20 \times 2–6$ μ. CONIDIA (phialoconidia) hyaline, 0–1 septate, elliptical, $8–16 \times 2–6$ μ (A. Sivanesan & C. Critchett, *CMI Descr.* 437, 1974).

Australian or Tryon's citrus scab has been reported from: Argentina, Australia, Comores Islands, Fiji, Hong Kong, Indonesia, Malagasy Republic, Malawi, Malaysia, New Caledonia, Papua New Guinea, Zimbabwe, Solomon Islands, Sri Lanka and Zambia (Map 161). The symptoms resemble those caused by common scab (*Elsinoë fawcetti* q.v.) on citrus, but the 2 fungi cause symptoms that differ in the shape and appearance of those individual scabs of $>1–1.25$ mm diam. Australian scab is typically discoid, smooth, flat or crateriform (appearing to the unaided eye as distinct, raised, brown, circular spots up to 2 mm diam.); and not regularly pulvinate to subglobose, or highly irregular warty and deeply fissured as in common scab. The conidia are longer and the conidiophores more robust and darker than those of *E. fawcettii*. *S. fawcettii* var. *scabiosa* attacks several *Citrus* spp. and vars. It can be particularly severe on rough lemon and some mandarin cvs may be fairly susceptible. The disease is not apparently severe on sour orange. The characteristics of the disease and its control are much as for *E. fawcettii* and *E. australis* (q.v., sweet orange scab.).

Jenkins, A. E. et al. 1953. *Revta argent. agron.* **20**: 230 (occurrence in Argentina; **33**: 292).
McCleery, F. C. 1930. *Agric. Gaz. N.S.W.* **41**: 27 (on lemon & control; **9**: 450).

Sphaceloma perseae Jenkins, *Phytopathology* **24**: 84, 1934.

Lesions generally brown to almost black up to 3 mm diam. CONIDIAL FRUCTIFICATIONS acervuli at first, becoming sporodochial, or more or less free conidiphore tufts, scattered to effuse, in mass dark olive or light brownish olive, 25–70 μ in length. CONIDIOPHORES at first 1–2 celled, often *c*. 12 μ high × 2–7 μ at base, tapering, or acute to truncate at apex, from hyaline intra-epidermal hyphae or prosenchymatous stroma, palisaded, on rupturing epidermis increasing in length by continued growth or conidia in situ, often 25–50 μ long (up to 100 μ), more or less divergent, continuous to several septate, straight or geniculate, usually simple, sometimes denticulate. CONIDIA acrogenous or pleurogenous, hyaline or coloured, spherical to cylindrical, 2–30 × 2–5 μ; hyaline conidia ovoid or oblong elliptical, often 5–8 × 3–4 μ, sometimes biguttulate, continuous, at least at first; elongate coloured conidia 1–6 celled, reaching 30–35 μ, often 1–2 celled, 12–20 μ long; sometimes greatly enlarged or swollen (and at times muriform); germination by hyaline sprout conidia or by germ tubes, often produced apically or subapically but also laterally (after A. E. Jenkins l.c.).

Avocado scab occurs throughout Central America and the West Indies, also in: Argentina, Brazil, Guinea, Guyana, Mexico, Morocco, Peru, Philippines, Zimbabwe, South Africa, Venezuela, Zambia and USA (Florida and Texas; Map 232). Jenkins investigated the disease and compared the fungus with *Elsinoë fawcettii*, the cause of common citrus scab. *S. perseae* is restricted to avocado. The corky, raised, brownish oval spots on the fruit may coalesce as they become older, thus giving a russetted appearance. The subsequent cracking may allow secondary organisms to penetrate the fruit. Scabby lesions form on the leaves (which may be deformed) and small branches. Infections on the lower leaf surface are mainly on the midrib and main veins. In the early stages of scab development there is a velvety covering of conidia and conidiophores, light to dark olive in mass; fruits *c*. 1 month old may show this symptom. As the fruit ages the sporulation virtually ceases. The youngest tissues are most susceptible; fruit do not become immune until they are over half normal size. Cvs vary in their degree of susceptibility and in regions where avocado scab is likely to be severe fungicides should be used. Sprays should be applied from the end of the main bloom period when many fruits have set. Copper is generally recommended but where build up of this element is excessive ferbam was as effective in USA.

Fresa, R. 1964. *Rev. Invest. agropec. B. Aires* Ser. 5 **7**: 97 (general; **44**, 2211).
Jenkins, A. E. 1934. *J. agric. Res.* **49**: 859 (general; **14**: 459).
Ramallo, N. E. V. de 1969. *Revta ind. agric. Tucuman* **46**: 27 (brief account, control with Cu; **49**, 2140).
Ruehle, G. D. et al. 1962. *Proc. Fla St. hort. Soc.* **75**: 363 (control with ferbam; **43**, 882e).

SPHACELOTHECA de Bary, *Vergl. Morph. Biol. Pilze*: 187, 1884.
Endothlapsis Sorok, 1890
Sorisporium Ehrenb. ex Fr., 1825.
(see *Mycotaxon* **6**: 421, 1978)
(Ustilaginaceae)

SORI in the inflorescence, predominantly in the ovaries, covered with a definite pseudomembrane enclosing a dusty spore mass and a central columella; pseudomembrane later flaking away, composed largely or entirely of definite sterile fungus cells; such cells hyaline or slightly tinted, sometimes more or less firmly bound together. SPORES free, single, developed in a somewhat centripetal manner, small to medium size; germinating as in *Ustilago* (from B. B. Mundkur & M. J. Thirumalachar, *Ustilaginales of India*: 9, 1952)

Mundkur et al. (l.c.) gave descriptions of 3 *Sphacelotheca* spp. on *Saccharum*: *Sphacelotheca erianthi* (Syd.) Mundkur (synonym: *Ustilago erianthi* Syd.), *S. sacchari* (Rabenh.) Cif. (synonyms: *U.*

sacchari Rabenh. & *U. sacchari-ciliaris* Bref.), and *S. schweinfurthiana* (Thüm.) Sacc. (synonym: *U. schweinfurthiana* Thüm.). The last sp. has been recorded on sugarcane and *Imperata cylindrica* (lalang). The spores are 11–12 μ diam., and a var. *minor* Zundel has spores 7–9 μ diam. *S. macrospora* Yen & Wang was described on sugarcane from Taiwan. *U. scitaminea* (q.v.) is the only smut of economic importance on this crop. *S. cordobensis* (Speg.) Jacks occurs (amongst other hosts) on *Paspalum notatum* (Bahia grass).

Goyal, K. N. et al. 1975. *Indian Sug.* **25**: 439 (*Sphacelotheca schweinfurthiana* on sugarcane; **56**, 2652).

Khanna, K. L. et al. 1946. *Curr. Sci.* **15**: 253 (*S. schweinfurthiana* on *Saccharum munja*; **26**: 80).

Mundkur, B. B. et al. 1942. *Kew Bull.* 1941, 3: 209 (*Saccharum* & *Erianthus* smuts; **21**: 305).

Sharma, S. L. et al. 1955. *Proc. Indian Acad. Sci.* Sect. B **41**: 16 (*Sphacelotheca* on *Saccharum munja*; **34**: 752).

Yen, W. Y. et al. 1955. *J. agric. Assoc. China* **9**: 1 (*Sphacelotheca macrospora* sp. nov.; **37**: 595).

Sphacelotheca cruenta (Kühn) Potter, *Phytopathology* **2**: 98, 1912.
Ustilago cruenta Kühn, 1872
Sphacelotheca chrysopogonis Clinton, 1904
S. holci Jackson, 1934.

SORI in the ovaries, globose to elongated, *c.* 2.5–20 mm in length covered by a thin peridium of fungal tissue which soon cracks, flakes away, and disintegrates into groups of thin walled 'sterile cells', *c.* 10–12 μ diam. SPORE MASS dark brown, surrounding a well developed, unbranched columella of host tissue. SPORES globose to subglobose, light yellowish brown, wall *c.* 1 μ thick, finely echinulate under oil immersion, 6–10 μ diam. Sporidial cultures may be obtained on Czapek agar; see Blackmon et al. On *Sorghum bicolor* (sorghum), *S. sudanense* (Sudan grass), *S. halepense* (Johnson grass), and other *S.* spp., and (in India fide B. L. Chona et al. *Curr. Sci.* **20**: 301, 1951) *Saccharum officinarum* (sugarcane). (G. C. Ainsworth, *CMI Descr.* 71, 1965; and see Viennot-Bourgin.)

Loose smut of sorghum is fairly widespread in tropical and temperate areas but has not yet been reported from Australia, Oceania and the Malaysian–Indonesian archipelago (Map 408). For some of its distinguishing features from the other 2 sorghum smuts and further information, see *Sphacelotheca sorghi* (of generally less importance than *S. reiliana* and *S. sorghi* and distinguished from them (amongst other characteristics) by the marked stunting caused in the host). Most spikelets on a head are infected and the diseased panicles become looser, more bushy and darker than healthy ones. The glumes may be >2.5 cm long, vegetative proliferation in the inflorescence may occur, there is premature heading and increased tillering. Yields, especially those of forage, are reduced considerably. Sori may occur elsewhere apart from the inflorescence.

Infection (primarily seedborne) mostly occurs in the germinating seed before emergence, but heads can be infected through airborne spores. In the latter case there is no subsequent systemic spread. Stubble can be infected. Spore germination is opt. at *c.* 25° and infection high at 20–25°. Survival of spores in soil is low and probably too short to ensure viable inoculum for a succeeding crop, except in dry soil. *S. cruenta* is heterothallic and hybridisation with *S. reiliana* and *S. sorghi* has been recorded. Three races have been described and race 3, primarily on Johnson grass, is considered by some to be a distinct sp., *S. holci*. Control is through resistance and standard seed treatment (Dean). Where crops are treated as perennials the spread of infection in the rhizome should be considered.

Blackmon, C. W. et al. 1956. *Phytopathology* **46**: 403 (in vitro culture requirements).

Dean, J. L. 1966. *Ibid* **56**: 1342 (infection & chemical seed treatment; **46**, 1261).

Faris, J. A. et al. 1925. *Mycologia* **17**: 50 (mode of infection; **4**: 537).

Leukel, R. W. et al. 1950. *Phytopathology* **40**: 1061 (mode of infection, seed & stubble; **30**: 324).

Luttrell, E. S. et al. 1964. *Ibid* **54**: 612 (effect on vegetative growth; **43**, 2898).

Melchers, L. E. 1933. *J. agric. Res.* **47**: 339 (races; **13**: 227).

—— et al. 1943. *Ibid* **66**: 145 (effect on host development; **22**: 302).

Potter, A. A. 1915. *Phytopathology* **5**: 149 (morphology & taxonomy).

Rodenhiser, H. A. 1937. *Ibid* **27**: 643 (echinulation of spores & races; **16**: 667).

Shih, L. 1938. *Arch. Mikrobiol.* **9**: 167 (heterothallism; **17**: 740).

Shukla, D. D. et al. 1969. *Nova Hedwigia* **18**: 241, 245 (spore germination, light & temp.; **49**, 2468).

Takasugi, H. et al. 1933 & 1937. *Res. Bull. S. Manchuria Rly. Co.* **11**: 1; **16**: 49 (spore viability, germination & temp.; **13**: 436; **16**: 597).
Viennot-Bourgin, G. 1978. *Rev. Mycol.* **42**: 253 (morphology & histology).

Sphacelotheca destruens (Schlecht.) Stevenson & A. G. Johnston, *Phytopathology*, **34**: 613, 1944.
Caeoma destruens Schlechtendahl, 1824
Uredo segetum Pers. subsp. *panici-miliacea* Persoon, 1801
Ustilago panici-miliacea (Pers.) Winter, 1884
Sorosporium panici-miliacea (Pers.) Takahashi, 1902
Sphacelotheca panici-miliacea (Pers.) Bubak, 1912.
(further synonymy in Fischer, *Manual of North American smut fungi*)

SORI in the inflorescences which are replaced by greyish white or cream coloured structures up to 5 cm or more in length and 0.5–2 cm in width, each limited by a tough membrane of fungal tissue which flakes away. SPORE MASS powdery, dark brown, surrounding numerous fibres (vascular strands) of host tissue. SPORES globose to subglobose, olivaceous brown wall *c.* 1 μ thick, smooth or under oil immersion sometimes slightly punctate, 7–10 μ diam. SPORIDIAL cultures may be obtained on malt or PDA (Martin; Martin et al.). On *Panicum miliaceum* (common millet) and other *P.* spp. and the weed *Echinochloa crus-galli*. Hybrid maize (*Zea mays*) has been experimentally infected (Shevchenko 1964). (G. C. Ainsworth, *CMI Descr.* 72, 1965.)

Head smut of common millet is widespread in the north temperate zone and also occurs in N. Africa, South Africa, Argentina, Australia, Philippines and Uganda (Map 219). The sorus remains within the leaf sheath covered by the grey white membrane until maturity. Infected plants produce more stalks and leaves, and the former are taller than those of healthy plants. Diseased plants also tend to remain green longer. On inoculated maize the first symptoms appeared on the early, young leaves and invasion of the inflorescence led to it forming leafy structures, and individual flowers were replaced by smutted swellings. No close limits for opt. spore germination have been reported, it occurs particularly at 22–30°. No races have been described. *S. destruens* is seedborne (the spores overwinter and are viable for up to 8 years) and is controlled by standard seed treatments (e.g., Sevryukova; Steblyuk).

Buchheim, A. 1930. *Z. Bot.* **23**: 245 (effect on host growth in USSR; **10**: 181).
——. 1935. *Phytopathol. Z.* **8**: 615 (as above; **15**: 360).
Martin, W. J. et al. 1941. *Phytopathology* **31**: 761 (in vitro culture & variability; **21**: 20).
——. 1943. *Ibid* **33**: 569 (genetics; **22**: 476).
Richter, H. et al. 1943. *Zentbl. Bakt. ParasitKde* Abt 2 **106**: 32 (symptoms & seed treatment; **23**: 131).
Sevryukova, L. F. 1974. *Trudȳ Khar'kov. S.-Kh. Inst.* 202: 81 (seed treatment with carboxin; **55**, 1781).
Shevchenko. I. S. 1964. *Kukuruza* **9**: 48 (infection of maize; **44**, 2126).
Steblyuk, M. I. 1977. *Zakhist Roslin* 24: 68 (seed treatment with benomyl & carboxin; **57**, 582).
Yen, W. Y. 1937. *Bull. Soc. mycol. Fr.* **53**: 339 (morphology of spore germination; **17**: 628).

Sphacelotheca reiliana (Kühn) Clinton, *J. Mycol.* **8**: 141, 1902.
Ustilago reiliana Kühn, 1875
Sorosporium reilianum (Kühn) McAlpine, 1910.
(further synonymy in Fischer, *Manual of North American smut fungi*)

SORI destroying the inflorescences, each at first covered by a peridium of fungal tissue which soon disintegrates into globose to subglobose, hyaline or pale yellowish 'sterile cells', 5–15 μ diam. SPORE MASS powdery, dark brown, quickly dispersed to expose a tangled mass of vascular strands of the host or, rarely, a single central columella. SPORES globose to subglobose, or somewhat angled, pale yellowish to dark reddish brown, wall *c.* 1 μ thick, abundantly echinulate, 9–14 μ diam. SPORIDIAL CULTURES grow well on malt or carrot agar and not infrequently produce smut spores (Potter 1914). On *Sorghum bicolor* (sorghum), *S. sudanense* (Sudan grass) and other *S.* spp., *Zea mays* (maize), *Euchlaena mexicana* (teosinte), and spp. of *Andropogon, Hackelochloa* and *Cleistochne*. (G. C. Ainsworth, *CMI Descr.* 73, 1965).

Head smut of sorghum and maize is widespread in both tropical and temperate areas (Map 69); for some of its distinguishing features from 2 other sorghum smuts and source for a general account see *Sphacelotheca sorghi*. Overall losses from *S. reiliana* are probably not as great as those from *S. sorghi*. In maize the individual flowers of the tassel are infected whilst the cob may be partly or completely smutted. Rudimentary cobs and side shoots may develop into twisted leafy shoots; sori can occur on the leaves and the stem apex becomes galled. In sorghum the first symptoms occur at heading, the young head being replaced by a whitish gall (up to 5 × 10–15 cm). Heads may be only partially infected and panicles may be sterile but not smutted. The complete replacement of the head, so that the original inflorescence is hardly apparent, distinguishes *S. reiliana* from *S. sorghi* and *S. cruenta*, where each affected grain is replaced by a small sorus. Halisky (1963) gave a general account of the disease including tabulated comparisons of it on maize, sorghum and Sudan grass, and also a tabular differentiation of *S. reiliana* and *Ustilago maydis* (q.v.) on maize.

Infection can take place in seedlings up to a few weeks old (up to 9 weeks in sorghum) from soilborne inoculum which remains viable for *c.* 8 months. Spore germination is opt. at 27–31° and growth in vitro best at 28–30°. Most infection of sorghum occurs at 28° but lower temps down to 21° allow good infection which is favoured by lower soil moisture values. Temp. and moisture are thought to act directly on the pathogen. Different types (3) of host and parasite interaction, depending on host genotype and in one case on temp., have been described; and the systemic development in Sudan grass studied. There is no evidence for direct floral infection.

S. reiliana is heterothallic and a virulent white mutant (probably the first report of heritable albinism in *Sphacelotheca*) has been described. Two groups of the pathogen, one on sorghum and one on maize, are recognised; it has been suggested that these should become vars or forma speciales but specificity is not absolute, the form on maize being more specialised. Up to 5 races have been recognised in the sorghum group but only one from that on maize (Al-Sohaily et al.; Frederiksen et al.; Mehta et al.). Control through resistance is probably most effective. But crop sanitation (early destruction of smutted heads before spore dispersal) and crop rotation are also recommended. Seed treatment (Fullerton et al.; Simpson et al.) may achieve only partial control since infection can occur after germination from infested soil, and over a relatively long period of time. Fenwick et al. described some control with in-furrow treatment. Sorghum lines resistant to *S. sorghi* (q.v.; Cadady 1962) are mostly also resistant to *S. reiliana*.

Al-Sohaily, I. A. et al. 1963. *Phytopathology* **53**: 723 (5 races in sorghum group; **43**, 82).

Christensen, J. J. 1926. *Ibid* **16**: 353 (effect of soil temp. & moisture; **5**: 606).

Fenwick, H. S. et al. 1967. *Pl. Dis. Reptr* **51**: 626 (control with quintozene; **47**, 159).

Frederiksen, R. A. et al. 1969. *Ibid* **53**: 171 (race pathogenic to feterita sorghums; **48**, 1677).

—— ——. 1970. *Phytopathology* **60**: 583 (3 races in sorghum group).

Fullerton, R. A. et al. 1974. *N.Z. Jl exp. Agric.* **2**: 177 (maize seed treatment with captan + benomyl or carboxin & yield; **54**, 1241).

Gallegos, H. M. L. 1967. *Fitopatol. Mex.* **1**: 9 (requirements for in vitro growth; **47**, 2995).

Halisky, P. M. et al. 1959. *Pl. Dis. Reptr* **43**: 1084 (prevalence in California, symptoms, inoculation & resistance; **39**: 228).

——. 1961. *Phytopathology* **51**: 407 (heritable albinism; **40**: 751).

—— et al. 1962. *Ibid* **52**: 541 (systemic development; **42**: 28).

——. 1962. *Ibid* **52**: 199 (infection on maize in California, temp.; **42**: 18).

——. 1963. *Hilgardia* **34**: 287 (general; **42**: 759).

Hanna, W. F. 1929. *Phytopathology* **19**: 415 (cytology & mating pattern; **8**: 715).

Kruger, W. 1962. *S. Afr. J. agric. Sci.* **5**: 43 (infection & control; **41**: 652).

——. 1969. *Phytopathol. Z.* **64**: 367 (infection, soil moisture & organic matter; **48**, 3475).

Leukel, R. W. 1956. *Pl. Dis. Reptr* **40**: 737 (infection & sorghum seedling age; **36**: 315).

Mehta, B. K. et al. 1967. *Phytopathology* **57**: 925 (races; **47**, 167).

Potter, A. A. 1914. *J. agric. Res.* **2**: 339 (general).

Reed, G. M. et al. 1927. *Bull. Torrey bot. Club* **54**: 295 (maize & sorghum forms; **6**: 548).

Simpson, W. R. 1966. *Pl. Dis. Reptr* **50**: 215 (general).

—— et al. 1971. *Ibid* **55**: 501 (maize seed treatment with carboxin; **51**: 299).

Stewart, R. B. et al. 1958. *Ibid* **42**: 1133 (resistant cvs in sorghum; **38**: 205).

Tóffano, W. B. 1977. *Arg. Inst. biol. S. Paulo* **44**: 1 (inter- & intraspecific crosses with *Ustilago scitaminea*; **47**, 3444).

Ujević, I. et al. 1966. *Ochr. Rost.* NS **2**: 315 (fine structure of spores; **46**, 3009).

Vaheeduddin, S. 1936. *Phytopathology* **26**: 111 (hybridisation between *Sphacelotheca reiliana* & *S. cruenta*).

Wilson, J. M. et al. 1970. *Ibid* **60**: 828 (histopathology in sorghum; **49**, 3254).

—— ——. 1970. *Ibid* **60**: 1365 (histopathology & resistant sorghum; **50**, 1206).

Sphacelotheca sorghi (Link) Clinton, *J. Mycol.* 8: 140, 1902.

Sporisorium sorghi Link, 1825

Ustilago sorghi (Link) Passerini, 1873.

(more synonyms, or possible synonyms, are in Zundel, *The Ustilaginales of the world*, Fischer, *Sphacelotheca reiliana* (q.v.), Tarr, *Diseases of sorghum, Sudan grass and broom corn*)

SORI in the ovaries, producing from between the glumes as tapering, more or less cylindrical or subconical and rather persistent structures concolorous with, or darker than, the glumes, limited by a firm peridium of fungal tissue which breaks up into hyaline, more or less globose 'sterile cells' 6–18 μ diam., frequently in chains. SPORE MASS dark brown, surrounding a central columella of host tissue. SPORES globose to subglobose, olivaceous brown, apparently smooth but under oil immersion sometimes minutely echinulate, 5–9 (mostly 6–7)μ diam. SPORIDIAL CULTURES may be obtained on dextrose agar (Tyler). On *Sorghum bicolor* (sorghum), *S. sudanense* (Sudan grass), *S. halepense* (Johnson grass). As the 3 common *Sphacelotheca* smuts of sorghum are liable to be confused a comparison of a few of their differential characteristics may be tabulated (G. C. Ainsworth *CMI Descr.* 74, 1965).

SMUT	HOST	SORUS		SPORES	
		Site	Membrane	Surface	av. diam.
S. sorghi (covered smut)	Not stunted, heading normal	Ovary	Rather permanent	Apparently smooth	6–7 μ
S. cruenta (loose smut)	Stunted, heading premature	Ovary	Ruptures easily	Minutely echinulate	7–8 μ
S. reiliana (head smut)	Not stunted, heading normal	Inflorescence	Ruptures easily	Conspicously echinulate	9–12 μ

Covered smut, probably the most important disease of sorghum, is worldwide in distribution (Map 220) and presumably co-extensive with its main host. A full discussion on the smuts of sorghum is given by Tarr (l.c.) The first clear signs of symptoms are the smutted heads since the pathogen causes little if any stunting and abnormal tillering. Primary heads of infected plants may be healthy but on cutting back, smutted, secondary heads appear.

Spore germination is very variable morphologically and has an opt. temp. of 20–23°. Heavy seedling infection occurs at 25° and temps rather below this tend to favour the disease. In Australia max. infection took place when spores were placed on ungerminated seed or on seed with coleoptiles <5 mm long. Little infection occurred once the primary leaf had emerged. *S. sorghi* is carried over on the seed and since all infection takes place between sowing and emergence any cultural operations or soil factors which increase early growth may decrease infection rates; although the practical importance of these factors seems in doubt. This smut is heterothallic with multiple mating factors, morphologically variable in vitro and hybridisation with both *S. cruenta* and *S. reiliana* (q.v.) can occur (Rodenhiser; Tyler; Tyler et al.).

The most important control is to ensure healthy seed by a standard fungicidal treatment. Tarr (l.c.) tabulated 8 races and their reaction on 10 differential hosts. In USA Casady stated that lines resistant to *S. sorghi* were also resistant to *S. reiliana* and that resistance to races 1, 2 and 3 of *S. sorghi* was controlled by single genes (the 3 being linked) which were incompletely dominant.

Casady, A. J. 1961. *Crop Sci.* **1**: 63 (inheritance of resistance to races 1, 2, & 3; **41**: 387).

—— et al. 1962. *Ibid* **2**: 519 (breeding for resistance; **42**: 381).

——. 1963. *Ibid* **3**: 535 (inheritance of blasting reaction to race 1; **44**, 1080).

Ciccarone, A. et al. 1950. *Riv. Agric. subtrop. trop.* **44**: 145 (general in Venezuela; **30**: 564).

El-Helaly, A. F. 1939. *Bull. Minist. Agric. Egypt* 233 (effect of sowing & temp.; **19**: 13).

—— et al. 1957. *Phytopathology* **47**: 620 (symptoms, infection & cultural factors; **37**: 164).

Ficke, C. H. et al. 1930. *Ibid* **20**: 241 (cultural characters; **9**: 644).

Gorter, G. J. M. A. 1961. *S. Afr. J. agric. Sci.* **4**: 251 (2 races; **41**: 31).

Harris, K. M. 1963. *Ann. appl. Biol.* **51**: 367 (losses in Nigeria; **42**: 759).

Hsi, C. H. 1958. *Phytopathology* **48**: 22 (infection, irrigation & temp.; **37**: 352).

Isenbeck, K. 1935. *Phytopathol. Z.* **8**: 165 (temp. for infection & variation in monosporidial lines; **14**: 439).

Jones, G. H. et al. 1940. *Ann. appl. Biol.* **27**: 35 (effect of sowing depth & moisture in Egypt; **19**: 391, 392).

McKnight, T. 1966. *Qd J. agric. Anim. Sci.* **23**: 605 (infection; **47**, 511).

Marcy, D. E. 1937. *Bull. Torrey bot. Club* **64**: 209, 245 (inheritance of resistance to *Sphacelotheca sorghi* & *S. cruenta*; **16**: 598, 666).

Melchers, L. E. et al. 1932. *J. agric. Res.* **44**: 1 (5 races; **11**: 448).

——. 1933. *Ibid* **47**: 339, 343 (races & distribution in host; **13**: 227).

Melchers, L. E. et al. 1938. *Am. J. Bot.* **25**: 17 (effect of soil temp. & moisture; **17**: 453).
Reed, G. M. et al. 1924. *Ibid* **11**: 518 (effect of soil environment; **4**: 158).
Rodenhiser, H. A. 1932. *J. agric. Res.* **45**: 287 (heterothallism & hybridisation between *S. sorghi* & *S. cruenta*; **12**: 89).
——. 1934. *Ibid* **49**: 1069 (origin of races in *S. sorghi* & *S. cruenta*; **14**: 438).
——. 1937. *Phytopathology* **27**: 643 (hybridisation between *S. sorghi* & *S. cruenta*; **16**: 667).
Swanson, A. F. et al. 1931. *J. Hered.* **22**: 51 (inheritance of resistance; **10**: 725).
Tyler, L. J. et al. 1935. *Phytopathology* **25**: 385 (hybridisation between *S. sorghi* & *S. reiliana*; **14**: 504).
——. 1938. *Tech. Bull. Minn. agric. Exp. Stn* 133 (heterothallism, characteristics in culture & variability in pathogenicity; **19**: 210).
Vaheeduddin, S. 1938. *Phytopathology* **28**: 656 (race 6; **18**: 18).
——. 1951. *Indian Phytopathol.* **3**: 162 (races 7 & 8, key to races; **31**: 380).

SPHAEROPSIS Sacc., nom. cons., *Michelia* **2**: 105, 1880.
(synonymy in B. C. Sutton, *Mycol. Pap.* 141: 198, 1977; there is no satisfactory description, Sutton, personal communication, February, 1978)
(Sphaerioidaceae)

Sphaeropsis tumefaciens Hedges, *Phytopathology*, **1**: 64, 1911.

PYCNIDIA, on 'knots' produced on branches and stems of host, subglobose, dark brown, ostiolate, solitary or gregarious, 180–220 μ diam. pycnidial wall composed of many layers of elongated cells, the outermost layer of cells thicker walled than the innermost layers. CONIDIOPHORES hyaline, obpyriform to cylindrical, $5–10 \times 3–5\ \mu$, arising from the innermost layer of cells lining the inside of the pycnidium. CONIDIA hyaline to faintly yellow, oblong to obovoid, rounded at the apex and flat at the base, $20–34 \times 6–10\ \mu$, aseptate, occasionally septate but spores with 1–2 septa occur in culture. MICROCONIDIA hyaline, ellipsoid to ovoid, $3–5 \times 1.5\ \mu$. (P. Holliday & E. Punithalingam, *CMI Descr.* 278, 1970).

Knot of *Citrus* spp. is an old but little known disease of limited and uncertain distribution. Historically it appears to have been a disease of lime (*C. aurantifolia*) but in later years infection of Valencia orange (budded on rough lemon) has occurred; and more recently infection of ornamentals. Serious damage has only been described from Jamaica, though *S. tumefaciens* has also been reported from Cameroon, Cuba, Egypt, Guyana, Hawaii, India, Indonesia, Morocco, Pakistan, Peru, Sri Lanka, Sudan, Trinidad, USA and Venezuela (Map 386). The so called knots are growths like galls, rounded (1–7 cm diam.), sometimes elongated, on the stems. These swellings are at first covered with normal bark which changes to a whitish, rough tissue (like cork); this extends in size, becoming fissured, with much enlarged woody tissue. The knots are firmly attached and may occur in large numbers over considerable lengths of stem which may be girdled and killed. The surface of the knot may become soft and crumbling; but it is hard inside where black streaking indicates the presence of mycelium. A gall may form up to 40 shoots, from multiple bud formation, some over 1 m long and often themselves bearing knots or galls (witches' broom effect). These abnormal shoots eventually die. Knots can occur on the trunk and severe infection leads to death of the tree.

It is not, apparently, clear how natural infection takes place. Inoculations on citrus tend to be more successful on buds (especially young ones) than on internodal areas where wounding may be necessary. This picture seems to have been confirmed on ornamentals. Marlatt et al. (1974) found that the branches of the bottlebrush tree (*Callistemon viminalis*) showed symptoms after 21–105 days. Infection could cause the death of branches. Pycnidia do not seem to be common in the field but they may form abundantly in the development of knots and, therefore, are not readily observed. Mycelium is abundant in the cortex and wood of the knot and it is by growth in the woody tissue that the knot increases in size. Where several knots occur on one branch, with no macroscopic symptoms between them, there are mycelial connections. The method of spread is not known. But persistence in the host may be lengthy; in one case the fungus was isolated from secondary knots 4 years and 8 months from the time of inoculation. Pycnidia have been found in lime shoots after 2–3 weeks and they are formed in culture. Since the life history is insufficiently known, control measures are restricted to destruction of infected material as soon as it is detected.

Blazquez, C. H. et al. 1966. *Proc. Fla St. hort. Soc.* **79**: 344 (on lime in Jamaica; **46**, 3435).

——. 1967. *Ibid* **80**: 46 (on rough lemon rootstock in Jamaica; **48**, 1452a).
Hedges, F. et al. 1912. *Bull. Bur. Pl. Ind. U.S. Dep. Agric.* 246 (original monograph).
Marlatt, R. B. et al. 1974. *Phytopathology* **64**: 1001 (on bottlebrush tree; **54**, 1768).
——. 1976. *Pl. Dis. Reptr* **60**: 842 (on *Carissa grandiflora*; **56**, 4074).
Moreau, C. 1947. *Rev. Mycol.* Suppl. Colon. 3, **12**: 84 (in Cameroon; **27**: 564).
Prasad, N. et al. 1961. *Curr. Sci.* **30**: 110 (reported from India; **40**: 533).
Ridings, W. H. et al. 1976. *Proc. Fla St. hort. Soc.* **89**: 302 (on *Nerium oleander*; **57**, 2991).

SPHAEROSTILBE Tul., *Sel. Fung. carp.* Vol. 1: 130, 1861.
(Hypocreaceae)

Dingley (*Nectria* q.v.) placed *Sphaerostilbe* as a synonym of *Nectria* since the structures of the perithecia are identical. Rogerson considered that *Sphaerostilbe* was a *Nectria*. The basis for the former was the association with a synnematous conidial state (*Stilbum*). The fungus also known as *Sphaerostilbe musarum* Ashby (described as causing the unimportant bonnygate disease of banana) is referred to under *Calostilbe striispora*.

Rogerson, C. T. 1970. *Mycologia* **62**: 865 (Hypocreales, key to genera of Hypocreaceae; **50**, 2758).

Sphaerostilbe repens Berk. & Br., *J. Linn. Soc. Bot.* **14**: 114, 1873.
Corallomyces elegans Berk. & Curt., 1853.

This sp., which occurs on a wide range of tropical crops and in tropical soils, is characterised by its rhizomorphs which spread over the surface or in the cortex of the host and through adjacent soil. Conidia are produced on synnemata which develop from the rhizomorphs or from the mycelium. PERITHECIA also develop superficially in pustules from the rhizomorphs or from lateral ramifications of a rhizomorph. When mature the perithecia are scarlet, red to reddish brown and pyriform with a prominent neck, or more globose with a papillate apex; 500–650 × 400–450 μ diam. ASCI cylindrical to clavate, 185–215 × 8–9 μ, with 8 uniseriate ascospores. Mature ASCOSPORES hyaline to light brown, rough walled, faintly striate and somewhat constricted at the central septum, 18–20 × 8–9 μ. Cultures develop sparse white aerial mycelium but with the onset of conidial formation they assume a powdery appearance. The first formed conidia are borne on simple lateral phialides on the aerial mycelium or from lateral branches which terminate in 3–5 apical phialides. These are subulate and are 28–32 × 4 μ at the base.

CONIDIA hyaline, smooth and somewhat variable in size and shape but predominantly pyriform, 14–26 × 6–9 μ. Rough walled, sterile hyphae occur as part of the aerial mycelium and these are similar to those which later form the synnemata. RHIZOMORPHS tend to develop radially in culture from the point of inoculation, light reddish brown with white tips and branch dichotomously; later becoming black, up to 2 mm diam. SYNNEMATA develop both at the point of inoculation and at the tips of the lateral branches of the rhizomorphs; up to 4 mm high, from 300 μ to 1 mm diam. and formed of rough walled hyphae which branch and anastomose repeatedly. It is the lateral branches of these hyphae that terminate in rough walled, pyriform cells that project along the stalk of the synnemata, giving it a hairy appearance. When mature the synnemata are orange with a pale cream coloured, globose head formed by the mass of conidia. The conidia are similar to those formed from the aerial hyphae. CHLAMYDOSPORES form abundantly in culture and are globose, light brown, 12–20 μ diam. (C. Booth & P. Holliday, *CMI Descr.* 391, 1973).

Stinking root rot (sometimes called red or violet root rot) is widespread in the wet tropics (Map 288), usually occurring on woody hosts (plurivorous) including: avocado, cacao, citrus, coffee, mango, papaw, rubber and tea. The symptoms were well illustrated by Hilton in *Maladies of* Hevea *in Malaya* (1959). The characteristic and conspicuous rhizomorphs are seen on stripping away the bark on the main roots (usually the proximal parts) and the lower part of the stem. The young parts of the rhizomorphs are whitish and then shade to dark brown or purplish black (as in tea). The pinkish orange sporodochia and reddish perithecia are readily found on exposed woody tissue. Secondary symptoms are the above ground chlorosis and collapse of the canopy, and a sickly sour smell. Although this largely saprophytic fungus has frequently caused severe losses in the early days of agricultural development in the tropics (e.g., on lime in Dominica), attacks on living roots only take place when predisposing conditions arise, the most important of which are factors which contribute to poor aeration and waterlogging in the soil. Control is thus cultural, to avoid poor physical conditions in the soil.

Baker, R. E. D. 1938. *Trop. Agric. Trin.* **15**: 105 (on lime; **17**: 671).
Bugnicourt, F. 1935. *Bull. Econ. Indochine* **38**: 471 (on *Aleurites montana*; **15**: 271).
Goos, R. D. 1962. *Am. J. Bot.* **49**: 19 (in soil & in culture; **41**: 442).
Guillaumin, J. J. 1970. *Cah. ORSTOM Biol.* **12**: 41; **13**: 41 (rhizomorph morphology & differentiation).
John, K. P. 1964. *Plrs' Bull. Rubb. Res. Inst. Malaya* **75**: 244 (minor diseases on rubber; **44**, 1222g).
Petch, T. 1910. *Circ. agric. J. R. bot. Gdns Peradeniya* **5**: 65 (on rubber).
Small, W. et al. 1929. *Ann. R. bot. Gdns Peradeniya* **11**: 189 (inoculations; **8**: 610).

SPHAEROTHECA Lév., *Ann. Sci. Nat.* III **15**: 133, 1851

Albigo Auct.
(Erysiphaceae)

ASCOCARPS (cleistothecia, perithecia) with simple hyphoid appendages like those of *Erysiphe* but with a single ASCUS like *Podosphaera* (not described). CONIDIOPHORES usually abundant, base straight (not swollen as in *E. graminis*). CONIDIA borne in chains, with fibrosin bodies. MYCELIUM superficial (partly from C. E. Yarwood, *The fungi* (edited by G. C. Ainsworth et al.) Vol. IV A: 71, 1973).

The imperfect state is *Oidium* Sacc. = *Acrosporium* Nees ex Gray; *Erysiphe* (q.v.) for some general comments and ref. on the powdery mildews. *S. macularis* (Wallr. ex Fr.) Lind causes powdery mildew of hop (*Humulus lupulus*) and strawberry (*Fragaria* spp.), Synonyms: *S. humuli* (DC.) Burr., *Alphitomorpha macularis* Wallr., *Erysiphe macularis* (Wallr.) Fr. The fungus differs from *S. fuliginea* (q.v.) in the bigger cleistothecia and asci, and the unbranched appendages. *S. mors-uvae* (Schw.) Berk & Curt. causes powdery mildew of *Ribes* spp. (American gooseberry mildew). Synonyms: *E. mors-uvae* Schw., *Albigo mors-uvae* (Schw.) Kuntze. *S. pannosa* (Wallr. ex Fr.) Lév., rose powdery mildew (on Rosaceae), has few hyaline appendages mostly smaller than the cleistothecium. Synonyms: *Alphitomorpha pannosa* Wallr., *E. pannosa* (Wallr.) Fr.

Mukerji, K. G. 1968. *C.M.I. Descr. pathog. Fungi Bact.* 188, 189 (*Sphaerotheca macularis* & *S. pannosa*).
Parmelee, J. A. 1975. *Fungi Canadenses* 63 (*S. macularis*).
Purnell, T. J. et al. 1970. *C.M.I. Descr. pathog. Fungi Bact.* 254 (*S. mors-uvae*).

Sphaerotheca fuliginea (Schlecht. ex Fr.) Poll., *Atti r. 1st Bot. Univ. Pavia* **2**(9): 8, 1905.

Alphitomorpha fuliginea Schlecht., 1819
Erysiphe fuliginea (Schlecht.) Fr., 1829
Sphaerotheca humuli (DC.) Burr. var. *fuliginea* Salm., 1900.

MYCELIUM hyaline, occasionally brown when old, usually evanescent but sometimes persistent, forming white circular to irregular patches on the upper leaf surfaces. CLEISTOTHECIA scattered to densely gregarious, 66–68 μ diam. usually <85 μ, walls usually >25 μ wide. APPENDAGES variable in number, usually as long as the diam. of the ascocarp, mycelioid, brown, tortuous, interwoven with mycelium, but sometimes long, nearly straight and dark brown. ASCUS broadly elliptic to subglobose, 50–80 × 30–60 μ. ASCOSPORES 8, ellipsoid to nearly spherical, 17–22 × 12–20 μ. CONIDIA in long chains, often with distinct fibrosin bodies, ellipsoid to barrel shaped, 25–37 × 14–25 μ (J. N. Kapoor, *CMI Descr.* 159, 1967, q.v. for further synonymy and hosts).

Junell gave a taxonomic revision of the sp. *S. fuligenea* occurs on many plants and causes the important worldwide powdery mildew of cucurbits. The disease was once erroneously considered to be caused by *Erysiphe cichoracearum* (q.v.) which also occurs on cucurbits. As in other powdery mildew fungi (especially in tropical regions) the perfect state is often absent and the imperfect state needs to be used for identification. Zaracovitis (*Erysiphe* q.v.), in describing attempts to identify these fungi by conidial characters, refers briefly to earlier work. K. Hirata is quoted for finding that *S. fuliginea* is unique in that the conidium produces a forked germ tube. The conidia of *Sphaerotheca* contain fibrosin bodies whilst those of *E. cichoracearum*, often confused with *S. fuliginea*, do not. Both spp. form conidia in long chains (see Alcorn; Schlösser).

Ballantyne reviewed the literature in an account of the distribution, host range and sources of resistance. Records of perithecia of both fungi are given with those of the *Oidium* state where it resembles that of *S. fuliginea*. The absence of cleistothecia in some regions may have a genetic rather than an environmental cause. Hashioka (1938) described the direct penetration from the germinating conidium on plants varying in reaction from susceptible to immune. Hashioka (1937) found most conidial germination and infection (and haustorial formation) at *c.* 28°. Kumar et al. found abundant cleistothecia at 10–18°. Conidia form abundantly at RH 76–93% but are much reduced at saturation. In Israel the disease is most severe in summer. Reuveni et al. found that the disease developed more on squash in growth chambers under dry rather than humid conditions. The humidity has differing effects during disease development. Dryness tends to favour host colonisation, fungal sporulation and dispersal; high humidities favour infection and conidial survival. These processes will occur under humidity conditions far from their respective optima. Therefore, *S. fuliginea* develops under widely differing conditions. The disease can be severe on melon (*Cucumis melo*), particularly in dry regions, and may be less severe on pumpkin, marrow, squash and cucumber (*C. sativus*; but it can be severe on glasshouse cucumber); watermelon (*Citrullus lanatus*) is not often affected but severe disease outbreaks can occur (Ballantyne). Of Cucurbitaceous plants tested Stone found that only *Luffa* was immune. Since the fungus infects different crops (and weeds), for example, okra, cowpea, horsegram, cluster bean and papaw, carry over can be readily explained (see Alcorn).

Extensive work on resistance in cucumber and melons has been done (see Ballantyne; Harwood et al.; Robinson et al.; Shanmugasundaram et al.). The first worker listed resistance sources used in both crops; in melon 5 genes for resistance have been designated; in cucumber the inheritance of resistance is complex. It is likely that the designated races (1 and 2) are each a complex of races (Ballantyne). Thomas described a third race in 1978 and gave the reactions of 5 melon cvs to the 3 races. Robinson et al. (1975a) found that out of 590 watermelon accessions from 42 countries only 1 line was highly susceptible. Benomyl and triarimol are effective fungicides; examples of fungicide control are given for: India, Iran, Sudan, USA and West Indies. Resistance in *S. fuliginea* to systemic fungicides has been reported (e.g., Burth; Kooistra et al.; Peterson).

Alcorn, J. L. 1969. *Aust. J. Sci.* **31**: 296 (characteristics of *Sphaerotheca fuliginea*, infection of cucurbits & non-cucurbits).

Ballantyne, B. 1975. *Proc. Linn. Soc. N.S.W.* **99**: 100 (review, 132 ref.; **54**, 5619).

Banihashemi, Z. et al. 1972. *Pl. Dis. Reptr* **56**: 206 (fungicides in Iran; **51**, 3677).

Burth, U. 1973. *Arch. Phytopathol. PflSchutz* **9**: 411 (resistance to fungicides; **53**, 4172).

Daly, P. 1973. *Agron. trop.* Ser. 1 **28**: 18 (fungicides in West Indies; **52**, 3112).

Harwood, R. R. et al. 1968. *J. Hered.* **59**: 213 (genetic survey of resistance in melon; *Pl. Breed. Abstr.* **39**, 3684).

Hashioka, Y. 1937 & 1938. *Trans. nat. Hist. Soc. Formosa* **27**: 129; **28**: 47 (effects of temp. & RH on germination, viability & infection, infection of plants with varying resistance; **17**: 93, 579).

Jhooty, J. S. et al. 1972. *Indian J. agric. Sci.* **42**: 505 (benomyl seed treatment **52**, 1714).

Junell, L. 1966. *Svensk bot. Tidskr.* **60**: 365 (taxonomic revision of *S. fuliginea*; **46**, 1229).

Kooistra, T. et al. 1972. *Gewasbescherming* **3**: 121 (resistance to fungicides; **55**, 962).

Kumar, A. et al. 1974. *Indian J. Mycol. Pl. Pathol.* **4**: 201 (temp. & cleistothecia; **55**, 3792).

Omer, M. E. H. 1972. *Expl Agric.* **8**: 265 (fungicides in Sudan; **51**, 4574).

Peterson, R. A. 1973. *APPS Newsl.* **2**: 27 (resistance to benomyl; **54**, 3552).

Reuveni, R. et al. 1974. *Phytoparasitica* **2**: 25 (effects of RH on disease; **54**, 1492).

Robinson, R. W. et al. 1975a. *J. Am. Soc. hort. Sci.* **100**: 328 (resistance of watermelon; **55**, 961).

—— ——. 1975b. *J. Hered.* **66**: 310 (inheritance of susceptibility in watermelon; **55**, 2018).

Rudich, J. et al. 1969. *Israel Jnl agric. Res.* **19**: 41 (2 races on melon; **48**, 1426).

Schlösser, E. 1976. *NachrBl. dt. PflSchutzdienst.* **28**: 65 (conidial differentiation in *S. fuliginea* & *Erysiphe cichoracearum*; **56**, 465).

Shanmugasundaram, S. et al. 1971. *Phytopathology* **61**: 1218 (inheritance of resistance in cucumber; **51**, 2108).

Sowell, G. et al. 1974. *HortScience* **9**: 398 (races & resistance in melon; **54**, 1491).

Staub, T. et al. 1974. *Phytopathology* **64**: 364 (light & scanning electron microscopy of *S. fuliginea* and *E. graminis*; **53**, 3872).

Stone, O. M. 1962. *Ann. appl. Biol.* **50**: 203 (hosts; **41**: 758).

Thayer, P. L. et al. 1972. *Pl. Dis. Reptr* **56**: 45 (fungicides in Florida, USA; **51**, 3013).

Thomas, C. E. 1978. *Ibid* **62**: 223 (races on melon; **57**, 5184).

Waraitch, K. S. et al. 1975. *Indian Phytopathol.* **28**: 556 (fungicides in India; **57**, 3167).

Sphaerulina oryzina

SPHAERULINA Sacc., *Michelia* 1: 399, 1878. (Dothidiaceae)

PSEUDOTHECIA separate, immersed in substrate. ASCI fasciculate, bitunicate. ASCOSPORES hyaline, phragmosporous; like *Mycosphaerella* but with multiseptate spores (from R. W. G. Dennis, *British ascomycetes*: 365, 1968; key by E. S. Luttrell in *The fungi* (edited by G. C. Ainsworth et al.) Vol. IV A: 175, 1973).

S. rehmiana Japp causes a leaf spot and canker of rose and *S. rubi* Demaree & Wilcox causes a leaf spot of raspberry. *Sphaerulina* spp. have been described from *Cucurbita*, eucalyptus, maize and *Musa* (*Rev. appl. Mycol.* **32**: 149; **11**: 268; **46** 3632a; *Rev. Pl. Pathol.* **51**, 509; **53**, 1475).

Boerema, G. M. 1963. *Neth. J. Pl. Pathol.* **69**: 76 (*Sphaerulina rehmiana*; **42**: 553).

Demaree, J. B. et al. 1943. *Phytopathology* **33**: 986 (*S. rubi* sp. nov. general & stat. conid.; **23**: 135).

Tsonkovski, K. 1972. *Rastitelna Zashchita* **29**(9): 16 (*S. rubi*; **52**, 1606).

Sphaerulina oryzina Hara, *Diseases of the rice plant*, 1918 (Japanese).

stat. conid. *Cercospora oryzae* Miyake, 1910.

PERITHECIA scattered or gregarious, globose or subglobose, black, cellulo-membraneous, with a papilliform ostiole, immersed in the epidermal tissue of the host plant, 60–100 μ diam. ASCI cylindrical or club shaped, round at the top, stipitate, $50–60 \times 10–13$ μ. ASCOSPORES biseriate, spindle shaped, straight or slightly curved, 3 septate, hyaline, $20–23 \times 4–5$ μ. STROMA absent or small, 15–20 μ diam., brown. CONIDIOPHORES amphigenous, solitary or in fascicles of 2–3 (up to 7, rarely 15), pale to medium brown or sometimes dark brown, multiseptate 3 or more, slightly attenuated, unbranched 1–2 geniculations, $8–140 \times 4–9$ (mostly 3–5.5) μ, conidial scars present. CONIDIA hyaline, cylindrical to obclavate, straight or mildly curved, 3–10 (mostly 1–4 or 5) septate, base long obconic, apex blunt or rounded, 15–60 (10.6–73) $\times$ 3–6.4 μ. (Perfect state from G. W. Padwick, *Manual of rice diseases*: 155, 1950, and see Deighton for the 3 *Sphaerulina* spp. on rice. Stat. conid. J. L. Mulder & P. Holliday, *CMI Descr.* 420, 1974. The perfect state needs confirmation.)

Narrow brown leaf spot of rice is widespread in the tropics; it also occurs in S. and S.E. USA (Map 71). Sridhar reported the fungus on *Panicum repens*. On rice the linear leaf lesions are 2–10 mm long, usually not $>$ 1–1.5 mm wide, the long axis parallel to that of the leaf; centre dark brown with the border fading as the outer margin is reached. Lesions on the sheath are similar to those on the leaf, whilst those on the glumes are shorter and tend to spread laterally. On susceptible cvs leaf lesions are wider and a paler brown, whereas on resistant ones they are narrow and a more uniform dark brown. Symptoms are usually extensive during the later stages of growth and should not be confused with those caused by *Ramularia oryzae* (*Ramularia* q.v.).

There is no recent work on this minor disease; but severe losses may occur if susceptible cvs are grown. Older leaves are more susceptible than young ones. The opt. temp. for in vitro growth is 25–28°. No studies on spread have been described. In USA 10 races have been reported; 5 were characterised on 4 differentials, whilst 3 more differential hosts were used to distinguish sub-races with 3 of these races. In some cases a single dominant factor controls resistance to one race but in others 2 or more factors are involved. Control is through resistance; see Ou (*Rice diseases*) for a discussion on susceptible and resistant cvs.

Adair, C. R. 1941. *Tech. Bull. U.S. Dep. Agric.* 772, 18 pp. (inheritance of resistance with *Cochliobolus miyabeanus*; **20**: 490).

Chilton, S. J. P. et al. 1946. *Phytopathology* **36**: 950 (race 6; **26**: 124).

Deighton, F. C. 1967. *Trans. Br. mycol. Soc.* **50**: 499 (*Sphaerulina* spp. on rice; **47**, 179).

Jodon, N. E. et al. 1944. *J. Am. Soc. Agron.* **36**: 497 (inheritance of resistance; **23**: 499).

—— ——. 1946. *Ibid* **38**: 864 (characters inherited independently of reaction to races; **26**: 80).

Ryker, T. C. et al. 1940. *Phytopathology* **30**: 1041 (inheritance of resistance; **20**: 132).

—— ——. 1942. *J. Am. Soc. Agron.* **34**: 836 (inheritance & linkage of factors for resistance; **22**: 37).

——. 1943. *Phytopathology* **33**: 70 (differentiation of 5 races and sub-races; **22**: 224).

——. 1947. *Ibid* **37**: 19 (races 7 & 8; **26**: 261).

Sridhar, R. 1970. *Pl. Dis. Reptr* **54**: 272 (on *Panicum repens*).

Tasugi, H. et al. 1956. *Bull. natn. Inst. agric. Sci. Tokyo* Ser. C **6**: 167 (morphology, culture & pathogenicity; **37**: 56).

Tullis, E. C. 1937. *Phytopathology* **27**: 1005 (general; **17**: 201).

Yoshida, M. 1948. *Horoshimanogyo* **1**: 4 (connection of perfect state).

STAGONOSPORA (Sacc.) Sacc., nom. cons., *Sylloge Fung*. **3**: 445, 1884.
(synonymy in B. C. Sutton, *Mycol. Pap*. 141: 199, 1977)
(Sphaerioidaceae)

PYCNIDIA immersed, separate, spheroid, ampulliform or ovate, unilocular, ostiolate, pigmented or hyaline; walls thin, simple or several cells thick, outer layer forming a distinct specialised wall. CONIDIOGENOUS CELLS annellidic, holoblastic. CONIDIA hyaline, multiseptate, cylindrical, navicular, fusiform, truncate at the base; annellides doliiform, hyaline; prominent basal frill and appendages absent (from key by B. C. Sutton in *The fungi* (edited by G. C. Ainsworth et al.) Vol. IV A: 562, 1973).

S. meliloti (Lasch) Petrak (perfect state *Leptosphaeria pratensis* Sacc. & Briand) causes a leaf spot and root rot on lucerne and other legumes. Its taxonomy, synonymy, morphology and the disease caused were described by Jones et al. The pycnidia have a characteristic neck or rostrum extending through the epidermis. The leaf spots (up to 1 cm diam.) are a light buff to almost white, many have a darker brown border; some spots have brown, more or less concentric, bands in the central grey area which bears numerous pycnidia. On the roots (the more important phase of the disease) the surface of the lesion is dark brown to black, at first smooth, becoming rough and cracking; beneath the surface the tissues are reddish, they become necrotic and death of the root or part of the crown follows. The fungus can probably penetrate the intact roots of young plants; but the disease develops very slowly. In culture the best growth occurs at 18°, the earliest conidial germination is at 21–27° with germ tube length the most at 24–27°; and 19° appears opt. for pycnidial formation (Erwin, who found that *S. meliloti* did not survive after 3 months in soil at a level sufficient to cause disease). Control measures appear to be largely cultural. *S. recedens* (C. Massal.) F. R Jones & Weimer (from *Trifolium*) was thought distinct from *S. meliloti* because of the absence of the characteristic pycnidial neck (Jones et al., who considered that *Ascochyta trifolii* Bond. & Truss. was the same as *S. recedens*). Sprague, *Diseases of cereals and grasses in North America*, gave a key to (and described) the *Stagonospora* spp. on Gramineae in Canada and USA. *S. curtisii* (Berk.) Sacc. (leaf scorch of *Narcissus*) occurs on some ornamentals. The pathogen persists in the dry tissues of the flower bulb apex and spreads to the leaves as the plant develops. The bulb itself is not attacked. Primary necrotic spotting occurs at the leaf tips and then there is the damaging secondary spread to further areas of the leaves and to the flowers. In UK zineb (in petroleum oil) sprays have given control.

Assawah, M. W. 1968. *Phytopathol. Mediterranea* **7**: 21 (*Stagonospora curtisii* in Egypt; **48**, 1205).

Beaumont, A. 1950. *Ann. appl. Biol*. **37**: 591 (*S. curtisii* in UK; **30**: 322).

Bergman, B. H. H. et al. 1965. *PrakMeded Lab. BloembollOnderz. Lisse* 14, 7 pp. (*S. curtisii* in Netherlands; **45**, 2867).

Creager, D. B. 1933. *Phytopathology* **23**: 770 (*S. curtisii* in USA; **13**: 167).

Erwin, D. C. 1954. *Ibid* **44**: 137 (crown rot of lucerne & associated fungi; **33**: 730).

Grigoriou, A. C. 1975. *Annls Inst. Phytopathol. Benaki* **11**: 109 (*S. boltshauseri* (Sacc.) comb. nov. on common bean; **55**, 6016).

Jones, F. R. et al. 1938. *J. agric. Res*. **57**: 791 (*S. meliloti*, general; **18**: 320).

Stagonospora sacchari T. T. Lo & K. C. Ling, *J. Sugcane Res. Taiwan* **4**: 323, 1950.

COLONIES on potato sucrose agar form characteristic radiating strands at the advancing margin, white at first, becoming pale cinnamon pink then dark olivaceous or black in the centre where the pycnidia form. PYCNIDIA immersed in the leaves, commoner on the upper surface; subspherical to spherical, dark brown, 150–228 μ diam.; wall 13.7–17.1 μ thick; ostiole slightly raised, protruding, 17.1–27.4 μ diam. CONIDIA hyaline, ellipsoid, apex tapering, basal end somewhat rounded or truncated, straight or slightly curved, 44.6–51.5 × 10.3 μ, 3 septate (rarely 1 or 4 septate), constricted slightly at the septa; guttulate when mature (after T. T. Lo, *Sugarcane diseases of the world* Vol. 1: 212, 1961).

Leaf scorch of sugarcane occurs in Argentina, Japan, Panama, Philippines, South Africa, Taiwan, Thailand and Vietnam (Map 418). Further and earlier ref. to the disease were given by Exconde and by Lo (l.c.); and see *Rev. appl. Mycol.* **34**: 256, 257; **36**: 425. Exconde reported *Miscanthus sinensis* to be very susceptible and gave other hosts. The very small initial lesions (especially on young leaves) are red or brown and are visible 2–3 days after infection. The spots elongate, have a chlorotic halo, and become spindle shaped streaks which coalesce to form lesions from 0.3×5 cm to 1×17 cm. The central area becomes straw coloured and there is a reddish margin. The whole of the lamina can be killed. On older leaves the spots do not usually develop into streaks. The waterborne conidia germinate rapidly (95% after 4 hours) under moist conditions (opt. temp. 20–25°) and may form white (becoming pale buff and pinkish) mycelial patches over the pycnidia. Mycelial growth is opt. at 28°. Germinating conidia form appressoria and penetration takes place mostly through the stomatal guard cells. The fungus remains viable in infected leaves for several months. Under wet conditions and in susceptible clones sugarcane leaf scorch can spread rapidly and cause severe losses. Control is through resistance and such selections have been developed in the Philippines and Taiwan.

Chen, C. C. et al. 1955. *Rep. Taiwan Sug. Exp. Stn* **13**: 81 (infection; also as special publication No. 3, Coll. Agric., Natn. Taiwan Univ.).

Exconde, O. R. 1963. *Philipp. Agric.* **47**: 271 (general; **44**, 828).

Leu, L. S. 1968. *Pl. Prot. Bull. Taiwan* **10**: 1 (resistance, field trials; **51**, 4329).

Ling, K. C. 1962. *Rep. Taiwan Sug. Exp. Stn* **29**: 43 (conidial biology; **43**, 556).

STEMPHYLIUM Wallr., *Fl. crypt. Germ.* Pars. post 300, 1833.

(Dematiaceae)

COLONIES effuse, grey, brown, olivaceous brown or black, velvety or cottony. MYCELIUM immersed or partly superficial. Stroma sometimes present. SETAE and HYPHOPODIA absent. CONIDIOPHORES macronematous, mononematous, scattered or caespitose, unbranched or occasionally loosely branched, straight or flexuous, usually nodose with vesicular swellings, pale to mid brown or olivaceous brown, smooth or in part verruculose. CONIDIOGENOUS CELLS monoblastic, integrated, terminal, percurrent at first clavate or subspherical with the wall at the apex thin, later often becoming calyciform by invagination. CONIDIA solitary, dry, acrogenous, rounded at the ends, ellipsoidal, obclavate or subspherical; some spp. with a pointed conical apex and one with lateral conical protrusions, pale to mid dark or olivaceous brown, smooth, verrucose or echinulate, muriform, often constricted at one or more of the septa, cicatrised at the base (M. B. Ellis, *Demat. Hyphom.*: 165, 1971 where a key to and descriptions of 5 spp. were given).

References to leaf diseases on legume crops caused by spp. of the genus are fairly frequent in the literature. These are mostly on temperate crops and are briefly mentioned under Leguminosae. Groves et al. described *Stemphylium* as a seedborne fungus, Simmons its perfect states in *Pleospora* (q.v.) and the conidial states, and Wiltshire the conceptions of the genus.

Brett, M. A. 1931. *Trans. Br. mycol. Soc.* **16**: 89 (cyclic saltation in vitro; **11**: 393).

Groves, J. W. et al. 1944. *Can. J. Res.* Sect. C **22**: 190 (seedborne *Stemphylium*; **24**: 42).

Moreau, C. et al. 1947. *Bull. Soc. mycol. Fr.* **63**: 58 (on *Alternaria* & *Stemphylium*; **27**: 46).

Simmons, E. G. 1967. *Mycologia* **59**: 67 (typification of *Alternaria, Stemphylium* & *Ulocladium*; **46**, 2394).

——. 1969. *Ibid* **61**: 1 (perfect states of *Stemphylium*; **48**, 2220).

Wiltshire, S. P. 1938. *Trans. Br. mycol. Soc.* **21**: 211 (original & modern conceptions of *Stemphylium*; **18**: 141).

LEGUMINOSAE

Several *Stemphylium* spp. have been described as causing various leaf spot diseases of leguminous crops. These include: *S. botryosum* (perfect state *Pleospora herbarum* q.v.), *S. globuliferum* (Vestergren) Simmonds, *S. loti* Graham, *S. trifolii* Graham and *S. sarciniforme* (Cav.) Wiltshire (see Ellis, *Demat. Hyphom.*, 1971). The crops involved include: blue lupin (*Lupinus angustifolius*), lucerne (*Medicago sativa*), clover (*Trifolium*), chick pea (*Cicer arietinum*), sainfoin (*Onobrychis viciifolia*) and bird's-foot trefoil (*Lotus corniculatus*). Some of these fungi infect >1 of these plants, with *S. botryosum* having the widest host range (Graham et al.); differences in pathogenicity in *S. botryosum* and *S. sarciniforme* are possibly due to host specialisation. Tate found that *S. botryosum* from blue lupin differed from lettuce isolates in host specificity and temp.

requirements. *S. botryosum*, *S. sarciniforme* and *S. trifolii* have all been described as attacking lucerne. Forms of *S. solani* more pathogenic to blue lupin than to tomato have also been described.

Das, G. N. et al. 1961. *Pl. Dis. Reptr* **45**: 979 (*Stemphylium sarciniforme* on chick pea; **41**: 497).

Edwardson, J. R. et al. 1961. *Ibid* **45**: 958 (effect on yield of lupin; **41**: 392).

Focke, I. 1967. *Reprium nov. Spec. Regni veg.* **75**: 61 (*S. botryosum* & *S. sarciniforme* on lucerne; **46**, 3121).

Forbes, I. et al. 1961. *Crop Sci.* **1**: 184 (inheritance of resistance in lupin).

———. 1964. *Phytopathology* **54**: 54 (as above; **43**, 2307).

———. 1965. *Ibid* **55**: 627 (as above; **44**, 3084).

Gentner, G. 1918. *Prakt. Bl. PflBau PflSchutz* **16**: 97 (*S. sarciniforme* on lucerne & clover).

Graham, J. H. 1953. *Phytopathology* **43**: 577 (*S. loti* sp. nov. on bird's-foot trefoil; **33**: 356).

——. 1957. *Ibid* **47**: 213 (*S. trifolii* sp. nov. on clover; **36**: 593).

—— et al. 1960. *Ibid* **50**: 757 (pathogenicity & morphology of some leguminicolous & related *Stemphylium* spp.; **40**: 230).

Hancock, J. G. et al. 1965. *Ibid* **55**: 346, 356 (cell degrading enzymes in *S. botryosum* inter alia; **44**, 2182a & b).

Pierre, R. E. et al. 1965. *Ibid* **55**: 909 (histology, *S. botryosum* on lucerne & other *Stemphylium* spp.; **45**, 129).

Tate, K. G. 1970. *N.Z. Jl agric. Res.* **13**: 710 (*S. botryosum* on lupin; **50**, 1294).

Wells, H. D. et al. 1956. *Pl. Dis. Reptr* **40**: 803 (*S. botryosum* & *S. solani* on lupin; **36**: 325).

———. 1961. *Ibid* **45**: 725 (as above; **41**: 157).

———. 1962. *Ibid* **46**: 333 (*S. loti* on lupin; **41**: 719).

Stemphylium solani Weber, *Phytopathology* **20**: 516, 1930.

CONIDIOPHORES pale to mid brown, up to 200 μ long, 4–7 μ thick, vesicular swellings 8–10 μ diam. CONIDIA pointed at the apex, with 3–6 transverse and several longitudinal septa, mostly constricted at the medium septum only, pale to mid-golden brown, smooth or minutely verruculose, $35–55 \times 18–28$ μ (length:breadth ratio 2:1) (M. B. Ellis, *Demat. Hyphom.*: 167, 1971; and see Ellis et al.)

Grey leaf spot of tomato is fairly widespread (Map 333) with most records in tropical and subtropical regions. Two other spp. of the genus have been described as causes of leaf diseases in this crop. *S. lycopersici* (Enjoji) Yamamoto, who considered *S. floridanum* Hannon & Weber (1955) to be synonymous, and *S. botryosum* f. sp. *lycopersici* (*Pleospora herbarum* q.v.). The former was described by Ellis et al. (synonyms: *Thyrospora lycopersici* Enjoji, *T. solani* Sawada and *S. floridanum*). *S. lycopersici* differs from *S. solani* in that the conidia are constricted at 3 major transverse septa and in the length:breadth ratio of 3:1 or more. The latter also infects *Solanum* spp. (including eggplant) and *Capsicum* (see Blazquez).

Symptoms are restricted to the foliage; in contrast to other tomato leaf spots the spots are small, more regular and evenly distributed, and do not enlarge rapidly or show any concentric zonation. The disease may appear first in a seedbed where defoliation results without conspicuous yellowing, but in the field yellowing can be conspicuous. Black, circular spots (1 mm diam.) appear first and these may be so numerous as to cover half the lamina. When *c.* 2 mm diam. there is a definite chlorotic halo to the spots which may (rarely) reach 4 mm diam. and coalesce, forming larger necrotic areas, cracking, shotholes, yellowing and shedding of the leaves. Detailed information on host leaf penetration seems to be lacking; it occurs through stomata and the cuticle, the former process is more frequent. The best growth in vitro has been reported at 23–24° and 24–28°. Sporulation occurs on both leaf surfaces and readily in culture. Paulus et al. found more infection at 26° than at 14, 18 and 22°. With a high inoculum potential disease was very severe at 16, 20 and 24° and markedly reduced at 28°. Namekata et al. proposed 3 races. Hendrix et al. found that oligogenic resistance was controlled in a dominant fashion. Resistance is available in several cv. types of tomato, for example, from USA (Florida): Floradel, Floralou, Immokalee, Tropic and Walter. Fungicides are effective, i.e. captafol, chlorothalonil, mancozeb, maneb, metiram, propineb and zineb.

Blazquez, C. H. 1969. *Pl. Dis. Reptr* **53**: 756 (red pepper, control with zineb; **49**, 319).

——. 1971. *Proc. Fla St. hort. Soc.* **84**: 171 (as above, with mancozeb & chlorothalonil; **52**, 2442).

Dennet, R. K. 1950. *Proc. Am. Soc. hort. Sci.* **56**: 353 (association of resistance with that to *Fusarium oxysporum* f. sp. *lycopersici*; **30**: 436).

Diener, U. L. 1955. *Phytopathology* **45**: 141, 654 (sporulation in vitro, penetration & histology; **34**: 679; **35**: 553).

Ellis, M. B. et al. 1975. *C.M.I. Descr. pathog. Fungi Bact.* 471, 472 (*Stemphylium lycopersici* & *S. solani*).
Hannon, C. I. et al. 1955. *Phytopathology* **45**: 11 (tomato leaf spot caused by *S. lycopersici*; **34**: 493).
Hendrix, J. W. et al. 1949. *Tech. Bull. Hawaii Exp. Stn* 8, 24 pp. (inheritance of resistance; **28**: 648).
Namekata, T. et al. 1967. *Anais Esc. Sup. Agric. 'Luis Queiroz'* **24**: 273 (variability; **49**, 3005).
Paulus, A. O. et al. 1955. *Phytopathology* **45**: 168 (effect of temp. with *Alternaria tomato*; **34**: 680).
Poe, S. L. et al. 1972. *J. econ. Ent.* **65**: 792 (compatibility of fungicides with insecticides; **52**, 253).
Sobers, R. K. 1965. *Phytopathology* **55**: 131 (form of *S. lycopersici* on *Kalanchoë*; **45**, 1403).
Sproston, T. et al. 1968. *Mycologia* **60**: 104 (ergosterol & substitutes for UV radiation requirement for conidia formation; **47**, 2405).
Weber, G. F. 1930. *Phytopathology* **20**: 513 (general; **9**: 812).
—— et al. 1932. *Bull. agric. Exp. Stn Univ. Fla* 249, 35 pp. (general; **12**: 60).
Yamamoto, W. 1960. *Trans. mycol. Soc. Japan* **2**: 88 (synonymous spp. of *Alternaria* & *Stemphylium* in Japan, including *S. lycopersici*).

SYNCHYTRIUM de Bary & Woron., *Ber. naturf. Ges. Freiburg. i.B* 3: 22, 1863.

(synonymy in Karling)
(Synchytriaceae)

THALLUS developing into either a group of zoosporangia or a resting spore; initial thallus functioning as a sorus and segmenting directly into a number of zoosporangia or gametangia, or as an evanescent prosorus from which the contents emerge through a pore and form an attached vesicular sporangial or gametangial sorus outside of the initial thallus but within the infected host cell, or developing into a resting spore. ZOOSPORANGIA and GAMETANGIA variously shaped but predominantly polyhedral with a hyaline wall and red, orange, yellow, reddish yellow, lemon coloured, grey or hyaline granular content; dehiscing by a tear, cleft or papilla. ZOOSPORES and GAMETES ovoid, ellipsoid, solid oblong, spherical or pyriform, usually with 1 and sometimes 2 conspicuous, refringent globules, and a single, posterior, whiplash flagellum. RESTING SPORES formed parthenogenetically (?) by encystment of the initial thallus or by fusion of isogametes, with the resulting zygotes infecting the host and developing into diploid resting spores; exospore thick, smooth, rough or ridged, hyaline, amber coloured, reddish brown or light to dark brown; endospore relatively thin and hyaline; content variously coloured with a few to numerous refringent globules; functioning as a sporangium or a prosorus in germination (J. S. Karling: 117).

S. endobioticum (Schilb.) Percival causes the important wart disease of Irish potato (see Karling for a 31 page bibliography; Hampson et al.). The pathogen is widespread in Europe and is also present in: Bolivia, Canada (Newfoundland), Chile, China, Ecuador, Falkland Islands, India, Mexico, Nepal, New Zealand (South Island, attempts at eradication), Peru, South Africa, Uruguay and USA (Maryland and Pennsylvania; Map 1). *S. lagenariae* Mhatre & Mundkur and *S. trichosanthidis* Mhatre & Mundkur occur on cucurbits: *Citrullus, Cucumis, Luffa* and *Trichosanthes* (Rao et al.). *S. sesamicola* causes a disease of sesame (*Sesamum indicum*).

Recent interest in the crop potential of winged bean (or Goa bean, *Psophocarpus tetragonolobus*) has led to the investigation of one of its diseases, false rust caused by *S. psophocarpi* (Racib.) Gäumann (synonyms: *Uromyces psophocarpi* H. & P. Syd., *Woroninella psophocarpi* Raciborski). SORI solitary in epidermal cells, ovoid to subspherical, 140–230 μ diam.; orange red; segmenting into many sporangia which form a powdery mass when the sorus membrane and galls rupture. SPORANGIA globose, 20–25 μ diam., subglobose, ovate, 19–34 × 17–26 μ, or elongate 16 × 50 μ; orange red contents. ZOOSPORES pyriform, 3–3.5 × 6.8 μ, flagellum 5–8(?) μ long (Karling: 335, and see Drinkall et al.).

S. psophocarpi has been reported from: Indonesia, Malaysia, Papua New Guinea and Philippines. All above ground parts of the plant bear semi-globular, yellow galls, 180–460 μ diam., slightly raised above the host surface, usually closely aggregated; open galls are crater shaped. Young, infected leaves become abnormal in size and shape; small, thickened and curled. Stem infection causes an arrest in growth with bunched and twisted side shoots. Sporangial germination occurs at 5–30° (best 10–25°) and RH 84–100%. Zoospores (in water) emerge after 40 minutes incubation and there is max. germination after 70 minutes. Sporangia (exposed) lost viability after 4 days.

Drinkall, M. J. 1978. *PANS* **24**: 160 (*Synchytrium psophocarpi*, general; **57**, 5790).
—— et al. 1979. *Trans. Br. mycol. Soc.* **72**: 91 (as above, morphology, spore germination & viability).

Hampson, M. C. et al. 1974. *Pl. Prot. Bull. F.A.O.* **22**: 53 (*S. endobioticum*, review, distribution & problems in Newfoundland, 96 ref.; **54**, 1851).
Karling, J. S. 1964. *Synchytrium*, Academic Press (monograph; **44**: 64).
Rao, N. N. R. et al. 1978. *Indian J. agric. Sci.* **48**: 76 (on cucurbits).
Vera-Chaston, H. P. De 1977. *Kalikasan* **6**: 183 (*S. psophocarpi*, histology; **57**, 5791).

TAPHRINA Fr., *Obs. Mycol.* Vol. 1: 217, 1815. (Taphrinaceae)

Parasitic on higher plants or ferns; MYCELIUM intercellular, subcuticular or within the epidermal wall; forming asci in a subcuticular layer or a wall locule; overwintering in the form of BLASTOSPORES derived from ASCOSPORES by budding, or in a few spp. as perennial mycelium. Infection (so far as known) is by blastospores. ASCI arise from rounded ascogenous cells (chlamydospores), either by elongation of the ascogenous cell or by bursting out from the ascogenous cell wall. In many spp. a stalk cell (basal cell) is cut off from the ascus proper. Budding of the ascospores to form blastospores may occur within the ascus and continue after spore expulsion. Grows readily in artificial media if cultures originate from ascospores or blastospores; behaving in media as yeasts. Cells formed in culture are blastospores, hyphae, ascogenous cells and (rarely) asci (after A. J. Mix).

There is a monograph by Mix, and Snider et al. used electrophoresis in taxonomic studies. The morphology was discussed by C. L. Kramer in *The fungi* (edited by G. C. Ainsworth et al.) Vol. IV A: 35). *Taphrina* (the only genus in the family) forms no ascocarps. The spp. cause: galls on leaves, stems and fruits; hypertrophied and deformed leaves and fruits; necrotic, limited leaf spots with little or no hypertrophy; blister-like leaf lesions; and witches' brooms. Most spp. mentioned in pathology literature occur on temperate broad-leaved trees and stone fruits. One of the most investigated is *T. deformans* (Berk.) Tul. (peach leaf curl; Map 192). Accounts of this sp. can be found in: Butler & Jones, *Plant pathology*, and Walker, *Plant pathology*. Kramer (1960) described the morphological development and nuclear behaviour in 5 spp. including that of *T. deformans*; 3 distinct patterns of nuclear behaviour were described.

Kern, H. et al. 1975. *Phytopathol. Z.* **83**: 193 (formation of auxins & cytokinins by *Taphrina* spp.; **55**, 1656).
Kramer, C. L. 1960. *Mycologia* **52**: 295 (morphological development & nuclear behaviour; **41**: 19).
Mix, A. J. 1949. *Kans. Univ. Sci. Bull.* **33**: 3–167 (taxonomic monograph; **28**: 548).
——. 1953 & 1954. *Mycologia* **45**: 649; **46**: 721 (spp. differentiation in culture, N & C compounds; **33**: 384; **34**: 608).
Salata, B. 1974. *Fungi (Mycota), Ascomycetes, Taphrinales* Vol. 6; Warsaw.-Kraków, Pol. Akad. Nauk. Inst. Bot. 87 pp. (descriptions of 36 *Taphrina* spp. with key; **54**, 63).
Schneider, A. et al. 1977. *Annls Sci. nat. Bot. Biol. Veg.* **18**: 207 (structure of ascus wall; **57**, 1683).
Snider, R. D. et al. 1974. *Mycologia* **66**: 743, 754 (electrophoresis & taxonomy; **54**, 3184, 3185).
Watson, J. L. et al. 1974. *Ibid* **66**: 773 (fatty acid ratios of lipid solvent fractions; **54**, 3186).

Taphrina maculans Butler, *Annls Mycol.* **9**: 39, 1911.

Leaf spots amphigenous, mostly epiphyllous, at first yellow becoming brownish yellow, crowded, 1–5 mm diam. MYCELIUM growing within the epidermal walls, and within radial walls between cells of subepidermal layers, sometimes forming plates of hyphae and occasionally penetrating the host cells by haustoria. ASCI epiphyllous, clavate, rounded or truncate above, 8 spored, but may become multispored due to budding of the ascospores within the ascus, $20–36 \times 6–10\ \mu$, with a 1-, 2- or 3-celled stalk. ASCOSPORES hyaline, aseptate, smooth, ovoid to almost oblong, $4–6.5 \times 2–3.5\ \mu$ (A. Sivanesan & I. A. S. Gibson, *CMI Descr.* 507, 1976).

T. maculans occurs on spp. in the Zingiberaceae, including *Curcuma* and *Zingiber*. It causes a leaf spot of turmeric (*C. domestica*) in India and has been reported from Burma and Japan. Mix (1949, *Taphrina* l.c.) examined material from Japan (on *Z. mioga*) although the fungus is not given in a list of plant diseases in this country (*Rev. appl. Mycol.* **46**, 1510). *T. maculans* has not, apparently, been recorded on ginger (*Z. officinale*), but it occurs on Indian arrowroot (*C. angustifolia*). Turmeric leaf spot was described by Butler in *Fungi and disease in plants*. The leaf spots are sometimes very numerous, mostly on the upper surface. They are at first pale yellow, becoming dirty yellow and deepening in colour to old gold or sometimes nearly a shade of brown; small, usually 1–2 mm diam., not sharply defined and coalesce freely; not limited by the leaf veins and not causing any leaf distortion. Infected leaves tend to turn yellow sooner than healthy ones. Plants are not killed but very severe spotting leads to a reduction in yield.

Ascospores, and probably blastospores, are forcibly discharged from the asci; the ascospores multiply by budding after discharge (Upadhyay et al. 1973). Pavgi et al. described the development of haustoria after host penetration by germinating blastospores. The fungus survives on infected leaf debris between crops through viable ascogenous cells, and through dried spores (discharged from asci during the crop season) on soil and leaf debris. Upadhyay et al. (1966) described secondary infections occurring from discharged ascospores which formed blastospores, and which required no dormant period; such infections caused profuse spotting. Leaves remain susceptible over a long period (Upadhyay et al. 1967b). Zineb gives some control and resistance has been described. Cvs differ in reaction to infection.

Ahmed, L. et al. 1968. *Mycopathol. Mycol. appl.* **35**: 325 (carry over & spread; **48**, 558).

Chiplonkar, A. 1969. *Caryologia* **22**: 197 (dikaryotisation; **49**, 694).

Kulkarni, N. B. et al. 1968. *Mycopathol. Mycol. appl.* **34**: 364 (histology & morphology, life cycle; **47**, 2801).

Nambiar, K, K. N. et al. 1977. *J. Pl. Crops* **5**(2): 124 (resistance; **57**, 4086).

Pavgi, M. S. et al. 1967. *Cytologia* **32**: 262 (haustoria; **48**, 889).

Reddy, G. S. et al. 1963. *Andhra agric. J.* **10**: 146 (resistance; **45**, 1123).

Upadhyay, R. et al. 1966. *Phytopathol. Z.* **56**: 151 (secondary infection; **46**, 401).

—— ——. 1967a. *Ibid* **59**: 136 (carry over; **46**, 3531).

—— ——. 1967b. *Ann. phytopathol. Soc. Japan* **33**: 176 (factors affecting incidence; **46**, 3161).

—— ——. 1967c. *Indian Phytopathol.* **20**: 29 (resistance; **47**, 870).

—— ——. 1973. *Mycopathol. Mycol. appl.* **50**: 109 (morphology, development & cytology; **53**, 74).

—— ——. 1974. *Riv. Patol. veg.* **10**: 153 (fungicides; **55**, 367).

THANATEPHORUS Donk, *Reinwardtia* 3: 376, 1956.

(Tulasnellaceae)

Typically parasitic on plant parts in or near soil but often saprobic in soil or on rotten wood, forming a rhizoctonia state and often forming sclerotia. FRUCTIFICATION resupinate, pruinose–pellicular, flaky to somewhat tufted or almost hypochnoid. HYPHAE wide (sometimes up to 17 μ), branching at a wide angle and often forming cruciform cells, monomitic, basal hyphae longer celled, often coloured and thick walled; ascending hyphae shorter celled, thin walled, barrel shaped, bearing basidia in discontinuous clusters of small asymmetrical cymes or less typically racemes; clamp connections absent. BASIDIA short, barrel shaped to subcylindrical or obovoid, not uniform or constricted about the middle, about the same diam. as the supporting hyphae; STERIGMATA (1–) 4 (–7), straight, stout reaching the same length as the metabasidia or longer, rarely becoming septate. BASIDIOSPORES capable of repetition, not amyloid, smooth, hyaline. No known conidial states (P. H. B. Talbot, *Persoonia* 3: 386, 1965).

Talbot (and in Parmeter, *T. cucumeris* q.v.) discussed the tortuous taxonomy of this genus and related genera (with descriptions and keys). *Pellicularia* Cooke was rejected (and see Donk 1954). Talbot considered that *Koleroga* Donk is a synonym of *Ceratobasidium* Rogers. Donk (1958) erected the genus *Koleroga* to accommodate the fungus causing black rot or koleroga of coffee in India. The disease is a web or thread blight of the leaves (Coleman et al.; Narasimhan; Mathew).

Anon. 1970. *Commonw. phytopathol. News* Pt 2: 4 (*Pellicularia* versus *Thanatephorus* for the perfect state of *Rhizoctonia solani*).

Coleman, L. C. et al. 1923. *Bull. Dep. Agric. Mysore* Mycol. Ser. 5, 12 pp. (koleroga of coffee; **3**: 397).

Donk, M. A. 1954. *Reinwardtia* **2**: 425 (notes on resupinate hymenomycetes I, on *Pellicularia* Cooke; **33**: 564).

——. 1958. *Fungus* **28**: 16 (as above V; **38**: 468).

Mathew, K. T. 1954. *Proc. Indian Acad. Sci.* Sect. B. **39**: **39**: 133 (koleroga of coffee; **34**: 721).

Narasimhan, M. J. 1933. *Phytopathology* **23**: 875 (as above; **13**: 230).

Talbot, P. H. B. 1965. *Persoonia* **3**: 371 (on '*Pellicularia*' and associated genera of Hymenomycetes; **45**: 388).

Thanatephorus cucumeris (Frank) Donk, *Reinwardtia* 3: 376, 1956.
Corticium solani (Prill. & Delacr.) Bourd. & Galz.
Pellicularia filamentosa (Pat.) Rogers
Sclerotial state *Rhizoctonia solani* Kühn, 1858.

COLONIES on PDA at first colourless, rapidly becoming brown. Aerial mycelium variable, giving a felted or mealy surface on which long, sparsely branched hyphae are frequently present. Some isolates show diurnal zonation. SCLEROTIA develop as crust radiating out from point of inoculation or as separate entities scattered over colony surface. Cells of hyphae at advancing edge of colony usually 5–12 μ wide and up to 250 μ long. Branches arise near distal end of cell, are constricted at point of origin and septate shortly above. Phase contrast microscopy shows the cells to be multinucleate (2–25, mostly 4–8) with conspicuous dolipore septa. Older mycelium shows great variation in hyphal dimensions and shorter cells owing to formation of secondary septa; the angle of branching approaches 90° and branches may arise at various points along the cell length. Some hyphae are differentiated into swollen monilioid cells often 30 μ or more in width, and similar cells derived from repeated branching of one or more hyphae form the homogeneous, prosoplectenchymatous sclerotia.

On host hyphae frequently follow lines of junction of epidermal cells before forming short branches on which the simple lobed appressoria or infection cushions and their associated infection pegs develop. Young hyphae on surfaces, appressoria, infection cushions and mycelium in host tissues are hyaline; the surface hyphae, which often colonise large areas beyond those penetrated, and sclerotia, gradually develop a brown pigment. HYMENIUM a thin sheet or collar commonly on stems or leaves just above soil surface or on soil particles, discontinuous, composed of barrel shaped to subcylindrical holobasidia, 10–25 μ long $\times$6–19 μ wide, arranged in imperfect cymes or racemes. STERIGMATA usually 4 (2–7) per basidium, stout, straight, 5.5–36.5 μ long, occasionally with adventitious septa. BASIDIOSPORES hyaline, oblong to broad ellipsoid and unilaterally flattened, prominently apiculate, smooth, thin walled, 6–14$\times$4–8 μ. Basidiospore germination by repetition occurs. Some spp. of *Ceratobasidium* Rogers have morphologically similar *Rhizoctonia* states; most of these are known to have binucleate mycelium (J. E. M. Mordue, *CMI Descr.* 406, 1974).

The work on this ubiquitous saprophyte and parasite is too extensive to be at all adequately treated here. A few general points are made and some of the more important crop diseases are described. But full discussions on all aspects of the biology of *T. cucumeris* were given in accounts based on a symposium (1965) held by the American Phytopathological Society (Parmeter). Tomkins provided a bibliography. Other recent, general accounts are by: Ogoshi; Saksena; Tu; Ui; see also Dodman; Flentje; Henis in *Root diseases and soil-borne pathogens* (Toussoun et al., editors) and Mordue (l.c.). The taxonomy (including synonymy) of the sclerotial and perfect states was discussed by Parmeter et al. and Talbot (in Parmeter), respectively.

T. cucumeris, an aggregated sp., is a soil inhabitant which has evolved different parasitic patterns, i.e. aerial, soil surface and subterranean. Isolates have also been grouped on the basis of: plants attacked, morphology, survival characteristics, temp. requirements, carbon dioxide tolerance, virulence, nutrition and anastomosis patterns (heterothallism). Disease symptoms caused are: seed decay, damping off, stem lesion and canker, root rot, above ground rot, leaf (web or thread) blight and storage rot. Talbot (l.c.) gave a translation of Frank's original description in 1883, a part of which reads: 'A fungus, easily observable with the naked eye . . ., which covers the lower parts of the plants. From soil level upwards a fibrous grey or brownish grey fungal membrane covers the upper parts of roots, stems and even the petioles of the lower leaves up to several centimetres from the soil surface, creeping upwards with a floccose or radiating outer margin. This membrane is the fungal mycelium, the older parts of which are covered with a hymenium in which more or less densely grouped basidia are to be found, each bearing a single, 1 celled, ovate, hyaline spore on each of the 4 sterigmata. The spores give the upper surface a powdery appearance. The mycelial mat loosely covers the above-named parts of the cucumber plant and can easily be removed, exposing fresh and healthy looking tissue although the fungus has invaded it at some place on the lower part. . . .'

The fungus penetrates youngish, intact cuticle or invades natural openings and wounds. Infection lessens with increasing host maturity. Direct infection from basidiospores probably occurs. Isolates vary from avirulent to highly virulent. Host specificity is found but is not as sharply limited as in

Fusarium oxysporum. The fungus grows through soil and survives as sclerotia, in crop residues and on weeds. The whole range of plant disease control measures have been used against *T. cucumeris*. Synergism with nematodes and other plant pathogens occurs. The diseases are described by plant family. Only ref. relating to aetiology and control are given, with some more recent ones; see Tomkins and the other general works for a full coverage of the literature.

Clark, C. A. et al. 1978. *Phytopathology* **68**: 1234 (elutriation procedures for quantitative soil assay).

Griesbach, E. 1975. *Zentbl. Bakt. ParasitKde* Abt. 2 **130**: 45, 64 (weeds & transmission; **55**, 1162, 1163).

Henis, Y. et al. 1978. *Phytopathology* **68**: 371 (soil sampler to study populations in soil).

Lewis, J. A. 1976. *Can. J. Microbiol.* **22**: 1300 (effect of volatiles from decomposing plant tissues; **56**, 1028).

Naiki, T. et al. 1975. *Soil Biol. Biochem.* **7**: 301 (fine structure of decayed sclerotia; **54**, 5302).

Ogoshi, A. 1975. *Rev. Pl. Prot. Res.* **8**: 93 (grouping & perfect states, 46 ref.; **55**, 5058).

Papavizas, G. C. et al. 1975. *Phytopathology* **65**: 871 (inoculum density in soil, morphology & pathogenicity of types; **55**, 600).

Parmeter, J. R. (editor). 1970. Rhizoctonia solani, *biology and pathology*. Univ. California Press (>1000 ref.; 52, 3153).

Puhalla, J. E. et al. 1976. *Phytopathology* **66**: 1348 (role of H locus in heterokaryosis; **56**, 2372).

Saksena, H. K. 1971. *Indian Phytopathol.* **24**: 1 (review of parasitism, 70 ref.; **51**, 3253).

Tomkins, C. M. 1975. *Contr. Reed Herb.* 24, 169 pp. (bibliography, & on *Pythium*; **55**, 5053).

Tu, C. C. 1968. *Pl. Prot. Bull. Taiwan* **10**(4): 39 (morphology, including perfect state; 67 ref.; **49**, 962).

—— et al. 1977. *Can. J. Bot.* **55**: 2419 (cytology & fine structure of *T. cucumeris* & the *Rhizoctonia* complex; **57**, 1634).

Ui, T. 1973. *Rev. Pl. Prot. Res.* **6**: 115 (diseases in Japan, 52 ref.; **55**, 80).

Weinhold, A. R. 1977. *Phytopathology* **67**: 566 (technique, soil population; **56**, 5469).

CRUCIFERAE

On cabbage, besides the common damping off, *T. cucumeris* causes bottom (and heart) rot. Basal leaves usually show the first symptoms of sunken, black lesions, with sparse web-like mycelium. There is a decay of the leaf base and chlorosis of the apical area; leaf fall may occur. The midribs of the outer leaves of the head are weakened by infection and signs of wilt appear. The stem and core are not attacked and the head remains upright. Part of the damping off syndrome (wirestem) results when more mature seedlings are attacked. A necrotic lesion envelops the lowest few cm of the stem; the seedlings appear tough, hardened and are smaller than usual. The general control methods against damping off apply and soil disinfestation by chemicals may be appropriate. If the disease persists short rotations with crucifers should be avoided. Good drainage, rich soil and weed control are essential. Williams et al. described the inheritance of resistance.

In pre-emergence damping off of radish Benson et al. found that the disease was most severe at 26° when germination was rapid. Greater inoculum densities were needed for significant disease as temps became less favourable. Survival of the fungus was longest in cool dry soil. Roots (scurfy root) are also attacked; the lesions are brown, superficial and infection causes stunting in roots and tops. Humayadan et al. found resistance in some breeding lines in USA but the popular red globe cvs were highly susceptible (and see Henis for control).

Benson, D. M. et al. 1974. *Phytopathology* **64**: 38, 957, 1163 (epidemiology in radish, effect of quintozene, inoculum potential & survival; **53**, 4631; **54**, 1484, 2526).

Henis, Y. et al. 1978. *Ibid* **68**: 900 (integrated control).

Humayadan, H. S. et al. 1976. *Pl. Dis. Reptr* **60**: 156 (resistance in radish with *Aphanomyces raphani*; **55**, 5389).

Weber, G. F. 1931. *Bull. Fla agric. Exp. Stn* 242, 31 pp. (general on cabbage; **11**: 417).

Wellman, F. L. 1932. *J. agric. Res.* **45**: 461 (as above; **12**: 133).

Williams, P. H. et al. 1966. *Phytopathology* **56**: 367 (inheritance of resistance in cabbage; **45**, 2279).

GRAMINEAE

The fungus causing sheath blight of rice is frequently called *Corticium sasakii* (Shirai) Matsumoto but it is now considered that the pathogen is a form of *T. cucumeris*. Kozaka gave a review of the disease. On the leaf sheath the spots are ellipsoid, at first greenish grey, 1–3 cm long or more; they become greyish white with a narrow dark brown margin. The readily detached sclerotia form in the spots which also show a hyaline mycelium. On the lamina the lesions tend to be irregular in shape; those on the stem are infrequent, occurring only when the disease is very severe. The spots are generally first seen near the waterline. Severe infection causes extensive killing of the leaves, stem break and lodging, the perfect state forms on the plants at about the booting stage.

Infection occurs from sclerotia which float to the surface of the water. Hashiba et al. (1975) described sclerotial morphology and buoyancy. Primary infection often occurs soon after transplanting. Before heading sheath blight tends to be limited to the lower leaves and sheaths. After heading it may spread as the upper parts of the plants become more susceptible. Once infection is established it spreads laterally and upwards. The disease is not usually seen until the temp. is 21–22° and the opt. temp. is 28–32°. Infection occurs more rapidly at 32° (*c*. 18 hours) than at 28° (24 hours). Factors which increase and maintain high RH and high nitrogen increase disease. Early and late cvs may be differentially infected because of temp. variation. In Japan late maturing cvs are less affected by sheath blight (disease escape) since temps are lower. Hashiba et al. (1974) found that isolates from high temp. regions grew well on PDA at 35° but those from low temp. ones did not. The rate of upward development of the disease at an av. 26–28° was greater with the isolates from the high temp. regions. The sclerotia overwinter in Japan in soil and early work showed that their survival time was several months in soil and in water (and see Prabhat et al.; Roy). Chin et al. reported that infested soil remained infective for 4 weeks in wet and 32 weeks in dry soil.

Cultural control measures advocated include destruction and deep ploughing in of infected residues, destruction of grass weeds, and avoidance of excessive nitrogenous fertilisers. Inherent high resistance appears not to have been found. Such resistance that has been reported is probably slight or should be explained in terms of disease escape. The antibiotics validamycin and kasugamycin (see Furuta) applied as foliage sprays are very effective. Organoarsenicals, benomyl, edifenphos and hymexazol (see Takahi et al.) have also been used. Where fungicides are uneconomic cultural measures and the use of less susceptible cvs must suffice.

Chin, K. M. et al. 1975. *MARDI Res. Bull.* 3(2): 19 (survival, spread & host range; **56**, 707).
Furuta, H. 1973. *Jap. Pestic. Inf.* 15: 28 (validamycin for control; **53**, 532).
Hashiba, T. et al. 1974. *Proc. Assoc. Pl. Prot. Hokuriku* 22: 8 (relation between mycelial growth & temps where isolates originated; **55**, 238).
——. 1975. *Phytopathology* **65**: 159 (development changes in sclerotia; **54**, 3915).
——. 1977. *Ann. phytopathol. Soc. Japan* **43**: 1 (quantitative N & starch changes in rice sheaths during vertical disease development; **56**, 5627).
Hashioka, Y. et al. 1971. *Res. Bull. Fac. Agric. Gifu Univ.* 31: 99 (fine structure of infection; **52**, 3316).
Kozaka, T. 1975. *Rev. Pl. Prot. Res.* **8**: 69 (review, 61 ref.; **55**, 5187).
Prabhat, C. A. M. et al. 1974. *Agric. Res. J. Kerala* **12**: 96 (sclerotial viability; **55**, 2714).
Roy, A. K. 1976. *Phytopathol. Z.* **86**: 270 (sclerotial survival; **56**, 1117).
Takahi, Y. et al. 1974. *Proc. Kansai Pl. Prot. Soc.* 16: 52 (hymexazol & control; **54**, 1738).
Varma, A. S. et al. 1977. *Madras agric. J.* **64**: 416 (fungicides; **57**, 2517).
Watanabe, T. et al. 1975. *Trans. mycol. Soc. Japan* **16**: 253 (on sugarcane, morphology; **55**, 5909).

LEGUMINOSAE

T. cucumeris infection of common bean has been extensively investigated; and important diseases of green gram (*Phaseolus aureus*) and lima bean (*P. lunatus*) also occur. Most forms of the disease (damping off, stem canker, soil rot and foliage web blight) have been described. The fungus causing web blight was called *Corticium microsclerotia* Weber (nomen nudum), a synonym of *T. cucumeris*. The blight begins as small, circular, watersoaked leaf spots; the mycelium then spreads over the upper parts of the plant. Lesions reach 3 cm diam. and the hyphae may bind the plant organs into a mat. Dodman et al. and Christou described penetration of the hypocotyl of common bean. Stem lesions are elongate, sunken, reddish brown and near the soil. Person, working with isolates from other hosts, described 4 groups based on pathogenicity to common bean. The pathogen was reported by Diaz Polanco et al. to be a main cause of lowering the plant population in bean fields in Venezuela. Their isolates were divided into 6 groups on the basis of morphology and pathogenicity. The work of Richards suggests that the disease is more likely to be found under cooler conditions (and see Henis et al. for inoculum potential). Other leguminous crops attacked include: lucerne (root canker, Smith); lentil (Shatla et al.); pea (tip blight and stem rot, Flentje et al.); soybean (Boosalis, aerial blight; Cardoso et al.; O'Neill et al.) and *Tephrosia vogelii* (Ruppel et al.).

On green gram in small scale tests, a single soil

drench with thiophanate methyl and 2 drenches of benomyl gave *c.* 90% control. More seedlings survived when this fungicide was used as a post-inoculation drench than when benomyl or chloroneb were so used (Kataria et al.). Control of root rot in common bean by quintozene (as a furrow spray on the seed) was described by Crossan et al. The black seeded cvs of this crop show resistance to seed infection and pre-emergence damping off. White seeded cvs tend to be susceptible. The seed coats of white cvs cracked readily before emergence whilst those of the black cvs did not do so. The resistance associated with coloured seed is inherited (Deakin et al.; Prasad et al.; see also Warren et al. for resistance in lima bean). Except where stated ref. are on common bean.

Bateman, D. F. et al. 1965 & 1967. *Phytopathology* **55**: 734; **57**: 127 (relation of calcium & pectic substances in hypocotyls of different ages to susceptibility, respiration in infected hypocotyls & lesion maturation; **45**, 292; **46**, 2135b).

Boosalis, M. G. 1950. *Ibid* **40**: 820 (on soybean seed & seedlings; **30**: 258).

Cardoso, J. E. et al. 1978. *Fitopat. Brasileira* **3**: 193, 205 (on soybean, inoculation & host reaction).

Christou, T. 1962. *Phytopathology* **52**: 381 (penetration & histology; **41**: 751).

Crossan, D. F. et al. 1963. *Pl. Dis. Reptr* **47**: 109 (fungicide; **42**: 505).

Deakin, J. R. et al. 1975. *HortScience* **10**: 269 (breeding for resistance; **54**, 5634).

Diaz Polanco, C. et al. 1973. *Agron. trop.* **23**: 47 (pathogenic & morphological groups; **53**, 4190).

Dodman, R. L. et al. 1968. *Phytopathology* **58**: 31, 1271 (penetration of common bean & radish; **47**, 2591; **48**, 332).

Flentje, N. T. et al. 1964. *Ibid* **54**: 788 (on pea; **44**, 292).

Henis, Y. et al. 1970. *Ibid* **60**: 1351 (effect of propagule size on saprophytic growth, infectivity & virulence; **50**, 1504).

Homma, Y. et al. 1975. *Mem. Fac. Agric. Hokkaido Univ.* **9**: 177 (cultural characters & growth of single basidiospore isolates & synthesised heterokaryons; **54**, 3565).

Kataria, H. R. et al. 1978. *Ann. appl. Biol.* **88**: 257 (fungicides on green gram; **57**, 4219).

Lisker, N. et al. 1976. *Ann. Bot.* **40**: 625 (fine structure of infection; **55**, 5419).

Maxwell, D. P. et al. 1967. *Phytopathology* **57**: 132 (changes in oxidases in infected hypocotyls in relation to lesion maturation; **46**, 2135c).

O'Neill, N. R. et al. 1977. *Pl. Dis. Reptr* **61**: 713 (on soybean; **57**, 1912).

Person, L. H. 1944. *Phytopathology* **34**: 1056 (pathogenicity & pathogenic groups; **24**: 261).

Prasad, K. et al. 1969. *Pl. Dis. Reptr* **53**: 350 (seed coat & resistance; **48**, 2662).

—— ——. 1976. *Phytopathology* **66**: 342 (as above; **55**, 6010).

Richards, B. L. 1923. *J. agric. Res.* **25**: 431 (soil temp. & infection of pea & common bean; **3**: 439).

Ruppel, E. G. et al. 1965. *Phytopathology* **55**: 612 (on *Tephrosia vogelii*; **44**, 3122).

Shatla, M. N. et al. 1974. *Arch. Phytopathol. PflSchutz* **10**: 333 (on lentil; **54**, 4706).

Smith, O. F. 1943, 1945 & 1946. *Phytopathology* **33**: 1081; **35**: 832; **36**: 638 (on lucerne, pathogenicity & effect of soil temp.; **23**: 109; **25**: 118; **26**: 15).

Van Etten, H. et al. 1967. *Ibid* **57**: 121 (lesion maturation, fungal development & cell degrading enzymes in hypocotyls; **46**, 2135a).

Warren, H. L. et al. 1972. *Pl. Dis. Reptr* **56**: 268 (resistance in lima bean; **51**, 4508).

——. 1973. *Phytopathology* **63**: 1024 (colonisation of host tissues & infectivity; **53**, 1631).

Weber, G. F. 1939. *Ibid* **29**: 559 (general, web blight; **19**: 3).

MALVACEAE

T. cucumeris causes the seedling stem canker disease of cotton called soreshin. Under favourable conditions for infection seedlings are blighted either before or after emergence. Under less favourable conditions lesions on the hypocotyl appear later; reddish brown, linear, cortical cankers form near the soil line, the stem may be girdled. Neal described a cotton leaf spot. The spots are at first light brown, irregular in outline, varying in size, and bordered by a purplish ring and a chlorotic outer zone. Shothole symptoms appeared later; for penetration of the hypocotyl see Khadga et al.; Nakayama; Weinhold et al. (1973). Basidiospore infection of the boll and bracts occurs and boll decay can result (Luke et al.; Pinckard et al.). In the Yemen Arab Republic Daniels described a diseased condition (Abyan root rot) of cotton in which the fungus was strongly implicated, but the syndrome was unusual. Symptoms on seedlings resembled those seen elsewhere but they persisted well beyond the seedling stage. The outer foot and root tissues rotted (lesions a few cm above soil level); there was a discolouration of the stele and a wilt. It was considered that penetration by the pathogen took place in early growth and was followed by continued growth in the host. The fungus survived for 4 years in air dried soil.

In USA Walker reported an opt. soil temp. of 17–23° for the death of cotton seedlings. Seedlings

12 days old are more resistant than 5–6-day-old ones (Hunter 1974; Veech). Infection can lead to a marked stunting of the plants in natural soil; there is less shoot and root growth in infected seedlings at 19° but not at 28°. Whilst field plants with lesions may grow as well as healthy ones they can yield less seed cotton (Brown et al.). An increase in soreshin can be caused when there is damage by *Meloidogyne incognita*. Opt. soil temps for disease caused by this pathogen complex are 18° and 21°. Increasing the coarse particle content of the soil increased this synergistic interaction (Carter; Reynolds et al.). Control can be given by seed dressings, by applying the fungicide in the furrow and mixing with the covering soil. Fungicides used include: benodanil, captan, chloroneb, etridiazole, fenaminosulf and quintozene. Crandall described a zonate leaf spot (zonate spot with a red halo) infection of kenaf (*Hibiscus cannabinus*) and roselle (*H. sabdariffa*); and see Adeniji for control on these crops with seed dressing fungicides, particularly phenylmercury acetate. Except where stated ref. are on cotton.

Adeniji, M. O. 1972. *Niger. J. Sci.* **5**: 155 (kenaf & roselle, seed fungicides; **54**, 3338).

Bell, D. K. et al. 1963. *Pl. Dis. Reptr* **47**: 1016 (effects of soil temp. & fungicide placement on damping off; **43**, 1653).

Brown, E. A. et al. 1976. *Phytopathology* **66**: 111 (effect of seedling disease on subsequent host growth & yield; **55**, 4734).

Carter, W. W. 1975. *J. Nematol.* **7**: 229, 234 (effects of soil temps, inoculum levels & soil texture on interaction with *Meloidogyne incognita*; **55**, 1812, 1813).

Cole, D. L. et al. 1977. *Rhod. Jnl agric. Res.* **15**: 45 (seed dressing fungicides; **56**, 5067).

Crandall, B. S. 1948. *Phytopathology* **38**: 503 (on kenaf & roselle; **28**: 65).

Daniels, J. 1965. *Emp. Cotton Grow. Rev.* **42**: 104 (aetiology in Yemen Arab Republic; **44**, 2147).

Hayman, D. S. 1970. In *Root diseases and soil-borne pathogens*: 99, Univ, California Press (effect of seed exudate on infection).

Hunter, R. E. 1974. *Physiol. Pl. Pathol.* **4**: 151 (inactivation of pectic enzymes by polyphenols in seedlings of different ages; **53**, 3982).

——. 1978. *Phytopathology* **68**: 1032 (effects of catechin on growth & polygalacturonase).

—— et al. 1978. *Ibid* **68**: 347 (terpenoid accumulation in ageing & infected hypocotyls).

Khadga, B. B. et al. 1963. *Phytopathology* **53**: 1331 (penetration & infection; **43**, 1028).

Luke, W. J. et al. 1974. *Ibid* **64**: 107 (basidiospore infection of bolls; **53**, 3499).

Nakayama, T. 1940. *Ann. phytopathol. Soc. Japan* **10**: 93 (penetration & infection; **20**, 461).

Neal, D. C. 1944. *Phytopathology* **34**: 599 (leaf symptoms; **23**: 485).

Neubauer, R. et al. 1973. *Ibid* **63**: 651 (effect of herbicide triflurin; **53**, 567).

Paulus, A. O. et al. 1976. *Calif. Agric.* **30**(10): 15 (seed dressing fungicides; **56**, 2077).

Pinckard, J. A. et al. 1967. *Pl. Dis. Reptr* **51**: 67 (basidiospore infection of bolls; **46**, 1592).

Reynolds, H. W. et al. 1957. *Phytopathology* **47**: 256 (infection with *M. incognita*; **36**: 697).

Shatla, M. M. et al. 1963. *Ibid* **53**: 1407 (resistance to quintozene; **43**, 1652).

Veech, J. A. 1976. *Ibid* **66**: 1072 (localisation of peroxidase in infected seedlings; **56**, 2076).

Walker, M. N. 1928. *Bull. Fla agric. Exp. Stn* 197: 345 (effects of soil temp. & moisture; **9**: 178).

Weinhold, A. R. et al. 1969. *Phytopathology* **59**: 1601 (virulence as affected by nutrition of pathogen; **49**, 1038).

—— ——. 1973. *Ibid* **63**: 157 (penetration & infection; **52**, 3706).

SOLANACEAE

Infection of tomato by *T. cucumeris* causes stem and branch canker, foot rot, damping off and fruit rot (Conover; Ramsey et al.; Schroeder; Verhoeff). Lesions on branches may be partial or girdling, sunken and show alternate bands of dark and light brown. The margin is sharply delimited. Branches may wilt or break. Seedlings are damped off or stunted with sunken lesions becoming dark brown at soil level; the disease occurs at all temps suitable for growth. Under wet conditions on fruit close to or touching the soil small watersoaked, then brown circular spots form. They enlarge to large irregular areas covering >half the fruit. On green fruit the lesions tend to have zonate markings but with ripening these disappear. The disease may be evident in transit. Gonzalez et al. reported an opt. temp. of 24° for soil rot of tomato fruit. Disease complexes involving *T. cucumeris* and *Meloidogyne incognita* on roots of this crop were described by Golden et al. and Hazarika et al. (and see Barksdale for resistance). Basidiospore infection of tobacco leaves, causing shotholes, was described by Vargas (and see Batten et al.; Jochems). Seed transmission occurs in red pepper, eggplant, tomato and many other crops (e.g., Baker).

Baker, K. F. 1947. *Phytopathology* **37**: 912 (seed transmission; **27**: 346).
Barksdale, T. H. 1974. *Pl. Dis. Reptr* **58**: 406 (resistance in tomato fruit; **53**, 4119).
Batten, C. K. et al. 1971. *J. Nematol.* **3**: 164 (on tobacco with *Meloidogyne incognita*; **51**, 1901).
Cole, D. L. et al. 1978. *Ann. appl. Biol.* **90**: 187 (field control on tobacco with benomyl & benodanil).
Conover, R. A. 1949. *Phytopathology* **39**: 950 (tomato canker; **29**: 388).
Golden, J. K. et al. 1975. *Ibid* **65**: 265 (on tomato & *Hibiscus esculentus* with *M. incognita*; **54**, 3810).
Gonzalez, L. C. et al. 1963. *Ibid* **53**: 82 (tomato soil rot; **42**: 574).
Hazarika, B. P. et al. 1974. *Indian J. Nematol.* **4**: 246 (on eggplant with *M. incognita*; **55**, 6048).
Jochems, S. C. J. 1926. *Bull. Deli Proefstn Medan* 21, 13 pp. (on tobacco; **6**, 4).
Ramsey, G. B. et al. 1929. *Phytopathology* **19**: 383 (tomato soil rot; **8**: 681).
Schroeder, W. T. et al. 1961. *Pl. Dis. Reptr* **45**: 160 (tomato fruit rot; **40**: 564).
Vargas, G. E. 1973. *Turrialba* **23**: 347 (basidiospore infection of tobacco; **53**, 1942).
Verhoeff, K. 1963. *Neth. J. Pl. Pathol.* **69**: 265 (tomato foot rot; **43**, 1147).

OTHER CROPS

Banded leaf blight of arrowroot (*Maranta arundinacea*) was described by Ramakrishnan et al. Infection results in transverse necrotic patches, light green to yellowish brown separated by healthy green or light green. The mycelium spreads over the whole leaf, with sclerotia on the lower surface. Bordeaux was said to be effective for control. A seed and root rot of avocado (*Persea americana*) occurred in USA. Invasion of the cotyledons and embryo prevents germination and girdling lesions are caused on the roots (Mircetich et al.). Infection of roots of betel (*Piper betle*) and leaves of black pepper (*P. nigrum*) were reported by Chowdhury and Vanderweyen, respectively. Carrot seedlings can be attacked soon after emergence and the stand is reduced. Some control is given by quintozene, spraying the soil surface and incorporating before sowing or spraying the soil immediately before sowing and incorporating by action of the planter (Shlevin et al.). Infection of celery causes crater spot. Brown, ovoid to elongated (>2.5 cm), sharply sunken lesions develop on either surface of the petiole. The first signs of the disease show on the outer petioles touching the soil (Houston et al.). Madarang described damping off in *Cinchona*.

Corticium areolatum Stahel (invalid) causes areolate leaf spot of citrus. It is uncertain whether the fungus is a sp. of *Ceratobasidium* or *Thanatephorus* (Talbot in Parmeter, *T. cucumeris* q.v.). In wet weather the initial leaf spots enlarge, the water-soaked rings which form are daily increments in the growth of the spot. The dark brown, narrow rings are ridges of gummed cells; they are more prominent on the lower surface. The areas between the ridges are light brown (10–20 rings are usual). The lower surface of the spots is covered with a white mycelium in wet weather. Stahel stated that control of common citrus scab (*Elsinoë fawcettii*) with Bordeaux will also control areolate leaf spot. Carpenter described target leaf spot of rubber (*Hevea brasiliensis*) in the Americas. In Peru it became serious in nurseries and reduced the vigour of young plants through reduction of the leaf area (including defoliation). Infection of rubber by basidiospores (discharged mostly at night) was demonstrated by Carpenter. Copper and other fungicides are effective. Tu et al. reported on basidiospore infection of jute (*Corchorus*), causing leaf spots on mature plants. The spots are circular to irregular, 1–2 mm diam., reddish brown with light brown centres, enlarging under humid conditions to form large necrotic areas; infected leaves fall prematurely. The hymenium develops on the stalk at soil level. Damping off also occurs. Bottom rot of lettuce (called black rot by Verhoeff et al.) was described by Townsend. The plant shows a basal decay and the necrosis spreads (usually from near soil level) from leaf to leaf until the entire head is destroyed. Cvs with an upright habit may escape infection. If the disease is checked before the head is killed the lesions dry out and become sunken, chocolate brown spots. Townsend in USA found most disease at mean daily temps of $>19.5°$ with a daily min. of $>10°$. Control is achieved by growing plants on raised beds and applying benomyl or chlorothalonil to the basal leaves. A collar rot of tea was reported from S. India on plants with well developed stems in the nursery and there was heavy callus above and below the injured area; foliage became chlorotic and withered. The disease was brought on by planting too deeply and injudicious mulching (Venkataramani et al.). Mundkur described infection on sweet potato, and Strider the control of *T. cucumeris* root rot of *Euphorbia pulcherrima* (poinsettia) with benomyl.

Carpenter, J. B. 1949. *Phytopathology* **39**: 980 (on rubber, basidiospore discharge; **29**: 382).
——. 1951. *Tech. Bull. U.S. Dep. Agric.* 1028, 34 pp. (on rubber; **31**: 294).
Chowdhury, S. 1944. *Proc. Indian Acad. Sci.* Sect. B. **20**: 229 (on betel; **24**: 248).
Duncan, E. J. 1974. *Trans. Br. mycol. Soc.* **63**: 115 (on vegetative hyphae of *Corticium areolatum*; **54**, 1278).
Houston, B. R. et al. 1949. *Phytopathology* **39**: 470 (on celery; **28**: 559).
Madarang, S. A. 1941. *Philipp. J. For.* **4**: 105 (on *Cinchona* seedlings; **21**: 302).
Mircetich, S. M. et al. 1964. *Phytopathology* **54**: 211 (on avocado; **43**, 2363).
Mundkur, B. B. 1936. *Indian J. agric. Sci.* **6**: 994 (on sweet potato; **16**: 122).
Pieczarka, D. J. et al. 1974. *Pl. Dis. Reptr* **58**: 837 (control on lettuce; **54**, 630).
Ramakrishnan, K. et al. 1948. *Indian Phytopathol.* **1**: 129 (on arrowroot; **29**: 194).
Shlevin, E. et al. 1975. *Pl. Dis. Reptr* **59**: 29 (control on carrot seedlings; **54**, 3032).
Stahel, G. 1940. *Phytopathology* **30**: 119 (on citrus; **19**: 340).
Strider, D. L. 1977. *Pl. Dis. Reptr* **61**: 746 (control on poinsettia; **57**, 1751).
Townsend, G. R. 1934. *Mem. Cornell Univ. agric. Exp. Stn* 158, 46 pp. (on lettuce; **14**: 73).
Tu, C. C. et al. 1977. *Pl. Dis. Reptr* **61**: 80 (on jute; **56**, 5071).
Vanderweyen, A. 1957. *Bull. agric. Congo Belge* **48**: 365 (on black pepper; **36**: 727).
Venkataramani, K. S. et al. 1959. *Phytopathology* **49**: 527 (on tea; **39**: 193).
Verhoeff, K. et al. 1963. *Neth. J. Pl. Pathol.* **69**: 294 (on lettuce; **43**, 865).

THIELAVIOPSIS Went, *Meded. Proefstn SuikRiet W. Java* 'Kagok', 5: 4, 1893.

Stilbochalara Ferdinandsen & Winge, 1910
Hughesiella Batista & Vital, 1956.
(Dematiaceae)

COLONIES effuse, grey, olivaceous, dark blackish brown or black, velvety or pulverulent. MYCELIUM partly superficial, partly immersed. STROMA none. SETAE and HYPHOPODIA absent. CONIDIOPHORES macronematous, mononematous, unbranched or irregularly branched, straight or flexuous, hyaline or pale brown, smooth. CONIDIOGENOUS CELLS fragmenting to form arthroconidia, terminal and intercalary, determinate, cylindrical. CONIDIA catenate, dry, schizogenous, often seceding with difficulty, simple, doliiform, ellipsoidal, obovoid or oblong, mid to very dark brown, smooth in most spp., aseptate. In *T. basicola* the short chains of conidia remain intact for a long time and resemble multiseptate conidia. There is also a phialidic *Chalara* state and often a *Ceratocystis* state (M. B. Ellis, *Demat. Hyphom.*: 30, 1971).

The sp. described (*T. basicola*) differs from the *Thielaviopsis* state of *Ceratocystis paradoxa* (q.v.) in that the conidia (arthroconidia) remain together for a long time (in chains), and they are short oblong, wider than long, compared with those of *C. paradoxa* where the conidia are doliiform, obovoid or ellipsoidal, longer than wide. The confusion in identity with *Thielavia* Zopf. (Eurotiaceae) is referred to under *Thielaviopsis basicola*, and see Härri for an account of the biology of the 2 genera. Booth et al. gave a key to *Thielavia* spp.

Booth, C. 1961. *Mycol. Pap.* 83, 15 pp. (*Thielavia* & notes on allied genera).
—— et al. 1966. *Trans. Br. mycol. Soc.* **49**: 665 (*Thielavia pilosa* sp. nov. & key to *Thielavia* spp.).
Härri, E. von 1959. *Phytopathol. Z.* **36**: 27 (*Thielavia* & *Thielaviopsis*; **39**: 156).

Thielaviopsis basicola (Berk. & Br.) Ferraris, *Flora Italica, Hyphales*: 233, 1912.

COLONIES grey or olivaceous, often velvety. CONIDIOPHORES up to 50×6–$9\ \mu$. ARTHROCONIDIA usually in chains of 4–8 remaining together for a long time and resembling large multiseptate conidia, finally seceding but with difficulty, oblong or short cylindrical, only the apical one in each chain rounded at the end, others truncate at both ends, mid to dark golden brown, smooth, 7–12 μ long, 10–17 μ wide. PHIALIDES up to 100 μ long, 5–8 μ thick in the broadest part, tapering to 3–4 μ. PHIALOCONIDIA cylindrical, truncate at the ends, hyaline, 7–17×2.5–$4.5\ \mu$ (M. B. Ellis, *Demat. Hyphom.*: 31, 1971).

Subramanian wrote: 'Berkeley and Broome (1850) found the fungus on peas and on *Nemophila auriculata* in England. They observed the chlamdospores and named the fungus *Torula basicola*. Zopf (*Sitzb. bot. Prov. Brandenburg* 18: 101–105, 1876) observed a fungus on *Senecio elegans* producing chlamydospores similar to those of *Torula basicola*; he also saw endoconidia and cleistothecia and, assuming a connection, named the fungus *Thielavia basicola* (Berk. & Br.) Zopf. It is now known that *Thielavia basicola* grows only in association with *Thielaviopsis basicola* which is a distinct fungus (McCormick, *Bull. Conn. agric. Exp. Stn* 269, 1925). *Thielavia basicola* occurs in and on tobacco roots only if they are infected with *Thielaviopsis basicola*.' The fungus has a wide host range attacking plants in over 15 families, primarily belonging to the Leguminosae (groundnut, soybean, *Lespedeza*, clover, lucerne, cowpea, lupin, sunn hemp and bean), Solanaceae and Cucurbitaceae.

The disease caused (black root rot) is widespread in both temperate and tropical regions (Map 218). The pathogen is a soil and root inhabitant, primitive and unspecialised sensu Garrett (*Biology of root infecting fungi*) who discussed its active saprophytic behaviour. The characteristics of the arthroconidia, usually referred to as chlamydospores, have been discussed by Tsao et al. (1970a). They found that these spore chains are produced in acropetal succession whereas those of the type sp., the *Thielaviopsis* state of *Ceratocystis paradoxa* (q.v.), are basipetal in development. They, therefore, considered that *T. basicola* should be transferred to a new genus. The phialoconidia are commonly called endoconidia (see Brierley). Spore fine structure has been described by Christias et al. (1970); Delvecchio et al.; Tsao et al. (1970b). Johnson, Berkner and Yarwood et al. gave some information on hosts, and isolation techniques were described by Lloyd et al.; Papavizas (1964); Tsao (1964); Yarwood (1946). Stover (1950, 1956, see under tobacco) described 2 distinct natural forms (brown and grey wild types) which could be distinghished in vitro. The grey form is slower growing, less pathogenic, more thiamine deficient and survives for shorter periods in a dormant state; see also Huang et al. (1971); Johnson et al. (1935); Rawlings.

The fungus invades the root cortex of young plants or of plants growing under sub-optimal conditions. Necrotic areas form in the cortex, and these can be so extensive that the whole root system becomes reduced and blackened. Apart from the host the arthroconidia are the main survival units; phialoconidia can also survive in soil for limited periods (Schippers; Tsao et al. 1966). Both conidial types germinate well at 20–33°. Soil populations of propagules increase markedly in the rhizosphere of host plants but non-host plants have little or no effect on these populations. When disease is well advanced some increase in the number of propagules also occurs in non-rhizosphere soil (Bateman). Arthroconidia germination is stimulated by host proximity, some organic amendments and soil moisture; drier soil and low temps increase their survival rates (Adams et al.; Linderman et al.; Papavizas et al. 1971; Patrick). Their survival in moist, warm soil was poor (Clough et al. 1976a). Control measures are mainly the use of sufficiently resistant or tolerant cvs and soil fumigation. Acid soils may reduce disease. The most studied of the diseases caused by *T. basicola* is that on tobacco; some account is also given of the black root rots of cotton, poinsettia and, collectively, of some other crops.

Adams, P. B. et al. 1969. *Phytopathology* **59**: 135 (soil fungistasis; **48**, 2179).

Bateman, D. F. 1963. *Ibid* **53**: 1174 (effect of hosts & non-hosts on soil populations; **43**, 935).

Berkner, F. 1937. *Pflanzenbau* **13**: 321 (on leguminous crops, *Lupinus*; **16**: 539).

Brierley, W. B. 1915. *Ann. Bot.* **29**: 483 (endoconidia).

Christias, C. et al. 1967. *Phytopathology* **57**: 1363 (chitinase & separation of arthroconidia; **47**, 1426).

—— ——. 1970. *Can. J. Bot.* **48**: 2305 (fine structure of arthroconidia; **50**, 3525).

Clough, K. S. et al. 1976a & b. *Soil Biol. Biochem.* **8**: 456, 473 (biotic factors & viability of arthroconidia; **56**, 3895, 3896).

Delvecchio, V. G. et al. 1969. *J. gen. Microbiol.* **58**: 23 (fine structure of hyphae & conidia; **49**, 967).

Hawes, C. R. et al. 1977. *Trans. Br. mycol. Soc.* **68**: 304 (conidium ontogeny; **56**, 4395).

Hawthorne, B. T. et al. 1970. *Phytopathology* **60**: 891 (effect of separation & culture age on germination of arthroconidia; **49**, 3647).

Huang, H. C. et al. 1971. *Can. J. Bot.* **49**: 1041 (variability in culture; **51**, 144).

—— ——. 1972. *Ibid* **50**: 2423 (nuclear distribution & behaviour; **52**, 2172).

—— ——. 1974. *Ibid* **52**: 2263 (karyology of conidiogenesis & conidial germination; **54**, 3721).

Johnson, E. M. et al. 1935. *Phytopathology* **25**: 1011 (as above; **15**: 263).

Johnson, J. 1916. *J. agric. Res.* **7**: 289 (hosts).

Linderman, R. G. et al. 1967. *Phytopathology* **57**: 729 (behaviour of conidia in soil; **46**: 3350).

—— ——. 1968. *Ibid* **58**: 1578 (pathogenesis in non-sterile soil; **48**, 1731).

Lloyd, A. B. et al. 1962. *Ibid* **52**: 1314 (isolation; **42**: 645).

Maier, C. R. 1967. *Bull. New Mex. agric. Exp. Stn* 517, 21 pp. (as a soilborne pathogen).

Mathre, D. E. et al. 1966. *Phytopathology* **56**: 337 (conidial germination; **45**, 2401).

Papavizas, G. C. 1964. *Ibid* **54**: 1475 (isolation; **44**, 1428).
—— et al. 1969. *Ibid* **59**: 371 (conidial survival in soil; **48**, 2180).
—— ——. 1971. *Ibid* **61**: 108 (as above; **50**: 2746).
—— ——. 1972. *Ibid* **62**: 417 (conidial survival & effects of volatile soil fungicides; **51**, 3797).
—— ——. 1972. *Ibid* **62**: 688 (stimulation of conidial germination by fatty acids from rhizosphere soil; **52**, 350).
Patrick, Z. A. 1965. *Ibid* **55**: 466 (germination of arthroconidia; **44**, 2608).
Rawlings, R. E. 1940. *Ann. Mo. bot. Gdn* **27**: 561 (culture & differences in pathogenicity; **21**: 431).
Sattler, F. 1936. *Phytopathol. Z.* **9**: 1 (general; **15**: 536).
Schippers, B. 1970. *Neth. J. Pl. Pathol.* **76**: 206 (survival of phialoconidia in soil; **49**, 3127).
Smith, J. G. 1960. *Acta bot. neerl.* **9**: 59 (effect of antagonistic fungi; **39**: 462).
Subramanian, C. V. 1968. *C.M.I. Descr. pathog. Fungi Bact.* 170 (general).
Tsao, P. H. 1964. *Phytopathology* **54**: 548 (isolation; **43**, 2829).
—— et al. 1964. *Ibid* **54**: 633 (methods for estimating soil populations; **43**, 3153).
—— ——. 1966. *Ibid* **56**: 1012 (arthroconidia as soil survival units; **46**, 260).
—— ——. 1970a. *Mycologia* **62**: 960 (acropetal development of arthroconidia; **50**, 2773).
Tsao, P. W. et al. 1970b. *Phytopathology* **60**: 613 (fine structure of arthroconidia; **49**, 3128).
Yarwood, C. E. 1946. *Mycologia* **38**: 346 (isolation; **25**: 486).
——. 1974. *Pl. Dis. Reptr* **58**: 54 (habitat distribution in USA, California; **53**, 2448).
—— et al. 1976. *Ibid* **60**: 347 (association with crops in USSR, Moscow region; **56**, 1064).

EUPHORBIA PULCHERRIMA

A general root rot of poinsettia in glasshouses in USA results from infection by several fungi. At the earlier stages of growth *Pythium ultimum* and *Thanatephorus cucumeris* (q.v.) cause most damage. A second phase of the rot is marked by *Thielaviopsis basicola* attacks which are more severe in the presence of the other 2 pathogens. Recently *Chalaropsis thielavioides* Peyronel was implicated as a further causal component (Perry); but it is a relatively weak pathogen, entering via wounds. Root destruction leads to plant stunting, wilt, leaf fall and stem lesions which are dark brown and may girdle. Necrotic areas on the larger leaf veins and petioles caused by *T. basicola* infection have also been described. Root development in poinsettia is best at *c.* 26°. Root rot caused by *T. basicola* is most extensive at 17° but can be severe at 13–26°. At 30° there is only a trace of the disease which, in general, is favoured by lower temps. Soil moisture also affects disease incidence. *T. basicola* causes appreciable damage at 36% (moisture holding capacity) and this increases up to 70%. *T. cucumeris* is most severe at <40% and severity decreases with increasing soil moisture. At >70% *P. ultimum* becomes a problem. The root rot is reduced in soils that are on the acid side (pH 5.5 or less); pH affects *T. basicola* and *P. ultimum* particularly.

Plant overcrowding should be avoided and suitable temps and soil moistures used; it may be necessary to sterilise soil (Grasso et al.). Benomyl, fenaminosulf and thiabendazole in the planting mix give control; although plants grown in treated mixes were not so large as those in untreated and uninfested mixes (Raabe et al. 1971). Manning et al. reported the effectiveness of benomyl soil drenches on newly potted rooted cuttings and established plants with root rot. Some variations in the degree of cv. susceptibility have been described.

Bateman, D. F. et al. 1959. *Phytopathology* **49**: 641 (effects of temp.; **39**: 316).
——. 1961. *Ibid* **51**: 445 (effects of soil moisture; **41**: 91).
——. 1962. *Ibid* **52**: 559 (effects of soil pH; **42**: 27).
Grasso, S. et al. 1969. *Tec. agric. Catania* **31**: 3 (control; **49**, 3335).
Keller, J. R. et al. 1955. *Phytopathology* **45**: 552 (general; **35**: 299).
Manning, W. J. et al. 1970. *Pl. Dis. Reptr* **54**: 328 (control with benomyl; **49**, 3336).
Perry, R. S. 1970. *Ibid* **54**: 451 (*Chalariopsis thielavioides*; **49**, 3337).
——. 1971. *Illinois nat. Hist. Surv. Bull.* **30**: 419 (comparison with *C. thielavioides*; **51**, 2570).
Raabe, R. D. et al. 1970. *Calif. Agric.* **24**(1): 9 (control with fungicides; **49**, 2886).
—— ——. 1971. *Pl. Dis. Reptr* **55**: 238 (control with benomyl & thiabendazole; **50**, 3832).
Shanks, J. B. et al. 1958. *Proc. Am. Soc. hort. Sci.* **71**: 522 (cultural factors & incidence; **38**: 7).
Tompkins, C. M. et al. 1950. *Hilgardia* **20**: 171 (*Pythium* spp. & *Thanatephorus cucumeris*; **30**: 611).

GOSSYPIUM

Infection of cotton (black root rot) seedlings can cause stunting, the leaves become small, pale green with marginal browning. Mature plants can be affected and show signs of a wilt. There is swelling in the collar region a few cm below soil level, blackening of the tap root and the typical general root decay. The disease is most serious in the very early growth stages. Mathre et al. (1966) found that root rot in seedlings could occur with inoculum densities as low as 100 propagules/g in soil. In older plants stunting showed only when inoculum was at least 10^3 propagules/g and at higher levels (10^4) some reduction in boll set took place. In California (USA) *T. basicola* did not seem to be important in the summer growing season. Blank et al. reported more disease at av. soil temps of 21° compared with those of 25–27°.

In USA *T. basicola* has frequently been described as only one factor in a seedling root rot complex (Maier et al. 1963; Staffeldt). Schnathorst described a characteristic sudden wilt syndrome caused by a combined infection with *Verticillium albo-atrum* (*V. dahliae*, cotton q.v.). Field studies by Mathre et al. (1967a) confirmed glasshouse ones, in that low to medium levels of *T. basicola* in the soil did not increase the severity of wilt caused by *V. dahliae* in wilt tolerant strs of *G. hirsutum*. *Alternaria* spp. have also been implicated in cotton seedling root rot (Maier 1965). In control build up of the pathogen in the soil should be avoided. Non-host plants can be grown and seed treatment (phenthiuram and maneb have been recommended) is effective. Mathre et al. (1967b) found resistance in strs of *G. arboreum* and *G. herbaceum*, in work with other fungal pathogens of cotton as well; and see Linderman et al.

Blank, L. M. et al. 1953. *Pl. Dis. Reptr* **37**: 473 (effect of soil temp.; **33**: 153).

King, C. J. et al. 1942. *Phytopathology* **32**: 752 (general; **22**: 21).

Linderman, R. G. et al. 1968. *Ibid* **58**: 1431 (breakdown of resistance by hydrocinnamic acid; **48**, 813).

——— ——. 1968. *Ibid* **58**: 1571 (predisposition by phytotoxins from decomposing residues; **48**, 814).

Maier, C. R. et al. 1960. *Pl. Dis. Reptr* **44**: 956 (cultural & pathogenic variability with *Thanatephorus cucumeris*; **40**: 363).

——— ——. 1963. *Bull. New Mex. agric. Exp. Stn* 474, 41 pp. (seedling disease complex).

——. 1965. *Pl. Dis. Reptr* **49**: 904 (*Alternaria* spp. in seedling disease complex; **45**, 1081).

Mathre, D. E. et al. 1966. *Phytopathology* **56**: 1213 (inoculum potential & effect on yield; **46**, 1009).

——— ——. 1967a. *Ibid* **57**: 604 (effect of infection by *Thielaviopsis basicola* on that by *Verticillium dahliae*; **46**, 2740).

——— ——. 1967b. *Pl. Dis. Reptr* **51**: 864 (sources of resistance to *T. basicola* inter alia; **47**, 527).

Schnathorst, W. C. 1964. *Ibid* **48**: 90 (infection with *V. dahliae*; **43**, 1933).

Staffeldt, E. E. 1959. *Ibid* **43**: 506 (*T. basicola* as part of seedling disease complex; **38**: 693).

Zdrozhevskaya, S. D. 1968. *Byull. vses. nauch.-issled. Inst. Zashch. Rast.* 2(14): 39 (seed treatment; **50**, 692).

NICOTIANA TABACUM

Tobacco black root rot is the most studied of the diseases caused by *T. basicola* (see Lucas, *Diseases of tobacco*, for an account). Although it is generally now much less of a problem, many acres of susceptible cvs are still grown in which black root rot can cause serious damage. In 1971 Gayed referred to it as the major disease of flue-cured tobacco in Canada. Infection may first appear in the seedbed and cause damping off. Larger seedlings which escape early infection become stunted and chlorotic, especially if soil fertility is low. In young plants this root pathogen can colonise the stem and first leaves. Seedlings at the early stage of stunting show the characteristic cortical rot of black root tips. Later the roots become almost completely decayed, appearing short and stubby, and any new roots formed are usually invaded in turn. The plant growth and appearance in infested seedbeds become uneven. Diseased plants put out in the field make slow growth and show varying degrees of stunting. In warm and otherwise favourable conditions for the host the plants may make some recovery. An infested field will show an uneven development in plant height, some chlorosis and premature flowering. The effects of the root rot on the leaf are both quantitative and qualitative. The characteristic conidia are abundant in the necrotic roots.

Stover (1959b, c, 1956) described the growth of the fungus over the root surface and the frequent penetration of cells and host colonisation by characteristically constricted hyphae, and conidial formation. He described the brown and grey forms of the pathogen (already referred to) and the cultural variants (and see Conant). The disease is favoured by relatively low soil temps (17–23°, max. in vitro growth at 22–28°) and by a soil pH that rises above *c.* 6; there is little or no disease at <pH 5.6 (at any temp.). Above 26° black root rot becomes less severe (Doran; Jewett; Johnson et al.; Lucas; Yamaguchi). Gayed (1971) described the disease as being usually more severe on fine textured soils and in poorly drained conditions. In transplanting experiments it

was found that, whilst there was a marked transplant shock in seedlings (unpotted) compared with those in peat pots, host resistance was not affected by the relative host vigour during the active phase of fungal infection. Black root rot can cause severe damage over a wide range of soil moistures. Gayed (1972) described the host range and persistence in tobacco soil. Patrick et al. considered that the phytotoxic effects of decomposing plant residues on tobacco may partly explain an apparent breakdown of resistance in the field; the effect of the toxic compounds being largely on the host. Infection by *T. basicola* may be associated with that of other organisms in tobacco. Stover (1950c) reported that the generally more pathogenic brown forms consisted of 2 races with a different geographical distribution in Canada. Race 2 and all grey forms were less pathogenic than race 1. He found that all cultural 'mutant' types were less pathogenic than the wild type cultures.

Control depends on resistance, seedbed sterilisation and cultural measures. Immunity occurs in *Nicotiana rustica, N. alata, N. paniculata, N. sanderae* and *N. debneyi*, that of the last (dominant and oligogenic) has been most used as a resistance source. Polygenic resistance also occurs in *N. tabacum*. Gayed (1969a) described a leaf disk inoculation technique for screening material. Besides those in USA (Clayton) resistant cvs have been described from Canada (White; Povilaitis et al.), Europe (Ivancheva-Gabrovska; Klarner; Klarner et al.; Oberthür), Japan (Nakamura; Ohashi et al.) and New Zealand (Wright et al.). Continuous cultivation of tobacco can lead to breakdown of some resistance through inoculum build up and long rotations with grass crops (legumes being generally susceptible) may be effective. Seedbed soil should be steam sterilised or chemically treated. Benomyl, allyl bromide, dazomet, methyl bromide, thiophanate methyl, metham sodium and methyl isothiocyanate have been used (Anon; Gayed 1967, 1969b, 1976; Hartill et al.; Lautz; Papavizas et al.; Saenger; Stover et al.). Gayed (1969b, 1970) demonstrated the systemic protection given by benomyl in leaf disks and that the fungicide checked the disease in the field. Soil pH should be kept below 5.6.

Anderson, P. J. et al. 1926. *Bull. Mass. agric. Exp. Stn* 229: 117 (effect of soil pH).

Anon. 1968. *Can. Tob. Grow.* **16**: 23 (control, steam, methyl isothiocyanate & dazomet; **50**, 2493).

Baltruschat, H. et al. 1975. *Phytopathol. Z.* **84**: 172 (effect of mycorrhiza on infection; **55**, 2358).

Clayton, E. E. 1969. *Tob. N.Y.* **168**: 30 (patterns & inheritance of resistance; **49**, 1483).

Conant, G. H. 1927. *Am. J. Bot.* **14**: 457 (host morphological response to invasion; **7**: 278).

Doran, W. L. 1929. *J. agric. Res.* **39**: 853 (effects of soil temp. & pH; **9**: 348).

——. 1931. *Bull. Mass. agric. Exp. Stn* 276: 118 (use of soil pH in control; **10**: 762).

Gayed, S. K. 1967. *Lighter* **37**: 10 (control with soil fumigants; **47**, 1285).

——. 1969a. *Phytopathology* **59**: 1596 (relation between leaf & root necrosis, bearing on resistance; **49**, 1139).

——. 1969b. *Can. Pl. Dis. Surv.* **49**: 70 (control with benomyl including field trials; **49**, 2994).

——. 1970. *Neth. J. Pl. Pathol.* **76**: 125 (systemic action of benomyl; **49**, 2993).

——. 1971. *Can. Pl. Dis. Surv.* **51**: 142 (transplanting effects on plant vigour & susceptibility; **51**, 4353).

——. 1972. *Can. J. Pl. Sci.* **52**: 869 (hosts & persistence in tobacco soil; **53**, 271).

—— et al. 1975. *Phytopathology* **65**: 1049 (levels of chlorogenic acid & infection).

——. 1976. *Phytoprotection* **57**: 109 (control, steam compared with fumigants; **56**, 3711).

Hartill, W. F. T. et al. 1972. *Pl. Dis. Reptr* **56**: 708 (control in seedbeds, benomyl & thiophanate methyl; **52**, 1260).

Ivancheva-Gabrovska, T. 1960. *Rastit. Zasht.* **8**: 15 (resistance of cvs & *Nicotiana* spp.; **40**: 248).

Jewett, F. L. 1938. *Bot. Gaz.* **100**: 276 (effects of soil temp. & pH; **18**: 350).

Johnson, J. et al. 1919. *J. agric. Res.* **17**: 41 (effects of soil temp. & other soil factors).

——. 1930. *Tech. Bull. U.S. Dep. Agric.* 175, 20 pp. (breeding for resistance).

Klarner, S. et al. 1963. *Roczn. Nauk roln.* Ser. A **88**: 143 (resistance of cvs; **43**, 1140).

——. 1968. *Ibid* **94**: 277 (as above; **48**, 595).

Koch, L. W. 1935. *Can. J. Res.* Sect. C. **13**: 174 (*Thielaviopsis basicola* inter alia; **15**: 178).

Lautz, W. 1957. *Pl. Dis. Reptr* **41**: 174 (control, vapam, chlorobromopropene & allyl bromide; **36**: 668).

Lucas, G. B. 1955. *Mycologia* **47**: 793 (effects of temp. & pH; **35**: 642).

Nakamura, A. 1967. *Bull. Hatano Tob. Exp. Stn* **59**: 1 (breeding for disease resistance; **47**, 3563).

Oberthür, K. 1955. *Züchter* **25**: 17 (resistance of cvs & *Nicotiana* spp.; **34**: 678).

Ohashi, Y. et al. 1959. *Bull. Hatano Tob. Exp. Stn* **44**: 65, 75, 83 (resistance of cvs & *Nicotiana* spp., variation in pathogenicity; **40**: 561).

Otani, Y. 1962. *Bull. Okayama Tob. Exp. Stn* 23, 118 pp. (general).

Papavizas, G. C. et al. 1971 & 1972. *Pl. Dis. Reptr* **55**: 352; **56**: 15 (control with fungicides & effect of organic amendments; **51**, 668, 3254).

Patrick, Z. A. et al. 1963. *Can. J. Bot.* **41**: 747 (adverse effect of phytotoxic substances from plant residues on resistance; **42**: 704).
Povilaitis, B. et al. 1964. *Can. J. Pl. Sci.* **44**: 126 (assessment procedure for resistance).
Saenger, H. L. 1970. *Pl. Dis. Reptr* **54**: 136 (control, benomyl inter alia; **49**, 2176).
Steinberg, R. A. 1951. *Bull. Torrey bot. Club* **78**: 227 (amino acid toxicity to cvs differing in resistance; **30**: 435).
Stover, R. H. 1950a. *Pl. Dis. Reptr* **34**: 387 (control; **31**: 259).
——. 1950b. *Can. J. Res.* Sect. C **28**: 445 (general; **30**: 200).
——. 1950c. *Ibid* **28**: 726 (brown & grey variants; **30**: 435).
—— et al. 1952. *Sci. Agric.* **32**: 411 (control, methyl bromide; **32**: 344).
——. 1956. *Can. J. Bot.* **34**: 459, 875 (effect of in vitro nutrition on variants, albinism; **36**: 205, 492).
Unbehaun, L. M. et al. 1970. *Phytopathology* **60**: 304 (pectic enzymes; **49**, 2644).
White, F. H. 1969. *Lighter* **39**: 10 (resistance in cvs; **48**, 1967).
Wright, D. S. C. et al. 1967. *Tob. N.Y.* **164**: 30 (as above; **47**, 892).
Yamaguchi, Y. 1957. *Ann. phytopathol. Soc. Japan* **22**: 204 (effects of temp., pH and N in culture; **38**: 36).

OTHER CROPS

T. basicola has been described as a citrus pathogen. Sweet orange grown in infested soil eventually shows foliar chlorosis, delay of top growth, reduction in weight and a necrotic root system. The roots are sparsely branched and covered with many dark lesions (specks, streaks or irregular patches), usually up to 1.5 mm long; there are also root cracks (4–5 mm long). Many citrus vars and hybrids are attacked but there are some differences in reaction between them. Infection has been associated with attacks by *Radopholus similis*. The growth of sweet orange seedlings in soil infested with *T. basicola* was reduced by 89–95% at 15° and 20°, 83–85% at 25° and 34–36% at 30° (Feder et al.; Tsao et al.).

Black root rot has also been studied in common bean. Christou described direct penetration of the root epidermis, without appressoria formation, and spread of the fungus in this crop. Root hairs are invaded. Using 6 isolates, varying from avirulent to highly virulent, Maier found that there was most disease at 15.6° and 18.9° compared with 23.3° and 26.7°. At the lower temps root rot in seedlings developed more rapidly 2–3 weeks after emergence, and at 3–4 weeks at the higher ones. In the field infection caused stunting and delayed plant maturity. Papavizas and co-workers reported that organic amendments and fungicides reduced root rot in the glasshouse. Some treatments reduced both inoculum density and viability of arthroconidia. Genetic control of resistance is partially recessive; *c.* 3 genes are involved and those controlling resistance to *T. basicola* and *Fusarium solani* f. sp. *phaseoli* (q.v.) appear not to be linked (Hassan et al.).

Lloyd investigated the effect of temp. on infection of pea (*Pisum sativum*), a low temp. organism (opt. 13–17°), comparing it with infection of other hosts with higher opt. In the latter black root rot tended to be most severe at 17–25°, i.e. below the opt. temps for host growth (26–31°). On sesame Adams called the disease red root since it resulted in reddish brown lesions on the roots and occasionally on the stems; lesions sometimes showed cracking. High temps (30° and 35°) controlled the disease in sesame. With a soil temp. regime of 10 hours at 25° and 14 at 30° there was less red root than at a constant 25°.

Other crops on which black root rot has been described include castor (*Ricinus communis*), ginseng (*Panax schinseng*), groundnut and soybean. Hsi found that black hull of groundnut was more severe in fields under monoculture than in those under a rotation with sorghum. The pathogen has been implicated in the specific replant disorder of cherry and plum in UK; the comparable disorder on apple was considered to be distinct (Pepin et al.; Sewell et al.).

Adams, P. B. 1971. *Phytopathology* **61**: 93 (effects of soil amendments & temp. in sesame; **50**, 3090).
Christou, T. 1962. *Ibid* **52**: 194 (penetration & spread in common bean; **41**: 751).
Feder, W. A. et al. 1963. *Pl. Dis. Reptr* **47**: 666 (on citrus with *Radopholus similis*; **42**: 761).
Hassan, A. A. et al. 1971. *J. Am. Soc. hort. Sci.* **96**: 628, 631 (genetics & inheritance of resistance in common bean; **51**, 2052, 2053).
Hsi, D. C. H. 1978. *Phytopathology* **68**: 1442 (effects of crop sequence, groundnut black hull & sampling time on soil populations).
Lee, M. et al. 1977. *Ibid* **67**: 1360 (effect of herbicide on disease in soybean; **57**, 3707).
Lloyd, A. B. et al. 1961. *Pl. Dis. Reptr* **45**: 422 (on pea; **41**: 73).
——. 1963. *Phytopathology* **53**: 329 (on pea, effects of soil temp., str. & cvs; **42**: 642).
Lockwood, J. L. et al. 1970. *Pl. Dis. Reptr* **54**: 849 (on soybean; **50**, 1519).

Lumsden, R. D. et al. 1968. *Phytopathology* **58**: 219 (phosphatide degrading enzymes in infection of common bean; **47**, 2005).
Maier, C. R. 1961. *Pl. Dis. Reptr* **45**: 804 (soil temps & infection of common bean; **41**: 346).
Papavizas, G. C. et al. 1961. *Phytopathology* **51**: 92 (isolation from common bean rhizosphere; **40**: 642).
——. 1968 *Ibid* **58**: 421 (effects of organic amendments, common bean; **47**, 2615b).
—— et al. 1970. *Pl. Dis. Reptr* **54**: 114 (as above & fungicides; **49**, 1972).
Pepin, H. S. et al. 1975. *Ann. appl. Biol.* **79**: 171 (effects of soil populations on cherry stocks growth; **54**, 2881).
Rosenbaum, J. 1912. *Phytopathology* **2**: 191 (on ginseng).
——. 1963. *Ibid* **53**: 738 (effect of soil moisture in citrus; **43**, 91).
——. 1963. *Pl. Dis. Reptr* **47**: 437 (relative susceptibility in types of citrus; **42**: 681).
Sewell, G. W. F. et al. 1975. *Ann. appl. Biol.* **79**: 149 (role in replant disorders in cherry & plum; **54**, 2880).
Sneh, B. et al. 1976. *Can. J. Bot.* **54**: 1499 (germination lysis, possible control in soybean; **56**, 928).
Thomas, C. A. et al. 1965. *Pl. Dis. Reptr* **49**: 256 (on castor & sesame; **44**, 1921).
Tsao, P. H. et al. 1960. *Phytopathology* **50**: 657 (effect of soil temps on infection in citrus; **40**: 223).
—— ——. 1962. *Ibid* **52**: 781 (on citrus; **42**: 263).

TILLETIA Tul., *Annls Sci. nat.* (Bot.) **3**(7): 112, 1847.
(Tilletiaceae)

SORI mostly in the ovaries, occasionally in vegetative tissues, forming powdery to agglutinated spore masses, often foetid. SPORES single, formed from intercalary cells of a sporogenous mycelium, or terminally on sporogenous hyphae, commonly encased in a hyaline to tinted, gelatinoid sheath, comparatively large and regular, exhibiting a wide range in size, variously sculptured (commonly reticulate, cerebriform, verrucose, spiny, tuberculate, or rarely smooth), pallid to opaque, germinating by a continuous promycelium, bearing terminal sporidia that usually fuse in situ, giving rise to secondary sporidia; immature spores few to copious, similar in morphology to the mature spores but pigmented to a lesser degree; 'sterile cells' single, present in varying numbers, hyaline to tinted, smooth to granular, naked or sheathed, variously shaped (from R. Duran & G. W. Fischer, *The genus* Tilletia: 17, 1961).

The genus was monographed by Duran et al. who gave analytical and host genus keys to the spp. which occur almost exclusively on Gramineae. There are several important pathogens on temperate cereals. The grass genera on which *Tilletia* spp. have been recorded include: *Aegilops* (1 spp.), *Brachiaria* (1), *Bromus* (2), *Coix* (1), *Digitaria* (3), *Eragrostis* (3), *Oryza* (1), *Panicum* (9), *Paspalum* (3), *Setaria* (5), *Themeda* (2) and *Vetiveria* (1). *T. ayresii* Berk. occurs on *Panicum maximum* (guinea grass) and *Setaria sphacelata* (golden timothy). Besides rice, *T. barclayana* (q.v.), hosts include *Pennisetum typhoides* (bulrush millet).

Duran, R. et al. 1961. *The genus* Tilletia, 138 pp., Washington State Univ. (monograph; **41**: 19).

Tilletia barclayana (Bref.) Sacc. & Syd. in Saccardo, *Sylloge Fung.* **14**: 422, 1899.
Neovossia barclayana Brefeld, 1895
Tilletia horrida Takahashi, 1896, fide Tullis & Johnson (1952)
N. horrida (Tak.) Padwick & Azmatullah Khan, 1944
T. pulcherrima Ellis & Galloway, 1904, fide Duran & Fisher (1961, *Tilletia* q.v.)
T. ajrekari Mundkar, 1939, fide Tullis & Johnson (1952).
(additional synonyms and discussions of synonymy in Tullis et al. and Duran et al. l.c.)

SORI in the ovaries of some, or more rarely all, spikelets, concealed by the glumes but sometimes bursting through at maturity. Spore mass granular or agglutinated, black. STERILE CELLS intermixed with the spores, few to many, mostly more or less globose but rather variable in shape, hyaline to yellowish tinted, smooth, wall 2–4 μ thick, 10–30 μ diam. SPORES globose to subglobose, at first light brown becoming dark brown at maturity, enclosed in a tinted or hyaline sheath with or without a short apiculus; exospore, when young ornamented with curved spines (1.5–4 μ in length) which become truncated scales at maturity, mostly *c.* 17–25 μ diam. but occasionally larger (up to 35 μ). SPORIDIA filiform or needle shaped, curved in various ways, 10–12 in number and 38–53 μ in length (fide Takahashi; G. C. Ainsworth, *CMI Descr.* 75, 1965).

Kernel smut is a minor disease of rice. It is widespread in the cultivated areas of Asia and USA; and in Central and S. America, Australia, Cuba, Fiji,

Sierra Leone, Togo and Trinidad; there is one record for Europe (Greece; Map 75). First symptoms are formed towards maturity as minute black pustules or streaks bursting through the glumes. Sometimes the whole grain becomes a smutted mass but frequently only a part is affected and this results in a characteristic twisting over and outward of the unaffected part. The spores require a 4–5 months dormant period and light for germination. They may form up to 60 sporidia and germination is high at *c*. 30°. The primary sporidia may form smaller, secondary sporidia (10–14 μ long) or, alternatively, a mycelial mat is formed first on the water surface and this develops the secondary sporidia which are forcibly discharged and infect the flowers. Air-dispersed sporidia also originate from spores in the soil. There is no evidence for systemic infection. The sporidia germinate and penetrate the ovary via the style. In India disease incidence is favoured by light, sandy loam soils, frequent light showers, high RH (85% or more) and temps of 25–30° at anthesis. Spores can remain viable for >2 years. A 3 year rotation was recommended in India (Singh 1975) but no extensive work on control seems to have been reported. High nitrogen may increase kernel smut and its incidence varies on different cvs.

Anderson, A. P. 1902. *Bull. Torrey bot. Club* **29**: 35 (morphology & taxonomy).
Chauha, L. S. et al. 1964. *Sci. Cult.* **30**: 201 (incidence in different cvs; **43**, 3213).
Chowdhury, S. 1951. *Indian Phytopathol.* **4**: 25 (general in Assam; **32**: 33).
Kameswar Row, K. V. S. R. 1962. *Sci. Cult.* **28**: 534 (incidence in different cvs; **42**: 462).
Khanna, A. et al. 1968. *Mycologia* **60**: 655 (spore morphology of *T. barclayana* inter alia; **47**, 3738).
Lin, C. K. 1936. *Bull. Coll. Agric. For. Univ. Nanking* **45**: 1 (spore germination).
——. 1955. *Acta phytopathol. sin.* **1**: 183 (light & spore germination; **37**: 56).
Padwick, G. W. et al. 1944. *Mycol. Pap.* **10**: 2 (morphology & taxonomy; **26**: 265).
Reyes, G. M. 1933. *Philipp. J. Agric.* **4**: 241 (general; **13**: 469).
Singh, R. A. et al. 1970. *Indian Phytopathol.* **23**: 51 (incidence in different cvs; **49**, 3270).
—— ——. 1972. *Riso* **21**: 259 (cytology of spore germination; **53**, 164).
—— ——. 1973. *Ibid* **22**: 243 (development of sorus; **53**, 4435).
—— ——. 1973. *Nova Hedwigia* **24**: 487 (culture; **54**, 4493).
—— ——. 1975. *Riso* **24**: 77 (epidemiology; **55**, 3176).
Templeton, G. E. et al. 1960. *Arkans. Fm Res.* **9**: 10 (general; **40**: 465).
—— ——. 1961. *Rice J.* **64**: 24 (infection & spore development; cf. *Phytopathology* **51**: 130; **40**: 532, 606).
——. 1963 & 1964. *Arkans. Fm Res.* **12**: 12 (effect of N; *Rice J.* **67**: 39; **43**, 2612).
Teng, S. C. 1931. *Contr. biol. Lab. Sci. Soc. China*, Bot. Ser. **6**: 111 (spore development; **11**: 324).
Tullis, E. C. et al. 1952. *Mycologia* **44**: 773 (synonymy; **32**: 512).

TOLYPOSPORIUM Woron., *Abh. senckenb. naturforsch. Ges.* **12**: 557, 1882.
(Saccardo (**7**: 501): Woron. 1882 in *Schroet. Pilzfl. Schles*: 276).
(Ustilaginaceae)

SORI in the inflorescence, usually the ovaries, forming a granular to agglutinated spore mass. SPORE BALLS of numerous spores bound together by ridged folds or thickenings of their outer walls: SPORES small to medium in size, often with a reticulate aspect due to the characteristic folds or thickenings which bind them together in the balls. Germination by 3–4-celled promycelium with sporidia borne at the septa (from G. W. Fischer, *Manual of North American smut fungi*: 196, 1953).

In a taxonomic revision Thirumalachar et al. erected *Tolyposporidium* Thirum. & Neergaard (gen. nov.) because the type of *Tolyposporium* is a *Sorosporium*. Transfers from the former genus to the new one and to other genera were made. They transferred *T. cymbopogonis* Mundkur and *T. senegalense* Speg. to *Sorosporium*; and *T. bullatum* (Schroet.) Schroet. was considered to belong to the new genus *Tolypoderma*. *Tolyposporium senegalense* was described on *Pennisetum typhoides* (bulrush millet), and *T. bullatum* from *Echinochloa crus-galli* (barnyard millet). Hirschhorn considered that the 2 fungi were the same. Fischer (l.c.) gave the synonymy of *T. bullatum* as *S. bullatum* and *T. senegalense*. Corbetta considered *T. bullatum* further. Two spp. have been described from *Cymbopogon*: *T. cymbopogonis* on *C. citratus* (lemon grass) and *T. christensenii* Raghunath on *C. flexuosus* (Malabar grass).

Corbetta, G. 1954. *Phytopathol. Z.* **22**: 275 (*Tolyposporium bullatum*; **34**: 230).
Hirschhorn, E. 1941. *Revta argent. Agron.* **8**: 384 (*T. bullatum* & *T. senegalense*; **24**: 62).

Mundkur, B. B. 1944. *Indian J. agric. Sci.* **14**: 49 (rare & new smuts; **24**: 224).
Raghunath, T. 1968. *Mycopathol. Mycol. appl.* **34**: 119 (*T. christensenii* sp. nov.; **47**, 2513).
——. 1970. *Phytopathol. Z.* **68**: 35 (mode of infection by *T. christensenii*; **50**, 139).
——. 1970. *J. Univ. Poona* Sect. Sci. Technol. **38**: 109 (teliospore germination & culture of *T. christensenii*; **51**, 554).
Thirumalachar, M. J. et al. 1977. *Friesia* **11**: 177 (on *Tolyposporium*, *Tolysporidium* gen. nov.; **57**, 4835).
Zambettakis, C. et al. 1973. *Bull. Soc. mycol. Fr.* **89**: 83 (numerical systematics; **54**, 2091).

Tolyposporium ehrenbergii (Kühn) Patouillard, *Bull. Soc. mycol. Fr.* **19**: 254, 1903.
Sorosporium ehrenbergii Kühn, 1877
Tolyposporium filiferum Busse, 1904, fide Mason (1926)
S. filiferum (Busse) Zundel, 1930.
(Thirumalachar et al., *Tolyposporium* q.v., retained *S. ehrenbergii* for this fungus. *T. ehrenbergii* Kühn var. *grandiglobum* Uppal & Patel (as *grandiglobosum*) was transferred to *Sorosporium*.)

SORI in the ovaries, irregularly distributed, more or less cylindrical with tapered ends, up to 4 (mostly 1–3) cm long, 0.5–1 cm wide, surrounded by a firm, false membrane of fungal tissue which ruptures from the apex. Spore mass granular, black, intermixed with shreds of host tissue. SPORE BALLS many spored, permanent, rather irregular in size and shape, globose to elongated and irregular, dark brown and opaque, 45 μ or less, up to >200 μ in max. diam. SPORES at surface of ball dark brown with free surface papillate, inner spores paler in colour and smooth, globose to subglobose or angular, 10–15 (12) μ diam. Sporidial cultures may be obtained on PDA or carrot agar (Kamat). On *Sorghum bicolor* and other *Sorghum* spp. (G. C. Ainsworth, *CMI Descr.* 76, 1965).

Long smut of sorghum is a relatively minor disease that is fairly widespread in Africa, present in parts of S. and central Asia but not reported from America and Australasia (Map 377). External symptoms are restricted to the fruiting head where usually not >10% of the flowers are affected. Most of the evidence of inoculations shows that infection takes place from airborne sporidia which invade the young flowers before the ear emerges from the boot leaf. Infection may also occur at anthesis. Reports of infection of germinating seedlings do not seem to have been confirmed (Hafiz; Hassan et al.; Manzo; Prasad; Ragab et al.; Ramakrishnan et al.; Vasudeva et al.). The opt. temp. for spore germination is 28–30° and for linear growth *c.* 25°. When inoculation is done into the boot enclosing the young inflorescence, symptoms occur in 2–3 weeks. No races have been found but a form on the wild grass *Sorghum purpureo-sericeum* was designated *T. ehrenbergii* var. *grandiglobum* Uppal & Patel; it is distinguished by the larger spore balls, 45–240 (107) μ and greater number of spores in each ball, i.e. >50. Until the life history is better known control measures will presumably remain uncertain, both seed treatment and rotation could reduce infection and some host tolerance may occur.

Hafiz, A. 1958. *Pakist. J. scient. Res.* **10**: 83 (morphology, incidence & infection; **38**: 82).
Hassan, S. F. et al. 1970. *J. agric. Res. Punjab* **8**: 411 (floral infection; **51**, 4818).
Jilani, M. B. et al. 1967. *W. Pakist. Jnl agric. Res.* **5**: 66 (effect of temp., pH, C & N sources in vitro; **47**: 2996).
Kamat, M. N. 1933. *Phytopathology* **23**: 985 (development in vitro & temp.; **13**: 300).
Kul'pinova, M. P. 1957. *Pl. Prot. Moscow* **4**: 38 (host reaction, culture & temp.; **34**: 475).
Manzo, S. K. 1976. *Pl. Dis. Reptr* **60**: 948 (inoculations of inflorescence & seed; **56**, 3006).
Mason, E. W. 1926. *Trans. Br. mycol. Soc.* **11**: 284 (taxonomy & synonymy; **6**: 190).
Prasad, N. 1945. *Curr. Sci.* **14**: 239 (infection; **25**: 32).
Qazi, A. Q. et al. 1957. *Pakist. J. scient. Res.* **9**: 117 (host tolerance; **37**: 584).
Ragab, M. A. et al. 1966. *Mycologia* **58**: 184 (infection & host susceptibility; **45**: 2822).
Ramakrishnan, T. S. et al. 1949. *Curr. Sci.* **18**: 418 (floral infection; **29**: 152).
Tamimi, S. A. 1970–1. *Mesopotamia J. Agric.* **5/6**: 47 (reaction of cvs; **51**, 4817).
Uppal, B. N. et al. 1943. *Indian J. agric. Sci.* **13**: 520 (*T. ehrenbergii* var. *grandiglobum*; **23**: 480).
Vasudeva, R. S. et al. 1950. *Curr. Sci.* **19**: 123 (floral infection; **29**: 463).
Zaprometov, N. G. 1964. *Trudȳ tashkent. sel'-khoz Inst.* 16: 283 (general & seed treatment; **44**: 2128).

Tolyposporium penicillariae Brefeld, *Unters-Gesammt. Mykol.* **12**: 154, 1895.
(the comb. nov. *Tolyposporidium penicillariae* (Bref.) Thirum. & Neergard was made, *Tolyposporium* q.v.)

SORI in the ovaries, irregularly distributed, pear shaped, 3–5 mm long, limited by a dark brown membrane of host tissue which disintegrates at maturity. Spore mass granular, black. SPORE BALLS many spored, permanent, globose to angled or elongate, opaque, irregular in size 50–150 μ (mostly 90–120 μ) diam. SPORES globose to subglobose, or ovoid, light brown, smooth, 6–12 μ (mostly *c.* 8–10 μ) diam. On *Pennisetum typhoides*. (G. C. Ainsworth, *CMI Descr.* 77, 1965).

This is a little known smut of bulrush millet which is widespread in Africa and India; it occurs also in Australia (N. Territory). Burma, Iraq, Pakistan, USA (Georgia) and Yemen (Map 150). Bhowmik et al. (1976) and Chauhan described losses. Primary infection is considered to occur in the inflorescence (before emergence) from germinating spores in the soil where survival has been reported to be *c.* 2 years. Symptoms are seen 15 days after inoculation. No dormant mycelium in the seed has been found. Intensity of the disease increases with successive crops. Evidence for control by seed treatment apparently conflicts but fungicidal control with head and foliage sprays may be worthwhile.

Ajrekar, S. L. et al. 1933. *Curr. Sci.* **1**: 215 (general; **12**: 367).

Ashraf, M. et al. 1961. *Biologia Lahore* **6**: 223 (heterothallism; **41**: 18).

Bhowmik, T. P. et al. 1971. *Pl. Dis. Reptr* **55**: 87 (control with systemic fungicides; **50**, 2265).

—— ——. 1976. *Indian J. agric. Sci.* **46**: 528 (effect of N & loss; **56**, 4501).

Bhatt, R. S. 1946. *J. Indian bot. Soc.* **25**: 163 (mode of infection; **26**: 200).

Chauhan, S. K. 1955. *Agra Univ. J. Res. Sci.* **4**: 367 (loss; **35**: 673).

Patel, M. H. et al. 1959. *Curr. Sci.* **28**: 28 (spore longevity; **38**: 576).

Wells, H. D. et al. 1963. *Pl. Dis. Reptr* **47**: 16 (in USA, comparison with *Tolyposporium bullatum*; **42**: 382).

——. 1967. *Ibid* **51**: 468 (fungicides; **46**, 3087).

Vasudeva, R. S. et al. 1950. *Curr. Sci.* **19**: 123 (inoculation of ears; **29**: 464).

Yadav, R. P. 1974. *Agra Univ. J. Res. Sci.* **23**: 37 (inheritance of resistance; **55**, 3162).

TRACHYSPHAERA Tabor & Bunting, *Ann. Bot.* **37**: 156, 1923.
(Pythiaceae)

CONIDIOPHORES variable in form; the simplest are upright hyphae terminating in a single conidium, but more usually the conidiophore bears a terminal vesicle to which a whorl of pedicellate conidia is attached. The more complex types may show a series of such enlargements, each with a whorl of conidia or with lateral fertile branches replacing some of the latter. CONIDIA spherical, strongly echinulate. ANTHERIDIA amphigynous. OOGONIA thick walled, characterised by irregular sac-like outgrowths which vary considerably in size and shape, from short rounded bosses to finger-like processes. OOSPORE thin walled with little or no epispore (from R. J. Tabor & R. H. Bunting l.c.; the genus is monotypic).

Trachysphaera fructigena Tabor & Bunting, *Ann. Bot.* **37**: 153, 1923.

HYPHAE aseptate, coarse, intercellular with branches of smaller diam. penetrating the cells. Subiculum forms beneath the epidermis which is ruptured on sporulation. CONIDIOPHORES erect, usually with a terminal vesicle; simple with one terminal conidium or complex with a whorl (or several whorls) of conidia on branches arising from the main hyphae. CONIDIA spherical, 13–48 (mean 35) μ, echinulate, thin walled, hyaline, pedicel 10–30 μ. Thick walled conidia (?chlamydospores) inside the host. ANTHERIDIA amphigynous, usually terminal. OOGONIA somewhat pyriform 24×40 μ, thick walled with sac-like outgrowths which vary from short rounded bosses to finger-like processes, curved and slightly forked. OOSPORE wall thin with little or no epispore. In mature oogonia the antheridia become detached with walls intact, and antheridial membrane surrounds the oogonial stalk (P. Holliday, *CMI Descr.* 229, 1970).

The fungus is a saprophyte which causes a few minor diseases; it has a limited distribution (W. and central Africa and Malagasy Republic, Map 249). But it may be found in banana ripening rooms where this fruit has been imported from regions where *T. fructigena* occurs naturally. The diseases are on cacao (mealy pod), coffee (berry rot) and banana (finger rot). Plant tissue needs to be wounded (or possibly moribund) before infection takes place.

On cacao pods a brownish rot spreads rapidly over the whole fruit. The dense, conidial masses then appear on the surface, at first white and becoming pinkish brown, covering the whole lesion. On bananas in the field infection can be restricted to the blossom end. On developing fruit infection shows the burnt appearance of the cigar-end syndrome (also caused by *Verticillium theobromae*). A wet rot can also occur. In the banana ripening room a generalised rot is found. The skin is thrown into folds, conidial masses appear, the pulp dries out and the finger becomes mummified. The banana rot can also occur in transport. Noviello et al. infected several fruit types (through wounds) and flowers sprayed with conidial suspensions.

The opt. temp. for growth in vitro is *c*. 24°. In banana ripening rooms infection can spread to healthy fruit from fruit that was infected before shipment. Lesions develop more rapidly at 21–26.5° (ripening temps) than at 14.5° (shipping temp.; Buxton et al.). Dakwa inoculated cacao pods in various ways. Maramba et al. considered that survival rates for conidia were generally low. On banana in Cameroon Tezenas du Montcel et al. reported the disease to be mainly in the highest areas; it appeared when max. temps were <27°. In Cameroon field control in banana through copper sprays, deflowering and bagging the inflorescences before opening of the bracts has been described (Beugnon et al.; Brun; Brun et al.; and see Mallamaire for coffee).

Beugnon, M. et al. 1970. *Fruits* **25**: 187 (banana, control; **50**, 2390).

Brun, J. 1954. *Ibid* **9**: (banana, climate & control; **34**, 657).

—— et al. 1955. *Ibid* **10**: 163 (control; **35**: 906).

Buxton, J. W. et al. 1963. *Rothamsted exp. Stn Rep.* 1962 (banana; **42**: 653).

Dakwa, J. T. 1976. *Turrialba* **26**: 279 (cacao, inoculation; **56**, 3489).

Goujon, M. M. 1965. *Rev. Gen. Bot.* **72**: 353 (in vitro; **44**, 3228).

Mallamaire, A. 1934. *Agron. Colon.* **23**: 114 (coffee, loss & control; **14**: 153).

Maramba, P. et al. 1974. *Trans. Br. mycol. Soc.* **63**: 391 (conidial survival; **54**, 1132).

Meredith, D. S. 1960. *Ibid* **43**: 100 (banana, general; **39**: 727).

Noviello, C. et al. 1967. *Annali Fac. Sci. agr. Univ. Napoli* Ser. 4 2: 489 (inoculation of various flowers & fruit; **50**, 3053).

Tezenas du Montcel, H. et al. 1977. *Fruits* **32**: 77 (banana, climate; **56**, 5122).

TUNSTALLIA Agnihothrudu, *Phytopathol. Z.* **40**: 280, 1961.
(Sphaeriales)

PERITHECIA clustered, immersed erumpent, strongly stromatic, carbonaceous, aggregated in groups up to 5 in number with a prominent spinescent ectostroma and a well formed blackened marginal zone. ASCI clavate to cylindrical, tetra- or octosporus, distinctly pedicillate with a truncate apex. ASCOSPORES elongate fusoid, cymbiform or anguilliform, hyaline to subhyaline, aseptate (V. Agnihothrudu, l.c.).

The genus was erected to accommodate the fungus which causes thorny stem blight of tea, *Tunstallia aculeata* (Petch) Agnihothrudu (synonyms: *Aglaospora aculeata* Petch; *Phragmodiaporthe aculeata* (Petch) Wehmeyer). The following morphology of the pathogen is taken from the author: FRUCTIFICATIONS on the bark surface, strongly pustulate, subspherical to acuminate swellings up to 2.5 mm diam. STROMA immersed erumpent, well developed, dark carbonaceous, of hyphae and cortical tissues with a well developed dark, marginal zone enclosing a fascicle of confluent perithecia up to 5 in number, with a common ostiole opening into the spinescent conical apex of the ectostroma up to 2 mm in height. PERITHECIA as irregular, flattened, shallow cavities up to 2500 by 600 μ diam., embedded in a black pseudoparenchymatous stroma. ASCI clavate to cylindrical, paraphysate, somewhat narrow above with a truncate apex, typically pedicillate, pedicel straight or often recurved; 4 spored, $148–240 \times 22–44\ \mu$. ASCOSPORES lying parallel in the ascus, elongate fusoid to cymbiform, aseptate, hyaline to subhyaline, guttulate, $80–110 \times 6–13\ \mu$. A var. *kesabii* was also described by Agnihothrudu from tea. It differs from *T. aculeata* in having 8-spored asci with anguilliform ascospores, $100–162 \times 6–12\ \mu$.

T. aculeata is a wound pathogen which typically affects stems and branches of tea in India and Sri Lanka. It has been largely studied in N.E. India (see reports of the Tocklai Experimental Station, Assam). The disease's name arises from the projecting conical ostioles which give the infected bark a rough, pustular surface. The area around the ostioles (or 'thorns') is slightly raised and cracked. Black fans of mycelium form on the wood surface beneath the bark. Black lines appear in the wood which is discoloured with black patches. Spread is through the ascospores. The fungus can spread down the main stems to the collar and death of the whole bush can occur. Thorny stem blight is generally a disease of weakened tea plants. The cuts arising from removing diseased stems should be treated with a fungicidal paste. Weak bushes, intended for medium pruning, should be rested for at least 2 months beforehand. The plants should be desnagged in the year after medium pruning and the bigger cuts treated with a fungicide.

Agnihothrudu, V. 1961. *Phytopathol. Z.* **40**: 277 (generic, sp. and var. morphology; symptoms; **40**: 560).

UROCYSTIS Rabenh. in Klotzsch, *Herb. Viv. mycol.* ed. II, No. 393, 1856.
(Tilletiaceae)

SORI in various parts of the host, most frequently in vegetative parts but in some spp. in reproductive organs, usually dark brown to black, powdery to granular; SPORES in spore balls composed of 1 to many spores, firmly adhering, and covered partially or completely by an investing cortex of smaller, lighter coloured, sterile cells also firmly adhering; spore germination by short to long promycelium with a terminal whorl of a few to several primary sporidia fusing in pairs, or with several branches in lieu of sporidia and developing directly into a mycelium (from G. W. Fischer, *Manual of the North American smut fungi*: 204, 1953).

Several spp. occur on temperate ornamentals and cereals, see the key by Fischer (l.c.). *U. brassicae* Mundkur was described in India on Indian mustard (*Brassica juncea*) and other *Brassica* spp. can be infected. The SORI are 25–58 × 20–45 (av. 38 × 32) μ with 1–5 (mostly 2–3) fertile cells; the deep brown SPORES are 13–25 × 9–20 (av. 20–16) μ, surrounded by numerous, bright brown, elongated, sterile cells, 5–15 × 3–10 (av. 9.9–6.1) μ, forming a continuous layer. The spores are formed in wart like, tuberculate, leaden grey galls, up to 4 cm diam., in the underground parts of the plant. *U. coralloides* Rostrup has also been described on Cruciferae including mustard.

Mitra, M. 1928. *Agric. J. India* **23**: 104 (*Urocystis coralloides*; 7: 554).
Mundkur, B. B. 1938. *Phytopathology* **28**: 134 (*U. brassicae* sp. nov.; 17: 430).

Urocystis cepulae Frost, *A. Rep. Sec. Mass. St. Bd Agric.*, 24: 175, 1877.
Tuburcina cepulae (Frost) Liro, 1922
Urocystis colchici (Schlecht) Rabenh. var. *cepulae* Cooke, 1877.

SORI present beneath epidermis in pustules or elongate streaks. Spore mass is dark brown, powdery. SPORE BALLS spherical to ellipsoid (14–22 μ diam.), composed of a single spore surrounded by a layer of slightly coloured sterile cells (4–6 μ diam.). SPORES spherical to ellipsoid, reddish brown, wall smooth (11–14 μ diam.). On spp. of *Allium* including *A. cepa*, *A. porrum* and *A. sativum* (J. L. Mulder & P. Holliday, *CMI Descr.* 298, 1971; Chupp considered that *U. cepulae* and *U. colchici* were distinct).

Onion smut (infection of other *Allium* spp. is much less serious) is widespread in Europe, W. Asia, and N. and Central America; also in Australia, Chile, Egypt, Japan, Korea, Morocco, Nepal, Philippines, Peru and Thailand (Map 12). In India measures for eradication were taken (Urs et al.). Symptoms are first seen at the cotyledon stage when a dark thickened area appears; this (when large) may cause a downward curvature. As growth occurs the lesions, which break open to reveal the dark spore masses, form at the base of the leaves. Most plants are killed in 3–4 weeks but if they survive the leaves become short, brittle, distorted and may bear lesions throughout their length. Spores develop on the bulb which may be undersize; and, although it does not rot in storage, resistance to secondary infections from other pathogens is low (see Anderson; Anderson et al.). The disease first became prominent in USA where it had probably been introduced from Europe, having been later found there on wild *Allium*. Grayson et al. described cytology.

Although *U. cepulae* occurs in the tropics it is unlikely to be serious in the lowlands where the high temps are above the opt. for growth of, and infection by, the pathogen; and also enable the host to grow more rapidly through the susceptible stage. The spores germinate to form a mycelium which is more extensive from fresh ones. No typical sporidia are formed; the hyphae break up into oidial fragments. Penetration takes place (mostly before emergence) through the cuticle of the cotyledon and without an appressorium. The cotyledon is susceptible for *c.* 3 weeks and if infection is near the meristematic zone, from which the leaves form, each successive leaf becomes infected as it passes through the short susceptible period.

Spores and oidia germinate at temps of 13–22°. In vitro on an organic agar medium the opt. was 24° but lower optima have been reported. Germination of spores begins 60–72 hours from plating and organic carbon and nitrogen are required for high germination. There is abundant infection at a soil temp. of 10–12° and this can occur up to *c.* 25°, above which there is a rapid reduction and no infection at 29° (Walker et al. 1921, 1926). Stienstra et al. (1972) found max, infection at 20°; Kochman et al. found this at 15°. Temp. acts both on the pathogen directly and on the host which, by growing more rapidly at a higher temp., escapes infection. The spores are of 1 or 2 mating types, although there seems to be no adequate explanation for this unusual condition. Both types are required for infection but the way this comes about and the position of meiosis is not clear. No physiologic specialisation has been reported. It has not been definitely established whether the pathogen is seedborne and this is not considered to be important. Setts and transplants form the main manner of spread and *U. cepulae* persists in the soil as dormant spores for an indefinite number of years.

Onion smut is controlled by chemical treatment of the seed. Originally this was with formalin trickled in with the seeder, but thiram, carboxin, captan, ferbam of folpet with a sticker or in a granular formulation are now used (Crete et al.; Croxall et al.; Stienstra et al. 1970). Methyl bromide is an effective eradicant (Harrow). No resistance has been found in onion but it occurs in other *Allium* spp., and *A. fistulosum* × *A. cepa* gives resistant progeny, especially on backcrossing to the resistant parent. Biochemical changes associated with resistance in seedlings have been described.

Anderson, P. J. et al. 1924. *Bull. Mass. agric. Exp. Stn* 221, 29 pp (history & general, 55 ref.).

——. 1925. *J. agric. Res.* **31**: 275 (*Urocystis* spp. on *Allium* spp.; **5**: 204).

Blizzard, A. W. 1926. *Bull. Torrey bot. Club* **53**: 77 (infection & cytology; **5**: 531).

Chupp, C. 1960. *Mycologia* **52**: 343 (*U. cepulae* & *U. colchici* distinct; **41**: 18).

Crete, R. et al. 1973. *Phytoprotection* **54**: 32 (3 years of chemical control; **52**, 4300).

Croxall, H. E. et al. 1953. *Ann. appl. Biol.* **40**: 176 (seed treatment control; **33**: 135).

Dow, R. L. et al. 1969. *Phytopathology* **59**: 1219 (growth in vitro; **49**, 613).

Evans, R. I. 1933 & 1937. *Am. J. Bot.* **20**: 255; **24**: 214 (penetration & development in seedling; **12**: 610; **16**: 652).

Grayson, R. L. et al. 1975. *Phytopathology* **65**: 994 (spore development & nuclear history; **55**, 2054).

Harrow, K. M. 1970. *N.Z. Comm. Grow.* **25**(14): 27 (use of methyl bromide; **51**, 941).

Kochman, J. et al. 1974. *Acta agrobot.* **27**: 85 (factors affecting fungicides; **54**, 2574).

Lacy, M. L. 1968. *Phytopathology* **58**: 1460 (germination in vitro; **48**: 1001).

Stienstra, W. C. et al. 1970. *Pl. Dis. Reptr* **54**: 290 (control & effect of planting depth; **49**, 3064).

—— ——. 1972. *Phytopathology* **62**: 282 (effect of inoculum density, planting depth & soil temp. on infection; **51**, 2989).

Szembel, S. J. 1926. *Pl. Prot. Leningrad* **2**: 524 (soil survival & on wild onion; **5**: 646).

Tachibana, H. et al. 1964. *Mycologia* **56**: 289 (mating types & infection; **43**, 2549).

—— ——. 1966. *Phytopathology* **56**: 136 (survival; **45**, 1600).

Urs, N. V. R. et al. 1963. *Curr. Sci.* **32**: 445 (outbreak in S. India, eradication; **43**, 1490).

Walker, J. C. et al. 1921. *J. agric. Res.* **22**: 235 (infection & temp.; **1**: 281).

—— ——. 1926. *Ibid* **32**: 133 (as above; **5**: 464).

—— ——. 1944. *Ibid* **69**: 1 (*A. fistulosum* as resistance source; **24**: 45).

Whitehead, T. 1921. *Trans. Br. mycol. Soc.* **7**: 65 (life history & morphology; **1**: 35).

UROMYCES (Link) Unger, *Einfl. d. Bodens*: 216, 1836.

(see G. F. Laundon, *Mycol. Pap.* 99, 1865)
(Pucciniaceae)

SPERMOGONIA deeply embedded in the tissues of the host, flask shaped with conical mouth and ostiolar filaments, and flexuous hyphae. AECIDIA usually with an evident, generally cup shaped, peridium; AECIDIOSPORES with indistinct pores. UREDOSPORES formed singly on their pedicels, with several, usually rather distinct pores, rarely accompanied by paraphyses. TELIOSPORES aseptate, on distinct pedicels, almost always with an apical pore. BASIDIOSPORES flattened on one side or kidney shaped. Autoecious or heteroecious (from M. Wilson & D. M. Henderson, *British rust fungi*: 310, 1966).

Uromyces (see Guyot) is distinguished from *Puccinia* by its aseptate teliospores only. The spp. on these crops are considered separately: broad bean, chick pea, cowpea, lentil, lucerne, pea, *Phaseolus*, sunn hemp and vetch. *U. setariae-italicae* Yosh. on *Setaria* spp. can cause considerable damage to *S. italica* (foxtail millet) in India. The rust attacks other gramineous genera including *Brachiaria, Panicum* and *Pennisetum*. The fungus is sometimes referred to as *U. leptodermus* H. Syd. & P. Syd. which is one of the synonyms listed by Cummins (*The rust fungi of cereals, grasses and bamboos*: 448) who described it: AECIDIA (*Aecidium brasiliense* Diet.) on *Cordia* spp., AECIDIOSPORES 20–27 × 18–23 μ, globoid or ellipsoid, wall 1 μ thick, verrucose. UREDIA amphigenous, cinnamon brown; UREDOSPORES (25–) 27–33 (–35) × (20–) 23–28 (–30) μ, broadly obovoid or ellipsoid, wall (1–) 1.5 (–2) μ thick, cinnamon brown, echinulate, germ pores 3, equatorial. TELIA amphigenous, covered by the epidermis, blackish, small and inconspicuous; TELIOSPORES (16–) 18–25 (–28) × (14–) 16–20 μ, variable, mostly angularly globoid or obovoid, wall uniformly 1–1.5 μ thick, clear chestnut brown, smooth; pedicels colourless, thin walled and collapsing, to 20 μ long but usually broken near the spore. Ramakrishnan (1949) reported a short life for the uredospores and found that the forms from *S. italica* and *S. glauca* did not infect the other host. Using *Panicum* spp. and *Brachiaria* spp. Sundaram also found host specialisation.

Firman has pointed out the confusion in the rusts on banana (*Musa*): *Uredo musae* Cummins and *Uromyces musae* P. Henn. The latter, known only from Africa, has finely echinulate uredospores, 20–28 × 17–24 μ, wall 2.5 μ thick. The former (Pacific region) has more crowded uredia whose spores have a thinner (1.5 μ) wall and more pronounced echinulations. Neither fungus is of any economic importance (see Mulder et al.).

Joshi et al. (1959) referred to severe outbreaks of indigo (*Indigofera*) rust in Delhi and elsewhere in India. They considered that there were 2 forms of *U. indigoferae* Diet. & Holw.; 1 on *I. linifolia* and the other on *I. tinctoria*. Also (1958) *U. orientalis* Syd. was considered to be a synonym of *U. indigoferae*. The rust on pigeon pea (*Cajanus cajan*) is frequently and erroneously (see Sivanesan 1970a) referred to as *U. dolicholi* Arth. which occurs only on *Rhynchosia* spp. The rust on pigeon pea is *Uredo cajani* Syd. (Anahosur et al.). The 2 rusts cannot be distinguished since the telia have not been reported for *U. cajani*. The other rust on *Rhynchosia* is *Uromyces rhynchosiae* Cooke. *U. mucunae* Rabenh. (velvet bean rust) occurs on *Stizolobium atterrimum* (Bengal bean). *Uredo mucunicola* P. Henn. and *U. mucunae* P. Henn. are said to be synonyms of *Uromyces mucunae* (Laundon et al.; Thirumalachar). Six *Uromyces* spp. on cassava are discussed in the 1976 annual report of the Centro Internacional de Agricultura Tropical (CIAT), Cali, Colombia. These include *U. manihotis* P. Henn., the most widely distributed in the Americas, and *U. jatrophae* Diet. & Holw. The rust on beet is *U. betae* Kickx (see Punithalingam 1968a); and amongst rusts on ornamentals are those on *Dianthus* (including *U. dianthi* (Pers.) Niessl. and *Geranium* (including *U. geranii* (DC.) Fr.).

Anahosur, K. H. et al. 1978. *C.M.I. Descr. pathog. Fungi Bact.* 590 (*Uredo cajani*).

Firman, I. D. 1976. *Fiji agric. J.* **38**: 85 (banana rust in the Pacific region; **57**, 2598).

Guyot, A. L. 1938, 1951, 1957. *Les Urédinées* (*ou Rouilles des végétaux*) I. II. III. *Uromyces* (*Encycl. Mycol.* **8**, **15**, **29**), 438 pp., 331 pp., 647 pp. (systematics, morphology, hosts & distribution; **19**: 239; **31**: 146; **36**: 789).

Hitatsuka, N. 1973. *Rep. Tottori mycol. Inst.* **10**: 1 (in Japan, systematics, hosts & distribution; **53**, 2883).

Joshi, L. M. et al. 1958. *Indian Phytopathol.* **11**: 59 (on *Indigofera*; **38**: 676).

—— ——. 1959. *Ibid* **12**: 25 (as above; **39**: 332).

Laundon, G. F. et al. 1971. *C.M.I. Descr. pathog. Fungi Bact.* 290 (*U. mucunae*).

Mulder, J. L. et al. 1971. *Ibid* 295 (*U. musae*).

Narasimhan, M. J. et al. 1964. *Mycologia* **56**: 555 (on *Setaria*; **44**, 128).

Patil, B. V. et al. 1969. *Indian Phytopathol.* **22**: 110 (as above; **49**, 748).

Punithalingam, E. 1968a & b. *C.M.I. Descr. pathog. Fungi Bact.* 177, 180 (*U. betae* & *U. dianthi*).

Ramakrishnan, K. 1949. *Indian Phytopathol.* **2**: 31 (on *Setaria*; **29**: 506).

Ramakrishnan, T. S. 1955. *Proc. Indian Acad. Sci.* Sect B **41**: 241 (as above **35**: 96).

Savile, D. B. O. 1973. *Rep. Tottori mycol. Inst.* **10**: 225 (aecidiospore types in *Puccinia* & *Uromyces* on Cyperaceae, Juncaceae & Poaceae; **53**, 4341).

Sivanesan, A. 1970a & b. *C.M.I. Descr. pathog. Fungi Bact.* 269, 270 (*U. dolicholi* & *U. geranii*).

Sundaram, N. V. 1964. *Indian J. agric. Sci.* **34**: 215 (on *Panicum* & *Brachiaria*).

Thirumalachar, M. J. 1945. *Mycologia* **37**: 295 (*U. mucunae* inter alia; **25**: 11).

Uromyces appendiculatus (Pers.) Unger, *Einfl. d. Bodens*: 216, 1836.
Aecidium caulicola P. Henn., 1907
A. nigro-cinctum Pat. & Har., 1906
A. vignae Cooke, 1879
Uredo vignae Bres., 1891
Uromyces dolichi Cooke, 1882
U. pazschkeanus P. Henn., 1893
U. phaseoli (Pers.) Wint., 1880
U. punctiformis Syd., 1901
U. vignae Barcl., 1891
(further synonyms in Guyot, *Encycl. Mycol.* **29**: 482, 1957)

PYCNIA mostly epiphyllous in small groups. AECIDIA mostly hypophyllous in crowded groups up to 8 mm diam. cupulate, 0.25–0.3 mm diam. AECIDIOSPORES angular ellipsoidal, 15–26 μ diam.; wall hyaline, verrucose 1–1.5 μ thick. UREDIA amphigenous irregularly scattered or in small groups, cinnamon, *c.* 1 mm diam. UREDOSPORES ellipsoidal to obovoidal, 20–30 × 20–26 μ; wall sienna, finely echinulate, 1–2 μ thick, pores 2, ± equatorial or superequatorial. TELIA like the uredia but almost black. TELIOSPORES ellipsoidal, umbonate at the apex, 28–38 × 20–36 μ; wall bay, umber to sienna with hyaline to sienna umbo over pore, smooth or sparsely warted or striate, 2.5–3.5 μ thick at the side, 4–8 μ thick above; pedicels fragile, hyaline to pale luteous, up to 20 μ long, sometimes swelling in water.

On *Dolichos, Phaseolus* and *Vigna* spp. Of the many rusts described on these 3 genera only 2 are well known and clearly distinct from *U. appendiculatus: Phakopsora pachyrhizi* Syd. (q.v.) has minute, peridiate uredia and sessile teliospores, and *Puccinia tristachyae* Doidge which is a heteroecious sp. with aecidia (*A. decipiens* Syd.) on *Vigna* in South Africa. Three other rusts are poorly known and may not be distinct: *U. azukicola* Hirata (on *Phaseolus*, Japan) has rather irregularly shaped teliospores with wall 1.5–2 μ thick; *U. kisantuensis* P. Henn (on *Dolichos*, Congo) has smaller, densely verrucose teliospores; *U. vignae-sinensis* Miura (on *Vigna*, Manchuria) has uredospores with 2–4 germ pores (G. F. Laundon & J. M. Waterston, *CMI Descr.* 57, 1965).

This autoecious, heterothallic, leguminous rust is worldwide (Map 290) and probably occurs wherever beans (*Phaseolus*) are grown. Groth et al. (1978) completed the life cycle on common bean. But only in certain areas and seasons does it become destructive, causing complete loss of crop where infection occurs in the early stages of growth. The pathogen is significantly restricted by high temps and low RH. Limits for severe infection are 20–25° and overcast conditions must prevail. At day and night temps of 32–33° and 26–27°, respectively, no symptoms develop. High temp. limits spread of the pathogen in the host and infection by uredospores. The aecidial stage (and sometimes the telial stage) is very rare and in warmer areas carry over between crops is through the uredospores; teliospores only becoming important at low temps. Host penetration is typical of the rusts; germinating uredospores are unaffected (in penetration) by closed stomata, and no differential effect of light and dark has been found. Storage of uredospores (race 33) at −18° has a survival rate of 16% after 600 days. There may be some spread by seed but crop debris (besides infected stands) is the most important inoculum source.

Considerable work on the physiology and biochemistry of this host and parasite relationship has been done: *Rev. appl. Mycol.* **36**: 159; **38**: 50, 306; **40**: 3291; **41**, 346; **42**: 65; **44**, 35, 296, 1727, 2663; **45**, 1256, 1956, 2007; **46**, 873, 1796, 2886; **48**, 309; *Rev. Pl. Pathol.* **49**, 1205, 1206, 2230; **50**, 2042; **51**, 3773; **52**, 2416, 3124; **53**, 1628; **54**, 4693; **55**, 4368, 4369, 4370, 4371, 5435, 6013; **56**, 1361, 2301, 2302, 5281; **58**, 438. For fine structure see *Rev. Pl. Pathol.* **50**, 386, 3290; **51**, 895, 2987; **52**, 530; **53**, 2747; **54**, 1053, 3051, 3053, 4250, 4703, 5657; **55**, 1541, 5418; **56**, 495; **57**, 4719; and for defence mechanism: **51**, 2449; **52**, 2417, 4281; **55**, 1546, 6014.

Several control measures may be necessary. Long rotations may be advisable. The fungicides used more recently include: carboxin, chlorothalonil, maneb, oxycarboxin, pyracarbolid, triforine and zineb (Dongo; Iamamoto et al.; Okioga et al.; Parambaramani et al.; So; Wimalajeewa et al.; Yoshii et al.). Resistant cvs should be used but the range of their effectiveness is lessened by variability in this rust; > 35 races probably occur. Coyne et al. pointed out that such vertical resistance afforded only temporary protection and discussed other forms of resistance. Races have been described in Australia (Ballantyne; Ogle et al.; Waterhouse); Brazil (Carvalho et al.; De Menezes; Dias); Colombia (Rey et al.; Zuniga de Rodriguez et al.); Costa Rica (Christen et al.); E. Africa (Howland et al.); Mexico (Crispin et al.); New Zealand (Yen et al.); Peru (Guerra et al.) and USA (Augustin et al.; Davison et al. 1963b, 1964; Fisher; Gay; Groth et al. 1977; Harter et al.; Sappenfield; Zaumeyer). The earlier infection grades in differentiating 20 races (Harter et al. 1941) were reduced to 5 and the integrity of 34 races was maintained (Davison et al. 1963b). For the genetics of resistance see Augustin et al.; Carvalho et al.; Wingard, Zaumeyer et al.

Andrus, D. F. 1931. *J. agric. Res.* **42**: 559 (heterothallism; **10**: 810).

——. 1933. *J. Wash. Acad. Sci.* **23**: 544 (heterothallism & cytology in the aecidial stage; **13**: 319).

Augustin, E. et al. 1972. *J. Am. Soc. hort. Sci.* **97**: 526 (inheritance of resistance; **52**, 288).

Ballantyne, B. 1974. *Proc. Linn. Soc. N.S.W.* **98**: 107 (resistance; **53**, 3248).

Brien, R. M. et al. 1954. *N.Z. Jl Sci. Technol.* Sect. A. **36**: 280 (general in New Zealand; **34**: 623).

Carvalho, L. P. et al. 1978. *Fitopat. Brasileira* **3**: 181 (inheritance of resistance).

Christen, R. et al. 1967. *Turrialba* **17**: 7 (races in Costa Rica; **46**, 2564).

Coyne, D. P. et al. 1975. *Euphytica* **24**: 795 (genetic & breeding strategy for resistance; **55**, 3805).

Crispin, A. et al. 1962. *Pl. Dis. Reptr* **46**: 411 (races in Mexico; **42**: 66).

Curtis, C. R. 1966. *Phytopathology* **56**: 1316 (effect of light on uredospore germination; **46**, 793).

Davison, A. D. et al. 1963a. *Ibid* **53**: 736 (uredospore longevity in storage; **43**, 299).

—— ——. 1963b. *Ibid* **53**: 456 (identification of races; **42**: 643).

—— ——. 1964. *Ibid* **54**: 336 (effect of uredospore inoculum concentration on race determination; **43**, 2451).

De Menezes, O. B. 1952. *Dusenia* **3**: 309 (races in Brazil; **32**: 357).

Dias, F. I. R. 1968. *Pesq. agropec. Bras.* **3**: 165 (races in Brazil; **49**, 3049).

Dongo, D. S. 1971. *Invest. agropec. Peru* **2**: 23 (fungicides; **52**, 2418).

Fisher, H. H. 1952. *Pl. Dis. Reptr* **36**: 103 (new races in USA; **31**: 467).

Fromme, F. D. et al. 1921. *J. agric. Res.* **21**: 385 (symptoms on resistant & susceptible cvs).

——. 1924. *Phytopathology* **14**: 67 (morphology on *Vigna*; **3**: 566).

Gay, J. D. 1971. *Pl. Dis. Reptr* **55**: 384 (new race on cowpea; **51**, 896).

Groth, J. V. et al. 1977. *Ibid* **61**: 756 (virulence; **57**, 1902).

—— ——. 1978. *Phytopathology* **68**: 1674 (completing life cycle on common bean).

Guerra, E. et al. 1972–3. *Invest. agropec. Peru* **3**: 92 (races in Peru; **56**, 477).

Harter, L. L. et al. 1935. *J. agric. Res.* **50**: 737 (general; **41**: 669).

—— ——. 1941. *Ibid* **62**: 717 (identification of races; **20**: 555).

Howland, A. K. et al. 1966. *E. Afr. agric. For. J.* **32**: 208 (races in E. Africa; **46**, 1148).

Iamamoto, T. et al. 1971. *Biológico* **37**: 266 (fungicides; **51**, 4484).

Maheshwari, R. et al. 1967. *Can. J. Bot.* **45**: 447 (cytology of uredospore infection; **46**, 2565).

Ogle, H. J. et al. 1974. *Qd J. agric. Anim. Sci.* **31**: 71 (races & control in Australia; **54**, 2547).

Okioga, D. M. et al. 1972. *Misc. Rep. Trop. Pestic. Res. Inst. Tanzania* 805, 13 pp. (fungicides; **52**, 2783).

Parambaramani, C. et al. 1971. *Madras agric. J.* **58**: 705 (fungicides; **51**, 2174).

Rey, G. J. V. et al. 1961. *Acta agron. Palmira* **11**: 147 (races in Colombia; **42**: 66).

Sappenfield, W. P. 1954. *Pl. Dis. Reptr* **38**: 282 (new race in USA; **33**: 698).

Schein, R. D. 1961. *Phytopathology* **51**: 674 (effect of temp. on infection & colonisation; **41**: 346).

——. 1962. *Ibid* **52**: 653 (storage of uredospores; **42**: 167).

—— et al. 1965. *Mycologia* **57**: 397 (effect of temp. & RH on uredospore viability; **44**, 2995).

——. 1965. *Phytopathology* **55**: 454 (leaf age & susceptibility; **44**, 2288).

So, V. 1971. *Agric. Sci. Hong Kong* **1**: 265 (fungicides; **53**, 3678).

Waterhouse, W. L. 1953. *Proc. Linn. Soc. N.S.W.* **78**: 226 (races in Australia; **34**: 425).

Wei, C. T. 1937. *Phytopathology* **27**: 1090 (factors affecting host reaction types; **17**: 287).

Wimalajeewa, D. L. S. et al. 1973. *Trop. Agric.* **129**: 61 (fungicides; **53**, 4184).

Wingard, S. A. 1933. *Bull. Va agric. Exp. Stn* 51, 40 pp. (inheritance of resistance; **13**: 205).

Wynn, W. K. 1976. *Phytopathology* **66**: 136 (appressorium formation; **55**, 4933).

Yen, D. E. et al. 1960. *N.Z. Jl Agric.* **3**: 358 (races & breeding in New Zealand; **39**: 757).

Yoshii, K. et al. 1976. *Fitopatologia* **11**: 66 (fungicides; **56**, 1813).

Zaumeyer, W. J. et al. 1941. *J. agric. Res.* **63**: 599 (inheritance of resistance; **21**: 181).

——. 1960. *Pl. Dis. Reptr* **44**: 459 (new race in USA; **39**: 758).

Zuniga de Rodriguez, J. E. et al. 1975. *Acta agron. Palmira* **25**: 75 (races in Colombia; **56**, 2747).

Uromyces ciceris-arietini Jacz. apud. Boy. & Jacz., *Annls Ec. natu. Agric. Montpellier* 8: 195, 1894.

Uredo ciceris arientini Grogn., 1863
Uromyces ciceris-arietini Jacz. apud Boy. & Jacz. var. *aetnensis* Scalia, 1899.
(as *U. ciceris-arietini* (Grog.) Jacz. var. *aetnensis* Scalia)

PYCNIA and AECIDIA not known. UREDIA hypophyllous, scattered, pulverulent, cinnamon brown. UREDOSPORES globose to subglobose, 20–28 μ diam., wall 3–4 μ thick, yellowish brown, echinulate, 4–8 germ pores with hyaline papillae over the pores. TELIA like uredia but dark brown. TELIOSPORES variable in shape, round, ovate, or angular, 18–30 × 18–24 μ, with an unthickened apex and paler over the pore; wall brown, 3–3.5 μ thick, verrucose, pedicel short and hyaline. On *Cicer arietinum* and *Trigonella polycerata*.

Other rust fungi recorded on *Cicer* include *U. soongaricae* Ahmad which has large teliospores (25.5 × 20–30 μ) with a thickened apex. *U. appendiculatus* (q.v.) reported on *Cicer* has uredospores with ±2 super equatorial pores and teliospores with a smooth or warted, sparsely striate wall. Jørstad (*Uredinales of the Canary Islands*: 101, 1958) listed *U. ciceris-arietini* under *U. anthyllidis* which has very similar teliospores. Pycnia and aecidia are known for the latter on *Euphorbia* but not for the former, and it has not been demonstrated that the rust on *Anthyllis* and *Cicer* can be cross-inoculated. For these reasons the fungus described is regarded as a distinct sp. (E. Punithalingam, *CMI Descr*. 178, 1968).

Coalescence of the chick pea rust pustules can cause premature defoliation and result in considerable reduction in yield. The pathogen is widely distributed in the Mediterranean region and S.E. Europe, and also occurs in Asia (Afghanistan, India, Iran, Israel, Nepal, USSR), E. Africa (Kenya, Malawi) and Mexico (Map 235). The rust has not been studied extensively but it has caused considerable losses in India and Mexico. In India it was calculated that a 10% increase in disease intensity caused a loss of yield of 0.38 g/plant. Good germination of uredospores takes place at 20–26°, with 18–20° being reported as opt.; there is little germination at 30°. At 30–40° viability was lost in 2–4 weeks but at 5–7° remained for 48 weeks. The incubation period for uredospore infection was 11–13 days at 20–24°; teliospores do not appear to have been germinated. The disease is most serious in India in the cooler months, temps are too high for survival in the lowlands between crops. But natural infection of *T. polycerata* occurs in the Himalayan foothills from whence spores are considered to spread to the lowlands each season. Resistance has been reported from India.

Asthana, R. P. 1956–7. *Nagpur agric. Coll. Mag.* **31**: 20A (an outbreak in India; **37**: 260).
Bahadur, P. et al. 1970. *Indian Phytopathol.* **23**: 626 (physiologic specialisation; **51**, 901).
Dalela, G. G. 1962. *Agra Univ. J. Res. Sci.* **11**: 117 (method for estimating loss; **42**: 716).
Mehta, P. R. et al. 1946. *Indian J. agric. Sci.* **16**: 186 (general; **27**: 110).
Saksena, H. K. et al. 1955. *Indian Phytopathol.* 8: 94 (general; **36**: 230).
Payak, M. M. 1962. *Curr. Sci.* **31**: 433 (natural occurrence on *Trigonella polycerata* in India; **42**: 251).

Uromyces decoratus Syd., *Annls Mycol.* **5**: 491, 1907

UREDIA hypophyllous, rarely epiphyllous, numerous, scattered, pulverulent and yellowish brown; uredia also present on other green parts of the plant. UREDOSPORES globose, to subglobose or ellipsoid, 21–26 μ diam., wall 1.5–2 μ thick, light brown, echinulate with 4–6 ± equatorial germ pores. TELIA hypophyllous, smaller than uredia, black and pulverulent. TELIOSPORES globose to oblong, 20–32 × 14–20 μ, wall 1.5–2 μ uniformly thick, verrucose, apex hyaline, 3–5 μ thick. Pedicel hyaline, slender and easily detached. On *Crotalaria juncea*, *C. maxillaris*, *C. medicaginea*, *C. retusa* and *C. vitatoni*. Also by inoculation on *C. pumila*, *C. shrica* and *C. verrucosa* (Joshi 1960).

Several other rust fungi have been recorded on *Crotalaria*. *Uromyces crotalariae* (Arth.) Baxter (synonyms: *Uropyxis crotalariae* Arth., *Haplopyxis crotalariae* (Arth.) Syd., *Uromyces harmsianus* Doidge and *Uredo harmsiana* P. Henn.) has uredospores with 4–6 scattered germ pores and teliospores with a bilaminate wall, the inner chestnut brown, 2 μ thick; the outer, hyaline, after swelling in water 5–10 μ thick. *Maravalia crotalariae* Syd. has hyaline, verruculose uredospores and oblong to ovate, hyaline, smooth walled teliospores, 20–25 × 9–16 μ with undifferentiated germ pores. *Phakospora crotalariae*

Uromyces pisi-sativi

(Diet.) Arth. (synonym: *Uredo crotalariae* Diet.) has paraphysate uredia and telia in crusts of largely adherent spores. *Uredo theresiae* Neger reported from Colombia is said to have hyaline uredospores. *Uredo crotalariicola* P. Henn. from Africa has larger uredospores (24–30 × 21–28 μ) with spore walls 5–7 μ thick. *Uredo crotalariae-vitellinae* Rangel, described from Brazil, has paraphysate uredia and pale yellow uredospores. Other rust fungi recorded are *Aecidium crotalariae* P. Henn., *A. crotalariicola* P. Henn., *A. crotalariicolum* P. Henn. and *A. dielsii* P. Henn. (E. Punithalingam, *CMI Descr.* 179, 1968).

Severe infection of sunn hemp by this rust (of limited distribution) has occurred in India. The distribution is: Africa (Ghana, Guinea, Ivory Coast), Asia (China, India, Japan, Pakistan, Sri Lanka, Taiwan, Thailand), America (Costa Rica, Venezuela; Map 472). Uredospores germinate well at 18–25° and teliospores also will do so in <48 hours at 21–25°. At 20–25° uredospores remained viable for 40–45 days (stored at 5° viability is 4 months). Neither spore type nor dormant mycelium is considered to survive the summer in the lowlands of N. India. Survival presumably occurs in the uplands on perennial *Crotalaria*. Attempts at infection of *Crotalaria* with basidiospores in India were unsuccessful. No investigations on control seem to have been reported.

Joshi, L. M. 1960. *Indian Phytopathol.* **13**: 90 (spore germination, viability & inoculation; **42**: 124).
——. 1964. *Ibid* **17**: 257 (formation of secondary sporidia; **44**, 2068).
Merny, J. 1962. *Rev. Mycol. Paris* **27**: 169 (outbreak in Africa, description; **43**, 777).

Uromyces pisi-sativi (Pers.) Liro, *Uredineae Fennicae*: 100, 1908.

Uredo pisi-sativi Pers., 1801
Uromyces pisi ([Pers.] DC.) Wint. 1884
U. lathyri Fuck., 1869.

PYCNIA mostly hypophyllous scattered among the aecidia. AECIDIA hypophyllous, scattered over the entire leaf surface, often covering most leaves of the infected plant, cupulate, 0.3–0.4 mm diam. AECIDIOSPORES broadly ellipsoidal, 17–22 μ diam., wall hyaline, finely verrucose, 1 μ thick. UREDIA mostly hypophyllous, irregularly scattered, cinnamon, up to 1 mm diam. UREDOSPORES broadly ellipsoidal, 22–28 × 19–24 μ; wall luteous to sienna, very finely echinulate, 1.5–2 μ thick; pores 3–6, scattered. TELIA like the uredia but chestnut. TELIOSPORES broadly ellipsoidal, slightly bullate at the apex, 22–28 × 17–22 μ; wall sienna, paler over the pore, finely warted, 2–3 μ thick at the sides, 4–5 μ thick at the apex; pedicels almost hyaline, fragile, short. Pycnia and aecidia on *Euphorbia*, uredia and telia on *Lathyrus, Orobus, Pisum* and *Vicia*.

Numerous other rusts have been recorded on these hosts; the most important of these is the autoecious *U. viciae-fabae* (q.v.) which differs in having uredospores with equatorial pores and smooth teliospores with greatly thickened apical wall. It is not uncommon for both *U. pisi-sativi* and *U. viciae-fabae* to infect the same leaf; see also Parmelee in *U. striatus* (q.v.) for a comparison. The malodorous pycnia, the reddish orange aecidia, and the abnormal thickness and shape of the host leaves serve to distinguish *U. pisi-sativi* from other rusts attacking *Euphorbia* (Moehrke; G. F. Laundon & J. M. Waterston, *CMI Descr.* 58, 1965).

This rust of pea, vetches and broad bean has only occasionally been reported as causing serious losses. It is largely a European pathogen but has also been reported from Africa (Canary Islands, Ethiopia, Libya, Morocco), Asia (Burma, China, Hong Kong, Iran, Israel, Lebanon, Nepal, Pakistan, Turkey, USSR) and S. America (Argentina, Brazil, Chile); it appears to be absent from N. and Central America (Map 404). Reference should be made to Guyot (1957, *Uromyces* q.v.) and to Wilson and Henderson (*British rust fungi*). Studies on *U. pisi-sativi* have been mostly confined to the various physiologic forms occurring on leguminous hosts and to inoculation of (and spread within) the main aecidial host *E. cyparissias*. Opt. temp. for aecidiospore germination was reported as 17–20° from USSR but no temp. requirements for the other spore stages appear to have been determined. Some races are restricted to *Pisum* and others occur on *Lathyrus* and *Vicia*. Sidenko (1966) described 2 forms; *U. pisi* ff. sp. *pisi* and *lathyri*. In Europe infection spreads to the leguminous crop from the aecidial host. Virtually no experimental work on control has been done.

Bucheim, A. 1922 & 1924. *Zentbl. Bakt. ParasitKde* Abt. 2. **55**: 507; **60**: 534 (host range on legumes & inoculations; **2**: 341; **3**: 486).
Hartwich, W. 1955. *Phytopathol. Z.* **24**: 73 (inoculation of & spread in *Euphorbia cyparissias*; **35**: 370).

Jørstad, I. 1948. *Nord. Jordbr. Forsk.*: 198 (physiologic forms; inoculation of *Pisum*; **29**: 242).
Moehrke, L. 1927. *Bot. Arch.* **18**: 347 (symptoms on *E. cyparissias*; **7**: 540).
Noffray, E. 1924. *C. r. hebd. Séanc. Acad. Agric. Fr.* **10**: 140 (cultural control; **3**: 498).
Sidenko, I. E. 1960. *Zashch. Rast. Mosk.* **5**: 30 (physiologic forms & control; **39**: 717).
——. 1966. *Nov. Sist. niz. Rast.*: 188 (physiologic forms; **46**: 1028).
Stilbach, K. 1932. *Dt. landwirt. Presse* **59**: 302 (inoculation & testing of *Pisum*; **11**: 760).

Uromyces striatus Schroeter, *Abh. Schles. Ges.* 1869–72; 11, 1870.
Uromyces medicaginis (Pass.) Thum., 1874
U. oblongus Vize, 1877
U. medicaginis-falcatae Wint., 1884.

PYCNIA and AECIDIA as for *U. pisi-sativi* (q.v.). UREDIA amphigenous and caulicolous, irregularly scattered, cinnamon, *c.* 0.5 mm diam. UREDOSPORES ellipsoidal, 18–26 × 16–22 μ; wall sienna to luteous, finely echinulate, 1.5–2 μ thick; pores 3–4 ± equatorial. TELIA like the uredia, slightly darker coloured. TELIOSPORES spheroidal to ellipsoidal, papillate over the apical pore, 18–24 × 16–20 μ; wall sienna, striate or warted striate, 1.5–2 μ thick; pedicels pale, fragile, short. Pycnia and aecidia on *Euphorbia*, uredia and telia on *Medicago* and *Trifolium* spp.

Several other rusts have been described on *Medicago*, most of which are probably synonymous with either the above spp. or *U. anthyllidis*. The latter differs in having uredospores with scattered pores and verrucose teliospores. *U. striatus* is considered by Jørstad (*Opera Bot.* 1: 93, 1953) to be a form of *U. pisi-sativi*. Parmelee, whilst acknowledging the close relationship, considered that the smaller number and equatorial position of the pores in the uredospores of *U. striatus*, and the striate wall sculpturing of the teliospore of this sp., in contrast to the punctate markings of *U. pisi-sativi*, are adequate criteria for separating them (G. F. Laundon & J. M. Waterston, *CMI Descr.* 59, 1965).

Like its host, lucerne rust occurs chiefly in temperate and subtropical areas, and in cooler parts of the tropics. It is very widespread, wherever *M. sativa* is grown (Map 342). Serious losses have been reported from Israel, South Africa, Sudan, Uganda and USSR (where in the middle Volga region very susceptible *Euphorbia* spp. were considered to cause severe spread). The rust perennates and is systemic in *E. cyparissias* but uredospores can overwinter, retaining viability and virulence for *c.* 6 months. For races in Europe see Guyot (1957, *Uromyces* q.v.); in Canada *Medicago* spp. were more susceptible than *Trifolium* spp. In Australia isolates from *Trigonella suavissima* and *M. sativa* (considered to be 2 races) were grouped into 4 by inoculations on *Medicago* with the 2 races, i.e. all *Medicago* spp. susceptible, susceptible to the *T. suavissima* race but not to the *M. sativa* one, resistant to the *T. suavissima* race but not to the other, and resistant to both races. There were 3 clear infection types: immune, susceptible and intermediate (flecks with small pustules). In USA repeated selection (8 cycles) showed a significant association between resistance to races 1 and 2. A new form of rust on *Melilotus italica* was described from Iowa. Resistant vars offer the best hope for control; frequent cutting or grazing has been recommended in Israel.

Hill, R. R. et al. 1963. *Phytopathology* **53**: 432 (selection for resistance in lucerne; **42**: 618).
Koepper, J. M. 1942. *Ibid* **32**: 1048 (inoculation and resistance; **22**: 209).
Leppik, E. E. 1960. *Pl. Dis. Reptr* **44**: 184 (new form on *Melilotus italica*; **39**: 474).
Lopatin, V. I. 1938. *Socialistic Grain Fmg Saratoff* **2**: 110 (susceptibility in aecidial & telial hosts; **18**: 397).
Novikov, V. A. 1937. *C. r. Acad. Sci. U.R.S.S.* **15**: 53 (interference with metabolism in lucerne; **16**: 754).
Parmelee, J. A. 1962. *Can. J. Bot.* **40**: 491 (review, 55 ref.; **41**: 584).
Waterhouse, W. L. 1953. *Proc. Linn. Soc. N.S.W.* **78**: 147 (races on lucerne & *Trigonella suavissima*; **34**: 373).

Uromyces viciae-fabae (Pers.) Schroet., *Hedwigia* **14**: 161, 1875.
Uredo viciae-fabae Pers., 1801
Uromyces fabae (Pers.) de Bary, 1863
U. orobi (Pers.) Fuckel, 1869
U. viciae Fuckel, 1869
U. polymorphus Peck & Clint., 1878
U. yoshinagai P. Henn., 1901.

PYCNIA amphigenous, in small groups associated with the aecia. AECIDIA amphigenous or hypophyllous, usually in groups surrounding the pycnia or sometimes scattered, cupulate, 0.3–0.4 mm diam. AECIDIOSPORES spheroidal, 18–26 μ diam.; wall hyaline, verrucose, 1 μ thick. UREDIA amphigenous and on the petioles and stems, scattered, cinnamon, 0.5–1 mm diam. UREDOSPORES ellipsoidal or obovoidal, 22–28 × 19–22 μ; wall luteous to sienna, very finely echinulate, 1–2.5 μ thick; pores 3–4, equatorial or occasionally scattered on *Lathyrus*. TELIA like the uredia but black and larger; 1–2 mm diam. TELIOSPORES ellipsoidal, obovoidal or cylindrical, rounded or subacute above, 25–40 × 18–26 μ; wall chestnut, smooth, 1–2 μ thick at the sides, 5–12 μ thick above; pedicels sienna to luteous, up to 100 μ long. On *Lathyrus, Lens, Orobus, Pisum* and *Vicia* spp.

Numerous other rusts have been described on these hosts. Often they differ in rather minor respects and identification should be undertaken with great caution. Of the most important spp. *Uromyces pisi-sativi* (q.v.) has uredospores with 3–6 scattered pores and verrucose teliospores, 22–28 × 17–22 μ; *U. heimerlianus* Magn. (synonym: *U. fischeri-eduardi* Magn.), on *Vicia*, has uredospores with 4–6 scattered pores and verrucose teliospores, 25–30 × 20–27 μ; *U. viciae-craccae* Const. (on *Lathyrus, Lens, Vicia*) has uredospores with 3 equatorial pores and smooth teliospores with a thickened apical wall. (G. F. Laundon & J. M. Waterston, *CMI Descr.* 60, 1965.)

A worldwide (Map 200), generally minor, autoecious, heterothallic rust on broad bean, lentil, pea and vetch. De Bary (1863) in historical, first experiments with rusts sowed teliospores from *Vicia faba* and *Pisum sativum* on to these hosts and obtained aecidia. The disease is restricted by high temps. In Yugoslavia and Japan uredospores germinated best at 16–22° and in inoculations to broad bean developed uredo sori at 14–24°, the number decreasing at 26–30°. Teliospores from lentil require no rest period before germination. They remain viable throughout the hot season in India (Indo-Gangetic plain) and germinate (at 12–22°) to cause fresh outbreaks of the disease in January. During the growing season in India spread is by the aecidiospores; aecidia form at < 25° whilst at 25° and above uredia develop. Uredospores have an opt. temp. for germination of 17–18° whilst that for aecidiospores is probably lower (Prasada et al.). In some areas the fungus overwinters as uredo mycelium. Races have been described on broad bean and pea. Some forms seem to be restricted to individual host spp., for example, in Yugoslavia isolates from broad bean did not infect lentil. Control is probably most effective through resistant vars although fungicides (notably sulphur compounds and dithiocarbamates) have been used successfully.

Accatino, P. 1963–4. *Agric. téc.* **23–24**: 7 (fungicides on lentil; **46**, 205).

Brown, A. M. 1940. *Can. J. Res.* Sect. C. **18**: 18 (heterothallism; **19**: 304).

De Bary, A. 1863. *Annls Sci. nat.* Ser. IV **20**: 74 (experimental formation of aecidia).

El-Helaly, A. F. 1939. *Bull. Minist. Agric. Egypt* 201 (incidence & control; **18**: 567).

Gäumann, E. 1934. *Annls Mycol.* **32**: 464 (physiologic forms in *Vicia*; **14**: 141).

Hiratsuka, N. 1933. *Jap. J. Bot.* **6**: 329 (physiologic forms & related *Uromyces* spp. **12**: 596).

——. 1934. *Bot. Mag. Tokyo* **48**: 309 (temp., life history & resistance in *Pisum*; **13**: 670).

Jordi, E. 1904. *Zentbl. Bakt. ParasitKde* Abt 2 **11**: 763 (physiologic forms).

Kapooria, R. G. et al. 1966. *Indian Phytopathol.* **19**: 229 (host range in *Vicia*; **46**, 2890).

——. 1971. *Neth. J. Pl. Pathol.* **77**: 91 (cytology of promycelia & chromosome number; **51**: 140).

Kišpatic, J. 1944. *Phytopathol. Z.* **14**: 475 (possible races on broad bean; **25**: 484).

——. 1949. *Annls Trav. agric. scient.* **1**(2), 61 pp. (general, life history, races & control; **29**: 5).

Mishra, R. P. et al. 1969. *Jnl appl. Sci. India* **1**: 54 (screening *Lathyrus* spp. for resistance; **49**, 2706).

Prasada, R. et al. 1948. *Indian Phytopathol.* **1**: 142 (life history on lentil; **29**: 193).

Raggi, V. 1966. *Annali fac. Univ. Perugia* **21**: 293 (effect on photosynthesis & respiration in broad bean; **47**, 956).

Reichert, I. et al. 1946. *Palest. J. Bot. Rehovot Ser.* **5**: 202 (development, temp. & control; **26**: 523).

Savile, D. B. O. 1939. *Am. J. Bot.* **26**: 585 (cytology; **19**: 167).

Sempio, C. et al. 1966. *Phytopathol. Z.* **55**: 117 (N metabolism in broad bean; **45**, 2666).

Steven, W. F. 1936. *J. Bot. Lond.* **74**: 79 (aecidial stage on broad bean; **15**: 552).

USTILAGINOIDEA Bref., *Unters. Mykol.* 12: 194, 1895.

(Dematiaceae)

FRUCTIFICATIONS in the ovaries of individual grains of grasses, transforming them into large, very dark olive green or sometimes orange, velvety masses; when cut open the inner part is seen to be bright orange towards the surface, almost white in the centre. MYCELIUM partly superficial, partly immersed. STROMA always present, prosenchymatous. SETAE and HYPHOPODIA absent. CONIDIOPHORES micronematous, mononematous, sparingly branched, flexuous, intertwined, hyaline, smooth. CONIDIOGENOUS CELLS polyblastic, integrated, intercalary, determinate, cylindrical, denticulate; denticles short, cylindrical. CONIDIA solitary, dry, pleurogenous, simple, spherical or subspherical, pale to dark olive green, verruculose, aseptate (M. B. Ellis, *Demat. Hyphom.*: 42, 1971).

Besides *U. virens* (described) there is another sp. (*U. ochracea* P. Henn) which occurs on many grasses, especially Paniceae in the tropics; it has larger conidia, 6–8 μ diam. (Ellis l.c.).

Ustilaginoidea virens (Cke) Takahashi, *Bot. Mag. Tokyo* **10**: 16, 1896.

Ustilago virens Cooke, 1878
Tilletia oryzae Patouillard, 1887
Ustilaginoidea oryzae (Pat.) Bref., 1895.

FRUCTIFICATION surrounds grain almost completely. Diam. *c.* equal to the length of the grain, changing from yellow to orange, olive green to black. Fleshy when young, hard later. The central mycelial tissue is hard, surrounded by 3 sporiferous layers, the inner pale yellow and the outermost olivaceous to greenish black. CONIDIA olivaceous, warty when mature, spherical to elliptical, $3–5 \times 4–6$ μ; younger spores pale and almost smooth; borne laterally on minute sterigmata. The perfect state is reported to be an apothecium arising from a sclerotium; numerous asci are formed (Hashioka et al. 1951; after J. L. Mulder & P. Holliday, *CMI Descr.* 299, 1971).

False smut of rice is a minor disease which may at times become locally serious; the fungus also occurs on maize. It is widespread in the tropics (Map 347). The symptoms are quite conspicuous olive green, velvety, globose masses, up to 1 cm diam., in some of the ears of the inflorescence. This spore ball, beneath the dark layer of mature spores, is orange yellow, paling inwards until it is almost white; as it ages it becomes black. The glumes are closely applied to the lower part of the spore ball which is at first covered with a membrane. Infection with similar symptoms has been reported on the male inflorescence of maize and on wild *Oryza* spp.

The life cycle appears to be incompletely known. Infection of rice by injection of conidia and ascospores at the booting stage and on coleoptiles 3–6 mm high has been reported from Japan where it is considered that most natural infection occurs just before flowering (Hashioka et al. 1951; Ikegami; Yoshino et al.). But other observations have suggested that it occurs either in the flowers before fertilisation or afterwards when the grain is formed. Kulkarni et al. got infection by inoculating ovaries. Relatively few ears form spore balls and there may be empty spikelets. In vitro the conidia form a mycelium and secondary conidia, smaller and hyaline; teritary conidia can also be formed. The opt. temp. for growth in vitro is *c.* 27°. There appears to be no evidence for seed transmission and the conidia are probably short lived; they have a diurnal periodicity with a peak at 2200 hours; numbers are very low between 0400 and 1600 hours (Sreeramulu et al.). The ascospores may also be air dispersed. The fine structure of conidia has been compared with that of cereal smut spores. Control can apparently be obtained, where warranted, by fungicide treatment before heading. Nitrogen tends to increase disease incidence and cv. differences have been reported.

Chou, T. G. 1967. *Pl. Prot. Bull. Taiwan* **9**: 51 (symptoms, cv. differences; **49**: 1021).

Hashioka, Y. et al. 1951. *Res. Bull. Saitama agric. Exp. Stn* 2, 20 pp. (conidial germination, infection & perfect state; **31**: 455).

———. 1966. *Res. Bull. Fac. Agric. Gifu Univ.* **22**: 40 (fine structure of conidia; **46**: 2015).

Haskell, R. J. et al. 1929. *Phytopathology* **19**: 589 (on maize; **8**: 716).

Ikegami, H. 1960. *Res. Bull. Fac. Agric. Gifu Univ.* **12**: 45 (ascospore inoculation; **40**: 466).

——. 1962. *Ibid* **16**: 45 (nutrition).

——. 1962. *Ann. phytopathol. Soc. Japan* **27**: 16 (conidial inoculation; **42**: 319).

——. 1963. *Res. Bull. Fac. Agric. Gifu Univ.* **18**: 47, 54 (sclerotia & infection).

Kulkarni, C. S. et al. 1975. *Curr. Sci.* **44**: 483 (infection & spread; **55**, 1249).

Lepori, L. 1951. *Notiz. Mal. Piante* **16**: 40 (observations in Italy; **31**: 456).

Rao, P. G. et al. 1955. *Indian Phytopathol.* **8**: 72 (severity in India & on wild rice).

Raychaudhuri, S. P. 1946. *J. Indian bot. Soc.* **25**: 145 (mode of infection; **26**: 125).

Revilla, V. A. 1955. *Boln Estac. exp. agric. La Molina* 61, 14 pp. (symptoms; **36**: 57).

Seth, L. N. 1945. *Indian J. agric. Sci.* **15**: 53 (in Burma, conidial behaviour in vitro; **25**: 139).

Sreeramulu, T. et al. 1966. *Trans. Br. mycol. Soc.* **49**: 443 (diurnal periodicity of, & natural infection by, conidia; **46**: 316).
Yoshino, M. et al. 1952. *Agric. Hort. Tokyo* **27**: 291 (infection by injection; **31**: 574).

USTILAGO (Pers.) Roussel, *Flore du Calvados*: 47, 1806.
(Ustilaginaceae)

SORI in various parts of the host, forming at maturity dusty, dark, spore masses; SPORES single, produced irregularly in fertile mycelial threads which later disappear through gelatinisation. Germination by septate promycelium producing infection threads or sporidia formed terminally and laterally near the septa; sporidia germinating in water to form infection hyphae or budding indefinitely in nutrient solutions (B. B. Mundkur & M. J. Thirumalachar, *Ustilaginales of India*: 24, 1952).

This is a large genus with most spp. on Gramineae. Keys for the N. American and Nordic spp. were given by Fischer (*Manual of the North American smut fungi*) and Lindberg, respectively; see also Ahmed for Pakistan (*Mycol. Pap.* 64, 1956), Mundkur et al. (l.c.), and Ainsworth and Sampson (*The British smut fungi*). The delimitations and subdivisions of *Ustilago* were discussed by J. A. Nannfeldt (see Lindberg: 148). Among the important pathogens on temperate cereals are *U. avenae* (Pers.) Rostrup (loose smut of oats and barley), *U. hordei* (Pers.) Lagerh. (covered smut of these crops) and *U. nuda* (Jensen) Rostrup (loose smut of wheat, barley and rye; see Punithalingam et al.).

U. coicis Bref. causes a smut of Job's tears (*Coix lachryma-jobi*). SORI in the ovaries completely destroying all of them, 9–13 mm long, 5–9 mm broad, blackish brown, pulverulent. SPORES held together by the hard floral glumes, liver brown, subglobose to ellipsoidal, 7–13 μ diam. (av. 9 μ); epispore of medium thickness with minute but clear echinulations giving the margin a serrate appearance; also with prominent circular pits on the surface (Mundkur et al. l.c.). Chowdhury reported serious damage in India (Assam). Most spore germination occurred at 30° after 24 hours. Seed treatments with copper carbonate and mercurials give control.

U. crus-galli Tracey & Earle causes a smut of Japanese barnyard millet (*Echinochloa frumentacea*); (synonyms: *Cintractia seymouriana* P. Magnus, *C. crus-galli* (Tracey & Earle) P. Magnus). SORI entirely destroying the inflorescence, also on stems, especially at nodes, on young shoots and in axils of older leaves; shoot infection causing considerable deformity resulting in a twisted mass of leafy shoots with sometimes aborted ears; sori large, on stems up to 1.3 cm diam., swollen, covered with a hispid, grey membrane of host tissue, exposing a pulverulent spore mass. SPORES spherical to slightly ellipsoidal, Mikado brown, 9–12 μ diam. (av. 10.5 μ); epispore thick, covered with blunt, dense echinulations, sometimes verruculose; germination by the formation of a typical septate promycelium with lateral and terminal sporidia (Mundkur et al. l.c.). Venkatarayan described *U. crus-galli* on *E. colona* (jungle rice). This smut is said to be seedborne and standard seed treatments should give some control.

Two other *Ustilago* spp. occur on *E. frumentacea*: *U. panici-frumentacei* Bref. and *U. paradoxa* Syd. & Butl. (see Mundkur 1943). In the first the SORI enlarge the ovaries ×2–3 their normal size; covered by the seed coat, opening at maturity by an apical pore; seed coat rendered hairy; spore masses pulverulent, deep black brown. SPORES spherical, subspherical, buckthorn brown, 6–10 μ diam. (av. 8 μ) epispore thick, minutely echinulate. In the second the SORI do not enlarge the ovaries; covered by a tough, slightly hairy, grey indusium which later wears away exposing a black, pulverulent spore mass. SPORES globoid to oval, tawny olive with granular contents, 8–11 μ diam. (av. 9.5 μ); epispore smooth, thin; germination by a long branched germ tube without forming a true promycelium or sporidia (from Mundkur et al. l.c.). Sharma described infection by *U. paradoxa*; the seedlings of Japanese barnyard millet are most susceptible when 2–3 days old (no infection after 6 days). About 10% of the spores germinated when 4 years old. Both smuts on this crop can be controlled by standard seed treatments. Sharma et al. found some resistance to *U. paradoxa*.

U. esculenta P. Henn. causes hypertrophy on the culms of *Zizania aquatica* (American wild rice) and *Z. latifolia*. On the latter the gall that is formed by the infection is important as a vegetable in Taiwan, E. China and other countries of S.E. Asia (Enkina; Su; Yang et al.). Infected stems are dwarfed with only 2–5 shortened internodes (normally *c.* 10), inflorescences are deformed and hypertrophied, and there is no seed. The spores are finely echinulate, 6–9.5 μ diam. according to Fischer (l.c.), but described as having a smooth wall, 5.5–8 × 5–7 μ, by Enkina (see also Yu). *U. vetiveriae* Padwick was described from the inflorescences of *Vetiveria zizanioides* (khuskus grass); spores are 6.5–11 μ diam. and the sori *c.* 1 cm long and 1.5 mm wide.

Bradford, L. S. et al. 1975. *Bot. Gaz.* **136**: 109 (electrophoretic survey of 14 *Ustilago* spp.; **55**, 586).

Chowdhury, S. 1946. *J. Indian bot. Soc.* **25**: 123 (*U. coicis*; **26**: 101).

Enkina, T. V. 1968. *Zashch. Rast. Mosk.* **13**: 40 (*U. esculenta*; **47**, 2714).

Hirschhorn, E. 1959. In *Omagiu lui T. Săvulescu*, Bucharest: 297 (*Ustilago* spp. on *Paspalum*; **39**: 667, 715).

Lindberg, B. 1959. *Symb. bot. upsal.* **16**(2): 148 (Ustilaginales of Sweden; **39**: 284).

Mundkur, B. B. 1943. *Indian J. agric. Sci.* **13**: 631 (*Ustilago* spp. on Japanese barnyard millet; **24**: 15).

Padwick, G. W. 1946. *Mycol. Pap.* 17, 12 pp. (Indian fungi including *U. vetiveriae* sp. nov.; **26**: 266).

Punithalingam, E. et al. 1970. *C.M.I. Descr. pathog. Fungi Bact.* 279, 280 (*U. avenae* & *U. nuda*).

Sharma, B. B. 1963. *Proc. natn. Acad. Sci. India* Sect. B. **33**: 618 (*U. paradoxa*; **43**, 2899).

Sharma, B. L. et al. 1973. *JNKVV Res. Jnl* 7: 108 (resistance to *U. paradoxa*; **53**, 3003).

Su, H. J. 1962. *Spec. Publ. Coll. Agric. Taiwan* 10: 139 (*U. esculenta*; **42**: 172).

Venkatarayan, S. V. et al. 1951. *Curr. Sci.* **20**: 329 (*U. crus-galli*; **31**: 493).

Yang, H. C. et al. 1978. *Phytopathology* **68**: 1572 (*U. esculenta*).

Yu, Y. N. 1974. *Phytotaxonomica Sinica* **12**: 317 (genus *Yenia* in Yeniaceae & *U. esculenta*; **54**, 2097).

Ustilago crameri Körnike in Fuckel, *Symb. Mycol.*, Nacht. **2**: 11, 1873.

SORI in all ovaries of infected spikes, more or less globose, 2–4 mm diam., limited by a thin membrane of host tissue which soon ruptures. SPORE MASS powdery, dark brown. SPORES globose, pale brown, smooth, 8–12 μ. Cultures may be obtained on malt agar or PDA. Single spore cultures regularly develop smut spores (Wang 1938, 1943). On *Setaria italica* (Italian or foxtail millet) and its vars and *S. viridis* (G. C. Ainsowrth, *CMI Descr.* 78, 1965).

Head smut of foxtail millet is an important disease which is widespread in Asia, N. America and S.E. Europe; it also occurs in Argentina, Australia, Kenya and South Africa (Map 239). Every shoot and grain of an infected plant usually becomes smutted. The spores can germinate without a rest period but a high percentage of them germinated after 3 years, and some did so after 64 years. Penetration (the promycelia functioning as infective hyphae) takes place in the 2-day-old seedling. In Europe spores placed in soil in Aug. germinated in 7 days or at least before winter if so placed in Sept. In both cases spores were non-infective after overwintering. Since the spores are carried, as a typical smut, on the seed, standard chemical treatments are effective. Six races on 5 differential hosts have been demonstrated. Resistant forms are known, have been investigated, and in 1 cross resistance was controlled by a single dominant factor.

Kühnel, W. 1963. *Zentbl. Bakt. ParasitKde* Abt. 2 **117**: 180 (spore survival in soil; **43**, 2858).

Melchers, L. E. 1927. *Phytopathology* **17**: 739 (seed treatment control; **7**: 238).

Porter, R. H. et al. 1928. *Ibid* **18**: 911 (as above; **8**: 234).

—— ——. 1930. *Ibid* **20**: 915 (selection for resistance in China; **10**: 238).

Sundararaman, S. 1912. *Bull. agric. Res. Inst. Pusa* 97 (general).

Tu, C. et al. 1935. *Phytopathology* **25**: 648 (selection for resistance in China; **14**: 691).

Wang, C. S. 1936. *Ibid* **26**: 1086 (spore longevity; **16**: 247).

——. 1938. *Ibid* **28**: 860 (spores & morphology in vitro; **18**: 174).

——. 1943. *Ibid* **33**: 1122 (cytology & host penetration; **23**: 174).

——. 1944. *Ibid* **34**: 1050 (races; **24**: 187).

Yu, T. F. et al. 1934. *Bull. Nanking Coll. Agric. For.* **14**: 18 (seed treatment control; **13**: 629).

—— ——. 1937. *Chin. J. exp. Biol.* **1**: 235 (selection for resistance in China; **16**: 741).

—— ——. 1942. *Sci. Rec. Chunking* **1**: 248 (inheritance of resistance; **23**: 481).

Ustilago cynodontis (Pass.) P. Henn., *Bot. Jb.* (Engler) **14**: 369, 1891.

Ustilago carbo (DC.) Tul. var. *cynodontis* Pass., 1870
U. cynodontis (Pass.) Curzi, 1927
U. carbo (DC.) Tul. f. *cynodontis-dactylonis* McOwan & Thum., 1877
U. segetum (Bull.) Ditton var. *cynodontis* P. Henn., 1891.

SORI in the inflorescence of host reducing it to dusty spore masses. SPORES reddish brown, globose to subglobose, 5–8 μ diam., wall smooth. On spp. of *Cynodon* particularly *C. dactylon* (J. L. Mulder & P. Holliday, *CMI Descr.* 297, 1971).

Bermuda or Bahama grass smut is most conspicuous on the inflorescence which may be wholly or partly destroyed. Spore formation can also occur on the axis below the inflorescence and on the leaves. Galls have been reported. The pathogen is widespread in both tropical and temperate areas. Zambettakis found 3 other smuts on *C. dactylon*: *U. hitchcockiana* Zundel, *U. paraguariensis* Speg. and *Sorosporium cynodontis* Ling, and reported 3 races of *U. cynodontis*. Spread of the disease probably takes place mainly by planting infected rhizomes. But original infections presumably occur at the seedling stage and also in the mature plant when this has been cut. In USA infected plants have been produced by spore inoculation of seed and a clipped stand; a healthy stand which was clipped and inoculated with a spore suspension in June developed smutted heads in Nov. In Morocco spore trapping showed that *U. cynodontis* accounted for 68.4% of the total count for *Ustilago*; the highest numbers occurring from April to Aug. The normal 2 mating types exist but sexuality may prove to be multipolar when collections from different geographical areas are examined. In planting rhizomes a smut-free source should be used.

Chabert, J. et al. 1966. *Bull. Soc. mycol. Fr.* **82**: 569 (aerobiology in Morocco; **46**: 2252).
Chevalier, L. 1960. *Naturalia monspel.* Sér. bot. **12**: 3 (field symptoms; **40**: 690).
Halisky, P. M. et al. 1963. *Nature Lond.* **197**: 919 (heterothallism).
McCain, A. H. et al. 1962. *Phytopathology* **52**: 742 (inoculation of seed & clipped stand; **42**: 555).
Mehta, K. C. 1923. *J. Indian bot. Soc.* **3**: 243 (infection & spread in rhizome; **3**: 336).
Singh, H. 1965. *Curr. Sci.* **34**: 159 (gall symptoms; **44**: 2171).
Zambettakis, C. 1963. *Rev. Mycol.* **28**: 312 (smuts on Bermuda grass; **43**, 2947).
Zundel, G. L. 1939. *Mycologia* **31**: 572 (as above inter alia; **19**: 120).

Ustilago maydis (DC.) Corda, *Icon. Fung.* **5**: 3, 1842.

Uredo maydis de Candolle, 1815
U. zeae Schweinitz, 1822
Ustilago zeae (Schw.) Unger, 1833
U. zeae-mays Magnus, 1895.

(further synonyms in Zundel, *The Ustilaginales of the world*: 179, 1953; Fischer, *Manual of the North American smut fungi*: 281, 1953; Stevenson & Johnson, *Pl. Dis. Reptr* **28**: 663, 1944)

SORI in the inflorescences, leaves and stems as irregular swellings 1 cm to > 10 cm in length, at first limited by a white, cream or greenish membrane of fungus and host tissue which soon ruptures. SPORE MASS powdery, very dark brown. SPORES globose or subglobose to ellipsoidal, prominently and bluntly echinulate, 8–12 (7–10) μ diam. SPORIDIAL CULTURES may be obtained on PDA, carrot or other media. A range of synthetic nutrient media was tested by Ranker, *J. agric. Res.* **41**: 435, 1930 (G. C. Ainsworth, *CMI Descr.* 79, 1965).

Blister (boil or common) smut of maize (also on *Euchlaena mexicana*) is worldwide (Map 93). But it is absent from New Zealand, was eradicated from Australia, New South Wales (Magee, *Rep. 5th Commonw. mycol. Conf.*: 65, 1954), and in Australasia and Oceania has only been reported from Fiji, Hawaii and Papua New Guinea. The literature on this disease, with its spectacular symptoms, is very extensive; see the monograph by Christensen and the bibliography by Fischer, *The smut fungi*. General accounts of *U. maydis* were given by Butler, *Fungi and disease in plants*; Dickson, *Diseases of field crops*; Walker, *Plant pathology*; and see Fischer & Holton, *Biology and control of the smut fungi*. The fungus has been very commonly used as a tool in genetical research (ref. omitted) and in studying general variation. Much of the more recent literature on the disease originates in E. Europe (including USSR). Ullstrup on maize diseases (*Agric. Handb. U.S. Dep. Agric.* 199, 1974) stated that blister smut rarely occurs in tropical climates. Only a small selection of the more recent ref. is given.

The external symptoms are the galls which can form on any part of the plant where there is meristematic tissue. They are commonest on ears, tassels, stems and node shoots. In size they vary from very small to very large. Main stem galls (mostly just above nodes) may be 20–30 cm diam. and attached for 2–5 cm. Galls are at first firm, light in colour (almost white), semi-glossy peridium; the enclosing membrane eventually cracks and dies, thus exposing the dry, powdery, almost black spore mass. The leaf galls are usually largest on the leaf sheath; on the lamina they are small but can be very numerous. Galls vary in shape and may be lobed and convoluted. Ovaries, glumes, husks and male flowers show galls. The whole tassel and stalk may become a large smut mass. Inoculated (injection) seedlings develop severe hypertrophy and die. In the field galls are seldom seen until the plant is 0.34–1 m tall. Korshunova reported galls on the roots but this seems to be rare.

Infection occurs only through young tissue above ground (directly or via wounds) and is by sporidia from germinating spores. The fungus is heterothallic but, besides the haploid sporidia, diploid sporidia occur. Haploid hyphae can penetrate the host but fusion is needed to form the dicaryon and initiate the parasitic phase. Diploid sporidia which are heterozygous for the 2 mating alleles are pathogenic (solopathogens); but those which are homozygous for this factor are not pathogenic. The young galls may be visible in a few days and mature spores can occur 7–9 days after infection. Under humid conditions a gall may form masses of aerial sporidia in chains. Infection of seedlings can be systemic but in the older plants all infections are localised. The spores (mostly uninucleate and diploid) are spread aerially (like the sporidia) and are released from the host, particularly at harvest. The fungus has a very characteristic saprophytic phase and persists for long periods in the soil, on crop debris and refuse. Spores may be carried on the seed but this is not a major source of inoculum. Spores require nutrients for germination which is most rapid at *c*. 30° and good at 20–30°. Sporidial growth is best at the same temps. Spores overwinter readily and viabilities of > 10 years have been reported (König). But viability in silage may be only a few weeks. Blister smut is sometimes sporadic in occurrence. A temp. of 25° is probably most favourable for infection. *U. maydis* shows extreme variability both in pathogenicity and many other characteristics; no specific races are recognised and breeding for resistance has been to select for the more general (rather than specific) form of this aspect of control. Although seed treatment has been often recommended (e.g., Crespy et al.), the likely contamination and infection via the soil may make this control method ineffective unless maize is being introduced to a smut-free area. Under continuous cultivation neither rotation nor sanitation are important measures. The most effective control is to avoid susceptible hybrids. Wilcoxson found less smut where the plant populations were 18 000 or more per acre, compared with 5000–7000.

Billett, E. E. et al. 1978. *Physiol. Pl. Pathol.* **12**: 93, 103 (effect of infection on growth, translocation of C labelled assimilates; **57**, 3925, 3926).

Bojanowski, J. 1969. *Theor. appl. Genet.* **39**: 32 (inheritance of host reaction; **48**, 2347).

Callow, J. A. et al. 1973. *Ibid* **3**: 489 (histology in seedlings; **53**, 1345).

Christensen, J. J. 1963. *Monogr. Am. phytopathol. Soc.* 2, 41 pp. (367 ref.).

Crespy, A. et al. 1973. *Phytiat. Phytopharm.* **22**: 33 (seed treatment, damage in W. France; **53**, 3461).

Hindorf, H. 1978. *Z. PflKrankh. PflSchutz* **85**: 169 (inoculation & control; **58**, 188).

König, K. 1971. *Bayer landw. Jb.* **48**: 696 (ecology & control; **51**, 3976).

Korshunova, A. F. 1962. *Zashch. Rast. Mosk.* **7**: 59 (root symptoms; **42**: 259).

Kuznetsov, L. V. 1963. *Vest. Mosk. Univ.* Ser. Biol. **18**: 30 (viability; **43**, 1625).

Perova, A. Ya 1972. *Vest. Mosk. Univ.* Ser. 6 **27**: 43 (fine structure in inoculated maize; **52**, 1520).

Ponte, J. J. Da 1967. *Sydowia* **21**: 159 (resistance of spores to high temps; **48**, 1661).

Slepyan, E. I. 1968. *Mikol. i Fitopatol.* **2**: 49 (gall morphology; **47**, 2473).

Urech, P. A. 1972. *Phytopathol. Z.* **73**: 1 (infection; **51**, 3975).

Wilcoxson, R. D. 1975. *Pl. Dis. Reptr* **59**: 678 (infection & host population; **55**, 196).

Wimalajeewa, D. L. S. et al. 1971. *Physiol. Pl. Pathol.* **1**: 523 (common antigen relationship between pathogen & host; **51**, 1413).

Ustilago scitaminea Sydow, *Ann. Mycol. Berl.* **22**: 281, 1924.

Ustilago sacchari Auct. (non Rabenh.; non Fischer de Waldh.).

SORI in the inflorescences which are replaced by whip-like structures, each at first covered by a thin silver grey membrane of host tissue which soon flakes away. SPORE MASS powdery, dark brown. SPORES globose to subglobose, reddish brown, typically smooth or punctate, 5–10 μ (mostly 6–8 μ; av. 7.5 μ, fide Mundkur, *Kew Bull. 1939*: 525, 1940) diam. SPORIDIAL CULTURES on malt or PDA frequently develop typical smut spores.

Ustilago scitaminea

Mundkur (l.c.) distinguished a smut on *Saccharum barberi* in India as *U. scitaminea* var. *sacchari-barberi* by its slightly smaller, 5.1–8.6 (6.7) μ, finely verrucose spores, and a form from India and elsewhere on *S. officinarum* and *S. spontaneum* as *U. scitaminea* var. *sacchari-officinarum* which has slightly larger, 6.5–11.3 (8.4) μ, coarsely echinulate spores. Hirschhorn distinguished six groups of *U. scitaminea* on spore morphology. The range of morphological variation in *U. scitaminea* needs critical re-examination (G. C. Ainsworth, *CMI Descr.* 80, 1965; and see Mundkur & Thirumalachar, *Ustilaginales of India*).

This culmicolous smut of sugarcane has been reported on *Saccharum* spp., *Imperata cylindrica* and *Erianthus saccharoides*. *U. scitaminea* is widespread (Map 79) with the crop but has not been reported from Australia, some West Indian islands, and Papua New Guinea. It was recently found in Florida, USA. There are several reviews on this smut disease, the latest is by Lee-Lovick (see also Antoine; Appalanarasiah; Arruda et al.; Vizioli). The disease, first reported in the 1870s in South Africa, has periodically caused severe losses but at other times has been of little importance. Epidemics presumably can arise where extremely susceptible cvs are being grown and climatic conditions become very favourable.

The most characteristic symptom is the often curved, whip-like structure, usually up to 90 cm long, at the apex of a cane. This has a central core surrounded by the spore masses which soon become free for wind dispersal. An infected stool, before production of the whip, shows these symptoms: early stimulation of tillering, shoots thin and grass like with long internodes, and growing rather faster than healthy cane, leaves short and stiff, and carried at a more acute angle. If an infected sett is planted all the subsequent shoots become smutted and develop whips (primary infection). If the sett is initially healthy new shoots arising from the stool can become infected or not and therefore not all canes, in this case, will produce whips (secondary infection). Smutted galls have been described from leaves. Byther et al. described some unusual symptoms (convoluted sori from lateral buds, stalk distortion, galls and multiple buds) from Hawaii.

Infection occurs in the buds and very young shoots (<4 cm long) at or below the soil surface. Young buds, or those beginning to develop, are most susceptible; infection occurring beneath the inner scales where the epidermis remains undifferentiated. Although sporidia can cause infection this mostly occurs through hyphae arising from the spore promycelium. Sporidia appear to be formed only under certain conditions. As the new shoot develops pockets of mycelium become established at each primordium which becomes the nodal bud. Mycelium is absent from the internodes. On development each infected bud will eventually form the characteristic sporulating whip after 4–8 months. As tillering becomes less with age so the disease tends to be checked. Until the bud develops the pathogen remains latent. The max. formation of appressoria occurs at 31° and 8 hours is adequate for infection. The growth of sporidia in vitro and of promycelia has a broad opt. of 27–31°. Free moisture is required for spore germination. Under moist conditions viability of spores was lost in 2–3 weeks, when dry they were viable for 105–128 days at 25°. Whip production was highest at 29° compared with 25° and 22°. High temperatures and excessive moisture favour *U. scitaminea*; it is more severe on irrigated sugarcane. Where infected setts have been used there are 2 flushes of whip production, the first arising from primary infection and the second from infection of new shoots as they emerge through the soil. Spores are dispersed mostly around noon.

Two races have been reported from Taiwan and 2 from Hawaii. In Taiwan a third race was formed from the hybridisation of races 1 and 2. These races differ not only in pathogenicity to cvs but also (1 and 2) in cultural characteristics on PDA (Comstock et al.; Hsieh et al.; Leu et al. 1972a, 1978). Mating types, cytology and spore viability have been described (Alexander et al.; James; Khan et al.; Luthra et al.; Muhammad et al.; Saxena et al.; Viswanathan).

The most satisfactory control measure is probably selection for resistance. Lee-Lovick gave a full discussion of control (roguing, healthy planting material, disinfection, fallow, rotation, burning at harvest and resistance). Simmonds, after an outbreak of sugarcane smut in Guyana, considered the relative costs of replacing susceptible cvs with resistant lines without disturbing the normal replanting cycle or a rapid replanting program. He concluded that the former strategy was preferable.

Alexander, K. C. et al. 1966. *Curr. Sci.* **35**: 603 (2 mating types; **46**, 1714).

Antoine, R. 1961. In *Sugarcane diseases of the world* Vol. 1: 327 (general, 88 ref.).
Appalanarasiah, P. 1961. *Indian J. Sugcane Res. Dev.* **6**: 34 (review; **41**: 615).
Arruda, S. C. et al. 1951. *Biológico* **17**: 155 (review, Brazil; **31**: 400).
Bock, K. R. 1964. *Trans. Br. mycol. Soc.* **47**: 403 (conditions for infection, resistance screening in Kenya; **44**, 229).
Byther, R. S. et al. 1974. *Pl. Dis. Reptr* **58**: 401 (unusual symptoms in Hawaii; **53**, 3606).
Chona, B. L. 1943. *Indian Fmg* **4**: 401 (general; **23**: 407).
Comstock, J. C. et al. 1977. *Sugcane Pathol. Newsl.* 19: 24 (new race in Hawaii; **57**, 3595).
Fawcett, G. L. 1944. *Boln Estac. exp. agric. Tucumán* 47, 15 pp. (general in Argentina; **24**: 73).
Hirschhorn, E. 1943. *Notic. Mus. La Plata* **8**: 23 (general & morphology in Argentina; **25**: 11).
Hsieh, W. H. et al. 1976. *Rep. Taiwan Sug. Exp. Stn* 73: 51 (compatibility & pathogenicity of 2 races; **56**, 3207).
—— ——. 1977. *Ibid* 76: 53 (culture of 2 races; **57**, 766).
James, G. L. 1969. *Sugcane Pathol. Newsl.* 3: 10 (viability of spores in soil; **49**, 549).
Khan, A. M. et al. 1964. *Indian Sugcane Jnl* **9**: 55 (temp., RH & viability of spores; **45**, 1497).
Lambat, A. K. et al. 1968. *Mycopathol. Mycol. appl.* **36**: 300 (morphology; **49**, 243).
Lee-Lovick, G. 1978. *Rev. Pl. Pathol.* **57**: 181 (review, 90 ref.).
Leu, L. S. 1971. *Pl. Prot. Bull. Taiwan* **13**: 6 (histology; **51**, 2820).
—— et al. 1972a. *Sugcane Pathol. Newsl.* 8: 12 (races; **51**, 4319).
—— ——. 1972b. *Rep. Taiwan Sug. Exp. Stn* 56: 37 (germination & storage of spores, compatibility; **52**, 3416).
—— ——. 1978. *Ann. phytopathol. Soc. Japan* **44**: 321 (new race by hybridisation).
Luthra, J. C. et al. 1938. *Indian J. agric. Sci.* **8**: 849 (life history & survival).
Muhammad, S. et al. 1962. *Biologia Lahore* **8**: 65 (mating types; **42**: 277).
Muthusamy, S. 1974. *S. Afr. Sug. J.* **58**: 14 (var. susceptibility & bud characters; **54**, 4129).
Saxena, K. M. S. et al. 1966. *Indian Phytopathol.* **19**: 286 (2 mating types; **47**, 304).
Saxena, S. K. et al. 1964. *J. Indian bot. Soc.* **43**: 61 (spore germination & RH; **44**, 1230).
Sharma, S. L. 1956. *Proc. int. Soc. Sugcane Technol.* 9th Congr.; 1134 (morphological change in host; **38**: 33).
Simmonds, N. W. 1976. *Int. Sug. J.* **78**: 329 (control strategy in W. Indies; **56**, 2651).
Sreeramulu, T. et al. 1972. *Trans. Br. mycol. Soc.* **58**: 301 (spore dispersal; **51**, 4316).
Subramanian, T. V. et al. 1951. *Proc. 1st Conf. Sugcane Res. Wkrs India* 2: 55 (infection & development in host; **33**: 633).
Viswanathan, T. S. 1964. *Mycopathol. Mycol. appl.* **23**: 203 (nuclear behaviour in spores; **44**, 1231).
Vizioli, J. 1953. *Brasil açucar* **42**: 535 (review; **33**: 502).
Waller, J. M. 1967. *E. Afr. agric. For. J.* **32**: 399 (testing for resistance; **46**, 3189).
——. 1969. *Trans. Br. mycol. Soc.* **52**: 139 (epidemiology in Kenya; **48**, 1941).
——. 1970. *Ibid* **54**: 405 (infection & resistance in Kenya; **50**, 258).
Wiehe, P. O. 1949. *Rev. agric. Ile Maurice* **28**: 7 (incidence & infection in Mauritius; **28**: 544).

USTULINA Tulasne, *Sel. Fung. carp.* 2: 23, 1863.

(Xylariaceae)

PERITHECIA globose to slightly flask shaped, ostiolate, clustered in a common stroma which may have a black, brown, reddish or even whitish surface and may vary in shape from a hemispherical cushion to a thin spreading, flat topped crust, not gelatinous, on dead wood or bark. ASCI with persistent wall, mostly cylindrical, unitunicate. ASCOSPORES elliptic fusiform to bean shaped, dark brown, aseptate, with a distinct colourless furrow down one side. Differs from *Hypoxylon* in its large perithecia and ascospores.

Ustulina deusta (Hoffm. ex Fr.) Lind, *Danish fungi*: 252, 1913.
Sphaeria deusta Hoffm., 1787
Hypoxylon deustum (Hoffm. ex Fr.) Grev., 1828
Ustulina vulgaris Tul., 1863
U. zonata (Lev.) Sacc., 1882
(further synonymy in *CMI Descr.* 360 and J. H. Miller, *A monograph of the world species of* Hypoxylon, 1961)

STROMATA massive, orbicular to effuse, forming discrete cushions erumpent through the periderm or confluent, undulate to pulvinate, arising on the site of the conidial state, greyish white and leathery at first, becoming reddish black then black and brittle with age, readily detached from the substratum (separating along the line of the perithecial bases), stromatic tissue internally whitish, composed of vertical fibrous masses which disintegrate in old stroma, crust usually 1–3.5 mm thick, but when perennial becoming thicker and concentrically zoned.

Ustulina deusta

PERITHECIA scattered in the stroma, immersed in its whitish flesh, with short projecting, black, papillate ostioles, very large, vertically orientated and elongated to 1.5 × 1 mm. ASCI cylindrical, very thin walled, often becoming gelatinised, 8 spored, 190–300 × 8–15 μ, stalk 50–60 μ long. PARAPHYSES numerous, branching from the base, filiform. ASCOSPORES uniseriate, dark brown to black, narrowly ellipsoid to fusiform, sides inequilateral, apices acute, with a longitudinal furrow which is sometimes indistinct, 26–40 × 6–13 μ. STAT. CONID. on host precedes the formation of the perfect state stromata, greyish white to greyish brown, forming an effuse continuous layer where the periderm has disappeared, consisting of vertically orientated compacted tufts of parallel conidiophores, usually *c.* 0.5 mm thick. CULTURES growing readily on PDA and malt agar at 23°, white to grey at first but becoming greyish brown with the formation of conidia on vertical tufts of conidiophores, becoming black in reverse after 2–3 weeks. CONIDIOPHORES hyaline to pale fuscous, sparsely branched. CONIDIA hyaline to pale fuscous, simple, narrowly ovate, smooth walled, 5–7 (–9) × 2–3.5 μ.

U. zonata has been considered to differ from *U. deusta* by some in having slightly wider ascospores, predominating in the tropics and in its parasitism. But it does not seem to be specifically distinct, although it may merit recognition as an infraspecific taxon. *U. deusta* is a readily identifiable sp. which is distinguished from spp. of *Hypoxylon* by its large perithecia and ascospores, and the way in which the massive carbonaceous stroma is easily removed from the host tissues. The stromata characteristically occur on the basal parts of the host while those of *Hypoxylon* spp. may often be found higher up the plant as well (D. L. Hawksworth, *CMI Descr.* 360, 1972).

U. deusta is very widespread (Map 417) and plurivorous on woody hosts. Wilkins gave an annotated list of early references (to 1932), a host index and described germination and infection (temperate hosts) for both spore types. Among the tropical tree crops on which it can cause losses are cacao, oil palm, rubber and tea. It is most important as a pathogen causing charcoal stump rot of tea in N.E. India and Sri Lanka (see Ram et al. for S. India). On oil palm the disease is called charcoal base rot. The root, collar and stem rots caused by this fungus must be distinguished from other fungi which attack the roots; for tea see the key by Hainsworth, *Tea pests and diseases and their control* and for rubber see Anon., *Plrs' Bull. Rubb. Res. Inst. Malaysia* 133: 111, 1974.

On tea (gradual and partial or sudden and entire death, with leaves remaining attached for a time) the characteristic fructifications, at first white then black, form on the bark and exposed wood at the collar and on exposed roots. White, fan shaped mycelium occurs between the bark and wood. The wood remains almost normal in colour and it shows irregular, single or double black bands or lines. On rubber the first signs of infection can be latex bleeding from the trunk or branches. The rotted wood is a uniform light brown (roots are brittle) and shows a network of black, usually double lines. The fructifications are at first velvety, greyish becoming black. On oil palm the characteristic symptoms of a dying tree can be seen when the stem is split. Its base shows a black, dry rot, extremely hard and extending as a band across the entire base; affected palms may fall. Tissues above the thin zone of basal blackening may exhibit a dry fibrous rot extending upwards for 45 cm. *U. deusta*, common on dead wood, spreads through root contact, from diseased woody material in the soil and by airborne spores. All cut woody surfaces, and wounded branches or trunks can be directly infected by spores. Infection of rubber roots may also take place through moribund root initials and lenticels; the fungus then spreads to the collar and stem. In Sri Lanka *U. deusta* infects exposed stump surfaces of *Grevillea robusta* and then spreads to adjoining tea. In tea, generally, shade trees may be a source of infection.

The principles of disease control for the plurivorous root, collar and stem fungal pathogens in these crops apply to *U. deusta*. Because of the importance of airborne inoculum special attention should be paid to protecting (or removing) freshly cut or otherwise damaged woody surfaces. In rubber use the recommended fungicidal paint. Tea bushes struck by lightning should be removed without delay as they are liable to be attacked quickly. Attacks on tea may arise from the cut stumps of shade trees which have become infected. In removing (thinning) healthy trees the stumps should also be removed. If this is not possible the trees should be ring barked in full leaf and only felled after complete defoliation. In Sri Lanka methyl bromide, used for the control of *Poria hypolateritia* (q.v.) effectively controls *U. deusta*. Shade trees infected with charcoal stump rot should be dug out completely.

Brooks, F. T. 1915. *New Phytol.* **14**: 152 (on rubber).
Ram, C. S. V. et al. 1974. *Bull. United Plrs' Assoc. S. India* 31: 11 (fumigation against *Armillariella mellea* & status of *Ustulina deusta*: **55**, 1944).
Rogers, J. D. 1968. *Mycopathol. Mycol. appl.* **35**: 249 (chromosome number & taxonomy; **48**, 730).
Sharples, A. 1918. *Ann. appl. Biol.* **4**: 153 (on rubber).
Thompson, A. 1936. *Malay. agric. J.* **24**: 222 (on oil palm; **15**: 647).
Varghese, G. 1971. *J. Rubb. Res. Inst. Malaya* **23**: 157 (infection of rubber; **51**: 2805).
Wallace, G. B. 1936. *E. Afr. agric. J.* **1**: 266 (on cacao; **15**: 427).
Wilkins, W. H. 1934. *Trans. Br. mycol. Soc.* **18**: 320 (literature & hosts; **13**: 597).
——. 1936. *Ibid* **20**: 133 (on *Tilia vulgaris*; **15**: 471).
——. 1938. *Ibid* **22**: 47 (ascospore germination & infection; **18**: 69).
——. 1939. *Ibid* **23**: 65, 171 (conidial germination & infection, on *Ulmus campestris*; **18**: 638; **19**: 49).
——. 1944. *Ibid* **26**: 169 (on *Fagus sylvatica*; **23**: 157).

VALSA Fr., *Summ. veg. Scand.* Sect. Post.: 410, 1849.

(Diaporthaceae)

PSEUDOSTROMATA without dark lines, often reduced to clypeus-like structures. ASCOMATA in the pseudostromata often arranged in circles. ASCI fusiform, clavate or nearly cylindrical, unitunicate, loosening from the ascogenous cells at an early stage and lying free in the ascoma. ASCOSPORES allantoid. On woody plants with conidial states in *Cytospora* q.v. (from key by E. Müller & J. A. von Arx, in *The fungi* (edited by G. C. Ainsworth et al.) Vol. IV A: 113, 1973).

Leucostoma (Nits.) Hohn. is sometimes united with *Valsa*. The former has a distinct (better developed) pseudostroma which is often delineated by dark lines in the bark. The genera were studied by Défago. *Valsa* needs revision (Dennis, *British Ascomycetes*). Most spp. occur on temperate, woody dicotyledons, and on conifers in the northern temperate regions. Examples of the latter category are: *V. abietis* (Fr.) Fr. (stat. conid. *Cytospora abietis* Sacc.), *V. curreyi* Nits. (stat. conid. *C. curreyi* Sacc.) and *V. kunzei* (Fr.) Fr. (stat. conid. *C. kunzei* Sacc.). These 3 fungi are weak parasites associated with canker and dieback.

Défago, G. 1944. *Phytopathol. Z.* **14**: 103 (characters & taxonomy; **22**: 40).
Kern, H. 1961. *Ibid* **40**: 303 (*Leucostoma*; **40**: 522).
Smerlis, E. 1971. *Phytoprotection* **52**: 28 (pathogenicity of *Valsa abietis* inter alia; **51**, 732).
Waterman, A. M. 1955. *Phytopathology* **45**: 686 (*V. kunzei* & conifer canker; **35**: 564).

Valsa eugeniae Nutman & Roberts, *Trans. Br. mycol. Soc.* 36: 229, 1953.

STROMATA ash grey, develop in the outer bark, initially in the phellogen but subsequently other layers develop, fused with the host tissue. ASCOCARPS black, sometimes in small groups of 4–12 but more frequently in groups of 100 or more, forming tiers at different levels in the stromata, the oldest externally. As the lower layers develop the necks elongate until their tops are about the same level. Globose or laterally compressed, 200–280 μ diam., with necks 600–900 μ (up to 5 mm in thick bark), 50–60 μ diam. with rounded apex. ASCI sessile, numerous, loosely aggregated, subclavate, wall not readily visible, 8 spored, $17–26 \times 4–7$ μ, eventually filling the lumen of the ascocarp. PARAPHYSES absent. ASCOSPORES small, allantoid, hyaline, aseptate, $4–5 \times 1–1.5$ μ, colourless or opalescent in mass. PYCNIDIA black, flattened at base, not associated with any definite stroma, unilocular, hyaline internally, with the conidia forming layer deeply convoluted, 200–300 μ diam. and 280–500 μ high. CONIDIOPHORES hyaline, $10–42 \times 1–1.5$ μ. PYCNIDIOSPORES allantoid, aseptate, $2–4 \times 0.4–1$ μ, extruded to form slimy yellow globules. In culture pycnidia may be multilocular (A. Sivanesan & P. Holliday, *CMI Descr.* 230, 1970).

V. eugeniae causes sudden death and a dieback of clove (*Eugenia caryophyllus*). It was also reported to have caused the death of cashew (*Anacardium occidentale*) in Tanzania. In Jamaica a dieback of pimento (*Pimenta dioica*), attributed to *Ceratocystis fimbriata* (q.v.), yielded isolates of *V. eugeniae*. Sudden death of clove should be a salutary example of the time it can take (and of the vicissitudes that result) to find the cause of a plant disease. It was known in Zanzibar and Pemba for nearly a century before its aetiology (possibly still not fully elucidated) was determined in 1951. Nutman et al. (1953b) gave a revealing account of the history of attempts to solve the problem. In 1949 it was estimated that half the clove trees in Zanzibar had been killed in the previous 10–12 years. Between 1939–40 3 plantations showed losses of 12.5, 9 and 36.6%.

The fungus also occurs in Malagasy Republic, Malaysia (W.) and Thailand.

The syndrome of sudden death is complicated, firstly by the differing effects on the plant by the pathogen and secondly by the presence, often on the same tree, of the also previously unknown *Endothia eugeniae* (q.v.) which causes a serious dieback of the same crop. The classical symptoms of sudden death occur only in mature trees. A barely discernible yellowing of the foliage is followed by a heavy, green leaf fall and wilt of the remaining leaves which wither to a russet colour. A few days after death a characteristic yellow stain in the wood is seen at the collar; this spreads up the trunk and becomes generally distributed in the tree 9 months later. The stain can be found much earlier in the roots. It is often separated from healthy wood by a black zone line and a narrow zone stained bluish grey. Tyloses and gum occur in the vessels. The pycnidia, which appear before the ascocarps, are abundant on cut, split or broken woody surfaces in moist conditions. The ascocarps appear through cracks in the bark at ground level. Fructification moves upwards in the tree and can continue for several years on the trunk and branches. Where new clove plantings have been made in areas previously killed off by sudden death the young trees (up to nearly 20 years old) show a slow decline. They appear unthrifty with yellowish and sparse foliage. Such trees have a slow but progressive root rot. Seedlings are immune to infection whilst plants up to 4 years old are very resistant. Finally, *V. eugeniae* can cause a branch dieback that is quite unconnected with any infection of the roots; here it functions as a wound parasite but spreads more slowly than *E. eugeniae*.

Before the aetiology was described the conditions in cloves known as sudden death and dieback had been confused. Damage due to the dieback, caused by both these fungi, has apparently been aggravated by bad harvesting methods, i.e. breaking and stripping branches to get at the crop. One of the disastrous side effects of this malpractice may have been to increase the amount of inoculum of *V. eugeniae* over the years, one result possibly being the sudden death type of symptom. In this type infection appears to begin in the absorbing roots, spreads centripetally, and so to the collar and trunk. Improvement in husbandry and fungicidal treatment of wounds above ground will presumably lead to control of the dieback disease. This could lead to a reduction of the root infection by decreasing inoculum in the soil. But no details on control measures appear to be available.

Nutman, F. J. et al. 1949. *Ann. appl. Biol.* **36**: 419 (symptoms & epidemiology; **29**: 277).

———. 1951. *Emp. J. exp. Agric.* **19**: 145 (loss; **30**: 580).

———. 1953a. *Nature Lond.* **171**: 128 (**32**: 278).

———. 1953b. *E. Afr. agric. For. J.* **18**: 146 (sudden death history in Zanzibar; **33**: 180).

———. 1953c. *Trans. Br. mycol. Soc.* **36**: 229 (*Valsa eugeniae* & *Cryptosporella eugeniae* spp. nov.; **33**: 179).

———. 1954. *Ann. appl. Biol.* **41**: 23 (aetiology of sudden death; **34**: 63).

———. 1971. *PANS* **17**: 147 (clove industry & diseases; **51**, 1751).

Wallace, G. B. et al. 1955. *E. Afr. agric. For. J.* **21**: 42 (on cashew; **35**: 496).

VERONAEA Cif. & Montem., *Atti Ist. bot. Univ. Lab. crittogam Pavia* Ser. 5 **15**: 68, 1958 (in separate, 1957).

Sympodina Subram. & Lodha, 1964.
(Dematiaceae)

COLONIES effuse, brown, greyish, brown or blackish brown, cottony hairy or velvety. Mycelium partly immersed, partly superficial. STROMA none. SETAE and HYPHOPODIA absent. CONIDIOPHORES macronematous, mononematous, unbranched or occasionally loosely branched, straight or flexuous, sometimes geniculate, pale to mid brown or olivaceous brown, smooth. CONIDIOGENOUS CELLS polyblastic, integrated, terminal often becoming intercalary, sympodial, cylindrical, cicatrised, scars usually small, flat. CONIDIA solitary, dry, acropleurogenous, simple, usually ellipsoidal or fusiform, sometimes cylindrical, rounded at the apex truncate at the base, usually colourless, pale brown or olivaceous brown, smooth or minutely verruculose, with 0, 1 or a few transverse septa (M. B. Ellis, *Demat. Hyphom.*: 245, 1971).

V. musae M. B. Ellis (synonym: *Chloridium musae* Stahel, nomen non rite publicatum) is the commonest of 3 dematiaceous fungi causing minor leaf spots on banana called speckle. The others are: *Periconiella musae* M. B. Ellis (synonym: *Ramichloridium musae* Stahel, nomen non rite publicatum), and *Cladosporium musae* Mason (see Martyn, *Cladosporium* q.v.). *P. musae* and *C. musae* are both described by Ellis l.c. pp. 296 and 317, respectively. *V. musae* is described by Ellis in *More Demat. Hyphom.*: 209, 1976. This fungus penetrates the stomata and, since spread in the leaf is restricted to an air chamber and the palisade cells immediately surrounding it, only a small pin point, brown or black lesion results. These speckles are more prominent on the upper surface; they are very numerous and become aggregated into large (up to 4 cm diam.) circular blotches which are chlorotic amongst the necrotic speckles. No control measures are required; spraying for control of sigatoka (*Mycosphaerella musicola*) reduces speckle diseases.

Ananthanarayanan, A. et al. 1964. *Madras Agric. J.* **51**: 294 (*Periconiella musae*).
Ellis, M. B. 1967. *Mycol. Pap.* 111, 46 pp. (*Periconiella* inter alia).
Frossard, P. 1963. *Fruits* **18**: 443 (*Cladosporium* sp.; **43**, 2686).
Stahel, G. 1937. *Trop. Agric. Trin.* **14**: 42 (*Veronaea musae* & *P. musae*; **16**: 476).

VERTICILLIUM Nees ex Wallr., *Fl. crypt. Germ.* Vol. 2. 301, 1833.

Acrostalagmus Corda, 1838.
(Moniliaceae)

MYCELIUM composed of creeping, septate, branching hyphae, hyaline or lightly coloured. CONIDIOPHORES erect, septate, simple or branched. Branches of the first order whorled, opposite or alternate; branches of second order whorled, dichotomous or trichotomous on the branches of the first order. Terminal elements on main stalk and branches phialides. PHIALIDES flask shaped, somewhat elongate, narrowed toward the apex. CONIDIA usually aseptate, formed singly from the tips of the phialides, often held together in slime to form false heads; globose, elliptical, oval, inverted egg shaped, short fusiform, hyaline or slightly coloured (after C. V. Subramanian, *Hyphomycetes*: 648, 1971; see S. J. Hughes *Mycol. Pap.* 45, 1951 for some illustrations).

There is no monographic treatment of this economically important genus or the diseases it causes. The reviews by Pegg (on pathology) and Isaac (1967, on speciation) should be consulted. The distinctions between the 2 major pathogens (*V. albo-atrum* and *V. dahliae* q.v.) have been and still are ignored or confused; or on which there is disagreement. Their differences are briefly summarised. The 2 spp. can be distinguished morphologically in that *V. dahliae* forms true microsclerotia, whilst *V. albo-atrum* forms only dark, resting mycelium which does not bud to form microsclerotia. Talboys (1960) described a culture medium as an aid in identification. *V. dahliae* shows some growth at 30° and tends to be the sp. isolated from warmer geographical areas. *V. albo-atrum* does not grow at this temp. The latter occurs on dicotyledons, mostly herbaceous, whilst the former occurs on woody hosts as well (Anon.; Engelhard; Meer; Rudolph). *V. albo-atrum* has been described as the more virulent of the two on several crops; thus a correct identification may well be of economic importance. Recent work on protein patterns (Hall; Milton et al.; Pelletier et al. 1971; Whitney et al.), fine structure (Griffiths et al. 1971) and hybridisation (Hastie 1973) supports the view that the 2 spp. are distinct.

Both fungi are vascular pathogens, although true wilt symptoms do not always occur. Both are considered to be root inhabitants (sensu S. D. Garrett). Isaac (1953b) could not recover the 2 spp. from soil after 6 months from the time of inoculation. But whilst survival of the non-microsclerotial *V. albo-atrum* declines rapidly (hence fallow or crop rotation exerts some control), *V. dahliae* persists in the soil as effective inoculum for considerably longer. In both cases non-pathogenic infection of weeds or crops (as in *Fusarium oxysporum*) may be factors in survival. Powelson reviewed soil behaviour, Talboys (1964) formed a concept of the host and parasite relationship, and Hastie (1970) reviewed genetical work. Ludbrook and Edgington et al. compared these 2 spp. with respect to temp. In vitro growth of *V. albo-atrum* being best at 22° and for *V. dahliae c.* 24°, the former showing considerably less growth at 28°. *V. dahliae* causes disease over a wide temp. range (up to 30°) and this can be severe at 28°. *V. albo-atrum* can cause severe disease at 24° but above this temp. symptom severity falls off. Also high air temps tend to reduce disease in this sp. even if soil temps are favourable. Selvaraj et al. found that conidia of *V. dahliae* germinated at $>30°$ whilst those of *V. albo-atrum* did not. Where temps are continually high in the tropics neither pathogen may be of much significance; in such areas forms of *F. oxysporum* are more important wilt pathogens. Examples of differences in geographical range within a restricted region are given by Aubé et al. for Canada and Smith for New Zealand. Since both the host range and distribution of *V. dahliae* and *V. albo-atrum* overlap to some extent, the temp. factor may be important in determining which sp. is the main pathogen in any given disease. For example Isaac (1949) found that *V. dahliae* from sainfoin was pathogenic to tomato at 21.5, 25 and 27°; whereas *V. albo-atrum* from cucumber did not induce wilt in sainfoin and tomato at 25° and 27° but did so at 21.5°. This worker and others have described the interactions between these 2 spp. and 3 other spp. which are only weakly pathogenic and have more marked saprophytic characteristics: *V. nigrescens* Pethybridge, *V. nubilum* Pethybridge and *V. tricorpus* Isaac (*CMI Descr.* 257, 258, 260). The spread of the diseases caused by these 5 spp., and the dormancy and germination of the resting structures have been described (Isaac 1953; Isaac et al. 1962, 1966). Host penetration and colonisation, and pathogenicity amongst the spp. have been studied by Griffiths et al. (1963, 1964, 1966); and see Wilhelm for resistance.

V. tricorpus can be distinguished from other *Verticillium* spp. by the orange yellow colour of the first formed prostrate hyphae, and the formation of chlamydospores and microsclerotia. It causes a wilt in tomato; Taylor (1968) described severe symptoms on hop, and mild ones on eggplant, tobacco and tomato. Fluctuations in soil populations during the year were considered to be due to the formation of short lived conidia by the microsclerotia (Taylor 1969). Hoes described the development of the chlamydospores in *V. nigrescens* and *V. nubilum*. Those of the latter are larger and produced in chains of several cells, and should not be confused with the microsclerotia of *V. dahliae*. *V. theobromae* (a non-vascular sp.) is described separately.

Anon. 1957. *Pl. Dis. Reptr* Suppl. 244: 24 (host index).

Aubé, C. et al. 1964. *Can. J. Pl. Sci.* **44**: 427 (on forage legumes; **44**, 738).

Berkeley, G. H. et al. 1931. *Sci. Agric.* **11**: 739 (taxonomy of *Verticillium albo-atrum* and *V. dahliae*; **11**: 130).

Devaux, A. L. et al. 1966. *Can. J. Bot.* **44**: 803 (*Verticillium* spp. on horticultural crops; **45**, 3082).

Edgington, L. V. et al. 1957. *Phytopathology* **47**: 594 (effect of temp. on infection of tomato; **37**: 186).

Ende, G. Van den 1958. *Acta bot. neerl.* **7**: 665 (biology of *V. albo-atrum* and *V. dahliae*; **38**: 312).

Engelhard, A. W. 1957. *Pl. Dis. Reptr* Suppl. 244: 23 (host index).

Griffiths, D. A. et al. 1963. *Hort. Res.* **2**: 104 (wilt of *Lupinus* & sunflower; **42**: 768).

— —. 1963. *Ann. appl. Biol.* **51**: 231 (infection of tomato leaves; **42**: 634).

— —. 1964. *Mycopathol. Mycol. appl.* **24**: 103 (mechanical resistance in root hairs; **44**, 2430).

— —. 1966. *Ann. appl. Biol.* **57**: 59 (colonisation of cellular membranes; **45**, 1675).

— —. 1966. *Ibid* **58**: 259 (host & parasite relationships in tomato; **46**, 438).

— —. 1971. *Can. J. Microbiol.* **17**: 1533 (fine structure of resting mycelium of *V. albo-atrum*; **51**, 3209).

Hall, R. 1969. *Can. J. Bot.* **47**: 2110 (protein patterns in *V. albo-atrum* & *V. dahliae*; **49**, 1994).

Hastie, A. C. 1970. In *Root diseases and soil-borne pathogens*: 55, Univ. California Press (review of genetics, 50 ref.).

—. 1973. *Trans. Br. mycol. Soc.* **60**: 511 (hybridisation of *V. albo-atrum* & *V. dahliae*; **53**, 431).

Heale, J. B. et al. 1965. *Ibid* **48**: 39 (environment & formation of dark resting structures; **44**, 2069).

Hoes, J. A. 1971. *Can. J. Bot.* **49**: 1863 (chlamydospore development in *V. nigrescens* & *V. nubilum*, **51**, 2246).

Isaac, I. 1949. *Trans. Br. mycol. Soc.* **32**: 137 (comparisons between pathogenic isolates; **29**: 333).

—. 1953a. *Ibid* **36**: 180 (as above; **33**: 185).

—. 1953b & c. *Ann. appl. Biol.* **40**: 623, 630 (interactions between & spread of *Verticillium* spp.; **33**: 544, 545).

— et al. 1962. *Nature Lond.* **195**: 826 (germination of resting structures; **42**: 97).

— —. 1966. *Trans. Br. mycol. Soc.* **49**: 669 (dormancy & germination of resting structures; **46**, 1212).

—. 1967. *A. Rev. Phytopathol.* **5**: 201 (speciation, 111 ref.).

Leal, J. A. et al. 1962. *Nature Lond.* **195**: 1328 (lack of pectic enzyme in non-pathogenic spp.; **42**: 183).

Ludbrook, W. V. 1933. *Phytopathology* **23**: 117 (temp. & comparisons between *V. albo-atrum* & *V. dahliae*; **12**: 470).

Meer, J. H. H. Van der 1925. *Meded. LandbHoogesch. Wageningen* 28: 82 pp. (*V. dahliae* & *V. albo-atrum*, morphology & temperate hosts; **4**, 495).

Milton, J. M. et al. 1971. *Trans. Br. mycol. Soc.* **56**: 61 (protein patterns in 5 spp.; **50**, 2136).

Pegg, G. F. 1974. *Rev. Pl. Pathol.* **53**: 157 (diseases, 142 ref.).

Pelletier, G. et al. 1970. *Can. J. Microbiol.* **16**: 231 (environment, conidial size & contents; **49**, 3137).

— —. 1971. *Can. J. Bot.* **49**: 1293 (protein patterns in 4 spp.; **51**, 1196).

Powelson, R. L. 1970. In *Root diseases and soil-borne pathogens*: 31, Univ. California Press (populations in soil, 52 ref.).

Reiss, J. 1969. *Z. PflKrankh. PflPath. PflSchutz* **75**: 480 (distinction between *V. albo-atrum* & *V. dahliae*; **48**, 3368).

Rudolph, B. A. 1931. *Hilgardia* **5**: 197 (survey & host list, 329 refs; **10**: 757).
Selvaraj, J. C. et al. 1975. *Indian Phytopathol.* **28**: 412 (temp. & conidial germination in *V. dahliae* & *V. albo-atrum*; **56**, 2367).
Sherbakoff, C. D. 1949. *Bot. Rev.* **15**: 377 (breeding for resistance to *Fusarium* & *Verticillium* **28**: 532).
Smith, H. C. 1965. *N.Z. Jl agric. Res.* **8**: 450 (morphology of *V. albo-atrum, V. dahliae* & *V. tricorpus*; **44**, 3007).
Talboys, P. W. 1960. *Pl. Pathol.* **9**: 57 (identification of *V. albo-atrum* & *V. dahliae* in culture; **40**: 90).
——. 1964. *Nature Lond.* **202**: 361 (a concept of host & parasite relationship; **43**, 2521).
—— et al. 1970. *Trans. Br. mycol. Soc.* **55**: 367 (pectic enzymes produced by *Verticillium* spp.; **50**, 2135).
Taylor, J. B. 1968. *N.Z. Jl agric. Res.* **11**: 521 (host range of *V. tricorpus*; **47**, 2658).
——. 1969. *Can. J. Bot.* **47**: 737 (soil populations of *V. tricorpus*; **48**, 3347).
Tolmsoff, W. J. 1972. *Phytopathology* **62**: 407 (genetic factors as sources of variation in *Verticillium* spp.; **51**, 4604c).
Whitney, P. J. et al. 1968. *J. exp. Bot.* **19**: 415 (protein patterns in *V. albo-atrum, V. dahliae* & *Fusarium oxysporum*; **47**, 3374).
Wilhelm, S. 1975. In *Biology and control of soilborne plant pathogens*, Am. Phytopathol. Soc.: 166 (resistance; 42 ref.).

Verticillium albo-atrum Reinke & Berthold, *Die Zersetzung der Kartoffel durch Pilze*: 75, 1879.
Verticillium albo-atrum var. *caespitosum* Wollenw., 1929
V. albo-atrum var. *caespitosum* f. *pallens* Wollenw., 1929
V. albo-atrum var. *tuberosum* Rudolph, 1931.

CULTURES growing rapidly on PDA and malt agar at 23°; the prostrate hyphae which are first produced are hyaline. MYCELIUM becoming flocculose and white to greyish, rather more densely compacted on PDA than MA, hyaline, whitish to cream in reverse after 1 week. After 2–3 weeks becoming brownish to cream and black centrally due to the formation of dark resting mycelium. White sectors formed frequently in the generally greyish colonies. CONIDIOPHORES abundant, more or less erect, hyaline, verticillately branched, 2–4 phialides arising at each node, phialides sometimes secondarily branched; characteristically darkened at the base when growing on plant tissue. PHIALIDES variable in size, mainly 20–30 (–50) $\times$ 1.4–3.2 μ. CONIDIA arise singly at the apices of the phialides, ellipsoidal to irregularly subcylindrical, hyaline, mainly simple, occasionally 1 septate, 3.5–10.5 (–12.5) $\times$ 2–4 μ. RESTING MYCELIUM appearing after 10–15 days, dark brown to blackish but sometimes hyaline in sectors, regularly septate, becoming swollen between the septa so as to appear somewhat torulose in parts, never budding to form structures like microsclerotia, hyphae 3–7 μ diam. Formation of dark resting mycelium tends to cease after prolonged subculturing. CHLAMYDOSPORES and MICROSCLEROTIA absent. May be distinguished from *V. dahliae* by the absence of microsclerotia, no growth at 30° and the dark resting mycelium. Also on plant tissues the bases of the conidiophores in *V. albo-atrum* tend to be brownish (D. L. Hawksworth & P. W. Talboys, *CMI Descr.* 255, 1970).

Since the presence or absence of microsclerotia is not always mentioned in the literature it is not always clear whether this sp. or *V. dahliae* is being reported. Where a paper refers to the presence of microsclerotia it has been assumed that the organism concerned is *V. dahliae* even though it may have been assigned by the author(s) to *V. albo-atrum*. *V. albo-atrum* is plurivorous on dicotyledonous plants, mostly herbaceous (*Verticillium* q.v. for host lists; Born; Himelick; Snyder et al.). It is a widespread (Map 365) organism associated with plant roots, frequently causing serious diseases in cooler regions on several temperate crops. One of the most studied of these host and parasite relationships is the disease on hop (*Humulus lupulus*). A brief account of which was given by H. H. Glasscock (in Western, *Diseases of crop plants*; see also the annual reports of the East Malling Research Station, UK, and papers by I. Isaac, W. G. Keyworth, G. W. F. Sewell, P. W. Talboys inter alia (not quoted). Panton gave a review of *V. albo-atrum* up to 1964; and see Himelick.

The pathogen invades the vascular tissue and causes wilt diseases; permanent wilt commonly occurs but not invariably. In hop there is chlorosis leading to marginal and interveinal necrosis, and progressive desiccation but without flaccidity. One-sided symptom development may occur where only a few vascular bundles are infected, and diseased xylem shows a brown discolouration which may be considerably delayed. McKeen, investigating isolates from several hosts in Canada, found high disease levels in the glasshouse at 21–27° (max. 24°) at an opt. soil moisture. At higher moisture levels the temp. range for severe disease was broader. The fungus persists in the soil in infected host debris and infects roots that come into contact with it. Sewell found that, while a very limited and transient extension of free mycelium into soil occurred from certain food bases, the fungus was largely confined to those tissues which it had initially invaded as a parasite. Sporulation took place on inoculum fragments after their deposition in the soil, and also on infected roots after death. Conidia may play a role in spread. Initially (during the sporulation phase) root contact with diseased residues may not be a prerequisite for infection.

Isolates from one host sp. often cause disease in other spp. without causing obvious symptoms. There may be some specificity. In hop some strs of the pathogen cause a form of the disease known as fluctuating. The progressive or more serious form of the disease is caused by a virulent str. In England legislation has aimed at limiting the spread of this str. which can kill hop plants in the first or second year and spread rapidly through a garden. Pegg and Stoddart et al. have described phytotoxins, and various workers the cell degrading enzymes.

Presumably, since microsclerotia are not formed, the spectrum of control measures is wider than for *V. dahliae*, and may include a rotation with non-dicotyledonous crops. Sewell et al., in the disease on hop, could not recover *V. albo-atrum* after 4 years of a weed-free grass cover. Soil infectivity was reduced to a low level under bare fallow. Under grass infectivity was low after 2 years and apparently nil after 3–5 years. Where weeds were allowed to develop infectivity declined after a single annual cultivation but finally increased. Resistant or tolerant cvs are often available and soil fumigation is effective. Systemic fungicides offer a possible further means of control. Two diseases, on lucerne and tomato, are described.

Aubé, C. et al. 1968. *Can. J. Microbiol.* **14**: 606 (intrahyphal spores, **47**, 2625).
Born, G. L. 1974. *Bull. Ill. nat. Hist. Surv.* 31: 209 (hosts 52 ref.; **54**, 2743).
Campbell, W. P. et al. 1974. *Can. J. Microbiol.* **20**: 163 (morphological variation; **53**, 2864).
Chaudhuri, H. 1923. *Ann. Bot.* **37**: 519 (growth in culture).
Gupta, D. P. et al. 1971. *J. gen. Microbiol.* **63**: 163 (induction of cellulase; **50**, 2692).
Heale, J. B. et al. 1968. *Can. J. Genet. Cytol.* **10**: 321 (nuclear division; **47**, 3743).
—— ——. 1971. *J. gen. Microbiol.* **63**: 175 (utilisation of cellobiose; **50**, 2693).
Himelick, E. B. 1969. *Biol. Notes nat. Hist. Surv. Div. St. Ill.* 66: 1 (review, tree & shrub hosts, 44 ref.; **49**, 855).
Kaiser, W. J. 1964. *Phytopathology* **54**, 481 (biotin requirement; **43**, 2535).
Keen, N. T. et al. 1970. *Physiol. Pl.* **23**: 691, 878 (induction & repression of betagalactosidase synthesis & properties of; **50**, 512, 1094).
—— ——. 1971. *Phytopathology* **61**: 1266 (dimorphism as affected by initial spore conc. & anti-sporulant chemicals; **51**, 2529).
McKeen, C. D. 1943. *Can. J. Res.* **21**: 95 (pathogenicity; **22**: 323).
Malca, I. et al. 1966. *Phytopathology* **56**: 401 (effects of pH, C & N on growth; **45**, 2721).
Mussel, H. W. et al. 1972. *Can. J. Biochem. Physiol.* **50**: 625 (2 polygalacturonases; **52**, 3223).
Panton, C. A. 1964. *Acta Agric. scand.* **14**: 97 (review, 83 ref.; **44**, 1008).
Patil, S. S. et al. 1967. *Phytopathology* **57**: 492 (inhibition of polygalacturonase; **46**, 2997).
Pegg, G. F. 1965. *Nature Lond.* **208**: 1228 (phytotoxin).
Roth, J. N. et al. 1964. *Phytopathology* **54**: 363 (nuclei; **43**, 2840).
—— ——. 1964. *Ibid* **54**: 1454 (environment & hereditary variation in monospore cultures; **44**, 1417).
Russel, S. 1975. *Phytopathol. Z.* **84**: 222 (cellulase; **55**, 2536).
Schreiber, L. R. et al. 1966. *Phytopathology* **56**: 1110 (anastomosis in soil; **46**, 261).
Sewell, G. W. F. 1959. *Trans. Br. mycol. Soc.* **42**: 312 (direct observation in soil; **39**: 502).
—— et al. 1966. *Ann. appl. Biol.* **58**: 241 (soil survival, effects of grass cover & bare fallow in hop cultivation; **46**, 398).
Snyder, W. C. et al. 1950. *Pl. Dis. Reptr* **34**: 26 (hosts; **29**: 448).
Stoddart, J. L. et al. 1966. *Ann. appl. Biol.* **58**: 81 (phytotoxins; **45**, 3487).
Vessey, J. C. et al. 1973. *Trans. Br. mycol. Soc.* **60**: 133 (autolysis & chitinase production in culture; **52**, 1431).
Watanabe, T. et al. 1973. *Ann. phytopathol. Soc. Japan* **39**: 344 (on Chinese cabbage; **53**, 3227).
Wilhelm, S. 1950. *Phytopathology* **40**: 368 (vertical distribution in soil; **29**: 540).
—— et al. 1953. *Ibid* **43**: 593 (soil fumigation; **33**: 508).

LYCOPERSICON ESCULENTUM

Bewley, when investigating tomato wilt diseases caused by *V. albo-atrum* and *Fusarium oxysporum* f. sp. *lycopersici*, referred to the condition as sleepy disease; he was concerned mostly with the former pathogen which is only likely to be serious in cooler regions. Since the ranges of both fungi overlap these two vascular wilts may be confused in the field. A tomato disease can also be caused by *V. dahliae* (q.v.). The first symptoms are usually seen in the lowermost leaves which begin to yellow. If conditions are somewhat dry the following wilt, spreading upwards, may take place fairly rapidly. Often, however, wilt may not occur; the plant becomes stunted with short internodes, small fruit and the leaf margins may curl upwards. Chlorotic blotches appear on the leaves which may eventually wither and fall. A brown or straw coloured symptom is seen in the vascular strands and this may extend to the apex. Host invasion takes place through the roots. Selman et al. (1959) found that rapid systemic infection only occurred when roots were damaged and considered that natural infection took place via wounding to the xylem. Bewley found that the disease developed at 15–24°, it was most active at 21–23° and at 25° there was little infection. The raising of temps to 25° in glasshouses is one of the recommended control measures (Ogilvie, *Diseases of vegetables*). Provvidenti et al. induced leaf infection experimentally and reported an opt. temp. for this similar to that for root infection. Rudolph considered that seed transmission was unimportant; Kadow reported that infection of the seed can occur. Nutrient imbalance in the host may increase infection, for example, high nitrogen with low potassium (Roberts), and see Walker et al. No races on tomato appear to have been characterised; Pegg (1957) described a hyaline variant that was pathogenic. In USA cvs resistant in California were susceptible when inoculated with an isolate from Ohio (Alexander).

Many papers have described the possible roles of toxins, growth substances, enzymes and vessel blockage by tyloses in producing the disease syndrome, but the relative action and importance of these several, interrelated factors is not clear. The behaviour of the pathogen has also been compared after infection of susceptible and resistant or tolerant cvs. Dixon et al. (1969) considered that tylosis did not constitute a specific resistance mechanism and that resistance was not related directly to the rate or degree of fungal colonisation. Pegg et al. (1969) compared the reactions of susceptible and resistant cvs with infection (at different inoculum potentials) by a str. causing severe wilt in tomato and strs which caused the progressive and fluctuating forms of wilt in hop (*Humulus lupulus*). Susceptibility to *V. albo-atrum* depends on host cv., fungal str. and initial inoculum conc. In some cv. and pathogen combinations susceptibility is directly proportional to the amount of mycelium in the vessels. In others a physiological resistance mechanism, independent of the degree of fungal colonisation, appeared to operate. In a third category increased disease development, rather than resistance, was associated with high levels of tyloses. In more recent work it was concluded that ethylene has a multiple role in this vascular pathogenesis—first as an inducer of resistance, and, secondly, as a toxin synergist when interacting with pathogenic metabolites after the establishment of the pathogen (Cronshaw et al.; Pegg 1976d; Pegg et al. 1976b & c).

Cell degrading enzymes have been examined by Blackhurst et al. (1963a); Cooper et al. (1975, 1978); Deese et al.; Pegg et al. (1973); Pegg (1976d); Russel; Wood. For phytoalexins see Pollock et al. (1976b); Tjamos et al. (1974) and for electron microscopy of the fungus in the xylem see Cooper et al. (1974); Pegg et al. (1976a). Jones et al. (1975b) found that symptoms developed more rapidly under a 4-hour photoperiod compared with one of 8, 12 or 16 hours.

Thanassoulopoulos reported an increase in the severity of symptoms 20–30 days earlier in the season when tomato was also infected with tobacco mosaic virus. Conroy et al. found that infection by *V. albo-atrum* increased at most inoculum densities when the stubby root nematode (*Trichodorus christiei*) was present. Infection also increased as the density of the nematode did so. No consistent interaction with *Meloidogyne incognita* was found (and see Jones et al. 1976).

In glasshouse culture raising the temp. to 25°, sanitation measures and reducing watering to a minimum should control an outbreak but if the disease is being caused by *V. dahliae* the measures would not be effective. Soil fumigation may have to be carried out and chloropicrin has given good results. In USA the use of nematicides improved control (Jones et al. 1971, 1978; Overman et al.). Resistant cvs are available (Denby et al.; Jones et al. 1973, 1975a). In Canada Denby et al. described the incorporation of the resistance gene from cv. Loran Blood in 45 established cvs (and see Schiable et al.; Tjamos et al. 1975).

Alexander, L. J. 1962. *Phytopathology* **52**: 998 (pathogenicity of different isolates; **42**, 344).
Bewley, W. F. 1922. *Ann. appl. Biol.* **9**: 116 (general, temp., hosts & *Fusarium oxysporum*; **2**: 148).
Blackhurst, F. M. 1963. *Ibid* **52**: 79 (induction of

symptoms in detached shoots from resistant & susceptible cvs; **43**, 238).
—— et al. 1963a. *Ibid* **52**: 89 (role of pectic & cellulytic enzymes; **43**, 239).
—— ——. 1963b. *Trans. Br. mycol. Soc.* **46**: 385 (spread in resistant & susceptible cvs; **43**, 1150).
Conroy, J. J. et al. 1974. *Phytopathology* **64**: 1118 (interactions with nematodes; **54**, 2469).
Cooper, R. M. et al. 1974. *Physiol. Pl. Pathol.* **4**: 443 (electron microscopy of xylem; **54**, 1437).
—— ——. 1975. *Ibid* **5**: 135 (regulation of synthesis of cell degrading enzymes with *F. oxysporum* f. sp. *lycopersici*; **54**, 3480).
—— ——. 1978. *Ibid* **13**: 101 (on polysaccharidases with *F. oxysporum* f. sp. *lycopersici*; **58**, 371).
Cronshaw, D. K. et al. 1976. *Ibid* **9**: 33 (ethylene as a toxin synergist; **56**, 849).
Deese, D. C. et al. 1962. *Phytopathol. Z.* **46**: 53 (pectic enzymes; **42**: 410).
Denby, L. G. et al. 1962. *Can. J. Pl. Sci.* **42**: 681 (development of resistant cvs; **42**: 279).
Dixon, G. R. et al. 1969. *Trans. Br. mycol. Soc.* **53**: 109 (infection, hyphal lysis & tylose formation; **48**, 3670).
—— ——. 1972. *Ann. Bot.* **36**: 147 (amino acid content of xylem sap after infection; **51**, 2860).
Green, R. J. 1954. *Phytopathology* **44**: 433 (in vitro toxin formation; **34**: 266).
Jones, J. P. et al. 1971. *Pl. Dis. Reptr* **55**: 26 (soil fumigants; **50**, 2503).
—— ——. 1973. *Ibid* **57**: 122 (effect of wilt on resistant, tolerant & susceptible cvs; **52**, 3430).
—— ——. 1975a. *Ibid* **59**: 3 (reaction of resistant, tolerant & susceptible cvs to wilt; **54**, 4156).
—— ——. 1975b. *Phytopathology* **65**: 647 (effect of photoperiod on wilt; **54**, 5554).
—— ——. 1976. *Pl. Dis. Reptr* **60**: 913 (tomato wilts, nematodes & yields as affected by soil reaction & a nematicide; **56**, 3719).
—— ——. 1978. *Ibid* **62**: 451 (soil fumigants; **58**, 372).
Kadow, K. J. 1934. *Phytopathology* **24**: 1265 (seed transmission; **14**: 283).
Kerr, A. 1961. *Trans. Br. mycol. Soc.* **44**: 365 (antagonism of root surface organisms; **41**: 102).
Overman, A. J. et al. 1970. *Proc. Fla St. hort. Soc.* **83**: 203 (soil fumigants; **51**, 3558).
Pegg, G. F. 1957. *Phytopathology* **47**: 57 (pathogenicity of a hyaline variant; **36**: 430).
—— et al. 1959. *Ann. appl. Biol.* **47**: 222 (host growth response & infection, production of growth substances; **39**: 49).
—— ——. 1969. *Ibid* **63**: 389 (4 categories of host & pathogen interaction, resistant & susceptible cvs; **48**, 3669).
—— ——. 1973. *Physiol. Pl. Pathol.* **3**: 207 (chitinase in host & relationship to in vitro lysis of mycelium; **52**, 4223).
—— ——. 1976a. *Ibid* **8**: 221 (electron microscopy of xylem; **55**, 5927).
—— ——. 1976b. *Ibid* **8**: 279 (ethylene formation in infected plants; **55**, 5929).
—— ——. 1976c. *Ibid* **9**: 215 (response of ethylene treated plants to infection; **56**, 3227).
——. 1976d. *J. exp. Bot.* **27**: 1093 (occurrence of 1,3 betaglucanase; **56**, 2217).
Pollock, C. J. et al. 1976a. *Phytopathol. Z.* **86**: 56 (phenolic compounds & resistance; **56**, 393).
—— ——. 1976b. *Ibid* **86**: 353 (growth in resistant & susceptible cvs; **56**, 1737).
Provvidenti, R. et al. 1959. *Pl. Dis. Reptr* **43**: 821 (foliage infection; **39**: 49).
Roberts, F. M. 1943. *Ann. appl. Biol.* **30**: 327 (soil nutrient factors in infection; **23**: 154).
——. 1944. *Ibid* **31**: 191 (leaf : shoot ratio & nutrient factors in infection; **24**: 125).
Rudolph, B. A. 1944. *Phytopathology* **34**: 622 (unimportance of seed dissemination; **24**: 79).
Russel, S. 1975. *Phytopathol. Z.* **82**: 35 (role of cellulase; **54**, 4640).
Scheffer, R. P. et al. 1956. *Phytopathology* **46**: 83 (physiology of wilt; **36**: 68).
Schiable, L. et al. 1951. *Ibid* **41**: 986 (inheritance of resistance).
Selman, I. W. et al. 1957. *Ann. appl. Biol.* **45**: 674 (young plant growth response to infection; **37**: 377).
—— ——. 1959. *Trans. Br. mycol. Soc.* **42**: 227 (factors affecting root invasion; **39**: 49).
Sinha, A. K. et al. 1967. *Ann. appl. Biol.* **59**: 143 (response to infection in susceptible & resistant host; **46**, 1741).
—— ——. 1967. *Ibid* **60**: 117 (effect of growth substances on wilt; **46**, 3570).
—— ——. 1968. *Ibid* **62**: 319 (nature of resistance; **48**, 603).
Thanassoulopoulos, C. C. 1976. *Phytoparasitica* **4**: 137 (effect of tobacco mosaic virus with *F. oxysporum* f. sp. *lycopersici*; **56**, 3713).
Threlfall, R. J. 1959. *Ann. appl. Biol.* **47**: 57 (physiology of wilt; **38**: 545).
Tigchelaar, E. C. et al. 1975. *HortScience* **10**: 623 (induced resistance from inoculation with *F. oxysporum* f. sp. *lycopersici*; **55**, 4276).
Tjamos, E. C. et al. 1974. *Physiol. Pl. Pathol.* **4**: 249 (phytoalexins; **53**, 4118).
—— ——. 1975. *Ibid* **6**: 215 (expression of resistance in monogenically resistant plants; **55**, 1469).
Walker, J. C. et al. 1954. *Am. J. Bot.* **41**: 760 (host nutrition & disease; **34**: 111).
Wilhelm, S. 1951. *Phytopathology* **41**: 684 (effect of soil amendments on inoculum potential; **31**: 35).
Wood, R. K. S. 1961. *Ann. appl. Biol.* **49**: 120 (role of pectic & cellulytic enzymes; **40**: 631).

MEDICAGO SATIVA

The forms of *V. albo-atrum* causing this severe lucerne wilt appear to be confined to Europe. Although *V. dahliae* attacks this host in the same areas it is of little or no economic importance (Isaac et al. 1959b). In warmer periods the upper leaves droop, isolated plants become pale and stunted. The lower leaves and shoots gradually become chlorotic, whitish, dried up and there is severe defoliation. The symptoms move upwards until the whole plant is affected; sometimes there is a fairly rapid, whole plant collapse. Any partial recovery from wilt is transient. The basal few cm of stem are frequently covered with conidiophores giving a superficial greyish appearance. The typical vascular discolouration can be traced from the small lateral roots, through roots, stem and petioles. In the second and third years the disease may have spread into large patches or strips often following the line of crop cutting. The disease generally becomes more severe as the age of the stand increases.

Penetration occurs through intact surfaces and wounds; both conidia and hyphae infect the wound surfaces of newly cut plants. Spread is through infected host debris (assisted by cutting), and there is some evidence that airborne conidia can also cause spread (Davies et al.). The pathogen spreads rapidly underground from diseased to healthy plants. Resting mycelium remains viable for 9 months in host pieces at 30 cm soil depth, 7 months at 15 cm and 5 months on the soil surface. At room temp. viability was maintained for 13 months but not 2 years. *V. dahliae* is only virulent in superphosphate rich soil. A survey in UK showed that less wilt occurred in a warm and partially dry year compared with 2 cooler and wetter ones (Roberts et al.). The fungal strs from lucerne have some host specificity (e.g., Knoll) and do not cause disease in other crops, although these crops can be infected (Heale et al. 1963), as also can weeds which show no external symptoms (Müller). Isolates from lucerne used to inoculate weeds remain infective to the crop when re-isolated. Müller found that sainfoin (*Onobrychis viciifolia*) developed severe symptoms after inoculation with isolates from lucerne, and see Richter et al. A. J. H. Carr (in Western, editor, *Diseases of crop plants*) stated that a specific str. of *V. dahliae* causes symptoms in sainfoin similar to those in lucerne, but that sainfoin can replace lucerne in badly affected fields.

Although the pathogen is apparently not borne in or on the seed, seed treatment (organo-mercurials or thiram) is advisable since disease spread from debris with seed can occur. Measures should be taken to prevent spread in cutting or other cultural operations. Some crops may be safely grown on badly infested fields, provided they do not result in a build up of the lucerne strs. Existing cvs are not adequately resistant. Breeding for resistance was described by Panton (1965, 1967a & b; Steuckardt); the genetic control is complex (Panton 1967c). Other *Medicago* spp. are being used, and see also the annual reports of the Plant Breeding Institute, Cambridge, UK. Dixon found that thiabendazole, as a soil drench in the glasshouse, was effective; also in a field trial there appeared to be a cumulative reduction in wilt development which took place more quickly in some of the cvs.

Aubé, C. 1967. *Can. J. Microbiol.* **13**: 227 (antagonism of soil fungi; **46**, 2055).

Davies, R. R. et al. 1958. *Nature Lond.* **181**: 649 (airborne spread of conidia; **37**: 497).

Dixon, G. R. 1972. *Pl. Pathol.* **21**: 129 (effects of thiabendazole; **52**, 1585).

Flood, J. et al. 1978. *Ann. appl. Biol.* **89**: 329 (phytoalexin).

Gupta, D. P. 1973. *Indian Phytopathol.* **26**: 90 (endopolygalacturonase stimulation; **53**, 4020).

Heale, J. B. et al. 1963. *Ann. appl. Biol.* **52**: 439 (pathogenicity, including *V. dahliae*, spread & survival in soil & weeds; **43**, 1681).

—— ——. 1972. *Trans. Br. mycol. Soc.* **58**: 19 (mechanism of vascular wilt; **51**, 3162).

Isaac, I. 1957. *Ann. appl. Biol.* **45**: 550 (general; **37**: 176).

——. 1959a. *N.A.A.S. q. Rev.* **46**: 75 (general; **39**: 592).

—— et al. 1959b. *Ann. appl. Biol.* **47**: 673 (seasonal effects, hosts & seed treatment; **39**: 420).

—— ——. 1961. *Ibid* **49**: 675 (viability & seed treatment; **41**: 393).

Khan, F. Z. et al. 1975. *Physiol. Pl. Pathol.* **7**: 179 (phytoalexin; **55**, 2285).

—— ——. 1978. *Ibid* **13**: 215 (as above).

Kiessig, R. et al. 1957. *Phytopathol. Z.* **31**: 185 (general; **37**: 416).

Knoll, F. A. 1972. *Zentbl. Bakt. ParasitKde* Abt. 2 **127**: 332 (distribution of *Verticillium* spp. in lucerne; **52**, 2670).

Michail, S. H. et al. 1966. *Trans. Br. mycol. Soc.* **49**: 133 (culture filtrates for resistance screening; **45**, 2157).

Müller, H. L. 1969. *Phytopathol. Z.* **65**: 69 (pathogenic strs, weed hosts & resistance testing; **48**, 3537).

Naumann, K. et al. 1971. *Arch. PflSchutz* **7**: 37 (fungal wilt diseases, infection & colonisation; **50**, 3902).

Noble, M. et al. 1953. *Pl. Pathol.* **2**: 31 (symptoms & isolation; **32**: 629).

Panton, C. A. 1965. *Acta Agric. scand.* **15**: 85 (screening methods & selection for resistance; **44**, 3081).
——. 1967a. *Ibid* **17**: 59 (toxicity of culture filtrate; **46**, 3119).
——. 1967b. *Hereditas* **57**: 115 (transgressive segregation in F_3 & effect on resistance in inbreeding; **47**, 244).
——. 1967c. *Ibid* **57**: 333 (inheritance of resistance; **47**, 556).
Richter, H. et al. 1938. *NachrBl. dt. PflSchutzdienst Berl.* **18**: 57 (hosts, on *Onobrychis viciifolia*, inoculation; **17**: 754).
Roberts, E. T. et al. 1963. *Pl. Pathol.* **12**: 47 (disease survey in UK; **42**: 689).
Schmiedeknecht, M. 1969. *Arch. PflSchutz* **5**: 143 (root penetration & colonisation; **49**, 1068).
Steuckardt, R. et al. 1976. *Arch. Züchtungforschung* **6**: 201 (breeding for resistance; **57**, 1765).
Whitney, P. et al. 1969. *J. gen. Microbiol.* **56**: 215 (carboxymethylcellulase production; **48**, 3324).
Whitney, P. J. et al. 1972. *J. exp. Bot.* **23**: 400 (protein changes in infected cvs; **51**, 4108).
Zaleski, A. 1957. *Pl. Pathol.* **6**: 137 (disease reactions amongst cvs; **37**: 297).

Verticillium dahliae Klebahn, *Mycol. Centralb*, **3**: 66, 1913.

Verticillium dahliae var. *longisporum* C. Stark, 1961
V. albo-atrum var. *medium* Wollenw., 1929
V. albo-atrum auct. pro parte.

CULTURES growing rapidly on PDA and malt agar at 23°; the prostrate hyphae which are first produced are hyaline. MYCELIUM becoming flocculose and white, rather more densely compacted on PDA than MA; hyaline, whitish to cream in reverse after 1 week, later becoming black with the formation of microsclerotia. Hyaline sectors arise very frequently in the generally white colonies. CONIDIOPHORES abundant, more or less erect, hyaline, verticillately branched, 3–4 phialides arising at each node, phialides sometimes secondarily branched. PHIALIDES variable in size, mainly $16–35 \times 1.25\ \mu$. CONIDIA arise singly at the apices of the phialides, ellipsoidal to irregularly subcylindrical, hyaline, mainly simple but occasionally 1 septate, $2.5–8 \times 1.4–3.2\ \mu$ in var. *dahliae*; $5–12.5 \times 1.6–3.4\ \mu$ in var. *longisporum*. Dark brown RESTING MYCELIUM only formed in association with microsclerotia. CHLAMYDOSPORES absent. MICROSCLEROTIA arising centrally in cultures, dark brown to black, torulose or botryoidal, consisting of swollen almost globular cells. Each microsclerotium arises from a single hypha by repeated budding; very variable in shape, elongate to irregularly spherical; very variable in size, 15–50 (–100) μ diam.

May be distinguished from *V. albo-atrum* (q.v.) by the presence of true microsclerotia, forming colonies which are entirely black in reverse with growth at 30°. On plant tissue the basal parts of the conidiophores in *V. albo-atrum* are often brownish. The var. *longisporum*, with conidia almost twice as long as *V. dahliae* and known only from *Armoracia rusticana*, appears to be a stable diploid (fide Ingram, *Trans. Br. mycol. Soc.* **51**: 339–341; D. L. Hawksworth & P. W. Talboys, *CMI Descr.* 256, 1970).

Whether microsclerotia are present or absent is not always mentioned in the literature; it is thus not always clear if this sp. or *V. albo-atrum* is being reported. Where a paper refers to the presence of microsclerotia it has been assumed that the organism concerned is *V. dahliae*, even though it may have been assigned by the author(s) to *V. albo-atrum*. *V. dahliae* is plurivorous on herbaceous and woody dicotyledons (*Verticillium* q.v. for host lists, and see Stark; Woolliams); it is a widespread organism associated with plant roots, frequently causing serious diseases of both temperate and tropical crops. With a distribution (Map 366) that extends into warmer regions it is more likely to be encountered in the tropics than *V. albo-atrum*. Some of the diseases described here under *V. dahliae* have been attributed to *V. albo-atrum* but with no indication (in some papers) as to the presence or absence of microsclerotia. They are given here partly for convenience and partly because on tropical or subtropical crops *V. dahliae* is the most likely sp.

Although wilt is used to describe the diseases, and often indicates their onset, actual wilt does not always occur. Permanent wilt, followed by chlorosis and necrosis, often shows a one-sided pattern that results from the infection of a few vascular bundles. Infected xylem commonly shows brown discolouration. Foliar chlorosis can lead to a progressive necrosis and defoliation. Woody plants frequently show leaf chlorosis and necrosis, followed by defoliation of individual branches, and the plant may recover. Isolates from many hosts can cause disease in several other spp. and genera, and may invade plants without causing obvious symptoms. Some host specificity has, however, been reported.

The fungus survives for long periods in the soil, at

first in host plant debris and then probably as microsclerotia whose germination (Ben-Yephet et al.; Gordee et al.; Wilhelm) and behaviour in soil (Green et al.; Menzies et al.; Schreiber et al.) have been investigated. Evans and Evans et al. (1973) described the effect of weed hosts on the ecology of *V. dahliae* in newly cultivated areas of the Namoi Valley in New South Wales, Australia. The evidence was that the fungus was not native to the valley but had been introduced (probably on weed seed) and had then built up on dicotyledonous weeds from which it had spread to crops, for example, cotton. *V. dahliae* was isolated from 26 weed spp. in naturally infested soils and 15 new hosts were recorded. Some spp. consistently yielded more colonies of the fungus than others, but susceptible ones were colonised no more frequently than those that were immune to systemic infection. Discrete colonies (*c.* 2 mm long) occurred at random along the root system, each colony apparently arising from a different sclerotium. Colonies appeared to be restricted to the rhizoplane or to superficial sites in the root cortex (and see Evans et al. 1974; Lacy et al.). For techniques for the assessment of microsclerotia in the soil see also *Rev. Pl. Pathol.* **52**, 1016, 2505; **53**, 4731; **54**, 1144, 1149; **56**, 1469; **57**, 983, 1624. Nadakavukaren et al. described a medium for selectively determining microsclerotia in the soil and Nelson et al. reported on the thermal death range.

Kaiser found that no microsclerotia formed in continuous blue light but they occurred under yellow, orange and red light, and in the dark. Blue light did not inhibit microsclerotial formation under an 18 hour dark and 6 hour light regime but sporulation was much less than with a 12 hour dark and light one. Brandt (1964a) attributed the delay in, or suppression of, the formation of microsclerotia by near UV light to the inhibition of a self-induced morphogenetic factor by UV. This factor affects the formation of microsclerotia and melanin, hyphal elongation and sporulation (Brandt 1964b; 1967; Brandt et al.). The fine structures of conidia (Buckley et al.), microsclerotia (Brown et al.; Griffiths 1970; Nadakavukaren), hyphae colonising cellophane (Griffiths 1971a), and growth in and on roots (Griffiths 1973; Levy et al.) have been described. Disease control depends largely on resistance cvs and soil fumigation (e.g., Henis et al.; McKeen et al.). In contrast to diseases caused by *V. albo-atrum* crop rotation methods are less satisfactory. Systemic fungicides offer further possibilities in the future. Jordan et al. described inoculum suppression as a possible aid to control by chemical treatment.

Ben-Yephet, Y. et al. 1977. *Phytoparasitica* **5**: 159 (germination of individual microsclerotia).
Beyma Thoe Kingma, F. H. van 1940. *Antonie van Leeuwenhoek* **6**: 34 (taxonomy; **19**: 367).
Brandt, W. H. 1964a. *Can. J. Bot.* **42**: 1017 (effect of light on microsclerotia & melanin; **44**, 350).
——. 1964b. *Am. J. Bot.* **51**: 820 (a self-induced, non-hereditary variation in colony form).
—— et al. 1964. *Ibid* **51**: 922 (a self-induced, diffusible morphogenetic factor; **44**, 995).
—— ——. 1965. *Phytopathology* **55**: 1200 (loss of melanin containing structures in culture; **45**, 1001).
——. 1967. *Mycologia* **59**: 736 (effect of near UV light on hyphal elongation; **46**, 3675).
Brown, M. F. et al. 1970. *Phytopathology* **60**: 538 (fine structure of microsclerotia; **49**, 2730).
Buckley, P. M. et al. 1969. *Mycologia* **61**: 240 (fine structure of conidia; **49**, 154).
Congly, H. et al. 1976. *Can. J. Bot.* **54**: 1214 (effects of osmotic potential on microsclerotial germination & colony growth; **55**, 5543).
Emmatty, D. A. et al. 1969. *Phytopathology* **59**: 1590 (fungistasis & microsclerotial behaviour in soil; **49**, 1289).
Evans, G. 1971. *Ann. appl. Biol* **67**: 169 (effect of weed hosts in ecology of cultivated areas in New South Wales; **50**, 2652).
—— et al. 1973. *Aust. J. biol. Sci.* **26**: 151 (origin & nature of colonisation of plant roots; **52**, 3615).
—— ——. 1974. *Can. J. Microbiol.* **20**: 119 (quantitative bioassay for low microsclerotial numbers in soil; **53**, 2451.
Farley, J. D. et al. 1971. *Phytopathology* **61**: 260 (germination & sporulation of microsclerotia in soil; **50**: 2747).
Galanopoulos, N. et al. 1974. *Trans. Br. mycol. Soc.* **63**: 85 (conidial survival; **54**, 340).
Gordee, R. S. et al. 1961. *Mycologia* **53**: 171 (structure, germination & physiology of microsclerotia; **41**: 612).
Green, R. J. et al. 1968. *Phytopathology* **58**: 567 (effects of C, C:N ratios & organic amendments on survival; **47**, 3023).
——. 1969. *Ibid* **59**: 874 (survival & inoculum potential in soil; **48**, 3668).
Griffiths, D. A. 1970. *Arch. Mikrobiol.* **74**: 207 (fine structure of developing microsclerotia; **50**, 2696).
—— et al. 1970. *Can. J. Microbiol.* **16**: 1132 (hyphal interaction in microsclerotial development; **50**: 1644).
——. 1971a. *Ibid* **17**: 79 (fine structure in colonising cellophane; **50**, 3404).

Griffiths, D. A. 1971b. *Ibid* **17**: 441 (development of lignitubers in roots after infection; **50**, 3405).
——. 1973. *Skokubutsu Byogai Kenkyu* **8**: 147 (fine structure of host reaction in roots; **53**, 2114).
Henis, Y. et al. 1973. *Phytoparasitica* **1**: 95 (sensitivity of microsclerotia to chloropicrin; **53**, 3320).
Ioannou, N. et al. 1977. *Phytopathology* **67**: 637, 645, 651 (effects of water potential, temp., O_2, CO_2, ethylene, flooding of soil & gas composition on microsclerotia; **57**, 424, 425, 426).
Jordan, V. W. L. et al. 1978. *Ann. appl. Biol.* **89**: 139 (inoculum suppression).
Kaiser, W. J. 1964. *Phytopathology* **54**: 765 (effects of light on growth & sporulation; **44**, 40).
Lacy, M. L. et al. 1966. *Ibid* **56**: 427 (behaviour in rhizosphere of plant roots; **45**, 2435).
Levy, J. et al. 1976. *Trans. Br. mycol. Soc.* **67**: 91 (fine structure of growth over & in roots; **56**, 1995).
McKeen, C. D. et al. 1964. *Can. J. Pl. Sci.* **44**: 466 (control by soil fumigation in glasshouse; **44**, 1263).
Menzies, J. D. et al. 1967. *Phytopathology* **57**: 703 (survival & saprophytic growth in uncropped soil; **46**, 3348).
Mountain, W. B. et al. 1962. *Nematologia* **7**: 261 (effect on population of *Pratylenchus penetrans*).
Nadakavukaren, M. J. et al. 1959. *Phytopathology* **49**: 527 (a selective medium for microsclerotia; **39**: 157).
——. 1963. *Can. J. Microbiol.* **9**: 411 (fine structure of microsclerotia; **43**: 40).
Nelson, P. E. et al. 1958. *Phytopathology* **48**: 613 (thermal death range; **38**: 210).
Schippers, B. et al. 1966. *Ibid* **56**: 549 (growth from *Senecio vulgaris* seed on agar & in soil; **45**, 3067).
Schnathorst, W. C. 1965. *Mycologia* **57**: 343 (origin of new growth in microsclerotia; **44**, 3008).
Schreiber, L. R. et al. 1962. *Phytopathology* **52**: 288 (comparative soil survival of mycelium, conidia & microsclerotia; **42**: 2).
—— ——. 1963. *Ibid* **53**: 260 (soil fungistasis, root exudates & germination of conidia & microsclerotia; **42**: 603).
Stark, C. 1961. *Gartenbauwissenschaft* **26**: 493 (occurrence on horticultural crops, new hosts; **42**: 181).
Talboys, P. W. et al. 1976. *Ann. appl. Biol.* **82**: 41 (benomyl resistance; **55**, 3899).
Wilhelm, S. 1954. *Phytopathology* **44**: 609 (aerial microsclerotia from conidial anastomosis; **34**: 308).
——. 1955. *Ibid* **45**: 180 (longevity; **34**: 679).
Woolliams, G. E. 1966. *Can. J. Pl. Sci.* **46**: 661 (host range & symptoms in economic, weed & native plants in British Colombia; **46**, 922).

GOSSYPIUM

Cotton wilt is largely caused by *V. dahliae* and ref. to it are placed here, although some papers refer to the pathogen as *V. albo-atrum*. The severe stunting of the plant following infection at the early stages of growth is a direct cause of loss. Later infection may still cause inferior fibre to form. Bugbee et al. recently investigated the effects of infection on yield, fibre characteristics and seed quality in USA (N. Dakota). Most damage to fibre and seed germination resulted from inoculations in August when bolls were in the early stage of maturation. Inoculations a month later caused most loss in seed weight. Cotton wilts caused by this pathogen and *Fusarium oxysporum* f. sp. *vasinfectum* (q.v.) may occur together, the latter has a higher opt. temp.; separation of the 2 diseases in the field may be difficult and isolations therefore necessary. The disease is serious in many parts of the world, including USA and USSR where it has been intensively studied. Perches et al. gave a general account of it in Mexico and Schnathorst described the more virulent str. of the fungus in Peru. In 1972 Isaac et al. referred to crop losses in India of 70–90%. Evans et al. (1967b) emphasised the importance of wilt in a part of Australia (Namoi Valley, New South Wales), and Wildermuth later said that the cotton cv. most commonly planted in Queensland (Deltapine Smoothleaf) was one of the most susceptible.

In young plants the first symptoms are a pronounced epinasty of the leaves, followed by marginal curling, irregular chlorotic mottling (often characteristically at the margins of the lamina and between the main veins) and necrosis. Stunting of the plant is common. Defoliation may occur, this being distinctive of the defoliating str. which is apparently restricted to parts of the Americas. On older plants symptoms are seen first on the lower leaves, spreading upwards; the chlorotic areas enlarge and become paler. These plants are not usually killed, even though leaf fall may be severe, and secondary branches or shooting from the base may develop. The entire vascular system shows a light brown discolouration (see Erwin et al.; Presley). Sporulation can occur over the stem surface for a short distance above soil level and microsclerotia form in the stem, leaves and roots. Senescent leaf tissue on moist soil may form abundant conidia (Evans et al. 1966b). Penetration takes place on the hypocotyl, root cap, root hairs, region of root elongation and maturation, and the torn areas where lateral roots emerge. The pathogen can be isolated from the tops of plants 115 cm high within 24 hours from immersion of the root systems in a conidial suspension (Garber et al. 1966; Presley et al. 1966). Ashworth et al. made field determinations on inoculum density and infection; they found that the infection threshold was *c.* 0.03 microsclerotia/g soil, and that essentially 100% infection resulted when inoculum was at least 3.5 microsclerotia/g soil.

Soil temps of 20–25° are favourable for disease development and Halisky et al. found no wilt symptoms at 35°. Microsclerotia are abundant in leaves at 18–30° and can develop at much lower temps; none form at 32° (Brinkerhoff). In USA (California) Garber et al. (1971) found that high temps can mask the disease. In a warm summer neither a susceptible cv. nor a tolerant one were seriously affected, but in a cool one the relative disease tolerances could be distinguished. At the higher temps of a warm summer the fungus is less readily isolated. A defoliating pathotype withstood high temps better than a non-defoliating one (Temple et al.).

Microsclerotia readily survive from one season to the next in soil, host debris or on other dicotyledonous hosts (including weeds) which are symptomless. But several factors affect this survival. Even 1 year of a cereal can reduce infection of cotton grown on infested soil. Using sorghum it was found that more wilt occurred in cotton if the previous sorghum crops (4 years) were weed infested (*Amaranthus* spp.) than if these crops were kept weed free. Build up of inoculum may occur on several weed spp. and other crops (Brown et al.; Butterfield et al.; Evans et al. 1967b; Minton). Benson et al. considered that survival on immune crop roots and crop residues was not very significant. Huisman et al., in concluding that short rotations were of little or no value in control, found that once the soil was infested the rate of inoculum decrease was very low even under immune crops. After 6 years (from cotton) microsclerotial populations had not dropped to a safe level. De Vay described the characteristics and conc. of fungal propagules in air-dried field soils. Karaca et al. and Shtok et al. reported penetration of cotton seed (and see Allen; Evans et al. 1966a; Rudolph et al.). Seed can be contaminated through plant debris and washings from commercially delinted seed may contain sclerotia.

Important differences in pathogenicity between strs have been described from USA, other countries in America and USSR. Isolates of these strs are divided into 2 groups: the virulent (severe), defoliating group and the mild, non-defoliating one. The virulent str. appears to have arisen in USA and was reported by Schnathorst et al. (1966a) to have been a major factor in cotton wilt in California since 1960. Using susceptible, tolerant and highly resistant cvs they were able to distinguish between the two strs; the degree of difference in virulence between them depending on host genotype. Temp. is an important factor in separating the strs and determining host reactions; it should not be >24.5°. Prior inoculation with the mild str. gave some protection against the virulent one (Barrow 1970b; Schnathorst et al. 1966b). Wyllie et al. found that the conidia from the defoliating str. had a higher germination rate, it sporulated more and formed microsclerotia over a wider range of carbon:nitrogen ratios compared with the non-defoliating one. Only conidia from the former germinated at 33°. The 2 strs differed antigenically. The defoliating one has a more restricted host range but causes wilt in olive (*Olea europaea*) (Mathre et al.; Schnathorst et al. 1971a). Isolates from Australia were all similar to the mild USA str. (Schnathorst et al. 1971b; and see Tjamos et al. for isolates in Greece). Infection of cotton with both *V. dahliae* and *F. oxysporum* f. sp. *vasinfectum* has been investigated (Al-Shukri). In small scale experiments prior infection with *Rhizoctonia solani* and *Meloidogyne incognita* var. *acrita* increased wilt (Khoury et al.). A disease complex with *Thielaviopsis basicola* (q.v.) has also been described. Phytoalexins, cell degrading enzymes and xylem occlusion have been investigated by several workers.

Control is largely through host resistance or tolerance. Some reduction in disease can be obtained by rotation with other crops (Butterfield et al.; Hinkle et al.; but see Huisman et al.). Seed treatment may also be advisable. Resistant forms of *Gossypium hirsutum* (Upland) have been developed in USA and USSR. Schnathorst et al. (1976) evaluated cvs from USSR which were tolerant of infection by *V. dahliae*. In the tolerant cvs of Upland the chance of disease lessens as host age increases. The inheritance of this tolerance is not well understood (Barrow 1970a; but see Roberts et al.). Resistance in Upland shows partial dominance; it is more marked in other cottons from *G. arboreum* (Asiatic), *G. barbadense* (Sea Island) and *G. herbaceum* (Asiatic; Verhalen et al.; Wiles). Phytoalexin conc. was higher (×20) in a resistant *G. barbadense* cv. compared with susceptible and tolerant *G. hirsutum* cvs when inoculated with the defoliating str. Crosses between these 2 spp. showed that the capacity for increased phytoalexin production was inherited largely in a dominant fashion. The increase in host resistance with temps appears to be due to the higher opt. temps for defence reactions (opt. for phytoalexin synthesis is 25–27.5°), rather than to any direct effects on the growth of the pathogen in vivo (Bell; Bell et al.; and see Bassett; Marani et al.; Wilhelm et al. 1974). Benomyl and thiabendazole have given control in field trials but their justification on economic grounds is in doubt (Booth et al.; Buchenauer et al.; Erwin; Ranney; and see *Rev. Pl. Pathol.* **52**, 4086; **53**, 1830, 3981, 4457, 4459; **56**, 1997). Wilhelm et al. (1972) described soil fumigation with methyl bromide and chloropicrin.

Allen, R. M. 1951. *Pl. Dis. Reptr* **35**: 11 (seed transmission; **30**: 367).

Al-Shukri, M. M. 1968. *Ibid* **52**: 910 (interaction with *Fusarium oxysporum* f. sp. *vasinfectum*; **48**, 1190).

——. 1969. *Ibid* **53**: 126 (survival; **48**, 1723).

Ashworth, L. J. et al. 1972. *Phytopathology* **62**: 901 (inoculum potential; **52**, 1154).

—— ——. 1974. *Ibid* **64**: 563 (free & bound microsclerotia in soil; **54**, 470).

Barrow, J. R. 1970a. *Ibid* **60**: 301 (inheritance of host tolerance; **49**, 2488).

——. 1970b. *Ibid* **60**: 559 (temp. & requirements for expression of host tolerance; **49**, 2865).

Bassett, D. M. 1974. *Crop Sci.* **14**: 864 (resistance & yield; **54**, 4944).

Bell, A. A. 1967. *Phytopathology* **57**: 759 (phytoalexin; **46**, 3449).

—— et al. 1969. *Ibid* **59**: 1119, 1141, 1147 (phytoalexin, effects of temp., formation in resistant & susceptible cvs; **49**, 153a, b, c).

Benken, A. A. et al. 1964. *Zasch. Rast. Mosk.* **9**: 15 (survival, microsclerotia in leaves; **44**, 1573).

Benson, D. M. et al. 1976. *Phytopathology* **66**: 883 (survival on nonsusceptible roots & residues in soil; **56**, 1136).

Booth, J. A. et al. 1971. *Pl. Dis. Reptr* **55**: 569 (field control with benomyl; **51**, 1534).

——. 1974. *Can. J. Bot.* **52**: 2219 (effect of root exudate on growth of, & pectolytic enzyme formation by, the pathogen; **54**, 2266).

Brinkerhoff, L. A. 1969. *Phytopathology* **59**: 805 (effect of factors on microsclerotial development in abscissed leaves; **48**, 3500).

Brown, F. H. et al. 1970. *Pl. Dis. Reptr* **54**: 508 (infection of cvs & weeds, symptomless hosts; **50**, 106).

Buchenauer, H. et al. 1971. *Phytopathology* **61**: 433 (control with benomyl & thiabendazole; **50**, 3795).

Bugbee, W. M. et al. 1970. *Crop Sci.* **10**: 649 (effect on yield, fibre & seed; **51**, 2528).

Butterfield, E. J. et al. 1978. *Phytopathology* **68**: 1217 (effect of crop sequences on disease incidence & pathogen populations in soil).

De Vay, J. E. et al. 1974. *Ibid* **64**: 22 (conc. & characteristics of pathogen propagules in relation to wilt; **53**, 4458).

Erwin, D. C. et al. 1965. *Ibid* **55**: 663 (stem inoculation & symptom severity; **44**, 3057).

——. 1969. *Wld Rev. Pest Control* **8**: 6 (review of systemic & fungitoxic properties of chemicals, benomyl & thiabendazole; **49**, 649).

Evans, G. et al. 1966a. *Phytopathology* **56**: 460 (seed dissemination & treatment; **45**, 2498).

—— ——. 1966b. *Ibid* **56**: 590 (inoculation, infection & sporulation in vivo; **45**, 3149).

—— ——. 1967a. *Ibid* **57**: 1250 (amounts of microsclerotia in soil; **47**, 824).

—— ——. 1967b. *J. Aust. Inst. agric. Sci.* **33**: 210 (incidence & distribution in Namoi Valley, New South Wales).

Garber, R. H. et al. 1966. *Phytopathology* **56**: 1121 (host penetration & colonisation; **46**, 331).

—— ——. 1967. *Ibid* **57**: 885 (resistance; **46**, 3450).

—— ——. 1971. *Ibid* **61**: 204 (air temp. & development of wilt in the field; **50**, 2973).

Hafez, A. A. R. et al. 1975. *Agron. J.* **67**: 359 (K uptake & wilt; **55**, 3194).

Halisky, P. M. et al. 1959. *Pl. Dis. Reptr* **43**: 584 (effect of soil temp.; **38**: 598).

Hinkle, D. A. et al. 1963. *Bull. Ark. agric. Exp. Stn* 674, 15 pp. (yield & effect of crop rotations on disease incidence; **43**, 1932).

Howell, C. R. et al. 1973. *Can. J. Microbiol.* **19**: 1367 (virulence to cotton & tolerance of sanguinarine in *Verticillium*; **53**, 3496).

—— ——. 1976. *Physiol. Pl. Pathol.* **8**: 181 (effect of ageing on flavonoid content & resistance; **55**, 5210).

—— ——. 1976. *Ibid* **9**: 279 (importance of pectolytic enzymes in symptom expression; **56**, 3049).

Huisman, O. C. et al. 1976. *Phytopathology* **66**: 978 (effect of crop rotation on survival; **56**, 1615).

Isaac, I. et al. 1972. *Trans. Br. mycol. Soc.* **59**: 313 (in Tamil Nadu, India; **52**, 1548).

Kamal, M. et al. 1956. *Ann. appl. Biol.* **44**: 322 (pectic enzymes, effect of culture filtrates; **36**: 27).

Karaca, I. et al. 1973. *J. Turkish Phytopathol.* **2**: 30 (in seed; **52**, 3333).

Keen, N. T. et al. 1971. *Phytopathology* **61**: 198 (endopolygalacturonase; **50**, 2974).

Khoury, F. Y. et al. 1973. *Ibid* **63**: 352, 485 (effects of *Rhizoctonia solani* & *Meloidogyne incognita* var. *acrita* on infection; **52**, 4089; **53**, 177).

Krassilnikov, N. A. et al. 1969. *J. gen. appl. Microbiol. Tokyo* **15**: 1 (toxins; **49**, 767).

Mace, M. E. et al. 1976. *Can. J. Bot.* **54**: 2095 (histochemistry & identification of disease induced terpenoid aldehydes; **56**, 1137).
——. 1978. *Physiol. Pl. Pathol.* **12**: 1 (tyloses, phytoalexins & resistance; **47**, 3987).
—— ——. 1978. *Ibid* **13**: 143 (histochemistry & identification of flavanols).
Marani, A. et al. 1976. *Crop Sci.* **16**: 392 (tolerance & yield; **56**, 1616).
Mathre, D. E. et al. 1966. *Pl. Dis. Reptr* **50**: 930 (mild & virulent strs & hosts; **46**, 1008).
Minton, E. B. 1972. *Phytopathology* **62**: 582 (weed control in sorghum & subsequent disease incidence; **52**, 137).
Misaghi, I. J. et al. 1978. *Can. J. Bot.* **56**: 339 (xylem occlusion & symptoms; **57**, 4483).
Mussell, H. W. 1973. *Phytopathology* **63**: 62 (endopolygalacturonase; **52**, 3705).
Nazirov, N. N. 1960. *Dokl. Akad. Nauk SSSR* **6**: 44 (resistance & susceptibility in aerial part & root system, grafting; **40**: 682).
Perches, E. S. et al. 1971. *Foll. tec. Inst. nac. Invest. agric. Mexico* 56, 30 pp. (general; **51**, 4042).
Presley, J. T. 1950. *Phytopathology* **40**: 497 (general, variation in pathogen; **29**: 560).
—— et al. 1966. *Ibid* **56**: 375 (movement of conidia in host; **45**, 2499).
—— ——. 1969. *Ibid* **59**: 253 (effect of vessel ontogeny on disease in young plants; **48**, 2409).
Puhalla, J. E. et al. 1975. *Physiol. Pl. Pathol.* **7**: 147 (endopolygalacturonase & symptom expression; **55**, 2737).
Ranney, C. D. 1971. *Phytopathology* **61**: 783 (field control with benomyl & thiabendazole; **51**, 384).
Roberts, C. L. et al. 1972. *Crop Sci.* **12**: 63 (inheritance of tolerance; **53**, 176).
Rudolph, B. A. et al. 1944. *Phytopathology* **34**: 849 (unimportance of seed dissemination; **24**: 99).
Schnathorst, W. C. et al. 1966a. *Ibid* **56**: 1155 (differentiation of virulent & mild strs; **46**, 332).
—— ——. 1966b. *Ibid* **56**: 1204 (cross protection in host; **46**, 333).
——. 1969. *Pl. Dis. Reptr* **53**: 149 (virulent str. in Peru; **48**, 1725).
—— ——. 1971a. *Ibid* **55**: 780 (strs & infection of *Gossypium* & *Olea europaea*; **51**, 1262).
—— ——. 1971b. *Ibid* **55**: 977 (comparative virulence of isolates from USA & Australia; **51**, 1533).
—— ——. 1975. *Ibid* **59**: 863 (strs in Nevada, USA; **55**, 3193).
—— ——. 1976. *Ibid* **60**: 211 (cvs from USSR & tolerance; **55**, 5759).
Shtok, D. A. et al. 1974. *Mikol. i Fitopatol.* **8**: 374 (in seed; **54**, 2264).
Soloveva, A. I. et al. 1940. *Tashkent agric. Publ. Dept* 63 pp. (general in USSR; **27**: 19).
Straumal, B. P. 1966. *Khlopkovodsto* **16**: 31 (breeding for resistance; **45**, 2500).
Temple, S. H. et al. 1973. *Phytopathology* **63**: 953 (temp. & pathogenicity of defoliating & non-defoliating pathotypes; **53**, 1401).
Tjamos, E. C. et al. 1978. *Pl. Dis. Reptr* **62**: 456 (virulence of strs in Greece; **58**, 244).
Verhalen, L. M. et al. 1971. *Crop Sci.* **11**: 407 (genetics of resistance; **53**, 564).
Wang, M. C. et al. 1970. *Archs Biochem. Biophys.* **141**: 749 (endopolygalacturonase; **50**, 2290).
Wiese, M. V. et al. 1970. *Phytopathology* **60**: 641 (polygalacturonase & cultural characteristics; **49**, 3303).
Wildermuth, G. B. 1971. *Aust. J. exp. Agric. Anim. Husb.* **11**: 365 (varietal resistance; **51**, 389).
Wiles, A. B. 1960. *Pl. Dis. Reptr* **44**: 419 (screening for resistance in *Gossypium* spp.; **40**: 48).
Wilhelm, S. et al. 1972. *Calif. Agric.* **26**(10): 4 (soil fumigation; **52**, 2953).
—— ——. 1974a & b. *Phytopathology* **64**: 924, 931 (techniques of identification, sources & inheritance of resistance; **54**, 2262, 2265).
Wyllie, T. D. et al. 1970. *Ibid* **60**: 907 (growth characteristics of virulent & mild strs, *Verticillium nigrescens*; **49**, 3304).
—— ——. 1970. *Ibid* **60**: 1682 (immunological comparisons of virulent & mild strs & *V. nigrescens*; **50**, 1804).
Zaki, A. I. et al. 1972. *Ibid* **62**: 1398, 1402 (phytoalexins; **52**, 3331, 3332).

HELIANTHUS ANNUUS

Infection of young sunflower by *V. dahliae* causes stunting, wilt and death. Plants attacked at this stage can also show the characteristic interveinal, chlorotic mottling on the leaves; these symptoms are typical on maturing plants. Yellow areas turn necrotic except at the edge where chlorosis persists. Tissue bordering the leaf veins remains green as the chlorosis spreads. The affected leaves may occur on one side of the plant only and on older plants the lower leaves are the first to show symptoms, these spread upwards; there may be premature ripening and less seed. There is a light brown vascular discolouration. In Canada the disease has been called leaf mottle. Sackston et al. (1973) in the cv. Sunrise found that symptoms appeared sooner and were more severe in plants under long days (16 hours light) or short days with an interrupted dark period (hours: 10 light, 6.5 dark, 1.5 light, 6 dark), than under short days (10 hours). Under the short day treatment healthy plants flowered in 45–47 days but long day plants only reached the yellow bud stage. Development of wilt did not depend on the initiation of flowering. Typical symptoms developed in plants sown in uninfested soil with seed from diseased plants. Sackston et al. (1959) found a lack of consistency in isolation of the pathogen from seed from diseased plants. Isolates from sunflower were mostly avirulent to Irish potato, tomato, safflower and cotton; and some susceptible sunflower cvs can be infected by *V. dahliae* strs parasitic on other hosts (Orellana). Oligogenic and monogenic resistance to the pathogen is known; highly resistant inbred lines and resistant hybrids have been developed.

Fick, G. N. et al. 1974. *Crop Sci.* **14**: 895 (inheritance of resistance; **54**, 5033).
Hoes, J. A. et al. 1973. *Phytopathology* **63**: 1517 (resistance in wild *Helianthus*; **53**, 3578).
Moser, P. E. et al. 1973. *Ibid* **63**: 1521 (inoculation methods; **53**, 3129).
Orellana, R. G. 1969. *Phytopathol. Z.* **65**: 183 (relative pathogenicity to sunflower & other hosts; **49**, 534).
Putt, E. D. 1964. *Crop Sci.* **4**: 177 (inheritance of resistance; **44**, 1197).
Robb, J. et al. 1975. *Can. J. Bot.* **53**: 2725 (fine structure of syndrome; **55**, 3244).
——. 1977. *Ibid* **55**, 139 (as above; **56**, 4144).
Sackston, W. E. et al. 1957. *Pl. Dis. Reptr* **41**: 337 (general; **37**: 52).
——. 1959. *Can. J. Bot.* **37**: 759 (seed transmission; **39**: 417).
——. 1973. *Ibid* **51**: 23 (effect of day length on host reaction; **52**, 2711).
Zimmer, D. E. et al. 1973. *Pl. Dis. Reptr* **57**: 624 (resistance, with *Puccinia helianthi*; **53**, 646).

NICOTIANA TABACUM

Tobacco wilt caused by *V. dahliae* (and to a lesser extent by *V. albo-atrum*) is an important disease in New Zealand where it limits tobacco production in certain areas (Wright et al. 1973). These workers induced severe symptoms in tobacco with an isolate from tomato in Canada, and drew attention to the potential threat of this disease to tobacco in the country. Sheppard et al. (1974) reported *V. nigrescens* as a cause of wilt. A very distinctive symptom is the orange colour of wilted leaves. The interveinal areas become orange and then necrotic, leaving an orange border between living and dead tissue. Wilt appears at the beginning of flowering. Symptoms are at first restricted to 1 or 2 leaves but eventually spread to all of them. The midrib shows a light brown vascular discolouration. There are no external root symptoms.

Mahanty considered that the living tissues of the root tip were not penetrated directly and that invasion is probably through wounded or moribund roots. Wright (1968b, 1969) described the invasion of intact roots at high inoculum levels. Conidia inoculated in the roots spread rapidly through the roots, stem and leaf midribs. Entry of roots of resistant and susceptible cvs was similar but the pathogen eventually became more restricted in a resistant cv. At the end of the season the fungus becomes more abundant near the top of stems than at the base. Toxins, not vessel plugging, have been considered to be the cause of wilt (McLeod et al. 1961). Survival of the pathogen presumably occurs in the characteristic way. The isolates from tobacco are considered to be a distinct str. Those from some other crops, and including *V. albo-atrum*, did not attack tobacco (Christie); also the tobacco str. of *V. dahliae* did not attack tomato or Irish potato. *Verticillium* cultures from solanaceous hosts were more pathogenic to tobacco than those from non-solanaceous ones (Sheppard et al. 1974). No differences in virulence between tobacco isolates appear to have been found.

Cvs with resistance (inheritance polygenic) and tolerance have been selected. Grafting experiments with cvs differing widely in the degree of resistance give data which suggest that most of the resistance factors are contributed by the roots at all growth stages. Hartill et al. described tolerance levels in cvs. Methyl bromide controlled wilt for 3 years and gave increases in root weight, yield, crop indices and quality; chloropicrin also reduced wilt for 3 years (Taylor et al. 1970). Hartill found that benomyl applied before planting markedly reduced leaf loss; it was less effective as a root dip and ineffective as a late spray. There was a yield increase due to the delayed appearance of the disease in treated plants. Less control was obtained in late than in early maturing cvs.

Canter-Visscher, T. W. 1967. *N.Z. Tob. Grow. J.* Mar.: 6 (survival).
Christie, T. 1956. *N.Z. Jl Sci. Technol.* Sect. A **38**: 17 (pathogenicity test with *Verticillium*; **35**: 927).
——. 1966. *N.Z. Jl agric. Res.* **9**: 149 (pathogenicity of *Verticillium* to tobacco; **45**, 1511).
Gibbins, L. N. et al. 1968. *Ibid* **11**: 789 (effects of roots & stems on symptoms; **48**, 1309a).
Hartill, W. F. T. 1971. *Pl. Dis. Reptr* **55**: 889 (control with benomyl; **51**, 1897).

—— et al. 1976. *N.Z. Jl agric. Res.* **19**: 377 (tolerance of cvs; **56**, 1265).
McLeod, A. G. et al. 1959. *Ibid* **2**: 792 (resistance trials).
—— ——. 1961. *Ibid* **4**: 123 (effect of culture filtrates; **41**: 171).
Mahanty, H. K. 1970. *Ibid* **13**: 699 (histology of colonisation, resistant & susceptible cvs; **50**, 971).
Sheppard, J. W. et al. 1974. *Can. Pl. Dis. Surv.* **54**: 57 (survey in Quebec, Canada; **54**, 4141).
—— ——. 1976. *Can. J. Pl. Sci.* **56**: 157 (chlorogenic acid & wilt; **55**, 4273).
Taylor, J. B. et al. 1970. *N.Z. Jl Sci.* **13**: 591 (soil fumigation; **50**, 3166).
Taylor, M. E. U. 1968. *Span* **11**: 96 (resistance; **47**, 3402).
Thompson, R. et al. 1959. *N.Z. Jl agric. Res.* **2**: 785 (general; **39**: 245).
Wright, D. S. C. 1968a. *Ibid* **11**: 655 (screening cvs for resistance; **47**, 3564).
——. 1968b. *Ibid* **11**: 803 (movement in host; **48**, 1309c).
—— et al. 1968. *Ibid* **11**: 797 (pathogenicity of isolates; **48**, 1309b).
——. 1969. *Ibid* **12**: 228 (symptoms in different cvs; **48**, 1969).
—— ——. 1970. *N.Z. Jl Bot.* **8**: 326 (fine structure in vivo & in vitro; **50**: 1967).
—— ——. 1973. *Can. J. Pl. Sci.* **53**: 391 (a potential disease in Canada; **53**, 1941).

SOLANUM MELONGENA

Infected eggplant shows chlorotic areas in the leaves which become flaccid; plants become stunted, wilt and die. If infection occurs at a young growth stage there is virtually no crop. Guba found most disease at soil temps of 21–25°; no wilt occurs at 35°. Vascular discolouration does not always show in all parts of the plant. In Canada nematode (*Pratylenchus penetrans*) damage on the roots increases the amount of fungal infection and reduction in the numbers of nematodes reduces disease caused by *V. dahliae*. It was also shown that root damage resulting from transplanting increases infection. Burton et al. found no evidence for seed transmission although it had been reported earlier. Ribeiro found that a virulent str. penetrated the seed but that a mild one did not. Wilt has been reduced and its onset delayed by soil fumigation with methyl isothiocyanate; both plant growth and yields improved (McKeen et al. 1967). Although differences in susceptibility between commercial cvs exist there seems to be no very high resistance to any of them. In small scale field tests susceptible cvs grafted on resistant tomato rootstocks gave a high level of resistance (Lockwood et al. 1970; and see Gindrat et al.). In India benomyl was said to be effective.

Burton, C. L. et al. 1958. *Pl. Dis. Reptr* **42**: 427 (seed transmission tests; **37**: 625).
Figueiredo, M. B. et al. 1968. *Arq. Inst. biol. S. Paulo* **35**: 9 (resistance in cvs; **48**, 323).
Gindrat, D. et al. 1976. *Rev. suisse Vitic. Arboric. Hort.* **8**: 71 (use of resistant tomato stocks; **55**, 5451).
Guba, E. F. 1934. *Phytopathology* **24**: 906 (general; **14**: 74).
Lockwood, J. L. et al. 1967. *Q. Bull. Mich. St. Univ. agric. Exp. Stn* **50**: 50 (breeding for resistance 1961–6, races; **47**, 722).
—— ——. 1970. *Pl. Dis. Reptr* **54**: 846 (use of resistant tomato stocks; **50**, 1546).
McKeen, C. D. et al. 1960. *Can. J. Bot.* **38**: 789 (synergism with *Pratylenchus penetrans*; **40**: 261).
—— ——. 1967. *Can. J. Pl. Sci.* **47**: 1 (soil fumigation; **46**, 1825).
Mountain, W. B. et al. 1965. *Can. J. Bot.* **43**: 619 (effect of transplant injury & *P. penetrans*).
Ribeiro, R. de L. D. 1972. *Arq. Univ. Fed. Rur. Rio de Janeiro* **2**: 17 (seed transmission; **53**, 4934).
Sivaprakasam, K. et al. 1974. *Indian Phytopathol.* **27**: 304 (use of benomyl; **55**, 3836).

OTHER CROPS

Diseases caused by *V. dahliae* (sometimes as *V. albo-atrum*) on 12 other crops are described briefly. Evans et al. described the effects of crops on microsclerotial numbers. On groundnut the lower leaves show chlorotic areas, usually towards flowering; plants are stunted and show the typical vascular discolouration. Isolates from other hosts (e.g., cotton, eggplant and red pepper) caused disease in groundnut. The Spanish-Valencia forms are more susceptible. Incidence of this groundnut wilt necessitates breeding for resistance in Israel where selection has been done (Frank et al.; Purss; Smith). Isaac described a wilt of Brussels sprout; inoculations suggested that the str. on this host is distinct. The pathogen is not seedborne but can be spread through infected tissues. Runner beans, broccoli and cauliflower are resistant. Isaac et al. reported, too, on a wilt of pea (*Pisum sativum*). The symptoms included acropetal progression of leaf chlorosis and necrosis and premature defoliation. These are indistinguishable from senescence and thus there can be confusion in the field. The pea isolate was pathogenic to Irish potato, sweet pea, antirrhinum and broad bean. Isolates from Irish potato, lucerne and sweet pea infected pea. Kendrick et al. found that isolates from red pepper differed from those of other hosts; only those from red pepper caused severe disease (stunting) in this crop. Red pepper isolates affected okra (*Hibiscus esculentus*) severely. Disease could be high at 15–30° soil temps (air temp. 24°) but decreased markedly at 35°. Woolliams et al. screened many cvs and lines but found no very high resistance, and noted that losses in Canada (British Colombia) could be of economic significance (see also Curzi; Elenkov; Snyder et al.). On okra the lower leaves show chlorotic areas; irregular, marginal and bounded by the veins; the leaf margins curl upwards. Shoots wilt, including those that form after the leaves fall, and there is extensive vascular necrosis.

V. dahliae was described from cowpea (*Vigna*) and it occurs (as microsclerotia) both on and in the seed of safflower (Carpenter; Goethal; Klisiewicz; Strobel). Conroy et al., in investigations on *V. dahliae* on tomato, described the interactions with the root lesion nematode (*Pratylenchus penetrans*). They found that at all inoculum potentials infection by the fungus increased when the nematode was present; it also became more frequent as the nematode population increased. When a resistant tomato cv. was inoculated with the root knot nematode (*Meloidogyne javanica*) and the fungus the resistance was not broken down. But the same inoculation of a susceptible cv. led to enhanced incidence of *V. dahliae* and symptoms (Orion et al.; and see Patrick et al.). In USA (Florida) infection of young mango trees caused death of the main branches; the necrotic leaves remained attached (Marlatt et al. 1970). Selections of castor (*Ricinus communis*) showed high tolerance of strs virulent on cotton (Brigham et al.). Avocado wilt was described from USA (California and Florida; Marlatt et al. 1969; Zentmyer). The sudden wilt of individual branches or whole trees and rapid death of leaves which remain attached are distinctive symptoms (cf. the much more serious root rot of the same crop caused by *Phytophthora cinnamomi* q.v.). Death of the tree, in which up to 75% of the foliage may wilt, is rare and some months after the initial collapse new shoots appear, leading to recovery. Some trees may show recurrent symptoms but complete recovery is more usual. The xylem shows a dark brown streaking in stems and roots. Mexican stocks, more resistant than Guatemalan ones, should be used. In severe disease cases fumigate with chloropicrin. Monocotyledonous hosts do not usually show disease symptoms but Sherrod et al. observed them in sorghum when inoculated with a cotton isolate. A wilt of cacao was described from Uganda by Leakey, Emechebe and Emechebe et al. In the field leaves droop, become dry and brittle but remain attached for >5 weeks; leaves and smaller branches may fall. The symptoms are similar to forms of dieback (which may have other causes) in cacao; there is internal vascular necrosis. Plants may apparently recover. The penetration of roots of cacao seedlings has been described. Deaths of field plants have been reported, following either a prolonged dry season or after heavy rain, and the condition called sudden death.

Brigham, R. D. et al. 1969. *Pl. Dis. Reptr* **53**: 262 (tolerance in castor; **48**, 2509).

Carpenter, C. W. 1918. *J. agric. Res.* **12**: 529 (on okra).

Conroy, J. J. et al. 1972. *Phytopathology* **62**: 362 (on tomato with *Pratylenchus penetrans*; **51**, 4373).

Curzi, M. 1925. *Riv. Patol. veg. Pavia* **15**: 145 (on red pepper; **5**: 206).

Elenkov, E. 1957. *Byul. Rast. Zashch. Sofia* **6**: 32 (on red pepper, testing cvs for resistance; **37**: 435).

Emechebe, A. M. et al. 1971. *Ann. appl. Biol.* **69**: 223 (on cacao; **51**, 2335).

—— ——. 1972. *Ibid* **70**: 157 (as above, infection of seedling roots; **51**, 3914).

—— ——. 1974. *E. Afr. agric. For. J.* **39**: 337 (as above, vessel blockage & host recovery; **54**, 3816).

——. 1974. *Ibid* **40**: 168 (as above, inoculation; **55**, 1171).

——. 1975. *Ibid* **40**: 271 (as above; **55**, 3115).

—— ——. 1975. *Ibid* **41**: 184 (as above, time & infection; **57**, 108).

Evans, G. et al. 1975. *Can. J. Pl. Sci.* **55**: 827, 857 (effect of crops on microsclerotial numbers & a str. pathogenic to red pepper; **55**, 597, 1000).

Frank, Z. R. et al. 1968. *Israel Jnl agric. Res.* **18**: 83 (on groundnut; testing cvs for resistance; **47**, 3286).

Goethal, M. 1971. *Al Awamia* **39**: 39 (on safflower; **53**, 4082).

Isaac, I. 1957. *Ann. appl. Biol.* **45**: 276 (on Brussels sprout; **36**: 742).

—— et al. 1974. *Ibid* **76**: 27 (on pea; **53**, 3680).

Kendrick, J. B. et al. 1959. *Phytopathology* **49**: 23 (on red pepper, soil temp.; **38**: 378).

Klisiewicz, J. M. 1974. *Pl. Dis. Reptr* **58**: 926 (assay in safflower seed; **54**, 952).

——. 1975. *Phytopathology* **65**: 696 (survival & spread in safflower seed; **55**, 355).

Leakey, C. L. A. 1965. *E. Afr. agric. For. J.* **31**: 21 (on cacao; **45**, 1742).

Locke, T. et al. 1976. *Pl. Pathol.* **25**: 59 (on tomato, resistance to benomyl; **55**, 4858).

Marlatt, R. B. et al. 1969. *Pl. Dis. Reptr* **53**: 583 (on avocado; **48**, 3595).
——. 1970. *Ibid* **54**: 569 (on mango; **50**, 196).
Orion, D. et al. 1976. *Phytoparasitica* **4**: 41 (on tomato with *Meloidogyne javanica*; **56**, 3226).
Patrick, T. W. et al. 1977. *Can. J. Bot.* **55**: 377 (cytokinin levels; **56**, 5198).
Purss, G. S. 1961. *Qd J. agric. Sci.* **18**: 453 (on groundnut; **41**: 689).
Sherrod, L. L. et al. 1967. *Phytopathology* **57**: 14 (on sorghum; **46**, 2009).
Smith, T. E. 1961. *Ibid* **51**: 411 (on groundnut; **41**: 76).
Snyder, W. C. et al. 1939. *Ibid* **29**: 359 (on red pepper; **18**: 570).
Strobel, J. W. 1961. *Proc. Fla St. hort. Soc.* 74: 171 (on okra & *Vigna*; **42**: 172).
Woolliams, G. E. et al. 1962. *Can. J. Pl. Sci.* **42**: 515 (on red peppper, testing cvs for resistance; **42**: 172).
Zentmyer, G. A. 1949. *Phytopathology* **39**: 677 (on avocado; **29**: 105).

Verticillium theobromae (Turc.) Mason & Hughes, apud Hughes, *Mycol. Pap.*, 45: 10, 1951.
Stachylidium theobromae Turc. 1920.

CULTURES growing rapidly on PDA and malt agar at 23°; the prostrate hyphae which are first produced are hyaline. MYCELIUM white flocculose, densely compacted or sparse, becoming olivaceous grey brown underneath after 1–2 weeks, never entirely black; hyaline sectors unknown. CONIDIOPHORES abundant, more or less erect, hyaline sometimes becoming brownish below, verticillately branched, 3–6 phialides arising at each node, up to 6 whorls of phialides per conidiophore, phialides sometimes branched. PHIALIDES variable in size, mainly $14–37 \times 1.5–5\ \mu$. CONIDIA arise singly at the apices of the phialides, attached by one of their poles, ellipsoidal to subcylindrical, hyaline, $3–8 \times 1.5–3\ \mu$. RESTING MYCELIUM pale brown septate hyphae, not torulose, 2–3.5 μ diam., often poorly developed to absent. CHLAMYDOSPORES and microsclerotia absent.

Differs from *V. albo-atrum* (q.v.) in the absence of torulose resting mycelium, the tendency of the conidiophores to be brownish below, the different colour of the cultures underneath and in the smaller conidia (D. L. Hawksworth & P. Holliday, *CMI Descr.* 259, 1970).

One of the fungi involved in cigar end of banana in the field, and in crown rot during the shipment of boxed fruit. On banana (on the plant) the symptoms begin as a necrosis at the pistillate end. The skin becomes folded and shrunk and the dead floral parts tend to become persistent. The powdery, greyish conidia form on the shrivelled black end of the fruit giving rise to the characteristic cigar end appearance. The internal rot which results is a dry one in contrast to the wet rot caused by *Trachysphaera fructigena* (q.v.). *V. theobromae* may also be associated with black pitting and spot of banana fruit. The fungus is widespread in the tropics (Map 146). It is presumably air dispersed and also through banana debris in the field, on which it is a common inhabitant.

It is not clear whether *V. theobromae* can function as a primary invader of bananas. The evidence suggests that it is secondary in its role, occurring frequently with *Deightoniella torulosa* (q.v.) in Jamaica and *T. fructigena* in W. Africa in the distal end rots. Damage in the floral region or the incomplete formation of the corky barrier at the base of the dying parts of the flower may increase the risk of infection. There is apparently no evidence that penetration through an intact host surface can occur. In crown rot the pathogen was isolated from 81% of ripe crowns and 41% of the isolates induced a rot when placed on a freshly cut surface. The rot is probably more extensive at 24–26° than at lower temps. In boxed fruit rotting caused primarily by *V. theobromae* was slight but increased when inoculations were done with *Fusarium roseum* 'Gibbosum' and *Giberella fujikuroi* (q.v.). Conidial germination is highest in a water film. At <RH 32% conidia may survive for 60 days, but at 50–83% there was only 10% survival after 20 days. Growth in vitro is best at 25°.

Attempted control is probably through the general measures taken against fruit rotting fungi in banana. For cigar end hand removal of floral remains and copper spraying may give some control. Treatment with thiabendazole has given good results against the general rot found in boxed banana hands.

Daudin, J. 1953. *Fruits* **8**: 488 (control by removing floral remains; **34**: 657).
El-Helaly, A. F. et al. 1955. *Alex. J. agric. Res.* **11**: 9 (general; **35**: 781).
Malan, E. F. 1953. *Fmg S. Afr.* **28**: 365 (control; **33**: 363).
Meredith, D. S. 1961. *Trans. Br. mycol. Soc.* **44**: 487 (*Verticillium theobromae* & *Deightoniella torulosa* finger rots; **41**: 321).
——. 1965. *Ibid* **48**: 327 (general, in Jamaica; **45**: 161).
Wardlaw, C. W. 1931. *Trop. Agric. Trin.* **8**: 293 (general; **11**: 312).

EPILOGUE

It could be said, perhaps paradoxically, that the fundamental problem of applied biology is its application. Nowhere is this more true than over large agricultural regions in the tropics and subtropics (lowland or upland). With inadequate help what is the intelligent but scientifically ignorant farmer to do? There is no doubt that there are innumerable instances where a few simple field experiments, and proper instructions and help, are all that is needed to assist such farmers. Where what is often needed is the proper application, without any extensive research superstructure, of knowledge gained in the agriculturally more fortunate countries. How effectively one wonders do the elaborate, expensive and bureaucratic, agricultural research organisations operate in the tropics? An administrator or politician can be more easily impressed by an electron microscope than by labourers using fungicide sprayers. But the latter are usually needed more.

In Sarawak we came across a really remarkable example of disease control used by the Chinese growers of black pepper. The crop in S. E. Asia has been traditionally cultivated (in small holdings) by these people in a marvellously intensive way for hundreds of years. One farmer was being employed by us to cultivate a field trial. When the vines were 90–120 cm high we noticed slight damage (such as might have been caused by an insect) on the young, green stems which were being tied to the supporting post. The cause of the markings on these stems became known after a lengthy interpretation from an obscure Chinese dialect. Our farmer was removing with his finger nails the first visible pustules (1 mm diam.) of the pink disease pathogen (*Corticium salmonicolor*). The amount of labour must have been staggering, with some 600 vines, 3 stems for each and doubtless several pustules per stem.

We had frequently wondered why black pepper seemed only to get pink disease at a late growth stage, perhaps at 3 years old and more. The reason was now plain. Up to about the time when the plant reached the top of its supporting post (nearly 3 m above ground level) the farmer was able to cope with this disease problem. But thereafter with larger plants, and the extra work of harvesting and fertilising, the task was beyond him. A few fungicide field trials by any efficient agronomist was probably all that was needed to spare the cultivator his tedious and extraordinary method of cultural control. Bordeaux was being used against pink disease on rubber in the same region more than 70 years ago. But, apparently, no one had thought to communicate this fact effectively to the grower of a nearby crop, even by the late 1950s!

APPENDIX: HOSTS and PATHOGENS

Each entry for a plant consists of up to 4 elements in this order:

1. Plant genus and species with common name(s).
2. Pathogens (not italicised) (and disease name) fully treated.
3. Other pathogens (genera or species) where the plant is mentioned in a briefer treatment of a disease.
4. Selected ref. to crop pathology.

Abies (fir)
(Rhizina undulata)

Abutilon
(Puccinia cacabata)

Acacia, *dealbata* (silver wattle), *decurrens* (green wattle), *kao* (koa acacia), *cibaria*, *longifolia*.
(Armillariella mellea, Calonectria, C. crotalariae, Colletotrichum truncatum, Glomerella cingulata, Phoma)

Agave, *amaniensis* (blue sisal), *angustifolia* (dwarf sisal), *sisalana* (sisal), *zapupe* (zapupe azul), *lespinassei*.
Phytophthora nicotianae var. parasitica (bole rot)
(Glomerella cingulata)
Bock, K. R. 1965. *Wld Crops* **17**: 64.

Aglaonema simplex (Chinese evergreen)
(Pythium splendens)

Agrostis
(Gloeocercospora sorghi, Pythium aphanidermatum, P. butleri, Sclerospora graminicola)

Albizia
(Fusarium oxysporum, F. solani, Gibberella, Neocosmospora, Phytophthora drechsleri, Pleiochaeta)

Aleurites (tung)
(Armillariella mellea, Botryosphaeria ribes, Cristulariella, Mycosphaerella)
Large, J. R. 1949. *Pl. Dis. Reptr* **33**: 22 (in S. USA).
Plakidas, A. G. 1937. *Bull. La agric. Exp. Stn* 282, 11 pp. (in Louisiana, USA).
Saccas, A. M. et al. 1951. *Agron. Trop.* **6**: 239 (in equatorial Africa).
Wiehe, P. O. 1952. *Pl. Dis. Reptr* suppl. 216: 189 (bibliography).

Allium, *ampeloprasum* (great-headed garlic, leek), *ascalonicum* (shallot), *bouddhae*, *fistulosum* (Welsh onion), *cepa* (onion, shallot), *sativum* (garlic), *schoenoprasum* (chive), *ursinum* (wild garlic)
Alternaria porri (purple blotch)
Botrytis allii (grey mould neck rot)
Colletotrichum circinans & C. dematium (smudge)
Fusarium oxysporum f. sp. cepae (wilt)
Peronospora destructor (downy mildew)
Phytophthora porri (white tip)
Puccinia allii (rust)
Pyrenochaeta terrestris (pink root)
Sclerotinia porri (neck rot)
S. squamosa (small sclerotial neck rot)
Sclerotium cepivorum (white rot)
Urocystis cepulae (smut)
(Botryodiplodia theobromae, Botrytis, Cladosporium, Fusarium solani, Glomerella cingulata, Penicillium, Pleospora herbarum, Pythium graminicola, P. mammillatum, Sclerotinia, S. fuckeliana)
Hughes, I. K. 1970. *Qd agric. J.* **96**: 607 (in Australia).
Jones, H. A. & Mann, L. K. 1963. *Onions and their allies*: 180, Leonard Hill.
Walker, J. C. et al. 1961. *Agric. Handb. U.S. Dep. Agric.* 208, 27 pp.

Alnus (alder)
(Microsphaera)

Alocasia, *macrorrhiza* (giant alocasia)
Phytophthora colocasiae (leaf spot)

Althaea rosea (hollyhock)
(Ascochyta gossypii, Phytophthora megasperma, Puccinia)

Alysum
(Peronospora parasitica)

Amaranthus
(Verticillium albo-atrum)

Amygdalus persica (peach)
(Rhizopus stolonifer, Taphrina)

Anacardium occidentale (cashew)
(Glomerella cingulata, Phomopsis, Sphaceloma, Valsa eugeniae)

Ananas comosus (pineapple)
Ceratocystis paradoxa (water blister)
(Gibberella fujikuroi, Nigrospora, Penicillium, Phytophthora cinnamoni, P. meadii, P. nicotianae var. parasitica, P. palmivora, Pythium aphanidermatum, P. arrhenomanes, P. graminicola, P. mamillatum, P. splendens)

Ananas comosus (*cont.*)
Collins, J. L. 1960. *The pineapple, botany and utilization*: 187, Leonard Hill.
Merny, G. 1949. *Fruits* **4**: 125, 288, 327.
Andropogon
(Curvularia, Sphacelotheca reiliana)
Anethum graveolens (dill)
(Alternaria, A. radicina, Fusarium oxysporum)
Annona squamosa (sugar apple)
(Cercospora, Glomerella cingulata, Phomopsis)
Purss, G. S. 1953. *Qd J. agric. Sci.* **10**: 247 (fruit rots).
Anthurium andreanum (arum lily)
(Glomerella cingulata)
Anthylis
(Uromyces ciceris-arietini)
Antirrhinum majus (snapdragon)
(Myrothecium roridum, Puccinia, Pythium ultimum)
Apium graveolens var. **dulce** (celery)
(Alternaria radicina, Fusarium oxysporum, Mycocentrospora acerina, Puccinia, Septoria, Thanatephorus cucumeris)
Arabis
(Albugo candida)
Arachis hypogaea (groundnut), *glabrata, marginata, morticola, nambyquarae, prostrata.*
Aspergillus flavus (aflaroot)
A. niger (crown rot)
Calonectria crotalariae (black rot)
Mycosphaerella arachidis (early leaf spot)
M. berkeleyi (late leaf spot)
Puccinia arachidis (rust)
(Botryodiplodia theobromae, Colletotrichum dematium, Corticium rolfsii, Cristulariella, Cylindrocladium scoparium, Fusarium solani, Gibberella, Guignardia, Leptosphaerulina trifolii, Oidium, Pythium myriotylum, Rhizopus stolonifer, Sclerotinia sclerotiorum, Sphaceloma, Verticillium dahliae)
Bell, D. K. 1974. *Phytopathology* **64**: 241 (seed decay in soil).
Feakin, S. D. (editor). 1973. *PANS Manual* 2, 197 pp. (disease control).
Frezzi, M. J. 1960. *Revta Invest. agric. B. Aires* **14**: 113 (in Argentina, 45 ref.).
Hanlin, R. T. 1969. *Mycopathol. Mycol. appl.* **38**: 93 (fungi in pods).
Jackson, C. R. et al. 1969. *Res. Bull. Coll. Agric. Exp. Stn Ga* 56, 137 pp. (in the USA, 480 ref.).
McDonald, D. 1969. *Rev. appl. Mycol.* **48**: 465 (pod diseases; 95 ref.).
Purss, G. S. 1962. *Qd agric. J.* **88**: 540 (in Australia).
Araucaria
(Cylindrocladium)
Arctium lappa (great burdock)
(Erysiphe cichoracearum)
Areca catechu (areca palm)
Phytophthora arecae (koleroga)
P. palmivora (bud rot)
(Ganoderma boninense, Glomerella cingulata)
Reddy, M. K. et al. 1978. *J. Pl. Crops* **6**(1): 28 (in India).
Armoracia rusticana (horse radish)
(Albugo candida, Verticillium dahliae)
Artemesia dracunculus (tarragon)
(Puccinia)
Artocarpus, *altilis* (breadfruit), *heterophyllus* (jackfruit)
(Rhizopus stolonifer)
Butani, D. K. 1978. *Fruits* **33**: 35 (34 ref.).
Reddy, D. B. (editor). 1970. *Inf. Lett. F.A.O. Pl. Prot. Comm. S.E. Asia Pacific Reg.* 79, 8 pp.
Asparagus officinalis (asparagus)
(Phytophthora megasperma, Puccinia, P. allii)
Avena (oat)
(Cochliobolus, Ustilago)
Bauhinia purpurea (camel's-foot tree)
(Rigidoporus zonalis)
Benincasa hispida (wax gourd)
(Fusarium oxysporum f. sp. melonis, Pseudoperonospora cubensis, Pythium aphanidermatum)
Bertholletia excelsa (Brazil nut)
(Phytophthora heveae)
Beta vulgaris (beet, sugar beet)
(Aphanomyces, Erysiphe cruciferarum, Peronospora, Physoderma, Pleospora, Phytophthora drechsleri, Pythium debaryanum, P. ultimum, Rhizopus oryzae, Uromyces)
Bixa orellana (annatto)
(Glomerella cingulata, Oidium)
Boehmeria nivea (ramie)
(Rosellinia necatrix)
Bougainvillea
(Colletotrichum dematium)
Bombax malabaricum (semul, silk cotton)
(Phomopsis juniperivora)
Borassus flabellifer (palmyra palm)
Phytophthora palmivora (bud rot)
Bouteloua
(Puccinia cacabata)
Brachiara
(Uromyces)
Brassica, *alba* (white mustard), *campestris* (field mustard), *chinensis* (Chinese mustard), *juncea* (Indian mustard), *napobrassica* (rutabaga, swede), *napus* (rape), *nigra* (black mustard), *oleracea* var. *botrytis* (broccoli, cauliflower), var. *capitata* cabbage), var. *gemmifera* (Brussels sprout), var. *gongyloides* (kohlrabi), *rapa* (turnip)
Albugo candida (white blister)
Alternaria brassicae (grey leaf spot)
A. brassicicola (black leaf spot)
A. raphani (black pod blotch)

Brassica (*cont.*)
Fusarium oxysporum f. sp. conglutinans (yellows)
Leptosphaeria maculans (black leg)
Mycosphaerella brassicicola (ringspot)
Peronospora parasitica (downy mildew)
Pseudocercosporella capsellae (white spot)
Sclerotinia sclerotiorum (cottony soft rot)
(Alternaria alternata, Aphanomyces, A. raphani, Cercospora, Colletotrichum, Erysiphe cruciferarum, Myrothecium roridum, Phytophthora megasperma, P. porri, Pythium mamillatum, Sclerotinia fuckeliana, Thanatephorus cucumeris, Urocystis, Verticillium dahliae)

Anon. 1964. *Bull. Minist. Agric. Fish. Fd Lond.* 131, 39 pp.
Richardson, M. J. 1970. *Proc. int. Seed Test. Assoc.* **35**: 207 (seedborne pathogens).
Sumner, D. R. 1974. *Phytopathology* **64**: 692 (ecology & control of seedling diseases).
Walker, J. C. et al. 1958. *Agric. Handb. U.S. Dep. Agric.* 144, 41 pp.

Bromus inermis (smooth brome)
(Pythium graminicola)

Cajanus cajan (pigeon pea)
Fusarium udum (wilt)
(Cercoseptoria, Mycovellosiella, Neocosmospora, Phytophthora, P. vignae, Pyrenochaeta, Uromyces)

Caladium
(Sclerotinia ricini)

Callirhoë
(Puccinia cacabata)

Callistemon viminalis (bottle brush tree)
(Sphaeropsis tumefaciens)

Calopogonium
(Mycosphaerella cruenta)

Camellia sinensis (tea)
Exobasidium vexans (blister blight)
Hypoxylon serpens (wood rot)
Pestalotiopsis theae (grey blight)
Phomopsis theae (collar and branch canker)
Poria hypobrunnea (root rot, branch canker)
Rosellinia arcuata (black root rot)
R. necatrix (white root rot)
Tunstallia aculeata (thorny stem blight)
Ustulina deusta (charcoal stump rot)
(Armillariella mellea, Calonectria, C. kyotensis, Cercoseptoria, Corticium salmonicolor, Cylindrocladium, Elsinoë, Glomerella, G. cingulata, Hypoxylon, Macrophoma, Phellinus noxius, Thanatephorus cucumeris)

Agnihothrudu, V. 1964. *J. Madras Univ.* Sect. B **34**: 155 (fungi on tea, 360 ref.).
Eden, T. 1976. *Tea*: 113, Longman.
Gadd, C. H. 1949. *Monogr. Tea prodn Ceylon* 2, 94 pp. (in Sri Lanka).
Hainsworth, E. 1952. *Tea pests and diseases and their control*, W. Heffer.
Kasai, K. 1972. *Jap. Pest. Inf.* **11**: 25 (control).
Mulder, D. 1963. *Pl. Prot. Bull. F.A.O.* **11**: 121 (developments in Sri Lanka).
Petch, T. 1923. *The diseases of the tea bush*, Macmillan.
Sarmah, K. C. 1960. *Mem. Tocklai exp. Stn* 26, 68 pp. (in N.E. India).
Shanmuganathan, N. 1969. *Tea Q.* **40**: 19 (root diseases in Sri Lanka).

Campanula
(Phytophthora porri)

Canavalia, *ensiformis* (jack bean), *gladiata* (sword bean)
(Elsinoë, E. phaseoli)

Canna
(Puccinia)

Cannabis sativa (hemp)
(Ascochyta, Didymella, Fusarium oxysporum, Pseudoperonospora, Septoria)

Barloy, J. et al. 1962. *Annls Epiphyt.* **13**: 117 (in France, 31 ref.).

Capsella
(Albugo candida, Peronospora parasitica)

Capsicum (chilli, red pepper), *annuum, frutescens* (bird pepper), *pendulum, pubescens.*
Colletotrichum capsici (anthracnose)
Leveillula taurica (powdery mildew)
Phytophthora capsici (blight)
(Alternaria, A. alternata, Ascochyta gossypii, Choanephora, Colletotrichum coccodes, Guignardia, Peronospora tabacina, Pestalotiopsis palmarum, Phaeoramularia, Pleospora herbarum, Puccinia, Pyrenochaeta lycopersici, Rhizopus stolonifer, Sclerotinia fuckeliana, Stemphylium solani, Thanatephorus cucumeris, Verticillium dahliae)

Boswell, V. R. et al. 1952. *Fmrs' Bull. U.S. Dep. Agric.* 2051, 30 pp.
Higgins, B. B. 1934. *Bull. Ga agric. Exp. Stn* 186, 20 pp. (in USA).
Weber, G. F. 1932. *Bull. Fla agric. Exp. Stn* 244, 46 pp. (in USA).

Cardamine
(Albugo candida)

Carica papaya (papaw, papaya)
Asperisporium caricae (black spot)
Phytophthora palmivora (root and stem rot, canker)
(Ascochyta, Calonectria crotalariae, Corynespora cassiicola, Drechslera rostrata, Fusarium solani, Glomerella cingulata, Mycosphaerella, Oidium, Phoma, Phomopsis, Pythium aphanidermatum, P. splendens, Rhizopus stolonifer, Sphaceloma)

Frossard, P. 1969. *Fruits* **24**: 473, 483 (82 ref.).
Hine, R. B. et al. 1965. *Bull. Hawaii agric. Exp. Stn* 136, 26 pp.

Carica papaya (*cont.*)
Hunter, J. E. et al. 1972. *Trop. Agric. Trin.* **49**: 61 (in Hawaii).
Krochmal, A. 1974. *Ceiba* **18**: 19 (48 ref.).
Simmonds, J. H. 1965. *Qd agric. J.* **91**: 666 (in Australia).
Srivastava, M. P. et al. 1971. *PANS* **17**: 51 (postharvest).

Carpentaria acuminata
(Cercospora elaeidis)

Carthamus tinctorius (safflower), *glaucus, lunatus, oxycantha, syriacus, tenuis.*
Alternaria carthami (leaf spot)
Fusarium oxysporum f. sp. carthami (wilt)
Phytophthora drechsleri (root rot)
Puccinia carthami (rust)
(Phytophthora cryptogea, P. palmivora, Pseudocercosporella, Pythium aphanidermatum, P. splendens, P. ultimum, Ramularia, Sclerotinia fuckeliana, S. sclerotiorum, Septoria, Verticillium dahliae)
Ashri, A. 1971. *Oléagineux* **26**: 559 (reaction to pathogens).
——. 1971. *Crop Sci.* **11**: 253 (world collection, reaction to pathogens).
Conners, I. L. 1943. *Pl. Dis. Reptr* **27**: 194 (rusts & other fungi).
——. 1943. *Phytopathology* **33**: 789 (rusts).
Foucart, G. 1954. *Bull. Agric. Congo Belge* **45**: 599.
Ikata, S. 1928. *Ann. phytopathol. Soc. Japan* **2**: 140.
Klisiewicz, J. M. 1965. *Circ. Calif. agric. Exp. Stn Ext. Serv.* 532: 33.
Robinson, R. A. 1963. *E. Afr. agric. For. J.* **28**: 164 (in Kenya).
Vasudeva, R. S. 1961. In *Niger and safflower*: 128, Indian Central Oilseeds Committee, Hyderabad.

Carya pecan (pecan)
(Cristulariella)

Cassia, *siamea* (djoowar), *tora* (sickle senna)
(Cochliobolus nodulosus, Fusarium oxysporum, F. oxysporum f. sp. medicaginis)

Castanae (chestnut)
(Endothia)

Cattleya
(Fusarium oxysporum, Pythium ultimum)

Caucalis tenella
(Alternaria dauci)

Ceiba pentandra (kapok)
(Rhizopus zonalis)

Celosia argentea
(Gibberella)

Cenchrus
(Cladosporium, Puccinia substriata)

Chamaerops humulis
(Pestalotiopsis palmarum)

Cheiranthus
(Leptosphaeria maculans, Peronospora parasitica)

Chloridion (=Stereochlaena) **cameronii**
(Puccinia substriata)

Chloris, *gayana* (rhodes grass), *elegans*
(Drechslera, Puccinia cacabata, Sclerospora)

Chondrilla juncea (skeleton weed)
(Puccinia)

Chrysanthemum, *cinerariaefolium* (pyrethrum), *morifolium* (chrysanthemum)
Didymella chrysanthemi (ray blight)
(Mycosphaerella, Puccinia, Ramularia, Septoria)

Cicer arietinum (chick pea)
Ascochyta rabiei (blight)
Fusarium oxysporum f. sp. ciceris (wilt)
Operculella padwickii (foot rot)
Uromyces ciceris-arietini (rust)
(Fusarium solani f. sp. pisi, Pleospora herbarum, Sclerotinia fuckeliana, Stemphylium)
Nene, Y. L. et al. 1978. *Inf. Bull. ICRISAT* 1, 43 pp. (bibliography).

Cichorium endiva (endive), *intybus* (chicory)
(Alternaria, Bremia lactucae)

Cinchona, *officinalis, pitayensis pubescens* (quinine)
(Phytophthora, P. cinnamomi, P. nicotianae var. parasitica, P. palmivora, Thanatephorus cucumeris)
Chevaugeon, J. et al. 1956. *J. agric. trop. Bot. appl.* **3**: 605 (in French Guiana).
Stoffels, E. H. J. 1945. *Publs Inst. natn. Etude agron. Congo Belge* Sér. tech. 24a, 57 pp.

Cinnamomum zeylanicum (cinnamon)
(Phytophthora cinnamomi)

Citrullus colocynthis (colocynth)
Fusarium oxysporum f. sp. niveum (wilt)

Citrullus lanatus (watermelon)
Alternaria cucumerina (leaf spot)
Didymella bryoniae (gummy stem blight)
Fusarium oxysporum f. sp. niveum (wilt)
Glomerella cingulata (anthracnose)
Phytophthora capsici (soft rot)
Pseudoperonospora cubensis (downy mildew)
Sphaerotheca fuliginea (powdery mildew)
(Glomerella, Puccinia, Pythium, Synchytrium)
Parris, G. K. 1952. *Bull. Fla agric. Exp. Stn* 491, 48 pp.
Schenk, N. C. 1968. *Phytopathology* **58**: 91 (airborne fungus spores)

Citrus, *aurantifolia* (lime), *aurantium* (sour orange), *grandis* (pummelo), *limon* (lemon), *medica* (citron), *paradisi* (grapefruit), *reticulata* (mandarin), *sinensis* (sweet orange)
Alternaria citri (black rot)
Botryodiplodia theobromae (stem end rot)
Deuterophoma tracheiphila (mal secco)
Diaporthe citri (melanose)
Elsinoë australis (sweet orange scab)
E. fawcettii (common scab)

Citrus (*cont.*)
Geotrichum candidum (sour rot)
Glomerella cingulata (anthracnose, withertip)
Guignardia citricarpa (black spot)
Mycosphaerella citri (greasy spot)
Penicillium digitatum (green mould)
P. italicum (blue mould)
Phytophthora citricola (brown rot)
P. citrophthora (brown rot, gummosis)
P. hibernalis (brown rot)
P. nicotianae var. parasitica (brown rot, gummosis)
Sphaceloma fawcettii var. scabiosa (Australian scab)
Sphaeropsis tumefaciens (knot)
(Alternaria alternata, Armillariella mellea, Aspergillus flavus, Botryosphaeria ribis, Cercospora, Corticium salmonicolor, Corynespora, Curvularia, Fusarium solani, Ganoderma, Gibberella, G. fujikuroi, G. stilboides, Oidium, Physoderma, Phytophthora palmivora, Pleospora herbarum, Rosellinia necatrix, Septoria, Sclerotinia sclerotiorum, Thanatephorus cucumeris, Thielaviopsis basicola)

Eckert, J. W. 1978. *Outl. Agric.* **9**: 225 (postharvest fruit diseases, 62 ref.).
Fawcett, H. S. 1936. *Citrus diseases and their control*, McGraw Hill.
Hanna, A. D. 1969. *PANS* **15**: 340 (control).
Jamoussi, B. 1955. *Rev. Mycol.* **20** Suppl. Colon. 1: 1 (wilts).
Klotz, L. J. et al. 1970. *Citrograph* **55**: 259 (control of postharvest decay).
——. 1973. *Color handbook of citrus diseases*, Univ. California, Riverside.
Knorr, L. C. et al. 1957. *Handbook of citrus diseases in Florida*, 157 pp. Univ. Florida, Gainesville.
——. 1973. *Citrus diseases and disorders* Univ. Florida, Gainesville.
——. 1973. *PANS* **19**: 441 (bibliography, 350 ref.).
Reuther, W. et al. (editors). 1978. *The citrus industry* Vol. IV, Crop protection, 362 pp. Univ. California, Berkeley.

Cleistachne
(Sphacelotheca reiliana)

Cocos nucifera (coconut)
Ceratocystis paradoxa (bleeding)
Drechslera incurvata (leaf spot)
Marasmiellus cocophilus (lethal bole rot)
Pestalotiopsis palmarum (leaf spot)
Phytophthora arecae (koleroga)
P. palmivora (bud rot)
(Curvularia, Drechslera gigantea, D. halodes, Ganoderma boninense, Guignardia)

Briton-Jones, H. R. 1940. *The diseases of the coconut palm* (revised by E. E. Cheesman), 176 pp., Baillière, Tindall & Cox.
Child, R. 1974. *Coconuts*: 213, Longman.
Dwyer, R. E. P. 1937. *New Guinea agric. Gaz.* **3**(1): 28.
——. 1939. *Ibid* **5**(3): 31 (association with soil conditions).
——. 1953. *Papua New Guinea agric. Gaz.* **8**(1): 24.
Martyn, E. B. 1945. *Trop. Agric. Trin.* **22**: 51, 69 (in Jamaica).
Shaw, D. E. 1965. *Papua New Guinea agric. J.* **17**: 67.

Coffea (coffee), *arabica* (arabica), *canephora* (robusta), *liberica* (liberica)
Ceratocystis fimbriata (canker)
Cercospora coffeicola (brown eyespot)
Gibberella stilboides (Storey's bark)
G. xylarioides (tracheomycosis)
Glomerella cingulata (berry disease)
Hemileia coffeicola (rust)
H. vastatrix (rust)
Mycena citricolor (cock's eye)
Trachysphaera fructigena (fruit rot)
(Acremonium, Armillariella mellea, Ascochyta, Calostilbe striispora, Corticium rolfsii, C. salmonicolor, Fusarium oxysporum, F. solani, Gibberella, Helicobasidium, Marasmius, Myrothecium roridum, Phoma, Rosellinia bunodes, R. necatrix, Sclerotinia fuckeliana)

Bitancourt, A. A. 1954. *Biológico* **20**: 205 (in W. Africa).
Haarer, A. E. 1956. *Modern coffee production*: 275, Leonard Hill.
Wellman, F. L. 1961. *Coffee*: 250, Leonard Hill.

Coix lachryma-jobi (Job's tears)
(Drechslera, Ustilago)

Colocasia esculenta (eddoe, cocoyam, dasheen, taro)
Phytophthora colocasiae (leaf spot)
(Cercoseptoria, Choanephora, Cladosporium, Corticium rolfsii, Pseudocercospora)

Gollifer, D. E. et al. 1973. *Ann. appl. Biol.* **73**: 349 (corm storage loss in the Solomon Islands).
Jackson, G. V. H. et al. 1975. *Ibid* **80**: 217 (as above).
—— ——, 1975. *PANS* **21**: 45 (in the Solomon Islands).
Parris, G. K. 1941. *Circ. Hawaii agric. Exp. Stn* 18, 29 pp.

Corchorus, *aestuans*, *capsularis*, *olitorius* (jute)
(Cercospora, Colletotrichum, Physoderma, Thanatephorus cucumeris)

Cordia, *alliodora* (cypre)
(Puccinia, Uromyces)

Coriandrum sativum (coriander)
Fusarium oxysporum f. sp. coriandrii (wilt)
Protomyces macrosporus (stem gall)
(Alternaria, Cercospora, Phoma, Ramularia, Sclerotinia sclerotiorum)

Corylus (hazel)
(Phyllactinia)

Crambe
(Alternaria brassicicola, Aphanomyces raphani)

Crotalaria, *anagyroides*, *intermedia*, *juncea* (sunn hemp), *mucronata*, *retusa*, *triquetra*

Crotalaria (*cont.*)
Calonectria crotalariae (collar rot)
Ceratocystis fimbriata (wilt)
Colletotrichum curvatum (see C. dematium, stem break)
Fusarium udum f. sp. crotalariae (wilt)
Uromyces decoratus (rust)
(Gibberella fujikuroi var. subglutinans, Neocosmospora, Pleiochaeta, Poria hypobrunnea, Protomycopsis)

Cucumis melo (melon)
Alternaria cucumerina (leaf spot)
Cladosporium cucumerinum (scab)
Didymella bryoniae (gummy stem blight)
Fusarium oxysporum f. sp. melonis (wilt)
F. solani f. sp. cucurbitae (foot rot)
Glomerella cingulata (anthracnose)
Phytophthora capsici (soft rot)
Pseudoperonospora cubensis (downy mildew)
Sphaerotheca fuliginea (powdery mildew)
(Botryodiplodia theobromae, Geotrichum candidum, Glomerella, Myrothecium roridum, Pleospora herbarum, Pythium ultimum, Sclerotinia sclerotiorum, Synchytrium)

Wiant, J. S. 1937. *Tech. Bull. U.S. Dep. Agric.* 563, 47 pp. (market diseases)

Cucumis sativus (cucumber)
Cladosporium cucumerinum (scab)
Didymella bryoniae (gummy stem blight)
Fusarium oxysporum f. sp. cucumerinum (wilt)
F. solani f. sp. cucurbitae (foot rot)
Glomerella cingulata (anthracnose)
Phytophthora capsici (soft rot)
Pseudoperonospora cubensis (downy mildew)
Sphaerotheca fuliginea (powdery mildew)
(Corynespora cassiicola, Gibberella, Myrothecium roridum, Phomopsis, Phytophthora, Pythium mamillatum, P. splendens, P. ultimum, Sclerotinia fuckeliana)

Munger, H. M. et al. 1953. *Phytopathology* **43**: 254 (breeding for disease resistance).

Cucurbita, *ficifolia* (Malabar gourd), *maxima* (pumpkin, squash), *moschata* (pumpkin), *pepo* (marrow)
Cladosporium cucumerinum (scab)
Didymella bryoniae (gummy stem blight)
Fusarium solani f. sp. cucurbitae (foot rot)
Glomerella cingulata (anthracnose)
Phytophthora capsici (soft rot)
Pseudoperonospora cubensis (downy mildew)
Sphaerotheca fuliginea (powdery mildew)
(Choanephora, Glomerella, Phomopsis, Septoria, Sclerotinia sclerotiorum, Sphaerulina)

Ellis, D. E. 1953. *Bull. N. Carol. agric. Exp. Stn* 380, 12 pp.

Middleton, J. T. et al. 1953. *Plant diseases, the yearbook of agriculture*: 483, U.S. Dep. Agric.

Rogers, I. S. 1964. *J. Dep. Agric. S. Aust.* **68**: 86.

Sitterlly, W. R. 1972. *A. Rev. Phytopathol.* **10**: 471 (breeding for disease resistance, 65 ref.).

Cuminum cyminum (cumin)
Alternaria burnsii (blight)
Erysiphe heraclei (powdery mildew)
Fusarium oxysporum f. sp. cumini (wilt)

Curcuma, *angustifolia* (Indian arrowroot), *domestica* (turmeric)
Taphrina maculans (leaf spot)
(Colletotrichum capsici, Corticium rolfsii, Pythium graminicola)

Pavgi, M. S. et al. 1968. *Sydowia* **21**: 100 (on turmeric in India).

Cyamopsis tetragonoloba (cluster bean, guar)
(Alternaria alternata, Cercospora kikuchii, Cochliobolus lunatus, Colletotrichum capsici, C. truncatum, Gibberella fujikuroi, Myrothecium roridum, Neocosmospora)

Hymowitz, T. 1963. *Processed Ser. Okla Univ. Exp. Stn* P468, 23 pp. (bibliography)

Cymbopogon, *citratus* (lemon grass), *flexuosus* (Malabar grass), *nardus* (citronella grass)
Drechslera sacchari (leaf spot)
(Curvularia, Drechslera rostrata, Puccinia purpurea, Tolyposporium)

Cynara scolymus (globe artichoke)
(Didymella chrysanthemi, Leveillula taurica)

Ciccarone, A. (1967) 1969. *Atti I Cong. internaz. Studi Carciofo*: 181 (62 ref.).

Cynodon, *dactylon* (Bermuda grass, star grass), *magennissi*, *transvaalensis* (masindi grass)
Drechslera gigantea (zonate eyespot)
Gloeocercospora sorghi (copper spot)
Puccinia cynodontis (rust)
Ustilago cynodontis (smut)
(Cochliobolus, C. spicifer, Drechslera rostrata, Léptosphaeria, Phyllachora, Physoderma)

Dactylis glomerata (cocksfoot)
(Rhynchosporium)

Dahlia
(Didymella chrysanthemi)

Dalbergia sissoo (sissoo)
(Fusarium solani, Nectria, Phyllactinia)

Datura stramonium (thorn apple)
Alternaria crassa (leaf spot)

Daucus carota (carrot)
Alternaria dauci (leaf blight)
A. radicina (black rot)
A. raphani (black pod blotch)
Erysiphe heraclei (powdery mildew)
Mycocentrospora acerina (licorice rot)
(Chalaropsis, Geotrichum candidum, Helicobasidium, Phytophthora megasperma, Pythium, P. debaryanum, Rhizoctonia, Rhizopus stolonifer, Sclerotinia fuckeliana, Thanatephorus cucumeris)

Daucus carota (*cont.*)
Årsvoll, K. 1969. *Meld. Norg Landbr Høisk.* **48**(2), 52 pp. (in Norway).
Crête, R. 1977. *Publs Canada Dep. Agric.* 1615, 21 pp. (in Canada).
KenKnight, G. et al. 1945. *Bull. Idaho agric. Exp. Stn* 262, 23 pp.
Mukula, J. 1957. *Acta Agric. scand.* Suppl. 2, 132 pp. (on stored carrots in Finland, 207 ref.).
Rader, W. E. 1952. *Bull. Cornell Univ. agric. Exp. Stn* 889, 64 pp. (on stored carrots, 100 ref.).
Derris
(Glomerella cingulata)
Desmodium (tick clovers)
(Colletotrichum capsici, C. truncatum, Elsinoë)
Dianthus caryophyllus (carnation)
(Phytophthora nicotianae var. parasitica, Uromyces)
Digitaria
Puccinia oahuensis (rust)
(Pyricularia grisea, Sclerophthora rayssiae)
Jailloux, F. et al. 1970. *Agron. trop.* **25**: 765 (in French West Indies)
Dioscorea (yam)
(Botryodiplodia theobromae, Cercoseptoria, Glomerella cingulata, Penicillium, Phaeoramularia, Pseudocercospora, Rhizopus oryzae)
Adeniji, M. O. 1970. *Phytopathology* **60**: 590, 1698 (storage rots & effects of moisture & temp.)
Bammi, R. K. et al. 1972. *Pl. Dis. Reptr* **56**: 990 (benomyl control of tuber rot).
Baudin, P. 1956. *Rev. Mycol.* **21** Suppl. Colon. 2: 87 (in Ivory Coast).
Ogundana, S. K. et al. 1970 & 1971. *Trans. Br. mycol. Soc.* **54**: 445; **56**: 73 (storage rots).
Diospyros kaki (Japanese persimmon)
(Pestalotiopsis theae)
Dolichos uniflorus (horse gram)
Uromyces appendiculatus (rust)
(Colletotrichum lindemuthianum, C. truncatum, Pyrenochaeta)
Draba
(Albugo candida)
Durio zibethinus (durian)
(Phytophthora palmivora)
Echinochloa, *colona* (jungle rice), *crus-galli* (barnyard millet), *frumentacea* (Japanese barnyard millet)
(Cochliobolus nodulosus, Drechslera, D. gigantea, Physoderma, Sclerospora graminicola, Sphacelotheca destruens, Tolyposporium, Ustilago)
Eichhornia crassipes (water hyacinth)
(Alternaria, Myrothecium roridum)
Elaeis guineensis (oil palm)
Ceratocystis paradoxa (dry basal rot)
Cercospora elaeidis (freckle)
Fusarium oxysporum f. sp. elaeidis (wilt)
Ganoderma boninense & other spp. (basal stem rot)
Phellinus noxius (upper stem rot)
(Armillariella mellea, Curvularia, Drechslera halodes, Gibberella fujikuroi var. subglutinans, Marasmius, Pestalotiopsis palmarum, Phomopsis, Leptosphaeria, Phytophthora palmivora, Pythium splendens, Ustulina deusta)
Aderungboye, F. O. 1977. *PANS* **23**: 305 (110 ref.).
Bull, R. A. 1954. *Jl W. Afr. Inst. Oil Palm Res.* **2**: 53 (disease list for Nigeria).
Hartley, C. W. S. 1977. *The oil palm*: 605, Longman.
Robertson, J. S. et al. 1968. *Jnl Niger. Inst. Oil Palm Res.* **4**: 381 (in W. Africa).
Turner, P. D. & Bull, R. A. 1967. *Diseases and disorders of the oil palm in Malaysia*, 247 pp., Incorporated Society of Planters, Kuala Lumpur (163 ref.).
——. 1971. *Phytopathol. Pap.* 14, 58 pp. (annotated list of micro-organisms, 345 ref.).
Wood, B. J. 1977. *PANS* **23**: 253 (economics).
Elettaria cardamomum (cardamom)
(Phakopsora, Phyllosticta, Sphaceloma)
Eleusine, *coracana* (finger millet), *indica* (fowl foot grass)
Cochliobolus nodulosus (leaf blight)
(Drechslera gigantea, Marasmiellus, Melanopsichium, Pyricularia, P. oryzae, Sclerospora)
Govindu, H. C. 1970. *Plant disease problems, Proc. 1st. int. Symp. Pl. Pathol. (1966–7)*, IARI, New Delhi: 415. (and see Ramakrishnan under *Pennisetum typhoides*)
Eragrostis tef (teff)
(Drechslera)
Erianthus
(Glomerella tucumanensis, Puccinia polysora, Sclerospora, Ustilago scitaminae)
Erythrina spp. (dadap), *glauca* (swamp immortelle), *poeppigiana* (mountain immortelle)
Calostilbe striispora (bark rot)
(Poria hypobrunnea)
Eucalyptus
(Armillariella mellea, Calonectria crotalariae, C. kyotensis, Corticium salmonicolor, Cylindrocladium, C. scoparium, Cytospora, Diaporthe, Endothia, Fusarium oxysporum Hypoxylon, Mycosphaerella, Phytophthora cinnamomi, P. drechsleri, P. heveae, Puccinia psidii, Pythium, Ramularia, Sclerotinia fuckeliana, Sphaerulina)
Euchlaena (teosinte), *luxurians*, *mexicana*, *perennis*
(Cochliobolus heterostrophus, Physoderma maydis, Puccinia polysora, P. sorghi, Sclerospora graminicola, S. maydis, S. philippinensis, S. sacchari, S. sorghi, Setosphaeria turcica, Sphacelotheca reiliana)
Eugenia, *caryophyllus* (clove), *cumini* (jambolan), *jambos* (rose apple), *malaccensis* (pomerac)
Endothia eugeniae (acute dieback)

Eugenia (*cont.*)
Valsa eugeniae (sudden death)
(Calonectria, Phomopsis, Phytophthora cinnamomi, Puccinia psidii, Rigidoporus zonalis)

Euphorbia
(Chalaropsis, Melampsora, Sclerotinia ricini, Sphaceloma, Thanatephorus cucumeris, Thielaviopsis basicola, Uromyces ciceris-arietini, U. pisi-sativi, U. striatus)

Fagus sylvatica (European beech)
(Hypoxylon serpens)

Ficus, *benghalensis* (Indian banyan), *carica* (fig)
(Cerotelium, Rigidoporus zonalis, Sphaceloma)

Foeniculum vulgare (fennel)
Erysiphe heraclei (powdery mildew)
(Alternaria, Ramularia)
Anahosur, K. H. et al. 1972. *Indian J. agric. Sci.* **42**: 990 (control of seed mycoflora)

Fortunella margarita (oval kumquat)
(Elsinoë fawcettii)

Fragaria
(Pyrenochaeta lycopersici, Rhizopus)

Gardenia
(Myrothecium roridum)

Geranium
(Uromyces)

Gerbera
(Phytophthora cryptogea)

Glycine max (soybean)
Cercospora kikuchi (purple seed stain)
Colletotrichum truncatum (anthracnose)
Diaporthe phaseolorum (stem canker)
Fusarium oxysporum f. sp. tracheiphilum (wilt)
Peronospora manshurica (downy mildew)
Phakopsora pachyrhizi (rust)
Phytophthora megasperma var. sojae (stalk rot)
Septoria glycines (brown spot)
(Aphanomyces euteiches, Ascochyta gossypii, Cercospora, Corticium rolfsii, Corynespora cassiicola, Fusarium oxysporum, F. oxysporum f. sp. vasinfectum, Melanopsichium, Microsphaera, Nematospora coryli, Neocosmospora, Phaeoisariopsis griseola, Pyrenochaeta, Pythium aphanidermatum, P. debaryanum, P. graminicola, P. myriotylum, P. ultimum, Sclerotinia sclerotiorum, Sphaceloma, Thanatephorus cucumeris, Thielaviopsis basicola)
Costa, A. S. 1977. *Summa Phytopathol.* 3: 3 (in Brazil, 113 ref.).
Hildebrand, A. A. et al. 1947. *Phytopathology* **37**: 111 (in Canada & seed treatment).
Kurata, H. 1960. *Bull. natn. Inst. agric. Sci. Tokyo* Ser. C 12: 1 (in Japan).
Ling, L. 1951. *Pl. Dis. Reptr* Suppl. 204: 111 (bibliography, 503 ref.).
Liu, S. T. 1948. *Bot. Bull. Acad. sin. Shanghai* **2**: 69 (seedborne diseases).
Sinclair, J. B. et al. (editors). 1975. *Am. Phytopathol. Soc.*, 69 pp. (compendium).
——— ———. 1975. INTSOY Ser. 7, Univ. Illinois, 280 pp. (annotated bibliography).

Gossypium (cotton)
Alternaria macrospora (leaf spot)
Ascochyta gossypii (wet weather blight)
Aspergillus flavus (boll rot)
Botryodiplodia theobromae (black boll rot)
Colletotrichum gossypii (anthracnose)
Fusarium oxysporum f. sp. vasinfectum (wilt)
Nematospora coryli (stigmatomycosis)
N. gossypii (stigmatomycosis)
Phakopsora gossypii (rust)
Puccinia cacabata (rust)
Ramularia gossypii (grey mildew)
Thielaviopsis basicola (black root rot)
Verticillium dahliae (wilt)
(Alternaria alternata, Cochliobolus spicifer, Colletotrichum, C. capsici, Corynespora cassiicola, Eremothecium, Gibberella fujikuroi, Myrothecium roridum, Pythium aphanidermatum, P. debaryanum, P. graminicola, P. mamillatum, P. ultimum, Rhizopus stolonifer, Thanatephorus cucumeris)
Ebbels, D. L. 1976. *Rev. Pl. Pathol.* **55**: 747 (in Africa, 155 ref.).
Hanna, A. D. 1969. *PANS* **15**: 186 (review).
Jakob, M. 1969. *Pfl'shutz Nachr. Bayer* **22**: 244 (on seedlings in Egypt).
Mostafa, M. A. 1959. *Egypt. Rev. Sci.* 3, 55 pp. (241 ref.).
Prentice, A. N. 1972. *Cotton with special reference to Africa*: 195, Longman.
Tarr, S. A. J. 1959. *Outl. Agric.* **2**: 168 (seed treatment).

Grevillea robusta (silky oak)
(Macrophomina phaseolina, Ustulina deusta)

Hackelochloa
(Sphacelotheca reiliana)

Helianthus, *annuus* (sunflower), *rigidus*, *tuberosus* (Jerusalem artichoke)
Plasmopora halstedii (false or downy mildew)
Puccinia helianthi (rust)
Sclerotinia sclerotiorum (cottony soft rot)
Septoria helianthi (leaf spot)
Verticillium dahliae (leaf mottle)
(Albugo, Alternaria, A. alternata, Didymella chrysanthemi, Gibberella fujikuroi, Phialophora, Phoma, Rhizopus oryzae, Sclerotinia fuckeliana)
Irwin, J. A. G. 1977. *Qd agric. J.* **103**: 516 (in Queensland, Australia).
Johnson, H. W. 1931. *J. agric. Res.* **43**: 337 (storage rots in Jerusalem artichoke).

Helianthus (*cont.*)
Martens, J. W. et al. 1970. *E. Afr. agric. For. J.* **35**: 389 (in Kenya).
Zimmer, D. E. et al. 1972. *Crop Sci.* **12**: 859 (diseases & achene & oil quality).
Heliconia
(Fusarium oxysporum f. sp. cubense)
Heliopsis
(Puccinia helianthi)
Herrania
(Crinipellis perniciosa)
Hevea brasiliensis (rubber)
Ceratocystis fimbriata (mouldy rot)
Colletotrichum dematium (leaf spot)
Corticium salmonicolor (pink disease)
Drechslera heveae (bird's eye spot)
Ganoderma philippii (red root rot)
Glomerella cingulata (secondary leaf fall)
Microcyclus ulei (South American leaf blight)
Odium heveae (secondary leaf fall)
Phellinus noxius (brown root rot)
Phytophthora botryosa (abnormal leaf fall)
P. meadii (abnormal leaf fall)
P. palmivora (abnormal leaf fall, black stripe, patch canker)
Rigidoporus lignosus (white root rot)
Ustulina deusta (collar and root rot)
(Armillariella mellea, Calonectria, Calostilbe striispora, Corynespora cassiicola, Elsinoë, Fusarium solani, Guignardia, Hypoxylon, Periconia, Pestalotiopsis palmarum, Phomopsis, Phyllachora, Phytopthora heveae, Rigidoporus zonalis, Thanatephorus cucumeris)
Chee, K. H. 1976 *Micro-organisms associated with rubber* (Hevea brasiliensis Mull. Arg), Rubb. Res. Inst. Malaysia, 78 pp. (list, 351 ref.).
Petch, T. 1911. *The physiology and diseases of* Hevea brasiliensis, Dulau.
——. 1921. *The diseases and pests of the rubber tree*, Macmillan.
Pichel, R. J. 1956. *Publs Inst. natn. Etude agron. Congo Belge* Sér. tech. 49, 480 pp. (in Zaire, 335 ref.).
Rao, B. S. 1975. *Maladies of* Hevea *in Malaysia*, Rubb. Res. Inst. Malaysia, 108 pp. (advisory).
Saccas, A. M. 1953. *Agron. trop.* **8**: 176, 229 (in equatorial Africa).
Sharples, A. 1936. *Diseases and pests of the rubber tree*, Macmillan.
Wastie, R. L. 1975. *PANS* **21**: 268 (diseases & control, 70 ref.).
Weir, J. R. 1926. *Bull. U.S. Dep. Agric.* 1380, 129 pp. (survey in Amazon valley).
Hibiscus, *cannabis* (kenaf), *esculentus* (okra), *panduraeformis*, *rosa-sinensis*, *sabdariffa* (roselle)
Glomerella cingulata (anthracnose)
Phytophthora nicotianae var. parasitica (foot rot)
(Ascochyta, A. gossypii, Choanephora, Cristulariella, Erysiphe cichoracearum, Fusarium oxysporum f. sp. vasinfectum, Myrothecium roridum, Phoma, Phytophthora palmivora, Pythium aphanidermatum, Sclerotinia sclerotiorum, Thanatephorus cucumeris, Verticillium dahliae)
Hordeum (barley)
(Rhynchosporium, Sclerophthora rayssiae)
Humulus lupulus (hop)
(Gibberella, Phytophthora citricola, Pseudoperonospora, Verticillium albo-atrum)
Hyacinthus
(Pythium ultimum)
Hyoscyamus, *bohemicus*, *niger* (henbane)
(Alternaria solani, Peronospora tabacina)
Hypocyrta
(Myrothecium roridum)
Iberis
(Peronospora parasitica)
Imperata cylindrica (lalang)
(Ustilago scitaminea)
Indigofera (indigo)
(Uromyces)
Ipomoea aquatica (water spinach)
Albugo ipomoeae-aquaticae
(Pseudocercosporella)
Ipomoea batatas (sweet potato), *biloba*, *horsfalliae*, *pentaphylla*
Albugo ipomoeae-panduratae (white rust)
Botryodiplodia theobromae (Java black rot)
Ceratocystis fimbriata (black rot)
Diaporthe phaseolorum (dry rot)
Fusarium oxysporum f. sp. batatas (wilt)
Monilochaetes infuscans (scurf)
Rhizopus stolonifer (whiskers)
(Coleosporium, Elsinoë, Helicobasidium, Phaeoisariopsis, Phoma, Pleospora herbarum, Pythium ultimum, Rhizopus oryzae, Septoria, Thanatephorus cucumeris)
Arene, O. B. et al. 1978. *PANS* **24**: 294 (in Nigeria, 71 ref.).
Harter, L. L. et al. 1929. *Tech. Bull. U.S. Dep. Agric.* 99, 117 pp. (monograph, 211 ref.).
Hildebrand, E. M. et al. 1959. *Fmrs' Bull. U.S. Dep. Agric.* 1059, 28 pp. (advisory).
Steinbauer, C. E. et al. 1971. *Agric. Handb. U.S. Dep. Agric.* 388: 49 (advisory).
Iris
(Puccinia)
Iva
(Puccinia helianthi)
Juglans regia (European walnut)
(Chalariopsis)
Juncellus serotinus
(Rhizoctonia oryzae-sativae)

Hosts and pathogens

Juniperus (juniper)
(Phomopsis juniperivora (blight))
Lablab niger, now *L. purpureus* (hyacinth bean)
(Cercospora canescens, Colletotrichum lindemuthianum, Elsinoë, E. phaseoli, Mycosphaerella cruenta)
Lactuca sativa (lettuce), *saligna*, *serriola* (wild lettuce)
Bremia lactucae (downy mildew)
Sclerotinia fuckeliana (grey mould)
S. minor (drop)
Septoria lactucae (leaf spot)
(Cercospora, Didymella chrysanthemi, Erysiphe cichoracearum, Mycocentrospora acerina, Pleospora herbarum, Sclerotinia sclerotiorum, Thanatephorus cucumeris)
Brien, R. M. et al. 1957. *Inf. Ser. Dep. scient. ind. Res. N.Z.* 14, 38 pp. (advisory).
Grogan, R. G. et al. 1955. *Circ. Calif. agric. Exp. Stn* 448, 28 pp. (advisory).
Lagenaria siceraria (bottle gourd)
(Fusarium oxysporum)
Larix (larch)
(Armillariella mellea, Rhizina undulata)
Lathyrus, *odoratus* (sweet pea), *sativus* (grass pea)
Uromyces pisi-sativi (rust)
U. viciae-fabae (rust)
(Aphanomyces euteiches, Ascochyta pisi, Erysiphe trifolii, Fusarium oxysporum, Mycosphaerella pinodes, Peronospora viciae, Physoderma, Ramularia)
Lens esculenta (lentil)
Uromyces viciae-fabae (rust)
(Erysiphe pisi, Fusarium oxysporum, Thanatephorus cucumeris)
Lespedeza
(Glomerella cingulata, Phyllachora)
Lilium longiflorum (Easter lily)
(Pythium splendens)
Linum usitatissimum (flax)
(Melampsora, Phoma)
Liquidambar formosana (Chinese sweet gum)
(Endothia)
Lolium
(Pythium aphanidermatum)
Lotus corniculatus (bird's-foot trefoil)
(Stemphylium)
Luffa
(Didymella bryoniae, Synchytrium)
Lupinus (lupin)
(Aphanomyces euteiches, Ascochyta gossypii, Diaporthe, Erysiphe pisi, Glomerella cingulata, Pleiochaeta, Stemphylium)
Lycopersicon esculentum (tomato), *glandulosum*, *hirsutum*, *peruvianum*, *pimpinellifolium* (currant tomato)
Alternaria solani (early blight)
A. tomato (nailhead spot)
Colletotrichum coccodes (anthracnose)
Didymella lycopersici (stem and fruit rot)
Fulvia fulva (leaf mould)
Fusarium oxysporum f. sp. lycopersici (wilt)
Leveillula taurica (powdery mildew)
Phytophthora capsici (blight)
P. infestans (late blight)
P. nicotianae var. parasitica (buckeye, foot rot)
Pleospora herbarum (foliage blight)
Pseudocercospora fuligena (leaf spot)
Pyrenochaeta lycopersici (corky root)
Sclerotinia fuckeliana (grey mould)
Septoria lycopersici (leaf spot)
Stemphylium solani (grey leaf spot)
Verticillium albo-atrum (wilt)
(Alternaria alternata, Ascochyta gossypii, Colletotrichum dematium, Corticium rolfsii, Corynespora cassiicola, Geotrichum candidum, Helminthosporium, Myrothecium roridum, Peronospora tabacina, Phoma, Phytophthora, P. arecae, P. cryptogea, P. hibernalis, P. palmivora, Puccinia, Pythium aphanidermatum, P. myriotylum, Rhizopus stolonifer, Sclerotinia sclerotiorum, Thanatephorus cucumeris, Verticillium, V. dahliae)
Barksdale, T. H. et al. 1977. *Agric. Handb. U.S. Dep. Agric.* 203, 109 pp. (advisory).
McKay, R. 1949. *Tomato diseases*, At the Sign of the Three Candles, Dublin.
McKeen, C. D. 1972. *Publ. Can. Dep. Agric.* 1479, 62 pp. (advisory).
Pelham, J. et al. 1968. *N.A.A.S. q. Rev.* **81**: 19 (use of resistance in glasshouse).
Walter, J. M. 1967. *A. Rev. Phytopathol.* **5**: 131 (hereditary resistance, 213 ref.).
Macadamia (macadamia nut)
(Phytophthora cinnamomi, Sclerotinia fuckeliana)
Maesopsis
(Fusarium solani, Nectria)
Malcolmia
(Peronospora parasitica)
Malus pumila (apple)
(Botryosphaeria ribis, Penicillium)
Mangifera indica (mango)
Botryodiplodia theobromae (postharvest rot)
Ceratocystis fimbriata (wilt)
Glomerella cingulata (anthracnose)
(Didymella, Elsinoë, Gibberella fujikuroi, G. fujikuroi var. subglutinans, Guignardia, Macrophoma, Oidium, Pestalotiopsis, Phoma, Phytophthora palmivora, Rigidoporus zonalis, Verticillium dahliae)
Lele, V. C. et al. 1975. *Progressive Hortic.* **7**: 39
Singh, L. B. 1968. *The mango*: 266, Leonard Hill.
Thompson, A. K. 1971. *Trop. Agric. Trin.* **48**: 63, 71 (West Indies, storage & transport).

Manihot esculenta (cassava)
(Botryodiplodia theobromae, Cercosporidium, Diaporthe, Glomerella cingulata, Gnomonia, Mycosphaerella, Periconia, Phaeoramularia, Phytophthora drechsleri, Sphaceloma, Uromyces)
Brekelbaum, T. et al. (editors). 1978. Proc. cassava protection workshop. CIAT, Cali, Colombia, 7–12 Nov. 1977, 244 pp.
Chevaugeon, J. 1956. *Encycl. Mycol.* **28**, 205 pp. (in W. Africa, 143 ref.).
Ingram, J. S. et al. 1972. *Trop. Sci.* **14**: 131 (in storage, 76 ref.).
Lozano, J. C. et al. 1974. *PANS* **20**: 30 (125 ref.).
Terry, E. R. et al. 1976. *Span* **29**: 116 (21 ref.).
Manilkara achras (sapodilla), *hexandra*
(Pestalotiopsis, P. palmarum, Phytophthora palmivora)
Maranta arundinacea (arrowroot)
(Pythium graminicola, Thanatephorus cucumeris)
Matthiola
(Alternaria raphani, Fusarium oxysporum f. sp. conglutinans, Peronospora parasitica)
Medicago sativa (alfalfa, lucerne)
Fusarium oxysporum f. sp. medicaginis (wilt)
Glomerella cingulata (anthracnose)
Leptosphaerulina trifolii (pepper spot)
Peronospora trifoliorum (downy mildew)
Phoma medicaginis (black stem)
Physoderma alfalfae (crown wart)
Phytophthora megasperma (root rot)
Pleospora herbarum (ringspot)
Uromyces striatus (rust)
Verticillium albo-atrum (wilt)
(Aphanomyces euteiches, Colletotrichum, Erysiphe pisi, Gibberella, Kabatiella, Leptosphaeria, Peronospora trifoliorum, Pythium debaryanum, P. graminicola, P. mamillatum, P. myriotylum, P. splendens, Rosellinia necatrix, Sclerotinia, S. fuckeliana, Stagonospora, Stemphylium, Thanatephorus cucumeris)
Chilton, S. J. P. et al. 1943. *Misc. Publ. U.S. Dep. Agric.* 499, 152 pp. (fungi on *Medicago*, *Melilotus* & *Trifolium*, 1733 ref.).
(see *A compendium of alfalfa* (*lucerne*) *diseases*, Am. Phytopathol. Soc., 1975)
Melilotus
(Aphanomyces euteiches, Pleospora herbarum, Uromyces striatus)
Mentha (mint)
(Puccinia)
Microcitrus australis (Australian wild lime)
(Phytophthora citrophthora)
Mikania
(Rigidoporus lignosus)
Miscanthus
(Sclerospora miscanthi, Stagonospora sacchari)
Momordica
(Didymella bryoniae)
Morus (mulberry)
(Gibberella)
Musa (banana, plantain)
Botryodiplodia theobromae (finger rot)
Colletotrichum musae (anthracnose)
Cordana musae (leaf blotch)
Deightoniella torulosa (black leaf spot and swamp spot)
Fusarium oxysporum f. sp. cubense (Panama wilt)
Guignardia musae (freckle)
Mycosphaerella fijiensis (black leaf streak)
M. fijiensis var. difformis (black sigatoka)
M. musicola (sigatoka)
Pyricularia grisea (pitting)
Trachysphaera fuctigena (fruit rot)
Verticillium theobromae (cigar end)
(Armillariella mellea, Calostilbe striispora, Ceratocystis paradoxa, Cercospora, Cladosporium, Cylindrocarpon, Drechslera, D. gigantea, Gibberella fujikuroi, Glomerella cingulata, Haplobasidion, Khuskia oryzae, Marasmiellus, Nigrospora, Pestalotiopsis palmarum, Phyllachora, Sphaerulina, Uromyces, Veronaea)
Anon. 1977. Pest control in bananas, *PANS Manual* 1, 126 pp.
Firman, I. D. 1972. *Trop. Agric. Trin.* **49**: 189 (susceptibility & leaf diseases in Fiji).
Meredith, D. S. 1970. *Rev. Pl. Pathol.* **49**: 539 (major diseases, 121 ref.).
——. 1971. *Trop. Agric. Trin.* **48**: 35 (transport & storage diseases, 82 ref.).
Noronha, A. do R. 1970. *Agron. Moçamb.* **4**: 161, 227 (in Mozambique).
Rowe, P. R. et al. 1975. *Bull. Trop. agric. Res. Serv.* (SIATSA) 2, 41 pp. (Honduras, breeding for resistance).
Stover, R. H. 1972. *Banana, plantain and abaca diseases*, Commonw. Mycol. Inst.
Wardlaw, C. W. 1972. *Banana diseases, including plantains and abaca*, Longman.
Musa textilis (abaca, Manilla hemp)
Deightoniella torulosa (brown leaf spot)
Fusarium oxysporum f. sp. cubense (Panama wilt)
(Colletotrichum musae, Marasmiellus)
(see Anon., Stover and Wardlaw above)
Myrciaria
(Puccinia psidii)
Narcissus
(Stagonospora)

Hosts and pathogens

Nicotiana tabacum (tobacco), *alata* (tobacco flower), *debneyi*, *exigua*, *goodspeedii*, *ingulba*, *langsdorfii*, *longiflora*, *paniculata*, *plumbaginifolia*, *repanda*, *sanderae*, *sylvestris*
Alternaria longipes (brown spot)
Ascochyta gossypii (leaf spot)
Cercospora nicotianae (frog eye)
Erysiphe cichoracearum (white mould)
Peronospora tabacina (blue mould)
Phytophthora nicotianae var. nicotianae (black shank)
Rhizopus oryzae (barn rot)
Thielaviopsis basicola (black root rot)
Verticillium dahliae (wilt)
(Aspergillus flavus, Cochliobolus spicifer, Colletotrichum, Corynespora cassiicola, Curvularia, Fusarium oxysporum, F. oxysporum f. sp. vasinfectum, Glomerella cingulata, Pyrenochaeta lycopersici, Pythium aphanidermatum, P. ultimum, Sclerotinia fuckeliana, S. sclerotiorum, Thanatephorus cucumeris)
Gayed, S. K. 1978. *Publ. Can. Dep. Agric.* 1641, 56 pp. (advisory).
Hopkins, J. C. F. 1956. *Tobacco diseases, with special reference to Africa*, Commonw. Mycol. Inst.
Lucas, G. B. 1975. *Diseases of tobacco*, Biological Consulting Associates, USA.
Nakamura, A. 1967. *Bull. Hatano Tob. Exp. Stn* 59, 86 pp. (breeding for resistance, 117 ref.).
Wolf, F. A. 1957. *Tobacco diseases and decays*, Duke Univ. Press.

Olea europaea (olive)
(Verticillium dahliae)

Onobrychis viciifolia (sainfoin)
(Erysiphe trifolii, Ramularia, Stemphylium, Verticillium albo-atrum)

Orbignya
(Pestalotiopsis palmarum)

Oryza sativa (rice)
Alternaria padwickii (stackburn)
Cochliobolus miyabeanus (brown spot)
Entyloma oryzae (leaf smut)
Gibberella fijikuroi (bakanae)
Magnaporthe salvinii (stem rot)
Pyricularia oryzae (blast)
Rhynchosporium oryzae (leaf scald)
Sarocladium oryzae (sheath rot)
Sclerophthora macrospora (downy mildew)
Sphaerulina oryzina (narrow brown leaf spot)
Tilletia barclayna (kernel smut)
Ustilaginoidea virens (false smut)
(Ascochyta, Cochliobolus, C. lunatus, Curvularia, Cylindrocladium scoparium, Drechslera rostrata, Gaeumannomyces, Gibberella zeae, Khuskia oryzae, Phomopsis, Phytophthora, Pyrenochaeta, Pythium graminicola, Ramularia, Rhizoctonia oryzae-sativae, Septoria, Thanatephorus cucumeris)
Anon. 1976. Pest control in rice, *PANS Manual* 3, 295 pp.
Atkins, J. G. 1974. *Agric. Handb. U.S. Dep. Agric.* 448, 106 pp. (review, diseases in America, 739 ref.).
Hashioka, Y. 1969–72. *Riso* **18**: 279; **19**: 11, 111, 309; **20**: 235; **21**: 11 (fungus diseases, 296 ref.).
Khush, G. S. 1977. In *Advances in Agronomy* **29**: 265 (review, resistance, 8 pp. ref.).
Mogi, S. 1976. *Jap. Pestic. Inf.* **27**: 5 (seed disinfection).
Ou, S. H. et al. 1969. *A. Rev. Phytopathol.* **7**: 383 (review, resistance, 166 ref.).
——. 1972. *Rice diseases*, Commonw. Mycol. Inst.
Padwick, G. W. 1950. *Manual of rice diseases*, Commonw. Mycol. Inst.
Ramakrishnan, T. S. 1971. *Diseases of rice*, ICAR, New Delhi.
Roger, L. 1941–42. *Bull. écon. Indochine* Fasc. 6, 1–5 302 pp. (sclerotial fungi).
Yamaguchi, T. 1977. *Rev. Pl. Prot. Res.* **10**: 49 (seed disinfection, 59 ref.).
——. 1978. *Outl. Agric.* **9**: 278 (in Japan).
(Several authors in *Proc. int. Seed Test. Assoc.* **37**, 723, 731, 985, 1972 (seed health testing).

Oxalis
(Puccinia purpurea, P. sorghi)

Panax schinseng (ginseng)
(Cylindrocarpon, Nectria, Thielaviopsis basicola)
(*Rev. appl. Mycol.* **40**: 120, 1961)

Panicum miliaceum (common millet), *maximum* (Guinea grass), *repens*
Sphacelotheca destruens (head smut)
(Cercospora, Claviceps, Cochliobolus nodulosus, C. setariae, Drechslera gigantea, Sclerospora graminicola, Sorosporium, Sphaerulina oryzina, Tilletia, Uromyces)
(see Ramakrishnan under *Pennisetum*)

Papaver somniferum (opium poppy)
(Peronospora, Pleospora, Pythium mamillatum)
Schmitt, C. G. et al. 1975. U.S. Dep. Agric. ARS-NE-62, 186 pp. (bibliography, 622 ref.).

Paspalum dilatum (dallis grass), *notatum* (Bahia grass), *scrobiculatum* (kodo millet)
(Claviceps, Sphacelotheca, Sorosporium)

Passiflora edulis (passion fruit), *quadrangularis* (giant granadilla)
Alternaria passiflorae (brown spot)
Fusarium oxysporum f. sp. passiflorae (wilt)
(Phytophthora cinnamomi, Pseudocercospora, Septoria)

Pastinaca sativa (parsnip)
(Alternaria radicina, Mycocentrospora acerina)

Pelargonium
(Puccinia)

Pennisetum clandestinum (kikuyu grass)
(Puccinia substriata)

Pennisetum purpureum (elephant grass)
Drechslera sacchari (leaf spot)
(Claviceps, Gloeocercospora sorghi, Puccinia substriata, Pyricularia)
Pennisetum typhoides (bulrush, pearl or spiked millet)
Cochliobolus setariae (leaf spot)
Puccinia substriata (rust)
Sclerospora graminicola (green ear)
Tolyposporium penicillariae (smut)
(Cochliobolus lunatus, Curvularia, Drechslera gigantea, D. rostrata, Gloeocercospora sorghi, Phakopsora, Plasmopora, Pyricularia, Ramulispora, Sclerospora sorghi, Tolyposporium, Uromyces)
Mathur, S. K. et al. 1973. *Seed Sci. Technol.* 1: 811 (seedborne fungi).
Ramakrishnan, T. S. 1963. *Diseases of millets*, ICAR, New Delhi.
Peperomia
(Phytophthora palmivora)
Persea americana (avocado pear)
Phytophthora cinnamomi (root rot)
Sphaceloma perseae (scab)
(Botryodiplodia theobromae, Botryosphaeria ribis, Cercoseptoria, Fusarium solani, Gibberella, G. fujikuroi, Glomerella cingulata, Guignardia, Physalospora, Phytophthora citricola, P. heveae, P. palmivora, Pseudocercospora, Thanatephorus cucumeris, Verticillium dahliae)
Horne, W. T. 1934. *Bull. Calif. agric. Exp. Stn* 585, 72 pp. (in California, USA).
Ruehle, G. D. 1958. *Bull. Fla agric. Exp. Stn* 602: 74.
Stevens, H. E. et al. 1941. *Circ. U.S. Dep. Agric.* 582, 46 pp. (in Florida, USA).
Zentmyer, G. A. 1961. *Ceiba* 9: 61 (in the Americas, 48 ref.).
—— et al. 1965. *Circ. Calif. agric. Exp. Stn* 534, 11 pp.
Petroselinum crispum (parsley)
Erysiphe heraclei (powdery mildew)
(Alternaria radicina)
Phaseolus aureus, now *Vigna radiata* (green gram, mung)
Cercospora canescens (leaf spot)
(Elsinoë, Erysiphe pisi, Mycosphaerella cruenta, Pythium aphanidermatum, Thanatephorus cucumeris)
Nath et al. 1970. *Proc. int. Seed Test. Assoc.* **35**: 225 (seedborne fungi).
Phaseolus coccineus (scarlet runner bean)
Fusarium oxysporum f. sp. phaseoli (yellows)
F. solani f. sp. phaseoli (dry rot)
Phaseolus lunatus (lima bean)
Diaporthe phaseolorum (pod blight)
Elsinoë phaseoli (scab)
Phaeoisariopsis griseola (angular leaf spot)
Phytophthora phaseoli (downy mildew)
(Colletotrichum truncatum, Cristulariella, Elsinoë, Fusarium solani f. sp. phaseoli, Sclerotinia fuckeliana, Thanatephorus cucumeris)
Phaseolus mungo, now *Vigna mungo* (black gram)
(Erysiphe pisi, Protomycopsis)
Phaseolus vulgaris (common bean)
Ascochyta phaseolorum (leaf spot)
Cercospora canescens (leaf spot)
Colletotrichum lindemuthianum (anthracnose)
Fusarium oxysporum f. sp. phaseoli (yellows)
F. solani f. sp. phaseoli (dry rot)
Mycovellosiella phaseoli (floury leaf spot)
Phaeoisariopsis griseola (angular leaf spot)
Sclerotinia sclerotiorum (cottony soft rot)
Uromyces appendiculatus (rust)
(Alternaria alternata, Ascochyta, Cercospora kikuchi, Choanephora, Colletotrichum truncatum, Corticium rolfsii, Cristulariella, Elsinoë phaseoli, Entyloma, Erysiphe pisi, Nematospora coryli, Pleiochaeta, Pythium, P. aphanidermatum, P. butleri, P. debaryanum, P. myriotylum, P. ultimum, Sclerotinia fuckeliana, Thanatephorus cucumeris, Thielaviopsis basicola)
Brien, R. M. et al. 1955. *Bull. N.Z. Dep. scient. ind. Res.* 114, 91 pp. (in New Zealand).
Hubbeling, N. 1955. *Meded. Inst. plziektenk. Onderz.* 83, 80 pp. (monograph).
Zaumeyer, W. J. et al. 1957. *Tech. Bull. U.S. Dep. Agric.* 868, 225 pp. (monograph, 1207 ref.).
—— ——. 1975. *A. Rev. Phytopathol.* **13**: 313 (resistance, 143 ref.).
Phaseolus spp.
Cercospora canescens (leaf spot)
Colletotrichum lindemuthianum (anthracnose)
(Aphanomyces euteiches, Ascochyta gossypii, Mycosphaerella cruenta, M. pinodes, Nematospora coryli, N. gossypii, Phialophora gregata, Phytophthora vignae)
Phoenix dactylifera (date palm)
(Chalaropsis)
Picea (spruce)
(Cylindrocladium scoparium, Rhizina undulata)
Pimenta dioica (pimento), *racemosa* (bay)
Ceratocystis fimbriata (canker)
Puccinia psidii (rust)
Pinus (pine), *canariensis* (Canary island), *caribaea* (Caribbean pitch), *contorta* (lodgepole), *elliotii* (slash), *nigra* (black), *palustris* (long leaf), *patula* (patula), *ponderosa* (ponderosa), *radiata* (radiata), *resinosa* (Canadian red), *sylvestris* (Scots), *taeda* (loblolly)
Diplodia pinea (shoot blight)
Cercoseptoria pini-densiflorae (blight)
Rhizina undulata (group dying)
Scirrhia acicola (brown spot needle blight)
S. pini (red band needle blight)

Pinus (*cont.*)
(Armillariella mellea, Botryosphaeria ribes, Calonectria kyotensis, Coleosporium, Colletotrichum, Cylindrocladium, C. scoparium, Gibberella, G. fujikuroi, Mycosphaerella, Phytophthora cinnamomi, P. cryptogea, P. drechsleri)
Gibson, I. A. S. 1970. *Commonw. For. Rev.* **49**: 267 (*Pinus patula*, review, 53 ref.).
——. 1979 *Diseases of forest trees widely planted as exotics in the tropics and southern hemisphere Pt II. The genus* Pinus, Commonw. Mycol. Inst. and Forestry Inst., 135 pp.

Piper, *betle* (betel), *methysticum* (yangona), *nigrum* (black pepper)
Phytophthora palmivora (foot rot)
P. nicotianae var. parasitica (foot rot)
(Colletotrichum dematium, Corticium salmonicolor, Elsinoë, Glomerella cingulata, Oidium, Thanatephorus cucumeris)
Mei, A. 1956. *Riv. Agric. subtrop. trop.* **50**: 407 (fungi on black pepper).
Singh, R. A. et al. 1971. *Mycopathol. Mycol. appl.* **43**: 109 (fungi on betel; and see *PANS* **25**: 150, 1979).

Pisum sativum (pea)
Aphanomyces euteiches (root rot)
Ascochyta pisi (leaf stem and pod spot)
Erysiphe pisi (powdery mildew)
Fusarium oxysporum f. sp. pisi (wilt)
F. solani f. sp. pisi (foot rot)
Mycosphaerella pinodes (foot rot)
Peronospora viciae (downy mildew)
Uromyces pisi-sativi (rust)
U. viciae-fabae (rust)
(Aspergillus, Cladosporium, Colletotrichum, Elsinoë phaseoli, Gibberella fujikuroi, Neocosmospora, Phoma medicaginis, Pythium aphanidermatum, P. debaryanum, P. graminicola, P. mamillatum, P. ultimum, Sclerotinia fuckeliana, Septoria, Thanatephorus cucumeris, Thielaviopsis basicola, Verticillium dahliae)
Zaumeyer, W. J. 1962. *Agric. Handb. U.S. Dep. Agric.* 228, 30 pp.

Plantago
(Puccinia cynodontis)

Platanus acerifolia (London plane)
(Ceratocystis fimbriata)

Poa pratensis (Kentucky blue grass)
(Pythium graminicola)

Poncirus trifoliata (trifoliate orange)
(Cylindrocarpon, Elsinoë fawcettii, Phytophthora citrophthora)

Populus (poplar)
(Hypoxylon, Melampsora)

Prunus domestica (plum)
(Phellinus)

Pseudotsuga menziesii (Douglas fir)
(Phomopsis juniperivora, Rhizina undulata)

Psidium guajava (guava)
Puccinia psidii (rust)
(Fusarium oxysporum, F. solani, Glomerella cingulata, Macrophoma, Pestalotiopsis, Physalospora, Sphaceloma)
Bose, S. K. 1969. *Progressive Hort.* **1**: 29 (in India).
Srivastava, M. P. et al. 1969. *Pl. Dis. Reptr* **53**: 206 (postharvest).
Williamson, D. 1975. In *Advances in mycology and plant pathology*: 179 (55 ref., editors, S. P. Raychaudhuri et al.).

Psophocarpus tetragonolobus (Goa or winged bean)
(Pseudocercospora, Synchytrium)

Quercus (oak)
(Ceratocystis, Hypoxylon, Phellinus)

Raphanus sativus (radish), *raphinastrum* (wild radish)
Albugo candida (white blister)
Alternaria raphani (black pod blotch)
Aphanomyces raphani (black root)
Fusarium oxysporum f. sp. conglutinans (yellows)
Peronospora parasitica (downy mildew)
Pseudocercosporella capsellae (white spot)
(Leptosphaeria maculans, Thanatephorus cucumeris)

Rauvolfia serpentina (Java devil pepper, sarpagandha)
(Cochliobolus lunatus, Colletotrichum dematium, Fusarium oxysporum)

Rhododendron
(Botryosphaeria ribes, Exobasidium, Phytophthora citricola)

Rhynchosia
(Uromyces)

Ribes
(Botryosphaeria ribis, Sphaerotheca, Microsphaera)

Ricinis communis (castor)
Alternaria ricini (capsule rot)
Sclerotinia ricini (grey capsule mould)
(Fusarium oxysporum, Melampsora, Phytophthora nicotianae var. parasitica, Sphaceloma, Thielaviopsis basicola, Verticillium dahliae)
Jain, J. P. et al. 1969. *Indian Phytopathol.* **22**: 209 (fungi on seed).

Ridolphia segatum
(Alternaria dauci)

Robinia
(Gibberella)

Rosa (rose)
(Chalaropsis, Sphaerulina)

Rubus
(Didymella, Sphaerulina)

Rudbeckia
(Didymella chrysanthemi)

Sabal causiarum (yaray palm)
(Phytophthora palmivora)

Saccharum (sugarcane)
Ceratocystis paradoxa (pineapple disease)
Cercospora koepkei (yellow spot)
C. longipes (brown spot)
Drechslera rostrata (leaf spot)
D. sacchari (eyespot)
D. stenospila (brown stripe)
Gibberella fujikuroi (pokkah boeng, sett rot)
G. fujikuroi var. subglutinans (wilt)
Glomerella tucumanensis (red rot)
Leptosphaeria michotii (leaf spot)
L. sacchari (ringspot)
Phaeocytostroma sacchari (rind disease)
Phytophthora megasperma (sett rot)
Puccinia erianthi (rust)
P. kuehnii (rust)
Pythium arrhenomanes (root rot)
P. graminicola (root rot)
Sclerophthora macrospora (downy mildew)
Sclerospora miscanthi (leaf splitting)
S. sacchari (cane dew, downy mildew)
Stagonospora sacchari (leaf scorch)
Ustilago scitaminea (culmicolous smut)
(Ceratocystis, Claviceps, Cochliobolus lunatus, C. nodulosus, Colletotrichum graminicola, Curvularia, Cytospora, Deightoneilla, Drechslera gigantea, D. halodes, Elsinoë, Gloeocercospora sorghi, Gnomonia, Leptosphaeria, Marasmius, Periconia, Phoma insidiosa, Phomopsis, Pyrenochaeta, Pythium aphanidermatum, P. mamillatum, Ramulispora, Sclerospora philippinensis, Sphacelotheca, S. cruenta)

Edgerton, C. W. 1958. *Sugarcane and its diseases*, Louisiana State Univ. Press.
Hughes, C. G., Abbot, E. V., Wismer, C. A. (editors). 1964. *Sugarcane diseases of the world* Vol. 2, Elsevier.
——. 1978. *PANS* **24**: 123 (review, 62 ref.).
Martin, J. P., Abbot, E. V., Hughes, C. G. (editors). 1961. *Sugarcane diseases of the world* Vol. 1, Elsevier.
Parthasarathy, S. V. 1972. *Sugarcane in India*: 683, K.C.P. Ltd, Madras.

Samanea saman (rain tree, saman)
(Fusarium solani)

Santalum
(Sphaceloma)

Sapindus trifoliatus (soap nut tree)
(Rigidoporus zonalis)

Secale (rye)
(Claviceps, Rhynchosporium, Ustilago)

Sesamum indicum (sesame)
Alternaria sesami (blight)
(Cercoseptoria, Cercospora, Corynespora cassiicola, Drechslera, Fusarium oxysporum, Myrothecium roridum, Oidium, Phytophthora nicotianae var. parasitica, Synchytrium, Thielaviopsis basicola)

Malaguti, G. 1973. *Revta Fac. Agron. Maracay* 7: 109 (leaf diseases).
Mathur, S. B. et al. 1975. *Seed Sci. Technol.* **3**: 655 (seedborne fungi).

Sesbania, *aculeata*, *grandiflora*
(Colletotrichum capsici, Protomycopsis)

Setaria italica (foxtail millet), *sphacelata* (golden timothy), *viridis*
Cochliobolus setariae (leaf spot)
Sclerospora graminicola (green ear)
Ustilago crameri (head smut)
(Cladosporium, Phakopsora, Phoma, Pyricularia, Pythium graminicola, Sclerospora graminicola, Tilletia, Uromyces)

Grewal, J. S. et al. 1965. *Indian Phytopathol.* **18**: 123 (seedborne fungi).
Mathur, S. B. et al. 1967. *Proc. int. Seed Test. Assoc.* **32**: 633 (as above).
(see Ramakrishnan under *Pennisetum*)

Severina buxifolia (box orange)
(Phytophthora citrophthora)

Shorea robusta (sal)
(Hypoxylon, Nectria, Rigidoporus zonalis)

Sinapis
(Peronospora parasitica)

Sisymbrium altissimum
(Leptosphaeria maculans)

Solanum melongena (eggplant), *carolinense* (Carolina horse nettle), *torvum* (soushumber), *Solanum* spp.
Leveillula taurica (powdery mildew)
Phomopsis vexans (tip over)
Phytophthora capsici (blight)
Verticillium dahliae (wilt)
(Alternaria, A. solani, Ascochyta, A. gossypii, Colletotrichum coccodes, C. dematium, Fusarium oxysporum, Mycovellosiella, Myrothecium roridum, Oidium, Peronospora tabacina, Phytophthora hibernalis, P. nicotianae var. parasitica, Puccinia substriata, Pythium aphanidermatum, Sclerotinia sclerotiorum, Septoria lycopersici, Stemphylium solani, Thanatephorus cucumeris)

Solanum tuberosum (Irish potato)
(Alternaria solani, Colletotrichum coccodes, C. dematium, Fusarium solani, Gibberella, Helminthosporium, Macrophomina phaseolina, Mycovellosiella, Phoma, Phytophthora cryptogea, P. drechsleri, P. infestans, Puccinia, Septoria lycopersici, Verticillium dahliae)

Sorghum bicolor (sorghum), *arundinaceum* (wild sorghum), *arundinaceum* var. *sudanensis* (Sudan grass), *dochna* (broomcorn), *halepense* (Johnson grass), *subglabrescens* (milo)
Cercospora sorghi (grey leaf spot)
Colletotrichum graminicola (anthracnose)

Sorghum bicolor (*cont.*)
Gibberella fujikuroi (stalk rot)
Gloeocercospora sorghi (zonate leaf spot)
Periconia circinata (milo)
Puccinia purpurea (rust)
Pythium arrhenomanes (root rot)
Ramulispora sorghi (sooty stripe)
Sclerospora graminicola (green ear)
S. sorghi (downy mildew)
Sphacelotheca cruenta (loose smut)
S. reiliana (head smut)
S. sorghi (covered smut)
Tolyposporium ehrenbergii (long smut)
(Ascochyta, Claviceps, Cochliobolus heterostrophus, C. lunatus, C. nodulosus, Curvularia, Didymella, Drechslera, D. rostrata, Gibberella fujikuroi var. subglutinans, Glomerella tucumanensis, Khuskia oryzae, Mycosphaerella, Phoma, Phomopsis, Phyllachora, Physalospora, Pythium graminicola, Ramulispora, Sclerospora, S. sacchari, Sorosporium, Verticillium dahliae)
Edmunds, L. K. et al. 1975. *Agric. Handb. U.S. Dep. Agric.* 468, 47 pp. (77 ref.).
Nishihara, N. 1973. *Bull. natn. Grassland Res. Inst.* 3: 134 (bibliography, 140 ref.).
Saccas, A. M. 1954. *Agron. trop. Nogent.* **9**: 135, 263, 647.
Tarr, S. A. J. 1962. *Diseases of sorghum, sudan grass and broomcorn*, Commonw. Mycol. Inst.
Spathodea campanulata (tulip tree)
(Poria hypobrunnea)
Spinacia oleracea (spinach)
(Albugo, Cladosporium, Fusarium oxysporum, Peronospora, Pythium ultimum)
Sumner, D. R. et al. 1976. *Phytopathology* **66**: 1267 (root diseases).
Stylosanthes
(Glomerella cingulata, Phoma)
Stizolobium, *aterrimum* (Bengal bean), *deeringianum* (Florida velvet bean)
(Elsinoë phaseoli, Mycosphaerella cruenta, Phytophthora drechsleri, Uromyces)
Tectona grandis (teak)
(Armillariella mellea, Fusarium solani, Helicobasidium, Nectria, Olivea, Rigidoporus zonalis, Sphaceloma)
Tephrosia (clover crop)
(Poria hypobrunnea, Thanatephorus cucumeris)
Theobroma cacao (cacao)
Botryodiplodia theobromae (brown pod, dieback)
Calonectria rigidiuscula (dieback)
Ceratocystis fimbriata (canker)
Crinipellis perniciosa (witches' broom)
Moniliophthora roreri (frosty pod rot)
Oncobasidium theobromae (vascular streak dieback)
Phytophthora palmivora (black pod, canker)
Trachysphaera fructigena (mealy pod)
(Armillariella mellea, Calostilbe striispora, Ceratocystis paradoxa, Colletotrichum dematium, Corticium salmonicolor, Glomerella cingulata, Phellinus noxius, Phytophthora arecae, P. heveae, Poria hypobrunnea, Ustulina deusta, Verticillium dahliae)
Briton-Jones, H. R. 1934. *The diseases and curing of cacao*, Macmillan.
Thorold, C. A. 1975. *Diseases of cocoa*, Clarendon Press.
Thlaspi arvense (penny cress)
(Leptosphaeria maculans)
Toona ciliata
(Rigidoporus zonalis)
Tragopogon porrifolius (salsify)
(Albugo)
Trapa bispinosa (water chestnut)
(Myrothecium roridum)
Trichosanthes cucumerina (snake gourd)
(Didymella bryoniae, Synchytrium)
Trifolium (clover)
(Aphanomyces euteiches, Erysiphe trifolii, Kabatiella, Myrothecium roridum, Peronospora trifoliorum, Phoma medicaginis, Physoderma, Pleospora herbarum, Pythium debaryanum, P. splendens, Sclerotinia, Stemphylium, Uromyces striatus)
Trigonella foenum-graecum (fenugreek), *polycerata*, *suavissima*
(Cercospora, Erysiphe pisi, Peronospora trifoliorum, Uromyces ciceris-arietini, U. striatus)
Tripsacum
(Physopella zeae, Puccinia polysora)
Triticum (wheat)
(Phoma insidiosa, Ustilago)
Tsuga
(Rhizina undulata)
Tulipa (tulip)
(Phytophthora cryptogea, P. porri)
Ulmus (elm)
(Ceratocystis, Chalaropsis, Rigidoporus)
Urena lobata (aramina)
(Sclerotinia fuckeliana)
Vaccinium
(Exobasidium)
Vanilla fragrans (vanilla)
(Fusarium oxysporum, Phytophthora palmivora)
Alconero, R. 1969. *Phyton B. Aires* **26**: 17 (on roots).
Bouriquet, G. 1933. *Annls Cryptog. exot.* **6**: 59 (in Malagasy Republic)
Petch, T. et al. 1927. *Ann. R. bot. Gdns Peradeniya* **10**: 181 (in Sri Lanka).
Vetivaria zizanioides (khuskhus grass)
(Ustilago)
Vicia faba (broad bean), *Vicia* spp.
Fusarium oxysporum f. sp. fabae (wilt)
Uromyces pisi-sativi (rust)

Vicia faba (*cont.*)
U. viciae-fabae (rust)
(Aphanomyces euteiches, Ascochyta, A. pisi, Botrytis, Cochliobolus spicifer, Colletotrichum lindemuthianum, Erysiphe pisi, Fusarium solani, Gibberella, Mycosphaerella pinodes, Peronospora viciae, Pleospora herbarum, Sclerotinia, S. fuckeliana)

Vigna unguiculata (cowpea), *Vigna* spp.
Cercospora canescens (leaf spot)
Fusarium oxysporum f. sp. tracheiphilum (wilt)
Mycosphaerella cruenta (leaf spot)
Phytophthora vignae (stem rot)
Uromyces appendiculatus (rust)
(Ascochyta gossypii, Choanephora, Cladosporium, Colletotrichum lindemuthianum, Corynespora cassiicola, Elsinoë phaseoli, Entyloma, Myrothecium roridum, Phaeoisariopsis griseola, Pleiochaeta, Protomycopsis, Verticillium dahliae)
Williams, R. J. 1977. *Trop. Agric. Trin.* **54**: 53 (resistance)

Vinca (periwinkle)
(Cercospora kikuchi, Myrothecium roridum)

Vitis (grapevine)
(Botryosphaeria, Elsinoë, Physopella, Plasmopora)

Voandzeia subterranea (bambara groundnut)
(Cercospora canescens, Fusarium oxysporum)

Xanthium
(Puccinia helianthi)

Xanthosoma (tannia)
(Phytophthora colocasiae)

Zea mays (maize)
Cochliobolus carbonum (leaf spot)
C. heterostrophus (southern leaf blight)
Colletotrichum graminicola (anthracnose)
Diplodia maydis (stalk rot)
Gibberella fujikuroi (stalk rot)
G. fujikuroi var. subglutinans (stalk rot)
G. zeae (stalk rot)
Mycosphaerella zeae-maydis (yellow leaf blight)
Physoderma maydis (brown spot)
Physopella zeae (rust)
Puccinia polysora (southern rust)
P. sorghi (rust)
Pythium arrhenomanes (root rot)
P. graminicola (root rot)
Sclerophthora macrospora (crazy top)
S. rayssiae (brown stripe)
Sclerospora maydis (Java downy mildew)
S. philippinensis (Philippine downy mildew)
S. sacchari (downy mildew)
S. sorghi (downy mildew)
Setosphaeria turcica (northern leaf blight)
Sphacelotheca reiliana (head smut)
Ustilago maydis (blister smut)
(Acremonium, Aphanomyces euteiches, Ascochyta, Aspergillus flavus, Cercospora, Claviceps, Cochliobolus nodulosus, Corticium rolfsii, Curvularia, Drechslera rostrata, Gibberella, Kabatiella, Khuskia oryzae, Marasmiellus, Marasmius, Phaeocytostroma, Phialophora, Phoma insidiosa, Phyllachora, Physalospora, Pythium butleri, P. splendens, P. ultimum, Rhizopus stolonifer, Sclerospora, S. graminicola, Septoria, Sphacelotheca destruens, Sphaerulina, Ustilaginoidea virens)
Anon. 1973. *A compendium of corn diseases*, 64 pp., Am. Phytopathol. Soc.
Fredericksen, R. A. 1976. *Kasetsart J.* **10**: 164 (downy mildews, 36 ref.).
Kenneth, R. G. 1976. *Ibid* **10**: 79 (downy mildews, 66 ref.).
Koehler, B. 1959 & 1960. *Bull. Ill. agric. exp. Stn* 639, 87 pp.; 658, 90 pp. (ear & stalk rots, 178 ref.).
Nishihara, N. 1972. *Bull. natn. Grassland Res. Inst.* 1: 59 (bibliography, Japan, 217 ref.).
Ullstrup, A. J. 1974. *Agric. Handb. U.S. Dep. Agric.* 199, 56 pp. (advisory).

Zingiber officinale (ginger)
(Fusarium oxysporum, Phyllosticta, Pythium aphanidermatum, P. graminicola, P. myriotylum, Taphrina maculans)
Pegg, K. G. et al. 1974. *Qd agric. J.* **100**: 611 (in Australia).
Sharma, N. D. et al. 1977. *PANS* **23**: 474 (check list & bibliography, 126 ref.).

Zinnia
(Didymella chrysanthemi)

Zizania, *aquatifolia* (American wild rice), *latifolia*
(Magnaporthe salvinii, Melanopsichium, Rhizoctonia oryzae-sativae, Ustilago)

LIST OF COMMON DISEASE NAMES
(with pathogens and hosts)

(This is not a complete list of all such names given in the text. But a name for most diseases is included, distinctive or otherwise. Undistinctive names, for example, anthracnose, rust, smut and wilt are given alphabetically by the main host's common name.)

Abnormal leaf fall, rubber (*Phytophthora botryosa*, *P. meadii*, *P. palmivora*)
Abyan root rot, cotton (*Thanatephorus cucumeris*)
Acute dieback, clove (*Endothia eugeniae*)
Aerial blight, soybean (*Thanatephorus cucumeris*)
Aflaroot, groundnut (*Aspergillus flavus*)
Anabe-roga, areca palm (see *Ganoderma boninense*)
Angular black leaf spot, black gram, cowpea (see *Protomycopsis*)
Angular leaf spot, common and lima beans (*Phaeoisariopsis griseola*)
  tung (see *Mycosphaerella*)
Anthracnose, acacia (*Glomerella cingulata*)
  annatto (*G. cingulata*)
  annona (*G. cingulata*)
  anthurium (*G. cingulata*)
  arecanut (*G. cingulata*)
  arum lily (*G. cingulata*)
  avocado (*G. cingulata*)
  banana (*Colletotrichum musae*, *G. cingulata*)
  betel (*G. cingulata*)
  black pepper (*G. cingulata*)
  brassicas (see *Colletotrichum*)
  cacao (*G. cingulata*)
  cashew (*G. cingulata*)
  cassava (*G. cingulata*)
  citrus (*G. cingulata*)
  cluster bean (*C. capsici*)
  common bean (*C. lindemuthianum*)
  cotton (see *Colletotrichum*, *G. cingulata*)
  cucurbits (see *Glomerella*, *G. cingulata*)
  derris (*G. cingulata*)
  guava (*G. cingulata*)
  jute (see *Colletotrichum*, *G. cingulata*)
  kenaf (*G. cingulata*)
  legumes (see *Colletotrichum*, *G. cingulata*)
  lucerne (*G. cingulata*)
  maize (*C. graminicola*)
  mango (*G. cingulata*)
  melon (*G. cingulata*)
  papaw (*G. cingulata*)
  pea (see *Colletotrichum*, *G. cingulata*)
  pine (see *Colletotrichum*)
  red pepper (*C. capsici*)
  silver wattle (*G. cingulata*)
  sisal (*G. cingulata*)
  sorghum (*C. graminicola*)
  soybean (*C. truncatum*)
  Stylosanthes (see *Colletotrichum*, *G. cingulata*)
  tick clovers (*C. capsici*)
  tobacco (*G. cingulata*)
  tomato (*G. cingulata*)
  turmeric (*C. capsici*)
  watermelon (*G. cingulata*)
  yam (*G. cingulata*)
Areolate leaf spot, citrus (*Thanatephorus cucumeris*)
Australian scab, citrus (*Sphaceloma fawcettii* var. *scabiosa*)
Bakanae, rice (*Gibberella fujikuroi*)
Banded leaf blight, arrowroot (*Thanatephorus cucumeris*)
Bark canker, cinnamon (*Phytophthora cinnamomi*)
Bark rot, immortelle (*Calostilbe striispora*)
Barn rot, tobacco (see *Rhizopus oryzae*)
Basal rot, onion (*Fusarium oxysporum* f. sp. *cepae*)
Basal stem rot, oil palm (see *Ganoderma boninense*)
Berry disease, coffee (*Glomerella cingulata*)
  rot, coffee (*Trachysphaera fructigena*)
Bird's eye rot, grapevine (see *Elsinoë*)
Bird's eye spot, rubber (*Drechslera heveae*)
  tea (see *Calonectria* and *Cercoseptoria*)
Black boll rot, cotton (*Botryodiplodia theobromae*)
Black bundle, maize (see *Acremonium*)
Black cross, banana (see *Phyllachora*)
Black crown rot, celery (*Mycocentrospora acerina*)
Black fruit rot, cucurbits (*Didymella bryoniae*)
Black head, banana (*Ceratocystis paradoxa*)
Black heart, banana (*Gibberella fujikuroi*)
Black hull, groundnut (*Thielaviopsis basicola*)
Black kernel, rice, cereals (*Cochliobolus lunatus*)
Black leaf spot, banana (*Deightoniella torulosa*)
  brassicas (*Alternaria brassicicola*)
Black leaf streak, banana (*Mycosphaerella fijiensis*)
Black leg, brassicas (*Leptosphaeria maculans*)
  sugar beet (see *Pleospora*)
Black pod, cacao (*Phytophthora palmivora*)
Black pod blotch, radish (*Alternaria raphani*)

Common disease names

Black root, radish (*Aphanomyces raphani*)
Black root rot, cotton (*Thielaviopsis basicola*)
  hop (*Phytophthora citricola*)
  tea (*Rosellinia arcuata*)
  tobacco (*T. basicola*)
  various (*R. arcuata*, *R. bunodes*, *R. pepo*)
Black rot, carrot (*Alternaria radicina*)
  citrus (*A. citri*)
  cucumber (see *Phomopsis*)
  groundnut (*Calonectria crotalariae*)
  sugarcane (see *Ceratocystis*)
  sweet potato (*C. fimbriata*)
Black rust, chrysanthemum (see *Puccinia*)
Black shank, tobacco (*Phytophthora nicotianae* var. *nicotianae*)
Black sigatoka, banana (see *Mycosphaerella fijiensis*)
Black spot, citrus (*Guignardia citricarpa*)
  papaw (*Asperisporium caricae*)
Black stem, lucerne (*Phoma medicaginis*)
Black stem rot, tobacco (*Pythium aphanidermatum*)
Black stripe, sugarcane (*Cercospora atrofiliformis*)
  rubber (*Phytophthora botryosa*, *P. palmivora*)
Black tip, banana (*Deightoniella torulosa*)
Blast, oil palm (*Pythium splendens*)
  rice (*Pyricularia oryzae*)
Bleeding, areca palm, coconut (*Ceratocystis paradoxa*)
Blight, castor (*Alternaria ricini*)
  chick pea (*Ascochyta rabei*)
  cluster bean (*Gibberella fujikuroi*)
  cummin (*Alternaria burnsii*)
  eggplant (*Phytophthora capsici*)
  finger millet (*Cochliobolus nodulosus*)
  juniper (*Phomopsis juniperivora*)
  macadamia nut (*Phytophthora nicotianae* var. *parasitica*)
  mango (see *Macrophoma*)
  pines (see *Cercoseptoria*)
  red pepper (*P. capsici*)
  sesame (*A. sesami*)
  tomato (*P. capsici*)
Blister blight, tea (*Exobasidium vexans*)
Blister smut, maize (*Ustilago maydis*)
Blossom blight, avocado pear (see *Physalospora*)
Blue mould, apple (see *Penicillium*)
  citrus (*P. italicum*)
  tobacco (*Peronospora tabacina*)
Bole rot, sisal (*Phytophthora nicotianae* var. *parasitica*)
Boll rot, cotton (several fungi)
Bonnygate, banana (*Calostilbe striispora*)
Bottom rot, cabbage, lettuce (*Thanatephorus cucumeris*)
Branch canker, tea (see *Macrophoma*, *Poria hypobrunnea*)
Brown blight, coffee (*Glomerella cingulata*)
  tea (*G. cingulata*)
Brown etch, pumpkin (*Didymella bryoniae*)
Brown eyespot, coffee (*Cercospora coffeicola*)
Brown leaf mould, red pepper (see *Phaeoramularia*)
Brown leaf spot, abaca (*Deightoniella torulosa*)
  cassava (*Cercosporidium henningsii*)
  safflower (see *Ramularia*)
  sesame (see *Cercoseptoria*)
Brown margin, lettuce (*Bremia lactucae*)
Brown pod, cacao (*Botryodiplodia theobromae*)
Brown root rot, rubber, various trees (*Phellinus noxius*)
Brown rot, citrus (*Phytophthora citricola*, *P. citrophthora*, *P. hibernalis*, *P. nicotianae* var. *parasitica*)
Brown spot, avocado (see *Pseudocercospora*)
  banana (see *Cercospora*)
  Crotalaria (see *Pleiochaeta*)
  maize (*Physoderma maydis*)
  passion fruit (*Alternaria passiflorae*)
  rice (*Cochliobolus miyabeanus*)
  soybean (*Septoria glycines*)
  sugarcane (*Cercospora longipes*)
  tobacco (*A. longipes*)
  various (see *Pleiochaeta*)
Brown spot needle blight, pine (*Scirrhia acicola*)
Brown stem rot, soybean (*Phialophora gregata*)
Brown stripe, maize (*Sclerophthora rayssiae*)
  sugarcane (*Drechslera stenospila*)
Buckeye, tomato (*Phytophthora nicotianae* var. *parasitica*)
Bud rot, palms (*Phytophthora palmivora*)
  pyrethrum (see *Ramularia*)
Bunch rot, oil palm (see *Marasmius*)
Calyx end rot, tomato (*Alternaria solani*)
Cane dew, sugarcane (*Sclerospora sacchari*)
Canker, cacao (*Ceratocystis fimbriata*)
  chestnut (see *Endothia*)
  coffee (*C. fimbriata*)
  Douglas fir (see *Phomopsis juniperivora*)
  eucalyptus (see *Cytospora*, *Diaporthe*, *Endothia*)
  mango (*Phytophthora palmivora*)
  parsnip (*Mycocentrospora acerina*)
  pimento (*C. fimbriata*)
  quinine (*P. nicotianae* var. *parasitica*)
Canker stain, London plane (*Ceratocystis fimbriata*)
Capsule rot, castor (*Alternaria ricini*)
Charcoal, oak (see *Hypoxylon*)
Charcoal base rot, oil palm (*Ustulina deusta*)
Charcoal rot, various (*Macrophomina phaseolina*)
Charcoal stump rot, tea (*Ustulina deusta*)
Chocolate spot, broad bean (see *Botrytis* and *Sclerotinia fuckeliana*)
Cigar end, banana (*Verticillium theobromae*)
Cock's eye, coffee (*Mycena citricolor*)
Collar and branch canker, tea (*Phomopsis theae*)
Collar canker, coffee (see *Helicobasidium*)
Collar and root rot, rubber (*Ustulina deusta*)
Collar rot, acacia (*Calonectria crotalariae*)
  papaw (*C. crotalariae*)
  quinine (*Phytophthora palmivora*)
  rice (see *Ascochyta*)
  safflower (*P. palmivora*)
  tea (*Thanatephorus cucumeris*)

Common leaf spot, sorghum (see *Ramulispora*)
Common scab, citrus (*Elsinoë fawcettii*)
Concentric canker, citrus (see *Ganoderma*)
Copper spot, turf grasses (*Gloeocercospora sorghi*)
Corky root, tomato (*Pyrenochaeta lycopersici*)
Cottony blight, grasses (*Pythium aphanidermatum*)
Cottony leak, cucurbits and others (*Pythium aphanidermatum*)
Cottony rot, cucurbits (*Pythium butleri*)
Cottony soft rot, sunflower and vegetables (*Slerotinia sclerotiorum*)
Covered smut, sorghum (*Sphacelotheca sorghi*)
Crater rot, carrot (see *Rhizoctonia*)
Crater spot, celery (*Thanatephorus cucumeris*)
Crazy top, maize (*Sclerophthora macrospora*)
Crown bud rot, lucerne (see *Gibberella*)
Crown rot, banana (*Deightoniella torulosa*)
  groundnut (*Aspergillus niger*)
  hollyhock (*Phytophthora megasperma*)
Crown sheath rot, rice (see *Gaeumannomyces*)
Crown wart, lucerne (*Physoderma alfalfae*)
Culmicolous smut, sugarcane (*Ustilago scitaminae*)
Diamond leaf spot, banana (see *Haplobasidion*)
Dieback, cacao (*Botryodiplodia theobromae*, *Calonectria rigidiuscula*)
  grapevine (see *Botryosphaeria*)
  rubber (see *Phomopsis*)
  silk cotton (see *P. juniperivora*)
Dollar spot, turf grasses (see *Sclerotinia*)
Downy mildew, beet (see *Peronospora*)
  clover (*P. trifoliorum*)
  common millet (*Sclerospora graminicola*)
  crucifers (*P. parasitica*)
  cucurbits (*Pseudoperonospora cubensis*)
  Gramineae (*S. graminicola*)
  hop (see *Pseudoperonospora*)
  lettuce (*Bremia lactucae*)
  lima bean (*Phytophthora phaseoli*)
  lucerne (*Peronospora trifoliorum*)
  maize (*S. sacchari*, *S. sorghi*)
  onion, Allium (*P. destructor*)
  pea (*P. viciae*)
  poppy (see *Peronospora*)
  rice (*Sclerophthora macrospora*)
  sorghum (*Sclerospora sorghi*)
  soybean (*P. manshurica*)
  sugarcane (*Sclerophthora macrospora*, *Sclerospora sacchari*)
  sunflower (*Plasmopora halstedii*)
Drop, lettuce (*Sclerotinia minor*)
Dry basal rot, oil palm (*Ceratocystis paradoxa*)
Dry rot, common bean (*Fusarium solani* f. sp. *phaseoli*)
  sweet potato (*Diaporthe phaseolorum*)
Dutch elm, elm (see *Ceratocystis*)
Ear and stalk rot, maize (see *Physalospora*)
Early blight, Irish potato (*Alternaria solani*)
  tomato (*A. solani*)
Early leaf spot, groundnut (*Mycosphaerella arachidis*)
Ergot, cereals and grasses (see *Claviceps*)
European canker, apple and pear (see *Nectria*)
Eyespot, banana and Bermuda grass (*Drechslera gigantea*)
  maize (see *Kabatiella*)
  sugarcane (*D. halodes*)
  temperate cereals and grasses (see *Pseudocercosporella*)
False mildew, sunflower (*Plasmopora halstedii*)
False rust, Goa bean (*Synchytrium psophocarpi*)
False smut, rice (*Ustilaginoidea virens*)
Finger rot, banana (*Botryodiplodia theobromae*, *Trachysphaera fructigena*)
Fireblight, pimento (*Ceratocystis fimbriata*)
Floury leaf spot, common bean (*Mycovellosiella phaseoli*)
Foliage blight, tomato (*Pleospora herbarum*)
Foliage web blight, common bean (*Thanatephorus cucumeris*)
Foliar blight, bougainvillea (*Phytophthora nicotianae* var. *parasitica*)
Foot rot, betel (*Phytophthora nicotianae* var. *parasitica*)
  black pepper (*P. palmivora*)
  chick pea (see *Operculella*)
  cucurbits (*Fusarium solani* f. sp. *cucurbitae*)
  kenaf, roselle (*P. nicotianae* var. *parasitica*)
  pea (*F. solani* f. sp. *pisi*, *Mycosphaerella pinodes*)
  sweet potato (see *Phoma*)
  tomato (*Phytophthora cryptogea*, *P. nicotianae* var. *parasitica*)
Freckle, banana (*Guignardia musae*)
  oil palm (*Cercospora elaeidis*)
Frog eye, soybean (see *Cercospora*)
  tobacco (*C. nicotianae*)
Frosty pod rot, cacao (*Moniliphthora roreri*)
Fruit canker, guava (see *Pestalotiopsis*)
Fruit rot, banana (*Trachysphaera fructigena*)
  coffee (*T. fructigena*)
  guava (see *Macrophoma*)
  papaw (see *Phomopsis*, *Phytophthora palmivora*)
Fruit stem end rot, banana (*Ceratocystis paradoxa*)
Fruitlet core rot, pineapple (*Gibberella fujikuroi*)
Gall, cacao (*Calonectria rigidiuscula*)
Gangrene, Irish potato (see *Phoma*)
Ghost spot, tomato (*Sclerotinia fuckeliana*)
Graft rot, rose (see *Chalaropsis*)
Greasy spot, citrus (*Mycosphaerella citri*)
  papaw (*Corynespora cassiicola*)
Green ear, bulrush and foxtail millets (*Sclerospora graminicola*)
  sorghum (*S. graminicola*)
Green mould, citrus (*Penicillium digitatum*)
Grey blight, tea (*Pestalotiopsis theae*)
Grey bulb rot, ornamentals (see *Rhizoctonia*)
Grey capsule mould, castor (*Sclerotinia ricini*)

Common disease names

Grey ear rot, maize (see *Physalospora*)
Grey leaf spot, brassicas (*Alternaria brassicae*)
cereals and grasses (*Pyricularia grisea*)
common bean (see *Cercospora vanderysti*)
maize (see *Cercospora*)
sorghum (*C. sorghi*)
tomato (*Stemphylium solani*)
Grey mildew, cotton (*Ramularia gossypii*)
Grey mould, various (*Sclerotinia fuckeliana*)
Grey mould neck rot, onion (*Botrytis allii*)
Group dying, conifers (*Rhizina undulata*)
Gummosis, citrus (*Phytophthora citrophthora*, *P. nicotianae* var. *parasitica*)
Gummy stem blight, cucurbits (*Didymella bryoniae*)
Head rot, sunflower (*Rhizopus oryzae*)
Head smut, common millet (*Sphacelotheca destruens*)
foxtail millet (*Ustilago crameri*)
maize and sorghum (*S. reiliana*)
Heart rot, pineapple (*Phytophthora nicotianae* var. *parasitica*)
plum (see *Phellinus*)
Iliau, sugarcane (see *Gnomonia*)
Interfruitlet corking, pineapple (see *Penicillium*)
Japanese blight, tea (see *Exobasidium vexans*)
Java black rot, sweet potato (*Botryodiplodia theobromae*)
Java downy mildew, maize (*Sclerospora maydis*)
Kernel smut, rice (*Tilletia barclayana*)
Knot, citrus (*Sphaeropsis tumefaciens*)
Koleroga, areca palm (*Phytophthora arecae*)
Late blight, Irish potato (*Phytophthora infestans*)
tomato (*P. infestans*)
Late leaf spot, groundnut (*Mycosphaerella berkeleyi*)
Late wilt, maize (see *Acremonium*)
Leaf blast, sugarcane (*Leptosphaeria michotii*)
Leaf blight, carrot (*Alternaria dauci*)
celery (see *Septoria*)
finger millet and fowl foot grass (*Cochliobolus nodulosus*)
sugarcane (see *Leptosphaeria*)
Leaf blotch, banana (*Cordana musae*)
sesame (see *Helminthosporium*)
Leaf curl, peach (see *Taphrina*)
Leaf gall, Sesbania (see *Protomycopsis*)
Leaf mottle, sunflower (*Verticillium dahliae*)
Leaf mould, tomato (*Fulvia fulva*)
Leaf, pod and stem spot, pea (*Ascochyta pisi*)
Leaf scald, barley and rice (see *Rhynchosporium*)
Leaf scorch, narcissus (see *Stagonospora*)
sugarcane (*S. sacchari*)
Leaf smut, rice (*Entyloma oryzae*)
Leaf speckle, banana (see *Cladosporium*, *Veronaea*)
Leaf splitting, sugarcane (*Sclerospora miscanthi*)
Leaf spot, bulrush millet (*Cochliobolus setariae*)
castor (*Phytophthora nicotianae* var. *parasitica*)
chick pea (*Pleospora herbarum*)
chicory (see *Alternaria*)
cluster bean (*A. alternata*)
coconut (*Drechslera incurvata*, *Pestalotiopsis palmarum*)
common bean (*Ascochyta phaseolorum*, *Cercospora canescens*)
cotton (*Alternaria macrospora*)
cowpea (*C. canescens*, see *Entyloma*, *Mycosphaerella cruenta*)
eggplant (*A. alternata*)
elephant grass (*D. sacchari*)
endive (see *Alternaria*)
foxtail millet (*Cochliobolus setariae*)
green gram (*Cercospora canescens*)
lemon grass (*D. sacchari*)
lettuce (*Mycocentrospora acerina*, *Pleospora herbarum*, *Septoria lactucae*)
maize (*Cochliobolus carbonum*)
melon (*A. cucumerina*, *P. herbarum*)
onion (*P. herbarum*)
pea (*Ascochyta pisi*)
red pepper (*P. herbarum*)
rubber (*Colletotrichum dematium*)
safflower (*Alternaria carthami*)
sesame (*Phytophthora nicotianae* var. *parasitica*)
sorghum (see *Phoma*, *Ramulispora*)
sugarcane (*D. rostrata*)
sunflower (*S. helianthi*)
sweet potato (*Pleospora herbarum*)
taro (*Phytophthora colocasiae*)
thorn apple (*A. crassa*)
tobacco (*Ascochyta gossypii*)
tomato (*Pseudocercospora fuligena*, *S. lycopersici*)
tumeric (*Taphrina maculans*)
watermelon (*Alternaria cucumerina*)
Leaf stripe, sugarcane (*Sclerospora sacchari*)
Leak, various (*Rhizopus stolonifer*)
Lethal bole rot, coconut (see *Marasmiellus*)
Licorice rot, carrot (*Mycocentrospora acerina*)
Light leaf spot, brassicas (see *Colletotrichum*)
Little broom, cotton (*Glomerella cingulata*)
Long smut, sorghum (*Tolyposporium ehrenbergii*)
Loose smut, barley and oat (*Ustilago avenae*)
sorghum (*Sphacelotheca cruenta*)
wheat and rye (*U. nuda*)
Malformation, mango (*Gibberella fujikuroi*)
Mal secco, citrus (*Deuterophoma tracheiphila*)
Marbled gall, lucerne (*Physoderma alfalfae*)
Mealy pod, cacao (*Trachysphaera fructigena*)
Melanose, citrus (*Diaporthe citri*)
Milo, sorghum (*Periconia circinata*)
Mottle necrosis, sweet potato (*Pythium ultimum*)
Mottle scab, tea (see *Elsinoë*)
Mouldy rot, rubber (*Ceratocystis fimbriata*)
Mycelial neck rot, onion (see *Botrytis*)
Nailhead spot, tomato (*Alternaria tomato*)
Narrow brown leaf spot, rice (*Sphaerulina oryzina*)

Neck rot, onion (see *Botrytis*, *B. allii*, *Sclerotinia porri*, *S. squamosa*)
Net blotch, groundnut (see *Mycosphaerella arachidis*)
Northern anthracnose, clover and lucerne (see *Kabatiella*)
Northern leaf blight, maize and sorghum (*Setosphaeria turcica*)
Panama wilt, abaca and banana (*Fusarium oxysporum* f. sp. *cubense*)
Patch canker, durian (*Phytophthora palmivora*)
  rubber (*P. palmivora*)
Pepper spot, lucerne and various (*Leptosphaerulina trifolii*)
Philippine downy mildew, maize (*Sclerospora philippinensis*)
Pineapple disease, sugarcane (*Ceratocystis paradoxa*)
Pink disease, rubber and various (*Corticium salmonicolor*)
Pink root, onion and Allium (*Pyrenochaeta terrestris*)
Pink rot, Irish potato (*Phytophthora cryptogea*)
Pitch canker, pines (see *Gibberella*)
Pitting, banana (*Pyricularia grisea*)
Pod blight, lima bean (*Diaporthe phaseolorum*)
Pokkah boeng, sugarcane (*Gibberella fujikuroi*)
Pollu, black pepper (*Glomerella cingulata*)
Powdery mildew, alder (see *Microsphaera*)
  annatto (see *Oidium*)
  betel (see *Oidium*)
  black gram (see *Erysiphe pisi*)
  carrot (*E. heraclei*)
  citrus (see *Oidium*)
  clover (see *E. pisi*)
  common bean (see *E. pisi*)
  crucifers (*E. cruciferarum*)
  cucurbits (*Sphaerotheca fuliginea*)
  cumin (*E. heraclei*)
  eggplant (*Leveillula taurica*, see *Oidium*)
  fennel (*E. heraclei*)
  fenugreek (see *E. pisi*)
  globe artichoke (*L. taurica*)
  gooseberry (see *Microsphaeria*, *Sphaerotheca*)
  green gram (see *E. pisi*)
  groundnut (see *Oidium*)
  hop (see *Sphaerotheca*)
  legumes (see *E. pisi*)
  lettuce (see *E. cichoracearum*)
  lilac (see *Microsphaera*)
  mango (see *Oidium*)
  okra (*E. cichoracearum*)
  papaw (see *Oidium*)
  parsley (*E. heraclei*)
  pea (*E. pisi*)
  red pepper (*L. taurica*)
  rose (see *Sphaerotheca*)
  rubber (*O. heveae*)
  sesame (see *Oidium*)
  sisam (see *Phyllactinia*)
  soybean (see *Microsphaera*)
  strawberry (see *Sphaerotheca*)
  sugar beet (see *E. cruciferarum*)
  tomato (*L. taurica*)
  umbellifers (*E. heraclei*)
Purple blotch, Allium (*Alternaria porri*)
Purple leaf spot, Sesbania (see *Protomycopsis*)
Purple seed stain, soybean (*Cercospora kikuchi*)
Ray blight, chrysanthemum and pyrethrum (*Didymella chrysanthemi*)
Red band needle blight, pines (*Scirrhia pini*)
Red leg, lettuce (*Sclerotinia fuckeliana*)
Red root, sesame (*Thielaviopsis basicola*)
Red root rot, rubber and various (*Ganoderma philippii*)
Red rot, sugarcane (*Glomerella tucumanensis*)
Red sheath rot, abaca (*Deightoniella torulosa*)
Red spot, sugarcane (*Cercospora vaginae*)
Rind disease, sugarcane (*Phaeocytostroma sacchari*)
Ringspot, brassicas (*Mycosphaerella brassicicola*)
  lucerne (*Pleospora herbarum*)
  sugarcane (*Leptosphaeria sacchari*)
Robinson dieback, tangerine (*Botryodiplodia theobromae*)
Root canker, lucerne (*Thanatephorus cucumeris*)
Root rot, avocado (*Phytophthora cinnamomi*)
  broad bean (see *Gibberella*)
  cabbage and cauliflower (see *Aphanomyces*)
  cassava (*P. drechsleri*)
  clove (*P. cinnamomi*)
  common bean (*Thanatephorus cucumeris*)
  date palm (see *Chalaropsis*)
  elm (see *Chalaropsis*)
  lucerne (*P. megasperma*)
  maize (*Pythium arrhenomanes*)
  papaw (*Phytophthora palmivora*)
  passion fruit (*P. cinnamomi*)
  pea (*A. euteiches*)
  poinsettia (*Thielaviopsis basicola*)
  quinine (*P. cinnamomi*)
  rubber (*Poria hypobrunnea*)
  safflower (*Phytophthora cryptogea*, *P. drechsleri*)
  sorghum (*Pythium arrhenomanes*)
  sugarcane (*P. arrhenomanes*, *P. graminicola*)
  tea (*Poria hypobrunnea*)
Rough leaf spot, sorghum (see *Ascochyta*)
Rust, antirrhinum (see *Puccinia*)
  asparagus (see *Puccinia*)
  banana (see *Uromyces*)
  beans (*U. appendiculatus*)
  Bermuda grass (*P. cynodontis*)
  bulrush millet (*P. substriata*)
  castor (see *Melampsora*)
  celery (see *Puccinia*)
  chick pea (*U. ciceris-arietini*)
  coffee (*Hemileia coffeicola*, *H. vastatrix*)
  cotton (*Phakopsora gossypii*, *Puccinia cacabata*)
  cowpea (*U. appendiculatus*)

Rust (*cont.*)
Digitaria (*P. oahuensis*)
fig (see *Cerotelium*)
flax (see *Melampsora*)
foxtail millet (see *Uromyces*)
garlic (*P. allii*)
geranium (see *Puccinia*)
grapevine (see *Physopella*)
grass pea (*U. pisi-sativae*, *U. viciae-fabae*)
groundnut (*Puccinia arachidis*)
guava (*P. psidii*)
hollyhock (see *Puccinia*)
horse gram (*U. appendiculatus*)
indigo (see *Uromyces*)
iris (see *Puccinia*)
Irish potato (see *Puccinia*)
lentil (*U. viciae-fabae*)
lucerne (*U. striatus*)
maize (*Physopella zeae*, *Puccinia polysora*, *P. sorghi*)
mint (see *Puccinia*)
onion and Allium (*P. allii*)
pangola grass (*P. oahuensis*)
pea (*U. pisi-sativi*, *U. viciae-fabae*)
pelargonium (see *Puccinia*)
pigeon pea (see *Uromyces*)
pimento (*P. psidii*)
poplar (see *Melampsora*)
safflower (*P. carthami*)
sorghum (*P. purpurea*)
soybean (*Pakopsora pachyrhizi*)
sugarcane (*Puccinia erianthi*, *P. kuehnii*)
sunflower (*P. helianthi*)
sunn hemp (*U. decoratus*)
sweet potato (see *Coleosporium*)
tarragon (see *Puccinia*)
teak (see *Olivea*)
tomato (see *Puccinia*)
velvet bean (see *Uromyces*)
Scab, avocado (*Sphaceloma perseae*)
black pepper (see *Elsinoë*)
cardamom (see *Sphaceloma*)
cashew (see *Sphaceloma*)
castor (see *Sphaceloma*)
cucurbits (*Cladosporium cucumerinum*)
green gram (see *Elsinoë*)
groundnut (see *Sphaceloma*)
guava (see *Sphaceloma*)
hyacinth bean (see *Elsinoë*)
jack bean (see *Elsinoë*)
lima bean (*E. phaseoli*)
mango (see *Elsinoë*)
papaw (see *Sphaceloma*)
poinsettia (see *Sphaceloma*)
rice (*Gibberella zeae*)
rubber (see *Elsinoë*)
sandalwood (see *Sphaceloma*)
soybean (see *Sphaceloma*)
sweet orange (*E. australis*)
sweet potato (see *Elsinoë*)
sword bean (see *Elsinoë*)
teak (see *Sphaceloma*)
Scorch, clover (see *Kabatiella*)
Scurf, sweet potato (*Monilochaetes infuscans*)
Scurfy root, radish (*Thanatephorus cucumeris*)
Secondary leaf fall, rubber (*Glomerella cingulata*, *Oidium heveae*)
Sett rot, sugarcane (*Gibberella fujikuroi*, *Phytophthora megasperma*)
Seven curls, onion (*Glomerella cingulata*)
Shanking, tulip (*Phytophthora cryptogea*)
Sheath blight, finger millet (see *Marasmiellus*)
rice (*Thanatephorus cucumeris*)
Sheath brown spot, rice (see *Pyrenochaeta*)
Sheath net blotch, rice (*Cylindrocladium scoparium*)
Sheath rot, abaca and banana (see *Marasmiellus*)
rice (*Sarocladium oryzae*)
sugarcane (see *Cytospora*)
Sheath spot, rice (see *Rhizoctonia oryzae-sativae*)
Shoot blight, eucalpytus (see *Ramularia*)
pine (*Diplodia pinea*)
Sigatoka, banana (*Mycosphaerella musicola*)
Silver scurf, Irish potato (see *Helminthosporium*)
Sleepy, maize (*Sclerospora maydis*, *S. philippinensis*)
tomato (*Verticillium albo-atrum*)
Small sclerotial neck rot, onion (*Sclerotinia squamosa*)
Smudge, onion (*Colletotrichum circinans*)
Smut, American wild rice (see *Ustilago*)
barnyard millet (see *Tolyposporium*)
Bermuda grass (*Ustilago cynodontis*)
bulrush millet (*T. penicillariae*)
finger millet (see *Melanopsichium*, *Tolyposporium*)
golden timothy (see *Tilletia*)
guinea grass (see *Tilletia*)
Indian mustard (see *Urocystis*)
Japanese barnyard millet (see *Ustilago*)
Job's tears (see *Ustilago*)
khuskus grass (see *Ustilago*)
kodo millet (see *Sorosporium*)
lemon grass (see *Tolyposporium*)
Malabar grass (see *Tolyposporium*)
onion (*Urocystis cepulae*)
soybean (see *Melanopsichium*)
Soft rot, carrot (*Phytophthora megasperma*)
cucurbits (*P. capsici*)
pineapple (*Ceratocystis paradoxa*)
various (see *Choanephora*)
Sooty stripe, sorghum (*Ramulispora sorghi*)
Soreshin, cotton (*Thanatephorus cucumeris*)
Sour rot, citrus (*Geotrichum candidum*)
South American leaf blight, rubber (*Microcyclus ulei*)
Southern anthracnose, Leguminosae (*Colletotrichum*, *Glomerella cingulata*)

Southern leaf blight, maize (*Cochliobolus heterostrophus*)
Southern rust, maize (*Puccinia polysora*)
Southern stem rot, various (*Corticium rolfsii*)
Spear rot, oil palm (*Gibberella fujikuroi* var. *subglutinans*)
Speckled blotch, rice (see *Septoria*)
Spot anthracnose, grapevine (see *Elsinoë*)
Spring black stem, lucerne and pea (*Phoma medicaginis*)
Spring dead spot, Bermuda grass (*Cochliobolus spicifer*)
Spur blight, raspberry (see *Didymella*)
Squirter, banana (see *Nigrospora*)
St John's, pea (see *Fusarium oxysporum* f. sp. *pisi*)
Stackburn, rice (*Alternaria padwickii*)
Stalk rot, maize (*Diplodia maydis*, *Gibberella fujikuroi*, *G. fujikuroi* var. *subglutinans*, *G. zeae*, inter alia)
  sorghum (*G. fujikuroi*)
  soybean (*Phytophthora megasperma* var. *sojae*)
Stem break, sunn hemp (see *Colletotrichum dematium*)
Stem canker, Albizia (*Phytophthora drechsleri*)
  macadamia (*P. cinnamomi*)
  soybean (*Diaporthe phaseolorum*)
Stem end rot, avocado (*Botryodiplodia theobromae*)
  citrus (*B. theobromae*)
  tomato (*Alternaria solani*)
Stem and fruit rot, tomato (*Didymella lycopersici*)
Stem gall, coriander (*Protomyces macrosporus*)
Stem rot, cowpea (*Phytophthora vignae*)
  rice (*Magnaporthe salvinii*)
Stigmatomycosis, cotton and various (*Eremothecium*, *Nematospora*, *N. coryli*, *N. gossypii*)
Stinking root rot, various (*Sphaerostilbe repens*)
Storey's bark, coffee (*Gibberella stilboides*)
Stripe canker, quinine (*Phytophthora nicotianae* var. *parasitica*)
Sudden death, clove (*Valsa eugeniae*)
Sugary disease (see ergot)
Superelongation, cassava (see *Sphaceloma*)
Swamp spot, banana (*Deightoniella torulosa*)
Tap root rot, sugar beet (*Phytophthora drechsleri*)
Tar spot, various (see *Phyllachora*)
Target blotch, sugarcane (*Drechslera halodes*)
Target leaf spot, rubber (*Thanatephorus cucumeris*)
Target spot, various (*Corynespora cassiicola*)
Tarry root rot, tea (see *Hypoxylon serpens*)
Terminal crook pine (see *Colletotrichum*)
Thorny stem blight, tea (see *Tunstallia*)
Tip blight, pea (*Thanatephorus cucumeris*)
Tip over, eggplant (*Phomopsis vexans*)
Tracheomycosis, coffee (*Gibberella xylarioides*)
Tuber rot, cassava (*Botryodiplodia theobromae*)
  yam (*B. theobromae*, *Rhizopus oryzae*)
Upper stem rot, oil palm (*Phellinus noxius*)
Vascular streak dieback, cacao (*Oncobasidium theobromae*)
Veneer blotch, sugarcane (see *Deightoniella*)
Violet root rot, carrot, teak and various (see *Helicobasidium*)
Wart, Irish potato (see *Synchytrium*)
Warty, coffee (*Sclerotinia fuckeliana*)
Water blister, pineapple (*Ceratocystis paradoxa*)
Web blight, lima bean (*Thanatephorus cucumeris*)
Wet weather blight, cotton (*Ascochyta gossypii*)
Whiskers, various (*Rhizopus stolonifer*)
White blister, crucifers (*Albugo candida*)
White fan blight, rubber (see *Marasmius*)
White heart rot, temperate, deciduous trees (see *Phellinus*)
White leaf spot, cassava (*Phaeoramularia manihotis*)
White leaf streak, rice (see *Ramularia*)
White mould, tobacco (*Erysiphe cichoracearum*)
White pocket rot, various (*Rigidoporus zonalis*)
White rash, sugarcane (see *Elsinoë*)
White root rot, rubber and various (*Rigidoporus lignosus*)
  various (*Rosellinia necatrix*)
White rot, Allium (*Sclerotiorum cepivorum*)
White rust, chrysanthemum (see *Puccinia*)
  salsify, spinach and sunflower (see *Albugo*)
  sweet potato (*A. ipomoeae-panduratae*)
  water spinach (*A. ipomoeae-aquaticae*)
White scab, tea (see *Elsinoë*)
White spot, crucifers (*Pseudocercosporella capsellae*)
White tip, *Allium* (*Phytophthora porri*)
White leaf spot, cassava (see *Cercosporidium*)
(for wilts caused by Fusarium oxysporum where the f. sp. is not given see the end of the section on this pathogen)
Wilt, Albizia (*Fusarium oxysporum*)
  avocado (*Verticillium dahliae*)
  bambara groundnut (*F. oxysporum*)
  bottle gourd (*F. oxysporum*)
  broad bean (*F. o.* f. sp. *fabae*)
  Brussels sprout (*V. dahliae*)
  cacao (*V. dahliae*)
  cassia (*F. oxysporum*)
  castor (*F. oxysporum*, *V. dahliae*)
  celery (*F. oxysporum*)
  chick pea (*F. o.* f. sp. *ciceris*)
  coffee (*F. oxysporum*)
  coriander (*F. o.* f. sp. *coriandrii*)
  cotton (*F. o.* f. sp. *vasinfectum*, *V. dahliae*)
  cowpea (*F. o.* f. sp. *tracheiphilum*, *V. dahliae*)
  cucumber (*F. o.* f. sp. *cucumerinum*)
  cumin (*F. o.* f. sp. *cumini*)
  dill (*F. oxysporum*)
  eggplant (*F. oxysporum*, *V. dahliae*)
  eucalyptus (*F. oxysporum*)
  ginger (*F. oxysporum*)
  grass pea (*F. oxysporum*)
  groundnut (*V. dahliae*)
  guava (*F. oxysporum*)
  hemp (*F. oxysporum*)
  lentil (*F. oxysporum*)
  lettuce (*F. oxysporum*)
  lucerne (*F. o.* f. sp. *medicaginis*, *V. albo-atrum*)

Wilt (*cont.*)
mango (*Ceratocystis fimbriata*, *V. dahliae*)
melon (*F. o.* f. sp. *melonis*)
oak (see *Ceratocystis*)
oil palm (*F. o.* f. sp. *elaeidis*)
onion (*F. o.* f. sp. *cepae*)
okra (*V. dahliae*)
orchid (*F. oxysporum*)
passion fruit (*F. o.* f. sp. *passiflorae*)
pea (*F. o.* f. sp. *pisi*, *V. dahliae*)
pigeon pea (*F. udum*)
pineapple (*Phytophthora cinnamomi*)
red pepper (*V. dahliae*)
safflower (*F. o.* f. sp. *carthami*, *V. dahliae*)
sarpagandha (*F. oxysporum*)
sesame (*F. oxysporum*)
soybean (*F. o.* f. sp. *tracheiphilum*)
spinach (*F. oxysporum*)
sugarcane (*Gibberella fujikuroi* var. *subglutinans*)
sunn hemp (*F. udum* f. sp. *crotalariae*)
sweet potato (*F. o.* f. sp. *batatas*)
tobacco (*F. oxysporum*, *V. dahliae*)
tomato (*F. o.* f. sp. *lycopersici*, *V. albo-atrum*, *V. dahliae*)
vanilla (*F. oxysporum*)
watermelon (*F. o.* f. sp. *niveum*)
yangona (*P. palmivora*)
Wirestem, cabbage (*Thanatephorus cucumeris*)
Witches' broom, cacao (*Crinipellis perniciosa*)
Withertip, lime (*Glomerella cingulata*)
Wood rot, tea (*Hypoxylon serpens*)
Yeast spot, common bean and soybean (*Nematospora corylii*)
Yellow leaf blight, maize (*Mycosphaerella zeae-maydis*)
Yellow leaf mould, Pueraria (see *Mycovellosiella*)
Yellow spot, sugarcane (*Cercospora koepkei*)
Yellow trunk rot, oak (see *Phellinus*)
Yellow wilt, rice (*Sclerophthora macrospora*)
Yellows, common bean and scarlet runner bean (*Fusarium oxysporum* f. sp. *phaseoli*)
crucifers (*F. o.* f. sp. *conglutinans*)
sunflower (see *Phialophora*)
Zebra leaf spot, sisal (*Phytophthora nicotianae* var. *parasitica*)
Zonal leaf spot, coffee (see *Acremonium*)
Zonate eyespot, Bermuda grass (*Drechslera gigantea*)
Zonate leaf spot, common bean, groundnut, kenaf, pecan and tung (see *Cristulariella*)
kenaf and roselle (*Thanatephorus cucumeris*)
sorghum (*Gloeocercospora sorghi*)

LIST OF COMMON HOST NAMES
(with botanical equivalents)

Abaca, *Musa textilis*
Adzuki bean, *Phaseolus angularis*
Alfalfa, see lucerne
American pitch pine, see long leaf pine
American wild rice, *Zizania aquatifolia*
Annatto, *Bixa orellana*
Apple, *Malus pumila*
Arabica coffee, *Coffea arabica*
Aramina, *Urena lobata*
Areca palm, *Areca catechu*
Arrowroot, *Maranta arundinacea*
Arum lily, *Anthurium andreanum*
Asparagus, *Asparagus officinalis*
Australian wild lime, *Microcitrus australis*
Avocado pear, *Persea americana*
Bambara groundnut, *Voandzeia subterranea*
Banana, *Musa*
Barley, *Hordeum*
Barnyard millet, *Echinochloa crus-galli*
Bay, *Pimenta racemosa*
Bean, see common bean
Beet, *Beta vulgaris*
Bengal bean, *Stizolobium aterrinum*
Bent grass, *Agrostis palustris*
Bermuda grass, *Cynodon dactylon*
Betel, *Piper betle*
Bird pepper, see red pepper
Bird's-foot trefoil, *Lotus corniculatus*
Black gram, *Phaseolus mungo* (now *Vigna mungo*)
Black mustard, *Brassica nigra*
Black pepper, *Piper nigrum*
Black pine, *Pinus nigra*
Blue sisal, *Agave amaniensis*
Bottle brush tree, *Callistemon viminalis*
Bottle gourd, *Lagenaria siceraria*
Box orange, *Severina buxifolia*
Brazil nut, *Bertholletia excelsa*
Breadfruit, *Artocarpus altilis*
Broad bean, *Vicia faba*
Broccoli, see cauliflower
Broomcorn, *Sorghum dochna*
Brussels sprout, *Brassica oleracea* var. *gemmifera*
Bulrush millet, *Pennisetum typhoides*
Cabbage, *Brassica oleracea* var. *capitata*
Cacao, *Theobroma cacao*
Camel's-foot tree, *Bauhinia purpurea*
Canadian red pine, *Pinus resinosa*
Canary island pine, *P. canariensis*
Cardamom, *Elettaria cardamomum*
Caribbean pitch pine, *Pinus caribaea*
Carnation, *Dianthus caryophyllus*
Carolina horse nettle, *Solanum caroliense*
Carrot, *Daucus carota*
Cashew, *Anacardium occidentale*
Cassava, *Manihot esculenta*
Castor, *Ricinus communis*
Cauliflower, *Brassica oleracea* var. *botrytis*
Celery, *Apium graveolens* var. *dulce*
Chestnut, *Castanea*
Chick pea, *Cicer arietinum*
Chicory, *Chichorium intybus*
Chilli, see red pepper
Chinese cabbage, *Brassica chinensis*
Chinese evergreen, *Aglaonema simplex*
Chinese mustard, *B. chinensis*
Chinese sweet gum, *Liquidambar formosana*
Chrysanthemum, *Chrysanthemum morifolium*
Cinnamon, *Cinnamomum zeylanicum*
Citron, *Citrus medica*
Citronella grass, *Cymbopogon nardus*
Clove, *Eugenia caryophyllus*
Clover, *Trifolium*
Cluster bean, *Cyamopsis tetragonoloba*
Cocksfoot, *Dactylis glomerata*
Coconut, *Cocos nucifera*
Cocoyam, *Colocasia esculenta*
Coffee, *Coffea*
Colocynth, *Citrullus colocynthis*
Common bean, *Phaseolus vulgaris*
Common millet, *Panicum miliaceum*
Coriander, *Coriandrum sativum*
Corn, see maize
Cotton, *Gossypium* spp.
Cowpea, *Vigna unguiculata*
Cucumber, *Cucumis sativus*
Cumin, *Cuminum cyminum*
Currant tomato, *Lycopersicon pimpinellifolium*
Cypre, *Cordia alliodora*
Dadap, *Erythrina*
Dahlia, *Dahlia rosea*
Dallis grass, *Paspalum dilatum*
Dasheen, see cocoyam
Date palm, *Phoenix dactylifera*
Derris, *Derris* spp.

Common host names

Dill, *Anethum graveolens*
Djoowar, *Cassia siamea*
Douglas fir, *Pseudotsuga menziesii*
Durian, *Durio zibethinus*
Dwarf sisal, *Agave angustifolia*
Easter lily, *Lilium longiflorum*
Eastern red cedar, *Juniperus virginiana*
Eddoe, see cocoyam
Eggplant, *Solanum melongena*
Elephant grass, *Pennisetum purpureum*
Elm, *Ulmus*
Endive, *Cichorium endivia*
Eucalyptus, *Eucalyptus* spp.
European beech, *Fagus sylvatica*
European walnut, *Juglans regia*
Fennel, *Foeniculum vulgare*
Fenugreek, *Trigonella foenum-graecum*
Field mustard, *Brassica campestris*
Fig, *Ficus carica*
Finger millet, *Eleusine coracana*
Flax, *Linum usitatissimum*
Florida velvet bean, *Stizolobium deeringianum*
Fowl foot grass, *Eleusine indica*
Foxtail millet, *Setaria italica*
French bean, see common bean
Garlic, *Allium sativum*
Giant alocasia, *Alocasia macrorrhiza*
Giant granadilla, *Passiflora quadrangularis*
Ginger, *Zingiber officinale*
Ginseng, *Panax schinseng*
Globe artichoke, *Cynara scolymus*
Goa bean, *Psophocarpus tetragonolobus*
Golden gram, see green gram
Golden timothy, *Setaria sphacelata*
Gooseberry, *Ribes grossularia*
Grapefruit, *Citrus paradisi*
Grapevine, *Vitis vinifera*
Grass pea, *Lathyrus sativus*
Great burdock, *Arctium lappa*
Great-headed garlic, *Allium ampeloprasum*
Green gram, *Phaseolus aureus* (now *Vigna radiata*)
Green wattle, *Acacia decurrens*
Groundnut, *Arachis hypogaea*
Guar, see cluster bean
Guava, *Psidium guajava*
Guinea grass, *Panicum maximum*
Haricot bean, see common bean
Hemp, *Cannabis sativa*
Henbane, *Hyoscyamus niger*
Hollyhock, *Althaea rosea*
Hop, *Humulus lupulus*
Horse bean, see broad bean
Horse gram, *Dolichos uniflorus*
Horse radish, *Armoracia rusticana*
Hyacinth, *Hyacinthus*
Hyacinth bean, *Lablab niger* (now *L. purpureus*)
Immortelle, *Erythrina*
Indian arrowroot, *Curcuma angustifolia*
Indian banyan, *Ficus benghalensis*
Indian mustard, *Brassica juncea*
Irish potato, *Solanum tuberosum*
Jack bean, *Canavalia ensiformis*
Jackfruit, *Artocarpus heterophyllus*
Jambolan, *Eugenia cuminii*
Japanese barnyard millet, *Echinochloa frumentacea*
Japanese persimmon, *Diospyros kaki*
Java devil pepper, see sarpagandha
Jerusalem artichoke, *Helianthus tuberosus*
Jimson weed, see thorn apple
Job's tears, *Coix lachryma-jobi*
Johnson grass, *Sorghum halepense*
Jungle rice, *Echinochloa colona*
Juniper, *Juniperus* spp.
Jute, *Corchorus* spp.
Kapok, *Ceiba pentandra*
Kenaf, *Hibiscus cannabinus*
Kentucky blue grass, *Poa pratensis*
Khuskhus grass, *Vetiveria zizanioides*
Kidney bean, see common bean
Kikuyu grass, *Pennisetum clandestinum*
Koa acacia, *Acacia koa*
Kodo millet, *Paspalum scrobiculatum*
Kohlrabi, *Brassica oleracea* var. *gongyloides*
Kudzu, *Pueraria thunbergiana*
Lalang, *Imperata cylindrica*
Larch, *Larix* spp.
Leek, *Allium ampeloprasum*
Lemon, *Citrus limon*
Lemon grass, *Cymbopogon citratus*
Lentil, *Lens esculenta*
Lettuce, *Lactuca sativa*
Liberica coffee, *Coffea liberica*
Lima bean, *Phaseolus lunatus*
Lime, *Citrus aurantifolia*
Loblolly pine, *Pinus taeda*
Lodgepole pine, *P. contorta*
London plane, *Platanus acerifolia*
Longleaf pine, *Pinus palustris*
Lucerne, *Medicago sativa*
Lupin, *Lupinus* spp.
Macadamia nut, *Macadamia* spp.
Maize, *Zea mays*
Malabar gourd, *Cucurbita ficifolia*
Malabar grass, *Cymbopogon flexuosus*
Mandarin, *Citrus reticulata*
Mango, *Mangifera indica*
Manila hemp, see abaca
Marrow, *Cucurbita pepo*
Masindi grass, *Cynodon transvaalensis*
Meadow grass, see Kentucky blue grass
Melon, *Cucumis melo*
Milo, *Sorghum subglabrescens*

Mint, *Mentha*
Mountain immortelle, *Erythrina poeppigiana*
Mulberry, *Morus* spp.
Mung, see green gram
Napier grass, see elephant grass
Oak, *Quercus* spp.
Oat, *Avena*
Oil palm, *Elaeis guineensis*
Okra, *Hibiscus esculentus*
Olive, *Olea europaea*
Onion, *Allium cepa*
Opium poppy, *Papaver somniferum*
Oval kumquat, *Fortunella margarita*
Palmyra palm, *Borassus flabellifer*
Pan, see betel
Pangola grass, *Digitaria decumbens*
Papaw, papaya, *Carica papaya*
Parsley, *Petroselinum crispum*
Parsnip, *Pastinaca sativa*
Passion fruit, *Passiflora edulis*
Patula pine, *Pinus patula*
Pea, *Pisum sativum*
Peach, *Amygdalus persica*
Peanut, see groundnut
Pearl millet, see bulrush millet
Pecan, *Carya pecan*
Penny cress, *Thlaspi arvense*
Pepper, see black and red
Periwinkle, *Vinca*
Pigeon pea, *Cajanus cajan*
Pimento, *Pimenta dioica*
Pine, *Pinus*
Pineapple, *Ananas comosus*
Plantain, *Musa*
Plum, *Prunus domestica*
Pomerac, *Eugenia malaccensis*
Ponderosa pine, *Pinus ponderosa*
Poplar, *Populus* spp.
Potato, see Irish and sweet
Pummelo, *Citrus grandis*
Pumpkin, *Cucurbita* spp.
Pyrethrum, *Chrysanthemum cinerariaefolium*
Quinine, *Cinchona* spp.
Radiata pine, *Pinus radiata*
Radish, *Raphanus sativus*
Rain tree, see saman
Ramie, *Boehmeria nivea*
Rape, *Brassica napus*
Raspberry, *Rubus*
Red gram, see pigeon pea
Red pepper, *Capsicum*
Rhodes grass, *Chloris gayana*
Rice, *Oryza sativa*
Robusta coffee, *Coffea canephora*
Rocky mountain juniper, *Juniperus scopulorum*
Rose, *Rosa*
Rose apple, *Eugenia jambos*
Roselle, *Hibiscus sabdariffa*
Rubber, *Hevea brasiliensis*
Rutabaga, *Brassica napobrassica*
Rye, *Secale*
Safflower, *Carthamus tinctorius*
Sainfoin, *Onobrychis viciifolia*
Sal, *Shorea robusta*
Salsify, *Tragopogon porrifolius*
Saman, *Samanea saman*
Sapodilla, *Manilkara achras*
Sarpagandha, *Rauvolfia serpentina*
Scarlet runner bean, *Phaseolus coccineus*
Scots pine, *Pinus sylvestris*
Semul, see silk cotton
Sesame, *Sesamum indicum*
Shallot, *Allium cepa*
Sickle senna, *Cassia tora*
Silk cotton, *Bombax malabaricum*
Silky oak, *Grevillea robusta*
Silver wattle, *Acacia dealbata*
Sisal, *Agave sisalana*
Sissoo, *Dalbergia sissoo*
Skeleton weed, *Chondrilla juncea*
Slash pine, *Pinus elliotii*
Smooth brome, *Bromus inermis*
Snake gourd, *Trichosanthes cucumerina*
Snapdragon, *Antirrhinum majus*
Soap nut tree, *Sapindus trifoliatus*
Sorghum, *Sorghum bicolor*
Sour orange, *Citrus aurantium*
Soushumber, *Solanum torvum*
Soybean, *Glycine max*
Spiked millet, see bulrush millet
Spinach, *Spinacia oleracea*
Squash, *Curcurbita* spp.
Star grass, see Bermuda grass
Sudan grass, *Sorghum arundinaceum* var. *sudanensis*
Sugar apple, *Annona squamosa*
Sugar beet, *Beta vulgaris*
Sugarcane, *Saccharum* spp.
Sunflower, *Helianthus annuus*
Sunn hemp, *Crotalaria juncea*
Swamp immortelle, *Erythrina glauca*
Swede, see rutabaga
Sweet orange, *Citrus sinensis*
Sweet pea, *Lathyrus odoratus*
Sweet potato, *Ipomoea batatas*
Sword bean, *Canavalia gladiata*
Tannia, *Xanthosoma* spp.
Taro, see cocoyam
Tarragon, *Artemesia dracunculus*
Tea, *Camellia sinensis*
Teak, *Tectona grandis*
Teff, *Eragrostis tef*

Teosinte, *Euchlaena* spp.
Thorn apple, *Datura stramonium*
Tick clovers, *Desmodium* spp.
Tobacco, *Nicotiana tabacum*
Tobacco flower, *N. alata*
Tomato, *Lycopersicon esculentum*
Trifoliate orange, *Poncirus trifoliata*
Tulip, *Tulipa*
Tulip tree, *Spathodea campanulata*
Tung, *Aleurites* spp.
Tumeric, *Curcuma domestica*
Turnip, *Brassica rapa*
Vanilla, *Vanilla fragrans*
Vetch, *Vicia*
Water chestnut, *Trapa bispinosa*
Water hyacinth, *Eichhornia crassipes*
Watermelon, *Citrullus lanatus*
Water spinach, *Ipomoea aquatica*
Wax gourd, *Benincasa hispida*
Welsh onion, *Allium fistulosum*
Wheat, *Triticum*
White mustard, *Brassica alba*
Wild garlic, *Allium ursinum*
Wild lettuce, *Lactuca saligna*, *L. serriola*
Wild radish, *Raphanus raphinastrum*
Wild sorghum, *Sorghum arundinaceum*
Winged bean, see Goa bean
Yam, *Dioscorea* spp.
Yangona, *Piper methysticum*
Yaray palm, *Sabal causiarum*
Zapupe azul, *Agave zapupe*
Zinnia, *Zinnia*

GENERAL WORKS

(1) *Pathology*

Agrios, G. N. 1978. *Plant Pathology*, 2nd edition, Academic Press.

Anon. 1978. *Pest control in tropical root crops, PANS Manual* No. 4, Centre for Overseas Pest Research, London.

Baker, K. F. & Snyder, W. C. (editors) 1965. *Ecology of soil-borne pathogens,* International Symposium Univ. California 1963, John Murray.

Boyce, J. S. 1961. *Forest pathology,* 3rd edition, McGraw Hill.

Brooks, F. T. 1953. *Plant diseases*, 2nd edition, Oxford Univ. Press.

Browne, F. G. 1968. *Pest and diseases of forest plantation trees*, Clarendon Press.

Bruehl, G. W. (editor) 1975. *Biology and control of soil-borne plant pathogens*, International Symposium Univ. Minnesota 1973, American Phytopathological Society.

Butler, E. J. 1918. *Fungi and disease in plants*, Thacker, Spink.

—— & Jones, S. G. 1949. *Plant pathology*, Macmillan.

Byrde, R. J. W. & Cutting, C. V. (editors) 1973. *Fungal pathogenicity and the plant's response*. Proceedings Symposium Long Ashton 1971, Academic Press.

Carefoot, G. L. & Sprott, E. R. 1969. *Famine on the wind*, Angus & Robertson.

Chupp, C. & Sherf, A. F. 1960. *Vegetable diseases*, Constable.

Couch, H. B. 1962. *Diseases of turf grasses*, Reinhold.

Dickson, J. G. 1956. *Diseases of field crops*, 2nd edition, McGraw Hill.

Fischer, G. W. & Holton, C. S. 1957. *Biology and control of the smut fungi*, Ronald Press.

Garrett, S. D. 1944. *Root disease fungi*, Chronica Botanica.

——. 1956. *Biology of root-infecting fungi*, Cambridge Univ. Press.

——. 1970. *Pathogenic root-infecting fungi*, Cambridge Univ. Press.

Graham, K. M. 1971. *Plant diseases of Fiji*, HMSO, London.

Heald, F. D. 1933. *Manual of plant diseases*, 2nd edition, McGraw Hill.

Hepting, G. H. 1971. *Diseases of forest and shade trees of the United States*, USDA Agriculture Handbook 386.

Holton, C. S. (editor) 1959. *Plant pathology, problems and progress*, Univ. Wisconsin Press.

Kranz, J., Schmutterer, H. & Koch, W. (editors) 1977. *Diseases, pests and weeds in tropical crops*, Paul Parey.

Large, E. C. 1940. *The advance of the fungi*, Jonathan Cape.

Leach, J. G. 1940. *Insect transmission of plant diseases*, McGraw Hill.

MacFarlane, H. O. (compiler) 1968. *Review of Applied Mycology, plant host–pathogen index to volumes 1–40 (1922–61)*, Commonwealth Mycological Institute.

Martin, H. (editor) 1972. *Pesticide manual*, 3rd edition, British Crop Protection Council. (Now a 5th edition, 1977.)

Neergaard, P. 1977. *Seed pathology*, 2 Volumes, Macmillan.

Nowell, W. 1923. *Diseases of crop plants in the lesser Antilles*, West India Committee.

Ogilvie, L. 1969. *Diseases of vegetables*, 6th edition, HMSO, London.

Peace, T. R. 1962. *Pathology of trees and shrubs*, Clarendon Press.

Richardson, M. J. 1979. *An annotated list of seed-borne diseases. Phytopathol. Pap.* 23 (also as ISTA Seed Health Testing Handbook Sect. 1.1.).

Sampson, K. & Western, J. H. 1954. *Diseases of British grasses and herbage legumes*, 2nd edition, Cambridge Univ. Press.

Sprague, R. 1950. *Diseases of cereals and grasses in North America*, Ronald Press.

Tarr, S. A. J. 1972. *Principles of plant pathology*, Macmillan.

Toussoun, T. A., Bega, R. V. & Nelson, P. E. (editors) 1970. *Root diseases and soil-borne pathogens*, Univ. California Press.

Walker, J. C. 1952. *Diseases of vegetable crops*, McGraw Hill.

——. 1969. *Plant pathology*, 3rd edition, McGraw Hill.

Western, J. H. (editor) 1971. *Diseases of crop plants*, Macmillan.

Wheeler, B. E. J. 1969. *An introduction to plant diseases*, John Wiley (2nd edition in preparation).

Wood, R. K. S. 1967. *Physiological plant pathology*, Blackwell.

(2) *Mycology*

Ainsworth, G. C., James, P. W. & Hawksworth, D. L. 1971. *Ainsworth & Bisby's Dictionary of the fungi*, 6th edition, Commonwealth Mycological Institute.

General works

Ainsworth, G. C. & Sampson, K. 1950. *British smut fungi*, Commonwealth Mycological Institute.

——, Sparrow, F. K. & Sussman, A. S. (editors) 1973. *The fungi* Volumes IVA & IVB, Academic Press.

Arthur, J. C. 1934. *Manual of the rusts in United States and Canada*, Purdue Research Foundation.

Bessey, E. A. 1950. *Morphology and taxonomy of fungi*, Blakiston.

Clements, F. E. & Shear, C. L. 1931. *The genera of fungi*, H. W. Wilson.

Cummins, G. B. 1971. *The rust fungi of cereals, grasses and bamboos*, Springer-Verlag.

——. 1978. *Rust fungi on legumes and composites in North America*, Univ. Arizona Press.

Dennis, R. W. G. 1968. *British ascomycetes*, Cramer (Now a revised edition 1978.)

Domsch, K. H. & Gams, W. 1972. *Fungi in agricultural soils*, English translation by P. S. Hudson, Longman.

Ellis, M. B. 1971. *Dematiaceous hyphomycetes*, Commonwealth Mycological Institute.

——. 1976. *More dematiaceous hyphomycetes*, Commonwealth Mycological Institute.

Fischer, G. W. 1951. *The smut fungi*, Ronald Press.

——. 1953. *Manual of North American smut fungi*, Ronald Press.

Fitzpatrick, H. M. 1930. *The lower fungi Phycomycetes*, McGraw Hill.

Hawksworth, D. L. 1974. *Mycologist's handbook*, Commonwealth Mycological Institute.

Moore, W. C. 1959. *British parasitic fungi*, Cambridge Univ. Press.

Smith, G. 1969. *An introduction to industrial mycology*, 6th edition, E. Arnold.

Subramanian, C. V. 1971. *Hyphomycetes*, ICAR, New Delhi.

Sutton, B. C. 1980. *The Coelomycetes*, Commonwealth Mycological Institute.

Talbot, P. H. B. 1971. *Principles of fungal taxonomy*, Macmillan.

Webster, J. 1970. *Introduction to fungi*, Cambridge Univ. Press.

Wilson, M. & Henderson, D. M. 1966. *British rust fungi*, Cambridge Univ. Press.

Wolf, F. A. & Wolf, F. T. 1947. *The fungi*, 2 Volumes, John Wiley.

(3) *Botany*

Airy-Shaw, H. K. 1966. *Willis' Dictionary of the flowering plants and ferns*, 7th edition, Cambridge Univ. Press (Now an 8th edition 1973).

Howes, F. N. 1974. *A dictionary of useful and everyday plants and their common names*, Cambridge Univ. Press.

MacMillan, H. F. 1943. *Tropical planting and gardening*, 5th edition, Macmillan.

Purseglove, J. W. 1968. *Tropical crops. Dicotyledons*, Volumes 1 & 2, Longman.

——. 1972. *Tropical crops. Monocotyledons*, Volumes 1 & 2, Longman.

Usher, G. 1970. *A dictionary of botany*, Constable.

——. 1974. *A dictionary of plants used by man*, Constable.

Willis, J. C. 1931. *Dictionary of flowering plants and ferns*, 6th edition, Cambridge Univ. Press.

ADDENDA

Booth, C. 1978. *C.M.I. Descr. pathog. Fungi Bact.* 571, 572, 573 (*Fusarium equiseti, F. heterosporum* & *F. semitectum*).

Bumbieris, M. 1979. *Aust. J. Bot.* **27**: 11 (on *Phytophthora cryptogea*).

Chen, W. H. et al. 1979. *Can. J. Bot.* **57**: 528 (*Sclerospora sacchari* in tissue culture).

Curren, T. 1969. *Can. J. Microbiol.* **15**: 1241 (cell degrading enzymes of *Alternaria radicina*).

Harrison, J. G. 1978. *Trans. Br. mycol. Soc.* **70**: 35 (*Botrytis fabae*, seedborne infection & epidemiology).

Hino, T. et al. 1978. *Tech. Bull. Trop. Agric. Res. Centre* 11, 130 pp. (*Cercospora* & related spp. in Brazil).

Jarvis, W. R. et al. 1978. *Phytopathology* **68**: 1679 (taxonomy of *Fusarium oxysporum* causing tomato foot & root rot).

Kaveriappa, K. M. et al. 1978. *Proc. Indian Acad. Sci.* Sect. B **87**: 303 (seedborne *Sclerospora sorghi*).

Kittle, D. R. et al. 1979. *Pl. Dis. Reptr* **63**: 231 (soil factors & pathogenicity of *Phytophthora megasperma* var. *sojae*).

Kochman, J. K. 1979. *Aust. J. agric. Res.* **30**: 273 (temp. & *Phakopsora pachyrhizi*).

Long, P. G. 1979. *Trans. Br. mycol. Soc.* **72**: 299 (*Mycosphaerella fijiensis* in W. Samoa).

Nachmias, A. et al. 1979. *Physiol. Pl. Pathol.* **14**: 135 (*Deuterophoma tracheiphila* & phytotoxin).

Raghavendra, S. et al. 1977. *Kavaka* **5**: 65 (*Sclerophthora macrospora*, oospore germination & viability).

Reyes, A. A. 1979. *Can. J. Microbiol.* **25**: 227 (*Fusarium oxysporum* f. sp. *spinaciae* in rhizosphere & soil).

Rossman, A. Y. 1979. *Mycotaxon* **8**: 321, 485 (*Calonectria* taxonomy).

Roth, D. A. et al. 1979. *Can. J. Microbiol.* **25**: 157 (low temp. & germination of *Cylindrocladium* microsclerotia).

Samson, R. A. 1979. *Stud. Mycol.* 18, 38 pp. (*Aspergillus* spp. described since 1965).

Shaw, C. G. et al. 1978. *Bull. Wash. St. Univ.* 867, 53 pp. (bibliography Peronosporaceae on Gramineae).

Skipton, W. A. 1979. *Trans. Br. mycol. Soc.* **72**: 161 (*Calonectria camelliae* sp. nov.).

Shit, S. K. et al. 1978. *Indian J. agric. Sci.* **48**: 629 (possible races in *Fusarium udum*).

Stamps, D. J. 1978. *C.M.I. Descr. pathog. Fungi Bact.* 600 (*Aphanomyces euteiches*).

Sun, E. J. et al. 1978. *Phytopathology* **68**: 1672 (*Fusarium oxysporum* f. sp. *cubense* race 4).

Vaartaja, O. et al. 1979. *Can. J. Pl. Sci.* **59**: 307 (*Phytophthora megasperma* var. *sojae*, chemical & biological control).

Venn, L. 1979. *Australasian Pl. Pathol.* **8**: 5 (genetic control of sexual compatibility in *Leptosphaeria maculans*).

Young, L. D. et al. 1979. *Phytopathology* **69**: 8 (soybean response to infection by *Septoria glycines*).

Zentmyer, G. A. et al. 1979. *Mycologia* **71**: 55 (temp., nutrition & formation of sexual state in *Phytophthora cinnamomi*).

INDEX OF FUNGI

(Accepted names; where there are perfect and imperfect state names the fungus is indexed, with few exceptions, under the former. For accepted imperfect state names of these see the next index.)

Acremonium 1–2, 185
furcatum 212
kiliense 1
strictum 1
terricola 212
zonatum 1
Aecidium
crotalariae 526
crotalariicola 526
crotalariicolum 526
dielsii 526
Albugo 2–3
bliti 2
candida 3–4
ipomoeae-aquaticae 4–5
ipomoeae-panduratae 5
occidentalis 2
tragopogonis 2–3
Alternaria 6–7
alternata 7–8, 15
brassicae 8–9, 19–20
brassicicola 8–10, 19–20
burnsii 10
carthami 10–11
cheiranthi 20
cichorii 6
citri 11–12
crassa 12
cucumerina 12–13
dauci 13–14, 379
eichhorniae 6
fasciculata 23
helianthi 6
longipes 15–16
longissima 6
macrospora 16
melongenae 6
padwickii 16–17
passiflorae 17–18
poonensis 6
porri 18, 381
radicina 13–14, 19
raphani 19–20
ricini 20–1
sesami 21
solani 21–4, 324
tenuissima 6, 24
tomato 23–4
umbellifericola 6
zinniae 6
Amphobotrys 467
Amyloflagellula pulchra 260

Aphanocladium 1
Aphanomyces 24–5
brassicae 25
cladogamus 25
cochlioides 25
euteiches 25–7, 195
helicoides 25
laevis 25
raphani 27
Armillariella 27
mellea 27–31, 436
montagnei 28
tabescens 28
Ascochyta 31–33
abelmoschi 32
caricae-papayae 32
fabae 32
gossypii 32–4
melongenae 32
oryzae 32
phaseolorum 32–4, 324
pisi 32, 34–6, 282, 326
prasadii 32
rabiei 36
sorghi 32
sorghina 32
tarda 32
trifolii 497
Aspergillus 37, 296
flavus 37–40
niger 40–1, 70, 98
ruber 37
Asperisporium 41
caricae 41
Aureobasidium 245
caulivorum 245

Blakeslea 76
Botryodiplodia 42
theobromae 42–4, 54–5, 107, 190, 333, 371
Botryosphaeria 45, 333
obtusa 45
ribes 45–6
Botryotinia 458
Botrytis 46–7, 457–8
allii 46–9, 98, 458, 474
byssoidea 46–7, 458, 474
fabae, 47, 462
Bremia 49
lactucae 49–51, 154, 461

Calonectria 51–2, 120–1
colhounii 51
crotalariae 52–3
ilicicola 121
kyotensis 51–4, 122
quinqueseptata 51
rigidiuscula 42, 54–5, 61
theae 51–3
Calostilbe 55
striispora 55–6, 493
Ceratobasidium 295, 502–3, 508
Ceratocytis 56–7, 509
adiposa 57
fagacearum 57
fimbriata 57–62, 537
paradoxa 57, 62–4, 107, 509–10
radicicola 75
ulmi 56–7
Cercoseptoria 64–5
cajanicola 65
pini-densiflorae 65
theae 65
sesame 65
Cercospora 65–7
apii 65–6, 72
angolensis 66
anonae 66
atrofiliformis 71
brassicicola 66
canescens 668
citri-grisea 276
coffeicola 68–9
corchori 66
elaeidis 69
fusimaculans 66
hayi 66, 107
kikuchi 69–71
koepkei 71
longipes 71–2
longissima 66
nicotianae 72–3
sesame 66
sojina 66
sorghi 73–4
traversiana 66
vaginae 71
vanderysti 66–7
zeae-maydis 66
Cercosporella 387
carthami 387
Cercosporidium 74
henningsii 74

Index of fungi

Cerotelium 75, 336
fici 75
Chaetoporus vinctus 383
Chalara 56, 509
Chalaropsis 56, 75–6
thielavioides 75–6, 511
Choanephora 76–7
cucurbitarum 76–7, 348
trispora 348
Cladosporium 77–8
allii-cepae 77
cladosporiodes 77
colocasiae 77
cucumerinum 78–9
herbarum 77
macrocarpum 77
musae 77, 538
oxysporum 77
pisicolum 77
sphaerospermum 77
spongiosum 77
variabile 77
vignae 77
Claviceps 79–80
fusiformis 80
gigantea 79
maximensis 79–80
microcephala 80
paspali 80
purpurea 79
sorghi 80
Cochliobolus 80–1, 139–40
carbonum 82–3, 141
chloridis 139
cymbopogonis 81, 119
cynodontis 81
geniculatus 81, 119
hawaiiensis 140
heterostrophus 82–6, 141, 208, 284, 483
lunatus 86–7
miyabeanus 87–9
nodulosus 89
sativus 81
setariae 89–90
spicifer 90–1
victoriae 81–2
Coleosporium 91
asterum 91
barclayense 91
helianthi 91, 397
ipomoeae 91
pinicola 91
tussilaginis 91
Colletotrichum 91–6, 217, 486
acutatum 92, 223
brassicae 93
capsici 93, 96–7, 218
circinans 97–8
coccodes 98–101, 218, 406
corchori 93
corchorum 93
crassipes 101
curvatum 101
dematium 92–4, 96–7, 101–2, 108, 218, 226; ff. sp. circinans 101; spinaciae 101; truncata 94, 101, 108
destructivum 94–5, 108
gloeosporioides var. minor 218
graminicola 101–4, 218, 231–2
higginsianum 93
indicum 93, 226
lindemuthianum 94, 104–6, 109, 218
musae 92, 106–8, 218–19, 291
obiculare 225
spinaciae 101
tabacum 93
trichellum 101
trifolii 94–5, 245
truncatum 94–5, 101, 108–9
Coniothyrium minitans 458, 471, 477
scirpi 250
Cordana 109
musae 109–10, 125
Corticium 110, 260
rolfsii 110–12
salmonicolor 112–13, 557
Corynespora 114
cassiicola 114–15
citricolor 114
Crinipellis 115, 260
actinophorus 260
perniciosa 43, 61, 115–18, 266, 371; var. citriniceps 116–17, ecuadoriensis 116–17, perniciosa 116–17
Cristulariella 118
pyramidalis 118
Cunninghamella 76
echinulata 77
Curvularia 81, 118–20
andropogonis 119
clavata 119
eragrostidis 119
pallescens 119
penniseti 119
senegalensis 87, 119
verruculosa 119
Cylindrocarpon 120–1
musae 120
panacis 121
tenue 121
tonkinense 121
Cylindrocladium 121–2
braziliensis 122
camelliae 121
clavatum 121–2
parvum 121
pteridis 121
scoparium 53, 122–3
Cytospora 123–4, 537
australiae 123
eucalypticola 123
sacchari 123
Deightoniella 124
papuana 124
torulosa 107, 110, 124–6, 555
Dendryphion 379
Deuterophoma 126, 323
tracheiphila 126–7
Diaporthe 127–8
citri 128–30, 235
cubensis 128, 149–50
eres 128
lokoyae 328
manihotis 128
medusaea 128
phaseolorum 130–1; var. batatatis 130, caulivora 130–1, phaseolorum 130, sojae 130–1
woodii 128
Didymella 131–2, 323
applanata 131
arcuata 131
bryoniae 43, 132–3
chrysanthemi 133–4
lycopersici 32, 131, 134–5, 324
mangiferae 131
rabiei 36
sorghina 131
Didymosphaeria arachidicola 271
Diplodia 135–6
macrospora 136
maydis 136–7, 208–9, 211, 214
pinea 137–9
sapinea 137
Drechslera 81, 139–40, 238, 482
australiensis 139
frumentacei 140
gigantea 141
halodes 141–2; var. elaeicola 141
heveae 142
holmii 482
incurvata 142–3
miyakei 140
monoceras 140
musae-sapientum 140
pedicillata 482
rostrata 87, 141, 143
sacchari 143–5, 250
sesami 140
sorghicola 140
stenospila 144–5

Elsinoë 145–6, 486
ampelina 146
australis 146–7, 487
batatas 145
canavaliae 145, 149
dolichi 145
fawcettii 146–8, 486–7, 508
heveae 146
iwatae 146
leucospila 146
mangiferae 146
phaseoli 145, 148–9; f. sp. vulgare 149
piperis 146

Elsinoë (*cont.*)
sacchari 146
theae 146
wisconsinensis 146
Emericellopsis 1
Endomyces geotrichum 203
Endothia 128, 149–50
eugeniae 150–1, 538
gyrosa 150
havanensis 149
parasitica 150
Entyloma 151, 385
oryzae 151–2
vignae 151
Eremothecium 152, 288
ashbyi 152
cymbalariae 152
Erysiphe 152–4, 494
betae 156
cichoracearum 154–6, 494
communis 156
cruciferarum 156
graminis 152, 494
heraclei 156–7
pisi 157–8
polygoni 158
trifolii 157
Eupenicillium 296
Exobasidium 159
reticulatum 159
vaccinii 159
vexans 159–61
Exserohilum 482–3

Fulvia 61
fulva 161–3, 387
Fusarium 163–4, 204
avenaceum 163
culmorum 163
oxysporum 164–89, 504, 539; ff. sp. anethi 188, apii 188, batatas 166–7, cannabis 188, carthami 167–8, cassiae 188, cattleyae 188, cepae 168, 407–8, ciceris 169, coffeae 188, conglutinans 169–70, coriandrii 170–1, cubense 165, 171–3, cucumerinum 173–4, cumini 174, elaeidis 174–5, eucalypti 188, fabae 175, glycines 188, lactucae 188, lagenariae 188, lathyri 188, lentis 188, lycopersici 165, 175–8, 542, medicaginis 178–9, melongenae 188, melonis 179–80, nicotianae 188, niveum 179–81, passiflorae 181, perniciosum 188, phaseoli 181, 184, pisi 165, 181–4, 194–5, psidii 188, raphani 170, 188, rauvolfiae 188, ricini 188, sesame 188, spinaciae 188, tracheiphilum 184, vanillae 188, vasinfectum 165, 176, 178, 185–7, 548–9, voandzeiae 188, zingiberi 188; var. redolens 165, 182
sacchari 211
semitectum 163
solani 165, 189–92, ff. sp. cucurbitae 191–2, fabae 175, phaseoli 192–5, 514, pisi 26, 182, 194–5, 425
udum 195–7; f. sp. crotalariae 197

Gaeumannomyces 197
cylindrosporus 198
graminis 1, 197–8; var. avenae 197, graminis 197–8, tritici 197
Ganoderma 198–9
applanatum 198–200, 320
boninense 199–201
chalceum 200
lucidum 199
neglectum 199
philippii 200–2, 320, 383, 436–7
tornatum 198–200
xylonoides 200
zonatum 199–200
Geotrichum 202
candidum 202–3; var. citri-aurantii 203
Gibberella 164, 203–5
acuminata 204
avenacea 204
baccata 204–5
cyanogena 204–5, 214
fujikuroi 107, 136–7, 205–11, 214, 555; var. subglutinans 136, 207–8, 211–12, 214
gordonia 204–5
intricans 204–5
pulicaris 204–5
stilboides 204, 212–13
xylariodes 213–14
zeae 136, 208–9, 211, 214–15
Gliocladium 296, 442
Gloeocercospora 215–16
inconspicua 216
sorghi 216, 427–8
Glomerella 92, 217
cingulata 32, 92, 94, 99, 104, 106, 217–31, 241, 293, 340; var. minor 218
graminicola 102
lindemuthianum 217
magna 217, 225
tucumanensis 101, 102, 207, 231–3, 316
Gnomonia 233
iliau 233
manihotis 233
Graphium 56, 442
Guignardia 233–4, 332
arachidis 234
capsici 234
citricarpa 234–5
cocoicola 234
heveae 234
mangiferae 234
musae Racib. 234, 236
perseae 234

Haplobasidion 236–7
musae 237
Helicobasidium 237
compactum 237
mompa 237
purpureum 237
Helminthosporium 139, 238, 482
carposaprum 238
sigmoideum var. irregulare 256
solani 238
velutinum 238
Hemileia 238–9
coffeicola 239
vastatrix 223, 239–43
Hendersonula 243
toruloidea 243
Hypoxylon 243, 536
mammatum 243–4
mediterraneum 243–4; var. microspora 244
nummularium 245
rubiginosum 243; var. tropica 244
serpens 243–5; var. effusum 244

Kabatiella 245
caulivora 95, 245
zeae 245
Khuskia 246
oryzae 101, 214, 246–7, 291

Leptodothiorella 234
Leptosphaeria 247–8, 323
elaeidis 247, 313
eustomoides 250
korrae 247
maculans 248–9
michotti 247, 249–50
narmari 247
pratensis 247, 497
sacchari van Breda de Haan 247, 250
spegazzinia 250
taiwanensis 247
Leptosphaerulina 251
americana 251
trifolii 251–2, 379
Leptothyrium theae 329
Leucostoma 537
Leveillula 153, 252, 331
taurica 152, 253

Index of fungi

Macrophoma 253–4
allahabadensis 254
mangifera 254
theicola 254
Macrophomina 254
phaseolina 136, 208–9, 214, 254–5
Magnaporthe 255–6
grisea 409–11
salvinii 198, 256–7
Marasmiellus 258–9
albus-corticis 258
cocophilus 258
dealbatus 258
inoderma 259
semiustus 259
stenophyllus 259, 261
troyanus 259
scandens 259
Marasmius 115, 258–61
crinisequi 260
cyphella 260
graminum var. brevispora 260–1
palmivorus 258, 260
plicatus 260–1
rigidichorda 260
sacchari 260–1; var. hawaiiensis 260–1
viegasii 261
Maravalia crotalariae 525
Massarina aquatica 409
Melampsora 261–2
lini 262
ricini Noronha 262, 394
Melanopsichium 262–3
eleusinis 262
esculentum 262
missouriense 262
Metasphaeria albescens 434
Microcyclus 263
ulei 228, 263–5
Microdiplodia 250
Microsphaera 153, 265–6
diffusa 266
grossulariae 266
penicillata 266
Monilinia 458
Moniliophthora 266
roreri 266–7, 371
Monilochaetes 267
infusans 267–8
Monochaetia karstenii 314
var. gallica 314
Monocillium 1
Mucor 431
Mycena 268
citricolor 268–9
Mycocentrospora 269
acerina 269–70
Mycosphaerella 270–1, 496
aleuritis 270
arachidis 111, 271–3, 393
areola 427
berkeleyi 111, 271–4, 393
brassicicola 274–5
caricae 270
citri 275–7
cruenta 277–8
fijiensis 237, 278–80; var. difformis 278
holci 270
horii 276
manihotis 270
maydis 283
minima 280
molleriana 271
musae 125, 280
musicola 66, 110, 270, 278–82, 539
nubilosa 271
pinicola 271
pinodes 32, 35, 95, 282–3, 326
puerariicola 271
websteri 270
zeae-maydis 283–4, 332
zeicola 283
Mycovellosiella 284–5
cajani 285
concors 284, 285
maclurae 284
phaseoli 285
puerariae 285
tarri 285
Myrothecium 285–6
advena 286
cinctum 286
roridum 286–7
verrucaria 286

Nectria 1, 51, 55, 120, 287–8, 493
dealbata 288
flavo-lanata 288
galligena 288
haematococca 189–90, 288
jungneri 121
lucida 120
ochroleuca 288
radicicola 288
Nematospora 152, 288–9
coryli 288–90
gossypii 289–90
lycopersici 288
phaseoli 288
Neocosmospora 290
africana 290
vasinfecta 290
Nigrospora 290–1
maydis 291
padwickii 291
panici 291
sacchari 291
sphaerica 107, 246, 291
vietnamensis 291

Oidiopsis 152, 252
Oidium 152–3, 266, 291–3, 494
arachidis 292
bixae 292
caricae 292
caricae-papayae 292
erysiphoides 292
heveae 152, 228–9, 293–4, 340
indicum 292
mangiferae 292
piperis 292
tingitaninum 292
Olivea 294
tectonae 294
Oncobasidium 294
theobromae 294–5
Operculella 295
padwickii 295–6
Orbilia obscura 255
Ovularia 457
Ovulariopsis papayae 292

Penicillium 296–7
corymbiferum 296
digitatum 297–9
expansum 296
funiculosum 210, 296–7
italicum 297–9
oxalicum 296
sclerotigenum 296
Peniophora 112, 438
Periconia 300
circinata 300–1, 418
macrospinosa 300–1
manihoticola 300
sacchari 300
shyamala 300
Periconiella musae 538
Peronosclerospora 449
Peronospora 301–2, 388
arborescens 302
destructor 302–3, 381
farinosa 302
hyoscyami ff. sp. hydrida 307; hyoscyami 307; tabacina 307; velutina 307
manshurica 303–4
parasitica 305–6
tabacina 306–11, 400
trifoliorum 311
viciae 311–12
Pestalotia 313
phoenicis 313
Pestalotiopsis 247, 312–13
funerea 313
guepini 313–14
mangiferae 313
palmarum 143, 313–14
papposa 313
psidii 313
theae 314–15
versicolor 313
Phacidiopycnis 295

Phaeocytostroma 315
ambiguum 315–16
sacchari 315–16; var. penniseti 316
Phaeoisariopsis 316
bataticola 316
griseola 316–17
Phaeoramularia 317–18
capsicicola 317
dioscorea 318
manihotis 74, 318
Phakopsora 318, 336
apoda 318, 404
crotalariae 525
elettariae 318
gossypii 318, 393
pachyrhizi 318–19, 523
setariae 318
Phanerochaete 112
Phellinus 319
ignarius 319
lamaensis 320
noxius 201, 319–21, 383, 436–7
pomaceus 319
robustus 319
Phialophora 197, 321–2
asteris 321; f. sp. helianthi 321
gregata 322–3
parasitica 321
radicicola 321
Phoma 31, 323–6, 332
destructiva 134, 324
exigua 32, 323–4; var. exigua 324, foveata 324, linicola 324
glomerata 323–4
herbarum 323–4
insidiosa 324
macrostoma 323–4
medicaginis 326: var. medicaginis 326–7, pinodella 35, 282, 326
pomorum 323–4
prunicola 324
viburni 324
Phomopsis 128, 327–8
anonacearum 327
carica-papayae 327
conorum 328
cucurbitae 328
heveae 327
juniperivora 328–9
occulta 328
salmalica 328
sclerotioides 328
theae 329–30
vexans 330
Phyllachora 330–1
cynodontis 331
huberi 331
lespedezae 331
maydis 331
musicola 331
sacchari 331
Phyllactinia 153, 331–2
dalbergiae 331–2
guttata 332
Phyllosticta 31, 234, 332
convallariae 332
elattariae 332
sorghina 250
zeae 283
zingiberi 332
zingiberis 332
Physalospora 332–3
persea 332–3
psidii 333
rhodina 42, 333
zeae 333
zeicola 333
Physoderma 333–4, 385
alfalfae 334–5
citri 334
corchori 334
cyndontis 334
echinochloae 334
lathyri 334
leproides (Trab.) Karl 334
maydis 335–6
trifolii 334
Physopella 318, 336
ampelopsidis 336
pallescens 337
zeae 318, 337, 399, 402
Phytophthora 337–9, 414, 446
arecae 339–40, 356, 363, 367
boehmeriae 367
botryosa 340–1, 356, 367, 369
cactorum 338, 346, 357
cajani 338
cambivora 338, 342
capsici 341–2, 365, 367
castaneae 367
cinnamomi 342–5, 349–50, 365, 554
citricola 345–6, 363
citrophthora 346–8, 353, 362–4, 367
colocasiae 76, 348–9
cryptogea 342, 349–50, 358
drechsleri 349–52, 365
erythroseptica 338, 349
heveae 352, 367, 369
hibernalis 346–7, 353, 363
infestans 203, 353–6, 364, 373
meadii 339–40, 356–7, 367, 369
megakarya 367
megasperma 357–60; var. sojae 357
melonis 338
mexicana 338, 367
nicotianae var. nictotianae 360–3, 367, parasitica 346–7, 353, 360, 362–8, 370
oryzae 338
palmivora 43, 54–5, 339–41, 347, 356, 360, 362–73
phaseoli 373–4
porri 374–5
quininea 338
syringae 353, 364
verrucosa 338
vignae 375
Plasmopora 302, 376, 388
halstedii 376–8
nivea 376
penniseti 376
viticola 376
Pleiochaeta 378
albizzae 378
setosa 378
Plenodomus destruens 325
Pleospora 378–9, 498
betae 379
herbarum 251, 379–82, 498; ff. sp. capsicum 381, lactucum 382, lycopersici 380–1
infectoria 379
papaveracea 379
Podosphaera 494
Poria 382, 435
hypobrunnea 382–4
hypolateritia 382, 536
punctata 383
versipora 383
vincta 383; var. cinerea 383–4; vincta 383
Protomyces 384–5
macrosporus 384–5
Protomycopsis 385–6
ajmeriensis 385–6
phaseoli 385–6
thirumalacharii 386
Pseudocercospora 386–7
colocasiae 386
contraria 386
fuligena 387
hiratsukana 386
psophocarpi 386
puerariae 386
purpurea 386
stahlii 386
ubi 386
Pseudocercosporella 387
capsellae 387–8
herpotrichoides 387
ipomoea 387
Pseudocochliobolus nisikadoi 140
Pseudoperonospora 388
cannabina 388
cubensis 389–90
humuli 388
Puccinia 390–1, 521
allii 391–2
antirrhini 391
apii 391
arachidis 392–3
asparagi 391–2
cacabata 318, 393–4
carduncelli 394

Puccinia (*cont.*)
carthami 394–5
cesatii 402
chondrillina 391
chrysanthemi 391
citrulli 391
cordiae 390–1
cynodontis 395–6
digitariae-vestitae 399
dracunculina 391
erianthi 396, 398
esclavensis 399
granularis 391
helianthi 397–8
horiana 391
iridis 391
jaagii 402
kuehnii 396, 398
leveillei 391
levis 399
malvacearum 391
massalis 397
melanocephala 396
menthae 391
minuscula 397
miscanthi 396
oahunensis 398–9
paulensis 391
pelargonii-zonalis 391
pitteriana 390
polysora 240, 337, 399–403
psidii 401
purpurea 401–2
schedonnardi 318, 393
sorghi 337, 399, 402–4
substriata var. indica 399, 404–5,
penicilliare 399, 404–5
tanaceti 391; var. dracunculina 391
thaliae 391
tristachyae 523
verruca 394
Pyrenochaeta 323, 405–6
cajani 405
dolichi 405
glycines 405
indica 405
lycopersici 100, 406–7
nipponica 405
oryzae 405
sacchari 405
terrestris 168, 406–8
Pyrenopeziza brassicae 92–3
Pyrenophora 139
Pyricularia 408–9
angulata 410
aquatica 409
didyma 409
grisea 107, 110, 409–10
oryzae 409–13
penniseti 409
setariae 409
Pyriculariopsis parasitica 410
Pythium 338, 413–15
aphanidermatum 368, 415–17, 419–20, 422, 424
arrhenomanes 136, 208, 301, 417–18, 421–2
butleri 136, 416, 418–19
debaryanum 26, 416, 419–20, 424–5
graminicola 417, 420–2
irregulare 419–20, 422, 424
mamillatum 421–2
myriotylum 190, 422–3
scleroteichum 424
splendens 420, 423–4
tardicrescens 421
ultimum 26, 182, 195, 416, 419–20, 422, 424–26, 511

Ramularia 426–7
bellunensis 426
carthami 426
carthamicola 426
coriandri 426
deusta 426
foeniculi 426
gossypii 427
onobrychidis 426
oryzae 426, 496
pitereka 426
Ramulispora 427–8
alloteropsis 428
sacchari 428
sorghi 216, 427–8
sorghicola 216, 427–8
Rhizina 429
undulata 429–30
Rhizoctonia 430
carotae 430
lamellifera 254, 423
oryzae 430
oryzae-sativae 430
tuliparum 430
Rhizopus 431
arrhizus 431
oryzae 431–2
microsporus 431
rhizopodiformis 431
sexualis 431
stolonifer 76, 431–4
Rhynchosporium 434–5
orthosporum 435
oryzae 434–5
secalis 434–5
Rigidoporus 435
lignosus 201–2, 320, 383, 435–8
ulmarius 435
zonalis 438–9
Rosellinia 439
aquila 439
arcuata 439–40
bunodes 439–42
necatrix 439, 441
pepo 439–40, 442
Sarocladium 442
attenuatum 443
oryzae 442–3
Schizopora paradoxa 383
Scirrhia 443
acicola 443–4
pini 65, 444–6
Sclerophthora 446, 449
cryophila 446
lolii 446
macrospora 446–8, 450, 454, 456
rayssiae 446, 448–9; var. zeae 448–9
Sclerospora 376, 446, 449–50
dichanthiicola 449–50
farlowii 450
graminicola 450–2, 456
iseilematis 450
maydis (Racib.) Butler 449, 452–3
miscanthi 449, 452–4
noblei 450
northii 450
philippinensis 448–9, 452–4, 456
sacchari 448–9, 453–6
sorghi 210, 449–51, 456–7
spontanea 449–50
westonii 449–50
Sclerotinia 46, 457–9
fuckeliana 47–8, 76, 459–66
globosa 47, 458
homoeocarpa 458
intermedia 458
minor 465–7
porri 47, 458, 467
ricini 467–8
sativa 458
sclerotiorum 110, 458, 464–6, 468–74
sphaerosperma 458
squamosa 46–8, 458, 474–5
trifoliorum 458; var. fabae 458
Sclerotium cepivorum 475–8
coffeicolum 111
delphinii 110
fumigatum 430
hydrophilum 430
oryzicola 430
Septoria 478–9
apiicola 478
bataticola 478
cannabis 478
carthami 478
chrysanthemella 478
citri 478
cucurbitacearum 478
depressa 478
fernandezii 481
glycines 479–80
helianthi 480
helianthicola 480
lactucae 480–1
lactucina Lobik 481
leucanthemi 478
limonum 478

Septoria (*cont*..)
ludoviciana 481
lycopersici 481–2
obesa 478
oryzae 478
passiflorae 478–9
pauper 480
pisi 479
schembelii 481
sikangensis 481
sojae 479
sojina 479
unicolor 481
Setosphaeria 139, 482–3
rostrata 482
turcica 208, 483–4
Sorosporium 484, 516
cenchri 485
cynodontis 532
digitariae 485
paspali-thunbergii 485
simii 485
Sphaceloma 145, 485–6
anarcardii 486
arachidis 486
cardamomi 486
fawcettii var. scabiosa 146–7, 486–7
glycines 486
manihoticola 486
papayae 486
perseae 487
poinsettiae 486
psidii 486
ricini 486
santali 486
tectonae 486
Sphacelotheca 485, 487–8
cordobensis 488
cruenta 488–91
destruens 489
erianthi 487
macrospora 488
reiliana 487–91
sacchari 487
schweinfurthiana 488
sorghi 488–92
Sphaeropis 492
tumefaciens 492–3
Sphaerostilbe 493
repens 493–4
Sphaerotheca 152–3, 494
fuliginea 154, 494–5
macularis 494
mors-uvae 153, 494
pannosa 494
Sphaerulina 496
oryzina 426, 496–7
rehmiana 496
rubi 496
Sporothrix 56
Stagonospora 497
curtisii 497
hortensis 32
recedens 497
sacchari 497–8
Stemphylium 6, 380, 498–9
globuliferum 498
loti 498
lycopersici 380–1, 499
sarciniforme 380, 498–9
solani 24, 380–1, 499–500
trifolii 498–9
Stromatinia gladioli 476
Syncephalis californica 432
Synchytrium 385, 500–1
endobioticum 500
lagenariae 500
psophocarpi 500
sesamicola 500
trichosanthidis 500

Taphrina 500
deformans 501
maculans 501–2
Thanatephorus 295, 502
cucumeris 100, 430, 503–9, 511
Thecaphora 485
Thielavia 509–10
basicola 510
Thielaviopsis 56, 509
basicola 100, 509–15, 549
Tilletia 515
ayresii 515
barclayana 151, 515–16
Tolypoderma 516
Tolyposporidium 516, 518
Tolyposporium 516–17
bullatum 516
christensenii 516
cymbopogonis 516
ehrenbergi 516; var. grandiglobum 516
penicillariae 518
Trachysphaera 518
fructigena 107, 125, 371, 518–19, 555
Trichoderma harzianum 111
viride 29, 111
Tunstallia 519
aculeata 519–20; var. kesabii 519

Ulocladium 6
Uredo cajani 522
crotalariae-vitellinae 526
crotalariicola 526
ficina 75
geniculata 402
musae 522
theresiae 526
Urocystis 520
brassicae 520
cepulae 520–1
coralloides 520
Uromyces 521–2
anthyllidis 525, 527
appendiculatus 523–5
azukicola 523
betae 522
ciceris-arietini 525
crotalariae 525
decoratus 525–6
dianthi 522
dolicholi 522
geranii 522
heimerlianus 528
indigoferae 522
jatrophae 522
junci 397
kisantuensis 523
manihotis 522
mucunae 522
musae 522
pegleriae 399
pisi-sativi 526–8
rhynchosiae 522
setariae-italicae 522
soongaricae 525
striatus 527
viciae-craccae 528
viciae-fabae 526–8
vignae-sinensis 523
Ustilaginoidea 528–9
ochracea 529
virens 529–30
Ustilago 485, 487, 530–1
avenae 530
coicis 530
crameri 531
crus-galli 530
cynodontis 531–2
esculenta 530–1
hitchcockiana 532
hordei 530
maydis 490, 532–3
nuda 530
panici-frumentacei 530
paradoxa 530
paraguariensis 532
scitaminea 488, 533–5; var. sacchari-barberi 534, sacchari-officinarum 534
vetiveriae 531
Ustulina 535
deusta 535–7

Valsa 123, 537
abietis 537
curreyi 537
eugeniae 61, 150, 537–8
kunzei 537
Veronaea 538–9
musae 538–9
Verticicladiella 56
Verticillium 442, 539–41
albo-atrum 539–48, 552, 555
dahliae 185, 321–2, 512, 539–43, 545–55
nigrescens 540, 552
nubilum 540
theobromae 107, 125, 519, 540, 555
tricorpus 540

INDEX OF FUNGUS SYNONYMS AND SOME CONIDIAL (IMPERFECT) STATES

(Accepted conidial state names, where there is a perfect state, are in bold face. This list also contains names which are confused, dubious and/or rejected; and some sclerotial names.)

Acrocylindrium oryzae 442
Acrosporium 266, 291, 494
heveae 293
Acrostalagmus 539
Acrotheca 426
Acrothecium
lunatum 86
penniseti 119
Aecidium
brasiliense 522
caulicola 523
decipiens 523
gossypii 393
helianthi-mollis 397
nigro-cinctum 523
oxalidis 402
plantaginus 395
vignae 523
Agaricus
albus-corticis 258
candidus 258
citricolor 268
melleus 27
Aglaospora aculeata 519
Albigo 494
mors-uvae 494
Allantospora 120
Alphitomorpha 152
fuliginea 494
guttata 332
heraclei 156
macularis 494
pannosa 494
pisi 157
Alternaria
brassicae var. dauci 13
carotae 13
circinans 9
cyamopsidis 13
daturae 12
fasciculata 23
longipedicellata 16
matthiolae 19
oleracea 9
tenuis 7
Ancylospora 386
Angiopoma 139
Angiopsora 336
ampelopsidis 336
zeae 337
Anatospora 269
macrospora 269
Aposphaeria ulei 263
Armillaria
fuscipes 28
mellea 27
Ascochyta
adzamethica 271
althaeina 32
arachidis 271
boltshauseri 32, 34
capsici 32
caricae 32
chrysanthemi 133
cucumis 132
imperfecta 326
lactucae 480
lycopersici 32, 134
nicotianae 32
pinodella 282, 326
pinodes 282
pisicola 34
Ashbia gossypii 289
Ashbya gossypii 289
Aspergillus circinatus 300
Asteroma brassicae 275
Asteromella brassicae 275

Bakerophoma tracheiphila 126
Basisporium 290
gallarum 246
Berkeleyna 300
Berteromyces 74
Bipolaris 139
rostrata 482
Botryobasidium salmonicolor 112
Botryodiplodia
ananassae 42
elasticae 42
gossypii 42
pinea 137
tubericola 42
Botryotinia
fuckeliana 459
porri 467
squamosa 474
Botrytis
bifurcata 467
cinerea 47, 459, 462
destructor 302
globosa 47, 458
infestans 353
parasitica 305
ricini 467
squamosa 47, 474
viciae 311
Brachycladium
penicillatum 379
spiciferum 90
Brachysporium
eragrostidis 119
sesami 119
torulosum 124
Bullaria carthami 394

Cadophora 321
Caeoma
destruens 489
helianthi 397
ricini 262
Calonectria
floridana 53
theae var. crotalariae 52
uniseptata 53
Calostilbella calostilbe 55
Candelospora 121
theae 51
Catacauma 330
Centrospora 269
acerina 269
Cephalosporium 1
acremonium 1
eichhorniae 1
fici 1
gregatum 322
maydis 1, 185
sacchari 1, 211
zonatum 1
Ceratophorum
albizziae 378
setosum 378
Ceratosphaeria grisea 409, 411
Ceratostomella
fimbriatum 57
paradoxa 62
Cercocladospora 386
Cercodeuterospora 284
trichophila 285
Cercospora
acerina 269
aleuritidis 270
arachidicola 271
arachidis 272
aranetae 386
bataticola 316
cajani 285

Cercospora (*cont.*)
capsici Marchal & Steyaert 317
capsici Unamuno 317
capsicola 317
caribaea 74
caricae 41
cassavae 74
coffeae 68
concors 285
crassa 12
cruenta 277
daturae 12
diazu 66
dioscorea 318
fijiensis 278
fuligena 387
henningsii 74
herrerana 68
heterosperma 285
indica 285
manihotis 74
melonis 114
musae 279
musarum 124
oryzae 496
passiflorae-foetidae 386
passiflorae-longipedis 386
personata 272
pini-densiflorae 65
psophocarpi 386
puerariae 386
purpurea 386
raciborski 72
sacchari 143
sesamicola 65
solimani 316
sorghi var. maydis 73
stahlii 386
taiwanensis 247
theae 65
unamunoi 317
vignicaulis 67
vignicola 114
Cercosporella
albomaculans 388
brassicae 388
gossypii 427
theae 52
Cercosporidium personatum 272
Cercosporina
aleuritidis 270
kikuchii 69
Cerotelium gossypii 318
Chaconia tectonae 294
Chaetodiplodia grisea 42
Chaetomium coccodes 98
Chloridium musae 538
Cintractia
crus-galli 530
seymouriana 530
Cladochytrium
alfalfae 334
maydis 335
Cladosporium
cucumeris 78
fulvum 161
personata 272
Clitocybe tabescens 28
Cochliobolus stenospilus 144
Colletotrichum
atramentarium 98, 101
caulicola 108
coffeanum 222–3
falcatum 101, 102, 218, 231
gloeosporiodes 217; var. minor 218
glycines 94, 108
gossypii 226
gossypii var. cephalosporioides 226
hibisci 229
lagenarium 225
phomoides 99
pisi 94–5
sativum 94
viciae 108
Collybia
troyana 259
sacchari 260
Coniothyrium sacchari 315
Coprotrichum 202
Corallomyces elegans 493
Corticium
areolatum 508
microsclerotia 505
sasaki 504–5
solani 503
Corynespora melonis 114
Crebrothecium 152
ashbyi 152
Creonectria 287
Cryptosporella eugeniae 150
Cryptosporium acicolum 443
Curvularia
coicis 139
cymbopogonis 119
geniculata 119
lunata 86, 119
maculans 119
sigmoidea 256
spicifera 90
Cylindrocarpon
destructans 120, 288
lucidum 120
victoriae 121
Cylindrocladium
crotalariae 52
floridanum 53, 122
ilicicola 121
macrosporum 121
pithecolobii 122
quinqueseptatum 51
scoparium var. braziliensis 122
theae 51
Cylindrosporium
capsellae 388
concentricum 92
sesame 65
Cyphella pulcher 260
Cystopus 2
Cytospora
eucalyptina 123
albietis 537
curreyi 537
kunzei 537
Cytosporina septospora 444

Dactylaria oryzae 410
Dactylella aquatica 409
Dematophora 439–40
necatrix 441
Dendryphion
papaveris 379
penicillatum 379
Dendryphium cajani 285
Diachora 330
Dialonectria 287
Diaporthe vexans 330
Dicaeoma
carthami 394
cynodontis 395
polysorum 399
purpureum 401
sorghi 402
Dichotomella 290
Dicladium graminicolum 102
Didymella
ligulicola 133
pinodes 282
rabiei 36
Didymellina pinodes 282
Didymosphaeria
bryoniae 132
taiwanensis 249
Didymotrichum 77
Diplocladium cylindrosporium 122
Diplodia
ananassae 42
cacaoicola 42
frumenti 333
gossypina 42
maydicola 136
natalensis 42
rapax 42
theobromae 42
tubericola 42
zeae 136
zeae-maydis 136
Diplodina
destructiva 324
lycopersici 134, 324
medicaginis 326
Dothidella ulei 263
Dothistroma pini 444
var. **keniensis** 445, **linearis** 445, **pini** 445
septospora 444

Index of fungus synonyms and conidial states

Drechslera
  chloridis 139
  coicis 139
  cynodontis 81
  hawaiiensis 87, 140
  maydis 83, 141
  nodulosa 89
  oryzae 87
  rostrata 482
  setariae 89
  sorokiniana 81
  turcica 483
  victoriae 81
  zeicola 82, 141

Endoconidiophora fimbriata 57
Endophyllachora 330
Endothlapsis 487
Epiclinium cumminsii 41
Erysibe 2
Erysiphe
  alni 332
  betulae 332
  fuliginea 494
  guttata 332
  macropus 157
  macularis 494
  martii 157
  mors-uvae 494
  pannosa 494
  scorzonerae 154
  taurica 253; var. andina 253;
    zygophyllii 253
  umbelliferarum 156
Exosporina fawcettii 243
Exotrichum 285

Fomes
  auberianus 435
  ignarius 319
  lignosus 435
  lucidus 199
  noxius 319
  philippii 201
  pseudoferreus 201
  robustus 319
  rugulosus 438
Fusarium
  acuminatum 204
  angustum 164
  avenaceum 204; f. sp. fabae 204
  batatatis 166
  bostrycoides 164
  bulbigenum 164; var. batatas 166,
    lycopersici 175, niveum 180,
    tracheiphilum 184
    conglutinans 164, 169
    cubense 171; var. inodoratum 171
    decemcellulare 54
  dianthi 164
  equiseti 204
  graminearum 214
  heterosporum 204
  javanicum 189, 191
  lateritium 204; ff. sp. cajani 195;
  cerealis 204, crotalariae 197,
  longum 212, mori 204, pini 204,
  xylarioides 213; forma ciceri 169;
  var. stilboides 212, uncinatum 195
  lini 164
  lycopersici 175
  martii var. pisi 194
  moniliforme 205; var.
    subglutinans 204–5, 207–8,
    211–12
  niveum 180
  orthoceras 164; var. pisi 181
  oxysporum var. cubense 171,
    orthoceras 169, 182; f. udum 195; f.
    xylariodes 213; f. 3 171; f. 4 171; f.
    8 182; ssp. lycopersici 175
  redolens 165; var. solani 165
  sambucinum 204; f. 6 204
  solani var. redolens 165
  spicaria-colorantis 54
  stilboides 212; var. minus 212
  sulphureum 204
  tracheiphilum 184
  udum var. cajani 195, crotalariae 197
  uncinatum 195
  vasinfectum 164, 185, 197; f.
    crotalaria 197; var. pisi 181
  xylariodes 213
Fusicladium
  caricae 41
  macrosporum 263
Fusisporium
  concors 285
  solani 189

Ganoderma
  miniatocintum 199
  noukahivense 199
  pseudoferreum 201
Geminospora 331
Geniculosporium serpens 244
Gibbera saubinettii 204, 214
Gibberella
  moniliforme 205
  saubinettii 204, 214
Giocladiopis sagariensis 121
Gliomastix 1
Gloeosporium 92, 218
  caulivorum 245
  concentricum 92
  limetticola 221
  lindemuthianum 104
  musarum 106
Glomerella
  folliicola 217, 221
  glycines 94, 108, 217
  gossypii 217, 226
  lagenarium 217
  major 217
  phomoides 99, 217
  psidii 217, 219
Gnomoniopsis 217
  cingulata 217
Guignardia
  heveae Fragoso & Cif. 234
  heveicola 234
  musae Stevens 236

Halstedia 331
Haplographium manihoticola 300
Haplopyxis crotalariae 525
Haplotheciella 131
Harpocephalum 300
Helicobasis 237
Helicomina 386
Helminthosporium
  acrothecioides 81
  australiense 139
  brassicicola 9
  californicum 81
  carbonum 82
  cassiicola 114
  cheiranthi 20
  coicis 139
  cymbopogonis 119
  cynodontis 81
  frumentaceum 140
  geniculatum 119
  giganteum 141
  halodes 141
  hawaiiense 140
  helianthi 6
  heveae 142
  inconspicuum 483
  incurvatum 142
  leucostylum 89
  maydis 83
  miyakei 140
  monoceras 140
  musae-sapientum 140
  nodosum 124
  nodulosum 89
  ocellum 143, 145
  oryzae 87
  papaveris 379
  papayae 114
  parasiticum 410
  rostratum 143, 482
  sacchari 143
  sativum 81
  sesami 140
  setariae 89
  sigmoideum 256
  sorghicola 140
  sorokinianum 81
  spiciferum 90
  stahlii 386
  stenospilum 144
  tenuissimum 6
  torulosum 124
  turcicum 483

Helminthosporium (*cont.*)
victoriae 81
vignae 114
Hemileiopsis 238
Hendersonia zeae 136
Heterosporium 77
Hughesiella 509
Hydatinophagus 24
Hypomyces 189
Hypoxylon deustum 535

Isariopsis griseola 316
Isotexis 145

Kawakamia macrospora 446
Koleroga 295, 502
Kuehneola
fici 75
gossypii 318

Lasiodiplodia
nigra 42
theobromae 42
triflorae 42
tubericola 42
Lasionectria 287
Lecanostica
acicola 443
pini 443
Leptosphaeria
sacchari Speg. 250
salvinii 256
trimera 249
Leptosphaerulina
arachidicola 251
argentinensis 251
australis 251
briosiana 251
crassiasca 252
vignae 251
Linocarpon oryzinum 197
Lisea fujikuroi 205

Macbridella striispora 55
Macrodiplodia
macrospora 136
zeae 136; var. macrospora 136
Macrophoma musae 236
phaseoli 254
phaseolina 254
pinea 137
vestita 42
zeae 333
Macrosporium 6
brassicae 8
carotae 13
cheiranthi 20; var circinans 9
cucumerinum 12
daturae 12
dauci 13
longipes 15
melophthorum 78
porri 18
ricini 20
sesami 21
solani 21
tennuisimum 6
tomato 23
Malustela 118
Manginia 485
Marasmius
actinophorus 260
byssicola 259
candidus 258
coronatus 260
equicrinis 260
graminum var. equicrinis 260
inoderma 259
perniciosus 115
pulcher 260
repens 260
scandens 259
stenophyllus 259
subsynodicus 259
troyanus 259
Melampsora ricini Pass. 262
Melampsorella ricini 262
Melanobasidium 485
Melanobasis 485
Melanophora 485
Melanopsammopsis 263
ulei 263
Metachora 330
Monilia roreri 266
Monotospora oryzae 246
Mucor
niger 432
stolonifer 432
Mycosphaerella
arachidicola W. A. Jenkins and Khokhryakov 271
argentinensis 271
dearnessii 443
ligulicola 133
melonis 132
rabiei 36
Myrotheciella 285
Myxocladium 77
Myxormia 285
Myxoxporium musae 106

Nakataea sigmoidea 256
Necator decretus 112
Nectria
rigidiuscula 54
striispora 55
Nectriopsis 287
Nematosporangium
aphanidermatum 415
Neovossia
barclayana 515
horrida 515
Nigrospora
musae 291
oryzae 246, 291

Oidiopsis taurica 253
Oidium citri-aurantii 203
Oligostroma acicola 443
Omphalia flavida 268
Oospora citri-aurantii 203
Ophiobolus
graminis 197
heterostrophus 83
oryzinus 197
Ophiochaeta graminis 197
Ophiostoma 56
coffeae 57
fimbriatum 57
paradoxa 62
Ovularia
indica 253
phaseoli 285
Ovulariopsis 331

Paecilomyces 1, 296
Passalora personata 272
Pellicularia 502
filamentosa 503
salmonicolor 112
Penicillium olivaceum 297
Periconia heveae 300
Peronoplasmopora 388
cubensis 389
Peronospora
brassicae 305
cubensis 389
dubia 307
effusa var. hyoscyami 307
graminicola 450
halstedii 376
infestans 353
maydis 452
nicotianae 307
pisi 311
schleideni 302
sojae 303
trifoliorum var. manshurica 303
viciae-sativae 311
Pestalotia guepini 313
palmarum 313
theae 314
Pestalozzia 313
Peyronellaea 323
Peziza
fuckeliana 459
sclerotiorum 468
Phaeocytosporella 315
zeae 315
Phaeocytostroma istrica 315
Phaeonectria 55
Phaeoramularia unamunoi 317
Phaeostagonosporopsis zeae 136
Phakopsora ampelopsidis 336

Index of fungus synonyms and conidial states

Phanerochaete salmonicolor 112
Phoma
arachidicola 271
brassicae 248
chrysanthemi 133
citricarpa 234
herbarum f. medicaginum 326
lingam 248; var. napobrassicae 248
lycopersici 134
musae 236
phaseoli 130
solani 330
subcircinata 130
terrestris 407
tracheiphila 126
trifolii 326
vexans 330
Phomopsis
californica 128
caribaea 128
citri 128
cytosporella 128
lekoyae 328
leptostromiformis 128
manihotis 128
phaseoli 130
Phragmodiaporthe aculeata 519
Phyllachora sorghi 331
Phyllactinia
corylea 332; var. subspiralis 332
subspiralis 332
suffulta 332
yarwoodii 332
Phyllosticta brassicae 248
cicerina 36
citricarpa 234
lycopersici 324
maydis 283
musarum 236
rabiei 36
sacchari 250
saccharicola 250
Phyllostictina 332
citricarpa 234
Physalospora
obtusa 45
tucumanensis 231
Physoderma
gibbosum 384
leproides (Trab.) Lagerh. 334
zeae-maydis 335
Physopella
fici 75
ficina 75
vitis 336
Phytophthora cactorum var. arecae 339,
applanata 345
erythroseptica var. drechsleri 350
faberi 367
hydrophila 341
infestans var. phaseoli 373
macrospora 446
melongena 360
omnivora var. arecae 339
parasitica 362; var. capsici 341,
piperina 365, rhei 360
sojae 357
terrestris 360
theobromae 367
Plasmopora
cubensis 389
helianthi 376
Plectodiscella 145
Plenodomus 323
Pleocyta 315
sacchari 315
Pleospora
allii 379
armeriae 379
australis 379
bjöerlingii 379
dianthi 379
evonymi 379
grossulariae 379
lycopersici 380
maculans 248
trifolii 251
Pleosporopsis 439
Polyporus
ignarius 319
lignosus 435
zonalis 438
Polystictus rigidus 438
Preussiaster 109
Protomyces
cari 384
graminicola 450
Protomycopsis
crotalariae 385
patelii 385
smithiae 385
Pseudocercospora
cruenta 277
fijiensis 278
musae 279
Pseudomelasmia 330
Pseudoplasmopora 376
Pseudoplea 251
trifolii 251
Pseudosphaeria trifolii 251
Pseudostemphylium radicinum 19
Puccinia
absinthii 391
allii-japonici 391
blasdalei 391
corticola 391
digitariae 398
digitariae-velutinae 398
granulispora 391
helianthorum 397
heliopsidis 397
kentrophylli 394
maydis 402
mixta 391
mutabilis 391.
penicillariae 404
penniseti 404
penniseti-spicati 404
porri 391
prunicolor 401
sanguinea 401
stakmanii 393
viguierae 397
xanthifoliae 397
zeae 402
Pucciniopsis caricae 41
Pyricularia musae 410
Pythiacystis citrophthora 346
Pythium teratosporon 350

Ragnihildiana 284
manihotis 74
Ramichloridium musae 538
Ramularia
areola 427
phaseoli 285
phaseolina 285
puerariae 285
rapae 388
scabiosa 486
Ramulispora andropogonis 428
Rhabdospora
helianthicola 480
hortensis 479
Rhamphospora 151
Rhaphidophora graminis 197
Rheosporangium aphanidermatum 415
Rhizina inflata 429
Rhizoctonia
bataticola 254
crocorum 237
solani 190, 430, 503, 549
Rhizopus
artocarpi 432
delemar 431
maydis 431
niger 432
nigricans 432
nodosus 431
stolonifer var. luxurians 432
Rhopalidium 6
Rhysotheca 376
halstedii 376
Rigidoporus surinamensis 438
Rosellinia
echinata 440
zingiberi 440
Rostrella coffeae 57

Scleropleela michotti 249
Sclerospora
andropogonis-sorghi 456
graminicola: var. andropogonis-
sorghi 456; setariae-italicae 450
indica 452–3
javanica 452

Sclerospora (*cont.*)
kriegeriana 446
macrospora 446
maydis Reinking 453
oryzae 446
sorghi-vulgaris 456
Sclerotinia libertiana 468
Sclerotium
oryzae 256
oryzae-sativae 430
rolfsii 110
tuliparum 430
Scoleconectria 51
Scolecotrichum
caricae 41
melophthorum 78
musae 109
Septogloeum arachidis 272
Septorella sorghi 428
Septoria
acicola 443
apii 478
apii-graveolentis 478
consimilis 480
lactucina Petrak 481
leguminum 479
petroselini var. apii 478
theae 65
Septoriopsis Stevens & Dalbey 64
Sorisporium 487
Sorosporium
bullatum 516
ehrenbergii 517
filiferum 517
panici-miliacea 489
paspali 485
reilianum 489
syntherismae 485
Sphacelia 79
sorghi 80
Sphaceloma
australis 146
fawcettii 147; var. viscosa 146
Sphacelotheca
chrysopogonis 488
holci 488
panici-miliacea 489
Sphaerella
brassicicola 274
bryoniae 132
chrysanthemi 133
michotii 249
pinodes 282
Sphaeria
brassicicola 274
bunodes 440
concava 34
cyanogena 204
dematium 101
deusta 535
herbarum 379
lingam 248
maculans 248
mammata 243
maydis 136
mediterranea 243
michotti 249
obtusa 45
phaseolorum 130
pinea 137
pinodes 282
rubiginosa 243
serpens 243
zeae 136, 214
Sphaeronaema
fimbriatum 57
lycopersici 134
Sphaeropsis
ambigua 315
ellisii 137
musarum 236
Sphaeropyxis 439
Sphaerostilbe
longiascus 55
musarum 493
Sphaerotheca
humuli 494; var. fuliginea 494
Sphaerulina trifolii 251
Spicaria colorans 54
Spondylocladium maculans 119
Sporidesmium
exitiosum var. dauci 13
longipedicellatum 16
Sporisorium sorghi 491
Sporocladium 77
Sporocybe 300
Sporodum 300
Sporotrichum
destructor 426
maydis 291
Stachylidium theobromae 555
Stagonopsis phaseoli 32
Stagonospora meliloti 497
Starkeyomyces 285
Stemphylium
botryosum 251, 379, 498–9
floridanum 380, 499
radicinum 19
Stenella sp. 275
Stenocarpella zeae 136
Sterigmatocystis niger 40
Stilbochalara 509
Stilbocrea 287
Stilbum flavidum 268
Stromatinia cepivorum 475
Symphyosira areola 427
Sympodina 538
Synchytrium phaseoli 385
Systremma acicola 443

Tetracytum 121
Thielaviopsis paradoxa 63
Thyrospora
lycopersici 499
radicina 19
solani 499
Tilletia
ajrekari 515
epiphylla 402
horrida 515
oryzae 529
pulcherrima 515
Titaeospora andropogonis 428
Tolyposporium
filiferum 517
senegalense 516
Torula
alternata 7
basicola 509
dimidiata 243
Trabutiella 331
Trichocephalum 300
Trichoconis padwickii 16
Trichometasphaeria turcica 483
Trullula sacchari 315
Tuburcina cepulae 520

Uredo
albugo 2
arachidis 392
chloridis-berroi 393
chloridis-polydactylidis 393
ciceris arietini 525
coffeicola 239
crotalariae 526
digitaria-ciliaris 398
digitariaecola 398
duplicata 398
fici 75
harmsiana 525
kuehnii 398
maydis 532
moricola 75
mucunae 522
mucunicola 522
pisi-sativi 526
ricini 262
segetum 489
sorghi 401
syntherismae 398
tectonae 294
vialae 336
viciae-fabae 527
vignae 523
vitis 336
zeae 532
Urocystis colchici var. cepulae 520
Uromyces
aemulus 391
ambiguus 391
arachidis 392
aterrimus 392
bicolor 392
ciceris-arietini var. aetnensis 525
dolichi 523
durus 391

Index of fungus synonyms and conidial states

Uromyces (*cont.*)
  fabae 527
  fischeri-eduardi 528
  harmsianus 525
  kuehnii 398
  lathyri 526
  leptodermus 522
  medicaginis 527
  medicaginis-falcatae 527
  oblongus 527
  orientalis 522
  orobi 527
  pazschkeanus 523
  phaseoli 523
  pisi 526
  polymorphus 527
  psophocarpi 500
  punctiformis 523
  viciae 527
  vignae 523
  yoshinagai 527
Urophlyctis 333–4
  alfalfae 334
Uropyxis crotalariae 525
Ustilaginoidea oryzae 529
Ustilago
  carbo var. cynodontis 531
  carbo f. cynodontis-dactylonis 531
  cruenta 488
  eleusinis 262
  erianthi 487
  panici-miliacea 489
  paspali-thunbergii 485
  reiliana 489
  sacchari 487–8, 533
  sacchari-ciliaris 488
  schweinfurthiana 488
  segetum var. cynodontis 531
  sorghi 491
  virens 529
  zeae 532
  zeae-mays 532
Ustulina
  vulgaris 535
  zonata 535–6

Vakrabeeja sigmoidea 256
Vellosiella 284
  cajani 285
Vermicularia 92
  atramentaria 98
  capsici 96
  circinans 97
  gloesoporioides 217
  truncata 108
Verticillium albo-atrum var.
  caespitosum 541, medium 546;
    tuberosum 541
  dahliae var. longisporum 546
Virgasporium 65

Whetzelina 458
  sclerotiorum 468
Woroninella psophocarpi 500

Zythia rabiei 36

INDEX OF OTHER ORGANISMS

Aceria mangiferae 210
Achatina fulica 371
Acyrthosiphon pisum 326
Aphelenchus avenae 100
Aphrophora parallela 138

Bean yellow mosaic virus 195, 425
Belonolaimus
  gracilis 185
  longicaudatus 185
Bradysia 178

Cacao swollen shoot viruses 54
Chalcodermus aeneus 76
Chlorochroa sayi 39
Corynebacterium insidiosum 178
Creontiades pallidus 433
Cucumber mosaic virus 425

Diabrotica longicornis 208
Diatraea saccharalis 207, 232
Distantiella theobroma 54
Drosophila melanogaster 203, 433
Dysdercus 289

Erwinia
  amylovora 45
  mangiferae 230

Helicotylenchus dihystera 421
Heterodera
  glycines 184, 357
  rostochiensis 100
Homoeosoma ellectellum 432
Hoplolaimus indicus 208

Limonia 202
Lygus hesperus 39

Maize dwarf mosaic virus 66
Meloidogyne
  arenaria 111
  hapla 357
  incognita 111, 176, 179, 185, 208, 227, 421, 507, 543, 549
  javanica 87, 184, 554

Nezara viridula 433

Ostrinia nubilalis 215
Oxya
  japonica 447
  velox 447

Pea common mosaic virus 195, 425
Pectinophora gossypiella 39, 433
Phyllocoptruta oleivora 276
Polyphagotarsonemus latus 228
Pratylenchus
  brachyurus 185, 361, 421
  coffeae 176
  penetrans 26, 176, 182, 193, 553–4
  sudanensis 185
  zeae 421
Pseudomonas
  cepacia 168
  glycinea 480
  solanacearum 171

Radopholus similis 120, 172, 514
Rotylenchus
  reniformis 185
  robustus 182

Sahlbergella singularis 54
Scirtothrips dorsalis 228
Sitona hispidulus 178
Spissistilus festinus 111
Squash mosaic virus 192
Sugarcane mosaic virus 421

Tobacco mosaic virus 543
Tobacco necrosis virus 225
Tobacco ringspot virus 70
Trichodorus christiei 543
Turnip mosaic virus 170
Tylenchorhynchus claytoni 424
Tylenchus
  agricola 424
  semipenetrans 190

Watermelon mosaic virus 192
Wild cucumber mosaic virus 192

Xanthomonas malvacearum 185—6, 227
Xyleborus
  affinis 61
  corniculatus 60
  ferrugineus 60